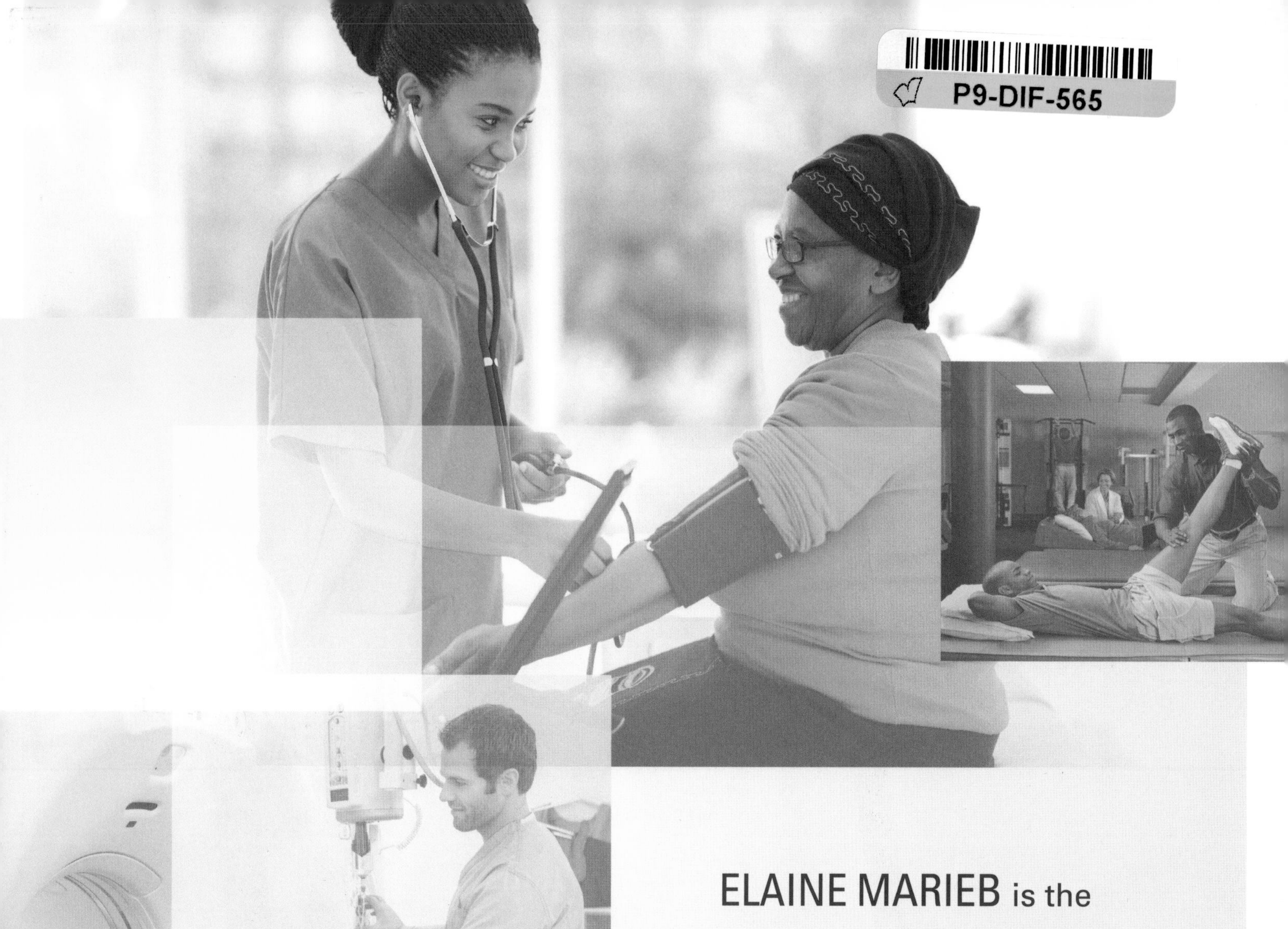

ELAINE MARIEB is the most trusted name in all of A&P. More than 3 million health care professionals started their careers with one of Elaine Marieb's Anatomy & Physiology texts.

Now, it's your turn.

READ ANYTIME, ANYWHERE

NEW! eText 2.0 brings your textbook to any web-enabled device.

- Now available on smartphones and tablets.
- Seamlessly integrated videos and other rich media.
- Accessible (screen-reader ready).
- Configurable reading settings, including resizable type and night reading mode.
- Instructor and student note-taking, highlighting, bookmarking, and search.

LEARN WHY THIS MATTERS

NEW! Chapter-opening Why This Matters videos describe how the material applies to your future career. **Scan the QR codes** to see brief videos of real health care professionals discussing how they use the chapter content every day in the field.

SEE WHERE YOU

NEW! Key concept organization presents the material in manageable chunks and helps you easily navigate the chapter. Each section header states the key concept of that section.

5 The Integumentary System

Overview of Key Concepts

KEY CONCEPTS

Would you be enticed by an ad for a coat that is waterproof, stretchable, washable, and air-conditioned, that automatically repairs small cuts, rips, and burns? How about one that's guaranteed to last a lifetime? Sounds too good to be true, but you already have such a coat—your skin.

The skin and its derivatives (sweat and oil glands, hairs, and nails) make up a complex set of organs that serves several functions, mostly protective. Together, these organs form the **integumentary system** (in-teg″u-men′tar-e).

Key Concept section header

5.1 The skin consists of two layers: the epidermis and dermis

Learning Objective

- [ ] **List the two layers of skin and briefly describe subcutaneous tissue.**

The skin receives little respect from its inhabitants, but architecturally it is a marvel. It covers the entire body, has a surface area of 1.2 to 2.2 square meters, weighs 4 to 5 kilograms (4–5 kg = 9–11 lb), and accounts for about 7% of total body weight in the average adult. Also called the integument ("covering"), the skin multitasks. Its functions go well beyond serving as a bag for body contents. Pliable yet tough, it takes constant punishment from external agents. Without our skin, we would quickly fall prey to bacteria and perish from water and heat loss.

Varying in thickness from 1.5 to 4.0 millimeters (mm) or more in different parts of the body, the skin is composed of two distinct layers (**Figure 5.1**):

- The *epidermis* (ep″ĭ-der′mis), composed of epithelial cells, is the outermost protective shield of the body (*epi* = upon).
- The underlying *dermis,* making up the bulk of the skin, is a tough, leathery layer composed mostly of dense connective tissue.

Only the dermis is vascularized. Nutrients reach the epidermis by diffusing through the tissue fluid from blood vessels in the dermis.

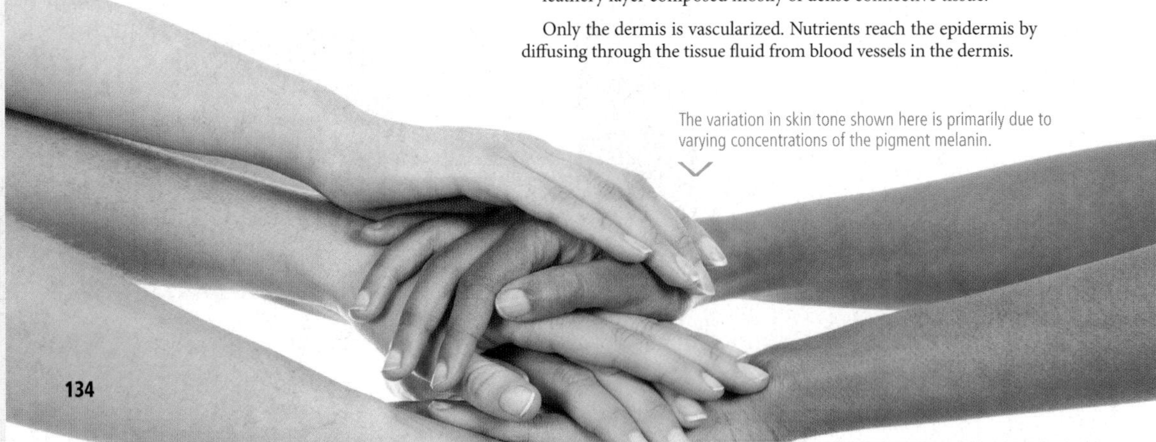

The variation in skin tone shown here is primarily due to varying concentrations of the pigment melanin.

ARE HEADED

Check Your Understanding questions end each section and allow you to assess your understanding of the concept before moving on.

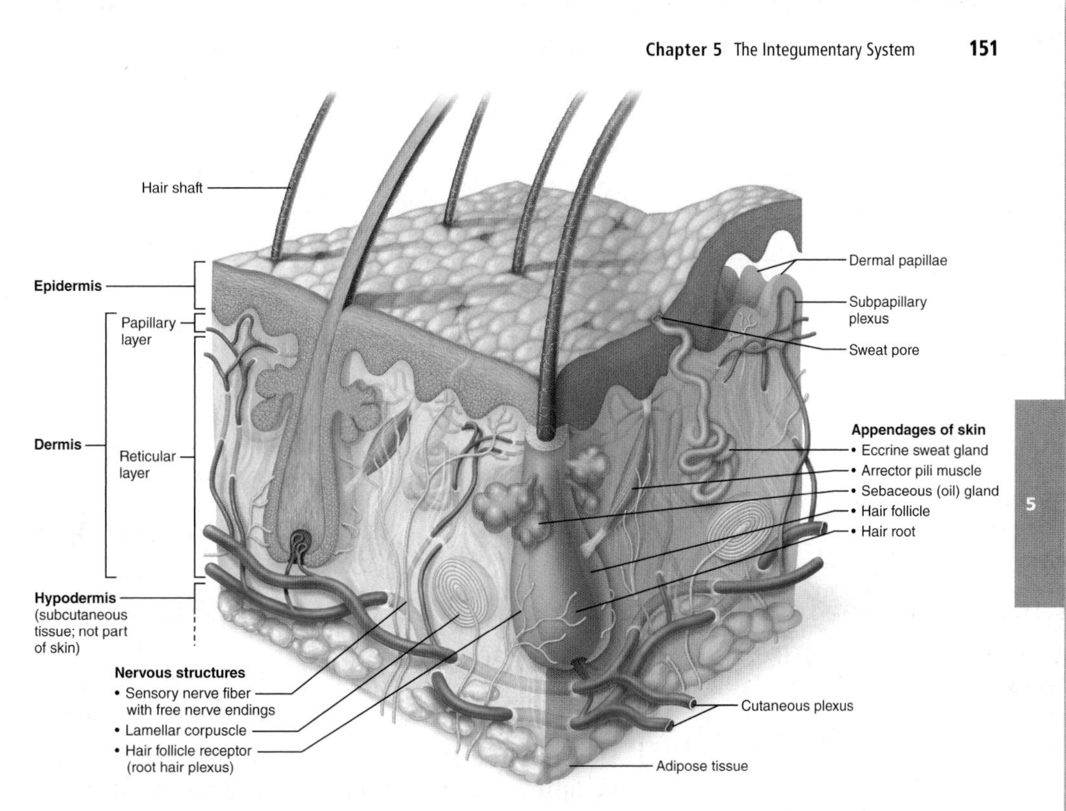

Figure 5.1 Skin structure. Three-dimensional view of the skin and underlying subcutaneous tissue. The epidermal and dermal layers have been pulled apart at the upper right corner to reveal the dermal papillae.

and accounts for about 7% of total body weight in the average adult. Also called the integument ("covering"), the skin multitasks. Its functions go well beyond serving as a bag for body contents. Pliable yet tough, it takes constant punishment from external agents. Without our skin, we would quickly fall prey to bacteria and perish from water and heat loss.

Varying in thickness from 1.5 to 4.0 millimeters (mm) or more in different parts of the body, the skin is composed of two distinct layers (**Figure 5.1**):

- The *epidermis* (ep″ĭ-der′mis), composed of epithelial cells, is the outermost protective shield of the body (*epi* = upon).
- The underlying *dermis,* making up the bulk of the skin, is a tough, leathery layer composed mostly of dense connective tissue.

Only the dermis is vascularized. Nutrients reach the epidermis by diffusing through the tissue fluid from blood vessels in the dermis.

The subcutaneous tissue just deep to the skin is known as the **hypodermis** (Figure 5.1). Strictly speaking, the hypodermis is not part of the skin, but it shares some of the skin's protective functions. The hypodermis, also called **superficial fascia** because it is superficial to the tough connective tissue wrapping (fascia) of the skeletal muscles, consists mostly of adipose tissue.

Besides storing fat, the hypodermis anchors the skin to the underlying structures (mostly to muscles), but loosely enough that the skin can slide relatively freely over those structures. Sliding skin protects us by ensuring that many blows just glance off our bodies. Because of its fatty composition, the hypodermis also acts as a shock absorber and an insulator that reduces heat loss.

☑ Check Your Understanding

1. Which layer of the skin—dermis or epidermis—is better nourished?

For answers, see Answers Appendix.

Check Your Understanding self-assessment

TOOLS TO HELP YOU

NEW! Find study tools online with references to MasteringA&P® in the book. Visit MasteringA&P for self-study modules, interactive animations, virtual lab tools, and more!

Practice art labeling
MasteringA&P®>Study Area>Chapter 10

Figure 10.26 Summary: Actions of muscles of the thigh and leg.

NEW! Easily find clinical examples to help you see how A&P concepts apply to your future career. The clinical content—including the Homeostatic Imbalance sections, clinical content modules, and the chapter-ending At the Clinic Case Study—has a unified new look and feel.

CLINICAL

12.9 Brain injuries and disorders have devastating consequences

→ Learning Objectives

- ☐ **Describe the cause (if known) and major signs and symptoms of cerebrovascular accidents, Alzheimer's disease, Parkinson's disease, and Huntington's disease.**
- ☐ **List and explain several techniques used to diagnose brain disorders.**

Brain dysfunctions are unbelievably varied and extensive. We have mentioned some of them already, but here we will focus on traumatic brain injuries, cerebrovascular accidents, and degenerative brain disorders.

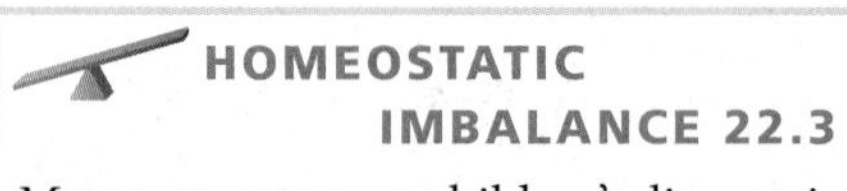

CLINICAL

HOMEOSTATIC IMBALANCE 22.3

Mumps, a common children's disease, is an inflammation of the parotid glands caused by the mumps virus (*myxovirus*), which spreads from person to person in saliva. If you check the location of the parotid glands in Figure 22.10a, you can understand why people with mumps complain that it hurts to open their mouth or chew. Other signs and symptoms include moderate fever and pain when swallowing acidic foods (pickles, grapefruit juice, etc.). Mumps in adult males carries a 25% risk of infecting the testes too, leading to sterility. ✚

ON YOUR JOURNEY

Stunning 3-D art with vibrant colors appears on every page to help you better visualize and understand key anatomical structures and their functions.

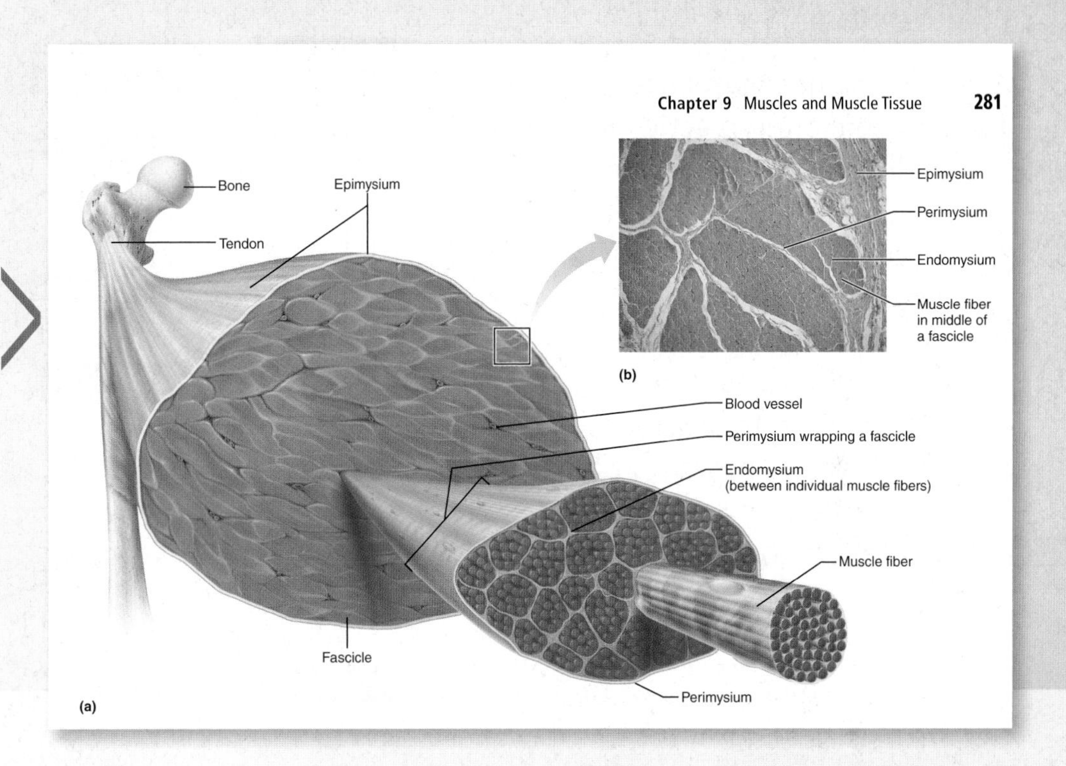

NEW! Making Connections questions in each chapter ask you to apply what you've learned across different body systems and chapters so that you build a cohesive understanding of the body.

☑ Check Your Understanding

21. What chemicals produced in the skin help provide barriers to bacteria? List at least three and explain how the chemicals are protective.

22. Which epidermal cells play a role in body immunity?

23. How is sunlight important to bone health?

24. MAKING **connections** When blood vessels in the dermis constrict or dilate to help maintain body temperature, which type of muscle tissue that you learned about (in Chapter 4) acts as the effector that causes blood vessel dilation or constriction?

For answers, see Answers Appendix.

PRACTICE MAKES PERFECT

NEW! Concept Maps are fun and challenging activities that help you solidify your understanding of a key course concept. These fully mobile activities allow you to combine key terms with linking phrases into a free-form map for topics such as protein synthesis, events in an action potential, and excitation-contraction coupling.

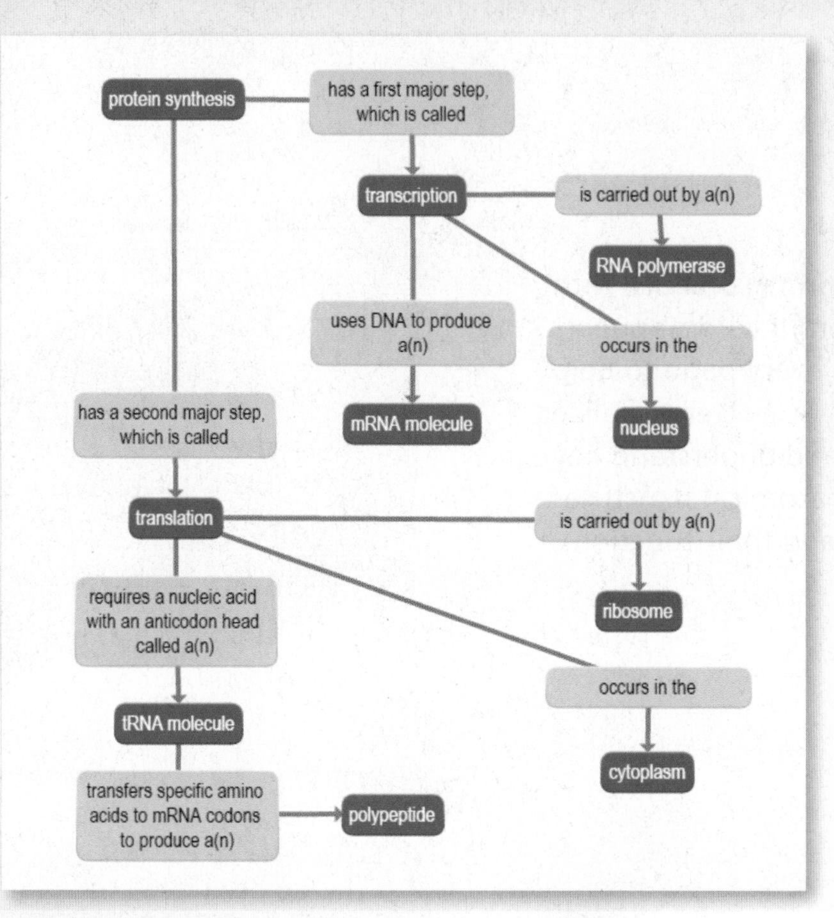

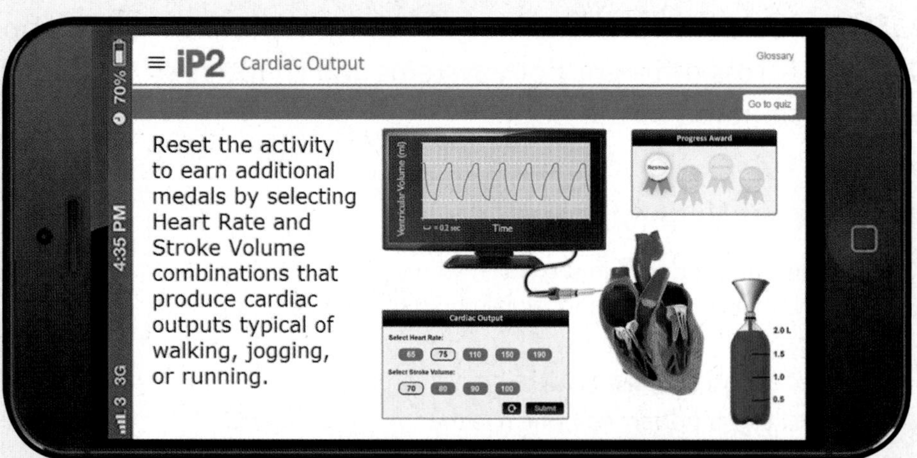

NEW! Interactive Physiology® 1.0 and 2.0 help you understand the hardest part of A&P: physiology. Fun, interactive tutorials, games, and quizzes give you additional explanations to help you grasp difficult concepts. IP 2.0 includes topics that have been updated for today's technology, such as **Resting Membrane Potential, Cardiac Output, Electrical Activity of the Heart, Factors Affecting Blood Pressure,** and **Cardiac Cycle**.

WITH MasteringA&P

A&P Flix™ are 3-D movie-quality animations with self-paced tutorials and gradable quizzes that help you master the toughest topics in A&P.

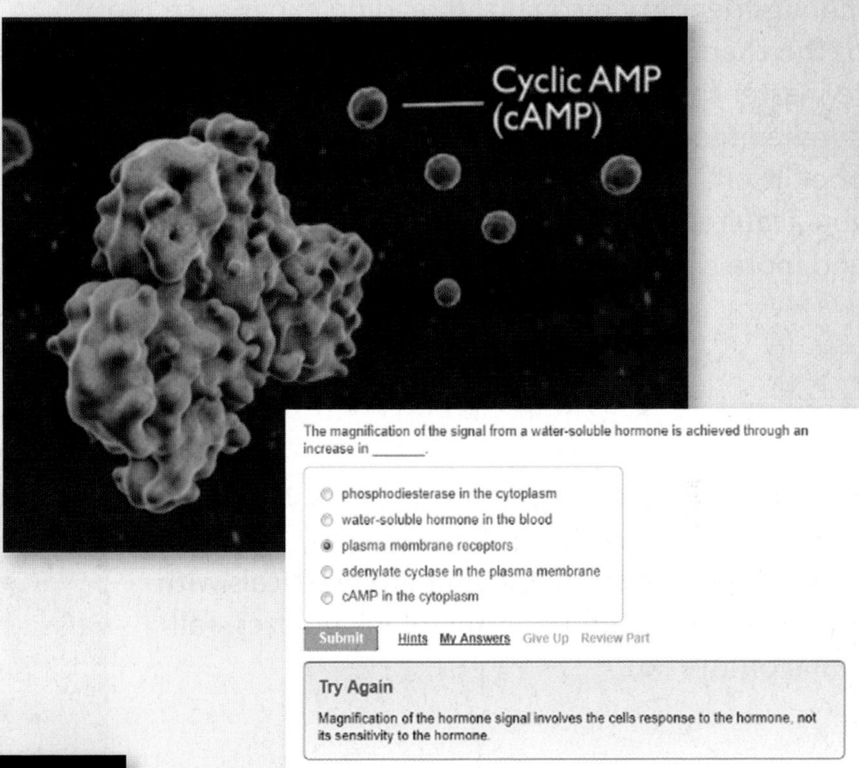

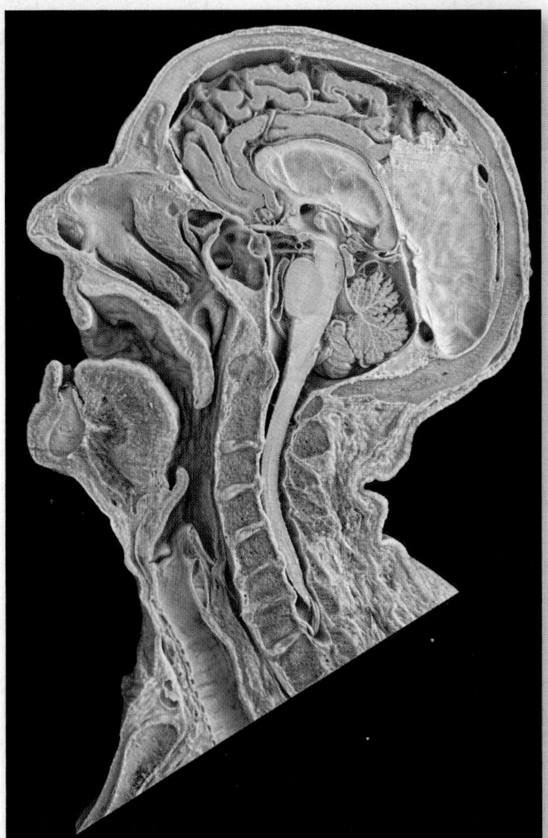

Practice Anatomy Lab™ (PAL™) 3.0 is a virtual anatomy study and practice tool that gives you 24/7 access to the most widely used lab specimens, including the human cadaver, anatomical models, histology, cat, and fetal pig. PAL 3.0 is easy to use and includes built-in audio pronunciations, rotatable bones, and simulated fill-in-the-blank lab practical exams.

STUDY ON THE GO WITH THESE MOBILE TOOLS

NEW! Dynamic Study Modules offer a mobile-friendly, personalized reading experience of the chapter content. As you answer questions to master the chapter content, you receive detailed feedback with text and art from the book itself. The Dynamic Study Modules help you acquire, retain, and recall information faster and more efficiently than ever before.

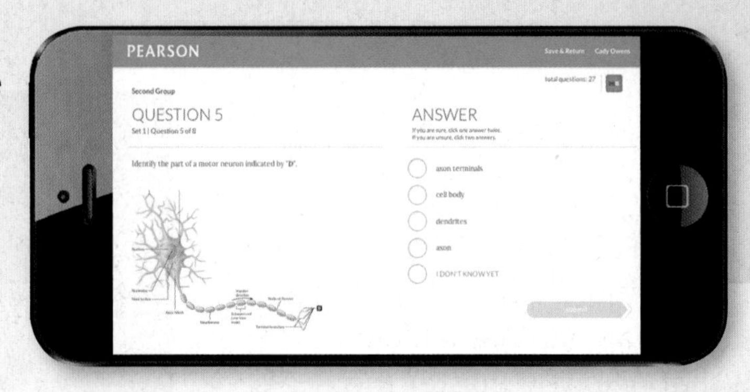

The PAL 3.0 App lets you access PAL 3.0 on your iPad or Android tablet. Enlarge images, watch animations, and study for your lab practicals with multiple-choice and fill-in-the-blank quizzes—all while on the go!

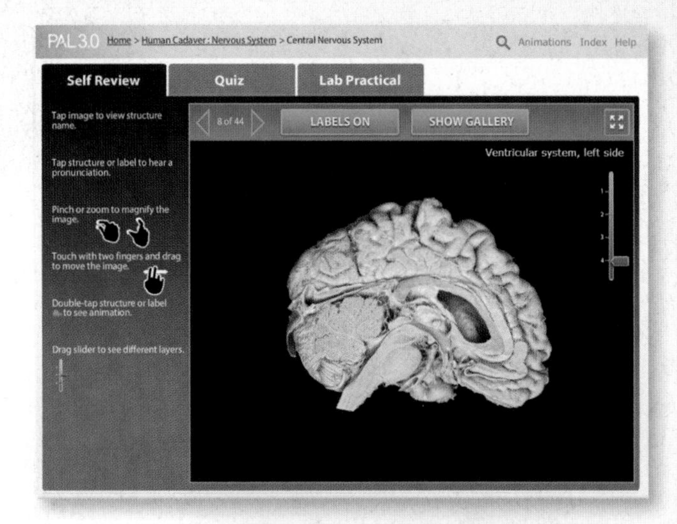

Learning Catalytics is a "bring your own device" (laptop, smartphone, or tablet) engagement, assessment, and classroom intelligence system. Use your device to respond to open-ended questions, and then discuss your answers in groups based on responses.

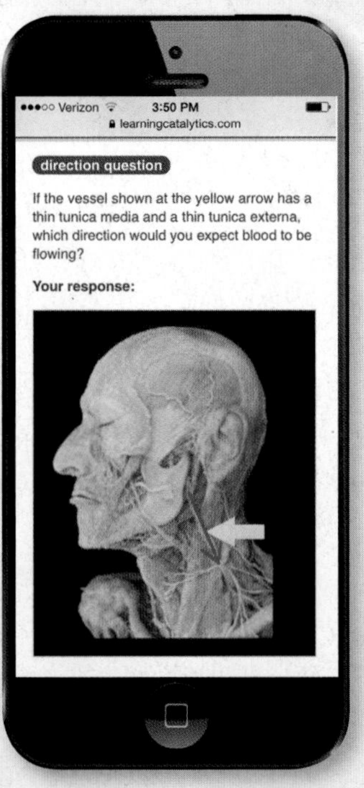

Anatomy & Physiology

Sixth Edition

Elaine N. Marieb, R.N., Ph.D.
Holyoke Community College

Katja Hoehn, M.D., Ph.D.
Mount Royal University

PEARSON

Editor-in-Chief: Serina Beauparlant
Sr. Acquisitions Editor: Brooke Suchomel
Production and Design Manager: Michele Mangelli
Program Manager: Tiffany Mok
Development Editor: Tanya Martin
Art Development Manager: Laura Southworth
Art Development Editors: Laura Southworth, Elisheva Marcus
Editorial Assistant: Nicky Montalvo
Director of Development: Barbara Yien
Program Management Team Lead: Michael Early
Project Management Team Lead: Nancy Tabor
Copyeditor: Anita Hueftle
Compositor: Cenveo® Publisher Services
Production Coordinator: David Novak
Art Coordinator: Jean Lake
Proofreader: Martha Ghent
Art Proofreader: Betsy Dietrich
Indexer: Sallie Steele
Cover and Interior Designer: Tandem Creative, Inc.
Illustrators: Imagineering STA Media Services Inc.
Text and Photo Permissions Management: Maya Gomez
Photo Researcher: Kristin Piljay
Sr. Procurement Specialist: Stacey Weinberger
Exec. Marketing Manager: Allison Rona
Sr. Anatomy & Physiology Specialist: Derek Perrigo

Cover Illustration: The plasma membrane, Imagineering STA Media Services/Precision Graphics

Acknowledgments of third party content appear on pages C-1 and C-2 , which constitutes an extension of this copyright page.

Library of Congress Cataloging-in-Publication Data

Marieb, Elaine Nicpon / Hoehn, Katja.
Anatomy & physiology / Elaine N. Marieb, R.N., Ph.D. Holyoke Community College, Katja Hoehn, M.D., Ph.D., Mount Royal University.
Anatomy and physiology
Sixth edition. / San Francisco : Pearson Education, Inc., [2017]
Includes index.
LCCN 2015035194 / ISBN 9780134156415
LCSH: Human physiology. / Human anatomy.
LCC QP34.5 .M454 2017 / DDC
612—dc23
LC record available at http://lccn.loc.gov/ 2015035194

www.pearsonhighered.com

ISBN 10: 0-13-415641-2; ISBN 13: 978-0-13-415641-5 (Student Edition)
ISBN 10: 0-13-421336-X; ISBN 13: 978-0-13-421336-1 (Instructor's Review Copy)
6 18

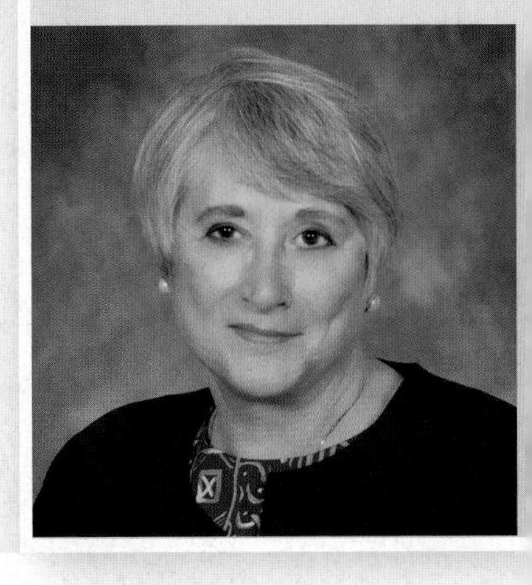

About the Authors

We dedicate this work to our students both present and past, who always inspire us to "push the envelope."

Elaine N. Marieb

For Elaine N. Marieb, taking the student's perspective into account has always been an integral part of her teaching style. Dr. Marieb began her teaching career at Springfield College, where she taught anatomy and physiology to physical education majors. She then joined the faculty of the Biological Science Division of Holyoke Community College in 1969 after receiving her Ph.D. in zoology from the University of Massachusetts at Amherst. While teaching at Holyoke Community College, where many of her students were pursuing nursing degrees, she developed a desire to better understand the relationship between the scientific study of the human body and the clinical aspects of the nursing practice. To that end, while continuing to teach full time, Dr. Marieb pursued her nursing education, which culminated in a Master of Science degree with a clinical specialization in gerontology from the University of Massachusetts. It is this experience that has informed the development of the unique perspective and accessibility for which her publications are known.

Dr. Marieb has partnered with Benjamin Cummings for over 30 years. Her first work was *Human Anatomy & Physiology Laboratory Manual* (*Cat Version*), which came out in 1981. In the years since, several other lab manual versions and study guides, as well as the softcover *Essentials of Human Anatomy & Physiology* textbook, have hit the campus bookstores. This textbook, now in its 10th edition, made its appearance in 1989 and is the latest expression of her commitment to the needs of students studying human anatomy and physiology.

Dr. Marieb has given generously to colleges both near and far to provide opportunities for students to further their education. She contributes to the New Directions, New Careers Program at Holyoke Community College by funding a staffed drop-in center and by providing several full-tuition scholarships each year for women who are returning to college after a hiatus or attending college for the first time and who would be unable to continue their studies without financial support. She funds the E. N. Marieb Science Research Awards at Mount Holyoke College, which promotes research by undergraduate science majors, and has underwritten renovation and updating of one of the biology labs in Clapp Laboratory at that college. Dr. Marieb also contributes to the University of Massachusetts at Amherst where she generously provided funding for reconstruction and instrumentation of a cutting-edge cytology research laboratory. Recognizing the severe national shortage of nursing faculty, she underwrites the Nursing Scholars of the Future Grant Program at the university.

In 1994, Dr. Marieb received the Benefactor Award from the National Council for Resource Development, American Association of Community Colleges, which recognizes her ongoing sponsorship of student scholarships, faculty teaching awards, and other academic contributions to Holyoke Community College. In May 2000, the science building at Holyoke Community College was named in her honor.

Dr. Marieb is an active member of the Human Anatomy and Physiology Society (HAPS) and the American Association for the Advancement of Science (AAAS). Additionally, while actively engaged as an author, Dr. Marieb serves as a consultant for the Benjamin Cummings *Interactive Physiology*® CD-ROM series.

When not involved in academic pursuits, Dr. Marieb is a world traveler and has vowed to visit every country on this planet. Shorter term, she serves on the scholarship committee of the Women's Resources Center and on the board of directors of several charitable institutions in Sarasota County. She is an enthusiastic supporter of the local arts and enjoys a competitive match of doubles tennis.

Katja Hoehn

Dr. Katja Hoehn is a professor in the Department of Biology at Mount Royal University in Calgary, Canada. Dr. Hoehn's first love is teaching. Her teaching excellence has been recognized by several awards during her 21 years at Mount Royal University. These include a PanCanadian Educational Technology Faculty Award (1999), a Teaching Excellence Award from the Students' Association of Mount Royal (2001), and the Mount Royal Distinguished Faculty Teaching Award (2004).

Dr. Hoehn received her M.D. (with Distinction) from the University of Saskatchewan, and her Ph.D. in Pharmacology from Dalhousie University. In 1991, the Dalhousie Medical Research Foundation presented her with the Max Forman (Jr.) Prize for excellence in medical research. During her Ph.D. and postdoctoral studies, she also pursued her passion for teaching by presenting guest lectures to first- and second-year medical students at Dalhousie University and at the University of Calgary.

Dr. Hoehn has been a contributor to several books and has written numerous research papers in Neuroscience and Pharmacology. She oversaw a recent revision of the Benjamin Cummings *Interactive Physiology*® CD-ROM series modules, and coauthored the newest module, *The Immune System*.

Following Dr. Marieb's example, Dr. Hoehn provides financial support for students in the form of a scholarship that she established in 2006 for nursing students at Mount Royal University.

Dr. Hoehn is also actively involved in the Human Anatomy and Physiology Society (HAPS) and is a member of the American Association of Anatomists. When not teaching, she likes to spend time outdoors with her husband and two sons, compete in triathlons, and play Irish flute.

Preface

As educators we continually make judgments about the enormous amount of information that besets us daily, so we can choose which morsels to pass on to our students. Yet even this refined information avalanche challenges the learning student's mind. What can we do to help students apply the concepts they are faced with in our classrooms? We believe that this new edition of our textbook addresses that question by building on the strengths of previous editions while using new, innovative ways to help students visualize connections between various concepts.

Unifying Themes

Three unifying themes that have helped to organize and set the tone of this textbook continue to be valid and are retained in this edition. These themes are:

Interrelationships of body organ systems. This theme emphasizes the fact that nearly all regulatory mechanisms have interactions with several organ systems. The respiratory system, for example, cannot carry out its role of gas exchange in the body if there are problems with the cardiovascular system that prevent the normal delivery of blood throughout the body.

Homeostasis. Homeostasis is the normal and most desirable condition of the body. Its loss is always associated with past or present pathology. This theme is not included to emphasize pathological conditions but rather to illustrate what happens in the body when homeostasis is lost.

Whenever students see a red balance beam symbol accompanied by an associated clinical topic, their understanding of how the body works to stay in balance is reinforced.

Complementarity of structure and function. This theme encourages students to understand the structure of some bodily part (cell, bone, lung, etc.) in order to understand the function of that structure. For example, muscle cells can produce movement because they are contractile cells.

Changes Past and Present

Many of the changes made to the 5th edition have been retained and are reinforced in this 6th edition.

- There are more step-by-step blue texts accompanying certain pieces of art (blue text refers to the instructor's voice).
- The many clinical features of the book have been clearly identified to help students understand why this material is important.
- The "Check Your Understanding" questions at the end of each module reinforce understanding throughout the chapter.
- We have improved a number of our Focus Figures. (Focus Figures are illustrations that use a "big picture" layout and dramatic art to walk the student through difficult processes in a step-by-step way.)
- MasteringA&P continues to provide text-integrated media of many types to aid learning. These include *Interactive Physiology* (IP) tutorials that help students to grasp difficult concepts, *A&PFlix* animations that help students visualize tough A&P topics, and the PAL (Practice Anatomy Lab) collection of virtual anatomy study and practice tools focusing on the most widely used lab specimens. These are by no means all of the helpful tools to which students have access. It's just a smattering.

New to the Sixth Edition

So, besides these tools, what is really new to this textbook this time around? Each chapter's first page has a "Why This Matters" icon and QR code that links to a video of a health-care professional telling us why the chapter's content is important for his or her work.

Other new features include (1) declarative headers at the beginning of each chapter module so that the student can quickly grasp the "big idea" for that module, (2) more modularization (chunking) of the text so that students can tackle manageable pieces of information as they read through the material, (3) increased readability of the text as a result of more bulleted lists and shorter paragraphs, (4) more summary tables to help students connect information, (5) improvements to many of the figures so that they teach even more effectively, and (6) "Making Connections" questions in each chapter that ask students to incorporate related information from earlier chapters or earlier modules in the same chapter, helping students to see the forest, not just the trees, as they study.

Chapter-by-Chapter Changes

Chapter 1 The Human Body: An Orientation

- Updated Figure 1.8 for better teaching effectiveness.

Chapter 2 Chemistry Comes Alive

- Updated Figure 2.18 for better teaching effectiveness.

Chapter 3 Cells: The Living Units

- Updated statistics on Tay-Sachs disease.
- Updated information about riboswitches and added information about small interfering RNAs (siRNAs).
- Added summary text to Figure 3.3 for better pedagogy.
- Updated Focus Figure 3.4.

Chapter 4 Tissue: The Living Fabric

- New photos of simple columnar epithelium, pseudostratified ciliated columnar epithelium, cardiac muscle tissue, and smooth muscle tissue (Figures 4.3c, d and 4.9b, c).

Chapter 5 The Integumentary System

- Added information about the role of tight junctions in skin.
- New photo of stretch marks (Figure 5.5).
- New photo of cradle cap (seborrhea) in a newborn (Figure 5.9).
- New photo of malignant melanoma (Figure 5.10).

Chapter 6 Bones and Skeletal Tissues

- Revised Figure 6.9 for improved teaching effectiveness.
- New X rays showing Paget's disease and normal bone (Figure 6.16).

Chapter 7 The Skeleton

- Illustrated the skull bone table to facilitate student learning (Table 7.1).
- Added three new Check Your Understanding figure questions asking students to make anatomical identifications.
- New photos of humerus, radius, and ulna (Figures 7.28 and 7.29).

Chapter 8 Joints

- Updated statistics for osteoarthritis.
- Updated figure showing movements allowed by synovial joints (Figure 8.5).
- New photos of special body movements (Figure 8.6).

Chapter 9 Muscles and Muscle Tissue

- Updated Table 9.2 information on sizes of skeletal muscle fiber types in humans.

Chapter 10 The Muscular System

- New photos showing surface anatomy of muscles used in seven facial expressions (Figure 10.7).

Chapter 11 Fundamentals of the Nervous System and Nervous Tissue

- Added overview figure of nervous system (Figure 11.2).
- Improved Focus Figure 11.2 (*Action Potential*) for better student understanding.
- New image of a motor neuron based on a computerized 3-D reconstruction of serial sections.
- Converted Figure 11.17 to tabular head style to teach better.

Chapter 12 The Central Nervous System

- Updated mechanisms of Alzheimer's disease to include propagation of misfolded proteins.
- Updated information about gender differences in the brain.
- Streamlined discussion of sleep, memory, and stroke.
- New figure to show distribution of gray and white matter (Figure 12.3).
- Functional neuroimaging of the cerebral cortex (Figure 12.6).
- Improved reticular formation figure with "author's voice" blue text (Figure 12.18).
- New figure showing decreased brain activity in Alzheimer's (Figure 12.26).

Chapter 13 The Peripheral Nervous System and Reflex Activity

- Updated description of cytostructure of human cochlear hair cells (they have no kinocilia).
- New data on the number of different odors that humans can detect.
- Reorganized discussion of sound transmission to the inner ear. New numbered text improves text-art correlation.
- New figure teaches the function of the basilar membrane (Figure 13.26).
- New figure on how the hairs on the cochlear hair cells transduce sound (Figure 13.27).
- New figure shows the structure and function of the macula (Figure 13.28).
- Updated and expanded description of axon regeneration (in Figure 13.31).

Chapter 14 The Autonomic Nervous System

- Improved teaching effectiveness of Figure 14.3 (differences in the parasympathetic and sympathetic nervous systems).
- New summary table for autonomic ganglia (Table 14.2).

Chapter 15 The Endocrine System

- New information on actions of vitamin D and location of its receptors.
- New summary table showing differences between water-soluble and lipid-soluble hormones (Table 15.1).

- New summary flowchart shows the signs and symptoms of diabetes mellitus (Figure 15.19).

Chapter 16 Blood

- Improved teaching effectiveness of Figure 16.14 (intrinsic and extrinsic clotting factors).

Chapter 17 The Cardiovascular System: The Heart

- Rearranged topics in this chapter for better flow.
- New section and summary table (Table 17.1) teach key differences between skeletal muscle and cardiac muscle.
- New Making Connections figure question (students compare three action potentials).
- Rearranged material so that all electrical events are presented in one module.
- Added tabular headers, a photo, and bullets to more effectively teach ECG abnormalities (Figure 17.18).
- Streamlined figure showing effects of norepinephrine on heart contractility (Figure 17.22).

Chapter 18 The Cardiovascular System: Blood Vessels

- New information about pericytes (now known to be stem cells and generators of scar tissue in the CNS).
- New information that the fenestrations in fenestrated capillaries are dynamic structures.
- Rearranged topics in the physiology section of this chapter for better flow.
- New micrograph of artery and vein (Figure 18.2).
- Revised Figure 18.3 (the structure of different types of capillaries), putting all of the information in one place.
- New figure summarizes the major factors determining mean arterial pressure to give a "big picture" view (Figure 18.9).
- New figure illustrating active hyperemia (Figure 18.15).
- Updated Focus Figure 18.1 (*Bulk Flow across Capillary Walls*).
- New Homeostatic Imbalance feature on edema relates it directly to the preceding Focus Figure 18.1) and incorporates information previously found in Chapter 25.
- New photos of pitting edema (Figure 18.18).

Chapter 19 The Lymphatic System and Lymphoid Organs and Tissues

- Updated statistics on survival of non-Hodgkin's lymphoma patients.
- Updated figure to improve teaching of primary and secondary lymphoid organs (Figure 19.4).

Chapter 20 The Immune System: Innate and Adaptive Body Defenses

- Updated information on aging and the immune system, particularly with respect to chronic inflammation.
- Added a new term, pattern recognition receptors, to help describe how our innate defenses recognize pathogens.
- Provided new research results updating the number of genes in the human genome to about 20,000.

Chapter 21 The Respiratory System

- New Check Your Understanding question with graphs reinforces concepts learned in Focus Figure 21.1 (*The Oxygen-Hemoglobin Dissociation Curve*).
- New figure illustrating pneumothorax (Figure 21.14).

Chapter 22 The Digestive System

- Updated information about the treatment of peptic ulcers.
- Updated information about the types and locations of epithelial cells of the small intestine.
- New information about roles of our intestinal flora.
- Updated hepatitis C treatment to include the new FDA-approved drug sofosbuvir.
- Added discussion of non-alcoholic fatty liver disease.
- New information about fecal transplants to treat antibiotic-associated diarrhea.
- Updated figure that compares and contrasts peristalsis and segmentation (Figure 22.3) for improved teaching effectiveness.
- Updated Figure 22.4 explaining the relationship between the peritoneum and the abdominal organs to improve teaching effectiveness.
- Enteric nervous system section rewritten and rearranged with new figure (Figure 22.6).
- Improved teaching effectiveness of Figure 22.14 (the steps of deglutition).
- Streamlined Figure 22.19 to enhance teaching of regulation of gastric secretion.
- Updated Figure 22.20 (the mechanism of HCl secretion by parietal cells) for improved teaching effectiveness.
- Improved the text flow by moving discussion of the liver, gallbladder, and pancreas before the small intestine.
- Improved teaching effectiveness of Figure 22.28 (mechanism promoting secretion and release of bile and pancreatic juice).
- Updated and revised sections about motility of the small and large intestines.
- Rearranged text to discuss digestion and absorption together for each nutrient. The figures for digestion and absorption of carbohydrates (Figure 22.35) and proteins (Figure 22.36) now parallel each other and appear together for easy comparison.
- Rearranged and rewrote lipid digestion and absorption text and updated Figure 22.37.

Chapter 23 Nutrition, Metabolism, and Energy Balance

- Chapter title changed from Nutrition, Metabolism, and Body Temperature Regulation in order to emphasize the concept of energy balance.
- Updated shape and mechanism of action of ATP synthase to reflect new research findings.
- Updated hypothalamic control of food intake per new research findings.
- Revised Figure 23.4 to enhance the ability of students to compare and contrast the mechanisms of phosphorylation that convert ADP to ATP.
- Revised figure describing ATP synthase structure and function (Figure 23.10).
- Revised Figure 23.13 to help students compare and contrast glycogenesis and glycogenolysis (Figure 23.12).
- Three new figures help students grasp the terms for key pathways in carbohydrate, protein, and fat metabolism (Figures 23.12, 23.14, and 23.18).

Chapter 24 The Urinary System

- New cadaver photo of urinary tract organs (Figure 24.2).
- New Check Your Understanding question for nephron labeling.
- Improved Focus Figure 24.1 (*Medullary Osmotic Gradient*) for better teaching effectiveness.
- Added new illustrations to improve teaching effectiveness of Figure 24.19 (the effects of ADH on the nephron).

Chapter 25 Fluid, Electrolyte, and Acid-Base Balance

- New Check Your Understanding figure question requires students to integrate information.

Chapter 26 The Reproductive System

- Updated screening recommendations for prostate cancer, as well as updated information on detection and treatment.
- Updated screening guidelines for cervical cancer.
- Updated breast cancer statistics.
- New Check Your Understanding figure labeling question.
- New figure teaches independent assortment (Figure 26.8).
- New photo of female pelvic organs (Figure 26.15c).
- New photos of mammograms showing normal and cancerous breast tissues (Figure 26.19).
- Revised Figure 26.23 to reflect recent research about follicular development in humans.
- Revised section describing the stages of follicle development to facilitate student learning and to incorporate recent research.

Appendices

- Added a table of the genetic code (Appendix B).

Acknowledgments

Each time we put this textbook to bed, we promise ourselves that the next time will be easier and will require less of our time. Now hear this! This is its 6th edition (and 30 years more or less) and fulfillment of this promise has yet to materialize. How could there be so much going on in physiology research and so many new medical findings? Winnowing through these findings to decide on the updates to include in this edition has demanded much of our attention. Many people at Pearson have labored with us to produce another fine text. Let's see if we can properly thank them.

As Katja and I worked on the first draft of the manuscript, Tanya Martin (our text Development Editor) worked tirelessly to improve the readability of the text, all the while trying to determine which topics could be shortened or even deleted in the 6th edition. After we had perused and acted on some of Tanya's suggestions, we forwarded the manuscript to Michele Mangelli who oversees everything having to do with getting a clean manuscript to production. Michele reviewed the entire revised manuscript. Nothing escaped her attention as she worked to catch every problem.

At the same time the text was in revision, the art program was going through a similar process. Laura Southworth, our superb Art Development Editor (aided briefly by Elisheva Marcus), worked tirelessly to make our Focus Figures and other art even better. Needing a handshake and a heartfelt "thank you" in the process are Kristin Piljay (Photo Researcher) and Jean Lake, who handled the administrative aspects of the art program. This team ensured that the artists at Imagineering had all the information they needed to produce beautiful final art products.

As the manuscript made the transition from Editorial to Production, Michele Mangelli, the Production and Design Manager, made her appearance known. The head honcho and skilled handler of all aspects of production, everyone answered to her from this point on. In all previous editions, the manuscript would simply go directly into production once the writing and editing phases were over, but our new modular design required extra steps to make the art-text correlation a reality—the electronic page layout. Working closely with Katja and her husband Larry Haynes, Michele's small but powerful team "yanked" the new design to attention, fashioning two-page spreads, each covering one or more topics with its supporting art or table. This was our Holy Grail for this edition and the ideal student coaching device. They made it look easy (which it was not). Thank you Katja, Larry, and Michele—you are the ideal electronic page layout team. This was one time I felt fortunate to be the elder author.

The remaining people who helped with Production include David Novak (our conscientious Production Supervisor), Martha Ghent (Proofreader), Betsy Dietrich (Art Proofreader), Sallie Steele (Indexer), Cynthia Mutheardy (Project Manager at Imagineering), and Tim Frelick (Compositor). Copyeditor Anita Hueftle (formerly Anita Wagner) is the unofficial third author of our book. We are absolutely convinced that she memorizes the entire text. She verified the spelling of new terms, checked the generic and popular names of drugs, confirmed our grammar, and is the person most responsible for the book's consistency and lack of typographical errors. We are grateful to Izak Paul for meticulously reading each chapter to find any remaining errors, and to Yvo Riezebos for his stunning design work on the cover, chapter opening pages, and the text.

Finally—what can we say about Brooke Suchomel, our Acquisitions Editor? She loved playing with the modular design and the chapter road maps and advising on Focus Figures, but most of her time was spent out in the field talking to professors, demonstrating the book's changes and benefits. She spent weeks on the road, smiling all the time—no easy task. Finally, we are fortunate to have the ongoing support and friendship of Serina Beauparlant, our Editor-in-Chief.

Other members of our team with whom we have less contact but who are nonetheless vital are: Barbara Yien (Director of Development), Michael Early (Program Manager Team Lead), Nancy Tabor (Project Manager Team Lead), Stacey Weinberger (our Senior Manufacturing Buyer), Allison Rona (our top-notch Executive Marketing Manager), and Derek Perrigo (Senior Anatomy & Physiology Specialist). We appreciate the hard work of our media production team headed by Laura Tomassi, Aimee Pavy, and Lauren Hill and also wish to thank Eric Leaver.

Kudos to our entire team. We feel we have once again prepared a superb textbook. We hope you agree.

There are many people who reviewed parts of this text—both professors and students, either individually or in focus groups, and we would like to thank them. Input from the following reviewers has contributed to the continued excellence and accuracy of this text:

Matthew Abbott, *Des Moines Area Community College*
Lynne Anderson, *Meridian Community College*
Martin W. Asobayire, *Essex Community College*
Yvonne Baptiste-Szymanski, *Niagara County Community College*
Claudia Barreto, *University of New Mexico–Valencia*
Diana Bourke, *Community College of Allegheny County*
Sherry Bowen, *Indian River State College*
Beth Braun, *Truman College*
C. Steven Cahill, *West Kentucky Community and Technical College*
Brandi Childress, *Georgia Perimeter College*
William Michael Clark, *Lone Star College–Kingwood*
Teresa Cowan, *Baker College of Auburn Hills*
Donna Crapanzano, *Stony Brook University*
Maurice M. Culver, *Florida State College at Jacksonville*
Smruti A. Desai, *Lone Star College–CyFair*
Karen Dunbar Kareiva, *Ivy Tech Community College*
Elyce Ervin, *University of Toledo*
Martha Eshleman, *Pulaski Technical College*
Juanita A. Forrester, *Chattahoochee Technical College*
Reza Forough, *Bellevue College*
Dean Furbish, *Wake Technical Community College*
Emily Getty, *Ivy Tech Community College*
Amy Giesecke, *Chattahoochee Technical College*
Abigail Goosie, *Walters State Community College*
Mary Beth Hanlin, *Des Moines Area Community College*
Heidi Hawkins, *College of Southern Idaho*
Martie Heath-Sinclair, *Hawkeye Community College*
Nora Hebert, *Red Rocks Community College*
Nadia Hedhli, *Hudson County Community College*
D.J. Hennager, *Kirkwood Community College*
Shannon K. Hill, *Temple College*
Mark Hollier, *Georgia Perimeter College*
H. Rodney Holmes, *Waubonsee Community College*
Mark J. Hubley, *Prince George's Community College*
Jason Hunt, *Brigham Young University–Idaho*
William Karkow, *University of Dubuque*
Suzanne Keller, *Indian Hills Community College*
Marta Klesath, *North Carolina State University*
Nelson H. Kraus, *University of Indianapolis*
Steven Lewis, *Metropolitan Community College–Penn Valley*
Jerri K. Lindsey, *Tarrant County College–Northeast*
Chelsea Loafman, *Central Texas College*
Paul Luyster, *Tarrant County College–South*
Abdallah M. Matari, *Hudson County Community College*
Bhavya Mathur, *Chattahoochee Technical College*
Tiffany Beth McFalls-Smith, *Elizabethtown Community and Technical College*
Todd Miller, *Hunter College of CUNY*
Regina Munro, *Chandler-Gilbert Community College*
Necia Nicholas, *Calhoun Community College*
Ellen Ott-Reeves, *Blinn College–Bryan*
Jessica Petersen, *Pensacola State College*
Sarah A. Pugh, *Shelton State Community College*
Rolando J. Ramirez, *The University of Akron*
Terrence J. Ravine, *University of South Alabama*
Laura H. Ritt, *Burlington County College*
Susan Rohde, *Triton College*
Brian Sailer, *Central New Mexico Community College*
Mark Schmidt, *Clark State Community College*
Amy Skibiel, *Auburn University*
Lori Smith, *American River College*
Ashley Spring-Beerensson, *Eastern Florida State College*
Justin R. St. Juliana, *Ivy Tech Community College*
Laura Steele, *Ivy Tech Community College*
Shirley A. Whitescarver, *Bluegrass Community and Technical College*
Patricia Wilhelm, *Johnson and Wales University*
Luann Wilkinson, *Marion Technical College*
Peggie Williamson, *Central Texas College*
MaryJo A. Witz, *Monroe Community College*
James Robert Yount, *Brevard Community College*

Interactive Physiology 2.0 Reviewers

Lynne Anderson, *Meridian Community College*
J. Gordon Betts, *Tyler Junior College*
Mike Brady, *Columbia Basin College*
Betsy Brantley, *Valencia College*
Tamyra Carmona, *Cosumnes River College*
Alexander G. Cheroske, *Mesa Community College at Red Mountain*
Sondra Dubowsky, *McLennan Community College*
Paul Emerick, *Monroe Community College*
Brian D. Feige, *Mott Community College*
John E. Fishback, *Ozarks Technical Community College*
Aaron Fried, *Mohawk Valley Community College*
Jane E. Gavin, *University of South Dakota*
Gary Glaser, *Genesee Community College*
Mary E. Hanlin, *Des Moines Area Community College*
Mark Hubley, *Prince George's Community College*
William Karkow, *University of Dubuque*
Michael Kielb, *Eastern Michigan University*
Paul Luyster, *Tarrant County College–South*

Louise Millis, *North Hennepin Community College*
Justin Moore, *American River College*
Maria Oehler, *Florida State College at Jacksonville*
Fernando Prince, *Laredo Community College*
Terrence J. Ravine, *University of South Alabama*
Mark Schmidt, *Clark State Community College*
Cindy Stanfield, *University of South Alabama*
Laura Steele, *Ivy Tech Community College*
George A. Steer, *Jefferson College of Health Sciences*
Shirley A. Whitescarver, *Bluegrass Community and Technical College*

Harvey Howell, my beloved husband and helpmate, died in August of 2013. He is sorely missed.

Katja would also like to acknowledge the support of her colleagues at Mount Royal University (Trevor Day, Sarah Hewitt, Tracy O'Connor, Izak Paul, Michael Pollock, Lorraine Royal, Karen Sheedy, Kartika Tjandra, and Margot Williams) and of Ruth Pickett-Seltner (Chair), Tom MacAlister (Associate Dean), and Jeffrey Goldberg (Dean). Thanks also to Katja's husband, Dr. Lawrence Haynes, who as a fellow physiologist has provided invaluable assistance to her during the course of the revision. She also thanks her sons, Eric and Stefan Haynes, who are an inspiration and a joy.

We would really appreciate hearing from you concerning your opinion—suggestions and constructive criticisms—of this text. It is this type of feedback that will help us in the next revision, and underlies the continued improvement of this text.

Elaine N. Marieb

Katja Hoehn

Elaine N. Marieb and Katja Hoehn
Anatomy and Physiology
Pearson Education
1301 Sansome Street
San Francisco, CA 94111

Contents

Although we use the reference values and common directional and regional terms to refer to all human bodies, you know from observing the faces and body shapes of people around you that we humans differ in our external anatomy. The same kind of variability holds for internal organs as well. In one person, for example, a nerve or blood vessel may be somewhat out of place, or a small muscle may be missing. Nonetheless, well over 90% of all structures present in any human body match the textbook descriptions. We seldom see extreme anatomical variations because they are incompatible with life.

Topics of Anatomy

Anatomy is a broad field with many subdivisions, each providing enough information to be a course in itself. **Gross**, or **macroscopic, anatomy** is the study of large body structures visible to the naked eye, such as the heart, lungs, and kidneys. Indeed, the term *anatomy* (from Greek, meaning "to cut apart") relates most closely to gross anatomy because in such studies preserved animals or their organs are dissected (cut up) to be examined.

Gross anatomy can be approached in different ways. In **regional anatomy**, all the structures (muscles, bones, blood vessels, nerves, etc.) in a particular region of the body, such as the abdomen or leg, are examined at the same time.

In **systemic anatomy** (sis-tem′ik),* body structure is studied system by system. For example, when studying the cardiovascular system, you would examine the heart and the blood vessels of the entire body.

Another subdivision of gross anatomy is **surface anatomy**, the study of internal structures as they relate to the overlying skin surface. You use surface anatomy when you identify the bulging muscles beneath a bodybuilder's skin, and clinicians use it to locate appropriate blood vessels in which to feel pulses and draw blood.

Microscopic anatomy deals with structures too small to be seen with the naked eye. For most such studies, exceedingly thin slices of body tissues are stained and mounted on glass slides to be examined under the microscope. Subdivisions of microscopic anatomy include **cytology** (si-tol′o-je), which considers the cells of the body, and **histology** (his-tol′o-je), the study of tissues.

Developmental anatomy traces structural changes that occur throughout the life span. **Embryology** (em″bre-ol′o-je), a subdivision of developmental anatomy, concerns developmental changes that occur before birth.

Some highly specialized branches of anatomy are used primarily for medical diagnosis and scientific research. For example, *pathological anatomy* studies structural changes caused by disease. *Radiographic anatomy* studies internal structures as visualized by X-ray images or specialized scanning procedures.

One essential tool for studying anatomy is a mastery of anatomical terminology. Others are observation, manipulation, and, in a living person, *palpation* (feeling organs with your hands) and *auscultation* (listening to organ sounds with a stethoscope). A simple example illustrates how some of these tools work together in an anatomical study.

Let's assume that your topic is freely movable joints of the body. In the laboratory, you will be able to *observe* an animal joint, noting how its parts fit together. You can work the joint (*manipulate* it) to determine its range of motion. Using *anatomical terminology*, you can name its parts and describe how they are related so that other students (and your instructor) will have no trouble understanding you. The list of word roots (at the back of the book) and the glossary will help you with this special vocabulary.

Although you will make most of your observations with the naked eye or with the help of a microscope, medical technology has developed a number of sophisticated tools that can peer into the body without disrupting it.

Topics of Physiology

Like anatomy, physiology has many subdivisions. Most of them consider the operation of specific organ systems. For example, **renal physiology** concerns kidney function and urine production. **Neurophysiology** explains the workings of the nervous system. **Cardiovascular physiology** examines the operation of the heart and blood vessels. While anatomy provides us with a static image of the body's architecture, physiology reveals the body's dynamic and animated workings.

Physiology often focuses on events at the cellular or molecular level. This is because the body's abilities depend on those of its individual cells, and cells' abilities ultimately depend on the chemical reactions that go on within them. Physiology also rests on principles of physics, which help to explain electrical currents, blood pressure, and the way muscles use bones to cause body movements, among other things. We present basic chemical and physical principles in Chapter 2 and throughout the book as needed to explain physiological topics.

Complementarity of Structure and Function

Although it is possible to study anatomy and physiology individually, they are really inseparable because function always reflects structure. That is, what a structure can do depends on its specific form. This key concept is called the **principle of complementarity of structure and function.**

For example, bones can support and protect body organs because they contain hard mineral deposits. Blood flows in one direction through the heart because the heart has valves that prevent backflow. Throughout this book, we accompany a description of a structure's anatomy with an explanation of its function, and we emphasize structural characteristics contributing to that function.

☑ Check Your Understanding

1. In what way does physiology depend on anatomy?
2. Would you be studying anatomy or physiology if you investigated how muscles shorten? If you explored the location of the lungs in the body?

For answers, see Answers Appendix.

*For the pronunciation guide rules, see the first page of the glossary in the back of the book.

1.2 The body's organization ranges from atoms to the entire organism

→ Learning Objectives

- ☐ **Name the different levels of structural organization that make up the human body, and explain their relationships.**
- ☐ **List the 11 organ systems of the body, identify their components, and briefly explain the major function(s) of each system.**

The human body has many levels of structural organization (**Figure 1.1**). The simplest level of the structural hierarchy is the **chemical level**, which we study in Chapter 2. At this level, *atoms*, tiny building blocks of matter, combine to form *molecules* such as water and proteins. Molecules, in turn, associate in specific ways to form *organelles*, basic components of the microscopic cells. *Cells* are the smallest units of living things. We examine the **cellular level** in Chapter 3. All cells have some common functions, but individual cells vary widely in size and shape, reflecting their unique functions in the body.

The simplest living creatures are single cells, but in complex organisms such as human beings, the hierarchy continues on to the **tissue level**. *Tissues* are groups of similar cells that have a common function. The four basic tissue types in the human body are epithelium, muscle, connective tissue, and nervous tissue.

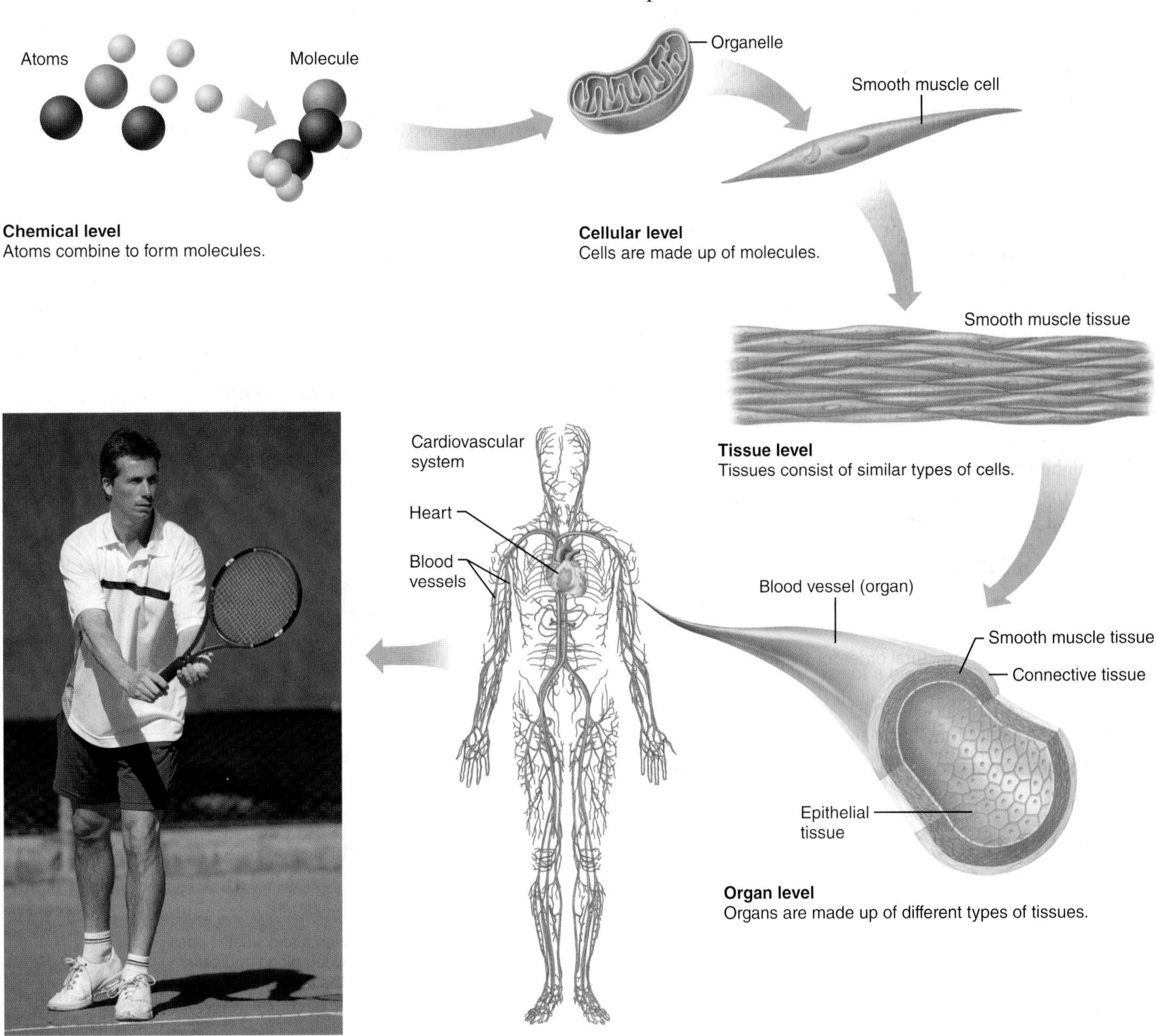

Figure 1.1 Levels of structural organization. Components of the cardiovascular system are used to illustrate the levels of structural organization in a human being.

Carbohydrates are the major energy fuel for body cells. Proteins, and to a lesser extent fats, are essential for building cell structures. Fats also provide a reserve of energy-rich fuel. Selected minerals and vitamins are required for the chemical reactions that go on in cells and for oxygen transport in the blood. The mineral calcium helps to make bones hard and is required for blood clotting.

Oxygen

All the nutrients in the world are useless unless **oxygen** is also available. Because the chemical reactions that release energy from foods are *oxidative* reactions that require oxygen, human cells can survive for only a few minutes without oxygen. Approximately 20% of the air we breathe is oxygen. The cooperative efforts of the respiratory and cardiovascular systems make oxygen available to the blood and body cells.

Water

Water accounts for 50–60% of our body weight and is the single most abundant chemical substance in the body. It provides the watery environment necessary for chemical reactions and the fluid base for body secretions and excretions. We obtain water chiefly from ingested foods or liquids. We lose it from the body by evaporation from the lungs and skin and in body excretions.

Normal Body Temperature

If chemical reactions are to continue at life-sustaining rates, **normal body temperature** must be maintained. As body temperature drops below 37°C (98.6°F), metabolic reactions become slower and slower, and finally stop. When body temperature is too high, chemical reactions occur at a frantic pace and body proteins lose their characteristic shape and stop functioning. At either extreme, death occurs. The activity of the muscular system generates most body heat.

Appropriate Atmospheric Pressure

Atmospheric pressure is the force that air exerts on the surface of the body. Breathing and gas exchange in the lungs depend on *appropriate* atmospheric pressure. At high altitudes, where atmospheric pressure is lower and the air is thin, gas exchange may be inadequate to support cellular metabolism.

• • •

The mere presence of these survival factors is not sufficient to sustain life. They must be present in the proper amounts. Too much and too little may be equally harmful. For example, oxygen is essential, but excessive amounts are toxic to body cells. Similarly, the food we eat must be of high quality and in proper amounts. Otherwise, nutritional disease, obesity, or starvation is likely. Also, while the needs listed here are the most crucial, they do not even begin to encompass all of the body's needs. For example, we can live without gravity if we must, but the quality of life suffers.

☑ Check Your Understanding

6. What separates living beings from nonliving objects?
7. What name is given to all chemical reactions that occur within body cells?
8. Why is it necessary to be in a pressurized cabin when flying at 30,000 feet?

For answers, see Answers Appendix.

1.4 Homeostasis is maintained by negative feedback

→ Learning Objectives

- ☐ **Define homeostasis and explain its significance.**
- ☐ **Describe how negative and positive feedback maintain body homeostasis.**
- ☐ **Describe the relationship between homeostatic imbalance and disease.**

When you think about the fact that your body contains trillions of cells in nearly constant activity, and that remarkably little usually goes wrong with it, you begin to appreciate what a marvelous machine your body is. Walter Cannon, an American physiologist of the early twentieth century, spoke of the "wisdom of the body," and he coined the word **homeostasis** (ho″me-o-sta′sis) to describe its ability to maintain relatively stable internal conditions even though the outside world changes continuously.

Although the literal translation of homeostasis is "unchanging," the term does not really mean a static, or unchanging, state. Rather, it indicates a *dynamic* state of equilibrium, or a balance, in which internal conditions vary, but always within relatively narrow limits. In general, the body is in homeostasis when its needs are adequately met and it is functioning smoothly.

Maintaining homeostasis is more complicated than it appears at first glance. Virtually every organ system plays a role in maintaining the constancy of the internal environment. Adequate blood levels of vital nutrients must be continuously present, and heart activity and blood pressure must be constantly monitored and adjusted so that the blood is propelled to all body tissues. Also, wastes must not be allowed to accumulate, and body temperature must be precisely controlled. A wide variety of chemical, thermal, and neural factors act and interact in complex ways—sometimes helping and sometimes hindering the body as it works to maintain its "steady rudder."

Homeostatic Control

Communication within the body is essential for homeostasis. Communication is accomplished chiefly by the nervous and endocrine systems, which use neural electrical impulses or bloodborne hormones, respectively, as information carriers. We cover the details of how these two great regulating systems operate in later chapters, but here we explain the basic characteristics of control systems that promote homeostasis.

Regardless of the factor or event being regulated—the **variable**—all homeostatic control mechanisms are processes involving at least three components that work together (**Figure 1.4**). The first component, the **receptor**, is some type of sensor that monitors the environment and responds to changes, called *stimuli*, by sending information (input) to the second component, the *control center*. Input flows from the receptor to the control center along the *afferent pathway*.

The **control center** determines the *set point*, which is the level or range at which a variable is to be maintained. It also analyzes the input it receives and determines the appropriate response. Information (output) then flows from the control center to the third component, the *effector*, along the *efferent pathway*. (To help you remember the difference between "afferent" and "efferent," note that information traveling along the afferent pathway *approaches* the control center and efferent information *exits* from the control center.)

The **effector** provides the means for the control center's response (output) to the stimulus. The results of the response then *feed back* to influence the effect of the stimulus, either reducing it so that the whole control process is shut off, or enhancing it so that the whole process continues at an even faster rate.

1

Negative Feedback Mechanisms

Most homeostatic control mechanisms are **negative feedback mechanisms**. In these systems, the output shuts off the original effect of the stimulus or reduces its intensity. These mechanisms cause the variable to change in a direction *opposite* to that of the initial change, returning it to its "ideal" value.

Let's start with an example of a nonbiological negative feedback system: a home heating system connected to a temperature-sensing thermostat. The thermostat houses both the receptor (thermometer) and the control center. If the thermostat is set at 20°C (68°F), the heating system (effector) is triggered ON when the house temperature drops below that setting. As the furnace produces heat and warms the air, the temperature rises, and when it reaches 20°C or slightly higher, the thermostat triggers the furnace OFF. This process results in a cycling of the furnace between "ON" and "OFF" so that the temperature in the house stays very near the desired temperature.

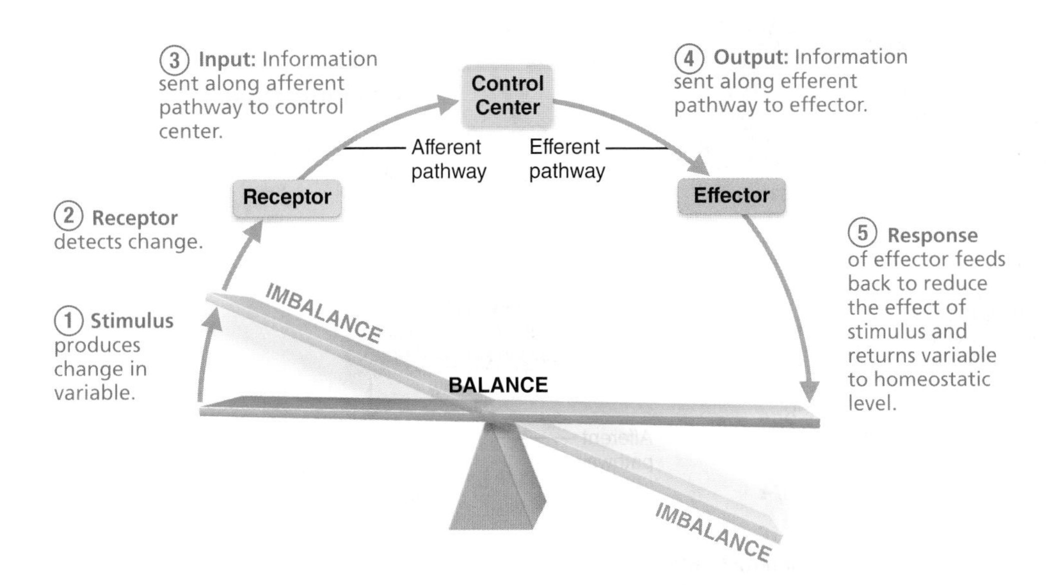

Figure 1.4 Interactions among the elements of a homeostatic control system maintain stable internal conditions.

Practice art labeling
MasteringA&P®>Study Area>Chapter 1

the palms face forward and the thumbs point away from the body. You can see the anatomical position in **Table 1.1** (top) and **Figure 1.7a**.

It is essential to understand the anatomical position because most of the directional terms used in this book refer to the body *as if it were in this position, regardless of its actual position.* Another point to remember is that the terms "right" and "left" refer to those sides of the person or the cadaver being viewed—not those of the observer.

Directional terms allow us to explain where one body structure is in relation to another. For example, we could describe the relationship between the ears and the nose by stating, "The ears are located on each side of the head to the right and left of the nose." Using anatomical terminology, this becomes "The ears are lateral to the nose." Using anatomical terms saves words and is less ambiguous.

Commonly used orientation and directional terms are defined and illustrated in Table 1.1. Many of these terms are also used in everyday conversation, but remember as you study them that their anatomical meanings are very precise.

Regional Terms

The two fundamental divisions of our body are its *axial* and *appendicular* (ap″en-dik′u-lar) parts. The **axial part**, which makes up the main *axis* of our body, includes the head, neck, and trunk. The **appendicular part** consists of the *appendages*, or *limbs*, which are attached to the body's axis. **Regional terms** used to designate specific areas within these major body divisions are indicated in Figure 1.7.

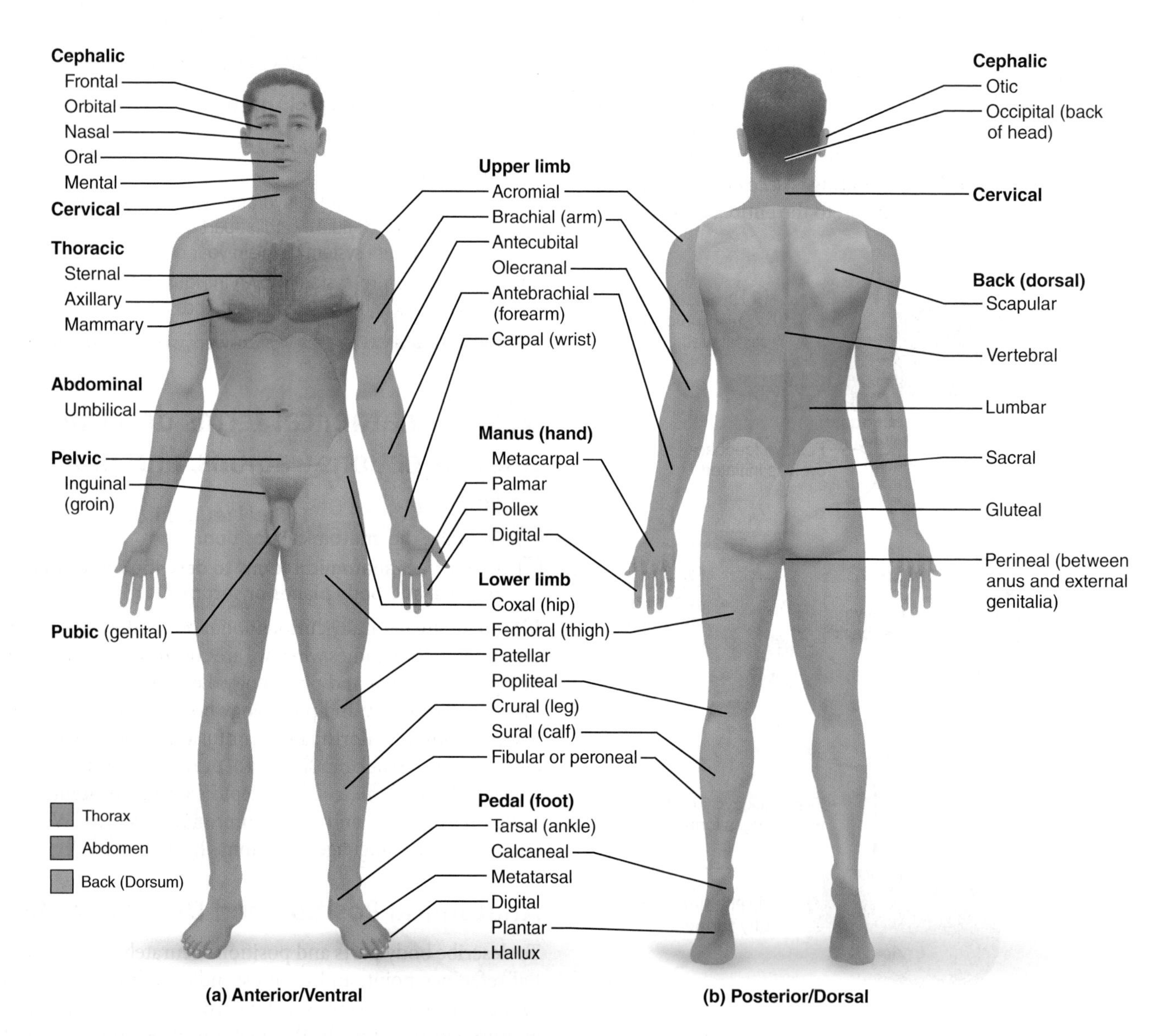

Figure 1.7 Regional terms used to designate specific body areas. Common terms for certain regions are shown in parentheses. **(a)** The anatomical position. **(b)** The heels are raised to show the plantar surface of the foot, which is actually on the inferior surface of the body.

Table 1.1 Orientation and Directional Terms

TERM	DEFINITION	EXAMPLE
Superior (cranial)	Toward the head end or upper part of a structure or the body; above	The head is superior to the abdomen.
Inferior (caudal)	Away from the head end or toward the lower part of a structure or the body; below	The navel is inferior to the chin.
Anterior (ventral)*	Toward or at the front of the body; in front of	The breastbone is anterior to the spine.
Posterior (dorsal)*	Toward or at the back of the body; behind	The heart is posterior to the breastbone.
Medial	Toward or at the midline of the body; on the inner side of	The heart is medial to the arm.
Lateral	Away from the midline of the body; on the outer side of	The arms are lateral to the chest.
Intermediate	Between a more medial and a more lateral structure	The collarbone is intermediate between the breastbone and shoulder.
Proximal	Closer to the origin of the body part or the point of attachment of a limb to the body trunk	The elbow is proximal to the wrist.
Distal	Farther from the origin of a body part or the point of attachment of a limb to the body trunk	The knee is distal to the thigh.
Superficial (external)	Toward or at the body surface	The skin is superficial to the skeletal muscles.
Deep (internal)	Away from the body surface; more internal	The lungs are deep to the skin.

*The terms *ventral* and *anterior* are synonymous in humans, but this is not the case in four-legged animals. *Anterior* refers to the leading portion of the body (abdominal surface in humans, head in a cat), but *ventral* specifically refers to the "belly" of a vertebrate animal, so it is the inferior surface of four-legged animals. Likewise, although the dorsal and posterior surfaces are the same in humans, the term *dorsal* specifically refers to an animal's back. Thus, the dorsal surface of four-legged animals is their superior surface.

The superior subdivision, the **thoracic cavity** (tho-ras′ik), is surrounded by the ribs and muscles of the chest. The thoracic cavity is further subdivided into lateral **pleural cavities** (ploo′ral), each enveloping a lung, and the medial **mediastinum** (me″de-ah-sti′num). The mediastinum contains the **pericardial cavity** (per″ĭ-kar′de-al), which encloses the heart, and it also surrounds the remaining thoracic organs (esophagus, trachea, and others).

The thoracic cavity is separated from the more inferior **abdominopelvic cavity** (ab-dom′ĭ-no-pel′vic) by the diaphragm, a dome-shaped muscle important in breathing. The abdominopelvic cavity, as its name suggests, has two parts. However, these regions are not physically separated by a muscular or membrane wall. Its superior portion, the **abdominal cavity**, contains the stomach, intestines, spleen, liver, and other organs. The inferior part, the **pelvic cavity**, lies in the bony pelvis and contains the urinary bladder, some reproductive organs, and the rectum. The abdominal and pelvic cavities are not aligned with each other. Instead, the bowl-shaped pelvis tips away from the perpendicular as shown in Figure 1.9a.

When the body is subjected to physical trauma (as in an automobile accident), the abdominopelvic organs are most vulnerable. Why? This is because the walls of the abdominal cavity are formed only by trunk muscles and are not reinforced by bone. The pelvic organs receive a somewhat greater degree of protection from the bony pelvis. ✚

Membranes in the Ventral Body Cavity

The walls of the ventral body cavity and the outer surfaces of the organs it contains are covered by a thin, double-layered membrane, the **serosa** (se-ro′sah), or **serous membrane**. The part of the membrane lining the cavity walls is called the **parietal serosa** (pah-ri′ĕ-tal; *parie* = wall). It folds in on itself to form the **visceral serosa**, covering the organs in the cavity.

You can visualize the relationship between the serosal layers by pushing your fist into a limp balloon (**Figure 1.10a**). The part of the balloon that clings to your fist can be compared to the visceral serosa clinging to an organ's external surface. The outer wall of the balloon represents the parietal serosa that lines the walls of the cavity. (However, unlike the balloon, the parietal serosa is never exposed but is always fused to the cavity wall.) In the body, the serous membranes are separated not by air but by a thin layer of lubricating fluid, called **serous fluid**, which is secreted by both membranes. Although there is a potential space between the two membranes, the barely present, slitlike cavity is filled with serous fluid.

The slippery serous fluid allows the organs to slide without friction across the cavity walls and one another as they carry out their routine functions. This freedom of movement is especially important for mobile organs such as the pumping heart and the churning stomach.

The serous membranes are named for the specific cavity and organs with which they are associated. For example, as shown in Figure 1.10b, the *parietal pericardium* lines the pericardial cavity and folds back as the *visceral pericardium*, which covers the heart. Likewise, the *parietal pleurae* (ploo′re) line the walls of the thoracic cavity, and the *visceral pleurae* cover the lungs. The *parietal peritoneum* (per″ĭ-to-ne′um) is associated with the walls of the abdominopelvic cavity, while the *visceral peritoneum* covers most of the organs within that cavity. (The pleural and peritoneal serosae are illustrated in Figure 4.11c on p. 130.)

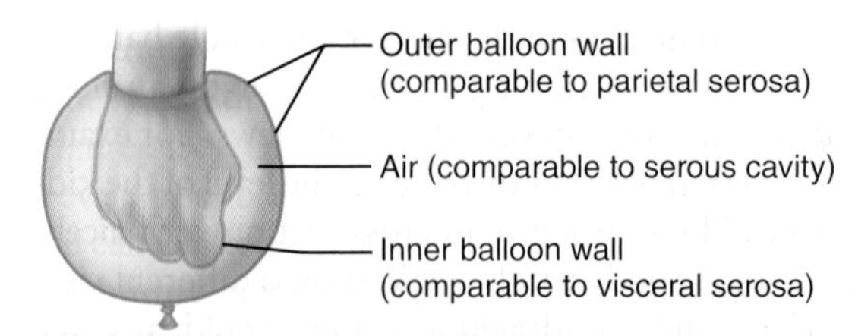

(a) A fist thrust into a flaccid balloon demonstrates the relationship between the parietal and visceral serous membrane layers.

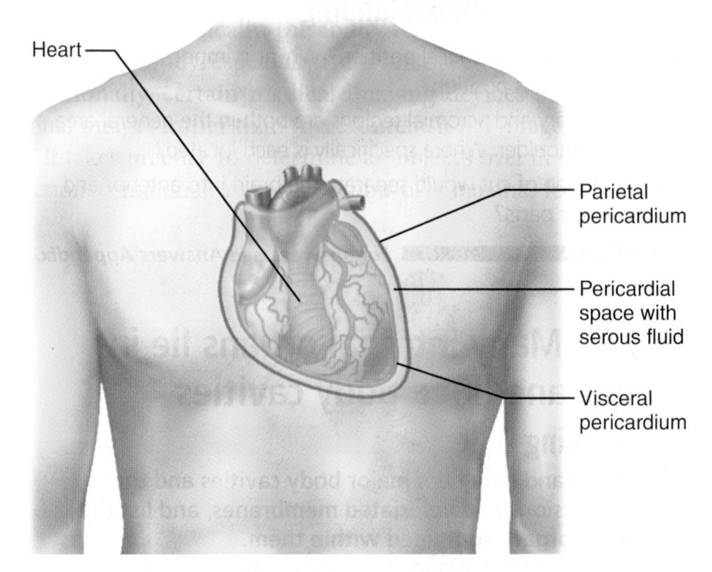

(b) The serosae associated with the heart.

Figure 1.10 Serous membrane relationships.

HOMEOSTATIC IMBALANCE 1.2 CLINICAL

When serous membranes are inflamed, their normally smooth surfaces become roughened. This roughness causes the membranes to stick together and drag across one another. Excruciating pain results, as anyone who has experienced *pleurisy* (inflammation of the pleurae) or *peritonitis* (inflammation of the peritoneums) knows. ✚

Abdominopelvic Regions and Quadrants

Because the abdominopelvic cavity is large and contains several organs, it helps to divide it into smaller areas for study. Medical personnel usually use a simple scheme to locate the abdominopelvic cavity organs (**Figure 1.11**). In this scheme, a transverse and a median plane pass through the umbilicus at right

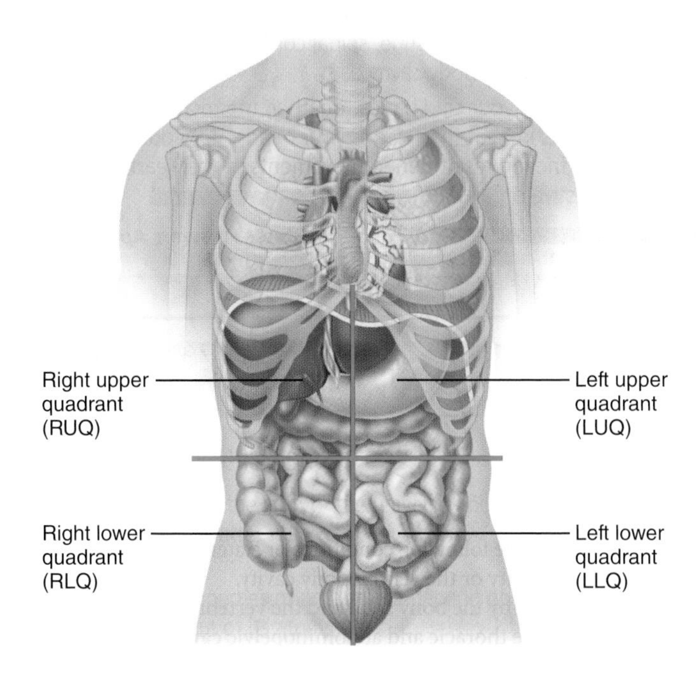

Figure 1.11 The four abdominopelvic quadrants. In this scheme, the abdominopelvic cavity is divided into four quadrants by two planes.

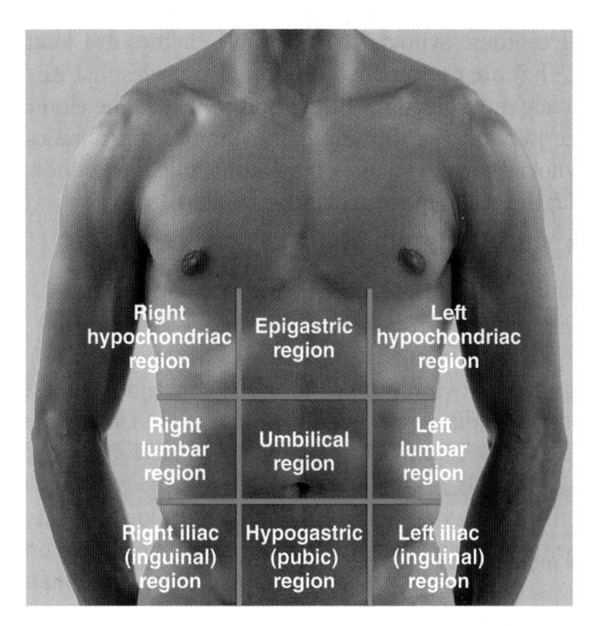

(a) Nine regions delineated by four planes

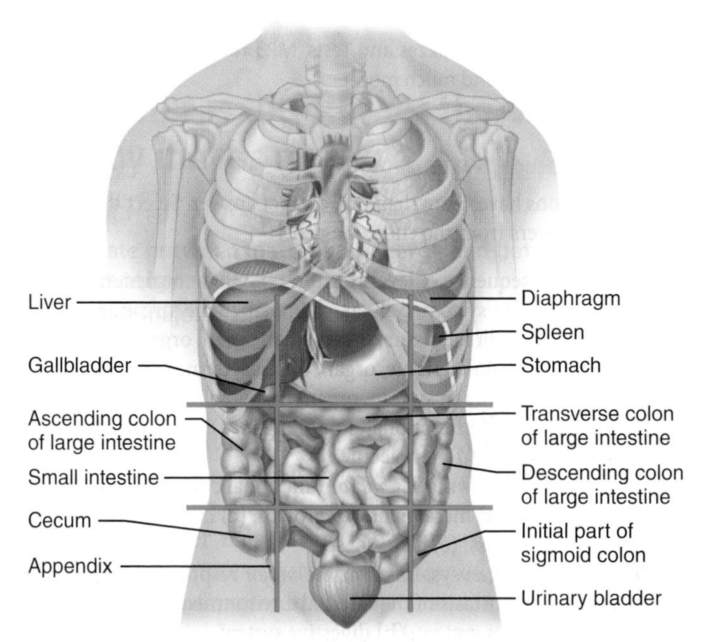

(b) Anterior view of the nine regions showing the superficial organs

Figure 1.12 The nine abdominopelvic regions. (a) The superior transverse plane is just inferior to the ribs; the inferior transverse plane is just superior to the hip bones; and the parasagittal planes lie just medial to the nipples.

angles. The four resulting quadrants are named according to their positions from the subject's point of view: the **right upper quadrant (RUQ)**, **left upper quadrant (LUQ)**, **right lower quadrant (RLQ)**, and **left lower quadrant (LLQ)**.

Another division method, used primarily by anatomists, uses two transverse and two parasagittal planes. These planes, positioned like a tic-tac-toe grid on the abdomen, divide the cavity into nine regions (**Figure 1.12**):

- The **umbilical region** is the centermost region deep to and surrounding the umbilicus (navel).
- The **epigastric region** is located superior to the umbilical region (*epi* = upon, above; *gastri* = belly).
- The **hypogastric (pubic) region** is located inferior to the umbilical region (*hypo* = below).
- The **right** and **left iliac**, or **inguinal**, **regions** (ing′gwĭ-nal) are located lateral to the hypogastric region (*iliac* = superior part of the hip bone).
- The **right** and **left lumbar regions** lie lateral to the umbilical region (*lumbus* = loin).
- The **right** and **left hypochondriac regions** lie lateral to the epigastric region and deep to the ribs (*chondro* = cartilage).

Other Body Cavities

In addition to the large closed body cavities, there are several smaller body cavities. Most of these are in the head and most open to the body exterior. Figure 1.7 provides the terms that will help you locate all but the last two cavities mentioned here.

- **Oral and digestive cavities.** The oral cavity, commonly called the mouth, contains the teeth and tongue. This cavity is part of and continuous with the cavity of the digestive organs, which opens to the body exterior at the anus.
- **Nasal cavity.** Located within and posterior to the nose, the nasal cavity is part of the respiratory system passageways.
- **Orbital cavities.** The orbital cavities (orbits) in the skull house the eyes and present them in an anterior position.
- **Middle ear cavities.** The middle ear cavities in the skull lie just medial to the eardrums. These cavities contain tiny bones that transmit sound vibrations to the hearing receptors in the inner ears.

Energy

Compared with matter, energy is less tangible. It has no mass, does not take up space, and we can measure it only by its effects on matter. **Energy** is defined as the capacity to do work, or to put matter into motion. The greater the work done, the more energy is used doing it. A baseball player who has just hit the ball over the fence uses much more energy than a batter who bunts the ball back to the pitcher.

Kinetic versus Potential Energy

Energy exists in two states, and each can be transformed to the other. **Kinetic energy** (ki-net′ik) is energy in action. We see evidence of kinetic energy in the constant movement of the tiniest particles of matter (atoms) as well as in larger objects (a bouncing ball). Kinetic energy does work by moving objects, which in turn can do work by moving or pushing on other objects. For example, a push on a swinging door sets it into motion.

Potential energy is stored energy, that is, inactive energy that has the *potential*, or capability, to do work but is not presently doing so. The batteries in an unused toy have potential energy, as does water confined behind a dam. Your leg muscles have potential energy when you sit still on the couch. When potential energy is released, it becomes kinetic energy and so is capable of doing work. For example, dammed water becomes a rushing torrent when the dam is opened, and that rushing torrent can move a turbine at a hydroelectric plant, or charge a battery.

Actually, energy is a topic of physics, but matter and energy are inseparable. Matter is the substance, and energy is the mover of the substance. All living things are composed of matter and they all require energy to grow and function. The release and use of energy by living systems gives us the elusive quality we call life. Now let's consider the forms of energy used by the body as it does its work.

Forms of Energy

- **Chemical energy** is the form stored in the bonds of chemical substances. When chemical reactions occur that rearrange the atoms of the chemicals in a certain way, the potential energy is unleashed and becomes kinetic energy, or energy in action.

  For example, some of the energy in the foods you eat is eventually converted into the kinetic energy of your moving arm. However, food fuels cannot be used to energize body activities directly. Instead, some of the food energy is captured temporarily in the bonds of a chemical called *adenosine triphosphate* (*ATP*; ah-den′o-sēn tri″fos′fāt). Later, ATP's bonds are broken and the stored energy is released as needed to do cellular work. Chemical energy in the form of ATP is the most useful form of energy in living systems because it is used to run almost all functional processes.
- **Electrical energy** results from the movement of charged particles. In your home, electrical energy is found in the flow of electrons along the household wiring. In your body, electrical currents are generated when charged particles called *ions* move along or across cell membranes. The nervous system uses electrical currents, called *nerve impulses*, to transmit messages from one part of the body to another. Electrical currents traveling across the heart stimulate it to contract (beat) and pump blood. (This is why a strong electrical shock, which interferes with such currents, can cause death.)
- **Mechanical energy** is energy *directly* involved in moving matter. When you ride a bicycle, your legs provide the mechanical energy that moves the pedals.
- **Radiant energy**, or **electromagnetic radiation** (e-lek″tro-mag-net′ik), is energy that travels in waves. These waves, which vary in length, are collectively called the *electromagnetic spectrum*. They include visible light, infrared waves, radio waves, ultraviolet waves, and X rays. Light energy, which stimulates the retinas of our eyes, is important in vision. Ultraviolet waves cause sunburn, but they also stimulate your body to make vitamin D.

Energy Form Conversions

With few exceptions, energy is easily converted from one form to another. For example, the chemical energy (in gasoline) that powers the motor of a speedboat is converted into the mechanical energy of the whirling propeller that makes the boat skim across the water.

Energy conversions are quite inefficient. Some of the initial energy supply is always "lost" to the environment as heat. It is not really lost because energy cannot be created or destroyed, but that portion given off as heat is at least partly *unusable*. It is easy to demonstrate this principle. Electrical energy is converted into light energy in a lightbulb. But if you touch a lit bulb, you will soon discover that some of the electrical energy is producing heat instead.

Likewise, all energy conversions in the body liberate heat. This heat helps to maintain our relatively high body temperature, which influences body functioning. For example, when matter is heated, the kinetic energy of its particles increases and they begin to move more quickly. The higher the temperature, the faster the body's chemical reactions occur. We will learn more about this later.

☑ Check Your Understanding

1. What form of energy is found in the food we eat?
2. What form of energy is used to transmit messages from one part of the body to another?
3. What type of energy is available when we are still? When we are exercising?

For answers, see Answers Appendix.

2.2 The properties of an element depend on the structure of its atoms

→ Learning Objectives

- ☐ Define chemical element and list the four elements that form the bulk of body matter.
- ☐ Define atom. List the subatomic particles, and describe their relative masses, charges, and positions in the atom.
- ☐ Define atomic number, atomic mass, atomic weight, isotope, and radioisotope.

All matter is composed of **elements**, unique substances that cannot be broken down into simpler substances by ordinary chemical methods. Among the well-known elements are oxygen, carbon, gold, silver, copper, and iron.

At present, 118 elements are recognized. Of these, 92 occur in nature. The rest are made artificially in particle accelerator devices.

Four elements—carbon, oxygen, hydrogen, and nitrogen—make up about 96% of body weight, and 20 others are present in the body, some in trace amounts. Table 2.1 on p. 22 lists those of importance to the body. An oddly shaped checkerboard called the **periodic table** provides a listing of the known elements and helps to explain the properties of each element.

Each element is composed of more or less identical particles or building blocks, called **atoms**. The smallest atoms are less than 0.1 nanometer (nm) in diameter, and the largest are only about five times as large. [1 nm = 0.0000001 (or 10^{-7}) centimeter (cm), or 40 billionths of an inch!]

Every element's atoms differ from those of all other elements and give the element its unique physical and chemical properties. *Physical properties* are those we can detect with our senses (such as color and texture) or measure (such as boiling point and freezing point). *Chemical properties* pertain to the way atoms interact with other atoms (bonding behavior) and account for the facts that iron rusts, animals can digest their food, and so on.

We designate each element by a one- or two-letter chemical shorthand called an **atomic symbol**, usually the first letter(s) of the element's name. For example, C stands for carbon, O for oxygen, and Ca for calcium. In a few cases, the atomic symbol is taken from the Latin name for the element. For example, sodium is indicated by Na, from the Latin word *natrium*.

Structure of Atoms

The word *atom* comes from the Greek word meaning "indivisible." However, we now know that atoms are clusters of even smaller particles called protons, neutrons, and electrons and that even those subatomic particles can be subdivided with high-technology tools. Still, the old idea of atomic indivisibility is useful because an atom loses the unique properties of its element when it is split into its subatomic particles.

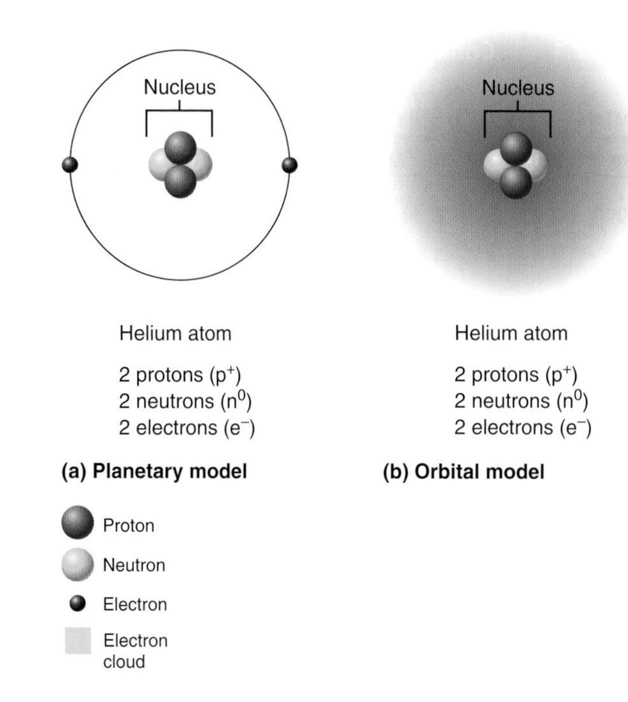

Figure 2.1 Two models of the structure of an atom.

An atom's subatomic particles differ in mass, electrical charge, and position in the atom. An atom has a central **nucleus** containing protons and neutrons tightly bound together. The nucleus, in turn, is surrounded by orbiting electrons (Figure 2.1). **Protons** (p^+) bear a positive electrical charge, and **neutrons** (n^0) are neutral, so the nucleus is positively charged overall. Protons and neutrons are heavy particles and have approximately the same mass, arbitrarily designated as 1 **atomic mass unit** (1 amu). Since all of the heavy subatomic particles are concentrated in the nucleus, the nucleus is fantastically dense, accounting for nearly the entire mass (99.9%) of the atom.

The tiny **electrons** (e^-) bear a negative charge equal in strength to the positive charge of the proton. However, an electron has only about 1/2000 the mass of a proton, and the mass of an electron is usually designated as 0 amu.

All atoms are electrically neutral because the number of protons in an atom is precisely balanced by its number of electrons (the + and − charges will then cancel the effect of each other). For example, hydrogen has one proton and one electron, and iron has 26 protons and 26 electrons. For any atom, the number of protons and electrons is always equal.

The **planetary model** of the atom (Figure 2.1a) is a simplified model of atomic structure. As you can see, it depicts electrons moving around the nucleus in fixed, generally circular orbits. But we can never determine the exact location of electrons at a particular time because they jump around following

Glucose is $C_6H_{12}O_6$, which indicates that it has 6 carbon atoms, 12 hydrogen atoms, and 6 oxygen atoms. To compute the molecular weight of glucose, you would look up the atomic weight of each of its atoms in the periodic table and compute its molecular weight as follows:

Atom	Number of Atoms		Atomic Weight		Total Atomic Weight
C	6	×	12.011	=	72.066
H	12	×	1.008	=	12.096
O	6	×	15.999	=	95.994
					180.156

Then, to make a *one-molar* solution of glucose, you would weigh out 180.156 grams (g), called a *gram molecular weight*, of glucose and add enough water to make 1 liter (L) of solution. In short, a one-molar solution (abbreviated 1.0 *M*) of a chemical substance is one gram molecular weight of the substance (or one gram atomic weight in the case of elemental substances) in 1 L (1000 milliliters) of solution.

The beauty of using the mole as the basis of preparing solutions is its precision. One mole of any substance always contains exactly the same number of solute particles, that is, 6.02×10^{23}. This number is called **Avogadro's number** (av″o-gad′rōz). So whether you weigh out 1 mole of glucose (180 g) or 1 mole of water (18 g) or 1 mole of methane (16 g), in each case you will have 6.02×10^{23} molecules of that substance.* This allows almost mind-boggling precision to be achieved.

Because solute concentrations in body fluids tend to be quite low, those values are usually reported in terms of millimoles (m*M*; 1/1000 mole).

Colloids

Colloids (kol′oidz), also called *emulsions*, are *heterogeneous* mixtures, which means that their composition is dissimilar in different areas of the mixture. Colloids often appear translucent or milky and although the solute particles are larger than those in true solutions, they still do not settle out. However, they do scatter light, so the path of a light beam shining through a colloidal mixture is visible.

Colloids have many unique properties, including the ability of some to undergo **sol-gel transformations**, that is, to change reversibly from a fluid (sol) state to a more solid (gel) state. Jell-O, or any gelatin product (Figure 2.4), is a familiar example of a nonliving colloid that changes from a sol to a gel when refrigerated (and that gel will liquefy again if placed in the sun). Cytosol, the semifluid material in living cells, is also a colloid, largely because of its dispersed proteins. Its sol-gel transformations underlie many important cell activities, such as cell division and changes in cell shape.

Suspensions

Suspensions are *heterogeneous* mixtures with large, often visible solutes that tend to settle out. An example of a suspension is a mixture of sand and water. So is blood, in which the living blood cells are suspended in the fluid portion of blood (blood plasma). If left to stand, the suspended cells will settle out unless some means—mixing, shaking, or circulation in the body—keeps them in suspension.

All three types of mixtures are found in both living and nonliving systems. In fact, living material is the most complex mixture of all, since it contains all three kinds of mixtures interacting with one another.

Distinguishing Mixtures from Compounds

Now let's zero in on how to distinguish mixtures and compounds from one another. Mixtures differ from compounds in several important ways:

- The chief difference between mixtures and compounds is that no chemical bonding occurs between the components of a mixture. The properties of atoms and molecules are not changed when they become part of a mixture. Remember they are only physically intermixed.
- Depending on the mixture, its components can be separated by physical means—straining, filtering, evaporation, and so on. Compounds, by contrast, can be separated into their constituent atoms only by chemical means (breaking bonds).
- Some mixtures are homogeneous, whereas others are heterogeneous. A bar of 100% pure (elemental) iron is homogeneous, as are all compounds. As already mentioned, heterogeneous substances vary in their makeup from place to place. For example, iron ore is a heterogeneous mixture that contains iron and many other elements.

☑ Check Your Understanding

7. What is the meaning of the term "molecule"?
8. Why is sodium chloride (NaCl) considered a compound, but oxygen gas is not?
9. Blood contains a liquid component and living cells. Would it be classified as a compound or a mixture? Why?

For answers, see Answers Appendix.

2.4 The three types of chemical bonds are ionic, covalent, and hydrogen

→ Learning Objectives

- ☐ **Explain the role of electrons in chemical bonding and in relation to the octet rule.**
- ☐ **Differentiate among ionic, covalent, and hydrogen bonds.**
- ☐ **Compare and contrast polar and nonpolar compounds.**

*The important exception to this rule concerns molecules that ionize and break up into charged particles (ions) in water, such as salts, acids, and bases (see pp. 34–35). For example, simple table salt (sodium chloride) breaks up into two types of charged particles. Therefore, in a 1.0 *M* solution of sodium chloride, 2 *moles* of solute particles are actually in solution.

As noted earlier, when atoms combine with other atoms, they are held together by **chemical bonds**. A chemical bond is not a physical structure like a pair of handcuffs linking two people together. Instead, it is an energy relationship between the electrons of the reacting atoms, and it is made or broken in less than a trillionth of a second.

The Role of Electrons in Chemical Bonding

Electrons forming the electron cloud around the nucleus of an atom occupy regions of space called **electron shells** that consecutively surround the atomic nucleus. The atoms known so far can have electrons in seven shells (numbered 1 to 7 from the nucleus outward), but the actual number of electron shells occupied in a given atom depends on the number of electrons the atom has. Each electron shell contains one or more orbitals. (Recall that *orbitals* are regions around the nucleus in which a given electron is likely to be found most of the time.)

It is important to understand that each electron shell represents a different **energy level**, because this prompts you to think of electrons as particles with a certain amount of potential energy. In general, the terms *electron shell* and *energy level* are used interchangeably.

How much potential energy does an electron have? The answer depends on the energy level that it occupies. The attraction between the positively charged nucleus and negatively charged electrons is greatest when electrons are closest to the nucleus and falls off with increasing distance. This statement explains why electrons farthest from the nucleus (1) have the greatest potential energy (it takes more energy for them to overcome the nuclear attraction and reach the more distant energy levels) and (2) are most likely to interact chemically with other atoms. (They are the least tightly held by their own nucleus and the most easily influenced by other atoms and molecules.)

Each electron shell can hold a specific number of electrons. Shell 1, the shell immediately surrounding the nucleus, accommodates only 2 electrons. Shell 2 holds a maximum of 8, and shell 3 has room for 18. Subsequent shells hold larger and larger numbers of electrons, and the shells tend to be filled with electrons consecutively. For example, shell 1 fills completely before any electrons appear in shell 2.

Which electrons are involved in chemical bonding? When we consider bonding behavior, the only electrons that are important are those in the atom's outermost energy level. Inner electrons usually do not take part in bonding because they are more tightly held by the nucleus.

When the outermost energy level of an atom is filled to capacity or contains 8 electrons, the atom is stable. Such atoms are *chemically inert*, that is, unreactive. A group of elements called the *noble gases*, which include helium and neon, typify this condition (**Figure 2.5a**). On the other hand, atoms in which the outermost energy level contains fewer than 8 electrons tend to gain, lose, or share electrons with other atoms to achieve stability (Figure 2.5b).

What about atoms that have more than 20 electrons, in which the energy levels beyond shell 2 can contain *more* than 8 electrons? The number of electrons that can participate in bonding is still limited to a total of 8. The term **valence shell** (va′lens) refers to an atom's outermost energy level *or that portion of it* containing the electrons that are chemically reactive. Hence, the key to chemical reactivity is the **octet rule** (ok-tet′), or **rule of eights**. Except for shell 1, which is full when it has 2 electrons, atoms tend to interact in such a way that they have 8 electrons in their valence shell.

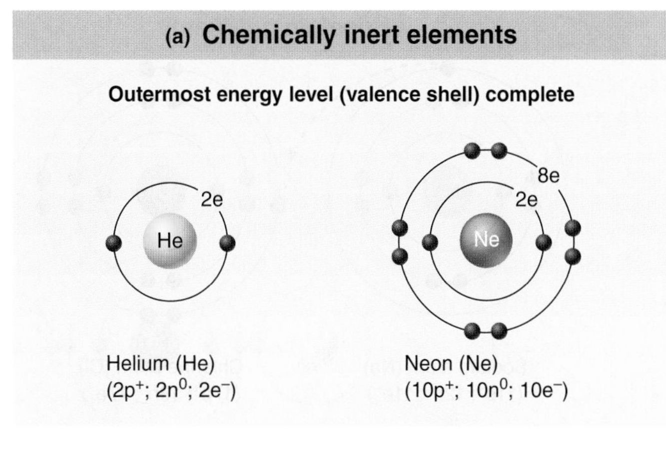

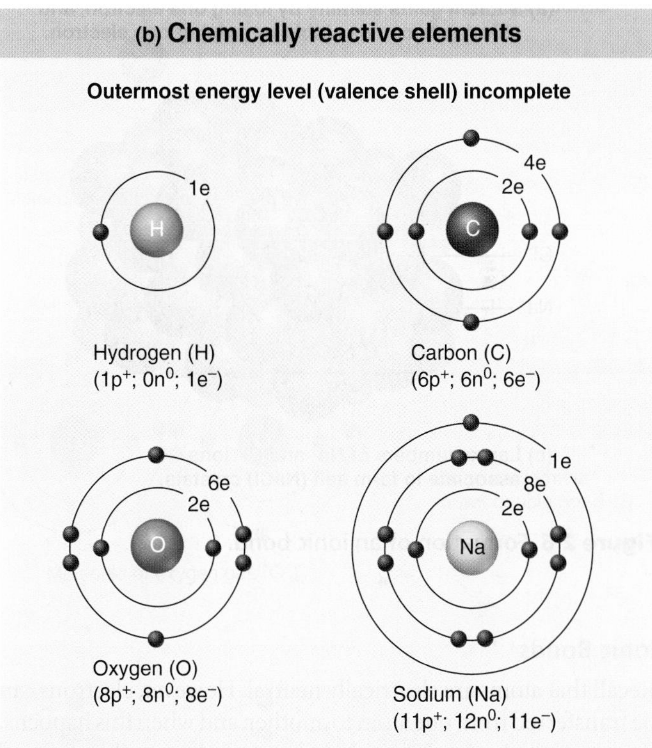

Figure 2.5 Chemically inert and reactive elements. (*Note*: For simplicity, each atomic nucleus is shown as a sphere with the atom's symbol; individual protons and neutrons are not shown.)

Types of Chemical Bonds

Three major types of chemical bonds—*ionic*, *covalent*, and *hydrogen bonds*—result from attractive forces between atoms.

In the previous equations, the reactants are atoms, as indicated by their atomic symbols (H, C). The product in each case is a molecule, as represented by its **molecular formula** (H_2, CH_4). The equation for the formation of methane may be read in terms of molecules or moles—as *either* "four hydrogen atoms plus one carbon atom yield one molecule of methane" *or* "four moles of hydrogen atoms plus one mole of carbon yield one mole of methane." Using moles is more practical because it is impossible to measure out one atom or one molecule of anything!

Types of Chemical Reactions

Most chemical reactions can be categorized as one of three types: *synthesis*, *decomposition*, or *exchange reactions*.

When atoms or molecules combine to form a larger, more complex molecule, the process is a **synthesis**, or **combination, reaction**. A synthesis reaction always involves bond formation. It can be represented (using arbitrary letters) as

$$A + B \rightarrow AB$$

Synthesis reactions are the basis of constructive, or **anabolic**, activities in body cells, such as joining small molecules called amino acids into large protein molecules (**Figure 2.11a**). Synthesis reactions are conspicuous in rapidly growing tissues.

A **decomposition reaction** occurs when a molecule is broken down into smaller molecules or its constituent atoms:

$$AB \rightarrow A + B$$

Essentially, decomposition reactions are reverse synthesis reactions: Bonds are broken. Decomposition reactions underlie all degradative, or **catabolic**, processes in body cells. For example, the bonds of glycogen molecules are broken to release simpler molecules of glucose (Figure 2.11b).

Exchange, or **displacement**, **reactions** involve both synthesis and decomposition. Bonds are both made and broken. In an exchange reaction, parts of the reactant molecules change partners, so to speak, producing different product molecules:

$$AB + C \rightarrow AC + B \quad \text{and} \quad AB + CD \rightarrow AD + CB$$

An exchange reaction occurs when ATP reacts with glucose and transfers its end phosphate group (indicated by a circled P in Figure 2.11c) to glucose, forming glucose-phosphate. At the same time, the ATP becomes ADP. This important reaction occurs whenever glucose enters a body cell, and it effectively traps the glucose fuel molecule inside the cell.

Another group of important chemical reactions in living systems is **oxidation-reduction reactions**, called **redox reactions** for short. Oxidation-reduction reactions are decomposition reactions in that they are the basis of all reactions in which food fuels are broken down for energy (that is, in which ATP is produced). They are also a special type of exchange reaction because electrons are exchanged between the reactants. The reactant losing the electrons is the *electron donor* and is said to be **oxidized**. The reactant taking up the transferred electrons is the *electron acceptor* and is said to become **reduced**.

Redox reactions also occur when ionic compounds are formed. Recall that in the formation of NaCl (see Figure 2.6), sodium loses an electron to chlorine. Consequently, sodium is oxidized and becomes a sodium ion, and chlorine is reduced and becomes a chloride ion. However, not all oxidation-reduction

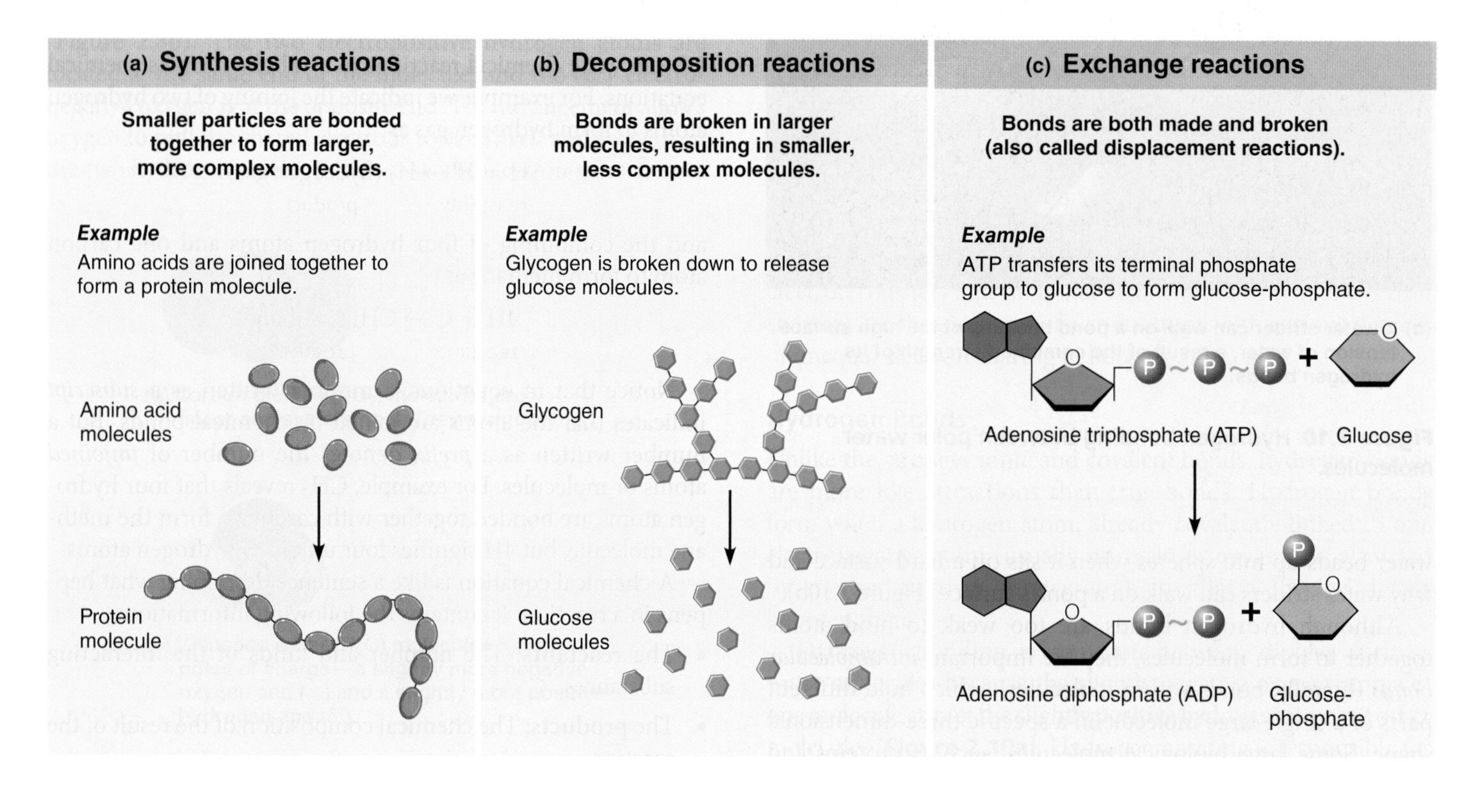

Figure 2.11 Types of chemical reactions.

reactions involve *complete transfer* of electrons—some simply change the pattern of electron sharing in covalent bonds. For example, a substance is oxidized both by losing hydrogen atoms and by combining with oxygen. The common factor in these events is that electrons that formerly "belonged" to the reactant molecule are lost. The electrons are lost either entirely (as when hydrogen is removed and takes its electron with it) or relatively (as the shared electrons spend more time in the vicinity of the very electronegative oxygen atom).

To understand the importance of oxidation-reduction reactions in living systems, take a look at the overall equation for *cellular respiration*, which represents the major pathway by which glucose is broken down for energy in body cells:

$$\underset{\text{glucose}}{C_6H_{12}O_6} + \underset{\text{oxygen}}{6O_2} \rightarrow \underset{\text{carbon dioxide}}{6CO_2} + \underset{\text{water}}{6H_2O} + \underset{\text{cellular energy}}{\text{ATP}}$$

As you can see, it is an oxidation-reduction reaction. Consider what happens to the hydrogen atoms (and their electrons). Glucose is oxidized to carbon dioxide as it loses hydrogen atoms, and oxygen is reduced to water as it accepts the hydrogen atoms. This reaction is covered in detail in Chapter 23.

Energy Flow in Chemical Reactions

Because all chemical bonds represent stored chemical energy, all chemical reactions ultimately result in net absorption or release of energy. Reactions that release energy are **exergonic reactions**. These reactions yield products with less energy than the initial reactants, along with energy that can be harvested for other uses. With a few exceptions, catabolic and oxidative reactions are exergonic.

In contrast, the products of energy-absorbing, or **endergonic**, reactions contain more potential energy in their chemical bonds than did the reactants. Anabolic reactions are typically endergonic reactions. Essentially exergonic and endergonic reactions add up to a case of "one hand washing the other"—the energy released when fuel molecules are broken down (oxidized) is captured in ATP molecules and then used to synthesize the complex biological molecules the body needs to sustain life.

Reversibility of Chemical Reactions

All chemical reactions are theoretically reversible. If chemical bonds can be made, they can be broken, and vice versa. Reversibility is indicated by a double arrow. When the arrows differ in length, the longer arrow indicates the major direction in which the reaction proceeds:

$$A + B \rightleftharpoons AB$$

In this example, the forward reaction (going to the right) predominates. Over time, the product (AB) accumulates and the reactants (A and B) decrease in amount.

When the arrows are of equal length, as in

$$A + B \rightleftarrows AB$$

neither the forward reaction nor the reverse reaction is dominant. In other words, for each molecule of product (AB) formed, one product molecule breaks down, releasing the reactants A and B. Such a chemical reaction is said to be in a state of **chemical equilibrium**.

Once chemical equilibrium is reached, there is no further *net change* in the amounts of reactants and products unless more of either are added to the mix. Product molecules are still formed and broken down, but the balance established when equilibrium was reached (such as greater numbers of product molecules) remains unchanged.

Chemical equilibrium is analogous to the admission scheme used by many nightclubs that restrict the number of patrons admitted to comply with safety regulations. To stay within their allowed capacity (for example, 300), once 300 people are inside no one else is admitted until others leave. Hence, when 10 leave, 10 more may go in. So there is a constant turnover but the number of patrons in the club remains at 300 throughout the night.

All chemical reactions are reversible, but many biological reactions show so little tendency to go in the reverse direction that they are irreversible for all practical purposes. Chemical reactions that release energy will not go in the opposite direction unless energy is put back into the system. For example, when our cells break down glucose during cellular respiration to yield carbon dioxide and water, some of the energy released is trapped in the bonds of ATP. Because the cells then use ATP's energy for various functions (and more glucose will be along with the next meal), this particular reaction is never reversed in our cells. Furthermore, if a product of a reaction is continuously removed from the reaction site, it is unavailable to take part in the reverse reaction. This situation occurs when the carbon dioxide that is released during glucose breakdown leaves the cell, enters the blood, and is eventually removed from the body by the lungs.

Factors Influencing the Rate of Chemical Reactions

What influences the speed of chemical reactions? For atoms and molecules to react chemically in the first place, they must *collide* with enough force to overcome the repulsion between their electrons. Interactions between valence shell electrons—the basis of bond making and breaking—cannot occur long distance. The force of collisions depends on how fast the particles are moving. Solid, forceful collisions between rapidly moving particles in which valence shells overlap are much more likely to cause reactions than are collisions in which the particles graze each other lightly.

Temperature Increasing the temperature of a substance increases the kinetic energy of its particles and the force of their collisions. For this reason, chemical reactions proceed more quickly at higher temperatures.

Concentration Chemical reactions progress most rapidly when the reacting particles are present in high numbers, because the chance of successful collisions is greater. As the concentration of the reactants declines, the reaction slows. Chemical equilibrium eventually occurs unless additional reactants are added or products are removed from the reaction site.

Particle Size Smaller particles move faster than larger ones (at the same temperature) and tend to collide more frequently and more forcefully. Hence, the smaller the reacting particles, the faster a chemical reaction goes at a given temperature and concentration.

Catalysts Many chemical reactions in nonliving systems can be speeded up simply by heating, but drastic increases in body temperature are life threatening because important biological molecules are destroyed. Still, at normal body temperatures, most chemical reactions would proceed far too slowly to maintain life were it not for the presence of catalysts. **Catalysts** (kat′ah-lists) are substances that increase the rate of chemical reactions without themselves becoming chemically changed or part of the product. Biological catalysts are called *enzymes* (en′zīmz). Later in this chapter we describe how enzymes work.

☑ Check Your Understanding

13. Which reaction type—synthesis, decomposition, or exchange—occurs when fats are digested in your small intestine?
14. Why are many reactions that occur in living systems irreversible for all intents and purposes?
15. What specific name is given to decomposition reactions in which food fuels are broken down for energy?

For answers, see Answers Appendix.

PART 2

BIOCHEMISTRY

Biochemistry is the study of the chemical composition and reactions of living matter. All chemicals in the body fall into one of two major classes: organic or inorganic compounds. **Organic compounds** contain carbon. All organic compounds are covalently bonded molecules, and many are large.

All other chemicals in the body are considered **inorganic compounds**. These include water, salts, and many acids and bases. Organic and inorganic compounds are equally essential for life. Trying to decide which is more valuable is like trying to decide whether the ignition system or the engine is more essential to run your car!

2.6 Inorganic compounds include water, salts, and many acids and bases

→ Learning Objectives

- ☐ **Explain the importance of water and salts to body homeostasis.**
- ☐ **Define acid and base, and explain the concept of pH.**

Water

Water is the most abundant and important inorganic compound in living material. It makes up 60–80% of the volume of most living cells. What makes water so vital to life? The answer lies in several properties:

- **High heat capacity.** Water has a high heat capacity. In other words, it absorbs and releases large amounts of heat before changing appreciably in temperature itself. This property of water prevents sudden changes in temperature caused by external factors, such as sun or wind exposure, or by internal conditions that release heat rapidly, such as vigorous muscle activity. As part of blood, water redistributes heat among body tissues, ensuring temperature homeostasis.
- **High heat of vaporization.** When water evaporates, or vaporizes, it changes from a liquid to a gas (water vapor). This transformation requires that large amounts of heat be absorbed to break the hydrogen bonds that hold water molecules together. This property is extremely beneficial when we sweat. As perspiration (mostly water) evaporates from our skin, large amounts of heat are removed from the body, providing efficient cooling.
- **Polar solvent properties.** Water is an unparalleled solvent. Indeed, it is often called the **universal solvent**. Biochemistry is "wet chemistry." Biological molecules do not react chemically unless they are in solution, and virtually all chemical reactions occurring in the body depend on water's solvent properties.

  Because water molecules are polar, they orient themselves with their slightly negative ends toward the positive ends of the solutes, and vice versa, first attracting the solute molecules, and then surrounding them. This polarity of water explains why ionic compounds and other small reactive molecules (such as acids and bases) *dissociate* in water, their ions separating from each other and becoming evenly scattered in the water, forming true solutions (**Figure 2.12**).

  Water also forms layers of water molecules, called **hydration layers**, around large charged molecules such as proteins, shielding them from the effects of other charged substances in the vicinity and preventing them from settling out of solution. Such protein-water mixtures are *biological colloids*. Blood plasma and cerebrospinal fluid (which surrounds the brain and spinal cord) are colloids.

  Water is the body's major transport medium because it is such an excellent solvent. Nutrients, respiratory gases, and metabolic wastes carried throughout the body are dissolved in blood plasma, and many metabolic wastes are excreted from the body in urine, another watery fluid. Lubricants (e.g., mucus) also use water as their dissolving medium.
- **Reactivity.** Water is an important *reactant* in many chemical reactions. For example, foods are broken down to their building blocks by adding a water molecule to each bond to be broken. We will discuss such decomposition reactions (called *hydrolysis reactions*) in the next module.
- **Cushioning.** By forming a resilient cushion around certain body organs, water helps protect them from physical trauma. The cerebrospinal fluid surrounding the brain exemplifies water's cushioning role.

Salts

A **salt** is an ionic compound containing cations other than H^+ and anions other than the hydroxyl ion (OH^-). As already noted, when salts are dissolved in water, they dissociate into their component ions (Figure 2.12). For example, sodium

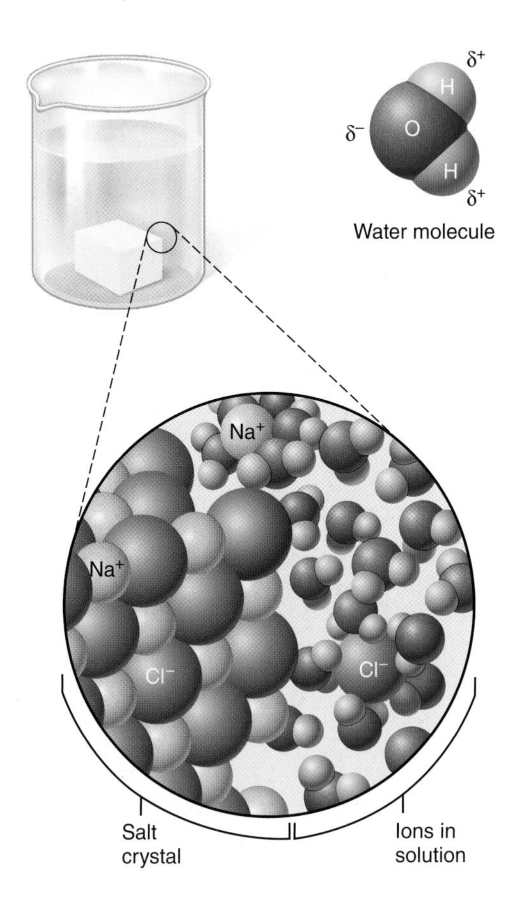

Figure 2.12 Dissociation of salt in water.

sulfate (Na_2SO_4) dissociates into two Na^+ ions and one SO_4^{2-} ion. It dissociates easily because the ions are already formed. All that remains is for water to overcome the attraction between the oppositely charged ions.

All ions are **electrolytes** (e-lek′tro-līts), substances that conduct an electrical current in solution. (Note that groups of atoms that bear an overall charge, such as sulfate, are called *polyatomic ions.*)

Salts commonly found in the body include NaCl, $CaCO_3$ (calcium carbonate), and KCl (potassium chloride). However, the most plentiful salts are the calcium phosphates that make bones and teeth hard. In their ionized form, salts play vital roles in body function. For instance, the electrolyte properties of sodium and potassium ions are essential for nerve impulse transmission and muscle contraction. Ionic iron forms part of the hemoglobin molecules that transport oxygen within red blood cells, and zinc and copper ions are important to the activity of some enzymes. Other important functions of the elements found in body salts are summarized in Table 2.1 on p. 22.

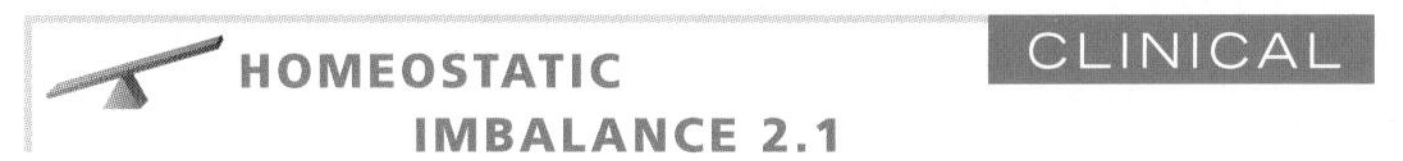

HOMEOSTATIC IMBALANCE 2.1

Maintaining proper ionic balance in our body fluids is one of the most crucial homeostatic roles of the kidneys. When this balance is severely disturbed, virtually nothing in the body works. Thousands of physiological activities are disrupted and grind to a stop. ✚

Acids and Bases

Like salts, acids and bases are electrolytes. They ionize and dissociate in water and can then conduct an electrical current.

Acids

Acids have a sour taste, can react with (dissolve) many metals, and "burn" a hole in your rug. But for our purposes the most useful definition of an acid is a substance that releases **hydrogen ions** (H^+) in detectable amounts. Because a hydrogen ion is just a hydrogen nucleus, or "naked" proton, acids are also defined as **proton donors**.

When acids dissolve in water, they release hydrogen ions (protons) and anions. It is the concentration of protons that determines the acidity of a solution. The anions have little or no effect on acidity. For example, hydrochloric acid (HCl), an acid produced by stomach cells that aids digestion, dissociates into a proton and a chloride ion:

$$HCl \rightarrow \underset{\text{proton}}{H^+} + \underset{\text{anion}}{Cl^-}$$

Other acids found or produced in the body include acetic acid ($HC_2H_3O_2$, commonly abbreviated as HAc), which is the acidic portion of vinegar; and carbonic acid (H_2CO_3). The molecular formula for an acid is easy to recognize because the hydrogen is written first.

Bases

Bases have a bitter taste, feel slippery, and are **proton acceptors**—that is, they take up hydrogen ions (H^+) in detectable amounts. Common inorganic bases include the *hydroxides* (hi-drok′sīds), such as magnesium hydroxide (milk of magnesia) and sodium hydroxide (lye). Like acids, hydroxides dissociate when dissolved in water, but in this case **hydroxyl ions (OH^-)** (hi-drok′sil) and cations are liberated. For example, ionization of sodium hydroxide (NaOH) produces a hydroxyl ion and a sodium ion, and the hydroxyl ion then binds to (accepts) a proton present in the solution. This reaction produces water and simultaneously reduces the acidity (hydrogen ion concentration) of the solution:

$$NaOH \rightarrow \underset{\text{cation}}{Na^+} + \underset{\text{hydroxyl ion}}{OH^-}$$

and then

$$OH^- + H^+ \rightarrow \underset{\text{water}}{H_2O}$$

Bicarbonate ion (HCO_3^-), an important base in the body, is particularly abundant in blood. **Ammonia (NH_3)**, a common waste product of protein breakdown in the body, is also a base. It has one pair of unshared electrons that strongly attracts protons. By accepting a proton, ammonia becomes an ammonium ion:

$$NH_3 + H^+ \rightarrow \underset{\text{ammonium ion}}{NH_4^+}$$

pH: Acid-Base Concentration

The more hydrogen ions in a solution, the more acidic the solution is. Conversely, the greater the concentration of hydroxyl ions (the lower the concentration of H^+), the more basic, or *alkaline* (al′kuh-līn), the solution becomes. The relative concentration of hydrogen ions in various body fluids is measured in concentration units called **pH units** (pe-āch′).

The idea for a pH scale was devised by a Danish biochemist and beer brewer named Sören Sörensen in 1909. He was searching for a convenient means of checking the acidity of his alcoholic product to prevent its spoilage by bacterial action. (Acidic conditions inhibit many bacteria.) The pH scale that resulted is based on the concentration of hydrogen ions in a solution, expressed in terms of moles per liter, or molarity. The pH scale runs from 0 to 14 and is *logarithmic*. In other words, each successive change of one pH unit represents a tenfold change in hydrogen ion concentration (**Figure 2.13**). The pH of a solution is thus defined as the negative logarithm of the hydrogen ion concentration $[H^+]$ in moles per liter, or $-\log[H^+]$. (Note that brackets [ ] indicate concentration of a substance.)

At a pH of 7 (at which $[H^+]$ is 10^{-7} *M*), the solution is *neutral*—neither acidic nor basic. The number of hydrogen ions exactly equals the number of hydroxyl ions (pH = pOH). Absolutely pure (distilled) water has a pH of 7.

Solutions with a pH below 7 are acidic—the hydrogen ions outnumber the hydroxyl ions. The lower the pH, the more acidic the solution. A solution with a pH of 6 has ten times as many hydrogen ions as a solution with a pH of 7.

Solutions with a pH higher than 7 are alkaline, and the relative concentration of hydrogen ions decreases by a factor of 10 with each higher pH unit. Thus, solutions with pH values of 8 and 12 have, respectively, 1/10 and 1/100,000 ($1/10 \times 1/10 \times 1/10 \times 1/10 \times 1/10$) as many hydrogen ions as a solution of pH 7.

The approximate pH of several body fluids and of a number of common substances appears in Figure 2.13. Notice that as the hydrogen ion concentration decreases, the hydroxyl ion concentration rises, and vice versa.

Neutralization

What happens when acids and bases are mixed? They react with each other in displacement reactions to form water and a salt. For example, when hydrochloric acid and sodium hydroxide interact, sodium chloride (a salt) and water are formed.

$$\underset{\text{acid}}{HCl} + \underset{\text{base}}{NaOH} \rightarrow \underset{\text{salt}}{NaCl} + \underset{\text{water}}{H_2O}$$

This type of reaction is called a **neutralization reaction**, because the joining of H^+ and OH^- to form water neutralizes the solution. Although the salt produced is written in molecular form (NaCl), remember that it actually exists as dissociated sodium and chloride ions when dissolved in water.

Buffers

Living cells are extraordinarily sensitive to even slight changes in the pH of the environment. Imagine what would happen to all those hydrogen bonds in biological molecules with large numbers of free H^+ running around. (Can't you just hear those molecules saying "Why share hydrogen when I can have my own?")

Homeostasis of acid-base balance is carefully regulated by the kidneys and lungs and by chemical systems (proteins and

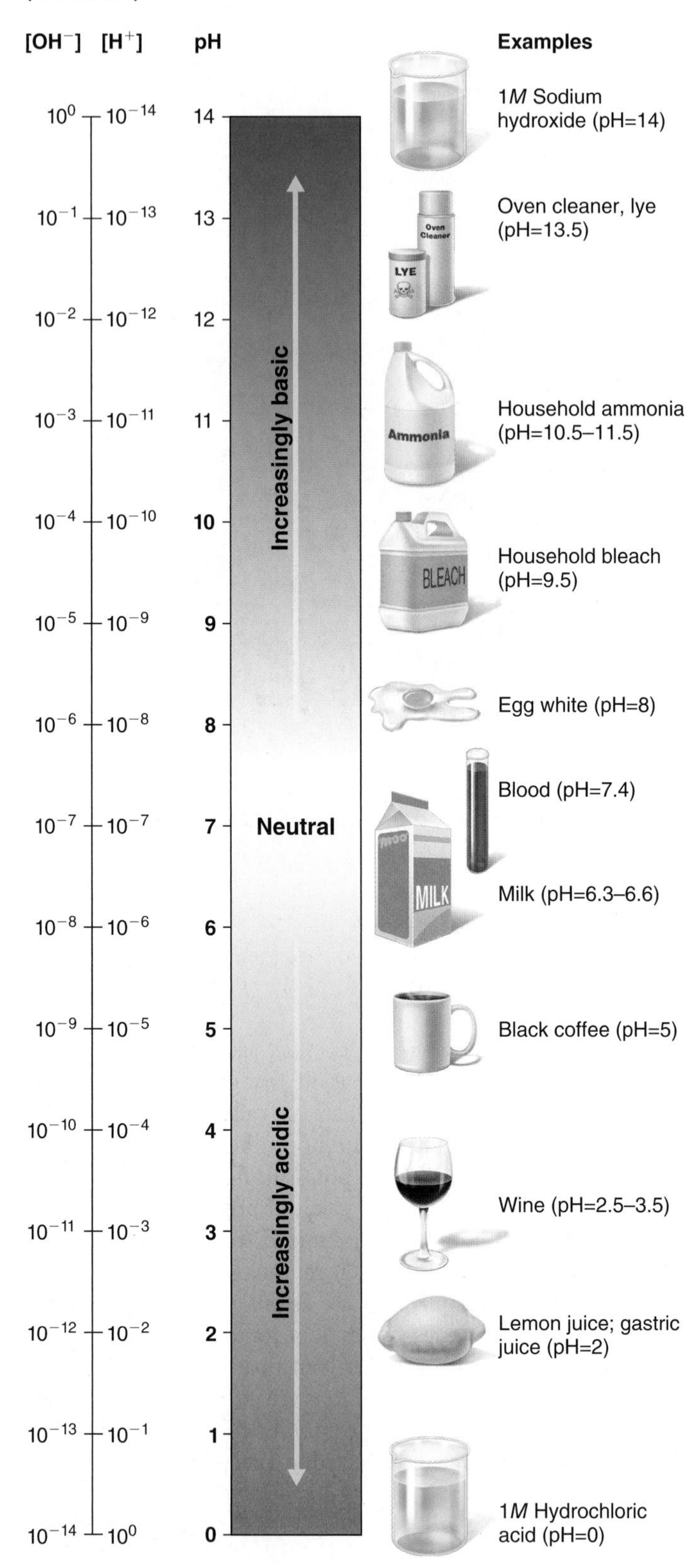

Figure 2.13 The pH scale and pH values of representative substances. The pH scale is based on the number of hydrogen ions in solution. The actual concentrations of hydrogen ions, $[H^+]$, and hydroxyl ions, $[OH^-]$, in moles per liter are indicated for each pH value noted. At a pH of 7, $[H^+] = [OH^-]$ and the solution is neutral.

other types of molecules) called **buffers**. Buffers resist abrupt and large swings in the pH of body fluids by releasing hydrogen ions (acting as acids) when the pH begins to rise and by binding hydrogen ions (acting as bases) when the pH drops. Because blood comes into close contact with nearly every body cell, regulating its pH is particularly critical. Normally, blood pH varies within a very narrow range (7.35 to 7.45). If the pH of blood varies from these limits by more than a few tenths of a unit, it may be fatal.

To comprehend how chemical buffer systems operate, you must thoroughly understand strong and weak acids and bases. The first important concept is that the acidity of a solution reflects *only* the free hydrogen ions, not those still bound to anions. Consequently, acids that dissociate completely and irreversibly in water are called **strong acids**, because they can dramatically change the pH of a solution. Examples are hydrochloric acid and sulfuric acid. If we could count out 100 molecules of hydrochloric acid and place them in 1 milliliter (ml) of water, we could expect to end up with 100 H^+, 100 Cl^-, and no undissociated hydrochloric acid molecules in that solution.

Acids that do not dissociate completely, like carbonic acid (H_2CO_3) and acetic acid (HAc), are **weak acids**. If we were to place 100 molecules of acetic acid in 1 ml of water, the reaction would be something like this:

$$100\ HAc \rightarrow 90\ HAc + 10\ H^+ + 10\ Ac^-$$

Because undissociated acids do not affect pH, the acetic acid solution is much less acidic than the HCl solution. Weak acids dissociate in a predictable way, and molecules of the intact acid are in dynamic equilibrium with the dissociated ions. Consequently, the dissociation of acetic acid may also be written as

$$HAc \rightleftharpoons H^+ + Ac^-$$

This viewpoint allows us to see that if H^+ (released by a strong acid) is added to the acetic acid solution, the equilibrium will shift to the left and some H^+ and Ac^- will recombine to form HAc. On the other hand, if a strong base is added and the pH begins to rise, the equilibrium shifts to the right and more HAc molecules dissociate to release H^+. This characteristic of weak acids allows them to play important roles in the chemical buffer systems of the body.

The concept of strong and weak bases is more easily explained. Remember that bases are proton acceptors. Thus, **strong bases** are those, like hydroxides, that dissociate easily in water and quickly tie up H^+. On the other hand, sodium bicarbonate (commonly known as baking soda) ionizes incompletely and reversibly. Because it accepts relatively few protons, its released bicarbonate ion is considered a **weak base**.

Now let's examine how one buffer system helps to maintain pH homeostasis of the blood. Although there are other chemical blood buffers, the **carbonic acid–bicarbonate system** is a major one. Carbonic acid (H_2CO_3) dissociates reversibly, releasing bicarbonate ions (HCO_3^-) and protons (H^+):

$$\underset{\substack{H^+ \text{ donor} \\ \text{(weak acid)}}}{H_2CO_3} \underset{\text{Response to drop in pH}}{\overset{\text{Response to rise in pH}}{\rightleftharpoons}} \underset{\substack{H^+ \text{ acceptor} \\ \text{(weak base)}}}{HCO_3^-} + \underset{\text{proton}}{H^+}$$

The chemical equilibrium between carbonic acid (a weak acid) and bicarbonate ion (a weak base) resists changes in blood pH by shifting to the right or left as H^+ ions are added to or removed from the blood. As blood pH rises (becomes more alkaline due to the addition of a strong base), the equilibrium shifts to the right, forcing more carbonic acid to dissociate. Similarly, as blood pH begins to drop (becomes more acidic due to the addition of a strong acid), the equilibrium shifts to the left as more bicarbonate ions begin to bind with protons. As you can see, strong bases are replaced by a weak base (bicarbonate ion) and protons released by strong acids are tied up in a weak one (carbonic acid). In either case, the blood pH changes much less than it would in the absence of the buffering system. We discuss acid-base balance and buffers in more detail in Chapter 25.

☑ Check Your Understanding

16. Salts are electrolytes. What does that mean?

17. Which ion is responsible for increased acidity?

18. To minimize the sharp pH shift that occurs when a strong acid is added to a solution, is it better to add a weak base or a strong base? Why?

19. MAKING **connections** We have learned about the complementarity of structure and function as it relates to anatomy and physiology (Chapter 1). See if you can extend your thinking about this principle to a simple molecule, and explain how the structure of a water molecule makes water an excellent solvent.

For answers, see Answers Appendix.

2.7 Organic compounds are made by dehydration synthesis and broken down by hydrolysis

→ Learning Objective

☐ **Explain the role of dehydration synthesis and hydrolysis in forming and breaking down organic molecules.**

Molecules unique to living systems—carbohydrates, lipids (fats), proteins, and nucleic acids—all contain carbon and hence are organic compounds. Organic compounds are generally distinguished by the fact that they contain carbon, and inorganic compounds are defined as compounds that lack carbon. You should be aware of a few exceptions to this generalization: Carbon dioxide and carbon monoxide, for example, contain carbon but are considered inorganic compounds.

For the most part, organic molecules are very large molecules, but their interactions with other molecules typically involve only small, reactive parts of their structure called *functional groups* (acid groups, amines, and others).

What makes carbon so special that "living" chemistry depends on its presence? To begin with, no other *small* atom is so precisely **electroneutral**. The consequence of its electroneutrality is that carbon never loses or gains electrons. Instead, it always shares them. Furthermore, with four valence shell electrons, carbon forms four covalent bonds with other elements, as well as with other carbon atoms. As a result, carbon can help form long, chainlike molecules (common in fats), ring structures (typical of carbohydrates and steroids), and many other structures that are uniquely suited for specific roles in the body.

Many biological molecules (carbohydrates and proteins for example) are polymers. **Polymers** are chainlike molecules made of many smaller, identical or similar units (**monomers**), which are joined together by a process called **dehydration synthesis** (Figure 2.14). During dehydration synthesis, a hydrogen atom is removed from one monomer and a hydroxyl group is removed from the monomer it is to be joined with. As a covalent bond unites the monomers, a water molecule is released. This removal of a water molecule at the bond site occurs each time a monomer is added to the growing polymer chain. The opposite reaction in which molecules are degraded is called **hydrolysis** (hi-drol′ĭ-sis; water splitting). In these reactions, a water molecule is added to each bond to be broken down, thereby releasing its building blocks or smaller molecules.

☑ Check Your Understanding

20. What is the result of hydrolysis reactions and how are these reactions accomplished in the body?

For answers, see Answers Appendix.

2.8 Carbohydrates provide an easily used energy source for the body

→ Learning Objective

☐ **Describe the building blocks, general structure, and biological functions of carbohydrates.**

Carbohydrates, a group of molecules that includes sugars and starches, represent 1–2% of cell mass. Carbohydrates contain

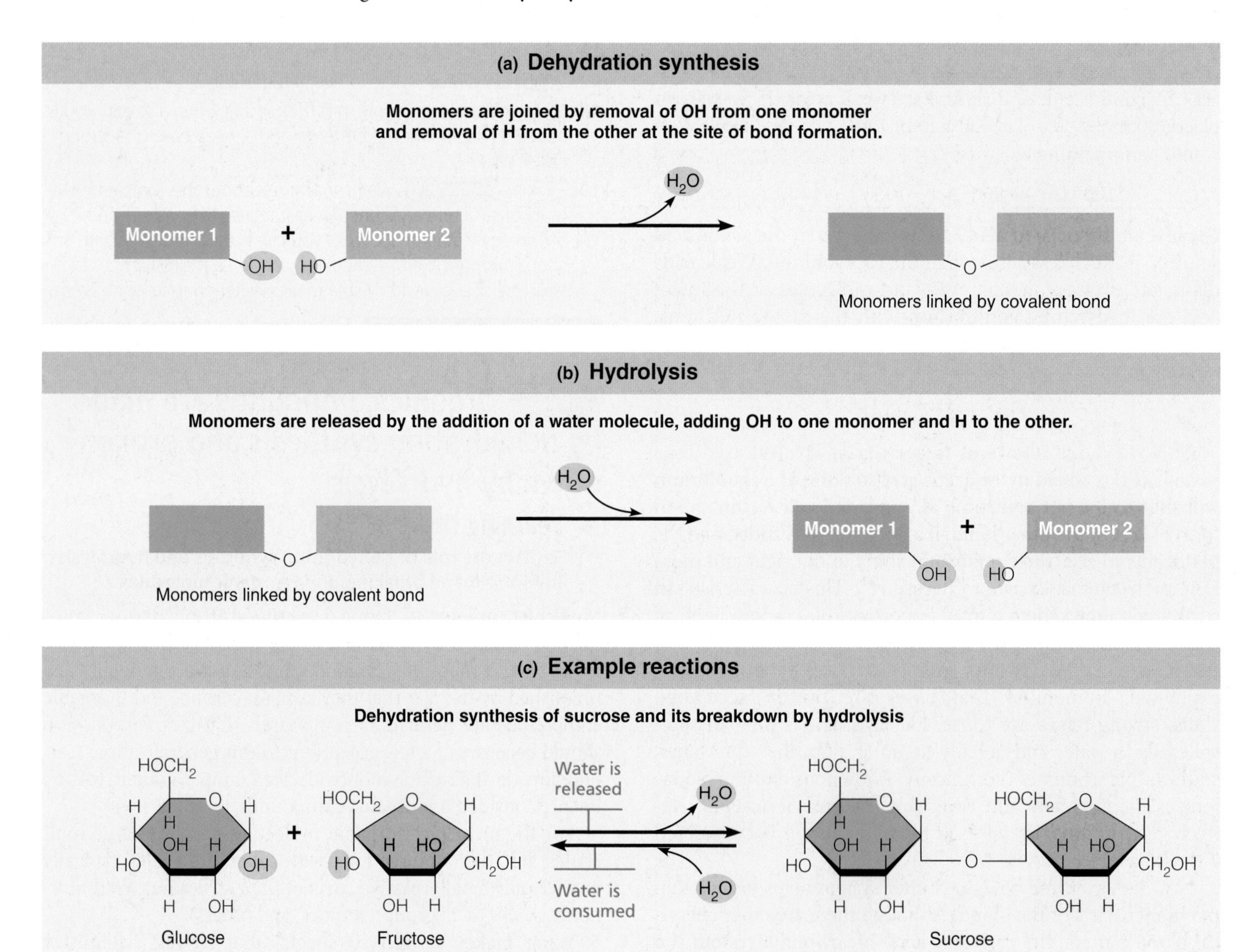

Figure 2.14 Dehydration synthesis and hydrolysis. Biological molecules are formed from their monomers, or units, by dehydration synthesis and broken down to the monomers by hydrolysis reactions.

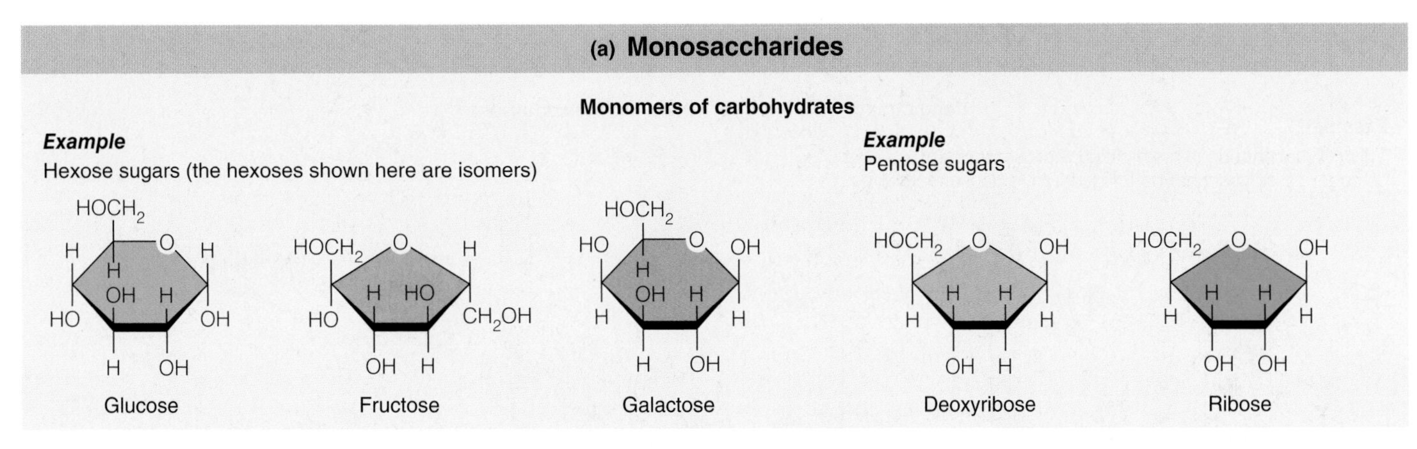

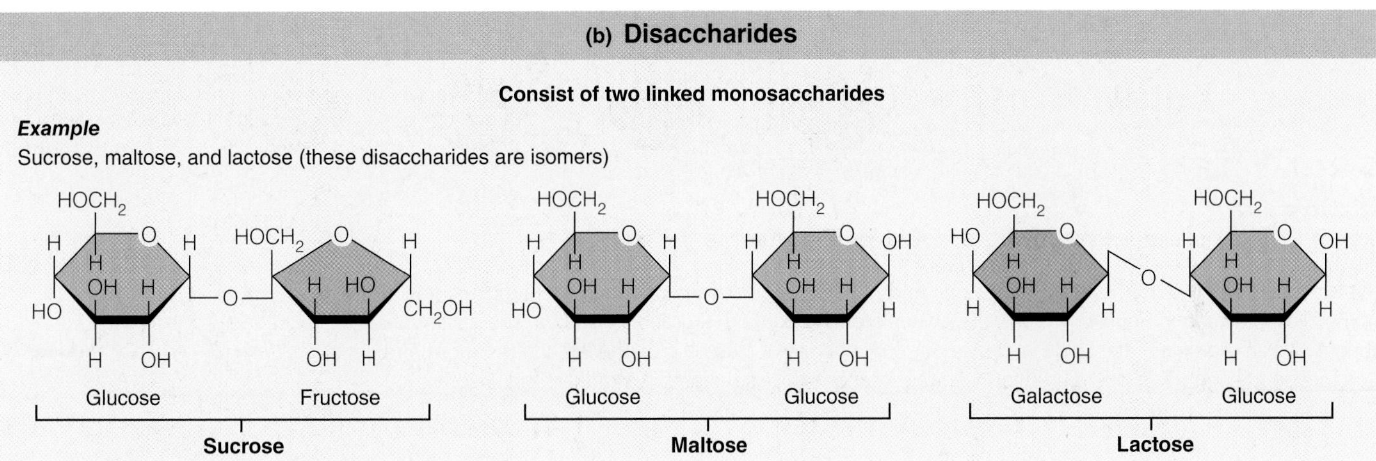

Figure 2.15 Carbohydrate molecules important to the body. *(Figure continues on p. 40.)*

carbon, hydrogen, and oxygen, and generally the hydrogen and oxygen atoms occur in the same 2:1 ratio as in water. This ratio is reflected in the word *carbohydrate* ("hydrated carbon").

A carbohydrate can be classified according to size and solubility as a monosaccharide ("one sugar"), disaccharide ("two sugars"), or polysaccharide ("many sugars"). Monosaccharides are the monomers, or building blocks, of the other carbohydrates. In general, the larger the carbohydrate molecule, the less soluble it is in water.

Monosaccharides

Monosaccharides (mon″o-sak′ah-rīdz), or *simple sugars*, are single-chain or single-ring structures containing from three to seven carbon atoms (**Figure 2.15a**). Usually the carbon, hydrogen, and oxygen atoms occur in the ratio 1:2:1, so a general formula for a monosaccharide is $(CH_2O)_n$, where *n* is the number of carbons in the sugar. Glucose, for example, has six carbon atoms, and its molecular formula is $C_6H_{12}O_6$. Ribose, with five carbons, is $C_5H_{10}O_5$.

Monosaccharides are named generically according to the number of carbon atoms they contain. Most important in the body are the pentose (five-carbon) and hexose (six-carbon) sugars. The pentose *deoxyribose* (de-ok″sĭ-ri′bōs) is part of DNA, and *glucose*, a hexose, is blood sugar.

Two other hexoses, *galactose* and *fructose*, are **isomers** (ī′so-mers) of glucose. That is, they have the same molecular formula ($C_6H_{12}O_6$), but their atoms are arranged differently, giving them different chemical properties (Figure 2.15a).

Disaccharides

A **disaccharide** (di-sak′ah-rīd), or *double sugar*, is formed when two monosaccharides are joined by *dehydration synthesis* (Figure 2.14a, c). In this synthesis reaction, a water molecule is lost as the bond is made, as illustrated by the synthesis of sucrose (soo′krōs):

$$\underset{\text{glucose + fructose}}{2C_6H_{12}O_6} \rightarrow \underset{\text{sucrose}}{C_{12}H_{22}O_{11}} + \underset{\text{water}}{H_2O}$$

Important disaccharides in the diet are *sucrose* (glucose + fructose), which is cane or table sugar; *lactose* (glucose + galactose), found in milk; and *maltose* (glucose + glucose), also called malt sugar (Figure 2.15b). Disaccharides are too large to be transported through cell membranes, so they must be digested by hydrolysis to their simple sugar units to be absorbed from the digestive tract into the blood (Figure 2.14b, c). A water molecule is added to each bond, breaking the bonds and releasing the simple sugar units.

(c) **Polysaccharides**

Long chains (polymers) of linked monosaccharides

Example
This polysaccharide is a simplified representation of glycogen, a polysaccharide formed from glucose molecules.

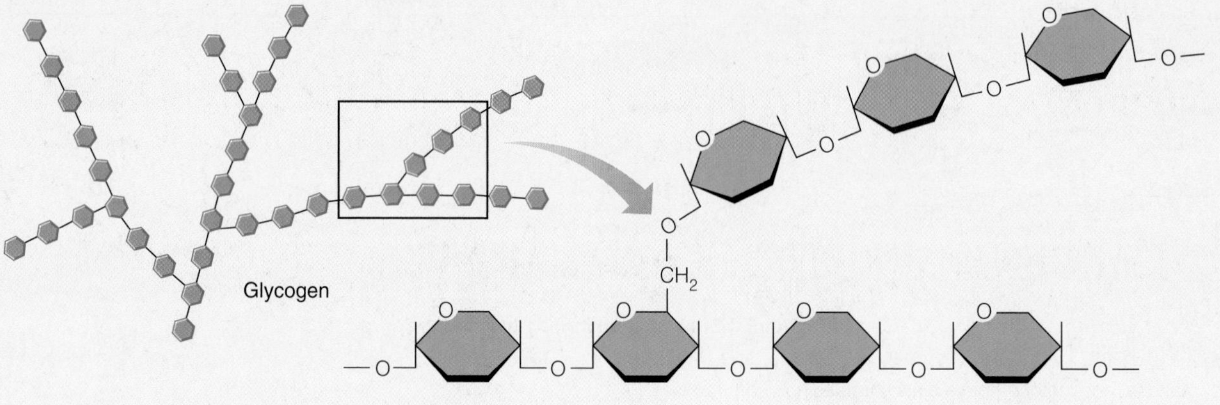

Glycogen

CH_2

Figure 2.15 *(continued)** **Carbohydrate molecules important to the body.**

*Notice that in Figure 2.15 the carbon (C) atoms present at the angles of the carbohydrate ring structures are not illustrated and in Figure 2.15c only the oxygen atoms and one CH_2 group are shown. The illustrations at right give an example of this shorthand style: The full structure of glucose is on the left and the shorthand structure on the right. This style is used for nearly all organic ringlike structures illustrated in this chapter.

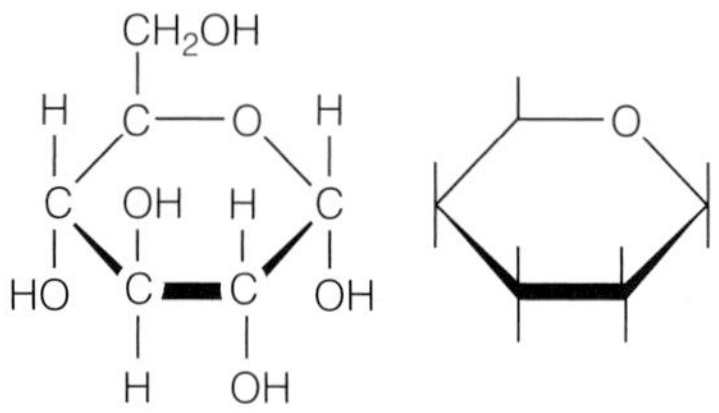

Polysaccharides

Polysaccharides (pol″e-sak′ah-rīdz) are polymers of simple sugars linked together by dehydration synthesis. Because polysaccharides are large, fairly insoluble molecules, they are ideal storage products. Another consequence of their large size is that they lack the sweetness of the simple and double sugars.

Only two polysaccharides are of major importance to the body: starch and glycogen. Both are polymers of glucose. Only their degree of branching differs.

Starch is the storage carbohydrate formed by plants. The number of glucose units composing a starch molecule is high and variable. When we eat starchy foods such as grain products and potatoes, the starch must be digested for its glucose units to be absorbed. We are unable to digest *cellulose*, another polysaccharide found in all plant products. However, it is important in providing the *bulk* (one form of fiber) that helps move feces through the colon.

Glycogen (gli′ko-jen), the storage carbohydrate of animal tissues, is stored primarily in skeletal muscle and liver cells. Like starch, it is highly branched and is a very large molecule (Figure 2.15c). When blood sugar levels drop sharply, liver cells break down glycogen and release its glucose units to the blood. Since there are many branch endings from which glucose can be released simultaneously, body cells have almost instant access to glucose fuel.

Carbohydrate Functions

The major function of carbohydrates in the body is to provide a ready, easily used source of cellular fuel. Most cells can use only a few types of simple sugars, and glucose is at the top of the "cellular menu." As described in our earlier discussion of oxidation-reduction reactions (pp. 32–33), glucose is broken down and oxidized within cells. During these reactions, electrons are transferred, releasing the bond energy stored in glucose. This energy is used to synthesize ATP. When ATP supplies are sufficient, dietary carbohydrates are converted to glycogen or fat and stored. Those of us who have gained weight from eating too many carbohydrate-rich snacks have personal experience with this conversion process!

Only small amounts of carbohydrates are used for structural purposes. For example, some sugars are found in our genes. Others are attached to the external surfaces of cells where they act as "road signs" to guide cellular interactions.

☑ Check Your Understanding

21. What are the monomers of carbohydrates called? Which monomer is blood sugar?

22. What is the animal form of stored carbohydrate called?

For answers, see Answers Appendix.

2.9 Lipids insulate body organs, build cell membranes, and provide stored energy

→ Learning Objective

- ☐ **Describe the building blocks, general structure, and biological functions of lipids.**

Lipids are insoluble in water but dissolve readily in other lipids and in organic solvents such as alcohol and ether. Like

(a) Triglyceride formation

Three fatty acid chains are bound to glycerol by dehydration synthesis.

Glycerol + 3 fatty acid chains → Triglyceride, or neutral fat + 3 water molecules ($3H_2O$)

(b) "Typical" structure of a phospholipid molecule

Two fatty acid chains and a phosphorus-containing group are attached to the glycerol backbone.

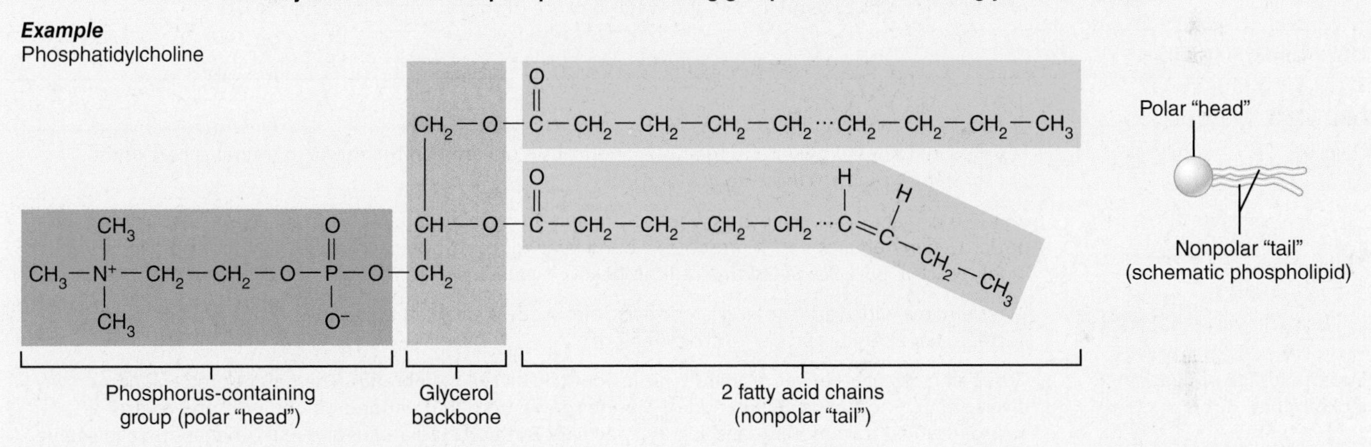

(c) Simplified structure of a steroid

Four interlocking hydrocarbon rings form a steroid.

Example
Cholesterol (cholesterol is the basis for all steroids formed in the body)

Figure 2.16 Lipids. The general structure of **(a)** triglycerides, or neutral fats, **(b)** phospholipids, and **(c)** cholesterol.

Practice art labeling
MasteringA&P®>Study Area>Chapter 2

carbohydrates, all lipids contain carbon, hydrogen, and oxygen, but the proportion of oxygen in lipids is much lower. In addition, phosphorus is found in some of the more complex lipids. Lipids include *triglycerides*, *phospholipids* (fos″fo-lip′idz), *steroids* (stĕ′roidz), and a number of other lipoid substances. Table 2.2 on p. 42 gives the locations and functions of some lipids found in the body.

Triglycerides (Neutral Fats)

Triglycerides (tri-glis′er-īdz), also called **neutral fats**, are commonly known as *fats* when solid or *oils* when liquid. Triglycerides are large molecules, often consisting of hundreds of atoms. They provide the body's most efficient and compact form of stored energy, and when they are oxidized, they yield large amounts of energy. A triglyceride is composed of two types of building blocks, **fatty acids** and **glycerol** (glis′er-ol), in a 3:1 ratio of fatty acids to glycerol (Figure 2.16a). Fatty acids are linear chains

Table 2.2 Representative Lipids Found in the Body

LIPID TYPE	LOCATION/FUNCTION
Triglycerides (Neutral Fats)	
	Fat deposits (in subcutaneous tissue and around organs) protect and insulate body organs, and are the major source of *stored* energy in the body.
Phospholipids (phosphatidylcholine; cephalin; others)	
	Chief components of cell membranes. Participate in the transport of lipids in plasma. Prevalent in nervous tissue.
Steroids	
Cholesterol	The structural basis for manufacture of all body steroids. A component of cell membranes.
Bile salts	These breakdown products of cholesterol are released by the liver into the digestive tract, where they aid fat digestion and absorption.
Vitamin D	Fat-soluble vitamin produced in the skin on exposure to UV radiation. Necessary for normal bone growth and function.
Sex hormones	Estrogen and progesterone (female hormones) and testosterone (a male hormone) are produced in the gonads. Necessary for normal reproductive function.
Adrenocortical hormones	Cortisol, a glucocorticoid, is a metabolic hormone necessary for maintaining normal blood glucose levels. Aldosterone helps to regulate salt and water balance of the body by targeting the kidneys.
Other Lipoid Substances	
Fat-soluble vitamins:	
A	Ingested in orange-pigmented vegetables and fruits. Converted in the retina to retinal, a part of the photoreceptor pigment involved in vision.
E	Ingested in plant products such as wheat germ and green leafy vegetables. Claims have been made (but not proved in humans) that it promotes wound healing, contributes to fertility, and may help to neutralize highly reactive particles called free radicals believed to be involved in triggering some types of cancer.
K	Prevalent in a wide variety of ingested foods; also made available to humans by the action of intestinal bacteria. Necessary for proper clotting of blood.
Eicosanoids (prostaglandins; leukotrienes; thromboxanes)	Group of molecules derived from fatty acids found in all cell membranes. The potent prostaglandins have diverse effects, including stimulation of uterine contractions, regulation of blood pressure, control of gastrointestinal tract motility, and secretory activity. Both prostaglandins and leukotrienes are involved in inflammation. Thromboxanes are powerful vasoconstrictors.
Lipoproteins	Lipoid and protein-based substances that transport fatty acids and cholesterol in the bloodstream. Major varieties are high-density lipoproteins (HDLs) and low-density lipoproteins (LDLs).
Glycolipids	Components of cell membranes. Lipids associated with carbohydrate molecules determine blood type, play a role in cell recognition, and in recognition of foreign substances by immune cells.

of carbon and hydrogen atoms (hydrocarbon chains) with an organic acid group (—COOH) at one end. Glycerol is a modified simple sugar (a sugar alcohol).

Fat synthesis involves attaching three fatty acid chains to a single glycerol molecule by dehydration synthesis. The result is an E-shaped molecule. The glycerol backbone is the same in all triglycerides, but the fatty acid chains vary, resulting in different kinds of fats and oils.

Their hydrocarbon chains make triglycerides nonpolar molecules. Because polar and nonpolar molecules do not interact (oil and water do not mix), digestion and absorption of fats is complicated and ingested fats and oils must be broken down to their building blocks.

Triglycerides are found mainly beneath the skin, where they insulate the deeper body tissues from heat loss and protect them from mechanical trauma. For example, women are usually more successful English Channel swimmers than men. Their success is due partly to their thicker subcutaneous fatty layer, which helps insulate them from the bitterly cold water of the Channel.

The length of a triglyceride's fatty acid chains and their degree of *saturation* with H atoms determine how solid the molecule is at a given temperature. Fatty acid chains with only single covalent bonds between carbon atoms are referred to as **saturated**. Their fatty acid chains are straight and, at room temperature, the molecules of a saturated fat are packed closely together, forming a solid. Fatty acids that contain one or more double bonds between carbon atoms are said to be **unsaturated** (**monounsaturated** or **polyunsaturated**, respectively). The double bonds cause the fatty acid chains to kink so that they cannot be packed closely enough to solidify. Hence, triglycerides with short fatty acid chains or unsaturated fatty acids are oils (liquid at room temperature) and are typical of plant lipids. Examples include olive and peanut oils (rich in monounsaturated fats) and corn, soybean, and safflower oils, which contain a high percentage of polyunsaturated fatty acids. Longer fatty acid chains and more saturated fatty acids are common in animal fats such as butterfat and the fat of meats, which are solid at room temperature. Of the two types of fatty acids, the unsaturated variety, especially olive oil, is said to be more "heart healthy."

Trans fats, common in many margarines and baked products, are oils that have been solidified by addition of H atoms at sites of carbon double bonds. They increase the risk of heart disease even more than the solid animal fats. Conversely, the **omega-3 fatty acids**, found naturally in cold-water fish, appear to decrease the risk of heart disease and some inflammatory diseases.

Phospholipids

Phospholipids are modified triglycerides. Specifically, they are diglycerides with a phosphorus-containing group and two, rather than three, fatty acid chains (Figure 2.16b). The phosphorus-containing group gives phospholipids their distinctive chemical properties. Although the hydrocarbon portion (the "tail") of the molecule is nonpolar and interacts only with nonpolar molecules, the phosphorus-containing part (the "head") is polar and attracts other polar or charged particles, such as water or ions. This unique characteristic of phospholipids allows them to be used as the chief material for building cellular membranes. Some biologically important phospholipids and their functions are listed in Table 2.2.

Steroids

Structurally, steroids differ quite a bit from fats and oils. **Steroids** are basically flat molecules made of four interlocking hydrocarbon rings. Like triglycerides, steroids are fat soluble and contain little oxygen. The single most important molecule in our steroid chemistry is *cholesterol* (ko-les′ter-ol) (Figure 2.16c). We ingest cholesterol in animal products such as eggs, meat, and cheese, and our liver produces some.

Cholesterol has earned bad press because of its role in atherosclerosis, but it is essential for human life. Cholesterol is found in cell membranes and is the raw material for synthesis of vitamin D, steroid hormones, and bile salts. Although steroid hormones are present in the body in only small quantities, they are vital to homeostasis. Without sex hormones, reproduction would be impossible, and a total lack of the corticosteroids produced by the adrenal glands is fatal.

Eicosanoids

The **eicosanoids** (i-ko′sah-noyds) are diverse lipids chiefly derived from a 20-carbon fatty acid (arachidonic acid) found in all cell membranes. Most important of these are the *prostaglandins* and their relatives, which play roles in various body processes including blood clotting, regulation of blood pressure, inflammation, and labor contractions (Table 2.2). Their synthesis and inflammatory actions are blocked by NSAIDs (nonsteroidal anti-inflammatory drugs).

☑ Check Your Understanding

23. How do triglycerides differ from phospholipids in body function and location?

For answers, see Answers Appendix.

2.10 Proteins are the body's basic structural material and have many vital functions

→ Learning Objectives

- ☐ **Describe the four levels of protein structure.**
- ☐ **Describe enzyme action.**

Protein composes 10–30% of cell mass and is the basic structural material of the body. However, not all proteins are construction materials. Many play vital roles in cell function. Proteins, which include enzymes (biological catalysts), hemoglobin of the blood, and contractile proteins of muscle, have the most varied functions of any molecules in the body. All proteins contain carbon, oxygen, hydrogen, and nitrogen, and many contain sulfur as well.

Amino Acids and Peptide Bonds

The building blocks of proteins are molecules called **amino acids**, of which there are 20 common types. As shown in the amino acid below, all amino acids have two important functional groups: a basic group called an *amine* (ah′mēn) *group* (—NH_2), and an organic *acid group* (—COOH).

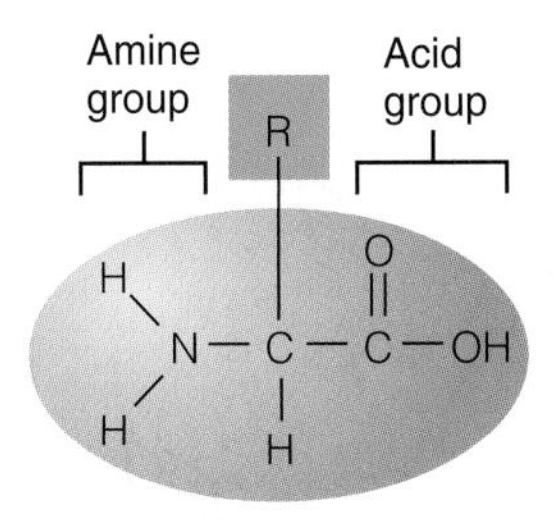

An amino acid may therefore act either as a base (proton acceptor) or an acid (proton donor). All amino acids are identical except for a single group of atoms called their *R group*. Hence, it is differences in the R group that make each amino acid chemically unique.

Proteins are long chains of amino acids joined together by dehydration synthesis, with the acid end of one amino acid linked to the amine end of the next. The resulting bond produces a characteristic arrangement of linked atoms called a **peptide bond** (Figure 2.17). Two united amino acids form a *dipeptide*, three a *tripeptide*, and ten or more a *polypeptide*. Although polypeptides containing more than 50 amino acids are called proteins, most proteins are **macromolecules**, large, complex molecules containing from 100 to over 10,000 amino acids.

Because each type of amino acid has distinct properties, the sequence in which they are bound together produces proteins that vary widely in both structure and function. We can think of the 20 amino acids as a 20-letter "alphabet" used in specific combinations to form "words" (proteins). Just as a change in one letter can produce a word with an entirely different meaning (flour → floor) or that is nonsensical (flour → flocr), changes in the kinds or positions of amino acids can yield proteins with different functions or proteins that are nonfunctional. Nevertheless, there are thousands of different proteins in the body, each with distinct functional properties, and all constructed from different combinations of the 20 common amino acids.

Structural Levels of Proteins

Proteins can be described in terms of four structural levels: primary, secondary, tertiary, and quaternery. The linear sequence of amino acids composing the polypeptide chain is the *primary structure* of a protein. This structure, which resembles a strand of amino acid "beads," is the backbone of the protein molecule (Figure 2.18a).

Proteins do not normally exist as simple, linear chains of amino acids. Instead, they twist or bend upon themselves to form a more complex *secondary structure*. The most common type of secondary structure is the **alpha (α)-helix**, which resembles a Slinky toy or a coiled spring (Figure 2.18b). The α-helix is formed by coiling of the primary chain and is stabilized by hydrogen bonds formed between NH and CO groups in amino acids in the primary chain which are about four amino acids apart. Hydrogen bonds in α-helices always link different parts of the *same* chain together.

In another type of secondary structure, the **beta (β)-pleated sheet**, the primary polypeptide chains do not coil, but are linked side by side by hydrogen bonds to form a pleated, ribbonlike structure that resembles an accordion's bellows (Figure 2.18b). Notice that in this type of secondary structure, the hydrogen bonds may link together *different polypeptide chains* as well as *different parts* of the same chain that has folded back on itself. A single polypeptide chain may exhibit both types of secondary structure at various places along its length.

Many proteins have *tertiary structure* (ter′she-a″re), the next higher level of complexity, which is superimposed on secondary structure and involves the amino acids' R-groups. Tertiary structure is achieved when α-helical or β-pleated regions of the polypeptide chain fold upon one another to produce a compact ball-like, or *globular*, molecule (Figure 2.18c). Hydrophobic R groups are on the inside of the molecule and hydrophilic R groups are on its outside. Their interactions plus those reinforced by covalent and hydrogen bonds help to maintain the unique tertiary shape.

When two or more polypeptide chains aggregate in a regular manner to form a complex protein, the protein has *quaternary structure* (kwah′ter-na″re). The transthyretin molecule with its four identical globular subunits represents this level of structure (Figure 2.18d). (Transthyretin transports thyroid hormone in the blood.)

How do these different levels of structure arise? Although a protein with tertiary or quaternary structure looks a bit like a clump of congealed pasta, the ultimate overall structure of any protein is very specific and is dictated by its primary structure. In other words, the types and relative positions of amino acids in the protein backbone determine where bonds can form to produce the complex coiled or folded structures that keep water-loving amino acids near the surface and water-fleeing amino acids buried in the protein's core.

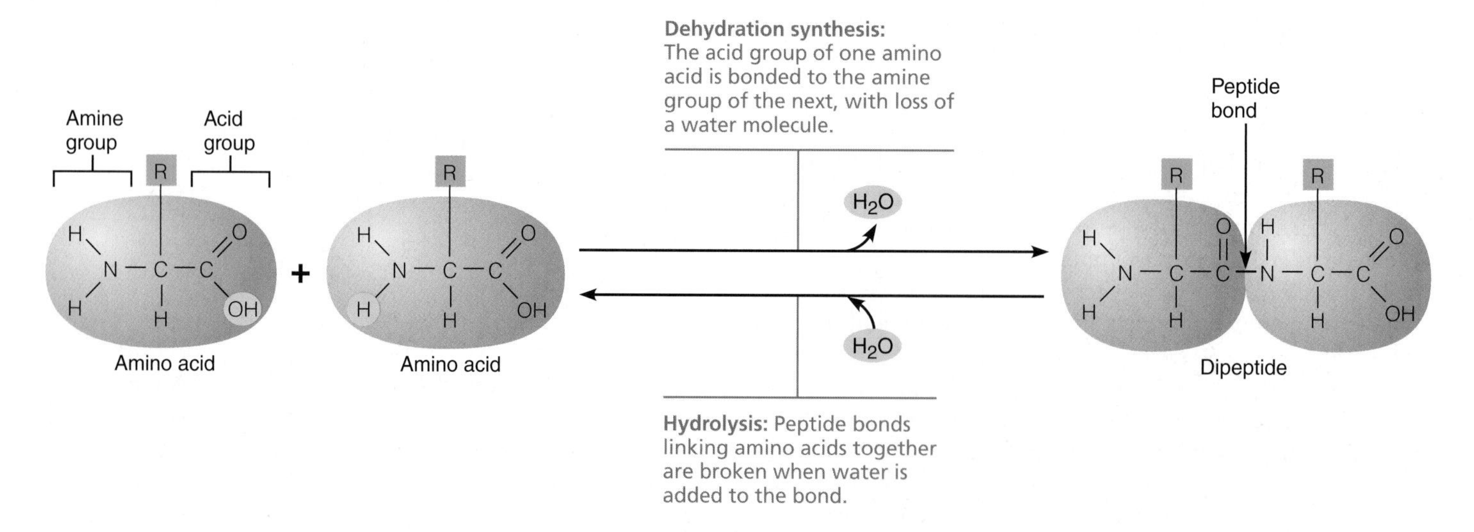

Figure 2.17 Amino acids are linked together by peptide bonds. Peptide bonds are formed by dehydration synthesis and broken by hydrolysis reactions.

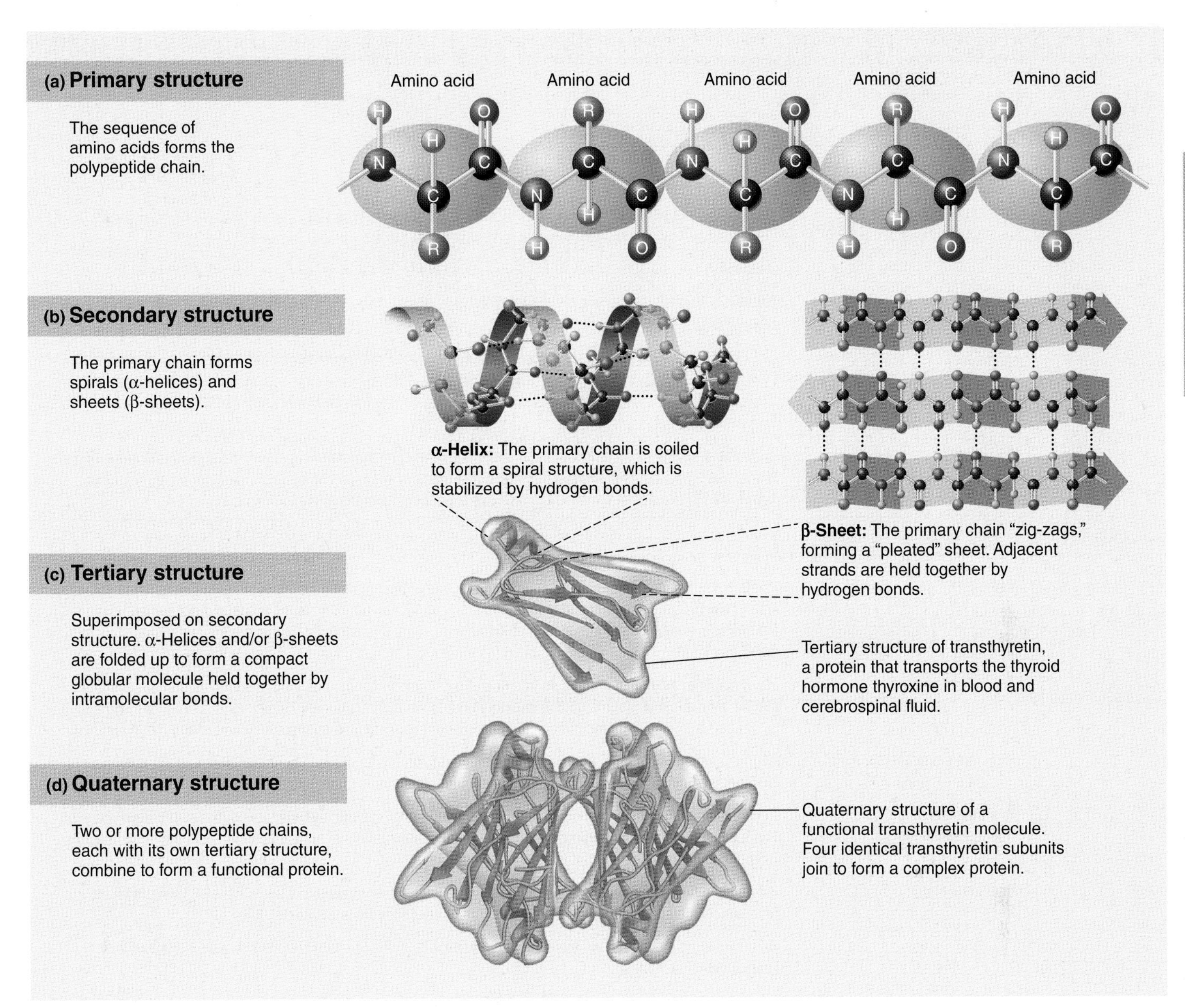

Figure 2.18 Levels of protein structure.

Fibrous and Globular Proteins

The overall structure of a protein determines its biological function. In general, proteins are classified according to their overall appearance and shape as either fibrous or globular.

Fibrous proteins, also known as **structural proteins**, are extended and strandlike. Some exhibit only secondary structure, but most have tertiary or even quaternary structure as well. For example, *collagen* (kol′ah-jen) is a composite of the helical tropocollagen molecules packed together side by side to form a strong ropelike structure. Fibrous proteins are insoluble in water, and very stable—qualities ideal for providing mechanical support and tensile strength to the body's tissues. Besides collagen, which is the single most abundant protein in the body, the fibrous proteins include keratin, elastin, and certain contractile proteins of muscle (Table 2.3 on p. 46).

Globular proteins, also called **functional proteins**, are compact, spherical proteins that have at least tertiary structure. Some also exhibit quaternary structure. The globular proteins are water-soluble, chemically active molecules, and they play crucial roles in virtually all biological processes. Some (antibodies) help to provide immunity, others (protein-based hormones) regulate growth and development, and still others (enzymes) are catalysts that oversee just about every chemical reaction in the body. The roles of these and selected other proteins found in the body are summarized in Table 2.3.

Protein Denaturation

Fibrous proteins are stable, but globular proteins are quite the opposite. The activity of a protein depends on its specific three-dimensional structure, and intramolecular bonds, particularly

Table 2.3 Representative Types of Proteins in the Body

CLASSIFICATION ACCORDING TO		
OVERALL STRUCTURE	**GENERAL FUNCTION**	**EXAMPLES FROM THE BODY**
Fibrous		
	Structural framework/ mechanical support	*Collagen*, found in all connective tissues, is the single most abundant protein in the body. It is responsible for the tensile strength of bones, tendons, and ligaments. *Keratin* is the structural protein of hair and nails and a water-resistant material of skin. *Elastin* is found, along with collagen, where durability and flexibility are needed, such as in the ligaments that bind bones together. *Spectrin* internally reinforces and stabilizes the plasma membrane of some cells, particularly red blood cells. *Dystrophin* reinforces and stabilizes the plasma membrane of muscle cells. *Titin* helps organize the intracellular structure of muscle cells and accounts for the elasticity of skeletal muscles.
	Movement	*Actin* and *myosin*, contractile proteins, are found in substantial amounts in muscle cells, where they cause muscle cell shortening (contraction); they also function in cell division in all cell types. Actin is important in intracellular transport, particularly in nerve cells.
Globular		
	Catalysis	Protein enzymes are essential to virtually every biochemical reaction in the body; they increase the rates of chemical reactions by at least a millionfold. Examples include *salivary amylase* (in saliva), which catalyzes the breakdown of starch, and *oxidase enzymes*, which act to oxidize food fuels.
	Transport	*Hemoglobin* transports oxygen in blood, and *lipoproteins* transport lipids and cholesterol. Other transport proteins in the blood carry iron, hormones, or other substances. Some globular proteins in plasma membranes are involved in membrane transport (as carriers or channels).
	Regulation of pH	Many plasma proteins, such as *albumin,* function reversibly as acids or bases, thus acting as buffers to prevent wide swings in blood pH.
	Regulation of metabolism	*Peptide* and *protein hormones* help to regulate metabolic activity, growth, and development. For example, *growth hormone* is an anabolic hormone necessary for optimal growth; *insulin* helps regulate blood sugar levels.
	Body defense	*Antibodies* (immunoglobulins) are specialized proteins released by immune cells that recognize and inactivate foreign substances (bacteria, toxins, some viruses). *Complement proteins,* which circulate in blood, enhance both immune and inflammatory responses.
	Protein management	*Molecular chaperones,* originally called "heat shock proteins," aid folding of new proteins in both healthy and damaged cells and transport of metal ions into and within the cell. They also promote breakdown of damaged proteins.

hydrogen bonds, are important in maintaining that structure. However, hydrogen bonds are fragile and easily broken by many chemical and physical factors, such as excessive acidity or temperature. Although individual proteins vary in their sensitivity to environmental conditions, hydrogen bonds begin to break when the pH drops or the temperature rises above normal (physiological) levels, causing proteins to unfold and lose their specific three-dimensional shape. In this condition, a protein is said to be **denatured**.

The disruption is reversible in most cases, and the "scrambled" protein regains its native structure when desirable conditions are restored. However, if the temperature or pH change is so extreme that protein structure is damaged beyond repair, the protein is *irreversibly denatured*. The coagulation of egg white (primarily albumin protein) that occurs when you boil or fry an egg is an example of irreversible protein denaturation. There is no way to restore the white, rubbery protein to its original translucent form.

When globular proteins are denatured, they can no longer perform their physiological roles because their function depends on the presence of specific arrangements of atoms, called *active sites*, on their surfaces. The active sites are regions that fit and interact chemically with other molecules of complementary shape and charge. Because atoms contributing to an active site may actually be far apart in the primary chain, disruption of intramolecular bonds separates them and destroys the active site. For example, hemoglobin becomes totally unable to bind and transport oxygen when blood pH is too acidic, because the structure needed for its function has been destroyed.

We will describe most types of body proteins in conjunction with the organ systems or functional processes to which they are closely related. However, one group of proteins—*enzymes*—is intimately involved in the normal functioning of all cells, so we will consider these incredibly complex molecules here.

Enzymes and Enzyme Activity

Enzymes are globular proteins that act as biological catalysts. *Catalysts* are substances that regulate and accelerate the rate of biochemical reactions but are not used up or changed in those reactions. More specifically, enzymes can be thought of as chemical traffic cops that keep our metabolic pathways flowing. Enzymes cannot force chemical reactions to occur between molecules that would not otherwise react. They can only increase the speed of reaction, and they do so by staggering amounts—from 100,000 to over 1 billion times the rate of an uncatalyzed reaction. Without enzymes, biochemical reactions proceed so slowly that for practical purposes they do not occur at all.

Characteristics of Enzymes

Some enzymes are purely protein. In other cases, the functional enzyme consists of two parts, collectively called a **holoenzyme**: an **apoenzyme** (the protein portion) and a **cofactor**. Depending on the enzyme, the cofactor may be an ion of a metal element such as copper or iron, or an organic molecule needed to assist the reaction in some way. Most organic cofactors are derived from vitamins (especially the B complex vitamins). This type of cofactor is more precisely called a **coenzyme**.

Each enzyme is chemically specific. Some enzymes control only a single chemical reaction. Others exhibit a broader specificity in that they can bind with molecules that differ slightly and thus regulate a small group of related reactions. The part of the enzyme where catalytic activity occurs is the **active site** and the substance on which an enzyme acts is called a **substrate**.

The presence of specific enzymes determines not only which reactions will be speeded up, but also which reactions will occur—no enzyme, no reaction. This also means that unwanted or unnecessary chemical reactions do not occur.

Most enzymes are named for the type of reaction they catalyze. *Hydrolases* (hi'druh-lās-es) add water during hydrolysis reactions and *oxidases* (ok'sĭ-dās-es) oxidize reactants by adding oxygen or removing hydrogen. You can recognize most enzyme names by the suffix *-ase*.

In many cases, enzymes are part of cellular membranes in a bucket-brigade type of arrangement. The product of one enzyme-catalyzed reaction becomes the substrate of the neighboring enzyme, and so on. Some enzymes are produced in an inactive form and must be activated in some way before they can function, often by a change in the pH of their surroundings. For example, digestive enzymes produced in the pancreas are activated in the small intestine, where they actually do their work. If they were produced in active form, the pancreas would digest itself.

Sometimes, enzymes are inactivated immediately after they have performed their catalytic function. This is true of enzymes that promote blood clot formation when the wall of a blood vessel is damaged. Once clotting is triggered, those enzymes are inactivated. Otherwise, you would have blood vessels full of solid blood instead of one protective clot. (Eek!)

Enzyme Action

How do enzymes perform their catalytic role? Every chemical reaction requires that a certain amount of energy, called **activation energy**, be absorbed to prime the reaction. The activation energy is needed to alter the bonds of the reactants so that they can be rearranged to become the product. It is present when kinetic energy pushes the reactants to an energy level where their random collisions are forceful enough to ensure interaction. Activation energy is needed regardless of whether the overall reaction is ultimately energy absorbing or energy releasing.

One way to increase kinetic energy is to increase the temperature, but higher temperatures denature proteins. (This is why a high fever can be a serious event.) Enzymes allow reactions to occur at normal body temperature by decreasing the amount of activation energy required (**Figure 2.19**).

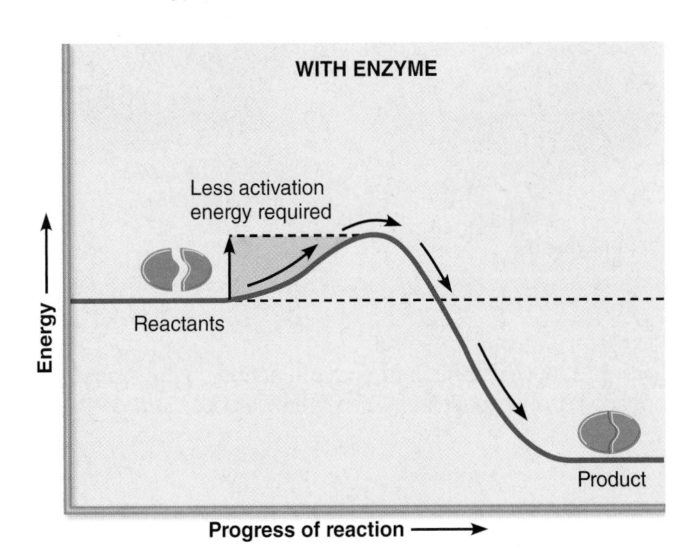

Figure 2.19 Enzymes lower the activation energy required for a reaction.

Exactly how do enzymes accomplish this remarkable feat? The answer is not fully understood. However, we know that, due to structural and electrostatic factors, they decrease the randomness of reactions by binding to the reacting molecules temporarily and presenting them to each other in the proper position for chemical interaction (bond making or breaking) to occur.

Three basic steps appear to be involved in enzyme action (**Figure 2.20**).

1. **Substrate(s) bind to the enzyme's active site, temporarily forming an enzyme-substrate complex.** Substrate binding causes the active site to change shape so that the substrate and the active site fit together precisely, and in an orientation that favors reaction. Although enzymes are specific for particular substrates, other (nonsubstrate) molecules may act as *enzyme inhibitors* if their structure is similar enough to occupy or block the enzyme's active site.
2. **The enzyme-substrate complex undergoes internal rearrangements that form the product(s).** This step shows the catalytic role of an enzyme.
3. **The enzyme releases the product(s) of the reaction.** If the enzyme became part of the product, it would be a reactant and not a catalyst. The enzyme is not changed and returns to its original shape, available to catalyze another reaction.

Because enzymes are unchanged by their catalytic role and can act again and again, cells need only small amounts of each enzyme. Catalysis occurs with incredible speed. Most enzymes can catalyze millions of reactions per minute.

☑ Check Your Understanding

24. What does the name "amino acid" tell you about the structure of this molecule?

25. What is the primary structure of proteins?

26. What are the two types of secondary structure in proteins?

27. How do enzymes reduce the amount of activation energy needed to make a chemical reaction go?

For answers, see Answers Appendix.

2.11 DNA and RNA store, transmit, and help express genetic information

→ Learning Objective

☐ **Compare and contrast DNA and RNA.**

The **nucleic acids** (nu-kle′ic), composed of carbon, oxygen, hydrogen, nitrogen, and phosphorus, are the largest molecules in the body. The nucleic acids include two major classes of molecules, **deoxyribonucleic acid (DNA)** (de-ok″sĭ-ri″bo-nu-kle′ik) and **ribonucleic acid (RNA)**.

The structural units of nucleic acids, called **nucleotides**, are quite complex. Each nucleotide consists of three components: a nitrogen-containing base, a pentose sugar, and a phosphate

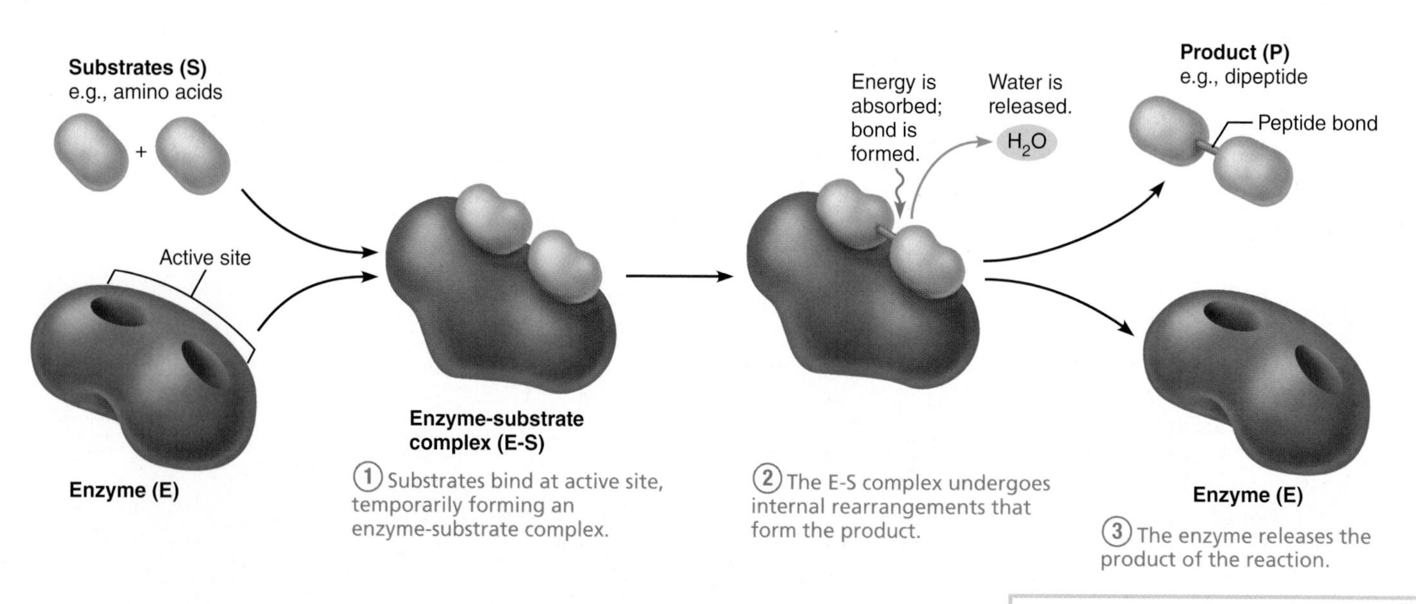

Figure 2.20 Mechanism of enzyme action. In this example, the enzyme catalyzes the formation of a dipeptide from specific amino acids. *Summary*: E + S → E-S → P + E

Practice art labeling
MasteringA&P®>Study Area>Chapter 2

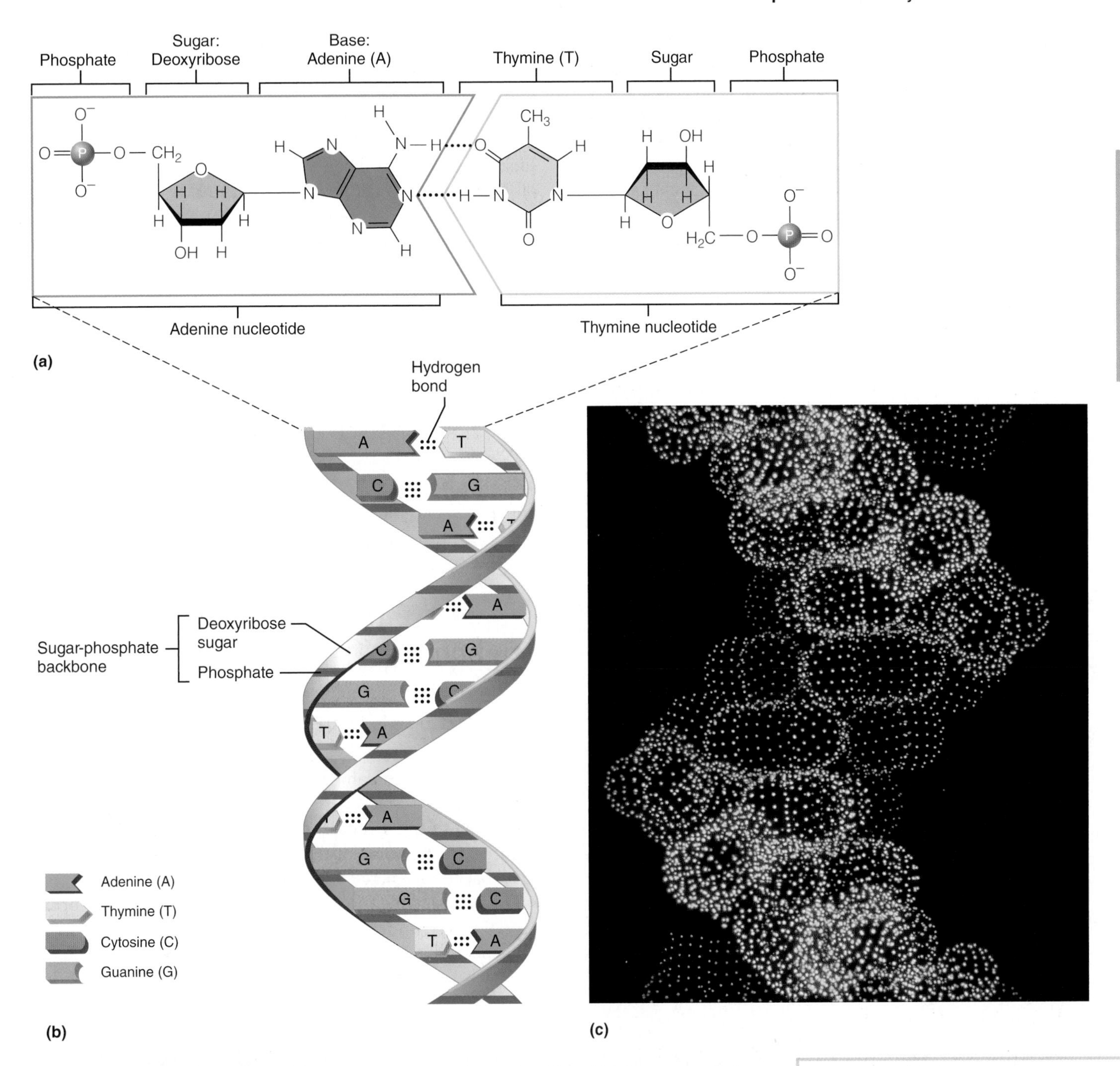

Figure 2.21 Structure of DNA. (a) The unit of DNA is the nucleotide, which is composed of a deoxyribose sugar molecule linked to a phosphate group, with a base attached to the sugar. Two nucleotides, linked by hydrogen bonds between their complementary bases, are illustrated. **(b)** DNA is a coiled double polymer of nucleotides (a double helix). The backbones of the ladderlike molecule are formed by alternating sugar and phosphate units. The rungs are formed by the binding together of complementary bases (A-T and G-C) by hydrogen bonds (shown by dotted lines). **(c)** Computer-generated image of a DNA molecule.

Practice art labeling
MasteringA&P®>Study Area>Chapter 2

group (Figure 2.21a). Five major varieties of nitrogen-containing bases can contribute to nucleotide structure: **adenine**, abbreviated **A** (ad′ĕ-nēn); **guanine**, **G** (gwan′ēn); **cytosine**, **C** (si′to-sēn); **thymine**, **T** (thi′mēn); and **uracil**, **U** (u′rah-sil). Adenine and guanine are large, two-ring bases (called purines), whereas cytosine, thymine, and uracil are smaller, single-ring bases (called pyrimidines).

The synthesis of a nucleotide involves the attachment of a base and a phosphate group to the pentose sugar.

Although DNA and RNA are both composed of nucleotides, they differ in many respects, as summarized in Table 2.4. Typically, DNA is found in the nucleus (control center) of the cell, where it constitutes the *genetic material*, also called the *genes*, or more recently the *genome*. DNA has two fundamental roles: It replicates (reproduces) itself before a cell divides, ensuring that the genetic information in the descendant cells is identical, and it provides the basic instructions for building every protein in the body. Although we have said that enzymes govern all chemical reactions, remember that enzymes, too, are proteins formed at the direction of DNA.

By providing the information for protein synthesis, DNA determines what type of organism you will be—frog, human, oak tree—and directs your growth and development. It also accounts for your uniqueness. DNA fingerprinting can help solve forensic mysteries (for example, verify one's presence at a crime scene), identify badly burned or mangled bodies after a disaster, and establish or disprove paternity. DNA fingerprinting analyzes tiny samples of DNA taken from blood, semen, or other body tissues and shows the results as a "genetic barcode" that distinguishes each of us from all others.

DNA is a long, double-stranded polymer—a double chain of nucleotides (Figure 2.21b and c). The bases in DNA are A, G, C, and T, and its pentose sugar is *deoxyribose* (as reflected in its name). Its two nucleotide chains are held together by hydrogen bonds between the bases, so that a ladderlike molecule is formed. Alternating sugar and phosphate components of each chain form the *backbones* or "uprights" of the "ladder," and the joined bases form the "rungs." The whole molecule is coiled into a spiral staircase–like structure called a **double helix**.

Bonding of the bases is very specific: A always bonds to T, and G always bonds to C. A and T are therefore called **complementary bases**, as are C and G. According to these base-pairing rules, ATGA on one DNA nucleotide strand would necessarily be bonded to TACT (a complementary base sequence) on the other strand.

RNA is located chiefly outside the nucleus and can be considered a "molecular slave" of DNA. That is, RNA carries out the orders for protein synthesis issued by DNA. [Viruses in which RNA (rather than DNA) is the genetic material are an exception to this generalization.]

RNA molecules are single strands of nucleotides. RNA bases include A, G, C, and U (U replaces the T found in DNA), and its sugar is *ribose* instead of deoxyribose. The three major varieties of RNA (messenger RNA, ribosomal RNA, and transfer RNA) are distinguished by their relative size and shape, and each has a specific role to play in carrying out DNA's instructions for protein synthesis. In addition to these three RNAs, there are several types of small RNA molecules, including *microRNAs*. MicroRNAs appear to control genetic expression by shutting down genes or altering their expression. We discuss DNA replication and the roles of DNA and RNA in protein synthesis in Chapter 3.

☑ Check Your Understanding

28. How do DNA and RNA differ in the bases and sugars they contain?

29. What are two important roles of DNA?

For answers, see Answers Appendix.

2.12 ATP transfers energy to other compounds

→ Learning Objective

☐ **Explain the role of ATP in cell metabolism.**

Glucose is the most important cellular fuel, but none of the chemical energy contained in its bonds is used directly to power cellular work. Instead, energy released during glucose catabolism is coupled to the synthesis of **adenosine triphosphate (ATP)**. In other words, some of this energy is captured and stored as small packets of energy in the bonds of ATP. ATP is the primary energy-transferring molecule in cells and it provides a form of energy that is immediately usable by all body cells.

Structurally, ATP is an adenine-containing RNA nucleotide to which two additional phosphate groups have been added

Table 2.4 Comparison of DNA and RNA

CHARACTERISTIC	DNA	RNA
Major cellular site	Nucleus	Cytoplasm (cell area outside the nucleus)
Major functions	Is the genetic material; directs protein synthesis; replicates itself before cell division	Carries out the genetic instructions for protein synthesis
Sugar	Deoxyribose	Ribose
Bases	Adenine, guanine, cytosine, thymine	Adenine, guanine, cytosine, uracil
Structure	Double strand coiled into a double helix	Single strand, straight or folded

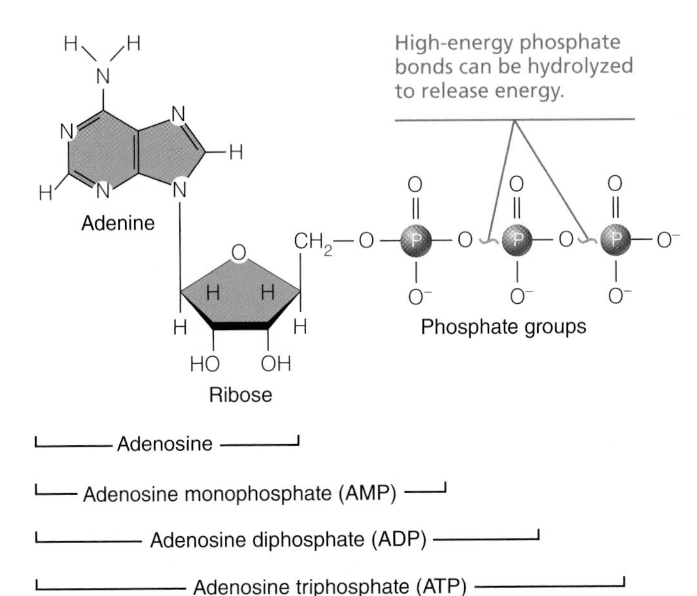

Figure 2.22 Structure of ATP (adenosine triphosphate). ATP is an adenine nucleotide to which two additional phosphate groups have been attached during breakdown of food fuels. When the terminal phosphate group is cleaved off, energy is released to do useful work and ADP (adenosine diphosphate) is formed. When the terminal phosphate group is cleaved off ADP, a similar amount of energy is released and AMP (adenosine monophosphate) is formed.

(**Figure 2.22**). Chemically, the triphosphate tail of ATP can be compared to a tightly coiled spring ready to uncoil with tremendous energy when the catch is released. Actually, ATP is a very unstable energy-storing molecule because its three negatively charged phosphate groups are closely packed and repel each other. When its terminal high-energy phosphate bonds are broken (hydrolyzed), the chemical "spring" relaxes and the molecule as a whole becomes more stable.

Cells tap ATP's bond energy during coupled reactions by using enzymes to transfer the terminal phosphate groups from ATP to other compounds. These newly *phosphorylated* molecules are said to be "primed" and temporarily become more energetic and capable of performing some type of cellular work. In the process of doing their work, they lose the phosphate group. The amount of energy released and transferred during ATP hydrolysis corresponds closely to that needed to drive most biochemical reactions. As a result, cells are protected from excessive energy release that might be damaging, and energy squandering is kept to a minimum.

Cleaving the terminal phosphate bond of ATP yields a molecule with two phosphate groups—*adenosine diphosphate (ADP)*—and an inorganic phosphate group, indicated by P_i, accompanied by a transfer of energy:

$$ATP \underset{H_2O}{\overset{H_2O}{\rightleftharpoons}} ADP + P_i + \text{energy}$$

As ATP is hydrolyzed to provide energy for cellular needs, ADP accumulates. Cleavage of the terminal phosphate bond of ADP liberates a similar amount of energy and produces adenosine monophosphate (AMP).

The cell's ATP supplies are replenished as glucose and other fuel molecules are oxidized and their bond energy is released. The same amount of energy that is liberated when ATP's terminal phosphates are cleaved off must be captured and used to reverse the reaction to reattach phosphates and re-form the energy-transferring phosphate bonds. Without ATP, molecules cannot be made or degraded, cells cannot transport substances across their membranes, muscles cannot shorten to tug on other structures, and life processes cease (**Figure 2.23**).

☑ Check Your Understanding

30. Glucose is an energy-rich molecule. So why do body cells need ATP?

31. What change occurs in ATP when it releases energy?

For answers, see Answers Appendix.

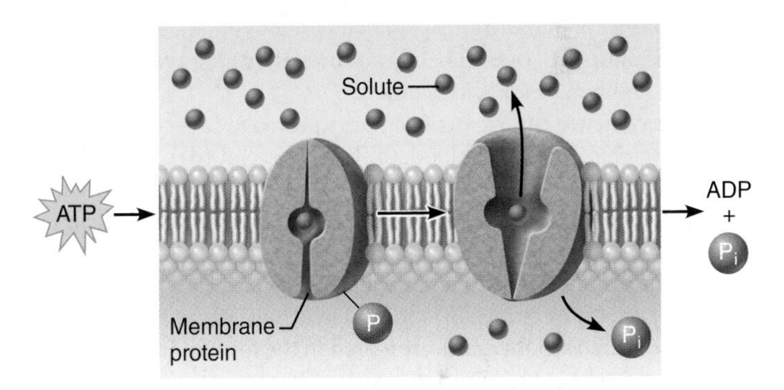

(a) Transport work: ATP phosphorylates transport proteins, activating them to transport solutes (ions, for example) across cell membranes.

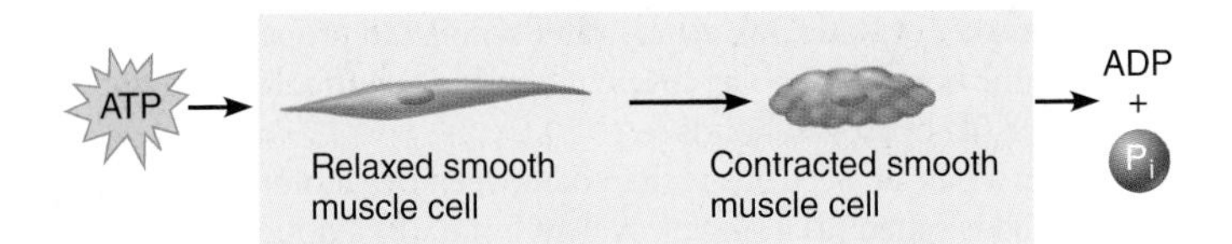

(b) Mechanical work: ATP phosphorylates contractile proteins in muscle cells so the cells can contract (shorten).

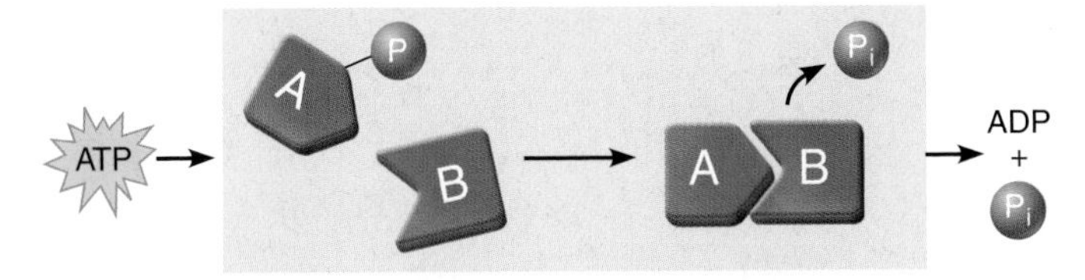

(c) Chemical work: ATP phosphorylates key reactants, providing energy to drive energy-absorbing chemical reactions.

Figure 2.23 Three examples of cellular work driven by energy from ATP.

REVIEW QUESTIONS

For more chapter study tools, go to the Study Area of MasteringA&P®.

There you will find:

- Interactive Physiology iP
- A&PFlix A&PFlix
- Practice Anatomy Lab PAL
- PhysioEx PEx
- Videos, Practice Quizzes and Tests, MP3 Tutor Sessions, Case Studies, and much more!

Multiple Choice/Matching

(Some questions have more than one correct answer. Select the best answer or answers from the choices given.)

1. Which of the following forms of energy is the stimulus for vision? **(a)** chemical, **(b)** electrical, **(c)** mechanical, **(d)** radiant.
2. All of the following are examples of the four major elements contributing to body mass except **(a)** hydrogen, **(b)** carbon, **(c)** nitrogen, **(d)** sodium, **(e)** oxygen.
3. The mass number of an atom is **(a)** equal to the number of protons it contains, **(b)** the sum of its protons and neutrons, **(c)** the sum of all of its subatomic particles, **(d)** the average of the mass numbers of all of its isotopes.
4. A deficiency in this element can be expected to reduce the hemoglobin content of blood: **(a)** Fe, **(b)** I, **(c)** F, **(d)** Ca, **(e)** K.
5. Which set of terms best describes a proton? **(a)** negative charge, massless, in the orbital; **(b)** positive charge, 1 amu, in the nucleus; **(c)** uncharged, 1 amu, in the nucleus.
6. The subatomic particles responsible for the chemical behavior of atoms are **(a)** electrons, **(b)** ions, **(c)** neutrons, **(d)** protons.
7. In the body, carbohydrates are stored in the form of **(a)** glycogen, **(b)** starch, **(c)** cholesterol, **(d)** polypeptides.
8. Which of the following does *not* describe a mixture? **(a)** properties of its components are retained, **(b)** chemical bonds are formed, **(c)** components can be separated physically, **(d)** includes both heterogeneous and homogeneous examples.
9. In a beaker of water, the water-water bonds can properly be called **(a)** ionic bonds, **(b)** polar covalent bonds, **(c)** nonpolar covalent bonds, **(d)** hydrogen bonds.
10. When a pair of electrons is shared between two atoms, the bond formed is called **(a)** a single covalent bond, **(b)** a double covalent bond, **(c)** a triple covalent bond, **(d)** an ionic bond.
11. Molecules formed when electrons are shared unequally are **(a)** salts, **(b)** polar molecules, **(c)** nonpolar molecules.
12. Which of the following covalently bonded molecules are polar?

H—Cl **(a)**

H—C(H)(H)—H **(b)**

Cl—C(H)(Cl)—Cl **(c)**

N≡N **(d)**

13. Identify each reaction as one of the following: **(a)** a synthesis reaction, **(b)** a decomposition reaction, **(c)** an exchange reaction.

_____ (1) $2Hg + O_2 \rightarrow 2HgO$

_____ (2) $HCl + NaOH \rightarrow NaCl + H_2O$

14. Factors that accelerate the rate of chemical reactions include all but **(a)** the presence of catalysts, **(b)** increasing the temperature, **(c)** increasing the particle size, **(d)** increasing the concentration of the reactants.
15. Which of the following molecules is an inorganic molecule? **(a)** sucrose, **(b)** cholesterol, **(c)** collagen, **(d)** sodium chloride.
16. Water's importance to living systems reflects **(a)** its polarity and solvent properties, **(b)** its high heat capacity, **(c)** its high heat of vaporization, **(d)** its chemical reactivity, **(e)** all of these.
17. Acids **(a)** release hydroxyl ions when dissolved in water, **(b)** are proton acceptors, **(c)** cause the pH of a solution to rise, **(d)** release protons when dissolved in water.
18. A chemist, during the course of an analysis, runs across a chemical composed of carbon, hydrogen, and oxygen in the proportion 1:2:1 and having a six-sided molecular shape. It is probably **(a)** a pentose, **(b)** an amino acid, **(c)** a fatty acid, **(d)** a monosaccharide, **(e)** a nucleic acid.
19. A triglyceride consists of **(a)** glycerol plus three fatty acids, **(b)** a sugar-phosphate backbone to which two amino groups are attached, **(c)** two to several hexoses, **(d)** amino acids that have been thoroughly saturated with hydrogen.
20. A chemical has an amine group and an organic acid group. It does not, however, have any peptide bonds. It is **(a)** a monosaccharide, **(b)** an amino acid, **(c)** a protein, **(d)** a fat.
21. The lipid(s) used as the basis of vitamin D, sex hormones, and bile salts is/are **(a)** triglycerides, **(b)** cholesterol, **(c)** phospholipids, **(d)** prostaglandin.
22. Enzymes are organic catalysts that **(a)** alter the direction in which a chemical reaction proceeds, **(b)** determine the nature of the products of a reaction, **(c)** increase the speed of a chemical reaction, **(d)** are essential raw materials for a chemical reaction that are converted into some of its products.

Short Answer Essay Questions

23. Define or describe energy, and explain the relationship between potential and kinetic energy.
24. Some energy is lost in every energy conversion. Explain the meaning of this statement. (Direct your response to answering the question: Is it really lost? If not, what then?)
25. Provide the atomic symbol for each of the following elements: (a) calcium, (b) carbon, (c) hydrogen, (d) iron, (e) nitrogen, (f) oxygen, (g) potassium, (h) sodium.
26. Consider the following information about three atoms:

$${}^{12}_{6}C \quad {}^{13}_{6}C \quad {}^{14}_{6}C$$

(a) How are they similar to one another? (b) How do they differ from one another? (c) What are the members of such a group of atoms called? (d) Using the planetary model, draw the atomic configuration of ${}^{12}_{6}C$ showing the relative position and numbers of its subatomic particles.

27. How many moles of aspirin, $C_9H_8O_4$, are in a bottle containing 450 g by weight? (*Note*: The approximate atomic weights of its atoms are C = 12, H = 1, and O = 16.)
28. Given the following types of atoms, decide which type of bonding, ionic or covalent, is most likely to occur: (a) two oxygen atoms; (b) four hydrogen atoms and one carbon atom; (c) a potassium atom (${}^{39}_{19}K$) and a fluorine atom (${}^{19}_{9}F$).
29. What are hydrogen bonds and how are they important in the body?

30. The following equation, which represents the oxidative breakdown of glucose by body cells, is a reversible reaction.

    Glucose + oxygen → carbon dioxide + water + ATP

    (a) How can you indicate that the reaction is reversible? (b) How can you indicate that the reaction is in chemical equilibrium? (c) Define chemical equilibrium.
31. Differentiate clearly between primary, secondary, and tertiary protein structure.
32. Dehydration and hydrolysis reactions are essentially opposite reactions. How are they related to the synthesis and degradation (breakdown) of biological molecules?
33. Describe the mechanism of enzyme action.
34. Explain why, if you pour water into a glass very carefully, you can "stack" the water slightly above the rim of the glass.

3 Cells: The Living Units

KEY CONCEPTS

Just as bricks and timbers are the structural units of a house, **cells** are the structural units of all living things, from one-celled "generalists" like amoebas to complex multicellular organisms such as humans, dogs, and trees. The human body has 50 to 100 trillion of these tiny building blocks.

This chapter focuses on structures and functions shared by all cells. We address specialized cells and their unique functions in later chapters.

3.1 Cells are the smallest unit of life

→ Learning Objectives

- ☐ **Define cell.**
- ☐ **Name and describe the composition of extracellular materials.**
- ☐ **List the three major regions of a generalized cell and their functions.**

The English scientist Robert Hooke first observed plant cells with a crude microscope in the late 1600s. Then, in the 1830s two German scientists, Matthias Schleiden and Theodor Schwann, proposed that all living things are composed of cells. German pathologist Rudolf Virchow extended this idea by contending that cells arise only from other cells.

Since the late 1800s, cell research has been exceptionally fruitful and provided us with four concepts collectively known as the **cell theory**:

- A *cell* is the basic structural and functional unit of living organisms. When you define cell properties, you define the properties of life.
- The activity of an organism depends on both the individual and the combined activities of its cells.
- According to the *principle of complementarity of structure and function*, the biochemical activities of cells are dictated by their shapes or forms, and by the relative number of the subcellular structures they contain.
- Cells can only arise from other cells.

Electron micrograph of a cancer cell undergoing cytokinesis at the end of mitosis.

We will expand on all of these concepts as we progress. Let us begin with the idea that the cell is the smallest living unit. Whatever its form, however it behaves, the cell is the microscopic package that contains all the parts necessary to survive in an ever-changing world. It follows then that loss of cellular homeostasis underlies virtually every disease.

The trillions of cells in the human body include over 250 different cell types that vary greatly in shape, size, and function (Figure 3.1). The disc-shaped red blood cells, branching nerve cells, and cubelike cells of kidney tubules are just a few examples of the shapes cells take. Cells also vary in length—ranging from 2 micrometers (1/12,000 of an inch) in the smallest cells to over a meter in the nerve cells that cause you to wiggle your toes. A cell's shape reflects its function. For example, the flat, tilelike epithelial cells that line the inside of your cheek fit closely together, forming a living barrier that protects underlying tissues from bacterial invasion.

Regardless of these differences, all cells have the same basic parts and some common functions. For this reason, it is possible to speak of a **generalized**, or **composite**, **cell** (Figure 3.2).

A human cell has three main parts:

- The *plasma membrane*: the outer boundary of the cell which acts as a selectively permeable barrier.
- The *cytoplasm* (si'to-plazm): the intracellular fluid packed with *organelles*, small structures that perform specific cell functions.
- The *nucleus* (nu'kle-us): an organelle that controls cellular activities. Typically the nucleus lies near the cell's center.

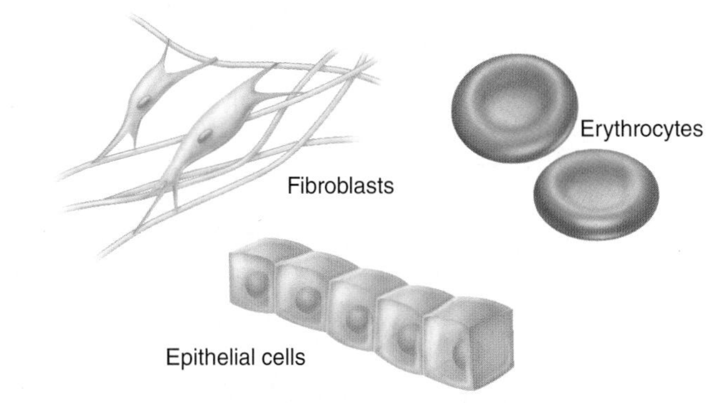

(a) Cells that connect body parts, form linings, or transport gases

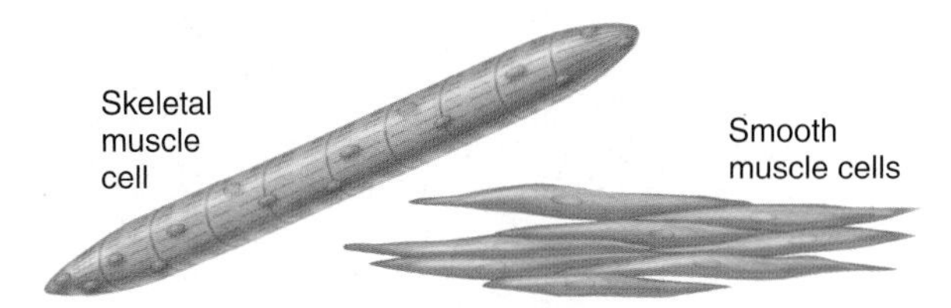

(b) Cells that move organs and body parts

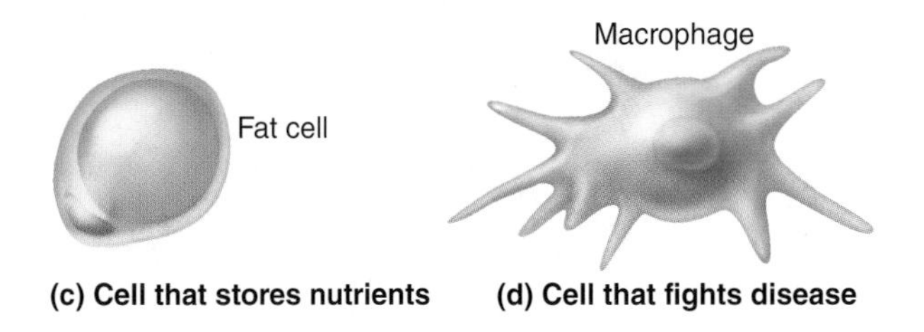

(c) Cell that stores nutrients **(d) Cell that fights disease**

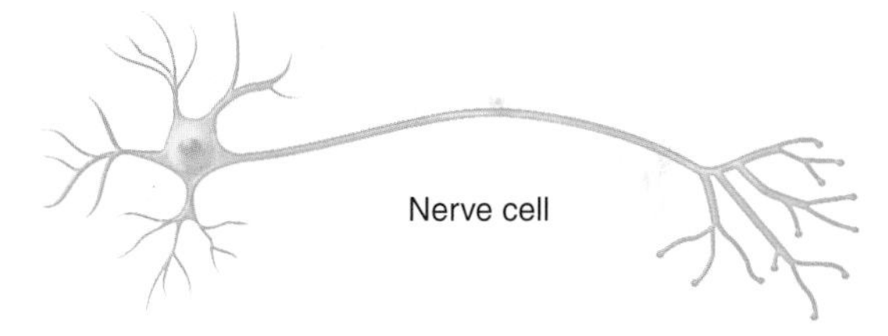

(e) Cell that gathers information and controls body functions

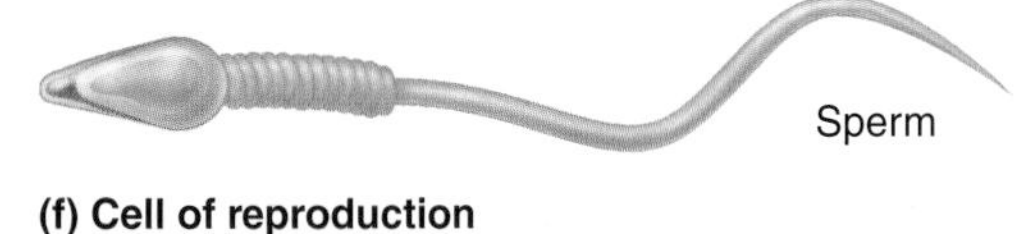

(f) Cell of reproduction

Figure 3.1 **Cell diversity.** (Note that cells are not drawn to the same scale.)

Extracellular Materials

Although we tend to think of the body as collections of cells—and it *is* that—it is impossible to discuss cells and their activities without saying something about extracellular materials. So—let's do that before going on to details about the generalized cell.

First of all, what are extracellular materials? **Extracellular materials** are substances contributing to body mass that are found outside the cells. Classes of extracellular materials include:

- *Body fluids*, also called extracellular fluids, include **interstitial fluid**, blood plasma, and cerebrospinal fluid. These fluids are important transport and dissolving media. Interstitial fluid is the fluid in tissues that bathes all of our cells, and has major and endless roles to play. Like a rich, nutritious "soup," interstitial fluid contains thousands of ingredients, including amino acids, sugars, fatty acids, regulatory substances, and wastes. To remain healthy, each cell must extract from this mix the exact amounts of the substances it needs depending on present conditions.
- *Cellular secretions* include substances that aid in digestion (intestinal and gastric fluids) and some that act as lubricants (saliva, mucus, and serous fluids).
- *The extracellular matrix* is the most abundant extracellular material. Most body cells are in contact with a jellylike substance composed of proteins and polysaccharides. Secreted by the cells, these molecules self-assemble into an organized

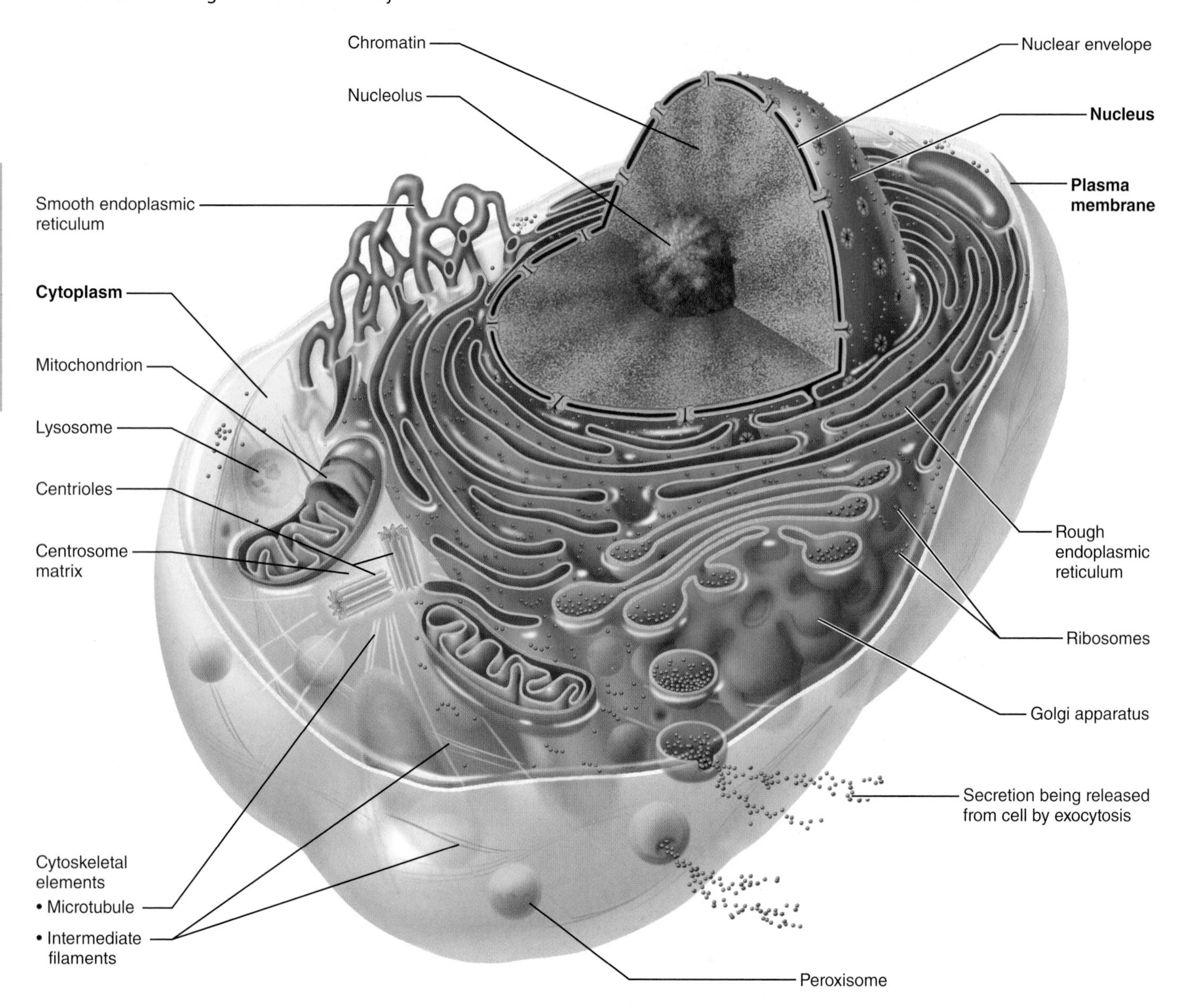

Figure 3.2 Structure of the generalized cell. No cell is exactly like this one, but this composite illustrates features common to many human cells. Note that not all of the organelles are drawn to the same scale in this illustration.

Practice art labeling
MasteringA&P®>Study Area>Chapter 3

mesh in the extracellular space, where they serve as a universal "cell glue" that helps to hold body cells together. As described in Chapter 4, the extracellular matrix is particularly abundant in connective tissues—in some cases so abundant that it (rather than living cells) accounts for the bulk of that tissue type. Depending on the structure to be formed, the extracellular matrix in connective tissue ranges from soft to rock-hard.

☑ Check Your Understanding

1. Summarize the four key points of the cell theory.
2. How would you explain the meaning of a "generalized cell" to a classmate?

For answers, see Answers Appendix.

PART 1

PLASMA MEMBRANE

The flexible **plasma membrane** separates two of the body's major fluid compartments—the *intracellular* fluid within cells and the *extracellular* fluid (ECF) outside cells. The term *cell membrane* is commonly used as a synonym for plasma membrane, but because nearly all cellular organelles are enclosed in a membrane, in this book we will always refer to the cell's surface, or outer limiting membrane, as the plasma membrane. The plasma membrane is much more than a passive envelope. As you will see, its unique structure allows it to play a dynamic role in cellular activities.

3.2 The fluid mosaic model depicts the plasma membrane as a double layer of phospholipids with embedded proteins

→ Learning Objectives

- ☐ **Describe the chemical composition of the plasma membrane and relate it to membrane functions.**
- ☐ **Compare the structure and function of tight junctions, desmosomes, and gap junctions.**

The **fluid mosaic model** of membrane structure depicts the plasma membrane as an exceedingly thin (7–10 nm) structure composed of a double layer, or bilayer, of lipid molecules with protein molecules "plugged into" or dispersed in it (**Figure 3.3**). The proteins, many of which float in the fluid *lipid bilayer*, form a constantly changing mosaic pattern. The model is named for this characteristic.

Membrane Lipids

The lipid bilayer forms the basic "fabric" of the membrane. It is constructed largely of *phospholipids*, with smaller amounts of *glycolipids* and *cholesterol*.

Phospholipids

Each lollipop-shaped phospholipid molecule has a polar "head" that is charged and is **hydrophilic** (*hydro* = water, *philic* = loving), and an uncharged, nonpolar "tail" that is made of two fatty acid chains and is **hydrophobic** (*phobia* = fear). The polar heads are attracted to water—the main constituent of both the intracellular and extracellular fluids—and so they lie on both the inner and outer surfaces of the membrane. The nonpolar tails, being hydrophobic, avoid water and line up in the center of the membrane.

The result is that all plasma membranes, indeed all biological membranes, share a sandwich-like structure: They consist of two parallel sheets of phospholipid molecules lying tail to tail, with their polar heads bathed in water on either side of the membrane or organelle. This self-orienting property of phospholipids encourages biological membranes to self-assemble into generally spherical structures and to reseal themselves when torn.

With a consistency similar to olive oil, the plasma membrane is a dynamic fluid structure in constant flux. Its lipid molecules move freely from side to side, parallel to the membrane surface, but because of their self-orienting properties, they do not flip-flop or move from one half of the bilayer to the other half. The inward-facing and outward-facing surfaces of the plasma membrane differ in the kinds and amounts of lipids they contain, and these variations help to determine local membrane structure and function.

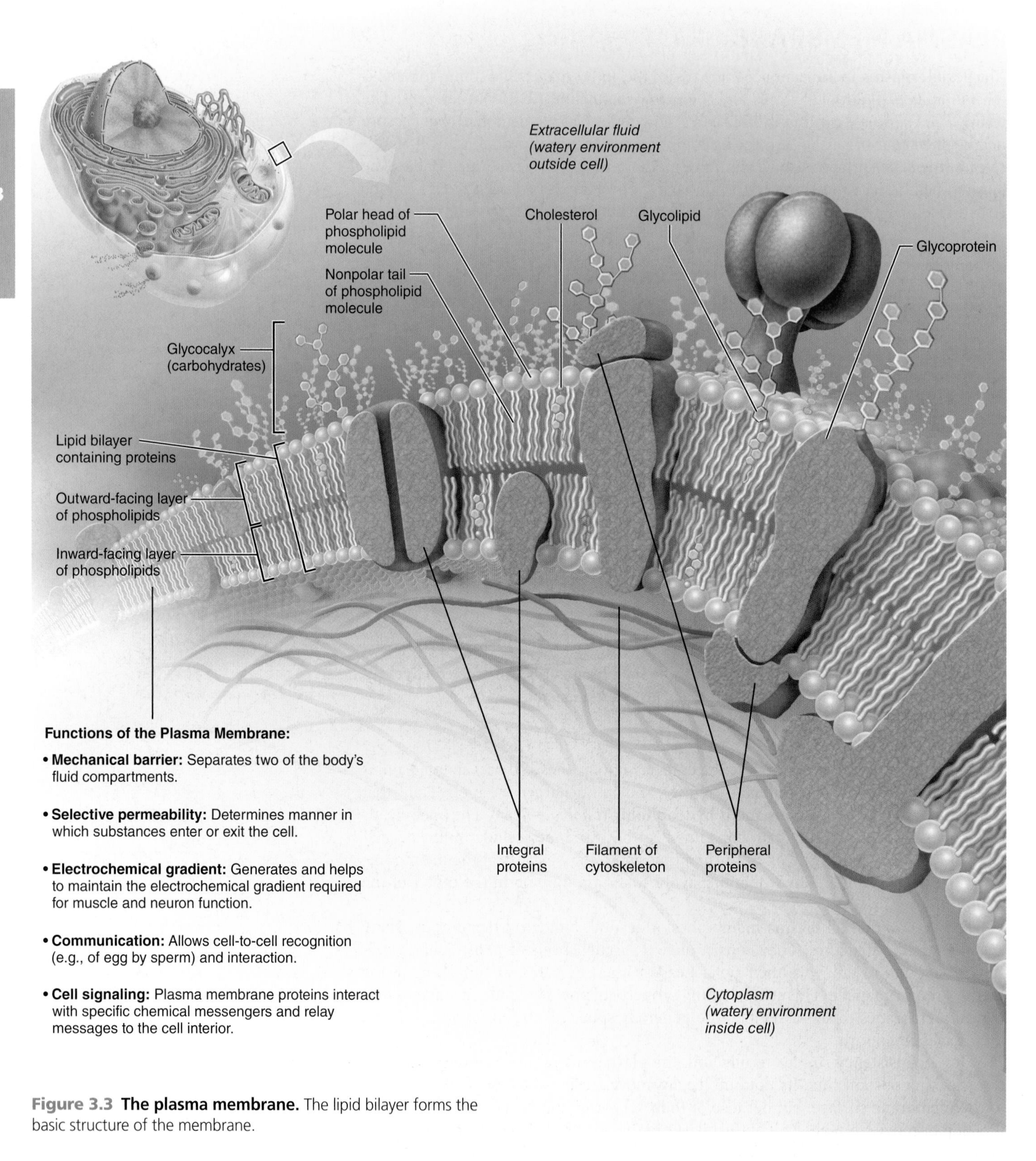

Figure 3.3 The plasma membrane. The lipid bilayer forms the basic structure of the membrane.

Glycolipids

Glycolipids (gli″ko-lip′idz) are lipids with attached sugar groups. Found only on the outer plasma membrane surface, glycolipids account for about 5% of total membrane lipids. Their sugar groups, like the phosphate-containing groups of phospholipids, make that end of the glycolipid molecule polar, whereas the fatty acid tails are nonpolar.

Cholesterol

Some 20% of membrane lipid is cholesterol. Like phospholipids, cholesterol has a polar region (its hydroxyl group) and a nonpolar region (its fused ring system). It wedges its platelike hydrocarbon rings between the phospholipid tails, which stabilize the membrane, while decreasing the mobility of the phospholipids and the fluidity of the membrane.

Membrane Proteins

A cell's plasma membrane bristles with proteins that allow it to communicate with its environment. Proteins make up about half of the plasma membrane by mass and are responsible for most of the specialized membrane functions. Some membrane proteins float freely. Others are "tethered" to intracellular or extracellular structures and are restricted in their movement.

There are two distinct populations of membrane proteins, integral and peripheral (Figure 3.3).

Integral Proteins

Integral proteins are firmly inserted into the lipid bilayer. Some protrude from one membrane face only, but most are *transmembrane proteins* that span the entire membrane and protrude on both sides. Whether transmembrane or not, all integral proteins have both hydrophobic and hydrophilic regions. This structural feature allows them to interact with both the nonpolar lipid tails buried in the membrane and the water inside and outside the cell.

Some transmembrane proteins are involved in transport, and cluster together to form *channels*, or pores. Small, water-soluble molecules or ions can move through these pores, bypassing the lipid part of the membrane. Others act as *carriers* that bind to a substance and then move it through the membrane (**Figure 3.4a**). Some transmembrane proteins are enzymes (Figure 3.4d). Still others are receptors for hormones or other chemical messengers and relay messages to the cell interior—a process called *signal transduction* (Figure 3.4b).

Peripheral Proteins

Unlike integral proteins, **peripheral proteins** (Figure 3.3) are not embedded in the lipid bilayer. Instead, they attach loosely

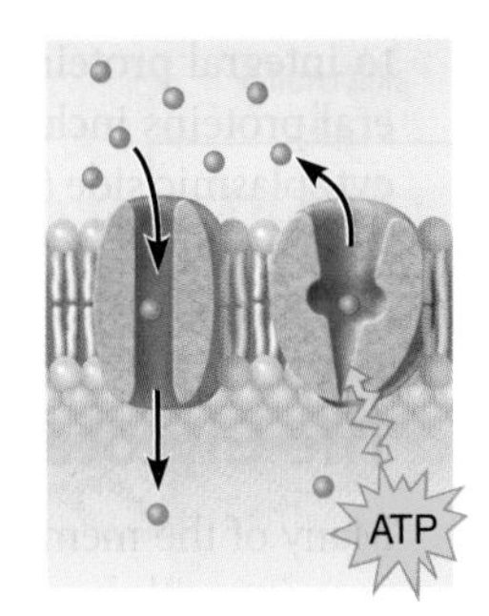

(a) Transport

- A protein (left) that spans the membrane may provide a hydrophilic channel across the membrane that is selective for a particular solute.
- Some transport proteins (right) hydrolyze ATP as an energy source to actively pump substances across the membrane.

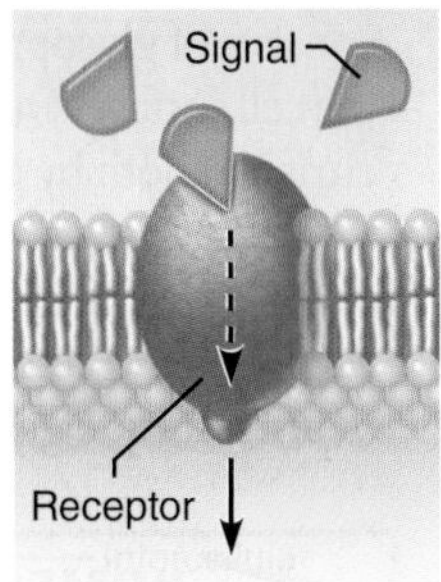

(b) Receptors for signal transduction

- A membrane protein exposed to the outside of the cell may have a binding site that fits the shape of a specific chemical messenger, such as a hormone.
- When bound, the chemical messenger may cause a change in shape in the protein that initiates a chain of chemical reactions in the cell.

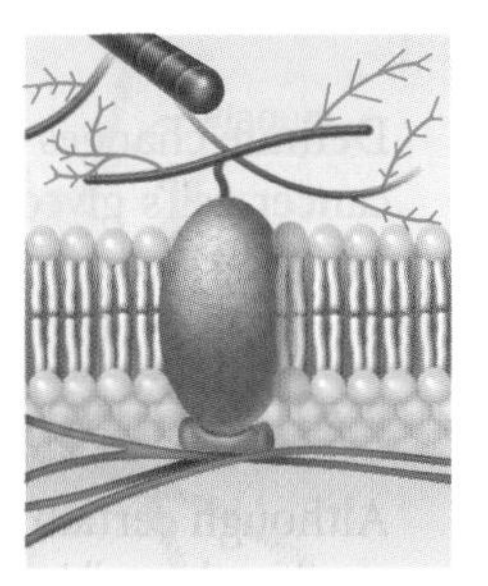

(c) Attachment to the cytoskeleton and extracellular matrix

- Elements of the cytoskeleton (cell's internal supports) and the extracellular matrix (fibers and other substances outside the cell) may anchor to membrane proteins, which helps maintain cell shape and fix the location of certain membrane proteins.
- Others play a role in cell movement or bind adjacent cells together.

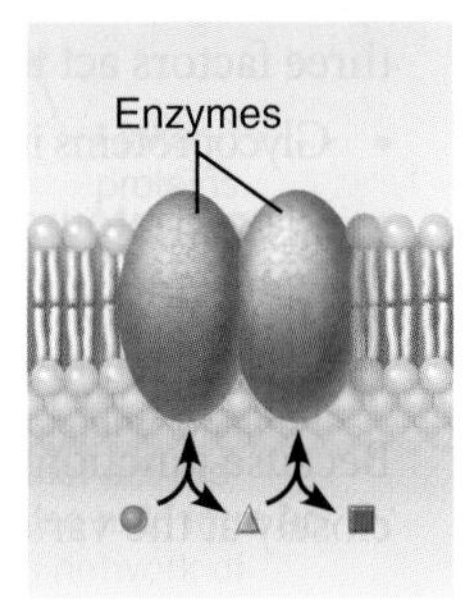

(d) Enzymatic activity

- A membrane protein may be an enzyme with its active site exposed to substances in the adjacent solution.
- A team of several enzymes in a membrane may catalyze sequential steps of a metabolic pathway as indicated (left to right) here.

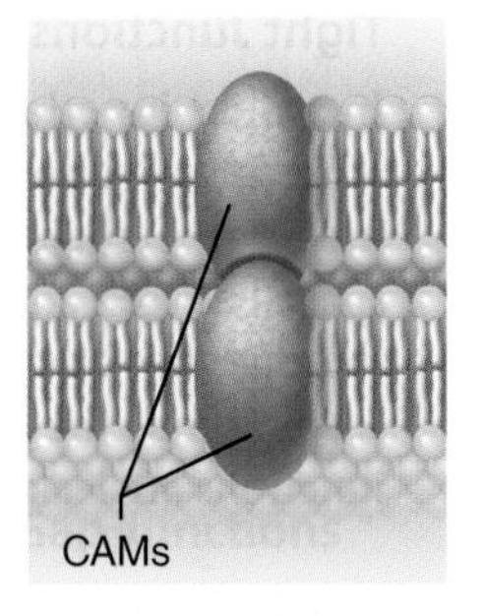

(e) Intercellular joining

- Membrane proteins of adjacent cells may be hooked together in various kinds of intercellular junctions.
- Some membrane proteins (cell adhesion molecules or CAMs) of this group provide temporary binding sites that guide cell migration and other cell-to-cell interactions.

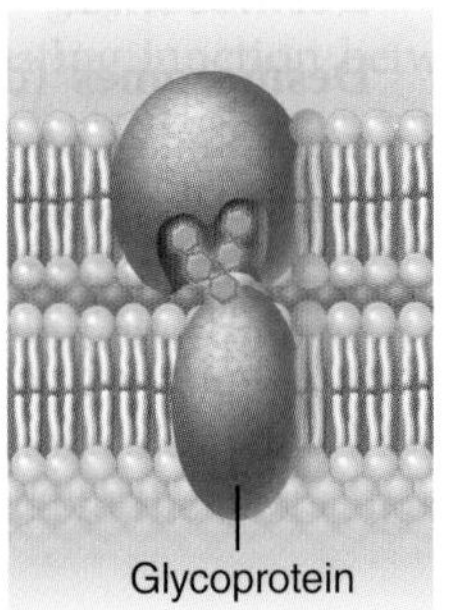

(f) Cell-cell recognition

- Some glycoproteins (proteins bonded to short chains of sugars which help to make up the glycocalyx) serve as identification tags that are specifically recognized by other cells.

Figure 3.4 Membrane proteins perform many tasks. A single protein may perform a combination of these functions.

FOCUS Primary Active Transport: The Na^+-K^+ Pump

Focus Figure 3.1 Primary active transport is the process in which solutes are moved across cell membranes against electrochemical gradients using energy supplied directly by ATP. The action of the Na^+-K^+ pump is an important example of primary active transport.

Watch full 3-D animations
MasteringA&P®>Study Area> ***A&PFlix***

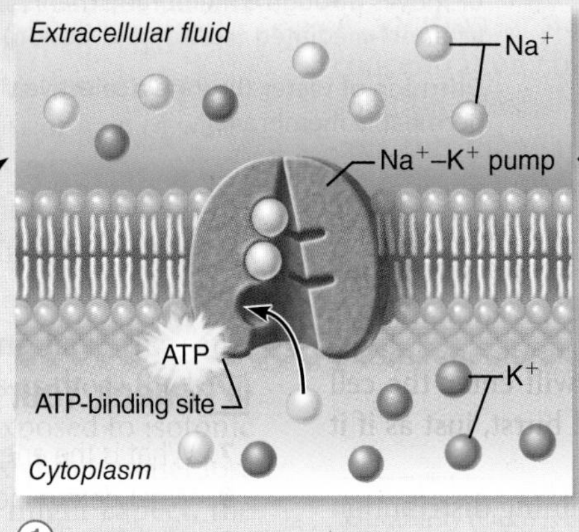

① Three cytoplasmic Na^+ bind to pump protein.

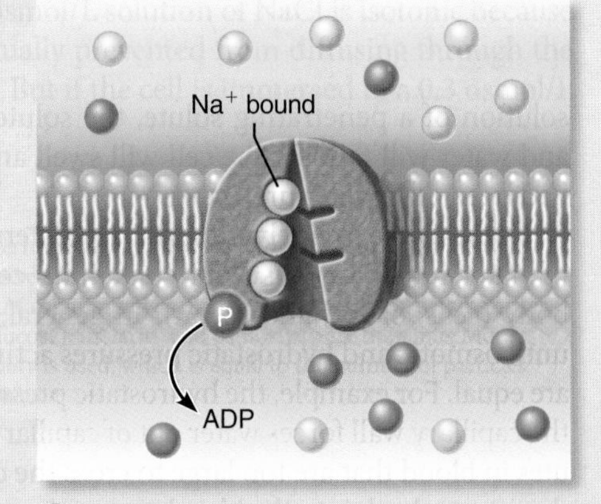

② Na^+ binding promotes hydrolysis of ATP. The energy released during this reaction phosphorylates the pump.

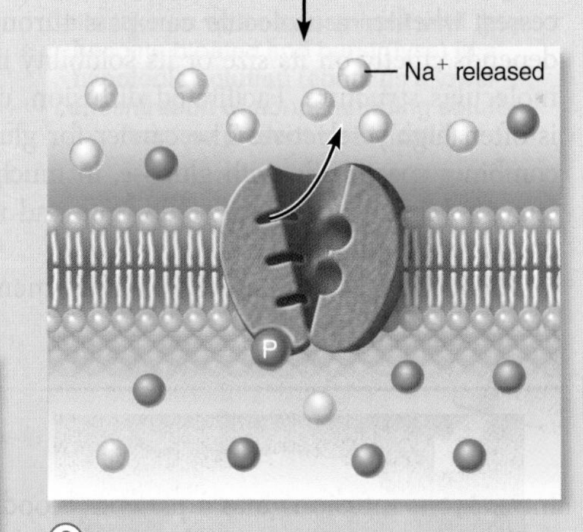

③ Phosphorylation causes the pump to change shape, expelling Na^+ to the outside.

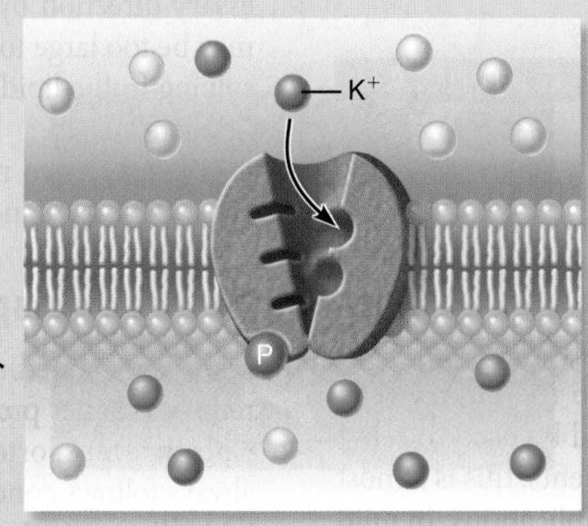

④ Two extracellular K^+ bind to pump.

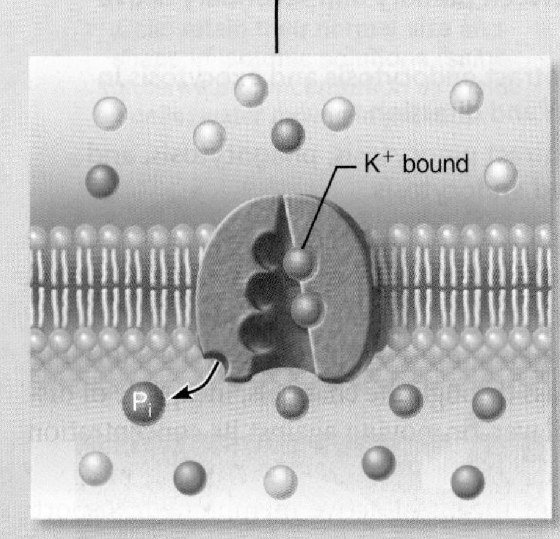

⑤ K^+ binding triggers release of the phosphate. The dephosphorylated pump resumes its original conformation.

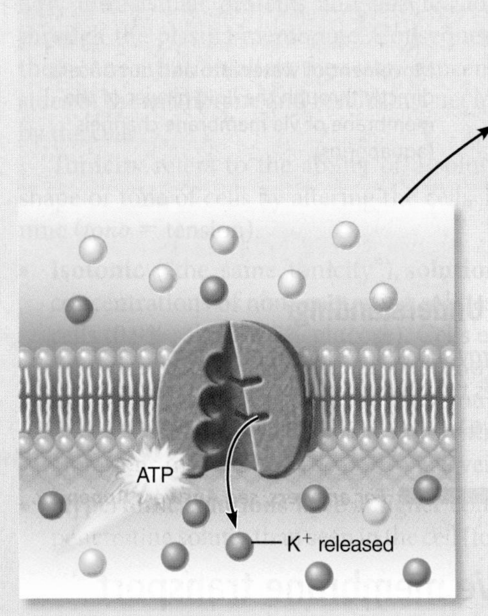

⑥ Pump protein binds ATP; releases K^+ to the inside, and Na^+ sites are ready to bind Na^+ again. The cycle repeats.

Active transport processes are distinguished according to their source of energy:

- In *primary active transport*, the energy to do work comes *directly from hydrolysis of ATP.*
- In *secondary active transport*, transport is driven indirectly *by energy stored in concentration gradients of ions* created by primary active transport pumps. Secondary active transport systems are all *coupled systems*; that is, they move more than one substance at a time.

In a **symport system,** the two transported substances move in the same direction (*sym* = same). In an **antiport system** (*anti* = opposite, against), the transported substances "wave to each other" as they cross the membrane in opposite directions.

Primary Active Transport

In **primary active transport**, hydrolysis of ATP results in the phosphorylation of the transport protein. This step causes the protein to change its shape in such a manner that it "pumps" the bound solute across the membrane.

Primary active transport systems include calcium and hydrogen pumps, but the most investigated example of a primary active transport system is the **sodium-potassium pump**, for which the carrier, or "pump," is an enzyme called **Na^+-K^+ ATPase. Focus Figure 3.1**, *Focus on Primary Active Transport: The Na^+-K^+ Pump*, describes the operation of the Na^+-K^+ pump, which moves Na^+ out of the cell and K^+ into the cell. As a result the concentration of K^+ inside the cell is some 10 times higher than that outside, and the reverse is true of Na^+. These ionic concentration differences are essential for excitable cells like muscle and nerve cells to function normally and for all body cells to maintain their normal fluid volume. Because Na^+ and K^+ leak slowly but continuously through leakage channels in the plasma membrane along their concentration gradient (and cross more rapidly in stimulated muscle and nerve cells), the Na^+-K^+ pump operates almost continuously. It simultaneously drives Na^+ out of the cell against a steep concentration gradient and pumps K^+ back in.

Earlier we said that solutes diffuse down their concentration gradients. This is true for uncharged solutes, but only partially true for ions. The negatively and positively charged faces of the plasma membrane can help or hinder diffusion of ions driven by a concentration gradient. It is more correct to say that ions diffuse according to **electrochemical gradients**, thereby recognizing the effect of both electrical and concentration (chemical) forces. Hence, the electrochemical gradients maintained by the Na^+-K^+ pump underlie most secondary active transport of nutrients and ions, and are crucial for cardiac, skeletal muscle, and neuron function.

Secondary Active Transport

A single ATP-powered pump, such as the Na^+-K^+ pump, can indirectly drive the **secondary active transport** of several other solutes. By moving sodium across the plasma membrane against its concentration gradient, the pump stores energy (in the ion gradient). Then, just as water held back by a dam can do work as it flows downward (to generate electricity, for instance), a substance pumped across a membrane can do work as it leaks back, propelled "downhill" along its concentration gradient. In this way, as sodium moves back into the cell with the help of a carrier protein, other substances are "dragged along," or cotransported, by the same carrier protein (**Figure 3.10**). This is a symport system.

For example, some sugars, amino acids, and many ions are cotransported via secondary active transport into cells lining the small intestine. Because the energy for this type of transport is the concentration gradient of the ion (in this case Na^+),

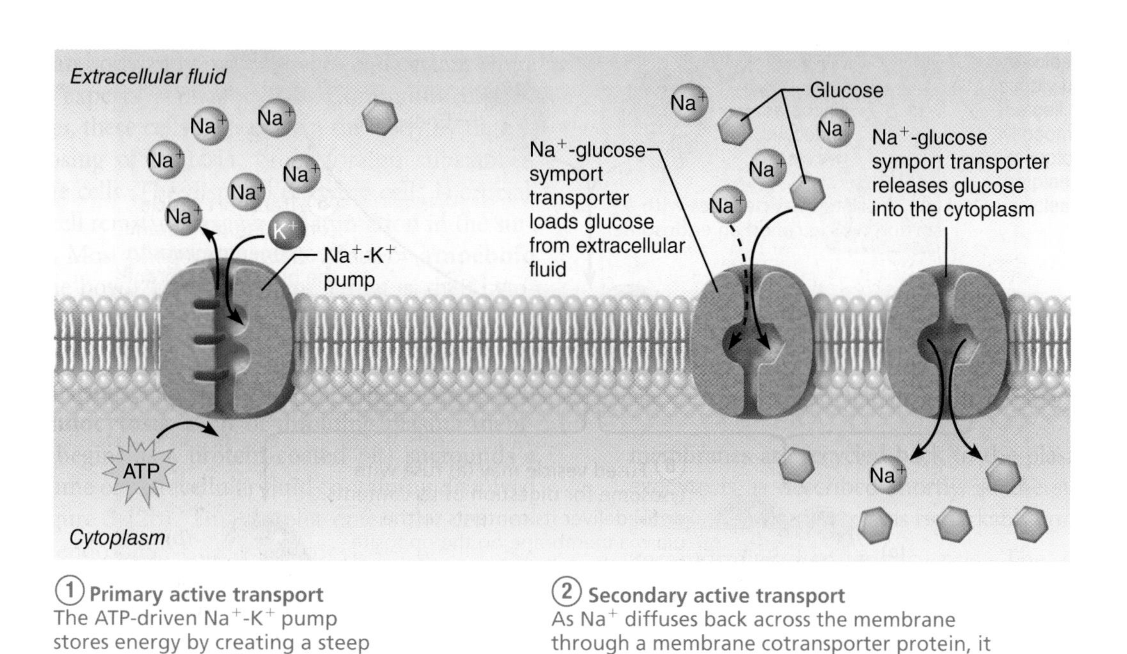

① **Primary active transport** The ATP-driven Na^+-K^+ pump stores energy by creating a steep concentration gradient for Na^+ entry into the cell.

② **Secondary active transport** As Na^+ diffuses back across the membrane through a membrane cotransporter protein, it drives glucose against its concentration gradient into the cell.

Figure 3.10 Secondary active transport is driven by the concentration gradient created by primary active transport.

K^+ Is the Key Player

Many kinds of ions are found both inside cells and in the extracellular fluid, but the resting membrane potential is determined mainly by the concentration gradient of potassium (K^+) and by the differential permeability of the plasma membrane to K^+ and other ions (**Figure 3.14**). K^+ and protein anions predominate inside body cells, and the extracellular fluid contains relatively more Na^+, which is largely balanced by Cl^-. The unstimulated plasma membrane is somewhat permeable to K^+ because of leakage channels, but impermeable to the protein anions. Consequently, as shown in Figure 3.14 ①, K^+ diffuses out of the cell along its concentration gradient but the protein anions are unable to follow, and this loss of positive charges makes the membrane interior more negative.

② As more and more K^+ leaves the cell, the negativity of the inner membrane face becomes great enough to attract K^+ back toward and even into the cell. ③ At a membrane voltage of −90 mV, potassium's concentration gradient is exactly balanced by the electrical gradient (membrane potential), and one K^+ enters the cell as one leaves.

In many cells, sodium (Na^+) also contributes to the resting membrane potential. Sodium is strongly attracted to the cell interior by its concentration gradient, bringing the resting membrane potential to −70 mV. However, K^+ still largely determines the resting membrane potential because the membrane is much more permeable to K^+ than to Na^+. Even though the membrane is permeable to Cl^-, in most cells Cl^- does not contribute to the resting membrane potential, because its concentration and electrical gradients exactly balance each other.

We may be tempted to believe that massive flows of K^+ ions are needed to generate the resting potential, but this is not the case. Surprisingly, the number of ions producing the membrane potential is so small that it does not change ion concentrations in any significant way.

In a cell at rest, very few ions cross its plasma membrane. However, Na^+ and K^+ are not at equilibrium and there is some net movement of K^+ out of the cell and of Na^+ into the cell. Na^+ is strongly pulled into the cell by both its concentration gradient and the interior negative charge. If only passive forces were at work, these ion concentrations would eventually become equal inside and outside the cell.

Active Transport Maintains Electrochemical Gradients

Now let's look at how active transport processes maintain the membrane potential that diffusion has established, with the result that the cell exhibits a *steady state*. The rate of active transport of Na^+ out of the cell is equal to, and depends on, the rate of Na^+ diffusion into the cell. If more Na^+ enters, more is pumped out. (This is like being in a leaky boat. The more water that comes in, the faster you bail!) The Na^+-K^+ pump couples sodium and potassium transport and, on average, each "turn" of the pump ejects $3Na^+$ out of the cell and carries $2K^+$ back in (see Focus Figure 3.1 on p. 68, *Focus on Primary Active Transport: The Na^+-K^+ Pump*). Because the membrane is about 25 times more permeable to K^+ than to Na^+, the ATP-dependent Na^+-K^+ pump maintains both the membrane potential (the charge separation) and the osmotic balance. Indeed, if Na^+ was not continuously removed from cells, so much would accumulate intracellularly that the osmotic gradient would draw water into the cells, causing them to burst.

As we described on p. 69, diffusion of charged particles across the membrane is affected not only by concentration gradients, but by the electrical charge on the inner and outer faces of the membrane. Together these gradients make up the *electrochemical gradient*. The diffusion of K^+ across the plasma membrane is aided by the membrane's greater permeability to it and by the ion's concentration gradient, but the negative charges on the cell interior resist K^+ diffusion. In contrast, a

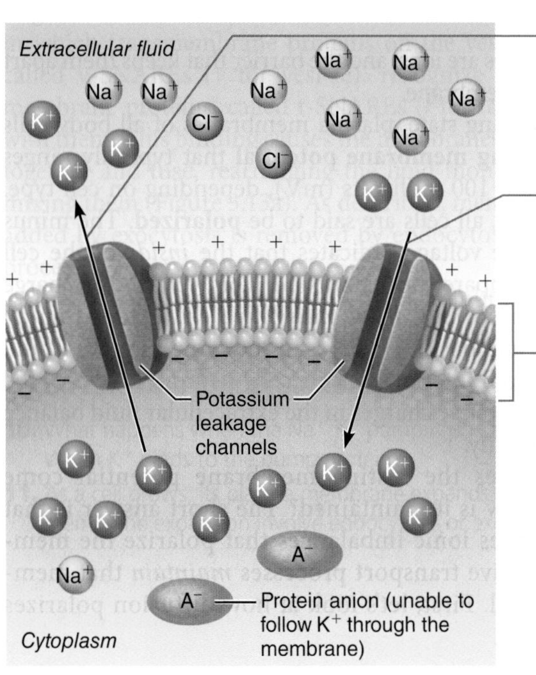

Figure 3.14 The key role of K^+ in generating the resting membrane potential. The resting membrane potential is largely determined by K^+ because at rest, the membrane is much more permeable to K^+ than Na^+. The active transport of sodium and potassium ions (in a ratio of 3:2) by the Na^+-K^+ pump maintains these conditions.

steep electrochemical gradient draws Na^+ into the cell, but the membrane's relative impermeability to it limits Na^+ diffusion.

The transient opening of gated Na^+ and K^+ channels in the plasma membrane "upsets" the resting membrane potential. As we describe in later chapters, this is a normal means of activating neurons and muscle cells.

☑ Check Your Understanding

14. What process establishes the resting membrane potential?

15. Is the inside of the plasma membrane negative or positive relative to its outside in a polarized membrane of a resting cell?

For answers, see Answers Appendix.

3.6 Cell adhesion molecules and membrane receptors allow the cell to interact with its environment

→ Learning Objectives

- ☐ Describe the role of the glycocalyx when cells interact with their environment.
- ☐ List several roles of membrane receptors and that of G protein–linked receptors.

Cells are biological minifactories and, like other factories, they receive and send orders from and to the outside community. But *how* does a cell interact with its environment, and *what* activates it to carry out its homeostatic functions?

Sometimes cells interact directly with other cells. However, in many cases cells respond to extracellular chemicals, such as hormones and neurotransmitters distributed in body fluids. Cells also interact with extracellular molecules that act as signposts to guide cell migration during development and repair.

Whether cells interact directly or indirectly, however, the glycocalyx is always involved. The best-understood glycocalyx molecules fall into two large families—cell adhesion molecules and plasma membrane receptors (see Figure 3.4).

Roles of Cell Adhesion Molecules (CAMs)

Thousands of **cell adhesion molecules (CAMs)** are found on almost every cell in the body. CAMs play key roles in embryonic development and wound repair (situations where cell mobility is important) and in immunity. These sticky glycoproteins (*cadherins* and *integrins*) act as:

- The molecular "Velcro" that cells use to anchor themselves to molecules in the extracellular space and to each other (see desmosome discussion on pp. 60–61)
- The "arms" that migrating cells use to haul themselves past one another
- SOS signals sticking out from the blood vessel lining that rally protective white blood cells to a nearby infected or injured area
- Mechanical sensors that respond to changes in local tension or fluid movement at the cell surface by stimulating synthesis or degradation of tight junctions
- Transmitters of intracellular signals that direct cell migration, proliferation, and specialization

Roles of Plasma Membrane Receptors

A huge and diverse group of integral proteins and glycoproteins that serve as binding sites are collectively known as **membrane receptors**. Some function in contact signaling, and others in chemical signaling. Let's take a look.

Contact Signaling

Contact signaling, in which cells come together and touch, is the means by which cells recognize one another. It is particularly important for normal development and immunity. Some bacteria and other infectious agents use contact signaling to identify their "preferred" target tissues.

Chemical Signaling

Most plasma membrane receptors are involved in *chemical signaling*. **Ligands** are chemicals that bind specifically to plasma membrane receptors. Ligands include most *neurotransmitters* (nervous system signals), *hormones* (endocrine system signals), and *paracrines* (chemicals that act locally and are rapidly destroyed).

Different cells respond in different ways to the same ligand. Acetylcholine, for instance, stimulates skeletal muscle cells to contract, but inhibits heart muscle. Why do different cells respond so differently? The reason is that a target cell's response depends on the internal machinery that the receptor is linked to, not the specific ligand that binds to it.

Though cell responses to receptor binding vary widely, there is a fundamental similarity: When a ligand binds to a membrane receptor, the receptor's structure changes, and cell proteins are altered in some way. For example, some membrane proteins respond to ligands by becoming activated enzymes, while others common in muscle and nerve cells respond by transiently opening or closing ion gates, which in turn changes the membrane potential of the cell.

G protein–linked receptors exert their effect *indirectly* through a **G protein**, a regulatory molecule that acts as a middleman or relay to activate (or inactivate) a membrane-bound enzyme or ion channel. This in turn generates one or more intracellular chemical signals, commonly called **second messengers**, which connect plasma membrane events to the internal metabolic machinery of the cell. Two important second messengers are **cyclic AMP** and ionic calcium, both of which typically activate *protein kinase enzymes*. These enzymes transfer phosphate groups from ATP to other proteins, activating a whole series of enzymes that bring about the desired cellular activity. Because a single enzyme can catalyze hundreds of reactions, the amplification effect of such a chain of events is tremendous, much like that stirred up by a chain letter. *Focus on G Proteins* (**Focus Figure 3.2**, p. 76) describes a G protein signaling system. Take a moment to study the figure carefully because this key signaling pathway is involved in neurotransmission, smell, vision, and hormone action (Chapters 11, 13, and 15).

Focus Figure 3.3 Mitosis is the process of nuclear division in which the chromosomes are distributed to two daughter nuclei. Together with cytokinesis, it produces two identical daughter cells.

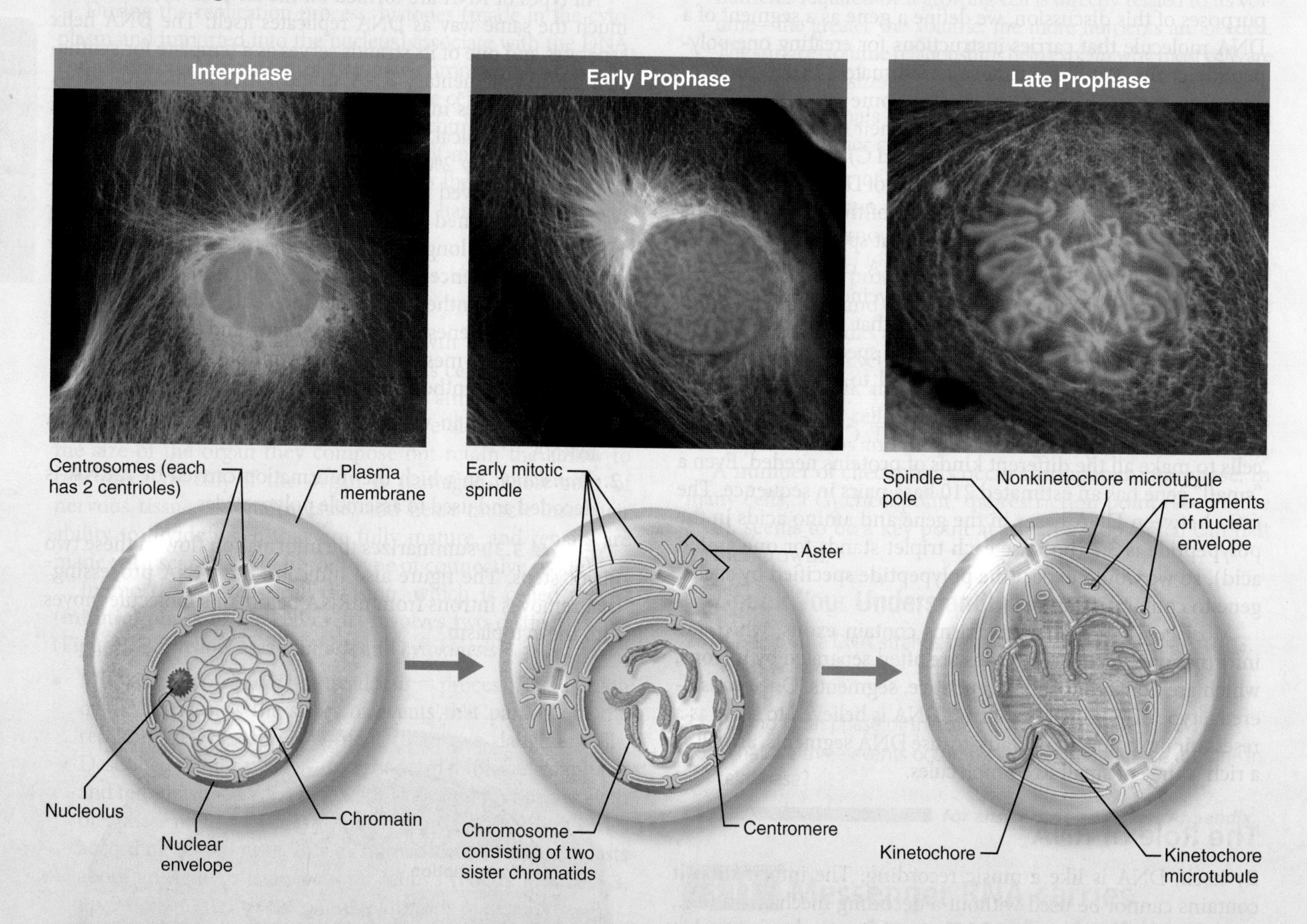

Interphase

Interphase

Interphase is the period of a cell's life when it carries out its normal metabolic activities and grows. Interphase is not part of mitosis.

- During interphase, the DNA-containing material is in the form of chromatin. The nuclear envelope and one or more nucleoli are intact and visible.
- There are three distinct periods of interphase: G_1, S, and G_2.

The light micrographs show dividing lung cells from a newt. The chromosomes appear blue and the microtubules green. (The red fibers are intermediate filaments.) The schematic drawings show details not visible in the micrographs. For simplicity, only four chromosomes are drawn.

Prophase—first phase of mitosis

Early Prophase

- The chromatin coils and condenses, forming barlike *chromosomes*.
- Each duplicated chromosome consists of two identical threads, called *sister chromatids*, held together at the *centromere*. (Later when the chromatids separate, each will be a new chromosome.)
- As the chromosomes appear, the nucleoli disappear, and the two centrosomes separate from one another.
- The centrosomes act as focal points for growth of a microtubule assembly called the *mitotic spindle*. As the microtubules lengthen, they propel the centrosomes toward opposite ends (poles) of the cell.
- Microtubule arrays called *asters* ("stars") extend from the centrosome matrix.

Late Prophase

- The nuclear envelope breaks up, allowing the spindle to interact with the chromosomes.
- Some of the growing spindle microtubules attach to *kinetochores* (ki-ne´ to-korz), special protein structures at each chromosome's centromere. Such microtubules are called *kinetochore microtubules*.
- The remaining (unattached) spindle microtubules are called *nonkinetochore microtubules*. The microtubules slide past each other, forcing the poles apart.
- The kinetochore microtubules pull on each chromosome from both poles in a tug-of-war that ultimately draws the chromosomes to the center, or equator, of the cell.

Watch full 3-D animations
MasteringA&P®>Study Area> ***A&PFlix***

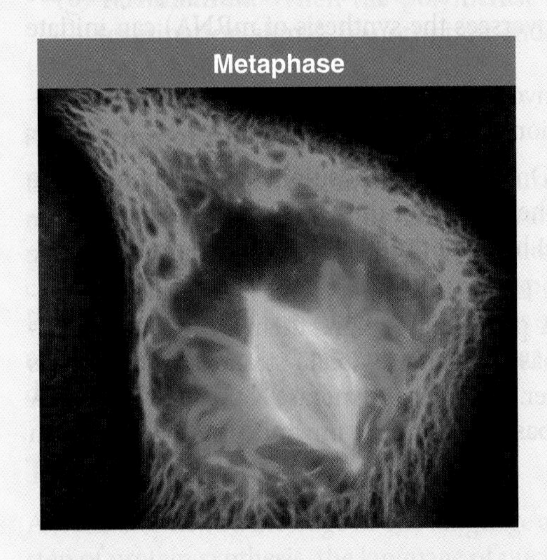

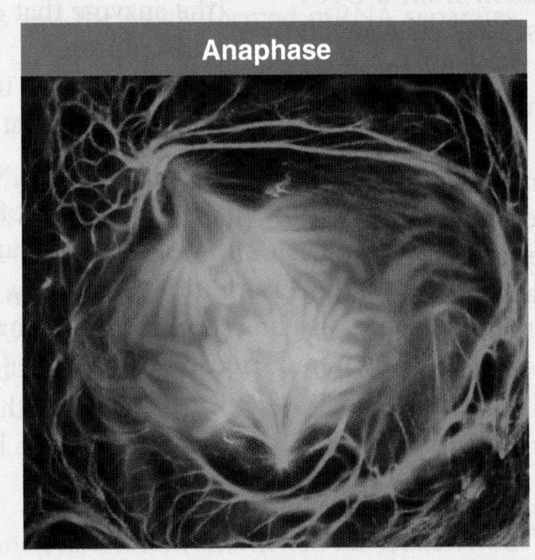

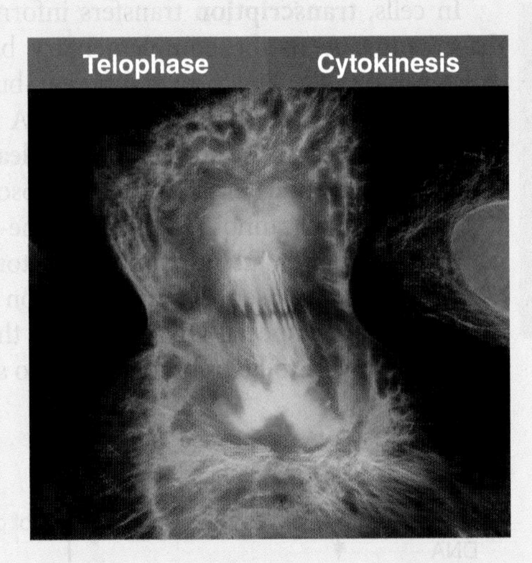

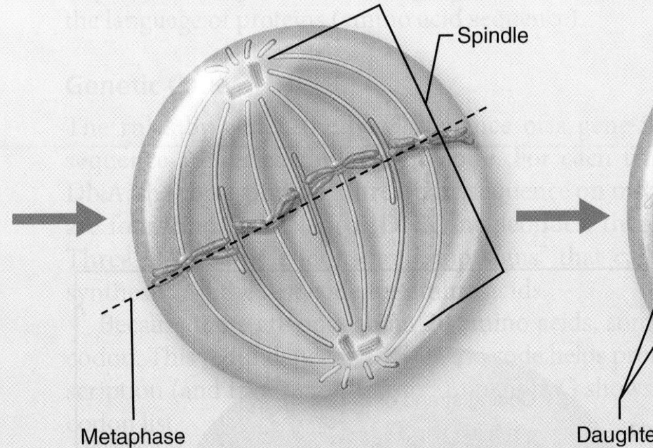

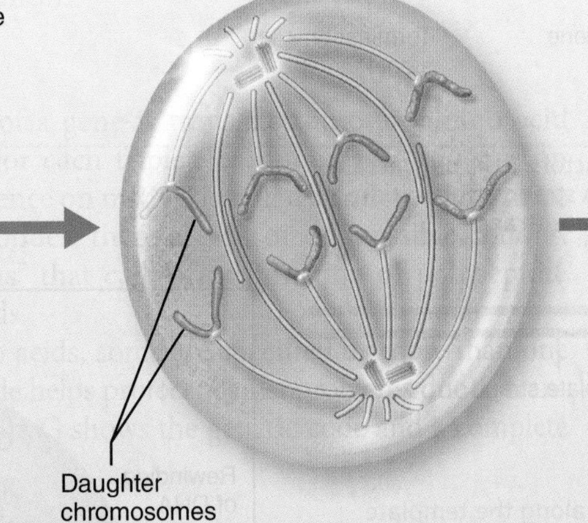

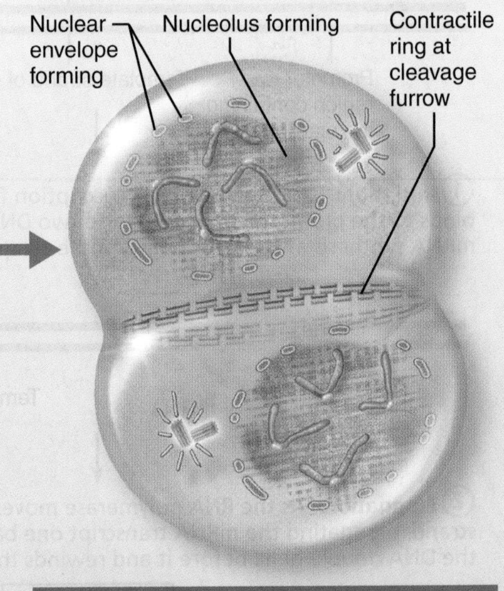

Metaphase—second phase of mitosis

- The two centrosomes are at opposite poles of the cell.
- The chromosomes cluster at the midline of the cell, with their centromeres precisely aligned at the spindle *equator*. This imaginary plane midway between the poles is called the *metaphase plate*.
- At the end of metaphase, enzymes that will act to separate the chromatids from each other are triggered.

Anaphase—third phase of mitosis

The shortest phase of mitosis, anaphase begins abruptly as the centromeres of the chromosomes split simultaneously. Each chromatid now becomes a chromosome in its own right.

- The kinetochore microtubules, moved along by motor proteins in the kinetochores, gradually pull each chromosome toward the pole it faces.
- At the same time, the nonkinetochore microtubules slide past each other, lengthen, and push the two poles of the cell apart.
- The moving chromosomes look V shaped. The centromeres lead the way, and the chromosomal "arms" dangle behind them.
- Moving and separating the chromosomes is helped by the fact that the chromosomes are short, compact bodies. Diffuse threads of chromatin would trail, tangle, and break, resulting in imprecise "parceling out" to the daughter cells.

Telophase—final phase of mitosis

Telophase
Telophase begins as soon as chromosomal movement stops. This final phase is like prophase in reverse.

- The identical sets of chromosomes at the opposite poles of the cell begin to uncoil and resume their threadlike chromatin form.
- A new nuclear envelope forms around each chromatin mass, nucleoli reappear within the nuclei, and the spindle breaks down and disappears.
- Mitosis is now ended. The cell, for just a brief period, is binucleate (has two nuclei) and each new nucleus is identical to the original mother nucleus.

Cytokinesis—division of cytoplasm

Cytokinesis begins during late anaphase and continues through and beyond telophase. A contractile ring of actin microfilaments forms the *cleavage furrow* and pinches the cell apart.

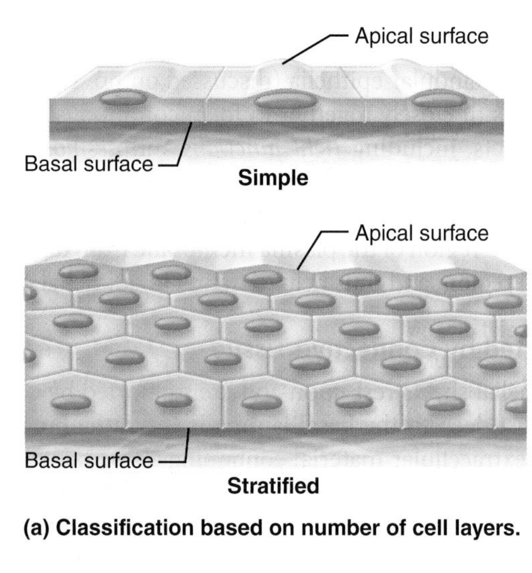

(a) Classification based on number of cell layers.

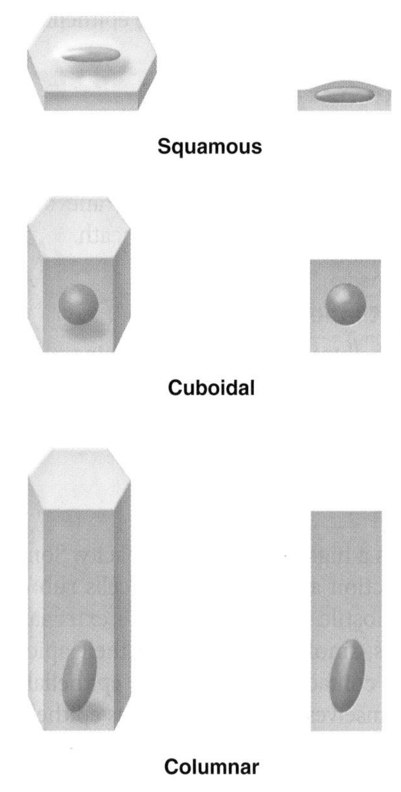

(b) Classification based on cell shape.

Figure 4.2 Classification of epithelia. Note that cell shape influences the shape of the nucleus.

looks like a honeycomb. This polyhedral shape allows the cells to be closely packed. However, epithelial cells vary in height, and on that basis, there are three common shapes of epithelial cells (Figure 4.2b):

- **Squamous cells** (skwa′mus) are flattened and scale-like (*squam* = scale).
- **Cuboidal cells** (ku-boi′dahl) are boxlike, approximately as tall as they are wide.
- **Columnar cells** (kŏ-lum′nar) are tall and column shaped.

In each case, the shape of the nucleus conforms to that of the cell. The nucleus of a squamous cell is a flattened disc; that of a cuboidal cell is spherical; and a columnar cell nucleus is elongated from top to bottom and usually located closer to the cell base. Keep nuclear shape in mind when you attempt to identify epithelial types.

Simple epithelia are easy to classify by cell shape because the cells usually have the same shape. In stratified epithelia, however, cell shape differs in the different layers. To avoid ambiguity, stratified epithelia are named according to the shape of the cells in the *apical* layer. This naming system will become clearer as we explore the specific epithelial types.

As you read about the epithelial classes, study **Figure 4.3**. Try to pick out the individual cells within each epithelium. This is not always easy, because the boundaries between epithelial cells often are indistinct. Furthermore, the nucleus of a particular cell may or may not be visible, depending on the plane of the cut made to prepare the tissue slides.

Simple Epithelia

The simple epithelia are most concerned with absorption, secretion, and filtration. Because they consist of a single cell layer and are usually very thin, protection is not one of their specialties.

Simple Squamous Epithelium The cells of a **simple squamous epithelium** are flattened laterally, and their cytoplasm is sparse (Figure 4.3a). In a surface view, the close-fitting cells resemble a tiled floor. When the cells are cut perpendicular to their free surface, they resemble fried eggs seen from the side, with their cytoplasm wisping out from the slightly bulging nucleus.

Thin and often permeable, simple squamous epithelium is found where filtration or the exchange of substances by rapid diffusion is a priority. In the kidneys, it forms part of the filtration membrane. In the lungs, it forms the walls of the air sacs across which gas exchange occurs (Figure 4.3a).

Two simple squamous epithelia in the body have special names that reflect their location.

- **Endothelium** (en″do-the′le-um; "inner covering") provides a slick, friction-reducing lining in lymphatic vessels and in all hollow organs of the cardiovascular system—blood vessels and the heart. Capillaries consist exclusively of endothelium, and its exceptional thinness encourages the efficient exchange of nutrients and wastes between the bloodstream and surrounding tissue cells.
- **Mesothelium** (mez″o-the′le-um; "middle covering") is the epithelium found in serous membranes, the membranes lining the ventral body cavity and covering its organs.

Simple Cuboidal Epithelium **Simple cuboidal epithelium** consists of a single layer of cells as tall as they are wide (Figure 4.3b). The generally spherical nuclei stain darkly. Important functions of simple cuboidal epithelium are secretion and absorption. This epithelium forms the walls of the smallest ducts of glands and of many kidney tubules.

Simple Columnar Epithelium **Simple columnar epithelium** is a single layer of tall, closely packed cells, aligned like

(a) Simple squamous epithelium

Description: Single layer of flattened cells with disc-shaped central nuclei and sparse cytoplasm; the simplest of the epithelia.

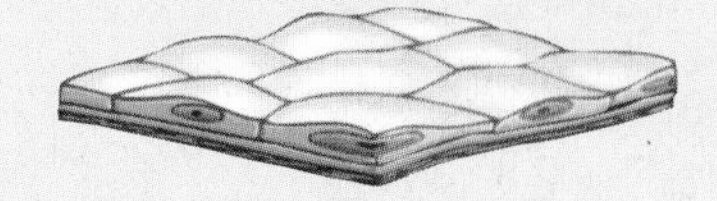

Function: Allows materials to pass by diffusion and filtration in sites where protection is not important; secretes lubricating substances in serosae.

Location: Kidney glomeruli; air sacs of lungs; lining of heart, blood vessels, and lymphatic vessels; lining of ventral body cavity (serosae).

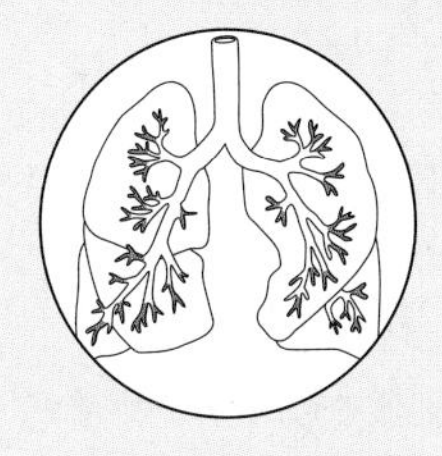

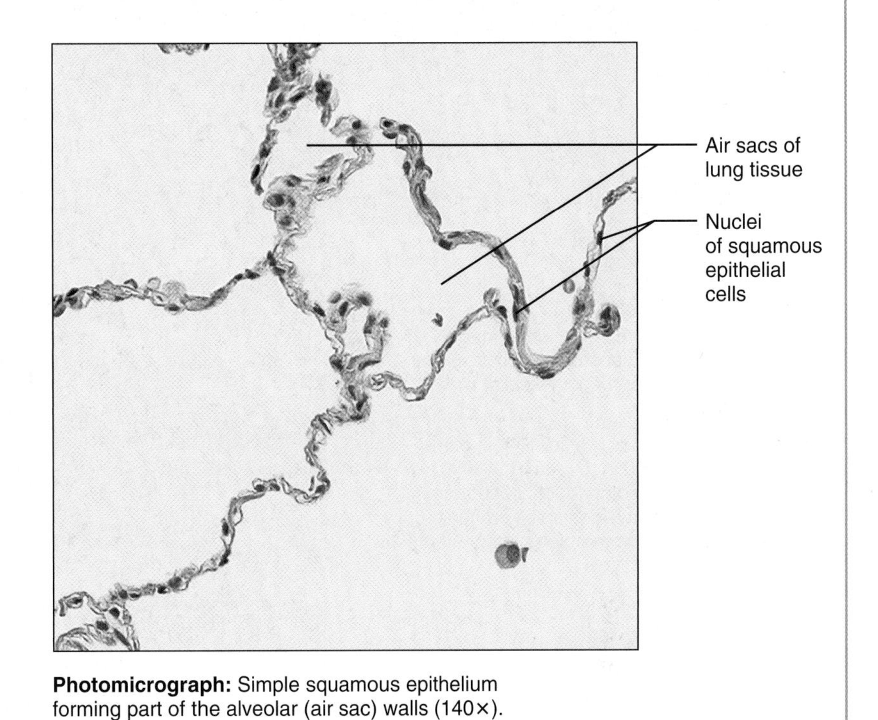

Photomicrograph: Simple squamous epithelium forming part of the alveolar (air sac) walls (140×).

(b) Simple cuboidal epithelium

Description: Single layer of cubelike cells with large, spherical central nuclei.

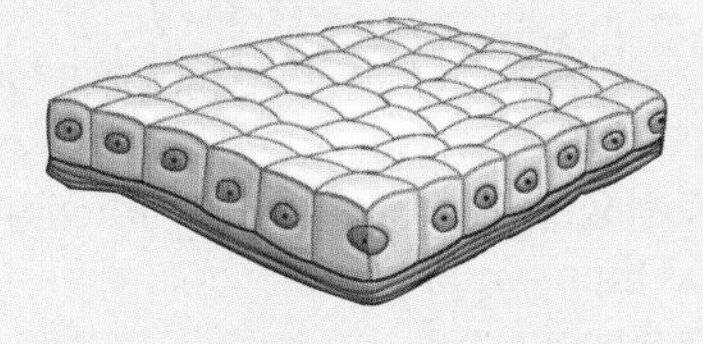

Function: Secretion and absorption.

Location: Kidney tubules; ducts and secretory portions of small glands; ovary surface.

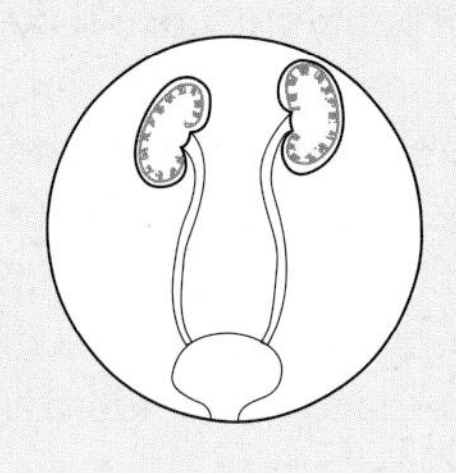

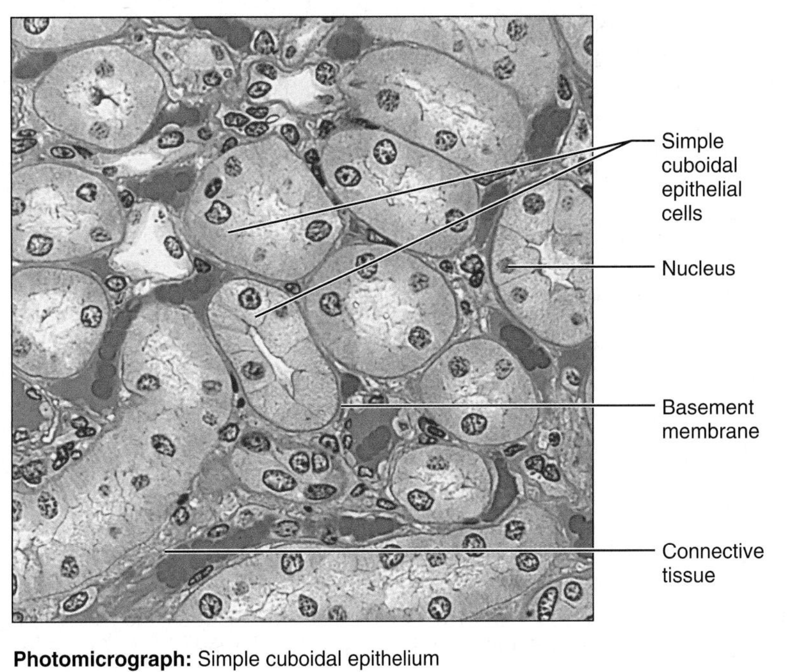

Photomicrograph: Simple cuboidal epithelium in kidney tubules (430×).

Figure 4.3 Epithelial tissues. (a) Simple epithelium. (For related images, see *A Brief Atlas of the Human Body,* Plates 1 and 2.) **(b)** Simple epithelium. (For a related image, see *A Brief Atlas of the Human Body,* Plate 3.)

View histology slides
MasteringA&P®>Study Area>PAL

(c) Simple columnar epithelium

Description: Single layer of tall cells with *round* to *oval* nuclei; many cells bear microvilli, some bear cilia; layer may contain mucus-secreting unicellular glands (goblet cells).

Function: Absorption; secretion of mucus, enzymes, and other substances; ciliated type propels mucus (or reproductive cells) by ciliary action.

Location: Nonciliated type lines most of the digestive tract (stomach to rectum), gallbladder, and excretory ducts of some glands; ciliated variety lines small bronchi, uterine tubes, and some regions of the uterus.

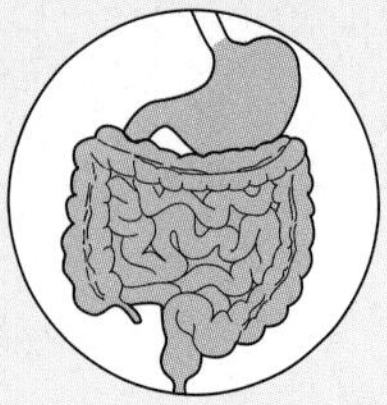

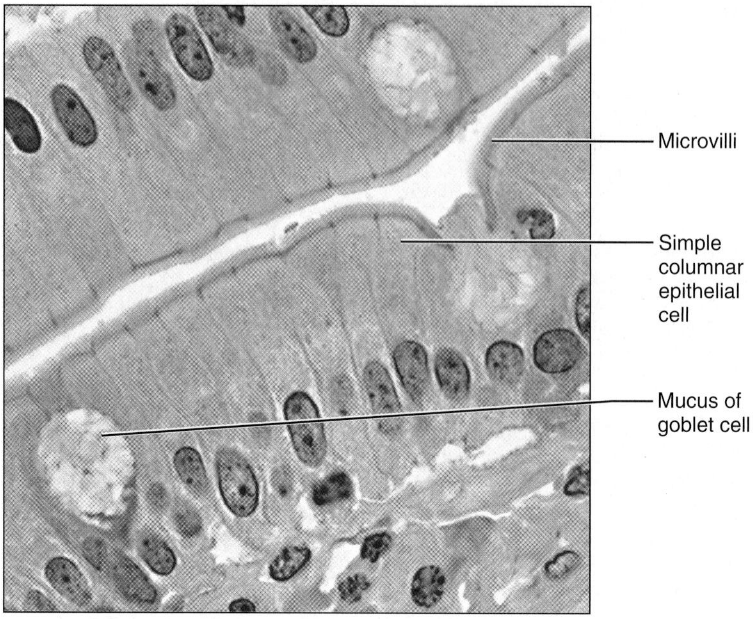

Microvilli

Simple columnar epithelial cell

Mucus of goblet cell

Photomicrograph: Simple columnar epithelium of the small intestine mucosa (640×).

Figure 4.3 *(continued)* **Epithelial tissues. (c)** Simple epithelium. (For related images, see *A Brief Atlas of the Human Body,* Plates 4 and 5.)

soldiers in a row (Figure 4.3c). It lines the digestive tract from the stomach through the rectum. Columnar cells are mostly associated with absorption and secretion, and the digestive tract lining has two distinct modifications that make it ideal for that dual function:

- Dense microvilli on the apical surface of absorptive cells
- Tubular glands made primarily of cells that secrete mucus-containing intestinal juice

Additionally, some simple columnar epithelia display cilia on their free surfaces, which help move substances or cells through an internal passageway.

Pseudostratified Columnar Epithelium The cells of **pseudostratified columnar epithelium** (soo″do-stră′tĭ-fīd) vary in height (Figure 4.3d). All of its cells rest on the basement membrane, but only the tallest reach the free surface of the epithelium. Because the cell nuclei lie at different levels above the basement membrane, the tissue gives the false (pseudo) impression that several cell layers are present; hence "pseudostratified." The short cells are relatively unspecialized and give rise to the taller cells. This epithelium, like the simple columnar variety, secretes or absorbs substances. A ciliated version containing mucus-secreting goblet cells lines most of the respiratory tract. Here the motile cilia propel sheets of dust-trapping mucus superiorly away from the lungs.

Stratified Epithelia

Stratified epithelia contain two or more cell layers. They regenerate from below; that is, the basal cells divide and push apically to replace the older surface cells. Stratified epithelia are considerably more durable than simple epithelia, and protection is their major (but not their only) role.

Stratified Squamous Epithelium **Stratified squamous epithelium** is the most widespread of the stratified epithelia (Figure 4.3e). Composed of several layers, it is thick and well suited for its protective role in the body. Its free surface cells are squamous, and cells of the deeper layers are cuboidal or columnar. This epithelium is found in areas subjected to wear and tear, and its surface cells are constantly being rubbed away and replaced by division of its basal cells. Because epithelium depends on nutrients diffusing from deeper connective tissue, the epithelial cells farther from the basement membrane are less viable and those at the apical surface are often flattened and atrophied.

To avoid memorizing all its locations, simply remember that this epithelium forms the external part of the skin and extends a short distance into every body opening that is directly continuous with the skin. The outer layer, or epidermis, of the skin is *keratinized* (ker′ah-tin″īzd), meaning its surface cells contain *keratin*, a tough protective protein. (We discuss the epidermis in Chapter 5.) The other stratified squamous epithelia of the body are *nonkeratinized*.

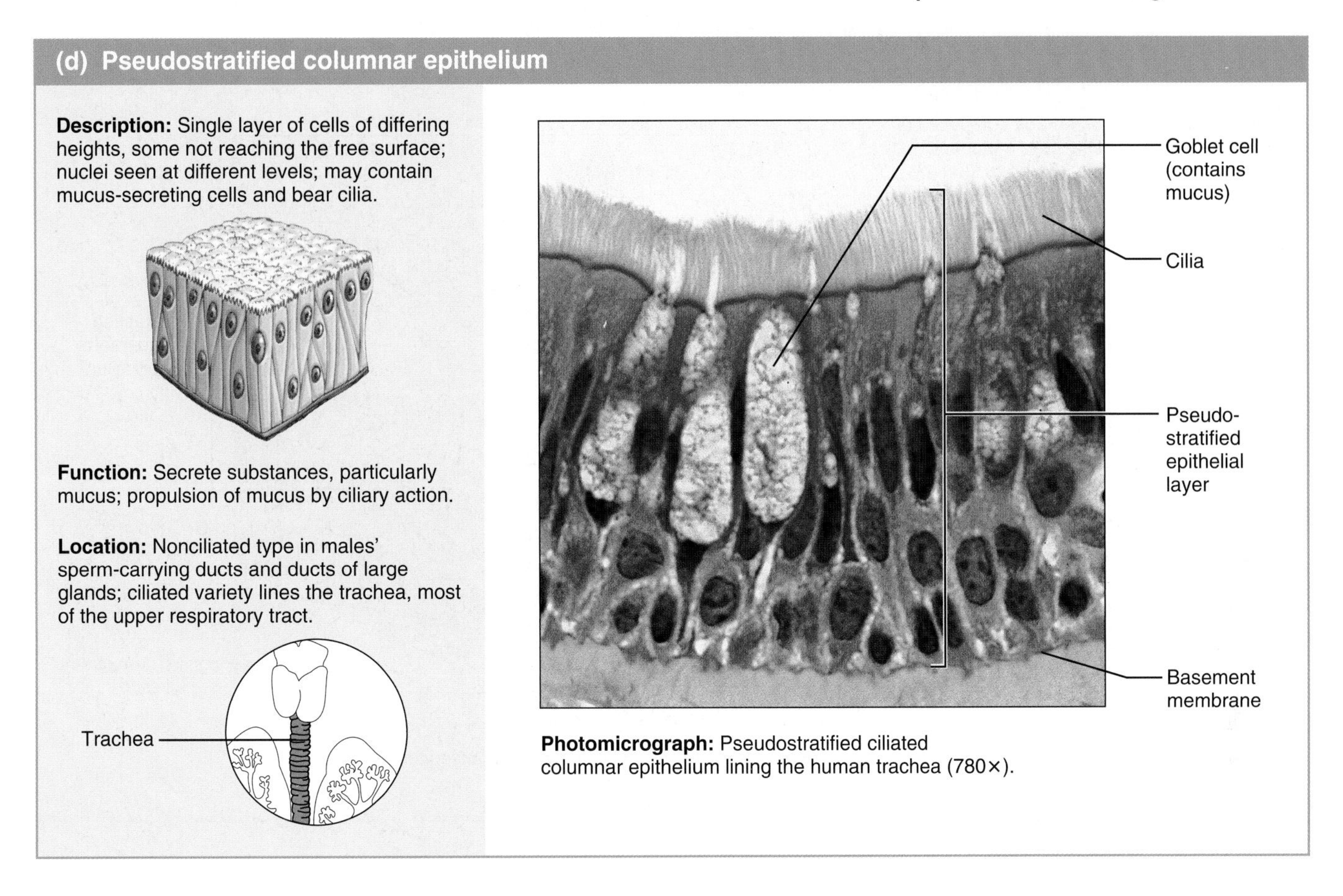

Figure 4.3 *(continued)* **(d)** Simple epithelium. (For a related image, see *A Brief Atlas of the Human Body,* Plate 6.)

Stratified Cuboidal and Columnar Epithelia **Stratified cuboidal epithelium** is quite rare in the body, mostly found in the ducts of some of the larger glands (sweat glands, mammary glands). It typically has two layers of cuboidal cells.

Stratified columnar epithelium also has a limited distribution in the body. Small amounts are found in the pharynx, the male urethra, and lining some glandular ducts. This epithelium also occurs at transition areas or junctions between two other types of epithelia. Only its apical layer of cells is columnar. Because of their relative scarcity in the body, Figure 4.3 does not illustrate these two stratified epithelia (but see *A Brief Atlas of the Human Body*, Plates 8 and 9).

Transitional Epithelium **Transitional epithelium** forms the lining of hollow urinary organs, which stretch as they fill with urine (Figure 4.3f). Cells of its basal layer are cuboidal or columnar. The apical cells vary in appearance, depending on the degree of distension (stretching) of the organ. When the organ is distended with urine, the transitional epithelium thins from about six cell layers to three, and its domelike apical cells flatten and become squamouslike. The ability of transitional cells to change their shape (undergo "transitions") allows a greater volume of urine to flow through a tubelike organ. In the bladder, it allows more urine to be stored.

Glandular Epithelia

A **gland** consists of one or more cells that make and secrete a particular product. This product, called a **secretion**, is an aqueous (water-based) fluid that usually contains proteins, but there is variation. For example, some glands release a lipid- or steroid-rich secretion.

Secretion is an active process. Glandular cells obtain needed substances from the blood and transform them chemically into a product that is then discharged from the cell. Notice that the term *secretion* can refer to both the gland's *product* and the *process* of making and releasing that product.

Glands are classified according to two sets of traits:

- Where they release their product—glands may be *endocrine* ("internally secreting") or *exocrine* ("externally secreting")
- Number of cells—glands may be *unicellular* ("one-celled") or *multicellular* ("many-celled")

Unicellular glands are scattered within epithelial sheets. By contrast, most multicellular epithelial glands form by invagination (inward growth) of an epithelial sheet into the underlying connective tissue. At least initially, most have *ducts*, tubelike connections to the epithelial sheets.

4

(e) Stratified squamous epithelium

Description: Thick membrane composed of several cell layers; basal cells are cuboidal or columnar and metabolically active; surface cells are flattened (squamous); in the keratinized type, the surface cells are full of keratin and dead; basal cells are active in mitosis and produce the cells of the more superficial layers.

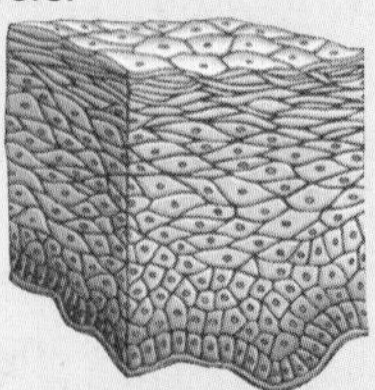

Function: Protects underlying tissues in areas subjected to abrasion.

Location: Nonkeratinized type forms the moist linings of the esophagus, mouth, and vagina; keratinized variety forms the epidermis of the skin, a dry membrane.

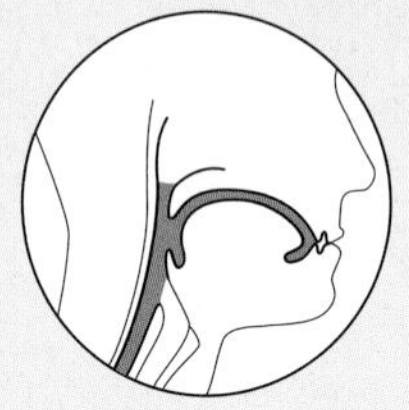

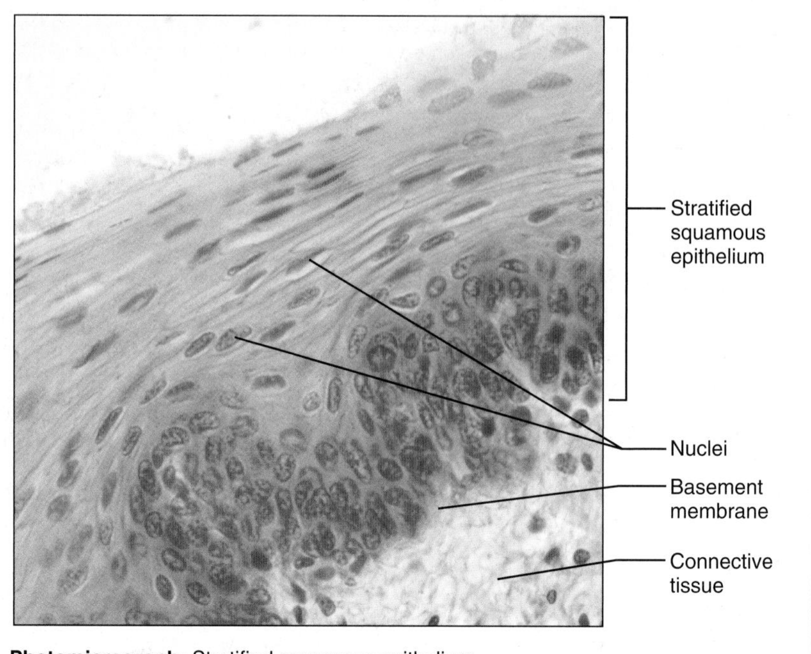

Photomicrograph: Stratified squamous epithelium lining the esophagus (285×).

(f) Transitional epithelium

Description: Resembles both stratified squamous and stratified cuboidal; basal cells cuboidal or columnar; surface cells dome shaped or squamouslike, depending on degree of organ stretch.

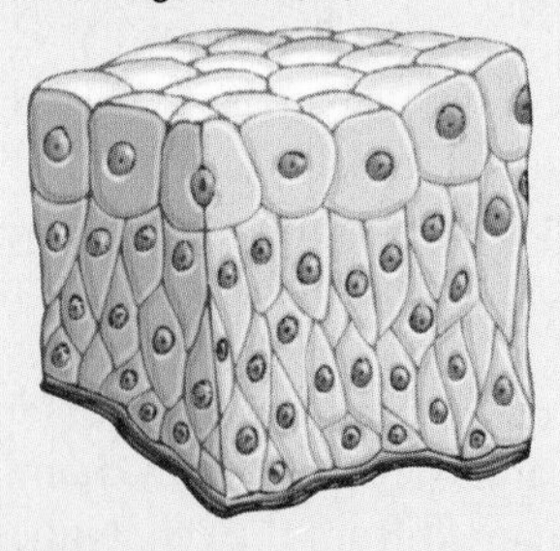

Function: Stretches readily, permits stored urine to distend urinary organ.

Location: Lines the ureters, bladder, and part of the urethra.

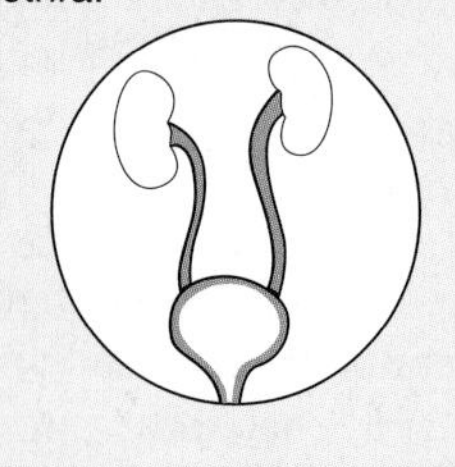

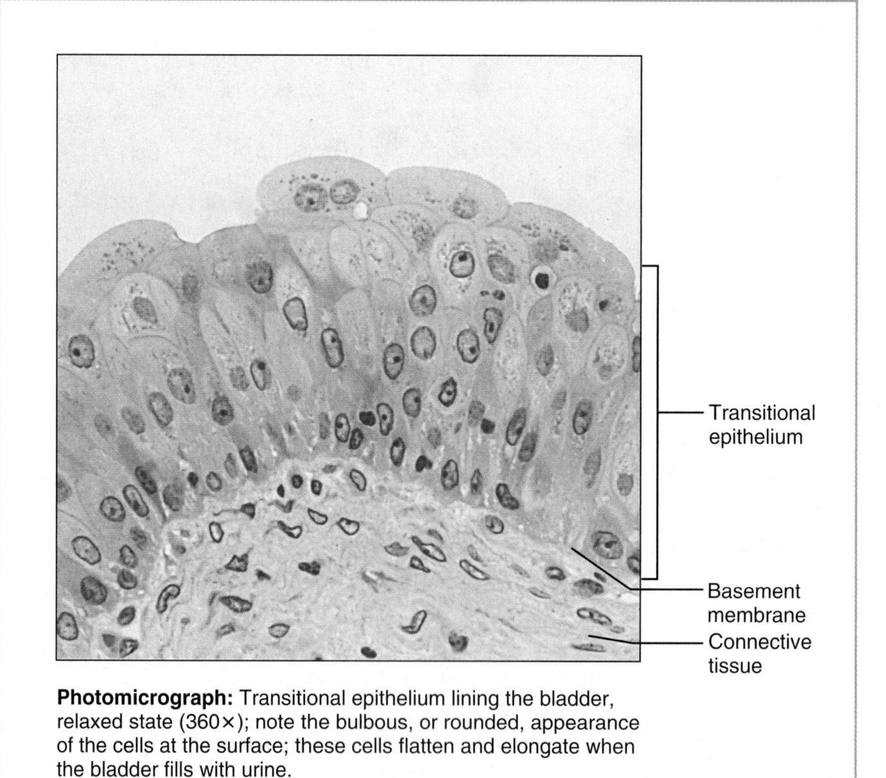

Photomicrograph: Transitional epithelium lining the bladder, relaxed state (360×); note the bulbous, or rounded, appearance of the cells at the surface; these cells flatten and elongate when the bladder fills with urine.

Figure 4.3 *(continued)* **Epithelial tissues. (e)** Stratified epithelium. (For a related image, see *A Brief Atlas of the Human Body,* Plate 7.) **(f)** Stratified epithelium. (For a related image, see *A Brief Atlas of the Human Body,* Plate 10.)

Endocrine Glands

Because **endocrine glands** eventually lose their ducts, they are often called *ductless glands*. They produce **hormones**, messenger chemicals that they secrete by exocytosis directly into the extracellular space. From there the hormones enter the blood or lymphatic fluid and travel to specific target organs. Each hormone prompts its target organ(s) to respond in some characteristic way. For example, hormones produced by certain intestinal cells cause the pancreas to release enzymes that help digest food in the digestive tract.

Endocrine glands are structurally diverse, so one description does not fit all. Most are compact multicellular organs, but some individual hormone-producing cells are scattered in the digestive tract lining (mucosa) and in the brain, giving rise to their collective description as the *diffuse endocrine system*. Endocrine secretions are also varied, ranging from modified amino acids to peptides, glycoproteins, and steroids. Not all endocrine glands are epithelial derivatives, so we defer their further consideration to Chapter 16.

Exocrine Glands

All **exocrine glands** secrete their products onto body surfaces (skin) or into body cavities. The unicellular glands do so directly (by exocytosis), whereas the multicellular glands do so via an epithelium-walled duct that transports the secretion to the epithelial surface. Exocrine glands are a diverse lot and many of their products are familiar. They include mucous, sweat, oil, and salivary glands, the liver (which secretes bile), the pancreas (which synthesizes digestive enzymes), and many others.

Unicellular Exocrine Glands The only important examples of **unicellular** (or one-celled) glands are *mucous cells* and *goblet cells*. Unicellular glands are sprinkled in the epithelial linings of the intestinal and respiratory tracts amid columnar cells with other functions (see Figure 4.3c).

In humans, all such glands produce **mucin** (mu′sin), a complex glycoprotein that dissolves in water when secreted. Once dissolved, mucin forms mucus, a slimy coating that protects and lubricates surfaces. In **goblet cells** the cuplike accumulation of mucin distends the top of the cell, making the cells look like a glass with a stem (thus "goblet" cell, **Figure 4.4**). This distortion does not occur in **mucous cells**.

Multicellular Exocrine Glands Compared to the unicellular glands, **multicellular exocrine glands** are structurally more complex. They have two basic parts: an epithelium-derived *duct* and a *secretory unit* (*acinus*) consisting of secretory cells. In all

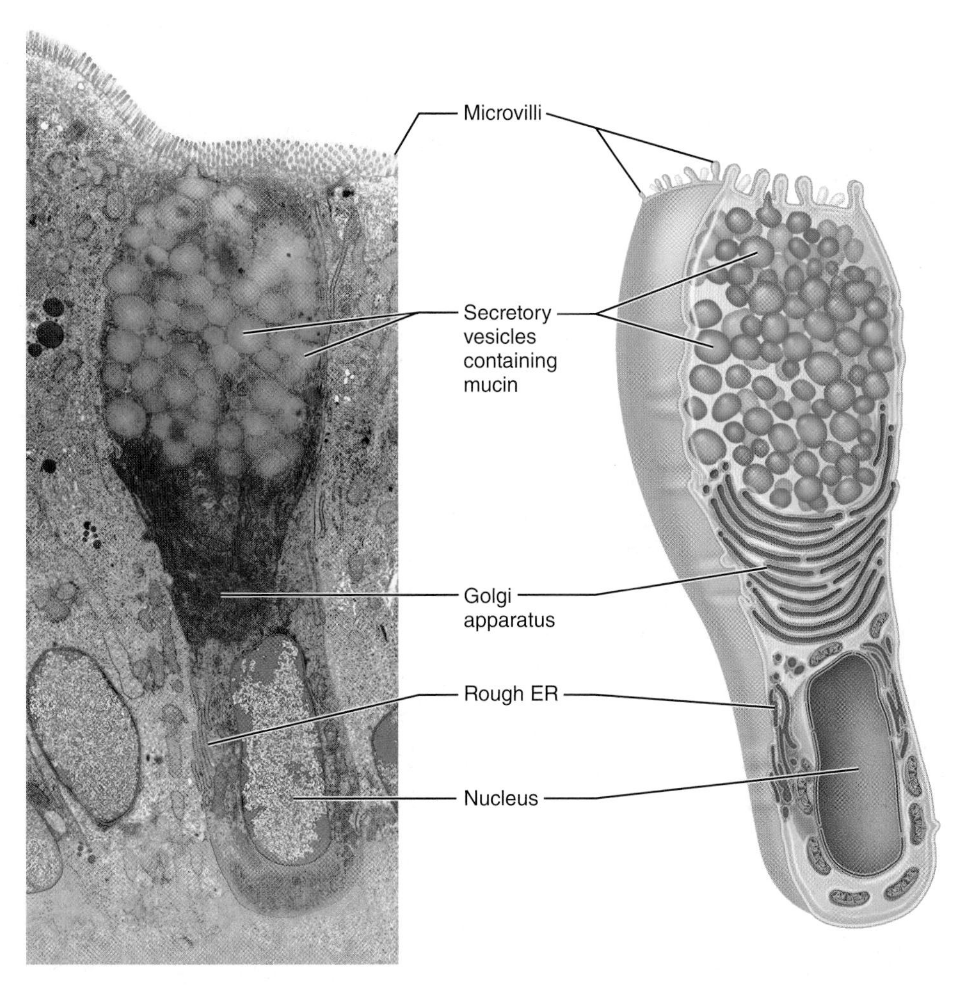

View histology slides
MasteringA&P®>Study Area>PAL

Figure 4.4 Goblet cell (unicellular exocrine gland). (a) Photomicrograph of a goblet cell in the simple columnar epithelium lining the small intestine (1640×). **(b)** Corresponding diagram. Notice the secretory vesicles and well-developed rough ER and Golgi apparatus.

but the simplest glands, *supportive connective tissue* surrounds the secretory unit, supplies it with blood vessels and nerve fibers, and forms a *fibrous capsule* that extends into the gland and divides it into *lobes*.

Multicellular exocrine glands can be classified by structure and by type of secretion.

- **Structural classification.** On the basis of their duct structures, multicellular exocrine glands are either simple or compound (**Figure 4.5**). **Simple glands** have an unbranched duct, whereas **compound glands** have a branched duct. The glands are further categorized by their secretory units as (1) **tubular** if the secretory cells form tubes; (2) **alveolar** (al-ve′o-lar) if the secretory cells form small, flasklike sacs (*alveolus* = "small hollow cavity"); or (3) **tubuloalveolar** if they have both types of secretory units. Note that many anatomists use the term **acinar** (as′ĭ-nar; "berry-like") interchangeably with alveolar.
- **Modes of secretion.** Multicellular exocrine glands secrete their products in different ways, so they can also be described functionally as *merocrine*, *holocrine*, or *apocrine* glands. Most are **merocrine glands** (mer′o-krin), which secrete their products by exocytosis as they are produced. The secretory cells are not altered in any way (so think "merely secrete" to remember their mode of secretion). The pancreas, most sweat glands, and salivary glands belong to this class (**Figure 4.6a**).

Secretory cells of **holocrine glands** (hol′o-krin) accumulate their products within them until they rupture. (They are replaced by the division of underlying cells.) Because holocrine gland secretions include the synthesized product plus dead cell fragments (*holo* = whole, all), you could say that their cells "die for their cause." Sebaceous (oil) glands of the skin are the only true example of holocrine glands (Figure 4.6b).

Although *apocrine glands* (ap′o-krin) are present in other animals, there is some controversy over whether humans have this gland type. Like holocrine glands, apocrine glands accumulate their products, but in this case only just beneath the free surface. Eventually, the apex of the cell pinches off (*apo* = from, off), releasing the secretory granules and a small amount of cytoplasm. The cell repairs its damage and the process repeats again and again. The best possibility in humans

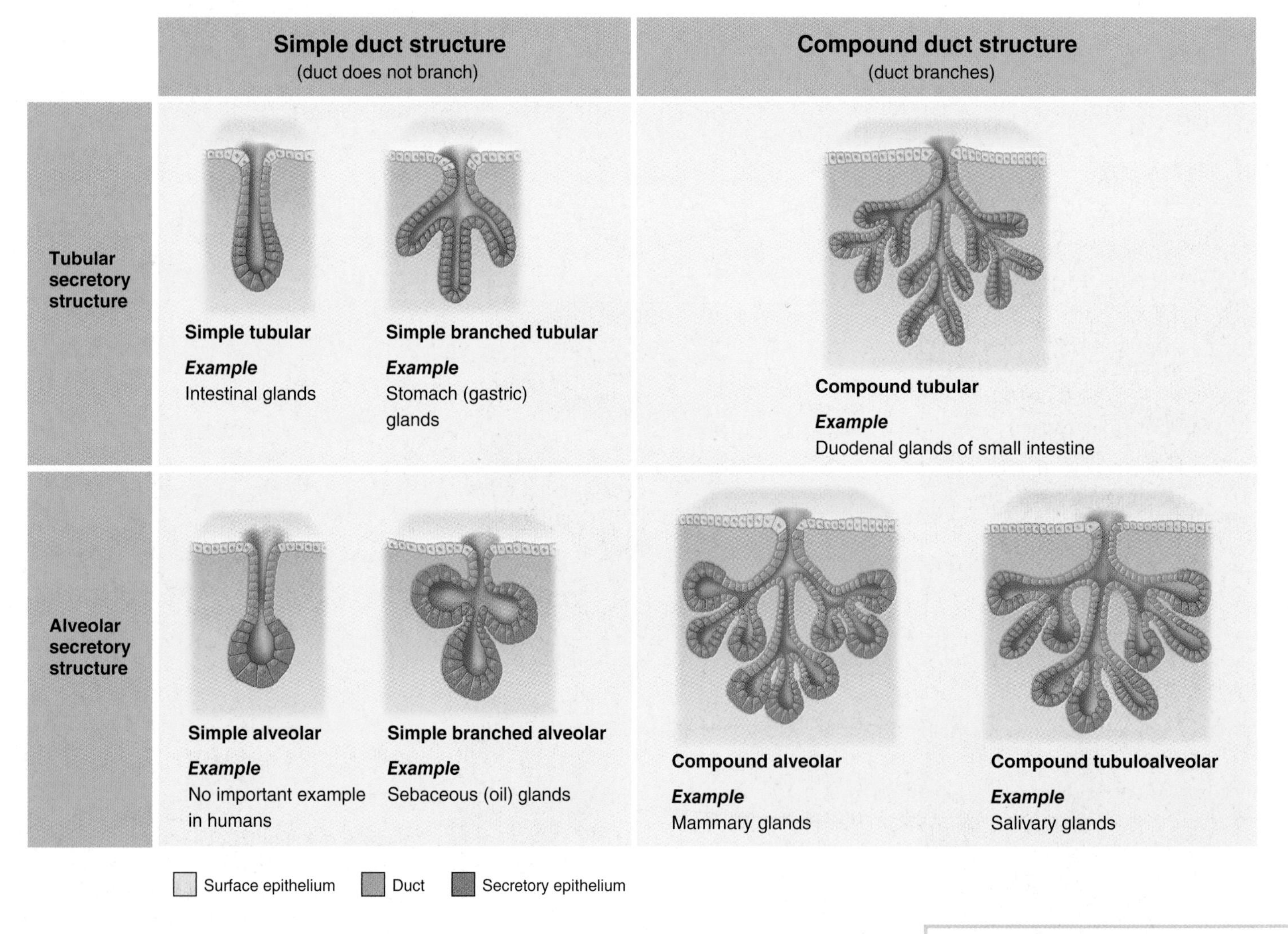

Figure 4.5 Types of multicellular exocrine glands. Multicellular glands are classified according to duct type (simple or compound) and the structure of their secretory units (tubular, alveolar, or tubuloalveolar).

Practice art labeling
MasteringA&P®>Study Area>Chapter 4

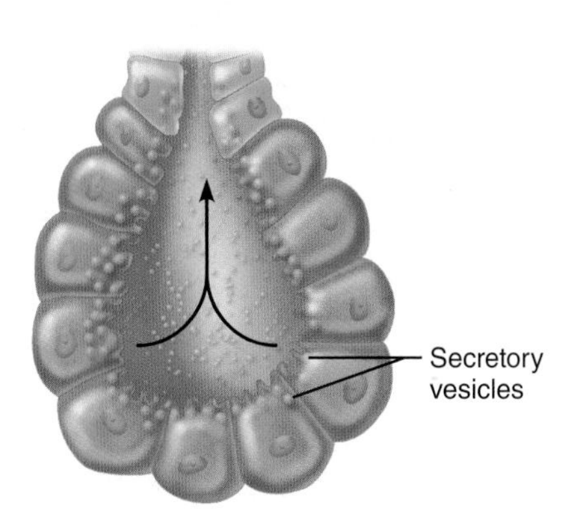

(a) Merocrine glands secrete their products by exocytosis.

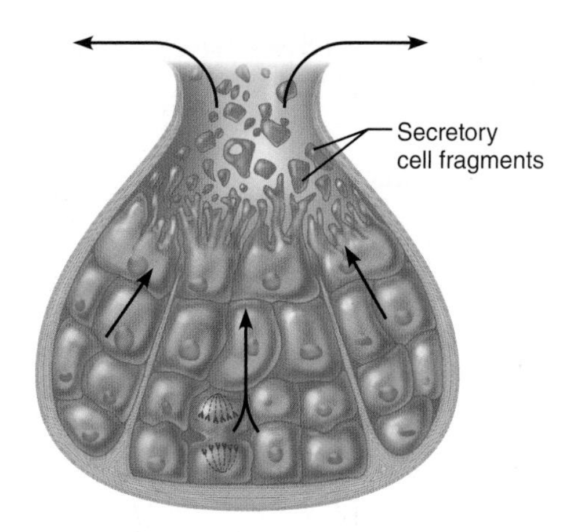

(b) In holocrine glands, the entire secretory cell ruptures, releasing secretions and dead cell fragments.

Figure 4.6 Chief modes of secretion in human exocrine glands.

is the release of lipid droplets by lactating mammary glands, but most histologists classify mammary glands as merocrine glands because this is the means by which milk proteins are secreted.

☑ Check Your Understanding

3. Epithelial tissue is the only tissue type that has polarity, that is, an apical and a basal surface. Why is this important?
4. Which gland type—merocrine or holocrine—would you expect to have the highest rate of cell division? Why?
5. Stratified epithelia are "built" for protection or to resist abrasion. What are the simple epithelia better at?
6. Some epithelia are pseudostratified. What does this mean?
7. Where is transitional epithelium found and what is its importance at those sites?

For answers, see Answers Appendix.

4.3 Connective tissue is the most abundant and widely distributed tissue in the body

→ Learning Objectives

- ☐ **Indicate common characteristics of connective tissue, and list and describe its structural elements.**
- ☐ **Describe the types of connective tissue found in the body, and indicate their characteristic functions.**

While **connective tissue** is prevalent in the body, its amount in particular organs varies. For example, skin consists primarily of connective tissue, while the brain contains very little.

There are four main classes of connective tissue and several subclasses (Table 4.1). These are (1) *connective tissue proper* (which includes fat and the fibrous tissue of ligaments), (2) *cartilage*, (3) *bone*, and (4) *blood*.

Connective tissue does much more than just *connect* body parts. Its major functions include (1) *binding and supporting*, (2) *protecting*, (3) *insulating*, (4) *storing* reserve fuel, and (5) *transporting* substances within the body. For example, bone and cartilage support and protect body organs by providing the hard underpinnings of the skeleton. Fat insulates and protects body organs and provides a fuel reserve. Blood transports substances inside the body.

Common Characteristics of Connective Tissue

Connective tissues share three characteristics that set them apart from other primary tissues:

- **Common origin.** All connective tissues arise from *mesenchyme* (an embryonic tissue).
- **Degrees of vascularity.** Connective tissues run the gamut of vascularity. Cartilage is avascular. Dense connective tissue is poorly vascularized, and the other types of connective tissue have a rich supply of blood vessels.
- **Extracellular matrix.** All other primary tissues are composed mainly of cells, but connective tissues are largely nonliving **extracellular matrix** (ma′triks; "womb"), which separates, often widely, the living cells of the tissue. Because of its matrix, connective tissue can bear weight, withstand great tension, and endure abuses, such as physical trauma and abrasion, that no other tissue can tolerate.

Structural Elements of Connective Tissue

Connective tissues have three main elements: *ground substance*, *fibers*, and *cells* (Table 4.1). Together ground substance and fibers make up the extracellular matrix. (Note that some authors use the term *matrix* to indicate the ground substance only.)

The composition and arrangement of these three elements vary tremendously. The result is an amazing diversity of connective tissues, each adapted to perform a specific function in

Table 4.1 Comparison of Classes of Connective Tissues

TISSUE CLASS AND EXAMPLE	SUBCLASSES	COMPONENTS: CELLS	COMPONENTS: MATRIX	GENERAL FEATURES
Connective Tissue Proper *Dense regular connective tissue*	1. Loose connective tissue • Areolar • Adipose • Reticular 2. Dense connective tissue • Regular • Irregular • Elastic	Fibroblasts Fibrocytes Defense cells Adipocytes	Gel-like ground substance All three fiber types: collagen, reticular, elastic	Six different types; vary in density and types of fibers Functions as a binding tissue Resists mechanical stress, particularly tension Provides reservoir for water and salts Nutrient (fat) storage
Cartilage *Hyaline cartilage*	1. Hyaline cartilage 2. Elastic cartilage 3. Fibrocartilage	Chondroblasts found in growing cartilage Chondrocytes	Gel-like ground substance Fibers: collagen, elastic fibers in some	Resists compression because of the large amounts of water held in the matrix Functions to cushion and support body structures
Bone Tissue *Compact bone*	1. Compact bone 2. Spongy bone	Osteoblasts Osteocytes	Gel-like ground substance calcified with inorganic salts Fibers: collagen	Hard tissue that resists both compression and tension Functions in support
Blood	(See Chapter 17 for details)	Erythrocytes or red blood cells (RBC) Leukocytes or white blood cells (WBC) Platelets	Plasma No fibers	A fluid tissue Functions to carry O_2, CO_2, nutrients, wastes, and other substances (such as hormones)

the body. For example, the matrix can be delicate and fragile to form a soft "packing" around an organ, or it can form "ropes" (tendons and ligaments) of incredible strength. Nonetheless, connective tissues have a common structural plan, and we use *areolar connective tissue* (ah-re′o-lar) as our *prototype*, or model (**Figure 4.7** and Figure 4.8a). All other subclasses are simply variants of this plan.

Ground Substance

Ground substance is the unstructured material that fills the space between the cells and contains the fibers. It is composed of *interstitial* (*tissue*) *fluid, cell adhesion proteins*, and *proteoglycans* (pro″te-o-gli′kanz). Cell adhesion proteins (*fibronectin, laminin*, and others) serve mainly as a connective tissue glue that allows connective tissue cells to attach to matrix elements. The proteoglycans consist of a protein core to which *glycosaminoglycans* (GAGs) (gli″kos-ah-me″no-gli′kanz) are attached. The strandlike GAGs, most importantly *chondroitin sulfate* and *hyaluronic acid* (hi″ah-lu-ron′ik), are large, negatively charged polysaccharides that stick out from the core protein like the fibers of a bottle brush. The proteoglycans tend to form huge aggregates in which the GAGs intertwine and trap water, forming a substance that varies from a fluid to a viscous gel. The higher the GAG content, the more viscous the ground substance.

The ground substance consists of large amounts of fluid and functions as a molecular sieve, or medium, through which nutrients and other dissolved substances can diffuse between the blood capillaries and the cells. The fibers embedded in the ground substance make it less pliable and hinder diffusion somewhat.

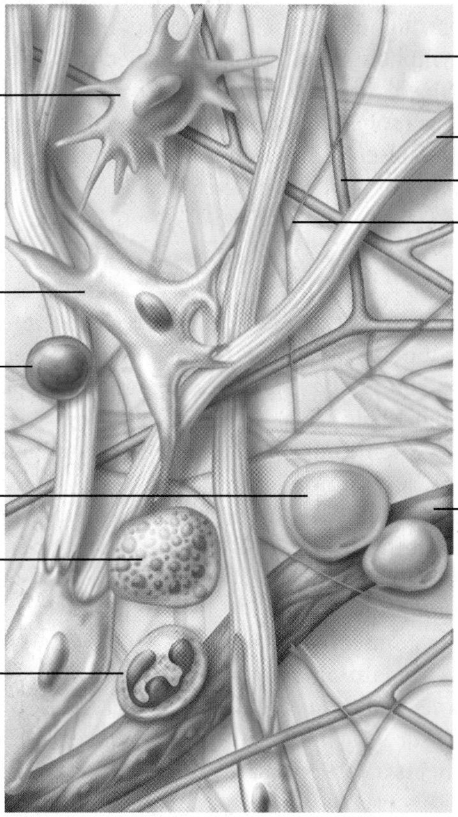

Practice art labeling
MasteringA&P®>Study Area>Chapter 4

Figure 4.7 Areolar connective tissue: A prototype (model) connective tissue. This tissue underlies epithelia and surrounds capillaries. Notice the various cell types and three classes of fibers (collagen, reticular, elastic) embedded in the ground substance. (See Figure 4.8a for a micrograph.)

Connective Tissue Fibers

The fibers of connective tissue are proteins that provide support. Three types of fibers are found in connective tissue matrix: collagen, elastic, and reticular fibers. Of these, collagen fibers are by far the strongest and most abundant.

Collagen Fibers These fibers are constructed primarily of the fibrous protein *collagen*. Collagen molecules are secreted into the extracellular space, where they assemble spontaneously into cross-linked fibrils, which in turn are bundled together into the thick collagen fibers seen with a microscope. Because their fibrils cross-link, collagen fibers are extremely tough and provide high tensile strength (the ability to resist being pulled apart) to the matrix. Indeed, stress tests show that collagen fibers are stronger than steel fibers of the same size!

Elastic Fibers Long, thin, elastic fibers form branching networks in the extracellular matrix. These fibers contain a rubberlike protein, *elastin*, that allows them to stretch and recoil like rubber bands. Connective tissue can stretch only so much before its thick, ropelike collagen fibers become taut. Then, when the tension lets up, elastic fibers snap the connective tissue back to its normal length and shape. Elastic fibers are found where greater elasticity is needed, for example, in the skin, lungs, and blood vessel walls.

Reticular Fibers These short, fine, collagenous fibers have a slightly different chemistry and form. They are continuous with collagen fibers, and they branch extensively, forming delicate networks (*reticul* = network) that surround small blood vessels and support the soft tissue of organs. They are particularly abundant where connective tissue is next to other tissue types, for example, in the basement membrane of epithelial tissues, and around capillaries, where they form fuzzy "nets" that allow more "give" than the larger collagen fibers.

Connective Tissue Cells

Each major class of connective tissue has a resident cell type that exists in immature (-blast) and mature (-cyte) forms (see Table 4.1). The immature cells, indicated by the suffix *-blast* (literally, "bud" or "sprout," but the suffix means "forming"), are actively mitotic cells. These cells secrete the ground substance and the fibers characteristic of their particular matrix. The primary blast cell types by connective tissue class are (1) connective tissue proper: **fibroblast**; (2) cartilage: **chondroblast** (kon′dro-blast″); and (3) bone: **osteoblast** (os′te-o-blast″). The *hematopoietic stem cell* (hem″ah-to-poy-et′ik) is the undifferentiated blast cell that produces blood cells. It is not included in Table 4.1 because it is not located in "its" tissue (blood) and does not make the fluid matrix (plasma) of that tissue. Blood formation is considered in Chapter 17.

Once they synthesize the matrix, the blast cells assume their mature, less active mode, indicated by the suffix *-cyte*. The mature cells maintain the health of the matrix. However, if the matrix is injured, they can easily revert to their more active state to repair and regenerate the matrix.

Additionally, connective tissue is home to an assortment of other cell types, such as:

- **Fat cells**, which store nutrients.
- **White blood cells** (neutrophils, eosinophils, lymphocytes) and other cell types that are concerned with tissue response to injury.
- **Mast cells**, which typically cluster along blood vessels. These oval cells detect foreign microorganisms (e.g., bacteria, fungi) and initiate local inflammatory responses against them. Mast cell cytoplasm contains secretory granules (*mast* = stuffed full of granules) with chemicals that mediate inflammation, especially in severe allergies. These chemicals include:
  - *Heparin* (hep′ah-rin), an anticoagulant chemical that prevents blood clotting when free in the bloodstream (but in human mast cells it appears to regulate the action of other mast cell chemicals)
  - *Histamine* (his′tah-mēn), a substance that makes capillaries leaky
  - *Proteases* (protein-degrading enzymes)
  - Other enzymes
- **Macrophages** (mak′ro-fāj″es; *macro* = large; *phago* = eat), large, irregularly shaped cells that avidly devour a broad variety of foreign materials, ranging from foreign molecules to entire bacteria to dust particles. These "big eaters" also dispose of dead tissue cells, and they are central actors in the immune system. Macrophages, which are peppered throughout loose connective tissue, bone marrow, and lymphoid tissue, may be attached to connective tissue fibers (fixed) or may migrate freely through the matrix. Some macrophages have selective appetites. For example, those of the spleen primarily dispose of aging red blood cells, but they will not turn down other "delicacies" that come their way.

Types of Connective Tissue

As noted, all classes of connective tissue consist of living cells surrounded by a matrix. Their major differences reflect cell type, and the types and relative amounts of fibers, as summarized in Table 4.1.

As mentioned earlier, mature connective tissues arise from a common embryonic tissue, called **mesenchyme** (meh′zin-kīm). Mesenchyme has a fluid ground substance containing fine sparse fibers and star-shaped *mesenchymal cells*. It arises during the early weeks of embryonic development and eventually differentiates (specializes) into all other connective tissue cells. However, some mesenchymal cells remain and provide a source of new cells in mature connective tissues.

Figure 4.8 illustrates the connective tissues that we describe in the next sections. Study this figure as you read along.

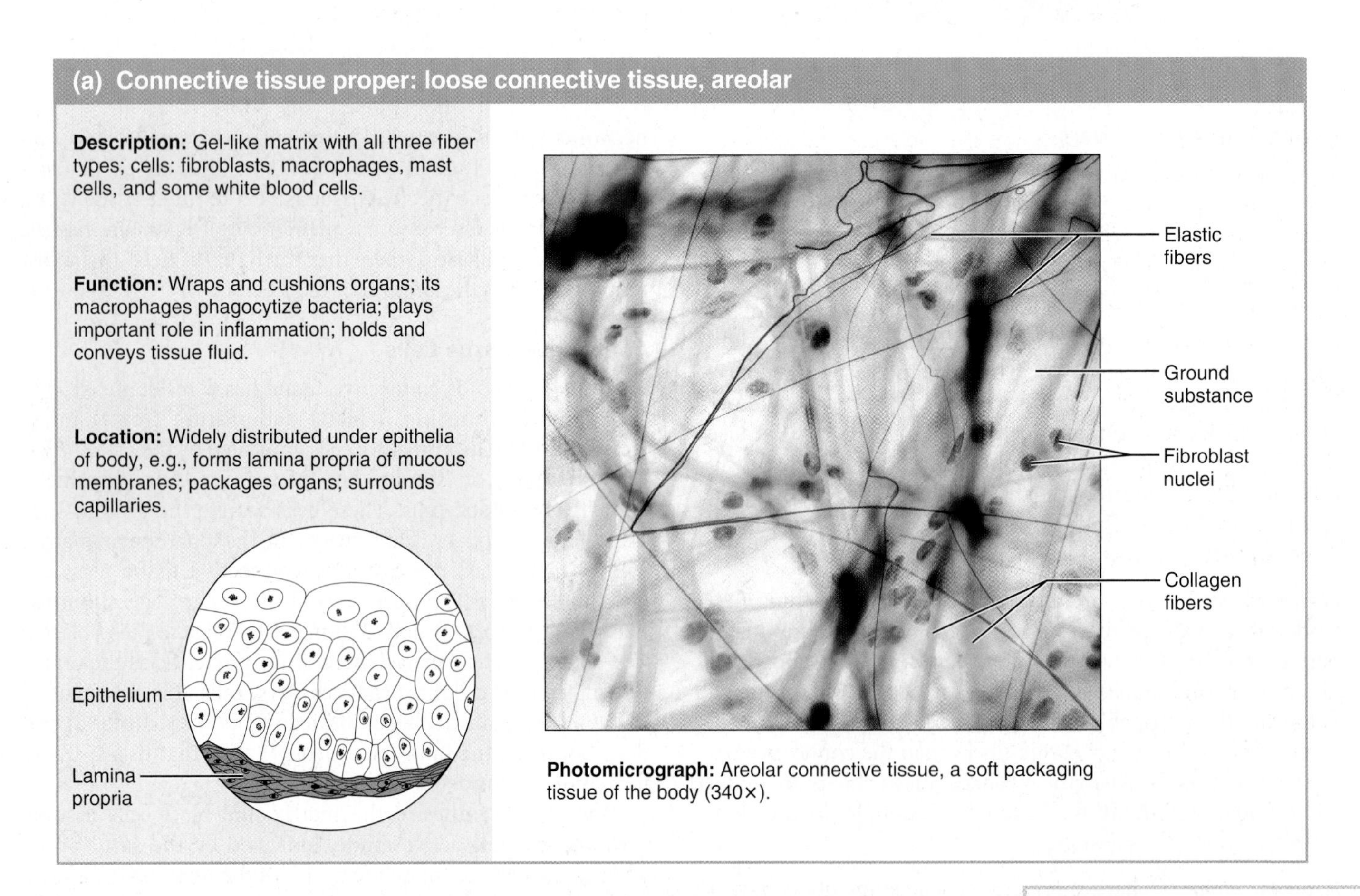

Figure 4.8 Connective tissues. (a) Connective tissue proper. (For a related image, see *A Brief Atlas of the Human Body*, Plate 11.)

View histology slides
MasteringA&P®>Study Area>PAL

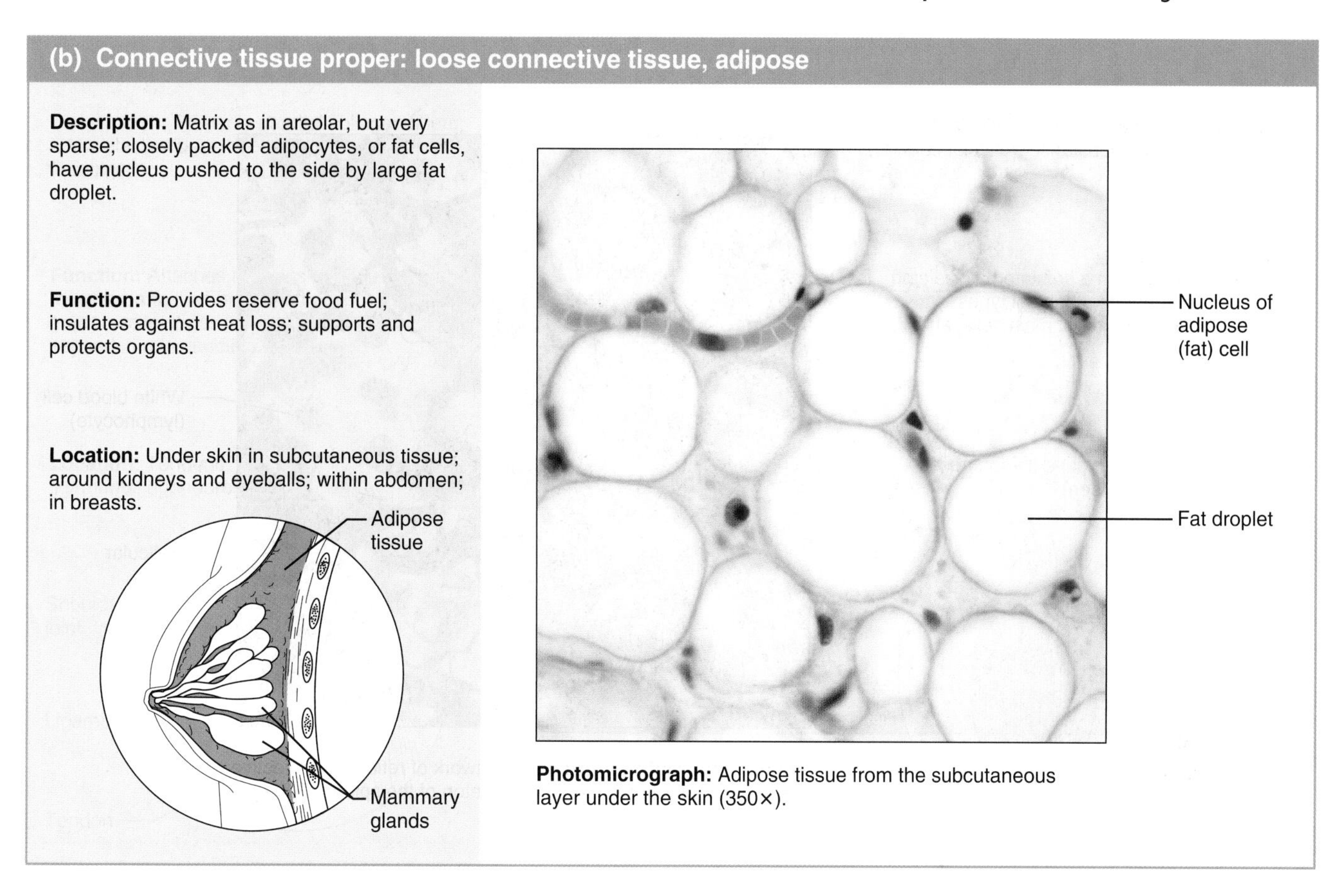

Figure 4.8 *(continued)* **(b)** Connective tissue proper. (For a related image, see *A Brief Atlas of the Human Body*, Plate 12.)

Connective Tissue Proper—Loose Connective Tissues

All mature connective tissues (except for bone, cartilage, and blood) are **connective tissue proper**. Connective tissue proper has two subclasses: **loose connective tissues** (areolar, adipose, and reticular) and **dense connective tissues** (dense regular, dense irregular, and elastic).

Areolar Connective Tissue The functions of **areolar connective tissue** (Figure 4.8a) include:

- Supporting and binding other tissues (the job of the fibers)
- Holding body fluids (the ground substance's role)
- Defending against infection (via the activity of white blood cells and macrophages)
- Storing nutrients as fat in adipocytes (fat cells)

Fibroblasts, flat branching cells that appear spindle shaped (elongated) in profile, predominate, but numerous macrophages are also seen and present a formidable barrier to invading microorganisms. Fat cells appear singly or in clusters, and occasional mast cells are identified easily by the large, darkly stained cytoplasmic granules that often obscure their nuclei. Other cell types are scattered throughout.

The most obvious structural feature of this tissue is the loose arrangement of its fibers. The rest of the matrix, occupied by ground substance, appears to be empty space when viewed through the microscope, and in fact, the Latin term *areola* means "a small open space." Because of its loose nature, areolar connective tissue provides a reservoir of water and salts for surrounding body tissues, always holding approximately as much fluid as there is in the entire bloodstream. Essentially all body cells obtain their nutrients from and release their wastes into this "tissue fluid."

The high content of hyaluronic acid makes its ground substance viscous, like molasses, which may hinder the movement of cells through it. Some white blood cells, which protect the body from disease-causing microorganisms, secrete the enzyme hyaluronidase to liquefy the ground substance and ease their passage. (Unhappily, some harmful bacteria have the same ability.)

Areolar connective tissue is the most widely distributed connective tissue in the body, and it serves as a universal packing material between other tissues. It binds body parts together while allowing them to move freely over one another; wraps small blood vessels and nerves; surrounds glands; and forms the subcutaneous tissue, which cushions and attaches the skin to underlying structures. It is the connective tissue that most epithelia rest on and is present in all mucous membranes as the *lamina propria.* (Mucous membranes line body cavities open to the exterior.)

Adipose (Fat) Tissue **Adipose tissue** (ad′ĭ-pōs) is similar to areolar tissue in structure and function, but its nutrient-storing ability is much greater. Consequently, **adipocytes** (ad′ĭ-po-sītz),

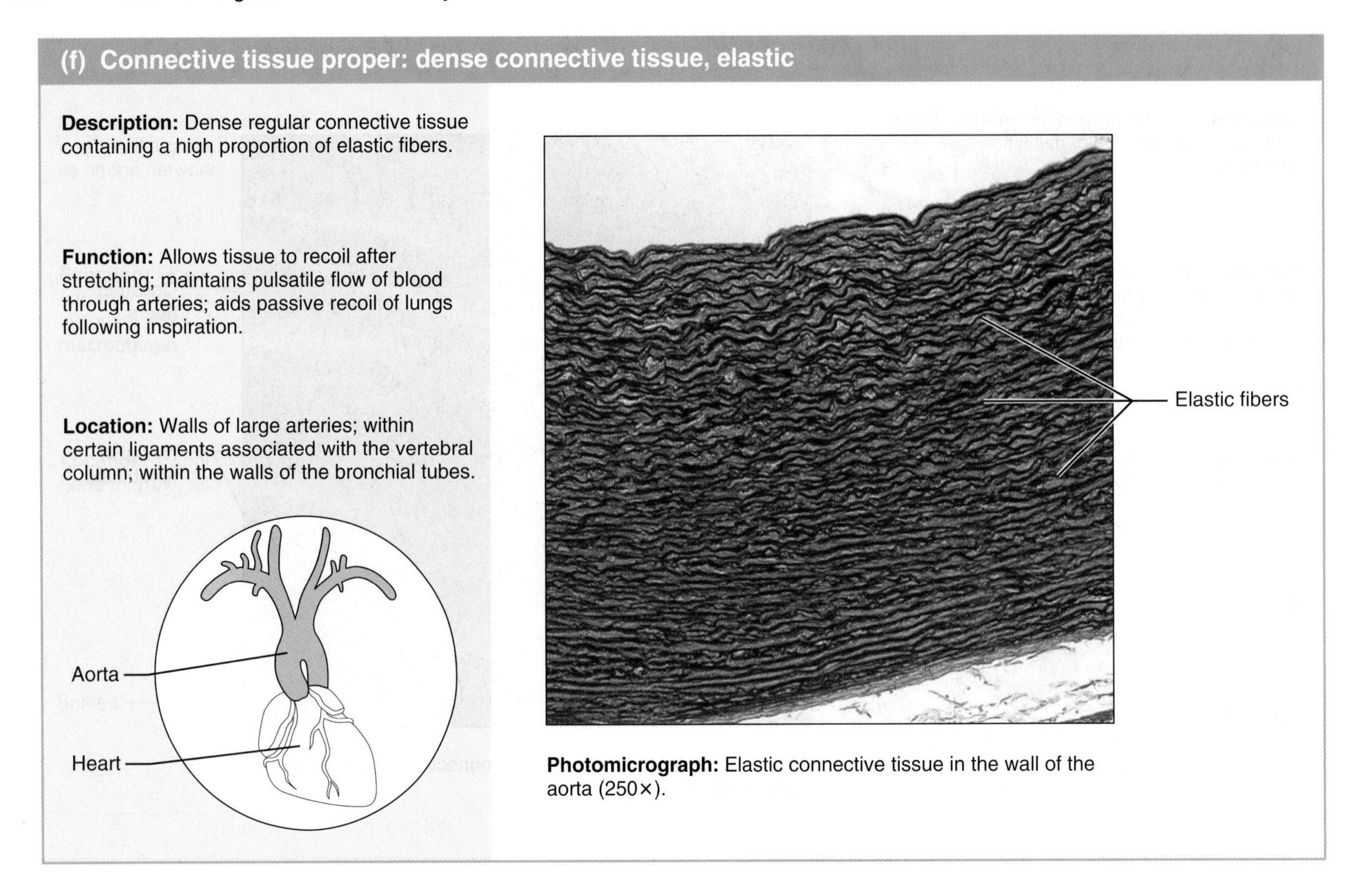

Figure 4.8 *(continued)* **Connective tissues. (f)** Connective tissue proper. (For a related image, see *A Brief Atlas of the Human Body,* Plate 16.)

Connective Tissue Proper—Dense Connective Tissues

The three varieties of dense connective tissue are dense regular, dense irregular, and elastic. Since all three have fibers as their prominent element, dense connective tissues are often called **fibrous connective tissues.**

Dense Regular Connective Tissue **Dense regular connective tissue** contains closely packed bundles of collagen fibers running in the same direction, parallel to the direction of pull (Figure 4.8d). This arrangement results in white, flexible structures with great resistance to tension (pulling forces) where the tension is exerted in a single direction. Crowded between the collagen fibers are rows of fibroblasts that continuously manufacture the fibers and scant ground substance.

Collagen fibers are slightly wavy (see Figure 4.8d). This allows the tissue to stretch a little, but once the fibers straighten out, there is no further "give" to this tissue. Unlike our model (areolar) connective tissue, this tissue has few cells other than fibroblasts and is poorly vascularized.

With its enormous tensile strength, dense regular connective tissue forms *tendons,* which are cords that attach muscles to bones; flat, sheetlike tendons called *aponeuroses* (ap″o-nu-ro′sēz) that attach muscles to other muscles or to bones; and the *ligaments* that bind bones together at joints. Ligaments contain more elastic fibers than tendons and are slightly more stretchy. Dense regular connective tissue also forms fascia (fash′e-ah; "a bond"), a fibrous membrane that wraps around muscles, groups of muscles, blood vessels, and nerves, binding them together like plastic wrap.

Dense Irregular Connective Tissue **Dense irregular connective tissue** has the same structural elements as the regular variety. However, the bundles of collagen fibers are much thicker and they are arranged irregularly; that is, they run in more than one plane (Figure 4.8e). This type of tissue forms sheets in body areas where tension is exerted from many different directions. It is found in the skin as the leathery *dermis,* and it forms fibrous joint capsules and the fibrous coverings that surround some organs (kidneys, bones, cartilages, muscles, and nerves).

Elastic Connective Tissue A few ligaments, such as those connecting adjacent vertebrae, are very elastic. The dense regular connective tissue in those structures is called **elastic connective tissue** (Figure 4.8f). Additionally, many of the larger arteries have stretchy sheets of elastic connective tissue in their walls.

Cartilage

Cartilage (kar′tĭ-lij) stands up to both tension *and* compression. Its qualities are between those of dense connective tissue and bone. It is tough but flexible, providing a resilient rigidity to the structures it supports.

(g) Cartilage: hyaline

Description: Amorphous but firm matrix; collagen fibers form an imperceptible network; chondroblasts produce the matrix and when mature (chondrocytes) lie in lacunae.

Function: Supports and reinforces; serves as resilient cushion; resists compressive stress.

Location: Forms most of the embryonic skeleton; covers the ends of long bones in joint cavities; forms costal cartilages of the ribs; cartilages of the nose, trachea, and larynx.

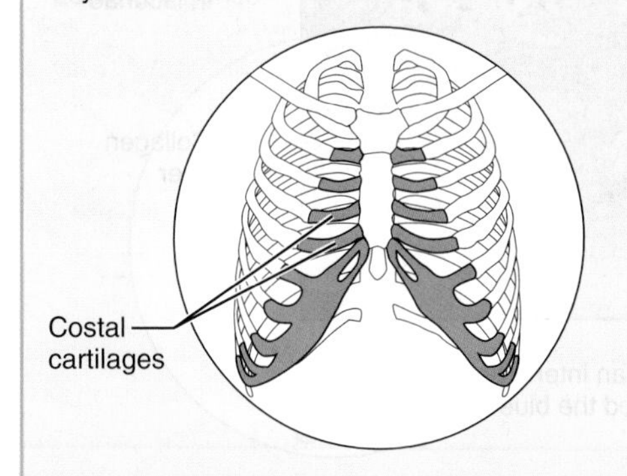

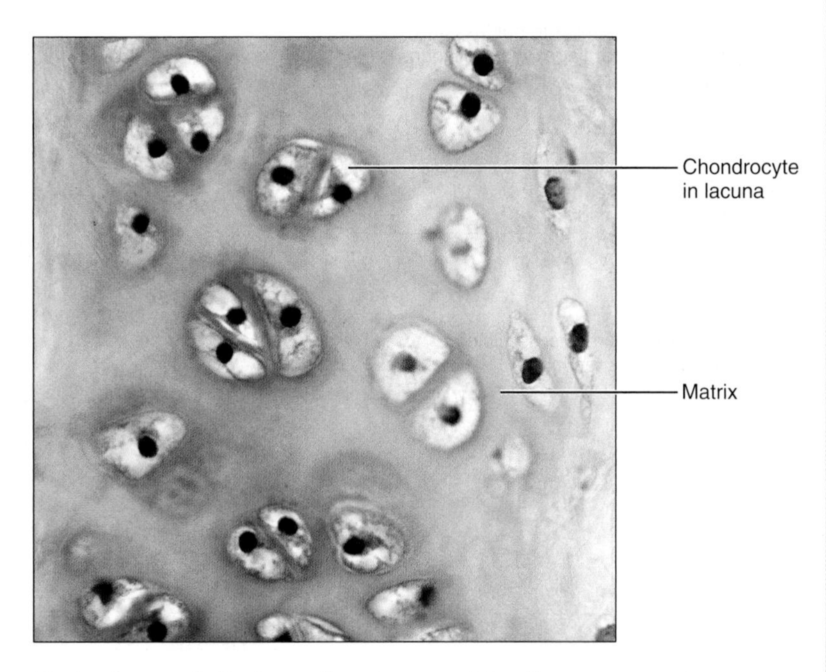

Photomicrograph: Hyaline cartilage from a costal cartilage of a rib (470×).

(h) Cartilage: elastic

Description: Similar to hyaline cartilage, but more elastic fibers in matrix.

Function: Maintains the shape of a structure while allowing great flexibility.

Location: Supports the external ear (pinna); epiglottis.

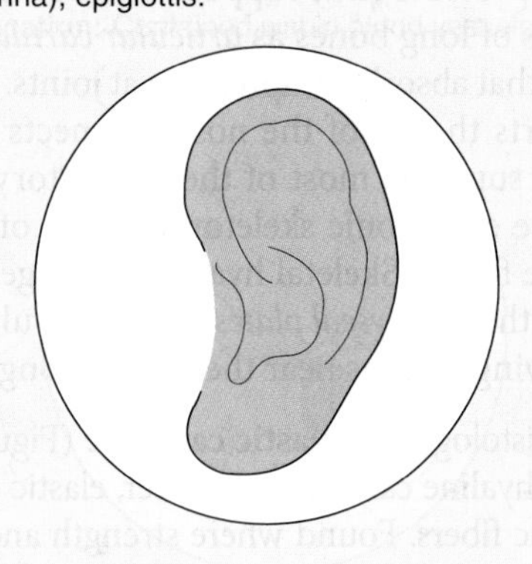

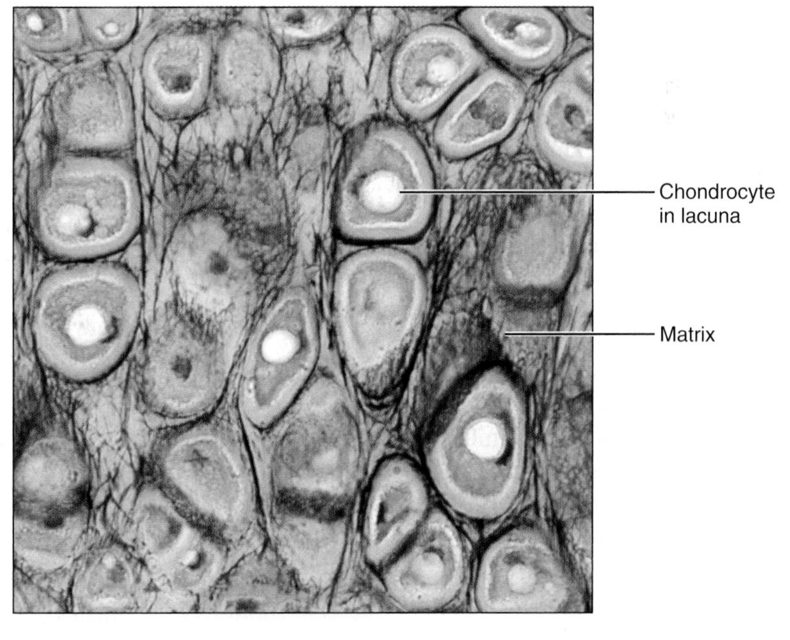

Photomicrograph: Elastic cartilage from the human ear pinna; forms the flexible skeleton of the ear (800×).

Figure 4.8 *(continued)* **(g)** and **(h)** Cartilage. (For related images, see *A Brief Atlas of the Human Body,* Plates 17 and 18.)

5 The Integumentary System

KEY CONCEPTS

Would you be enticed by an ad for a coat that is waterproof, stretchable, washable, and air-conditioned, that automatically repairs small cuts, rips, and burns? How about one that's guaranteed to last a lifetime? Sounds too good to be true, but you already have such a coat—your skin.

The skin and its derivatives (sweat and oil glands, hairs, and nails) make up a complex set of organs that serves several functions, mostly protective. Together, these organs form the **integumentary system** (in-teg″u-men′tar-e).

5.1 The skin consists of two layers: the epidermis and dermis

→ Learning Objective

☐ **List the two layers of skin and briefly describe subcutaneous tissue.**

The skin receives little respect from its inhabitants, but architecturally it is a marvel. It covers the entire body, has a surface area of 1.2 to 2.2 square meters, weighs 4 to 5 kilograms (4–5 kg = 9–11 lb), and accounts for about 7% of total body weight in the average adult. Also called the integument ("covering"), the skin multitasks. Its functions go well beyond serving as a bag for body contents. Pliable yet tough, it takes constant punishment from external agents. Without our skin, we would quickly fall prey to bacteria and perish from water and heat loss.

Varying in thickness from 1.5 to 4.0 millimeters (mm) or more in different parts of the body, the skin is composed of two distinct layers (**Figure 5.1**):

- The *epidermis* (ep″ĭ-der′mis), composed of epithelial cells, is the outermost protective shield of the body (*epi* = upon).
- The underlying *dermis,* making up the bulk of the skin, is a tough, leathery layer composed mostly of dense connective tissue.

Only the dermis is vascularized. Nutrients reach the epidermis by diffusing through the tissue fluid from blood vessels in the dermis.

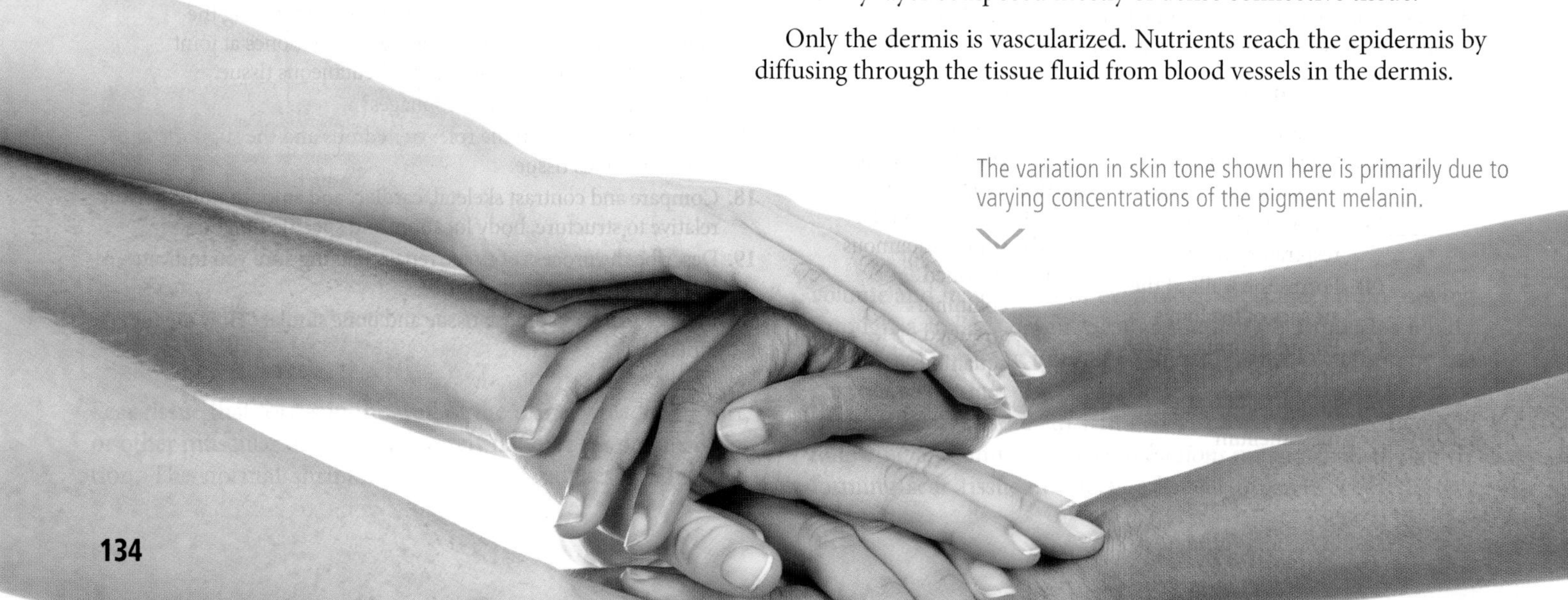

The variation in skin tone shown here is primarily due to varying concentrations of the pigment melanin.

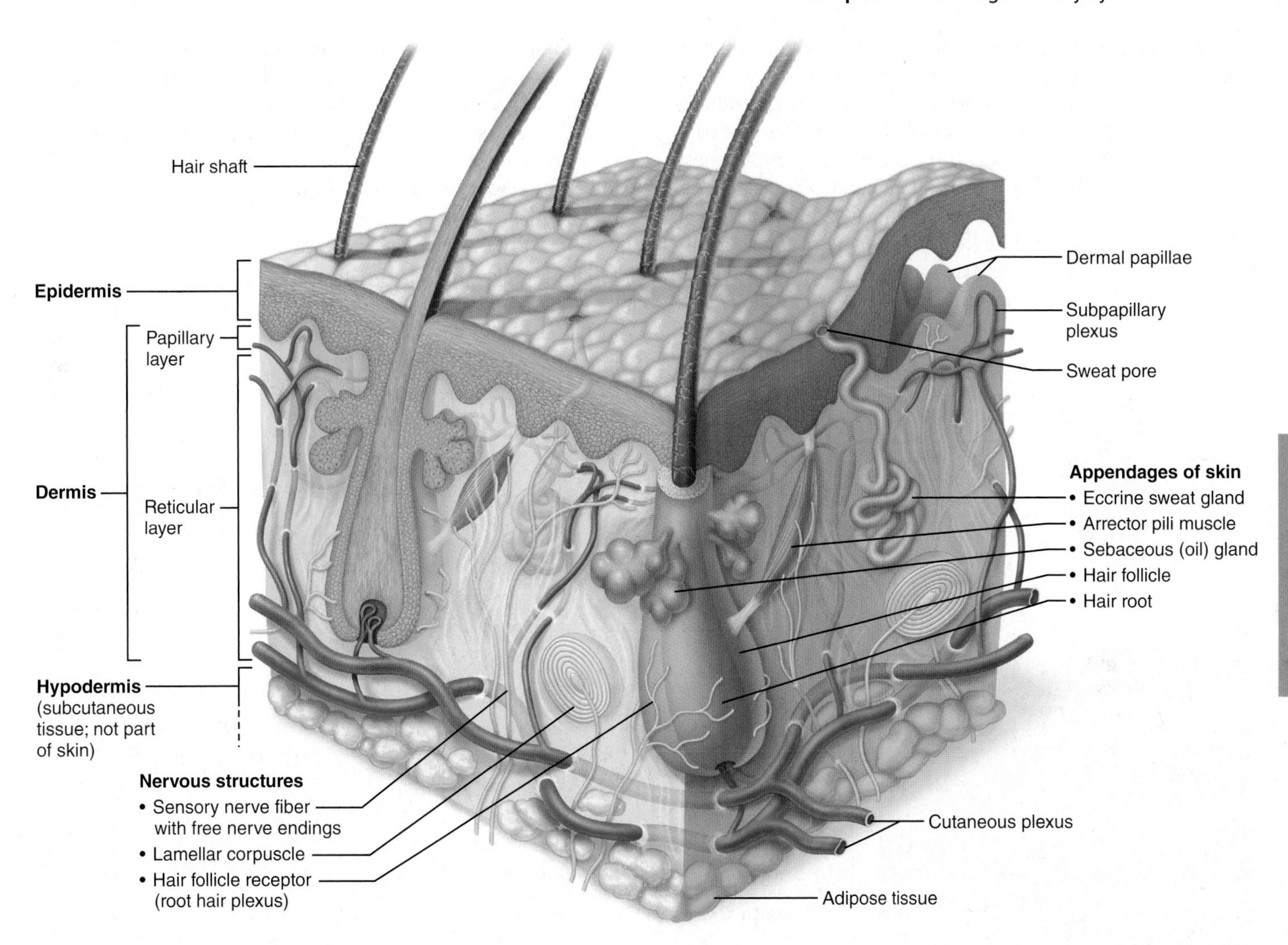

Figure 5.1 Skin structure. Three-dimensional view of the skin and underlying subcutaneous tissue. The epidermal and dermal layers have been pulled apart at the upper right corner to reveal the dermal papillae.

The subcutaneous tissue just deep to the skin is known as the **hypodermis** (Figure 5.1). Strictly speaking, the hypodermis is not part of the skin, but it shares some of the skin's protective functions. The hypodermis, also called **superficial fascia** because it is superficial to the tough connective tissue wrapping (fascia) of the skeletal muscles, consists mostly of adipose tissue.

Besides storing fat, the hypodermis anchors the skin to the underlying structures (mostly to muscles), but loosely enough that the skin can slide relatively freely over those structures. Sliding skin protects us by ensuring that many blows just glance off our bodies. Because of its fatty composition, the hypodermis also acts as a shock absorber and an insulator that reduces heat loss.

☑ Check Your Understanding

1. Which layer of the skin—dermis or epidermis—is better nourished?

For answers, see Answers Appendix.

5.2 The epidermis is a keratinized stratified squamous epithelium

→ Learning Objective

☐ **Name the tissue type composing the epidermis. List its major layers and describe the functions of each layer.**

The **epidermis** consists of four distinct cell types and four or five distinct layers.

Cells of the Epidermis

The cells populating the epidermis include *keratinocytes*, *melanocytes*, *dendritic cells*, and *tactile cells*.

Keratinocytes

The chief role of **keratinocytes** (kĕ-rat′ĭ-no-sītz″; "keratin cells") is to produce **keratin**, the fibrous protein that helps give the epidermis its protective properties (Greek *kera* = horn) (**Figure 5.2b**, orange cells). Most epidermal cells are keratinocytes.

Tightly connected by desmosomes, the keratinocytes arise in the deepest part of the epidermis from a cell layer called the stratum basale. These cells undergo almost continuous mitosis in response to prompting by epidermal growth factor, a peptide produced by various cells throughout the body. As these cells are pushed upward by the production of new cells beneath them, they make the keratin that eventually dominates their cell contents. By the time the keratinocytes reach the skin surface, they are dead, scale-like structures that are little more than keratin-filled plasma membranes.

Millions of dead keratinocytes rub off every day, giving us a totally new epidermis every 25 to 45 days, but cell production and keratin formation are accelerated in body areas regularly subjected to friction, such as the hands and feet. Persistent friction (from a poorly fitting shoe, for example) causes a thickening of the epidermis called a *callus*.

Melanocytes

Melanocytes (mel′ah-no-sītz), the spider-shaped epithelial cells that synthesize the pigment **melanin** (mel′ah-nin; *melan* = black), are found in the deepest layer of the epidermis (Figure 5.2b, gray cell). As melanin is made, it accumulates in membrane-bound granules called *melanosomes* that

View histology slides
MasteringA&P®>Study Area>PAL

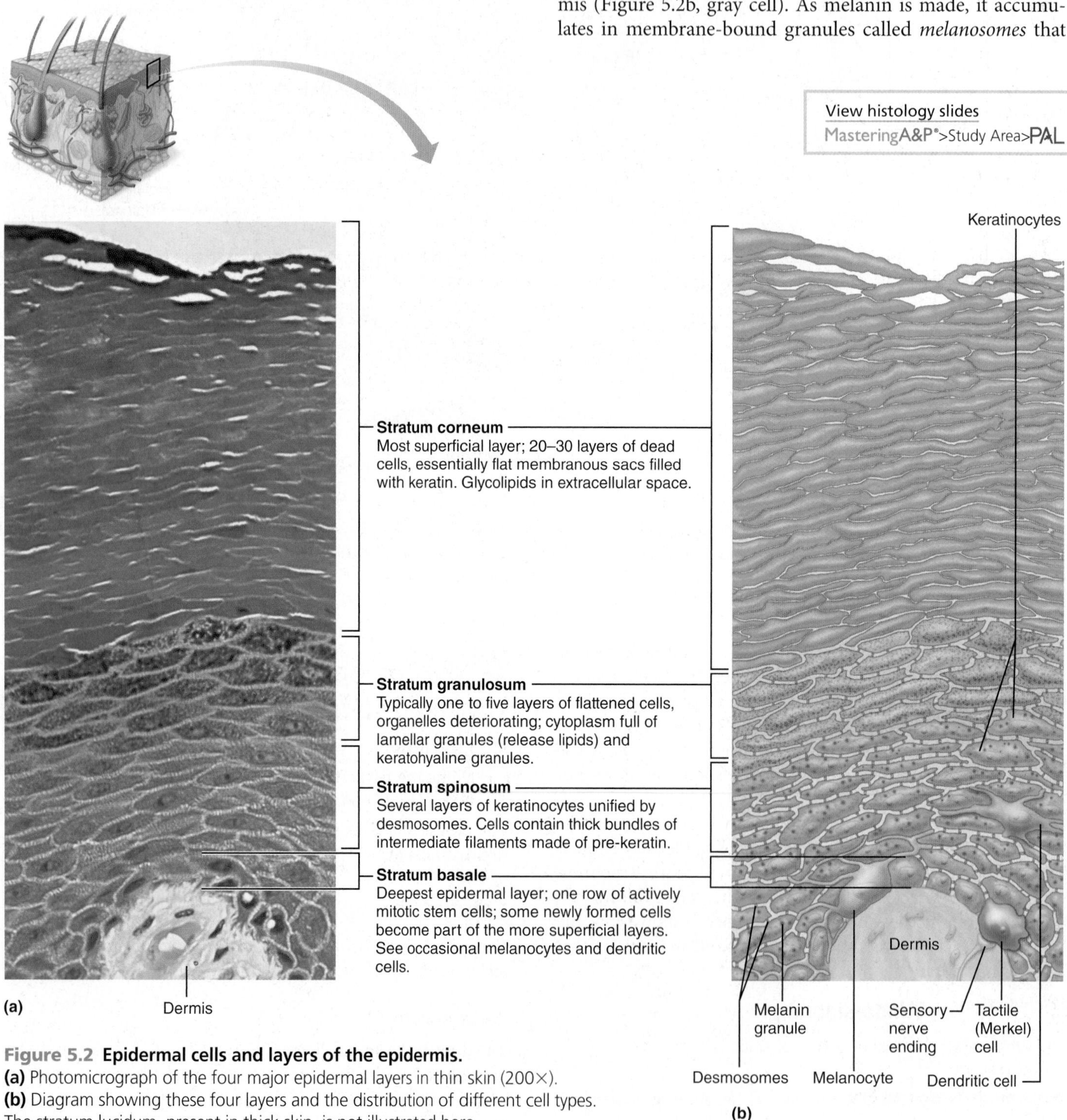

Figure 5.2 Epidermal cells and layers of the epidermis.
(a) Photomicrograph of the four major epidermal layers in thin skin (200×).
(b) Diagram showing these four layers and the distribution of different cell types. The stratum lucidum, present in thick skin, is not illustrated here.

motor proteins move along actin filaments to the ends of the melanocyte's processes (the "spider arms"). From there they are transferred to a number of nearby keratinocytes (4 to 10 depending on body area). The melanin granules accumulate on the superficial, or "sunny," side of the keratinocyte nucleus, forming a pigment shield that protects the nucleus from the damaging effects of ultraviolet (UV) radiation in sunlight.

Dendritic Cells

The star-shaped **dendritic cells** arise from bone marrow and migrate to the epidermis. Also called *Langerhans cells* (lahng′er-hanz) after a German anatomist, they ingest foreign substances and are key activators of our immune system, as described later in this chapter. Their slender processes extend among the surrounding keratinocytes, forming a more or less continuous network (Figure 5.2b, purple cell).

Tactile Cells

Occasional **tactile (Merkel) cells** are present at the epidermal-dermal junction. Shaped like a spiky hemisphere (Figure 5.2b, blue cell), each tactile cell is intimately associated with a disclike sensory nerve ending. The combination, called a *tactile* or *Merkel disc*, functions as a sensory receptor for touch.

Layers of the Epidermis

Variation in epidermal thickness determines if skin is *thick* or *thin*. In **thick skin**, which covers areas subject to abrasion—the palms, fingertips, and soles of the feet—the epidermis consists of five layers, or strata (stra′tah; "bed sheets"). From deep to superficial, these layers are stratum basale, stratum spinosum, stratum granulosum, stratum lucidum, and stratum corneum. In thin skin, which covers the rest of the body, the stratum lucidum appears to be absent and the other strata are thinner (Figure 5.2a, b).

Note that the terms "thick skin" and "thin skin" are really misnomers because they refer to the epidermis only. Indeed, the thickest skin in the body is on the upper back.

Stratum Basale (Basal Layer)

The **stratum basale** (stra′tum bah-sa′le), the deepest epidermal layer, is attached to the underlying dermis along a wavy borderline that resembles corrugated cardboard. For the most part, it consists of a single row of stem cells—a continually renewing cell population—representing the youngest keratinocytes. The many mitotic nuclei seen in this layer reflect the rapid division of these cells and account for its alternate name, **stratum germinativum** (jer′mĭ-nă″tiv-um; "germinating layer"). Each time one of these basal cells divides, one daughter cell is pushed into the cell layer just above to begin its specialization into a mature keratinocyte. The other daughter cell remains in the basal layer to continue the process of producing new keratinocytes.

Some 10–25% of the cells in the stratum basale are melanocytes, and their branching processes extend among the surrounding cells, reaching well into the more superficial stratum spinosum layer.

Stratum Spinosum (Prickly Layer)

The **stratum spinosum** (spi′no-sum; "prickly") is several cell layers thick. These cells contain a weblike system of intermediate filaments, mainly tension-resisting bundles of pre-keratin filaments, which span their cytosol to attach to desmosomes. Looking like tiny versions of the spiked iron balls used in medieval warfare, the keratinocytes in this layer appear to have spines, causing them to be called *prickle cells*. The spines do not exist in the living cells; they arise during tissue preparation when these cells shrink but their numerous desmosomes hold tight. Scattered among the keratinocytes are melanin granules and dendritic cells, which are most abundant in this epidermal layer.

Stratum Granulosum (Granular Layer)

The thin **stratum granulosum** (gran″u-lo′sum) consists of one to five cell layers in which keratinocyte appearance changes drastically, and the process of **keratinization** (in which the cells fill with keratin) begins. These cells flatten, their nuclei and organelles begin to disintegrate, and they accumulate two types of granules. The *keratohyaline granules* (ker″ah-to-hi′ah-lin) help to form keratin in the upper layers, as we will see.

The *lamellar granules* (lam′el-ar; "a small plate") contain a water-resistant glycolipid that is spewed into the extracellular space. Together with tight junctions, the glycolipid plays a major part in slowing water loss across the epidermis. The plasma membranes of these cells thicken as cytosol proteins bind to the inner membrane face and lipids released by the lamellar granules coat their external surfaces. These events produce an epidermal water barrier and make the cells more resistant to destruction. So, you might say that keratinocytes "toughen up" to make the outer strata the strongest skin region.

Like all epithelia, the epidermis relies on capillaries in the underlying connective tissue (the dermis in this case) for its nutrients. Above the stratum granulosum, the epidermal cells are too far from the dermal capillaries and the glycolipids coating their external surfaces cut them off from nutrients, so they die. This is a normal sequence of events.

Stratum Lucidum (Clear Layer)

Through the light microscope, the **stratum lucidum** (loo′sid-um; "light"), visible only in thick skin, is a thin translucent band just above the stratum granulosum. Considered by some to be a subdivision of the superficial stratum corneum, it consists of two or three rows of clear, flat, dead keratinocytes with indistinct boundaries. Here, or in the stratum corneum above, the gummy substance of the keratohyaline granules clings to the keratin filaments in the cells, causing them to aggregate in large, parallel arrays of intermediate filaments called *tonofilaments*.

Stratum Corneum (Horny Layer)

An abrupt transition occurs between the nucleated cells of the stratum granulosum and the flattened, anucleate cells of the **stratum corneum** (kor′ne-um). This outermost epidermal layer is a broad zone 20 to 30 cell layers thick that accounts for up to three-quarters of the epidermal thickness. Keratin and the thickened plasma membranes of cells in this stratum protect the skin against abrasion and penetration, and the

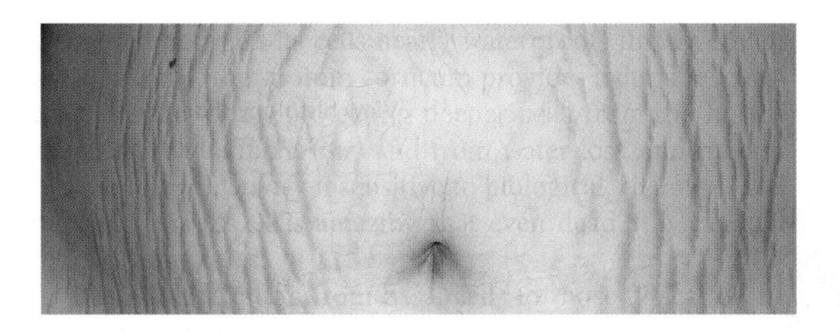

Figure 5.5 Stretch marks (striae).

Check Your Understanding

5. Which layer of the dermis is responsible for producing fingerprint patterns?
6. Which tissue of the hypodermis makes it a good shock absorber?
7. You have just gotten a paper cut. It is very painful, but it doesn't bleed. Has the cut penetrated into the dermis or just the epidermis?

For answers, see Answers Appendix.

5.4 Melanin, carotene, and hemoglobin determine skin color

Learning Objectives

- [ ] **Describe the factors that normally contribute to skin color.**
- [ ] **Briefly describe how changes in skin color may be used as clinical signs of certain disease states.**

Melanin, carotene, and hemoglobin determine skin color. Of these, only melanin is made in the skin.

Melanin is a polymer made of tyrosine amino acids. Its two forms range in color from reddish yellow to brownish black. Its synthesis depends on an enzyme in melanocytes called tyrosinase (ti-ro′sĭ-nās) and, as noted earlier, it passes from melanocytes to the basal keratinocytes. Eventually, lysosomes break down the melanosomes, so melanin pigment is found only in the deeper layers of the epidermis.

Human skin comes in different colors. However, distribution of those colors is not random—populations of darker-skinned people tend to be found nearer the equator (where greater protection from the sun is needed), and those with the lightest skin are found closer to the poles. Since all humans have the same relative number of melanocytes, differences in skin coloring reflect the kind and amount of melanin made and retained. Melanocytes of black- and brown-skinned people produce many more and darker melanosomes than those of fair-skinned individuals, and their keratinocytes retain it longer. *Freckles* and *pigmented nevi* (moles) are local accumulations of melanin.

When we expose our skin to sunlight, keratinocytes secrete chemicals that stimulate melanocytes. Prolonged sun exposure causes a substantial melanin buildup, which helps protect the DNA of viable skin cells from UV radiation by absorbing the rays and dissipating the energy as heat. Indeed, the initial signal for speeding up melanin synthesis seems to be a faster repair rate of DNA that has suffered photodamage (*photo* = light). In all but the darkest-skinned people, this defensive response causes skin to darken visibly (tanning occurs).

HOMEOSTATIC IMBALANCE 5.2 — CLINICAL

Despite melanin's protective effects, excessive sun exposure eventually damages the skin. It causes elastic fibers to clump, which results in leathery skin; temporarily depresses the immune system; and can alter the DNA of skin cells, leading to skin cancer. The fact that dark-skinned people get skin cancer less often than fair-skinned people and get it in areas with less pigment—the soles of the feet and nail beds—attests to melanin's effectiveness as a natural sunscreen.

Ultraviolet radiation has other consequences as well, such as destroying the body's folic acid that is necessary for DNA synthesis. This can have serious consequences, particularly in pregnant women because the deficit may impair the development of the embryo's nervous system.

Many chemicals induce photosensitivity; that is, they increase the skin's sensitivity to UV radiation and can cause an unsightly skin rash. Such substances include some antibiotic and antihistamine drugs, and many chemicals in perfumes and detergents. Small, itchy blisters erupt all over the body. Then the peeling begins—in sheets! +

Carotene (kar′o-tēn) is a yellow to orange pigment found in certain plant products such as carrots. It tends to accumulate in the stratum corneum and in fatty tissue of the hypodermis. Its color is most obvious in the palms and soles, where the stratum corneum is thickest, and most intense when large amounts of carotene-rich foods are eaten. In the body, carotene can be converted to vitamin A, a vitamin that is essential for normal vision, as well as for epidermal health.

The pinkish hue of fair skin reflects the crimson color of the oxygenated pigment **hemoglobin** (he′mo-glo″bin) in the red blood cells circulating through the dermal capillaries. Because Caucasian skin contains only small amounts of melanin, the epidermis is nearly transparent and allows hemoglobin's color to show through.

HOMEOSTATIC IMBALANCE 5.3 — CLINICAL

When hemoglobin is poorly oxygenated, both the blood and the skin of Caucasians appear blue, a condition called *cyanosis* (si″ah-no′sis; *cyan* = dark blue). Skin often becomes cyanotic during heart failure and severe respiratory disorders. In dark-skinned individuals, the skin does not appear cyanotic because of the masking effects of melanin, but cyanosis is apparent in the mucous membranes and nail beds.

Alterations in skin color can indicate certain disease states or even emotional states:

- *Redness*, or *erythema* (er″ĭ-the′mah): Reddened skin may indicate embarrassment (blushing), fever, hypertension, inflammation, or allergy.

- *Pallor*, or *blanching*: During fear, anger, and certain other types of emotional stress, some people become pale. Pale skin may also signify anemia or low blood pressure.
- *Jaundice* (jawn′dis), or *yellow cast*: An abnormal yellow skin tone usually signifies a liver disorder, in which yellow bile pigments accumulate in the blood and are deposited in body tissues.
- *Bronzing*: A bronze, almost metallic appearance of the skin is a sign of Addison's disease, in which the adrenal cortex produces inadequate amounts of its steroid hormones; or a sign of pituitary gland tumors that inappropriately secrete melanocyte-stimulating hormone (MSH).
- *Black-and-blue marks*, or *bruises*: Black-and-blue marks reveal where blood escaped from the circulation and clotted beneath the skin. Such clotted blood masses are called hematomas (he″mah-to′mah; "blood swelling"). +

☑ Check Your Understanding

8. Melanin and carotene are two pigments that contribute to skin color. What is the third and where is it found?
9. What is cyanosis and what does it indicate?
10. Which alteration in skin color may indicate a liver disorder?

For answers, see Answers Appendix.

Along with the skin itself, the integumentary system includes several derivatives of the epidermis. These **skin appendages** include hair and hair follicles, nails, sweat glands, and sebaceous (oil) glands. Each plays a unique role in maintaining body homeostasis. We will examine them in the next three modules.

5.5 Hair consists of dead, keratinized cells

→ Learning Objectives

- ☐ **List the parts of a hair follicle and explain the function of each part. Also describe the functional relationship of arrector pili muscles to the hair follicles.**
- ☐ **Name the regions of a hair and explain the basis of hair color. Describe the distribution, growth, replacement, and changing nature of hair during the life span.**

Millions of hairs are distributed over our entire skin surface except our palms, soles, lips, nipples, and parts of the external genitalia (such as the head of the penis). Although hair helps to keep other mammals warm, our sparse body hair is far less luxuriant and useful. Its main function in humans is to sense insects on the skin before they bite or sting us. Hair on the scalp guards the head against physical trauma, heat loss, and sunlight. Eyelashes shield the eyes, and nose hairs filter large particles like lint and insects from the air we inhale.

Structure of a Hair

Hairs, or **pili** (pi′li), are flexible strands produced by hair follicles and consist largely of dead, keratinized cells. The *hard keratin* that dominates hairs and nails has two advantages over the *soft keratin* found in typical epidermal cells: (1) It is tougher and more durable, and (2) its individual cells do not flake off.

The chief regions of a hair are the *shaft*, the portion in which keratinization is complete, and the *root*, where keratinization is still ongoing. The shaft, which projects from the skin, extends about halfway down the portion of the hair embedded in the skin (**Figure 5.6**). The root is the remainder of the hair deep within the follicle. If the shaft is flat and ribbonlike in cross section, the hair is kinky; if it is oval, the hair is silky and wavy; if it is perfectly round, the hair is straight and tends to be coarse.

A hair has three concentric layers of keratinized cells: the medulla, cortex, and cuticle (Figure 5.6a, b).

- The *medulla* (mĕ-dul′ah; "middle"), its central core, consists of large cells and air spaces. The medulla, the only part of the hair that contains soft keratin, is absent in fine hairs.
- The *cortex*, a bulky layer surrounding the medulla, consists of several layers of flattened cells.
- The outermost **cuticle** is formed from a single layer of cells overlapping one another like shingles on a roof. This arrangement helps separate neighboring hairs so the hair does not mat. (Hair conditioners smooth out the rough surface of the cuticle and make hair look shiny.) The most heavily keratinized part of the hair, the cuticle provides strength and helps keep the inner layers tightly compacted.

  Because it is subjected to the most abrasion, the cuticle tends to wear away at the tip of the hair shaft, allowing keratin fibrils in the cortex and medulla to frizz, creating "split ends."

Hair pigment is made by melanocytes at the base of the hair follicle and transferred to the cortical cells. Various proportions of melanins of different colors (yellow, rust, brown, and black) combine to produce hair color from blond to pitch black. Additionally, red hair is colored by a pigment called *pheomelanin*. When melanin production decreases (mediated by delayed-action genes) and air bubbles replace melanin in the hair shaft, hair turns gray or white.

Structure of a Hair Follicle

Hair follicles (*folli* = bag) fold down from the epidermal surface into the dermis. In the scalp, they may even extend into the hypodermis. The deep end of the follicle, located about 4 mm (1/6 in.) below the skin surface, expands to form a **hair bulb** (Figure 5.6c, d). A knot of sensory nerve endings called a **hair follicle receptor**, or **root hair plexus**, wraps around each hair bulb (see Figure 5.1). Bending the hair stimulates these endings. Consequently, our hairs act as sensitive touch receptors.

A *papilla of a hair follicle*, or more simply a *hair papilla*, is a nipple-like bit of dermal tissue that protrudes into the hair bulb. This papilla contains a knot of capillaries that supplies nutrients to the growing hair and signals it to grow. Except for its location, this papilla is similar to the dermal papillae underlying other epidermal regions.

The wall of a hair follicle is composed of an outer **peripheral connective tissue sheath** (or *fibrous sheath*), derived from the dermis; a thickened basal lamina called the *glassy membrane*;

5

6 Bones and Skeletal Tissues

KEY CONCEPTS

All of us have heard the expressions "bone tired" and "bag of bones"—rather unflattering and inaccurate images of one of our most phenomenal tissues and our main skeletal elements. Our brains, not our bones, convey feelings of fatigue. As for "bag of bones," they are indeed more prominent in some of us, but without bones to form our internal supporting skeleton, we would all creep along the ground like slugs, lacking any definite shape or form. Along with its bones, the skeleton contains resilient cartilages, which we briefly discuss in this chapter. However, our major focus is the structure and function of bone tissue and the dynamics of its formation and remodeling throughout life.

Electron micrograph of bone mineral crystals.

6.1 Hyaline, elastic, and fibrocartilage help form the skeleton

→ **Learning Objectives**

- ☐ **Describe the functional properties of the three types of cartilage tissue.**
- ☐ **Locate the major cartilages of the adult skeleton.**
- ☐ **Explain how cartilage grows.**

The human skeleton is initially made up of cartilages and fibrous membranes, but bone soon replaces most of these early supports. The few cartilages that remain in adults are found mainly in regions where flexible skeletal tissue is needed.

Basic Structure, Types, and Locations

A **skeletal cartilage** is made of some variety of *cartilage tissue* molded to fit its body location and function. Cartilage consists primarily of water, which accounts for its resilience, that is, its ability to spring back to its original shape after being compressed.

The cartilage, which contains no nerves or blood vessels, is surrounded by a layer of dense irregular connective tissue, the *perichondrium* (per″ĭ-kon′dre-um; "around the cartilage"). The perichondrium acts like a girdle to resist outward expansion when the cartilage is compressed. Additionally, the perichondrium contains the blood vessels from which nutrients diffuse through the matrix to reach the cartilage cells internally. This mode of nutrient delivery limits cartilage thickness.

As we described in Chapter 4, the three types of cartilage tissue are hyaline, elastic, and fibrocartilage. All three types have the same basic components—cells called *chondrocytes*, encased in small cavities (lacunae) within an *extracellular matrix* containing a jellylike ground substance and fibers.

Hyaline Cartilages

Hyaline cartilages, which look like frosted glass when freshly exposed, provide support with flexibility and resilience. They are the most abundant skeletal cartilages. Their chondrocytes are spherical (see Figure 4.8g, on p. 123), and the only fiber type in their matrix is fine collagen fibers (which are undetectable microscopically). Colored blue in **Figure 6.1**, skeletal hyaline cartilages include:

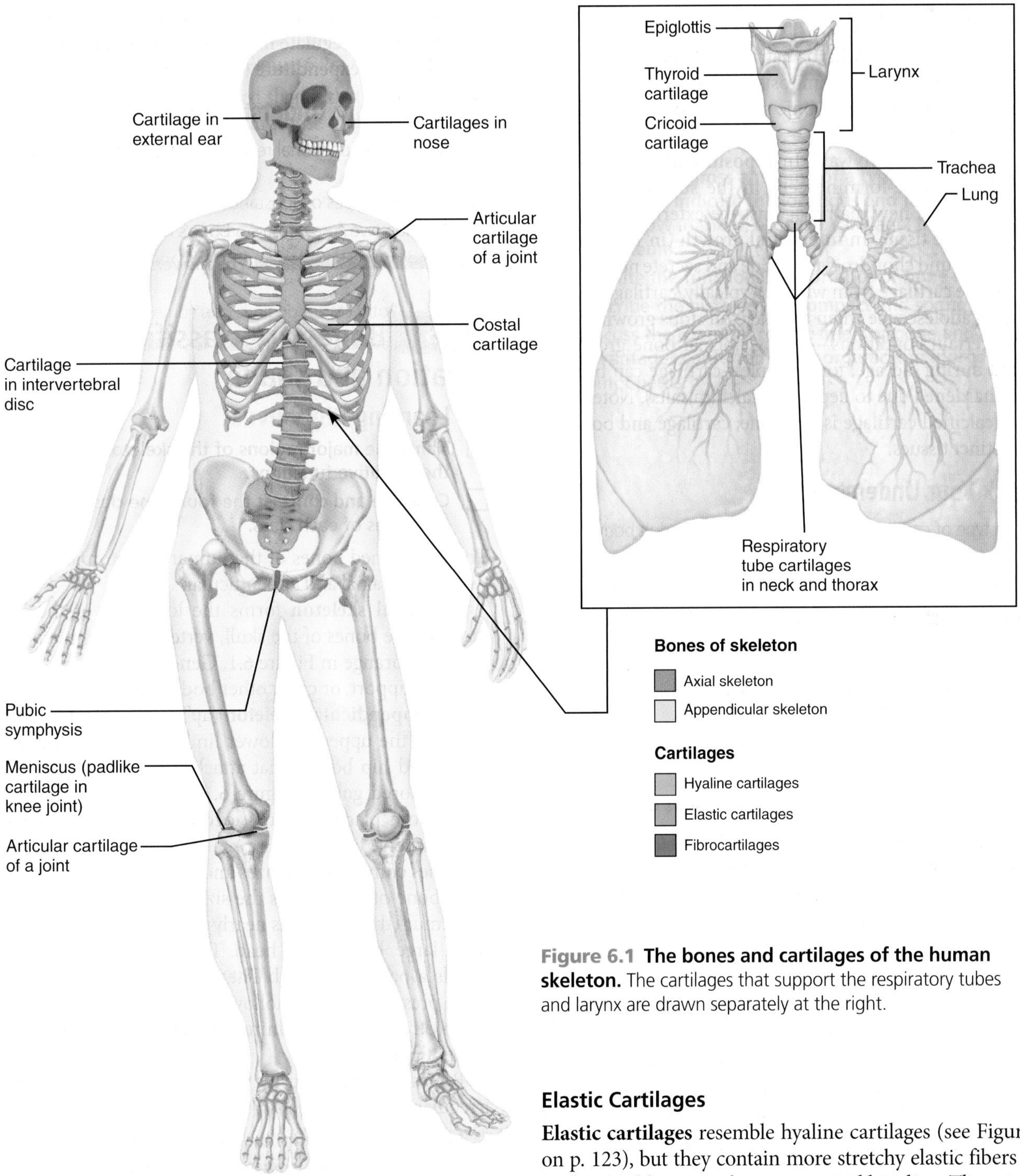

Figure 6.1 The bones and cartilages of the human skeleton. The cartilages that support the respiratory tubes and larynx are drawn separately at the right.

- *Articular cartilages* (*artic* = joint, point of connection), which cover the ends of most bones at movable joints
- *Costal cartilages*, which connect the ribs to the sternum (breastbone)
- *Respiratory cartilages*, which form the skeleton of the *larynx* (*voice box*) and reinforce other respiratory passageways
- *Nasal cartilages*, which support the external nose

Elastic Cartilages

Elastic cartilages resemble hyaline cartilages (see Figure 4.8h, on p. 123), but they contain more stretchy elastic fibers and so are better able to stand up to repeated bending. They are found in only two skeletal locations, shown in green in Figure 6.1—the external ear and the epiglottis (the flap that bends to cover the opening of the larynx each time we swallow).

Fibrocartilages

Highly compressible with great tensile strength, **fibrocartilages** consist of roughly parallel rows of chondrocytes alternating with thick collagen fibers (see Figure 4.8i, on p. 124). Fibrocartilages occur in sites that are subjected to both pressure and stretch, such as the padlike cartilages (menisci) of the knee and the discs between vertebrae, colored red in Figure 6.1.

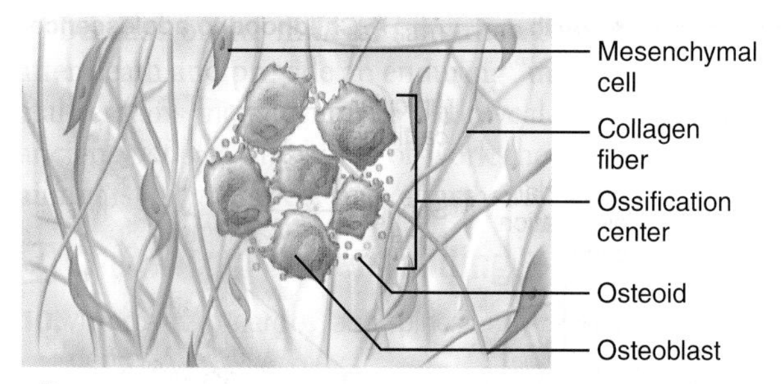

① **Ossification centers appear in the fibrous connective tissue membrane.**
- Selected centrally located mesenchymal cells cluster and differentiate into osteoblasts, forming an ossification center that produces the first trabeculae of spongy bone.

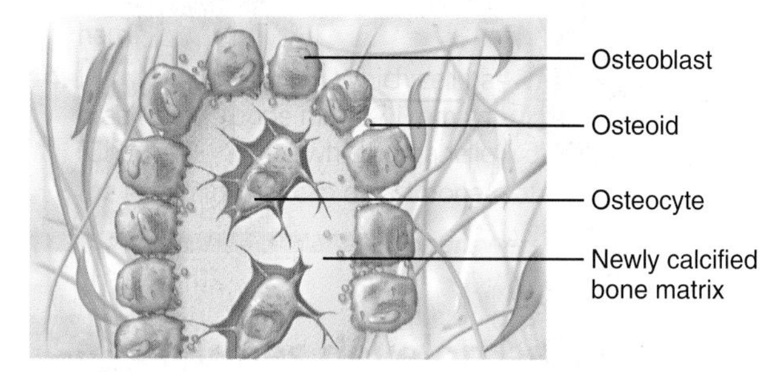

② **Osteoid is secreted within the fibrous membrane and calcifies.**
- Osteoblasts continue to secrete osteoid, which calcifies in a few days.
- Trapped osteoblasts become osteocytes.

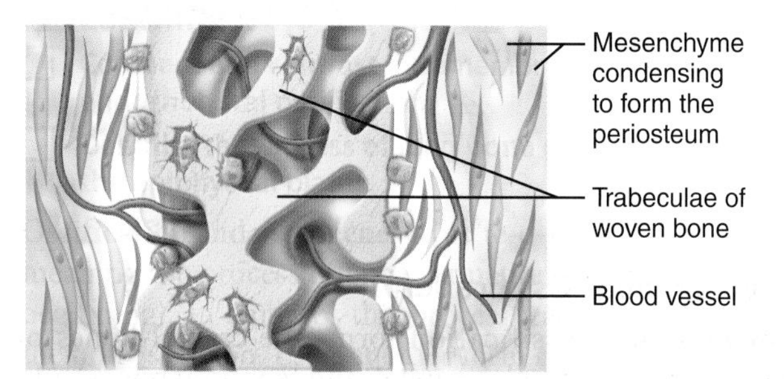

③ **Woven bone and periosteum form.**
- Accumulating osteoid is laid down between embryonic blood vessels in a manner that results in a network (instead of concentric lamellae) of trabeculae called woven bone.
- Vascularized mesenchyme condenses on the external face of the woven bone and becomes the periosteum.

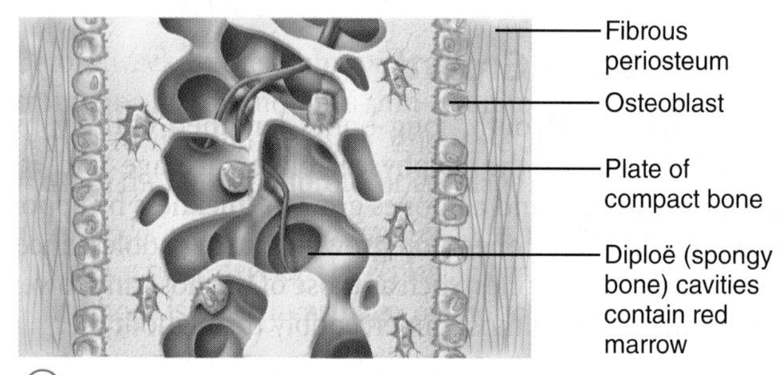

④ **Lamellar bone replaces woven bone, just deep to the periosteum. Red marrow appears.**
- Trabeculae just deep to the periosteum thicken. Mature lamellar bone replaces them, forming compact bone plates.
- Spongy bone (diploë), consisting of distinct trabeculae, persists internally and its vascular tissue becomes red marrow.

Figure 6.9 Intramembranous ossification. Diagrams ① and ② represent much greater magnification than diagrams ③ and ④.

(**Figure 6.10**). But the epiphyseal plate cartilage next to the diaphysis organizes into a pattern that allows fast, efficient growth. The cartilage cells here form tall columns, like coins in a stack.

① **Proliferation zone:** The cells at the "top" (epiphysis-facing) side of the stack next to the resting zone comprise the *proliferation* or *growth zone.* These cells divide quickly, pushing the epiphysis away from the diaphysis and lengthening the entire long bone.

② **Hypertrophic zone:** Meanwhile, the older chondrocytes in the stack, which are closer to the diaphysis (*hypertrophic*

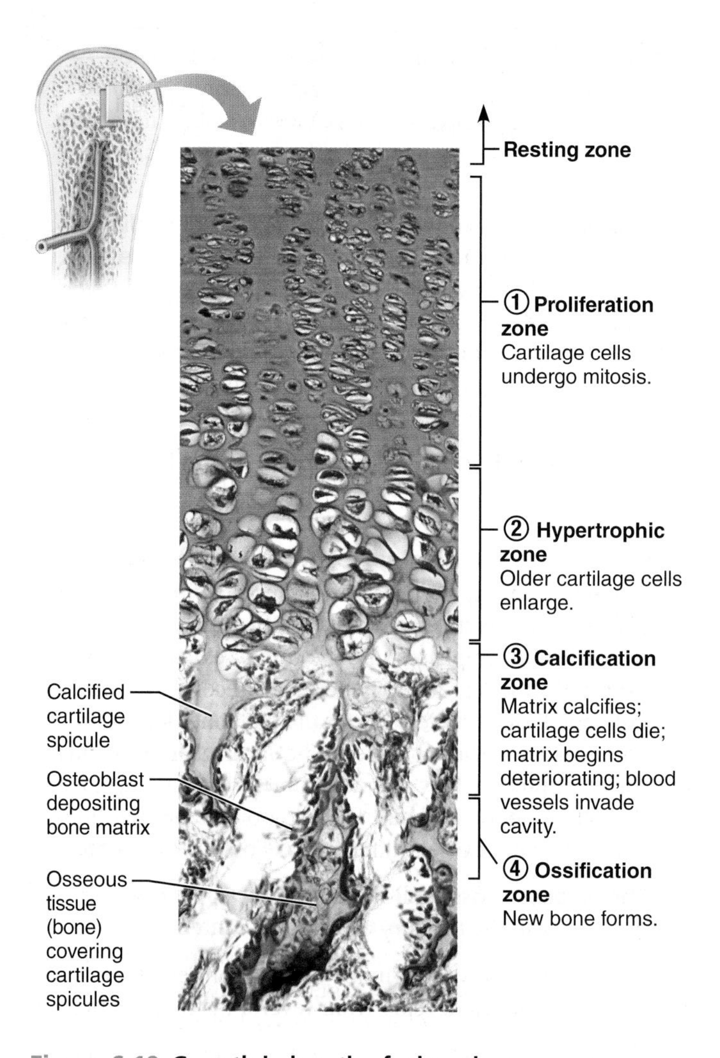

Figure 6.10 Growth in length of a long bone occurs at the epiphyseal plate. The side of the epiphyseal plate facing the epiphysis contains resting cartilage cells. The cells of the epiphyseal plate proximal to the resting cartilage area are arranged in four zones from the region of the earliest stage of growth ① to the region where bone is replacing the cartilage ④ (115×).

View histology slides
MasteringA&P®>Study Area>PAL

zone in Figure 6.10), hypertrophy, and their lacunae erode and enlarge, leaving large interconnecting spaces.

③ **Calcification zone:** Subsequently, the surrounding cartilage matrix calcifies and these chondrocytes die and deteriorate, producing the *calcification zone*.

④ **Ossification zone:** This leaves long slender spicules of calcified cartilage at the epiphysis-diaphysis junction, which look like stalactites hanging from the roof of a cave. These calcified spicules ultimately become part of the *ossification* or *osteogenic zone*, and are invaded by marrow elements from the medullary cavity. Osteoclasts partly erode the cartilage spicules, then osteoblasts quickly cover them with new bone. Ultimately spongy bone replaces them. Eventually as osteoclasts digest the spicule tips, the medullary cavity also lengthens.

During growth, the epiphyseal plate maintains a constant thickness because the rate of cartilage growth on its epiphysis-facing side is balanced by its replacement with bony tissue on its diaphysis-facing side.

Longitudinal growth is accompanied by almost continuous remodeling of the epiphyseal ends to maintain the proportion between the diaphysis and epiphyses. Bone remodeling involves both new bone formation and bone resorption (**Figure 6.11**).

As adolescence ends, the chondroblasts of the epiphyseal plates divide less often. The plates become thinner and thinner until they are entirely replaced by bone tissue. Longitudinal bone growth ends when the bone of the epiphysis and diaphysis fuses. This process, called *epiphyseal plate closure*, happens at about 18 years of age in females and 21 years of age in males. Once this has occurred, only the articular cartilage remains in bones. However, an adult bone can still widen by appositional growth if stressed by excessive muscle activity or body weight.

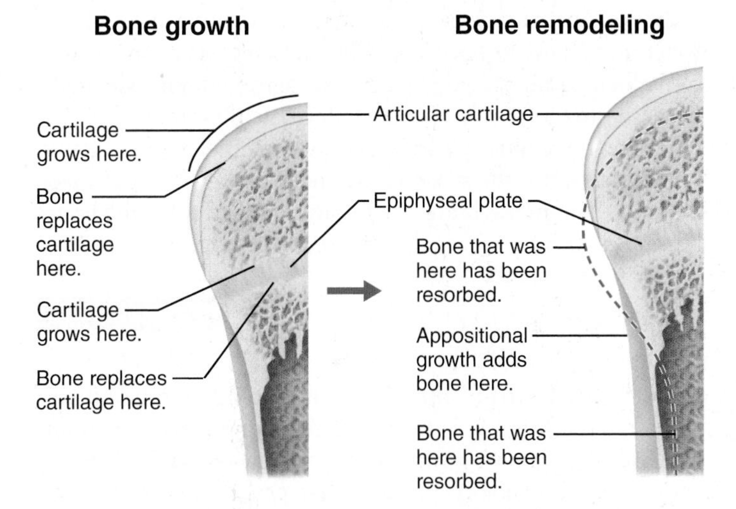

Figure 6.11 Long bone growth and remodeling during youth. Left: endochondral ossification occurs at the articular cartilages and epiphyseal plates as the bone lengthens. Right: bone remodeling during growth maintains proper bone proportions. The red dashes outline the area shown in the left view.

Growth in Width (Thickness)

Growing bones widen as they lengthen. As with cartilages, bones increase in thickness or, in the case of long bones, diameter, by appositional growth. Osteoblasts beneath the periosteum secrete bone matrix on the external bone surface as osteoclasts on the endosteal surface of the diaphysis remove bone (Figure 6.11). Normally there is slightly more building up than breaking down. This unequal process produces a thicker, stronger bone but prevents it from becoming too heavy.

Hormonal Regulation of Bone Growth

The bone growth that occurs until young adulthood is exquisitely controlled by a symphony of hormones. During infancy and childhood, the single most important stimulus of epiphyseal plate activity is *growth hormone* released by the anterior pituitary gland. Thyroid hormones modulate the activity of growth hormone, ensuring that the skeleton has proper proportions as it grows.

At puberty, sex hormones (testosterone in males and estrogens in females) are released in increasing amounts. Initially these sex hormones promote the growth spurt typical of adolescence, as well as the masculinization or feminization of specific parts of the skeleton. Later the hormones induce epiphyseal closure, ending longitudinal bone growth.

Excesses or deficits of any of these hormones can result in abnormal skeletal growth. For example, hypersecretion of growth hormone in children results in excessive height (gigantism), and deficits of growth hormone or thyroid hormone produce characteristic types of dwarfism.

☑ Check Your Understanding

15. Bones don't begin with bone tissue. What do they begin with?

16. When describing endochondral ossification, some say "bone chases cartilage." What does that mean?

17. Where is the primary ossification center located in a long bone? Where is (are) the secondary ossification center(s) located?

18. As a long bone grows in length, what is happening in the hypertrophic zone of the epiphyseal plate?

For answers, see Answers Appendix.

6.6 Bone remodeling involves bone deposit and removal

→ Learning Objectives

☐ **Compare the locations and remodeling functions of the osteoblasts, osteocytes, and osteoclasts.**

☐ **Explain how hormones and physical stress regulate bone remodeling.**

Bones appear to be the most lifeless of body organs, and may even summon images of a graveyard. But as you have just learned, bone is a dynamic and active tissue, and small-scale changes in bone architecture occur continually. Every week we recycle 5–7% of our bone mass, and as much as half a gram of calcium may enter or leave the adult skeleton each day! Spongy bone is replaced every three to four years; compact bone, every ten years or so. This is fortunate because when bone remains in place for long periods, more of the calcium salts crystallize and the bone becomes more brittle—ripe conditions for fracture.

In the adult skeleton, bone deposit and bone resorption occur at the surfaces of both the periosteum and the endosteum. Together, the two processes constitute **bone remodeling**. "Packets" of adjacent osteoblasts and osteoclasts called *remodeling units* coordinate bone remodeling (with help from the stress-sensing osteocytes).

In healthy young adults, total bone mass remains constant, an indication that the rates of bone deposit and resorption are essentially equal. Remodeling does not occur uniformly, however. For example, the distal part of the femur, or thigh bone, is fully replaced every five to six months, whereas its shaft is altered much more slowly.

Bone Deposit

An *osteoid seam*—an unmineralized band of gauzy-looking bone matrix 10–12 micrometers (μm) wide—marks areas of new matrix deposits by osteoblasts. Between the osteoid seam and the older mineralized bone, there is an abrupt transition called the *calcification front*. Because the osteoid seam is always of constant width and the change from unmineralized to mineralized matrix is sudden, it seems that the osteoid must mature for about a week before it can calcify.

The precise trigger for calcification is still controversial, but mechanical signals are definitely involved. One critical factor is the product of the local concentrations of calcium and phosphate (P_i) ions (the $Ca^{2+} \times P_i$ product) in the endosteal cavity. When the $Ca^{2+} \times P_i$ product reaches a certain level, tiny crystals of hydroxyapatite form spontaneously and catalyze further crystallization of calcium salts in the area. Other factors involved are matrix proteins that bind and concentrate calcium, and the enzyme *alkaline phosphatase* (shed in *matrix vesicles* by the osteoblasts), which is essential for mineralization. Once proper conditions are present, calcium salts are deposited all at once and with great precision throughout the "matured" matrix.

Bone Resorption

As noted earlier, the giant **osteoclasts** accomplish **bone resorption**. Osteoclasts move along a bone surface, digging depressions or grooves as they break down the bone matrix. The ruffled border of the osteoclast clings tightly to the bone, sealing off the area of bone destruction and secreting *protons* (H^+) and *lysosomal enzymes* that digest the organic matrix. The resulting acidic brew in the resorption bay converts the calcium salts into soluble forms that pass easily into solution. Osteoclasts may also phagocytize the demineralized matrix and dead osteocytes. The digested matrix end products, growth factors, and dissolved minerals are then endocytosed, transported across the osteoclast (by transcytosis), and released at the opposite side. There they enter the interstitial fluid and then the blood.

When resorption of a given area of bone is completed, the osteoclasts undergo apoptosis. There is much to learn about osteoclast activation, but parathyroid hormone and proteins secreted by T cells of the immune system appear to be important.

Control of Remodeling

Remodeling goes on continuously in the skeleton, regulated by genetic factors and two control loops that serve different "masters." One is a negative feedback hormonal loop that maintains Ca^{2+} homeostasis in the blood. The other involves responses to mechanical and gravitational forces acting on the skeleton.

The hormonal feedback becomes much more meaningful when you understand calcium's importance in the body. Ionic calcium is necessary for an amazing number of physiological processes, including transmission of nerve impulses, muscle contraction, blood coagulation, secretion by glands and nerve cells, and cell division.

The human body contains 1200–1400 g of calcium, more than 99% present as bone minerals. Most of the remainder is in body cells. Less than 1.5 g is present in blood, and the hormonal control loop normally maintains blood Ca^{2+} within the narrow range of 9–11 mg per dl (100 ml) of blood. Calcium is absorbed from the intestine under the control of vitamin D metabolites.

Hormonal Controls

The hormonal controls primarily involve **parathyroid hormone (PTH)**, produced by the parathyroid glands. To a much lesser extent **calcitonin** (kal″sĭ-to′nin), produced by parafollicular cells (C cells) of the thyroid gland, may be involved.

When blood levels of ionic calcium decline, PTH is released (**Figure 6.12**). The increased PTH level stimulates osteoclasts to resorb bone, releasing calcium into blood. Osteoclasts are no respecters of matrix age: When activated, they break down both old and new matrix. As blood concentrations of calcium rise, the stimulus for PTH release ends. The decline of PTH reverses its effects and causes blood Ca^{2+} levels to fall.

In humans, calcitonin appears to be a hormone in search of a function because its effects on calcium homeostasis are negligible. When administered at pharmacological (abnormally high) doses, it does lower blood calcium levels temporarily.

These hormonal controls act to preserve blood calcium homeostasis, not the skeleton's strength or well-being. In fact, if blood calcium levels are low for an extended time, the bones become so demineralized that they develop large holes.

Minute changes from the homeostatic range for blood calcium can lead to severe neuromuscular problems. For example, *hypocalcemia* (hi″po-kal-se′me-ah; low blood Ca^{2+} levels) causes hyperexcitability. In contrast, *hypercalcemia* (hi″per-kal-se′me-ah; high blood Ca^{2+} levels) causes nonresponsiveness and inability to function. In addition, sustained high blood levels of Ca^{2+} can lead to the formation of kidney stones or undesirable deposits of calcium salts in other organs, which may hamper their function. ✚

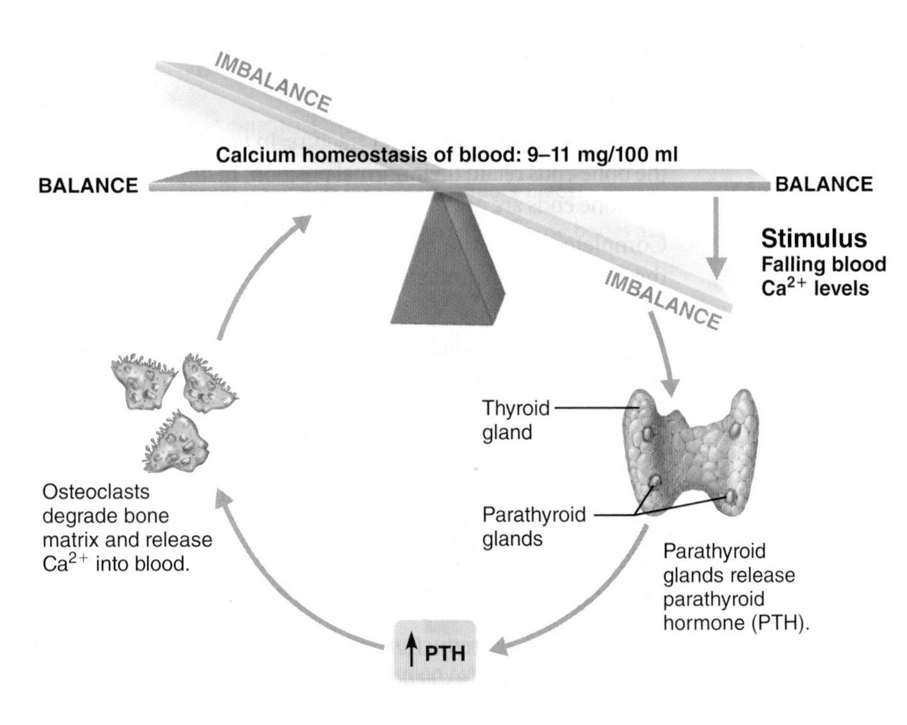

Figure 6.12 Parathyroid hormone (PTH) control of blood calcium levels.

Other hormones are also involved in modifying bone density and bone turnover. For example, *leptin*, a hormone released by adipose tissue, plays a role in regulating bone density. Best known for its effects on weight and energy balance, leptin may also inhibit osteoblasts through a brain (hypothalamus) pathway that activates sympathetic nerves serving bones. However, the full scope of leptin's bone-modifying activity in humans is still being worked out.

It is also evident that the brain, intestine, and skeleton have ongoing conversations that help regulate the balance between bone formation and destruction, with serotonin serving as a hormonal go-between. *Serotonin* is better known as a neurotransmitter that regulates mood and sleep, but most of the body's serotonin is made in the gut (intestine). The role of gut serotonin is still poorly understood. What is known is that when we eat, serotonin is secreted and circulated via the blood to the bones where it interferes with osteoblast activity. Reduction of bone turnover after eating may lock calcium in bone when new calcium is flooding into the bloodstream.

Response to Mechanical Stress

The second set of controls regulating bone remodeling, bone's response to mechanical stress (muscle pull) and gravity, keeps the bones strong where stressors are acting.

Wolff's law holds that a bone grows or remodels in response to the demands placed on it. The first thing to understand is that a bone's anatomy reflects the common stresses it encounters. For example, a bone is loaded (stressed) whenever weight bears down on it or muscles pull on it. This loading is usually off center and tends to bend the bone. Bending compresses the bone on one side and subjects it to tension (stretching) on the other (**Figure 6.13**).

As a result of these mechanical stressors, long bones are thickest midway along the diaphysis, exactly where bending stresses are greatest (bend a stick and it will split near the middle). Both compression and tension are minimal toward the center of the bone (they cancel each other out), so a bone can "hollow out" for lightness (using spongy bone instead of compact bone) without jeopardy.

Wolff's law also explains several other observations:

- Handedness (being right or left handed) results in the bones of one upper limb being thicker than those of the less-used limb. Vigorous exercise of the most-used limb leads to large increases in bone strength.

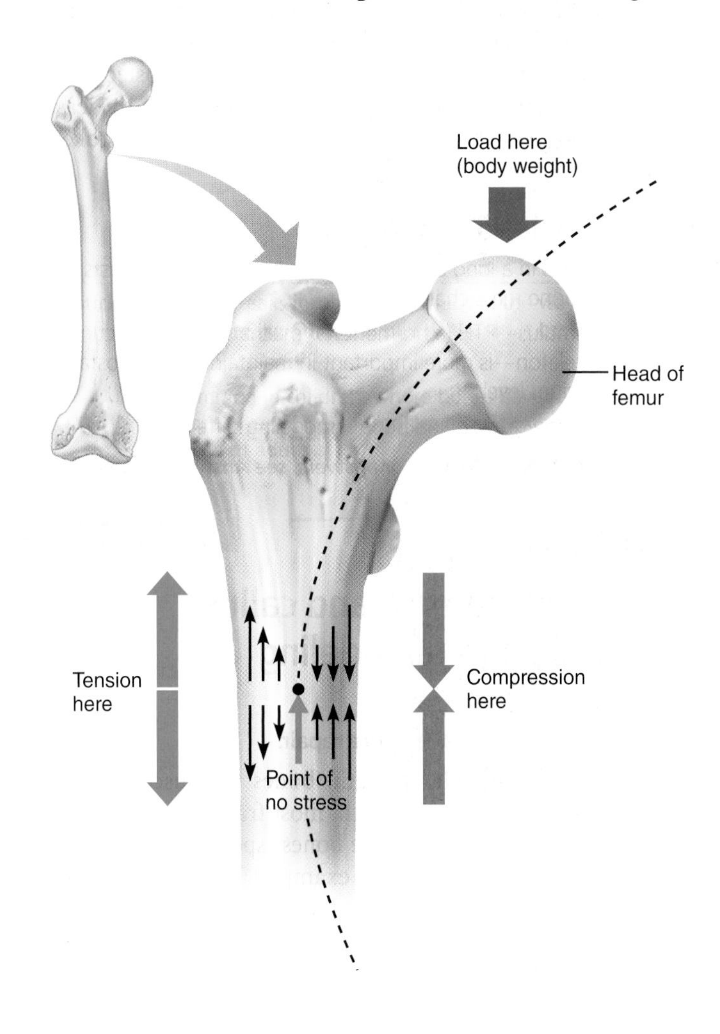

Figure 6.13 Bone anatomy and bending stress. Body weight transmitted to the head of the femur (thigh bone) threatens to bend the bone along the indicated arc, compressing it on one side (converging arrows on right) and stretching it on the other side (diverging arrows on left). Because these two forces cancel each other internally, much less bone material is needed internally than superficially.

7 The Skeleton

KEY CONCEPTS

The word *skeleton* comes from the Greek word meaning "dried-up body" or "mummy," a rather unflattering description. Nonetheless, the human skeleton is a triumph of design and engineering that puts most skyscrapers to shame. It is strong, yet light, and almost perfectly adapted for the protective, locomotor, and manipulative functions it performs.

The **skeleton**, or **skeletal system**, composed of bones, cartilages, joints, and ligaments, accounts for about 20% of body mass (about 30 pounds in a 160-pound person). Bones make up most of the skeleton. Cartilages occur only in isolated areas, such as the nose, parts of the ribs, and the joints. Ligaments connect bones and reinforce joints, allowing required movements while restricting motions in other directions. Joints provide for the remarkable mobility of the skeleton. We discuss joints and ligaments separately in Chapter 8.

PART 1 THE AXIAL SKELETON

As described in Chapter 6, the skeleton is divided into *axial* and *appendicular* portions (see Figures 6.1 and 7.1). The **axial skeleton** is structured from 80 bones segregated into three major regions: the *skull*, *vertebral column*, and *thoracic cage* (**Figure 7.1**). This part of the skeleton (1) forms the longitudinal axis of the body, (2) supports the head, neck, and trunk, and (3) protects the brain, spinal cord, and the organs in the thorax. As we will see later in this chapter, the bones of the appendicular skeleton, which allow us to interact with and manipulate our environment, are appended to the axial skeleton.

7.1 The skull consists of 8 cranial bones and 14 facial bones

→ **Learning Objectives**

☐ **Name, describe, and identify the skull bones. Identify their important markings.**

☐ **Compare and contrast the major functions of the cranium and the facial skeleton.**

☐ **Define the bony boundaries of the orbits, nasal cavity, and paranasal sinuses.**

The **skull** is the body's most complex bony structure. It is formed by *cranial* and *facial bones*, 22 in all. The cranial bones, or **cranium** (kra′ne-um), enclose and protect the fragile brain and furnish attachment sites for head and neck muscles. The facial bones:

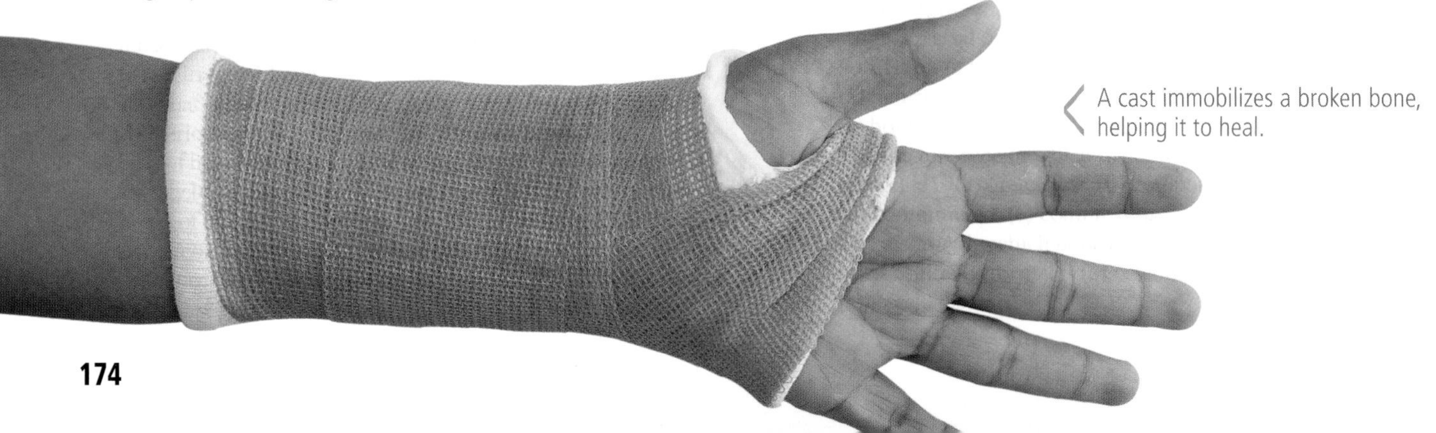

A cast immobilizes a broken bone, helping it to heal.

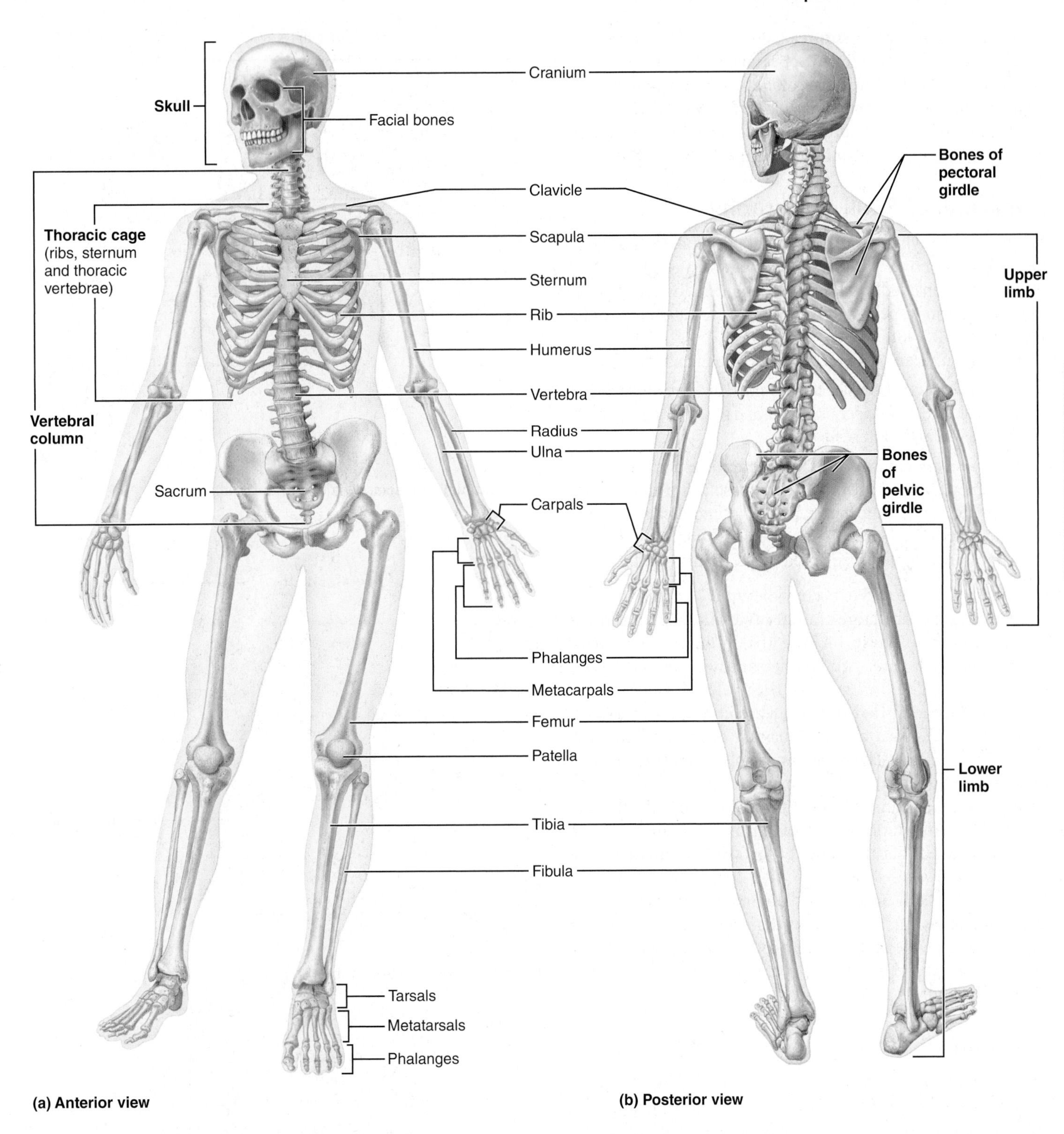

Figure 7.1 The human skeleton. Bones of the axial skeleton are colored green. Bones of the appendicular skeleton are gold.

- Form the framework of the face
- Contain cavities for the special sense organs of sight, taste, and smell
- Provide openings for air and food passage
- Secure the teeth
- Anchor the facial muscles of expression, which we use to show our feelings

Most skull bones are flat bones. Except for the mandible, which is connected to the rest of the skull by freely movable joints, all bones of the adult skull are firmly united by interlocking joints called **sutures** (soo′cherz). The suture lines have a saw-toothed or serrated appearance.

The major skull sutures, the *coronal*, *sagittal*, *squamous*, and *lambdoid sutures*, connect cranial bones (Figures 7.2a, 7.4b, and 7.5a). Most other skull sutures connect facial bones and are named according to the bones they connect.

Overview of Skull Geography

It is worth surveying basic skull "geography" before describing the individual bones. With the lower jaw removed, the skull resembles a lopsided, hollow, bony sphere. The facial bones form its anterior aspect, and the cranium forms the rest of the skull (**Figure 7.2a**).

The cranium can be divided into a vault and a base.

- The *cranial vault*, also called the *calvaria* (kal-va′re-ah; "bald part of skull"), forms the superior, lateral, and posterior aspects of the skull, as well as the forehead.
- The *cranial base* forms the skull's inferior aspect. Internally, prominent bony ridges divide the base into three distinct "steps" or fossae—the *anterior*, *middle*, and *posterior cranial fossae* (Figure 7.2b and c). The brain sits snugly in these cranial fossae, completely enclosed by the cranial vault.

Overall, the brain is said to occupy the *cranial cavity*.

In addition to the large cranial cavity, the skull has many smaller cavities. These include the middle and internal ear cavities and, anteriorly, the nasal cavity and the orbits (**Figure 7.3**). The *orbits* house the eyeballs. Several skull bones contain air-filled sinuses, which lighten the skull.

The skull also has about 85 named openings (foramina, canals, fissures, etc.). The most important of these provide passageways for the spinal cord, the major blood vessels serving the brain, and the 12 pairs of cranial nerves (numbered I through XII) that transmit information to and from the brain.

As you read about the bones of the skull, locate each bone on the skull views in **Figures 7.4**, **7.5** (pp. 178–180), and **7.6** (p. 181). The skull bones and their important markings are also summarized in **Table 7.1** at the end of the skull section (pp. 191–192). Note that the color-coded boxes before a bone's name in the text and in Table 7.1 correspond to the color of that bone in the figures.

Cranium

The eight cranial bones are the paired parietal and temporal bones and the unpaired frontal, occipital, sphenoid, and ethmoid bones. Together, these construct the brain's protective bony "helmet." Because its superior aspect is curved, the cranium is self-bracing. This allows the bones to be thin, and, like an eggshell, the cranium is remarkably strong for its weight.

Frontal Bone

The shell-shaped **frontal bone** (Figures 7.4a, 7.5, and 7.7) forms the anterior cranium. It articulates posteriorly with the paired parietal bones via the prominent *coronal suture*.

The most anterior part of the frontal bone is the vertical *squamous part*, commonly called the *forehead*. The frontal squamous

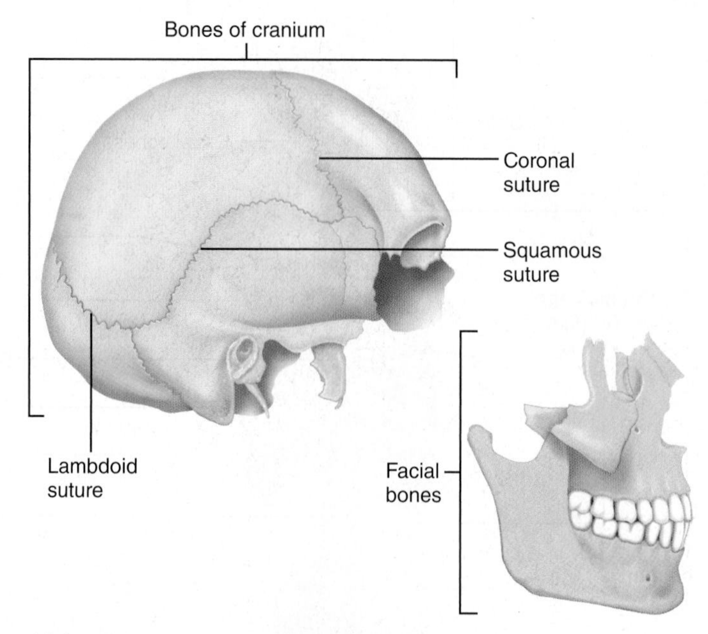

(a) Cranial and facial divisions of the skull

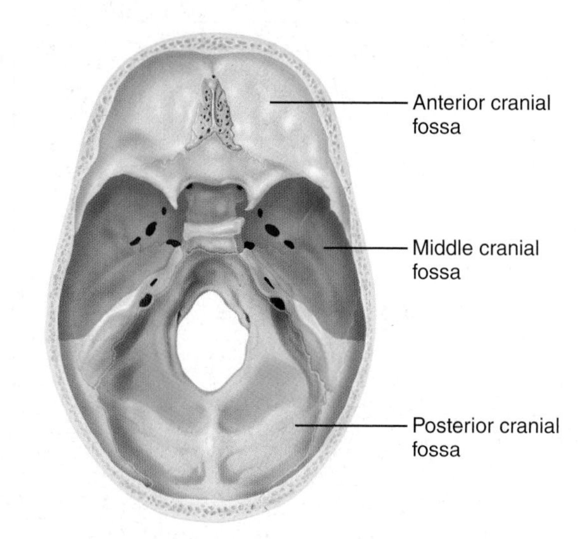

(b) Superior view of the cranial fossae

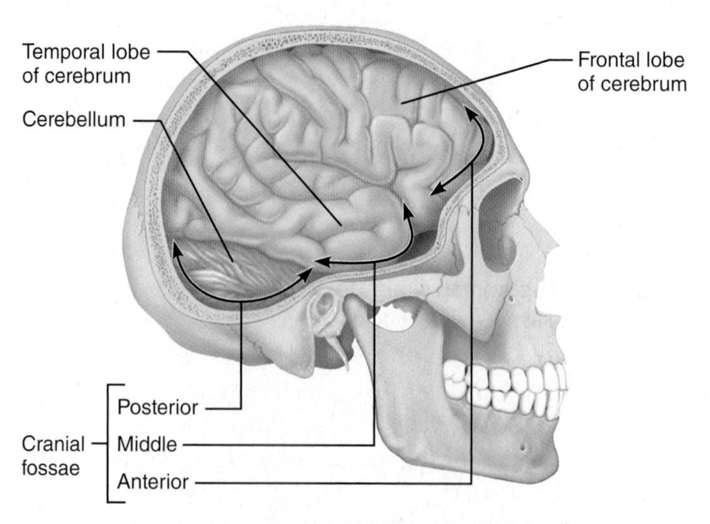

(c) Lateral view of cranial fossae showing the contained brain regions

Figure 7.2 The skull: Cranial and facial divisions and fossae.

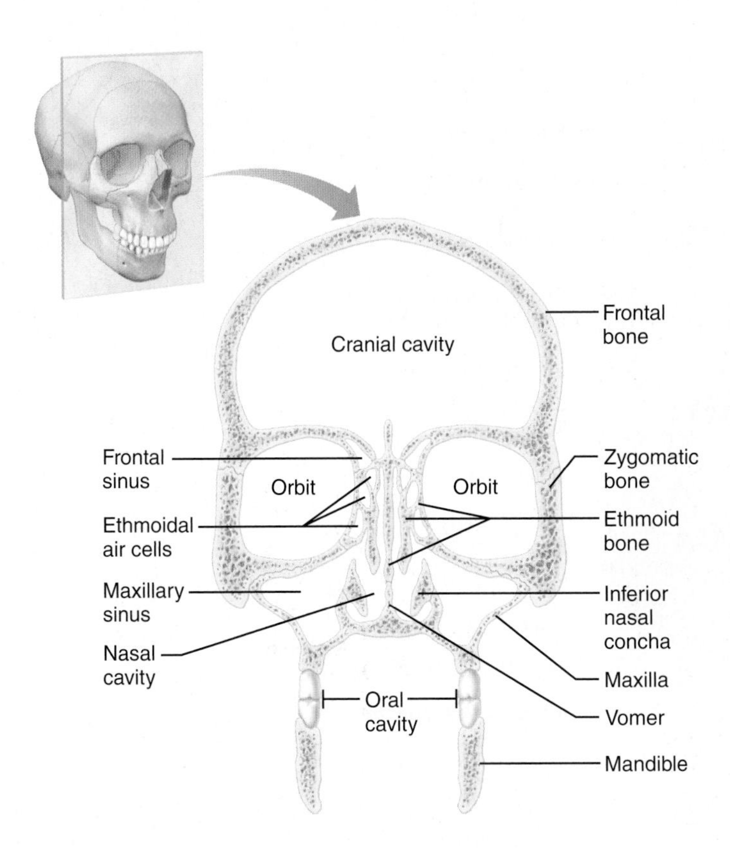

Figure 7.3 Major cavities of the skull, frontal section.

region ends inferiorly at the **supraorbital margins**, the thickened superior margins of the orbits that lie under the eyebrows. From here, the frontal bone extends posteriorly, forming the superior wall of the *orbits* and most of the **anterior cranial fossa** (**Figure 7.7a** and **b**, p. 182). This fossa supports the frontal lobes of the brain. Each supraorbital margin is pierced by a **supraorbital foramen (notch)**, which allows the supraorbital artery and nerve to pass to the forehead (Figure 7.4a).

The smooth portion of the frontal bone between the orbits is the **glabella** (glah-bel′ah). Just inferior to this the frontal bone meets the nasal bones at the *frontonasal suture* (Figure 7.4a). The areas lateral to the glabella contain sinuses, called the **frontal sinuses** (Figures 7.5c and 7.3).

Parietal Bones and the Major Sutures

The two large **parietal bones** are curved, rectangular bones that form most of the superior and lateral aspects of the skull; as such, they form the bulk of the cranial vault. The four largest sutures occur where the parietal bones articulate (form a joint) with other cranial bones:

- The **coronal suture** (kŏ-ro′nul), where the parietal bones meet the frontal bone anteriorly (Figures 7.2a and 7.5)
- The **sagittal suture**, where the parietal bones meet superiorly at the cranial midline (Figure 7.4b)
- The **lambdoid suture** (lam′doid), where the parietal bones meet the occipital bone posteriorly (Figures 7.2a, 7.4b, and 7.5)
- The **squamous suture** (one on each side), where a parietal and temporal bone meet on the lateral aspect of the skull (Figures 7.2a and 7.5)

Occipital Bone

The **occipital bone** (ok-sip′ĭ-tal) forms most of the skull's posterior wall and base. It articulates anteriorly with the paired parietal and temporal bones via the *lambdoid* and *occipitomastoid sutures*, respectively (Figure 7.5). The basilar part of the occipital bone also joins with the sphenoid bone in the cranial base (Figure 7.6a).

Internally, the occipital bone forms the walls of the **posterior cranial fossa** (Figures 7.7 and 7.2c), which supports the cerebellum of the brain. In the base of the occipital bone is the **foramen magnum** ("large hole") through which the inferior part of the brain connects with the spinal cord. The foramen magnum is flanked laterally by two occipital condyles (Figure 7.6). The rockerlike **occipital condyles** articulate with the first vertebra of the spinal column in a way that permits a nodding ("yes") motion of the head. Hidden medially and superiorly to each occipital condyle is a **hypoglossal canal** (Figure 7.7a), through which a cranial nerve (XII) passes.

Just superior to the foramen magnum is a median protrusion called the **external occipital protuberance** (Figures 7.4b, 7.5c and d, and 7.6). You can feel this knoblike projection just below the most bulging part of your posterior skull. A number of inconspicuous ridges, the *external occipital crest* and the *superior* and *inferior nuchal lines* (nu′kal), mark the occipital bone near the foramen magnum. The external occipital crest secures the *ligamentum nuchae* (lig″ah-men′tum noo′ke; *nucha* = back of the neck), a sheetlike elastic ligament that connects the vertebrae of the neck to the skull. The nuchal lines, and the bony regions between them, anchor many neck and back muscles. The superior nuchal line marks the upper limit of the neck.

Temporal Bones

The two **temporal bones** are best viewed on the lateral skull surface (Figure 7.5). They lie inferior to the parietal bones and meet them at the squamous sutures. The temporal bones form the inferolateral aspects of the skull and parts of the cranial base. The use of the terms *temple* and *temporal*, from the Latin word *temporum*, meaning "time," came about because gray hairs, a sign of time's passing, usually appear first at the temples.

Each temporal bone has a complicated shape (**Figure 7.8**, p. 183) and is described in terms of its three major parts, the *squamous*, *tympanic*, and *petrous parts*. The flaring **squamous part** ends at the squamous suture. Its barlike **zygomatic process** meets the zygomatic bone of the face anteriorly. Together, these two bony structures form the **zygomatic arch**, which you can feel as the projection of your cheek (*zygoma* = cheekbone). The small, oval **mandibular fossa** (man-dib′u-lar) on the inferior surface of the zygomatic process receives the condylar process of the mandible (lower jawbone), forming the freely movable *temporomandibular joint*.

The **tympanic part** (tim-pan′ik; "eardrum") (Figure 7.8) of the temporal bone surrounds the **external acoustic meatus**, or external ear canal (*meatus* = passage). The external acoustic meatus and the eardrum at its deep end are part of the *external ear*. In a dried skull, the eardrum has been removed and part of the middle ear cavity deep to the external meatus can also be seen.

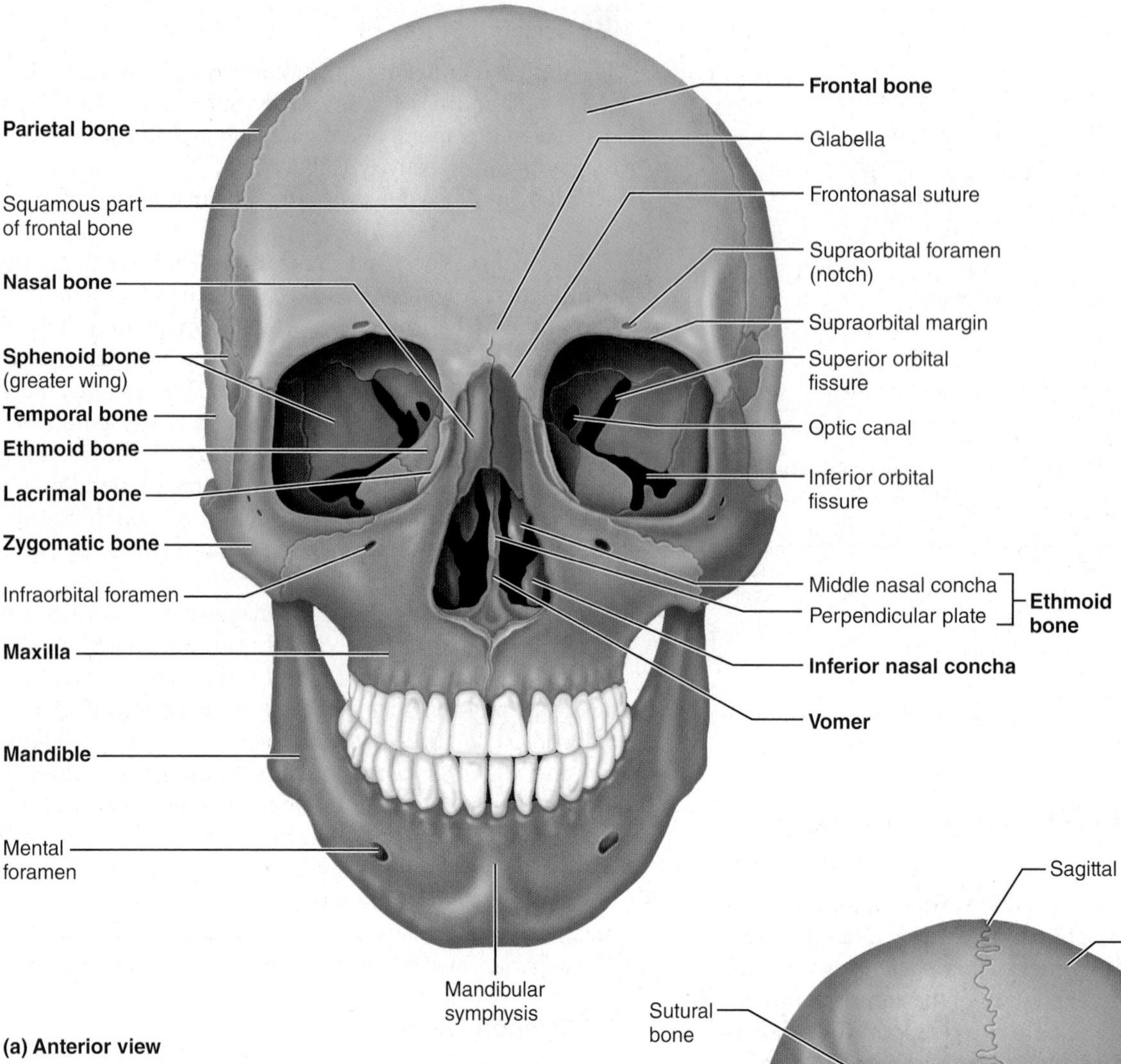

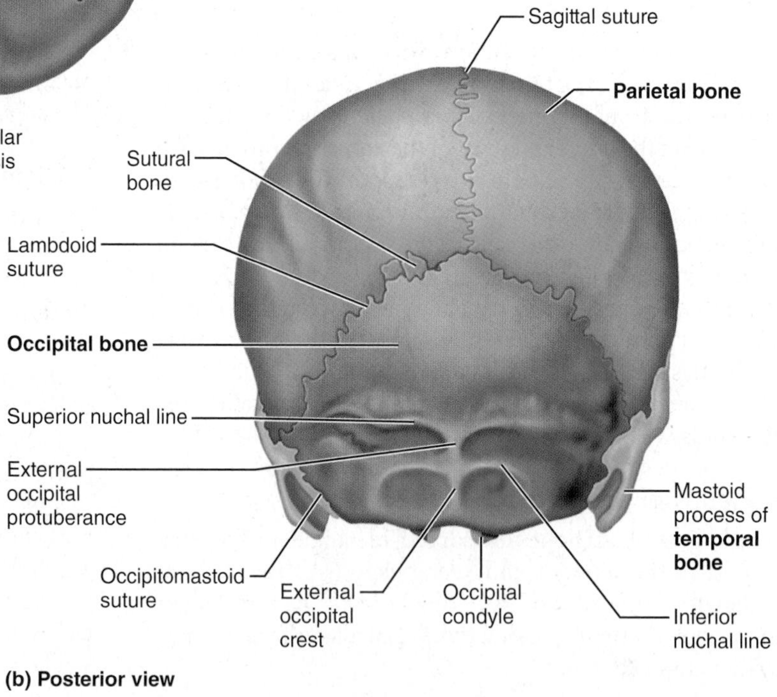

Practice art labeling
MasteringA&P®>Study Area>Chapter 7

Figure 7.4 Anterior and posterior views of the skull. (For related images, see *A Brief Atlas of the Human Body*, Figures 1 and 7.)

The thick **petrous part** (pet′rus) of the temporal bone houses the *middle* and *internal ear cavities*, which contain sensory receptors for hearing and balance. Extending from the occipital bone posteriorly to the sphenoid bone anteriorly, it contributes to the cranial base (Figures 7.6 and 7.7). In the floor of the cranial cavity, the petrous part of the temporal bone looks like a miniature mountain ridge (*petrous* = rocky). The posterior slope of this ridge lies in the posterior cranial fossa; the anterior slope is in the middle cranial fossa. Together, the sphenoid bone and the petrous portions of the temporal bones construct the **middle cranial fossa** (Figures 7.7 and 7.2b), which supports the temporal lobes of the brain.

Several foramina penetrate the bone of the petrous region (Figure 7.6). The large **jugular foramen** at the junction of the

(Text continues on p. 183.)

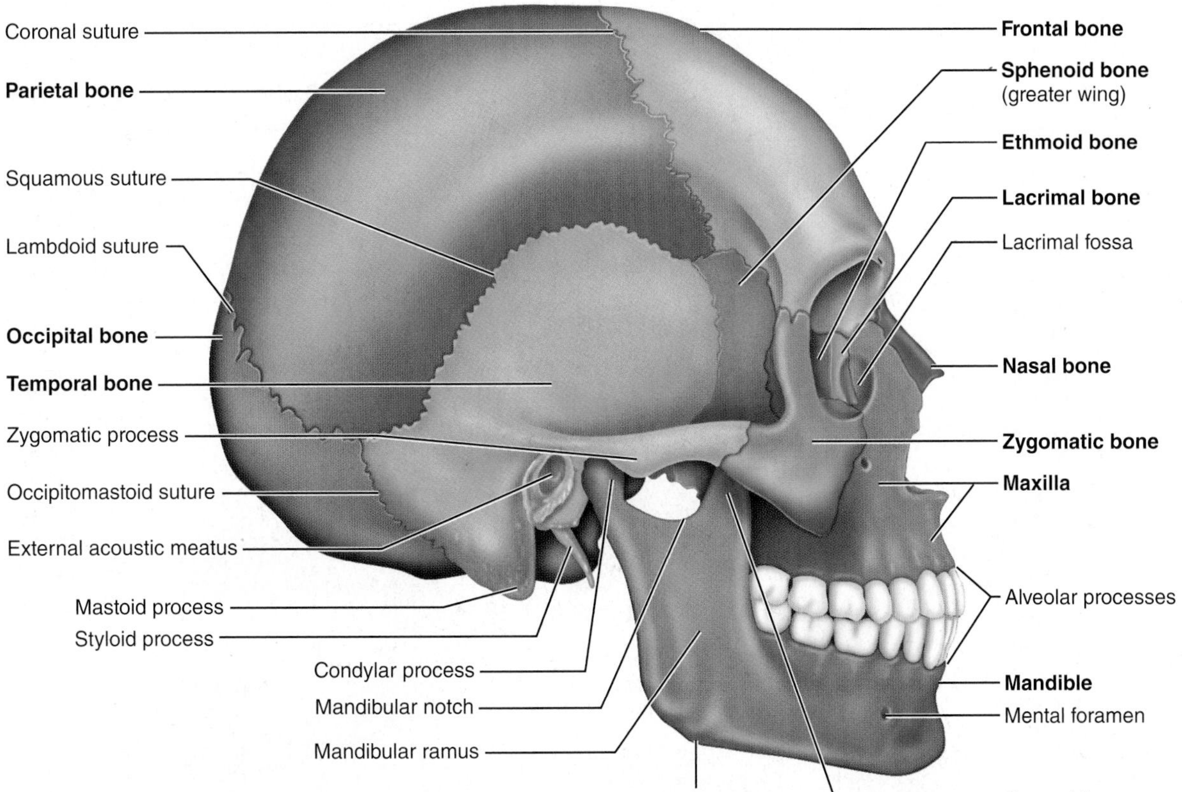

Coronal suture
Parietal bone
Squamous suture
Lambdoid suture
Occipital bone
Temporal bone
Zygomatic process
Occipitomastoid suture
External acoustic meatus
Mastoid process
Styloid process
Condylar process
Mandibular notch
Mandibular ramus
Mandibular angle
Frontal bone
Sphenoid bone (greater wing)
Ethmoid bone
Lacrimal bone
Lacrimal fossa
Nasal bone
Zygomatic bone
Maxilla
Alveolar processes
Mandible
Mental foramen
Coronoid process

(a) External anatomy of the right side of the skull

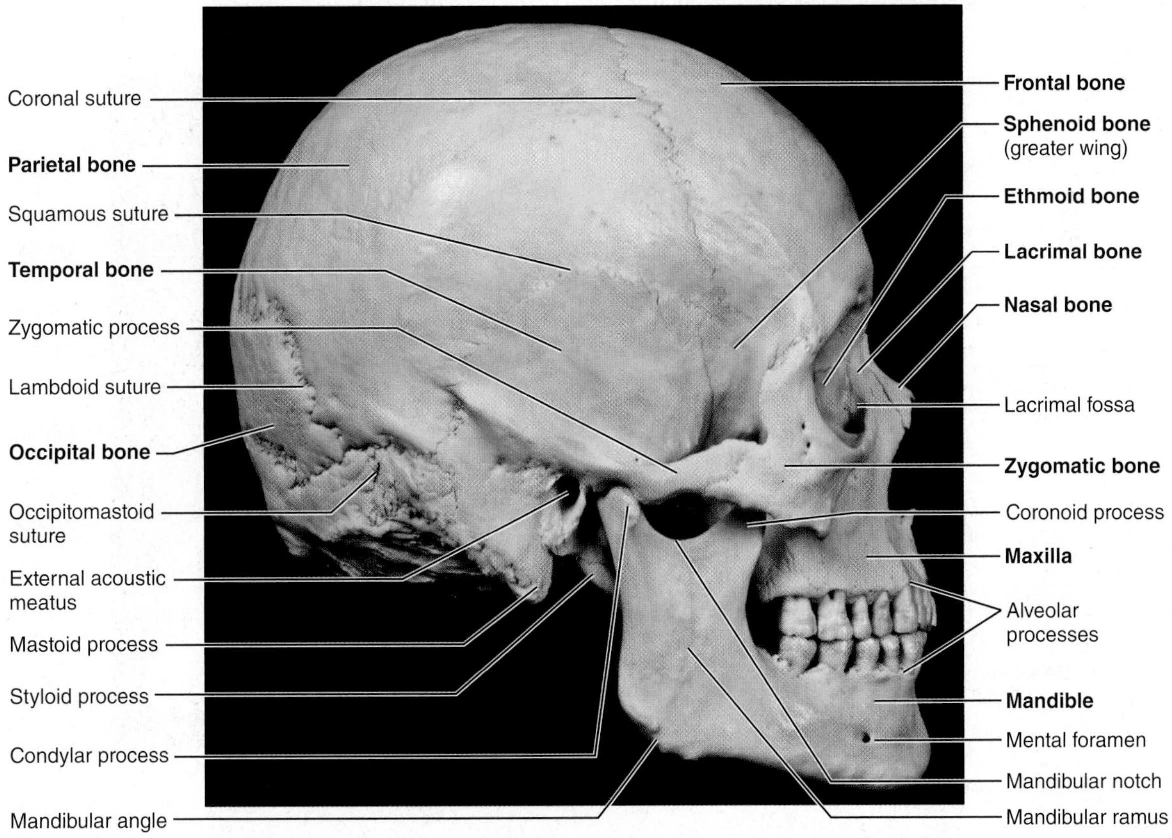

Coronal suture
Parietal bone
Squamous suture
Temporal bone
Zygomatic process
Lambdoid suture
Occipital bone
Occipitomastoid suture
External acoustic meatus
Mastoid process
Styloid process
Condylar process
Mandibular angle
Frontal bone
Sphenoid bone (greater wing)
Ethmoid bone
Lacrimal bone
Nasal bone
Lacrimal fossa
Zygomatic bone
Coronoid process
Maxilla
Alveolar processes
Mandible
Mental foramen
Mandibular notch
Mandibular ramus

(b) Photograph of right side of skull

Figure 7.5 Bones of the lateral aspect of the skull, external and internal views. (For related images, see *A Brief Atlas of the Human Body,* Figures 2 and 3.)

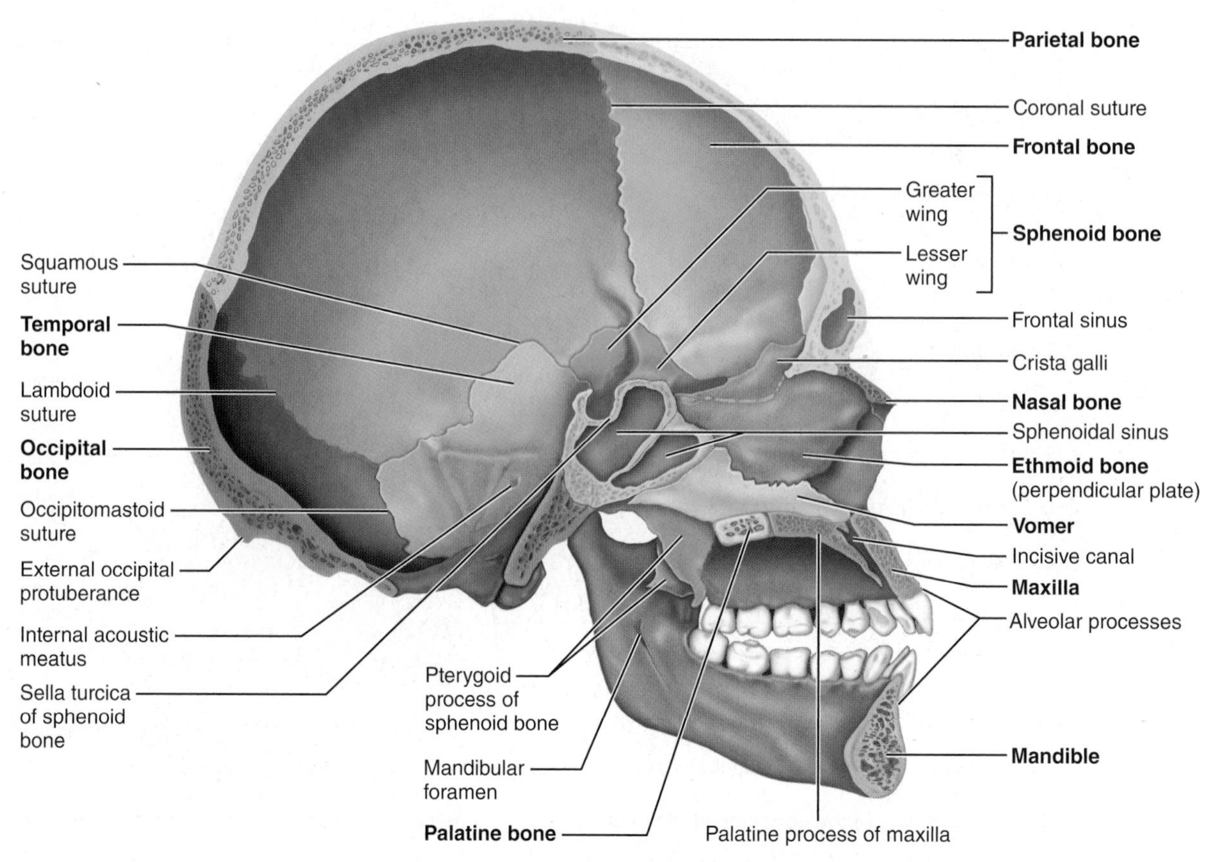

(c) Midsagittal section showing the internal anatomy of the left half of skull

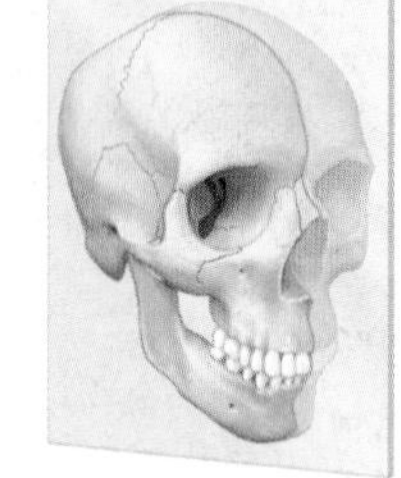

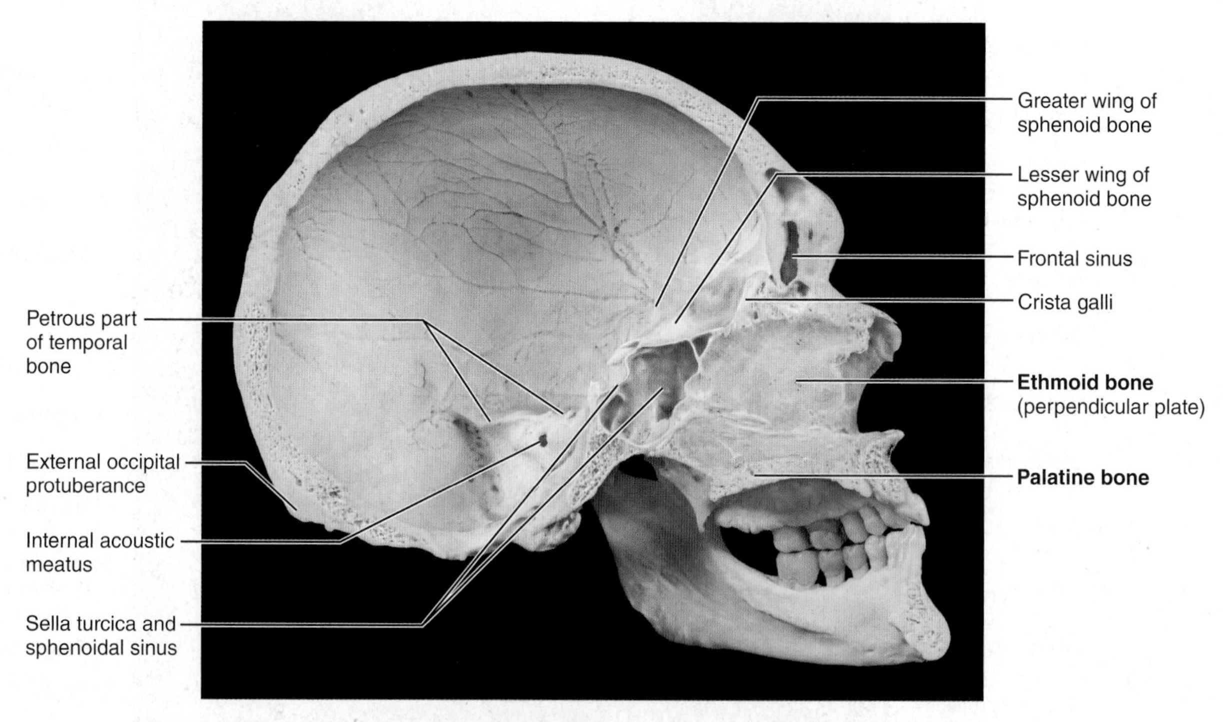

(d) Photo of skull cut through the midline, same view as in (c)

Figure 7.5 *(continued)* **Bones of the lateral aspect of the skull, external and internal views.**

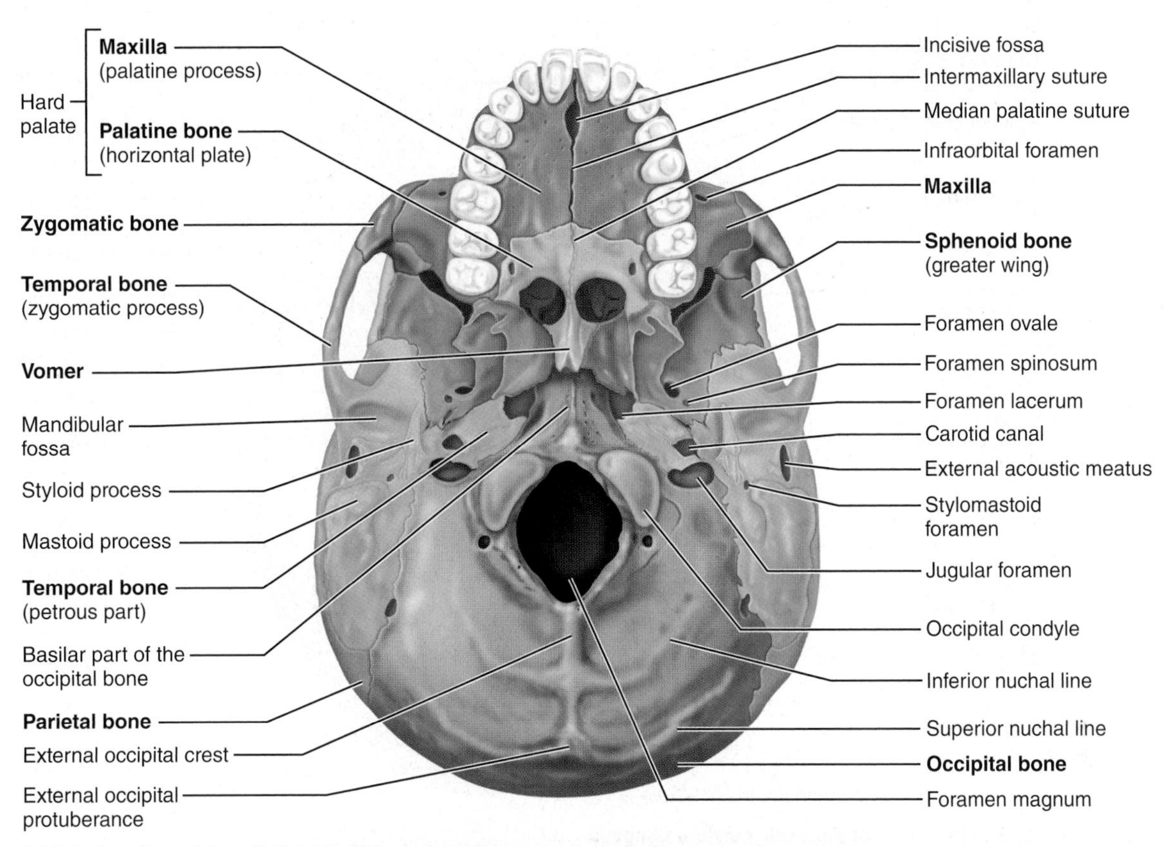

(a) Inferior view of the skull (mandible removed)

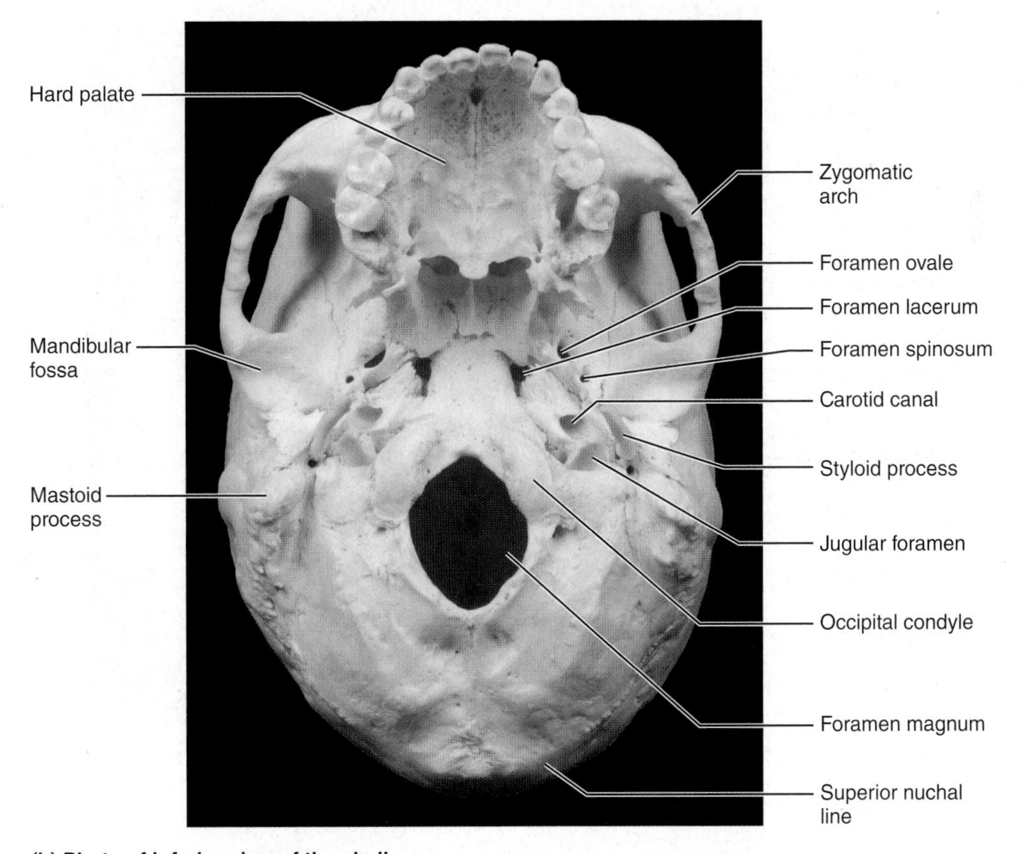

(b) Photo of inferior view of the skull

Figure 7.6 Inferior aspect of the skull, mandible removed. (For related images, see *A Brief Atlas of the Human Body*, Figure 4.)

Practice art labeling
MasteringA&P®>Study Area>Chapter 7

7

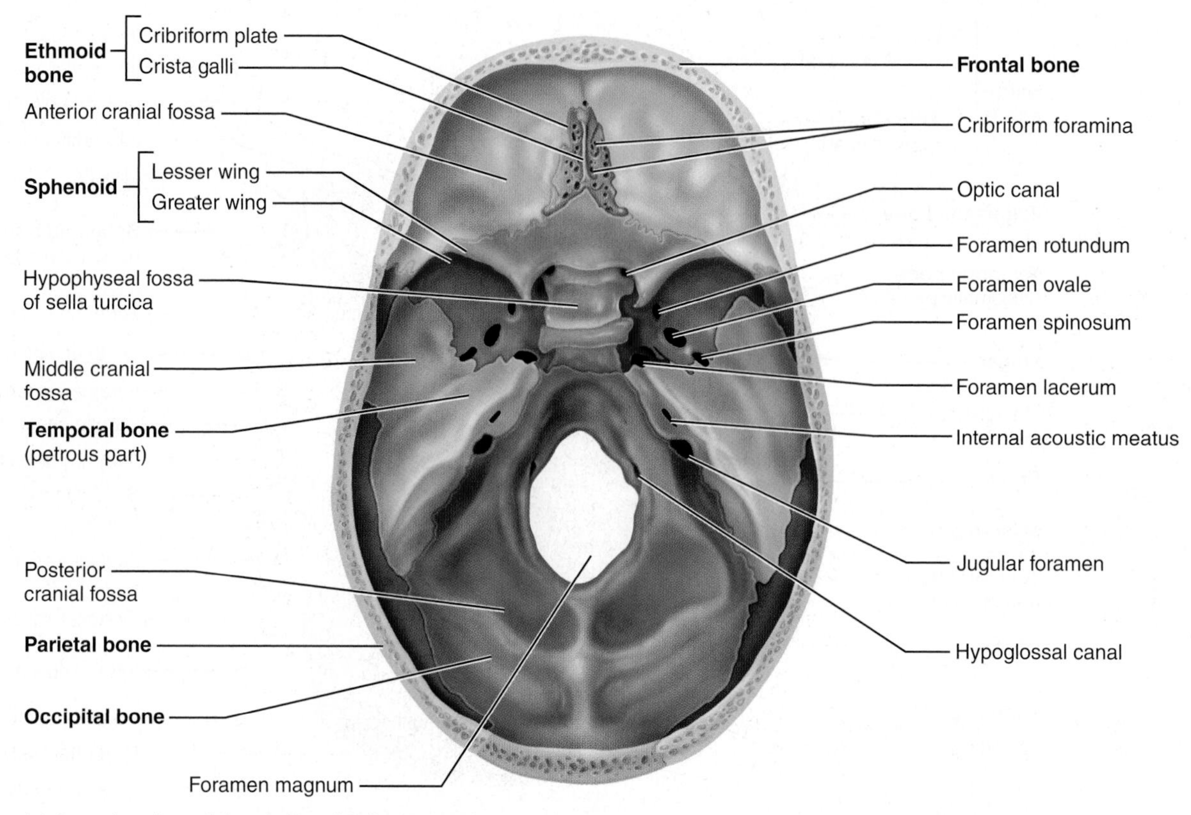

(a) Superior view of the skull, calvaria removed

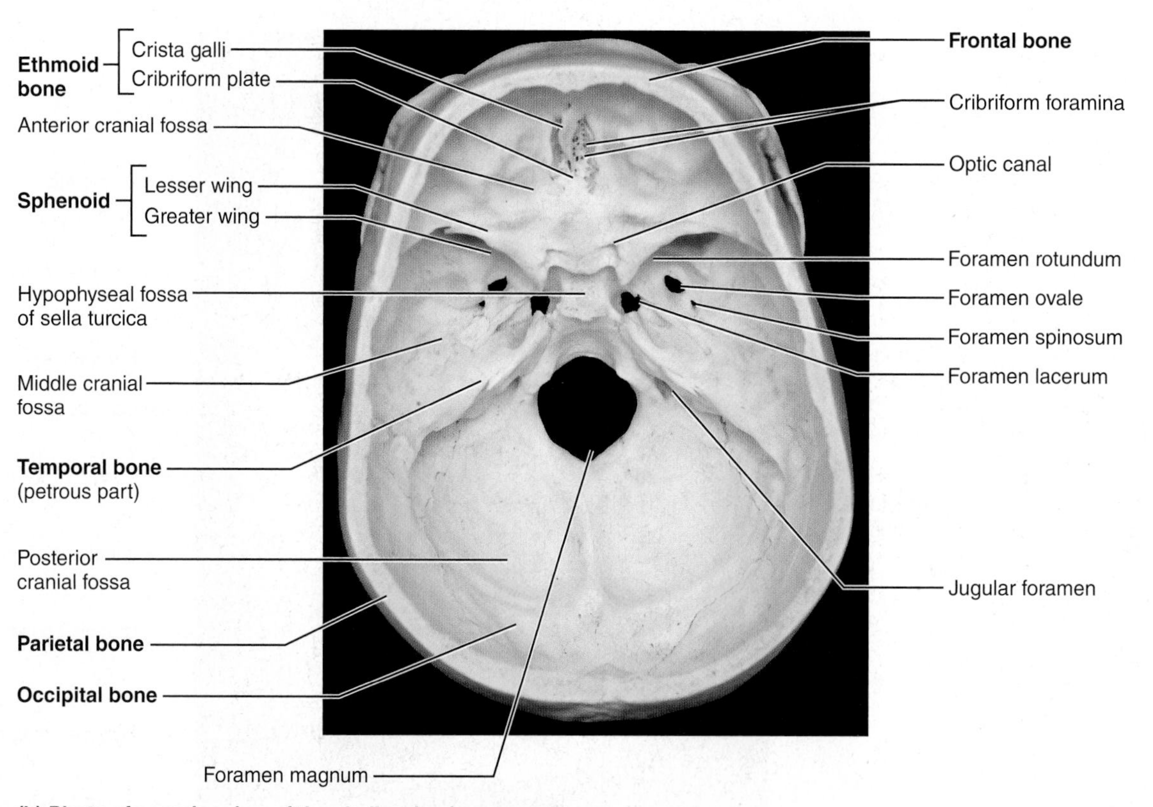

(b) Photo of superior view of the skull, calvaria removed

Figure 7.7 The base of the cranial cavity. (For related images, see *A Brief Atlas of the Human Body*, Figure 5.)

Practice art labeling
MasteringA&P®>Study Area>Chapter 7

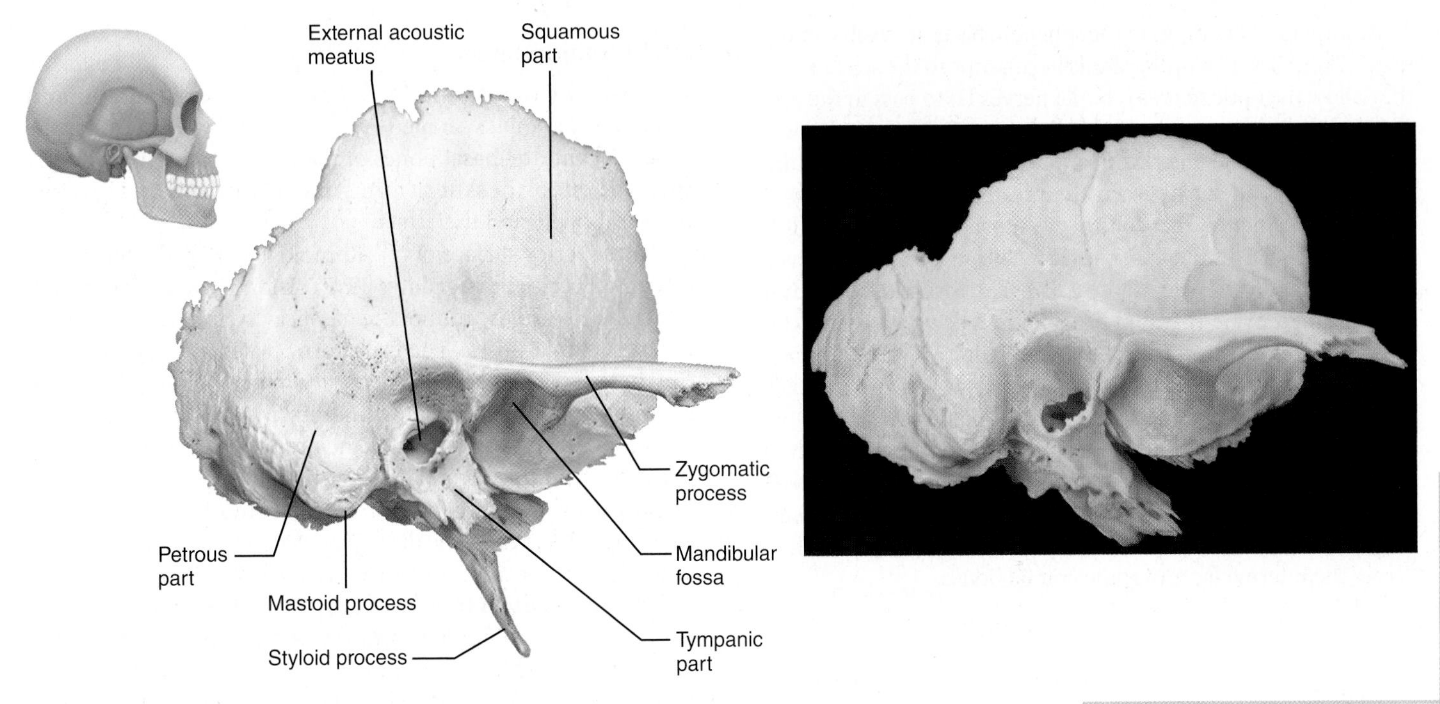

Figure 7.8 The temporal bone. Right lateral view. (For related images, see *A Brief Atlas of the Human Body*, Figures 2 and 8.)

Explore human cadaver
MasteringA&P®>Study Area>PAL

occipital and petrous temporal bones allows passage of the internal jugular vein and three cranial nerves (IX, X, and XI). The **carotid canal** (kah-rot′id), just anterior to the jugular foramen, transmits the internal carotid artery into the cranial cavity. The two internal carotid arteries supply blood to over 80% of the cerebral hemispheres of the brain; their closeness to the internal ear cavities explains why, during excitement or exertion, we may hear our rapid pulse as a thundering sound. The **foramen lacerum** (la′ser-um) is a jagged opening (*lacerum* = torn or lacerated) between the petrous temporal bone and the sphenoid bone. It is almost completely closed by cartilage in a living person, but it is conspicuous in a dried skull, and students usually ask its name. The **internal acoustic meatus**, positioned superolateral to the jugular foramen (Figures 7.5c and d, and 7.7), transmits cranial nerves VII and VIII.

A conspicuous feature of the petrous part of the temporal bone is the **mastoid process** (mas′toid; "breast"), which acts as an anchoring site for some neck muscles (Figures 7.5, 7.6, and 7.8). This process can be felt as a lump just posterior to the ear. The needle-like **styloid process** (sti′loid; "stakelike") is an attachment point for several tongue and neck muscles and for a ligament that secures the hyoid bone of the neck to the skull (see Figure 7.12). The **stylomastoid foramen**, between the styloid and mastoid processes, allows cranial nerve VII (the facial nerve) to leave the skull (Figure 7.6).

HOMEOSTATIC IMBALANCE 7.1 — CLINICAL

The mastoid process is full of air cavities (sinuses) called **mastoid air cells**. Their position adjacent to the middle ear cavity (a high-risk area for infections spreading from the throat) puts them at risk for infection themselves. A mastoid sinus infection, or *mastoiditis*, is notoriously difficult to treat. Because the mastoid air cells are separated from the brain by only a very thin bony plate, mastoid infections may spread to the brain as well. ✚

Sphenoid Bone

The bat-shaped **sphenoid bone** (sfe′noid; *sphen* = wedge) spans the width of the middle cranial fossa (Figure 7.7). The sphenoid is considered the keystone of the cranium because it forms a central wedge that articulates with all other cranial bones. It is a challenging bone to study because of its complex shape. As shown in **Figure 7.9**, it consists of a central body and three pairs of processes: the greater wings, lesser wings, and pterygoid processes (ter′ĭ-goid). Within the **body** of the sphenoid are the paired **sphenoidal sinuses** (see Figures 7.5c and d, and 7.14).

The superior surface of the body bears a saddle-shaped prominence, the **sella turcica** (sel′ah ter′sĭ-kah), meaning "Turk's saddle." The seat of this saddle, called the **hypophyseal fossa**, forms a snug enclosure for the pituitary gland (hypophysis).

The **greater wings** project laterally from the sphenoid body, forming parts of (1) the middle cranial fossa (Figures 7.7 and 7.2b), (2) the posterior walls of the orbits (Figure 7.4a), and (3) the external wall of the skull, where they are seen as flag-shaped, bony areas medial to the zygomatic arch (Figure 7.5). The hornlike **lesser wings** form part of the floor of the anterior cranial fossa (Figure 7.7) and part of the medial walls of the orbits. The trough-shaped **pterygoid processes** project inferiorly from the junction of the body and greater wings (Figure 7.9b). They anchor the pterygoid muscles, which are important in chewing.

A number of openings in the sphenoid bone are visible in Figures 7.7 and 7.9. The **optic canals** lie anterior to the sella turcica; they allow the optic nerves (cranial nerves II) to pass to the eyes. On each side of the sphenoid body is a crescent-shaped row of four openings. The anteriormost of these, the **superior orbital fissure**, is a long slit between the greater and lesser wings. It allows cranial nerves that control eye movements (III, IV, VI) to enter the orbit. This fissure is most obvious in an anterior view of the skull (Figure 7.4 and Figure 7.9b). The **foramen rotundum** and **foramen ovale** (o-va′le) provide passageways for branches of cranial nerve V to reach the face (Figure 7.7). The foramen rotundum is in the medial part of the greater wing and is usually oval, despite its name meaning "round opening." The foramen ovale, a large, oval foramen posterior to the foramen rotundum, is also visible in an inferior view of the skull (Figure 7.6). Posterolateral to the foramen ovale is the small **foramen spinosum** (Figure 7.7); it transmits the *middle meningeal artery*, which serves the internal faces of some cranial bones.

Ethmoid Bone

Like the temporal and sphenoid bones, the delicate **ethmoid bone** has a complex shape (**Figure 7.10**). Lying between the sphenoid and the nasal bones of the face, it is the most deeply situated bone of the skull. It forms most of the bony area between the nasal cavity and the orbits.

The superior surface of the ethmoid is formed by the paired horizontal **cribriform plates** (krib′rĭ-form) (see Figure 7.7), which help form the roof of the nasal cavity and the floor of the anterior cranial fossa. The cribriform plates are punctured by tiny holes (*cribr* = sieve) called *cribriform foramina* that allow the filaments of the olfactory nerves to pass from the smell receptors in the nasal cavity to the brain. Projecting superiorly between the cribriform plates is a triangular process called the **crista galli** (kris′tah gah′le; "rooster's comb"). The outermost covering of the brain (the dura mater) attaches to the crista galli and helps secure the brain in the cranial cavity.

The **perpendicular plate** of the ethmoid bone projects inferiorly in the median plane and forms the superior part of the nasal septum, which divides the nasal cavity into right and left halves (Figure 7.5c and d). Flanking the perpendicular plate on each side is a **lateral mass** riddled with sinuses called **ethmoidal air cells** (Figures 7.10 and 7.15), for which the bone itself is

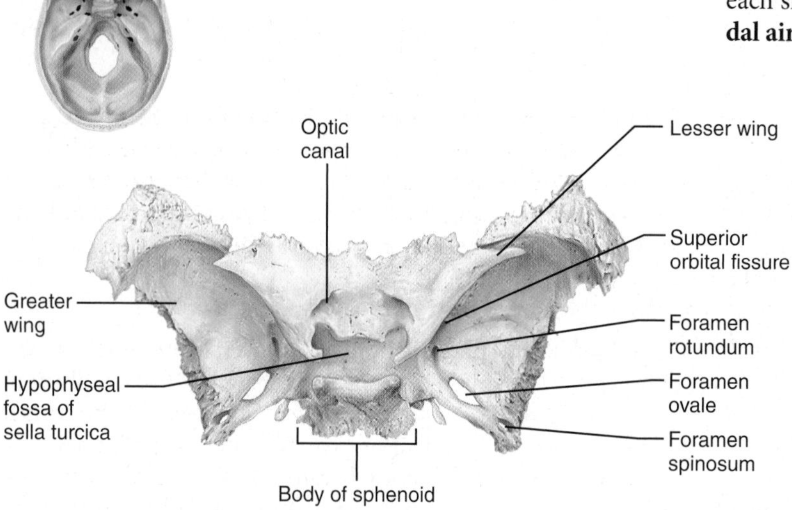

(a) Superior view

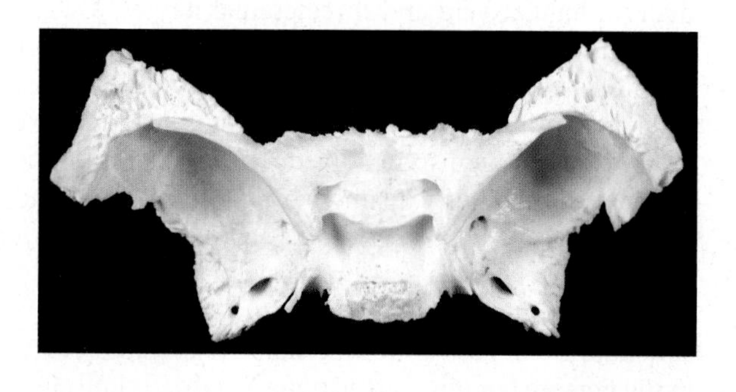

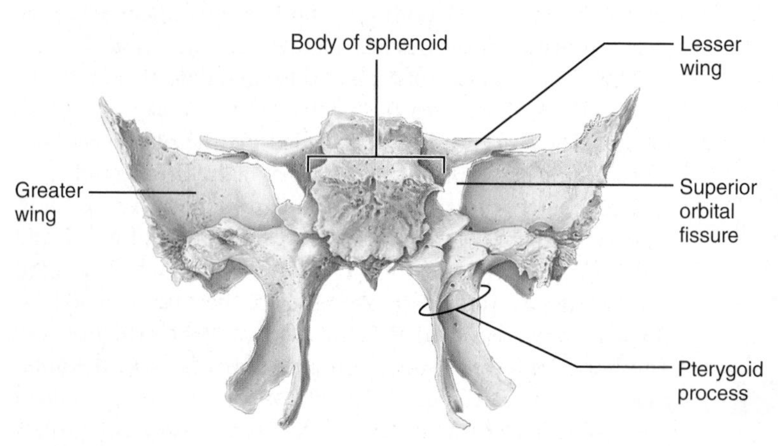

(b) Posterior view

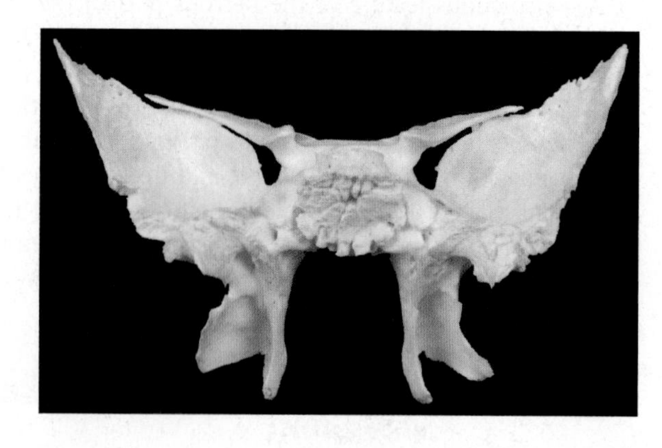

Figure 7.9 The sphenoid bone. (For related images, see *A Brief Atlas of the Human Body*, Figures 5 and 9).

Explore human cadaver MasteringA&P®>Study Area>PAL

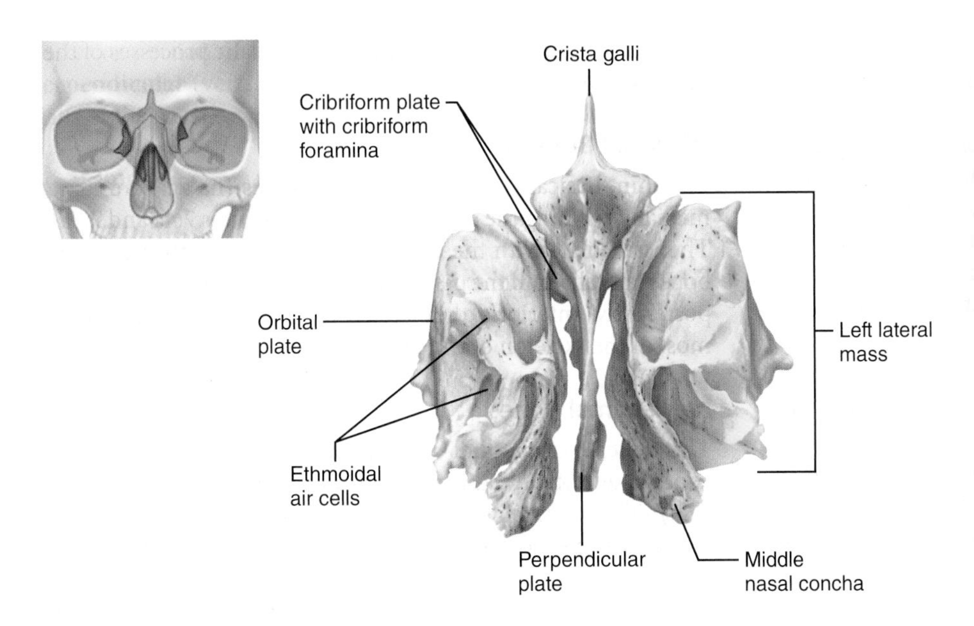

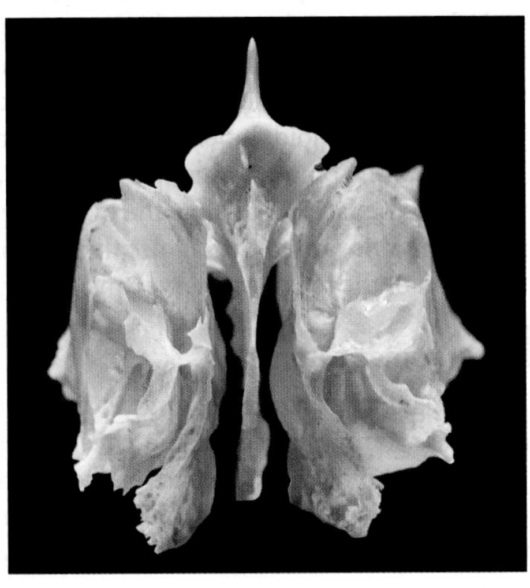

Figure 7.10 The ethmoid bone. Anterior view. (For related images, see *A Brief Atlas of the Human Body*, Figures 3 and 10.)

Explore human cadaver
MasteringA&P®>Study Area>PAL

named (*ethmos* = sieve). Extending medially from the lateral masses, the delicately coiled **superior** and **middle nasal conchae** (kong′ke; *concha* = shell), named after the conch shells found on warm ocean beaches, protrude into the nasal cavity (Figures 7.10 and 7.14a). The lateral surfaces of the ethmoid's lateral masses are called **orbital plates** because they contribute to the medial walls of the orbits.

Sutural Bones

Sutural bones are tiny, irregularly shaped bones or bone clusters that occur within sutures, most often in the lambdoid suture (Figure 7.4b). Not everyone has these bones and their significance is unknown.

Facial Bones

The facial skeleton is made up of 14 bones (see Figures 7.4a and 7.5a), of which only the mandible and the vomer are unpaired. The maxillae, zygomatics, nasals, lacrimals, palatines, and inferior nasal conchae are paired bones. As a rule, the facial skeleton of men is more elongated than that of women. Women's faces tend to be rounder and less angular.

Mandible

The U-shaped **mandible** (man′dĭ-bl), or lower jawbone (Figures 7.4a and 7.5, and **Figure 7.11a**), is the largest, strongest bone of the face. It has a body, which forms the chin, and two upright *rami* (*rami* = branches). Each ramus meets the body posteriorly at a **mandibular angle**. At the superior margin of each ramus are two processes separated by the **mandibular notch**. The anterior **coronoid process** (kor′o-noid; "crown-shaped") is an insertion point for the large temporalis muscle that elevates the lower jaw during chewing. The posterior **condylar process** articulates with the mandibular fossa of the temporal bone, forming the *temporomandibular joint* on the same side.

The mandibular **body** anchors the lower teeth. Its superior border, called the **alveolar process** (al-ve′o-lar), contains the sockets (*dental alveoli*) in which the teeth are embedded. In the midline of the mandibular body is a slight ridge, the **mandibular symphysis** (sim′fih-sis), indicating where the two mandibular bones fused during infancy (Figure 7.4a).

Large **mandibular foramina**, one on the medial surface of each ramus, permit the nerves responsible for tooth sensation to pass to the teeth in the lower jaw. Dentists inject lidocaine into these foramina to prevent pain while working on the lower teeth. The **mental foramina**, openings on the lateral aspects of the mandibular body, allow blood vessels and nerves to pass to the skin of the chin (*ment* = chin) and lower lip.

Maxillary Bones

The **maxillary bones**, or **maxillae** (mak-sil′le; "jaws") (Figures 7.4, 7.5, 7.6, and 7.11b and c), are fused medially. They form the upper jaw and the central portion of the facial skeleton. All facial bones except the mandible articulate with the maxillae. For this reason, the maxillae are considered the keystone bones of the facial skeleton.

The maxillae carry the upper teeth in their **alveolar processes**. Just inferior to the nose the maxillae meet medially, forming the pointed **anterior nasal spine** at their junction. The **palatine processes** (pă′lah-tīn) of the maxillae project posteriorly from the alveolar processes and fuse medially at the *intermaxillary suture*, forming the anterior two-thirds of the hard palate, or bony roof of the mouth (Figures 7.5c and d and 7.6). Just posterior to the teeth, a midline foramen called the **incisive fossa** leads into the **incisive canal**, a passageway for blood vessels and nerves.

The **frontal processes** extend superiorly to the frontal bone, forming part of the lateral aspects of the bridge of the nose (Figures 7.11b and 7.4a). The regions that flank the nasal cavity laterally contain the **maxillary sinuses** (see Figure 7.15), the largest of the paranasal sinuses. They extend from the orbits to

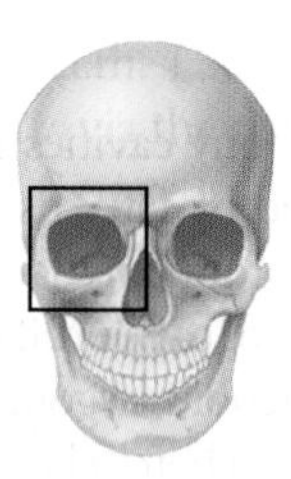

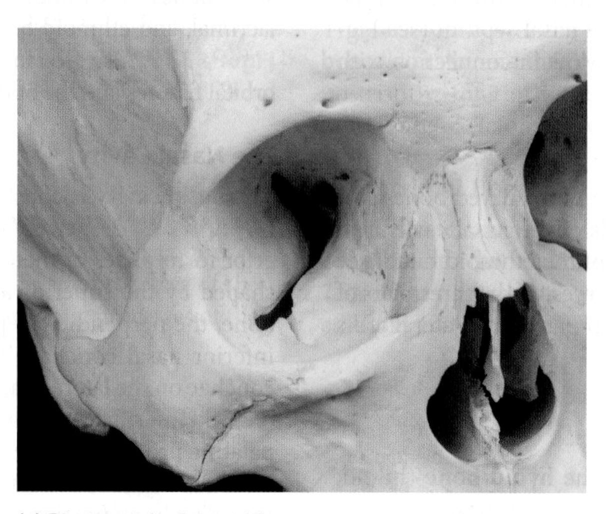

(a) Photograph, right orbit

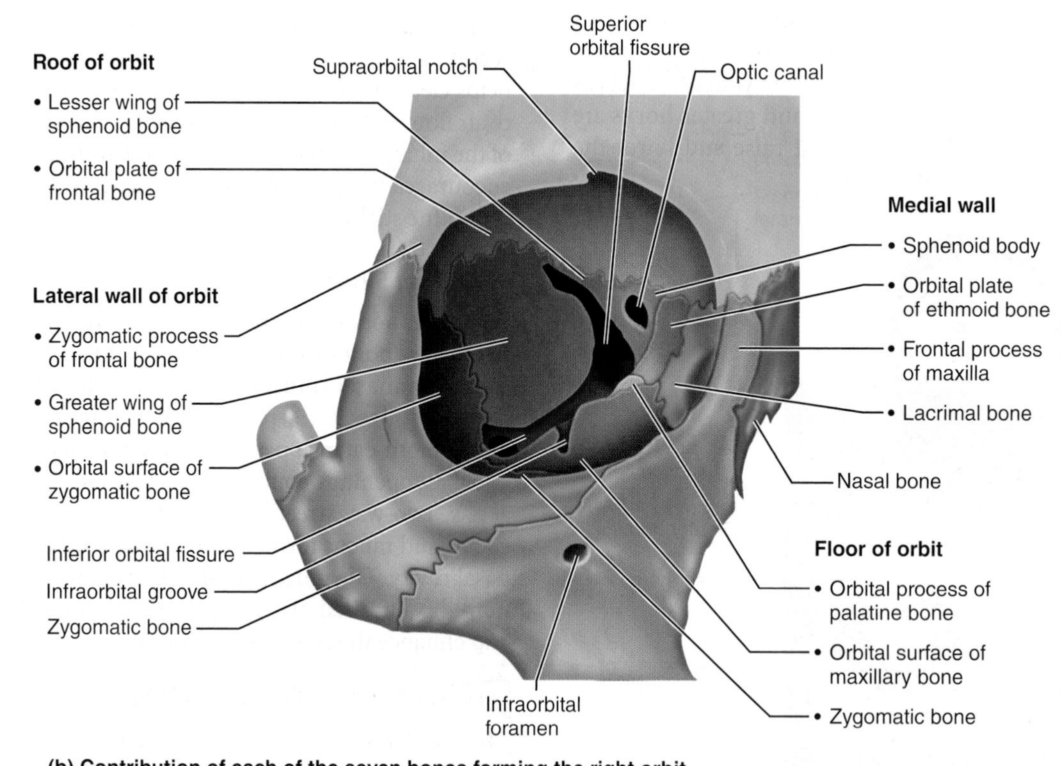

(b) Contribution of each of the seven bones forming the right orbit

Figure 7.13 Bones that form the orbits. (For a related image, see *A Brief Atlas of the Human Body,* Figure 14.)

Explore human cadaver
MasteringA&P®>Study Area>PAL

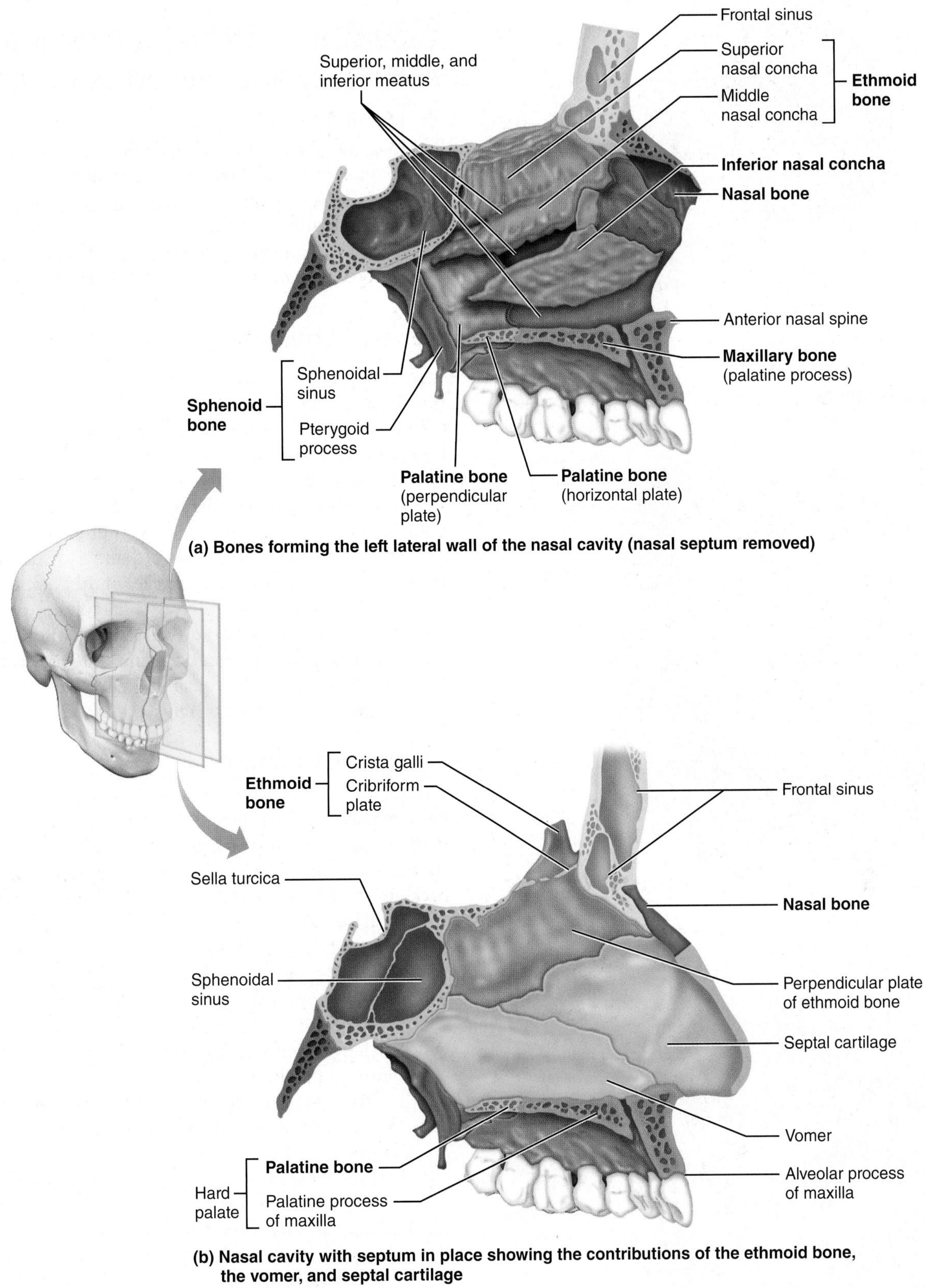

(a) Bones forming the left lateral wall of the nasal cavity (nasal septum removed)

(b) Nasal cavity with septum in place showing the contributions of the ethmoid bone, the vomer, and septal cartilage

Figure 7.14 Bones of the nasal cavity. (For a related image, see *A Brief Atlas of the Human Body*, Figure 15.)

Practice art labeling
MasteringA&P®>Study Area>Chapter 7

Name the process that leads to formation of most of the skull bones. Name the embryonic connective tissue that is converted to bone during this process.

For answers, see Answers Appendix.

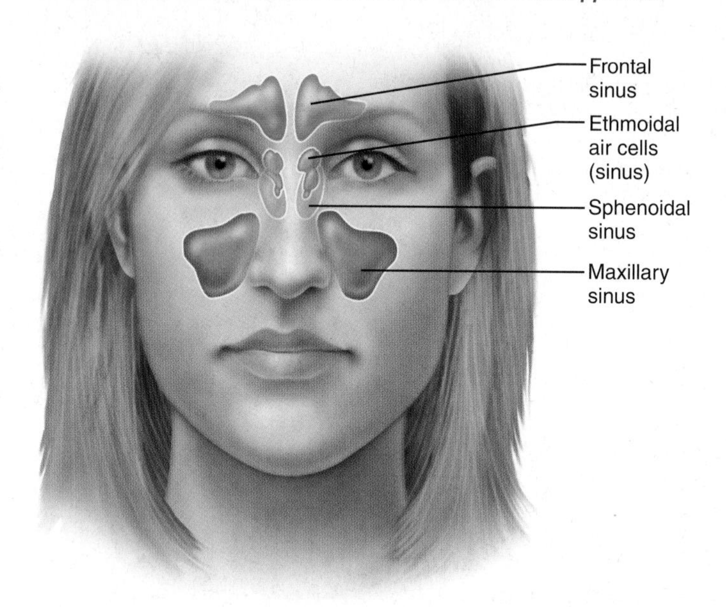

(a) Anterior aspect

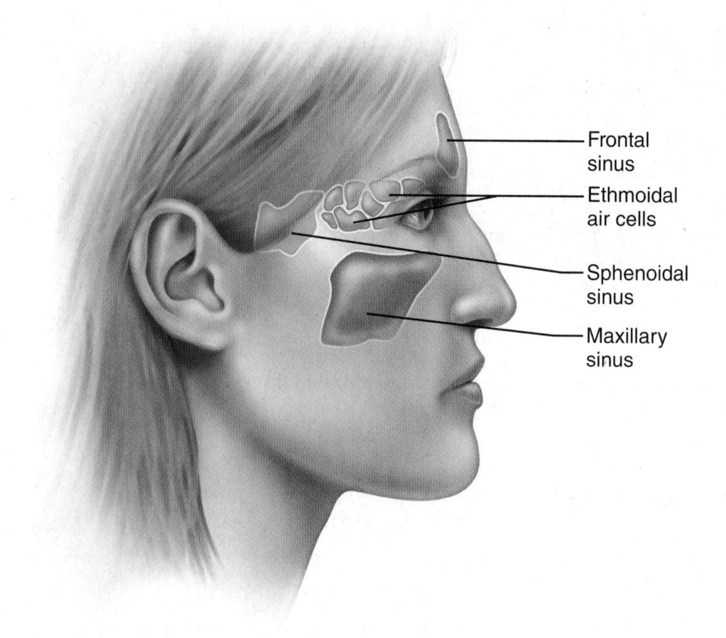

(b) Medial aspect

Figure 7.15 Paranasal sinuses.

7.2 The vertebral column is a flexible, curved support structure

→ Learning Objectives

- ☐ **Describe the structure of the vertebral column, list its components, and describe its curvatures.**
- ☐ **Indicate a common function of the spinal curvatures and the intervertebral discs.**
- ☐ **Discuss the structure of a typical vertebra and describe regional features of cervical, thoracic, and lumbar vertebrae.**

General Characteristics

Some people think of the **vertebral column** as a rigid supporting rod, but this is inaccurate. Also called the **spine** or **spinal column**, the vertebral column consists of 26 irregular bones connected in such a way that a flexible, curved structure results (**Figure 7.16**, p. 193).

The spine extends from the skull to the pelvis, where it transmits the weight of the trunk to the lower limbs. It also surrounds and protects the delicate spinal cord and provides attachment points for the ribs and for the muscles of the back and neck.

In the fetus and infant, the vertebral column consists of 33 separate bones, or **vertebrae** (ver′tĕ-bre). Inferiorly, nine of these eventually fuse to form two composite bones, the sacrum and the tiny coccyx. The remaining 24 bones persist as individual vertebrae separated by intervertebral discs.

Regions and Curvatures

The adult vertebral column is about 70 cm (28 inches) long and has five major regions (Figure 7.16). The seven vertebrae of the neck are the **cervical vertebrae** (ser′vĭ-kal), the next 12 are the **thoracic vertebrae** (tho-ras′ik), and the five supporting the lower back are the **lumbar vertebrae** (lum′bar). Remembering common meal times—7 AM, 12 noon, and 5 PM—will help you recall the number of bones in these three regions of the spine. The vertebrae become progressively larger from the cervical to the lumbar region, as they must support greater and greater weight.

Inferior to the lumbar vertebrae is the **sacrum** (sa′krum), which articulates with the hip bones. The terminus of the vertebral column is the tiny **coccyx** (kok′siks).

All of us have the same number of cervical vertebrae. Variations in numbers of vertebrae in other regions occur in about 5% of people.

When you view the vertebral column from the side, you can see the four curvatures that give it its S, or sinusoid, shape. The **cervical** and **lumbar curvatures** are concave posteriorly; the **thoracic** and **sacral curvatures** are convex posteriorly. These curvatures increase the resilience and flexibility of the spine, allowing it to function like a spring rather than a rigid rod.

Ligaments

Like a tall, tremulous TV transmitting tower, the vertebral column cannot stand upright by itself. It must be held in place by an elaborate system of cable-like supports. In the case of the vertebral column, straplike ligaments and the trunk muscles assume this role.

Table 7.1 Bones of the Skull

VIEW OF SKULL	BONE WITH COMMENTS*	IMPORTANT MARKINGS
	Cranial bones	
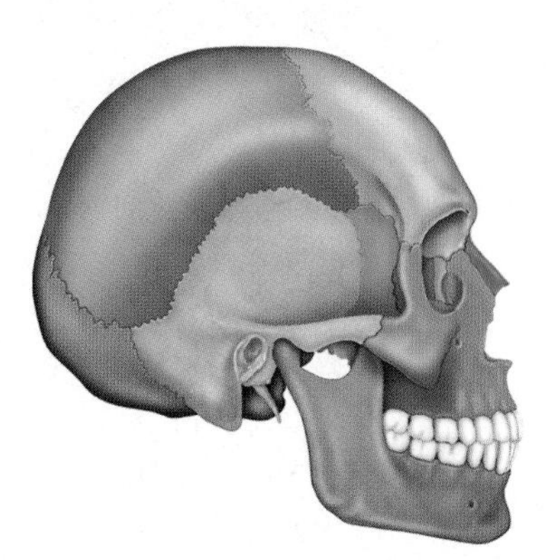 *Lateral view of skull (Figure 7.5)*	**Frontal (1)** Forms forehead, superior part of orbits, and most of the anterior cranial fossa; contains sinuses	**Supraorbital foramina (notches):** passageway for the supraorbital arteries and nerves
	Parietal (2) Form most of the superior and lateral aspects of the skull	
	Occipital (1) Forms posterior aspect and most of the base of the skull	**Foramen magnum:** allows passage of the spinal cord from the brain stem to the vertebral canal **Hypoglossal canals:** passageway for the hypoglossal nerve (cranial nerve XII) **Occipital condyles:** articulate with the atlas (first vertebra) **External occipital protuberance** and **nuchal lines:** sites of muscle attachment **External occipital crest:** attachment site of ligamentum nuchae
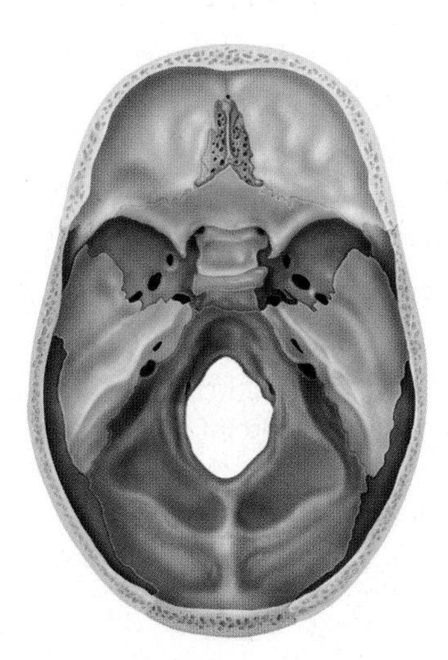 *Superior view of skull, calvaria removed (Figure 7.7)*	**Temporal (2)** Form inferolateral aspects of the skull and contribute to the middle cranial fossa; have squamous, tympanic, and petrous parts	**Zygomatic process:** contributes to the zygomatic arch, which forms the prominence of the cheek **Mandibular fossa:** articular point for the condylar process of the mandible **External acoustic meatus:** canal leading from the external ear to the eardrum **Styloid process:** attachment site for several neck and tongue muscles and for a ligament to the hyoid bone **Mastoid process:** attachment site for several neck muscles **Stylomastoid foramen:** passageway for cranial nerve VII (facial nerve) **Jugular foramen:** passageway for the internal jugular vein and cranial nerves IX, X, and XI **Internal acoustic meatus:** passageway for cranial nerves VII and VIII **Carotid canal:** passageway for the internal carotid artery
	Sphenoid (1) Keystone of the cranium; contributes to the middle cranial fossa and orbits; main parts are the body, greater wings, lesser wings, and pterygoid processes	**Sella turcica:** hypophyseal fossa portion is the seat of the pituitary gland **Optic canals:** passageway for cranial nerve II and the ophthalmic arteries **Superior orbital fissures:** passageway for cranial nerves III, IV, VI, part of V (ophthalmic division), and ophthalmic vein **Foramen rotundum (2):** passageway for the maxillary division of cranial nerve V **Foramen ovale (2):** passageway for the mandibular division of cranial nerve V **Foramen spinosum (2):** passageway for the middle meningeal artery
	Ethmoid (1) Small contribution to the anterior cranial fossa; forms part of the nasal septum and the lateral walls and roof of the nasal cavity; contributes to the medial wall of the orbit	**Crista galli:** attachment point for the falx cerebri, a dural membrane fold **Cribriform plates:** passageways for filaments of the olfactory nerves (cranial nerve I) **Superior and middle nasal conchae:** form part of lateral walls of nasal cavity; increase turbulence of air flow

➤

Table 7.1 Bones of the Skull *(continued)*

VIEW OF SKULL	BONE WITH COMMENTS*	IMPORTANT MARKINGS
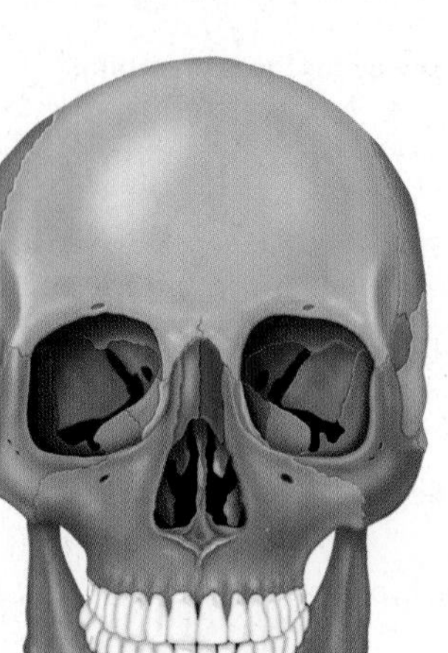 *Anterior view of skull (Figure 7.4)*	**Facial bones**	
	Nasal (2) Form the bridge of the nose	
	Lacrimal (2) Form part of the medial orbit wall	**Lacrimal fossa:** houses the lacrimal sac, which helps to drain tears into the nasal cavity
	Zygomatic (2) Form the cheek and part of the orbit	
	Inferior nasal concha (2) Form part of the lateral walls of the nasal cavity	
	Mandible (1) The lower jaw	**Coronoid processes:** insertion points for the temporalis muscles **Condylar processes:** articulate with the temporal bones to form the jaw (temporomandibular) joints **Mandibular symphysis:** medial fusion point of the mandibular bones **Dental alveoli:** sockets for the teeth **Mandibular foramina:** passageway for the inferior alveolar nerves **Mental foramina:** passageway for blood vessels and nerves to the chin and lower lip
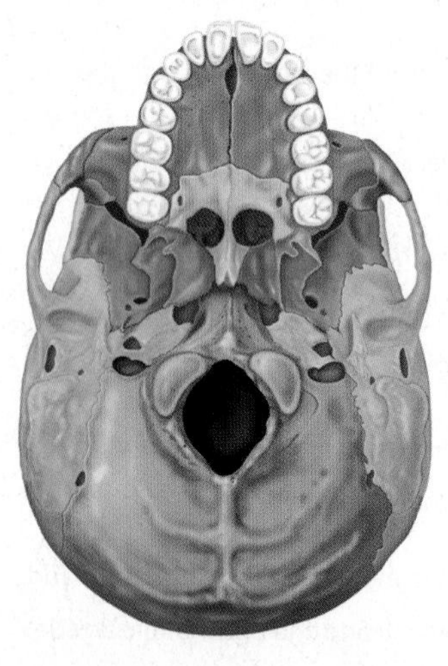 *Inferior view of skull, mandible removed (Figure 7.6)*	**Maxilla (2)** Keystone bones of the face; form the upper jaw and parts of the hard palate, orbits, and nasal cavity walls	**Dental alveoli:** sockets for teeth **Zygomatic process:** helps form the zygomatic arches **Palatine process:** forms the anterior hard palate; the two processes meet medially in the intermaxillary suture **Frontal process:** forms part of lateral aspect of bridge of nose **Incisive fossa** and **incisive canal:** passageway for blood vessels and nerves through anterior hard palate (fused palatine processes) **Inferior orbital fissure:** passageway for maxillary branch of cranial nerve V, the zygomatic nerve, and blood vessels **Infraorbital foramen:** passageway for infraorbital nerve to skin of face
	Palatine (2) Form posterior part of the hard palate and a small part of nasal cavity and orbit walls	
	Vomer (1) Inferior part of the nasal septum	
	Auditory ossicles (malleus, incus, and stapes) (2 each) Found in middle ear cavity; involved in sound transmission (see Chapter 13)	

*The color code beside each bone name corresponds to the bone's color in the illustrations (see Figures 7.4 to 7.14). The number in parentheses ( ) following the bone name indicates the total number of such bones in the body.

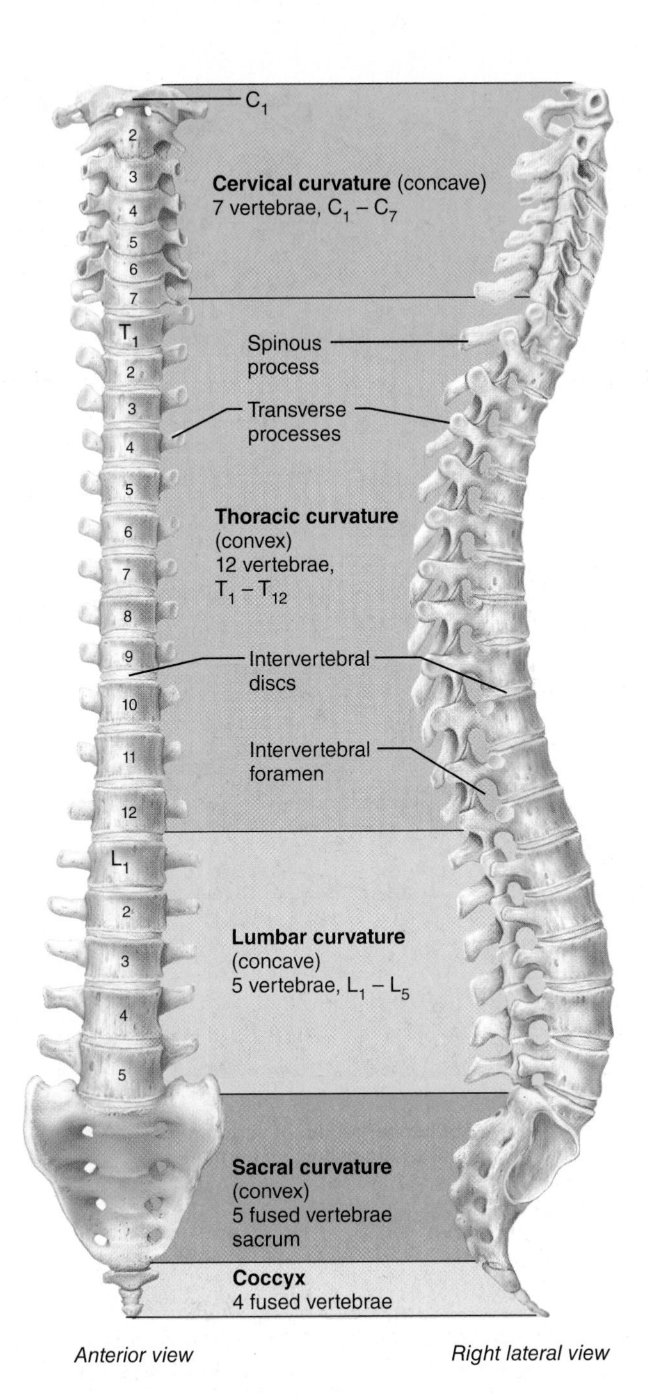

Figure 7.16 The vertebral column. Notice the curvatures in the lateral view. (The terms convex and concave refer to the curvature of the posterior aspect of the vertebral column.) (For a related image, see *A Brief Atlas of the Human Body,* Figure 17.)

The major supporting ligaments are the **anterior** and **posterior longitudinal ligaments** (Figure 7.17). These run as continuous bands down the front and back surfaces of the vertebrae from the neck to the sacrum. The broad anterior ligament is strongly attached to both the bony vertebrae and the discs. Along with its supporting role, it prevents hyperextension of the spine (bending too far backward). The posterior ligament, which resists hyperflexion of the spine (bending too far forward), is narrow and relatively weak. It attaches only to the discs. However, the **ligamentum flavum**, which connects adjacent vertebrae, contains elastic connective tissue and is especially strong. It stretches as we bend forward and then recoils when we resume an erect posture. Short ligaments connect each vertebra to those immediately above and below.

Intervertebral Discs

Each **intervertebral disc** is a cushionlike pad composed of two parts. The inner gelatinous **nucleus pulposus** (pul-po′sus; "pulp") acts like a rubber ball, giving the disc its elasticity and compressibility. Surrounding the nucleus pulposus is a strong collar composed of collagen fibers superficially and fibrocartilage internally, the **anulus fibrosus** (an′u-lus fi-bro′sus; "ring of fibers") (Figure 7.17a, c). The anulus fibrosus limits the expansion of the nucleus pulposus when the spine is compressed. It also acts like a woven strap to bind successive vertebrae together, withstands twisting forces, and resists tension in the spine.

Sandwiched between the bodies of neighboring vertebrae, the intervertebral discs act as shock absorbers during walking, jumping, and running. They allow the spine to flex and extend, and to a lesser extent to bend laterally. At points of compression, the discs flatten and bulge out a bit between the vertebrae. The discs are thickest in the lumbar and cervical regions, which enhances the flexibility of these regions.

Collectively the discs account for about 25% of the height of the vertebral column. They flatten somewhat during the course of the day, so we are always a few millimeters shorter at night than when we awake in the morning.

HOMEOSTATIC IMBALANCE 7.2 — CLINICAL

Severe or sudden physical trauma to the spine—for example, from bending forward while lifting a heavy object—may result in herniation of one or more discs. A **herniated (prolapsed) disc** (commonly called a *slipped disc*) usually involves rupture of the anulus fibrosus followed by protrusion of the spongy nucleus pulposus through the anulus (Figure 7.17c, d). If the protrusion presses on the spinal cord or on spinal nerves exiting from the cord, numbness or excruciating pain may result.

Herniated discs are generally treated with moderate exercise, massage, heat therapy, and painkillers. If this fails, the protruding disc may be removed surgically and a bone graft done to fuse the adjoining vertebrae. Another option is an outpatient procedure called percutaneous laser disc decompression, which involves vaporizing part of the disc with a laser. If necessary, tears in the anulus can be sealed by electrothermal means at the same time. ✚

HOMEOSTATIC IMBALANCE 7.3 — CLINICAL

There are several types of abnormal spinal curvatures (Figure 7.18). Some are congenital (present at birth); others result from disease, poor posture, or unequal muscle pull on the spine. *Scoliosis* (sko″le-o′sis), literally, "twisted disease," is an abnormal rotation of the spine that results in a *lateral* curvature, most often in the thoracic region. It is quite common during late childhood, particularly in girls. Other, more severe

Supraspinous ligament
Transverse process
Sectioned spinous process
Ligamentum flavum
Interspinous ligament
Inferior articular process
Intervertebral disc
Anterior longitudinal ligament
Intervertebral foramen
Posterior longitudinal ligament
Anulus fibrosus
Nucleus pulposus
Sectioned body of vertebra

(a) Median section of three vertebrae

Posterior longitudinal ligament
Anterior longitudinal ligament
Body of a vertebra
Intervertebral disc

(b) Anterior view of part of the spinal column

Vertebral spinous process (posterior aspect of vertebra)
Spinal nerve root
Transverse process
Herniated portion of disc
Anulus fibrosus of disc
Spinal cord
Nucleus pulposus of disc

(c) Superior view of a herniated intervertebral disc

Nucleus pulposus of intact disc
Herniated nucleus pulposus

(d) MRI of lumbar region of vertebral column in sagittal section showing herniated disc

Figure 7.17 Ligaments and fibrocartilage discs uniting the vertebrae.

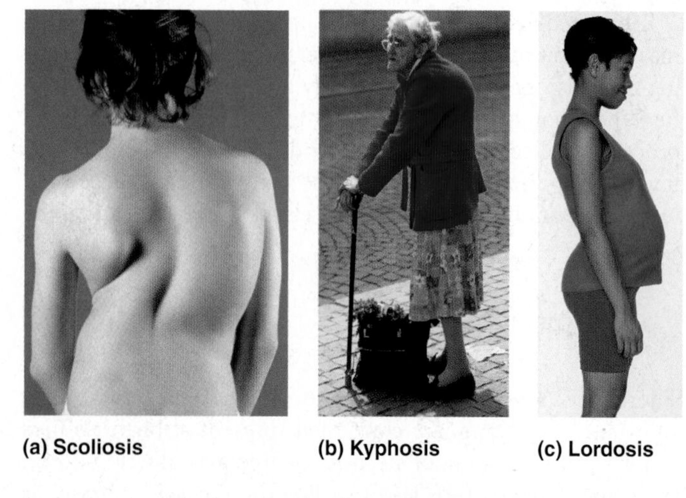

(a) Scoliosis **(b) Kyphosis** **(c) Lordosis**

Figure 7.18 Abnormal spinal curvatures.

cases result from abnormal vertebral structure, lower limbs of unequal length, or muscle paralysis. If muscles on one side of the body are nonfunctional, those of the opposite side exert an unopposed pull on the spine and force it out of alignment. Scoliosis is treated (with body braces or surgically) before growth ends to prevent permanent deformity and breathing difficulties due to a compressed lung.

Kyphosis (ki-fo′sis), or hunchback, is a *dorsally* exaggerated *thoracic* curvature. It is particularly common in elderly people because of osteoporosis, but may also reflect tuberculosis of the spine, rickets, or osteomalacia.

Lordosis, or swayback, is an accentuated *lumbar* curvature. It, too, can result from spinal tuberculosis or osteomalacia. Temporary lordosis is common in those carrying a large load up front, such as men with "potbellies" and pregnant women. In an attempt to maintain their center of gravity, these individuals automatically throw back their shoulders, accentuating their lumbar curvature. ✚

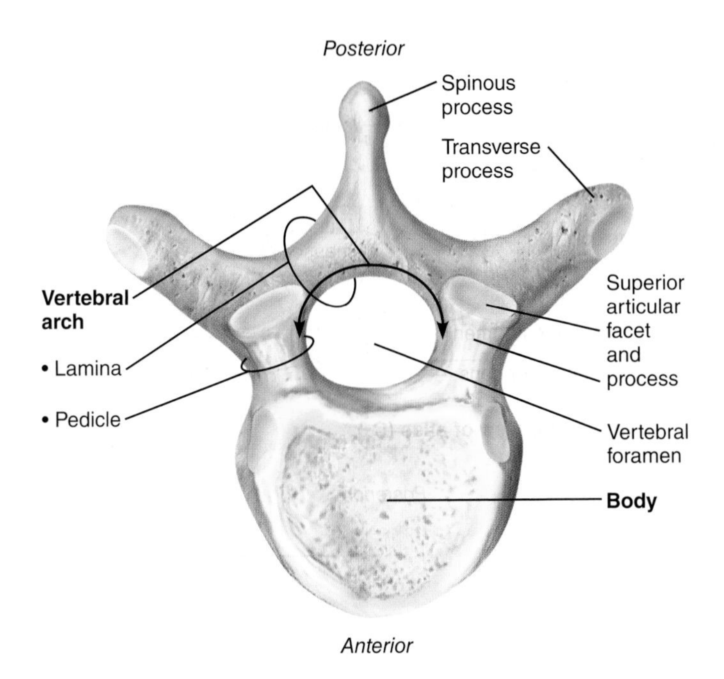

Figure 7.19 Typical vertebral structures. Superior view of a thoracic vertebra. Only bone features are illustrated in this and subsequent bone figures in this chapter. Articular cartilage is not depicted.

General Structure of Vertebrae

All vertebrae have a common structural pattern (**Figure 7.19**). Each vertebra consists of a **body**, or *centrum*, anteriorly and a **vertebral arch** posteriorly. The disc-shaped body is the weight-bearing region. Together, the body and vertebral arch enclose an opening called the **vertebral foramen**. Successive vertebral foramina of the articulated vertebrae form the long **vertebral canal**, through which the spinal cord passes.

The vertebral arch is a composite structure formed by two pedicles and two laminae. The **pedicles** (ped′ĭ-kelz; "little feet"), short bony pillars projecting posteriorly from the vertebral body, form the sides of the arch. The **laminae** (lam′ĭ-ne), flattened plates that fuse in the median plane, complete the arch posteriorly. The pedicles have notches on their superior and inferior borders, providing lateral openings between adjacent vertebrae called **intervertebral foramina** (see Figure 7.16). The spinal nerves issuing from the spinal cord pass through these foramina.

Seven processes project from the vertebral arch. The **spinous process** is a median posterior projection arising at the junction of the two laminae. A **transverse process** extends laterally from each side of the vertebral arch. The spinous and transverse processes are attachment sites for muscles that move the vertebral column and for ligaments that stabilize it. The paired **superior** and **inferior articular processes** protrude superiorly and inferiorly, respectively, from the pedicle-lamina junctions. The smooth joint surfaces of the articular processes, called *facets* ("little faces"), are covered with hyaline cartilage. The inferior articular processes of each vertebra form movable joints with the superior articular processes of the vertebra immediately below. Thus, successive vertebrae join both at their bodies and at their articular processes.

Regional Vertebral Characteristics

Beyond their common structural features, vertebrae exhibit variations that allow different regions of the spine to perform slightly different functions and movements. In general, movements that can occur between vertebrae are (1) flexion and extension (anterior bending and posterior straightening of the spine), (2) lateral flexion (bending the *upper body* to the right or left), and (3) rotation (in which vertebrae rotate on one another in the longitudinal axis of the spine). The regional vertebral characteristics are illustrated and summarized in **Table 7.2** on p. 198.

Cervical Vertebrae

The seven cervical vertebrae, identified as C_1–C_7, are the smallest, lightest vertebrae (see Figure 7.16). The "typical" cervical vertebrae (C_3–C_7) have the following distinguishing features (see Figure 7.21 and Table 7.2):

- The body is oval—wider from side to side than in the anteroposterior dimension.
- Except in C_7, the spinous process is short, projects directly back, and is *bifid* (bi′fid), or split at its tip.
- The vertebral foramen is large and generally triangular.
- Each transverse process contains a **transverse foramen** through which the vertebral arteries pass to service the brain.

The spinous process of C_7 is not bifid and is much larger than those of the other cervical vertebrae (see Figure 7.21a). Because its spinous process is palpable through the skin, C_7 can be used as a landmark for counting the vertebrae and is called the **vertebra prominens** ("prominent vertebra").

The first two cervical vertebrae, the atlas and the axis, are somewhat more robust than the typical cervical vertebra. They have no intervertebral disc between them, and they are highly modified, reflecting their special functions. The **atlas** (C_1) has no body and no spinous process (**Figure 7.20a** and **b**). Essentially, it is a ring of bone consisting of *anterior* and *posterior arches* and a *lateral mass* on each side. Each lateral mass has articular facets on both its superior and inferior surfaces. The superior articular facets receive the occipital condyles of the skull—they "carry" the skull, just as Atlas supported the heavens in Greek mythology. These joints allow you to nod your head "yes." The inferior articular facets form joints with the axis (C_2) below.

The **axis**, which has a body and the other typical vertebral processes, is not as specialized as the atlas. In fact, its only unusual feature is the knoblike **dens** (denz; "tooth") projecting superiorly from its body. The dens is actually the "missing" body of the atlas, which fuses with the axis during embryonic development. Cradled in the anterior arch of the atlas by the transverse ligament (**Figure 7.21a**), the dens acts as a pivot for the rotation of the atlas. Hence, this joint allows you to rotate your head from side to side to indicate "no."

Thoracic Vertebrae

Of the 12 thoracic vertebrae (T_1–T_{12}), the first looks much like C_7, and the last four show a progression toward lumbar vertebral structure (see Table 7.2, Figure 7.16, and Figure 7.21b). The

Table 7.2 Regional Characteristics of Cervical, Thoracic, and Lumbar Vertebrae

CHARACTERISTIC	CERVICAL (3–7)	THORACIC	LUMBAR
Body	Small, oval, wide side to side	Larger than cervical; heart shaped; bears two costal facets	Massive; kidney shaped
Spinous process	Short; bifid (except C_7); projects directly posterior	Long; sharp; projects inferiorly	Short; blunt; rectangular; projects directly posteriorly
Vertebral foramen	Triangular, large	Circular	Triangular
Transverse processes	Contain foramina	Bear facets for ribs (except T_{11} and T_{12})	Thin and tapered
Superior and inferior articular processes	Superior facets directed superoposteriorly	Superior facets directed posteriorly	Superior facets directed posteromedially (or medially)
	Inferior facets directed inferoanteriorly	Inferior facets directed anteriorly	Inferior facets directed anterolaterally (or laterally)
Movements allowed	Flexion and extension; lateral flexion; rotation; the spine region with the greatest range of movement	Rotation; lateral flexion possible but restricted by ribs; flexion and extension limited	Flexion and extension; some lateral flexion; rotation prevented

Superior View

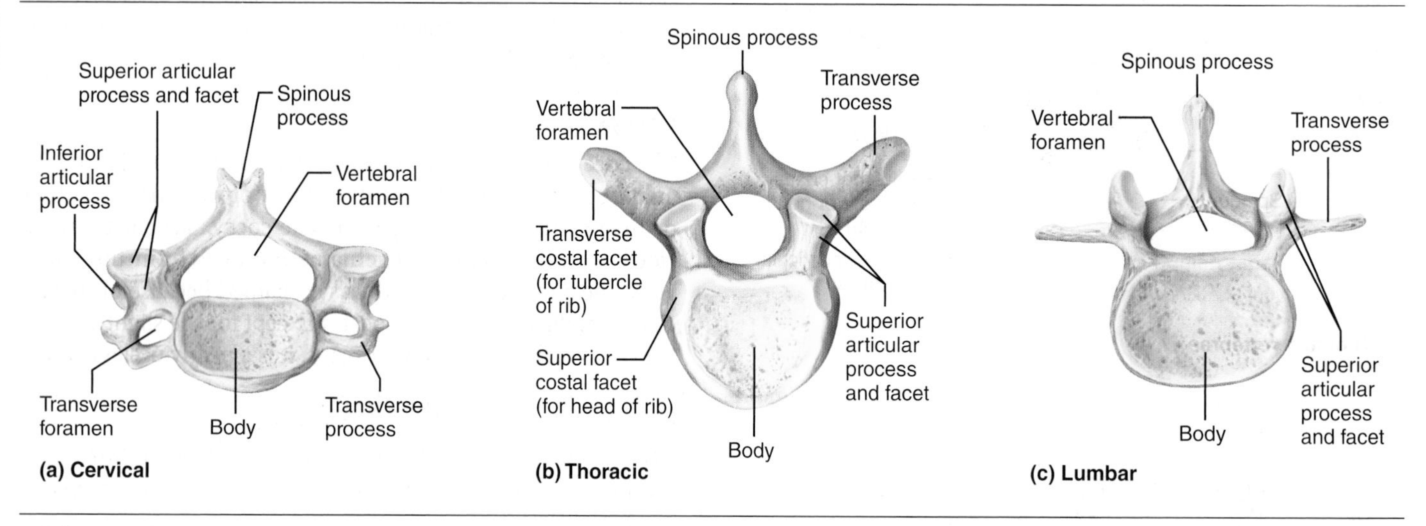

(a) Cervical **(b) Thoracic** **(c) Lumbar**

Right Lateral View

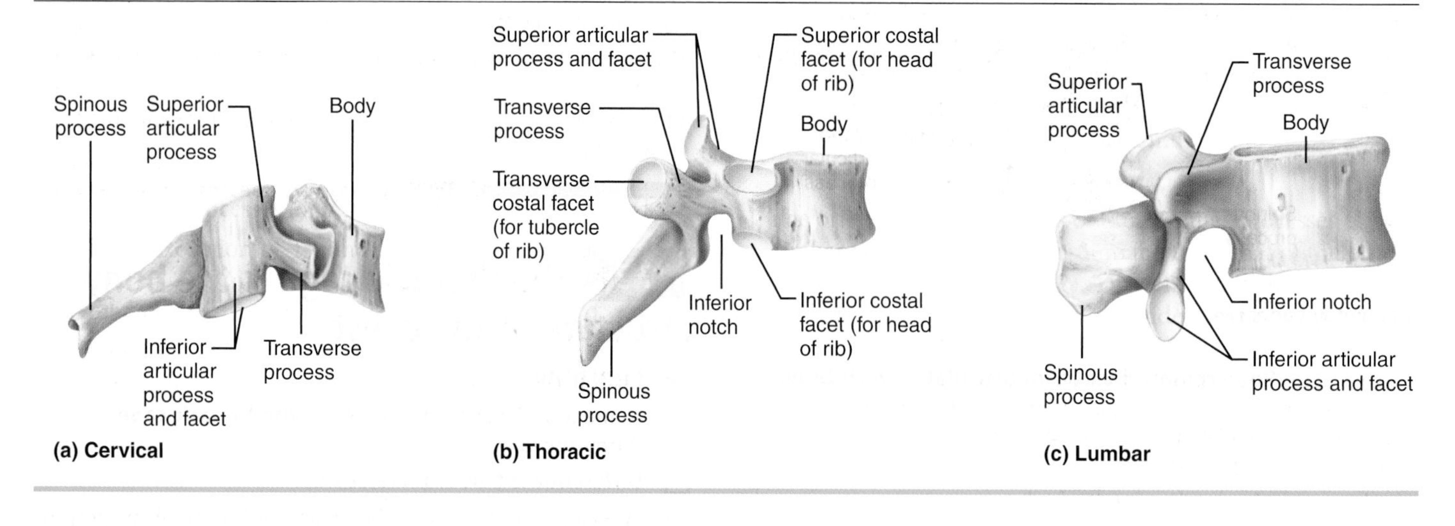

(a) Cervical **(b) Thoracic** **(c) Lumbar**

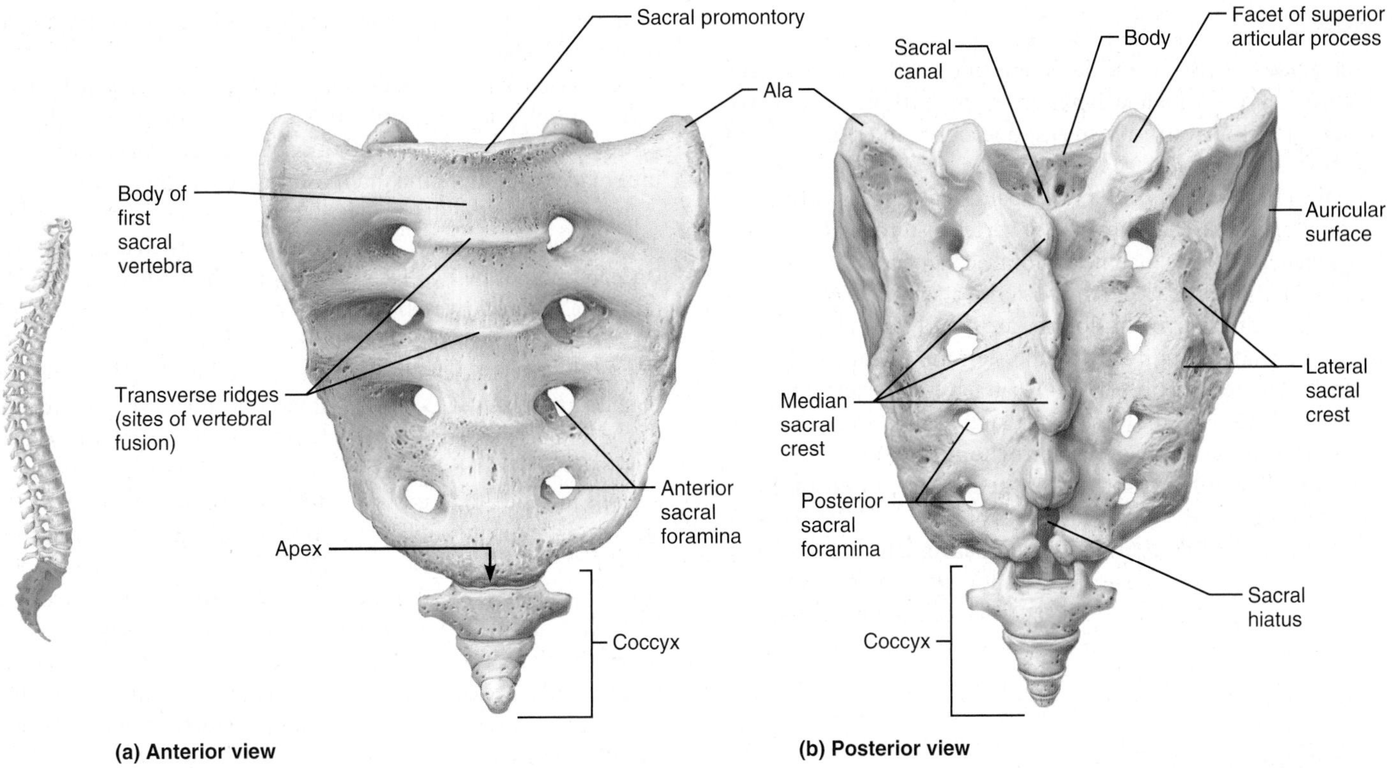

(a) Anterior view

(b) Posterior view

Figure 7.22 The sacrum and coccyx. (For a related image, see *A Brief Atlas of the Human Body*, Figure 22.)

the ribs laterally, and the sternum and costal cartilages anteriorly. The costal cartilages secure the ribs to the sternum (**Figure 7.23a**).

Roughly cone shaped with its broad dimension positioned inferiorly, the bony thorax forms a protective cage around the vital organs of the thoracic cavity (heart, lungs, and great blood vessels), supports the shoulder girdles and upper limbs, and provides attachment points for many muscles of the neck, back, chest, and shoulders. The *intercostal spaces* between the ribs are occupied by the intercostal muscles, which lift and then depress the thorax during breathing.

Sternum

The **sternum** (breastbone) lies in the anterior midline of the thorax. Vaguely resembling a dagger, it is a flat bone approximately 15 cm (6 inches) long, resulting from the fusion of three bones: the manubrium, the body, and the xiphoid process. The *manubrium* (mah-nu′bre-um; "knife handle") is the superior portion, which is shaped like the knot in a necktie. The manubrium articulates via its **clavicular notches** (klah-vik′u-lar) with the clavicles (collarbones) laterally, and just below this, it also articulates with the first two pairs of ribs. The *body*, or midportion, forms the bulk of the sternum. The sides of the body are notched where it articulates with the costal cartilages of the second to seventh ribs. The *xiphoid process* (zif′oid; "swordlike") forms the inferior end of the sternum. This small, variably shaped process is a plate of hyaline cartilage in youth, but it is usually ossified in adults over the age of 40. The xiphoid process articulates only with the sternal body and serves as an attachment point for some abdominal muscles.

The sternum has three important anatomical landmarks: the jugular notch, the sternal angle, and the xiphisternal joint (Figure 7.23). The easily palpated **jugular** (*suprasternal*) **notch** is the central indentation in the superior border of the manubrium. If you slide your finger down the anterior surface of your neck, it will land in the jugular notch. The jugular notch is generally in line with the disc between the second and third thoracic vertebrae and the point where the left common carotid artery issues from the aorta (Figure 7.23b).

The **sternal angle** is felt as a horizontal ridge across the front of the sternum, where the manubrium joins the sternal body. This cartilaginous joint acts like a hinge, allowing the sternal body to swing anteriorly when we inhale. The sternal angle is in line with the disc between the fourth and fifth thoracic vertebrae and at the level of the second pair of ribs. It is a handy reference point for finding the second rib and thus for counting the ribs during a physical examination and for listening to sounds made by specific heart valves.

The **xiphisternal joint** (zif″ĭ-ster′nul) is the point where the sternal body and xiphoid process fuse. It lies at the level of the ninth thoracic vertebra. The heart lies on the diaphragm just deep to this joint.

In some people, the xiphoid process projects posteriorly. In such cases, **chest trauma** consisting of a blow at the level of the xiphoid process can push the process into the underlying heart or liver, causing massive hemorrhage. +

Ribs

Twelve pairs of **ribs** form the flaring sides of the thoracic cage (Figure 7.23a). All ribs attach posteriorly to the thoracic vertebrae (bodies and transverse processes) and curve inferiorly toward the anterior body surface. The superior seven rib pairs attach directly to the sternum by individual costal cartilages (bars of hyaline cartilage). These are **true** or **vertebrosternal ribs** (ver″tě-bro-ster′nal). (Notice that the anatomical name indicates the two attachment points of a rib—the posterior attachment given first.)

The remaining five pairs of ribs are called **false ribs** because they either attach indirectly to the sternum or entirely lack a sternal attachment. Rib pairs 8–10 attach to the sternum indirectly, each joining the costal cartilage immediately above it. These ribs are also called **vertebrochondral ribs** (ver″tě-bro-kon′dral). The inferior margin of the rib cage, or **costal margin**, is formed by the costal cartilages of ribs 7–10. Rib pairs 11 and 12 are called **vertebral ribs** or **floating ribs** because they have no anterior attachments. Instead, their costal cartilages lie embedded in the muscles of the lateral body wall.

The ribs increase in length from pair 1 to pair 7, then decrease in length from pair 8 to pair 12. Except for the first rib, which lies deep to the clavicle, the ribs are easily felt in people of normal weight.

A typical rib is a bowed flat bone (**Figure 7.24**). The bulk of a rib is simply called the *shaft*. Its superior border is smooth, but its inferior border is sharp and thin and has a *costal groove* on its inner face that lodges the intercostal nerves and blood vessels.

In addition to the shaft, each rib has a head, neck, and tubercle. The wedge-shaped *head*, the posterior end, articulates with the vertebral bodies by two facets: One joins the body of the same-numbered thoracic vertebra, the other articulates with the body of the vertebra immediately superior. The *neck* is the constricted portion of the rib just beyond the head. Lateral to this, the knoblike *tubercle* articulates with the costal facet of the transverse process of the same-numbered thoracic vertebra. Beyond the tubercle, the shaft angles sharply forward (at the angle of the rib) and then extends to attach to its costal cartilage anteriorly. The costal cartilages provide secure but flexible rib attachments to the sternum.

The first pair of ribs is atypical. They are flattened superiorly to inferiorly and are quite broad, forming a horizontal table that supports the subclavian blood vessels that serve the upper limbs. There are also other exceptions to the typical rib pattern. Rib 1 and ribs 10–12 articulate with only one vertebral body, and ribs 11 and 12 do not articulate with a vertebral transverse process.

☑ Check Your Understanding

12. How does a true rib differ from a false rib?
13. What is the sternal angle and what is its clinical importance?
14. Besides the ribs and sternum, there is a third group of bones making up the thoracic cage. What is it?

For answers, see Answers Appendix.

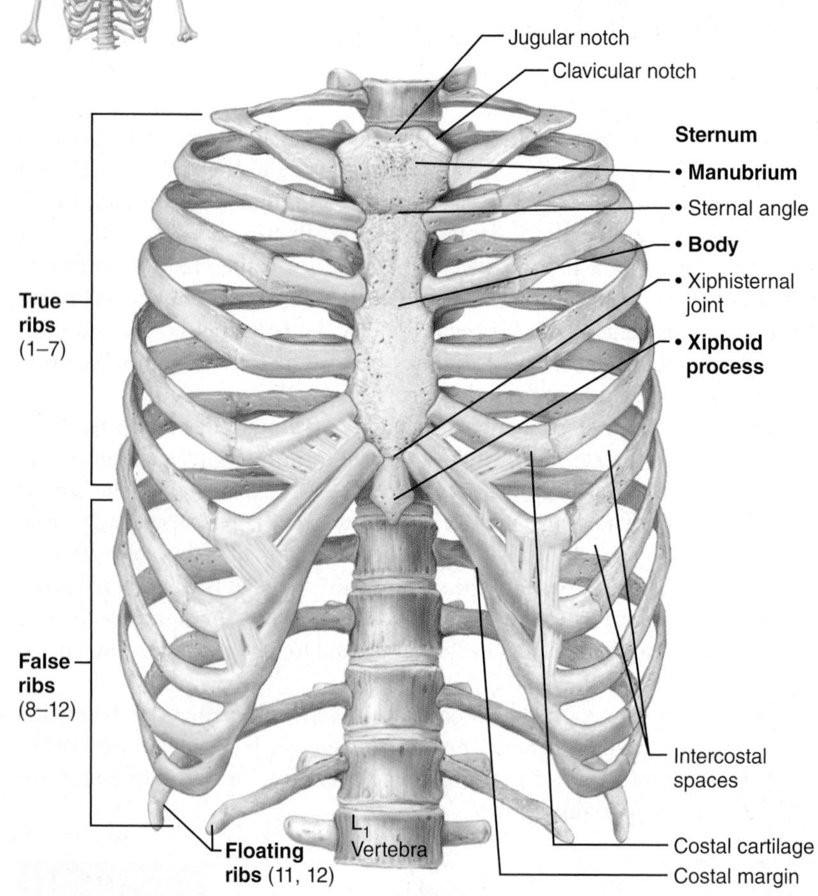

(a) Skeleton of the thoracic cage, anterior view

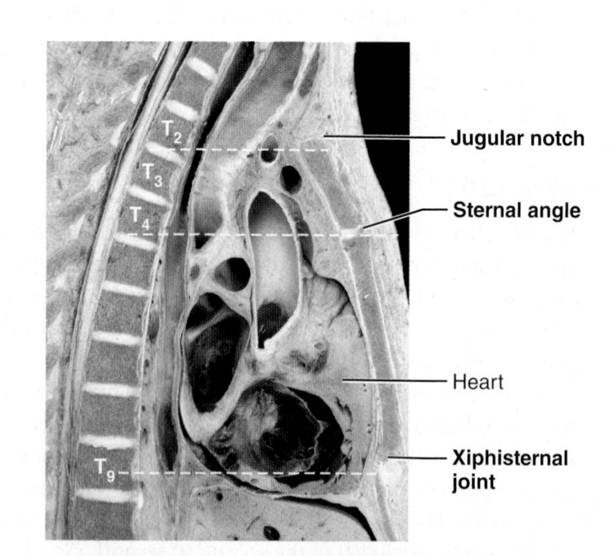

(b) Midsagittal section through the thorax, showing the relationship of surface anatomical landmarks of the thorax to the vertebral column

Figure 7.23 The thoracic cage. (For a related image, see *A Brief Atlas of the Human Body*, Figure 23a–d.)

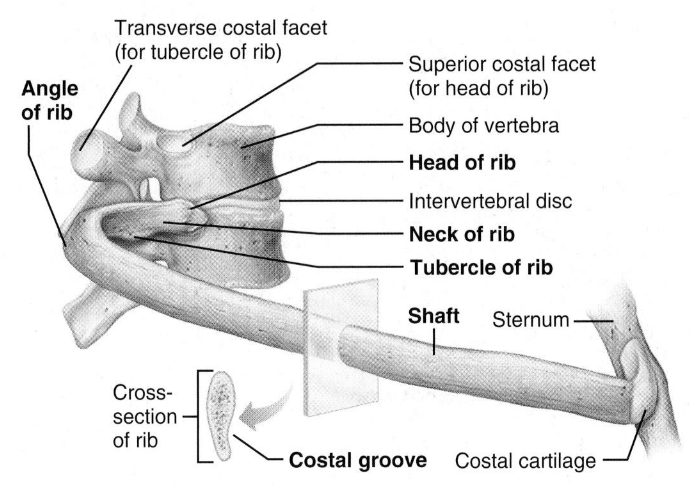

(a) Vertebral and sternal articulations of a typical true rib

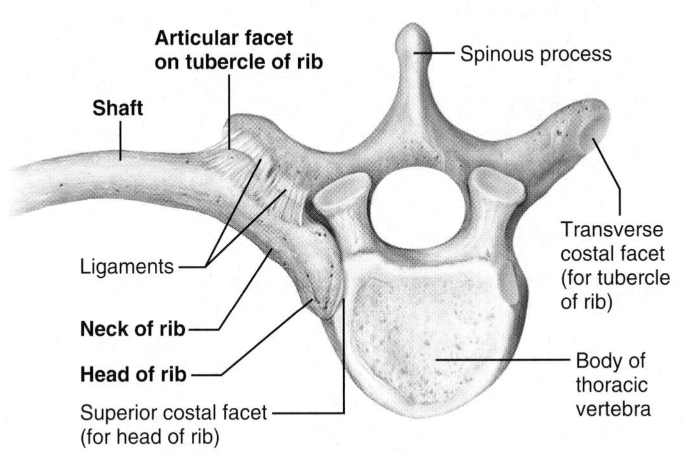

(b) Superior view of the articulation between a rib and a thoracic vertebra

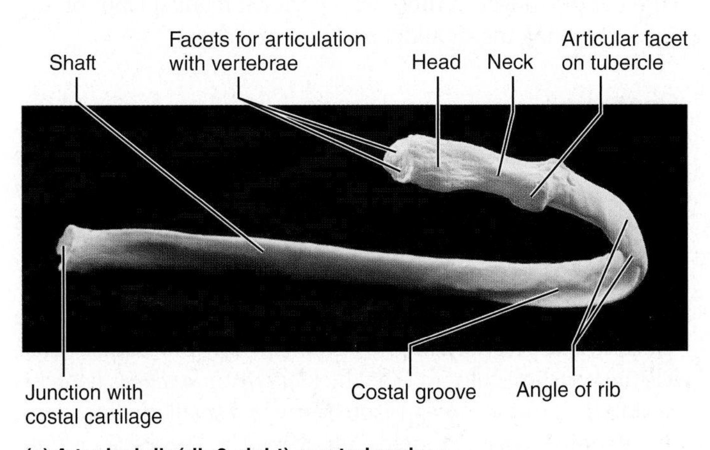

(c) A typical rib (rib 6, right), posterior view

Figure 7.24 Ribs. All ribs illustrated in this figure are right ribs. (For a related image, see *A Brief Atlas of the Human Body*, Figure 23e and f.)

Explore human cadaver
MasteringA&P®>Study Area>PAL

PART 2

THE APPENDICULAR SKELETON

Bones of the limbs and their girdles are collectively called the **appendicular skeleton** because they are appended to the axial skeleton (see Figure 7.1). The yokelike *pectoral girdles* (pek′tor-al; "chest") attach the upper limbs to the body trunk. The more sturdy *pelvic girdle* secures the lower limbs. Although the bones of the upper and lower limbs differ in their functions and mobility, they have the same fundamental plan: Each limb is composed of three major segments connected by movable joints.

The appendicular skeleton enables us to carry out the movements typical of our freewheeling and manipulative lifestyle. Each time we take a step, throw a ball, or pop a caramel into our mouth, we are making good use of our appendicular skeleton.

7.4 Each pectoral girdle consists of a clavicle and a scapula

→ **Learning Objectives**

- ☐ **Identify bones forming the pectoral girdle and relate their structure and arrangement to the function of this girdle.**
- ☐ **Identify important bone markings on the pectoral girdle.**

The **pectoral girdle**, or **shoulder girdle**, consists of the *clavicle* (klav′ĭ-kl) anteriorly and the *scapula* (skap′u-lah) posteriorly (**Figure 7.25** and Table 7.3 on p. 210). The paired pectoral girdles and their associated muscles form your shoulders. Although the term *girdle* usually signifies a beltlike structure encircling the body, a single pectoral girdle, or even the pair, does not quite satisfy this description. Anteriorly, the medial end of each clavicle joins the sternum; the distal ends of the clavicles meet the scapulae laterally. However, the scapulae fail to complete the ring posteriorly, because their medial borders do not join each other or the axial skeleton. Instead, the scapulae are attached to the thorax and vertebral column only by the muscles that clothe their surfaces.

The pectoral girdles attach the upper limbs to the axial skeleton and provide attachment points for many of the muscles that move the upper limbs. These girdles are very light and allow the upper limbs a degree of mobility not seen anywhere else in the body. This mobility is due to the following factors:

- Because only the clavicle attaches to the axial skeleton, the scapula can move quite freely across the thorax, allowing the arm to move with it.
- The socket of the shoulder joint (the scapula's glenoid cavity) is shallow and poorly reinforced, so it does not restrict the movement of the humerus (arm bone). Although this arrangement is good for flexibility, it is bad for stability—shoulder dislocations are fairly common.

Clavicles

The **clavicles** ("little keys"), or collarbones, are slender, S-shaped bones that can be felt along their entire course as they extend

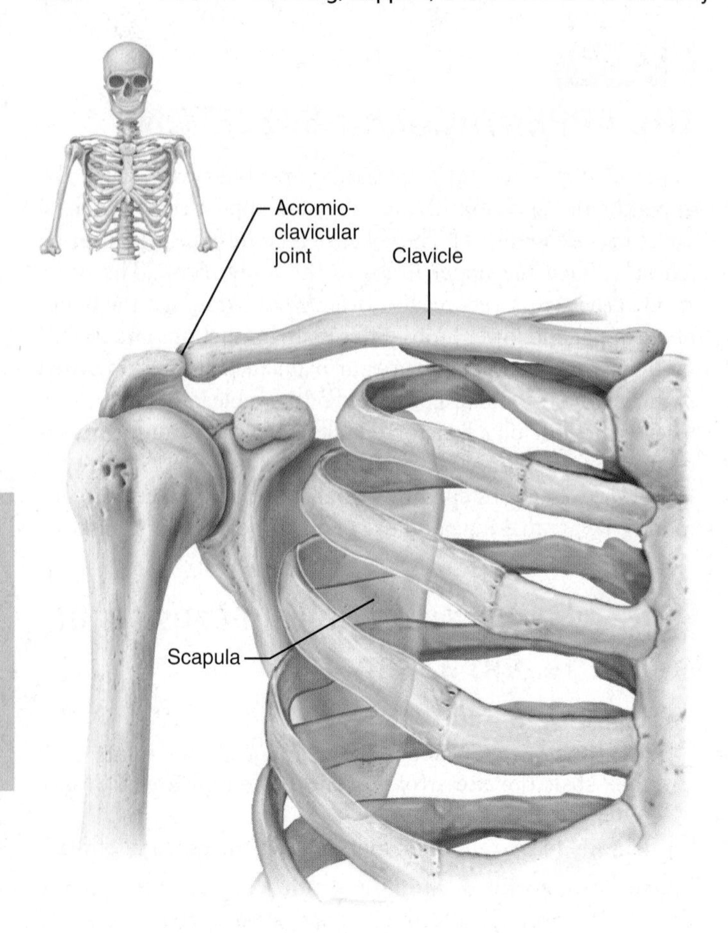

Figure 7.25 **The pectoral girdle with articulating bones.**

Practice art labeling
MasteringA&P®>Study Area>Chapter 7

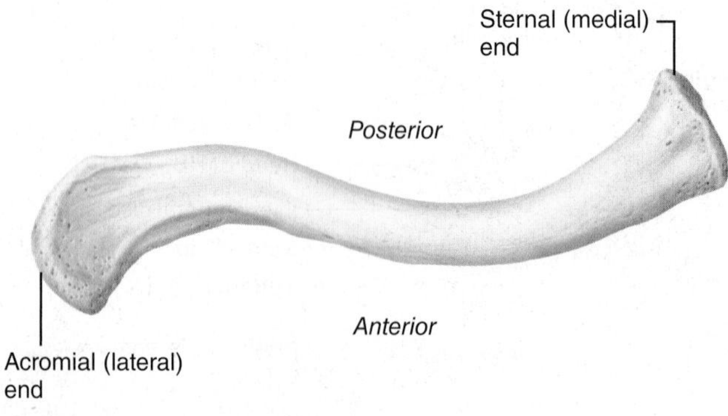

(a) Right clavicle, superior view

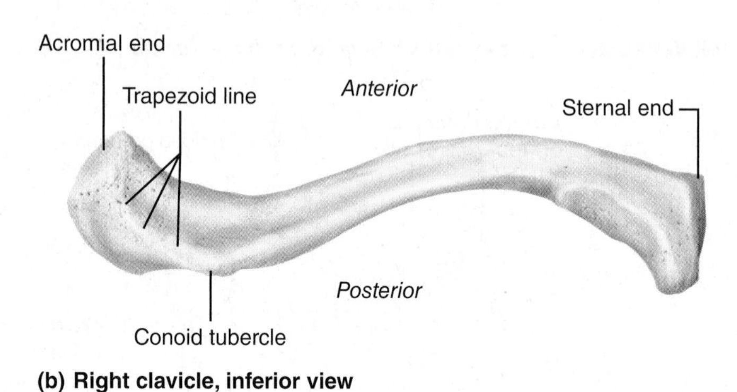

(b) Right clavicle, inferior view

Figure 7.26 **The clavicle.** (For a related image, see *A Brief Atlas of the Human Body*, Figure 24d.)

horizontally across the superior thorax (Figure 7.25). Besides anchoring many muscles, the clavicles act as braces: They hold the scapulae and arms out laterally, away from the narrower superior part of the thorax. This bracing function becomes obvious when a clavicle is fractured: The entire shoulder region collapses medially. The clavicles also transmit compression forces from the upper limbs to the axial skeleton, for example, when someone pushes a car to a gas station.

Each clavicle is cone shaped at its medial **sternal end**, which attaches to the sternal manubrium, and flattened at its lateral **acromial end** (ah-kro′me-al), which articulates with the scapula (**Figure 7.26**). The medial two-thirds of the clavicle is convex anteriorly; its lateral third is concave anteriorly. Its superior surface is fairly smooth, but the inferior surface is ridged and grooved by ligaments and by the action of the muscles that attach to it. The *trapezoid line* and the *conoid tubercle*, for example, are anchoring points for a ligament that connects the clavicle to the scapula.

The clavicles are not very strong and are likely to fracture, for example, when a person uses outstretched arms to break a fall. The curves in the clavicle ensure that it usually fractures anteriorly (outward). If it were to collapse posteriorly (inward), bone splinters would damage the subclavian artery, which passes just deep to the clavicle to serve the upper limb. The clavicles are exceptionally sensitive to muscle pull and become noticeably larger and stronger in those who perform manual labor or athletics involving the shoulder and arm muscles.

Scapulae

The **scapulae**, or *shoulder blades*, are thin, triangular flat bones (Figure 7.25 and **Figure 7.27**). Interestingly, their name derives from a word meaning "spade" or "shovel," for ancient cultures made spades from the shoulder blades of animals. The scapulae lie on the dorsal surface of the rib cage, between ribs 2 and 7.

Each scapula has three borders. The *superior border* is the shortest, sharpest border. The *medial*, or *vertebral*, *border* parallels the vertebral column. The thick *lateral*, or *axillary*, *border* is next to the armpit and ends superiorly in a small, shallow fossa, the **glenoid cavity** (gle′noid; "pit-shaped"). This cavity articulates with the humerus of the arm, forming the shoulder joint.

Like all triangles, the scapula has three corners or *angles*. The superior scapular border meets the medial border at the *superior angle* and the lateral border at the *lateral angle*. The medial and lateral borders join at the *inferior angle*. The inferior angle moves extensively as the arm is raised and lowered, and is an important landmark for studying scapular movements.

The anterior, or costal, surface of the scapula is concave and relatively featureless. Its posterior surface bears a prominent **spine** that is easily felt through the skin. The spine ends laterally in an enlarged, roughened triangular projection called the

Acromion
Suprascapular notch
Superior border
Coracoid process
Glenoid cavity
Superior angle
Lateral border
Subscapular fossa
Medial border
Inferior angle

(a) Right scapula, anterior aspect

Coracoid process
Suprascapular notch
Acromion
Superior angle
Supraspinous fossa
Spine
Glenoid cavity at lateral angle
Infraspinous fossa
Medial border
Lateral border

(b) Right scapula, posterior aspect

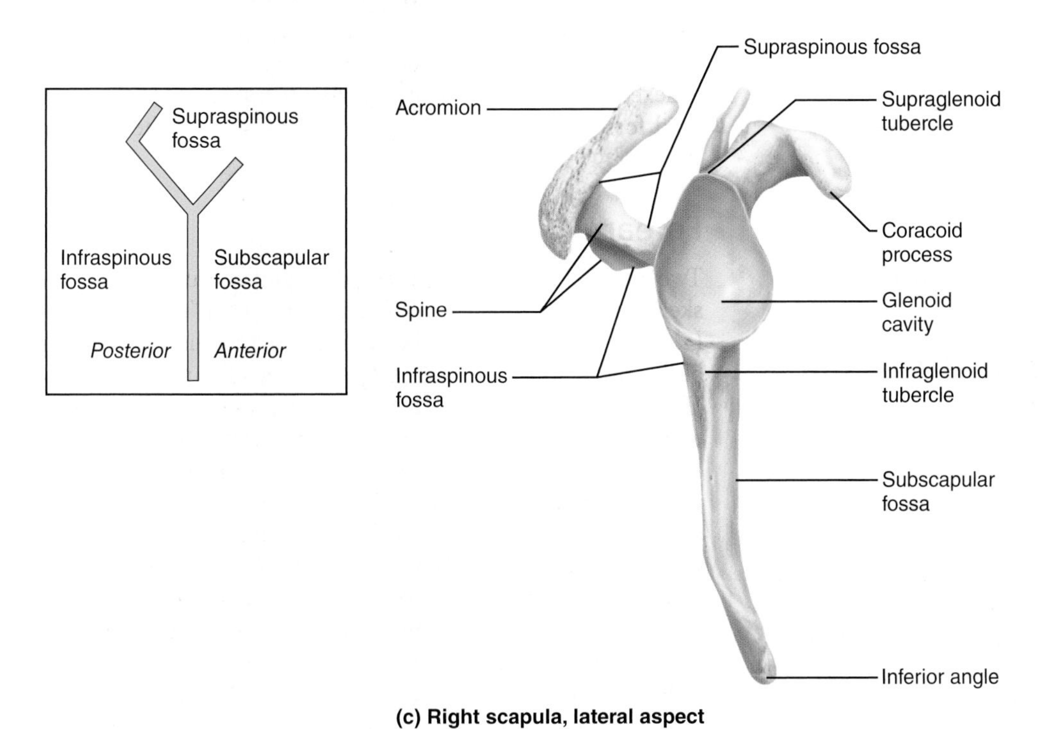

(c) Right scapula, lateral aspect

Figure 7.27 The scapula. View (c) is accompanied by a schematic representation of its orientation. (For a related image, see *A Brief Atlas of the Human Body*, Figure 24a–c, e.)

Practice art labeling
MasteringA&P®>Study Area>Chapter 7

Table 7.4 Comparison of the Male and Female Pelves

CHARACTERISTIC	FEMALE	MALE
General structure and functional modifications	Tilted forward; adapted for childbearing; true pelvis defines the birth canal; cavity of the true pelvis is broad, shallow, and has a greater capacity	Tilted less far forward; adapted for support of a male's heavier build and stronger muscles; cavity of the true pelvis is narrow and deep
Bone thickness	Less; bones lighter, thinner, and smoother	Greater; bones heavier and thicker, and markings are more prominent
Acetabula	Smaller; farther apart	Larger; closer
Pubic arch/subpubic angle	Broader (80° to 90°); more rounded	Angle is more acute (50° to 60°)
Anterior view	Pelvic brim Pubic arch	
Sacrum	Wider; shorter; sacral curvature is accentuated	Narrow; longer; sacral promontory more ventral
Coccyx	More movable; projects inferiorly	Less movable; projects anteriorly
Greater sciatic notch	Wide and shallow	Narrow and deep
Left lateral view	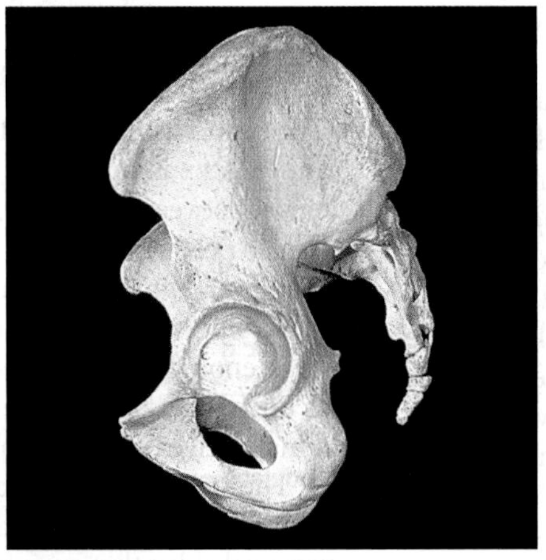	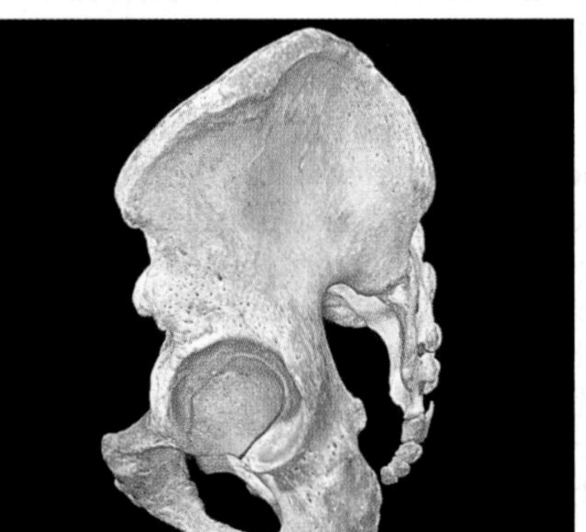
Pelvic inlet (brim)	Wider; oval from side to side	Narrow; basically heart shaped
Pelvic outlet	Wider; ischial tuberosities shorter, farther apart and everted	Narrower; ischial tuberosities longer, sharper, and point more medially
Posteroinferior view	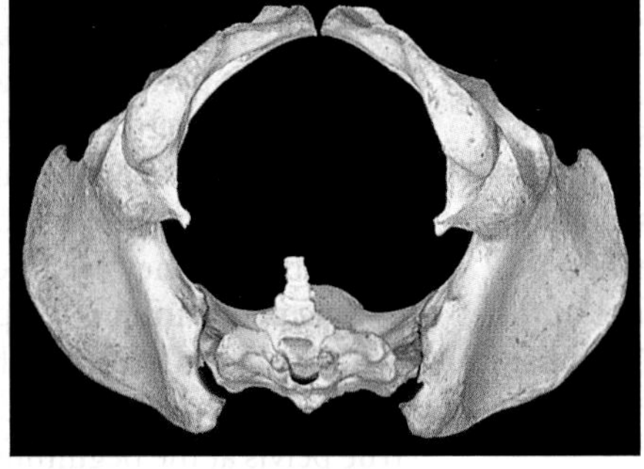	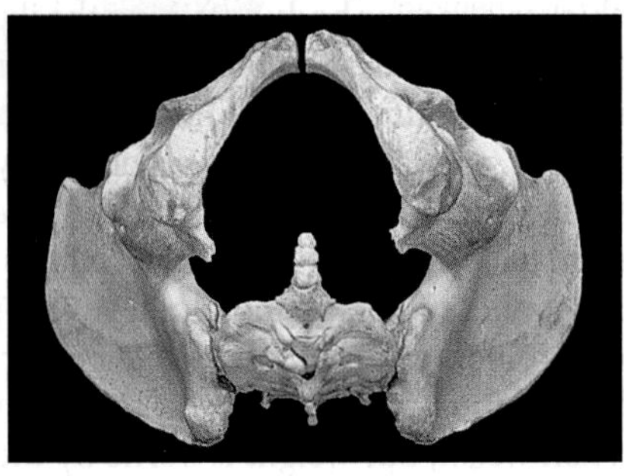

The **pelvic outlet**, illustrated in the photos at the bottom of Table 7.4, is the inferior margin of the true pelvis. It is bounded anteriorly by the pubic arch, laterally by the ischia, and posteriorly by the sacrum and coccyx. Both the coccyx and the ischial spines protrude into the outlet opening, so a sharply angled coccyx or unusually large spines can interfere with delivery. The largest dimension of the outlet is the anteroposterior diameter.

☑ Check Your Understanding

22. The ilium and pubis help to form the hip bone. What other bone is involved in forming the hip bone?
23. The pelvic girdle is a heavy, strong girdle. How does its structure reflect its function?
24. Which of the following terms or phrases refer to the female pelvis? Wider, shorter sacrum; cavity narrow and deep; narrow heart-shaped inlet; more movable coccyx; long ischial spines.

For answers, see Answers Appendix.

7.7 The lower limb consists of the thigh, leg, and foot

→ Learning Objective

☐ **Identify the lower limb bones and their important markings.**

The lower limbs carry the entire weight of the erect body and are subjected to exceptional forces when we jump or run. Thus, it is not surprising that the bones of the lower limbs are thicker and stronger than comparable bones of the upper limbs. The three segments of each lower limb are the thigh, the leg, and the foot (see Table 7.5, p. 217).

Thigh

The **femur** (fe′mur; "thigh"), the single bone of the thigh (Figure 7.33), is the largest, longest, strongest bone in the body. Its durable structure reflects the fact that the stress on the femur during vigorous jumping can reach 280 kg/cm^2 (about 2 tons per square inch)! The femur is surrounded by bulky muscles that prevent us from palpating its course down the length of the thigh. Its length is roughly one-quarter of a person's height.

Proximally, the femur articulates with the hip bone and then courses medially as it descends toward the knee. This arrangement allows the knee joints to be closer to the body's center of gravity and provides for better balance. The medial course of the two femurs is more pronounced in women because of their wider pelvis, a situation that may contribute to the greater incidence of knee problems in female athletes.

The ball-like **head** of the femur has a small central pit called the **fovea capitis** (fo′ve-ah kă′pĭ-tis; "pit of the head"). The short *ligament of the head of the femur* runs from this pit to the acetabulum, where it helps secure the femur. The head is carried on a *neck* that angles *laterally* to join the shaft. This arrangement reflects the fact that the femur articulates with the lateral aspect (rather than the inferior region) of the pelvis. The neck is the weakest part of the femur and is often fractured, an injury commonly called a broken hip.

At the junction of the shaft and neck are the lateral **greater trochanter** (tro-kan′ter) and posteromedial **lesser trochanter**. These projections serve as sites of attachment for thigh and buttock muscles. The two trochanters are connected by the **intertrochanteric line** anteriorly and by the prominent **intertrochanteric crest** posteriorly.

Inferior to the intertrochanteric crest on the posterior shaft is the **gluteal tuberosity**, which blends into a long vertical ridge, the **linea aspera** (lin′e-ah as′per-ah; "rough line"), inferiorly. Distally, the linea aspera diverges, forming the **medial** and **lateral supracondylar lines**. All of these markings are sites of muscle attachment. Except for the linea aspera, the femur shaft is smooth and rounded.

Distally, the femur broadens and ends in the wheel-like **lateral** and **medial condyles**, which articulate with the tibia of the leg. The **medial** and **lateral epicondyles** (sites of muscle attachment) flank the condyles superiorly. On the superior part of the medial epicondyle is a bump, the **adductor tubercle**. The smooth **patellar surface**, between the condyles on the anterior femoral surface, articulates with the *patella* (pah-tel′ah), or kneecap (see Figure 7.33 and Table 7.5). Between the condyles on the posterior aspect of the femur is the deep, U-shaped **intercondylar fossa**.

The **patella** ("small pan") is a triangular sesamoid bone enclosed in the (quadriceps) tendon that secures the anterior thigh muscles to the tibia. It protects the knee joint anteriorly and improves the leverage of the thigh muscles acting across the knee.

Leg

Two parallel bones, the tibia and fibula, form the skeleton of the leg, the region of the lower limb between the knee and the ankle (Figure 7.34). These two bones are connected by an *interosseous membrane* and articulate with each other both proximally and distally. Unlike the joints between the radius and ulna of the forearm, the *tibiofibular joints* (tib″e-o-fib′u-lar) of the leg allow essentially no movement. The bones of the leg thus form a less flexible but stronger and more stable limb than those of the forearm. The medial tibia articulates proximally with the femur to form the modified hinge joint of the knee and distally with the talus bone of the foot at the ankle. The fibula, in contrast, does not contribute to the knee joint and merely helps stabilize the ankle joint.

Tibia

The **tibia** (tib′e-ah; "shinbone") receives the weight of the body from the femur and transmits it to the foot. It is second only to the femur in size and strength. At its broad proximal end are the concave **medial** and **lateral condyles**, which look like two huge checkers lying side by side. These are separated by an irregular projection, the **intercondylar eminence**. The tibial condyles articulate with the corresponding condyles of the femur. The inferior region of the lateral tibial condyle bears a facet that indicates the site of the *superior tibiofibular joint.* Just inferior to the condyles, the tibia's anterior surface displays the rough **tibial tuberosity**, to which the patellar ligament attaches.

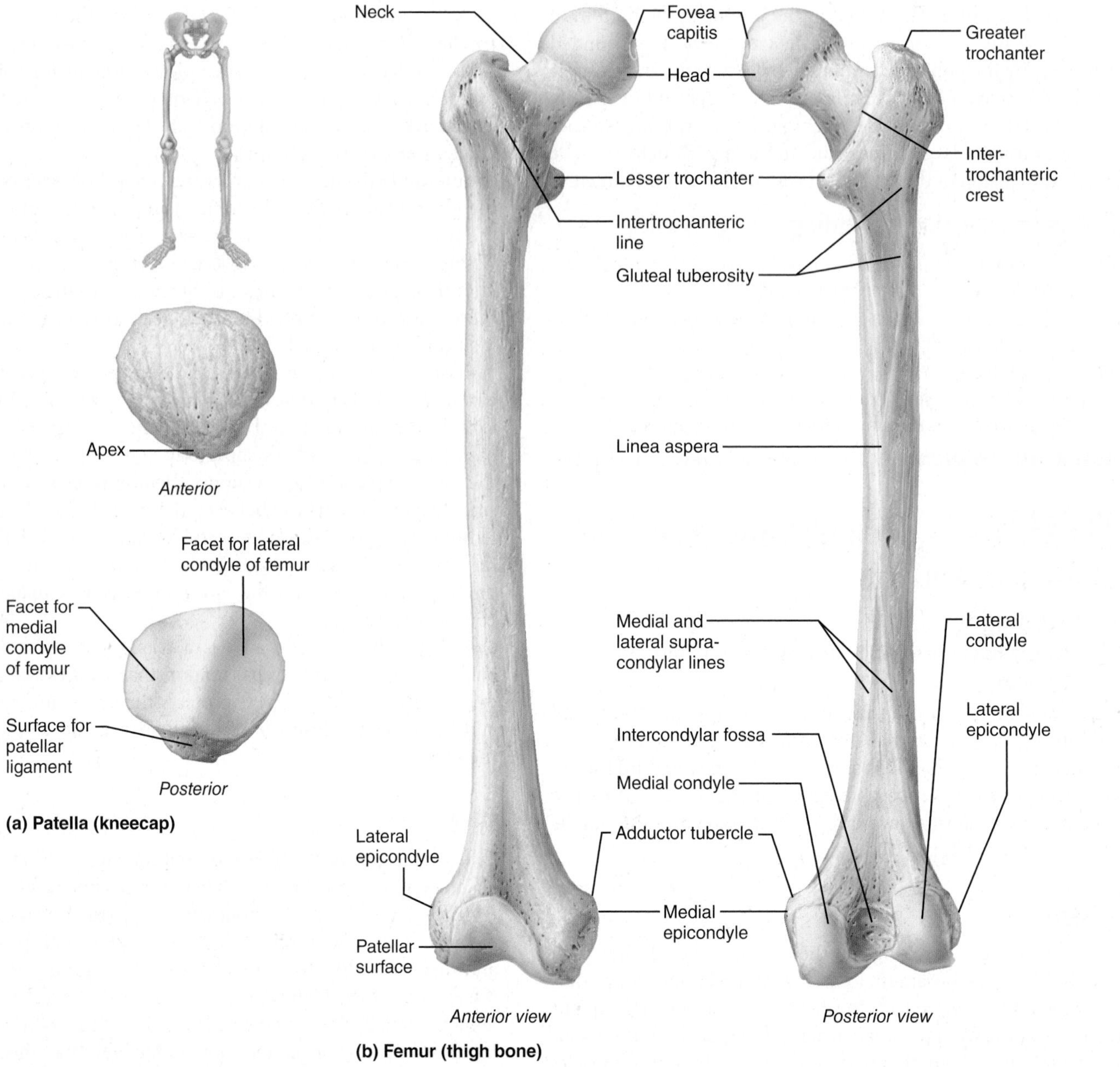

(a) Patella (kneecap)

(b) Femur (thigh bone)

Figure 7.33 Bones of the right knee and thigh. (For a related image, see *A Brief Atlas of the Human Body*, Figure 29a–k.)

Practice art labeling
MasteringA&P®>Study Area>Chapter 7

The tibial shaft is triangular in cross section. Neither the tibia's sharp **anterior border** nor its medial surface is covered by muscles, so they can be felt just deep to the skin along their entire length. The anguish of a "bumped" shin is an experience familiar to nearly everyone. Distally the tibia is flat where it articulates with the talus bone of the foot. Medial to that joint surface is an inferior projection, the **medial malleolus** (mah-le′o-lus; "little hammer"), which forms the medial bulge of the ankle. The **fibular notch**, on the lateral surface of the tibia, participates in the *inferior tibiofibular joint*.

Fibula

The **fibula** (fib′u-lah; "pin") is a sticklike bone with slightly expanded ends. It articulates proximally and distally with the lateral aspects of the tibia. Its proximal end is its **head**; its distal end is the **lateral malleolus**. The lateral malleolus forms the conspicuous lateral ankle bulge and articulates with the talus. The fibular shaft is heavily ridged and appears to have been twisted a quarter turn. The fibula does not bear weight, but several muscles originate from it.

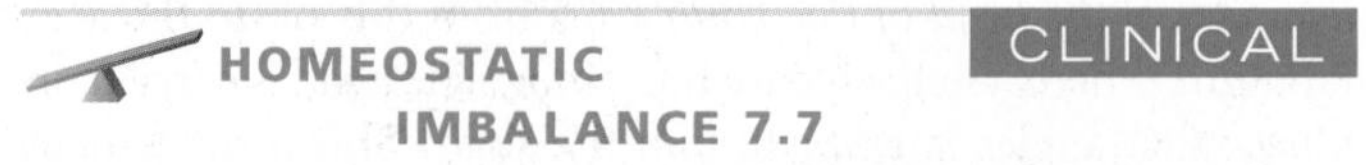

HOMEOSTATIC IMBALANCE 7.7 — CLINICAL

A *Pott's fracture* occurs at the distal end of the fibula, the tibia, or both. It is a common sports injury (Figure 7.34c). ✚

Foot

The skeleton of the foot includes the bones of the *tarsus*, the bones of the *metatarsus*, and the *phalanges*, or toe bones (**Figure 7.35**). The foot has two important functions: It supports our body weight, and it acts as a lever to propel the body forward when we walk and run. A single bone could serve both purposes, but it would adapt poorly to uneven ground. Segmentation makes the foot flexible, avoiding this problem.

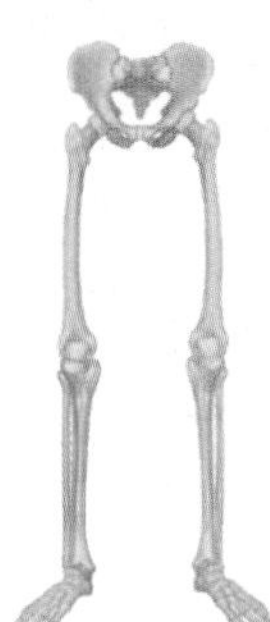

Tarsus

The **tarsus** is made up of seven bones called **tarsals** (tar′salz) that form the posterior half of the foot. It corresponds to the carpus of the hand. Body weight is carried primarily by the two largest, most posterior tarsals: the **talus** (ta′lus; "ankle"), which articulates with the tibia and fibula superiorly, and the strong **calcaneus** (kal-ka′ne-us; "heel bone"), which forms the heel of the foot and carries the talus on its superior surface. The thick *calcaneal*, or *Achilles*, *tendon* of the calf muscles attaches to the posterior surface of the calcaneus. The part of the calcaneus that touches the ground is the **calcaneal tuberosity**, and its shelflike projection that supports part of the talus is the **sustentaculum tali** (sus″ten-tak′u-lum ta′le; "supporter of the talus") or **talar shelf**. The tibia articulates with the talus at the *trochlea* of the talus. The remaining tarsals are the lateral **cuboid**, the medial **navicular** (nah-vik′u-lar), and the anterior **medial**, **intermediate**, and **lateral cuneiform bones** (ku-ne′i-form; "wedge-shaped"). The cuboid and cuneiform bones articulate with the metatarsal bones anteriorly.

Metatarsus

The **metatarsus** consists of five small, long bones called **metatarsals**. These are numbered I to V beginning on the medial (great toe) side of the foot. The first metatarsal, which plays an important role in

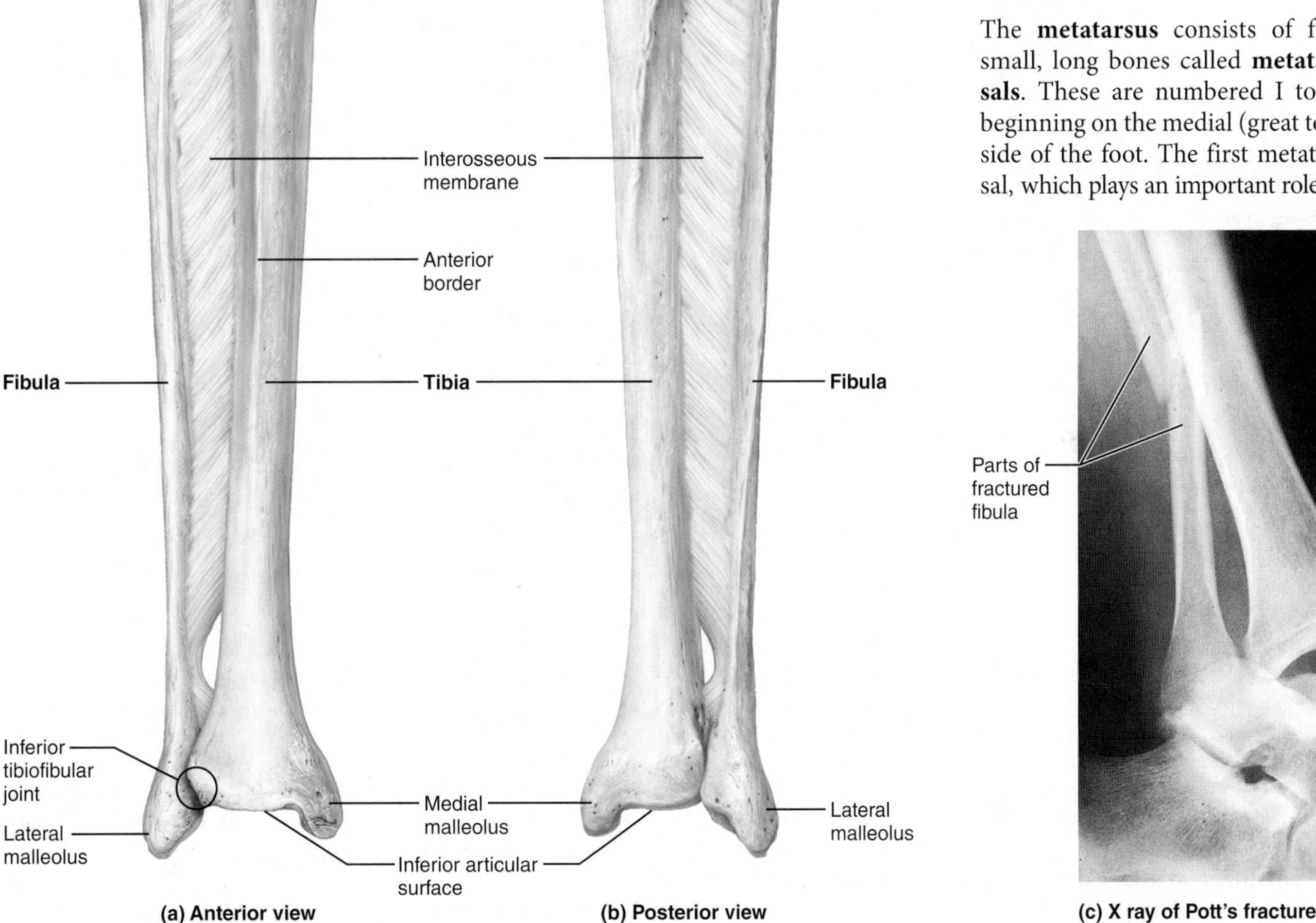

Figure 7.34 The tibia and fibula of the right leg. (For a related image, see *A Brief Atlas of the Human Body*, Figure 30a–j.)

Lateral condyle
Tibial tuberosity
Lateral condyle
Fibula articulates here
Line for soleus muscle

(d) Anterior view, proximal tibia
(e) Posterior view, proximal tibia

Figure 7.34 *(continued)* **The tibia and fibula of the right leg.**

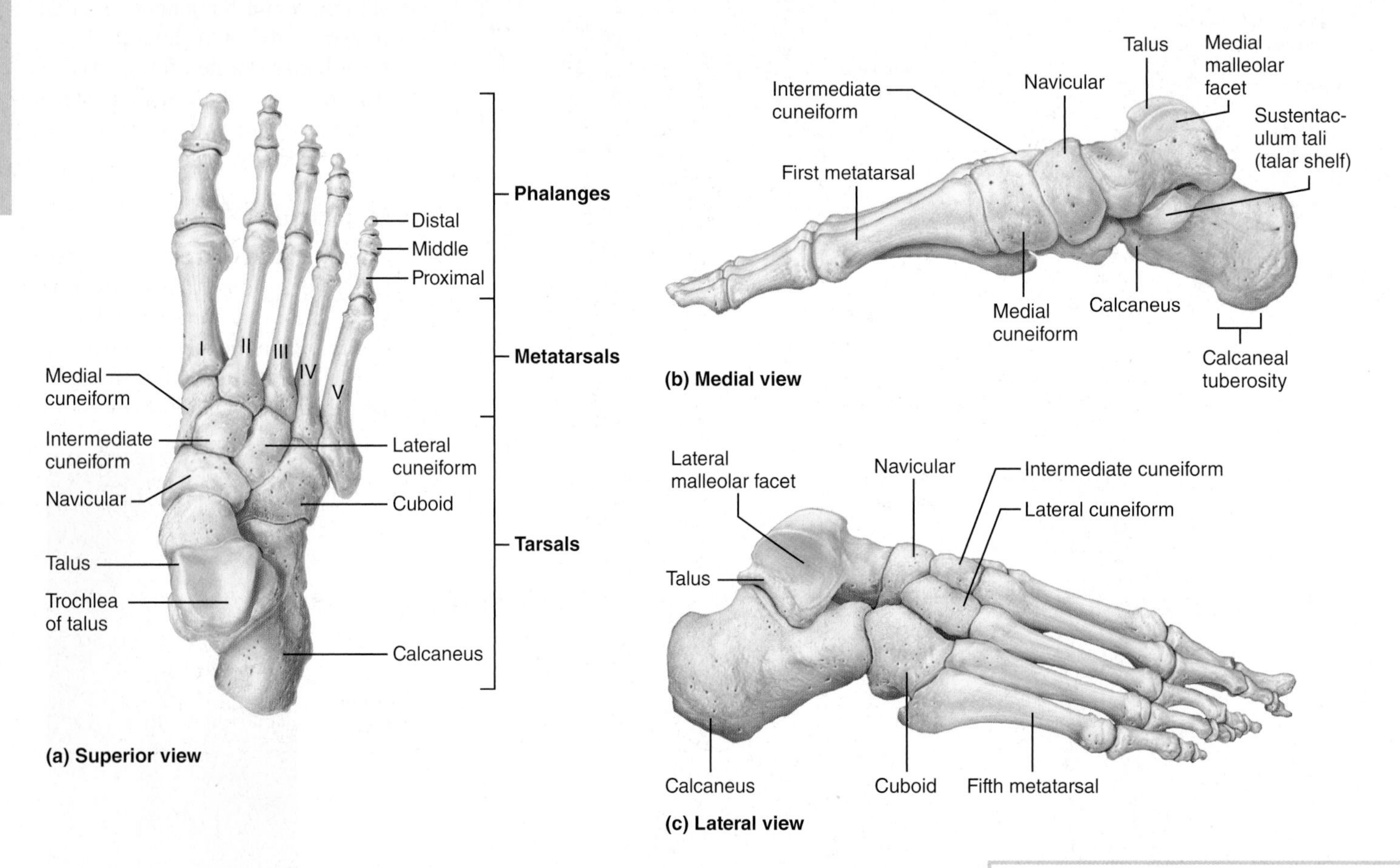

Figure 7.35 Bones of the right foot. (For a related image, see *A Brief Atlas of the Human Body,* Figure 31a, c, and d.)

Practice art labeling
MasteringA&P®>Study Area>Chapter 7

supporting body weight, is short and thick. The arrangement of the metatarsals is more parallel than that of the metacarpals of the hands. Distally, where the metatarsals articulate with the proximal phalanges of the toes, the enlarged head of the first metatarsal forms the "ball" of the foot.

Phalanges (Toes)

The 14 phalanges of the toes are a good deal smaller than those of the fingers and so are less nimble. But their general structure and arrangement are the same. There are three phalanges in each digit except for the great toe, the **hallux**. The hallux has only two, proximal and distal.

Arches of the Foot

A segmented structure can support weight only if it is arched. The foot has three arches: two *longitudinal arches* (*medial* and *lateral*) and one *transverse arch* (**Figure 7.36**), which account for its awesome strength. These arches are maintained by the

Table 7.5 Bones of the Appendicular Skeleton, Part 2: Pelvic Girdle and Lower Limb

BODY REGION	BONES*	ILLUSTRATION	LOCATION	MARKINGS
Pelvic girdle (Figures 7.31, 7.32)	**Coxal** (2) (hip)		Each hip (coxal) bone is formed by the fusion of an ilium, ischium, and pubis; the hip bones articulate anteriorly at the pubic symphysis and form sacroiliac joints with the sacrum posteriorly; girdle consisting of both hip bones and the sacrum is basinlike	Iliac crest; anterior and posterior iliac spines; auricular surface; greater and lesser sciatic notches; obturator foramen; ischial tuberosity and spine; acetabulum; pubic arch; pubic crest; pubic tubercle
Lower limb Thigh (Figure 7.33)	**Femur** (2)		Femur is the sole bone of thigh; between hip joint and knee; largest bone of the body	Head; greater and lesser trochanters; neck; lateral and medial condyles and epicondyles; gluteal tuberosity; linea aspera
Kneecap (Figure 7.33)	**Patella** (2)		Patella is a sesamoid bone formed within the tendon of the quadriceps (anterior thigh) muscles	
Leg (Figure 7.34)	**Tibia** (2)		Tibia is the larger and more medial bone of leg; between knee and foot	Medial and lateral condyles; tibial tuberosity; anterior border; medial malleolus
	Fibula (2)		Fibula is the lateral bone of leg; sticklike	Head; lateral malleolus
Foot (Figure 7.35)	7 **Tarsals** (14): talus, calcaneus, navicular, cuboid, lateral cuneiform, intermediate cuneiform, medial cuneiform		Tarsals are seven bones forming the proximal part of the foot; the talus articulates with the leg bones at the ankle joint; the calcaneus, the largest tarsal, forms the heel	
	5 **Metatarsals** (10)		Metatarsals are five bones numbered I–V	
	14 **Phalanges** (28): distal, middle, proximal		Phalanges form the toes; three in digits II–V, two in digit I (the great toe)	

* The number in parentheses ( ) following the bone name denotes the total number of such bones in the body.

interlocking shapes of the foot bones, by strong ligaments, and by the pull of some tendons during muscle activity. The ligaments and tendons provide a certain amount of springiness. In general, the arches "give," or stretch slightly, when weight is applied to the foot and spring back when the weight is removed, which makes walking and running more economical in terms of energy use than would otherwise be the case.

If you examine your wet footprints, you will see that the medial margin from the heel to the head of the first metatarsal leaves no print. This is because the **medial longitudinal arch** curves well above the ground. The talus is the keystone of this arch, which originates at the calcaneus, rises toward the talus, and then descends to the three medial metatarsals.

The **lateral longitudinal arch** is very low. It elevates the lateral part of the foot just enough to redistribute some of the weight to the calcaneus and the head of the fifth metatarsal (to the ends of the arch). The cuboid is the keystone bone of this arch.

The two longitudinal arches serve as pillars for the **transverse arch**, which runs obliquely from one side of the foot to the other, following the line of the joints between the tarsals and metatarsals. Together, the arches of the foot form a half-dome that distributes about half of a person's standing and walking weight to the heel bones and half to the heads of the metatarsals.

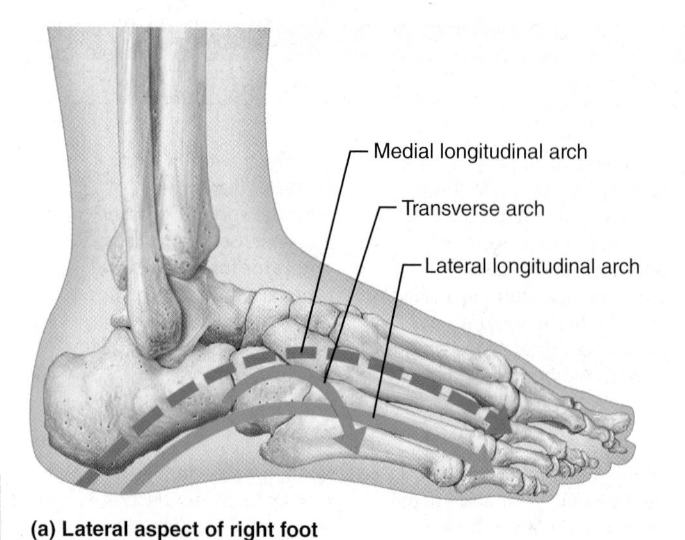

(a) Lateral aspect of right foot

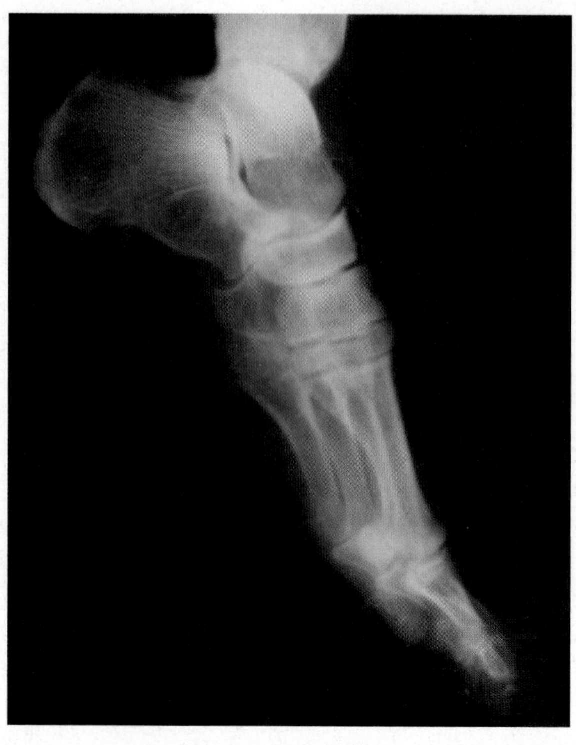

(b) X ray, medial aspect of right foot

Figure 7.36 Arches of the foot.

CLINICAL

HOMEOSTATIC IMBALANCE 7.8

Standing immobile for extended periods places excessive strain on the tendons and ligaments of the feet (because the muscles are inactive) and can result in fallen arches, or "flat feet," particularly if a person is overweight. Running on hard surfaces can also cause arches to fall unless the runner wears shoes that give proper arch support. +

Check Your Understanding

25. What lower limb bone is the second largest bone in the body?

26. The image below shows the posterior aspects of two bones. Name the bones. Are they from the right or left side of the body? Name the structures labeled a–c.

27. Which of the following sites is not a site of muscle attachment? Greater trochanter, lesser trochanter, gluteal tuberosity, lateral condyle.

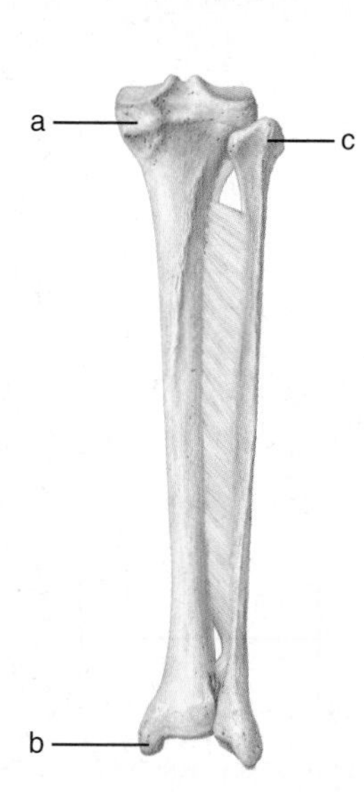

28. Besides supporting our weight, what is a major function of the arches of the foot?

29. What are the two largest tarsal bones in each foot, and which one forms the heel of the foot?

For answers, see Answers Appendix.

Our skeleton is a marvelous substructure, to be sure, but it is much more than that. It is a protector and supporter of other body systems, and without it (and the joints considered in Chapter 8), our muscles would be almost useless.

REVIEW QUESTIONS

For more chapter study tools, go to the Study Area of MasteringA&P®.

There you will find:

- Interactive Physiology iP
- A&PFlix A&PFlix
- Practice Anatomy Lab PAL
- PhysioEx PEx
- Videos, Practice Quizzes and Tests, MP3 Tutor Sessions, Case Studies, and much more!

Multiple Choice/Matching

(Some questions have more than one correct answer. Select the best answer or answers from the choices given.)

1. Match the bones in column B with their description in column A. (Note that some descriptions require more than a single choice.)

Column A	Column B
____ (1) connected by the coronal suture	(a) ethmoid
____ (2) keystone bone of cranium	(b) frontal
____ (3) keystone bone of the face	(c) mandible
____ (4) form the hard palate	(d) maxillary
____ (5) allows the spinal cord to pass	(e) occipital
____ (6) forms the chin	(f) palatine
____ (7) contain paranasal sinuses	(g) parietal
____ (8) contains mastoid sinuses	(h) sphenoid
	(i) temporal

2. Match the key terms with the bone descriptions that follow.

Key:
(a) clavicle (d) pubis (f) scapula
(b) ilium (e) sacrum (g) sternum
(c) ischium

____ (1) bone of the axial skeleton to which the pectoral girdle attaches
____ (2) markings include glenoid cavity and acromion
____ (3) features include the ala, crest, and greater sciatic notch
____ (4) doubly curved; acts as a shoulder strut
____ (5) hip bone that articulates with the axial skeleton
____ (6) the "sit-down" bone
____ (7) anteriormost bone of the pelvic girdle
____ (8) part of the vertebral column

3. Use key choices to identify the bone descriptions that follow.

Key:
(a) carpals (d) humerus (g) tibia
(b) femur (e) radius (h) ulna
(c) fibula (f) tarsals

____ (1) articulates with the acetabulum and the tibia
____ (2) forms the lateral aspect of the ankle
____ (3) bone that "carries" the hand
____ (4) the wrist bones
____ (5) end shaped like a monkey wrench
____ (6) articulates with the capitulum of the humerus
____ (7) largest bone of this "group" is the calcaneus

Short Answer Essay Questions

4. Name the cranial and facial bones and compare and contrast the functions of the cranial and facial skeletons.
5. How do the relative proportions of the cranium and face of a fetus compare with those of an adult skull?
6. Name and diagram the normal vertebral curvatures. Which are primary and which are secondary curvatures?
7. List at least two specific anatomical characteristics each for typical cervical, thoracic, and lumbar vertebrae that would allow anyone to identify each type correctly.
8. What is the function of the intervertebral discs?
9. Distinguish between the anulus fibrosus and nucleus pulposus regions of a disc. Which provides durability and strength? Which provides resilience? Which part is involved in a "slipped" disc?
10. What is a true rib? A false rib?
11. The major function of the shoulder girdle is flexibility. What is the major function of the pelvic girdle? Relate these functional differences to anatomical differences seen in these girdles.
12. List three important differences between the male and female pelves.
13. Peter Howell, a teaching assistant in the anatomy class, picked up a hip bone and pretended it was a telephone. He held the big hole in this bone right up to his ear and said, "Hello, obturator, obturator (operator, operator)." Name the structure he was helping the students to learn.

7

AT THE CLINIC

Clinical Case Study

Skeleton

Kayla Tanner, a 45-year-old mother of four, was a passenger on the bus involved in an accident on Route 91. When paramedics arrived on the scene, they found Mrs. Tanner lying on her side in the aisle. Upon examination, they found that her right thigh appeared shorter than her left thigh. They also noticed that even slight hip movement caused considerable pain. Suspecting a hip dislocation, they stabilized and transported her.

In the emergency department, doctors discovered a decreased ability to sense light touch in her right foot, and she was unable to move her toes or ankle. Dislocation of her right hip was confirmed by X ray. Mrs. Tanner was sedated to relax the muscles around the hip, and then doctors placed her in the supine position and performed a closed reduction ("popped" the femur back in place).

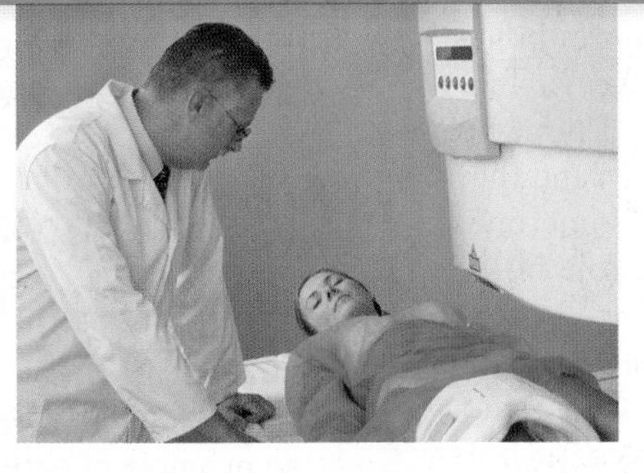

1. Mrs. Tanner's hip bone contains a hemispherical socket at the point where her femur attaches. Name this structure.
2. Name the structure on the femur that forms the "ball" that fits into the "socket" you named in question 1.
3. Three bones fuse together at a point within the structure that you named in question 1. Name those three bones.
4. Mrs. Tanner suffered an injury to the hip joint, but if you were asked to rest your hands on your hips, you would not actually touch this joint. What structure in the pelvic girdle would your hands be resting on?
5. The sedation that Mrs. Tanner was given was to relax the large muscles of the thigh and buttocks that attach to the proximal end of the femur. Name the structures on the femur where these muscles attach.
6. Mrs. Tanner's injury caused damage to the sciatic nerve that passes across the hip and down into the thigh, leg, and foot. Name the pelvic structure that this nerve passes through as it travels into the upper thigh.

For answers, see Answers Appendix.

8 Joints

KEY CONCEPTS

The graceful movements of ballet dancers and the rough-and-tumble grapplings of football players demonstrate the great variety of motion allowed by **joints**, or **articulations**—the sites where two or more bones meet. Our joints have two fundamental functions: They give our skeleton mobility, and they hold it together, sometimes playing a protective role in the process.

Joints are the weakest parts of the skeleton. Nonetheless, their structure resists various forces, such as crushing or tearing, that threaten to force them out of alignment.

A physiotherapist works on a hip joint.

8.1 Joints are classified into three structural and three functional categories

→ Learning Objectives

- ☐ **Define joint or articulation.**
- ☐ **Classify joints by structure and by function.**

Joints are classified by structure and by function. The *structural classification* focuses on the material binding the bones together and whether or not a joint cavity is present. Structurally, there are *fibrous*, *cartilaginous*, and *synovial joints* (Table 8.1). Only synovial joints have a joint cavity.

The *functional classification* is based on the amount of movement allowed at the joint. On this basis, there are **synarthroses** (sin″ar-thro′sēz; *syn* = together, *arthro* = joint), which are immovable joints; **amphiarthroses** (am″fe-ar-thro′sēz; *amphi* = on both sides), slightly movable joints; and **diarthroses** (di″ar-thro′sēz; *dia* = through, apart), or freely movable joints. Freely movable joints predominate in the limbs. Immovable and slightly movable joints are largely restricted to the axial skeleton. This localization of functional joint types makes sense because the less movable the joint, the more stable it is likely to be.

In general, fibrous joints are immovable, and synovial joints are freely movable. However, cartilaginous joints have both rigid and slightly movable examples. Since the structural categories are more clear-cut, we will use the structural classification in this discussion, indicating functional properties where appropriate.

☑ Check Your Understanding

1. What functional joint class contains the least-mobile joints?
2. How are joint mobility and stability related?

For answers, see Answers Appendix.

8.2 In fibrous joints, the bones are connected by fibrous tissue

→ Learning Objective

- ☐ **Describe the general structure of fibrous joints. Name and give an example of each of the three common types of fibrous joints.**

In **fibrous joints**, the bones are joined by the collagen fibers of connective tissue. No joint cavity is present. The amount of movement allowed depends on the length of the connective tissue fibers. Most fibrous joints are immovable, although a few are slightly movable. The three types of fibrous joints are *sutures*, *syndesmoses*, and *gomphoses*.

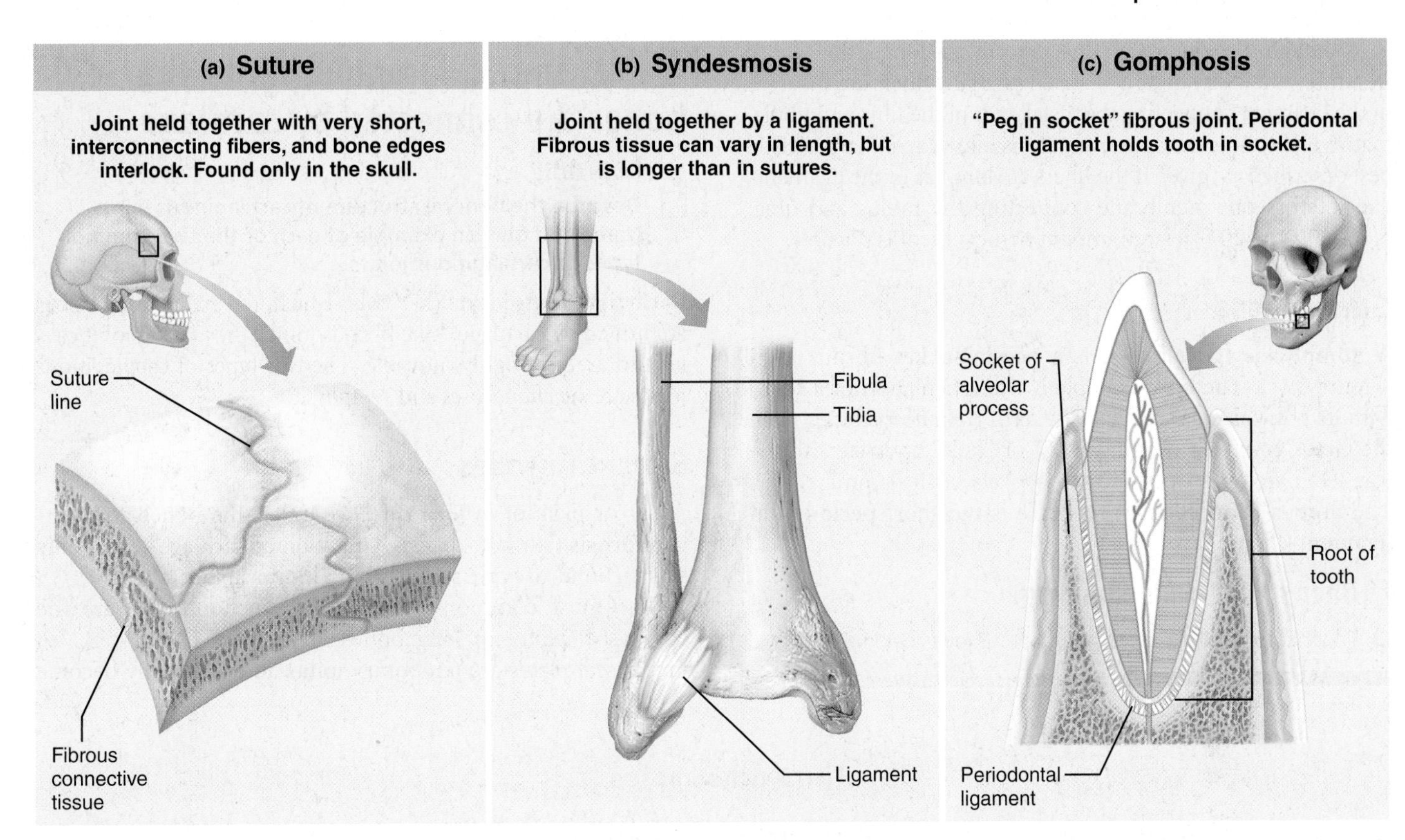

Figure 8.1 Fibrous joints.

Sutures

Sutures, literally "seams," occur only between bones of the skull (**Figure 8.1a**). The wavy articulating bone edges interlock, and the junction is completely filled by a minimal amount of very short connective tissue fibers that are continuous with the periosteum. The result is nearly rigid splices that knit the bones together, yet allow the skull to expand as the brain grows during youth. During middle age, the fibrous tissue ossifies and the skull bones fuse into a single unit. At this stage, the closed sutures are more precisely called **synostoses** (sin″os-to′sēz), literally, "bony junctions." Because movement of the cranial bones would damage the brain, the immovable nature of sutures is a protective adaptation.

Syndesmoses

In **syndesmoses** (sin″des-mo′sēz), the bones are connected exclusively by *ligaments* (*syndesmos* = ligament), cords or bands of fibrous tissue. The amount of movement allowed at a syndesmosis depends on the length of the connecting fibers.

Table 8.1 Summary of Joint Classes

STRUCTURAL CLASS	STRUCTURAL CHARACTERISTICS	TYPES	MOBILITY
Fibrous	Adjoining bones united by collagen fibers	Suture (short fibers)	Immobile (synarthrosis)
		Syndesmosis (longer fibers)	Slightly movable (amphiarthrosis) and immobile
		Gomphosis (periodontal ligament)	Immobile
Cartilaginous	Adjoining bones united by cartilage	Synchondrosis (hyaline cartilage)	Immobile
		Symphysis (fibrocartilage)	Slightly movable
Synovial	Adjoining bones covered with articular cartilage, separated by a joint cavity, and enclosed within an articular capsule lined with synovial membrane	• Plane • Condylar • Hinge • Saddle • Pivot • Ball-and-socket	Freely movable (diarthrosis; movements depend on design of joint)

Although the connecting fibers are always longer than those in sutures, they vary quite a bit in length. If the fibers are short (as in the ligament connecting the distal ends of the tibia and fibula, Figure 8.1b), little or no movement is allowed, a characteristic best described as "give." If the fibers are long (as in the ligament-like interosseous membrane connecting the radius and ulna, Figure 7.29, p. 207), a large amount of movement is possible.

Gomphoses

A **gomphosis** (gom-fo′sis) is a peg-in-socket fibrous joint (Figure 8.1c). The only example is the articulation of a tooth with its bony alveolar socket. The term *gomphosis* comes from the Greek *gompho*, meaning "nail" or "bolt," and refers to the way teeth are embedded in their sockets (as if hammered in). The fibrous connection in this case is the short **periodontal ligament** (Figure 22.12, p. 757).

☑ Check Your Understanding

3. To what functional class do most fibrous joints belong?

For answers, see Answers Appendix.

8.3 In cartilaginous joints, the bones are connected by cartilage

→ Learning Objective

☐ **Describe the general structure of cartilaginous joints. Name and give an example of each of the two common types of cartilaginous joints.**

In **cartilaginous joints** (kar″tĭ-laj′ĭ-nus), the articulating bones are united by cartilage. Like fibrous joints, they lack a joint cavity and are not highly movable. The two types of cartilaginous joints are *synchondroses* and *symphyses.*

Synchondroses

A bar or plate of *hyaline cartilage* unites the bones at a **synchondrosis** (sin″kon-dro′sis; "junction of cartilage"). Virtually all synchondroses are synarthrotic (immovable).

The most common examples of synchondroses are the epiphyseal plates in long bones of children (**Figure 8.2a**). Epiphyseal plates are temporary joints and eventually become

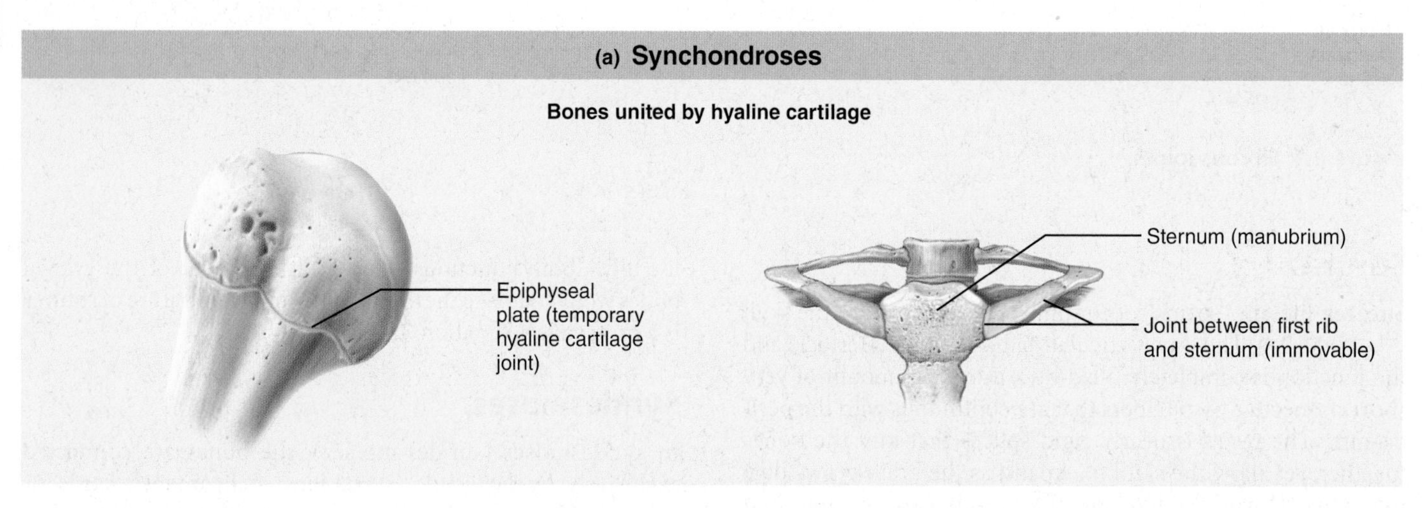

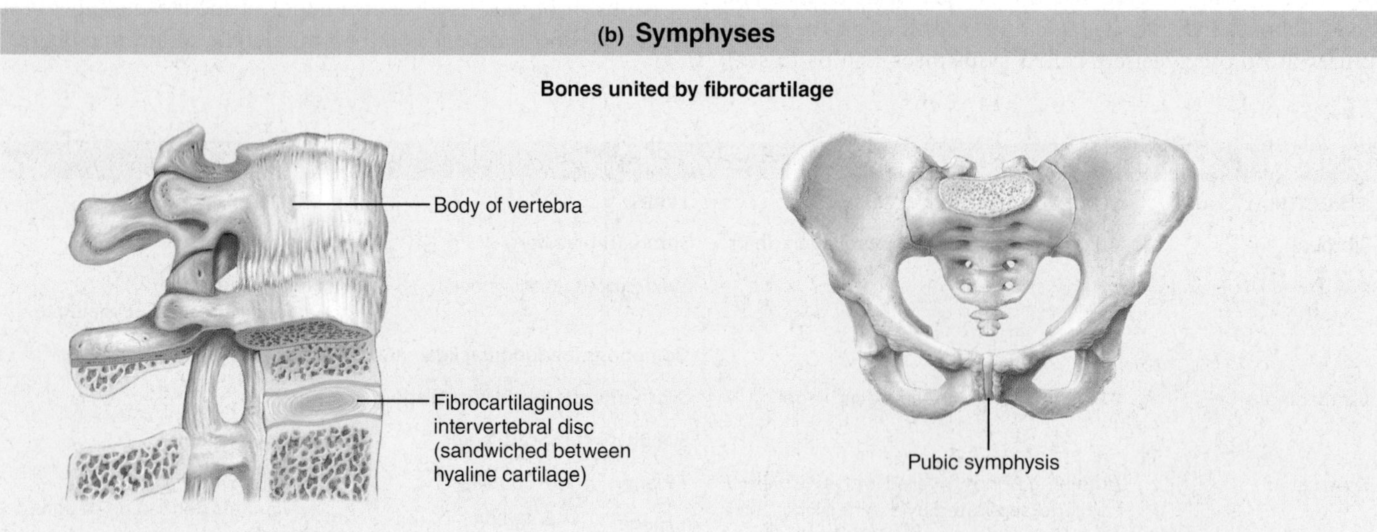

Figure 8.2 Cartilaginous joints.

synostoses. Another example of a synchondrosis is the immovable joint between the costal cartilage of the first rib and the manubrium of the sternum (Figure 8.2a).

Symphyses

A joint where *fibrocartilage* unites the bones is a **symphysis** (sim′fih-sis; "growing together"). Since fibrocartilage is compressible and resilient, it acts as a shock absorber and permits a limited amount of movement at the joint. Even though fibrocartilage is the main element of a symphysis, hyaline cartilage is also present in the form of articular cartilages on the bony surfaces. Symphyses are amphiarthrotic joints designed for strength with flexibility. Examples include the intervertebral joints and the pubic symphysis of the pelvis (Figure 8.2b, and see Table 8.2).

☑ Check Your Understanding

4. MAKING connections Evan is 25 years old. Would you expect to find synchondroses at the ends of his femur? Explain. (Hint: See Chapter 6.)

For answers, see Answers Appendix.

8.4 Synovial joints have a fluid-filled joint cavity

→ Learning Objectives

- ☐ Describe the structural characteristics of synovial joints.
- ☐ Compare the structures and functions of bursae and tendon sheaths.
- ☐ List three natural factors that stabilize synovial joints.
- ☐ Name and describe (or perform) the common body movements.
- ☐ Name and provide examples of the six types of synovial joints based on the movement(s) allowed.

Synovial joints (si-no′ve-al; "joint eggs") are those in which the articulating bones are separated by a fluid-containing joint cavity. This arrangement permits substantial freedom of movement, and all synovial joints are freely movable diarthroses. Nearly all joints of the limbs—indeed, most joints of the body—fall into this class.

General Structure

Synovial joints have six distinguishing features (Figure 8.3):

- **Articular cartilage.** Glassy-smooth hyaline cartilage covers the opposing bone surfaces as **articular cartilage**. These thin (1 mm or less) but spongy cushions absorb compression placed on the joint and thereby keep the bone ends from being crushed.
- **Joint (articular) cavity.** A feature unique to synovial joints, the joint cavity is really just a potential space that contains a small amount of synovial fluid.
- **Articular capsule.** The joint cavity is enclosed by a two-layered **articular capsule**, or *joint capsule*. The tough external **fibrous layer** is composed of dense irregular connective tissue that is continuous with the periostea of the articulating

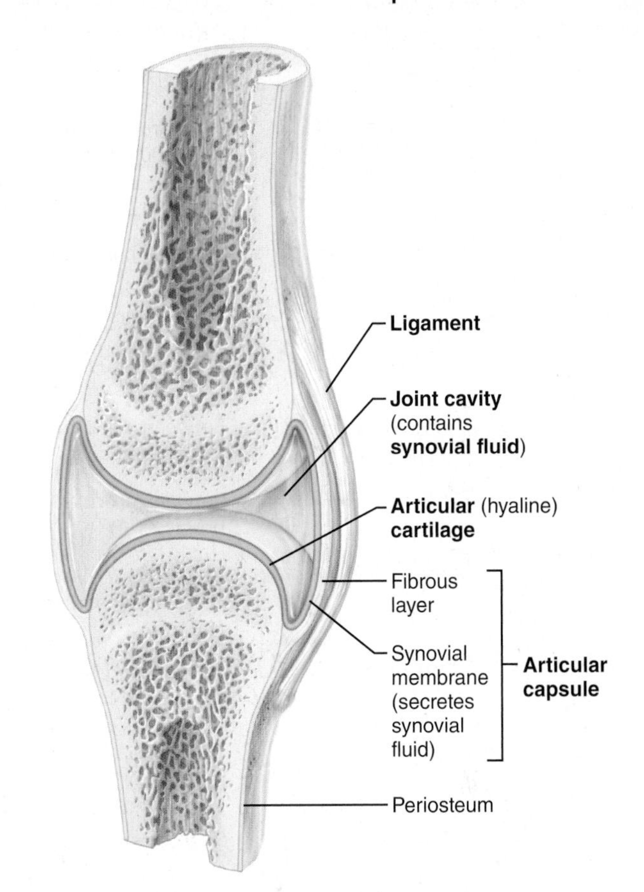

Figure 8.3 General structure of a synovial joint.

bones. It strengthens the joint so that the bones are not pulled apart. The inner layer of the joint capsule is a **synovial membrane** composed of loose connective tissue. Besides lining the fibrous layer internally, it covers all internal joint surfaces that are not hyaline cartilage. The synovial membrane's function is to make synovial fluid.

- **Synovial fluid.** A small amount of slippery **synovial fluid** occupies all free spaces within the joint capsule. This fluid is derived largely by filtration from blood flowing through the capillaries in the synovial membrane. Synovial fluid has a viscous, egg-white consistency (*ovum* = egg) due to hyaluronic acid secreted by cells in the synovial membrane, but it thins and becomes less viscous during joint activity.

Synovial fluid, which is also found *within* the articular cartilages, provides a slippery, weight-bearing film that reduces friction between the cartilages. Without this lubricant, rubbing would wear away joint surfaces and excessive friction could overheat and destroy the joint tissues. The synovial fluid is forced from the cartilages when a joint is compressed; then as pressure on the joint is relieved, synovial fluid seeps back into the articular cartilages like water into a sponge, ready to be squeezed out again the next time the joint is loaded (put under pressure). This process, called *weeping lubrication*, lubricates the free surfaces of the cartilages and nourishes their cells. (Remember, cartilage is avascular.) Synovial fluid also contains phagocytic cells that rid the joint cavity of microbes and cellular debris.

Table 8.2 Structural and Functional Characteristics of Body Joints

ILLUSTRATION	JOINT	ARTICULATING BONES	STRUCTURAL TYPE*	FUNCTIONAL TYPE; MOVEMENTS ALLOWED
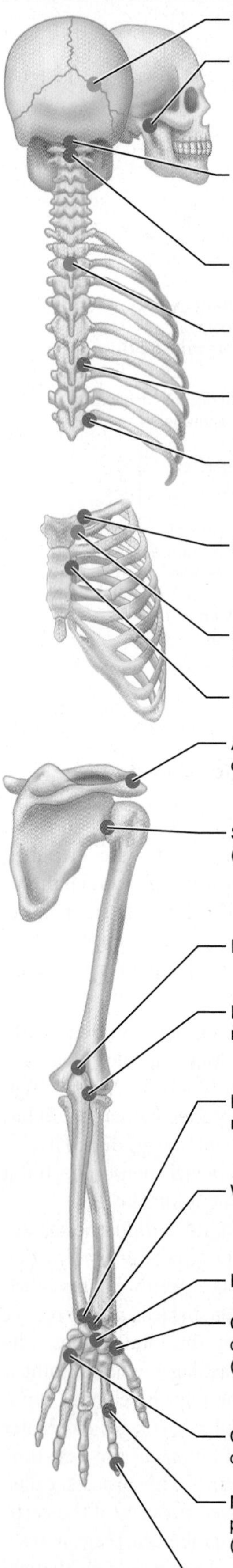	Skull	Cranial and facial bones	Fibrous; suture	Synarthrotic; no movement
	Temporo-mandibular	Temporal bone of skull and mandible	Synovial; modified hinge[†] (contains articular disc)	Diarthrotic; gliding and uniaxial rotation; slight lateral movement, elevation, depression, protraction, and retraction of mandible
	Atlanto-occipital	Occipital bone of skull and atlas	Synovial; condylar	Diarthrotic; biaxial; flexion, extension, lateral flexion, circumduction of head on neck
	Atlantoaxial	Atlas (C_1) and axis (C_2)	Synovial; pivot	Diarthrotic; uniaxial; rotation of the head
	Intervertebral	Between adjacent vertebral bodies	Cartilaginous; symphysis	Amphiarthrotic; slight movement
	Intervertebral	Between articular processes	Synovial; plane	Diarthrotic; gliding
	Costovertebral	Vertebrae (transverse processes or bodies) and ribs	Synovial; plane	Diarthrotic; gliding of ribs
	Sternoclavicular	Sternum and clavicle	Synovial; shallow saddle (contains articular disc)	Diarthrotic; multiaxial (allows clavicle to move in all axes)
	Sternocostal (first)	Sternum and rib I	Cartilaginous; synchondrosis	Synarthrotic; no movement
	Sternocostal	Sternum and ribs II–VII	Synovial; double plane	Diarthrotic; gliding
	Acromio-clavicular	Acromion of scapula and clavicle	Synovial; plane (contains articular disc)	Diarthrotic; gliding and rotation of scapula on clavicle
	Shoulder (glenohumeral)	Scapula and humerus	Synovial; ball-and-socket	Diarthrotic; multiaxial; flexion, extension, abduction, adduction, circumduction, rotation of humerus
	Elbow	Ulna (and radius) with humerus	Synovial; hinge	Diarthrotic; uniaxial; flexion, extension of forearm
	Proximal radioulnar	Radius and ulna	Synovial; pivot	Diarthrotic; uniaxial; pivot (convex head of radius rotates in radial notch of ulna)
	Distal radioulnar	Radius and ulna	Synovial; pivot (contains articular disc)	Diarthrotic; uniaxial; rotation of radius around long axis of forearm to allow pronation and supination
	Wrist	Radius and proximal carpals	Synovial; condylar	Diarthrotic; biaxial; flexion, extension, abduction, adduction, circumduction of hand
	Intercarpal	Adjacent carpals	Synovial; plane	Diarthrotic; gliding
	Carpometacarpal of digit I (thumb)	Carpal (trapezium) and metacarpal I	Synovial; saddle	Diarthrotic; biaxial; flexion, extension, abduction, adduction, circumduction, opposition of metacarpal I
	Carpometacarpal of digits II–V	Carpal(s) and metacarpal(s)	Synovial; plane	Diarthrotic; gliding of metacarpals
	Metacarpo-phalangeal (knuckle)	Metacarpal and proximal phalanx	Synovial; condylar	Diarthrotic; biaxial; flexion, extension, abduction, adduction, circumduction of fingers
	Interphalangeal (finger)	Adjacent phalanges	Synovial; hinge	Diarthrotic; uniaxial; flexion, extension of fingers

Table 8.2 (continued)

ILLUSTRATION	JOINT	ARTICULATING BONES	STRUCTURAL TYPE*	FUNCTIONAL TYPE; MOVEMENTS ALLOWED
	Sacroiliac	Sacrum and coxal bone	Synovial; plane in childhood, increasingly fibrous in adult	Diarthrotic in child; amphiarthrotic in adult; (more movement during pregnancy)
	Pubic symphysis	Pubic bones	Cartilaginous; symphysis	Amphiarthrotic; slight movement (enhanced during pregnancy)
	Hip (coxal)	Hip bone and femur	Synovial; ball-and-socket	Diarthrotic; multiaxial; flexion, extension, abduction, adduction, rotation, circumduction of thigh
	Knee (tibiofemoral)	Femur and tibia	Synovial; modified hinge† (contains articular discs)	Diarthrotic; biaxial; flexion, extension of leg, some rotation allowed in flexed position
	Knee (femoropatellar)	Femur and patella	Synovial; plane	Diarthrotic; gliding of patella
	Superior tibiofibular	Tibia and fibula (proximally)	Synovial; plane	Diarthrotic; gliding of fibula
	Inferior tibiofibular	Tibia and fibula (distally)	Fibrous; syndesmosis	Synarthrotic; slight "give" during dorsiflexion
	Ankle	Tibia and fibula with talus	Synovial; hinge	Diarthrotic; uniaxial; dorsiflexion, and plantar flexion of foot
	Intertarsal	Adjacent tarsals	Synovial; plane	Diarthrotic; gliding; inversion and eversion of foot
	Tarsometatarsal	Tarsal(s) and metatarsal(s)	Synovial; plane	Diarthrotic; gliding of metatarsals
	Metatarso-phalangeal	Metatarsal and proximal phalanx	Synovial; condylar	Diarthrotic; biaxial; flexion, extension, abduction, adduction, circumduction of great toe
	Interpha-langeal (toe)	Adjacent phalanges	Synovial; hinge	Diarthrotic; uniaxial; flexion, extension of toes

***Fibrous joints** indicated by orange circles (•); **cartilaginous joints** by blue circles (•); **synovial joints** by purple circles (•).
† These modified hinge joints are structurally bicondylar.

- **Reinforcing ligaments.** Synovial joints are reinforced and strengthened by a number of bandlike **ligaments**. Most often, these are **capsular ligaments**, which are thickened parts of the fibrous layer. In other cases, they remain distinct and are found outside the capsule (as **extracapsular ligaments**) or deep to it (as **intracapsular ligaments**). Since intracapsular ligaments are covered with synovial membrane, they do not actually lie *within* the joint cavity.

  People said to be double-jointed amaze the rest of us by placing both heels behind their neck. However, they have the normal number of joints. It's just that their joint capsules and ligaments are more stretchy and loose than average.

- **Nerves and blood vessels.** Synovial joints are richly supplied with sensory nerve fibers that innervate the capsule. Some of these fibers detect pain, as anyone who has suffered joint injury is aware, but most monitor joint position and stretch. Monitoring joint stretch is one of several ways the nervous system senses our posture and body movements (see p. 436).

  Synovial joints are also richly supplied with blood vessels, most of which supply the synovial membrane. There, extensive capillary beds produce the blood filtrate that is the basis of synovial fluid.

Besides the basic components just described, certain synovial joints have other structural features. Some, such as the hip and knee joints, have cushioning **fatty pads** between the fibrous layer and the synovial membrane or bone. Others have discs or wedges of fibrocartilage separating the articular surfaces. Where present, these **articular discs**, or **menisci** (mě-nis′ki; "crescents"), extend inward from the articular capsule and partially or completely divide the synovial cavity in two (see the menisci of the knee in Figure 8.7a, b, e, and f). Articular discs improve the fit between articulating bone ends, making the joint more stable and minimizing wear and tear on the joint surfaces. Besides the knees, articular discs occur in the jaw and a few other joints (see notations in the Structural Type column in Table 8.2).

Bursae and Tendon Sheaths

Bursae and tendon sheaths are not strictly part of synovial joints, but they are often found closely associated with them (**Figure 8.4**). Essentially bags of lubricant, they act as "ball bearings" to reduce friction between adjacent structures during joint activity. **Bursae** (ber′se; "purse") are flattened fibrous sacs lined with synovial membrane and containing a thin film of synovial fluid. They occur where ligaments, muscles, skin, tendons, or bones rub together.

A **tendon sheath** is essentially an elongated bursa that wraps completely around a tendon subjected to friction, like a bun around a hot dog. They are common where several tendons are crowded together within narrow canals (in the wrist, for example).

Factors Influencing the Stability of Synovial Joints

Because joints are constantly stretched and compressed, they must be stabilized so that they do not dislocate (come out of alignment). The stability of a synovial joint depends chiefly on three factors: the shapes of the articular surfaces; the number and positioning of ligaments; and muscle tone.

Articular Surfaces

The shapes of articular surfaces determine what movements are possible at a joint, but surprisingly, articular surfaces play only a minor role in joint stability. Many joints have shallow sockets or noncomplementary articulating surfaces ("misfits") that actually hinder joint stability. But when articular surfaces are large and fit snugly together, or when the socket is deep, stability is vastly improved. The ball and deep socket of the hip joint provide the best example of a joint made extremely stable by the shape of its articular surfaces.

Ligaments

The capsules and ligaments of synovial joints unite the bones and prevent excessive or undesirable motion. As a rule, the more ligaments a joint has, the stronger it is. However, when other stabilizing factors are inadequate, undue tension is placed on the ligaments and they stretch. Stretched ligaments stay stretched, like taffy, and a ligament can stretch only about 6% of its length before it snaps. Thus, when ligaments are the major means of bracing a joint, the joint is not very stable.

Muscle Tone

For most joints, the muscle tendons that cross the joint are the most important stabilizing factor. These tendons are kept under tension by the tone of their muscles. (*Muscle tone* is defined as low levels of contractile activity in relaxed muscles that keep the muscles healthy and ready to react to stimulation.) Muscle tone is extremely important in reinforcing the shoulder and knee joints and the arches of the foot.

Movements Allowed by Synovial Joints

Every skeletal muscle of the body is attached to bone or other connective tissue structures at no fewer than two points. The muscle's **origin** is attached to the immovable (or less movable) bone. Its other end, the **insertion**, is attached to the movable bone. Body movement occurs when muscles contract across joints and their insertion moves toward their origin. The movements can be described in directional terms relative to the lines, or axes, around which the body part moves and the planes of space along which the movement occurs, that is, along the transverse, frontal, or sagittal plane. (See Chapter 1 to review these planes.)

Range of motion allowed by synovial joints varies from **nonaxial movement** (slipping movements only) to **uniaxial**

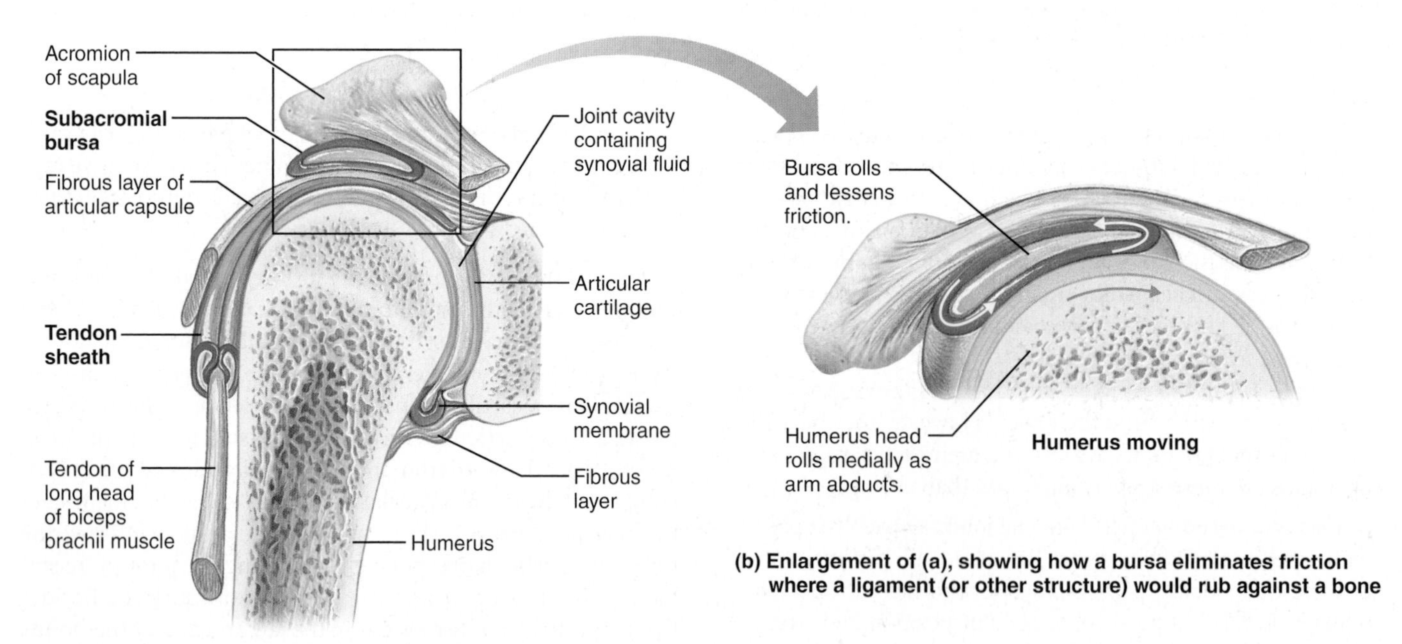

(a) Frontal section through the right shoulder joint

(b) Enlargement of (a), showing how a bursa eliminates friction where a ligament (or other structure) would rub against a bone

Figure 8.4 Bursae and tendon sheaths.

movement (movement in one plane) to **biaxial movement** (movement in two planes) to **multiaxial movement** (movement in or around all three planes of space and axes). Range of motion varies greatly. In some people, such as trained gymnasts or acrobats, range of joint movement may be extraordinary. The ranges of motion at the major joints are given in the far right column of Table 8.2.

There are three general types of movements: *gliding*, *angular movements*, and *rotation*. The most common body movements allowed by synovial joints are described next and illustrated in **Figure 8.5**.

Gliding Movements

Gliding occurs when one flat, or nearly flat, bone surface glides or slips over another (back-and-forth and side-to-side; Figure 8.5a) without appreciable angulation or rotation. Gliding occurs at the intercarpal and intertarsal joints, and between the flat articular processes of the vertebrae (Table 8.2).

Angular Movements

Angular movements (Figure 8.5b–e) increase or decrease the angle between two bones. These movements may occur in any plane of the body and include flexion, extension, hyperextension, abduction, adduction, and circumduction.

Flexion **Flexion** (flek′shun) is a bending movement, usually along the sagittal plane, that *decreases the angle* of the joint and brings the articulating bones closer together. Examples include bending the head forward on the chest (Figure 8.5b) and bending the body trunk or the knee from a straight to an angled position (Figure 8.5c and d). As a less obvious example, the arm is flexed at the shoulder when the arm is lifted in an anterior direction (Figure 8.5d).

Extension **Extension** is the reverse of flexion and occurs at the same joints. It involves movement along the sagittal plane that *increases the angle* between the articulating bones and typically straightens a flexed limb or body part. Examples include straightening a flexed neck, body trunk, elbow, or knee (Figure 8.5b–d). Continuing such movements beyond the anatomical position is called **hyperextension** (Figure 8.5b–d).

Abduction **Abduction** ("moving away") is movement of a limb *away* from the midline or median plane of the body, along the frontal plane. Raising the arm or thigh laterally is an example of abduction (Figure 8.5e). For the fingers or toes, abduction means spreading them apart. In this case the "midline" is the third finger or second toe. Notice, however, that lateral bending of the trunk away from the body midline in the frontal plane is called lateral flexion, not abduction.

Adduction **Adduction** ("moving toward") is the opposite of abduction, so it is the movement of a limb *toward* the body midline or, in the case of the digits, toward the midline of the hand or foot (Figure 8.5e).

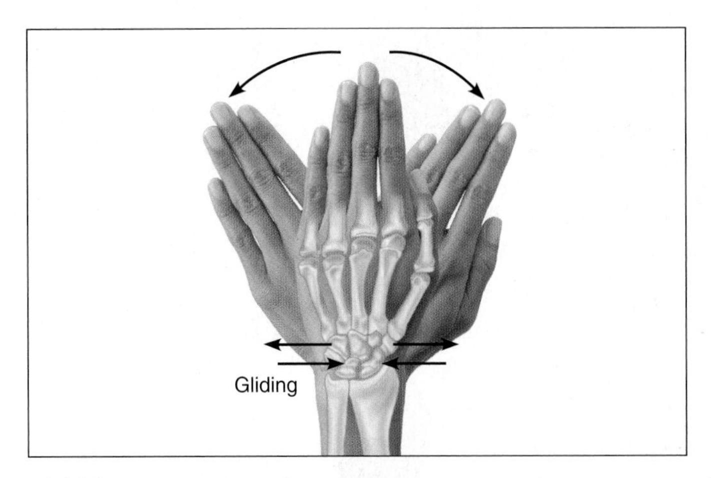

(a) Gliding movements at the wrist

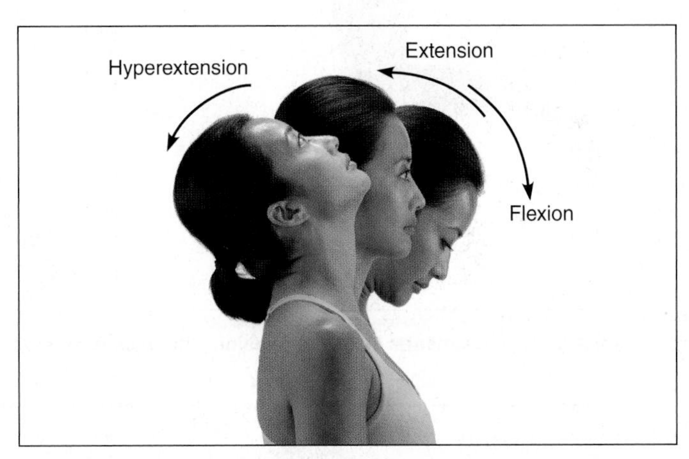

(b) Angular movements: flexion, extension, and hyperextension of the neck

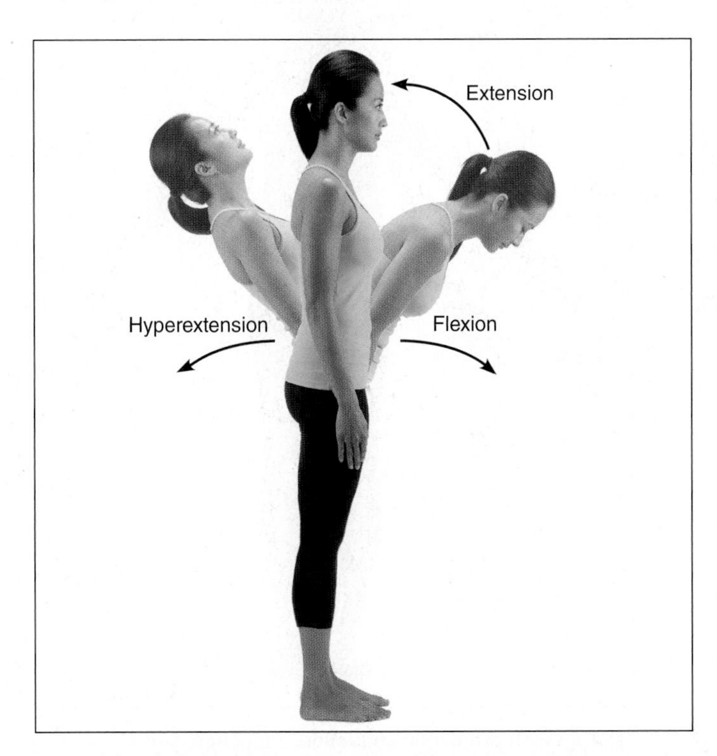

(c) Angular movements: flexion, extension, and hyperextension of the vertebral column

Figure 8.5 Movements allowed by synovial joints.

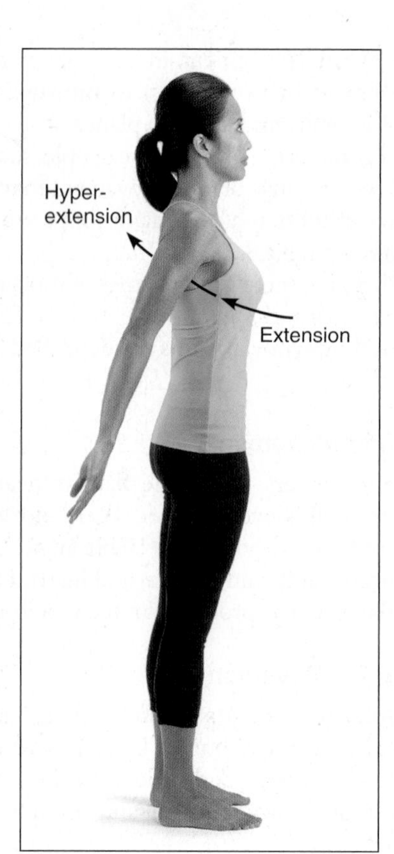

(d) Angular movements: flexion, extension, and hyperextension at the shoulder and knee

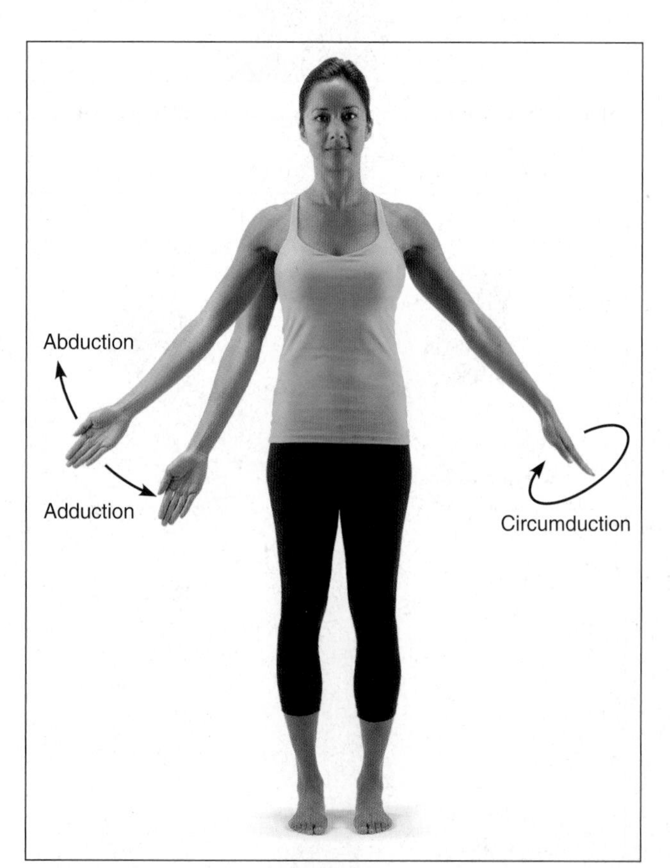

(e) Angular movements: abduction, adduction, and circumduction of the upper limb at the shoulder

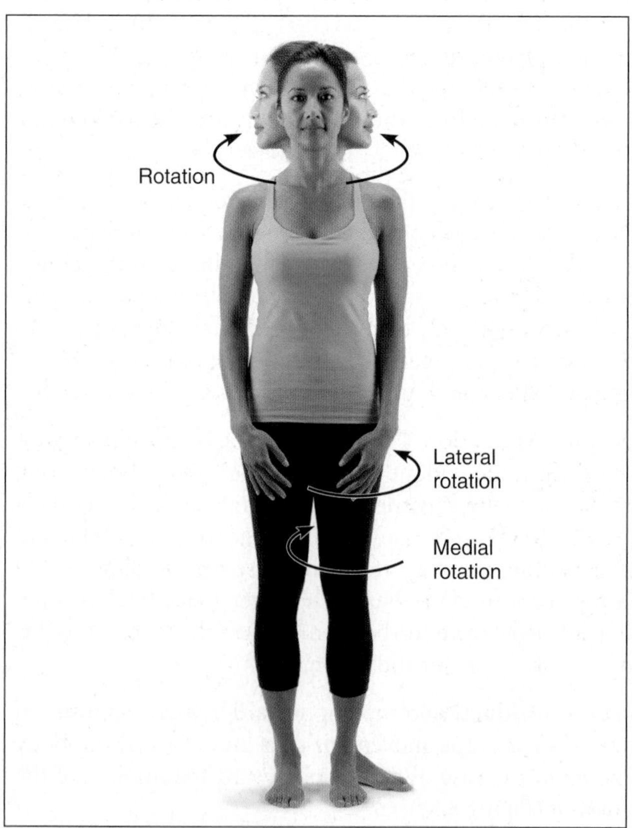

(f) Rotation of the head, neck, and lower limb

Figure 8.5 *(continued)* **Movements allowed by synovial joints.**

Circumduction **Circumduction** (Figure 8.5e) is moving a limb so that it describes a cone in space (*circum* = around; *duco* = to draw). The distal end of the limb moves in a circle, while the point of the cone (the shoulder or hip joint) is more or less stationary. A pitcher winding up to throw a ball is actually circumducting his or her pitching arm. Because circumduction consists of flexion, abduction, extension, and adduction performed in succession, it is the quickest way to exercise the many muscles that move the hip and shoulder ball-and-socket joints.

Rotation

Rotation is the turning of a bone around its own long axis. It is the only movement allowed between the first two cervical vertebrae and is common at the hip (Figure 8.5f) and shoulder joints. Rotation may be directed toward the midline or away from it. For example, in *medial rotation* of the thigh, the femur's anterior surface moves toward the median plane of the body; *lateral rotation* is the opposite movement.

Special Movements

Certain movements do not fit into any of the above categories and occur at only a few joints. Some of these special movements are illustrated in **Figure 8.6** on p. 232.

Supination and Pronation The terms **supination** (soo″pĭ-na′shun; "turning backward") and **pronation** (pro-na′shun; "turning forward") refer to the movements of the radius around the ulna (Figure 8.6a). Rotating the forearm laterally so that the palm faces anteriorly or superiorly is supination. In the anatomical position, the hand is supinated and the radius and ulna are parallel.

In pronation, the forearm rotates medially and the palm faces posteriorly or inferiorly. Pronation moves the distal end of the radius across the ulna so that the two bones form an X. This is the forearm's position when we are standing in a relaxed manner. Pronation is a much weaker movement than supination.

A trick to help you keep these terms straight: A *pro* basketball player pronates his or her forearm to dribble the ball.

Dorsiflexion and Plantar Flexion of the Foot The up-and-down movements of the foot at the ankle are given more specific names (Figure 8.6b). Lifting the foot so that its superior surface approaches the shin is **dorsiflexion** (corresponds to wrist extension), whereas depressing the foot (pointing the toes) is **plantar flexion** (corresponds to wrist flexion).

Inversion and Eversion **Inversion** and **eversion** are special movements of the foot (Figure 8.6c). In inversion, the sole of the foot turns medially. In eversion, the sole faces laterally.

Protraction and Retraction Nonangular anterior and posterior movements in a transverse plane are called **protraction** and **retraction**, respectively (Figure 8.6d). The mandible is protracted when you jut out your jaw and retracted when you bring it back.

Elevation and Depression **Elevation** means lifting a body part superiorly (Figure 8.6e). For example, the scapulae are elevated when you shrug your shoulders. Moving the elevated part inferiorly is **depression**. During chewing, the mandible is alternately elevated and depressed.

Opposition The saddle joint between metacarpal I and the trapezium allows a movement called **opposition** of the thumb (Figure 8.6f). This movement is the action taken when you touch your thumb to the tips of the other fingers on the same hand. It is opposition that makes the human hand such a fine tool for grasping and manipulating objects.

Types of Synovial Joints

Although all synovial joints have structural features in common, they do not have a common structural plan. Based on the shape of their articular surfaces, which in turn determine the movements allowed, synovial joints can be classified further into six major categories—plane, hinge, pivot, condylar (or ellipsoid), saddle, and ball-and-socket joints. The properties of these joints are summarized in *Focus on Types of Synovial Joints* (**Focus Figure 8.1**) on pp. 230–231.

☑ Check Your **Understanding**

5. How do bursae and tendon sheaths improve joint function?
6. Generally speaking, what factor is most important in stabilizing synovial joints?
7. John bent over to pick up a dime. What movement was occurring at his hip joint, at his knees, and between his index finger and thumb?
8. On the basis of movement allowed, which of the following joints are uniaxial? Hinge, condylar, saddle, pivot.

For answers, see Answers Appendix.

8.5 Five examples illustrate the diversity of synovial joints

→ Learning Objective

☐ **Describe the knee, shoulder, elbow, hip, and jaw joints in terms of articulating bones, anatomical characteristics of the joint, movements allowed, and joint stability.**

In this section, we examine five joints in detail: knee, shoulder, elbow, hip, and temporomandibular (jaw) joints. All have the six distinguishing characteristics of synovial joints, and we will not discuss these common features again. Instead, we will emphasize the unique structural features, functional abilities, and, in certain cases, functional weaknesses of each of these joints.

Knee Joint

The knee joint is the largest and most complex joint in the body (**Figure 8.7**, pp. 233–234). Despite its single joint cavity, the knee consists of three joints in one: an intermediate one between the patella and the lower end of the femur (the **femoropatellar joint**), and lateral and medial joints (collectively known as the **tibiofemoral joint**) between the femoral condyles above and the C-shaped **menisci**, or *semilunar cartilages*, of the tibia below (Figure 8.7b and e). Besides deepening the shallow tibial articular surfaces, the menisci help prevent side-to-side rocking of the femur on the tibia and absorb shock transmitted to

(Text continues on p. 233.)

Focus Figure 8.1 Six types of synovial joint shapes determine the movements that can occur at a joint.

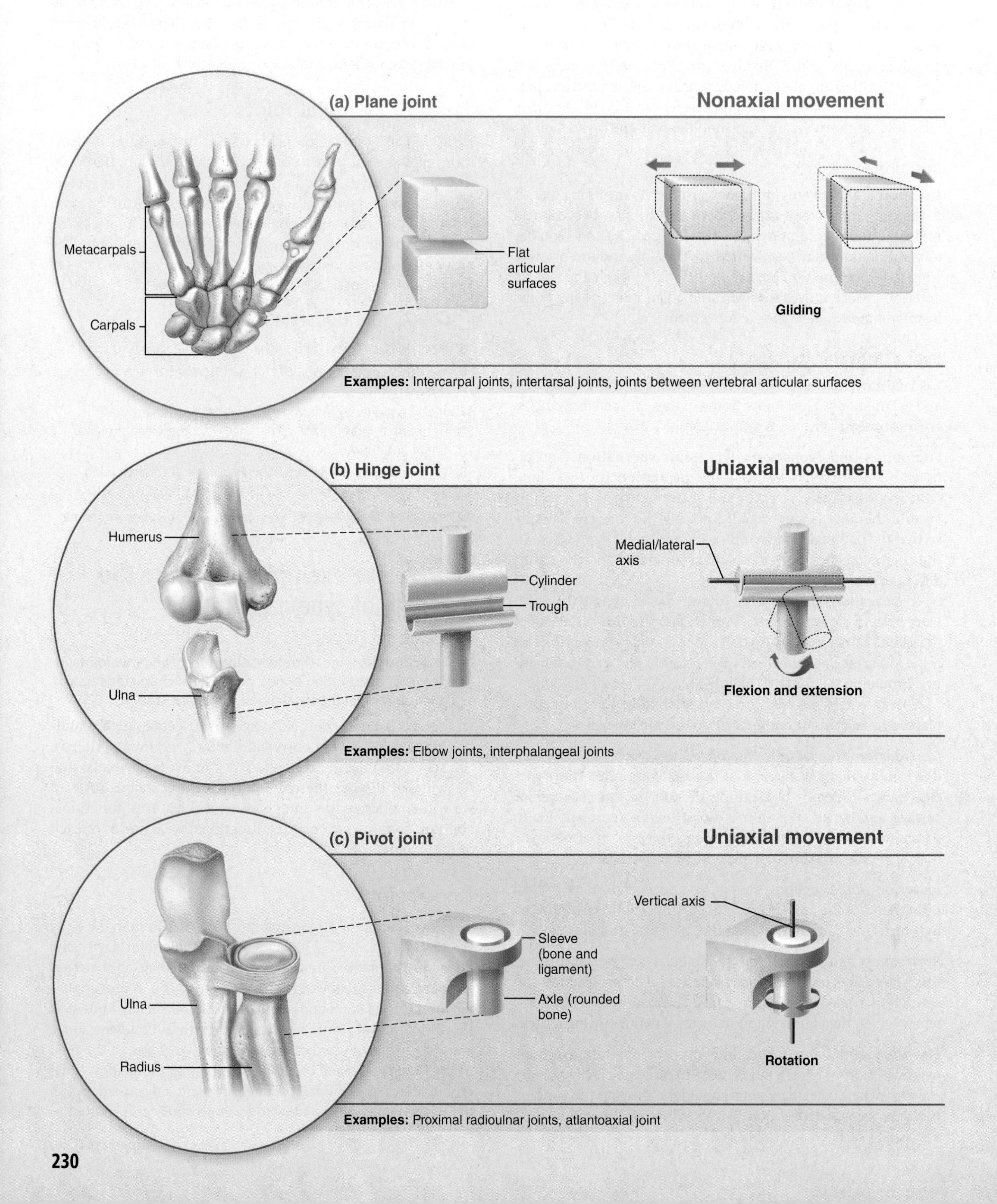

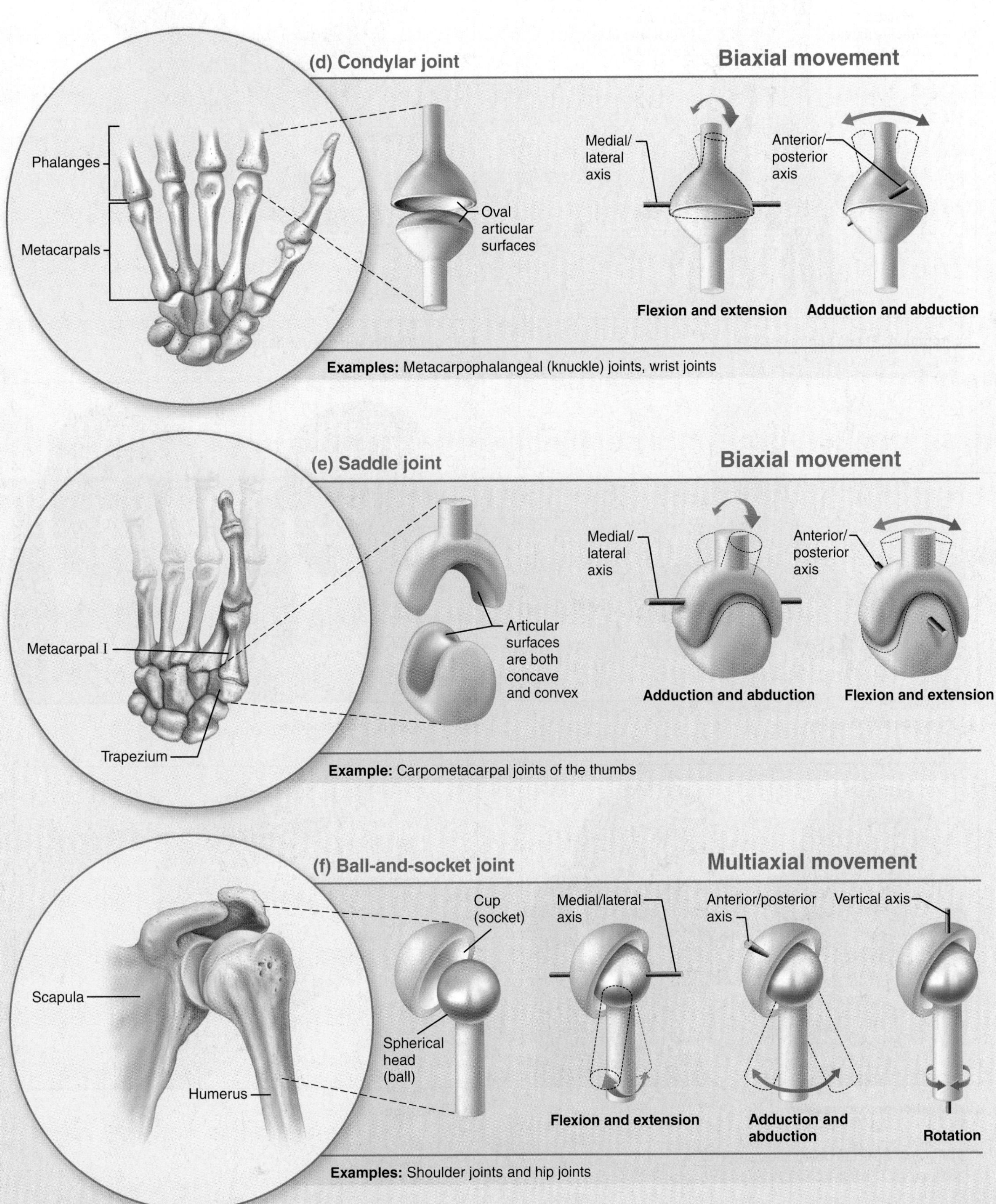
(d) Condylar joint
Biaxial movement
Phalanges
Metacarpals
Oval articular surfaces
Medial/ lateral axis
Anterior/ posterior axis
Flexion and extension
Adduction and abduction
Examples: Metacarpophalangeal (knuckle) joints, wrist joints
(e) Saddle joint
Biaxial movement
Metacarpal I
Trapezium
Articular surfaces are both concave and convex
Medial/ lateral axis
Anterior/ posterior axis
Adduction and abduction
Flexion and extension
Example: Carpometacarpal joints of the thumbs
(f) Ball-and-socket joint
Multiaxial movement
Scapula
Humerus
Cup (socket)
Spherical head (ball)
Medial/lateral axis
Anterior/posterior axis
Vertical axis
Flexion and extension
Adduction and abduction
Rotation
Examples: Shoulder joints and hip joints

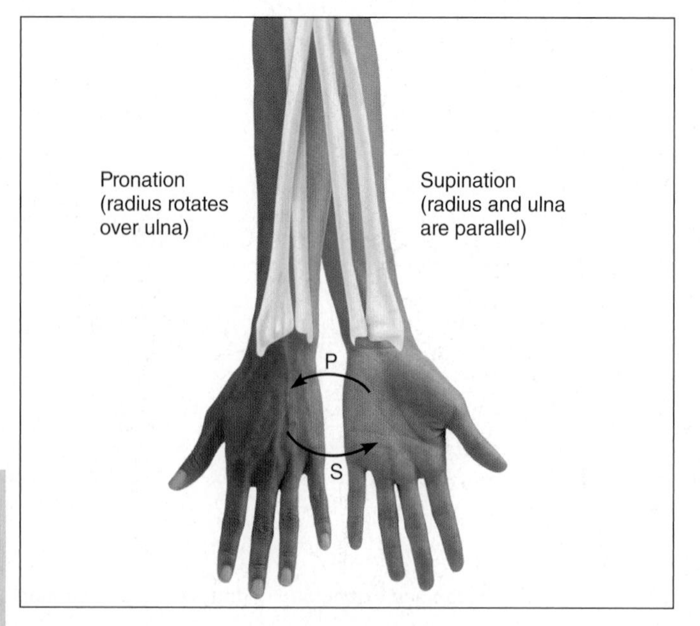

(a) Pronation (P) and supination (S)

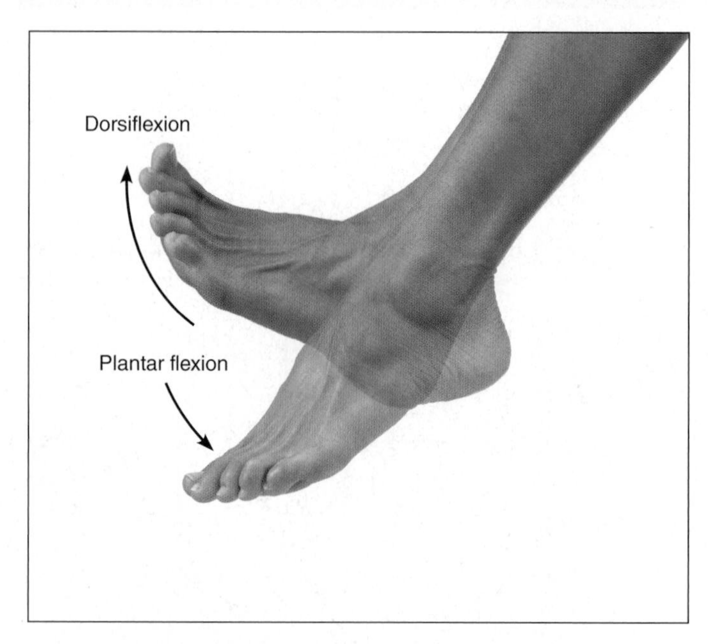

(b) Dorsiflexion and plantar flexion

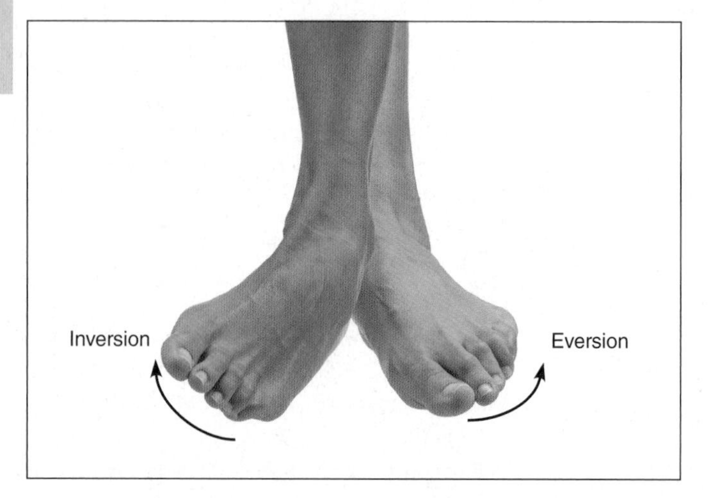

(c) Inversion and eversion

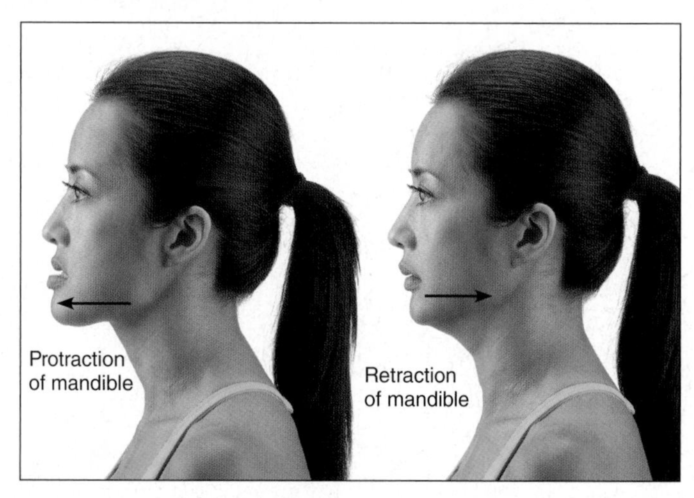

(d) Protraction and retraction

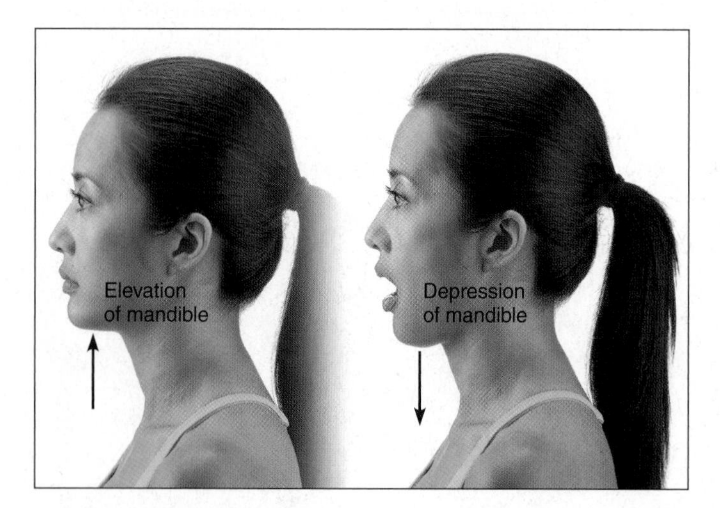

(e) Elevation and depression

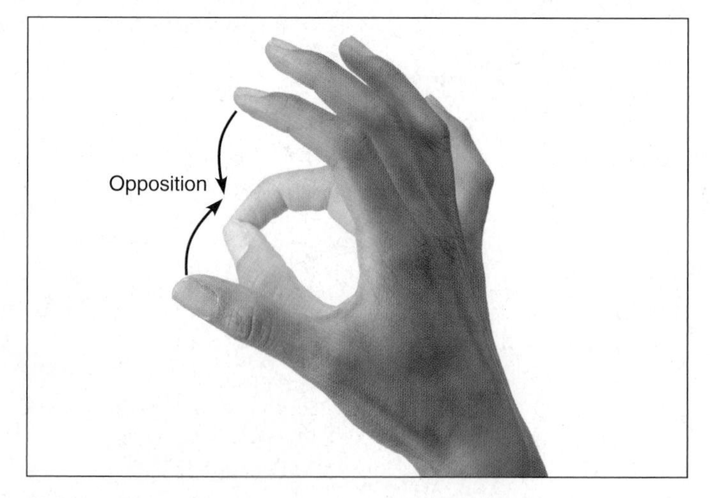

(f) Opposition

Figure 8.6 Special body movements.

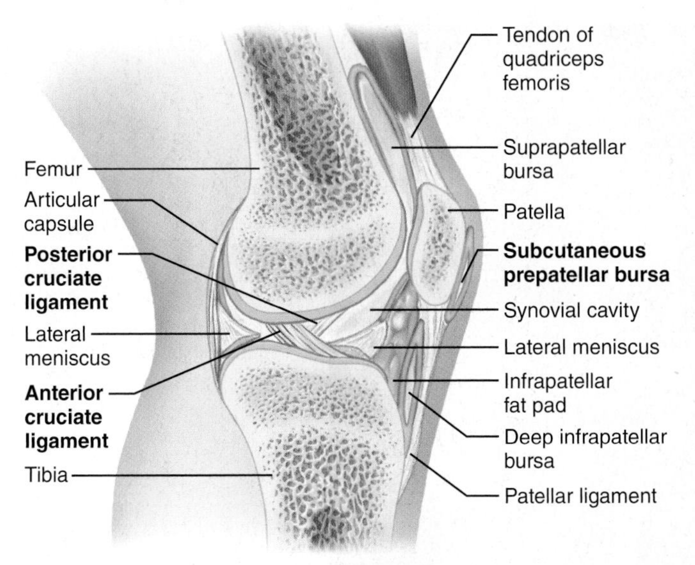

(a) Sagittal section through the right knee joint

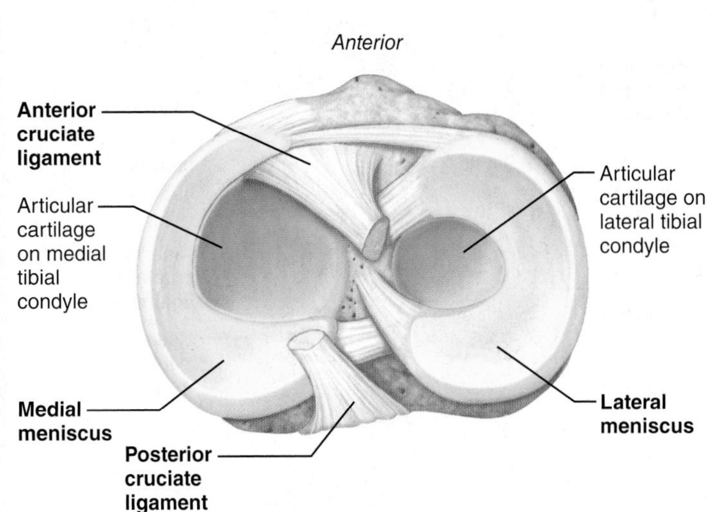

(b) Superior view of the right tibia in the knee joint, showing the menisci and cruciate ligaments

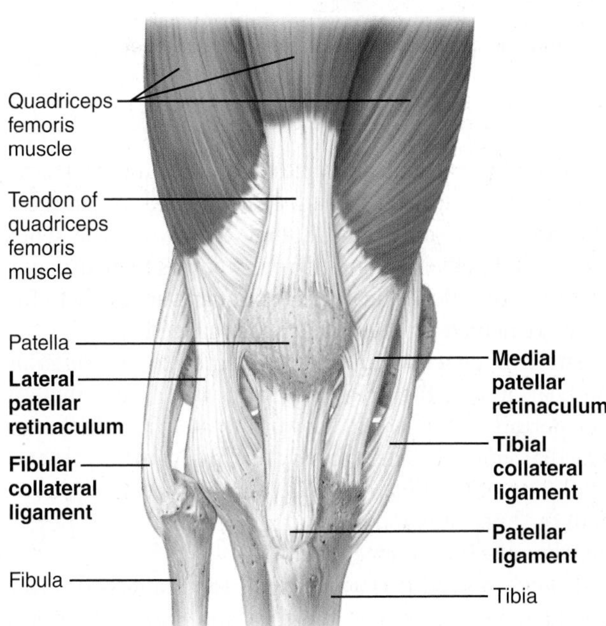
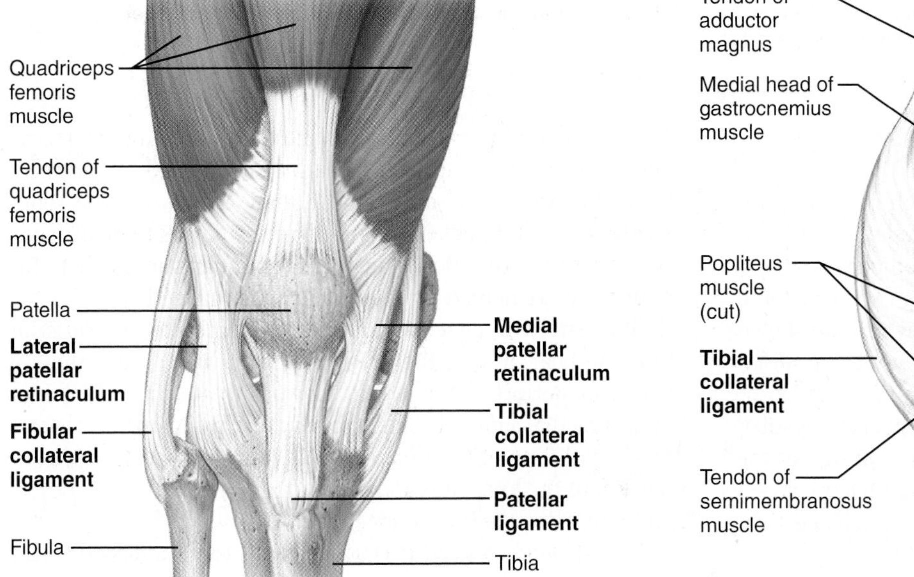

(c) Anterior view of right knee

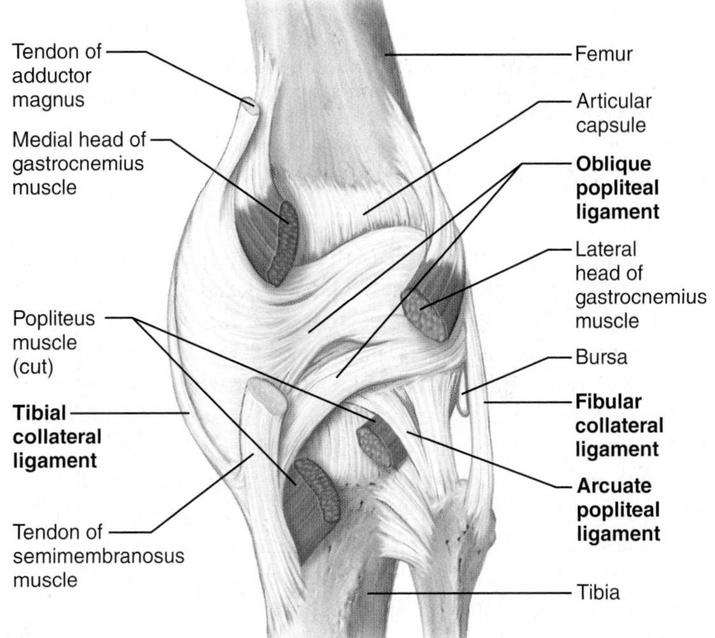

(d) Posterior view of the joint capsule, including ligaments

Figure 8.7 The knee joint.

Explore human cadaver
MasteringA&P®>Study Area>PAL

the knee joint. However, the menisci are attached only at their outer margins and are frequently torn free.

The tibiofemoral joint acts primarily as a hinge, permitting flexion and extension. However, structurally it is a bicondylar joint. Some rotation is possible when the knee is partly flexed, and when the knee is extending. But, when the knee is fully extended, side-to-side movements and rotation are strongly resisted by ligaments and the menisci. The femoropatellar joint is a plane joint, and the patella glides across the distal end of the femur during knee flexion.

The knee joint is unique in that its joint cavity is only partially enclosed by a capsule. The relatively thin articular capsule is present only on the sides and posterior aspects of the knee, where it covers the bulk of the femoral and tibial condyles. Anteriorly, where the capsule is absent, three broad ligaments run from the patella to the tibia below. These are the **patellar ligament** flanked by the **medial** and **lateral patellar retinacula** (ret″ĭ-nak′u-lah; "retainers"), which merge imperceptibly into the articular capsule on each side (Figure 8.7c). The patellar ligament and retinacula are actually continuations of the tendon

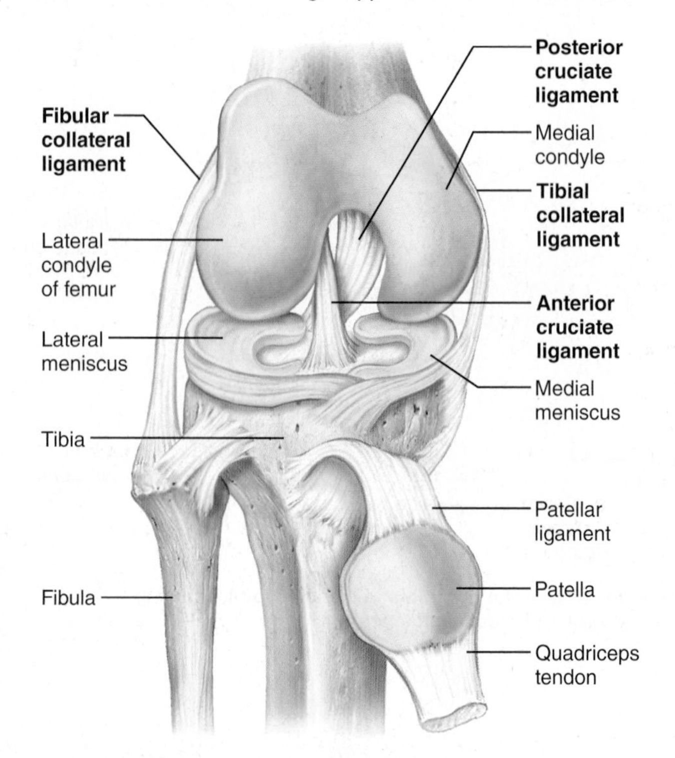

(e) Anterior view of flexed knee, showing the cruciate ligaments (articular capsule removed, and quadriceps tendon cut and reflected distally)

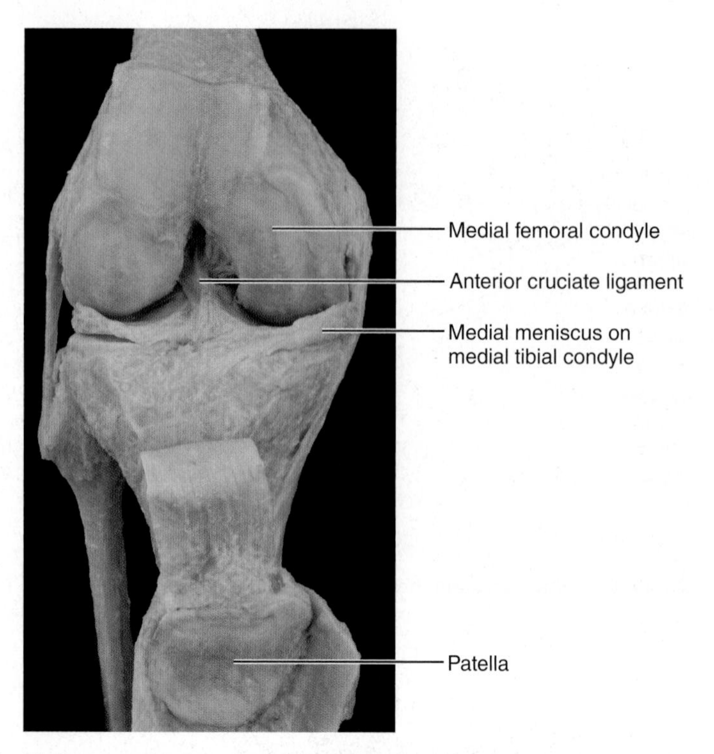

(f) Photograph of an opened knee joint; view similar to (e)

Figure 8.7 *(continued)* **The knee joint.**

of the bulky quadriceps muscle of the anterior thigh. Physicians tap the patellar ligament to test the knee-jerk reflex.

The synovial cavity of the knee joint has a complicated shape, with several extensions that lead into "blind alleys." At least a dozen bursae are associated with this joint, some of which are shown in Figure 8.7a. For example, notice the *subcutaneous prepatellar bursa*, which is often injured when the knee is bumped anteriorly.

All three types of joint ligaments (extracapsular, capsular, and intracapsular) stabilize and strengthen the capsule of the knee joint. All of the capsular and extracapsular ligaments act to prevent hyperextension of the knee and are stretched tight when the knee is extended. These include:

- The extracapsular **fibular** and **tibial collateral ligaments** are also critical in preventing lateral or medial rotation when the knee is extended. The broad, flat tibial collateral ligament runs from the medial epicondyle of the femur to the medial condyle of the tibial shaft below and is fused to the medial meniscus (Figure 8.7c–e).
- The **oblique popliteal ligament** (pop″lĭ-te′al) is actually part of the tendon of the semimembranosus muscle that fuses with the joint capsule and helps stabilize the posterior aspect of the knee joint (Figure 8.7d).
- The **arcuate popliteal ligament** arcs superiorly from the head of the fibula over the popliteus muscle and reinforces the joint capsule posteriorly (Figure 8.7d).

The knee's *intracapsular ligaments* are called *cruciate ligaments* (kroo′she-āt) because they cross each other, forming an X (*cruci* = cross) in the notch between the femoral condyles. They act as restraining straps to help prevent anterior-posterior displacement of the articular surfaces and to secure the articulating bones when we stand (Figure 8.7a, b, e). Although these ligaments are in the joint capsule, they are *outside* the synovial cavity, and synovial membrane nearly covers their surfaces. Note that the two cruciate ligaments both run superiorly to the femur and are named for their *tibial* attachment site.

The **anterior cruciate ligament** attaches to the *anterior* intercondylar area of the tibia (Figure 8.7b, e). From there it passes posteriorly, laterally, and upward to attach to the femur on the medial side of its lateral condyle. This ligament prevents forward sliding of the tibia on the femur and checks hyperextension of the knee. It is somewhat lax when the knee is flexed, and taut when the knee is extended.

The stronger **posterior cruciate ligament** is attached to the *posterior* intercondylar area of the tibia and passes anteriorly, medially, and superiorly to attach to the femur on the lateral side of the medial condyle (Figure 8.7a, b, e). This ligament prevents backward displacement of the tibia or forward sliding of the femur.

The knee capsule is heavily reinforced by muscle tendons. Most important are the strong tendons of the quadriceps muscles of the anterior thigh and the tendon of the semimembranosus muscle posteriorly (Figure 8.7c and d). The greater the strength and tone of these muscles, the less the chance of knee injury.

The knees have a built-in locking device that provides steady support for the body in the standing position. As we begin to stand up, the wheel-shaped femoral condyles roll like ball bearings across the tibial condyles and the flexed leg begins to extend at the knee. Because the lateral femoral condyle stops rolling before the medial condyle stops, the femur *spins* (rotates) medially on the tibia, until the cruciate and collateral ligaments of the knee are twisted and taut and the menisci are

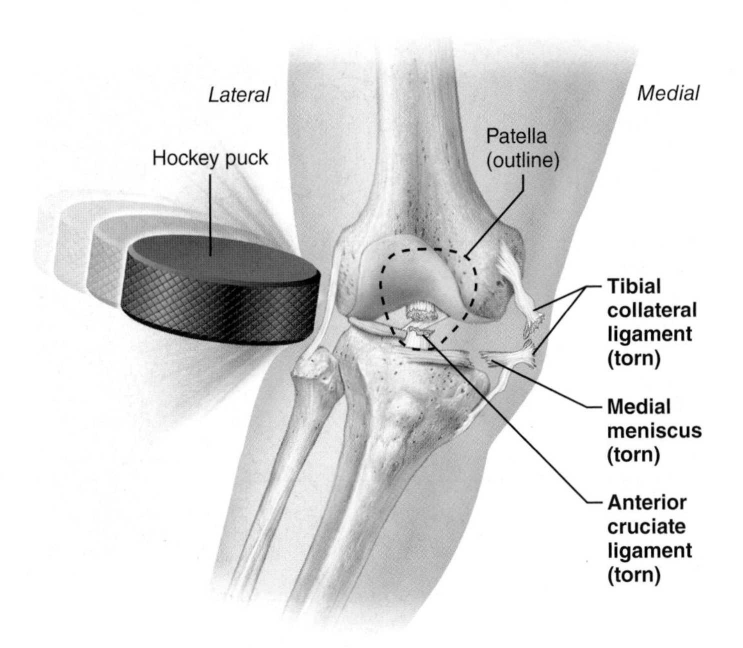

Figure 8.8 The "unhappy triad:" ruptured ACL, ruptured tibial collateral ligament, and torn meniscus. A common injury in hockey, soccer, and American football.

compressed. The tension in the ligaments effectively locks the joint into a rigid structure that cannot be flexed again until it is unlocked. This unlocking is accomplished by the popliteus muscle (see Figure 8.7d and Table 10.15, pp. 331–336). It rotates the femur laterally on the tibia, causing the ligaments to become untwisted and slack.

HOMEOSTATIC IMBALANCE 8.1 CLINICAL

Of all body joints, the knees are most susceptible to sports injuries because of their high reliance on nonarticular factors for stability and the fact that they carry the body's weight. The knee can absorb a vertical force equal to nearly seven times body weight. However, it is very vulnerable to *horizontal* blows, such as those that occur during blocking and tackling in football and in ice hockey.

When thinking of **common knee injuries**, remember the 3 Cs: collateral ligaments, cruciate ligaments, and cartilages (menisci). Most dangerous are *lateral* blows to the extended knee. These forces tear the tibial collateral ligament and the medial meniscus attached to it, as well as the anterior cruciate ligament (ACL) (**Figure 8.8**). It is estimated that 50% of all professional football players have serious knee injuries during their careers.

Although less devastating than the injury just described, injuries that affect only the anterior cruciate ligament are becoming more common, particularly as women's sports become more vigorous and competitive. Most ACL injuries occur when a runner changes direction quickly, twisting a hyperextended knee. A torn ACL heals poorly, so repair usually requires a graft taken from either the patellar ligament, the hamstring tendon, or the calcaneal tendon. ✚

Shoulder (Glenohumeral) Joint

In the shoulder joint, stability has been sacrificed to provide the most freely moving joint of the body. The shoulder joint is a ball-and-socket joint. The large hemispherical head of the humerus fits in the small, shallow glenoid cavity of the scapula (**Figure 8.9**), like a golf ball sitting on a tee. Although the glenoid cavity is slightly deepened by a rim of fibrocartilage, the **glenoid labrum** (*labrum* = lip), it is only about one-third the size of the humeral head and contributes little to joint stability (Figure 8.9d).

The articular capsule enclosing the joint cavity (from the margin of the glenoid cavity to the anatomical neck of the humerus) is remarkably thin and loose, qualities that contribute to this joint's freedom of movement. The few ligaments reinforcing the shoulder joint are located primarily on its anterior aspect. The superiorly located **coracohumeral ligament** (kor′ah-ko-hu′mer-ul) provides the only strong thickening of the capsule and helps support the weight of the upper limb (Figure 8.9c). Three **glenohumeral ligaments** (glē″no-hu′mer-ul) strengthen the front of the capsule somewhat but are weak and may even be absent (Figure 8.9c, d).

Muscle tendons that cross the shoulder joint contribute most to this joint's stability. The "superstabilizer" is the tendon of the long head of the biceps brachii muscle of the arm (Figure 8.9c). This tendon attaches to the superior margin of the glenoid labrum, travels through the joint cavity, and then runs within the intertubercular sulcus of the humerus. It secures the head of the humerus against the glenoid cavity.

Four other tendons (and the associated muscles) make up the **rotator cuff**. This cuff encircles the shoulder joint and blends with the articular capsule. The muscles include the subscapularis, supraspinatus, infraspinatus, and teres minor. (The rotator cuff muscles are illustrated in Figure 10.15, pp. 312–313.) The rotator cuff can be severely stretched when the arm is vigorously circumducted; this is a common injury of baseball pitchers. As noted in Chapter 7, shoulder dislocations are fairly common. Because the shoulder's reinforcements are weakest anteriorly and inferiorly, the humerus tends to dislocate in the forward and downward direction.

Elbow Joint

Our upper limbs are flexible extensions that permit us to reach out and manipulate things in our environment. Besides the shoulder joint, the most prominent of the upper limb joints is the elbow. The elbow joint provides a stable and smoothly operating hinge that allows flexion and extension only (**Figure 8.10**). Within the joint, both the radius and ulna articulate with the condyles of the humerus, but it is the close gripping of the trochlea by the ulna's trochlear notch that forms the "hinge" and stabilizes this joint (Figure 8.10a). A relatively lax articular capsule extends inferiorly from the humerus to the ulna and radius, and to the **anular ligament** (an′u-lar) surrounding the head of the radius (Figure 8.10b, c).

Anteriorly and posteriorly, the articular capsule is thin and allows substantial freedom for elbow flexion and extension. However, side-to-side movements are restricted by two strong capsular ligaments: the **ulnar collateral ligament** medially, and the **radial collateral ligament**, a triangular ligament on the lateral

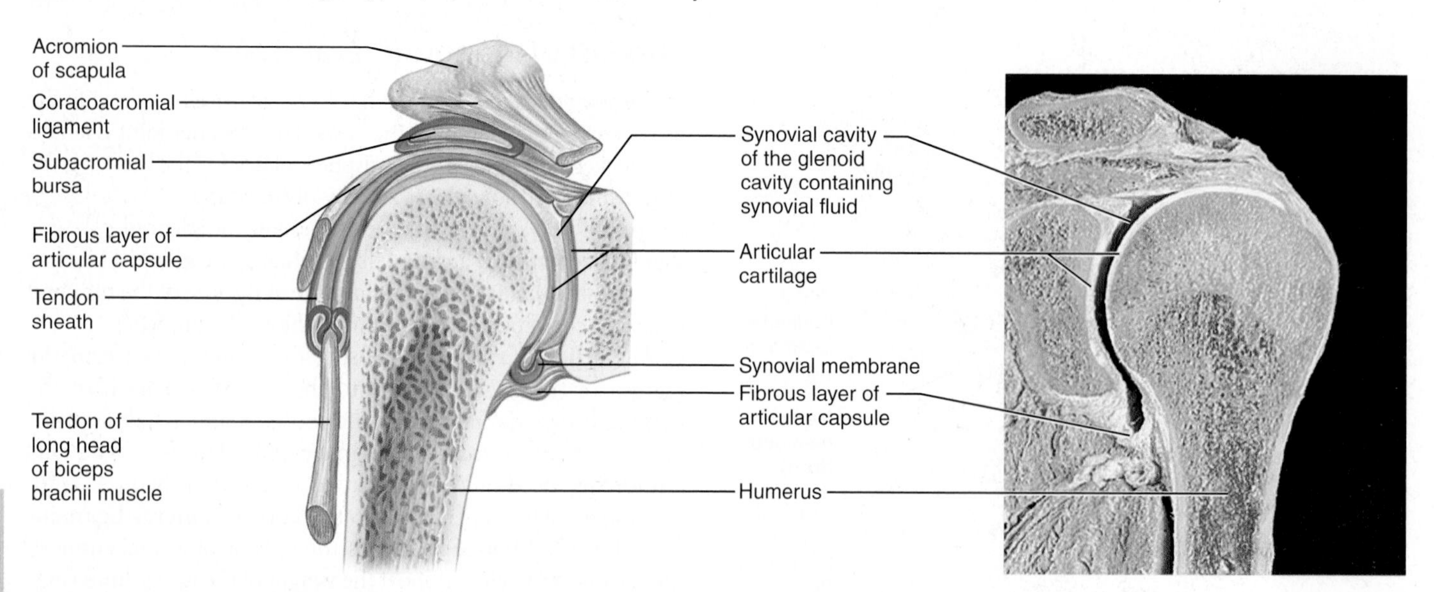

(a) Frontal section through right shoulder joint

(b) Cadaver photo corresponding to (a)

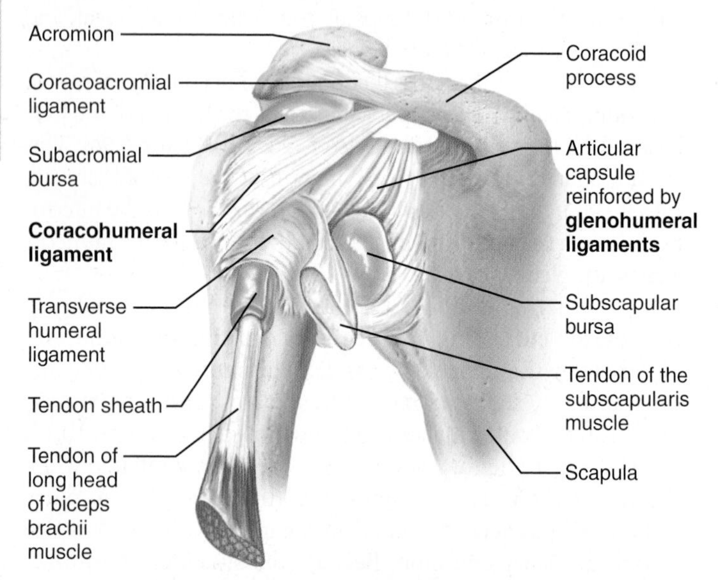

(c) Anterior view of right shoulder joint capsule

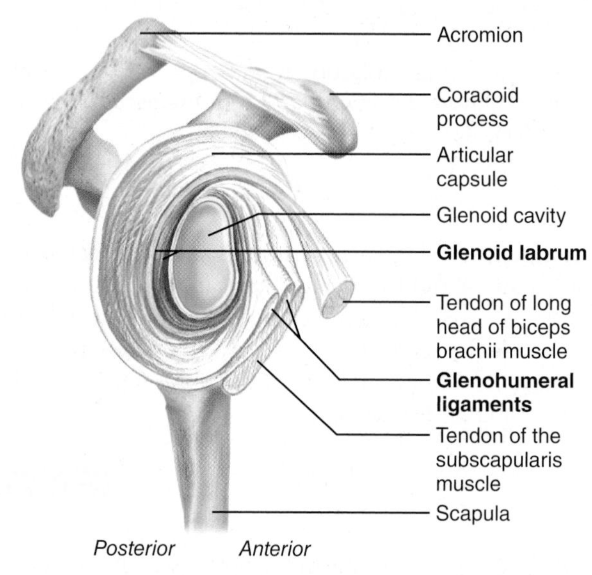

(d) Lateral view of socket of right shoulder joint, humerus removed

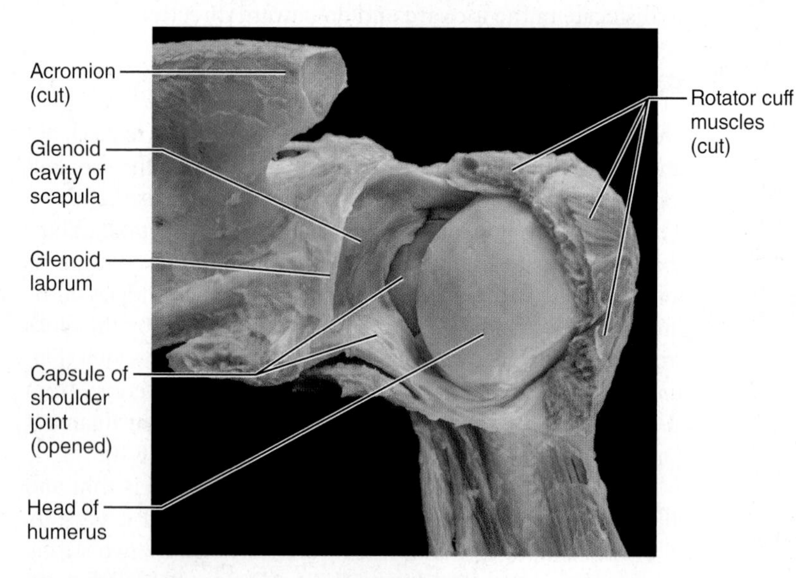

(e) Posterior view of an opened right shoulder joint

Explore human cadaver
MasteringA&P®>Study Area>PAL

Figure 8.9 The shoulder joint.

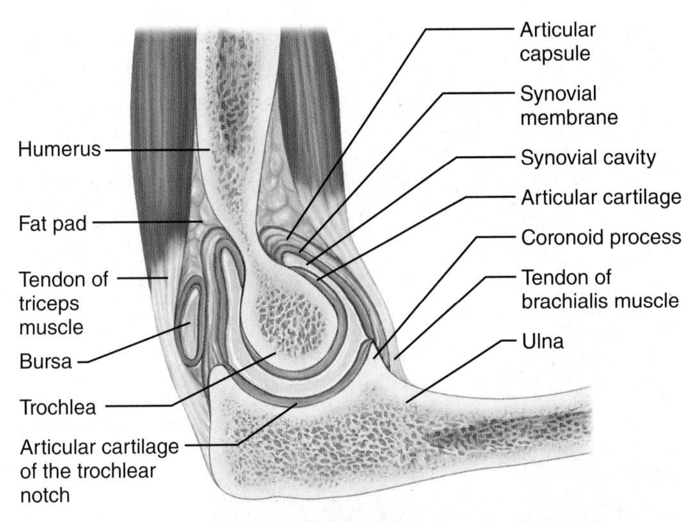

(a) Median sagittal section through right elbow (lateral view)

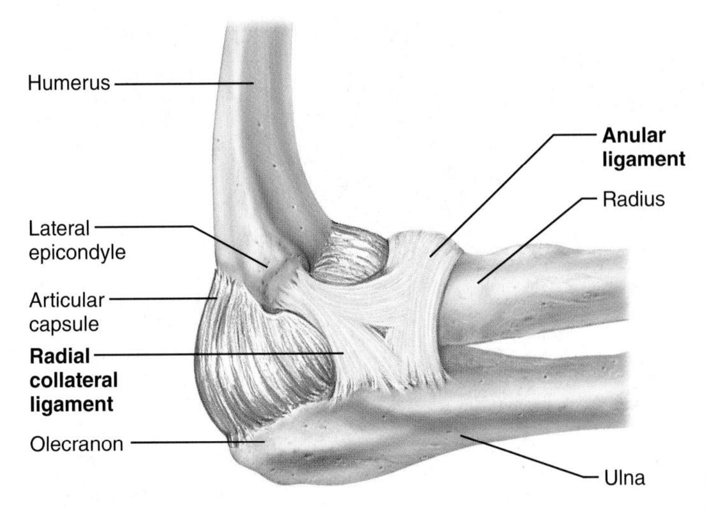

(b) Lateral view of right elbow joint

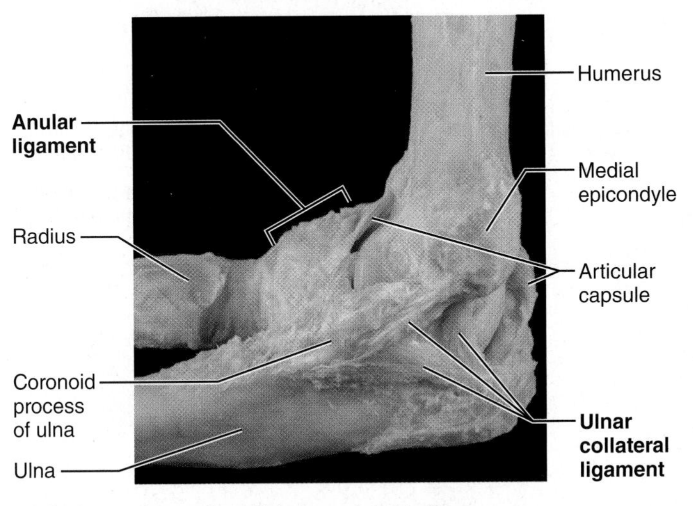

(c) Cadaver photo of medial view of right elbow

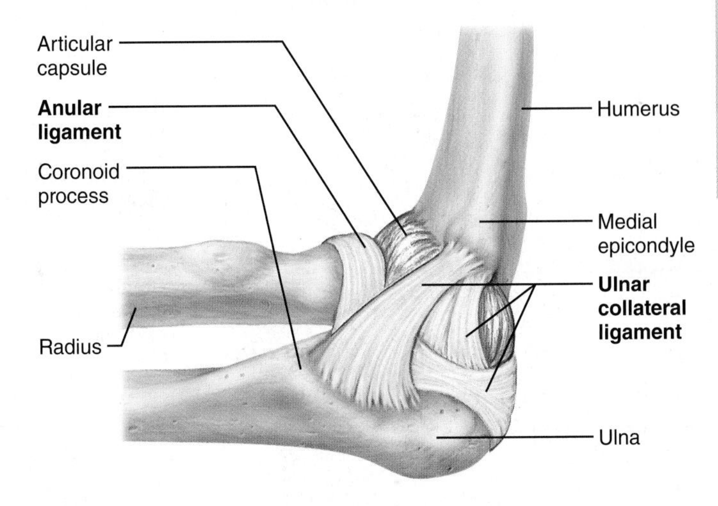

(d) Medial view of right elbow

Figure 8.10 The elbow joint.

Explore human cadaver MasteringA&P®>Study Area>PAL

side (Figure 8.10b, c, and d). Additionally, tendons of several arm muscles, such as the biceps and triceps, cross the elbow joint and provide security.

The radius is a passive "onlooker" in the angular elbow movements. However, its head rotates within the anular ligament during supination and pronation of the forearm.

Hip Joint

The **hip** (coxal) **joint**, like the shoulder joint, is a ball-and-socket joint. It has a good range of motion, but not nearly as wide as the shoulder's range. Movements occur in all possible planes but are limited by the joint's strong ligaments and its deep socket.

The hip joint is formed by the articulation of the spherical head of the femur with the deeply cupped acetabulum of the hip bone (**Figure 8.11**). The depth of the acetabulum is enhanced by a circular rim of fibrocartilage called the **acetabular labrum** (as″ĕ-tab′u-lar) (Figure 8.11a, b). The labrum's diameter is less than that of the head of the femur, and these articular surfaces fit snugly together, so hip joint dislocations are rare.

The thick articular capsule extends from the rim of the acetabulum to the neck of the femur and completely encloses the joint. Several strong ligaments reinforce the capsule of the hip joint. These include the **iliofemoral ligament** (il″e-o-fem′o-ral), a strong V-shaped ligament anteriorly; the **pubofemoral ligament** (pu″bo-fem′o-ral), a triangular thickening of the inferior part of the capsule; and the **ischiofemoral ligament** (is″ke-o-fem′o-ral), a spiraling posterior ligament (Figure 8.11c, d). These ligaments are arranged in such a way that they "screw" the femur head into the acetabulum when a person stands up straight, thereby providing stability.

The **ligament of the head of the femur**, also called the *ligamentum teres*, is a flat intracapsular band that runs from the femur head to the lower lip of the acetabulum (Figure 8.11a, b). This ligament is slack during most hip movements, so it is not important in stabilizing the joint. In fact, its mechanical function (if any) is unclear, but it does contain an artery that helps supply the head of the femur. Damage to this artery may lead to severe arthritis of the hip joint.

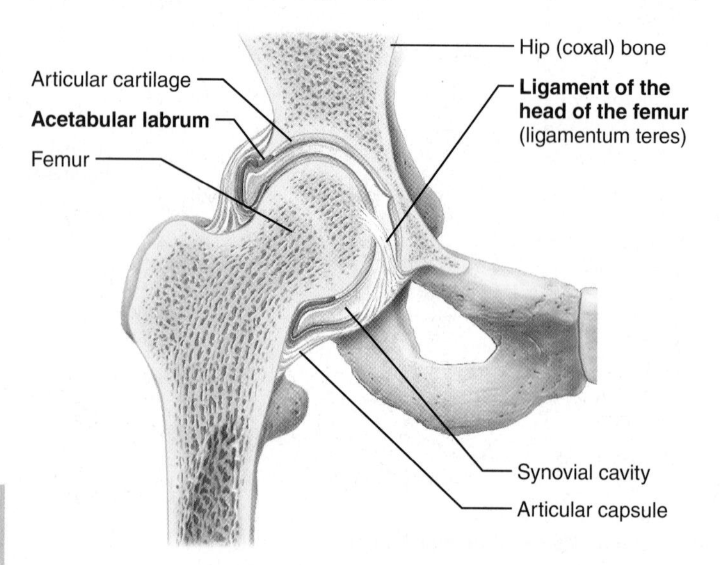

(a) Frontal section through the right hip joint

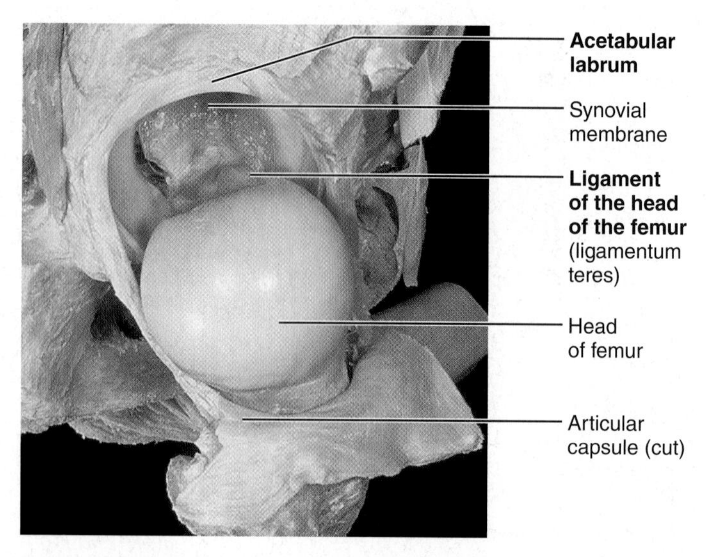

(b) Photo of the interior of the hip joint, lateral view

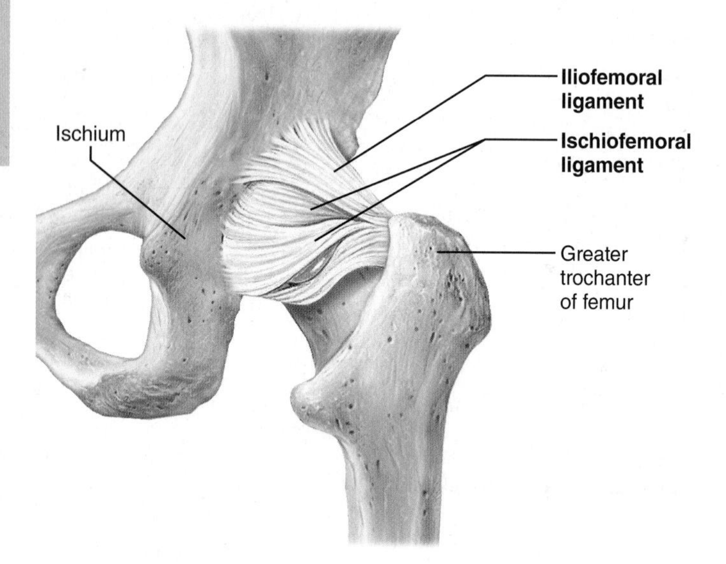

(c) Posterior view of right hip joint, capsule in place

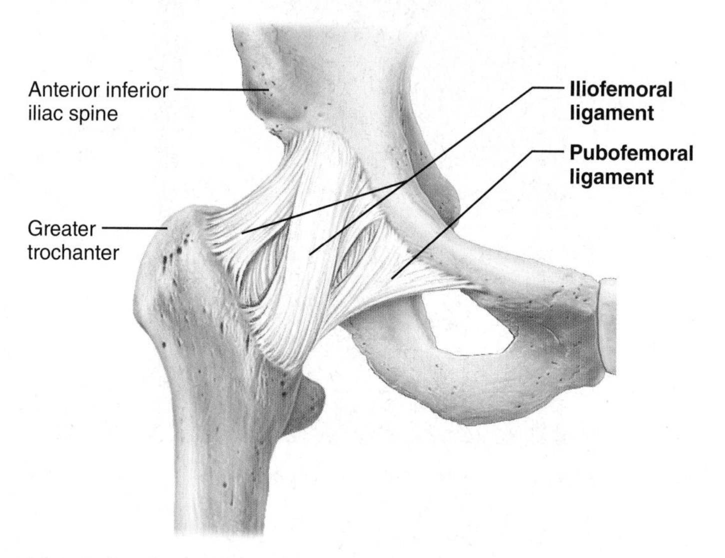

(d) Anterior view of right hip joint, capsule in place

Figure 8.11 The hip joint.

Explore human cadaver MasteringA&P®>Study Area>PAL

Muscle tendons that cross the joint and the bulky hip and thigh muscles that surround it contribute to its stability and strength. In this joint, however, stability comes chiefly from the deep socket that securely encloses the femoral head and the strong capsular ligaments.

Temporomandibular Joint

The **temporomandibular joint** (TMJ), or jaw joint, is a modified hinge joint. It lies just anterior to the ear (**Figure 8.12**). At this joint, the condylar process of the mandible articulates with the inferior surface of the squamous part of the temporal bone. The mandible's condylar process is egg shaped, whereas the articular surface of the temporal bone has a more complex shape. Posteriorly, it forms the concave **mandibular fossa**; anteriorly it forms a dense knob called the **articular tubercle**. The lateral aspect of the loose articular capsule that encloses the joint is thickened into a **lateral ligament**. Within the capsule, an articular disc divides the synovial cavity into superior and inferior compartments (Figure 8.12a, b).

Two distinct kinds of movement occur at the TMJ. First, the concave inferior disc surface receives the condylar process of the mandible and allows the familiar hingelike movement of depressing and elevating the mandible while opening and closing the mouth (Figure 8.12a, b). Second, the superior disc surface glides anteriorly along with the condylar process when the mouth is opened wide. This anterior movement braces the condylar process against the articular tubercle, so that the mandible is not forced through the thin roof of the mandibular fossa when one bites hard foods such as nuts or hard candies. The superior compartment also allows this joint to glide from side to side. As the posterior teeth are drawn into occlusion during grinding, the mandible moves with a side-to-side movement called *lateral excursion* (Figure 8.12c). This lateral jaw movement is unique to mammals and it is readily apparent in horses and cows as they chew.

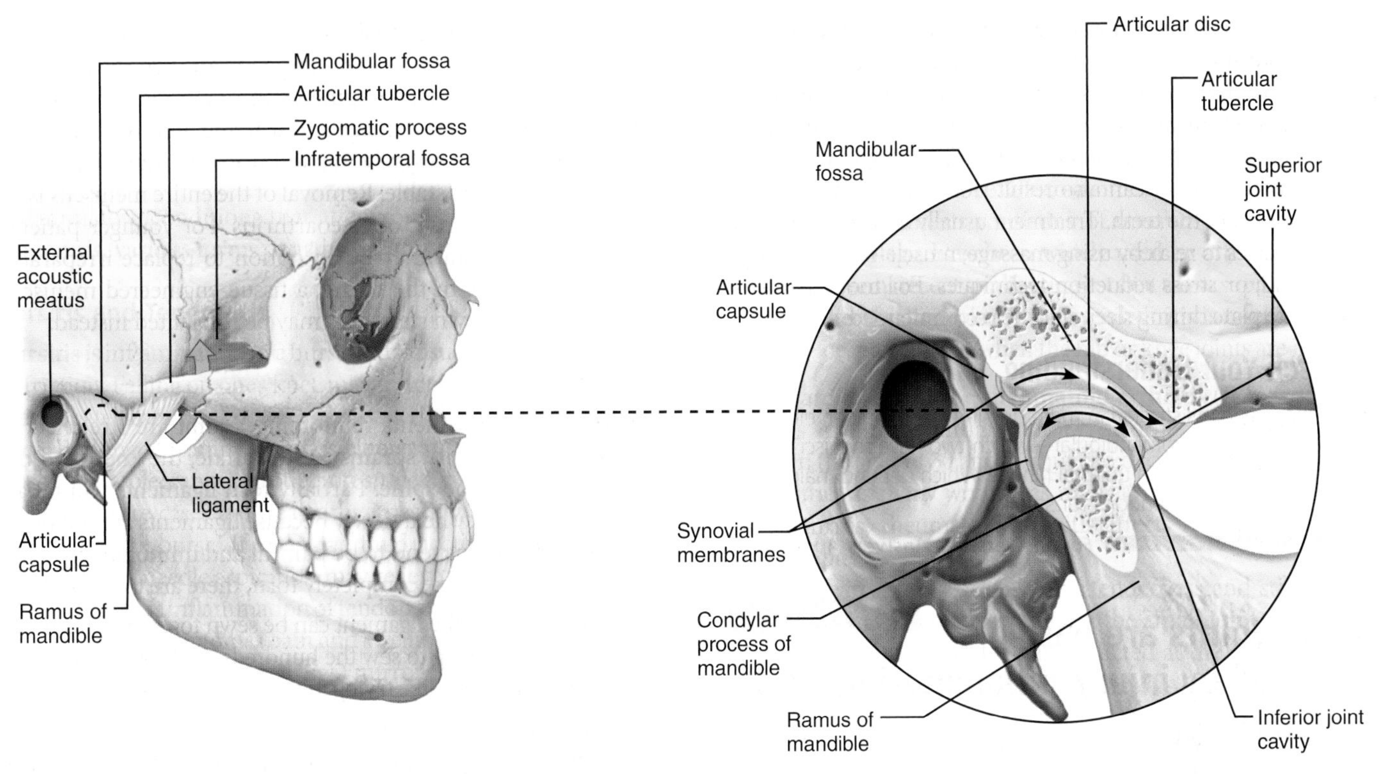

(a) Location of the joint in the skull

(b) Enlargement of a sagittal section through the joint

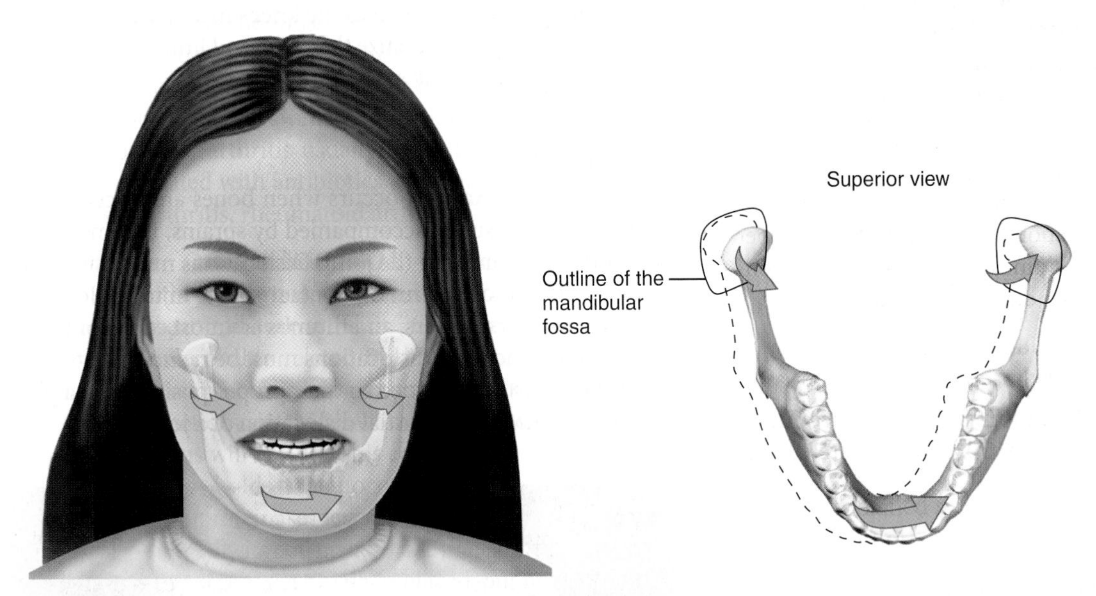

(c) Lateral excursion: lateral (side-to-side) movements of the mandible

Figure 8.12 The temporomandibular (jaw) joint. In **(b)**, note that the two parts of the joint cavity allow different movements, indicated by arrows. The inferior compartment of the joint cavity allows the condylar process of the mandible to rotate in opening and closing the mouth. The superior compartment lets the condylar process move forward to brace against the articular tubercle when the mouth opens wide, and also allows lateral excursion of this joint **(c)**.

HOMEOSTATIC IMBALANCE 8.2

CLINICAL

Dislocations of the TMJ occur more readily than any other joint dislocation because of the shallow socket in the joint. Even a deep yawn can dislocate it. This joint almost always dislocates anteriorly, the condylar process of the mandible ending up in a skull region called the *infratemporal fossa* (Figure 8.12a). In such cases, the mouth remains wide open. To realign a dislocated TMJ, the physician places his or her thumbs in the patient's mouth between the lower molars and the cheeks, and then pushes the mandible inferiorly and posteriorly.

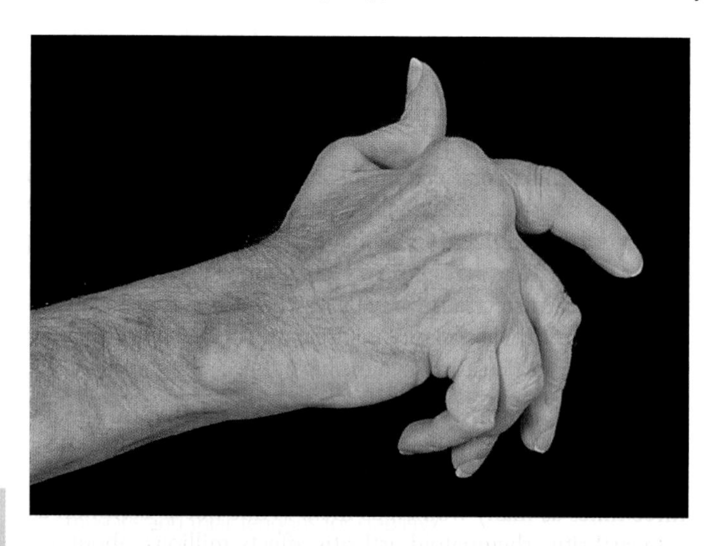

Figure 8.14 A hand deformed by rheumatoid arthritis.

An important target of many of these agents is an inflammatory chemical called *tumor necrosis factor*. Together, these drugs can dramatically slow the course of RA. As a last resort, replacing the joint with a joint prosthesis (artificial joint) may be an option to restore function. Indeed, some RA sufferers have over a dozen artificial joints.

Gouty Arthritis Uric acid, a normal waste product of nucleic acid metabolism, is ordinarily excreted in urine without any problems. However, when blood levels of uric acid rise excessively (due to its excessive production or slow excretion), it may be deposited as needle-shaped urate crystals in the soft tissues of joints. An inflammatory response follows, leading to an agonizingly painful attack of **gouty arthritis** (gow′te), or **gout**. The initial attack typically affects one joint, often at the base of the great toe.

Gout is far more common in men than in women because men naturally have higher blood levels of uric acid (perhaps because estrogens increase the rate of its excretion). Because gout seems to run in families, genetic factors are definitely implicated.

Untreated gout can be very destructive; the articulating bone ends fuse and immobilize the joint. Fortunately, several drugs (colchicine, nonsteroidal anti-inflammatory drugs, glucocorticoids, and others) that terminate or prevent gout attacks are available. Patients are advised to drink plenty of water and to avoid excessive alcohol consumption (which promotes uric acid overproduction) and foods high in purine-containing nucleic acids, such as liver, kidneys, and sardines.

Lyme Disease

Lyme disease is an inflammatory disease caused by spirochete bacteria transmitted by the bite of ticks that live on mice and deer. It often results in joint pain and arthritis, especially in the knees, and is characterized by a skin rash, flu-like symptoms, and foggy thinking. If untreated, neurological disorders and irregular heartbeat may ensue.

Because symptoms vary from person to person, the disease is hard to diagnose. Antibiotic therapy is the usual treatment, but it takes a long time to kill the infecting bacteria.

☑ Check Your Understanding

10. What does the term "arthritis" mean?

11. How would you determine by looking at someone suffering from arthritis if he or she has OA or RA?

12. What is the cause of Lyme disease?

For answers, see Answers Appendix.

The importance of joints is obvious: The skeleton's ability to protect other organs and to move smoothly reflects their presence. Now that we are familiar with joint structure and with the movements that joints allow, we are ready to consider how the muscles attached to the skeleton cause body movements by acting across its joints.

REVIEW QUESTIONS

Multiple Choice/Matching

(Some questions have more than one correct answer. Select the best answer or answers from the choices given.)

1. Match the key terms to the appropriate descriptions.

Key: **(a)** fibrous joints **(b)** cartilaginous joints **(c)** synovial joints

____ **(1)** exhibit a joint cavity
____ **(2)** types are sutures and syndesmoses
____ **(3)** bones connected by collagen fibers
____ **(4)** types include synchondroses and symphyses
____ **(5)** all are diarthrotic
____ **(6)** many are amphiarthrotic
____ **(7)** bones connected by a disc of hyaline cartilage or fibrocartilage
____ **(8)** nearly all are synarthrotic
____ **(9)** shoulder, hip, jaw, and elbow joints

2. Freely movable joints are **(a)** synarthroses, **(b)** diarthroses, **(c)** amphiarthroses.

3. Anatomical characteristics shared by all synovial joints include all except **(a)** articular cartilage, **(b)** a joint cavity, **(c)** an articular capsule, **(d)** presence of fibrocartilage.

4. Factors that influence the stability of a synovial joint include **(a)** shape of articular surfaces, **(b)** presence of strong reinforcing ligaments, **(c)** tone of surrounding muscles, **(d)** all of these.
5. The description "Articular surfaces deep and secure; capsule heavily reinforced by ligaments and muscle tendons; extremely stable joint" best describes **(a)** the elbow joint, **(b)** the hip joint, **(c)** the knee joint, **(d)** the shoulder joint.
6. Ankylosis means **(a)** twisting of the ankle, **(b)** tearing of ligaments, **(c)** displacement of a bone, **(d)** immobility of a joint due to fusion of its articular surfaces.
7. An autoimmune disorder in which joints are affected bilaterally and which involves pannus formation and gradual joint immobilization is **(a)** bursitis, **(b)** gout, **(c)** osteoarthritis, **(d)** rheumatoid arthritis.

Short Answer Essay Questions

8. Define joint.
9. Discuss the relative value (to body homeostasis) of immovable, slightly movable, and freely movable joints.
10. Compare the structure, function, and common body locations of bursae and tendon sheaths.
11. Joint movements may be nonaxial, uniaxial, biaxial, or multiaxial. Define what each of these terms means.
12. Compare and contrast the paired movements of flexion and extension with adduction and abduction.
13. How does rotation differ from circumduction?
14. Name two types of uniaxial, biaxial, and multiaxial joints.
15. What is the specific role of the menisci of the knee? Of the anterior and posterior cruciate ligaments?
16. The knee has been called "a beauty and a beast." Provide several reasons that might explain the negative (beast) part of this description.
17. Why are sprains and cartilage injuries a particular problem?
18. List the functions of the following elements of a synovial joint: fibrous layer of the capsule, synovial fluid, articular cartilage.

AT THE CLINIC

Clinical Case Study
Joints

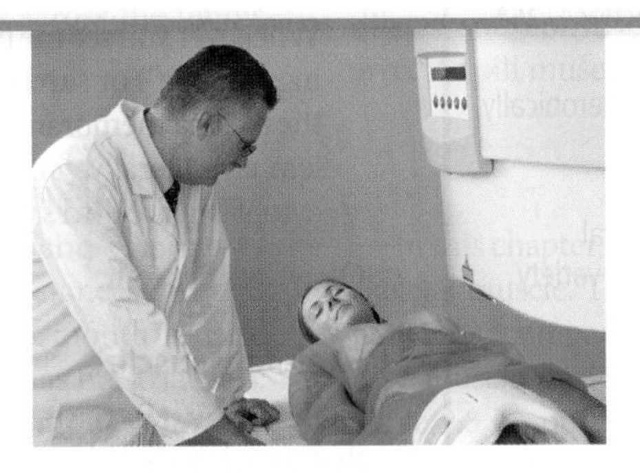

In the previous chapter, you met Kayla Tanner, a 45-year-old mother of four who suffered a dislocated right hip in the bus accident on Route 91. Prior to the closed reduction, the doctors noted that her right thigh was flexed at the hip, adducted, and medially rotated. After the reduction, the hip was put through a gentle range of motion (ROM) to assess the joint. A widened joint space in the postreduction X ray showed that the reduction was not complete, but no bone fragments were visible in the joint space. Mrs. Tanner was scheduled for immediate surgery.

The surgeons discovered that the acetabular labrum was detached from the rim of the acetabulum and was lying deep within the joint space. The detached portion of the labrum was excised, and the hip was surgically reduced. During the early healing phase (first two weeks), Mrs. Tanner was kept in traction with the hip abducted.

1. Joints can be classified by structure and by function. How would you structurally and functionally classify the joint involved in the injury in this case?
2. Name the six distinguishing features that define the structural classification of the joint involved in this injury.
3. The doctors noted that there were no bone fragments in the joint space. What is normally found in this space?
4. Surgeons had to remove a portion of Mrs. Tanner's acetabular labrum. What is this structure and what function does it supply at this joint?
5. The doctors noted that Mrs. Tanner's thigh was flexed at the hip, adducted, and medially rotated. Describe what this means in terms of the position of her leg.
6. Hip dislocations can be classified as anterior or posterior depending on which direction the head of the femur is facing after it dislocates. Based on the description you provided in question 5, which type of dislocation did Mrs. Tanner suffer?
7. In order to assess the joint as part of Mrs. Tanner's rehabilitation, clinicians would want to assess all of the movements that normally occur at the hip. List all the movements that the clinicians will need to assess.

For answers, see Answers Appendix.

Table 9.1 Structure and Organizational Levels of Skeletal Muscle

STRUCTURE AND ORGANIZATIONAL LEVEL	DESCRIPTION	CONNECTIVE TISSUE WRAPPINGS
Muscle (organ) 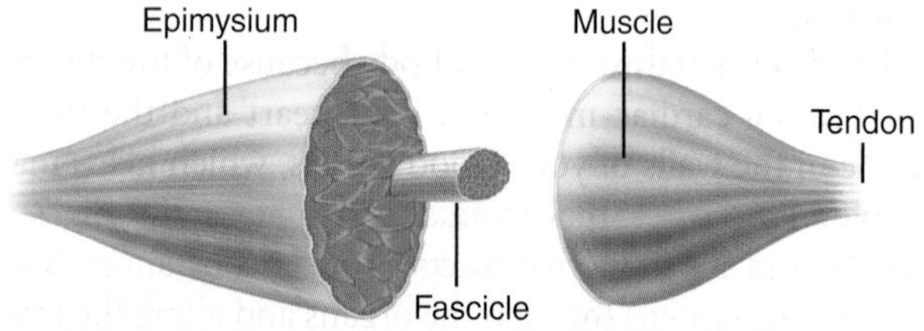	A muscle consists of hundreds to thousands of muscle cells, plus connective tissue wrappings, blood vessels, and nerve fibers.	Covered externally by the epimysium
Fascicle (a portion of the muscle) 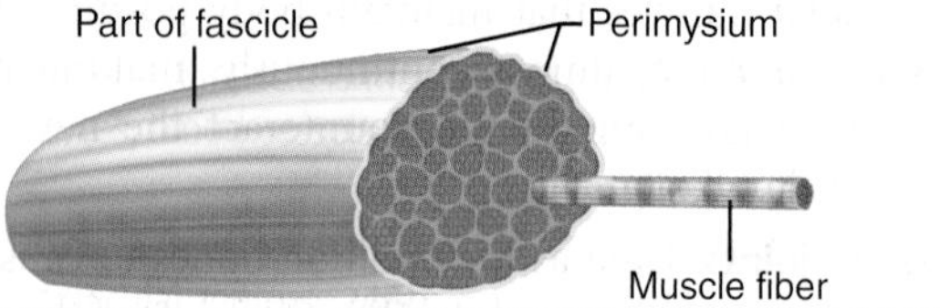	A fascicle is a discrete bundle of muscle cells, segregated from the rest of the muscle by a connective tissue sheath.	Surrounded by perimysium
Muscle fiber (cell) 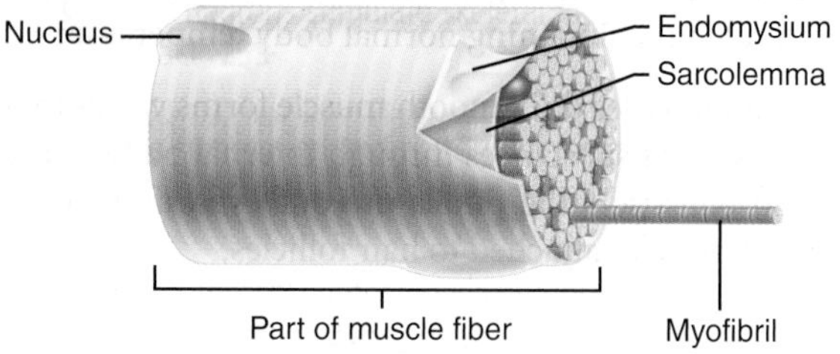	A muscle fiber is an elongated multinucleate cell; it has a banded (striated) appearance.	Surrounded by endomysium
Myofibril, or fibril (complex organelle composed of bundles of myofilaments) 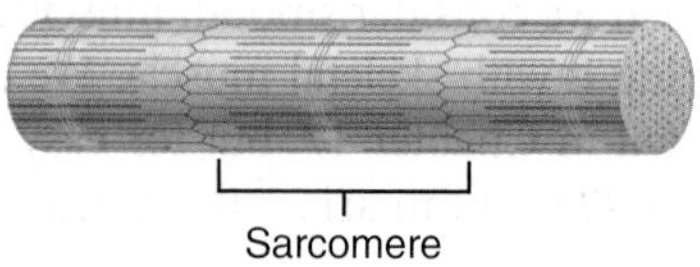	Myofibrils are rodlike contractile elements that occupy most of the muscle cell volume. Composed of sarcomeres arranged end to end, they appear banded, and bands of adjacent myofibrils are aligned.	—
Sarcomere (a segment of a myofibril) 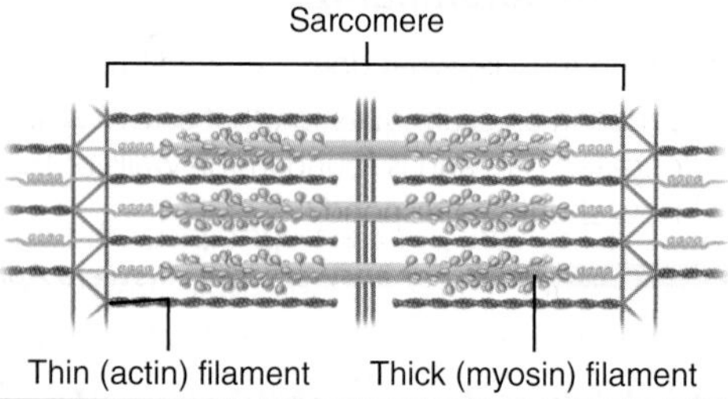	A sarcomere is the contractile unit, composed of myofilaments made up of contractile proteins.	—
Myofilament, or filament (extended macromolecular structure) 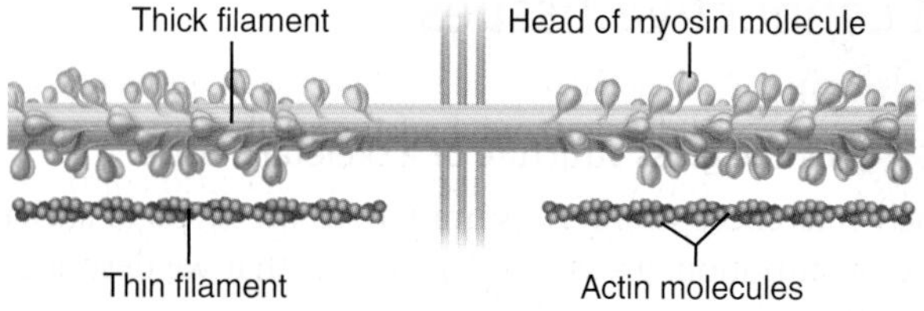	Contractile myofilaments are of two types—thick and thin. Thick filaments contain bundled myosin molecules; thin filaments contain actin molecules (plus other proteins). The sliding of the thin filaments past the thick filaments produces muscle shortening. Elastic filaments (not shown here) maintain the organization of the A band and provide elastic recoil when tension is released.	—

Each **skeletal muscle** is a discrete organ, made up of several kinds of tissues. Skeletal muscle fibers predominate, but blood vessels, nerve fibers, and substantial amounts of connective tissue are also present. We can easily examine a skeletal muscle's shape and its attachments in the body without a microscope.

Nerve and Blood Supply

In general, one nerve, one artery, and one or more veins serve each muscle. These structures all enter or exit near the central part of the muscle and branch profusely through its connective tissue sheaths (described below). Unlike cells of cardiac and smooth muscle tissues, which can contract without nerve stimulation, every skeletal muscle fiber is supplied with a nerve ending that controls its activity.

Skeletal muscle has a rich blood supply. This is understandable because contracting muscle fibers use huge amounts of energy and require almost continuous delivery of oxygen and nutrients via the arteries. Muscle cells also give off large amounts of metabolic wastes that must be removed through veins if contraction is to remain efficient. Muscle capillaries, the smallest of the body's blood vessels, are long and winding and have numerous cross-links, features that accommodate changes in muscle length. They straighten when the muscle stretches and contort when the muscle contracts.

Connective Tissue Sheaths

In an intact muscle, several different connective tissue sheaths wrap individual muscle fibers. Together these sheaths support each cell and reinforce and hold together the muscle, preventing the bulging muscles from bursting during exceptionally strong contractions.

Let's consider these connective tissue sheaths from external to internal (see **Figure 9.1** and the top three rows of Table 9.1).

- **Epimysium.** The **epimysium** (ep″ĭ-mis′e-um; "outside the muscle") is an "overcoat" of dense irregular connective tissue that surrounds the whole muscle. Sometimes it blends with the deep fascia that lies between neighboring muscles or the superficial fascia deep to the skin.
- **Perimysium and fascicles.** Within each skeletal muscle, the muscle fibers are grouped into **fascicles** (fas′ĭ-klz; "bundles")

9

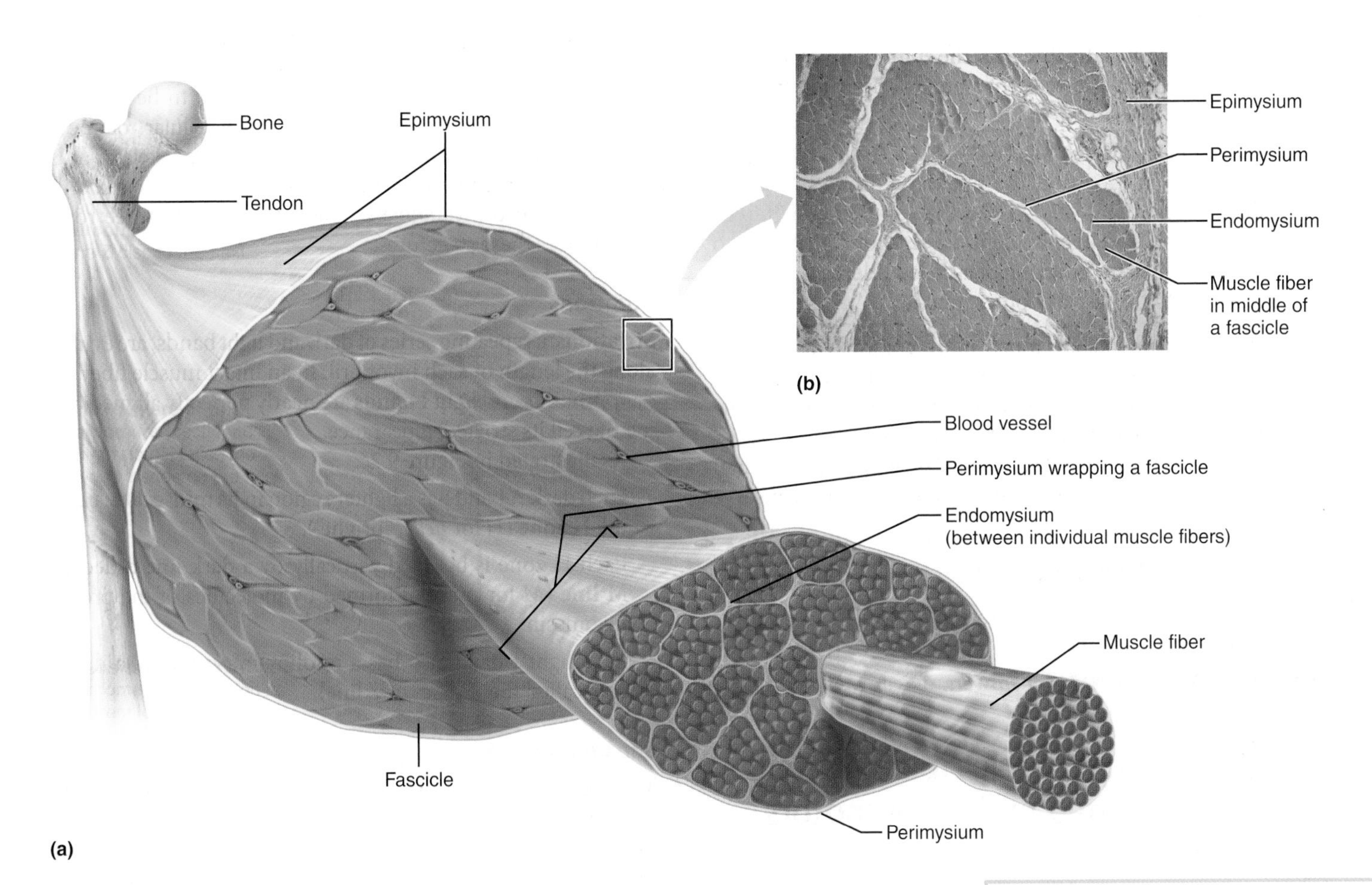

Figure 9.1 Connective tissue sheaths of skeletal muscle: epimysium, perimysium, and endomysium. (b) Photomicrograph of a cross section of part of a skeletal muscle (30×). (For a related image, see *A Brief Atlas of the Human Body*, Plate 29.)

Practice art labeling
MasteringA&P®>Study Area>Chapter 9

FOCUS Excitation-Contraction Coupling

Focus Figure 9.2 Excitation-contraction (E-C) coupling is the sequence of events by which transmission of an action potential along the sarcolemma leads to the sliding of myofilaments.

Watch full 3-D animations
MasteringA&P®>Study Area> *A&PFlix*

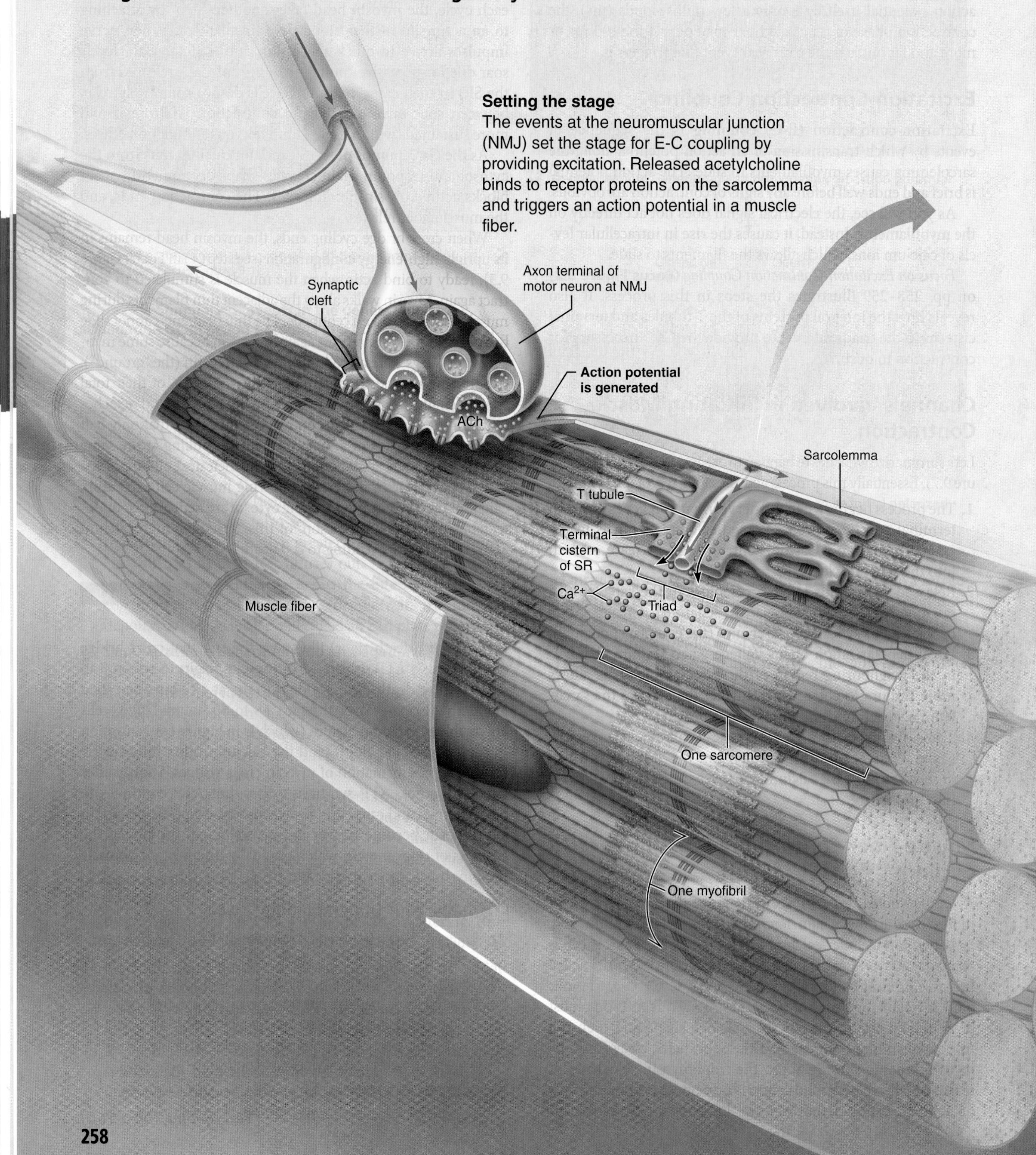

Steps in E-C Coupling:

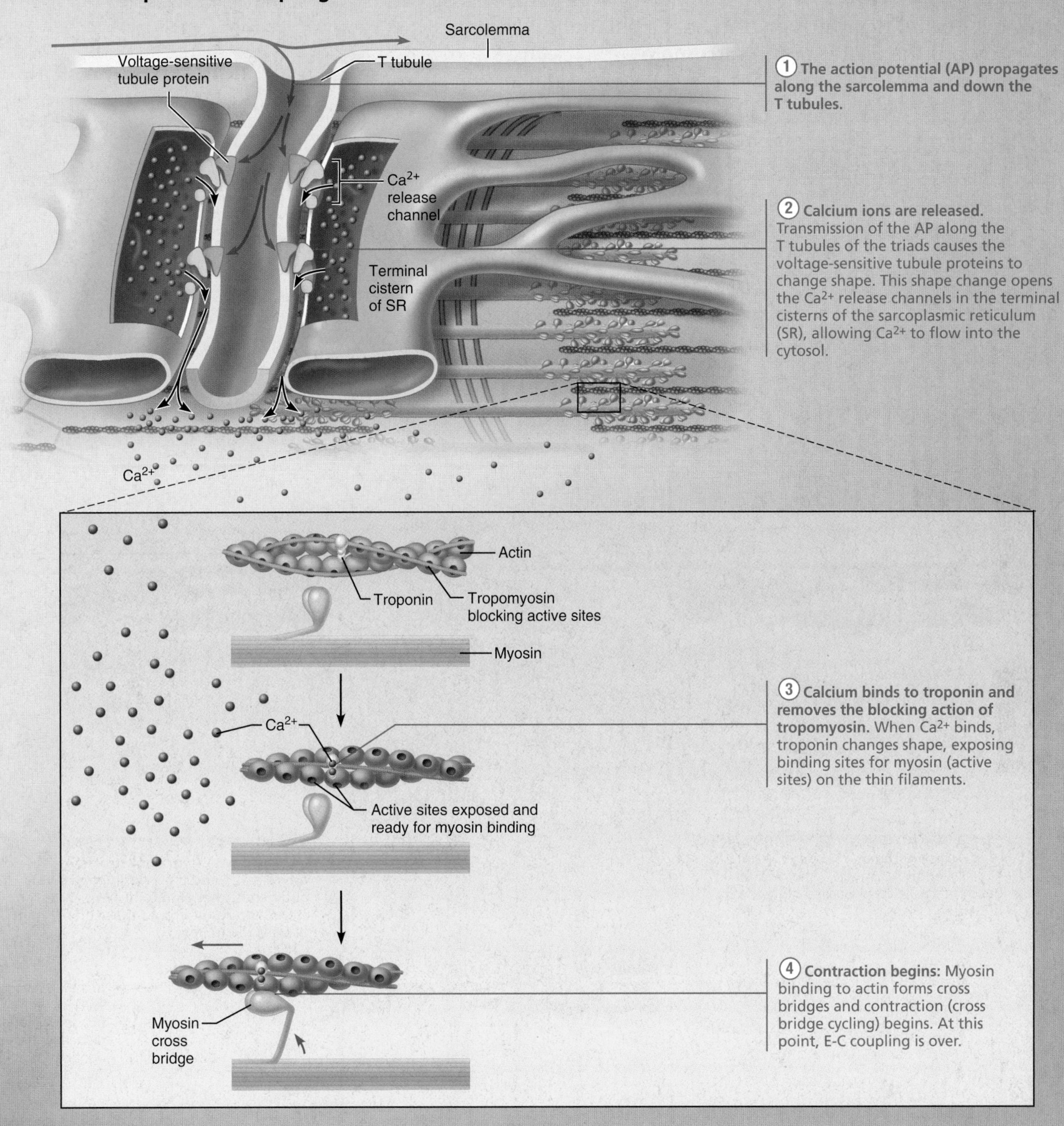

The aftermath

When the muscle AP ceases, the voltage-sensitive tubule proteins return to their original shape, closing the Ca^{2+} release channels of the SR. Ca^{2+} levels in the sarcoplasm fall as Ca^{2+} is continually pumped back into the SR by active transport. Without Ca^{2+}, the blocking action of tropomyosin is restored, myosin-actin interaction is inhibited, and relaxation occurs. Each time an AP arrives at the neuromuscular junction, the sequence of E-C coupling is repeated.

Focus Figure 9.3 The cross bridge cycle is the series of events during which myosin heads pull thin filaments toward the center of the sarcomere.

Watch full 3-D animations
MasteringA&P®>Study Area> *A&PFlix*

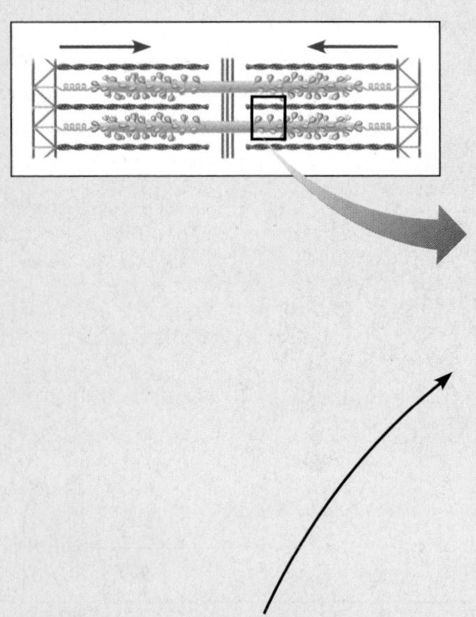

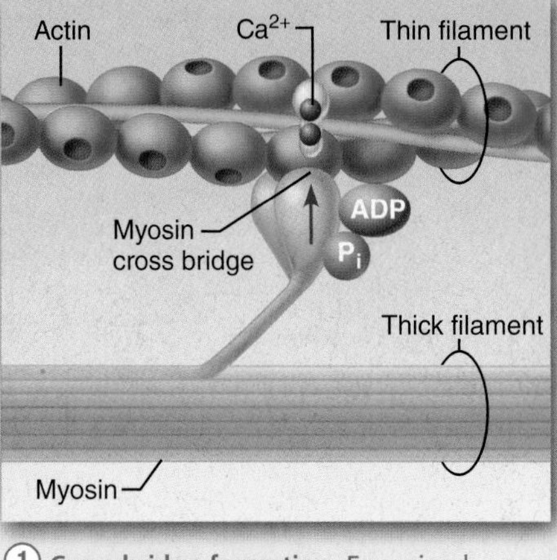

① **Cross bridge formation.** Energized myosin head attaches to an actin myofilament, forming a cross bridge.

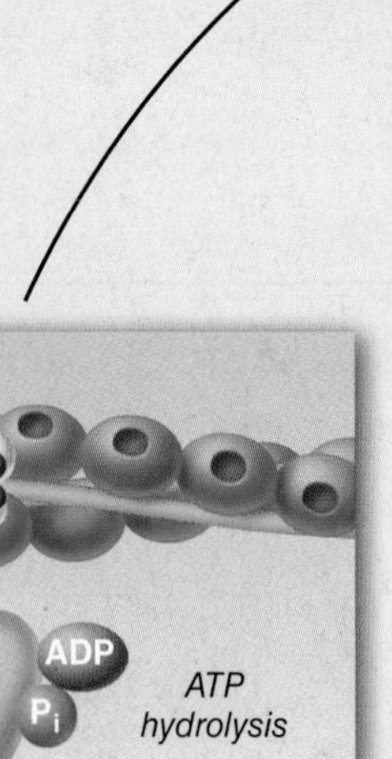

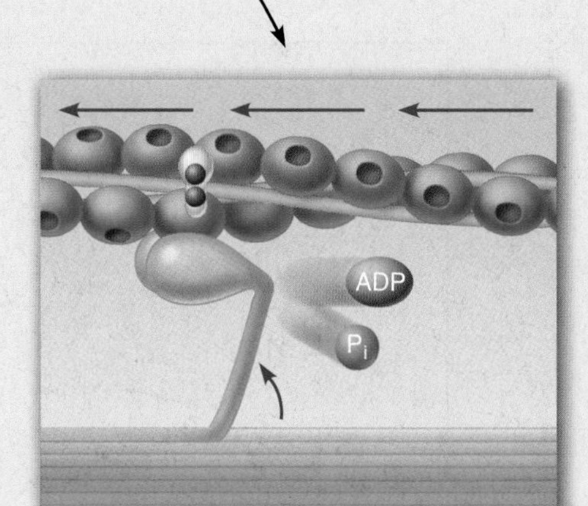

② **The power (working) stroke.** ADP and P_i are released and the myosin head pivots and bends, changing to its bent low-energy state. As a result it pulls the actin filament toward the M line.

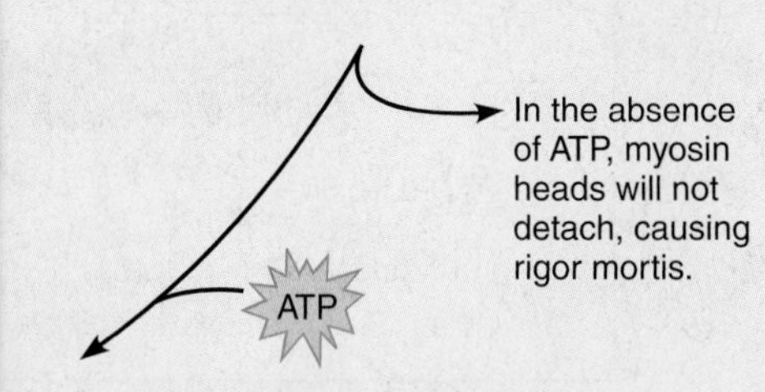

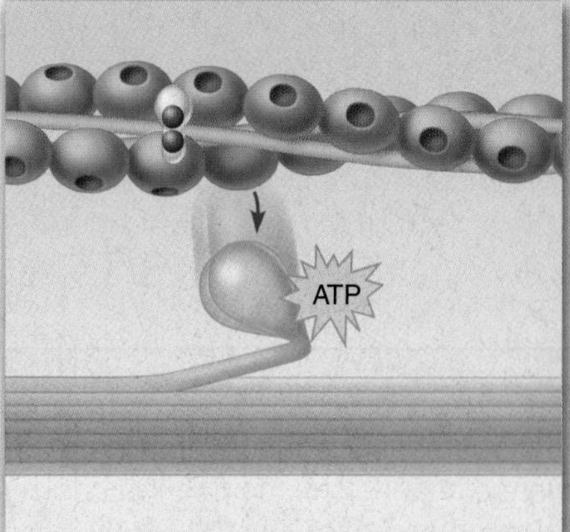

③ **Cross bridge detachment.** After ATP attaches to myosin, the link between myosin and actin weakens, and the myosin head detaches (the cross bridge "breaks").

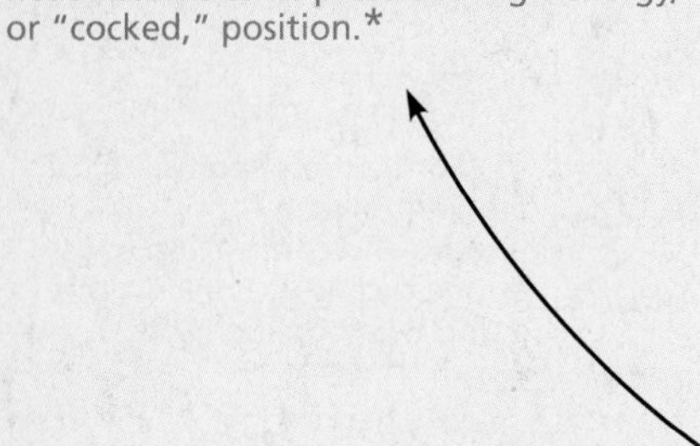

④ **Cocking of the myosin head.** As ATP is hydrolyzed to ADP and P_i, the myosin head returns to its prestroke high-energy, or "cocked," position.*

*This cycle will continue as long as ATP is available and Ca^{2+} is bound to troponin. If ATP is not available, the cycle stops between steps ② and ③.

9.5 Wave summation and motor unit recruitment allow smooth, graded skeletal muscle contractions

→ Learning Objectives

- ☐ Define motor unit and muscle twitch, and describe the events occurring during the three phases of a muscle twitch.
- ☐ Explain how smooth, graded contractions of a skeletal muscle are produced.
- ☐ Differentiate between isometric and isotonic contractions.

In its relaxed state, a muscle is soft and unimpressive, not what you would expect of a prime mover of the body. However, within a few milliseconds, it can contract to become a hard elastic structure with dynamic characteristics that intrigue not only biologists but engineers and physicists as well.

Before we consider muscle contraction on the organ level, let's note a few principles of muscle mechanics.

- The principles governing contraction of a single muscle fiber and of a skeletal muscle consisting of a large number of fibers are pretty much the same.
- The force exerted by a contracting muscle on an object is called **muscle tension**. The opposing force exerted on the muscle by the weight of the object to be moved is called the **load**.
- A contracting muscle does not always shorten and move the load. If muscle tension develops but the load is not moved, the contraction is called *isometric* ("same measure")—think of trying to lift a 2000-lb car. If the muscle tension developed overcomes the load and muscle shortening occurs, the contraction is *isotonic* ("same tension"), as when you lift a 5-lb sack of sugar. We will describe isometric and isotonic contractions in detail, but for now the important thing to remember when reading the accompanying graphs is this: *Increasing muscle tension* is measured for isometric contractions, whereas the *amount of muscle shortening* is measured for isotonic contractions.
- A skeletal muscle contracts with varying force and for different periods of time in response to our need at the time. To understand how this occurs, we must look at the nerve-muscle functional unit called a *motor unit*.

The Motor Unit

Each muscle is served by at least one *motor nerve*, and each motor nerve contains axons (fibrous extensions) of up to hundreds of motor neurons. As an axon enters a muscle, it branches into a number of endings, each of which forms a neuromuscular junction with a single muscle fiber. A **motor unit** consists of one motor neuron and all the muscle fibers it innervates, or supplies (**Figure 9.10**). When a motor neuron fires (transmits an action potential), all the muscle fibers it innervates contract.

The number of muscle fibers per motor unit may be as high as several hundred or as few as four. Muscles that exert fine control (such as those controlling the fingers and eyes) have small motor units. By contrast, large, weight-bearing muscles, whose movements are less precise (such as the hip muscles), have large motor units. The muscle fibers in a single motor unit are not clustered together but are spread throughout the muscle. As a result, stimulation of a single motor unit causes a weak contraction of the *entire* muscle.

The Muscle Twitch

Muscle contraction is easily investigated in the laboratory using an isolated muscle. The muscle is attached to an apparatus that produces a **myogram**, a recording of contractile activity. The line recording the activity is called a *tracing*.

A **muscle twitch** is a motor unit's response to a single action potential of its motor neuron. The muscle fibers contract quickly and then relax. Every twitch myogram has three distinct phases (**Figure 9.11a**).

1. **Latent period.** The **latent period** is the first few milliseconds following stimulation when excitation-contraction coupling is occurring. During this period, cross bridges begin to cycle but muscle tension is not yet measurable and the myogram does not show a response.
2. **Period of contraction.** During the period of contraction, cross bridges are active, from the onset to the peak of tension development, and the myogram tracing rises to a peak. This period lasts 10–100 ms. If the tension becomes great enough to overcome the resistance of the load, the muscle shortens.
3. **Period of relaxation.** This final phase, lasting 10–100 ms, is initiated by reentry of Ca^{2+} into the SR. Because the number of active cross bridges is declining, contractile force is declining. Muscle tension decreases to zero and the tracing returns to the baseline. If the muscle shortened during contraction, it now returns to its initial length. Notice that a muscle contracts faster than it relaxes, as revealed by the asymmetric nature of the myogram tracing.

As you can see in Figure 9.11b, twitch contractions of some muscles are rapid and brief, as with the muscles controlling eye movements. In contrast, the fibers of fleshy calf muscles (gastrocnemius and soleus) contract more slowly and remain contracted for much longer periods. These differences between muscles reflect variations in enzymes and metabolic properties of the myofibrils.

Graded Muscle Responses

Muscle twitches—like those single, jerky contractions provoked in a laboratory—may result from certain neuromuscular problems, but this is *not* the way our muscles normally operate. Instead, healthy muscle contractions are relatively smooth and vary in strength as different demands are placed on them. These variations, needed for proper control of skeletal movement, are referred to as **graded muscle responses**.

In general, muscle contraction can be graded in two ways: by changing the frequency of stimulation, and by changing the strength of stimulation.

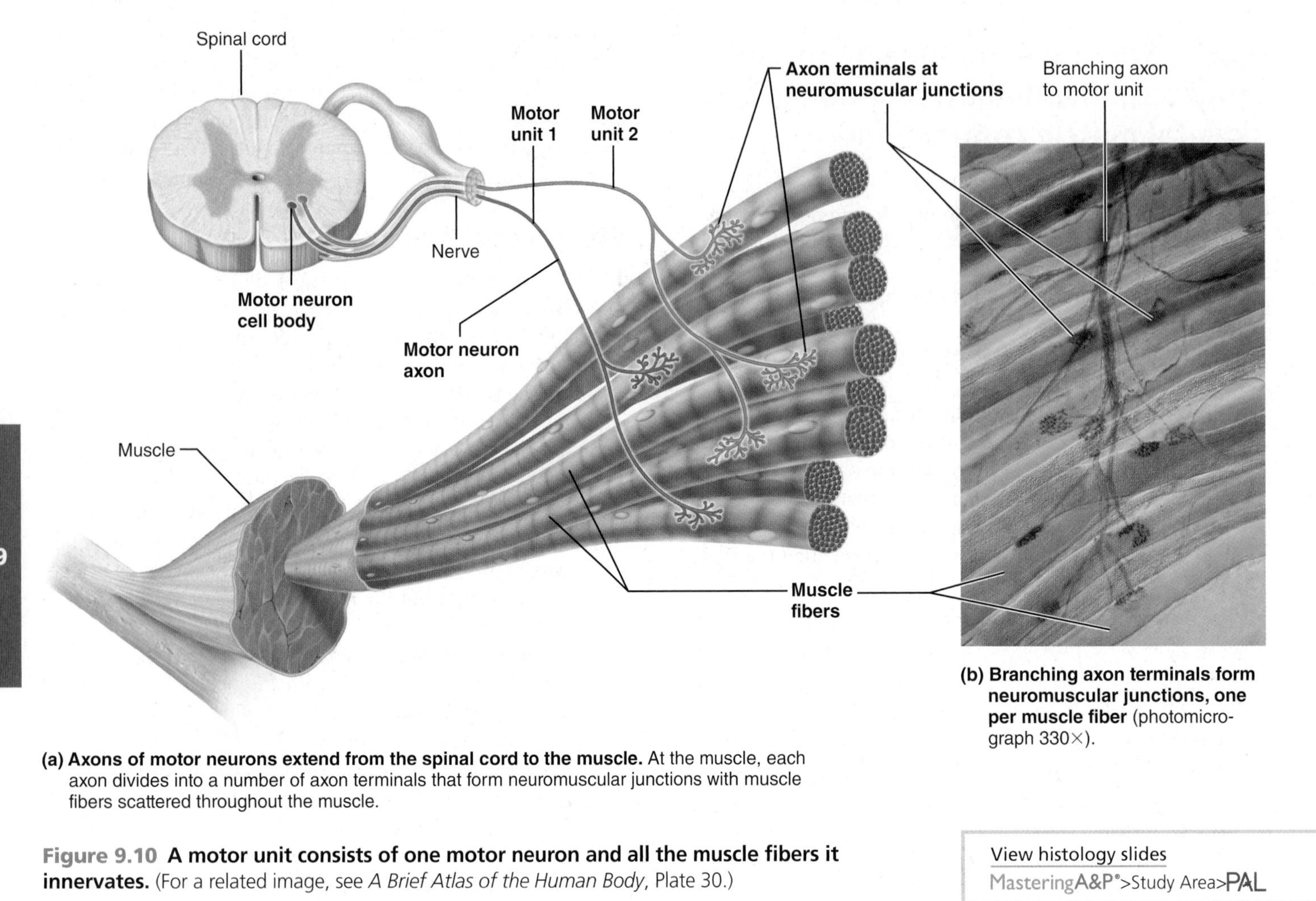

(a) Axons of motor neurons extend from the spinal cord to the muscle. At the muscle, each axon divides into a number of axon terminals that form neuromuscular junctions with muscle fibers scattered throughout the muscle.

(b) Branching axon terminals form neuromuscular junctions, one per muscle fiber (photomicrograph 330×).

Figure 9.10 A motor unit consists of one motor neuron and all the muscle fibers it innervates. (For a related image, see *A Brief Atlas of the Human Body*, Plate 30.)

View histology slides
MasteringA&P®>Study Area>PAL

Muscle Response to Changes in Stimulus Frequency

The nervous system achieves greater muscular force by increasing the firing rate of motor neurons. For example, if two identical stimuli (electrical shocks or nerve impulses) are delivered to a muscle in rapid succession, the second twitch will be stronger than the first. On a myogram the second twitch will appear to ride on the shoulders of the first (**Figure 9.12a, b**).

This phenomenon, called **wave** or **temporal summation**, occurs because the second contraction occurs before the muscle has completely relaxed. Because the muscle is already partially contracted and more calcium is being squirted into the cytosol to replace that being reclaimed by the SR, muscle tension produced during the second contraction causes more shortening than the first. In other words, the contractions are added together. (However, the refractory period is always honored. Thus, if a second stimulus arrives before repolarization is complete, no wave summation occurs.)

If the muscle is stimulated at an increasingly faster rate:

- The relaxation time between twitches becomes shorter and shorter.
- The concentration of Ca^{2+} in the cytosol rises higher and higher.
- The degree of wave summation becomes greater and greater, progressing to a sustained but quivering contraction referred to as **unfused** or **incomplete tetanus** (Figure 9.12b).
- Finally, as the stimulation frequency continues to increase, muscle tension increases until it reaches maximal tension. At this point all evidence of muscle relaxation disappears and the contractions fuse into a smooth, sustained contraction plateau called **fused** or **complete tetanus** (tet′ah-nus; *tetan* = rigid, tense) (Figure 9.12c).

In the real world, fused tetanus happens infrequently, for example, when someone shows superhuman strength by lifting a fallen tree limb off a companion. (Note that the term *tetanus* also describes a bacterial disease.)

Vigorous muscle activity cannot continue indefinitely. Prolonged tetanus inevitably leads to muscle fatigue. The muscle can no longer contract and its tension drops to zero.

Muscle Response to Changes in Stimulus Strength

Wave summation contributes to contractile force, but its primary function is to produce smooth, continuous muscle contractions by rapidly stimulating a specific number of muscle cells. **Recruitment**, also called **multiple motor unit**

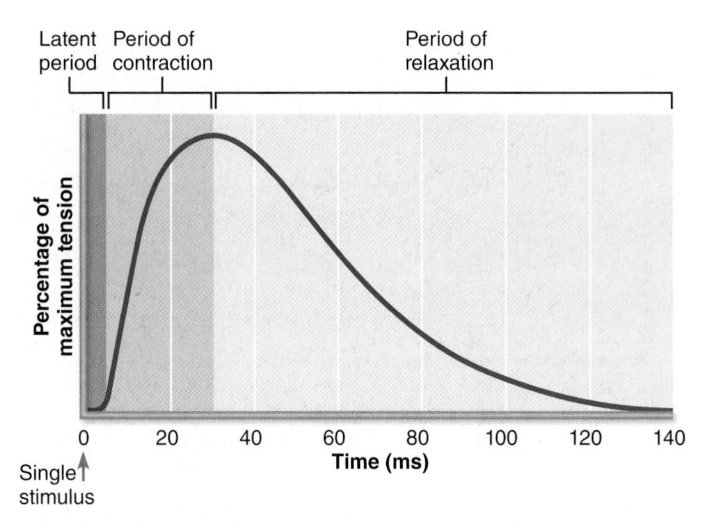

(a) Myogram showing the three phases of an isometric twitch

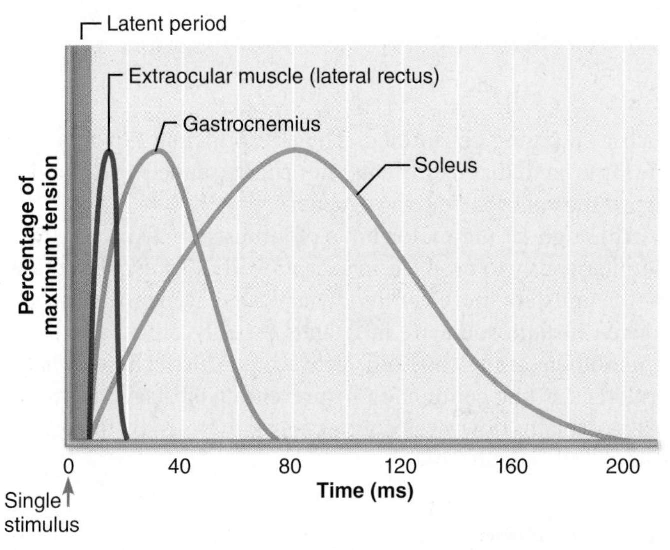

(b) Comparison of the relative duration of twitch responses of three muscles

Figure 9.11 The muscle twitch.

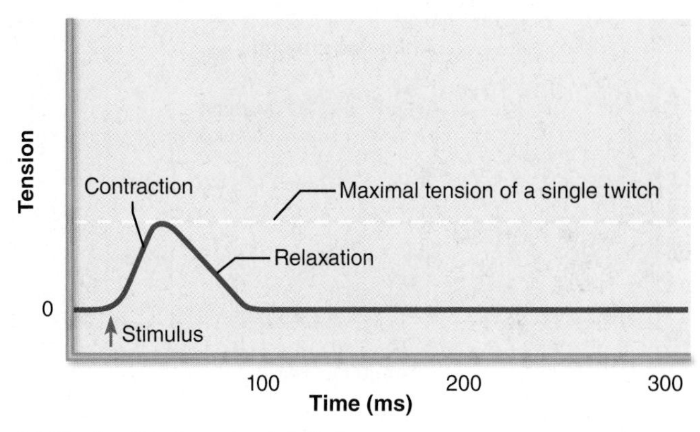

(a) Single stimulus: single twitch.
A single stimulus is delivered. The muscle contracts and relaxes.

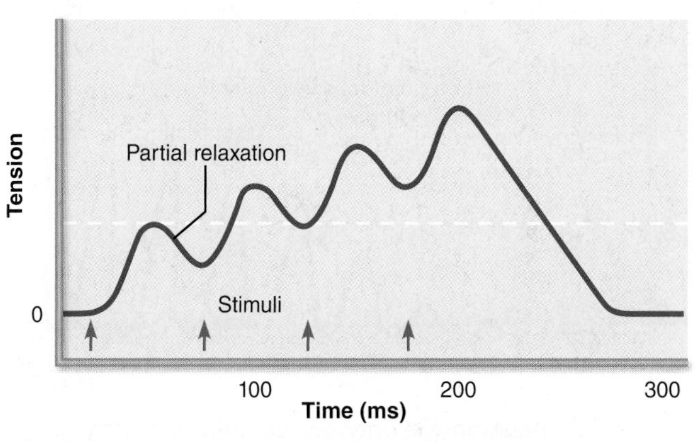

(b) Low stimulation frequency: unfused (incomplete) tetanus.
If another stimulus is applied before the muscle relaxes completely, then more tension results. This is wave (or temporal) summation and results in unfused (or incomplete) tetanus.

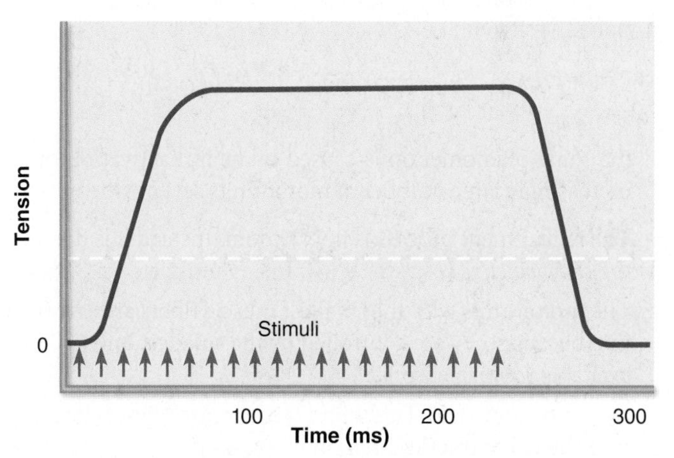

(c) High stimulation frequency: fused (complete) tetanus.
At higher stimulus frequencies, there is no relaxation at all between stimuli. This is fused (complete) tetanus.

Figure 9.12 A muscle's response to changes in stimulation frequency. (Note that tension is measured in grams.)

summation, controls the force of contraction more precisely. In the laboratory, recruitment is achieved by delivering shocks of increasing voltage to the muscle, calling more and more muscle fibers into play.

- Stimuli that produce no observable contractions are **subthreshold stimuli**.
- The stimulus at which the first observable contraction occurs is called the **threshold stimulus** (Figure 9.13). Beyond this point, the muscle contracts more vigorously as the stimulus strength increases.
- The **maximal stimulus** is the strongest stimulus that increases contractile force. It represents the point at which all the muscle's motor units are recruited. In the laboratory, increasing the stimulus intensity beyond the maximal stimulus does not produce a stronger contraction. In the body,

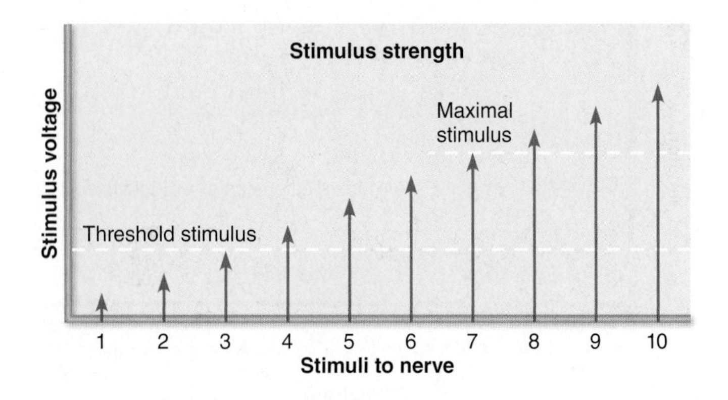

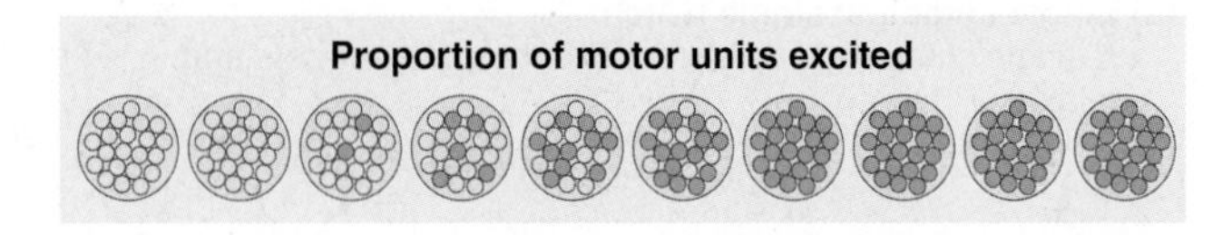

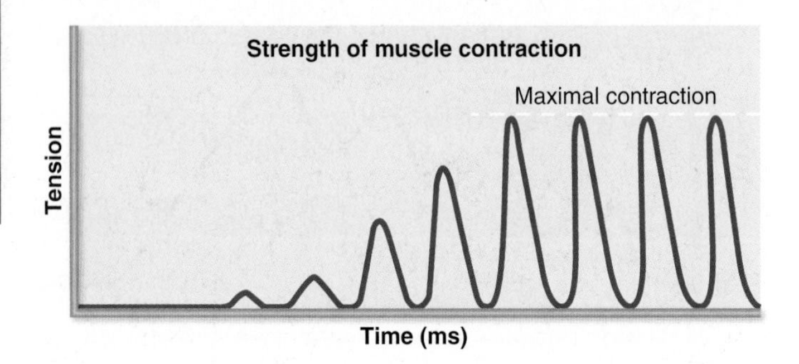

Figure 9.13 Relationship between stimulus intensity (graph at top) and muscle tension (tracing below). Below threshold voltage, the tracing shows no muscle response (stimuli 1 and 2). Once threshold (3) is reached, increases in voltage excite (recruit) more and more motor units until the maximal stimulus is reached (7). Further increases in stimulus voltage produce no further increase in contractile strength.

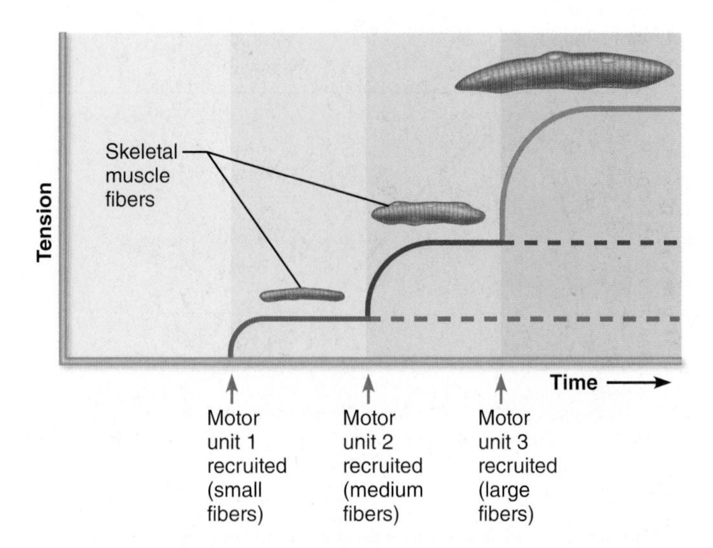

Figure 9.14 The size principle of recruitment. Recruitment of motor neurons controlling skeletal muscle fibers is orderly and follows the size principle.

the same phenomenon is caused by neural activation of an increasingly large number of motor units serving the muscle.

The recruitment process is not random. Instead it is dictated by the *size principle* (**Figure 9.14**). In any muscle:

- The motor units with the smallest muscle fibers are activated first because they are controlled by the smallest, most highly excitable motor neurons.
- As motor units with larger and larger muscle fibers begin to be excited, contractile strength increases.
- The largest motor units, containing large, coarse muscle fibers, are controlled by the largest, least excitable (highest-threshold) neurons and are activated only when the most powerful contraction is necessary.

Why is the size principle important? It allows the increases in force during weak contractions (for example, those that maintain posture or slow movements) to occur in small steps, whereas gradations in muscle force are progressively greater when large amounts of force are needed for vigorous activities such as jumping or running. The size principle explains how the same hand that lightly pats your cheek can deliver a stinging slap at the volleyball during a match.

Although *all* the motor units of a muscle may be recruited simultaneously to produce an exceptionally strong contraction, motor units are more commonly activated asynchronously. At a given instant, some are in tetanus (usually unfused tetanus) while others are resting and recovering. This technique helps prolong a strong contraction by preventing or delaying fatigue. It also explains how weak contractions promoted by infrequent stimuli can remain smooth.

Muscle Tone

Skeletal muscles are described as voluntary, but even relaxed muscles are almost always slightly contracted, a phenomenon called **muscle tone**. Muscle tone is due to spinal reflexes that activate first one group of motor units and then another in response to activated stretch receptors in the muscles. Muscle tone does not produce active movements, but it keeps the muscles firm, healthy, and ready to respond to stimulation. Skeletal muscle tone also helps stabilize joints and maintain posture.

Isotonic and Isometric Contractions

As noted earlier, there are two main categories of contractions—*isotonic* and *isometric*. In **isotonic contractions** (*iso* = same; *ton* = tension), muscle length changes and moves a load. Once sufficient tension has developed to move the load, the tension remains relatively constant through the rest of the contractile period (**Figure 9.15a**).

Isotonic contractions come in two "flavors"—*concentric* and *eccentric*. **Concentric contractions** are those in which the muscle shortens and does work, such as picking up a book or kicking a ball. Concentric contractions are probably more

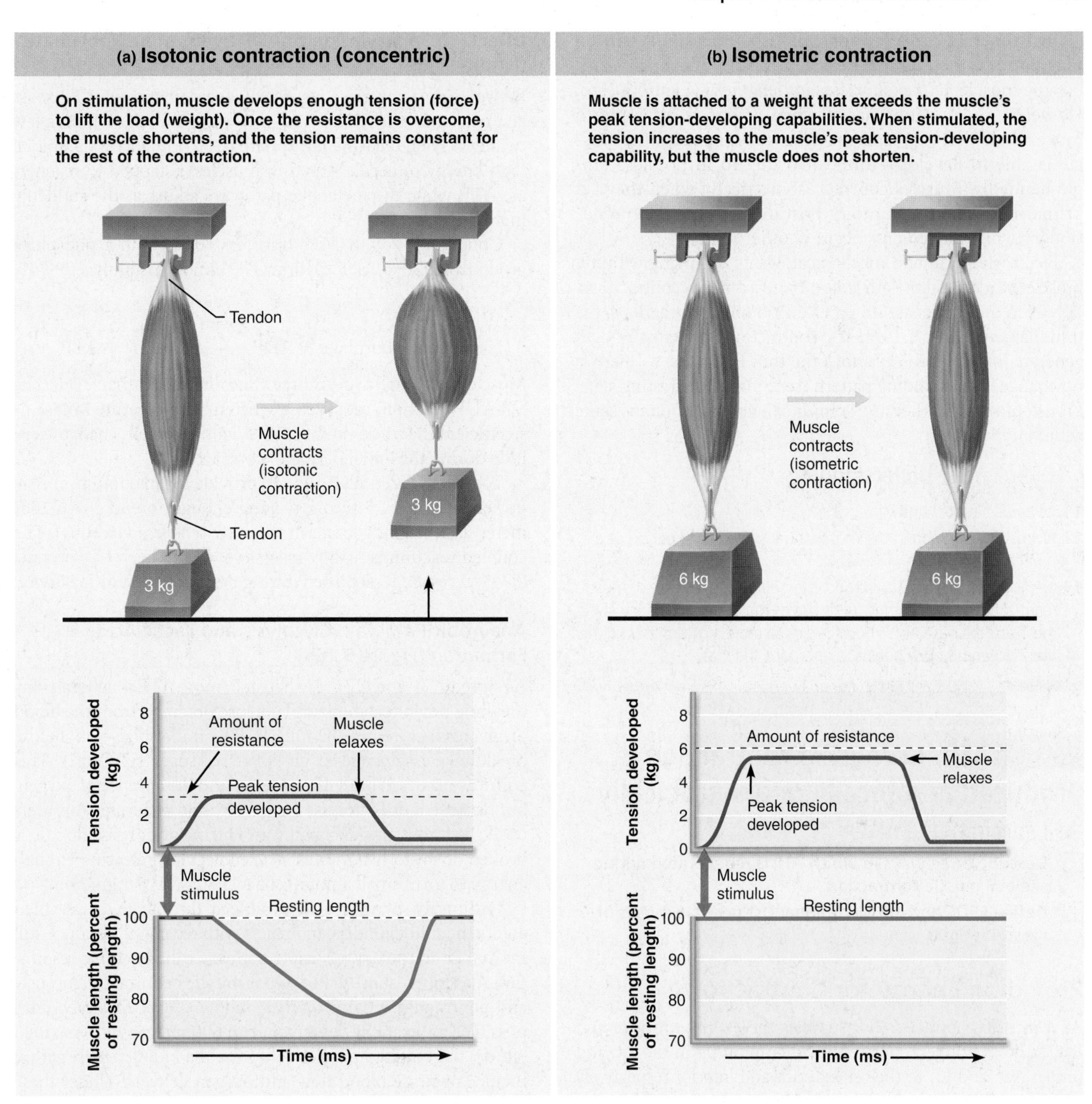

Figure 9.15 Isotonic (concentric) and isometric contractions.

familiar, but **eccentric contractions**, in which the muscle generates force as it lengthens, are equally important for coordination and purposeful movements.

Eccentric contractions occur in your calf muscle, for example, as you walk up a steep hill. Eccentric contractions are about 50% more forceful than concentric ones at the same load and more often cause delayed-onset muscle soreness. (Consider how your calf muscles *feel* the day after hiking up that hill.) The reason is unclear, but it may be that the muscle stretching that occurs during eccentric contractions causes microtears in the muscles.

Biceps curls provide a simple example of how concentric and eccentric contractions work together in our everyday activities. When you flex your elbow to draw a weight toward your shoulder, the biceps muscle in your arm is contracting concentrically. When you straighten your arm to return the weight to the bench, the isotonic contraction of your biceps is eccentric. Basically, eccentric contractions put the body in position to contract concentrically. All jumping and throwing activities involve both types of contraction.

In **isometric contractions** (*metric* = measure), tension may build to the muscle's peak tension-producing capacity, but the muscle *neither shortens nor lengthens* (Figure 9.15b). Isometric contractions occur when a muscle attempts to move a load that is greater than the force (tension) the muscle is able to develop—think of trying to lift a piano single-handedly. Muscles contract isometrically when they act primarily to maintain upright posture or to hold joints stationary while movements occur at other joints.

Electrochemical and mechanical events occurring within a muscle are identical in both isotonic and isometric contractions. However, the results are different. In isotonic contractions, the thin filaments slide. In isometric contractions, the cross bridges generate force but do *not* move the thin filaments, so there is no change in the banding pattern from that of the resting state. (You could say that they are "spinning their wheels" on the same actin binding sites.)

☑ Check Your Understanding

11. What is a motor unit?

12. What is happening in the muscle during the latent period of a twitch contraction?

13. Jacob is competing in a chin-up competition. What type of muscle contractions are occurring in his biceps muscles immediately after he grabs the bar? As his body begins to move upward toward the bar? When his body begins to approach the mat?

For answers, see Answers Appendix.

9.6 ATP for muscle contraction is produced aerobically or anaerobically

→ Learning Objectives

- ☐ **Describe three ways in which ATP is regenerated during skeletal muscle contraction.**
- ☐ **Define EPOC and muscle fatigue. List possible causes of muscle fatigue.**

Providing Energy for Contraction

As a muscle contracts, ATP supplies the energy to move and detach cross bridges, operate the calcium pump in the SR, and return Na^+ and K^+ to the cell exterior and interior respectively after excitation-contraction coupling. Surprisingly, muscles store very limited reserves of ATP—4 to 6 seconds' worth at most, just enough to get you going. Because ATP is the *only* energy source used directly for contractile activities, it must be regenerated as fast as it is broken down if contraction is to continue.

Fortunately, after ATP is hydrolyzed to ADP and inorganic phosphate in muscle fibers, it is regenerated within a fraction of a second by one or more of the three pathways summarized in **Figure 9.16**: (a) direct phosphorylation of ADP by creatine phosphate, (b) anaerobic glycolysis, which converts glucose to lactic acid, and (c) aerobic respiration. All body cells use glycolysis and aerobic respiration to produce ATP, so we touch on them here but describe them in detail later, in Chapter 23.

Direct Phosphorylation of ADP by Creatine Phosphate (Figure 9.16a)

As we begin to exercise vigorously, the demand for ATP soars and consumes the ATP stored in working muscles within a few twitches. Then **creatine phosphate (CP)** (kre′ah-tin), a unique high-energy molecule stored in muscles, is tapped to regenerate ATP while the metabolic pathways adjust to the suddenly higher demand for ATP.

Coupling CP with ADP transfers energy and a phosphate group from CP to ADP to form ATP almost instantly:

$$\text{Creatine phosphate} + \text{ADP} \xrightarrow{\text{creatine kinase}} \text{creatine} + \text{ATP}$$

Muscle cells store two to three times more CP than ATP. The CP-ADP reaction, catalyzed by the enzyme **creatine kinase**, is so efficient that the amount of ATP in muscle cells changes very little during the initial period of contraction.

Together, stored ATP and CP provide for maximum muscle power for about 15 seconds—long enough to energize a 100-meter dash (slightly longer if the activity is less vigorous). The coupled reaction is readily reversible, and to keep CP "on tap," CP reserves are replenished during periods of rest or inactivity.

Anaerobic Pathway: Glycolysis and Lactic Acid Formation (Figure 9.16b)

As stored ATP and CP are exhausted, more ATP is generated by breaking down (catabolizing) glucose obtained from the blood or glycogen stored in the muscle. The initial phase of glucose breakdown is **glycolysis** (gli-kol′ĭ-sis; "sugar splitting"). This pathway occurs in both the presence and the absence of oxygen, but because it does not use oxygen, it is an anaerobic (an-a′er-ōb-ik; "without oxygen") pathway. During glycolysis, glucose is broken down to two *pyruvic acid* molecules, releasing enough energy to form small amounts of ATP (2 ATP per glucose).

Ordinarily, pyruvic acid produced during glycolysis then enters the mitochondria and reacts with oxygen to produce still more ATP in the oxygen-using pathway called aerobic respiration, described shortly. But when muscles contract vigorously and contractile activity reaches about 70% of the maximum possible (for example, when you run 600 meters with maximal effort), the bulging muscles compress the blood vessels within them, impairing blood flow and oxygen delivery. Under these anaerobic conditions, most of the pyruvic acid is converted into **lactic acid**, and the overall process is referred to as **anaerobic glycolysis**. Thus, during oxygen deficit, lactic acid is the end product of cellular metabolism of glucose.

Most of the lactic acid diffuses out of the muscles into the bloodstream. Subsequently, the liver, heart, or kidney cells pick up the lactic acid and use it as an energy source. Additionally, liver cells can reconvert it to pyruvic acid or glucose and release it back into the bloodstream for muscle use or convert it to glycogen for storage.

The anaerobic pathway harvests only about 5% as much ATP from each glucose molecule as the aerobic pathway, but it produces ATP about 2½ times faster. For this reason, even when large amounts of ATP are needed for moderate periods (30–40 seconds)

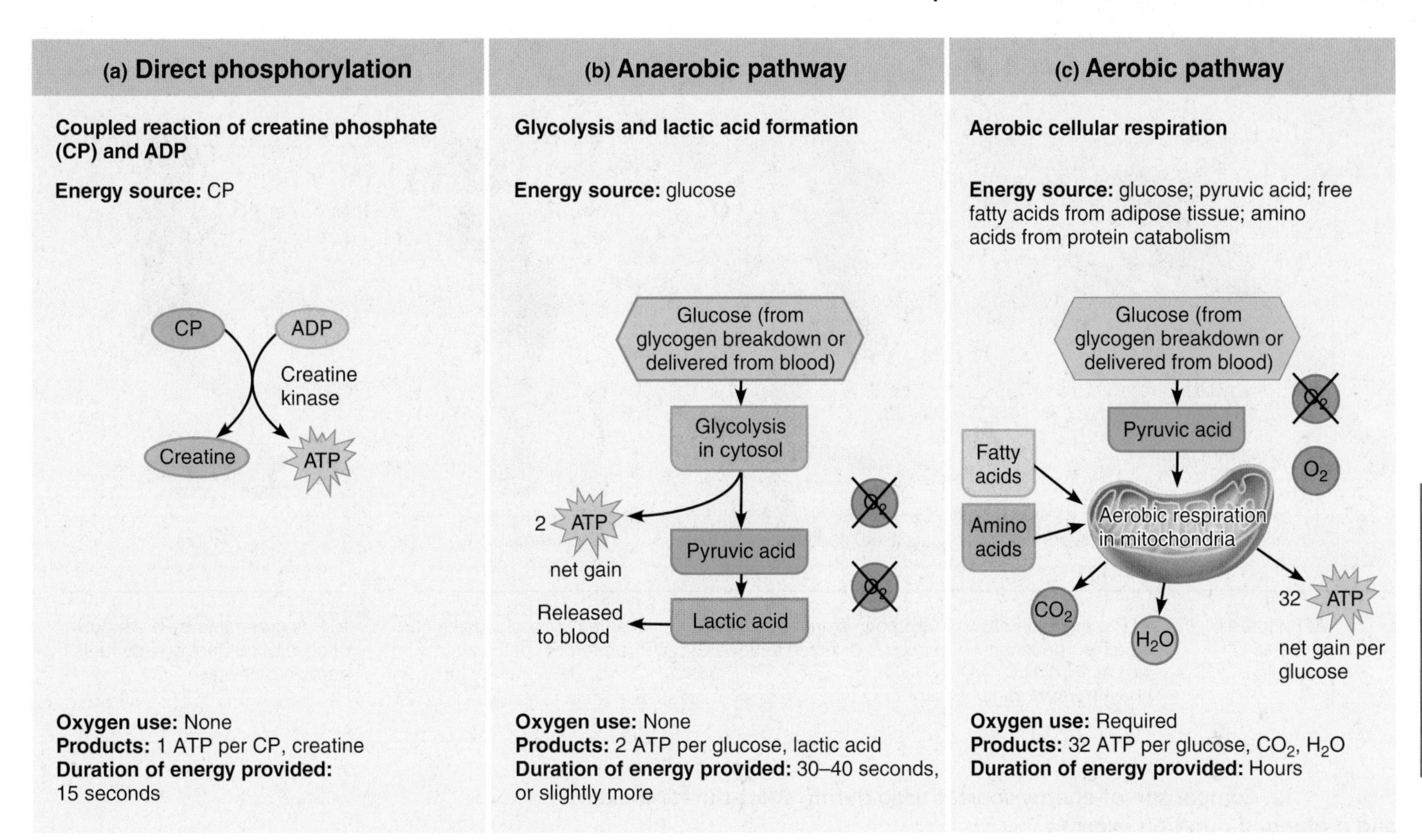

Figure 9.16 Pathways for regenerating ATP during muscle activity. The fastest pathway is direct phosphorylation **(a)**, and the slowest is aerobic respiration **(c)**.

of strenuous muscle activity, glycolysis can provide most of this ATP. Together, stored ATP and CP and the glycolysis–lactic acid pathway can support strenuous muscle activity for nearly a minute.

Although anaerobic glycolysis readily fuels spurts of vigorous exercise, it has shortcomings. Huge amounts of glucose are used to produce relatively small harvests of ATP, and the accumulating lactic acid is partially responsible for muscle soreness during intense exercise.

Aerobic Respiration (Figure 9.16c)

Because the amount of creatine phosphate is limited, muscles must metabolize nutrients to transfer energy from foodstuffs to ATP. During rest and light to moderate exercise, even if prolonged, 95% of the ATP used for muscle activity comes from aerobic respiration. **Aerobic respiration** occurs in the mitochondria, requires oxygen, and involves a sequence of chemical reactions that break the bonds of fuel molecules and release energy to make ATP.

Aerobic respiration, which includes glycolysis and the reactions that take place in the mitochondria, breaks down glucose entirely. Water, carbon dioxide, and large amounts of ATP are its final products.

$$\text{Glucose} + \text{oxygen} \longrightarrow \text{carbon dioxide} + \text{water} + \text{ATP}$$

The carbon dioxide released diffuses out of the muscle tissue into the blood, to be removed from the body by the lungs.

As exercise begins, muscle glycogen provides most of the fuel. Shortly thereafter, bloodborne glucose, pyruvic acid from glycolysis, and free fatty acids are the major sources of fuels. After about 30 minutes, fatty acids become the major energy fuels. Aerobic respiration provides a high yield of ATP (about 32 ATP per glucose), but it is slow because of its many steps and it requires continuous delivery of oxygen and nutrient fuels to keep it going.

Energy Systems Used during Exercise

Which pathways predominate during exercise? As long as a muscle cell has enough oxygen, it will form ATP by the aerobic pathway. When ATP demands are within the capacity of the aerobic pathway, light to moderate muscular activity can continue for several hours in well-conditioned individuals (**Figure 9.17**). However, when exercise demands begin to exceed the ability of the muscle cells to carry out the necessary reactions quickly enough, anaerobic pathways begin to contribute more and more of the total ATP generated. The length of time a muscle can continue to contract using aerobic pathways is called **aerobic endurance**, and the point at which muscle metabolism converts to anaerobic glycolysis is called **anaerobic threshold**.

Activities that require a surge of power but last only a few seconds, such as weight lifting, diving, and sprinting, rely entirely on ATP and CP stores. The slightly longer bursts of activity in tennis, soccer, and a 100-meter swim appear to be fueled almost

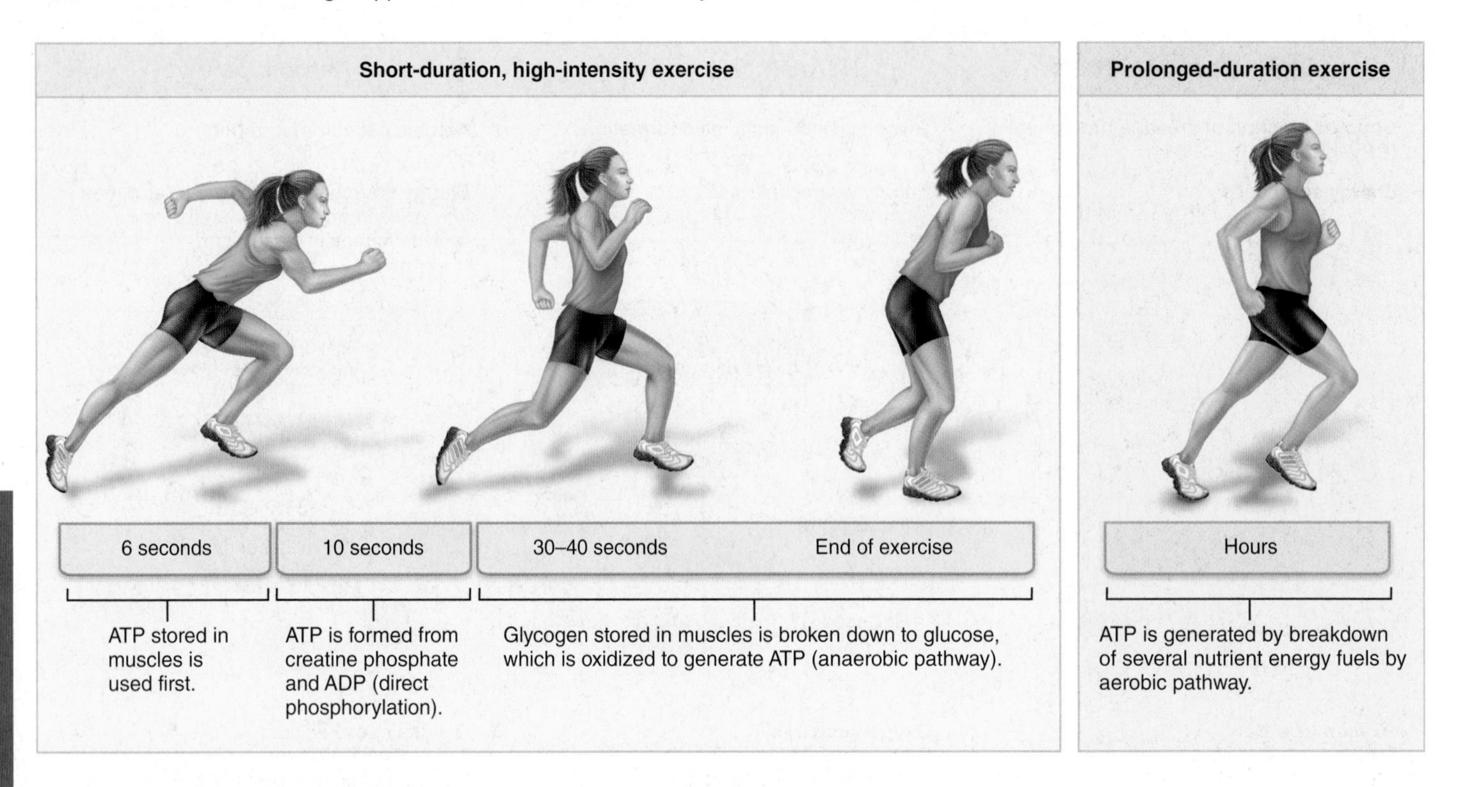

Figure 9.17 Comparison of energy sources used during short-duration exercise and prolonged-duration exercise.

entirely by anaerobic glycolysis (Figure 9.17). Prolonged activities such as marathon runs and jogging, where endurance rather than power is the goal, depend mainly on aerobic respiration using both glucose and fatty acids as fuels. Levels of CP and ATP don't change much during prolonged exercise because ATP is generated at the same rate as it is used—a "pay as you go" system. Compared to anaerobic energy production, aerobic generation of ATP is relatively slow, but the ATP harvest is enormous.

Muscle Fatigue

Muscle fatigue is a state of *physiological inability to contract* even though the muscle still may be receiving stimuli. Although many factors appear to contribute to fatigue, its specific causes are not fully understood. Most experimental evidence indicates that fatigue is due to a problem in excitation-contraction coupling or, in rare cases, problems at the neuromuscular junction. Availability of ATP declines during contraction, but it is abnormal to see major declines in ATP unless the muscles are severely stressed. So, lack of ATP is not a fatigue-producing factor in moderate exercise.

Several ionic imbalances contribute to muscle fatigue. As action potentials are transmitted, potassium is lost from the muscle cells, and accumulates in the fluids of the T tubules. This ionic change disturbs the membrane potential of the muscle cells and halts Ca^{2+} release from the SR.

Theoretically, in short-duration exercise, an accumulation of inorganic phosphate (P_i) from CP and ATP breakdown may interfere with calcium release from the SR. Alternatively, it may interfere with the release of P_i from myosin and thus hamper myosin's power strokes. Lactic acid has long been assumed to be a major cause of fatigue, and excessive intracellular accumulation of lactic acid raises the concentration of H^+ and alters contractile proteins. However, pH is normally regulated within normal limits in all but the greatest degree of exertion. Additionally, extracellular lactic acid actually counteracts the high K^+ levels that lead to muscle fatigue.

In general, intense exercise of short duration produces fatigue rapidly via ionic disturbances that alter E-C coupling, but recovery is also rapid. In contrast, the slow-developing fatigue of prolonged low-intensity exercise may require several hours for complete recovery. It appears that this type of exercise damages the SR, interfering with Ca^{2+} regulation and release, and therefore with muscle activation.

Excess Postexercise Oxygen Consumption (EPOC)

Whether or not fatigue occurs, vigorous exercise alters a muscle's chemistry dramatically. For a muscle to return to its preexercise state, the following must occur:

- Its oxygen reserves in myoglobin must be replenished.
- The accumulated lactic acid must be reconverted to pyruvic acid.
- Glycogen stores must be replaced.
- ATP and creatine phosphate reserves must be resynthesized.

The use of these muscle stores during anaerobic exercise simply defers when the oxygen is consumed, because replacing them requires oxygen uptake and aerobic metabolism after exercise ends. Additionally, the liver must convert any lactic acid persisting in blood to glucose or glycogen. Once exercise stops, the repayment process begins.

The extra amount of oxygen that the body must take in for these restorative processes is called the **excess postexercise oxygen consumption (EPOC)**, formerly called the oxygen debt. EPOC represents the difference between the amount of oxygen needed for totally aerobic muscle activity and the amount actually used. All anaerobic sources of ATP used during muscle activity contribute to EPOC.

☑ Check Your Understanding

14. When Eric returned from jogging, he was breathing heavily, sweating profusely, and complained that his legs ached and felt weak. His wife poured him a sports drink and urged him to take it easy until he could "catch his breath." On the basis of what you have learned about muscle energy metabolism, respond to the following questions: Why is Eric breathing heavily? Which ATP-generating pathway have his working muscles been using that makes him breathless? What metabolic products might account for his sore muscles and muscle weakness?

For answers, see Answers Appendix.

9.7 The force, velocity, and duration of skeletal muscle contractions are determined by a variety of factors

→ Learning Objectives

- ☐ **Describe factors that influence the force, velocity, and duration of skeletal muscle contraction.**
- ☐ **Describe three types of skeletal muscle fibers and explain the relative value of each type.**

Force of Muscle Contraction

The force of muscle contraction depends on the number of myosin cross bridges that are attached to actin. This in turn is affected by four factors (**Figure 9.18**):

- **Number of muscle fibers recruited.** The more motor units recruited, the greater the force.
- **Size of muscle fibers.** The bulkier the muscle and the greater the cross-sectional area, the more tension it can develop. The large fibers of large motor units produce the most powerful movements. Regular resistance exercise increases muscle force by causing muscle cells to *hypertrophy* (increase in size).
- **Frequency of stimulation.** When a muscle is stimulated more frequently, contractions are summed, becoming more vigorous and ultimately producing tetanus. So, the higher the frequency of muscle stimulation, the greater the force the muscle exerts.
- **Degree of muscle stretch.** If a muscle is stretched to various lengths and stimulated tetanically, the tension the muscle can generate varies with length. The ideal **length-tension relationship** occurs when the muscle is slightly stretched and the thin and thick filaments overlap optimally, because this permits sliding along nearly the entire length of the thin filaments (Figure 9.18 and **Figure 9.19**). If a muscle is stretched so much that the filaments do not overlap, the myosin heads have nothing to attach to and cannot generate tension. On the other hand, if the sarcomeres are so compressed that the thin filaments interfere with one another, little or no further shortening can occur. In the body, skeletal muscles are maintained near their optimal length by the way they are attached to bones. Our joints normally prevent bone movements that would stretch attached muscles beyond their optimal range.

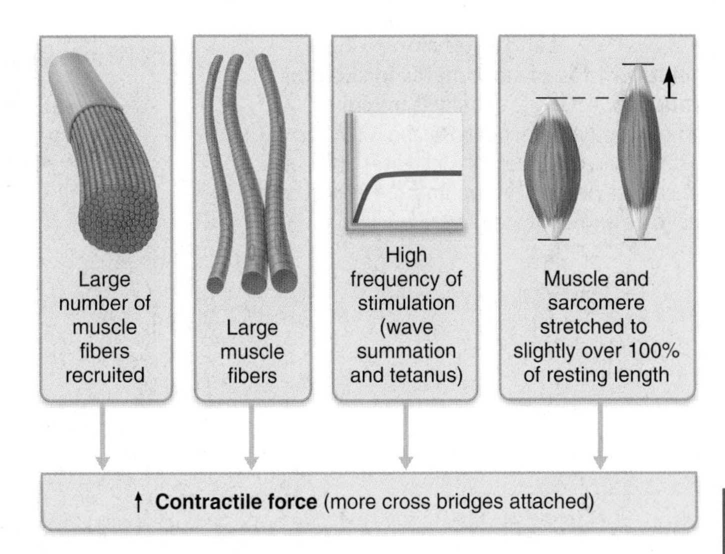

Figure 9.18 Factors that increase the force of skeletal muscle contraction.

Velocity and Duration of Contraction

Muscles vary in how fast they can contract and how long they can continue to contract before they fatigue. These characteristics are influenced by muscle fiber type, load, and recruitment (**Figure 9.20**).

Muscle Fiber Type

There are several ways of classifying muscle fibers, but learning about these classes will be easier if you pay attention to just two functional characteristics:

- **Speed of contraction.** On the basis of speed (velocity) of fiber shortening, there are **slow fibers** and **fast fibers**. The difference reflects how fast their myosin ATPases split ATP, and the pattern of electrical activity of their motor neurons. Contraction duration also varies with fiber type and depends on how quickly Ca^{2+} moves from the cytosol into the SR.
- **Major pathways for forming ATP.** The cells that rely mostly on the oxygen-using aerobic pathways for ATP generation are **oxidative fibers**. Those that rely more on anaerobic glycolysis and creatine phosphate are **glycolytic fibers**.

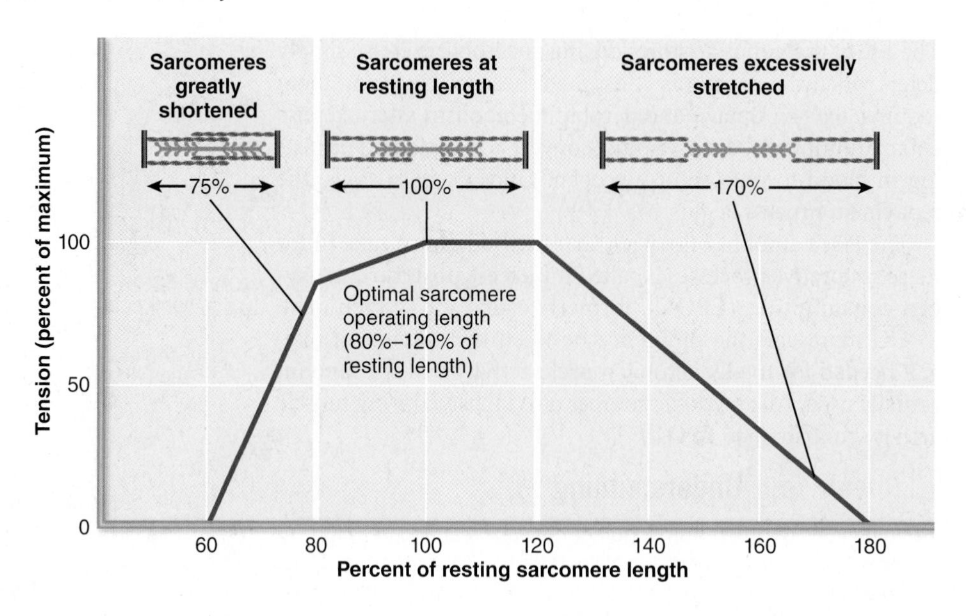

Figure 9.19 Length-tension relationships of sarcomeres in skeletal muscles. A muscle generates maximum force when it is between 80 and 120% of its optimal resting length. Increases and decreases beyond this optimal range reduce its force and ability to generate tension.

Using these two criteria, we can classify skeletal muscle cells as: **slow oxidative fibers**, **fast oxidative fibers**, or **fast glycolytic fibers**.

Table 9.2 gives details about each group, but a word to the wise: Do not approach this information by rote memorization—you'll just get frustrated. Instead, start with what you know for any category and see how the characteristics listed support that.

For example, think about a *slow oxidative fiber* (Table 9.2, first column, and Figure 9.20, right side). We can see that it:

- Contracts *slowly* because its myosin ATPases are slow (a criterion)
- Depends on *oxygen* delivery and aerobic pathways (its major pathways for forming ATP give it *high oxidative capacity*—a criterion)
- Resists fatigue and has high endurance (typical of fibers that depend on aerobic metabolism)
- Is thin (a large amount of cytoplasm impedes diffusion of O_2 and nutrients from the blood)
- Has relatively little power (a thin cell can contain only a limited number of myofibrils)
- Has many mitochondria (actual sites of oxygen use)
- Has a rich capillary supply (the better to deliver bloodborne O_2)
- Is red (its color stems from an abundant supply of myoglobin, muscle's oxygen-binding pigment that stores O_2 reserves in the cell and helps O_2 diffuse through the cell)

Add these features together and you have a muscle fiber best suited to endurance-type activities.

Now think about a *fast glycolytic fiber* (Table 9.2, third column, and Figure 9.20, left side). In contrast, it:

- Contracts *rapidly* due to the activity of fast myosin ATPases
- Uses little oxygen
- Depends on plentiful *glycogen* reserves for fuel rather than on blood-delivered nutrients
- Tires quickly because glycogen reserves are short-lived, making it a fatigable fiber
- Has a relatively large diameter, indicating the plentiful myofilaments that allow it to contract powerfully before it "poops out"
- Has few mitochondria, little myoglobin, and few capillaries (making it white), and is thicker than slow oxidative fibers (because it doesn't depend on continuous oxygen and nutrient diffusion from the blood)

For these reasons, a fast glycolytic fiber is best suited for short-term, rapid, intense movements (moving furniture across the room, for example).

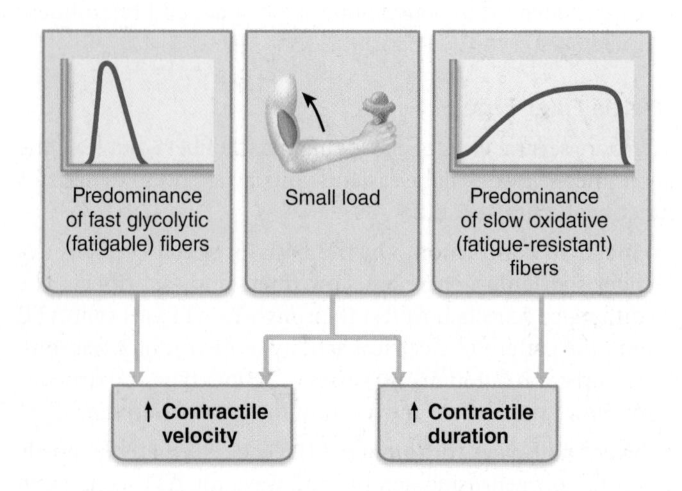

Figure 9.20 Factors influencing velocity and duration of skeletal muscle contraction.

Table 9.2 Structural and Functional Characteristics of the Three Types of Skeletal Muscle Fibers

	SLOW OXIDATIVE FIBERS	FAST OXIDATIVE FIBERS	FAST GLYCOLYTIC FIBERS
Metabolic Characteristics			
Speed of contraction	Slow	Fast	Fast
Myosin ATPase activity	Slow	Fast	Fast
Primary pathway for ATP synthesis	Aerobic	Aerobic (some anaerobic glycolysis)	Anaerobic glycolysis
Myoglobin content	High	High	Low
Glycogen stores	Low	Intermediate	High
Recruitment order	First	Second	Third
Rate of fatigue	Slow (fatigue-resistant)	Intermediate (moderately fatigue-resistant)	Fast (fatigable)
Activities Best Suited For			
	Endurance-type activities—e.g., running a marathon; maintaining posture (antigravity muscles)	Sprinting, walking	Short-term intense or powerful movements, e.g., hitting a baseball
Structural Characteristics			
Fiber diameter	Small	Large*	Intermediate
Mitochondria	Many	Many	Few
Capillaries	Many	Many	Few
Color	Red	Red to pink	White (pale)

*In animal studies, fast glycolytic fibers were found to be the largest, but not in humans.

Finally, consider the less common intermediate muscle fiber types, called *fast oxidative fibers* (Table 9.2, middle column). They have many characteristics intermediate between the other two types (fiber diameter and power, for example). Like fast glycolytic fibers, they contract quickly, but like slow oxidative fibers, they are oxygen dependent and have a rich supply of myoglobin and capillaries.

Some muscles have a predominance of one fiber type, but most contain a mixture of fiber types, which gives them a range of contractile speeds and fatigue resistance. But, as might be expected, all muscle fibers in a particular *motor unit* are of the same type.

Although everyone's muscles contain mixtures of the three fiber types, some people have relatively more of one kind. These differences are genetically initiated, but can be modified by exercise and no doubt determine athletic capabilities, such as endurance versus strength, to a large extent. For example, muscles of marathon runners have a high percentage of slow oxidative fibers (about 80%), while those of sprinters contain a higher percentage (about 60%) of fast oxidative and glycolytic fibers. Interconversion between the "fast" fiber types occurs as a result of specific exercise regimes, as we'll describe below.

Load and Recruitment

Because muscles are attached to bones, they are always pitted against some resistance, or load, when they contract. As you might expect, they contract fastest when there is no added load on them. A greater load results in a longer latent period, slower shortening, and a briefer duration of shortening (**Figure 9.21**).

In the same way that many hands on a project can get a job done more quickly and can keep working longer, the more motor units that are contracting, the faster and more prolonged the contraction.

☑ Check Your Understanding

15. List two factors that influence contractile force and two that influence velocity of contraction.

16. Jordan called several friends to help him move. Would he prefer to have those with more slow oxidative muscle fibers or those with more fast glycolytic fibers as his helpers? Why?

For answers, see Answers Appendix.

9

Figure 9.21 Influence of load on duration and velocity of muscle shortening.

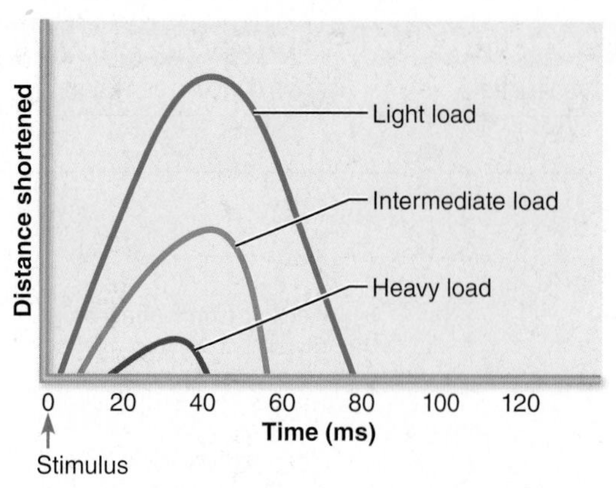

(a) The greater the load, the briefer the duration of muscle shortening.

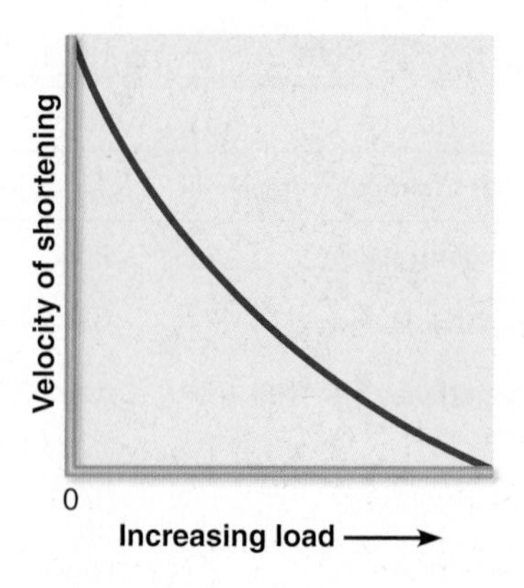

(b) The greater the load, the slower the muscle shortening.

9.8 How does skeletal muscle respond to exercise?

Learning Objective

- [ ] Compare and contrast the effects of aerobic and resistance exercise on skeletal muscles.

The amount of work a muscle does is reflected in changes in the muscle itself. When used actively or strenuously, muscles may become larger or stronger, or more efficient and fatigue resistant. Exercise gains are based on the overload principle. Forcing a muscle to work hard increases its strength and endurance. As muscles adapt to greater demand, they must be overloaded to produce further gains. Inactivity, on the other hand, *always* leads to muscle weakness and **atrophy** (at′ro-fe; a decrease in size).

Aerobic (Endurance) Exercise

Aerobic, or **endurance**, **exercise** such as swimming, jogging, fast walking, and biking results in several recognizable changes in skeletal muscles:

- The number of capillaries surrounding the muscle fibers increases.
- The number of mitochondria within the muscle fibers also increases.
- The fibers synthesize more myoglobin.

These changes occur in all fiber types, but are most dramatic in slow oxidative fibers, which depend primarily on aerobic pathways. The changes result in more efficient muscle metabolism and in greater endurance, strength, and resistance to fatigue. Regular endurance exercise may convert fast glycolytic fibers into fast oxidative fibers.

Resistance Exercise

The moderately weak but sustained muscle activity required for endurance exercise does not promote significant skeletal muscle hypertrophy, even though the exercise may go on for hours. Muscle hypertrophy—think of the bulging biceps of a professional weight lifter—results mainly from high-intensity **resistance exercise** (typically under anaerobic conditions) such as weight lifting or isometric exercise, which pits muscles against high-resistance or immovable forces. Here strength, not stamina, is important, and a few minutes every other day is sufficient to allow a proverbial weakling to put on 50% more muscle within a year.

The additional muscle bulk largely reflects the increased size of individual muscle fibers (particularly the fast glycolytic variety) rather than an increased number of muscle fibers. [However, some of the bulk may result from longitudinal splitting of the fibers and subsequent growth of these "split" cells, or from the proliferation and fusion of satellite cells (which help repair injured fibers).] Vigorously stressed muscle fibers also contain more mitochondria, form more myofilaments and myofibrils, store more glycogen, and develop more connective tissue between muscle cells.

Collectively these changes promote significant increases in muscle strength and size. Resistance activities can also convert fast oxidative fibers to fast glycolytic fibers. However, if the specific exercise routine is discontinued, the converted fibers revert to their original metabolic properties.

HOMEOSTATIC IMBALANCE 9.3 — CLINICAL

To remain healthy, muscles must be active. Immobilization due to enforced bed rest or loss of neural stimulation results in *disuse atrophy* (degeneration and loss of mass), which begins almost as soon as the muscles are immobilized. Under such conditions, muscle strength can decline at the rate of 5% per day!

Even at rest, muscles receive weak intermittent stimuli from the nervous system. When totally deprived of neural stimulation, a paralyzed muscle may atrophy to one-quarter of its initial size. Fibrous connective tissue replaces the lost muscle tissue, making muscle rehabilitation impossible. +

Check Your Understanding

17. How do aerobic and resistance exercise differ in their effects on muscle size and function?

For answers, see Answers Appendix.

9.9 Smooth muscle is nonstriated involuntary muscle

→ **Learning Objectives**

- ☐ **Compare the gross and microscopic anatomy of smooth muscle cells to that of skeletal muscle cells.**
- ☐ **Compare and contrast the contractile mechanisms and the means of activation of skeletal and smooth muscles.**
- ☐ **Distinguish between unitary and multi unit smooth muscle structurally and functionally.**

Except for the heart, which is made of cardiac muscle, the muscle in the walls of all the body's hollow organs is almost entirely smooth muscle. The chemical and mechanical events of contraction are essentially the same in all muscle tissues, but smooth muscle is distinctive in several ways, as summarized in Table 9.3 on pp. 276–277.

Microscopic Structure of Smooth Muscle Fibers

Smooth muscle fibers are spindle-shaped cells of variable size, each with one centrally located nucleus (Figure 9.22b). Typically, they have a diameter of 5–10 μm and are 30–200 μm long. Skeletal muscle fibers are up to 10 times wider and thousands of times longer.

Smooth muscle lacks the coarse connective tissue sheaths seen in skeletal muscle. However, a small amount of fine connective tissue (endomysium), secreted by the smooth muscles themselves and containing blood vessels and nerves, is found between smooth muscle fibers.

Most smooth muscle is organized into sheets of closely apposed fibers. These sheets occur in the walls of all but the smallest blood vessels and in the walls of hollow organs of the respiratory, digestive, urinary, and reproductive tracts. In most cases, there are two sheets of smooth muscle with their fibers oriented at right angles to each other, as in the intestine.

- In the *longitudinal layer*, the muscle fibers run parallel to the long axis of the organ. Consequently, when these fibers contract, the organ shortens.
- In the *circular layer*, the fibers run around the circumference of the organ. Contraction of this layer constricts the lumen (cavity inside) of the organ.

The alternating contraction and relaxation of these layers mixes substances in the lumen and squeezes them through the organ's internal pathway. This propulsive action is called **peristalsis** (per″ĭ-stal′sis; "around contraction"). Contraction of smooth muscle in the rectum, urinary bladder, and uterus helps those organs to expel their contents. Smooth muscle contraction also accounts for the constricted breathing of asthma and for stomach cramps.

Smooth muscle lacks the highly structured neuromuscular junctions of skeletal muscle. Instead, the innervating nerve fibers, which are part of the autonomic (involuntary) nervous system, have numerous bulbous swellings, called **varicosities** (Figure 9.23). The varicosities release neurotransmitter into a wide synaptic cleft in the general area of the smooth muscle cells. Such junctions are called **diffuse junctions**. Comparing the neural input to skeletal and smooth muscles, you could say that skeletal muscle gets priority mail while smooth muscle gets bulk mailings.

The sarcoplasmic reticulum of smooth muscle fibers is much less developed than that of skeletal muscle and lacks a specific pattern relative to the myofilaments. T tubules are absent, but

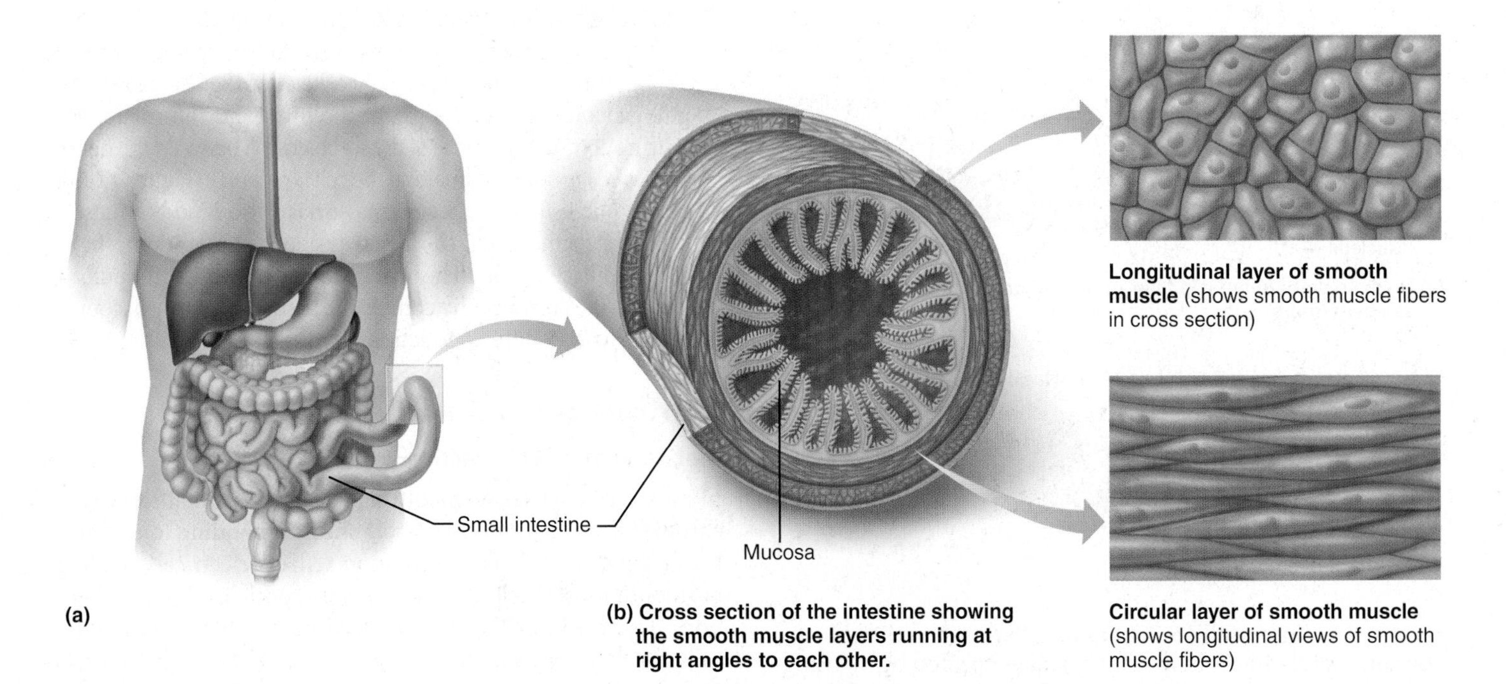

Figure 9.22 Arrangement of smooth muscle in the walls of hollow organs.

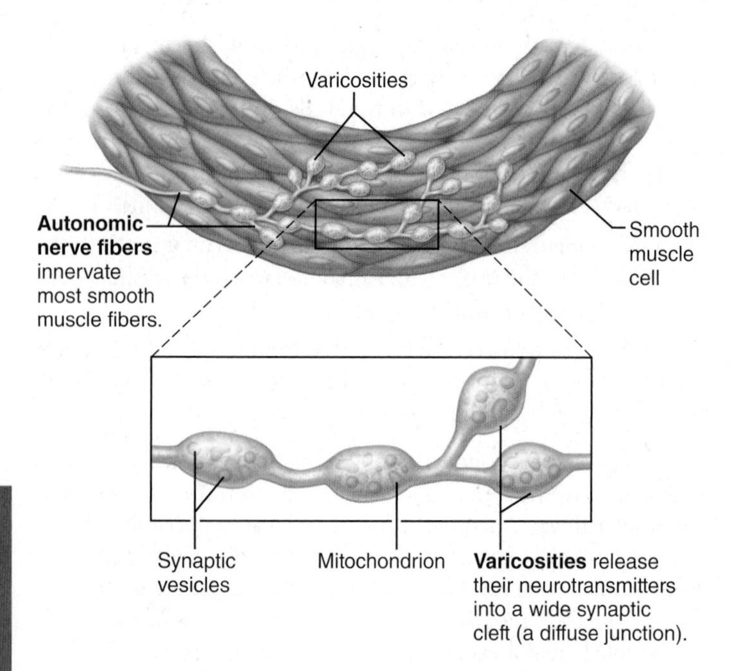

Figure 9.23 Innervation of smooth muscle.

the sarcolemma has multiple **caveolae**, pouchlike infoldings containing large numbers of Ca^{2+} channels (**Figure 9.24a**). Consequently, when calcium channels in the caveolae open, Ca^{2+} influx occurs rapidly. Although the SR *does* release some of the calcium that triggers contraction, most Ca^{2+} enters through calcium channels directly from the extracellular space. This situation is quite different from what we see in skeletal muscle, which does not depend on extracellular Ca^{2+} for excitation-contraction coupling. Contraction ends when cytoplasmic calcium is actively transported into the SR and out of the cell.

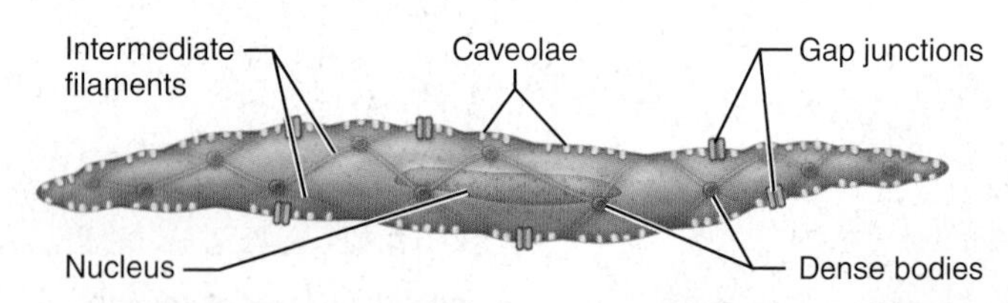

(a) Relaxed smooth muscle fiber (note that gap junctions connect adjacent fibers)

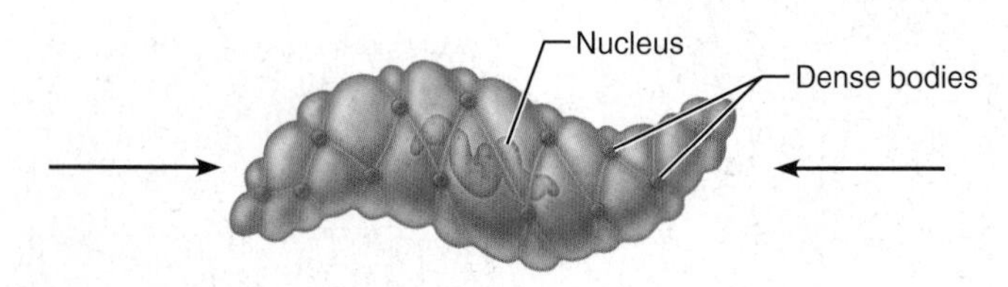

(b) Contracted smooth muscle fiber

Figure 9.24 Intermediate filaments and dense bodies of smooth muscle fibers harness the pull generated by myosin cross bridges. Intermediate filaments attach to dense bodies throughout the sarcoplasm.

There are no striations in smooth muscle, as its name indicates, and therefore no sarcomeres. Smooth muscle fibers do contain interdigitating thick and thin filaments, but the myosin filaments are a lot shorter than the actin filaments and the type of myosin contained differs from skeletal muscle. The proportion and organization of smooth muscle myofilaments differ from skeletal muscle in the following ways:

- **Thick filaments are fewer but have myosin heads along their entire length.** The ratio of thick to thin filaments is much lower in smooth muscle than in skeletal muscle (1:13 compared to 1:2). However, thick filaments of smooth muscle contain actin-gripping myosin heads along their *entire length,* a feature that makes smooth muscle as powerful as a skeletal muscle of the same size. Also, in smooth muscle the myosin heads are oriented in one direction on one side of the filament and in the opposite direction on the other side.
- **No troponin complex in thin filaments.** As in skeletal muscle, tropomyosin mechanically stabilizes the thin filaments, but smooth muscle has no calcium-binding troponin complex. Instead, a protein called *calmodulin* acts as the calcium-binding site.
- **Thick and thin filaments arranged diagonally.** Bundles of contractile proteins crisscross within the smooth muscle cell so they spiral down the long axis of the cell like the stripes on a barber pole. Because of this diagonal arrangement, the smooth muscle cells contract in a twisting way so that they look like tiny corkscrews (Figure 9.24b).
- **Intermediate filament–dense body network.** Smooth muscle fibers contain a lattice-like arrangement of noncontractile *intermediate filaments* that resist tension. They attach at regular intervals to cytoplasmic structures called dense bodies (Figure 9.24). The **dense bodies**, which are also tethered to the sarcolemma, act as anchoring points for thin filaments and therefore correspond to Z discs of skeletal muscle.

  The intermediate filament–dense body network forms a strong, cable-like intracellular cytoskeleton that harnesses the pull generated by the sliding of the thick and thin filaments. During contraction, areas of the sarcolemma between the dense bodies bulge outward, making the cell look puffy (Figure 9.24b). Dense bodies at the sarcolemma surface also bind the muscle cell to the connective tissue fibers outside the cell (endomysium) and to adjacent cells. This arrangement transmits the pulling force to the surrounding connective tissue and partly accounts for the synchronous contractions of most smooth muscle.

Contraction of Smooth Muscle

Mechanism of Contraction

In most cases, adjacent smooth muscle fibers exhibit slow, synchronized contractions, the whole sheet responding to a stimulus in unison. This synchronization reflects electrical coupling of smooth muscle cells by *gap junctions*, specialized cell connections described in Chapter 3. Skeletal muscle fibers are electrically isolated from one another, each stimulated to contract by its own neuromuscular junction. By contrast, gap junctions allow smooth muscles to transmit action potentials from fiber to fiber.

Some smooth muscle fibers in the stomach and small intestine are *pacemaker cells*: Once excited, they act as "drummers" to set the pace of contraction for the entire muscle sheet. These pacemakers depolarize spontaneously in the absence of external stimuli. However, neural and chemical stimuli can modify both the rate and the intensity of smooth muscle contraction.

Contraction in smooth muscle is like contraction in skeletal muscle in the following ways:

- Actin and myosin interact by the sliding filament mechanism.
- The final trigger for contraction is a rise in the intracellular calcium ion level.
- ATP energizes the sliding process.

During excitation-contraction coupling, the tubules of the SR release Ca^{2+}, but Ca^{2+} also moves into the cell from the extracellular space via membrane channels. In all striated muscle types, calcium ions activate myosin by binding to troponin. In smooth muscle, calcium activates myosin by interacting with a regulatory molecule called **calmodulin**, a cytoplasmic calcium-binding protein. Calmodulin, in turn, interacts with a kinase enzyme called **myosin kinase** or **myosin light chain kinase** which phosphorylates the myosin, activating it (**Figure 9.25**).

As in skeletal muscle, smooth muscle relaxes when intracellular Ca^{2+} levels drop—but getting smooth muscle to stop contracting is more complex. Events known to be involved include calcium detachment from calmodulin, active transport of Ca^{2+} into the SR and extracellular fluid, and dephosphorylation of myosin by a phosphorylase enzyme, which reduces the activity of the myosin ATPases.

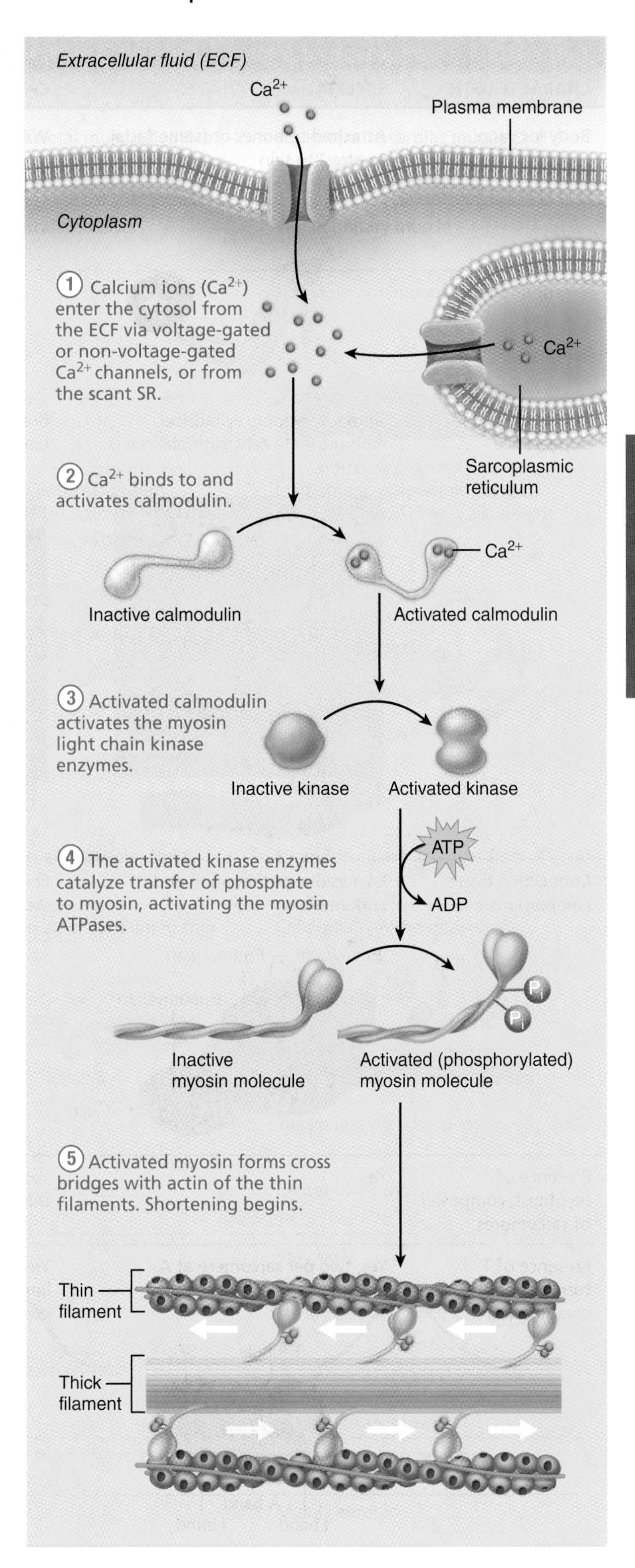

Figure 9.25 Sequence of events in excitation-contraction coupling of smooth muscle.

Energy Efficiency of Smooth Muscle Contraction

Smooth muscle takes 30 times longer to contract and relax than does skeletal muscle, but it can maintain the same contractile tension for prolonged periods at less than 1% of the energy cost. If skeletal muscle is like a speedy windup car that quickly runs down, then smooth muscle is like a steady, heavy-duty engine that lumbers along tirelessly.

Part of the striking energy economy of smooth muscle is the sluggishness of its ATPases compared to those in skeletal muscle. Moreover, smooth muscle myofilaments may latch together during prolonged contractions, saving energy in that way as well.

The smooth muscle in small arterioles and other visceral organs routinely maintains a moderate degree of contraction, called *smooth muscle tone*, day in and day out without fatiguing. Smooth muscle has low energy requirements, and as a rule, it makes enough ATP via aerobic pathways to keep up with the demand.

Regulation of Contraction

The contraction of smooth muscle can be regulated by nerves, hormones, or local chemical changes. Let's briefly consider each of these methods.

(Text continues on p. 278.)

Neural Regulation In some cases, the activation of smooth muscle by a neural stimulus is identical to that in skeletal muscle: Neurotransmitter binding generates an action potential, which is coupled to a rise in calcium ions in the cytosol. However, some types of smooth muscle respond to neural stimulation with graded potentials (local electrical signals) only.

Recall that all somatic nerve endings, that is, nerve endings that excite skeletal muscle, release the neurotransmitter acetylcholine. However, different autonomic nerves serving the smooth muscle of visceral organs release different neurotransmitters, each of which may excite or inhibit a particular group of smooth muscle cells.

The effect of a specific neurotransmitter on a smooth muscle cell depends on the type of receptor molecules on the cell's sarcolemma. For example, when acetylcholine binds to ACh receptors on smooth muscle in the bronchioles (small air passageways of the lungs), the response is strong contraction that narrows the bronchioles. When norepinephrine, released by a different type of autonomic nerve fiber, binds to norepinephrine receptors on the *same* smooth muscle cells, the effect is inhibitory—the muscle relaxes, which dilates the bronchioles. However, when norepinephrine binds to smooth muscle in the walls of most blood vessels, it stimulates the smooth muscle cells to contract and constrict the vessel.

Hormones and Local Chemical Factors Some smooth muscle layers have no nerve supply at all. Instead, they depolarize spontaneously or in response to chemical stimuli that bind to G protein–linked receptors. Other smooth muscle cells respond to both neural and chemical stimuli.

Several chemical factors cause smooth muscle to contract or relax without an action potential by enhancing or inhibiting Ca^{2+} entry into the sarcoplasm. They include certain hormones, histamine, excess carbon dioxide, low pH, and lack of oxygen. The direct response to these chemical stimuli alters smooth muscle activity according to local tissue needs and probably is most responsible for smooth muscle tone. For example, the hormone gastrin stimulates stomach smooth muscle to contract so it can churn foodstuffs more efficiently. We will consider activation of smooth muscle in specific organs as we discuss each organ in subsequent chapters.

Special Features of Smooth Muscle Contraction

Smooth muscle is intimately involved in the functioning of most hollow organs and has a number of unique characteristics. We have already considered some of these—smooth muscle tone, slow prolonged contractions, and low energy requirements. But smooth muscle also responds differently to stretch and can lengthen and shorten more than other muscle types. Let's take a look.

Response to Stretch Up to a point, when skeletal muscle is stretched, it responds with more vigorous contractions. Stretching of smooth muscle also provokes contraction, which automatically moves substances along an internal tract. However, the increased tension persists only briefly, and soon the muscle adapts to its new length and relaxes, while still retaining the ability to contract on demand.

This **stress-relaxation response** allows a hollow organ to fill or expand slowly to accommodate a greater volume without causing strong contractions that would expel its contents. This is an important attribute, because organs such as the stomach and intestines must store their contents long enough to digest and absorb the nutrients. Likewise, your urinary bladder must be able to store the continuously made urine until it is convenient to empty your bladder, or you would spend all your time in the bathroom.

Length and Tension Changes Smooth muscle stretches much more and generates more tension than skeletal muscles stretched to a comparable extent. The irregular, overlapping arrangement of smooth muscle filaments and the lack of sarcomeres allow them to generate considerable force, even when they are substantially stretched. The total length change that skeletal muscles can undergo and still function efficiently is about 60% (from 30% shorter to 30% longer than resting length), but smooth muscle can contract when it is anywhere from half to twice its resting length—a total range of 150%. This capability allows hollow organs to tolerate tremendous changes in volume without becoming flabby when they empty.

Types of Smooth Muscle

The smooth muscle in different body organs varies substantially in its (1) fiber arrangement and organization, (2) innervation, and (3) responsiveness to various stimuli. For simplicity, however, smooth muscle is usually categorized into two major types: *unitary* and *multi unit*.

Unitary Smooth Muscle

Unitary smooth muscle, commonly called **visceral muscle** because it is in the walls of all hollow organs except the heart, is far more common. All the smooth muscle characteristics described so far pertain to unitary smooth muscle.

For example, the cells of unitary smooth muscle:

- Are arranged in opposing (longitudinal and circular) sheets
- Are innervated by varicosities of autonomic nerve fibers and often exhibit rhythmic spontaneous action potentials
- Are electrically coupled by gap junctions and so contract as a unit (for this reason recruitment is not an option in unitary smooth muscle)
- Respond to various chemical stimuli

Multi Unit Smooth Muscle

The smooth muscles in the large airways to the lungs and in large arteries, the arrector pili muscles attached to hair follicles, and the internal eye muscles that adjust pupil size and allow the eye to focus visually are all examples of **multi unit smooth muscle**.

In contrast to unitary muscle, gap junctions and spontaneous depolarizations are rare. Like skeletal muscle, multi unit smooth muscle:

- Consists of muscle fibers that are structurally independent of one another
- Is richly supplied with nerve endings, each of which forms a motor unit with a number of muscle fibers
- Responds to neural stimulation with graded contractions that involve recruitment

However, skeletal muscle is served by the somatic (voluntary) division of the nervous system. Multi unit smooth muscle, like unitary smooth muscle, is innervated by the autonomic (involuntary) division and also responds to hormones.

☑ Check Your Understanding

18. Compare the structures of skeletal and smooth muscle fibers.

19. Calcium is the trigger for contraction of all muscle types. How does its binding site differ in skeletal and smooth muscle fibers?

20. How does the stress-relaxation response suit the role of smooth muscle in hollow organs?

21. MAKING **connections** Intracellular calcium performs other important roles in the body in addition to triggering muscle contraction. What are these roles? (Hint: See Chapter 3.)

For answers, see Answers Appendix.

In this chapter we have covered muscle anatomy from gross to molecular levels and have considered muscle physiology in some detail. Chapter 10 explains how skeletal muscles interact with bones and with each other, and describes the individual skeletal muscles that make up the muscular system.

REVIEW QUESTIONS

Multiple Choice/Matching

(Some questions have more than one correct answer. Select the best answer or answers from the choices given.)

1. The connective tissue covering that encloses the sarcolemma of an individual muscle fiber is called the **(a)** epimysium, **(b)** perimysium, **(c)** endomysium, **(d)** periosteum.

2. A fascicle is a **(a)** muscle, **(b)** bundle of muscle fibers enclosed by a connective tissue sheath, **(c)** bundle of myofibrils, **(d)** group of myofilaments.

3. Thick and thin myofilaments have different compositions. For each descriptive phrase, indicate whether the filament is **(a)** thick or **(b)** thin.

____ **(1)** contains actin
____ **(2)** contains ATPases
____ **(3)** attaches to the Z disc
____ **(4)** contains myosin
____ **(5)** contains troponin
____ **(6)** does not lie in the I band

4. The function of the T tubules in muscle contraction is to **(a)** make and store glycogen, **(b)** release Ca^{2+} into the cell interior and then pick it up again, **(c)** transmit the action potential deep into the muscle cells, **(d)** form proteins.

5. The sites where the motor nerve impulse is transmitted from the nerve endings to the skeletal muscle cell membranes are the **(a)** neuromuscular junctions, **(b)** sarcomeres, **(c)** myofilaments, **(d)** Z discs.

6. Contraction elicited by a single brief stimulus is called **(a)** a twitch, **(b)** wave summation, **(c)** multiple motor unit summation, **(d)** fused tetanus.

7. A smooth, sustained contraction resulting from very rapid stimulation of the muscle, in which no evidence of relaxation is seen, is called **(a)** a twitch, **(b)** wave summation, **(c)** multiple motor unit summation, **(d)** fused tetanus.

8. Characteristics of isometric contractions include all but **(a)** shortening, **(b)** increased muscle tension throughout the contraction phase, **(c)** absence of shortening, **(d)** used in resistance training.

9. During muscle contraction, ATP is provided by **(a)** a coupled reaction of creatine phosphate with ADP, **(b)** aerobic respiration of glucose, and **(c)** anaerobic glycolysis.

____ **(1)** Which provides ATP fastest?
____ **(2)** Which does (do) not require that oxygen be available?
____ **(3)** Which provides the highest yield of ATP per glucose molecule?
____ **(4)** Which results in the formation of lactic acid?
____ **(5)** Which has carbon dioxide and water products?
____ **(6)** Which is most important in endurance sports?

10. The neurotransmitter released by somatic motor neurons is **(a)** acetylcholine, **(b)** acetylcholinesterase, **(c)** norepinephrine.

11. The ions that enter the skeletal muscle cell during the generation of an action potential are **(a)** calcium ions, **(b)** chloride ions, **(c)** sodium ions, **(d)** potassium ions.

12. Myoglobin has a special function in muscle tissue. It **(a)** breaks down glycogen, **(b)** is a contractile protein, **(c)** holds a reserve supply of oxygen in the muscle.

13. Aerobic exercise results in all of the following except **(a)** more capillaries surrounding muscle fibers, **(b)** more mitochondria in muscle cells, **(c)** increased size and strength of existing muscle cells, **(d)** more myoglobin.

14. The smooth muscle type found in the walls of digestive and urinary system organs and that exhibits gap junctions and pacemaker cells is **(a)** multi unit, **(b)** unitary.

Short Answer Essay Questions

15. Name and describe the four special functional abilities of muscle that are the basis for muscle response.

16. Distinguish between (a) direct and indirect muscle attachments and (b) a tendon and an aponeurosis.

MUSCLE GALLERY

Table 10.8 Superficial Muscles of the Anterior and Posterior Thorax: Movements of the Scapula and Arm (Figure 10.14) *(continued)*

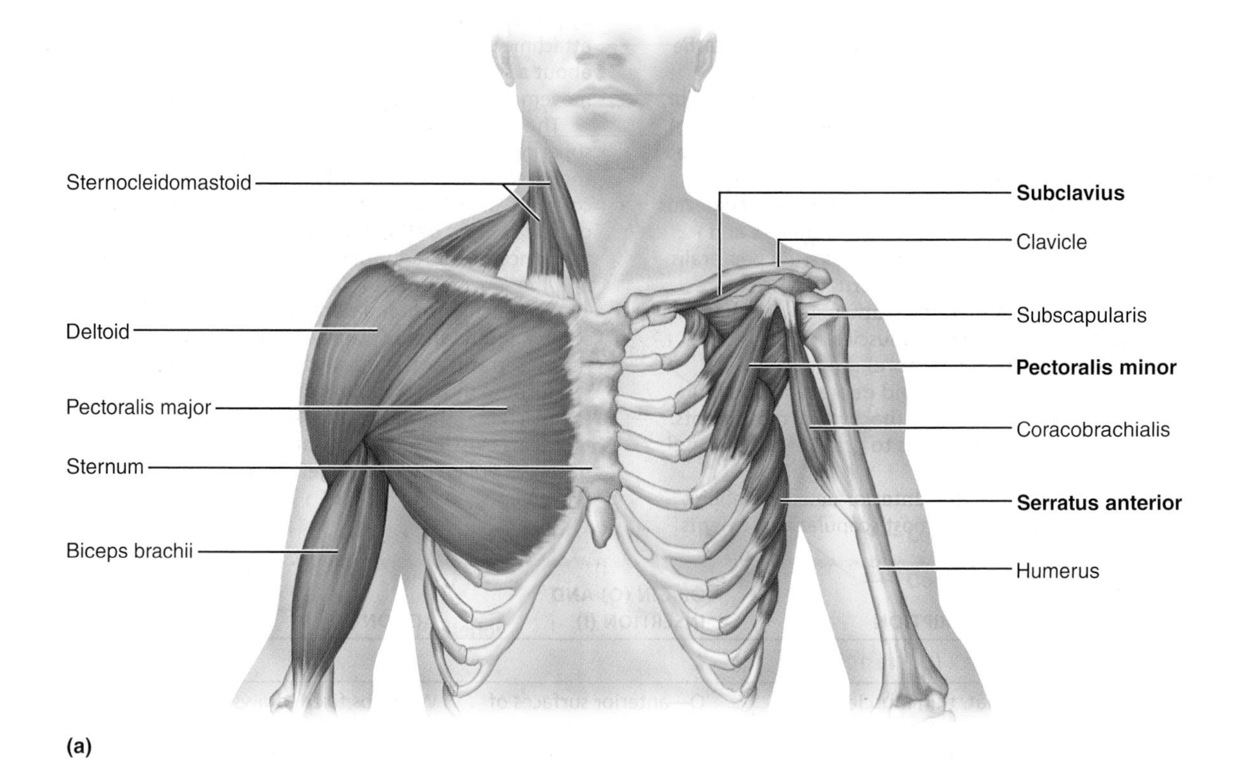

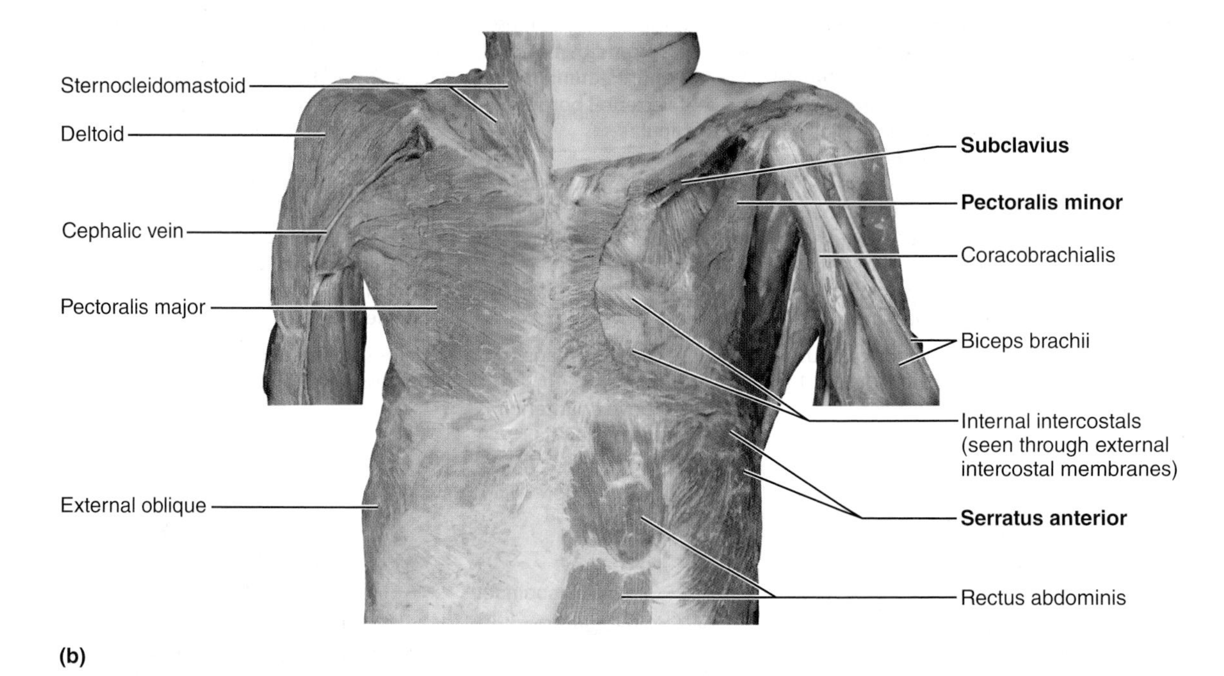

Figure 10.14 Superficial muscles of the thorax and shoulder acting on the scapula and arm. (a) Anterior view. The superficial muscles, which effect arm movements, are shown on the left side of the illustration. These muscles are removed on the right to show the muscles that stabilize or move the pectoral girdle. **(b)** Cadaver dissection, anterior view of superficial muscles of anterior thorax.

Explore human cadaver
MasteringA&P®>Study Area>PAL

MUSCLE GALLERY

Table 10.8 *(continued)*

MUSCLE	DESCRIPTION	ORIGIN (O) AND INSERTION (I)	ACTION	NERVE SUPPLY
MUSCLES OF THE POSTERIOR THORAX (Figure 10.14c–e)				
Trapezius (trah-pe′ze-us) (*trapezion* = irregular four-sided figure)	Most superficial muscle of posterior thorax; flat and triangular in shape; upper fibers run inferiorly to scapula; middle fibers run horizontally to scapula; lower fibers run superiorly to scapula	O—occipital bone, ligamentum nuchae, and spinous processes of C_7 and all thoracic vertebrae I—a continuous insertion along acromion and spine of scapula and lateral third of clavicle	Stabilizes, elevates, retracts, and rotates scapula; middle fibers retract (adduct) scapula; superior fibers elevate scapula (as in shrugging the shoulders) or help extend head with scapula fixed; inferior fibers depress scapula (and shoulder)	Accessory nerve (cranial nerve XI); C_3 and C_4
Levator scapulae (skap′u-le) (*levator* = raises)	Located at back and side of neck, deep to trapezius; thick straplike muscle	O—transverse processes of C_1–C_4 I—medial border of scapula, superior to spine	Elevates/adducts scapula in synergy with superior fibers of trapezius; tilts glenoid cavity downward when scapula is fixed, flexes neck to same side	Cervical spinal nerves and dorsal scapular nerve (C_3–C_5)
Rhomboids (rom′boidz)—major and minor (*rhomboid* = diamond shaped)	Two roughly diamond-shaped muscles lying deep to trapezius and inferior to levator scapulae; rhomboid minor is the more superior muscle	O—spinous processes of C_7 and T_1 (minor) and spinous processes of T_2–T_5 (major) I—medial border of scapula	Stabilize scapula; act together (and with middle trapezius fibers) to retract (adduct) scapula, thus "squaring shoulders"; rotate scapula so that glenoid cavity is downward (as when lowering arm against resistance; e.g., paddling a canoe)	Dorsal scapular nerve (C_4 and C_5)

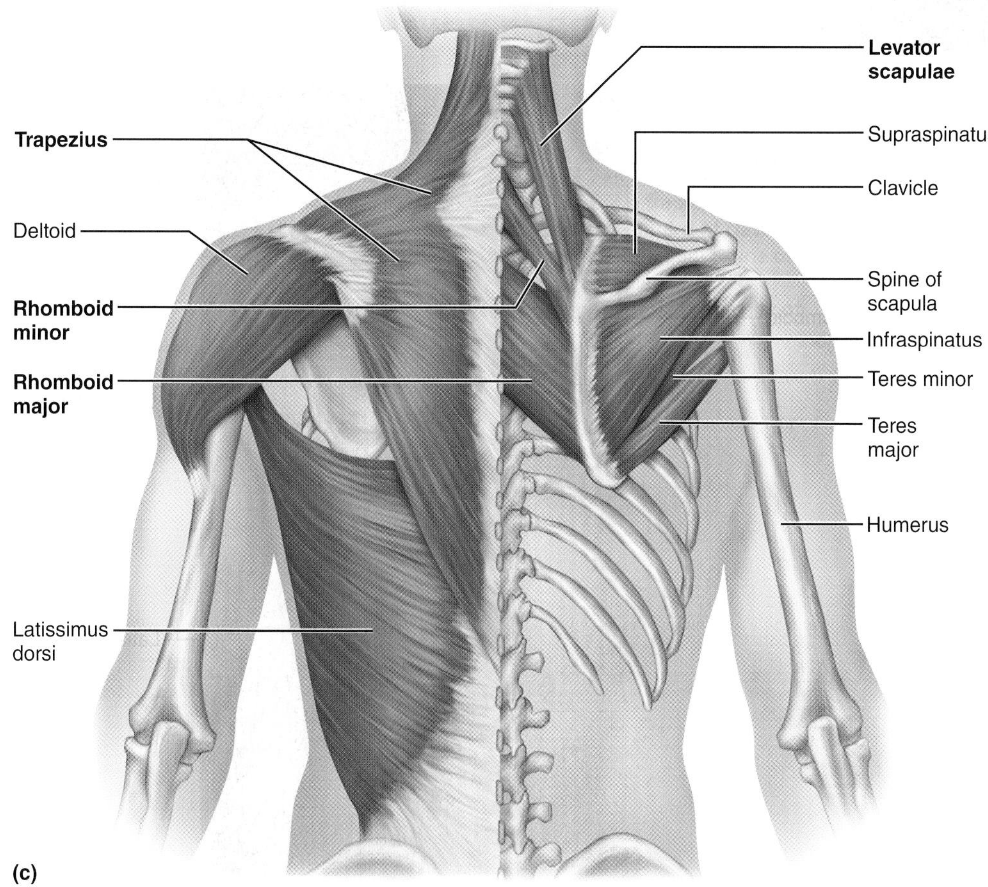

Figure 10.14 *(continued)*
(c) Posterior view. The superficial muscles are shown on the left side of the illustration. Superficial muscles are removed on the right side to reveal the deeper muscles acting on the scapula, and the rotator cuff muscles that help stabilize the shoulder joint.

10

MUSCLE GALLERY

Table 10.11 Muscles of the Forearm: Movements of the Wrist, Hand, and Fingers (Figures 10.16 and 10.17) *(continued)*

MUSCLE	DESCRIPTION	ORIGIN (O) AND INSERTION (I)	ACTION	NERVE SUPPLY
Flexor digitorum superficialis (dĭ″jĭ-tor′um soo″per-fish″e-al′is) (*digit* = finger, toe; *superficial* = close to surface)	Two-headed muscle; more deeply placed (therefore, actually forms an intermediate layer); overlain by muscles above but visible at distal end of forearm	O—medial epicondyle of humerus, coronoid process of ulna; shaft of radius I—by four tendons into middle phalanges of second to fifth fingers	Flexes wrist and middle phalanges of second to fifth fingers; the important finger flexor for speed and flexion against resistance	Median nerve (C_7, C_8, and T_1)
Deep Muscles				
Flexor pollicis longus (pah′lĭ-sis) (*pollix* = thumb)	Partly covered by flexor digitorum superficialis; parallels flexor digitorum profundus laterally	O—anterior surface of radius and interosseous membrane I—distal phalanx of thumb	Flexes distal phalanx of thumb	Branch of median nerve (C_8,T_1)
Flexor digitorum profundus (pro-fun′dus) (*profund* = deep)	Extensive origin; overlain entirely by flexor digitorum superficialis	O—coronoid process, anteromedial surface of ulna, and interosseous membrane I—by four tendons into distal phalanges of second to fifth fingers	Flexes distal interphalangeal joints; slow-acting flexor of any or all fingers; helps flex wrist	Medial half by ulnar nerve; lateral half by median nerve

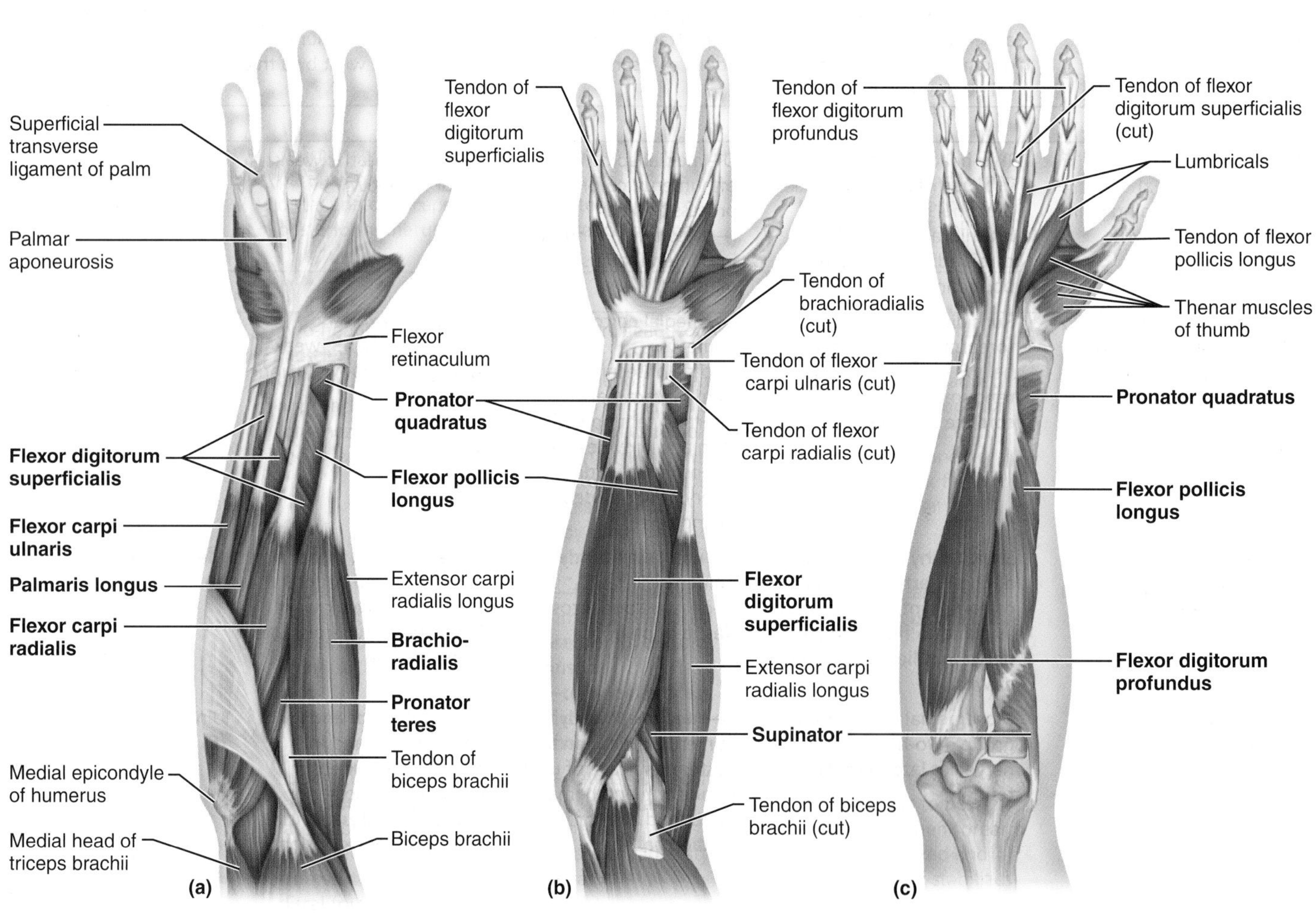

Figure 10.16 Muscles of the anterior fascial compartment of the forearm acting on the right wrist and fingers. **(a)** Superficial view. **(b)** The brachioradialis, flexors carpi radialis and ulnaris, and palmaris longus muscles have been removed to reveal the flexor digitorum superficialis. **(c)** Deep muscles of the anterior compartment. The lumbricals and thenar muscles) (intrinsic hand muscles) are also illustrated. (For a related image, *see A Brief Atlas of the Human Body*, Figure 37a.)

MUSCLE GALLERY

Table 10.11 *(continued)*

MUSCLE	DESCRIPTION	ORIGIN (O) AND INSERTION (I)	ACTION	NERVE SUPPLY
Pronator quadratus (kwod-ra'tus) (*quad* = square, four-sided)	Deepest muscle of distal forearm; passes downward and laterally; only muscle that arises solely from ulna and inserts solely into radius	O—distal portion of anterior ulnar shaft I—distal surface of anterior radius	Prime mover of forearm pronation; acts with pronator teres; also helps hold ulna and radius together	Median nerve (C_8 and T_1)
PART II: POSTERIOR MUSCLES (Figure 10.17)	These muscles of the posterior fascial compartment are listed from the lateral to the medial aspect. They are all innervated by the radial nerve or its branches. More than half of the posterior compartment muscles arise from a common extensor origin tendon attached to the posterior surface of the lateral epicondyle of the humerus and adjacent fascia. The extensor tendons are held in place at the posterior aspect of the hand by the *extensor retinaculum,* which prevents "bowstringing" of these tendons when the wrist is hyperextended. The *extensor* muscles of the fingers end in a broad hood over the dorsal side of the digits, the extensor expansion.			
Superficial Muscles				
Brachioradialis	See Table 10.10	See Table 10.10	See Table 10.10	See Table 10.10
Extensor carpi radialis longus (ek-sten'sor) (*extend* = increase angle between two bones)	Parallels brachioradialis on lateral forearm, and may blend with it	O—lateral supracondylar ridge of humerus I—base of second metacarpal	Extends hand in conjunction with extensor carpi ulnaris and abducts hand in conjunction with flexor carpi radialis	Radial nerve (C_6 and C_7)
Extensor carpi radialis brevis (brĕ'vis) (*brevis* = short)	Shorter than extensor carpi radialis longus and lies deep to it	O—lateral epicondyle of humerus I—base of third metacarpal	Extends and abducts hand; acts synergistically with extensor carpi radialis longus to steady wrist during finger flexion	Deep branch of radial nerve
Extensor digitorum	Lies medial to extensor carpi radialis brevis; a detached portion of this muscle, called *extensor digiti minimi,* extends little finger	O—lateral epicondyle of humerus I—by four tendons into extensor expansions and distal phalanges of second to fifth fingers	Prime mover of finger extension; extends hand; can abduct (flare) fingers	Posterior interosseous nerve, a branch of radial nerve (C_5 and C_6)
Extensor carpi ulnaris	Most medial of superficial posterior muscles; long, slender muscle	O—lateral epicondyle of humerus and posterior border of ulna I—base of fifth metacarpal	Extends hand in conjunction with extensor carpi radialis and adducts hand in conjunction with flexor carpi ulnaris	Posterior interosseous nerve
Deep Muscles				
Supinator (soo″pĭ-na'tor) (*supination* = turning palm anteriorly or upward)	Deep muscle at posterior aspect of elbow; largely concealed by superficial muscles	O—lateral epicondyle of humerus; proximal ulna I—proximal end of radius	Assists biceps brachii to forcibly supinate forearm; works alone in slow supination; antagonist of pronator muscles	Posterior interosseous nerve
Abductor pollicis longus (ab-duk'tor) (*abduct* = movement away from median plane)	Lateral and parallel to extensor pollicis longus; just distal to supinator	O—posterior surface of radius and ulna; interosseous membrane I—base of first metacarpal and trapezium	Abducts and extends thumb	Posterior interosseous nerve
Extensor pollicis brevis and longus	Deep muscle pair with a common origin and action; overlain by extensor carpi ulnaris	O—dorsal shaft of radius and ulna; interosseous membrane I—base of proximal (brevis) and distal (longus) phalanx of thumb	Extends thumb	Posterior interosseous nerve

MUSCLE GALLERY

Table 10.11 Muscles of the Forearm: Movements of the Wrist, Hand, and Fingers (Figures 10.16 and 10.17) *(continued)*

MUSCLE	DESCRIPTION	ORIGIN (O) AND INSERTION (I)	ACTION	NERVE SUPPLY
Extensor indicis (in′dĭ-sis) (*indicis* = index finger)	Tiny muscle arising close to wrist	O—posterior surface of distal ulna; interosseous membrane I—extensor expansion of index finger; joins tendon of extensor digitorum	Extends index finger (digit II) and assists in extending wrist	Posterior interosseous nerve

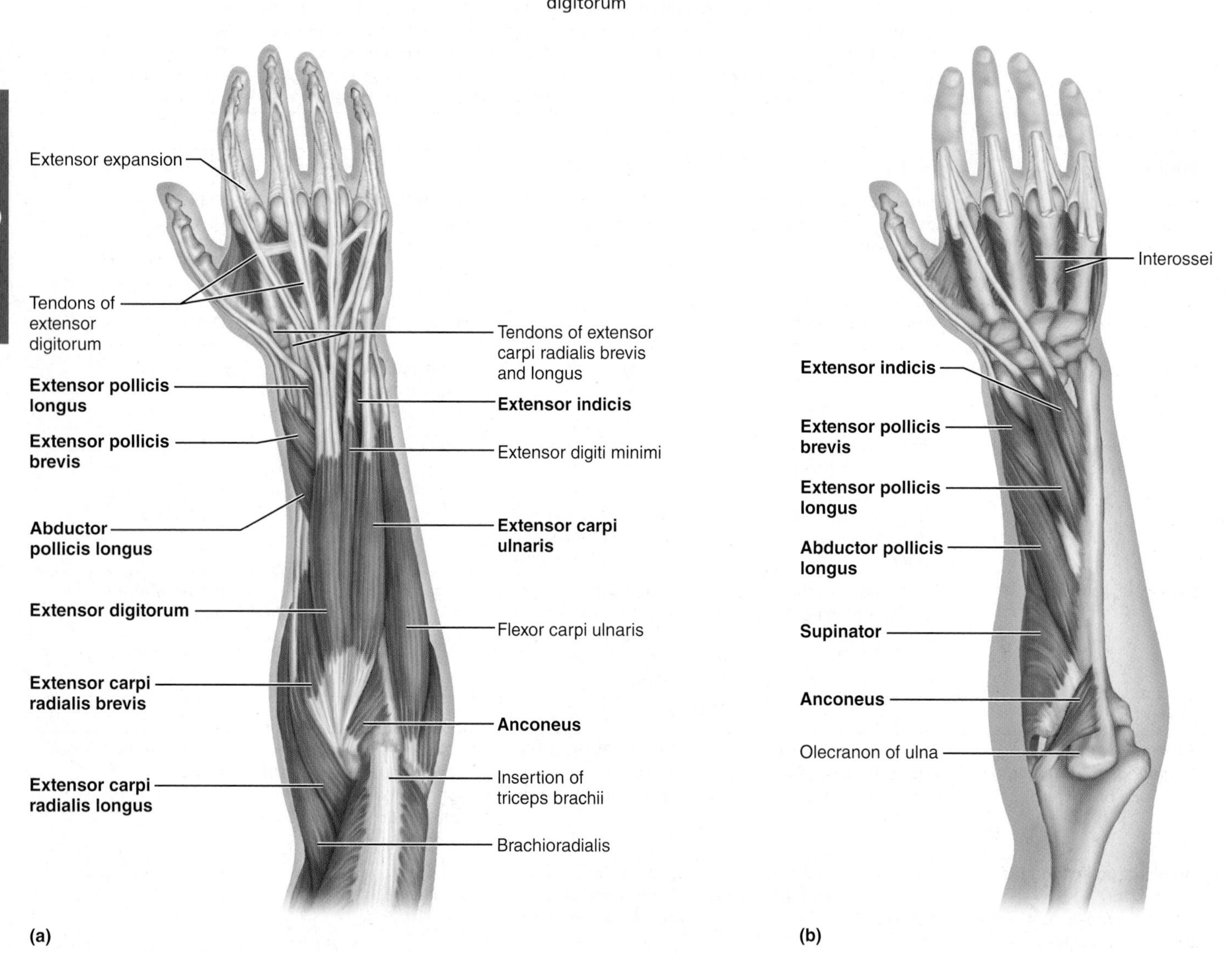

Figure 10.17 Muscles of the posterior fascial compartment of the right forearm acting on the wrist and fingers. (a) Superficial muscles, posterior view. (For a related image, see *A Brief Atlas of the Human Body*, Figure 37b.) **(b)** Deep posterior muscles, superficial muscles removed. The interossei, the deepest layer of intrinsic hand muscles, are also illustrated.

MUSCLE GALLERY

Table 10.12 Summary: Actions of Muscles Acting on the Arm, Forearm, and Hand (Figure 10.18)

PART I: Muscles Acting on the Arm (Humerus) (PM = Prime Mover)	Actions at the Shoulder					
	Flexion	Extension	Abduction	Adduction	Medial Rotation	Lateral Rotation
Pectoralis major	×			× (PM)	×	
Latissimus dorsi		× (PM)		× (PM)	×	
Teres major		×		×	×	
Deltoid	× (PM) (anterior fibers)	× (PM) (posterior fibers)	× (PM)		× (anterior fibers)	× (posterior fibers)
Subscapularis					× (PM)	
Supraspinatus			×			
Infraspinatus						× (PM)
Teres minor				× (weak)		× (PM)
Coracobrachialis	×			×		
Biceps brachii	× (weak)					
Triceps brachii				×		

PART II: Muscles Acting on the Forearm	Actions on the Forearm			
	Elbow Flexion	Elbow Extension	Pronation	Supination
Biceps brachii	× (PM)			×
Brachialis	× (PM)			
Triceps brachii		× (PM)		
Anconeus		×		
Pronator teres	× (weak)		×	
Pronator quadratus			× (PM)	
Supinator				×
Brachioradialis	×			

PART III: Muscles Acting on the Wrist and Fingers	Actions on the Wrist				Actions on the Fingers	
	Flexion	Extension	Abduction	Adduction	Flexion	Extension
Anterior Compartment						
Flexor carpi radialis	× (PM)		×			
Palmaris longus	× (weak)					
Flexor carpi ulnaris	× (PM)			×		
Flexor digitorum superficialis	× (PM)				×	
Flexor pollicis longus					× (thumb)	
Flexor digitorum profundus	×				×	
Posterior Compartment						
Extensor carpi radialis longus and brevis		×	×			
Extensor digitorum		×				× (PM, and abducts)
Extensor carpi ulnaris		×		×		
Abductor pollicis longus			×		(abducts thumb)	
Extensor pollicis longus and brevis						× (thumb)
Extensor indicis		× (weak)				× (index finger)

10

MUSCLE GALLERY

Table 10.12 Summary: Actions of Muscles Acting on the Arm, Forearm, and Hand (Figure 10.18) *(continued)*

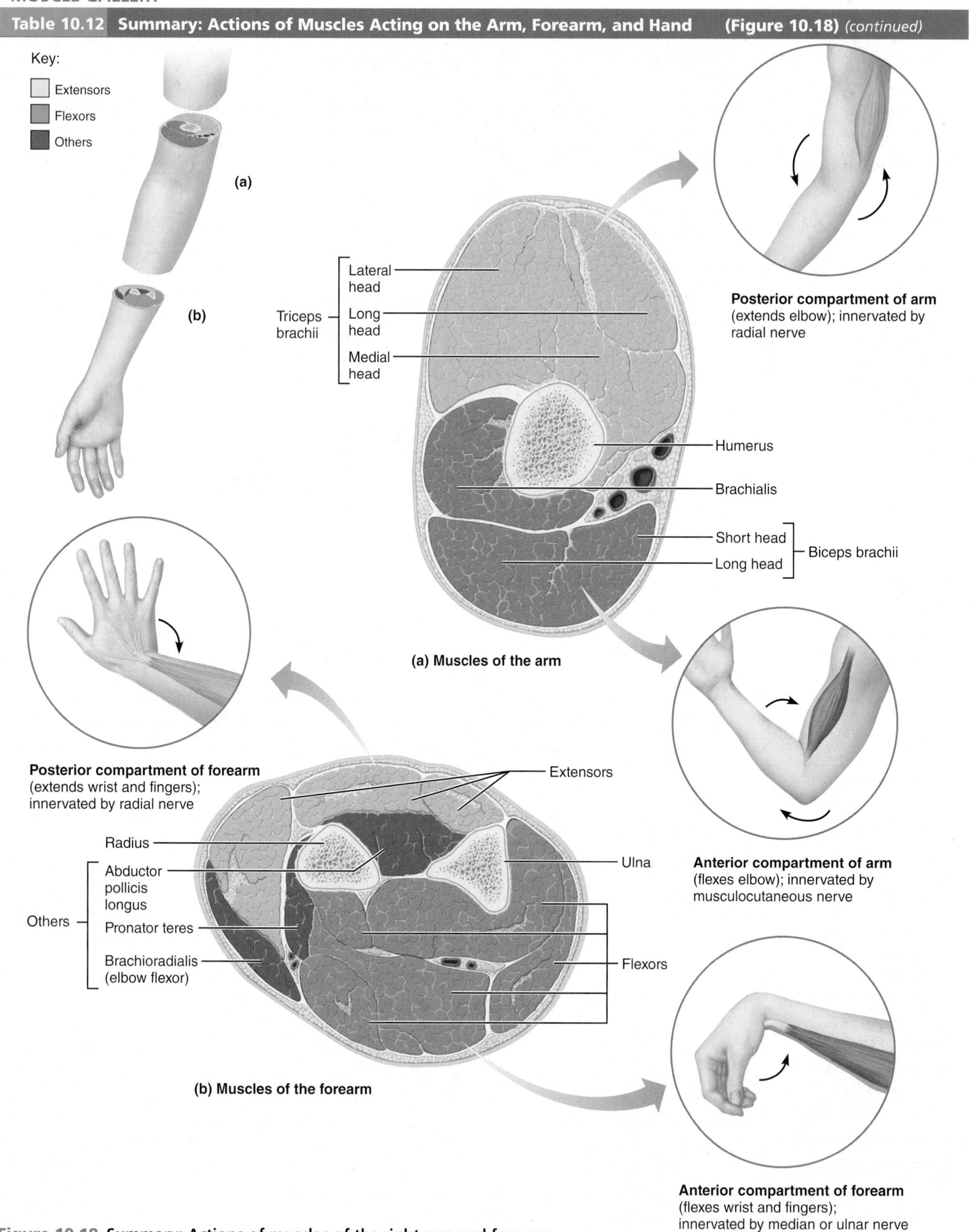

Figure 10.18 Summary: Actions of muscles of the right arm and forearm.

MUSCLE GALLERY

Table 10.13 Intrinsic Muscles of the Hand: Fine Movements of the Fingers (Figure 10.19)

In this table we consider the small muscles that lie entirely in the hand. All are in the palm, none on the hand's dorsal side. All move the metacarpals and fingers. Small, weak muscles, they mostly control precise movements (such as threading a needle), leaving the powerful movements of the fingers ("power grip") to the forearm muscles.

The intrinsic muscles include the main abductors and adductors of the fingers, as well as muscles that produce the movement of opposition—moving the thumb toward another digit of the same hand—that enables you to grip objects. Many palm muscles are specialized to move the thumb, and surprisingly many move the little finger.

Thumb movements are defined differently from movements of other fingers because the thumb lies at a right angle to the rest of the hand. The thumb flexes by bending medially along the palm, not by bending anteriorly, as do the other fingers. (To demonstrate this difference, start with your hand in the anatomical position or this will not be clear!) The thumb extends by pointing laterally (as in hitchhiking), not posteriorly, as do the other fingers. To abduct the fingers is to splay them laterally, but to abduct the thumb is to point it anteriorly. Adduction of the thumb brings it back posteriorly.

The intrinsic muscles of the palm are divided into three groups, those in:

- The *thenar eminence* (ball of the thumb)
- The *hypothenar eminence* (ball of the little finger)
- The midpalm

Thenar and hypothenar muscles are almost mirror images of each other, each containing a small flexor, an abductor, and an opponens muscle. The midpalmar muscles, called **lumbricals** and **interossei**, extend our fingers at the interphalangeal joints. The interossei are also the main finger abductors and adductors.

MUSCLE	DESCRIPTION	ORIGIN (O) AND INSERTION (I)	ACTION	NERVE SUPPLY
THENAR MUSCLES IN BALL OF THUMB (the′nar) (*thenar* = palm)				
Abductor pollicis brevis (*pollex* = thumb)	Lateral muscle of thenar group; superficial	O—flexor retinaculum and nearby carpals I—lateral base of thumb's proximal phalanx	Abducts thumb (at carpometacarpal joint)	Median nerve (C_8, T_1)
Flexor pollicis brevis	Medial and deep muscle of thenar group	O—flexor retinaculum and nearby carpals I—lateral side of base of proximal phalanx of thumb	Flexes thumb (at carpometacarpal and metacarpophalangeal joints)	Median (or occasionally ulnar) nerve (C_8, T_1)
Opponens pollicis (o-pōn′enz) (*opponens* = opposition)	Deep to abductor pollicis brevis, on metacarpal I	O—flexor retinaculum and trapezium I—whole anterior side of metacarpal I	Opposition: moves thumb to touch tip of another finger of the same hand	Median (or occasionally ulnar) nerve
Adductor pollicis	Fan-shaped with horizontal fibers; distal to other thenar muscles; oblique and transverse heads	O—capitate bone and bases of metacarpals II–IV; front of metacarpal III I—medial side of base of proximal phalanx of thumb	Adducts and helps to oppose thumb	Ulnar nerve (C_8, T_1)
HYPOTHENAR MUSCLES IN BALL OF LITTLE FINGER				
Abductor digiti minimi (dĭ′jĭ-ti min′ĭ-mi) (*digiti minimi* = little finger)	Medial muscle of hypothenar group; superficial	O—pisiform bone I—medial side of proximal phalanx of little finger	Abducts little finger at metacarpophalangeal joint	Ulnar nerve
Flexor digiti minimi brevis	Lateral deep muscle of hypothenar group	O—hamate bone and flexor retinaculum I—same as abductor digiti minimi	Flexes little finger at metacarpophalangeal joint	Ulnar nerve
Opponens digiti minimi	Deep to abductor digiti minimi	O—same as flexor digiti minimi brevis I—most of length of medial side of metacarpal V	Helps in opposition: brings metacarpal V toward thumb to cup the hand	Ulnar nerve

10

Table 10.13 Intrinsic Muscles of the Hand: Fine Movements of the Fingers (Figure 10.19) *(continued)*

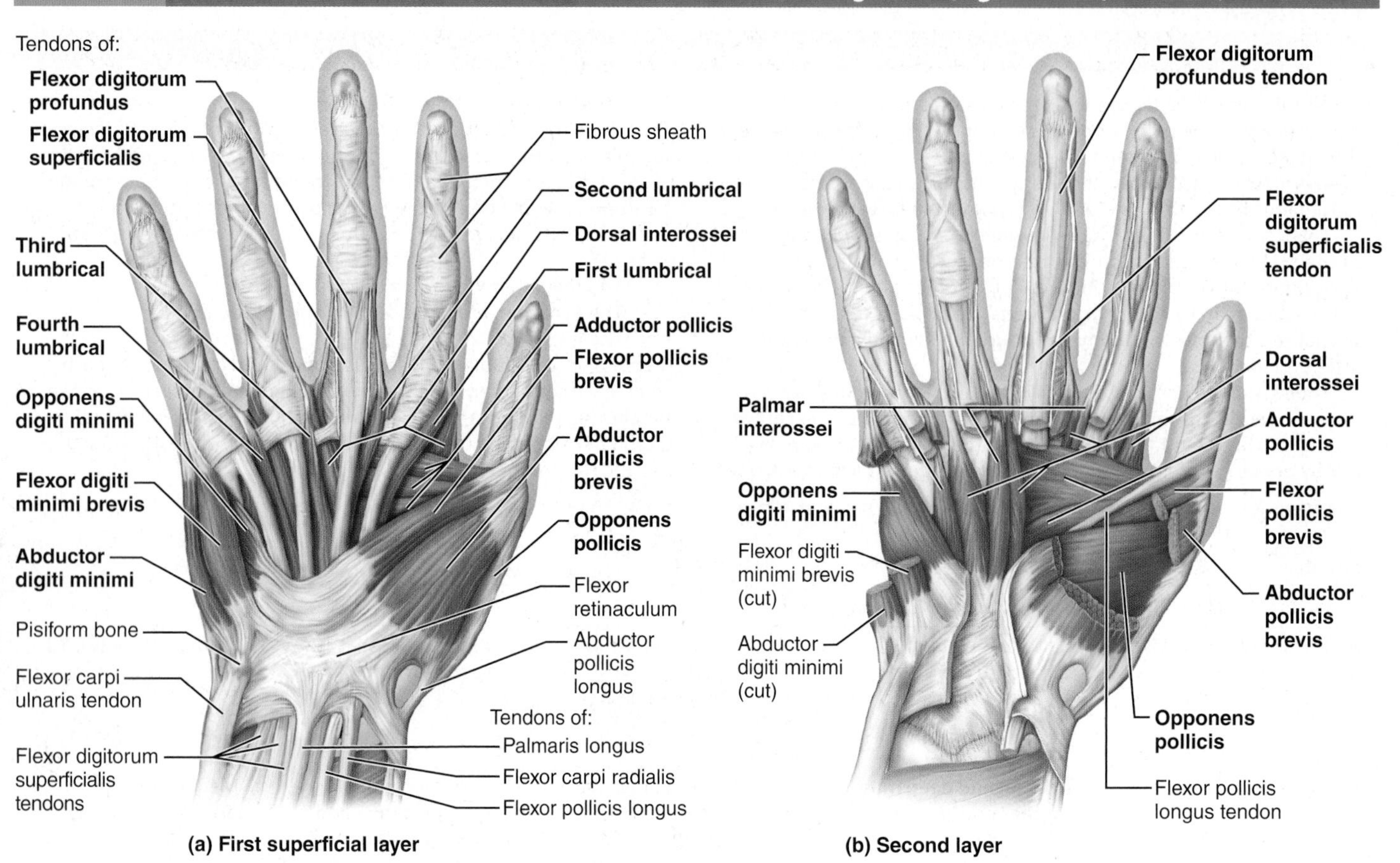

(a) First superficial layer

(b) Second layer

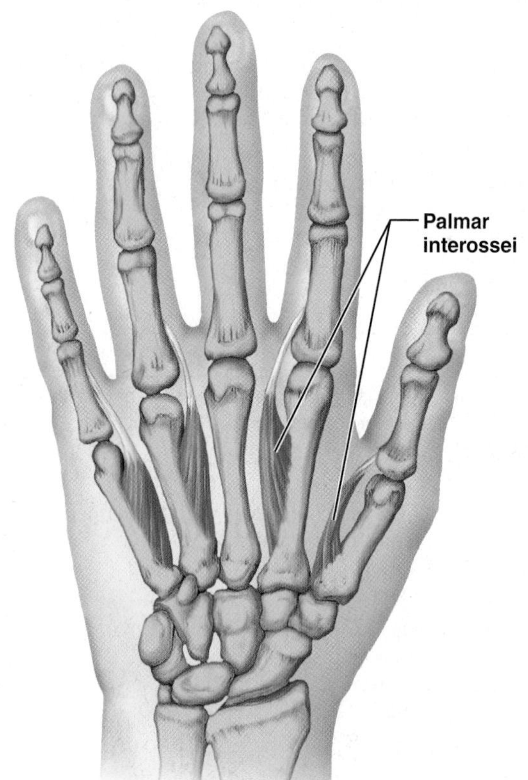

(c) Palmar interossei (isolated)

Dorsal interossei

(d) Dorsal interossei (isolated)

Figure 10.19 Hand muscles, ventral views of right hand.

Practice art labeling
MasteringA&P®>Study Area>Chapter 10

MUSCLE GALLERY

Table 10.13 *(continued)*

MUSCLE	DESCRIPTION	ORIGIN (O) AND INSERTION (I)	ACTION	NERVE SUPPLY
MIDPALMAR MUSCLES				
Lumbricals (lum′brĭ-klz) (*lumbric* = earthworm)	Four worm-shaped muscles in palm, one to each finger (except thumb); unusual because they originate from the tendons of another muscle	O—lateral side of each tendon of flexor digitorum profundus in palm I—lateral edge of extensor expansion on proximal phalanx of second to fifth fingers	Flex fingers at metacarpophalangeal joints but extend fingers at interphalangeal joints	Median nerve (lateral two) and ulnar nerve (medial two)
Palmar interossei (in″ter-os′e-i) (*interossei* = between bones)	Four long, cone-shaped muscles in the spaces between the metacarpals; lie ventral to the dorsal interossei	O—the side of each metacarpal that faces the midaxis of the hand (metacarpal III) but absent from metacarpal III I—extensor expansion on first phalanx of each finger (except third finger), on side facing midaxis of hand	Adduct fingers: pull fingers in toward third digit; act with lumbricals to extend fingers at interphalangeal joints and flex them at metacarpophalangeal joints	Ulnar nerve
Dorsal interossei	Four bipennate muscles filling spaces between the metacarpals; deepest palm muscles, also visible on dorsal side of hand (Figure 10.17b)	O—sides of metacarpals I—extensor expansion over proximal phalanx of second to fourth fingers on side opposite midaxis of hand (third finger), but on *both* sides of third finger	Abduct (diverge) fingers; extend fingers at interphalangeal joints and flex them at metacarpophalangeal joints	Ulnar nerve

MUSCLE GALLERY

Table 10.14 Muscles Crossing the Hip and Knee Joints: Movements of the Thigh and Leg (Figures 10.20 and 10.21)

The muscles fleshing out the thigh are difficult to segregate into groups on the basis of action. Some thigh muscles act only at the hip joint, others only at the knee, while still others act at both joints. However, *most anterior* muscles of the hip and thigh flex the femur at the hip and extend the leg at the knee, producing the foreswing phase of walking. The *posterior* muscles of the hip and thigh, by contrast, mostly extend the thigh and flex the leg—the backswing phase of walking. A third group of muscles in this region, the *medial*, or *adductor*, muscles, all adduct the thigh; they have no effect on the leg.

In the thigh, the anterior, posterior, and adductor muscles are separated by walls of fascia into *anterior*, *posterior*, and *medial compartments* (see Figure 10.26a). The deep fascia of the thigh, the *fascia lata*, surrounds and encloses all three groups of muscles like a support stocking.

Movements of the thigh

Movements of the thigh (occurring at the hip joint) are accomplished largely by muscles anchored to the pelvic girdle. Like the shoulder joint, the hip joint is a ball-and-socket joint permitting flexion, extension, abduction, adduction, circumduction, and rotation. Muscles effecting these movements are among the most powerful muscles of the body.

For the most part, the thigh *flexors* pass in front of the hip joint. The most important thigh flexors are the **iliopsoas** (the prime mover), **tensor fascia lata**, and **rectus femoris** (Figure 10.20a). They are assisted in this action by the **adductor muscles** of the medial thigh and the straplike **sartorius**.

Thigh *extension* is effected primarily by the massive **hamstring muscles** of the posterior thigh (Figure 10.21a). During forceful extension, the **gluteus maximus** of the buttocks is called into play. Buttock muscles that lie lateral to the hip joint (**gluteus medius** and **minimus**) *abduct* the thigh (Figure 10.21c).

Thigh adduction is the role of the adductor muscles of the medial thigh. Abduction and adduction of the thighs are extremely important during walking to shift the trunk from side to side and balance the body's weight over the limb that is on the ground. Many different muscles bring about medial and lateral rotation of the thigh.

Movements of the leg

At the knee joint, flexion and extension are the main movements. The sole knee *extensor* is the **quadriceps femoris** muscle of the anterior thigh, the most powerful muscle in the body (Figure 10.20a). The quadriceps is antagonized by the hamstrings of the posterior compartment, which are the prime movers of knee flexion.

Table 10.17 (Part I) summarizes the actions of these muscles.

MUSCLE	DESCRIPTION	ORIGIN (O) AND INSERTION (I)	ACTION	NERVE SUPPLY
PART I: ANTERIOR AND MEDIAL MUSCLES (Figure 10.20)				
Origin on the Pelvis or Spine				
Iliopsoas (il″e-o-so′us)	Iliopsoas is a composite of two closely related muscles (iliacus and psoas major) whose fibers pass under the inguinal ligament (see Figure 10.12) to insert via a common tendon on the femur.			
• **Iliacus** (il-e-ak′us) (*iliac* = ilium)	Large, fan-shaped, more lateral muscle	O—iliac fossa and crest, ala of sacrum I—lesser trochanter of femur via iliopsoas tendon	Iliopsoas is the prime mover for flexing thigh, or for flexing trunk on thigh as during a bow	Femoral nerve (L_2 and L_3)
• **Psoas major** (so′us) (*psoa* = loin muscle; *major* = larger)	Longer, thicker, more medial muscle of the pair (butchers refer to this muscle as the tenderloin in animals)	O—by fleshy slips from transverse processes, bodies, and discs of lumbar vertebrae and T_{12} I—lesser trochanter of femur via iliopsoas tendon	As above; also flexes vertebral column laterally; important postural muscle	Ventral rami (L_1–L_3)
Sartorius (sar-tor′e-us) (*sartor* = tailor)	Straplike superficial muscle running obliquely across anterior surface of thigh to knee; longest muscle in body; crosses both hip and knee joints	O—anterior superior iliac spine I—winds around medial aspect of knee and inserts into medial aspect of proximal tibia	Flexes, abducts, and laterally rotates thigh; a weak knee flexor; helps produce the cross-legged position	Femoral nerve

Table 10.14 *(continued)*

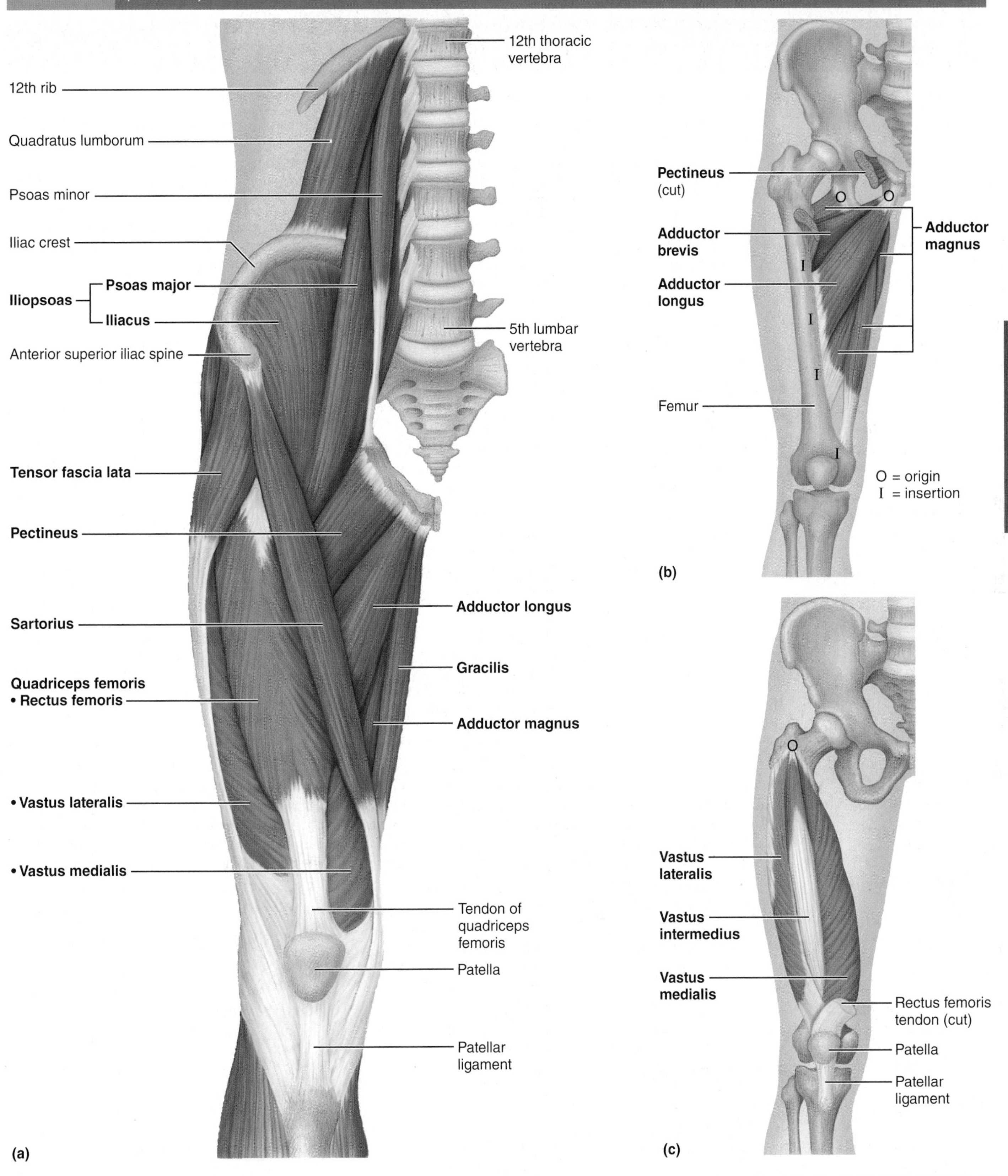

Figure 10.20 Anterior and medial muscles promoting movements of the thigh and leg. (For a related image, see *A Brief Atlas of the Human Body*, Figure 40.) **(a)** Anterior view of the deep muscles of the pelvis and superficial muscles of the right thigh. **(b)** Adductor muscles of the medial compartment of the thigh, isolated. **(c)** Vastus muscles of the quadriceps group, isolated.

Practice art labeling
MasteringA&P®>Study Area>Chapter 10

MUSCLE GALLERY

Table 10.14 Muscles Crossing the Hip and Knee Joints: Movements of the Thigh and Leg (Figures 10.20 and 10.21) *(continued)*

Muscle	Description	Origin (O) and Insertion (I)	Action	Nerve Supply
Muscles of the Medial Compartment of the Thigh				
Adductors (ah-duk′torz)	Large muscle mass consisting of three muscles (magnus, longus, and brevis) forming medial aspect of thigh; arise from inferior part of pelvis and insert at various levels on femur. All are used in movements that press thighs together, as when astride a horse; important in pelvic tilting movements that occur during walking and in fixing the hip when the knee is flexed and the foot is off the ground. Obturator nerve innervates entire group. Strain or stretching of this muscle group is called a "pulled groin."			
• **Adductor magnus** (mag′nus) (*adduct* = move toward midline; *magnus* = large)	Triangular muscle with a broad insertion; a composite muscle that is part adductor and part hamstring in action	O—ischial and pubic rami and ischial tuberosity I—linea aspera and adductor tubercle of femur	Anterior part adducts and medially rotates and flexes thigh; posterior part is a synergist of hamstrings in thigh extension	Obturator nerve and sciatic nerve (L_2–L_4)
• **Adductor longus** (*longus* = long)	Overlies middle aspect of adductor magnus; most anterior of adductor muscles	O—pubis near pubic symphysis I—linea aspera	Adducts, flexes, and medially rotates thigh	Obturator nerve (L_2–L_4)
• **Adductor brevis** (*brevis* = short)	In contact with obturator externus muscle; largely concealed by adductor longus and pectineus	O—body and inferior pubic ramus I—linea aspera above adductor longus	Adducts, flexes, and medially rotates thigh	Obturator nerve
Pectineus (pek-tin′e-us) (*pecten* = comb)	Short, flat muscle; overlies adductor brevis on proximal thigh; abuts adductor longus medially	O—pubis (and superior ramus) I—from lesser trochanter inferior to the linea aspera on posterior aspect of femur	Adducts, flexes, and medially rotates thigh	Femoral and sometimes obturator nerve
Gracilis (grah-sĭ′lis) (*gracilis* = slender)	Long, thin, superficial muscle of medial thigh	O—inferior ramus and body of pubis and adjacent ischial ramus I—medial surface of tibia just inferior to its medial condyle	Adducts thigh, flexes and medially rotates leg, especially during walking	Obturator nerve
Muscles of the Anterior Compartment of the Thigh				
Quadriceps femoris (kwod′rĭ-seps fem′o-ris)	Arises from four separate heads (*quadriceps* = four heads) that form the flesh of front and sides of thigh. These heads (rectus femoris, and lateral, medial, and intermediate vasti muscles) have a common insertion tendon, the *quadriceps tendon*, which inserts into the patella and then via the *patellar ligament* into tibial tuberosity. The quadriceps is a powerful knee extensor used in climbing, jumping, running, and rising from seated position. The tone of quadriceps plays an important role in strengthening the knee joint. Femoral nerve innervates the group.			
• **Rectus femoris** (rek′tus) (*rectus* = straight; *femoris* = femur)	Superficial muscle of anterior thigh; runs straight down thigh; longest head and only muscle of group to cross hip joint	O—anterior inferior iliac spine and superior margin of acetabulum I—patella and tibial tuberosity via patellar ligament	Extends leg and flexes thigh at hip	Femoral nerve (L_2–L_4)
• **Vastus lateralis** (vas′tus lat″er-a′lis) (*vastus* = large; *lateralis* = lateral)	Largest head of the group, forms lateral aspect of thigh; a common intramuscular injection site, particularly in infants (who have poorly developed buttock and arm muscles)	O—greater trochanter, intertrochanteric line, linea aspera I—as for rectus femoris	Extends and stabilizes leg	Femoral nerve
• **Vastus medialis** (me″de-a′lis) (*medialis* = medial)	Forms inferomedial aspect of thigh	O—linea aspera, intertrochanteric and medial supracondylar lines I—as for rectus femoris	Extends leg	Femoral nerve

10

MUSCLE GALLERY

Table 10.14 *(continued)*

MUSCLE	DESCRIPTION	ORIGIN (O) AND INSERTION (I)	ACTION	NERVE SUPPLY
• **Vastus intermedius** (in″ter-me′de-us) (*intermedius* = intermediate)	Obscured by rectus femoris; lies between vastus lateralis and vastus medialis on anterior thigh	O—anterior and lateral surfaces of proximal femur shaft I—as for rectus femoris	Extends leg	Femoral nerve
Tensor fascia lata (ten′sor fă′she-ah la′tah) (*tensor* = to make tense; *fascia* = band; *lata* = wide)	Enclosed between fascia layers of anterolateral aspect of thigh; functionally associated with medial rotators and flexors of thigh	O—anterior aspect of iliac crest and anterior superior iliac spine I—iliotibial tract*	Steadies the leg and trunk on thigh by making iliotibial tract taut; flexes and abducts thigh; rotates thigh medially	Superior gluteal nerve (L_4 and L_5)
PART II: POSTERIOR MUSCLES (Figure 10.21)				
Gluteal Muscles—Origin on Pelvis				
Gluteus maximus (gloo′te-us mak′sĭ-mus) (*glutos* = buttock; *maximus* = largest)	Largest and most superficial gluteus muscle; forms bulk of buttock mass; fibers are thick and coarse; site of intramuscular injection (dorsal gluteal site); overlies large sciatic nerve; covers ischial tuberosity only when standing; when sitting, moves superiorly, leaving ischial tuberosity exposed in the subcutaneous position	O—dorsal ilium, sacrum, and coccyx I—gluteal tuberosity of femur; iliotibial tract	Major extensor of thigh; complex, powerful, and most effective when thigh is flexed and force is necessary, as in rising from a forward flexed position and in thrusting the thigh posteriorly in climbing stairs and running; generally inactive during standing and walking; laterally rotates and abducts thigh	Inferior gluteal nerve (L_5, S_1, and S_2)
Gluteus medius (me′de-us) (*medius* = middle)	Thick muscle largely covered by gluteus maximus; important site for intramuscular injections (ventral gluteal site); considered safer than dorsal gluteal site because less chance of injuring sciatic nerve	O—between anterior and posterior gluteal lines on lateral surface of ilium I—by short tendon into lateral aspect of greater trochanter of femur	Abducts and medially rotates thigh; steadies pelvis; its action is extremely important in walking; e.g., muscle of limb planted on ground tilts or holds pelvis in abduction so that pelvis on side of swinging limb does not sag and foot of swinging limb can clear the ground	Superior gluteal nerve (L_5, S_1)
Gluteus minimus (mĭ′nĭ-mus) (*minimus* = smallest)	Smallest and deepest gluteal muscle	O—between anterior and inferior gluteal lines on external surface of ilium I—anterior border of greater trochanter of femur	As for gluteus medius	Superior gluteal nerve (L_5, S_1)
Lateral Rotators				
Piriformis (pir′ĭ-form-is) (*piri* = pear; *forma* = shape)	Pyramidal muscle located on posterior aspect of hip joint; inferior to gluteus minimus; issues from pelvis via greater sciatic notch	O—anterolateral surface of sacrum (opposite greater sciatic notch) I—superior border of greater trochanter of femur	Rotates extended thigh laterally; because inserted above head of femur, can also help abduct thigh when hip is flexed; stabilizes hip joint	S_1 and S_2, L_5

*The iliotibial tract is a thickened lateral portion of the *fascia lata* (the fascia that ensheathes all the muscles of the thigh). It extends as a tendinous band from the iliac crest to the knee (see Figure 10.21a).

MUSCLE GALLERY

Table 10.14 Muscles Crossing the Hip and Knee Joints: Movements of the Thigh and Leg (Figures 10.20 and 10.21) *(continued)*

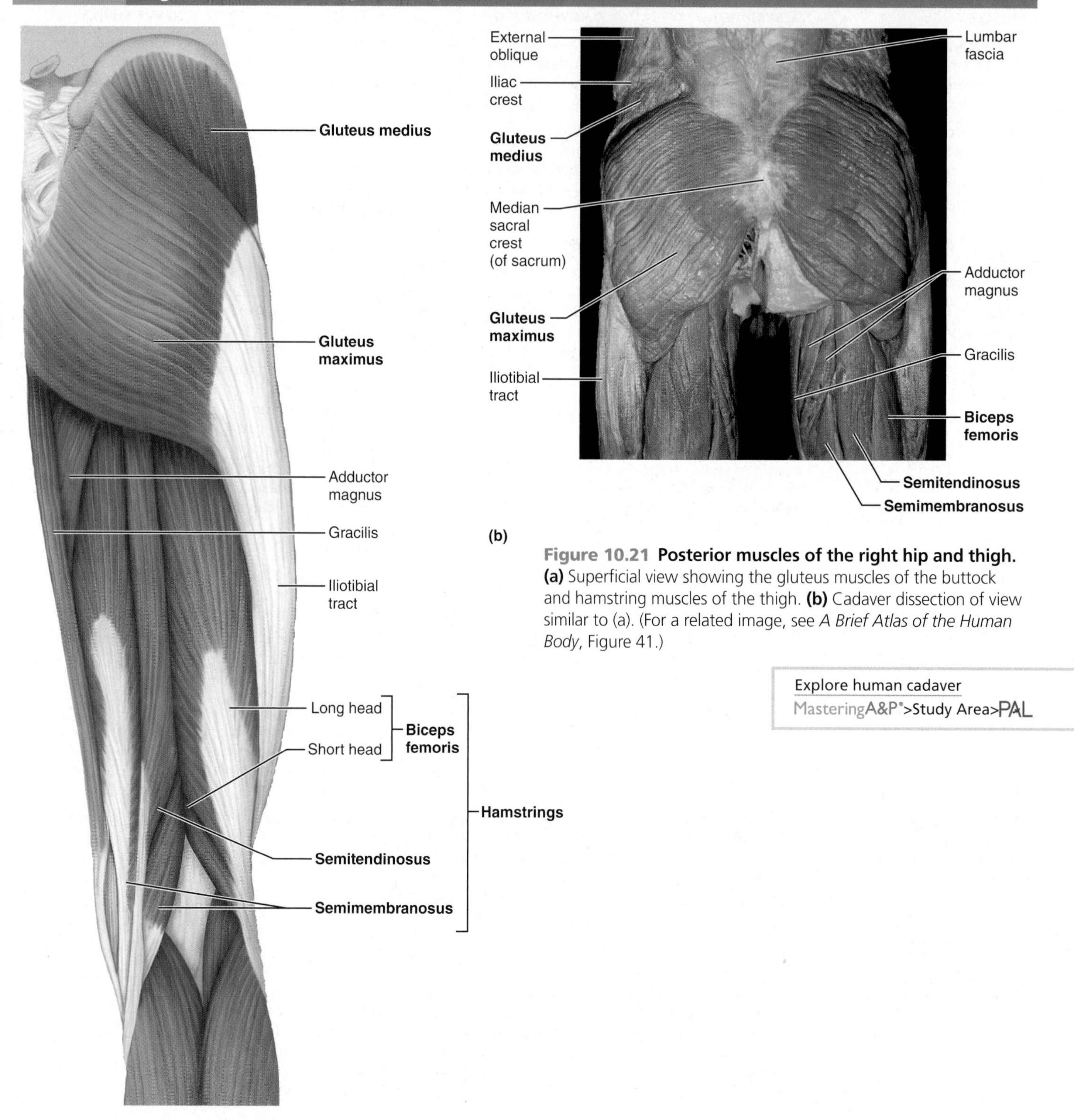

Figure 10.21 Posterior muscles of the right hip and thigh. **(a)** Superficial view showing the gluteus muscles of the buttock and hamstring muscles of the thigh. **(b)** Cadaver dissection of view similar to (a). (For a related image, see *A Brief Atlas of the Human Body*, Figure 41.)

Explore human cadaver
MasteringA&P®>Study Area>PAL

MUSCLE GALLERY

Table 10.14 *(continued)*

MUSCLE	DESCRIPTION	ORIGIN (O) AND INSERTION (I)	ACTION	NERVE SUPPLY
Obturator externus (ob″tu-ra′tor ek-ster′nus) (*obturator* = obturator foramen; *externus* = outside)	Flat, triangular muscle deep in superomedial aspect of thigh	O—outer surfaces of obturator membrane, pubis, and ischium, margins of obturator foramen I—by a tendon into trochanteric fossa of posterior femur	As for piriformis	Obturator nerve
Obturator internus (in-ter′nus) (*internus* = inside)	Surrounds obturator foramen within pelvis; leaves pelvis via lesser sciatic notch and turns acutely forward to insert on femur	O—inner surface of obturator membrane, greater sciatic notch, and margins of obturator foramen I—greater trochanter in front of piriformis	As for piriformis	L_5 and S_1
Gemellus (jĕ-mĕ′lis)—superior and inferior (*gemin* = twin, double; *superior* = above; *inferior* = below)	Two small muscles with common insertions and actions; considered extrapelvic portions of obturator internus	O—ischial spine (superior); ischial tuberosity (inferior) I—greater trochanter of femur	As for piriformis	L_5 and S_1
Quadratus femoris (*quad* = four-sided square)	Short, thick muscle; most inferior lateral rotator muscle; extends laterally from pelvis	O—ischial tuberosity I—intertrochanteric crest of femur	Rotates thigh laterally and stabilizes hip joint	L_5 and S_1

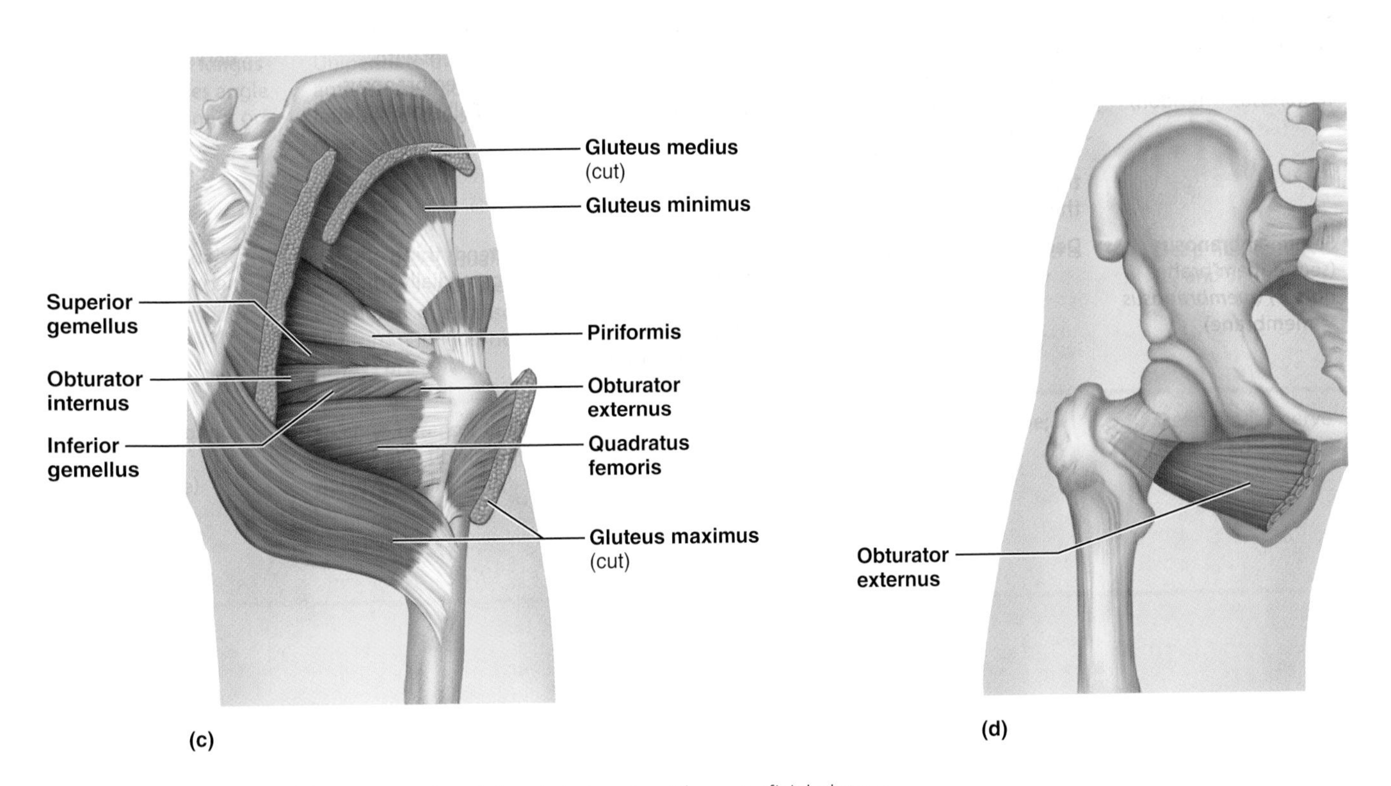

Figure 10.21 *(continued)* **(c)** Deep muscles of the gluteal region. The superficial gluteus maximus and medius have been removed. **(d)** Anterior view of the isolated obturator externus muscle.

Table 11.1 Comparison of Structural Classes of Neurons *(continued)*

NEURON TYPE		
MULTIPOLAR	**BIPOLAR**	**UNIPOLAR (PSEUDOUNIPOLAR)**
Functional Class: Neuron Type According to Direction of Impulse Conduction		
1. Most multipolar neurons are **interneurons** that conduct impulses within the CNS, integrating sensory input or motor output. May be one of a chain of CNS neurons, or a single neuron connecting sensory and motor neurons. 2. Some multipolar neurons are **motor neurons** that conduct impulses along the efferent pathways from the CNS to an effector (muscle/gland).	Essentially all bipolar neurons are **sensory neurons** that are located in some special sense organs. For example, bipolar cells of the retina are involved with transmitting visual inputs from the eye to the brain (via an intermediate chain of neurons).	Most unipolar neurons are **sensory neurons** that conduct impulses along afferent pathways to the CNS for interpretation. (These sensory neurons are called primary or first-order sensory neurons.)
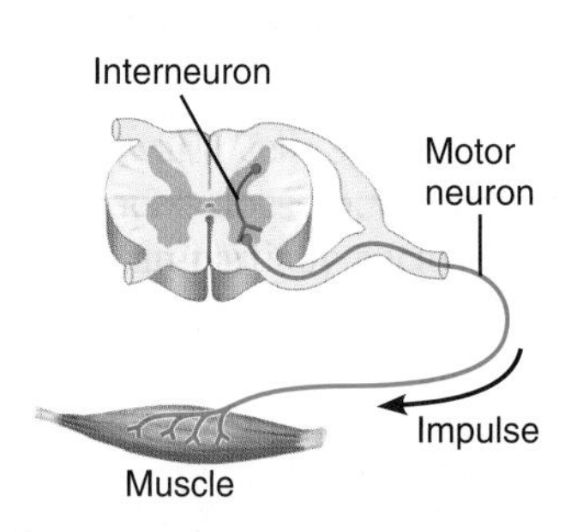	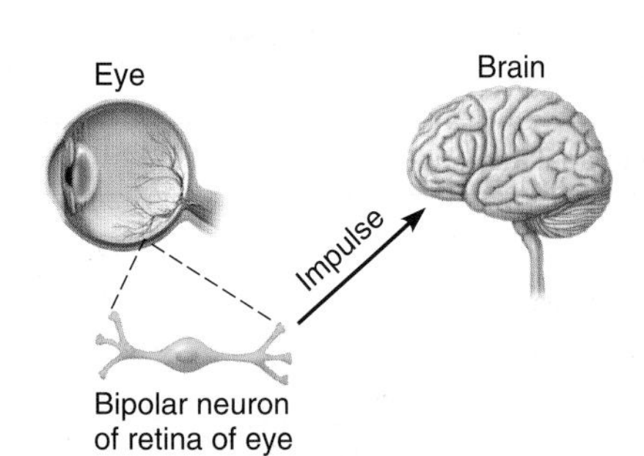	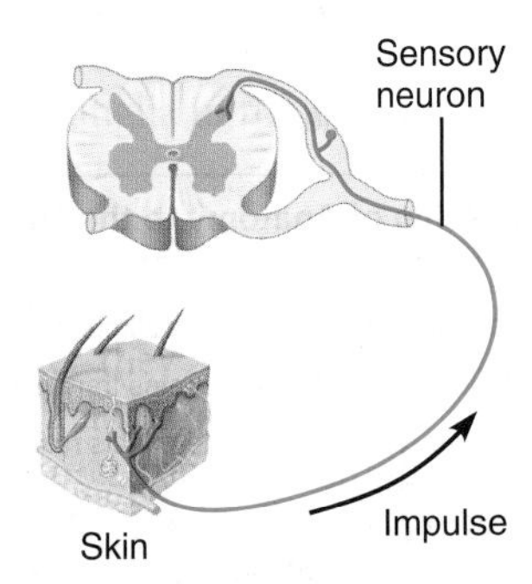

Motor, or **efferent**, **neurons** carry impulses *away from* the CNS to the effector organs (muscles and glands) of the body. Motor neurons are multipolar. Except for some neurons of the autonomic nervous system, their cell bodies are located in the CNS.

Interneurons, or *association neurons*, lie between motor and sensory neurons in neural pathways and shuttle signals through CNS pathways where integration occurs. Most interneurons are confined within the CNS. They make up over 99% of the neurons of the body, including most of those in the CNS.

Almost all interneurons are multipolar, but there is considerable diversity in size and fiber-branching patterns. The Purkinje and pyramidal cells illustrated in Table 11.1 are just two examples of their variety.

☑ Check Your Understanding

5. How does a nucleus within the brain differ from a nucleus within a neuron?
6. How is a myelin sheath formed in the CNS, and what is its function?
7. Which structural and functional type of neuron is activated first when you burn your finger? Which type is activated last to move your finger away from the source of heat?
8. MAKING connections Which part of the neuron is its fiber? How do nerve fibers differ from the fibers of connective tissue (see Chapter 4) and the fibers in muscle (see Chapter 9)?

For answers, see Answers Appendix.

11.4 The resting membrane potential depends on differences in ion concentration and permeability

→ Learning Objectives

- ☐ **Describe the relationship between current, voltage, and resistance.**
- ☐ **Identify different types of membrane ion channels.**
- ☐ **Define resting membrane potential and describe its electrochemical basis.**

Like all cells, neurons have a *resting membrane potential*. However, unlike most other cells, neurons can rapidly change their membrane potential. This ability underlies the function of neurons throughout the nervous system. In order to understand how neurons work, let's first explore some basic priciples of electricity and revisit the resting membrane potential.

Basic Principles of Electricity

The human body is electrically neutral—it has the same number of positive and negative charges. However, there are regions where one type of charge predominates, making those regions positively or negatively charged. Because opposite charges attract, energy must be used (work must be done) to separate them. On the other hand, the coming together of opposite charges liberates energy that can be used to do work. For this reason, situations in which there are separated electrical charges of opposite sign have potential energy.

Some Definitions: Voltage, Resistance, Current

Voltage, the measure of potential energy generated by separated electrical charges, is measured in either *volts* (V) or *millivolts* (1 mV = 0.001 V). Voltage is always measured between two points and is called the **potential difference** or simply the **potential** between the points. The greater the difference in charge between two points, the higher the voltage.

The flow of electrical charge from one point to another is a **current**, and it can be used to do work—for example, to power a flashlight. The amount of charge that moves between the two points depends on two factors: voltage and resistance. **Resistance** is the hindrance to charge flow provided by substances through which the current must pass. Substances with high electrical resistance are *insulators*, and those with low resistance are *conductors*.

Ohm's law gives the relationship between voltage, current, and resistance:

$$\text{Current } (I) = \frac{\text{voltage } (V)}{\text{resistance } (R)}$$

Ohm's law tells us three things:

- Current (I) is directly proportional to voltage: The greater the voltage (potential difference), the greater the current.
- There is no net current flow between points that have the same potential.
- Current is inversely related to resistance: The greater the resistance, the smaller the current.

In the body, electrical currents reflect the flow of ions across cellular membranes. (Unlike the electrons flowing along your house wiring, there are no free electrons "running around" in a living system.) Recall that there is a slight difference in the numbers of positive and negative ions on the two sides of cellular plasma membranes (a charge separation), so there is a potential across those membranes. The plasma membranes provide the resistance to current flow.

Role of Membrane Ion Channels

Recall that plasma membranes are peppered with a variety of membrane proteins that act as *ion channels*. Each of these channels is selective as to the type of ion (or ions) it allows to pass. For example, a potassium ion channel allows only potassium ions to pass.

Membrane channels are large proteins, often with several subunits. Some channels, **leakage** or **nongated channels**, are always open. Other channels are *gated*: Part of the protein forms a molecular "gate" that changes shape to open and close the channel in response to specific signals. There are three main types of gated channels:

- **Chemically gated channels**, also known as **ligand-gated channels**, open when the appropriate chemical (in this case a neurotransmitter) binds (**Figure 11.7a**).
- **Voltage-gated channels** open and close in response to changes in the membrane potential (Figure 11.7b).
- **Mechanically gated channels** open in response to physical deformation of the receptor (as in sensory receptors for touch and pressure).

When gated ion channels open, ions diffuse quickly across the membrane. Ions move along chemical *concentration gradients* when they diffuse passively from an area of their higher concentration to an area of lower concentration. They move

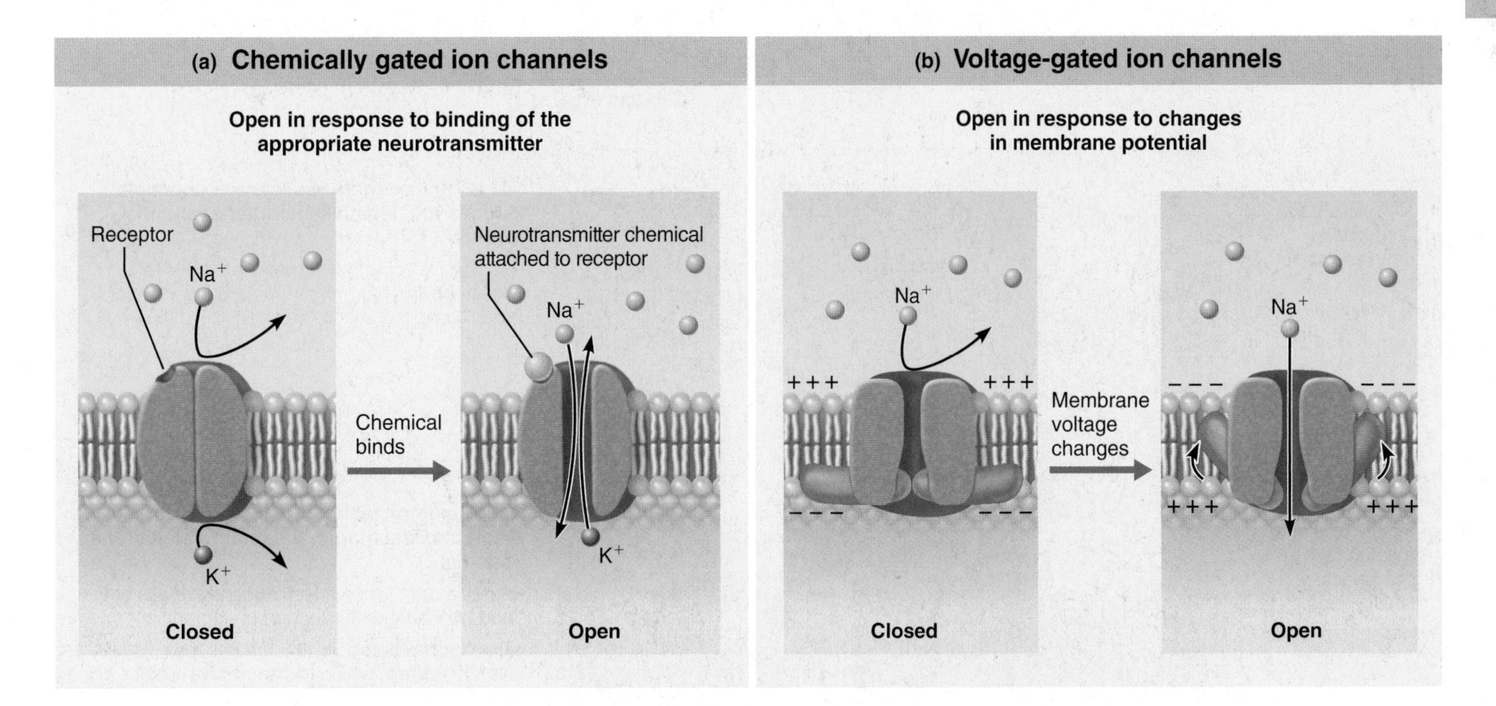

Figure 11.7 Operation of gated channels. (a) A chemically gated channel permeable to both Na^+ and K^+, and **(b)** a voltage-gated Na^+ channel.

Focus Figure 11.1 Generating a resting membrane potential depends on (1) differences in K^+ and Na^+ concentrations inside and outside cells, and (2) differences in permeability of the plasma membrane to these ions.

Interact with physiology
MasteringA&P®>Study Area>iP2

The concentrations of Na^+ and K^+ on each side of the membrane are different.

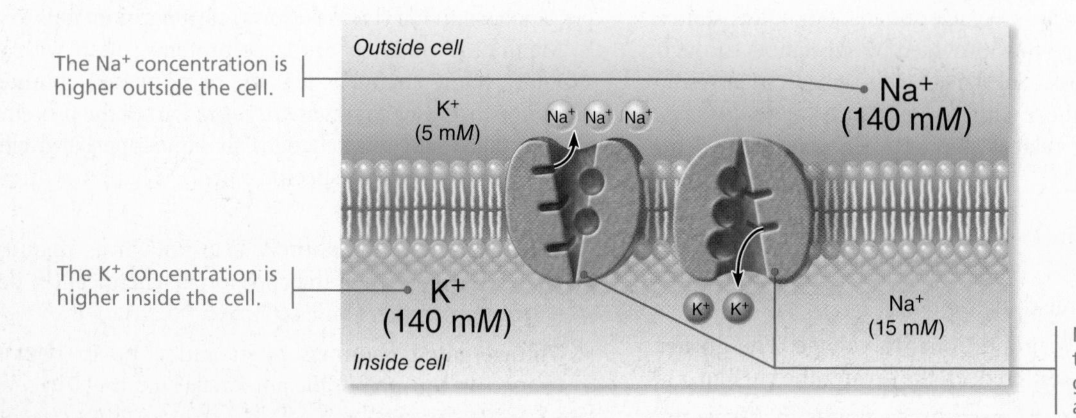

The permeabilities of Na^+ and K^+ across the membrane are different. In the next three panels, we will build the resting membrane potential step by step.

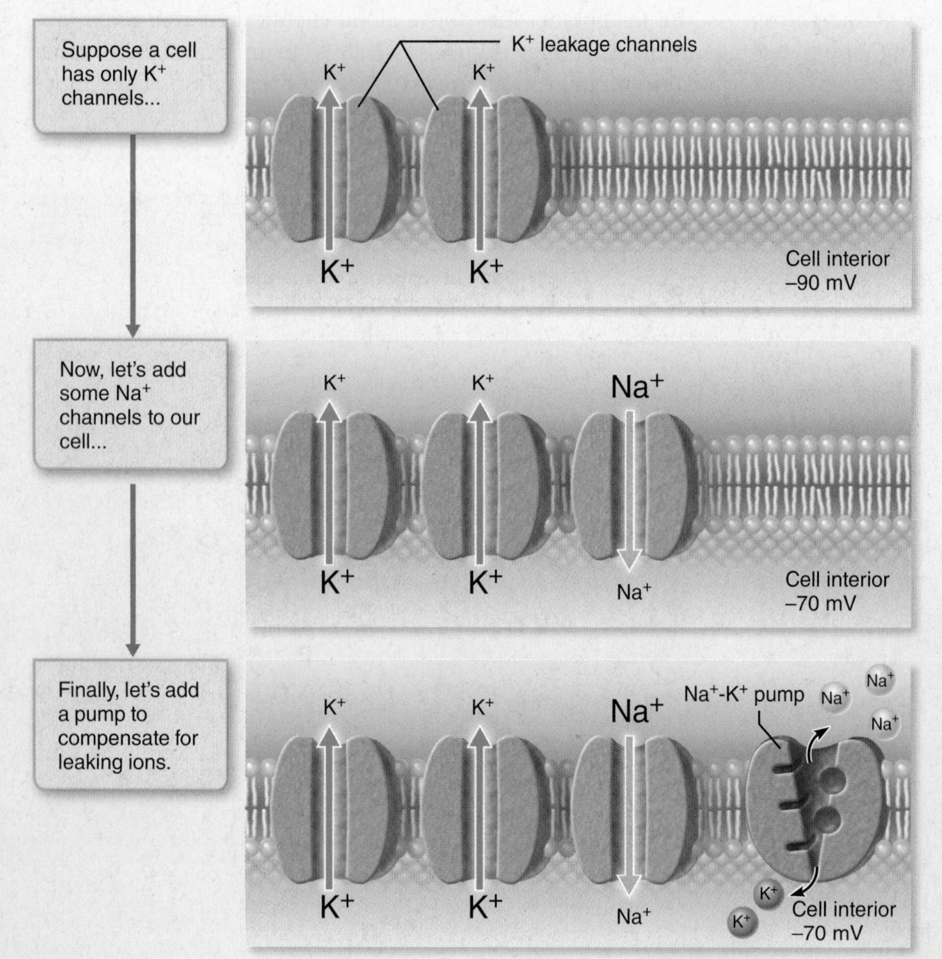

K^+ loss through abundant leakage channels establishes a negative membrane potential.

- The membrane is highly permeable to K^+, so K^+ flows down its concentration gradient.
- As positive K^+ leaks out, a negative voltage (electrical gradient) develops on the membrane interior. This electrical gradient pulls K^+ back in.
- At –90 mV, the concentration and electrical gradients for K^+ are balanced.

Na^+ entry through a few leakage channels reduces the negative membrane potential slightly.

- Adding Na^+ channels creates a small Na^+ permeability that brings the membrane potential to –70 mV.

Na^+-K^+ pumps maintain the concentration gradients, resulting in the resting membrane potential.

- A cell at rest is like a leaky boat: K^+ leaks out and Na^+ leaks in through open channels.
- The "bailing pump" for this boat is the **Na^+-K^+ pump**, which transports Na^+ out and K^+ in.

along *electrical gradients* when they move toward an area of opposite electrical charge. Together, electrical and concentration gradients constitute the **electrochemical gradient** that determines which way ions flow. Ions flowing along electrochemical gradients underlie all electrical events in neurons. Flowing ions create electrical currents and voltage changes across the membrane. These voltage changes are described by the rearranged Ohm's law equation:

$$\text{Voltage } (V) = \text{current } (I) \times \text{resistance } (R)$$

Generating the Resting Membrane Potential

A voltmeter is used to measure the potential difference between two points. When one microelectrode of the voltmeter is inserted into a neuron and the other is in the extracellular fluid, it records a voltage across the membrane of approximately −70 mV (**Figure 11.8**). The minus sign indicates that the cytoplasmic side (inside) of the membrane is negatively charged relative to the outside. This potential difference in a resting neuron (V_r) is called the **resting membrane potential**, and the membrane is said to be **polarized**. The value of the resting membrane potential varies (from −40 mV to −90 mV) in different types of neurons.

The resting potential exists only across the membrane; the bulk solutions inside and outside the cell are electrically neutral. Two factors generate the resting membrane potential: differences in the ionic composition of the intracellular and extracellular fluids, and differences in the permeability of the plasma membrane to those ions.

Differences in Ionic Composition

First, let's compare the ionic makeup of the intracellular and extracellular fluids, as shown in *Focus on Resting Membrane Potential* (**Focus Figure 11.1**). The cell cytosol contains a lower concentration of Na^+ and a higher concentration of K^+ than the extracellular fluid. Negatively charged (anionic) proteins (not shown) help to balance the positive charges of intracellular cations (primarily K^+). In the extracellular fluid, the positive charges of Na^+ and other cations are balanced chiefly by chloride ions (Cl^-). Although there are many other solutes (glucose, urea, and other ions) in both fluids, potassium (K^+) plays the most important role in generating the membrane potential.

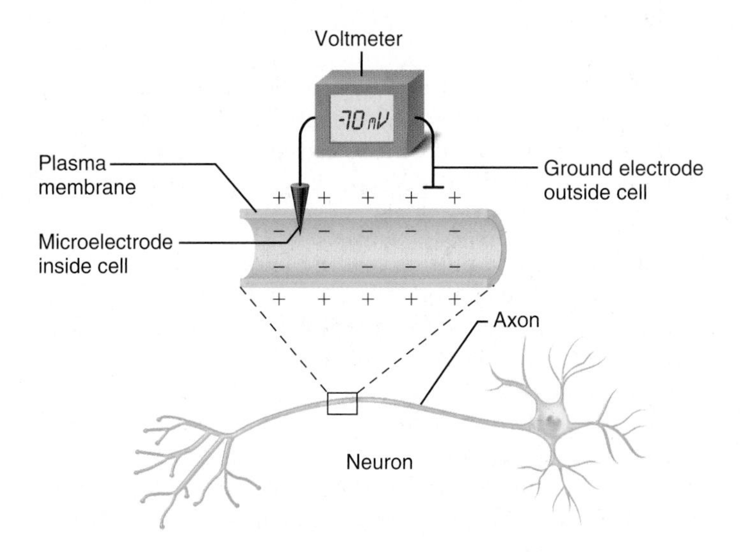

Figure 11.8 Measuring membrane potential in neurons. The potential difference between an electrode inside a neuron and the ground electrode in the extracellular fluid is approximately −70 mV (inside negative).

Differences in Plasma Membrane Permeability

Next, let's consider the differential permeability of the membrane to various ions (Focus Figure 11.1, bottom). At rest the membrane is impermeable to the large anionic cytoplasmic proteins, very slightly permeable to sodium, approximately 25 times more permeable to potassium than to sodium, and quite permeable to chloride ions. These resting permeabilities reflect the properties of the leakage ion channels in the membrane. Potassium ions diffuse out of the cell along their *concentration gradient* much more easily than sodium ions can enter the cell along theirs. K^+ flowing out of the cell causes the cell to become more negative inside. Na^+ trickling into the cell makes the cell just slightly more positive than it would be if only K^+ flowed. Therefore, at resting membrane potential, the negative interior of the cell is due to a much greater ability for K^+ to diffuse out of the cell than for Na^+ to diffuse into the cell.

Because some K^+ is always leaking out of the cell and some Na^+ is always leaking in, you might think that the concentration gradients would eventually "run down," resulting in equal concentrations of Na^+ and K^+ inside and outside the cell. This does not happen because the ATP-driven sodium-potassium pump first ejects three Na^+ from the cell and then transports two K^+ back into the cell. In other words, the **sodium-potassium pump (Na^+-K^+ ATPase)** stabilizes the resting membrane potential by maintaining the concentration gradients for sodium and potassium (Focus Figure 11.1, bottom).

Changing the Resting Membrane Potential

Neurons use changes in their membrane potential as signals to receive, integrate, and send information. A change in membrane potential can be produced by (1) anything that alters ion concentrations on the two sides of the membrane, or (2) anything that changes membrane permeability to any ion. However, only permeability changes (changes in the number of open channels) are important for transferring information.

Changes in membrane potential can produce two types of signals:

- *Graded potentials*—usually incoming signals operating over short distances
- *Action potentials*—long-distance signals of axons

The terms *depolarization* and *hyperpolarization* describe changes in membrane potential *relative to resting membrane potential.*

11

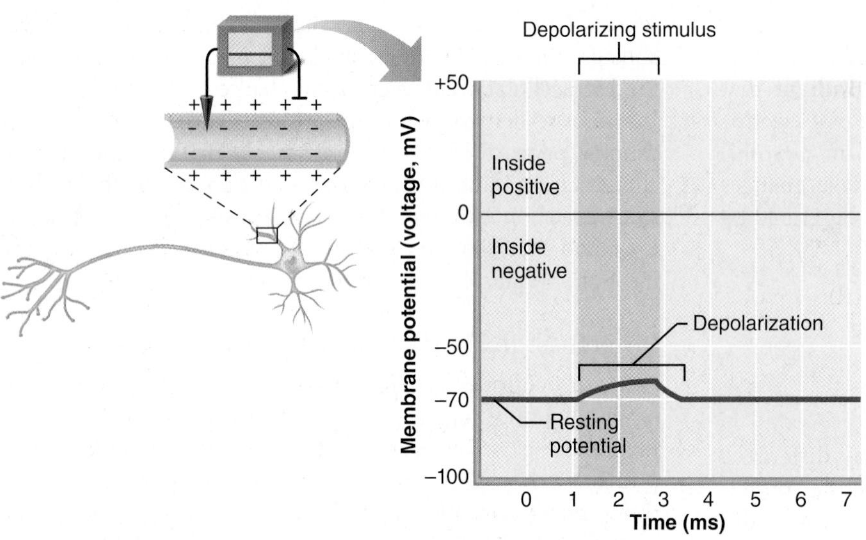

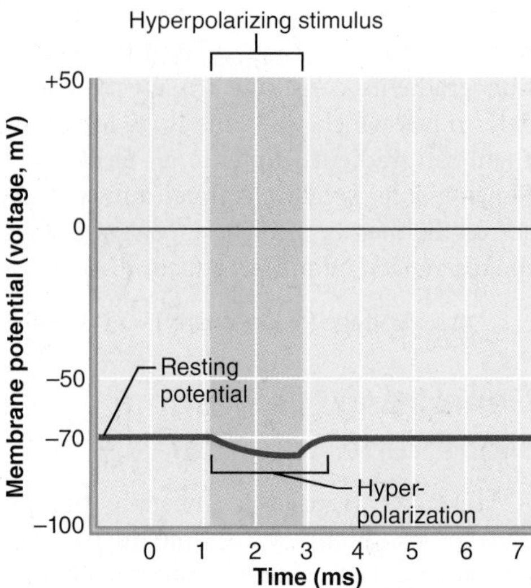

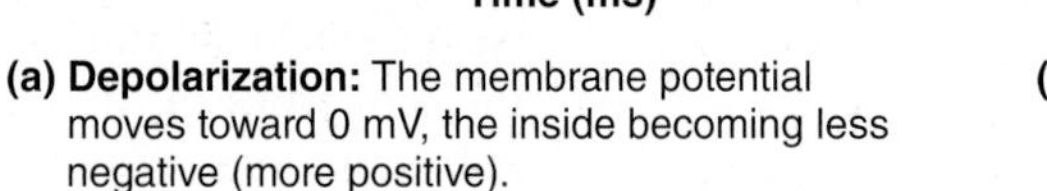

(a) Depolarization: The membrane potential moves toward 0 mV, the inside becoming less negative (more positive).

(b) Hyperpolarization: The membrane potential increases, the inside becoming more negative.

Figure 11.9 Depolarization and hyperpolarization of the membrane. The resting membrane potential is approximately −70 mV (inside negative) in neurons.

Depolarization is a decrease in membrane potential: The inside of the membrane becomes *less negative* (moves closer to zero) than the resting potential. For instance, a change in resting potential from −70 mV to −65 mV is a depolarization (**Figure 11.9a**). Depolarization also includes events in which the membrane potential reverses and moves above zero to become positive.

Hyperpolarization is an increase in membrane potential: The inside of the membrane becomes *more negative* (moves further from zero) than the resting potential. For example, a change from −70 mV to −75 mV is hyperpolarization (Figure 11.9b). As we will describe shortly, depolarization increases the probability of producing nerve impulses, whereas hyperpolarization reduces this probability.

☑ Check Your Understanding

9. For an open channel, what factors determine in which direction ions will move through that channel?

10. For which cation are there the largest number of leakage channels in the plasma membrane?

For answers, see Answers Appendix.

11.5 Graded potentials are brief, short-distance signals within a neuron

→ Learning Objective

☐ **Describe graded potentials and name several examples.**

Graded potentials are short-lived, localized changes in membrane potential. They can be either depolarizations or hyperpolarizations. These changes cause current flows that decrease in magnitude with distance. Graded potentials are called "graded" because their magnitude varies directly with stimulus strength. The stronger the stimulus, the more the voltage changes and the farther the current flows.

Graded potentials are triggered by some change (a stimulus) in the neuron's environment that opens gated ion channels. Graded potentials are given different names, depending on where they occur and the functions they perform.

- When the receptor of a sensory neuron is excited by some form of energy (heat, light, or other), the resulting graded potential is called a *receptor potential* or *generator potential.* We will consider these types of graded potentials in Chapter 13.
- When the stimulus is a neurotransmitter released by another neuron, the graded potential is called a *postsynaptic potential* because the neurotransmitter is released into a fluid-filled gap called a synapse and influences the neuron beyond the synapse.

Fluids inside and outside cells are fairly good conductors, and current, carried by ions, flows through these fluids whenever voltage changes. Suppose a stimulus depolarizes a small area of a neuron's plasma membrane (**Figure 11.10a**). Current (ions) flows on both sides of the membrane between the depolarized (active) membrane area and the adjacent polarized (resting) areas. Positive ions migrate toward more negative areas (the direction of cation movement is the direction of current flow), and negative ions simultaneously move toward more positive areas (Figure 11.10b).

For our patch of plasma membrane, positive ions (mostly K^+) inside the cell move away from the depolarized area and accumulate on the neighboring membrane areas, where they neutralize negative ions. Meanwhile, positive ions on the outside of the membrane move toward the depolarized region,

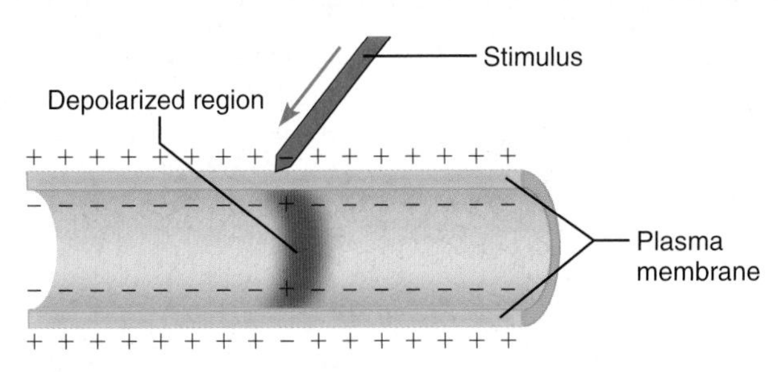

(a) Depolarization: A small patch of the membrane (red area) depolarizes.

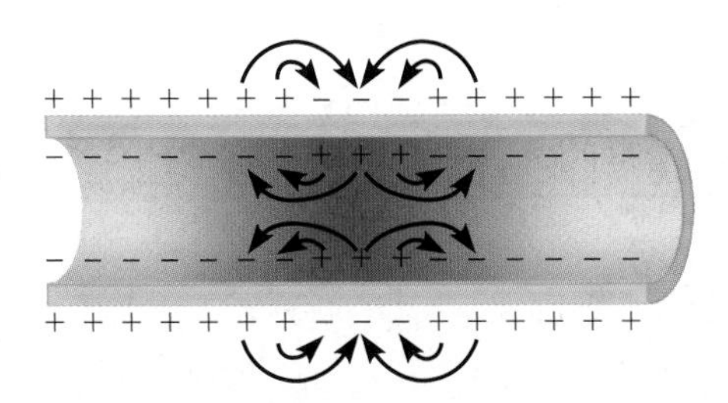

(b) Depolarization spreads: Opposite charges attract each other. This creates local currents (black arrows) that depolarize adjacent membrane areas, spreading the wave of depolarization.

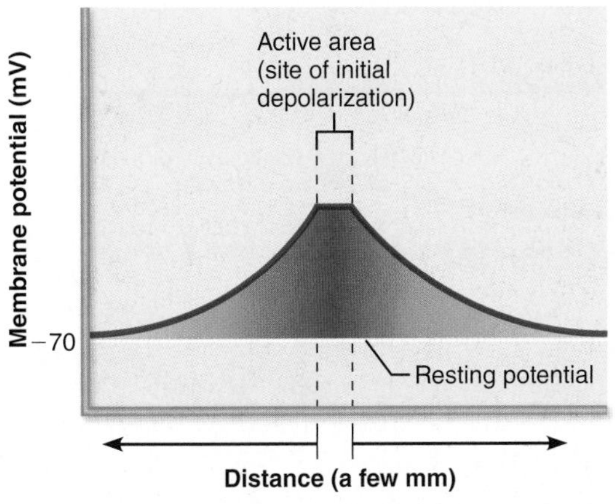

(c) Membrane potential decays with distance: Because current is lost through the "leaky" plasma membrane, the voltage declines with distance from the stimulus (the voltage is *decremental*). Consequently, graded potentials are short-distance signals.

Figure 11.10 The spread and decay of a graded potential.

which is momentarily less positive. As these positive ions move, their "places" on the membrane become occupied by negative ions (such as Cl^- and HCO_3^-), sort of like ionic musical chairs. In this way, at regions next to the depolarized region, the inside becomes less negative and the outside becomes less positive. The depolarization spreads as the neighboring membrane patch is, in turn, depolarized.

As just explained, the flow of current to adjacent membrane areas changes the membrane potential there as well. However, the plasma membrane is permeable like a leaky garden hose, and most of the charge is quickly lost through leakage channels. Consequently, the current dies out within a few millimeters of its origin and is said to be *decremental* (Figure 11.10c).

Because the current dissipates quickly and decays (declines) with increasing distance from the site of initial depolarization, graded potentials can act as signals only over very short distances. Nonetheless, they are essential in initiating action potentials, the long-distance signals.

☑ Check Your Understanding

11. What determines the size of a graded potential?

For answers, see Answers Appendix.

11.6 Action potentials are brief, long-distance signals within a neuron

→ Learning Objectives

- ☐ **Compare and contrast graded potentials and action potentials.**
- ☐ **Explain how action potentials are generated and propagated along neurons.**
- ☐ **Define absolute and relative refractory periods.**
- ☐ **Define saltatory conduction and explain how it differs from continuous conduction.**

The principal way neurons send signals over long distances is by generating and propagating (transmitting) action potentials. Only cells with *excitable membranes*—neurons and muscle cells—can generate action potentials.

An **action potential (AP)** is a brief reversal of membrane potential with a total amplitude (change in voltage) of about 100 mV (from –70 mV to +30 mV). Depolarization is followed by repolarization and often a short period of hyperpolarization. The whole event is over in a few milliseconds. Unlike graded potentials, action potentials do not decay with distance.

In a neuron, an AP is also called a **nerve impulse**, and is typically generated *only in axons*. A neuron generates a nerve impulse only when adequately stimulated. The stimulus changes the permeability of the neuron's membrane by opening specific voltage-gated channels on the axon.

These channels open and close in response to changes in the membrane potential. They are activated by local currents (graded potentials) that spread toward the axon along the dendritic and cell body membranes.

In many neurons, the transition from local graded potential to long-distance action potential takes place at the axon hillock. In sensory neurons, the action potential is generated by the peripheral (axonal) process just proximal to the receptor region. However, for simplicity, we will just use the term axon in our discussion. We'll look first at the generation of an action potential and then at its propagation.

Generating an Action Potential

Focus on an Action Potential (**Focus Figure 11.2**) on pp. 360–361 describes how an action potential is generated. Let's start with a neuron in the resting (polarized) state.

① **Resting state: All gated Na^+ and K^+ channels are closed.** Only the leakage channels are open, maintaining resting

Focus Figure 11.2 The action potential (AP) is a brief change in membrane potential in a patch of membrane that is depolarized by local currents.

Interact with physiology
MasteringA&P®>Study Area>iP2

The big picture

What does this graph show? During the course of an action potential (below), voltage changes over time at a given point within the axon.

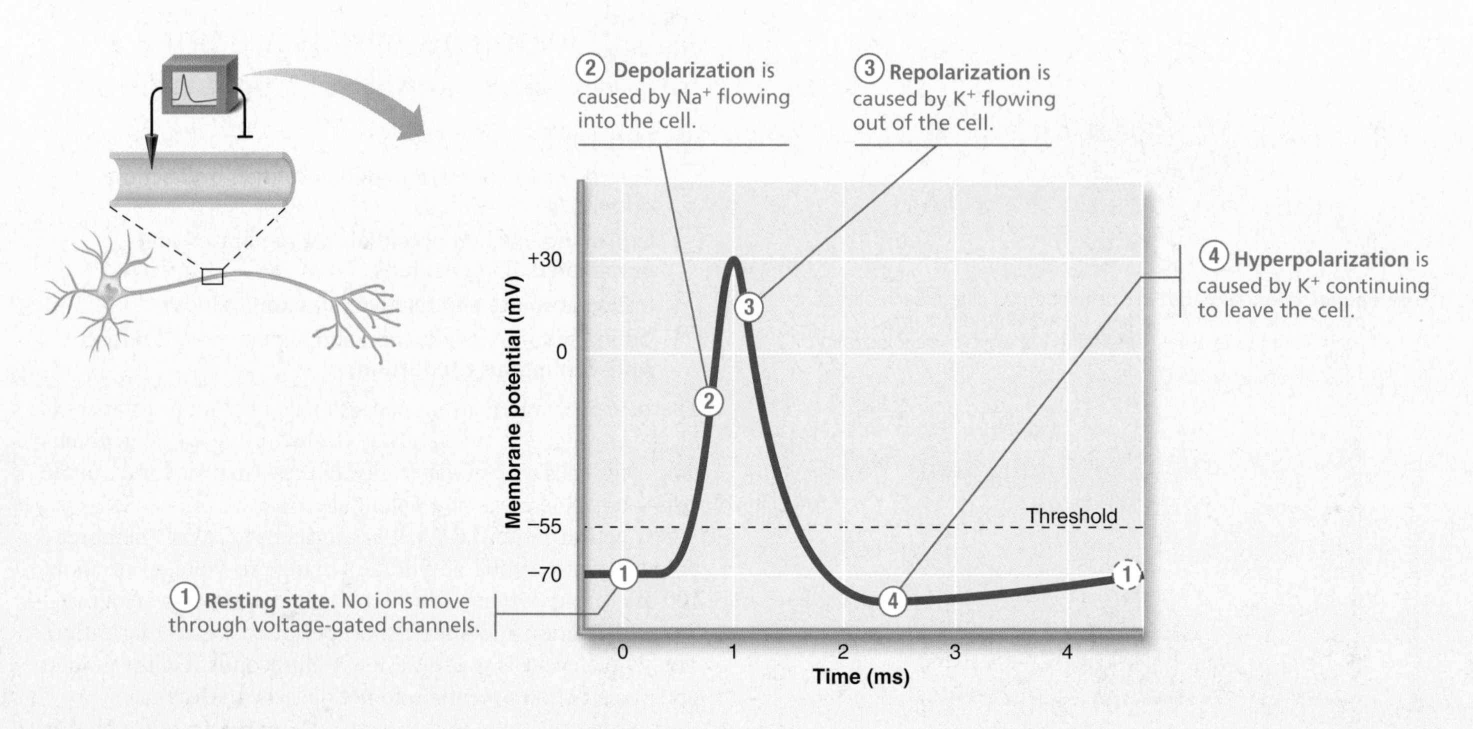

The key players

Voltage-gated Na^+ channels have two gates and alternate between three different states.

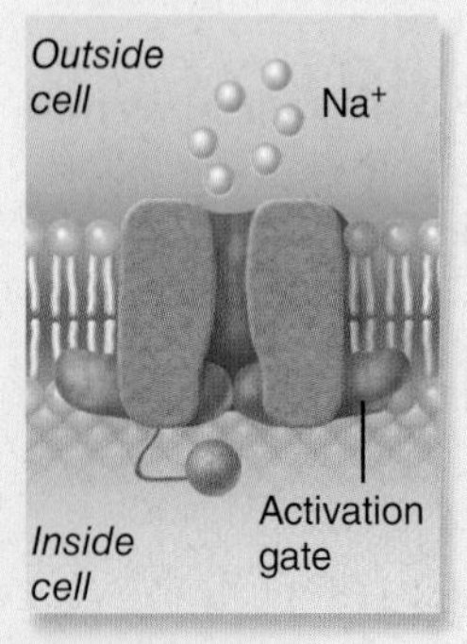

Closed at the resting state, so no Na^+ enters the cell through them

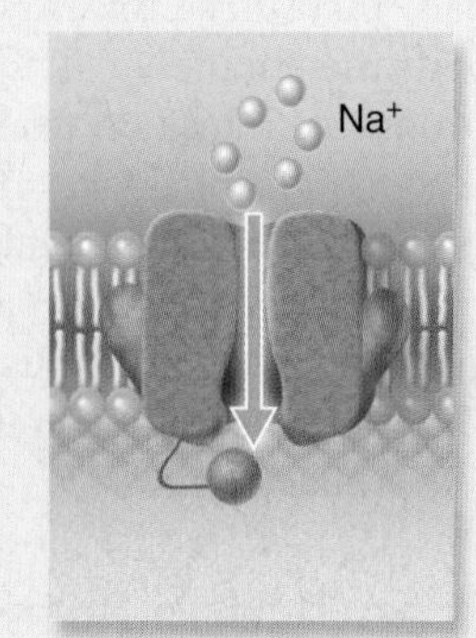

Opened by depolarization, allowing Na^+ to enter the cell

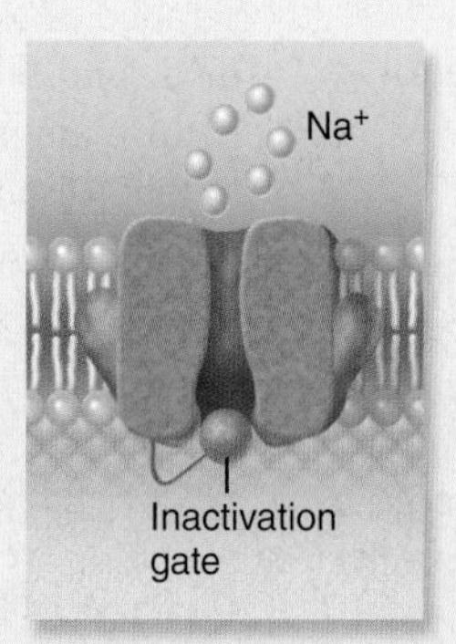

Inactivated— channels automatically blocked by inactivation gates soon after they open

Voltage-gated K^+ channels have one gate and two states.

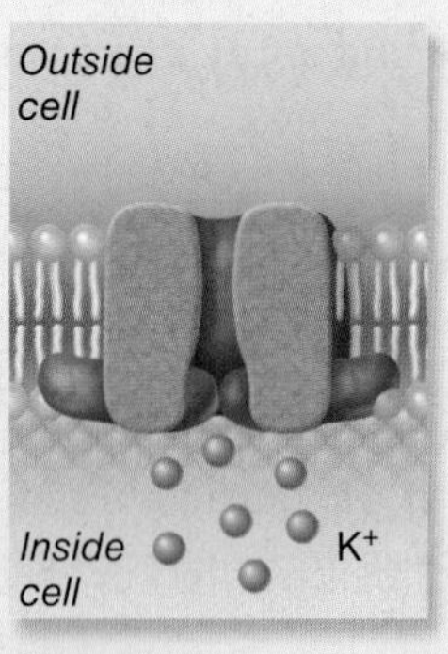

Closed at the resting state, so no K^+ exits the cell through them

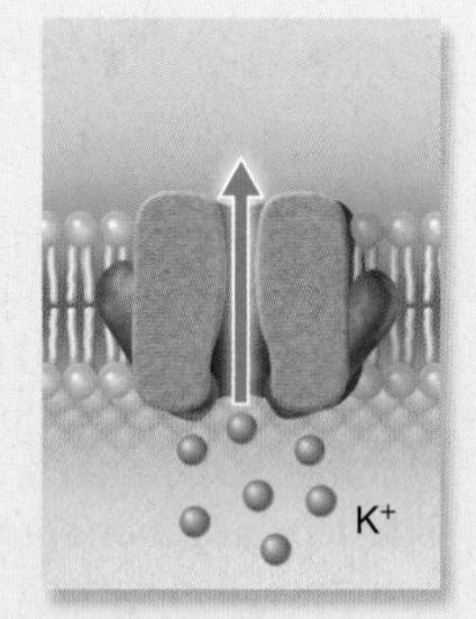

Opened by depolarization, after a delay, allowing K^+ to exit the cell

The events

Each step corresponds to one part of the AP graph.

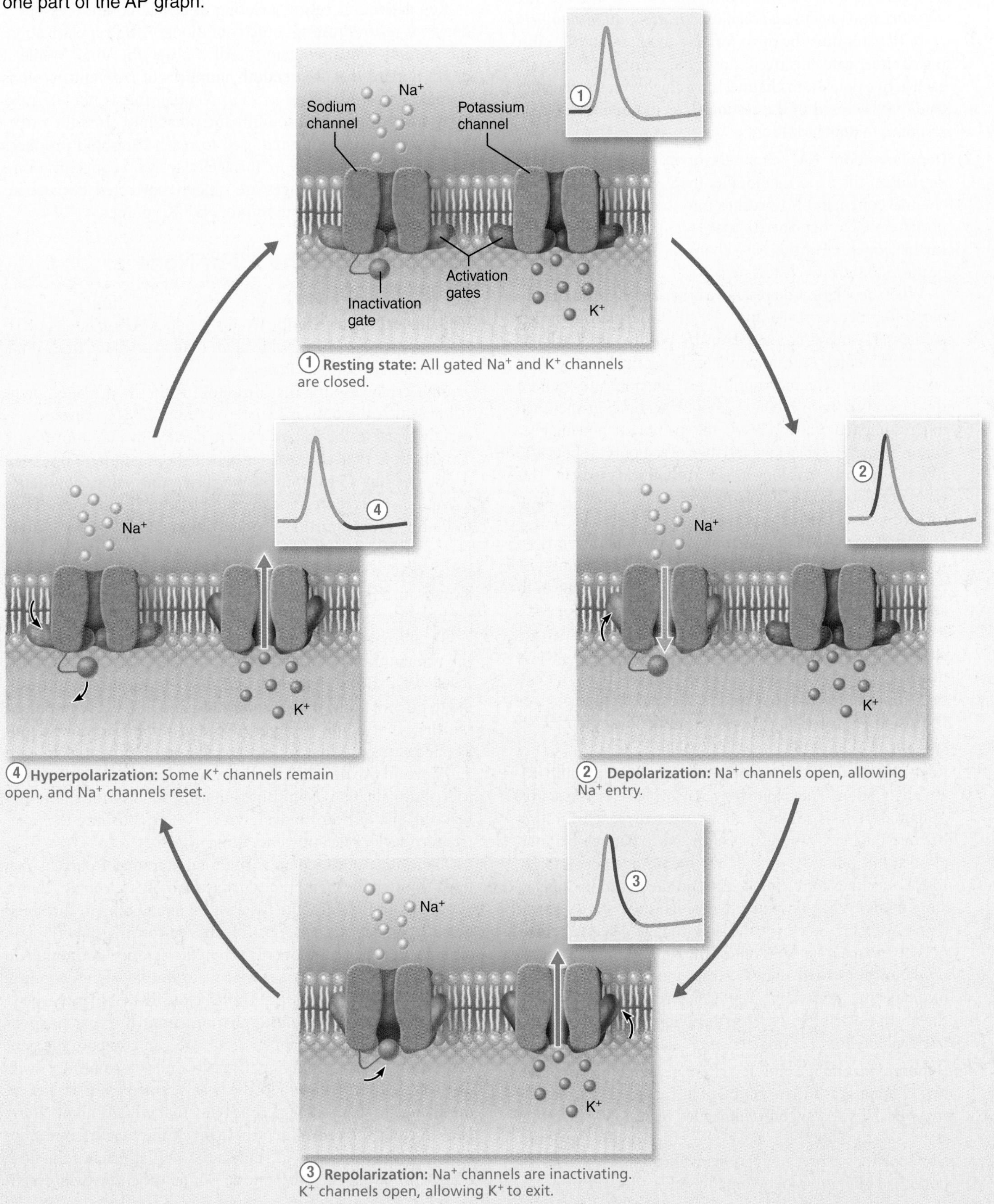

① **Resting state:** All gated Na^+ and K^+ channels are closed.

② **Depolarization:** Na^+ channels open, allowing Na^+ entry.

③ **Repolarization:** Na^+ channels are inactivating. K^+ channels open, allowing K^+ to exit.

④ **Hyperpolarization:** Some K^+ channels remain open, and Na^+ channels reset.

membrane potential. Each Na^+ channel has two gates: a voltage-sensitive *activation gate* that is closed at rest and responds to depolarization by opening, and an *inactivation gate* that blocks the channel once it is open. Thus, *depolarization opens and then inactivates sodium channels.*

Both gates must be open for Na^+ to enter, but the closing of *either* gate effectively closes the channel. In contrast, each active potassium channel has a single voltage-sensitive gate that is closed in the resting state and opens slowly in response to depolarization.

② **Depolarization: Na^+ channels open.** As local currents depolarize the axon membrane, the voltage-gated sodium channels open and Na^+ rushes into the cell. This influx of positive charge depolarizes that local patch of membrane further, opening more Na^+ channels so the cell interior becomes progressively less negative.

When depolarization reaches a critical level called **threshold** (often between −55 and −50 mV), depolarization becomes self-generating, urged on by positive feedback. As more Na^+ enters, the membrane depolarizes further and opens still more channels until all Na^+ channels are open. At this point, Na^+ permeability is about 1000 times greater than in a resting neuron. As a result, the membrane potential becomes less and less negative and then overshoots to about +30 mV as Na^+ rushes in along its electrochemical gradient. This rapid depolarization and polarity reversal produces the sharp upward *spike* of the action potential (Focus Figure 11.2).

Earlier, we stated that membrane potential depends on membrane permeability, but here we say that membrane permeability depends on membrane potential. Can both statements be true? Yes, because these two relationships establish a *positive feedback cycle*: Increasing Na^+ permeability due to increased channel openings leads to greater depolarization, which increases Na^+ permeability, and so on. This explosive positive feedback cycle is responsible for the rising (depolarizing) phase of an action potential—it puts the "action" in the action potential.

③ **Repolarization: Na^+ channels are inactivating, and K^+ channels open.** The explosively rising phase of the action potential persists for only about 1 ms. It is self-limiting because the inactivation gates of the Na^+ channels begin to close at this point. As a result, the membrane permeability to Na^+ declines to resting levels, and the net influx of Na^+ stops completely. Consequently, the AP spike stops rising.

As Na^+ entry declines, the slow voltage-gated K^+ channels open and K^+ rushes out of the cell, following its electrochemical gradient. This restores the internal negativity of the resting neuron, an event called **repolarization**. Both the abrupt decline in Na^+ permeability and the increased permeability to K^+ contribute to repolarization.

④ **Hyperpolarization: Some K^+ channels remain open, and Na^+ channels reset.** The period of increased K^+ permeability typically lasts longer than needed to restore the resting state. As a result of the excessive K^+ efflux before the potassium channels close, a hyperpolarization is seen on the AP curve as a slight dip following the spike. Also at this point, the Na^+ channels begin to reset to their original position by changing shape to reopen their inactivation gates and close their activation gates.

Repolarization restores resting electrical conditions, but it does *not* restore resting ionic conditions. After repolarization, the *sodium-potassium pump* redistributes the ions. While it might appear that tremendous numbers of Na^+ and K^+ ions change places during an action potential, this is not the case. Only small amounts of sodium and potassium cross the membrane. (The Na^+ influx required to reach threshold produces only a 0.012% change in intracellular Na^+ concentration.) These small ionic changes are quickly corrected because an axon membrane has thousands of Na^+-K^+ pumps.

Threshold and the All-or-None Phenomenon

Not all local depolarization events produce APs. The depolarization must reach threshold values if an axon is to "fire." What determines the *threshold point*?

One explanation is that threshold is the membrane potential at which the outward current created by K^+ movement is exactly equal to the inward current created by Na^+ movement. Threshold is typically reached when the membrane has been depolarized by 15 to 20 mV from the resting value. This depolarization status represents an unstable equilibrium state. If one more Na^+ enters, further depolarization occurs, opening more Na^+ channels and allowing more Na^+ to enter. If, on the other hand, one more K^+ leaves, the membrane potential is driven away from threshold, Na^+ channels close, and K^+ continues to diffuse outward until the potential returns to its resting value.

Recall that local depolarizations are graded potentials and their magnitude increases when stimuli become more intense. Brief weak stimuli (*subthreshold stimuli*) produce subthreshold depolarizations that are not translated into nerve impulses. On the other hand, stronger *threshold stimuli* produce depolarizing currents that push the membrane potential toward and beyond the threshold voltage. As a result, Na^+ permeability rises to such an extent that entering sodium ions "swamp" (exceed) the outward movement of K^+, establishing the positive feedback cycle and generating an AP.

The critical factor here is the total amount of current that flows through the membrane during a stimulus (electrical charge × time). Strong stimuli depolarize the membrane to threshold quickly. Weaker stimuli must be applied for longer periods to provide the crucial amount of current flow. Very weak stimuli do not trigger an AP because the local current flows they produce are so slight that they dissipate long before threshold is reached.

An AP is an **all-or-none phenomenon**: It either happens completely or doesn't happen at all. We can compare the generation of an AP to lighting a match under a small dry twig. The changes occurring where the twig is heated are analogous to the change in membrane permeability that initially allows more Na^+ to enter the cell. When that part of the twig becomes hot enough (when enough Na^+ enters the cell), it reaches the flash point (threshold) and the flame consumes the entire twig, even if

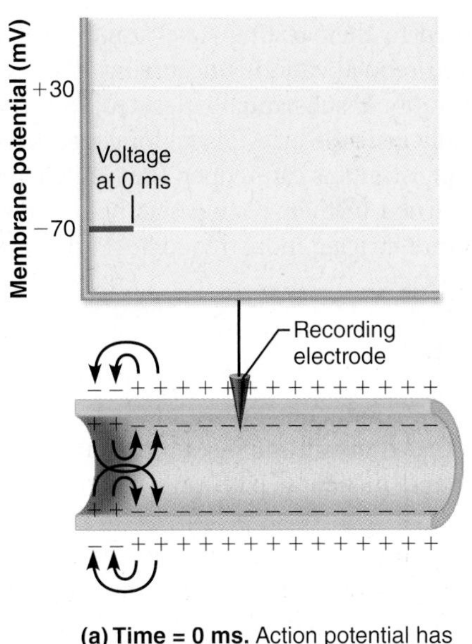

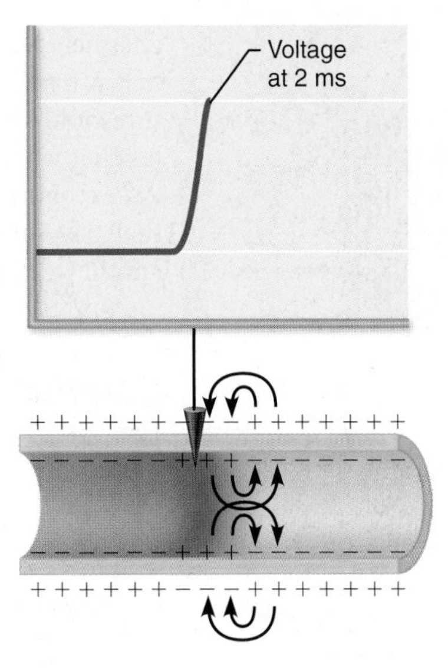

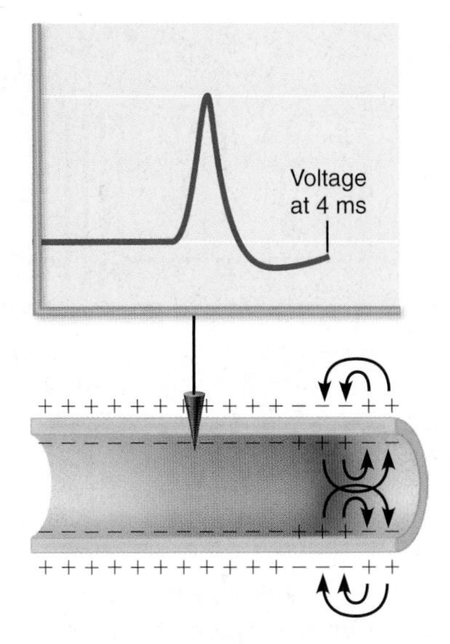

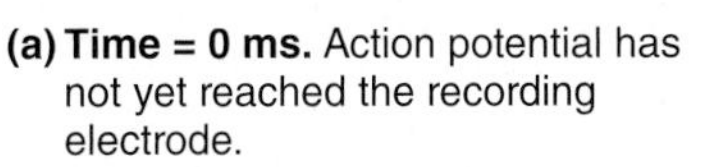

(a) Time = 0 ms. Action potential has not yet reached the recording electrode.

(b) Time = 2 ms. Action potential peak reaches the recording electrode.

(c) Time = 4 ms. Action potential peak has passed the recording electrode. Membrane at the recording electrode is still hyperpolarized.

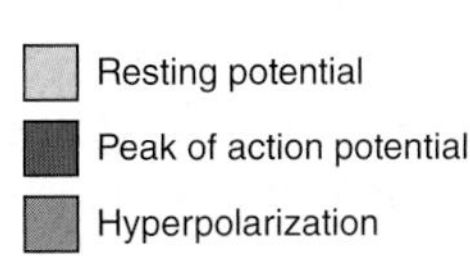

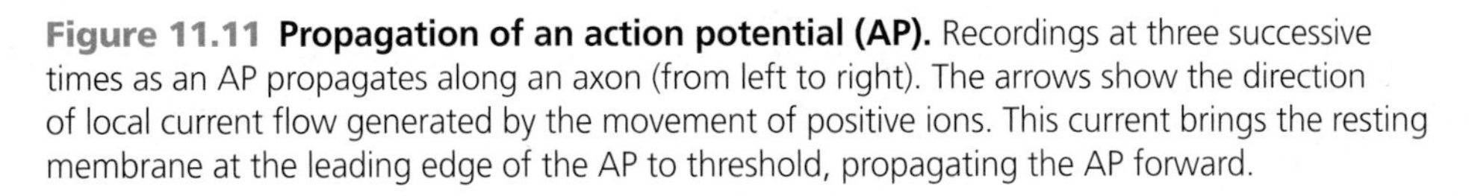

Figure 11.11 Propagation of an action potential (AP). Recordings at three successive times as an AP propagates along an axon (from left to right). The arrows show the direction of local current flow generated by the movement of positive ions. This current brings the resting membrane at the leading edge of the AP to threshold, propagating the AP forward.

you blow out the match. Similarly, the AP is generated and propagated whether or not the stimulus continues. But if you blow out the match before the twig reaches the threshold temperature, ignition will not take place. Likewise, if the number of Na^+ ions entering the cell is too low to achieve threshold, no AP will occur.

Propagation of an Action Potential

If it is to serve as the neuron's signaling device, an AP must be **propagated** along the axon's entire length. As we have seen, the AP is generated by the influx of Na^+ through a given area of the membrane. This influx establishes local currents that depolarize adjacent membrane areas in the forward direction (away from the origin of the nerve impulse), which opens voltage-gated channels and triggers an action potential there (**Figure 11.11**).

Because the area where the AP originated has just generated an AP, the sodium channels in that area are inactivated and no new AP is generated there. For this reason, the AP propagates away from its point of origin. (If an *isolated* axon is stimulated by an electrode, the nerve impulse will move away from the point of stimulus in both directions along the axon.) In the body, APs are initiated at one end of the axon and conducted away from that point toward the axon's terminals. Once initiated, an AP is *self-propagating* and continues along the axon at a constant velocity—something like a domino effect.

Following depolarization, each segment of axon membrane repolarizes, restoring the resting membrane potential in that region. Because these electrical changes also set up local currents, the repolarization wave chases the depolarization wave down the length of the axon.

The propagation process we have just described occurs on nonmyelinated axons. On p. 364, we will describe propagation along myelinated axons.

Although the phrase *conduction of a nerve impulse* is commonly used, nerve impulses are not really conducted in the same way that an insulated wire conducts current. In fact, neurons are fairly poor conductors, and as noted earlier, local current flows decline with distance because the charges leak through the membrane. The expression *propagation of a nerve impulse* is more accurate, because the AP is *regenerated anew* at each membrane patch, and every subsequent AP is identical to the one that was generated initially.

Coding for Stimulus Intensity

Once generated, all APs are independent of stimulus strength, and all APs are alike. So how can the CNS determine whether a particular stimulus is intense or weak—information it needs to initiate an appropriate response?

The answer is really quite simple: Strong stimuli generate nerve impulses more *often* in a given time interval than do weak stimuli. Stimulus intensity is coded for by the number of impulses per second—that is, by the *frequency of action potentials*—rather than by increases in the strength (amplitude) of the individual APs (**Figure 11.12**).

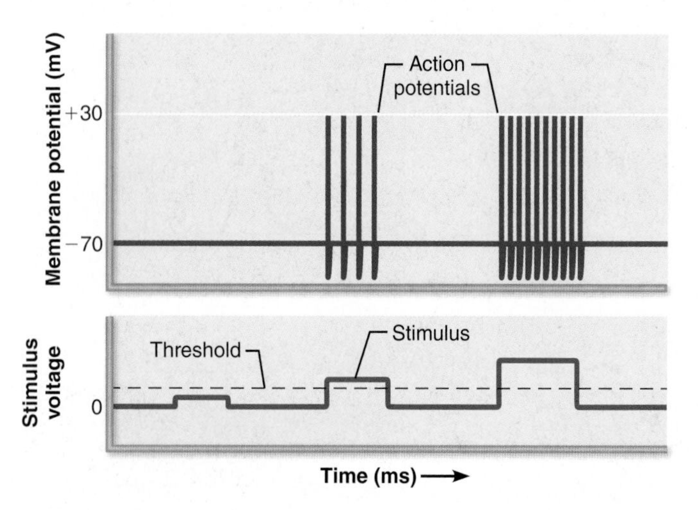

Figure 11.12 Relationship between stimulus strength and action potential frequency. APs are shown as vertical lines in the upper trace. The lower trace shows the intensity of the applied stimulus. A subthreshold stimulus does not generate an AP, but once threshold voltage is reached, the stronger the stimulus, the more frequently APs are generated.

Refractory Periods

When a patch of neuron membrane is generating an AP and its voltage-gated sodium channels are open, the neuron cannot respond to another stimulus, no matter how strong. This period, from the opening of the Na^+ channels until the Na^+ channels begin to reset to their original resting state, is called the **absolute refractory period** (Figure 11.13). It ensures that each AP is a separate, *all-or-none event* and enforces one-way transmission of the AP.

The **relative refractory period** follows the absolute refractory period. During the relative refractory period, most Na^+ channels have returned to their resting state, some K^+ channels are still open, and repolarization is occurring. The axon's threshold for AP generation is substantially elevated, so a stimulus that would normally generate an AP is no longer sufficient. An exceptionally strong stimulus can reopen the Na^+ channels that have already returned to their resting state and generate another AP. Strong stimuli trigger more frequent APs by intruding into the relative refractory period.

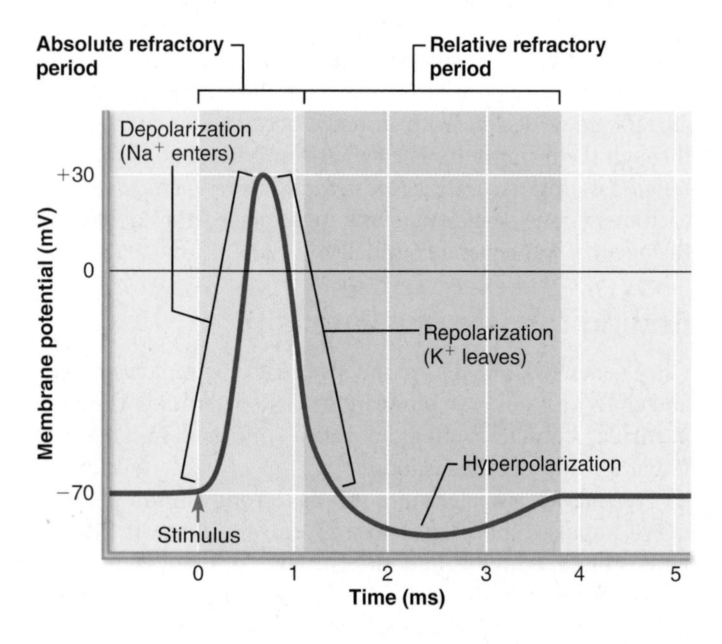

Figure 11.13 Absolute and relative refractory periods in an AP.

Conduction Velocity

How fast do APs travel? Conduction velocities of neurons vary widely. Nerve fibers that transmit impulses most rapidly (100 m/s or more) are found in neural pathways where speed is essential, such as those that mediate postural reflexes. Axons that conduct impulses more slowly typically serve internal organs (the gut, glands, blood vessels), where slower responses are not a handicap. The rate of impulse propagation depends largely on two factors:

- **Axon diameter.** As a rule, the larger the axon's diameter, the faster it conducts impulses. Larger axons conduct more rapidly because they offer less resistance to the flow of local currents, bringing adjacent areas of the membrane to threshold more quickly.
- **Degree of myelination.** Action potentials propagate because they are regenerated by voltage-gated channels in the membrane (Figure 11.14a, b). In **continuous conduction**, AP propagation involving nonmyelinated axons, these channels are immediately adjacent to each other. Continuous conduction is relatively slow.

  The presence of a myelin sheath dramatically increases the rate of AP propagation. By acting as an insulator, myelin prevents almost all charge from leaking from the axon and allows the membrane voltage to change more rapidly. Current can pass through the membrane of a myelinated axon *only* at the myelin sheath gaps, where there is no myelin sheath and the axon is bare. Nearly all the voltage-gated Na^+ channels are concentrated in these gaps.

  When an AP is generated in a myelinated fiber, the local depolarizing current does not dissipate through the adjacent membrane regions, which are nonexcitable. Instead, the current is maintained and moves rapidly to the next myelin sheath gap, a distance of approximately 1 mm, where it triggers another AP. Consequently, APs are triggered only at the gaps, a type of conduction called **saltatory conduction** (*saltare* = to leap) because the electrical signal appears to jump from gap to gap along the axon (Figure 11.14c). Saltatory conduction is about 30 times faster than continuous conduction.

The importance of myelin to nerve transmission is painfully clear to people with demyelinating diseases such as **multiple sclerosis (MS)**. This autoimmune disease is a result of the

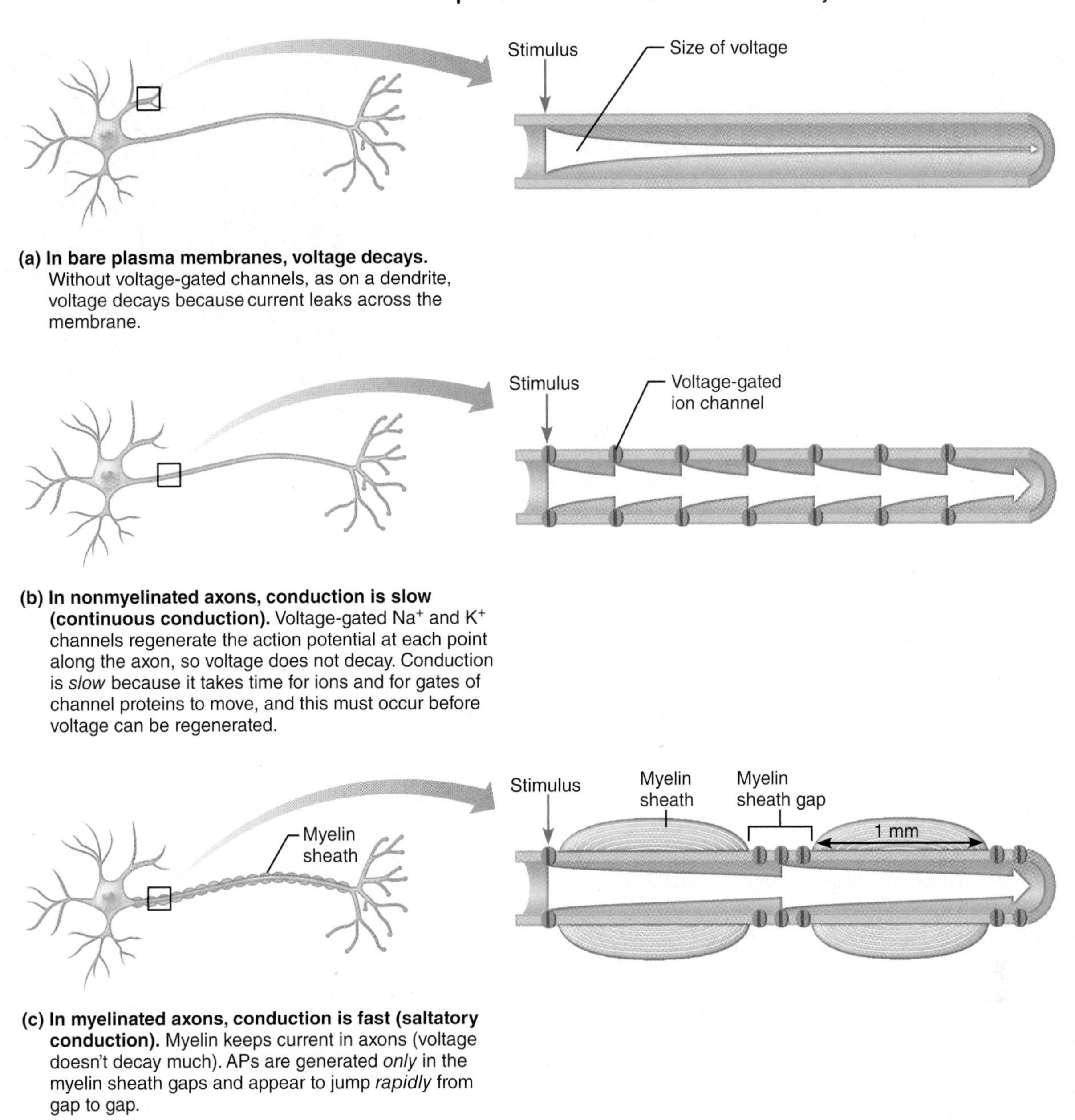

(a) In bare plasma membranes, voltage decays. Without voltage-gated channels, as on a dendrite, voltage decays because current leaks across the membrane.

(b) In nonmyelinated axons, conduction is slow (continuous conduction). Voltage-gated Na^+ and K^+ channels regenerate the action potential at each point along the axon, so voltage does not decay. Conduction is *slow* because it takes time for ions and for gates of channel proteins to move, and this must occur before voltage can be regenerated.

(c) In myelinated axons, conduction is fast (saltatory conduction). Myelin keeps current in axons (voltage doesn't decay much). APs are generated *only* in the myelin sheath gaps and appear to jump *rapidly* from gap to gap.

Figure 11.14 Action potential propagation in nonmyelinated and myelinated axons.

immune system's attack on myelin proteins and affects mostly young adults.

Multiple sclerosis gradually destroys myelin sheaths in the CNS, reducing them to nonfunctional hardened lesions called *scleroses*. The loss of myelin shunts and short-circuits the current so that successive gaps are excited more and more slowly, and eventually impulse conduction ceases. However, the axons themselves are not damaged and growing numbers of Na^+ channels appear spontaneously in the demyelinated fibers. This may account for the remarkably variable cycles of remission (symptom-free periods) and relapse typical of this disease. Common symptoms are visual disturbances (including blindness), problems controlling muscles (weakness, clumsiness, and ultimately paralysis), speech disturbances, and urinary incontinence.

The advent of drugs that modify the immune system's activity will continue to improve the lives of people with MS. These drugs seem to hold symptoms at bay, reducing complications and disability. Recent studies show that high blood levels of vitamin D reduce the risk of developing MS. +

Nerve fibers may be classified according to diameter, degree of myelination, and conduction speed.

- **Group A fibers** are mostly somatic sensory and motor fibers serving the skin, skeletal muscles, and joints. They have the largest diameter, thick myelin sheaths, and conduct impulses at speeds up to 150 m/s (over 300 miles per hour).
- **Group B fibers** are lightly myelinated fibers of intermediate diameter. They transmit impulses at an average rate of 15 m/s (about 30 mi/h).

- **Group C fibers** have the smallest diameter. They are nonmyelinated, so they are incapable of saltatory conduction and conduct impulses at a leisurely pace—1 m/s (2 mi/h) or less.

The B and C fiber groups include autonomic nervous system motor fibers serving the visceral organs; visceral sensory fibers; and the smaller somatic sensory fibers that transmit sensory impulses from the skin (such as pain and small touch fibers).

What happens when an action potential arrives at the end of a neuron's axon? That is the subject of the next section.

HOMEOSTATIC IMBALANCE 11.3 — CLINICAL

Impaired impulse propagation is caused by a number of chemical and physical factors. Local anesthetics like those used by your dentist act by blocking voltage-gated Na^+ channels. As we have seen, no Na^+ entry—no AP.

Cold and continuous pressure interrupt blood circulation, hindering the delivery of oxygen and nutrients to neuron processes and impairing their ability to conduct impulses. For example, your fingers get numb when you hold an ice cube for more than a few seconds, and your foot "goes to sleep" when you sit on it. When you remove the cold object or pressure, impulses are transmitted again, leading to an unpleasant prickly feeling. +

☑ Check Your Understanding

12. Which is bigger, a graded potential or an action potential? Which travels further? Which initiates the other?
13. An action potential does not get smaller as it propagates along an axon. Why not?
14. Why does a myelinated axon conduct action potentials faster than a nonmyelinated axon?
15. If an axon receives two stimuli close together in time, only one AP occurs. Why?

For answers, see Answers Appendix.

11.7 Synapses transmit signals between neurons

→ Learning Objectives

- ☐ **Define synapse.**
- ☐ **Distinguish between electrical and chemical synapses by structure and by the way they transmit information.**

The operation of the nervous system depends on the flow of information through chains of neurons functionally connected by synapses (**Figure 11.15**). A **synapse** (sin′aps), from the Greek *syn*, "to clasp or join," is a junction that mediates information transfer from one neuron to the next or from a neuron to an effector cell—it's where the action is.

The neuron conducting impulses toward the synapse is the **presynaptic neuron**, and the neuron transmitting the electrical signal away from the synapse is the **postsynaptic neuron.** At a given synapse, the presynaptic neuron sends the information, and the postsynaptic neuron receives the information. As you might anticipate, most neurons function as both presynaptic and postsynaptic neurons. Neurons have anywhere from 1000 to 10,000 axon terminals making synapses and are stimulated by an equal number of other neurons. Outside the central nervous system, the postsynaptic cell may be either another neuron or an effector cell (a muscle cell or gland cell).

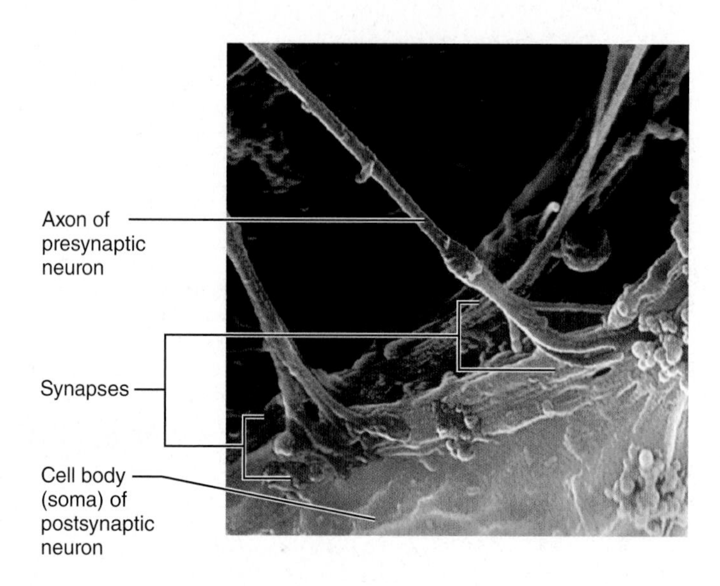

Figure 11.15 Synapses. Scanning electron micrograph (5300×).

Synapses between the axon endings of one neuron and the dendrites of other neurons are **axodendritic synapses.** Those between axon endings of one neuron and the cell body (soma) of another neuron are **axosomatic synapses** (**Figure 11.16**). Less common (and far less understood) are synapses between axons (*axoaxonal*), between dendrites (*dendrodendritic*), or between cell bodies and dendrites (*somatodendritic*).

There are two types of synapses: *chemical* and *electrical.*

Chemical Synapses

Chemical synapses are the most common type of synapse. They are specialized to allow the release and reception of chemical messengers known as *neurotransmitters.* A typical chemical synapse is made up of two parts:

- A knoblike *axon terminal* of the presynaptic neuron, which contains many tiny, membrane-bound sacs called **synaptic vesicles**, each containing thousands of neurotransmitter molecules
- A neurotransmitter *receptor region* on the postsynaptic neuron's membrane, usually located on a dendrite or the cell body

Although close to each other, presynaptic and postsynaptic membranes are separated by the **synaptic cleft**, a fluid-filled space approximately 30 to 50 nm (about one-millionth of an inch) wide.

Because the current from the presynaptic membrane dissipates in the fluid-filled cleft, chemical synapses prevent a nerve impulse from being *directly* transmitted from one neuron to another. Instead, an impulse is transmitted via a *chemical event* that depends on the release, diffusion, and receptor binding of

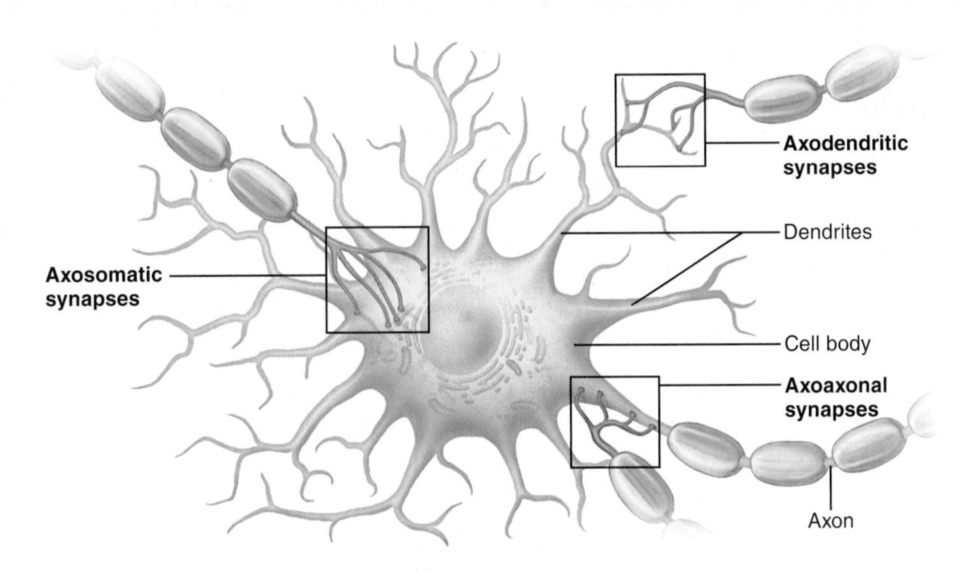

Figure 11.16 **Axodendritic, axosomatic, and axoaxonal synapses.**

neurotransmitter molecules and results in *unidirectional* communication between neurons.

In short, transmission of nerve impulses along an axon and across electrical synapses is a purely electrical event. However, chemical synapses convert the electrical signals to chemical signals (neurotransmitters) that travel across the synapse to the postsynaptic cells, where they are converted back into electrical signals.

Information Transfer across Chemical Synapses

In Chapter 9 we introduced a specialized chemical synapse called a neuromuscular junction (p. 255). The chain of events that occurs at the neuromuscular junction is simply one example of the general process that we will discuss next and show in *Focus on a Chemical Synapse* (**Focus Figure 11.3**):

① **Action potential arrives at axon terminal.** Neurotransmission begins with the arrival of an AP at the presynaptic axon terminal.

② **Voltage-gated Ca^{2+} channels open and Ca^{2+} enters the axon terminal.** Depolarization of the membrane by the action potential opens not only Na^+ channels but voltage-gated Ca^{2+} channels as well. During the brief time the Ca^{2+} channels are open, Ca^{2+} floods down its electrochemical gradient from the extracellular fluid into the terminal.

③ **Ca^{2+} entry causes synaptic vesicles to release neurotransmitter by exocytosis.** The surge of Ca^{2+} into the axon terminal acts as an intracellular messenger. A Ca^{2+}-sensing protein (*synaptotagmin*) binds Ca^{2+} and interacts with the SNARE proteins that control membrane fusion (see Figure 3.13 on p. 72). As a result, synaptic vesicles fuse with the axon membrane and empty their contents by exocytosis into the synaptic cleft. Ca^{2+} is then quickly removed from the terminal—either taken up into the mitochondria or ejected from the neuron by an active Ca^{2+} pump.

For each nerve impulse reaching the presynaptic terminal, many vesicles (perhaps 300) empty into the synaptic cleft. The higher the impulse frequency (that is, the more intense the stimulus), the greater the number of synaptic vesicles that fuse and spill their contents, and the greater the effect on the postsynaptic cell.

④ **Neurotransmitter diffuses across the synaptic cleft and binds to specific receptors on the postsynaptic membrane.**

⑤ **Binding of neurotransmitter opens ion channels, creating graded potentials.** When a neurotransmitter binds to the receptor protein, this receptor changes its shape. This change in turn opens ion channels and creates graded potentials. Postsynaptic membranes often contain receptor proteins and ion channels packaged together as chemically gated ion channels. Depending on the receptor protein to which the neurotransmitter binds and the type of channel the receptor controls, the postsynaptic neuron may be either excited or inhibited.

⑥ **Neurotransmitter effects are terminated.** Binding of a neurotransmitter to its receptor is reversible. As long as it is bound to a postsynaptic receptor, a neurotransmitter continues to affect membrane permeability and block reception of additional signals from presynaptic neurons. For this reason, some means of "wiping the postsynaptic slate clean" is necessary. The effects of neurotransmitters generally last a few milliseconds before being terminated in one of three ways, depending on the particular neurotransmitter:

- *Reuptake* by astrocytes or the presynaptic terminal, where the neurotransmitter is stored or destroyed by enzymes, as with norepinephrine
- *Degradation* by enzymes associated with the postsynaptic membrane or present in the synaptic cleft, as with acetylcholine
- *Diffusion* away from the synapse

Synaptic Delay

An impulse may travel at speeds of up to 150 m/s (300 mi/h) down an axon, but neural transmission across a chemical synapse is comparatively slow. It reflects the time required for neurotransmitter to be released, diffuse across the synaptic cleft, and bind to receptors. Typically, this **synaptic delay** lasts 0.3–5.0 ms, making transmission across the chemical synapse the *rate-limiting* (slowest) step of neural transmission. Synaptic delay helps explain why transmission along neural pathways involving only two or three neurons occurs rapidly, but transmission along multisynaptic pathways typical of higher mental functioning occurs much more slowly. However, in practical terms these differences are not noticeable.

Electrical Synapses

Electrical synapses are much less common than chemical synapses. They consist of gap junctions like those found between certain other body cells. Their channel proteins (connexons)

Focus Figure 11.3 Chemical synapses transmit signals from one neuron to another using neurotransmitters.

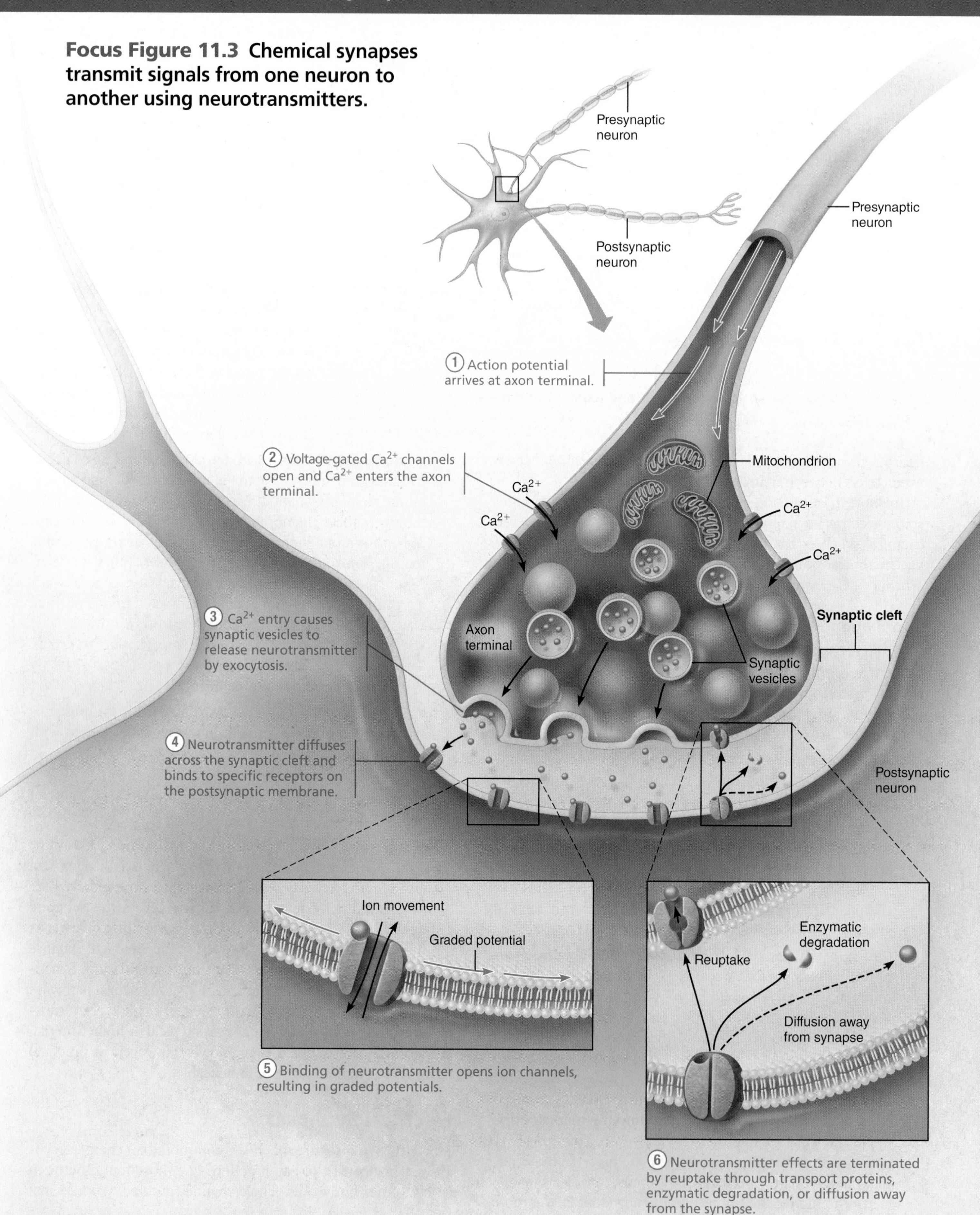

⑤ Binding of neurotransmitter opens ion channels, resulting in graded potentials.

⑥ Neurotransmitter effects are terminated by reuptake through transport proteins, enzymatic degradation, or diffusion away from the synapse.

connect the cytoplasm of adjacent neurons and allow ions and small molecules to flow directly from one neuron to the next. These neurons are *electrically coupled*, and transmission across these synapses is very rapid. Depending on the nature of the synapse, communication may be unidirectional or bidirectional.

Let's take a moment to compare the two methods of communication between neurons. The synaptic cleft of a chemical synapse is like a lake that the two neurons shout across. An electrical synapse, on the other hand, is like a doorway: Messages (ions) can move directly from one room (neuron) to another.

Electrical synapses between neurons provide a simple means of synchronizing the activity of all interconnected neurons. In adults, electrical synapses are found in regions of the brain responsible for certain stereotyped movements, such as the normal jerky movements of the eyes. They also occur in axoaxonal synapses in the hippocampus, a brain region involved in emotions and memory.

Electrical synapses are far more abundant in embryonic nervous tissue, where they permit exchange of guiding cues during early neuronal development so that neurons can connect properly with one another. As the nervous system develops, chemical synapses replace some electrical synapses and become the vast majority of all synapses. For this reason, we will focus on chemical synapses from now on.

☑ Check Your Understanding

16. Events at a chemical synapse usually involve opening both voltage-gated ion channels and chemically gated ion channels. Where are these ion channels located and what causes each to open?
17. What structure joins two neurons at an electrical synapse?

For answers, see Answers Appendix.

11.8 Postsynaptic potentials excite or inhibit the receiving neuron

→ Learning Objectives

- ☐ **Distinguish between excitatory and inhibitory postsynaptic potentials.**
- ☐ **Describe how synaptic events are integrated and modified.**

Many receptors on postsynaptic membranes at chemical synapses are specialized to open ion channels, in this way converting chemical signals to electrical signals. Unlike the voltage-gated ion channels responsible for APs, these chemically gated channels are relatively insensitive to changes in membrane potential. Consequently, channel opening at postsynaptic membranes cannot become self-amplifying or self-generating. Instead, neurotransmitter receptors mediate graded potentials—local changes in membrane potential that are *graded* (vary in strength) based on the amount of neurotransmitter released and how long it remains in the area. Table 11.2 compares graded potentials and action potentials.

Chemical synapses are either excitatory or inhibitory, depending on how they affect the membrane potential of the postsynaptic neuron.

Excitatory Synapses and EPSPs

At excitatory synapses, neurotransmitter binding depolarizes the postsynaptic membrane. In contrast to what happens on axon membranes, *chemically gated* ion channels open on postsynaptic membranes (those of dendrites and neuronal cell bodies). Each channel allows Na^+ and K^+ to diffuse *simultaneously* through the membrane but in opposite directions.

Although this two-way cation flow may appear to be self-defeating when depolarization is the goal, remember that the electrochemical gradient for sodium is much steeper than that for potassium. As a result, Na^+ influx is greater than K^+ efflux, and *net* depolarization occurs.

If enough neurotransmitter binds, depolarization of the postsynaptic membrane can reach 0 mV—well above an axon's threshold (about −50 mV) for firing an AP. However, unlike axons which have voltage-gated channels that make an AP possible, *postsynaptic membranes generally do not generate APs.* The dramatic polarity reversal seen in axons never occurs in membranes containing *only* chemically gated channels because the opposite movements of K^+ and Na^+ prevent excessive positive charge from accumulating inside the cell. For this reason, instead of APs, local graded depolarization events called **excitatory postsynaptic potentials (EPSPs)** occur at excitatory postsynaptic membranes (Figure 11.17a).

Each EPSP lasts a few milliseconds and then the membrane returns to its resting potential. The only function of EPSPs is to help trigger an AP distally at the axon hillock of the postsynaptic neuron. Although currents created by individual EPSPs decline with distance, they can and often do spread all the way to the axon hillock. If currents reaching the hillock are strong enough to depolarize the axon to threshold, axonal voltage-gated channels open and an AP is generated.

11

Inhibitory Synapses and IPSPs

Binding of neurotransmitters at inhibitory synapses *reduces* a postsynaptic neuron's ability to generate an AP. Most inhibitory neurotransmitters hyperpolarize the postsynaptic membrane by making the membrane more permeable to K^+ or Cl^-. Sodium ion permeability is not affected.

If K^+ channels open, K^+ moves out of the cell. If Cl^- channels open, Cl^- moves in. In either case, the charge on the inner face of the membrane becomes more negative. As the membrane potential increases and is driven farther from the axon's threshold, the postsynaptic neuron becomes *less and less likely* to "fire," and larger depolarizing currents are required to induce an AP. Hyperpolarizing changes in potential are called **inhibitory postsynaptic potentials (IPSPs)** (Figure 11.17b).

Integration and Modification of Synaptic Events

Summation by the Postsynaptic Neuron

A single EPSP cannot induce an AP in the postsynaptic neuron (Figure 11.18a, p. 372). But if thousands of excitatory axon terminals fire on the same postsynaptic membrane, or if a small number

Table 11.2 Comparison of Graded Potentials and Action Potentials

	GRADED POTENTIAL (GP)	ACTION POTENTIAL (AP)
Location of event	Cell body and dendrites, typically	Axon hillock and axon
Distance traveled	Short distance—typically within cell body to axon hillock (0.1–1.0 mm)	Long distance—from trigger zone at axon hillock through entire length of axon (a few mm to over a meter)
Amplitude (size)	Various sizes (graded); decays with distance	Always the same size (all-or-none); does not decay with distance
Stimulus for opening ion channels	Chemical (neurotransmitter) or sensory stimulus (e.g., light, pressure, temperature)	Voltage (depolarization, triggered by GP reaching threshold)
Positive feedback cycle	Absent	Present
Repolarization	Voltage independent; occurs when stimulus is no longer present	Voltage regulated; occurs when Na^+ channels inactivate and K^+ channels open
Summation	Stimulus responses can summate to increase amplitude of graded potential Temporal: increased frequency of stimuli Spatial: stimuli from multiple sources	Does not occur; an all-or-none phenomenon

of terminals deliver impulses rapidly, the probability of reaching threshold soars. EPSPs can add together, or **summate**, to influence the activity of a postsynaptic neuron. Otherwise, nerve impulses would never result.

Two types of summation occur: temporal and spatial.

- **Temporal summation** (*temporal* = time) occurs when one or more presynaptic neurons transmit impulses in rapid-fire order and bursts of neurotransmitter are released in quick succession. The first impulse produces a small EPSP, and before it dissipates, successive impulses trigger more EPSPs. These summate, causing the postsynaptic membrane to depolarize much more than it would from a single EPSP (Figure 11.18b).
- **Spatial summation** occurs when the postsynaptic neuron is stimulated simultaneously by a large number of terminals from one or, more commonly, many presynaptic neurons. Huge numbers of its receptors bind neurotransmitter and simultaneously initiate EPSPs, which summate and dramatically enhance depolarization (Figure 11.18c).

Table 11.2 *(continued)*

	GRADED POTENTIAL (GP)		ACTION POTENTIAL (AP)
	POSTSYNAPTIC POTENTIAL (A TYPE OF GP)		
	EXCITATORY (EPSP)	INHIBITORY (IPSP)	
Function	Short-distance signaling; depolarization that spreads to axon hillock; moves membrane potential *toward* threshold for generating an AP	Short-distance signaling; hyperpolarization that spreads to axon hillock; moves membrane potential *away* from threshold for generating an AP	Long-distance signaling; constitutes the nerve impulse
Initial effect of stimulus	Opens chemically gated channels that allow simultaneous Na^+ and K^+ fluxes	Opens chemically gated K^+ or Cl^- channels	Opens voltage-gated channels; first opens Na^+ channels, then K^+ channels
Peak membrane potential	Depolarizes; moves toward 0 mV	Hyperpolarizes; moves toward –90 mV	+30 to +50 mV

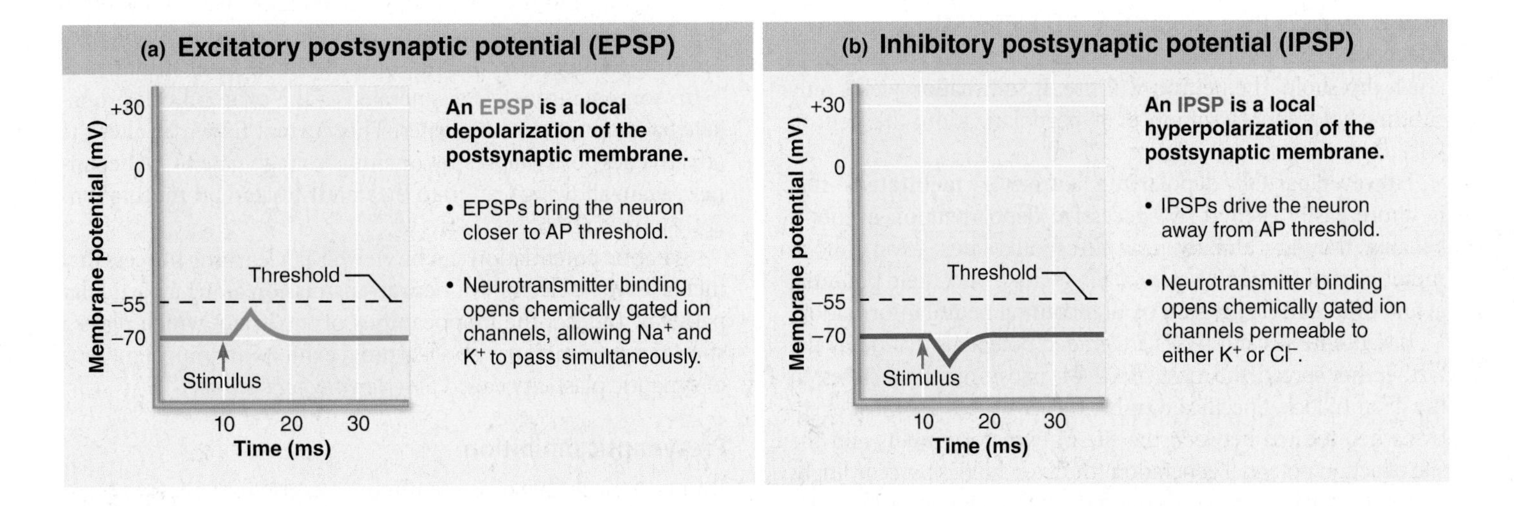

Figure 11.17 **Postsynaptic potentials can be excitatory or inhibitory.**

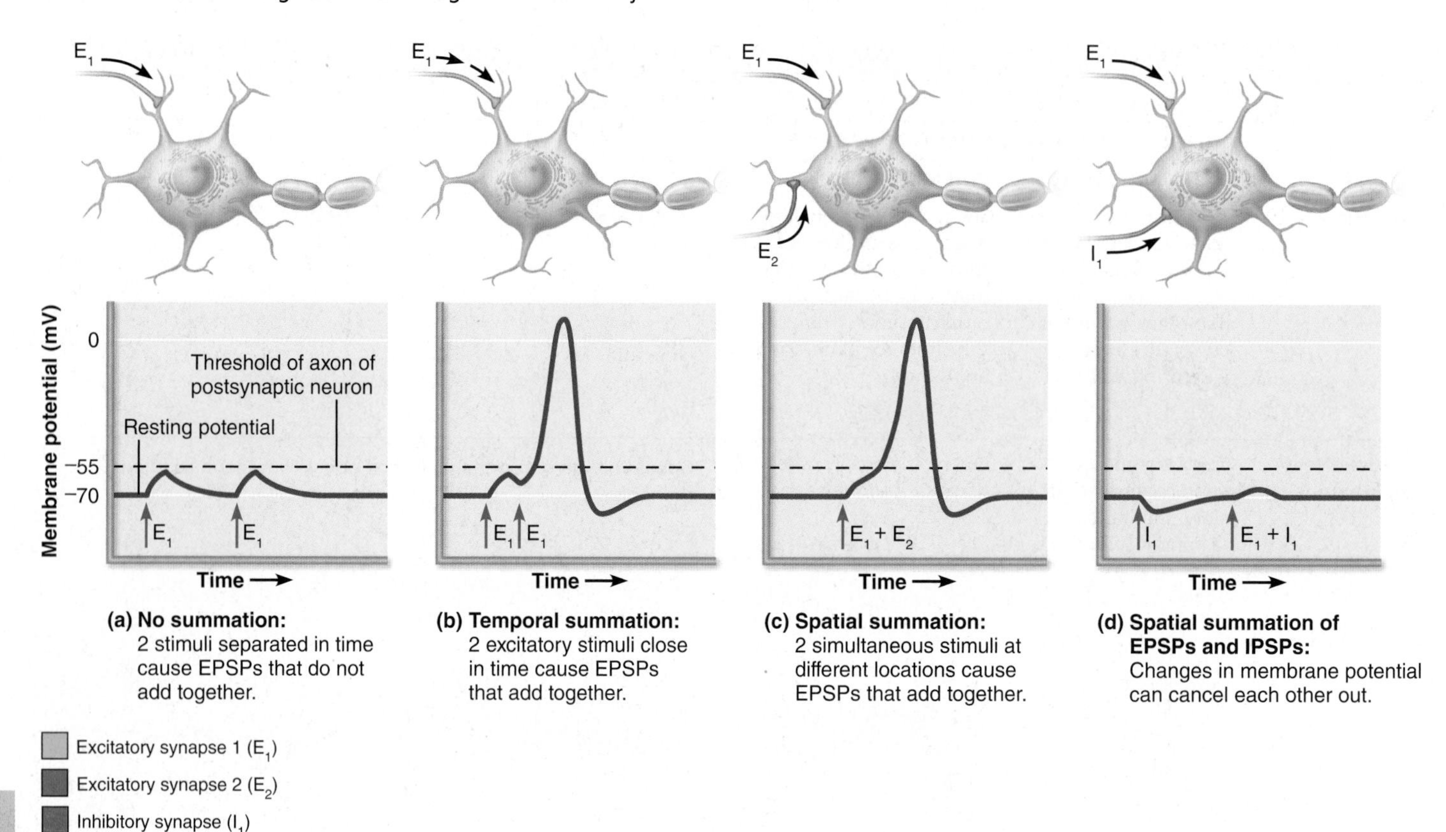

Figure 11.18 Neural integration of EPSPs and IPSPs.

Although we have focused on EPSPs, IPSPs also summate, both temporally and spatially. In this case, the postsynaptic neuron is inhibited to a greater degree.

Most neurons receive both excitatory and inhibitory inputs from thousands of other neurons. Additionally, the same axon may form different types of synapses (in terms of biochemical and electrical characteristics) with different types of target neurons. How is all this conflicting information sorted out?

Each neuron's axon hillock keeps a running account of all the signals it receives. Not only do EPSPs summate and IPSPs summate, but also EPSPs summate with IPSPs. If the stimulatory effects of EPSPs dominate the membrane potential enough to reach threshold, the neuron will fire. If summation yields only subthreshold depolarization or hyperpolarization, the neuron fails to generate an AP (Figure 11.18d).

However, partially depolarized neurons are **facilitated**—that is, more easily excited by successive depolarization events—because they are already near threshold. Thus, axon hillock membranes function as *neural integrators*, and their potential at any time reflects the sum of all incoming neural information.

Because EPSPs and IPSPs are graded potentials that decay the farther they spread, the most effective synapses are those closest to the axon hillock. Specifically, inhibitory synapses are most effective when located between the site of excitatory inputs and the site of action potential generation (the axon hillock). Accordingly, inhibitory synapses occur most often on the cell body and excitatory synapses occur most often on the dendrites (Figure 11.18d).

Synaptic Potentiation

Repeated or continuous use of a synapse (even for short periods) enhances the presynaptic neuron's ability to excite the postsynaptic neuron, producing larger-than-expected EPSPs. This phenomenon is **synaptic potentiation**. The presynaptic terminals at such synapses contain relatively high Ca^{2+} concentrations, a condition that triggers the release of more neurotransmitter, which in turn produces larger EPSPs.

Synaptic potentiation also brings about Ca^{2+} influx via dendritic spines into the postsynaptic neuron. As Ca^{2+} floods into the cell, it activates certain kinase enzymes that promote changes resulting in more effective responses to subsequent stimuli.

In some neurons, APs generated at the axon hillock propagate back up into the dendrites. This current flow may alter the effectiveness of synapses by opening voltage-gated Ca^{2+} channels, again allowing Ca^{2+} into the dendrites and promoting synaptic potentiation.

Synaptic potentiation can be viewed as a learning process that increases the efficiency of neurotransmission along a particular pathway. Indeed, the hippocampus of the brain, which plays a special role in memory and learning, exhibits an important type of synaptic plasticity called *long-term potentiation* (LTP).

Presynaptic Inhibition

Events at the presynaptic membrane can also influence postsynaptic activity. **Presynaptic inhibition** occurs when the release of excitatory neurotransmitter by one neuron is inhibited by the activity of another neuron via an axoaxonal synapse. More than one mechanism is involved, but the end result is that less neurotransmitter is released and bound, forming smaller EPSPs.

In contrast to postsynaptic inhibition by IPSPs, which decreases the excitability of the postsynaptic neuron, presynaptic inhibition decreases the excitatory stimulation of the postsynaptic neuron. In this way, presynaptic inhibition is like a functional synaptic "pruning."

☑ Check Your Understanding

18. Which ions flow through chemically gated channels to produce IPSPs? EPSPs?

19. What is the difference between temporal summation and spatial summation?

For answers, see Answers Appendix.

11.9 The effect of a neurotransmitter depends on its receptor

→ Learning Objectives

- ☐ **Define neurotransmitter and classify neurotransmitters by chemical structure and by function.**
- ☐ **Describe the action of neurotransmitters at channel-linked and G protein–linked receptors.**

Neurotransmitters, along with electrical signals, are the "language" of the nervous system—the means by which neurons communicate to process and send messages to the rest of the body. Sleep, thought, rage, hunger, memory, movement, and even your smile reflect the actions of these versatile molecules. Most factors that affect synaptic transmission do so by enhancing or inhibiting neurotransmitter release or destruction, or by blocking their binding to receptors. Just as speech defects may hinder interpersonal communication, anything that interferes with neurotransmitter activity may short-circuit the brain's "conversations" or internal talk.

More than 50 neurotransmitters or neurotransmitter candidates have been identified. Although some neurons produce and release only one kind of neurotransmitter, most make two or more and may release any one or all of them at a given time. It appears that in most cases, different neurotransmitters are released at different stimulation frequencies. This avoids producing a jumble of nonsense messages. However, co-release of two neurotransmitters from the same vesicles has been documented. The coexistence of more than one neurotransmitter in a single neuron makes it possible for that cell to exert several different influences.

Neurotransmitters are classified chemically and functionally. **Table 11.3** provides a detailed overview and key groups are discussed in the following sections.

Classification of Neurotransmitters by Chemical Structure

Neurotransmitters are grouped into several classes based on molecular structure.

Acetylcholine

Acetylcholine (ACh) (as″ĕ-til-ko′lēn), the first neurotransmitter identified, is still the best understood because it is released at neuromuscular junctions, which are much easier to study than synapses buried in the CNS.

ACh is synthesized from acetic acid (as acetyl CoA) and choline by the enzyme *choline acetyltransferase*, then transported into synaptic vesicles for later release. Once released by the presynaptic terminal, ACh binds briefly to the postsynaptic receptors. It is then released and degraded to acetic acid and choline by the enzyme **acetylcholinesterase (AChE)**, located in the synaptic cleft and on postsynaptic membranes. Presynaptic terminals recapture the released choline and reuse it to synthesize more ACh.

ACh is released by all neurons that stimulate skeletal muscles and by many neurons of the autonomic nervous system. ACh-releasing neurons are also found in the CNS.

Biogenic Amines

The **biogenic amines** (bi″o-jen′ik) include the **catecholamines** (kat″ĕ-kol′ah-mēnz), such as dopamine, norepinephrine (NE), and epinephrine, and the **indolamines**, which include sero-tonin and histamine. *Dopamine* and *NE* are synthesized from the amino acid tyrosine in a common pathway. The epinephrine-releasing cells of the brain and adrenal medulla use the same pathway. *Serotonin* is synthesized from the amino acid tryptophan. *Histamine* is synthesized from the amino acid histidine.

Biogenic amine neurotransmitters are broadly distributed in the brain, where they play a role in emotional behavior and help regulate the biological clock. Additionally, some motor neurons of the autonomic nervous system release catecholamines, particularly NE. Imbalances of these neurotransmitters are associated with mental illness. For example, overactive dopamine signaling occurs in schizophrenia. Additionally, certain psychoactive drugs (LSD and mescaline) can bind to biogenic amine receptors and induce hallucinations.

Amino Acids

It is difficult to prove a neurotransmitter role when the suspect is an amino acid, because amino acids occur in all cells of the body and are important in many biochemical reactions. The amino acids for which a neurotransmitter role is certain include **glutamate**, **aspartate**, **glycine**, and **gamma (γ)-aminobutyric acid (GABA)**, and there may be others.

Peptides

The **neuropeptides**, essentially strings of amino acids, include a broad spectrum of molecules with diverse effects. For example, a neuropeptide called **substance P** is an important mediator of pain signals. By contrast, **endorphins**, which include **beta endorphin**, **dynorphin**, and **enkephalins** (en-kef′ah-linz), act as natural opiates, reducing our perception of pain under stressful conditions. Enkephalin activity increases dramatically in pregnant women in labor. Endorphin release is enhanced when an athlete gets a so-called second wind and is probably responsible for the "runner's high." Additionally, some researchers claim that

(Text continues on p. 376.)

Table 11.3 Neurotransmitters and Neuromodulators

NEUROTRANSMITTER	FUNCTIONAL CLASSES	SITES WHERE SECRETED	COMMENTS
Acetylcholine (ACh)			
• At *nicotinic ACh receptors* (on skeletal muscles, autonomic ganglia, and in the CNS) • At *muscarinic ACh receptors* (on visceral effectors and in the CNS)	Excitatory Direct action Excitatory or inhibitory depending on subtype of muscarinic receptor Indirect action via second messengers	CNS: widespread throughout cerebral cortex, hippocampus, and brain stem PNS: all neuromuscular junctions with skeletal muscle; some autonomic motor endings (all preganglionic and parasympathetic postganglionic fibers)	Effects prolonged when AChE blocked by nerve gas or organophosphate insecticides (malathion), leading to tetanic muscle spasms. Release inhibited by botulinum toxin; binding to nicotinic ACh receptors inhibited by curare (a muscle paralytic agent) and to muscarinic ACh receptors by atropine. ACh levels decrease in certain brain areas in Alzheimer's disease; nicotinic ACh receptors destroyed in myasthenia gravis.
Biogenic Amines			
Norepinephrine (NE)	Excitatory or inhibitory depending on receptor type bound Indirect action via second messengers	CNS: brain stem, particularly in the locus coeruleus of the midbrain; limbic system; some areas of cerebral cortex PNS: main neurotransmitter of postganglionic neurons in the sympathetic nervous system	A "feel good" neurotransmitter. Release enhanced by amphetamines; removal from synapse blocked by tricyclic antidepressants and cocaine. Brain levels reduced by reserpine (an antihypertensive drug), leading to depression.
Dopamine	Excitatory or inhibitory depending on the receptor type bound Indirect action via second messengers	CNS: substantia nigra of midbrain; hypothalamus; the principal neurotransmitter of indirect motor pathways PNS: some sympathetic ganglia	A "feel good" neurotransmitter. Release enhanced by L-dopa and amphetamines; reuptake blocked by cocaine. Deficient in Parkinson's disease; dopamine neurotransmission increases in schizophrenia.
Serotonin (5-HT)	Mainly inhibitory Indirect action via second messengers; direct action at 5-HT_3 receptors	CNS: brain stem, especially midbrain; hypothalamus; limbic system; cerebellum; pineal gland; spinal cord	Plays a role in sleep, appetite, nausea, migraine headaches, and regulating mood. Drugs that block its uptake relieve anxiety and depression. Activity blocked by LSD and enhanced by ecstasy.
Histamine	Excitatory or inhibitory depending on receptor type bound Indirect action via second messengers	CNS: hypothalamus	Involved in wakefulness, appetite control, and learning and memory.
Amino Acids			
GABA (γ-aminobutyric acid)	Generally inhibitory Direct and indirect actions via second messengers	CNS: cerebral cortex, hypothalamus, Purkinje cells of cerebellum, spinal cord, granule cells of olfactory bulb, retina	Principal inhibitory neurotransmitter in the brain; important in presynaptic inhibition at axoaxonal synapses. Inhibitory effects augmented by alcohol, antianxiety drugs of the benzodiazepine class, and barbiturates. Substances that block its synthesis, release, or action induce convulsions.
Glutamate	Generally excitatory Direct action	CNS: spinal cord; widespread in brain where it represents the major excitatory neurotransmitter	Important in learning and memory. The "stroke neurotransmitter": excessive release produces excitotoxicity—neurons literally stimulated to death; most commonly caused by ischemia (oxygen deprivation, usually due to a blocked blood vessel).

Table 11.3 *(continued)*

NEUROTRANSMITTER	FUNCTIONAL CLASSES	SITES WHERE SECRETED	COMMENTS
Amino Acids, continued			
Glycine	Generally inhibitory Direct action	CNS: spinal cord and brain stem, retina	Principal inhibitory neurotransmitter of the spinal cord. Strychnine blocks glycine receptors, resulting in uncontrolled convulsions and respiratory arrest.
Peptides			
Endorphins, e.g., beta endorphin, dynorphin, enkephalins	Generally inhibitory Indirect action via second messengers	CNS: widely distributed in brain (hypothalamus; limbic system; pituitary) and spinal cord	Natural opiates; inhibit pain by inhibiting substance P. Effects mimicked by morphine, heroin, and methadone.
Tachykinins: substance P, neurokinin A (NKA)	Excitatory Indirect action via second messengers	CNS: basal nuclei, midbrain, hypothalamus, cerebral cortex PNS: certain sensory neurons of dorsal root ganglia (pain afferents), enteric neurons	Substance P mediates pain transmission in the PNS. In the CNS, tachykinins are involved in respiratory and cardiovascular controls and in mood.
Somatostatin	Generally inhibitory Indirect action via second messengers	CNS: widely distributed in brain (hypothalamus, basal nuclei, hippocampus, cerebral cortex) Pancreas	Often released with GABA. A gut-brain peptide. Inhibits growth hormone release.
Cholecystokinin (CCK)	Generally excitatory Indirect action via second messengers	Throughout CNS Small intestine	Involved in anxiety, pain, memory. A gut-brain peptide hormone. Inhibits appetite.
Purines			
ATP	Excitatory or inhibitory depending on receptor type bound Direct and indirect actions via second messengers	CNS: basal nuclei, induces Ca^{2+} wave propagation in astrocytes PNS: dorsal root ganglion neurons	ATP released by sensory neurons (as well as that released by injured cells) provokes pain sensation.
Adenosine	Generally inhibitory Indirect action via second messengers	Throughout CNS and PNS	Caffeine stimulates by blocking brain adenosine receptors. May be involved in sleep-wake cycle and terminating seizures. Dilates arterioles, increasing blood flow to heart and other tissues as needed.
Gases and Lipids			
Nitric oxide (NO)	Excitatory or inhibitory Indirect action via second messengers	CNS: brain, spinal cord PNS: adrenal gland; nerves to penis	Its release potentiates stroke damage. Some types of male impotence treated by enhancing NO action [e.g., with sildenafil (Viagra)].
Carbon monoxide (CO)	Excitatory or inhibitory Indirect action via second messengers	Brain and some neuromuscular and neuroglandular synapses	
Endocannabinoids, e.g., 2-arachidonoylglycerol, anandamide	Inhibitory Indirect action via second messengers	Throughout CNS	Involved in memory (as a retrograde messenger), appetite control, nausea and vomiting, neuronal development. Receptors activated by THC, the principal active ingredient of cannabis.

the placebo effect is due to endorphin release. These painkilling neurotransmitters remained undiscovered until investigators began to ask why morphine and other opiates reduce anxiety and pain. They found that these drugs attach to the same receptors that bind natural opiates, producing similar but stronger effects.

Some neuropeptides, known as **gut-brain peptides**, are also produced by nonneural body tissues and are widespread in the gastrointestinal tract. Examples include somatostatin and cholecystokinin (CCK).

Purines

Purines are nitrogen-containing chemicals (such as guanine and adenine) that are breakdown products of nucleic acids. **Adenosine triphosphate (ATP)**, the cell's universal form of energy, is now recognized as a major neurotransmitter (perhaps the most primitive one) in both the CNS and PNS. Like the receptors for glutamate and acetylcholine, certain receptors produce fast excitatory responses when ATP binds, while other ATP receptors trigger slow, second-messenger responses. Upon binding to receptors on astrocytes, ATP mediates Ca^{2+} influx.

In addition to the neurotransmitter action of extracellular ATP, **adenosine**, a part of ATP, also acts outside of cells on adenosine receptors. Adenosine is a potent inhibitor in the brain. Caffeine's well-known stimulatory effects result from blocking these adenosine receptors.

11

Gases and Lipids

Not so long ago, it would have been scientific suicide to suggest that small, short-lived, toxic gas molecules might be neurotransmitters. Nonetheless, the discovery of these unlikely messengers has opened up a new chapter in the story of neurotransmission.

Gasotransmitters These gases—the so-called "gasotransmitters" nitric oxide, carbon monoxide, and hydrogen sulfide—defy all the classical descriptions of neurotransmitters. Rather than being stored in vesicles and released by exocytosis, they are synthesized on demand and diffuse out of the cells that make them. Instead of attaching to surface receptors, they zoom through the plasma membrane of nearby cells to bind with intracellular receptors.

Both **nitric oxide (NO)** and **carbon monoxide (CO)** activate *guanylate cyclase*, the enzyme that makes the second messenger *cyclic GMP*. NO and CO are found in different brain regions and appear to act in different pathways, but their mode of action is similar. NO participates in a variety of processes in the brain, including the formation of new memories by increasing the strength of certain synapses.

Excessive release of NO is thought to contribute to the brain damage seen in stroke patients. In the PNS, NO causes blood vessels and intestinal smooth muscle to relax.

Less is known about **hydrogen sulfide (H_2S)**, the most recently discovered gasotransmitter. Unlike NO and CO, it appears to act directly on ion channels and other proteins to alter their function.

Endocannabinoids Just as there are natural opiate neurotransmitters in the brain, our brains make **endocannabinoids** (en″do-kă-nă′bĭ-noids) that act at the same receptors as tetrahydrocannabinol (THC), the active ingredient in marijuana. Their receptors, the *cannabinoid receptors*, are the most common G protein–linked receptors in the brain. Like the gasotransmitters, the endocannabinoids are lipid soluble and are synthesized on demand, rather than stored and released from vesicles. Like NO, they are thought to be involved in learning and memory. We are only beginning to understand the many other processes these neurotransmitters may be involved in, which include neuronal development, controlling appetite, and suppressing nausea.

Classification of Neurotransmitters by Function

In this text we can only sample the incredible diversity of functions that neurotransmitters mediate. We limit our discussion here to two broad ways of classifying neurotransmitters according to function, adding more details in subsequent chapters.

The important idea to keep in mind is this: The function of a neurotransmitter is determined by the receptor to which it binds.

Effects: Excitatory versus Inhibitory

Some neurotransmitters are excitatory (cause depolarization). Some are inhibitory (cause hyperpolarization). Others exert both effects, depending on the specific receptor types with which they interact.

For example, the amino acids GABA and glycine are usually inhibitory, whereas glutamate is typically excitatory (Table 11.3). On the other hand, ACh and NE each bind to at least two receptor types that cause opposite effects. For example, acetylcholine is excitatory at neuromuscular junctions in skeletal muscle and inhibitory in cardiac muscle.

Actions: Direct versus Indirect

Neurotransmitters that act *directly* are those that bind to and open ion channels. These neurotransmitters provoke rapid responses in postsynaptic cells by altering membrane potential. ACh and the amino acid neurotransmitters are typically direct-acting neurotransmitters.

Neurotransmitters that act *indirectly* promote broader, longer-lasting effects by acting through intracellular *second-messenger* molecules, typically via G protein pathways (see *Focus on G Proteins*, Focus Figure 3.2 on p. 76). In this way their action is similar to that of many hormones. The biogenic amines, neuropeptides, and dissolved gases are indirect neurotransmitters.

Neuromodulator is a term used to describe a chemical messenger released by a neuron that does not directly cause EPSPs or IPSPs but instead affects the strength of synaptic transmission. A neuromodulator may act presynaptically to influence the synthesis, release, degradation, or reuptake of neurotransmitter. Alternatively, it may act postsynaptically by altering the sensitivity of the postsynaptic membrane to neurotransmitter.

Receptors for neuromodulators are not necessarily found at a synapse. Instead, a neuromodulator may be released from one cell to act at many cells in its vicinity, similar to paracrines (chemical messengers that act locally and are quickly destroyed). The distinction between neurotransmitters and neuromodulators is fuzzy, but chemical messengers such as NO, adenosine, and a number of neuropeptides are often referred to as neuromodulators.

Neurotransmitter Receptors

In Chapter 3, we introduced the various receptors involved in cell signaling. Now we are ready to pick up that thread again as we examine the action of receptors that bind neurotransmitters. For the most part, neurotransmitter receptors are either channel-linked receptors, which mediate fast synaptic transmission, or G protein–linked receptors, which oversee slow synaptic responses.

Channel-Linked Receptors

Channel-linked receptors (*ionotropic receptors*) are ligand-gated ion channels that mediate direct neurotransmitter action. They are composed of several protein subunits in a "rosette" around a central pore. As the ligand binds to one (or more) receptor subunits, the proteins change shape. This event opens the central channel and allows ions to pass (**Figure 11.19**). As a result, the membrane potential of the target cell changes.

Channel-linked receptors are always located precisely opposite sites of neurotransmitter release, and their ion channels open instantly upon ligand binding and remain open 1 ms or less while the ligand is bound. At excitatory receptor sites (nicotinic ACh channels and receptors for glutamate, aspartate, and ATP), the channel-linked receptors are cation channels that allow small cations (Na^+, K^+, Ca^{2+}) to pass, but Na^+ entry contributes most to membrane depolarization. Channel-linked receptors that respond to GABA and glycine, and allow Cl^- to pass, mediate fast inhibition (hyperpolarization).

G Protein–Linked Receptors

Unlike responses to neurotransmitter binding at channel-linked receptors, which are immediate, simple, and brief, the

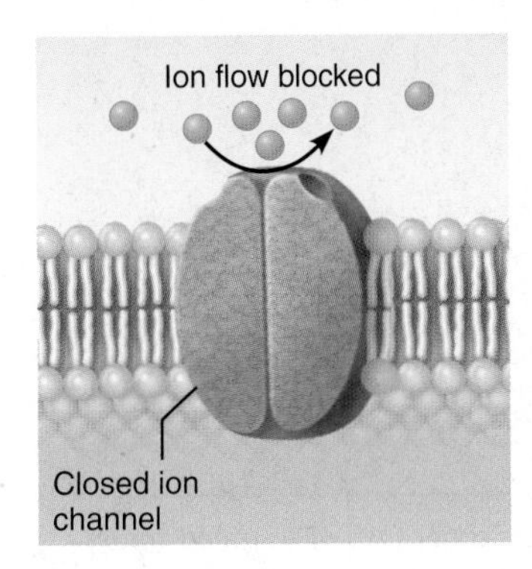

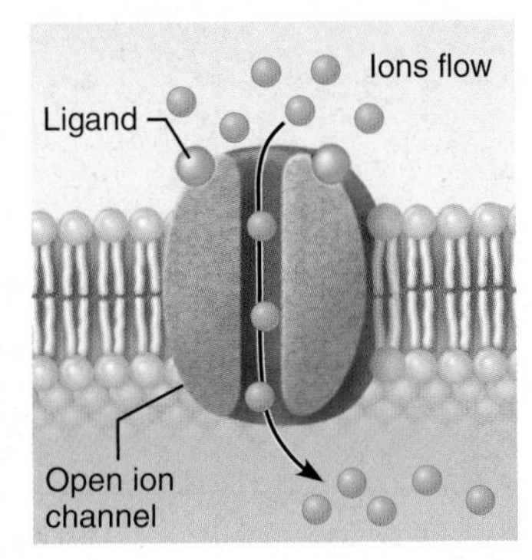

Figure 11.19 Channel-linked receptors cause rapid synaptic transmission. Ligand binding *directly* opens chemically gated ion channels.

activity mediated by **G protein–linked receptors** is indirect, complex, slow (hundreds of milliseconds or more), and often prolonged—ideal as a basis for some types of learning. Receptors in this class are transmembrane protein complexes. They include muscarinic ACh receptors and those that bind the biogenic amines and neuropeptides. Because their effects tend to bring about widespread metabolic changes, G protein–linked receptors are commonly called *metabotropic receptors.*

When a neurotransmitter binds to a G protein–linked receptor, the G protein is activated (**Figure 11.20**). (To orient yourself, refer back to the simpler G protein explanation in *Focus on G Proteins*, Focus Figure 3.2 on p. 76.) Activated G proteins typically work by controlling the production of second messengers such as **cyclic AMP**, **cyclic GMP**, **diacylglycerol**, or $\mathbf{Ca^{2+}}$.

These second messengers, in turn, act as go-betweens to regulate the opening or closing of ion channels or activate kinase enzymes that initiate a cascade of reactions in the target cells. Some second messengers modify (activate or inactivate) other proteins, including channel proteins, by attaching phosphate groups to them. Others interact with nuclear proteins that activate genes and induce synthesis of new proteins in the target cell.

☑ Check Your Understanding

20. ACh excites skeletal muscle and yet it inhibits heart muscle. How can this be?

21. Why is cyclic AMP called a second messenger?

For answers, see Answers Appendix.

11.10 Neurons act together, making complex behaviors possible

→ Learning Objectives

- ☐ **Describe common patterns of neuronal organization and processing.**
- ☐ **Distinguish between serial and parallel processing.**

Until now, we have concentrated on the activities of individual neurons. However, neurons function in groups, and each group contributes to still broader neural functions. In this way, the organization of the nervous system is hierarchical.

Any time you have a large number of *anything*—people included—there must be *integration*. In other words, the parts must be fused into a smoothly operating whole.

In this module, we move to the first level of **neural integration**: *neuronal pools* and their patterns of communicating with other parts of the nervous system. In Chapter 12 we discuss the highest levels of neural integration—how we think and remember. With this understanding of the basics and of the larger picture, in Chapter 13 we examine how sensory inputs interface with motor activity.

Organization of Neurons: Neuronal Pools

The billions of neurons in the CNS are organized into **neuronal pools**. These functional groups of neurons integrate

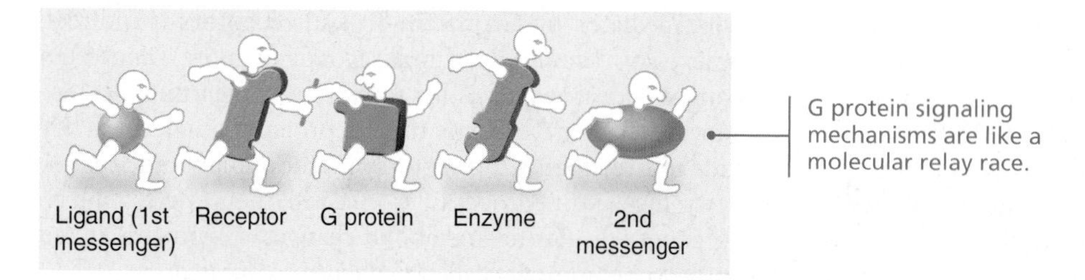

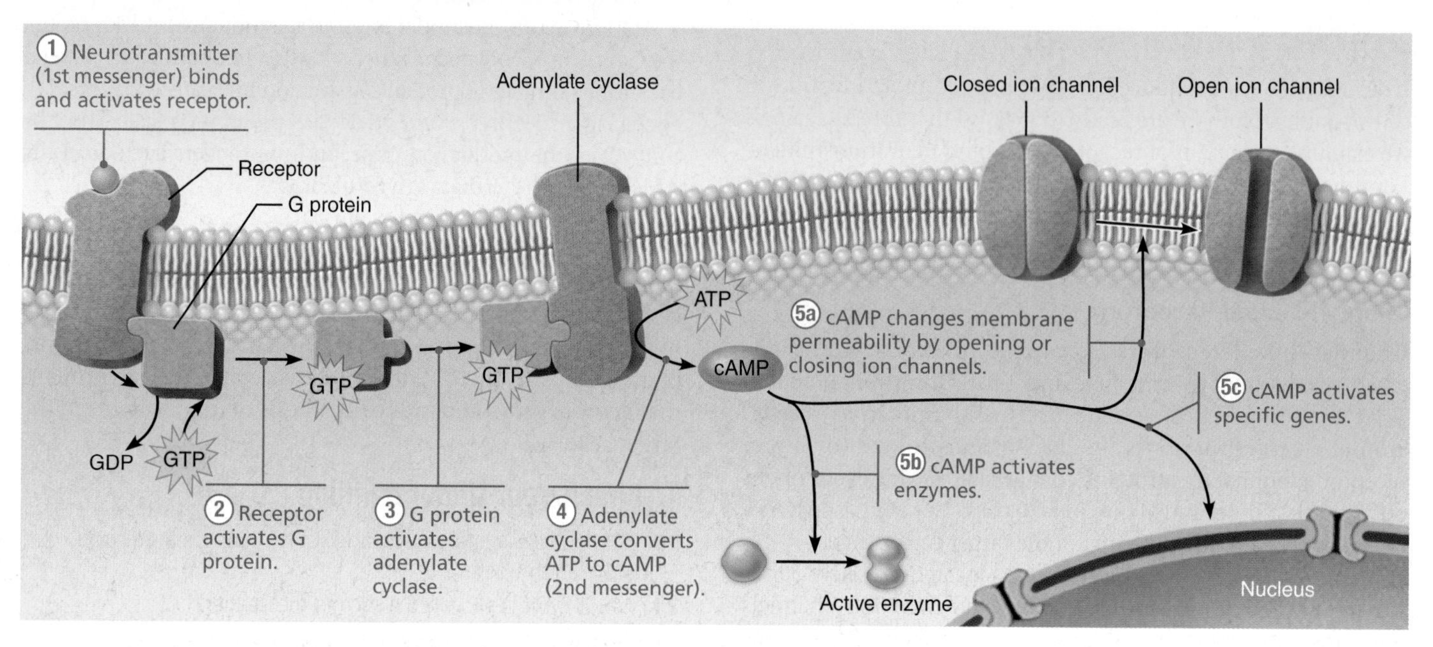

Figure 11.20 G protein–linked receptors cause the formation of intracellular second messengers. The neurotransmitter acts indirectly—via the second messenger cyclic AMP (cAMP) in this example—to bring about the cell's response. (For the basics of G protein signaling mechanisms, see Focus Figure 3.2 on p. 76.)

incoming information from receptors or different neuronal pools and then forward the processed information to other destinations.

In a simple type of neuronal pool (**Figure 11.21**), one incoming presynaptic fiber branches profusely as it enters the pool and then synapses with several different neurons in the pool. When the incoming fiber is excited, it will excite some postsynaptic neurons and facilitate others. Neurons most likely to generate impulses are those closely associated with the incoming fiber, because they receive the bulk of the synaptic contacts. Those neurons are in the *discharge zone* of the pool.

Neurons farther from the center are not usually excited to threshold, but they are facilitated and can easily be brought to threshold by stimuli from another source. For this reason, the periphery of the pool is the *facilitated zone.* Keep in mind, however, that our figure is a gross oversimplification. Most neuronal pools consist of thousands of neurons and include inhibitory as well as excitatory neurons.

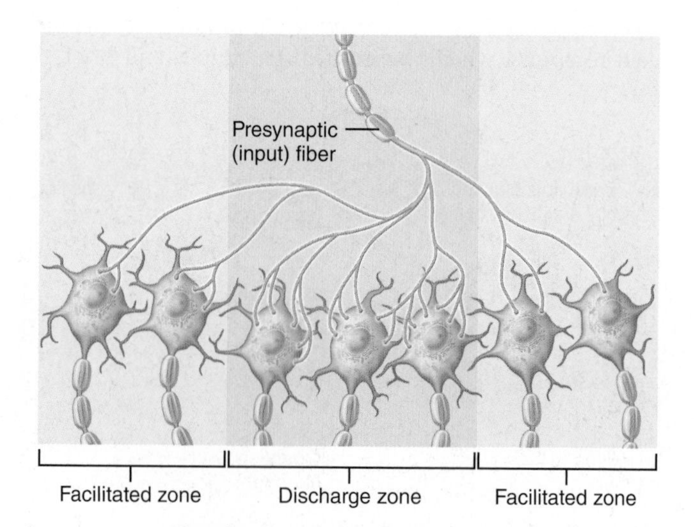

Figure 11.21 Simple neuronal pool. Postsynaptic neurons in the discharge zone receive more synapses and are more likely to discharge (generate APs). Postsynaptic neurons in the facilitated zone receive fewer synapses and are facilitated (brought closer to threshold).

Patterns of Neural Processing

Input processing is both *serial* and *parallel.* In serial processing, the input travels along one pathway to a specific destination. In parallel processing, the input travels along several different

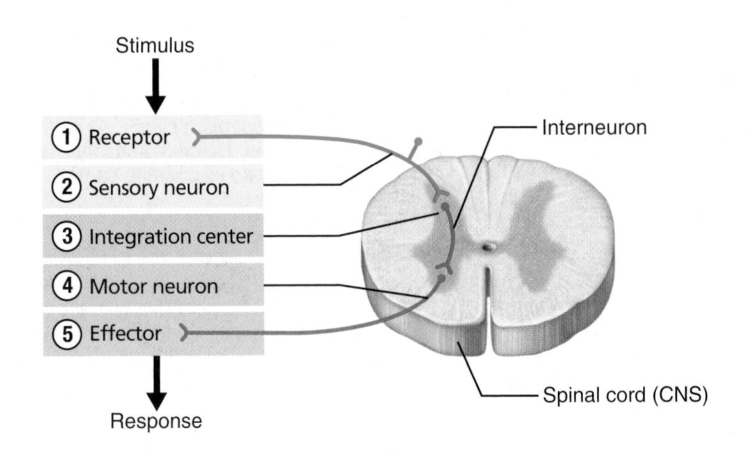

Figure 11.22 A simple reflex arc. Receptors detect a change in the internal or external environment that elicits a rapid stereotyped response. Effectors are muscles or glands.

pathways to be integrated in different CNS regions. Each mode has unique advantages, but as an information processor, the brain derives its power from its ability to process in parallel.

Serial Processing

In **serial processing**, the whole system works in a predictable all-or-nothing manner. One neuron stimulates the next, which stimulates the next, and so on, eventually causing a specific, anticipated response. The most clear-cut examples of serial processing are spinal reflexes. Straight-through sensory pathways from receptors to the brain are also examples. Because reflexes are the functional units of the nervous system, it is important that you understand them early on.

Reflexes are rapid, automatic responses to stimuli, in which a particular stimulus always causes the same response. Reflex activity, which produces the simplest behaviors, is stereotyped and dependable. For example, if you touch a hot object you jerk your hand away, and an object approaching your eye triggers a blink. Reflexes occur over neural pathways called **reflex arcs** that have five essential components—receptor, sensory neuron, CNS integration center, motor neuron, and effector (**Figure 11.22**).

Parallel Processing

In **parallel processing**, inputs are segregated into many pathways, and different parts of the neural circuitry deal simultaneously with the information delivered by each pathway. For example, smelling a pickle (the input) may cause you to remember picking cucumbers on a farm; or it may remind you that you don't like pickles or that you must buy some at the market; or perhaps it will call to mind *all* these thoughts. For each person, parallel processing triggers unique pathways. The same stimulus—pickle smell, in our example—promotes many responses beyond simple awareness of the smell. Parallel processing is not repetitious because the pathways do different things with the information. Each pathway or "channel" is decoded in relation to all the others to produce a total picture.

Think, for example, about what happens when you step on a sharp object while walking barefoot. The serially processed withdrawal reflex causes you to withdraw your foot immediately. At the same time, pain and pressure impulses are speeding up to your brain along parallel pathways that allow you to decide whether to simply rub the hurt spot or seek first aid.

Parallel processing is extremely important for higher-level mental functioning—for putting the parts together to understand the whole. For example, you can recognize a dollar bill in a split second. This task takes a serial-based computer a fairly long time, but your recognition is rapid because you use parallel processing. A single neuron sends information along several pathways instead of just one, so you process a large amount of information much more quickly.

Types of Circuits

Individual neurons in a neuronal pool both send and receive information, and synaptic contacts may cause either excitation or inhibition. The patterns of synaptic connections in neuronal pools, called **circuits**, determine the pool's functional capabilities. **Figure 11.23** illustrates four basic circuit patterns and their properties: diverging, converging, reverberating, and parallel after-discharge circuits.

☑ Check Your Understanding

22. Which types of neural circuits would give a prolonged output after a single input?

23. What pattern of neural processing occurs when your finger accidentally touches a hot grill? What is this response called?

24. What pattern of neural processing occurs when we smell freshly baked apple pie and remember Thanksgiving at our grandparents' house, the odor of freshly cooked turkey, sitting by the fire, and other such memories?

For answers, see Answers Appendix.

In this chapter, we have examined how the amazingly complex neurons, via electrical and chemical signals, serve the body in a variety of ways. Some serve as "lookouts," others process information for immediate use or for future reference, and still others stimulate the body's muscles and glands into activity. With this background, we are ready to study the most sophisticated mass of neural tissue in the entire body—the brain (and its continuation, the spinal cord), the focus of Chapter 12.

Input

Many outputs

(a) Diverging circuit
- One input, many outputs
- An *amplifying* circuit
- **Example:** A single neuron in the brain can activate 100 or more motor neurons in the spinal cord and thousands of skeletal muscle fibers

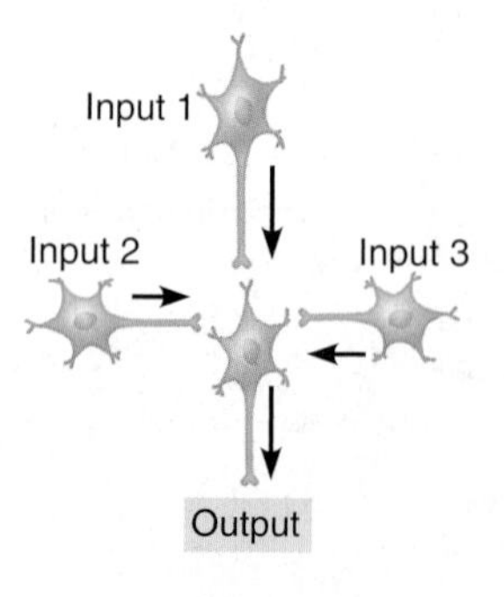

(b) Converging circuit
- Many inputs, one output
- A *concentrating* circuit
- **Example:** Different sensory stimuli can all elicit the same memory

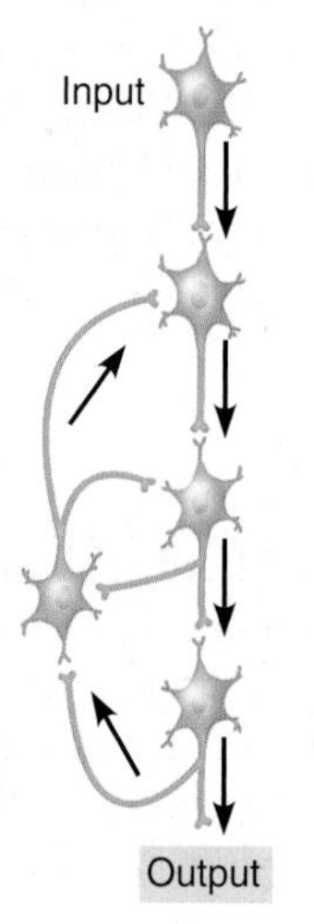

(c) Reverberating circuit
- Signal travels through a chain of neurons, each feeding back to previous neurons
- An *oscillating* circuit
- Controls rhythmic activity
- **Example:** Involved in breathing, sleep-wake cycle, and repetitive motor activities such as walking

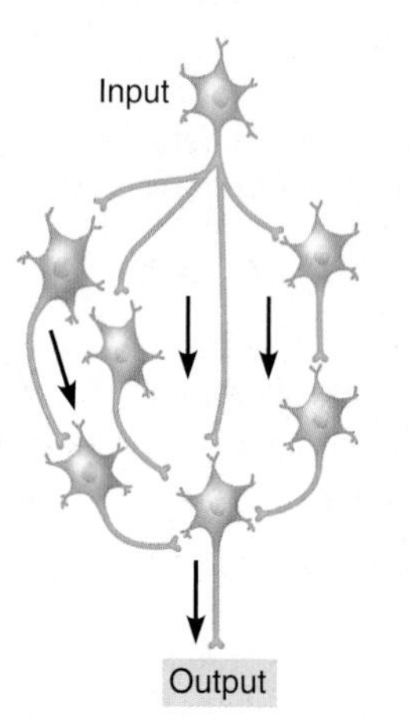

(d) Parallel after-discharge circuit
- Signal stimulates neurons arranged in parallel arrays that eventually converge on a single output cell
- Impulses reach output cell at different times, causing a burst of impulses called an *after-discharge*
- **Example:** May be involved in exacting mental processes such as mathematical calculations

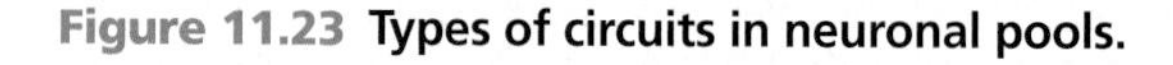

Figure 11.23 **Types of circuits in neuronal pools.**

11

REVIEW QUESTIONS

MAP For more chapter study tools, go to the Study Area of MasteringA&P®.
There you will find:
- Interactive Physiology iP
- A&PFlix A&PFlix
- Interactive Physiology 2.0 iP2
- PhysioEx PEx
- Practice Anatomy Lab PAL
- Videos, Practice Quizzes and Tests, MP3 Tutor Sessions, Case Studies, and much more!

Multiple Choice/Matching

(Some questions have more than one correct answer. Select the best answer or answers from the choices given.)

1. Which of the following structures is not part of the central nervous system? **(a)** the brain, **(b)** a nerve, **(c)** the spinal cord, **(d)** a tract.
2. Match the names of the supporting cells found in column B with the appropriate descriptions in column A.

Column A	Column B
____ **(1)** myelinates nerve fibers in the CNS	**(a)** astrocyte
____ **(2)** lines brain cavities	**(b)** ependymal cell
____ **(3)** myelinates nerve fibers in the PNS	**(c)** microglial cell
____ **(4)** CNS phagocyte	**(d)** oligodendrocyte
____ **(5)** helps regulate the ionic composition of CNS extracellular fluid	**(e)** satellite cell
	(f) Schwann cell

3. What type of current flows through the axolemma during the steep phase of repolarization? **(a)** chiefly a sodium current, **(b)** chiefly a potassium current, **(c)** sodium and potassium currents of approximately the same magnitude.
4. Assume that an EPSP is being generated on the dendritic membrane. Which will occur? **(a)** specific Na^+ channels will open, **(b)** specific K^+ channels will open, **(c)** a single type of channel will open, permitting simultaneous flow of Na^+ and K^+, **(d)** Na^+ channels will open first and then close as K^+ channels open.
5. The velocity of nerve impulse conduction is greatest in **(a)** heavily myelinated, large-diameter fibers, **(b)** myelinated, small-diameter fibers, **(c)** nonmyelinated, small-diameter fibers, **(d)** nonmyelinated, large-diameter fibers.
6. Chemical synapses are characterized by all of the following except **(a)** the release of neurotransmitter by the presynaptic membranes, **(b)** postsynaptic membranes bearing receptors that bind neurotransmitter, **(c)** ions flowing through protein channels from the presynaptic to the postsynaptic neuron, **(d)** a fluid-filled gap separating the neurons.

7. Biogenic amine neurotransmitters include all but **(a)** norepinephrine, **(b)** acetylcholine, **(c)** dopamine, **(d)** serotonin.
8. The neuropeptides that act as natural opiates are **(a)** substance P, **(b)** somatostatin and cholecystokinin, **(c)** tachykinins, **(d)** enkephalins.
9. Inhibition of acetylcholinesterase by poisoning blocks neurotransmission at the neuromuscular junction because **(a)** ACh is no longer released by the presynaptic terminal, **(b)** ACh synthesis in the presynaptic terminal is blocked, **(c)** ACh is not degraded, hence prolonged depolarization is enforced on the postsynaptic cell, **(d)** ACh is blocked from attaching to the postsynaptic ACh receptors.
10. The anatomical region of a multipolar neuron where the AP is initiated is the **(a)** soma, **(b)** dendrites, **(c)** axon hillock, **(d)** distal axon.
11. An IPSP is inhibitory because **(a)** it hyperpolarizes the postsynaptic membrane, **(b)** it reduces the amount of neurotransmitter released by the presynaptic terminal, **(c)** it prevents calcium ion entry into the presynaptic terminal, **(d)** it changes the threshold of the neuron.
12. Identify the neuronal circuits described by choosing the correct response from the key.

    Key: **(a)** converging **(b)** diverging **(c)** parallel after-discharge **(d)** reverberating

    ____ **(1)** Impulses continue around and around the circuit until one neuron stops firing.
    ____ **(2)** One or a few inputs ultimately influence large numbers of neurons.
    ____ **(3)** Many neurons influence a few neurons.
    ____ **(4)** May be involved in exacting types of mental activity.

Short Answer Essay Questions

13. Explain both the anatomical and functional divisions of the nervous system. Include the subdivisions of each.
14. (a) Describe the composition and function of the cell body. (b) How are axons and dendrites alike? In what ways (structurally and functionally) do they differ?
15. (a) What is myelin? (b) How does the myelination process differ in the CNS and PNS?
16. (a) Contrast unipolar, bipolar, and multipolar neurons structurally. (b) Indicate where each is most likely to be found.
17. What is the polarized membrane state? How is it maintained? (Note the relative roles of both passive and active mechanisms.)
18. Describe the events that must occur to generate an AP. Relate the sequence of changes in permeability to changes in the ion channels, and explain why the AP is an all-or-none phenomenon.
19. Since all APs generated by a given nerve fiber have the same magnitude, how does the CNS "know" whether a stimulus is strong or weak?
20. (a) Explain the difference between an EPSP and an IPSP. (b) What specifically determines whether an EPSP or IPSP will be generated at the postsynaptic membrane?
21. Since at any moment a neuron is likely to have thousands of neurons releasing neurotransmitters at its surface, how is neuronal activity (to fire or not to fire) determined?
22. The effects of neurotransmitter binding are very brief. Explain.
23. During a neurobiology lecture, a professor repeatedly refers to group A and group B fibers, absolute refractory period, and myelin sheath gaps. Define these terms.
24. Distinguish between serial and parallel processing.

AT THE CLINIC

Clinical Case Study

Nervous System

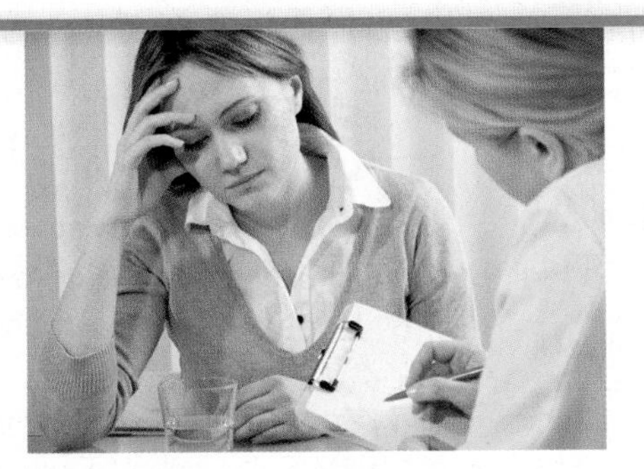

Elaine Sawyer, 35, was on her way to the local elementary school with her three children when the accident on Route 91 occurred. As Mrs. Sawyer swerved to avoid the bus, the right rear corner of her minivan struck the side of the bus, causing the minivan to tip over and slide on its side. Her children were shaken but unhurt. Mrs. Sawyer, however, suffered a severe head injury that caused post-traumatic seizures.

The drugs initially prescribed for her treatment were insufficient to control these seizures. Her doctor additionally prescribed Valium (diazepam), but suggested that she use it only for a month because Valium induces tolerance (loses its effectiveness). After a month of Valium treatment, Mrs. Sawyer no longer had seizures and gradually reduced and eliminated her use of Valium. After being seizure-free for another year, restrictions on her driver's license were lifted.

1. Seizures reflect uncontrolled electrical activity of groups of neurons in the brain. Valium is described as a drug that can "quiet the nerves," which means that it inhibits the ability of neurons to generate electrical signals. What are these electrical signals called, and what is happening at the level of the cell when they are generated?
2. Valium enhances inhibitory postsynaptic potentials (IPSPs). What is an IPSP? How does it affect action potential generation?
3. Valium enhances the natural effects of the neurotransmitter GABA [gamma (γ)-aminobutyric acid]. What chemical class of neurotransmitters does GABA belong to? What are some of the other neurotransmitters that fall into this same class?
4. Theoretically, there are a number of possible ways that a drug such as Valium could act to enhance the action of GABA. What are three such possibilities?
5. Valium actually works postsynaptically to promote binding of GABA to its receptor, thereby enhancing the influx of Cl^- ions into the postsynaptic cell (the natural effect produced by GABA). Why would this effect reduce the likelihood that this cell would be able to produce an electrical signal?

For answers, see Answers Appendix.

12 The Central Nervous System

KEY CONCEPTS

Interconnecting and directing a dizzying number of incoming and outgoing calls, the central switchboard of a telephone system historically seemed an apt comparison for the **central nervous system (CNS)**—the brain and spinal cord. Today, many people compare the CNS to a "cloud" of networked computers. These analogies may explain some workings of the spinal cord, but neither does justice to the fantastic complexity of the human brain. Whether we view the brain as an evolved biological organ, an impressive network of computers, or simply a miracle, it is one of the most amazing things known.

During the course of animal evolution, **cephalization** (sĕ″fah-lĭ-za′shun) has occurred. That is, there has been an elaboration of the *rostral* ("toward the snout"), or anterior, portion of the CNS, along with an increase in the number of neurons in the head. This phenomenon reaches its highest level in the human brain.

In this chapter, we examine the structure of the CNS and the functions associated with its various regions. We also touch on complex integrative functions, such as sleep-wake cycles and memory.

The unimpressive appearance of the human **brain** gives few hints of its remarkable abilities. It is about two good fistfuls of quivering pinkish gray tissue, wrinkled like a walnut, with a consistency somewhat like cold oatmeal. The average adult human brain has a mass of about 1500 g (3.3 lb).

12.1 Folding during development determines the complex structure of the adult brain

→ Learning Objectives

- ☐ **Describe how space constraints affect brain development.**
- ☐ **Name the major regions of the adult brain.**
- ☐ **Name and locate the ventricles of the brain.**

We begin with an introduction to brain embryology, as the terminology used for the structural divisions of the adult brain is easier to follow when you understand brain development.

The brain and spinal cord begin as an embryonic structure called the **neural tube** (**Figure 12.1a**). As soon as the neural tube forms, its anterior (rostral) end begins

Diffusion tensor MRI reveals myelinated fiber tracts in the brain.

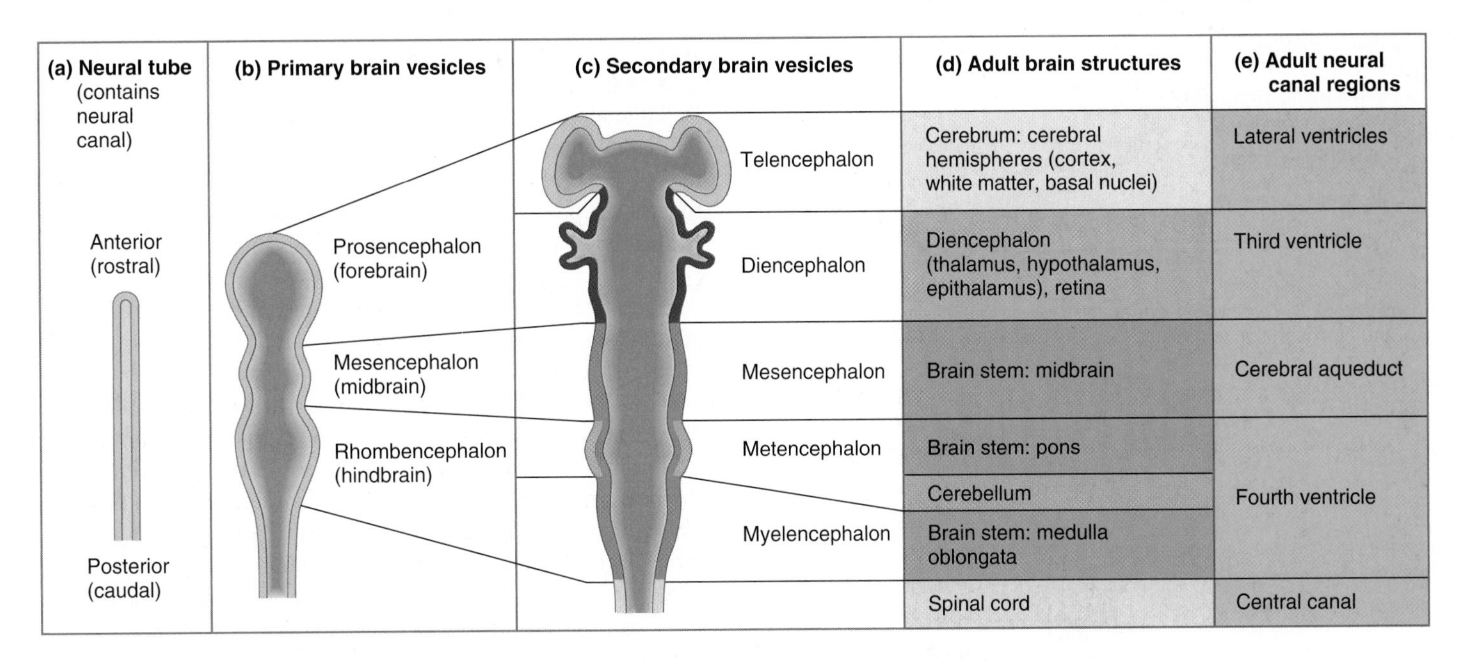

Figure 12.1 Embryonic development of the human brain. (a) Formed by week 4, the neural tube quickly subdivides into **(b)** the primary brain vesicles, which subsequently form **(c)** the secondary brain vesicles by week 5. These five vesicles differentiate into **(d)** the adult brain structures. **(e)** The adult structures derived from the neural canal.

to expand and constrictions appear that mark off the three **primary brain vesicles** (Figure 12.1b):

- **Prosencephalon** (pros″en-sef′ah-lon), or **forebrain**
- **Mesencephalon** (mes″en-sef′ah-lon), or **midbrain**
- **Rhombencephalon** (romb″en-sef′ah-lon), or **hindbrain**

(*Encephalo* means "brain.") The remaining *caudal* ("toward the tail"), or posterior, portion of the neural tube becomes the spinal cord, which we will discuss later in the chapter.

The primary vesicles give rise to the **secondary brain vesicles** (Figure 12.1c). The forebrain divides into the **telencephalon** ("endbrain") and **diencephalon** ("interbrain"), and the hindbrain constricts, forming the **metencephalon** ("afterbrain") and **myelencephalon** ("spinal brain"). The midbrain remains undivided.

Each of the five secondary vesicles then develops rapidly to produce the major structures of the adult brain (Figure 12.1d). The greatest change occurs in the telencephalon, which sprouts two lateral swellings that look like Mickey Mouse's ears. These become the two *cerebral hemispheres*, referred to collectively as the **cerebrum** (ser′ĕ-brum). The diencephalon specializes to form the *hypothalamus* (hi″po-thal′ah-mus), *thalamus*, *epithalamus*, and *retina* of the eye. Less dramatic changes occur in the mesencephalon, metencephalon, and myelencephalon as these regions transform into the *midbrain*, the *pons* and *cerebellum*, and the *medulla oblongata*, respectively. All these midbrain and hindbrain structures, except the cerebellum, form the **brain stem**.

The central cavity of the neural tube remains continuous and enlarges in four areas to form the fluid-filled *ventricles* (*ventr* = little belly) of the brain (Figure 12.1e). We will describe the ventricles shortly.

Because the brain grows more rapidly than the membranous skull that contains it, it folds up to occupy the available space. The *midbrain* and *cervical flexures* move the forebrain toward the brain stem (**Figure 12.2a**). The cerebral hemispheres are forced to take a horseshoe-shaped course and grow posteriorly and laterally (indicated by black arrows in Figure 12.2b). As a result, they grow back over and almost completely envelop the diencephalon and midbrain. By week 26, the continued growth of the cerebral hemispheres causes their surfaces to crease and fold into *convolutions* (Figure 12.2c), which increases their surface area and allows more neurons to occupy the limited space.

Brain Regions and Organization

Some textbooks discuss brain anatomy in terms of the *embryonic scheme* (see Figure 12.1c), but in this text, we will consider the brain in terms of the medical scheme and the four adult brain regions shown in Figure 12.2c: (1) cerebral hemispheres, (2) diencephalon, (3) brain stem (midbrain, pons, and medulla oblongata), and (4) cerebellum.

Gray and white matter have a unique distribution in the brain. The gray matter of the CNS consists of short, nonmyelinated neurons and neuron cell bodies. The white matter is composed of myelinated and nonmyelinated axons. The basic pattern of the CNS is a central cavity surrounded by gray matter, external to which is white matter. As shown in **Figure 12.3**:

① The spinal cord exhibits this basic pattern. This pattern changes with ascent into the brain stem.

② The brain stem has additional gray matter nuclei scattered within the white matter.

③ The cerebral hemispheres and the cerebellum have an outer layer or "bark" of gray matter called a *cortex*.

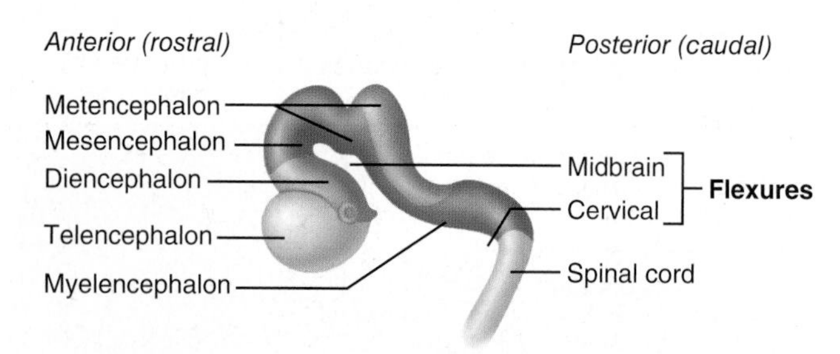

(a) Week 5: Two major flexures form, causing the telencephalon and diencephalon to angle toward the brain stem.

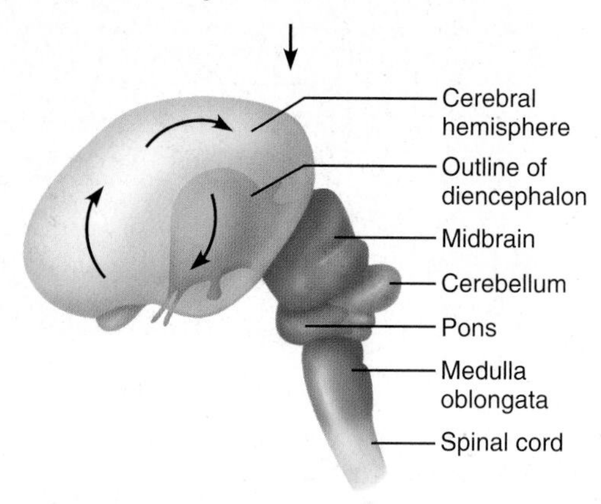

(b) Week 13: Cerebral hemispheres develop and grow posterolaterally to enclose the diencephalon and the rostral brain stem.

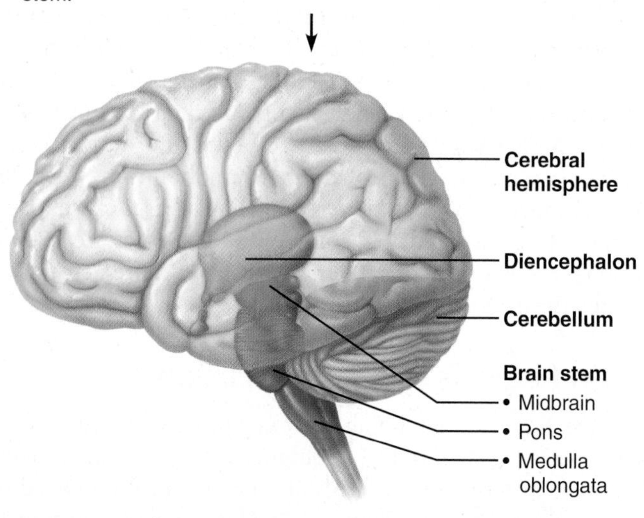

(c) Birth: Shows adult pattern of structures and convolutions.

Figure 12.2 Brain development. Initially, the cerebral surface is smooth. Folding begins in month 6, and convolutions become more obvious as development continues. See-through view in (b) and (c).

Knowledge of the basic pattern of the CNS will help you explore the brain, moving from the most rostral region (cerebrum) to the most caudal (brain stem). But first, let's explore the central hollow cavities that lie deep within the brain—the ventricles.

Ventricles

The brain **ventricles** are continuous with one another and with the central canal of the spinal cord (**Figure 12.4**). The hollow ventricular chambers are filled with cerebrospinal fluid and lined by *ependymal cells*, a type of neuroglia (see Figure 11.4c on p. 348).

The paired **lateral ventricles**, one deep within each cerebral hemisphere, are large C-shaped chambers that reflect the pattern of cerebral growth. Anteriorly, the lateral ventricles lie close together, separated only by a thin median membrane called the **septum pellucidum** (pě-lu′sid-um; "transparent wall"). (See Figure 12.11a, p. 395.)

Each lateral ventricle communicates with the narrow **third ventricle** in the diencephalon via a channel called an **interventricular foramen.**

The third ventricle is continuous with the **fourth ventricle** via the canal-like **cerebral aqueduct** that runs through the midbrain. The fourth ventricle lies in the hindbrain dorsal to the pons and superior medulla. It is continuous with the central canal of the spinal cord inferiorly. Three openings mark the walls of the fourth ventricle: the paired **lateral apertures** in its side walls and the **median aperture** in its roof. These apertures connect the ventricles to the *subarachnoid space* (sub″ah-rak′noid), a fluid-filled space surrounding the brain.

☑ Check Your Understanding

1. Which ventricle is surrounded by the diencephalon?
2. Which two areas of the adult brain have an outside layer of gray matter in addition to central gray matter and surrounding white matter?
3. What is the function of convolutions of the brain?

For answers, see Answers Appendix.

12.2 The cerebral hemispheres consist of cortex, white matter, and the basal nuclei

→ Learning Objectives

- ☐ **List the major lobes, fissures, and functional areas of the cerebral cortex.**
- ☐ **Explain lateralization of cortical function.**
- ☐ **Differentiate between commissures, association fibers, and projection fibers.**
- ☐ **Describe the general function of the basal nuclei (basal ganglia).**

The **cerebral hemispheres** form the superior part of the brain (**Figure 12.5**). The most conspicuous parts of an intact brain, together they account for about 83% of total brain mass. Picture how a mushroom cap covers the top of its stalk, and you have a good idea of how the paired cerebral hemispheres cover and obscure the diencephalon and the top of the brain stem (see Figure 12.2c).

Elevated ridges of tissue called **gyri** (ji′ri; singular: *gyrus;* "twisters")—separated by shallow grooves called **sulci** (sul′ki; singular: *sulcus;* "furrows")—mark nearly the entire surface of the cerebral hemispheres. Deeper grooves, called **fissures**, separate large regions of the brain (Figure 12.5c).

The more prominent gyri and sulci are important anatomical landmarks that are similar in all humans. The median **longitudinal**

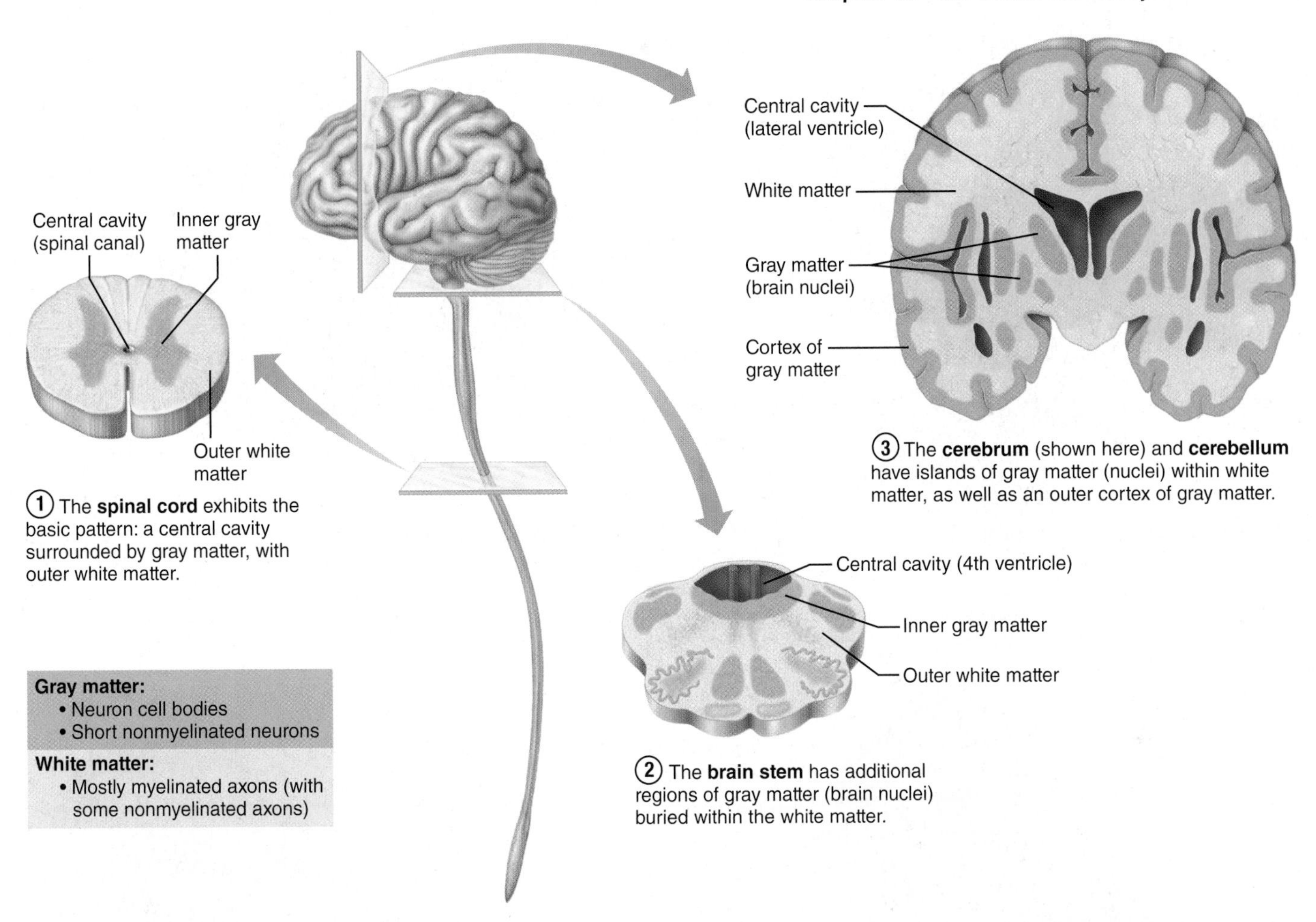

Figure 12.3 Pattern of distribution of gray and white matter in the CNS.

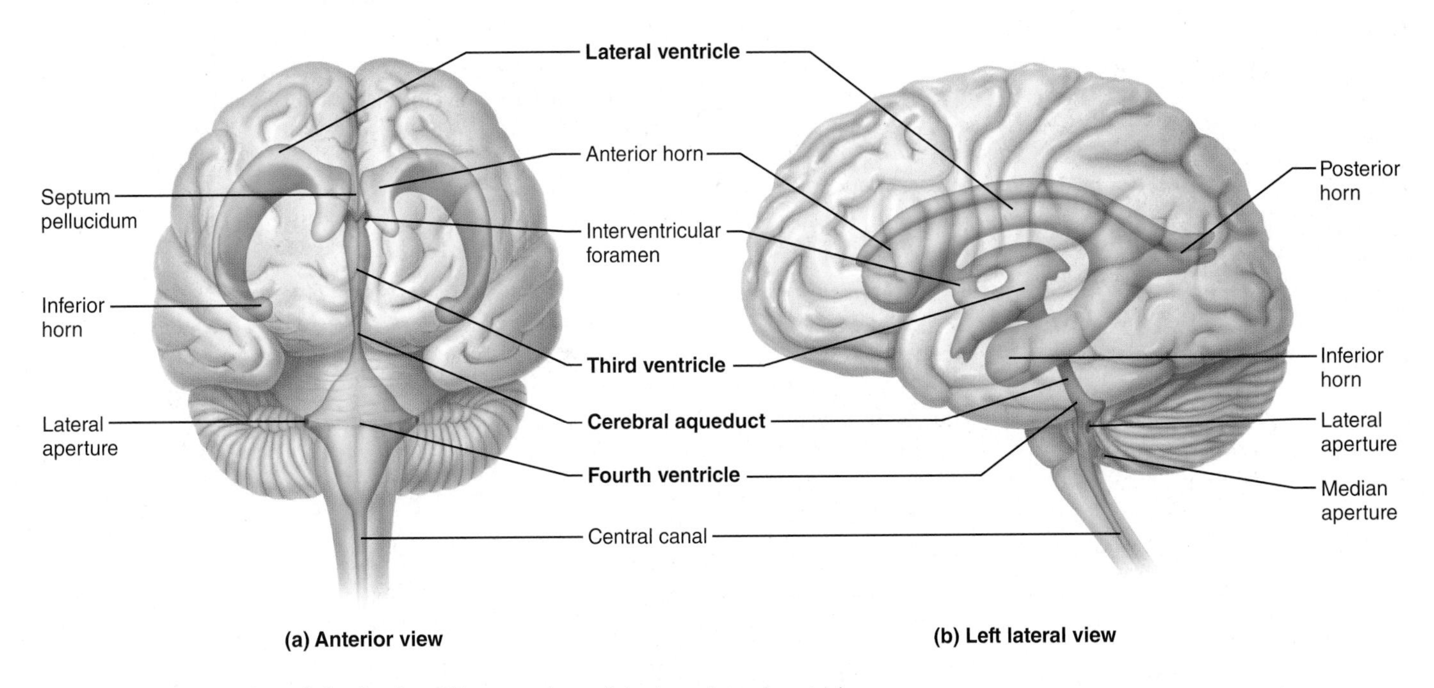

Figure 12.4 Ventricles of the brain. Different regions of the large lateral ventricles are labeled anterior horn, posterior horn, and inferior horn.

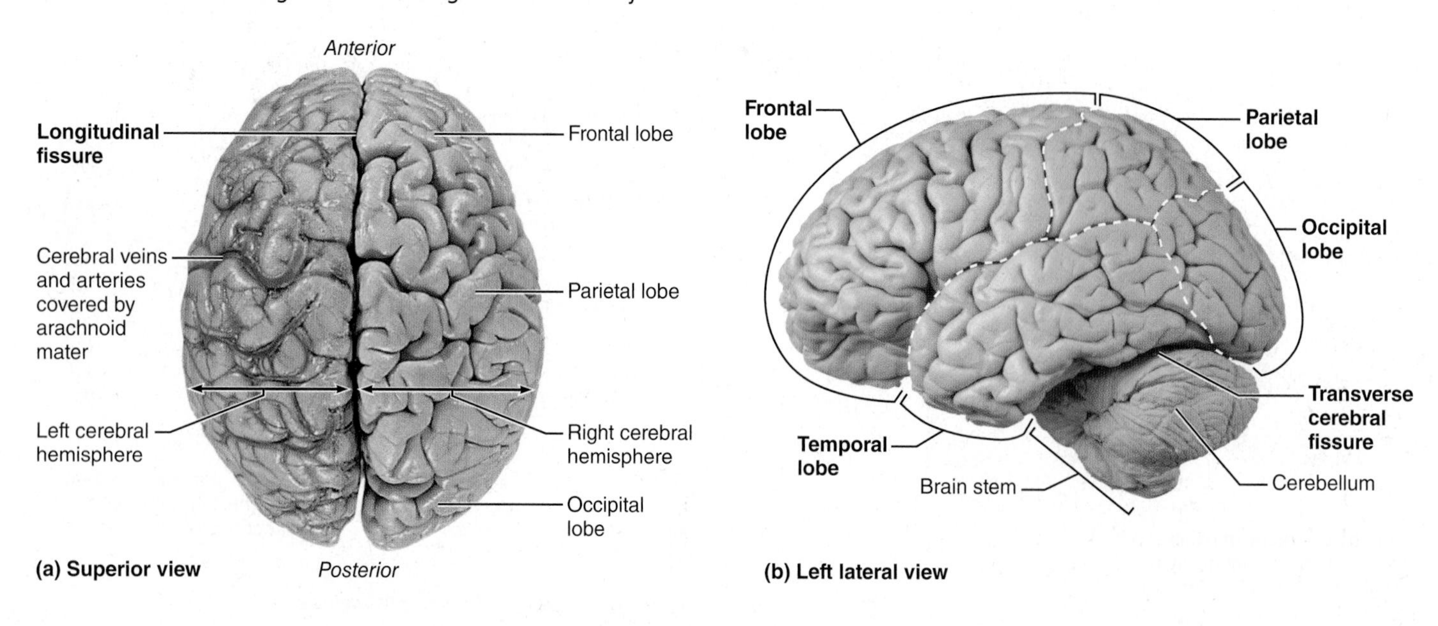

(a) Superior view

(b) Left lateral view

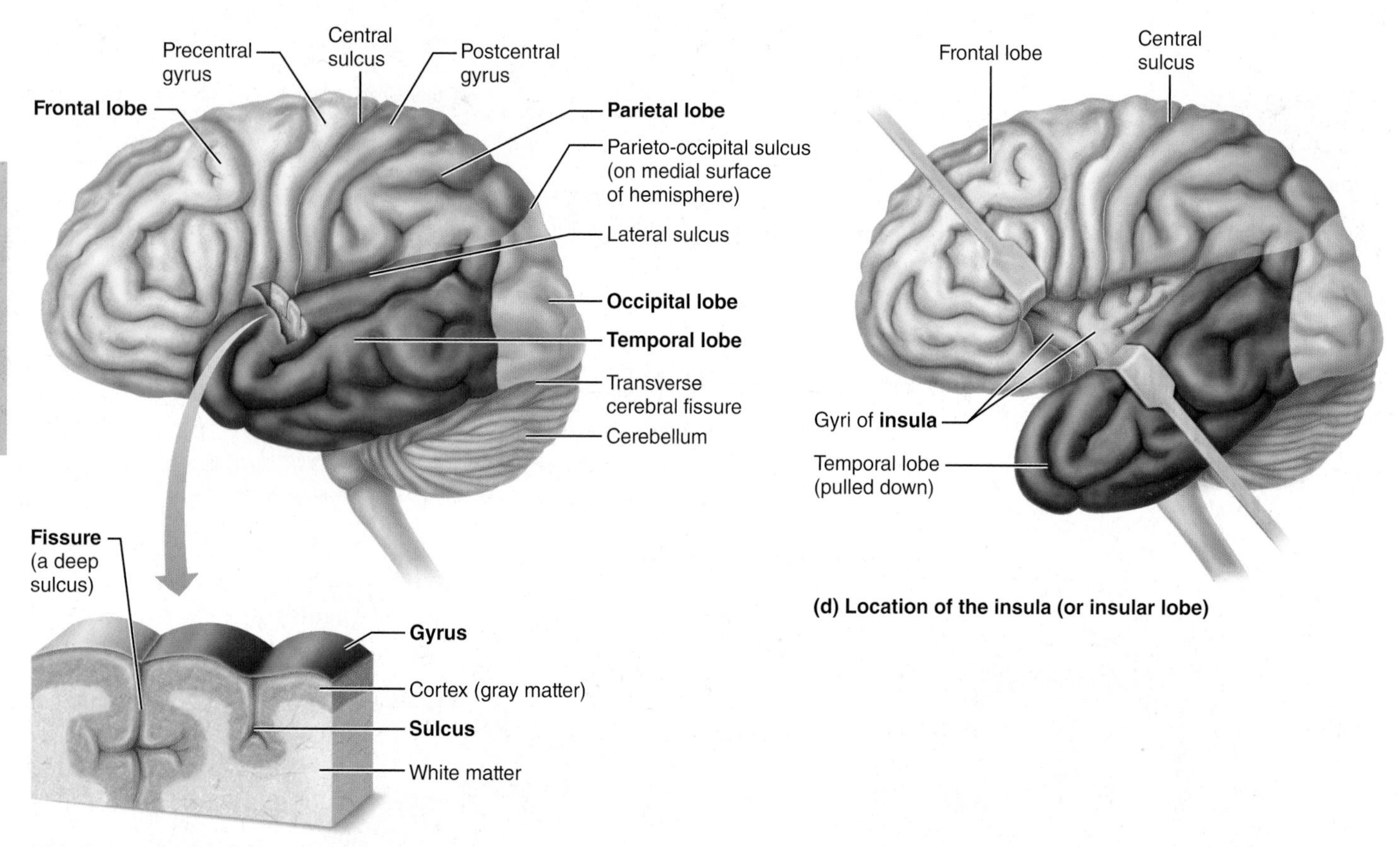

(c) Lobes and sulci of the cerebrum

(d) Location of the insula (or insular lobe)

Figure 12.5 Lobes, sulci, and fissures of the cerebral hemispheres.

Practice art labeling
MasteringA&P®>Study Area>Chapter 12

fissure separates the cerebral hemispheres (Figure 12.5a). Another large fissure, the **transverse cerebral fissure**, separates the cerebral hemispheres from the cerebellum below (Figure 12.5b, c).

Several sulci divide each hemisphere into five lobes—frontal, parietal, temporal, occipital, and insula (Figure 12.5c, d). All but the last are named for the cranial bones that overlie them (see Figure 7.5, pp. 179–180). The **central sulcus**, which lies in the frontal plane, separates the **frontal lobe** from the **parietal lobe**. Bordering the central sulcus are the **precentral gyrus** anteriorly and the **postcentral gyrus** posteriorly. The **parieto-occipital sulcus** (pah-ri″ĕ-to-ok-sip′ĭ-tal), located more posteriorly on the medial surface of the hemisphere, separates the **occipital lobe** from the parietal lobe.

The deep **lateral sulcus** outlines the flaplike **temporal lobe** and separates it from the parietal and frontal lobes. A fifth lobe of the cerebral hemisphere, the **insula** (in′su-lah; "island"), is buried deep within the lateral sulcus and forms part of its floor (Figure 12.5d). The insula is covered by portions of the temporal, parietal, and frontal lobes.

The cerebral hemispheres fit snugly in the skull. The frontal lobes lie in the anterior cranial fossa, and the anterior parts of the temporal lobes fill the middle cranial fossa (see Figure 7.2b, c, p. 176). The posterior cranial fossa, however, houses the brain stem and cerebellum. The occipital lobes are located well superior to that cranial fossa.

Each of the cerebral hemispheres has three basic regions (Figure 12.3 ③):

- A superficial *cerebral cortex* of gray matter, which looks gray in fresh brain tissue
- Internal *white matter*
- *Basal nuclei*, islands of gray matter situated deep within the white matter

We consider these regions next.

Cerebral Cortex

The **cerebral cortex** is the "executive suite" of the nervous system, where our *conscious mind* is found. It enables us to be aware of ourselves and our sensations, to communicate, remember, understand, and initiate voluntary movements.

The cerebral cortex is composed of gray matter: neuron cell bodies, dendrites, associated glia and blood vessels, but no fiber tracts. It contains billions of neurons arranged in six layers. Although only 2–4 mm (about 1/8 inch) thick, it accounts for roughly 40% of total brain mass. Its many convolutions effectively triple its surface area.

Modern imaging techniques allow us to see the brain in action—PET scans show maximal metabolic activity in the brain, and functional MRI scans reveal blood flow (**Figure 12.6**). They show that specific motor and sensory functions are localized in discrete cortical areas called *domains*. However, many higher mental functions, such as memory and language, appear to be spread over large areas of the cortex in overlapping domains.

Before we examine the functional regions of the cerebral cortex, let's consider four generalizations:

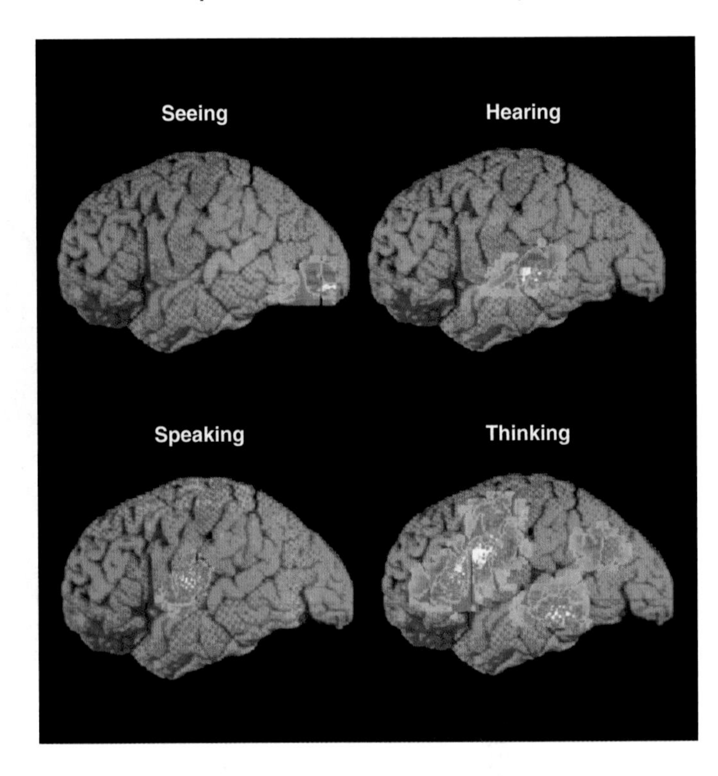

Figure 12.6 Functional neuroimaging (fMRI) of the cerebral cortex. Red and orange areas show increased blood flow in the cortical regions responsible for carrying out each activity.

- The cerebral cortex contains three kinds of functional areas: *motor areas*, *sensory areas*, and *association areas*. As you read about these areas, do not confuse the sensory and motor areas of the cortex with sensory and motor neurons. All neurons in the cortex are interneurons.
- Each hemisphere is chiefly concerned with the sensory and motor functions of the **contralateral** (opposite) side of the body.
- Although largely symmetrical in structure, the two hemispheres are not entirely equal in function. Instead, there is lateralization (specialization) of cortical functions.
- And finally, keep in mind that our approach is a gross oversimplification. No functional area of the cortex acts alone, and conscious behavior involves the entire cortex in one way or another.

Motor Areas

The following **motor areas** of the cortex, which control voluntary movement, lie in the posterior part of the frontal lobes: primary motor cortex, premotor cortex, Broca's area, and the frontal eye field (**Figure 12.7a**, dark and light red areas).

Primary Motor Cortex The **primary (somatic) motor cortex** is located in the precentral gyrus of the frontal lobe of each hemisphere (Figure 12.7, dark red area). Large neurons, called **pyramidal cells**, in these gyri allow us to consciously control the precise or skilled voluntary movements of our skeletal muscles. Their long axons, which project to the spinal cord, form the massive voluntary motor tracts called *pyramidal tracts* or

- **Various other centers.** Additional centers regulate such activities as vomiting, hiccuping, swallowing, coughing, and sneezing.

Notice that many functions listed above are also attributed to the hypothalamus (p. 397). The overlap is easily explained. The hypothalamus controls many visceral functions by relaying its instructions through medullary reticular centers, which carry them out.

☑ Check Your Understanding

10. What are the pyramids of the medulla? What is the result of decussation of the pyramids?

11. Which region of the brain stem is associated with the cerebral peduncles and the superior and inferior colliculi?

For answers, see Answers Appendix.

12.5 The cerebellum adjusts motor output, ensuring coordination and balance

→ Learning Objective

☐ **Describe the structure and function of the cerebellum.**

The cauliflower-like **cerebellum** (ser″ĕ-bel′um; "small brain"), exceeded in size only by the cerebrum, accounts for about 11% of total brain mass. The cerebellum is located dorsal to the pons and medulla (and to the intervening fourth ventricle). It protrudes under the occipital lobes of the cerebral hemispheres, from which it is separated by the transverse cerebral fissure (see Figure 12.5b).

By processing inputs received from the cerebral motor cortex, various brain stem nuclei, and sensory receptors, the cerebellum provides the precise timing and appropriate patterns of skeletal muscle contraction for smooth, coordinated movements and agility needed for our daily living—driving, typing, and for some of us, playing the tuba. Cerebellar activity occurs subconsciously—we have no awareness of it.

Cerebellar Anatomy

The cerebellum is bilaterally symmetrical. The wormlike **vermis** connects its two apple-sized **cerebellar hemispheres** medially (**Figure 12.16**). Its surface is heavily convoluted, with fine, transversely oriented pleatlike gyri known as **folia** ("leaves"). Deep fissures subdivide each hemisphere into **anterior**, **posterior**, and **flocculonodular lobes** (flok″u-lo-nod′u-lar). The small propeller-shaped flocculonodular lobes, situated deep to the vermis and posterior lobe, cannot be seen in a surface view.

Like the cerebrum, the cerebellum has a thin outer cortex of gray matter, internal white matter, and small, deeply situated, paired masses of gray matter, the most familiar of which are the *dentate nuclei*. Several types of neurons populate the cerebellar cortex, including **Purkinje cells** (see Table 11.1 on p. 353). These large cells, with their extensively branched dendrites, are the only cortical neurons that send axons through the white matter to synapse with the central nuclei of the cerebellum. The distinctive pattern of white matter in the cerebellum resembles a branching tree, a pattern fancifully called the **arbor vitae** (ar′bor vi′te; "tree of life") (Figure 12.16a, b).

The anterior and posterior lobes of the cerebellum, which coordinate body movements, have three sensory maps of the entire body as indicated by the homunculi in Figure 12.16d. The part of the cerebellar cortex that receives sensory input from a body region influences motor output to that region. The medial portions influence the motor activities of the trunk and girdle muscles. The intermediate parts of each hemisphere influence the distal parts of the limbs and skilled movements. The lateral-most parts of each hemisphere integrate information from the association areas of the cerebral cortex and appear to play a role in planning movements rather than executing them. The flocculonodular lobes receive inputs from the equilibrium apparatus of the inner ears, and adjust posture to maintain balance.

Cerebellar Peduncles

As noted earlier, three paired fiber tracts—the cerebellar peduncles—connect the cerebellum to the brain stem (see Figures 12.13 and 12.16b). Unlike the contralateral fiber distribution to and from the cerebral cortex, virtually all fibers entering and leaving the cerebellum are **ipsilateral** (*ipsi* = same)—from and to the same side of the body.

- The **superior cerebellar peduncles** connecting cerebellum and midbrain carry instructions from neurons in the deep cerebellar nuclei to the cerebral motor cortex via thalamic relays. Like the basal nuclei, the cerebellum has no *direct* connections to the cerebral cortex.
- The **middle cerebellar peduncles** carry one-way communications from the pons to the cerebellum, advising the cerebellum of voluntary motor activities initiated by the motor cortex (via relays in the pontine nuclei).
- The **inferior cerebellar peduncles** connect medulla and cerebellum. These peduncles convey sensory information to the cerebellum from (1) muscle proprioceptors throughout the body, and (2) the vestibular nuclei of the brain stem, which are concerned with equilibrium and balance.

Cerebellar Processing

Cerebellar processing fine-tunes motor activity as follows:

1. The motor areas of the cerebral cortex, via relay nuclei in the brain stem, notify the cerebellum of their intent to initiate voluntary muscle contractions.
2. At the same time, the cerebellum receives information from proprioceptors throughout the body (regarding tension in the muscles and tendons, and joint position) and from visual and equilibrium pathways. This information enables the cerebellum to evaluate body position and momentum—where the body is and where it is going.
3. The cerebellar cortex calculates the best way to coordinate the force, direction, and extent of muscle contraction to prevent overshoot, maintain posture, and ensure smooth, coordinated movements.

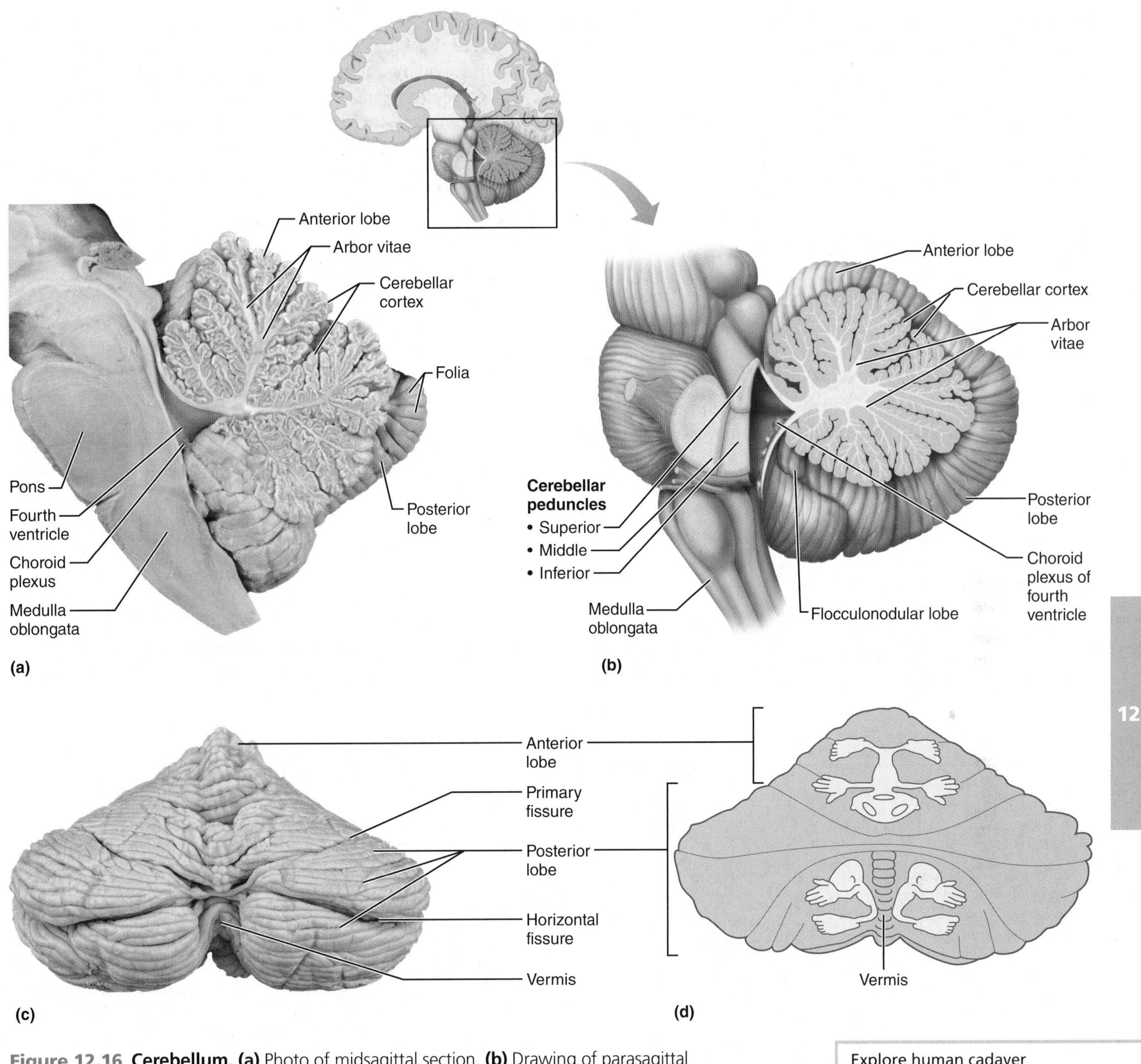

Figure 12.16 Cerebellum. (a) Photo of midsagittal section. **(b)** Drawing of parasagittal section. **(c)** Photograph of the posterior view of the cerebellum. **(d)** Three body maps of the cerebellar cortex (in the form of homunculi).

Explore human cadaver
MasteringA&P®>Study Area>PAL

4. Then, via the superior peduncles, the cerebellum dispatches to the cerebral motor cortex its "blueprint" for coordinating movement. Cerebellar fibers also send information to brain stem nuclei, which in turn influence motor neurons of the spinal cord.

Just as an automatic pilot compares a plane's instrument readings with the planned course, the cerebellum continually compares the body's performance with the higher brain's intention and sends out messages to initiate appropriate corrective measures. Cerebellar injury results in loss of muscle tone and clumsy, uncertain movements.

Cognitive Functions of the Cerebellum

Neuroanatomy, imaging studies, and observations of patients with cerebellar injuries suggest that the cerebellum also plays a role in thinking, language, and emotion. As in the motor system, the cerebellum may compare the actual output of these systems with the expected output and adjust accordingly. Much still remains to be discovered about the precise role of the cerebellum in nonmotor functions.

☑ Check Your Understanding

12. In what ways are the cerebellum and the cerebrum similar? In what ways are they different?

For answers, see Answers Appendix.

12.6 Functional brain systems span multiple brain structures

→ Learning Objective

☐ **Locate the limbic system and the reticular formation, and explain the role of each functional system.**

Functional brain systems are networks of neurons that work together but span relatively large distances in the brain, so they cannot be localized to specific regions. The *limbic system* and the *reticular formation* are excellent examples. Table 12.1 on pp. 406–407 summarizes their functions, as well as those of the cerebral hemispheres, diencephalon, brain stem, and cerebellum.

The Limbic System

The **limbic system** is a group of structures located on the medial aspect of each cerebral hemisphere and diencephalon. Its cerebral structures encircle (*limbus* = ring) the upper part of the brain stem (Figure 12.17). The limbic system includes the **amygdaloid body** (ah-mig′dah-loid), an almond-shaped nucleus that sits on the tail of the caudate nucleus, and other parts of the rhinencephalon (*cingulate gyrus, septal nuclei,* the C-shaped *hippocampus, dentate gyrus,* and *parahippocampal gyrus*). In the diencephalon, the main limbic structures are the *hypothalamus* and the *anterior thalamic nuclei.* The **fornix** ("arch") and other fiber tracts link these limbic system regions together.

The limbic system is our *emotional,* or *affective* (feelings), *brain.* The amygdaloid body and the anterior part of the **cingulate gyrus** seem especially important in emotions. The amygdaloid body is critical for responding to perceived threats (such as angry or fearful facial expressions) with fear or aggression. The cingulate gyrus plays a role in expressing our emotions through gestures and in resolving mental conflicts when we are frustrated.

Odors often trigger emotional reactions and memories. These responses reflect the origin of much of the limbic system in the primitive "smell brain" (rhinencephalon). Our reactions to odors are rarely neutral (a skunk smells *bad* and repulses us), and odors often recall memories of emotion-laden events.

Extensive connections between the limbic system and lower and higher brain regions allow the system to integrate and respond to a variety of environmental stimuli. Most limbic system output is relayed through the hypothalamus. Because the hypothalamus is the neural clearinghouse for both autonomic (visceral) function and emotional response, it is not surprising

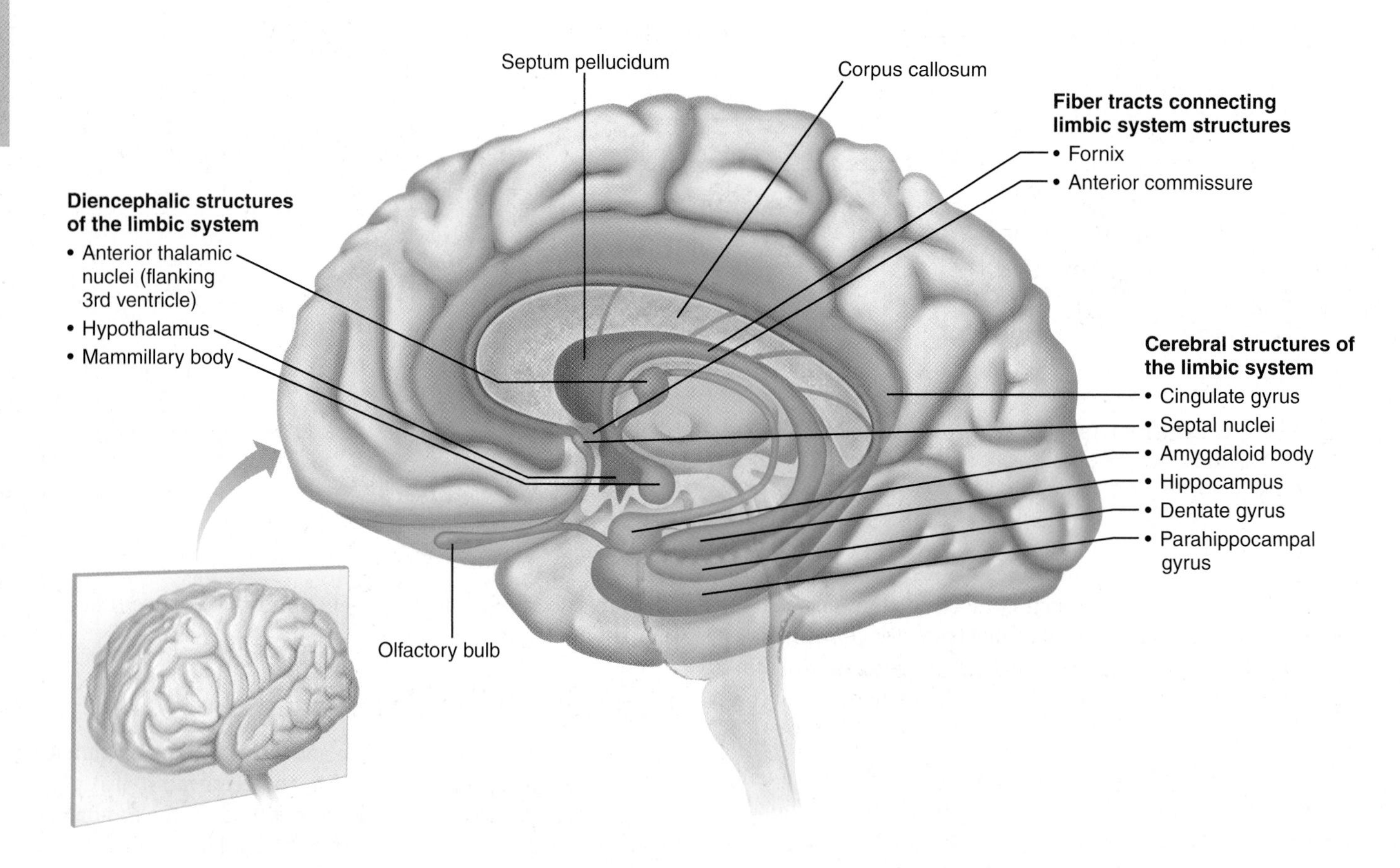

Figure 12.17 The limbic system. Medial view of the cerebrum and diencephalon illustrating the structures of the limbic system, the emotional-visceral brain.

that some people under acute or unrelenting emotional stress fall prey to visceral illnesses, such as high blood pressure and heartburn. Disorders with physical symptoms that originate from emotional causes are known as **psychosomatic illnesses**.

Because the limbic system interacts with the prefrontal lobes, there is an intimate relationship between our feelings (mediated by the emotional brain) and our thoughts (mediated by the cognitive brain). As a result, we (1) react emotionally to things we consciously understand to be happening, and (2) are consciously aware of the emotional richness of our lives. Communication between the cerebral cortex and limbic system explains why emotions sometimes override logic and, conversely, why reason can stop us from expressing our emotions inappropriately. Particular limbic system structures—the **hippocampus** and amygdaloid body—also play a role in memory.

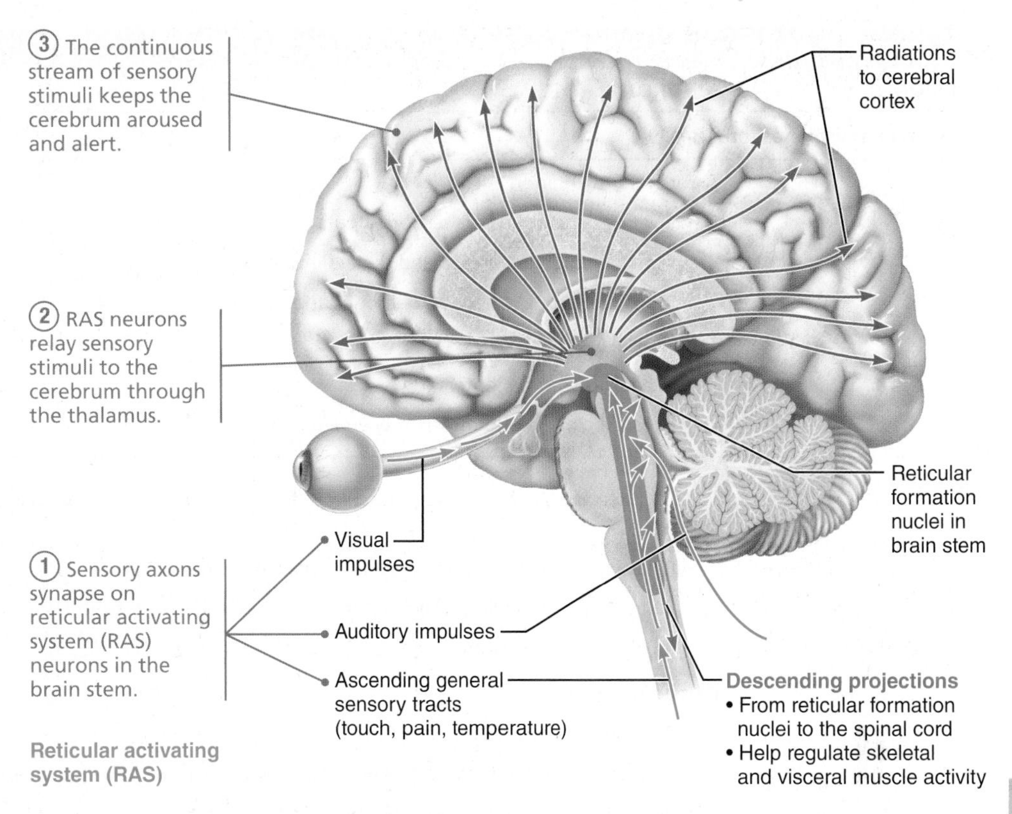

Figure 12.18 The reticular formation. This functional brain system (purple) extends the length of the brain stem. Part of this formation, the reticular activating system (RAS), maintains alert wakefulness of the cerebral cortex.

The Reticular Formation

The **reticular formation** extends through the central core of the medulla oblongata, pons, and midbrain (**Figure 12.18**). It is composed of loosely clustered neurons in what is otherwise white matter. These neurons form three broad columns along the length of the brain stem (Figure 12.15c): (1) the midline **raphe nuclei** (ra′fe; *raphe* = seam or crease), which are flanked laterally by (2) the **medial (large cell) group of nuclei** and (3) the **lateral (small cell) group of nuclei**.

The outstanding feature of the reticular neurons is their far-flung axonal connections. Individual reticular neurons project to the hypothalamus, thalamus, cerebral cortex, cerebellum, and spinal cord, making reticular neurons ideal for governing the arousal of the brain as a whole.

For example, unless inhibited by other brain areas, the neurons of the part of the reticular formation known as the **reticular activating system (RAS)** send a continuous stream of impulses to the cerebral cortex, keeping the cortex alert and conscious and enhancing its excitability. Impulses from all the great ascending sensory tracts synapse with RAS neurons, keeping them active and enhancing their arousing effect on the cerebrum. (This may explain why some students, stimulated by a bustling environment, like to study in a busy coffeehouse.)

The RAS also filters this flood of sensory inputs. Repetitive, familiar, or weak signals are filtered out, but unusual, significant, or strong impulses do reach consciousness. For example, you are probably unaware of your watch encircling your wrist, but would immediately notice it if the clasp broke. Between them, the RAS and the cerebral cortex disregard perhaps 99% of all sensory stimuli as unimportant. If this filtering did not occur, the sensory overload would drive us crazy. The drug LSD interferes with these sensory dampers, promoting an often overwhelming sensory overload.

- Take a moment to become aware of all the stimuli in your environment. Notice all the colors, shapes, odors, sounds, and so on. How many of these sensory stimuli are you usually aware of?

The RAS is inhibited by sleep centers located in the hypothalamus and other neural regions, and is depressed by alcohol, sleep-inducing drugs, and tranquilizers. Severe injury to this system, as might follow a knockout punch that twists the brain stem, results in permanent unconsciousness (irreversible *coma*). Although the RAS is central to wakefulness, some of its nuclei are also involved in sleep, which we will discuss later in this chapter.

The reticular formation also has a *motor* arm. Some of its motor nuclei project to motor neurons in the spinal cord via the *reticulospinal tracts*, and help control skeletal muscles during coarse limb movements. Other reticular motor nuclei, such as the vasomotor, cardiac, and respiratory centers of the medulla, are autonomic centers that regulate visceral motor functions.

Table 12.1 Functions of Major Brain Regions

REGION	FUNCTION
Cerebral Hemispheres (pp. 384–394)	
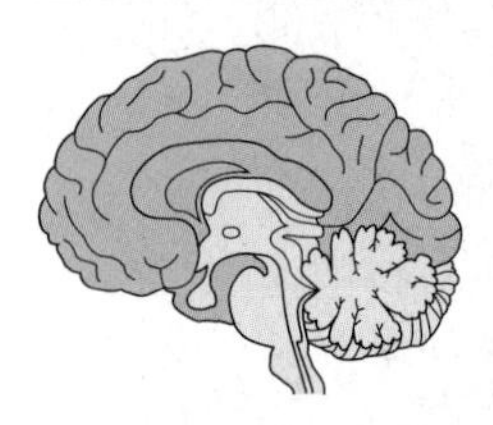	**Cortical gray matter:** • Localizes and interprets sensory inputs • Controls voluntary and skilled skeletal muscle activity • Functions in intellectual and emotional processing **Basal nuclei (ganglia):** • Subcortical motor centers • Help control skeletal muscle movements
Diencephalon (pp. 395–397)	
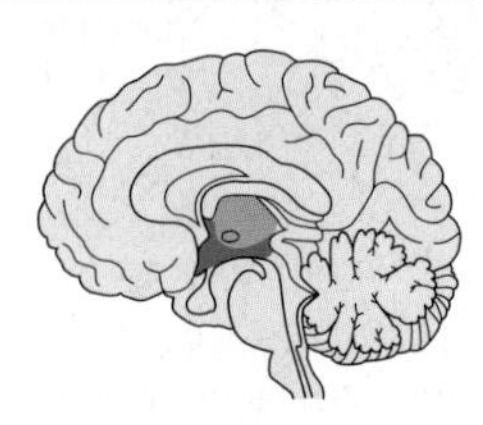	**Thalamus:** • Relays sensory impulses to cerebral cortex for interpretation • Relays impulses between cerebral motor cortex and lower (subcortical) motor centers, including cerebellum • Involved in memory processing **Hypothalamus:** • Chief integration center of autonomic (involuntary) nervous system • Regulates body temperature, food intake, water balance, thirst, and biological rhythms and drives • Regulates hormonal output of anterior pituitary gland • Acts as an endocrine organ, producing posterior pituitary hormones ADH and oxytocin
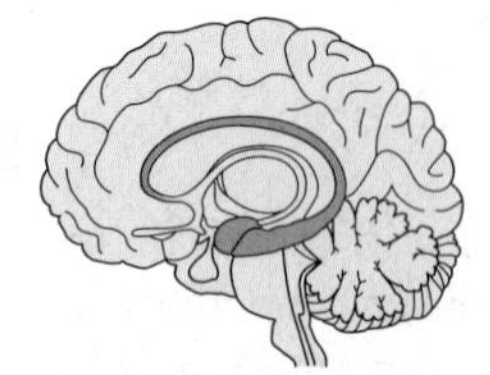	**Limbic system** (pp. 404–405)—A functional system: • Includes cerebral and diencephalon structures (e.g., hypothalamus and anterior thalamic nuclei) • Mediates emotional response • Involved in memory processing

☑ Check Your Understanding

13. The limbic system is sometimes called the emotional-visceral brain. Which part of the limbic system is responsible for the visceral connection?

14. When Taylor begins to feel drowsy while driving, she opens her window, turns up the volume of the car stereo, and sips ice-cold water. How do these actions keep her awake?

For answers, see Answers Appendix.

12.7 The interconnected structures of the brain allow higher mental functions

→ Learning Objectives

- ☐ **Identify the brain areas involved in language and memory.**
- ☐ **Identify factors affecting the formation of long-term memories.**
- ☐ **Define EEG and distinguish between alpha, beta, theta, and delta brain waves.**
- ☐ **Describe consciousness clinically.**
- ☐ **Compare and contrast the events and importance of slow-wave and REM sleep.**

During the last four decades, an exciting exploration of our "inner space," or what we commonly call the *mind*, has been going on. But researchers in the field of cognition are still struggling to understand how the mind's qualities spring from living tissue and electrical impulses. Souls and synapses are hard to reconcile!

Our ability to discuss "souls and synapses" is one of the things that makes us human. We will begin by looking at the mental functions of language and memory that give us this ability. Then we will explore brain waves, which reflect the underlying electrical activity of mental functions, followed by the related topics of consciousness and sleep.

Language

Language is such an important function of the brain that it involves practically all of the association cortex on the left side in one way or another. Pioneering studies of patients with *aphasias* (the loss of language abilities due to damage to specific areas of the brain)

Table 12.1 *(continued)*

REGION	FUNCTION
Brain Stem (pp. 398–402)	
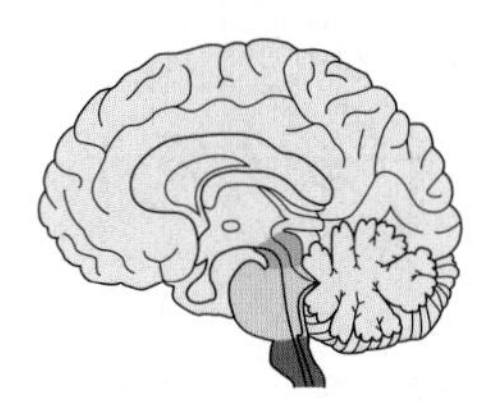	**Midbrain:** • Contains visual (superior colliculi) and auditory (inferior colliculi) reflex centers • Contains subcortical motor centers (substantia nigra and red nuclei) • Contains nuclei for cranial nerves III and IV • Contains projection fibers (e.g., fibers of the pyramidal tracts)
	Pons: • Relays information from the cerebrum to the cerebellum • Cooperates with the medullary respiratory centers to control respiratory rate and depth • Contains nuclei of cranial nerves V–VII • Contains projection fibers
	Medulla oblongata: • Relays ascending sensory pathway impulses from skin and proprioceptors through nuclei cuneatus and gracilis • Contains visceral nuclei controlling heart rate, blood vessel diameter, respiratory rate, vomiting, coughing, etc. • Relays sensory information to the cerebellum through inferior olivary nuclei • Contains nuclei of cranial nerves VIII–X and XII • Contains projection fibers • Site of decussation of pyramids
	Reticular formation (p. 405)—A functional system: • Maintains cerebral cortical alertness (reticular activating system) • Filters out repetitive stimuli • Helps regulate skeletal and visceral muscle activity
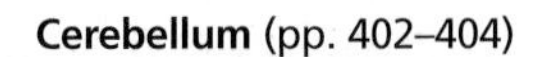 **Cerebellum** (pp. 402–404)	
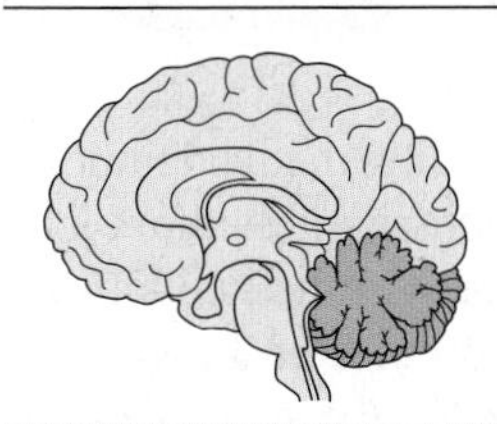	**Cerebellum:** • Processes information from cerebral motor cortex, proprioceptors, and visual and equilibrium pathways • Provides "instructions" to cerebral motor cortex and subcortical motor centers, resulting in smooth, coordinated skeletal muscle movements • Responsible for balance and posture

pointed to two critically important regions, Broca's area and Wernicke's area (see areas outlined by dashes in Figure 12.7a).

Patients with lesions involving **Broca's area** can understand language but have difficulty speaking (and sometimes cannot write or type or use sign language). On the other hand, patients with lesions involving **Wernicke's area** are able to speak but produce a type of nonsense often referred to as "word salad." They also have great difficulty understanding language.

In fact, this picture is clinically useful, but oversimplified. Broca's and Wernicke's areas together with the basal nuclei form a single language implementation system that analyzes incoming and produces outgoing word sounds and grammatical structures. A surrounding set of cortical areas forms a bridge between this system and the regions of cortex that hold concepts and ideas, which are distributed throughout the remainder of the association cortices.

The corresponding areas in the right or non-language-dominant hemisphere are involved in "body language"—the nonverbal emotional components of language. These areas allow the lilt or tone of our voice and our gestures to express our emotions when we speak, and permit us to comprehend the emotional content of what we hear. For example, a soft, melodious response to your question conveys quite a different meaning than a sharp reply.

Memory

Memory is the storage and retrieval of information. Memories are essential for learning and incorporating our experiences into behavior and are part and parcel of our consciousness. Stored somewhere in your 3 pounds of wrinkled brain are zip codes, the face of your grandfather, and the taste of yesterday's pizza. Your memories reflect your lifetime.

There are different kinds of memory: *Declarative (fact) memory* (names, faces, words, and dates), *procedural (skills) memory* (piano playing), *motor memory* (riding a bike), and *emotional memory* (your pounding heart when you hear a rattlesnake nearby). Let's focus on declarative memory.

Declarative memory storage involves two distinct stages: short-term memory and long-term memory (**Figure 12.19**).

Short-term memory (STM), also called *working memory*, is the preliminary step, as well as the power that lets you look up a telephone number, dial it, and then never think of it again. STM is limited to seven or eight chunks of information, such as the digits of a telephone number or the sequence of words in an elaborate sentence.

In contrast, **long-term memory (LTM)** seems to have a limitless capacity. Although our STM cannot recall numbers much longer than a telephone number, we can remember scores of telephone numbers by committing them to LTM. However, long-term memories can be forgotten, and so our memory bank continually changes with time. Furthermore, our ability to store and retrieve information declines with aging.

We do not remember or even consciously notice much of what is going on around us. As sensory inputs flood into our cerebral cortex, they are processed (yellow box in Figure 12.19). Some 5% of this information is selected for transfer to STM (light green box). STM serves as a temporary holding bin for data that we may or may not want to retain.

Information is then transferred from STM to LTM (dark green box). Many factors can influence this transfer, including:

- **Emotional state.** We learn best when we are alert, motivated, surprised, and aroused. For example, when we witness shocking events, transferral is almost immediate. Norepinephrine, a neurotransmitter involved in memory processing of emotionally charged events, is released when we are excited or "stressed out," which helps to explain this phenomenon.
- **Rehearsal.** Rehearsing or repeating the material enhances memory.
- **Association.** Tying "new" information to "old" information already stored in LTM appears to be important in remembering facts.
- **Automatic memory.** Not all impressions that become part of LTM are consciously formed. A student concentrating on a lecturer's speech may record an automatic memory of the pattern of the lecturer's tie.

Memories transferred to LTM take time to become permanent. The process of **memory consolidation** apparently involves fitting new facts into the categories of knowledge already stored in the cerebral cortex. The hippocampus and surrounding temporal cortical areas play a major role in memory consolidation by communicating with the thalamus and the prefrontal cortex.

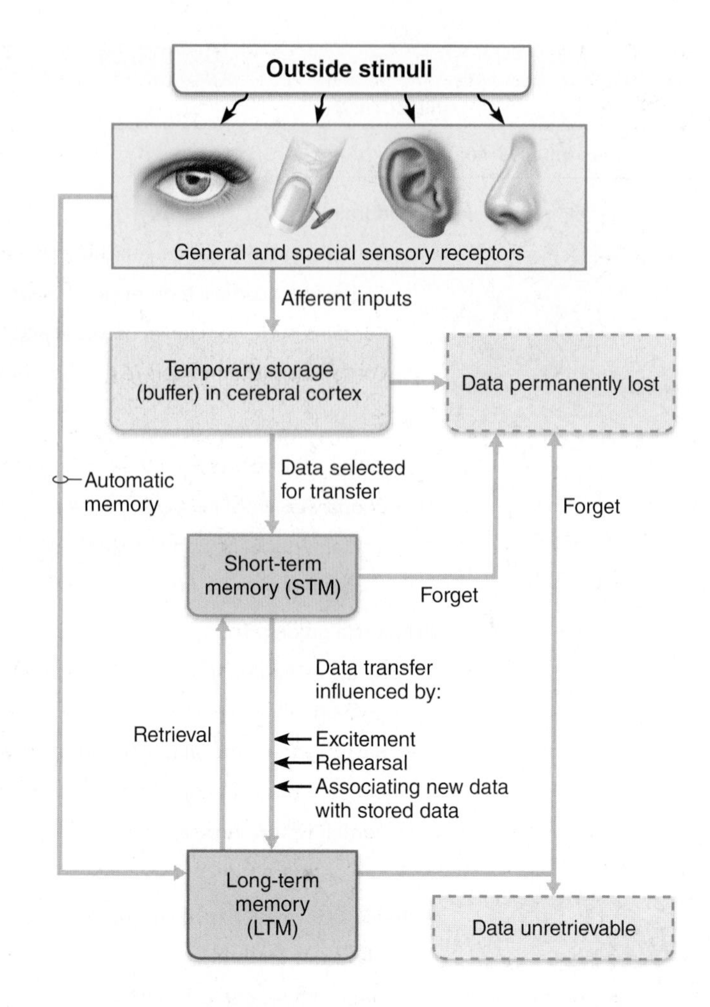

Figure 12.19 Memory processing.

Damage to the hippocampus and surrounding medial temporal lobe structures on either side results in only slight memory loss, but bilateral destruction causes widespread amnesia. Consolidated memories are not lost, but new sensory inputs cannot be associated with old, and the person lives in the here and now from that point on. This condition is called *anterograde amnesia* (an′ter-o-grād″), in contrast to *retrograde amnesia*, which is the loss of memories formed in the distant past. You could carry on an animated conversation with a person with anterograde amnesia, excuse yourself, return five minutes later, and that person would not remember you. +

Specific pieces of each memory are thought to be stored near regions of the brain that need them so new inputs can be quickly associated with the old. Accordingly, visual memories are stored in the occipital cortex, memories of music in the temporal cortex, and so on.

Brain Wave Patterns and the EEG

Brain waves reflect the electrical activity on which higher mental functions are based because normal brain function involves continuous electrical activity of neurons. An **electroencephalogram** (e-lek″tro-en-sef′ah-lo-gram), or **EEG**, records some aspects of this activity. EEGs are used for diagnosing epilepsy and sleep disorders, in research on brain function, and to determine brain death.

An EEG is made by placing electrodes on the scalp and connecting the electrodes to an apparatus that measures voltage differences between various cortical areas (**Figure 12.20a**). The patterns of neuronal electrical activity recorded, called **brain waves**, are generated by synaptic activity at the surface of the cortex, rather than by action potentials in the white matter.

Each of us has a brain wave pattern that is as unique as our fingerprints. For simplicity, however, we can group brain waves into the four frequency classes shown in Figure 12.20b. Each wave is a continuous train of peaks and troughs, and the wave frequency, expressed in hertz (Hz), is the number of peaks in one second. A frequency of 1 Hz means that one peak occurs each second.

The amplitude or intensity of any wave is represented by how high the wave peaks rise and how low the troughs dip. The amplitude of brain waves reflects the synchronous activity of many neurons and not the degree of electrical activity of individual neurons. Usually, brain waves are complex and low amplitude. During some stages of sleep, neurons tend to fire synchronously, producing similar, high-amplitude brain waves.

- **Alpha waves** (8–13 Hz) are relatively regular and rhythmic, low-amplitude, synchronous waves. In most cases, they indicate a brain that is "idling"—a calm, relaxed state of wakefulness.
- **Beta waves** (14–30 Hz) are also rhythmic, but less regular than alpha waves and with a higher frequency. Beta waves occur when we are mentally alert, as when concentrating on some problem or visual stimulus.
- **Theta waves** (4–7 Hz) are still more irregular. Though common in children, theta waves are uncommon in awake adults but may appear when concentrating.
- **Delta waves** (4 Hz or less) are high-amplitude waves seen during deep sleep and when the reticular activating system is suppressed, such as during anesthesia. In awake adults, they indicate brain damage.

Brain waves whose frequency is too high or too low suggest problems with cerebral cortical functions, and unconsciousness occurs at both extremes. Because spontaneous brain waves are always present, even during unconsciousness and coma, their absence—a "flat EEG"—is clinical evidence of *brain death*.

HOMEOSTATIC IMBALANCE 12.6

CLINICAL

Almost without warning, a person with *epilepsy* may lose consciousness and fall stiffly to the ground, wracked by uncontrollable jerking. These **epileptic seizures** reflect a torrent of electrical discharges by groups of brain neurons, and during their uncontrolled activity no other messages can get through.

Epilepsy, manifested by one out of 100 of us, is not associated with, nor does it cause, intellectual impairment. Genetic factors induce some cases, but epilepsy can also result from brain injuries caused by blows to the head, stroke, infections, or tumors.

Epileptic seizures vary tremendously in their expression and severity.

- *Absence seizures*, formerly known as *petit mal*, are mild forms in which the expression goes blank for a few seconds as consciousness disappears. These are typically seen in young children and usually disappear by age 10.

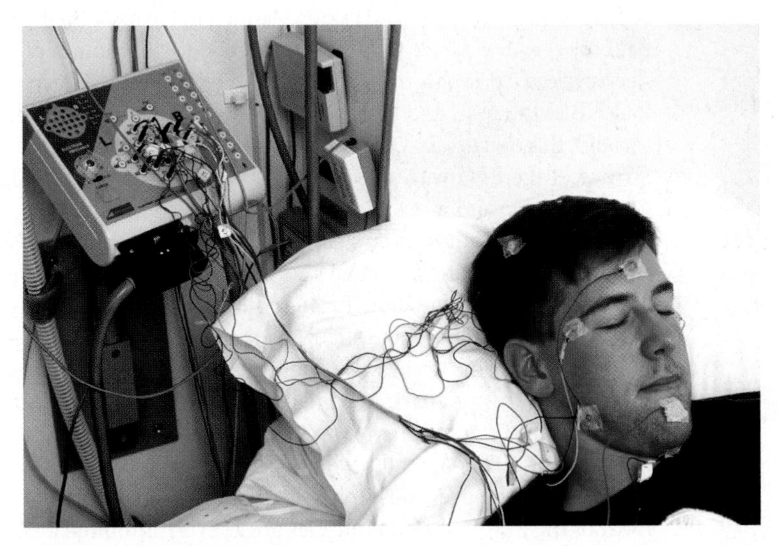

(a) Scalp electrodes are used to record brain wave activity.

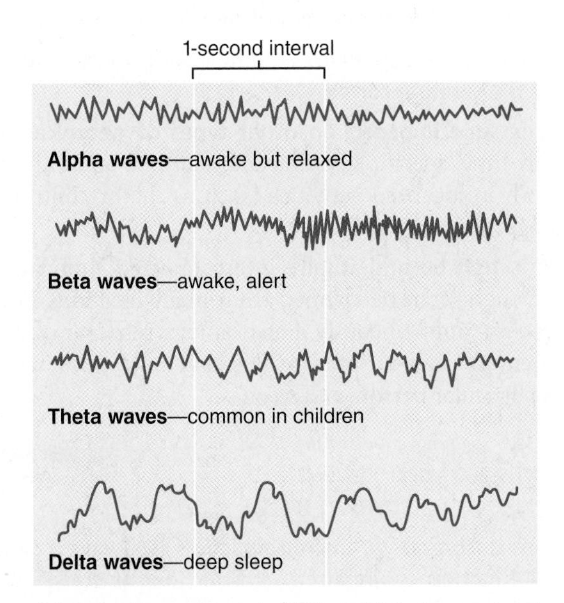

(b) Brain waves shown in EEGs fall into four general classes.

Figure 12.20 Electroencephalography (EEG) and brain waves.

- *Tonic-clonic seizures*, formerly called *grand mal*, are the most severe, convulsive form of epileptic seizures. The person loses consciousness, often breaking bones during the intense convulsions, showing the incredible strength of these muscle contractions. Loss of bowel and bladder control and severe biting of the tongue are common. The seizure lasts for a few minutes, then the muscles relax and the person awakens but remains disoriented for several minutes.

Many seizure sufferers experience a sensory hallucination, such as a taste, smell, or flashes of light, just before the seizure begins. This phenomenon, called an **aura**, is helpful because it gives the person time to lie down and avoid falling.

Epilepsy can usually be controlled by anticonvulsive drugs. If drugs fail to control the seizures, a *vagus nerve stimulator* or *deep brain stimulator* can be implanted. These devices deliver pulses to the vagus nerve or directly to the brain at predetermined intervals to stabilize the brain's electrical activity. +

Consciousness

Consciousness encompasses perception of sensations, voluntary initiation and control of movement, and capabilities associated with higher mental processing (memory, logic, judgment, perseverance, and so on). Clinically, consciousness is defined on a continuum that grades behavior in response to stimuli as (1) *alertness*, (2) *drowsiness* or *lethargy* (which proceeds to sleep), (3) *stupor*, and (4) *coma*. Alertness is the highest state of consciousness and cortical activity, and coma the most depressed.

12

Consciousness is difficult to define. And to be frank, reducing our response to a Key West sunset to a series of interactions between dendrites, axons, and neurotransmitters does not capture what makes that event so special. A sleeping person obviously lacks something that he or she has when awake, and we call this "something" consciousness.

Current suppositions about consciousness are as follows:

- **Consciousness involves simultaneous activity of large areas of the cerebral cortex.**
- **It is superimposed on other types of neural activity.** At any time, specific neurons and neuronal pools are involved both in localized activities (such as motor control) and in cognition.
- **It is holistic and totally interconnected.** Information for "thought" can be claimed from many locations in the cerebrum simultaneously. For example, retrieval of a specific memory can be triggered by several routes—a smell, a place, a particular person, and so on.

HOMEOSTATIC IMBALANCE 12.7 CLINICAL

Except during sleep, unconsciousness is always a signal that brain function is impaired. A brief loss of consciousness is called **fainting** or **syncope** (sing′ko-pe; "cut short"). Most often, syncope indicates inadequate cerebral blood flow due to low blood pressure, as might follow hemorrhage or sudden emotional stress.

Significant unresponsiveness to sensory stimuli for an extended period is called **coma**. Coma is *not* deep sleep. During sleep, the brain remains active and oxygen consumption resembles that of the waking state. In coma patients, oxygen use is always below normal resting levels.

Factors that can induce coma include: (1) blows to the head that cause widespread cerebral or brain stem trauma, (2) tumors or infections that invade the brain stem, (3) metabolic disturbances such as hypoglycemia (abnormally low blood sugar levels), (4) drug overdose, (5) liver or kidney failure. Strokes rarely cause coma unless they are massive and accompanied by extreme swelling of the brain, or are located in the brain stem.

When the brain has suffered irreparable damage, irreversible coma occurs. The result is **brain death**, a dead brain in an otherwise living body. Because life support can be removed only after death, physicians must determine whether a patient in an irreversible coma is legally alive or dead. +

Sleep and Sleep-Wake Cycles

Sleep is defined as a state of partial unconsciousness from which a person can be aroused by stimulation. This distinguishes sleep from coma, a state of unconsciousness from which a person *cannot* be aroused by even the most vigorous stimuli.

For the most part, cortical activity is depressed during sleep, but brain stem functions continue, such as control of respiration, heart rate, and blood pressure. Even environmental monitoring continues to some extent, as illustrated by the fact that strong stimuli ("things that go bump in the night") immediately arouse us.

Types of Sleep

The two major types of sleep, which alternate through most of the sleep cycle, are **non-rapid eye movement (NREM) sleep** and **rapid eye movement (REM) sleep**, defined in terms of their EEG patterns (**Figure 12.21**). During the first 30 to 45 minutes of the sleep cycle, we pass through the first two stages of NREM sleep and into NREM stages 3 and 4, also called **slow-wave sleep**. As we pass through these stages and slip into deeper and deeper sleep, the frequency of the EEG waves declines, but their amplitude increases. Blood pressure and heart rate also progressively decrease.

About 90 minutes after sleep begins, after reaching NREM stage 4, the EEG pattern changes abruptly. It becomes very irregular and backtracks quickly through the stages until alpha waves (more typical of the awake state) reappear, indicating the onset of REM sleep. This brain wave change is coupled with increases in heart rate, respiratory rate, and blood pressure and a decrease in gastrointestinal motility. Oxygen use by the brain is tremendous during REM—greater than during the awake state.

Although the eyes move rapidly under the lids during REM, most of the body's skeletal muscles are actively inhibited and go limp. Most dreaming occurs during REM sleep, and this temporary paralysis prevents us from acting out our dreams.

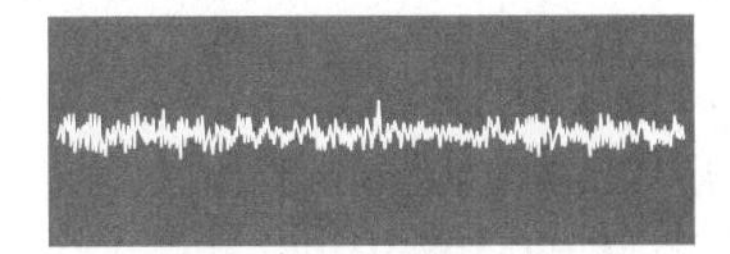

Awake

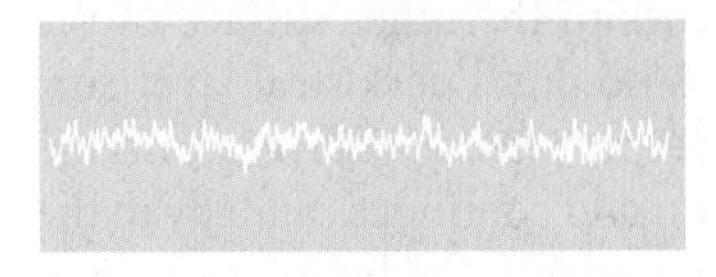

REM: Skeletal muscles (except ocular muscles and diaphragm) are actively inhibited; most dreaming occurs.

NREM stage 1: Relaxation begins; EEG shows alpha waves; arousal is easy.

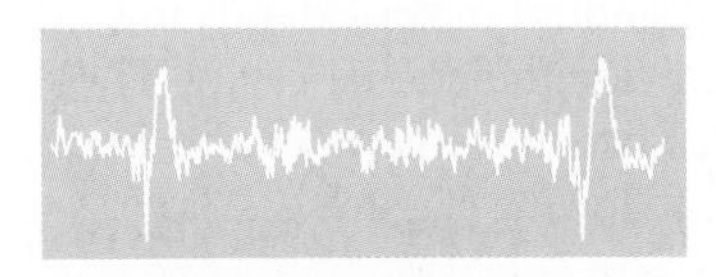

NREM stage 2: Irregular EEG with sleep spindles (short high-amplitude bursts); arousal is more difficult.

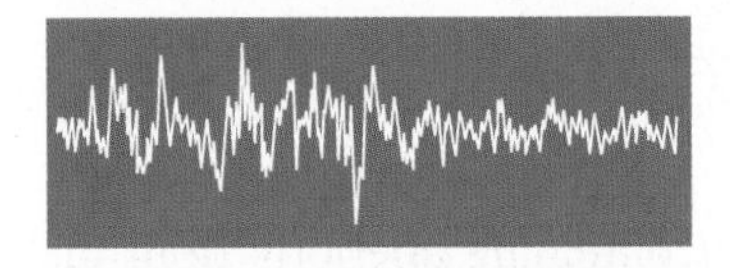

NREM stage 3: Sleep deepens; theta and delta waves appear; vital signs decline.

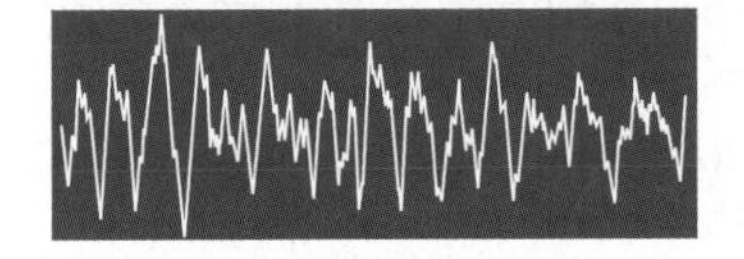

NREM stage 4: EEG is dominated by delta waves; arousal is difficult; bed-wetting, night terrors, and sleepwalking may occur.

(a) Typical EEG patterns

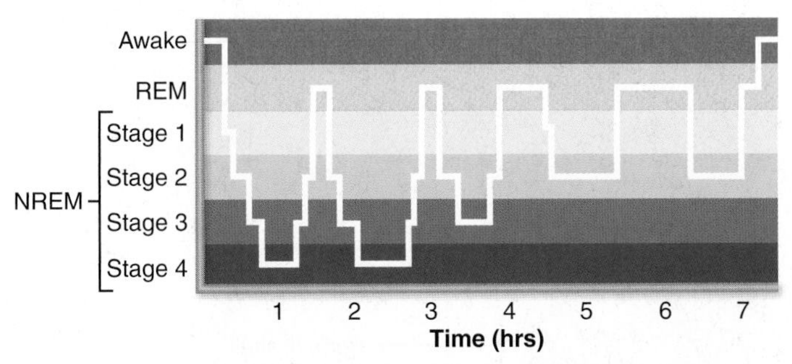

(b) Typical progression of an adult through one night's sleep stages

Figure 12.21 Types and stages of sleep. The four stages of non–rapid eye movement (NREM) sleep and rapid eye movement (REM) sleep are shown.

How Sleep Is Regulated

The alternating cycles of sleep and wakefulness reflect a natural *circadian*, or 24-hour, *rhythm*. The hypothalamus is responsible for the timing of the sleep cycle. Its *suprachiasmatic nucleus* (a biological clock) regulates its *preoptic nucleus* (a sleep-inducing center). By inhibiting the brain stem's reticular activating system (RAS; see Figure 12.18), the preoptic nucleus puts the cerebral cortex to sleep. However, sleep is much more than simply turning off the arousal system. RAS centers not only help maintain the awake state but also mediate some sleep stages, especially dreaming sleep.

Just before we wake, hypothalamic neurons release peptides called *orexins*, which act as "wake-up" chemicals. As a result, certain neurons of the brain stem reticular formation fire at maximal rates, arousing the sleepy cortex. A large number of chemical substances in the body cause sleepiness, but the relative importance of these various sleep-inducing substances is not known.

Importance of Sleep

Why do we sleep? Slow-wave (NREM stages 3 and 4) and REM sleep seem to be important in different ways. Slow-wave sleep is presumed to be restorative—the time when most neural activity can wind down to basal levels. When deprived of sleep, we spend more time than usual in slow-wave sleep during the next sleep episode.

A person persistently deprived of REM sleep becomes moody and depressed, and exhibits various personality disorders. REM sleep may (1) give the brain an opportunity to analyze the day's events and work through emotional problems in dream imagery, or (2) eliminate unneeded synaptic connections—in other words, we dream to forget.

HOMEOSTATIC IMBALANCE 12.8 CLINICAL

People with **narcolepsy** lapse abruptly into REM sleep from the awake state. These sleep episodes last about 15 minutes, can occur without warning, and are often triggered by a pleasurable event—a good joke, a game of poker.

Cells in the hypothalamus that secrete peptides called orexins (mentioned above as a wake-up chemical; also called hypocretins) are selectively destroyed in patients with narcolepsy, probably by the patient's own immune system. Replacing the orexins may be a key to future treatments.

Conversely, drugs that block the actions of orexin and promote sleep may treat **insomnia**, a chronic inability to obtain the *amount* or *quality* of sleep needed to function adequately during the day. Sleep requirements vary from four to nine hours a day in healthy people, so there is no way to determine the "right" amount.

True insomnia often reflects normal age-related changes, but perhaps the most common cause is psychological disturbance. We have difficulty falling asleep when we are anxious or upset, and depression is often accompanied by early awakening. +

☑ Check Your Understanding

15. Name three factors that can enhance transfer of information from STM to LTM.

16. When would you see delta waves in an EEG?

17. Which two states of consciousness are between alertness and coma?

18. During which sleep stage are most skeletal muscles actively inhibited?

For answers, see Answers Appendix.

12

12.8 The brain is protected by bone, meninges, cerebrospinal fluid, and the blood brain barrier

→ Learning Objectives

- ☐ **Describe how meninges, cerebrospinal fluid, and the blood brain barrier protect the CNS.**
- ☐ **Explain how cerebrospinal fluid is formed and describe its circulatory pathway.**

Nervous tissue is soft and delicate, and even slight pressure can injure neurons. However, the brain is protected by bone (the skull, discussed in Chapter 7), membranes (the meninges), and a watery cushion (cerebrospinal fluid). Furthermore, the blood brain barrier protects the brain from harmful substances in the blood.

Meninges

The **meninges** (mĕ-nin′jēz; *mening* = membrane) are three connective tissue membranes that lie just external to the CNS organs. The meninges:

- Cover and protect the CNS
- Protect blood vessels and enclose venous sinuses
- Contain cerebrospinal fluid
- Form partitions in the skull

From external to internal, the meninges (singular: **meninx**) are the dura mater, arachnoid mater, and pia mater (**Figure 12.22**).

Dura Mater

The leathery **dura mater** (du′rah ma′ter), meaning "tough mother," is the strongest meninx. Where it surrounds the brain, it is a two-layered sheet of fibrous connective tissue. The more superficial *periosteal layer* attaches to the inner surface of the skull (the periosteum). (There is no dural periosteal layer surrounding the spinal cord.) The deeper *meningeal layer* forms the true external covering of the brain and continues caudally in the vertebral canal as the spinal dura mater. The brain's two dural layers are fused together except in certain areas, where they separate to enclose **dural venous sinuses** that collect venous blood from the brain and direct it into the internal jugular veins of the neck (**Figure 12.23**).

In several places, the meningeal dura mater extends inward to form flat partitions that subdivide the cranial cavity. These **dural septa**, which limit excessive movement of the brain within the cranium, include the following (Figure 12.23a):

- **Falx cerebri** (falks ser′ĕ-bri). A large sickle-shaped (*falx* = sickle) fold that dips into the longitudinal fissure between the cerebral hemispheres. Anteriorly, it attaches to the crista galli of the ethmoid bone.
- **Falx cerebelli** (ser″ĕ-bel′i). Continuing inferiorly from the posterior falx cerebri, this small midline partition runs along the vermis of the cerebellum.
- **Tentorium cerebelli** (ten-to′re-um; "tent"). Resembling a tent over the cerebellum, this nearly horizontal dural fold extends into the transverse fissure between the cerebral hemispheres (which it helps to support) and the cerebellum.

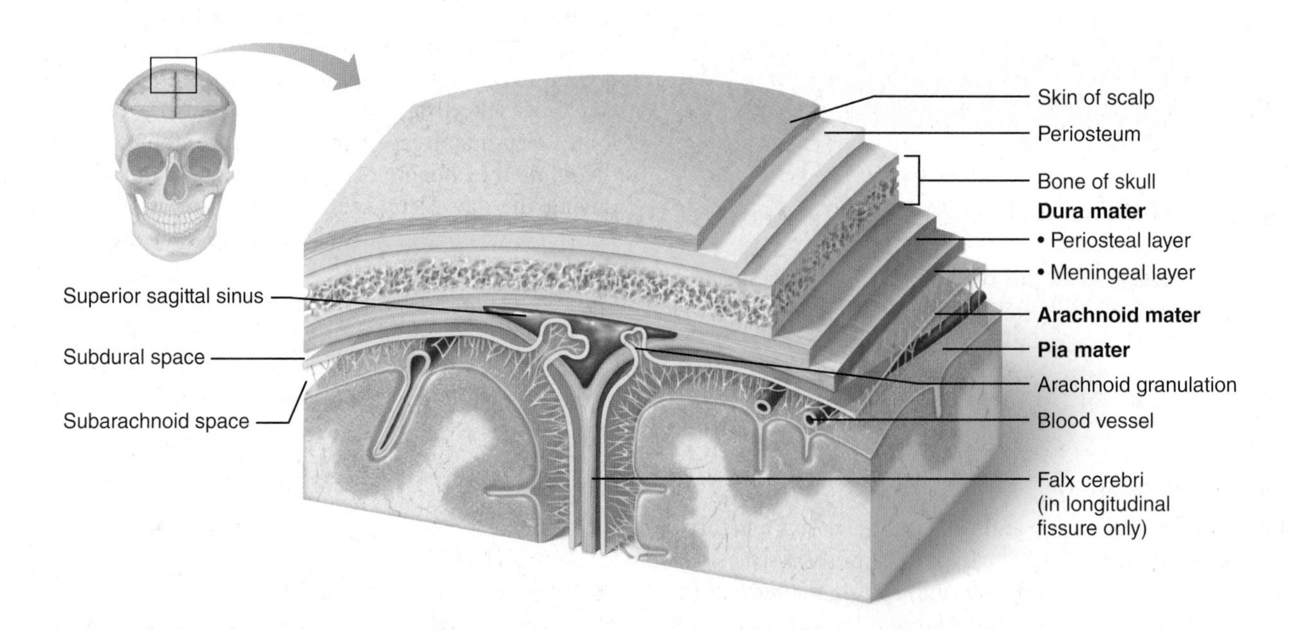

Figure 12.22 Meninges: dura mater, arachnoid mater, and pia mater. The meningeal dura forms the falx cerebri fold. A dural sinus, the superior sagittal sinus, is enclosed by the dural membranes superiorly. Arachnoid granulations return cerebrospinal fluid to the dural sinus. (Frontal section.)

Practice art labeling
MasteringA&P®>Study Area>Chapter 12

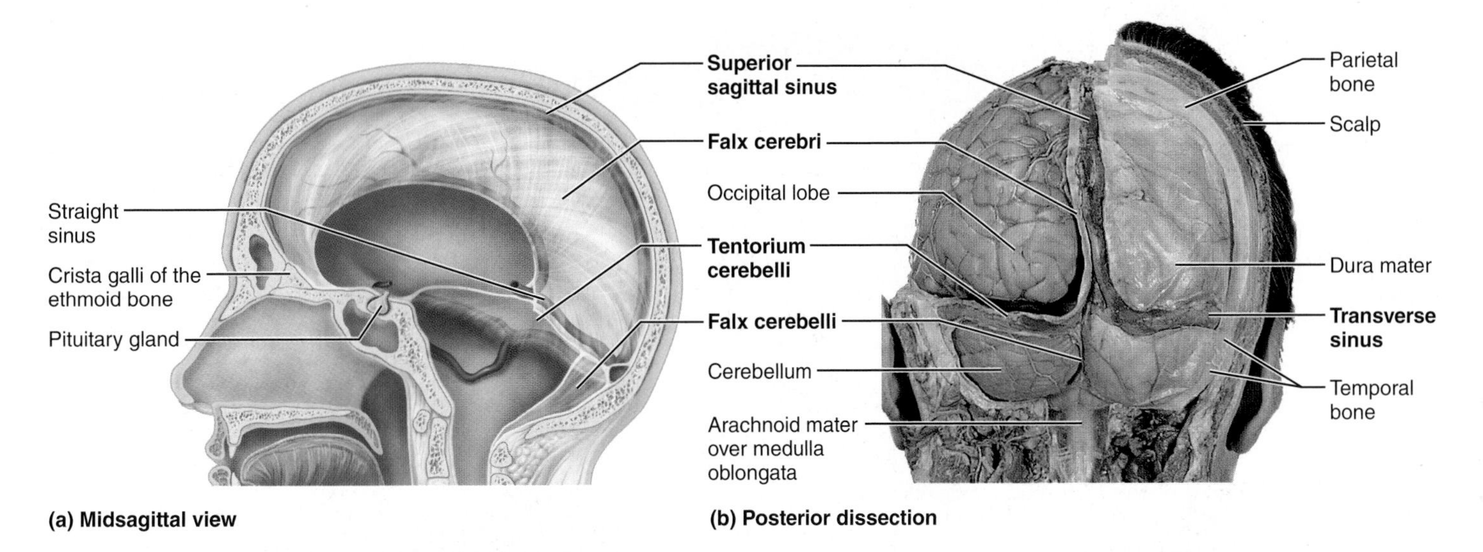

Figure 12.23 Dural septa and dural venous sinuses. (a) Dural septa are partitioning folds of dura mater in the cranial cavity. **(b)** Dural venous sinuses (injected with blue latex) are spaces between the periosteal and meningeal dura containing venous blood.

Explore human cadaver
MasteringA&P®>Study Area>PAL

Arachnoid Mater

The middle meninx, the **arachnoid mater** (ah-rak′noid), forms a loose brain covering, never dipping into the sulci at the cerebral surface. It is separated from the dura mater by a narrow serous cavity, the **subdural space**, which contains a film of fluid. Beneath the arachnoid membrane is the wide **subarachnoid space**. Spiderweb-like extensions span this space and secure the arachnoid mater to the underlying pia mater (*arachnida* means "spider"). The subarachnoid space is filled with cerebrospinal fluid and also contains the largest blood vessels serving the brain. Because the arachnoid mater is fine and elastic, these blood vessels are poorly protected.

Knoblike projections of the arachnoid mater called **arachnoid granulations** protrude superiorly through the dura mater and into the superior sagittal sinus (Figure 12.22). These granulations absorb cerebrospinal fluid into the venous blood of the sinus.

Pia Mater

The **pia mater** (pi′ah), meaning "gentle mother," is composed of delicate connective tissue and richly invested with tiny blood vessels. It is the only meninx that clings tightly to the brain like plastic wrap, following its every convolution. Small arteries entering the brain tissue carry ragged sheaths of pia mater inward with them for short distances.

HOMEOSTATIC IMBALANCE 12.9 — CLINICAL

Meningitis, inflammation of the meninges, is a serious threat to the brain because a bacterial or viral meningitis may spread to the CNS. Brain inflammation is called *encephalitis* (en-sef″ah-li′tis). Meningitis is usually diagnosed by obtaining a sample of cerebrospinal fluid via a lumbar tap (see Figure 12.28, p. 419) and examining it for microbes. ✚

Cerebrospinal Fluid (CSF)

Cerebrospinal fluid, found in and around the brain and spinal cord, forms a liquid cushion that gives buoyancy to CNS structures. By floating the jellylike brain, the CSF effectively reduces brain weight by 97% and prevents the delicate brain from crushing under its own weight. CSF also protects the brain and spinal cord from blows and other trauma. Additionally, although the brain has a rich blood supply, CSF helps nourish the brain, and there is some evidence that it carries chemical signals (such as hormones and sleep- and appetite-inducing molecules) from one part of the brain to another.

CSF is a watery "broth" similar in composition to blood plasma, from which it is formed. However, it contains less protein than plasma and its ion concentrations are different. For example, CSF contains more Na^+, Cl^-, and H^+ than does blood plasma, and less Ca^{2+} and K^+.

The **choroid plexuses** that hang from the roof of each ventricle form CSF. These plexuses are frond-shaped clusters of broad, thin-walled capillaries (*plex* = interwoven) enclosed first by pia mater and then by a layer of ependymal cells lining the ventricles (**Figure 12.24b**). These capillaries are fairly permeable, and tissue fluid filters continuously from the bloodstream. However, the choroid plexus ependymal cells are joined by tight junctions, and they have ion pumps that allow them to modify this filtrate by actively transporting only certain ions across their membranes into the CSF pool. This careful regulation of CSF composition is important because CSF mixes with the extracellular fluid bathing neurons and influences the composition of this fluid. Ion pumping also sets up ionic gradients that cause water to diffuse into the ventricles.

In adults, the total CSF volume of about 150 ml (about half a cup) is replaced every 8 hours or so. About 500 ml of CSF is formed daily. The choroid plexuses also help cleanse the CSF by removing waste products and unnecessary solutes.

12

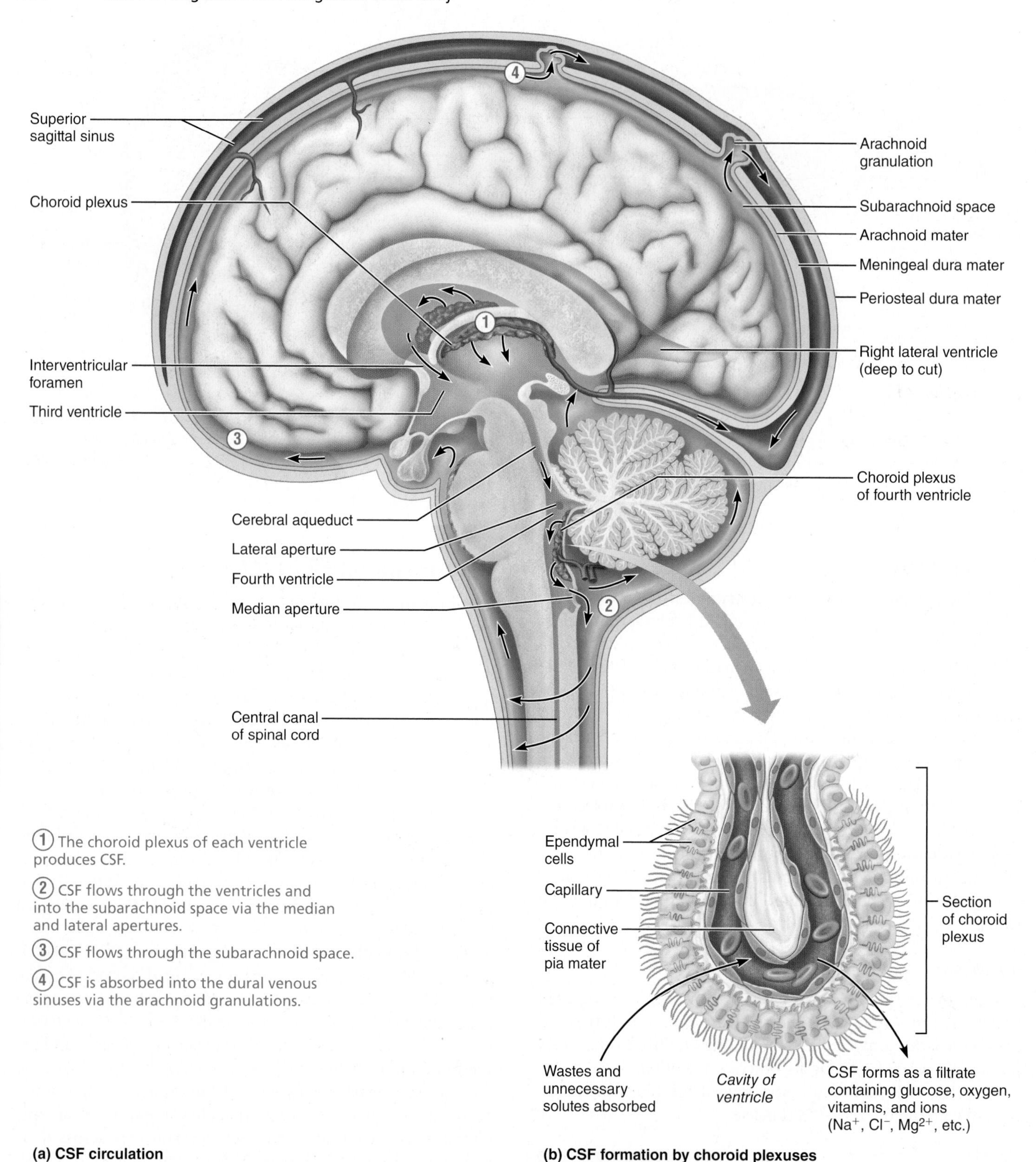

Figure 12.24 Formation, location, and circulation of CSF. (a) Location and circulatory pattern of cerebrospinal fluid (CSF). Arrows indicate the direction of flow. **(b)** Each choroid plexus consists of a knot of porous capillaries surrounded by a single layer of ependymal cells joined by tight junctions and bearing long cilia. Fluid leaking from porous capillaries is processed by the ependymal cells to form the CSF in the ventricles.

Practice art labeling
MasteringA&P®>Study Area>Chapter 12

Once produced, CSF moves freely through the ventricles. CSF enters the subarachnoid space via the lateral and median apertures in the walls of the fourth ventricle (Figure 12.24a). The long cilia of the ependymal cells lining the ventricles help keep the CSF in constant motion. In the subarachnoid space, CSF bathes the outer surfaces of the brain and spinal cord and then returns to the blood in the dural sinuses via the arachnoid granulations.

HOMEOSTATIC IMBALANCE 12.10 — CLINICAL

Ordinarily, CSF is produced and drained at a constant rate. However, if something (such as a tumor) obstructs its circulation or drainage, CSF accumulates and exerts pressure on the brain. This condition is called *hydrocephalus* ("water on the brain").

In a newborn baby with hydrocephalus, the head enlarges because the skull bones have not yet fused (**Figure 12.25**). In adults, however, the skull is rigid and hard, and hydrocephalus is likely to damage the brain because accumulating fluid compresses blood vessels and crushes the soft nervous tissue. Hydrocephalus is treated by inserting a shunt into the ventricles to drain excess fluid into the abdominal cavity. +

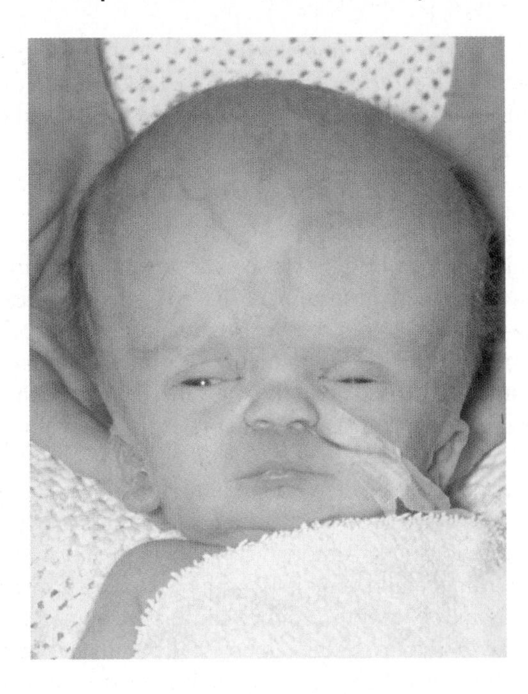

Figure 12.25 **Hydrocephalus in a newborn.**

Blood Brain Barrier

No other body tissue is so absolutely dependent on a constant internal environment as is the brain. In other body regions, the extracellular concentrations of hormones, amino acids, and ions are in constant flux, particularly after eating or exercise. If the brain were exposed to such chemical variations, its neurons would fire uncontrollably, because some hormones and amino acids serve as neurotransmitters and certain ions (particularly K^+) modify the threshold for neuronal firing.

The **blood brain barrier** is the protective mechanism that helps maintain the brain's stable environment. Exceptionally impermeable *tight junctions* between capillary endothelial cells are its major component.

Bloodborne substances in the brain's capillaries must pass through three layers before they reach the neurons: (1) the endothelium of the capillary wall, (2) a relatively thick basal lamina surrounding the external face of each capillary, and (3) the bulbous "feet" of the astrocytes clinging to the capillaries.

The astrocyte "feet" are not themselves the barrier, but play a role in its regulation: They supply required signals to the endothelial cells, causing them to make tight junctions. These tight junctions seamlessly join together the endothelial cells, making these the least permeable capillaries in the body.

The blood brain barrier is selective, not absolute. Nutrients such as glucose, essential amino acids, and some electrolytes move passively by facilitated diffusion through the endothelial cell membranes. Bloodborne metabolic wastes, proteins, certain toxins, and most drugs are denied entry. Small nonessential amino acids and potassium ions not only are prevented from entering the brain, but also are actively pumped from the brain across the capillary endothelium.

The barrier is ineffective against fats, fatty acids, oxygen, carbon dioxide, and other fat-soluble molecules that diffuse easily through all plasma membranes. This explains why bloodborne alcohol, nicotine, and anesthetics can affect the brain.

The structure of the blood brain barrier is not completely uniform. In some brain areas surrounding the third and fourth ventricles, the barrier is entirely absent and the capillary endothelium is quite permeable, allowing bloodborne molecules easy access to the neural tissue. One such region is the vomiting center of the brain stem, which monitors the blood for poisonous substances. Another is in the hypothalamus, which regulates water balance, body temperature, and many other metabolic activities. Lack of a blood brain barrier here is essential to allow the hypothalamus to sample the chemical composition of the blood. The barrier is incomplete in newborn and premature infants, and potentially toxic substances can enter the CNS and cause problems not seen in adults.

Injury to the brain, whatever the cause, may result in a localized breakdown of the blood brain barrier. Most likely, this breakdown reflects some change in the capillary endothelial cells or their tight junctions.

☑ Check Your Understanding

19. What is CSF? Where is it produced? What are its functions?

20. A brain surgeon is about to make an incision. Name all the tissue layers that she cuts through from the skin to the brain.

For answers, see Answers Appendix.

CLINICAL

12.9 Brain injuries and disorders have devastating consequences

→ Learning Objectives

- ☐ **Describe the cause (if known) and major signs and symptoms of cerebrovascular accidents, Alzheimer's disease, Parkinson's disease, and Huntington's disease.**
- ☐ **List and explain several techniques used to diagnose brain disorders.**

12

Brain dysfunctions are unbelievably varied and extensive. We have mentioned some of them already, but here we will focus on traumatic brain injuries, cerebrovascular accidents, and degenerative brain disorders.

Traumatic Brain Injuries

Head injuries are a leading cause of accidental death in North America. Consider, for example, what happens if you forget to fasten your seat belt and then rear-end another car. Your head is moving and then stops suddenly as it hits the windshield. Brain damage is caused not only by localized injury at the site of the blow, but also by the ricocheting effect as the brain hits the opposite end of the skull.

A **concussion** is an alteration in brain function, usually temporary, following a blow to the head. The victim may be dizzy or lose consciousness. Although typically mild and short-lived, even a seemingly mild concussion can be damaging, and multiple concussions over time produce cumulative damage.

More serious concussions can bruise the brain and cause permanent neurological damage, a condition called a **contusion**. In cortical contusions, the individual may remain conscious. Severe brain stem contusions always cause coma, lasting from hours to a lifetime because of injury to the reticular activating system.

Following a head injury, death may result from **subdural** or **subarachnoid hemorrhage** (bleeding from ruptured vessels into those spaces). Individuals who are initially lucid and then begin to deteriorate neurologically are, in all probability, hemorrhaging intracranially. Blood accumulating in the skull increases intracranial pressure and compresses brain tissue. If the pressure forces the brain stem inferiorly through the foramen magnum, control of blood pressure, heart rate, and respiration is lost. Intracranial hemorrhages are treated by surgically removing the hematoma (localized blood mass) and repairing the ruptured vessels.

Another consequence of traumatic head injury is **cerebral edema**, swelling of the brain. At best, cerebral edema aggravates the injury. At worst, it can be fatal in and of itself.

Cerebrovascular Accidents (CVAs)

The single most common nervous system disorder and the third leading cause of death in North America are **cerebrovascular accidents (CVAs)** (ser″ĕ-bro-vas′ku-lar), also called *strokes*. CVAs occur when blood circulation to a brain area is blocked and brain tissue dies of **ischemia** (is-ke′me-ah), a reduction of blood supply that impairs the delivery of oxygen and nutrients.

The most common cause of CVA is a blood clot that blocks a cerebral artery. A clot can originate outside the brain (from the heart, for example) or form on the roughened interior wall of a brain artery narrowed by atherosclerosis. Less frequently, strokes are caused by bleeding, which compresses brain tissue.

Many who survive a CVA are paralyzed on one side of the body (*hemiplegia*). Others commonly exhibit sensory deficits or have difficulty understanding or vocalizing speech. Even so, the picture is not hopeless. Some patients recover at least part of their lost faculties, because undamaged neurons sprout new branches that spread into the injured area and take over some lost functions. Physical therapy should start as soon as possible to prevent muscle contractures (abnormally shortened muscles due to differences in strength between opposing muscle groups).

Not all strokes are "completed." Temporary episodes of reversible cerebral ischemia, called **transient ischemic attacks (TIAs)**, are common. TIAs last from 5 to 50 minutes and are characterized by temporary numbness, paralysis, or impaired speech. These deficits are not permanent, but TIAs do constitute "red flags" that warn of impending, more serious CVAs.

A CVA is like an undersea earthquake. It's not the initial temblor that does most of the damage, it's the tsunami that floods the coast later. Similarly, the initial vascular blockage during a stroke is not usually disastrous because there are many blood vessels in the brain that can pick up the slack. Rather, it's the neuron-killing events outside the initial ischemic zone that wreak the most havoc.

Experimental evidence indicates that the main culprit is *glutamate*, an excitatory neurotransmitter. Glutamate plays a key role in learning and memory, as well as other critical brain functions. However, after brain injury, neurons totally deprived of oxygen begin to disintegrate, unleashing the cellular equivalent of "buckets" of glutamate. Under these conditions, glutamate acts as an *excitotoxin*, literally exciting surrounding cells to death.

At present, the most successful treatment for stroke is tissue plasminogen activator (tPA), which dissolves blood clots in the brain. Alternatively, a mechanical device can drill into a blood clot and pull it from a blood vessel like a cork from a bottle.

Degenerative Brain Disorders

Alzheimer's Disease

Alzheimer's disease (AD) (altz′hi-merz), a progressive degenerative disease of the brain, ultimately results in **dementia** (mental deterioration). Alzheimer's patients represent nearly half of the people living in nursing homes. Between 5 and 15% of people over 65 develop this condition, and for up to half of those over 85 it is a major contributing cause in their deaths.

Its victims exhibit memory loss (particularly for recent events), shortened attention span, disorientation, and eventual language loss. Over a period of several years, formerly good-natured people may become irritable, moody, and confused. Hallucinations may ultimately occur.

Examinations of brain tissue reveal senile plaques littering the brain like shrapnel between the neurons. The plaques consist of extracellular aggregations of *beta-amyloid peptide*, which is cut from a normal membrane precursor protein (APP) by enzymes. One form of Alzheimer's disease is caused by an inherited mutation in the gene for APP, which suggests that too much beta-amyloid may be toxic. Current clinical trials focus on using the immune system to clear away beta-amyloid peptide.

Another hallmark of Alzheimer's disease is the presence of *neurofibrillary tangles* inside neurons. These tangles involve a protein called tau, which functions like railroad ties to bind microtubule "tracks" together. In the brains of AD victims, tau abandons its microtubule-stabilizing role and grabs on to other tau molecules, forming spaghetti-like neurofibrillary tangles, which kill the neurons by disrupting their transport mechanisms.

Both plaques and tangles come about because the proteins that comprise them have misfolded. These misfolded proteins

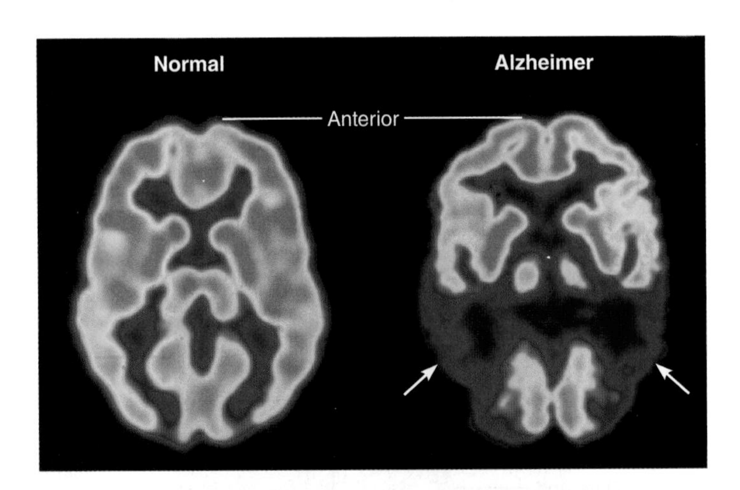

Figure 12.26 **Brain activity is decreased by Alzheimer's disease.** In these PET scans, high neural activity is indicated by reds and yellows. Arrows point to large decreases in parietal cortex activity.

clump together, and also catalyze the misfolding of normally folded copies of the same proteins. Recent evidence suggests that this protein misfolding spreads in a predictable way from one region of the brain to another. This helps explain both the different types of dementia and their progression, as neurons in more and more brain regions die.

As the brain cells die, their functions are lost and the brain shrinks. Particularly vulnerable brain areas include the hippocampus, the basal forebrain, and association areas of the cortex, all regions involved in thinking and memory (**Figure 12.26**). Loss of neurons in the basal forebrain is associated with a shortage of the neurotransmitter acetylcholine, and drugs that inhibit breakdown of acetylcholine slightly enhance cognitive function in AD patients.

Parkinson's Disease

Typically striking people in their 50s and 60s, **Parkinson's disease** results from a degeneration of the dopamine-releasing neurons of the substantia nigra. As those neurons deteriorate, the dopamine-deprived basal nuclei they target become overactive. Afflicted individuals have a persistent tremor at rest (exhibited by "pill-rolling" movements of the fingers and wrist), a forward-bent walking posture and shuffling gait, and a stiff facial expression. They are slow initiating and executing movement.

The cause of Parkinson's disease is still unknown, but multiple factors may interact to destroy dopamine-releasing neurons. Recent evidence points to abnormalities in certain mitochondrial proteins and protein degradation pathways. The drug *L-dopa* helps to alleviate some symptoms. It passes through the blood brain barrier and is then converted into dopamine.

However, as more and more neurons die off, L-dopa becomes ineffective. Mixing L-dopa with drugs that inhibit the breakdown of dopamine can prolong its effectiveness. Early in the disease, these drugs alone slow the neurological deterioration to some extent and delay the need to administer L-dopa.

Deep brain stimulation via implanted electrodes shuts down abnormal brain activity and can alleviate tremors. This treatment (for patients who no longer respond to drug therapy) is expensive and risky, but it works. Another possible future treatment is to use gene therapy to insert into adult brain cells the genes that would cause them to secrete the inhibitory neurotransmitter GABA. GABA then inhibits the abnormal brain activity just as the electrical stimulation does. Replacing dead or damaged cells by implanting stem cells is promising, but results to date are no better than more conventional treatments.

Huntington's Disease

Huntington's disease, a fatal hereditary disorder, strikes during middle age. Mutant *huntingtin* protein accumulates in brain cells and the tissue dies, leading to massive degeneration of the basal nuclei and later of the cerebral cortex. Its initial symptoms in many are wild, jerky, almost continuous "flapping" movements called *chorea* (Greek for "dance"). Although the movements appear to be voluntary, they are not. Late in the disease, marked mental deterioration occurs. Huntington's disease is progressive and usually fatal within 15 years.

The hyperkinetic manifestations of Huntington's disease are essentially the opposite of those of Parkinson's disease (overstimulation rather than inhibition of the motor drive). Huntington's is usually treated with drugs that block, rather than enhance, dopamine's effects.

Diagnostic Procedures for Assessing CNS Dysfunction

If you have had a routine physical examination, you are familiar with the reflex tests done to assess neural function. A tap with a reflex hammer stretches your quadriceps tendon and your anterior thigh muscles contract. This produces the knee-jerk response, which shows that your spinal cord and upper brain centers are functioning normally.

Abnormal responses to a reflex test may indicate such serious disorders as intracranial hemorrhage, multiple sclerosis, or hydrocephalus. They indicate that more sophisticated neurological tests are needed to identify the problem.

New imaging techniques have revolutionized the diagnosis of brain lesions. Together, various *CT* and *MRI scanning techniques* allow quick identification of most tumors, intracranial lesions, multiple sclerosis plaques, and areas of dead brain tissue (infarcts). PET scans can localize brain lesions that generate seizures (epileptic tissue), and new radiotracer dyes that bind to beta-amyloid promise earlier, more reliable diagnosis of Alzheimer's disease.

Take, for example, a patient arriving in the emergency room with a stroke. A race against time begins to save the affected area of the patient's brain. The first step is to determine if the stroke is due to a clot or a bleed by imaging the brain, most often with CT. If the stroke is due to a clot, the clot-busting drug tPA can be used, but only within the first hours. tPA is usually given intravenously, but a longer time window is possible if tPA is applied directly to the clot using a catheter guided into position. To visualize the location of the clot and the catheter, dye is injected to make arteries stand out in an X ray, a procedure called *cerebral angiography*. Cerebral angiography can also help patients who have had a warning stroke, or TIA.

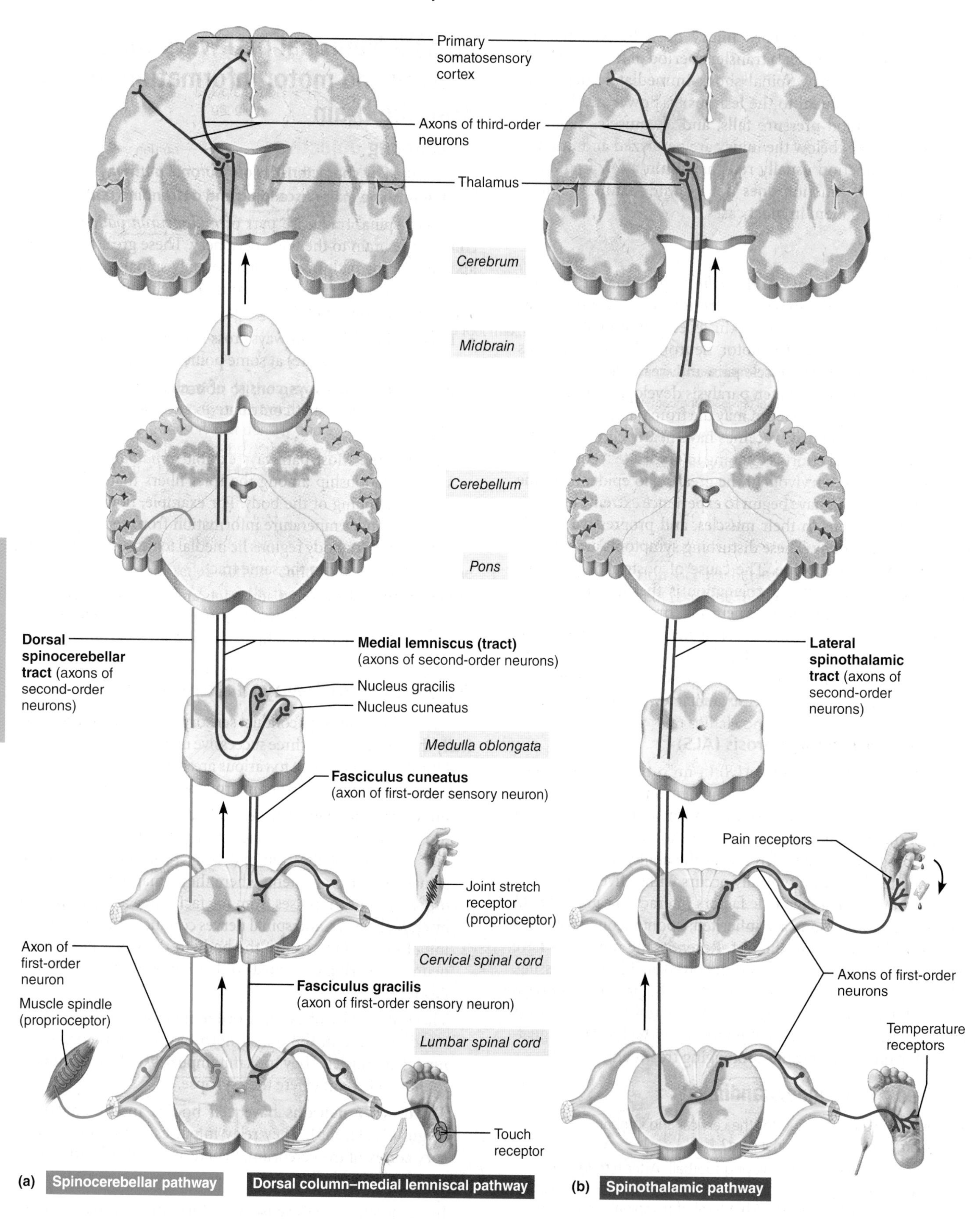

Figure 12.32 Pathways of selected ascending spinal cord tracts. Cross sections up to the cerebrum, which is shown in frontal section. **(a)** The spinocerebellar pathway (left) transmits proprioceptive information only to the cerebellum, and so is subconscious. The dorsal column–medial lemniscal pathway transmits discriminative touch and conscious proprioception signals to the cerebral cortex. **(b)** The lateral spinothalamic pathway transmits pain and temperature. See also Table 12.2.

spinothalamic pathways) transmit impulses via the thalamus to the sensory cortex for conscious interpretation. Collectively the inputs of these sister tracts provide *discriminative touch* and *conscious proprioception*. Both pathways decussate—the first in the medulla and the second in the spinal cord.

The third pathway, the *spinocerebellar pathway*, terminates in the cerebellum, and does not contribute to sensory perception. Let's examine these pathways more closely.

- **Dorsal column–medial lemniscal pathways.** The **dorsal column–medial lemniscal pathways** (lem-nis′kul; "ribbon") mediate precise, straight-through transmission of inputs from a single type (or a few related types) of sensory receptor that can be localized precisely on the body surface, such as discriminative touch and vibrations. These pathways are formed by the paired tracts of the **dorsal white column** of the spinal cord—**fasciculus cuneatus** and **fasciculus gracilis**—and the **medial lemniscus**.

  The medial lemniscus arises in the medulla and terminates in the thalamus (Figure 12.32a and **Table 12.2**). From the thalamus, impulses are forwarded to specific areas of the somatosensory cortex.

Table 12.2 Major Ascending (Sensory) Pathways and Spinal Cord Tracts

SPINAL CORD TRACT	LOCATION (FUNICULUS)	ORIGIN	TERMINATION	FUNCTION
Dorsal Column–Medial Lemniscal Pathways				
Fasciculus cuneatus and fasciculus gracilis (dorsal white column)	Dorsal	Central axons of sensory (first-order) neurons enter dorsal root of the spinal cord and branch. Branches enter dorsal white column on same side without synapsing.	By synapse with second-order neurons in nucleus cuneatus and nucleus gracilis in medulla. Fibers of medullary neurons cross over and ascend in medial lemniscus to thalamus, where they synapse with third-order neurons. Thalamic neurons then transmit impulses to somatosensory cortex.	Both tracts transmit sensory impulses from general sensory receptors of skin and proprioceptors, which are interpreted as discriminative touch, pressure, and "body sense" (limb and joint position) in opposite somatosensory cortex. Cuneatus transmits afferent impulses from upper limbs, upper trunk, and neck. Gracilis carries impulses from lower limbs and inferior body trunk.
Spinothalamic Pathways				
Lateral spinothalamic	Lateral	Interneurons (second-order neurons) in dorsal horns. Fibers cross to opposite side before ascending.	By synapse with third-order neurons in thalamus. Thalamic neurons then convey impulses to somatosensory cortex.	Transmits impulses concerned with pain and temperature to opposite side of brain for interpretation by somatosensory cortex.
Ventral spinothalamic	Ventral	Interneurons (second-order neurons) in dorsal horns. Fibers cross to opposite side before ascending.	By synapse with third-order neurons in thalamus. Thalamic neurons eventually convey impulses to somatosensory cortex.	Transmits impulses concerned with crude touch and pressure to opposite side of brain for interpretation by somatosensory cortex.
Spinocerebellar Pathways				
Dorsal spinocerebellar*	Lateral (dorsal part)	Interneurons (second-order neurons) in dorsal horn on same side of cord. Fibers ascend without crossing.	By synapse in cerebellum	Transmits impulses from trunk and lower limb proprioceptors on one side of body to same side of cerebellum for subconscious proprioception.
Ventral spinocerebellar*	Lateral (ventral part)	Interneurons (second-order neurons) of dorsal horn. Contains crossed fibers that cross back to the opposite side in the pons.	By synapse in cerebellum	Transmits impulses from the trunk and lower limb on the same side of body to cerebellum for subconscious proprioception.

*These spinocerebellar tracts carry information from the lower limbs and trunk only. The corresponding tracts for the upper limb and neck (rostral spinocerebellar and others) are beyond the scope of this book.

12

Table 12.3 Major Descending (Motor) Pathways and Spinal Cord Tracts

SPINAL CORD TRACT	LOCATION (FUNICULUS)	ORIGIN	TERMINATION	FUNCTION
Direct (Pyramidal) Pathways				
Lateral corticospinal	Lateral	Pyramidal cells of motor cortex of the cerebrum. Cross over in pyramids of medulla.	By synapse with ventral horn interneurons that influence motor neurons and occasionally with ventral horn motor neurons directly.	Transmits motor impulses from cerebrum to spinal cord motor neurons (which activate skeletal muscles on opposite side of body). A voluntary motor tract.
Ventral corticospinal	Ventral	Pyramidal cells of motor cortex. Fibers cross over at the spinal cord level.	Ventral horn (as above).	Same as lateral corticospinal tract.
Indirect Pathways				
Tectospinal	Ventral	Superior colliculus of midbrain of brain stem (fibers cross to opposite side of cord).	Ventral horn (as above).	Turns neck so eyes can follow a moving object.
Vestibulospinal	Ventral	Vestibular nuclei in medulla of brain stem (fibers descend without crossing).	Ventral horn (as above).	Transmits motor impulses that maintain muscle tone and activate ipsilateral limb and trunk extensor muscles and muscles that move head. Helps maintain balance during standing and moving.
Rubrospinal	Lateral	Red nucleus of midbrain of brain stem (fibers cross to opposite side just inferior to the red nucleus).	Ventral horn (as above).	In animals, transmits motor impulses concerned with muscle tone of distal limb muscles (mostly flexors) on opposite side of body. In humans, functions are largely assumed by corticospinal tracts.
Reticulospinal (medial and lateral)	Medial and lateral	Reticular formation of brain stem (medial nuclear group of pons and medulla). Both crossed and uncrossed fibers.	Ventral horn (as above).	Transmits impulses concerned with muscle tone and many visceral motor functions. May control most unskilled movements.

12

REVIEW QUESTIONS

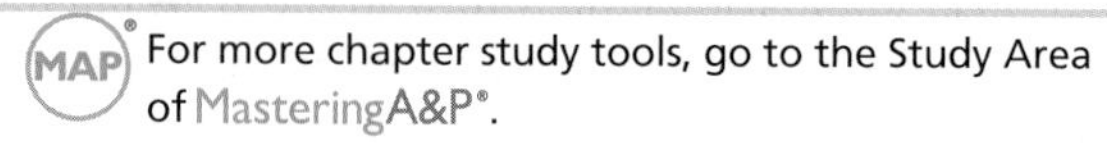

For more chapter study tools, go to the Study Area of MasteringA&P®.

There you will find:

- Interactive Physiology iP
- A&PFlix A&PFlix
- Practice Anatomy Lab PAL
- PhysioEx PEx
- Videos, Practice Quizzes and Tests, MP3 Tutor Sessions, Case Studies, and much more!

Multiple Choice/Matching

(Some questions have more than one correct answer. Select the best answer or answers from the choices given.)

1. The primary motor cortex, Broca's area, and the premotor cortex are located in which lobe? **(a)** frontal, **(b)** parietal, **(c)** temporal, **(d)** occipital.
2. The innermost layer of the meninges, delicate and adjacent to the brain tissue, is the **(a)** dura mater, **(b)** corpus callosum, **(c)** arachnoid mater, **(d)** pia mater.
3. Cerebrospinal fluid is formed by **(a)** arachnoid granulations, **(b)** dura mater, **(c)** choroid plexuses, **(d)** all of these.
4. A patient has suffered a cerebral hemorrhage that has caused dysfunction of the precentral gyrus of his right cerebral cortex. As a result, **(a)** he cannot voluntarily move his left arm or leg, **(b)** he feels no sensation on the left side of his body, **(c)** he feels no sensation on his right side.
5. Choose the correct term from the key to respond to the statements describing various brain areas.

Key:

(a) cerebellum	**(d)** striatum	**(g)** midbrain
(b) corpora quadrigemina	**(e)** hypothalamus	**(h)** pons
(c) corpus callosum	**(f)** medulla	**(i)** thalamus

____ (1) basal nuclei involved in fine control of motor activities
____ (2) region where there is a gross crossover of fibers of descending pyramidal tracts
____ (3) control of temperature, autonomic nervous system reflexes, hunger, and water balance
____ (4) houses the substantia nigra and cerebral aqueduct
____ (5) relay stations for visual and auditory stimuli input; found in midbrain
____ (6) houses vital centers for control of the heart, respiration, and blood pressure
____ (7) brain area through which all the sensory input is relayed to get to the cerebral cortex
____ (8) brain area most concerned with equilibrium, body posture, and coordination of motor activity

6. Which of the following tracts convey vibration and other specific sensations that can be precisely localized? **(a)** pyramidal tract, **(b)** medial lemniscus, **(c)** lateral spinothalamic tract, **(d)** reticulospinal tract.
7. Destruction of the ventral horn cells of the spinal cord results in loss of **(a)** integrating impulses, **(b)** sensory impulses, **(c)** voluntary motor impulses, **(d)** all of these.
8. Fiber tracts that allow neurons within the same cerebral hemisphere to communicate are **(a)** association fibers, **(b)** commissures, **(c)** projection fibers.
9. A number of brain structures are listed below. If an area is primarily gray matter, write **a** in the answer blank; if mostly white matter, respond with **b**.

____ (1) cerebral cortex
____ (2) corpus callosum and corona radiata
____ (3) red nucleus
____ (4) medial and lateral nuclear groups
____ (5) medial lemniscus
____ (6) cranial nerve nuclei
____ (7) spinothalamic tract
____ (8) fornix
____ (9) cingulate and precentral gyri

10. A professor unexpectedly blew a loud horn in his anatomy and physiology class. The students looked up, startled. The reflexive movements of their eyes were mediated by the **(a)** cerebral cortex, **(b)** inferior olives, **(c)** raphe nuclei, **(d)** superior colliculi, **(e)** nucleus gracilis.
11. Identify the stage of sleep described by using choices from the key. (Note that responses a–d refer to NREM sleep.)

Key: **(a)** stage 1 **(b)** stage 2 **(c)** stage 3 **(d)** stage 4 **(e)** REM

____ (1) the stage when blood pressure and heart rate reach their lowest levels
____ (2) indicated by movement of the eyes under the lids; dreaming occurs
____ (3) when sleepwalking may occur
____ (4) when the sleeper is very easily awakened; EEG shows alpha waves

12. All of the following descriptions refer to dorsal column–medial lemniscal ascending pathways except one: **(a)** they include the fasciculus gracilis and fasciculus cuneatus; **(b)** they include a chain of three neurons; **(c)** their connections are diffuse and poorly localized; **(d)** they are concerned with precise transmission of one or a few related types of sensory input.

Short Answer Essay Questions

13. Make a diagram showing the three primary (embryonic) brain vesicles. Name each and then use clinical terminology to name the resulting adult brain regions.
14. (a) What is the advantage of having a cerebrum that is highly convoluted? (b) What term is used to indicate its grooves? Its outward folds? (c) Which groove divides the cerebrum into two hemispheres? (d) What divides the parietal from the frontal lobe? The parietal from the temporal lobe?
15. (a) Make a rough drawing of the lateral aspect of the left cerebral hemisphere. (b) You may be thinking, "But I just can't draw!" So, name the hemisphere involved with most people's ability to draw. (c) On your drawing, locate the following areas and provide the major function of each: primary motor cortex, premotor cortex, somatosensory association cortex, primary somatosensory cortex, visual and auditory areas, prefrontal cortex, Wernicke's and Broca's areas.
16. (a) What does lateralization of cortical functioning mean? (b) Why is the term cerebral dominance a misnomer?
17. (a) What is the function of the basal nuclei? (b) Which basal nuclei form the striatum? (c) Which arches over the diencephalon?
18. Explain how the cerebellum is physically connected to the brain stem.
19. Describe the role of the cerebellum in maintaining smooth, coordinated skeletal muscle activity.
20. (a) Where is the limbic system located? (b) Which structures make up this system? (c) How is the limbic system important in behavior?
21. (a) Localize the reticular formation in the brain. (b) What does RAS mean, and what is its function?
22. What is an aura?
23. Describe the stages of sleep and outline the order in which we progress through these stages during a typical night's sleep.
24. Compare and contrast short-term memory (STM) and long-term memory (LTM) relative to storage capacity and duration of the memory.
25. Define memory consolidation.
26. List four ways in which the CNS is protected.
27. (a) How is cerebrospinal fluid formed and drained? Describe its pathway within and around the brain. (b) What happens if CSF does not drain properly? Why is this consequence more harmful in adults?
28. What constitutes the blood brain barrier?
29. (a) Define concussion and contusion. (b) Why does severe brain stem injury result in unconsciousness?
30. Describe the spinal cord, depicting its extent, its composition of gray and white matter, and its spinal roots.
31. How do the types of motor activity controlled by the direct (pyramidal) and indirect systems differ?
32. Describe the functional problems that would be experienced by a person in which these fiber tracts have been cut: (a) lateral spinothalamic, (b) ventral and dorsal spinocerebellar, (c) tectospinal.
33. Differentiate between spastic and flaccid paralysis.
34. How do the conditions paraplegia, hemiplegia, and quadriplegia differ?
35. (a) Define cerebrovascular accident or CVA. (b) Describe its possible causes and consequences.

AT THE CLINIC

Clinical Case Study

Central Nervous System

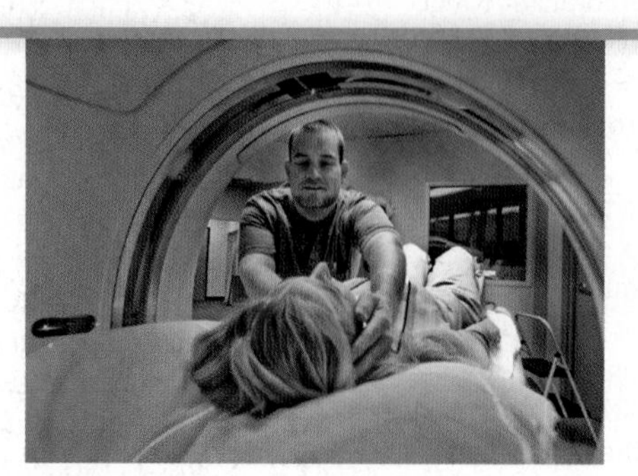

Margaret Bryans, a 39-year-old female, was a passenger on the bus that crashed on Route 91. When paramedics arrived on the scene, she was unconscious, with cuts on her arms, face, and scalp. She regained consciousness en route to the hospital and appeared agitated and combative. Paramedics observed that she had a right hemiparesis (muscle weakness), with a near complete paresis of her right upper extremity and a partial paresis of the right lower extremity. A head CT scan revealed an acute subdural hematoma and an extensive subarachnoid hemorrhage. Doctors noted that she was able to follow commands from medical personnel. With difficulty, she could speak haltingly, using only simple words.

Surgery to remove large clots from the subarachnoid space was performed immediately. Two weeks after the surgery, she showed significant improvement in her speech and motor function.

1. The adult brain can be broken down into four functional regions (see Table 12.1). Based on the observed signs in this case, which of these four brain regions is involved? What evidence did you use to determine this?
2. Which side of the brain is involved in Mrs. Bryans's injury? What evidence did you use to determine this?
3. What specific parts of the region of the brain you identified in question 1 are being affected by the injury to cause the muscle weakness and language problems?
4. What are the three membranes that make up the meninges? Describe their positions relative to the brain.
5. Relative to the meninges, describe the location of the bleeding revealed on the CT scan.

For answers, see Answers Appendix.

13 The Peripheral Nervous System and Reflex Activity

KEY CONCEPTS

→ Learning Objective

☐ **Define peripheral nervous system and list its components.**

The human brain, for all its sophistication, would be useless without its links to the outside world. Consider one experiment that illustrates this point. Volunteers hallucinated when they were deprived of sensory input by being blindfolded and suspended in warm water in a sensory deprivation tank. One saw charging pink and purple elephants. Another heard a chorus; still others had taste hallucinations. Our very sanity depends on a continuous flow of information from the outside.

Equally important to our well-being are the orders sent from the CNS to voluntary muscles and other effectors of the body, which allow us to move and take care of our own needs. The **peripheral nervous system (PNS)** provides these links from and to the world outside our bodies. Ghostly white nerves thread through virtually every part of the body, enabling the CNS to receive information and carry out its decisions.

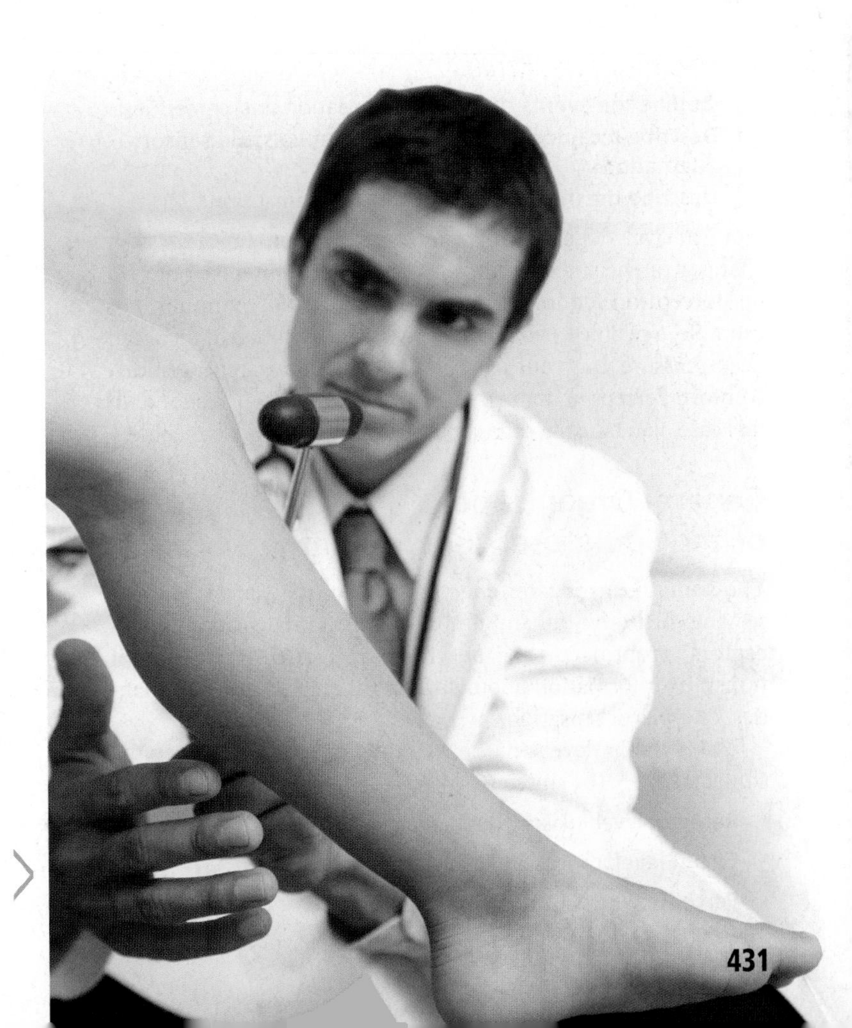

Striking the patellar tendon elicits the knee-jerk (patellar) reflex, providing information about the integrity of the nervous system.

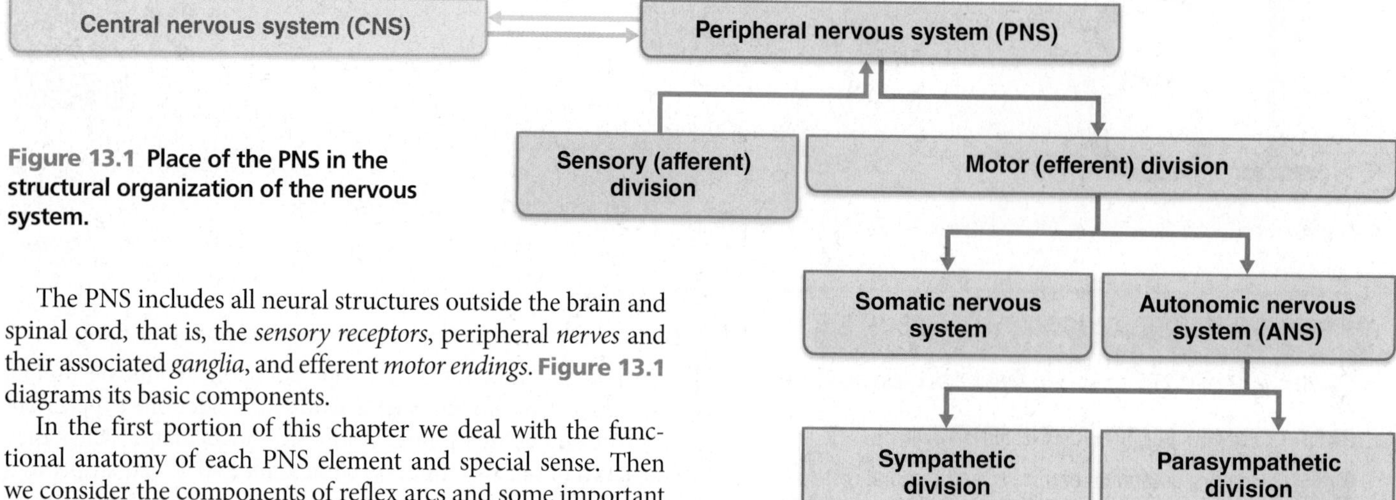

Figure 13.1 Place of the PNS in the structural organization of the nervous system.

The PNS includes all neural structures outside the brain and spinal cord, that is, the *sensory receptors*, peripheral *nerves* and their associated *ganglia*, and efferent *motor endings*. **Figure 13.1** diagrams its basic components.

In the first portion of this chapter we deal with the functional anatomy of each PNS element and special sense. Then we consider the components of reflex arcs and some important somatic reflexes, played out almost entirely in PNS structures, that help maintain homeostasis.

PART 1

SENSORY RECEPTORS AND SENSATION

13.1 Receptors, ascending pathways, and cerebral cortex process sensory information

→ **Learning Objectives**

- ☐ **Outline the events that lead to sensation and perception.**
- ☐ **Describe receptor and generator potentials and sensory adaptation.**
- ☐ **Describe the main aspects of sensory perception.**

Our survival depends not only on **sensation** (awareness of changes in the internal and external environments) but also on **perception** (conscious interpretation of those stimuli). For example, a pebble in your shoe causes the *sensation* of localized deep pressure, but your *perception* of it is an awareness of discomfort. Perception in turn determines how we will respond: In this case, you take off your shoe to get rid of the pesky pebble.

General Organization of the Somatosensory System

The **somatosensory system**—the part of the sensory system serving the body wall and limbs—receives inputs from exteroceptors, proprioceptors, and interoceptors. Consequently, it transmits information about several different sensory modalities, or types of sensation.

Three main levels of neural integration operate in the somatosensory (or any sensory) system:

- **Receptor level:** sensory receptors
- **Circuit level:** processing in ascending pathways
- **Perceptual level:** processing in cortical sensory areas

Sensory input is generally relayed toward the head, but note that it is also processed along the way. Let's examine this further.

Processing at the Receptor Level

Generating a Signal For sensation to occur, a stimulus must excite a receptor and action potentials must reach the CNS. For this to happen:

- The stimulus energy must match the *specificity* of the receptor. For example, a touch receptor may be sensitive to mechanical pressure, stretch, and vibration, but not to light energy (which is the province of receptors in the eye). The more complex the sensory receptor, the more specific it is.
- The stimulus must be applied within a sensory receptor's *receptive field*—the area the receptor monitors. Typically, the smaller the receptive field, the greater the ability of the brain to accurately localize the stimulus site.
- The stimulus energy must be converted into the energy of a *graded potential*, a process called **transduction**. This graded potential may be depolarizing or hyperpolarizing, similar to the EPSPs or IPSPs generated at postsynaptic membranes in response to neurotransmitter binding (see p. 369).

  Receptors can produce one of two types of graded potentials. When the receptor region is part of a sensory neuron (as with free dendrites or the encapsulated receptors of most general sense receptors), the graded potential is called a **generator potential** because it generates action potentials in a sensory neuron.

  When the receptor is a separate cell (as in most special senses), the graded potential is called a **receptor potential** because it occurs in a separate receptor cell. The receptor potential changes the amount of neurotransmitter released by the receptor cell onto the sensory neuron. The neurotransmitters then generate graded potentials in the sensory neuron.
- Graded potentials in the first-order sensory neuron must reach *threshold* so that voltage-gated sodium channels on

the axon are opened and nerve impulses are generated and propagated to the CNS.

Adaptation Information about a stimulus—its strength, duration, and pattern—is encoded in the frequency of nerve impulses: the greater the frequency, the stronger the stimulus. Many but not all sensory receptors exhibit **adaptation**, a change in sensitivity (and nerve impulse generation) in the presence of a constant stimulus. For example, when you step into bright sunlight from a darkened room, your eyes are initially dazzled, but your photoreceptors rapidly adapt, allowing you to see both bright areas and dark areas in the scene.

Phasic receptors are *fast adapting*, often giving bursts of impulses at the beginning and the end of the stimulus. Phasic receptors report *changes* in the internal or external environment. Receptors for light touch are phasic, so we are not constantly aware of our clothes touching our skin. On the other hand, **tonic receptors** provide a sustained response with little or no adaptation. Receptors for pain and proprioception (position sense) are tonic receptors because of the protective importance of their information.

Processing at the Circuit Level

At the second level of integration, the circuit level, the task is to deliver impulses to the appropriate region of the cerebral cortex for localization and perception of the stimulus.

Recall from Chapter 12 that ascending sensory pathways typically consist of a chain of three neurons called first-, second-, and third-order sensory neurons. The axons of first-order sensory neurons, whose cell bodies are in the dorsal root or cranial ganglia, link the receptor and circuit levels of processing. Central processes of first-order neurons branch diffusely when they enter the spinal cord. Some branches take part in local spinal cord reflexes. Others synapse with second-order sensory neurons, which then synapse with the third-order sensory neurons that take the message to the cerebral cortex.

Processing at the Perceptual Level

Sensory input is interpreted in the cerebral cortex. The ability to identify and appreciate sensations depends on the location of the target neurons in the sensory cortex, not on the nature of the message (which is, after all, just an action potential). Each sensory fiber is analogous to a "labeled line" that tells the brain "who" is calling—a taste bud or a pressure receptor—and from "where." The brain always interprets the activity of a specific sensory receptor ("who") as a specific sensation, no matter how it is activated.

For example, pressing on your eyeball activates photoreceptors, but what you "see" is light. The exact point in the cortex that is activated always refers to the same "where," regardless of how it is activated, a phenomenon called **projection**. Electrically stimulating a particular spot in the visual cortex causes you to "see" light in a particular place.

Let's examine the major features of sensory perception.

- **Perceptual detection** is the ability to detect that a stimulus has occurred. This is the simplest level of perception. As a general rule, inputs from several receptors must be summed for perceptual detection to occur.
- **Magnitude estimation** is the ability to detect how *intense* the stimulus is. Perceived intensity increases as stimulus intensity increases because of frequency coding (see Figure 11.12).
- **Spatial discrimination** allows us to identify the site or pattern of stimulation. A common tool for studying this quality in the laboratory is the **two-point discrimination** test. The test determines how close together two points on the skin can be and still be perceived as two points rather than as one. This test provides a crude map of the density of tactile receptors in the various regions of the skin. The distance between perceived points varies from less than 5 mm on highly sensitive body areas (tip of the tongue) to more than 50 mm on less sensitive areas (the back).
- **Feature abstraction** is the mechanism by which a neuron or circuit is tuned to one feature, or property, of a stimulus in preference to others. Sensation usually involves an interplay of several stimulus features.

  For example, one touch tells us that velvet is warm, compressible, and smooth but not completely continuous, each a feature that contributes to our perception of "velvet." Feature abstraction enables us to identify more complex aspects of a sensation.
- **Quality discrimination** is the ability to differentiate the submodalities of a particular sensation. Each sensory modality has several **qualities**, or submodalities. For example, taste is a sensory modality and its submodalities include sweet and bitter.
- **Pattern recognition** is the ability to take in the scene around us and recognize a familiar pattern, an unfamiliar one, or one that has special significance for us. For example, we can look at an image made of dots and recognize it as the portrait of a familiar face. We can listen to music and hear a melody, not just a string of notes.

Perception of Pain

Everyone has suffered pain—the cruel persistence of a headache, the smart of a bee sting or a cut finger. Although we may not appreciate it at the time, pain is invaluable because it warns us of actual or impending tissue damage and motivates us to take protective action. Managing a patient's pain can be difficult because pain is an intensely personal experience that cannot be measured objectively.

Pain receptors are activated by extremes of pressure and temperature as well as a veritable soup of chemicals released from injured tissue. Histamine, K^+, ATP, acids, and bradykinin are among the most potent pain-producing chemicals. All of these chemicals act on small-diameter fibers.

When you cut your finger, you may have noticed that you first felt a sharp pain followed some time later by burning or aching. Sharp pain is carried by the smallest of the myelinated sensory fibers, the A delta fibers, while burning pain is carried more slowly by small nonmyelinated C fibers. Both types of fibers release the neurotransmitters *glutamate* and *substance P*, which activate second-order sensory neurons. Axons from these second-order neurons ascend to the brain via the spinothalamic tract and other pathways.

If you cut your finger while fighting off an attacker, you might not notice the cut at all. How can that be? The brain has its own pain-suppressing analgesic systems in which the endogenous opioids such as *endorphins* and *enkephalins* (p. 373) play a key role. Various nuclei in the brain stem, including the periaqueductal gray matter of the midbrain, relay descending cortical and hypothalamic pain-suppressing signals. Descending fibers activate interneurons in the spinal cord, which release enkephalins. Enkephalins are inhibitory neurotransmitters that quash the pain signals generated by nociceptors.

HOMEOSTATIC IMBALANCE 13.1 CLINICAL

Normally the body maintains a steady state that correlates injury and pain. Long-lasting or very intense pain inputs, such as limb amputation, can disrupt this system, leading to **hyperalgesia** (pain amplification), chronic pain, and **phantom limb pain**. Intense or long-lasting pain activates *NMDA receptors*, the same receptors that strengthen neural connections during certain kinds of learning. Essentially, the spinal cord *learns* hyperalgesia. In light of this, it is crucial that health professionals effectively manage pain early to prevent chronic pain from becoming established.

Phantom limb pain (pain perceived in tissue that is no longer present) is a curious example of hyperalgesia. Until recently, surgical limb amputations were conducted under general anesthesia only and the spinal cord still experienced the pain of amputation. Epidural anesthetics block neurotransmission in the spinal cord, and using them during surgery greatly reduces the incidence of phantom limb pain. +

Visceral and Referred Pain

Visceral pain results from noxious stimulation of receptors in the organs of the thorax and abdominal cavity. Like deep somatic pain, it is usually a vague sensation of dull aching, gnawing, or burning. Important stimuli for visceral pain are extreme stretching of tissue, ischemia (low blood flow), irritating chemicals, and muscle spasms.

The fact that visceral pain afferents travel along the same pathways as somatic pain fibers helps explain the phenomenon of **referred pain**, in which pain stimuli arising in one part of the body are perceived as coming from another part. For example, a person experiencing a heart attack may feel pain that radiates along the medial aspect of the left arm. Because the same spinal segments (T_1–T_5) innervate both the heart and arm, the brain interprets these inputs as coming from the more common somatic pathway. **Figure 13.2** shows cutaneous areas to which visceral pain is commonly referred.

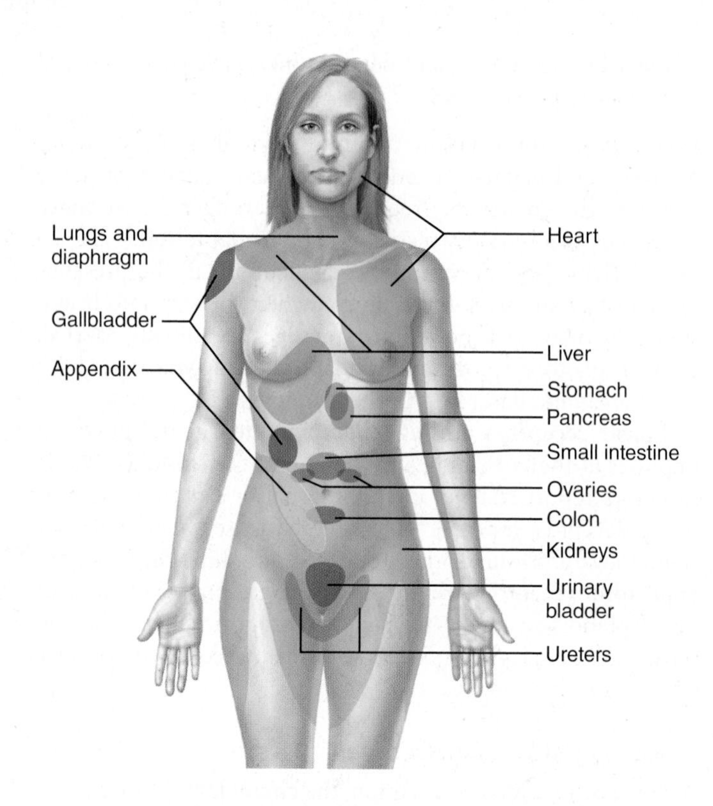

Figure 13.2 Map of referred pain. This map shows the anterior skin areas to which pain is referred from certain visceral organs.

Check Your Understanding

1. What are the three levels of sensory integration?
2. What is the key difference between tonic and phasic receptors? Why are pain receptors tonic?
3. Your cortex decodes incoming action potentials from sensory pathways. How does it tell the difference between hot and cold? Between cool and cold? Between ice on your finger and ice on your foot?

For answers, see Answers Appendix.

13.2 Sensory receptors are activated by changes in the internal or external environment

Learning Objective

☐ Classify general sensory receptors by stimulus detected, body location, and structure.

Sensory receptors are specialized to respond to changes in their environment, which are called **stimuli**. Typically, activation of a sensory receptor by an adequate stimulus results in graded potentials that in turn trigger nerve impulses along the afferent PNS fibers coursing to the CNS. Also, we have just learned that both sensation and perception of that stimulus occur in the brain. Now, let's examine how sensory receptors are classified. Basically, there are three ways to classify sensory receptors: (1) by the type of stimulus they detect; (2) by their body location; and (3) by their structural complexity.

Classification by Stimulus Type

These categories are easy to remember because the name usually indicates the stimulus that activates the receptor.

- **Mechanoreceptors** respond to mechanical force such as touch, pressure (including blood pressure), vibration, and stretch.

- **Thermoreceptors** respond to temperature changes.
- **Photoreceptors**, such as those of the retina of the eye, respond to light.
- **Chemoreceptors** respond to chemicals in solution (molecules smelled or tasted, or changes in blood or interstitial fluid chemistry).
- **Nociceptors** (no″se-sep′torz; *noci* = harm) respond to potentially damaging stimuli that result in pain. For example, searing heat, extreme cold, excessive pressure, and inflammatory chemicals are all interpreted as painful. These signals stimulate subtypes of thermoreceptors, mechanoreceptors, and chemoreceptors.

Classification by Location

Receptors can be grouped into three receptor classes according to either their location or the location of the activating stimulus: exteroceptors, interoceptors, and proprioceptors.

- **Exteroceptors** (ek″ster-o-sep′torz) are sensitive to stimuli arising outside the body (*extero* = outside), so most exteroceptors are near or at the body surface. They include touch, pressure, pain, and temperature receptors in the skin and most receptors of the special senses (vision, hearing, equilibrium, smell, and taste).
- **Interoceptors** (in″ter-o-sep′torz), also called *visceroceptors*, respond to stimuli within the body (*intero* = inside), such as from the internal viscera and blood vessels. Interoceptors monitor a variety of stimuli, including chemical changes, tissue stretch, and temperature. Sometimes their activity causes us to feel pain, discomfort, hunger, or thirst. However, we are usually unaware of their workings.
- **Proprioceptors** (pro″pre-o-sep′torz), like interoceptors, respond to internal stimuli. However, their location is much more restricted. Proprioceptors occur in skeletal muscles, tendons, joints, and ligaments and in connective tissue coverings of bones and muscles. (Some authorities include the equilibrium receptors of the inner ear in this class.) Proprioceptors constantly advise the brain of our body movements (*propria* = one's own) by monitoring how much the organs containing these receptors are stretched.

Classification by Receptor Structure

The overwhelming majority of sensory receptors belong to the **general senses** and are simply the modified dendritic endings of sensory neurons. They are found throughout the body and monitor most types of general sensory information.

Receptors for the **special senses** (vision, hearing, equilibrium, smell, and taste) are housed in complex **sense organs**. For example, the sense organ we know as the eye is composed not only of sensory neurons but also of nonneural cells that form its supporting wall, lens, and other associated structures.

Simple Receptors of the General Senses

The widely distributed general sensory receptors are involved in tactile sensation (a mix of touch, pressure, stretch, and vibration), temperature monitoring, and pain, as well as the "muscle sense" provided by proprioceptors. As you read about these receptors, notice that there is no perfect "one-receptor–one-function" relationship. Instead, one type of receptor can respond to several different kinds of stimuli. Likewise, different types of receptors can respond to similar stimuli. Anatomically, general sensory receptors are nerve endings that are either *nonencapsulated* (*free*) or *encapsulated*. Table 13.1 illustrates the general sensory receptors.

Nonencapsulated (Free) Nerve Endings Present nearly everywhere in the body, **nonencapsulated (free) nerve endings** of sensory neurons are particularly abundant in epithelia and connective tissues. Most of these sensory fibers are nonmyelinated, small-diameter group C fibers, and their distal endings (the sensory terminals) usually have small knoblike swellings.

Free nerve endings respond chiefly to temperature and painful stimuli, but some respond to tissue movements caused by pressure as well. Nerve endings that respond to cold (10–40°C, or 50–104°F) are located in the superficial dermis. Those responding to heat (32–48°C, or 90–120°F) are deeper in the dermis.

Heat or cold outside the range of thermoreceptors activates nociceptors and is perceived as painful. Nociceptors also respond to pinch and chemicals released from damaged tissue.

Another sensation mediated by free nerve endings is itch. Located in the dermis, the *itch receptor* escaped detection for years because of its thin diameter. A number of chemicals—notably histamine—present at inflamed sites activate these nerve endings.

Other nonencapsulated nerve endings include:

- **Tactile (Merkel) discs**, which lie in the deepest layer of the epidermis, function as light touch receptors. Certain free nerve endings associate with enlarged, disc-shaped epidermal cells (*tactile* or *Merkel cells*) to form tactile discs.
- **Hair follicle receptors**, free nerve endings that wrap basketlike around hair follicles, are light touch receptors that detect bending of hairs. The tickle of a mosquito landing on your skin is mediated by hair follicle receptors.

Encapsulated Nerve Endings All **encapsulated nerve endings** consist of one or more fiber terminals of sensory neurons enclosed in a connective tissue capsule. Virtually all encapsulated receptors are mechanoreceptors, but they vary greatly in shape, size, and distribution in the body. They include tactile corpuscles, lamellar corpuscles, bulbous corpuscles, muscle spindles, tendon organs, and joint kinesthetic receptors (Table 13.1):

- **Tactile corpuscles** or *Meissner's corpuscles* are small receptors in which a few spiraling sensory terminals are surrounded by Schwann cells and then by a thin egg-shaped connective tissue capsule. Tactile corpuscles are found just beneath the epidermis in the dermal papillae and are especially numerous in sensitive and hairless skin areas such as the nipples, fingertips, and soles of the feet. They are receptors for discriminative touch, and apparently play the same role in sensing light touch in hairless skin that hair follicle receptors do in hairy skin.
- **Lamellar corpuscles**, also called *Pacinian corpuscles*, are scattered deep in the dermis, and in subcutaneous tissue

Eyebrows

The **eyebrows** are short, coarse hairs that overlie the supraorbital margins of the skull (**Figure 13.3**). They help shade the eyes from sunlight and prevent perspiration trickling down the forehead from reaching the eyes.

Eyelids

Anteriorly, the eyes are protected by the **eyelids** or **palpebrae** (pal′pĕ-bre). The eyelids are separated by the **palpebral fissure** ("eyelid slit") and meet at the medial and lateral angles of the eye—the **medial** and **lateral commissures** (*canthi*) (Figure 13.3a).

The medial commissure sports a fleshy elevation called the **lacrimal caruncle** (kar′ung-kl; "a bit of flesh"). The caruncle contains sebaceous and sweat glands and produces the whitish, oily secretion (fancifully called the Sandman's eye-sand) that sometimes collects at the medial commissure, especially during sleep. In most Asian peoples, a vertical fold of skin called the *epicanthic fold* commonly appears on both sides of the nose and sometimes covers the medial commissure.

The eyelids are thin, skin-covered folds supported internally by connective tissue sheets called **tarsal plates** (Figure 13.3b). The tarsal plates also anchor the **orbicularis oculi** and **levator palpebrae superioris** muscles that run within the eyelid. The orbicularis muscle encircles the eye, and the eye closes when it contracts.

The eyelid muscles are activated reflexively to cause blinking every 3–7 seconds and to protect the eye from foreign objects. Each time we blink, accessory structure secretions (oil, mucus, and saline solution) spread across the eyeball surface, keeping the eyes moist.

Projecting from the free margin of each eyelid are the **eyelashes**. The follicles of the eyelash hairs are richly innervated by nerve endings (hair follicle receptors), and anything that touches the eyelashes (even a puff of air) triggers reflex blinking.

Several types of glands are associated with the eyelids. The **tarsal glands** are embedded in the tarsal plates (Figure 13.3b), and their ducts open at the eyelid edge just posterior to the eyelashes. These modified sebaceous glands produce an oily secretion that lubricates the eyelid and the eye and prevents the eyelids from sticking together. A number of smaller, more typical sebaceous glands are associated with the eyelash follicles. Modified sweat glands called ciliary glands lie between the hair follicles (*cilium* = eyelash).

Conjunctiva

The **conjunctiva** (kon″junk-ti′vah; "joined together") is a transparent mucous membrane. It lines the eyelids as the **palpebral conjunctiva** and folds back over the anterior surface of the eyeball as the **bulbar conjunctiva** (Figure 13.3b). The bulbar conjunctiva covers only the white of the eye, not the cornea. The bulbar conjunctiva is very thin, and blood vessels are clearly visible beneath it. (They are even more visible in irritated "bloodshot" eyes.)

When the eye is closed, a slitlike space occurs between the conjunctiva-covered eyeball and eyelids. This so-called

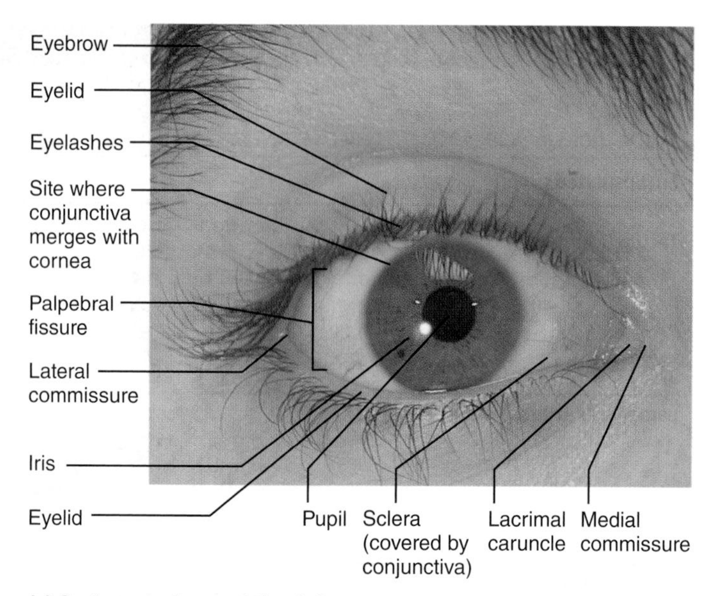

(a) Surface anatomy of the right eye

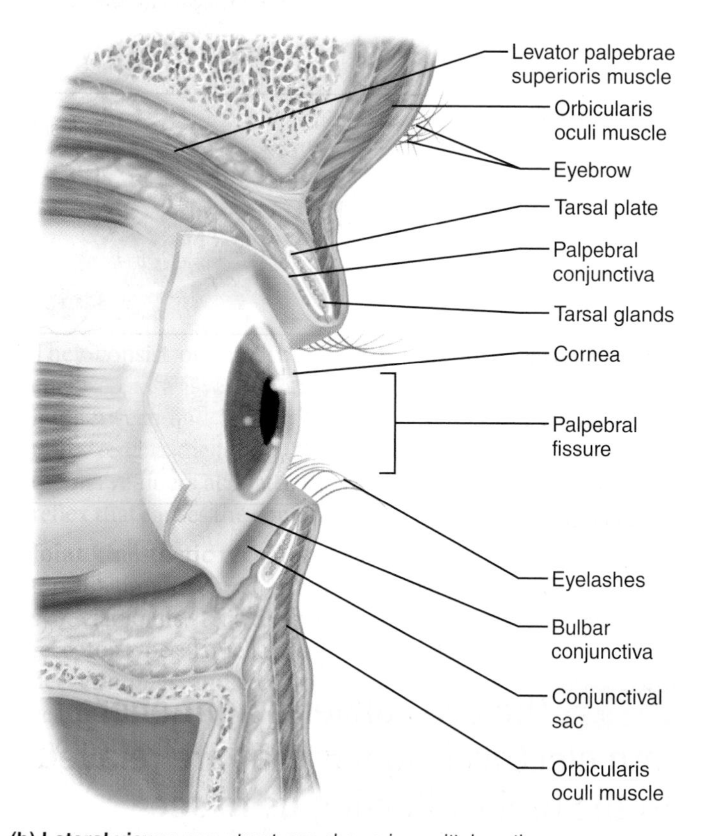

(b) Lateral view; some structures shown in sagittal section

Figure 13.3 The eye and accessory structures.

Practice art labeling
MasteringA&P®>Study Area>Chapter 13

conjunctival sac is where a contact lens lies, and eye medications are often administered into its inferior recess. The major function of the conjunctiva is to produce a lubricating mucus that prevents the eyes from drying out.

HOMEOSTATIC IMBALANCE 13.2

CLINICAL

Inflammation of the conjunctiva, called *conjunctivitis*, results in reddened, irritated eyes. *Pinkeye*, a conjunctival infection caused by bacteria or viruses, is highly contagious. +

Lacrimal Apparatus

The **lacrimal apparatus** (lak′rĭ-mal; "tear") consists of the lacrimal gland and the ducts that drain lacrimal secretions into the nasal cavity (**Figure 13.4**). The **lacrimal gland** lies in the orbit above the lateral end of the eye and is visible through the conjunctiva when the lid is everted. It continually releases a dilute saline solution called **lacrimal secretion**—or, more commonly, **tears**—into the superior part of the conjunctival sac through several small excretory ducts.

Blinking spreads the tears downward and across the eyeball to the medial commissure, where they enter the paired **lacrimal canaliculi** via two tiny openings called **lacrimal puncta** (literally, "prick points"), visible as tiny red dots on the medial margin of each eyelid. From the lacrimal canaliculi, the tears drain into the **lacrimal sac** and then into the **nasolacrimal duct**, which empties into the nasal cavity at the inferior nasal meatus.

Lacrimal fluid contains mucus, antibodies, and **lysozyme**, an enzyme that destroys bacteria. Thus, it cleanses and protects the eye surface as it moistens and lubricates it. When lacrimal secretion increases substantially, tears spill over the eyelids and fill the nasal cavities, causing congestion and the "sniffles." This spillover (tearing) happens when the eyes are irritated or when we are emotionally upset. In the case of eye irritation, enhanced tearing washes away or dilutes the irritating substance. The importance of emotionally induced tears is poorly understood.

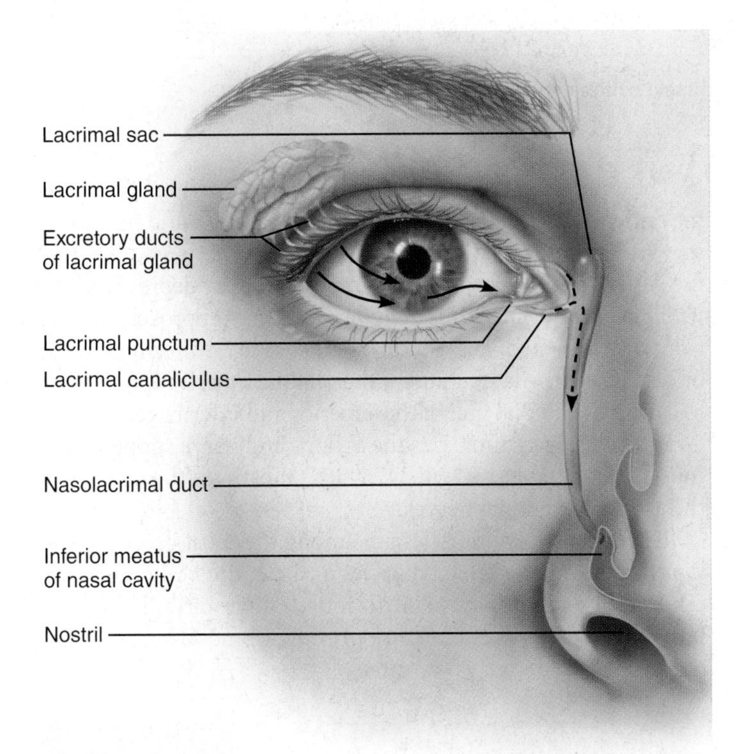

Figure 13.4 The lacrimal apparatus. Arrows indicate the flow of lacrimal fluid (tears) from the lacrimal gland to the nasal cavity.

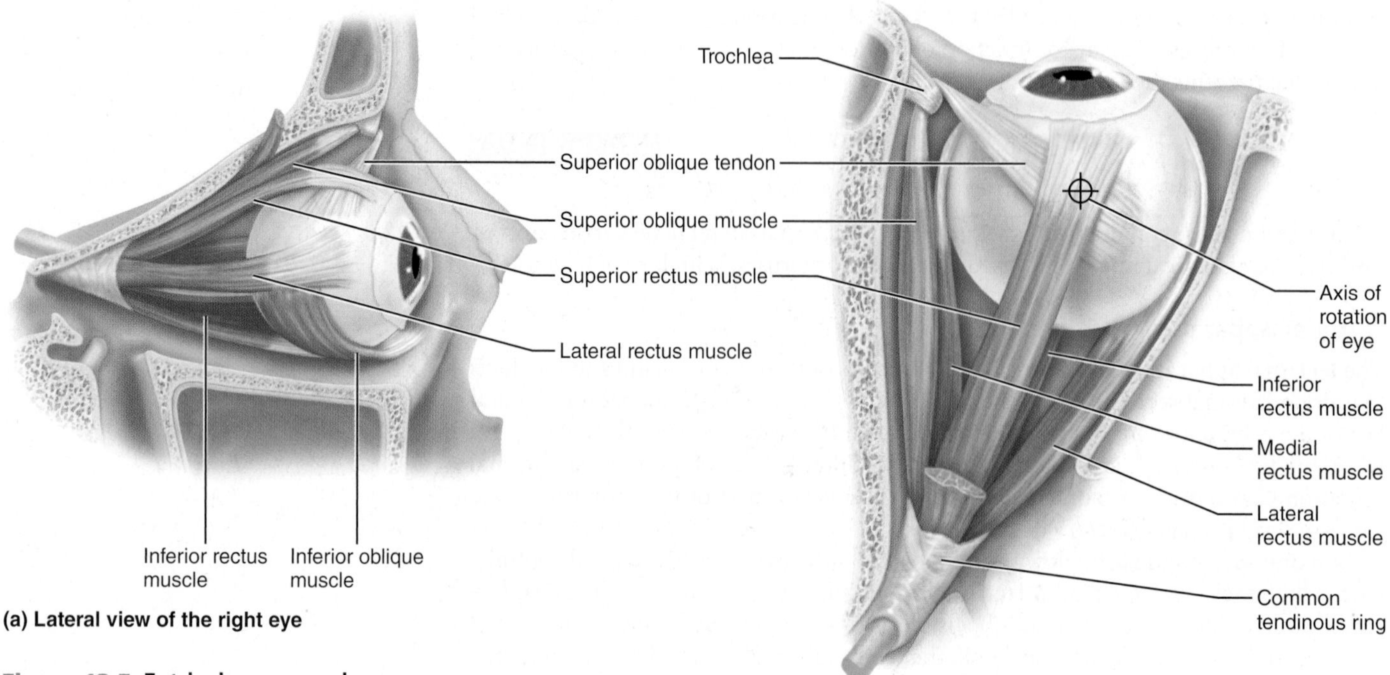

(a) Lateral view of the right eye

(b) Superior view of the right eye

Muscle	Action	Controlling cranial nerve
Lateral rectus	Moves eye laterally	VI (abducens)
Medial rectus	Moves eye medially	III (oculomotor)
Superior rectus	Elevates eye and turns it medially	III (oculomotor)
Inferior rectus	Depresses eye and turns it medially	III (oculomotor)
Inferior oblique	Elevates eye and turns it laterally	III (oculomotor)
Superior oblique	Depresses eye and turns it laterally	IV (trochlear)

(c) Summary of muscle actions and innervating cranial nerves

Figure 13.5 Extrinsic eye muscles.

HOMEOSTATIC IMBALANCE 13.3 CLINICAL

Because the nasal cavity mucosa is continuous with that of the lacrimal duct system, a cold or nasal inflammation often causes the lacrimal mucosa to swell. This swelling constricts the ducts and prevents tears from draining, causing "watery" eyes. +

Extrinsic Eye Muscles

How do our eyes move? Six straplike **extrinsic eye muscles** control the movement of each eyeball. These muscles originate from the walls of the orbit and insert into the outer surface of the eyeball (**Figure 13.5**). They allow the eyes to follow a moving object, help maintain the shape of the eyeball, and hold it in the orbit.

The four *rectus muscles* originate from the **common tendinous ring** (annular ring) at the back of the orbit and run straight to their insertion on the eyeball. Their locations and the movements that they promote are clearly indicated by their names: **superior, inferior, lateral**, and **medial rectus muscles**.

The actions of the two *oblique muscles* are less easy to deduce because they take rather strange paths through the orbit. They move the eye in the vertical plane when the eyeball is already turned medially by the rectus muscles. The **superior oblique muscle** originates in common with the rectus muscles, runs along the medial wall of the orbit, and then makes a right-angle turn and passes through a fibrocartilaginous loop called the **trochlea** (trok′le-ah; "pulley") suspended from the frontal bone before inserting on the superolateral aspect of the eyeball. It rotates the eye downward and somewhat laterally.

The **inferior oblique muscle** originates from the medial orbit surface and runs laterally and obliquely to insert on the inferolateral eye surface. It rotates the eye up and laterally.

The four rectus muscles would seem to provide all the eye movements we require—medial, lateral, superior, and inferior—so why the two oblique muscles? The simplest answer is that the superior and inferior recti cannot elevate or depress the eye *without also turning it medially* because they approach the eye from a posteromedial direction. For an eye to be *directly* elevated or depressed, the lateral pull of the oblique muscles is necessary to cancel the medial pull of the superior and inferior recti.

Figure 13.5c summarizes the actions and nerve supply of the muscles, and Table 13.4 (pp. 476–482) illustrates the courses of the associated cranial nerves.

The extrinsic eye muscles are among the most precisely and rapidly controlled skeletal muscles in the entire body. This precision reflects their high axon-to-muscle-fiber ratio: The motor units of these muscles contain only 8 to 12 muscle cells and in some cases as few as two or three.

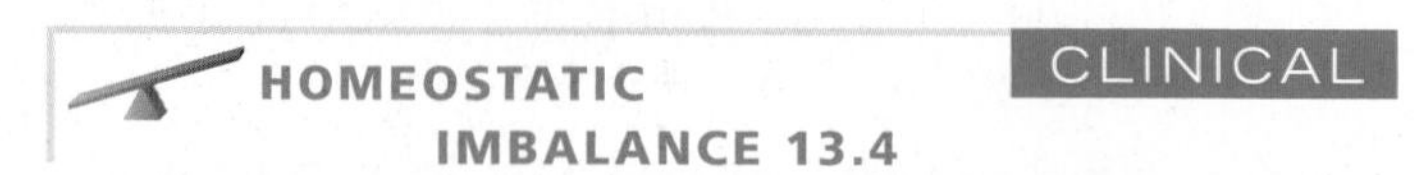
HOMEOSTATIC IMBALANCE 13.4 CLINICAL

When movements of the external muscles of the two eyes are not perfectly coordinated, a person cannot properly focus the

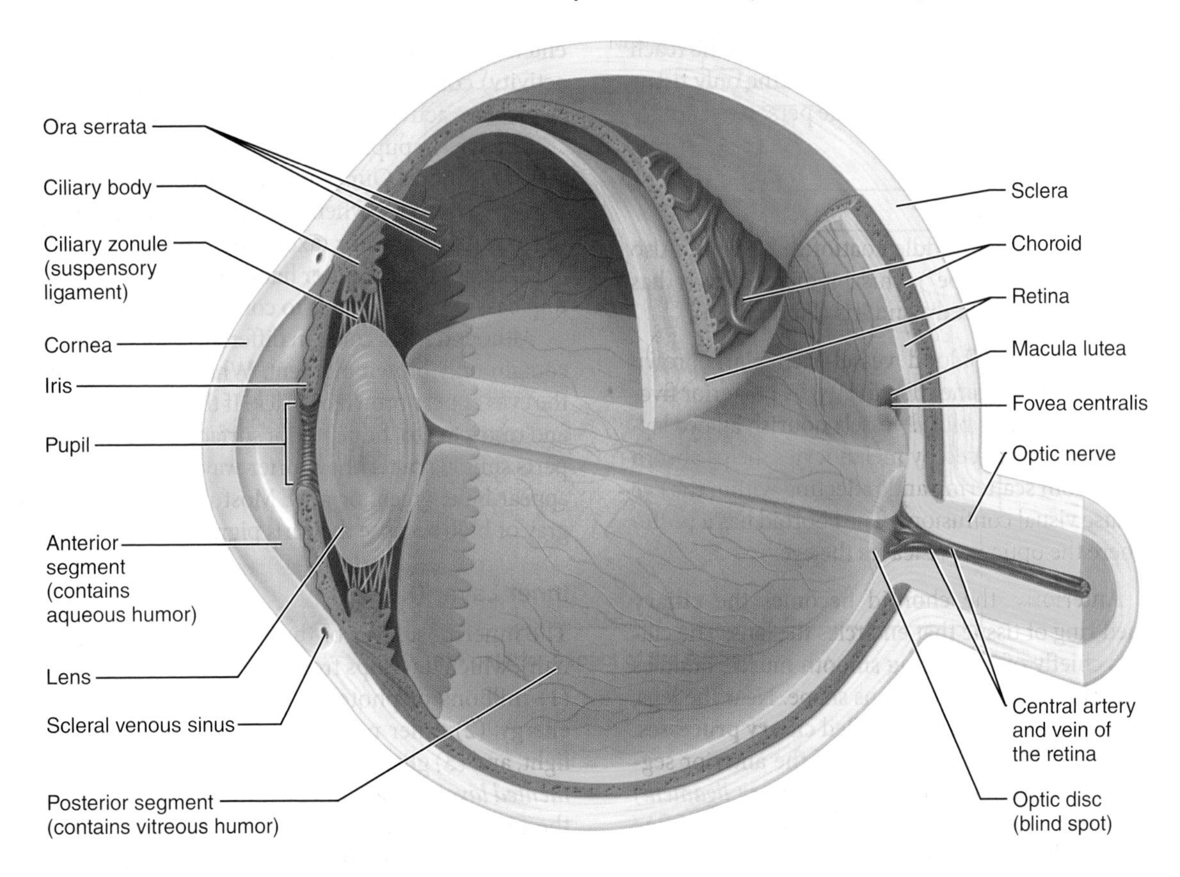

Figure 13.6 Internal structure of the eye (sagittal section). The vitreous humor is illustrated only in the bottom part of the eyeball.

Practice art labeling
MasteringA&P®>Study Area>Chapter 13

images of the same area of the visual field from each eye and so sees two images instead of one. This condition is called **diplopia** (dĭ-plo′pe-ah), or *double vision*. It can result from paralysis or weakness of certain extrinsic muscles, or neurological disorders.

Congenital weakness of the external eye muscles may cause **strabismus** (strah-biz′mus; "cross-eyed"), in which the affected eye rotates medially or laterally. To compensate, the eyes may alternate in focusing on objects. In other cases, only the controllable eye is used, and the brain begins to disregard inputs from the deviant eye, which (unless treated early) can then become functionally blind. +

Structure of the Eyeball

The eye itself, commonly called the **eyeball**, is a slightly irregular hollow sphere (**Figure 13.6**). Its wall is composed of three layers: the fibrous, vascular, and inner layers. Its internal cavity is filled with fluids called *humors* that help to maintain its shape. The lens, the adjustable focusing apparatus of the eye, is supported vertically within the eyeball and divides it into *anterior* and *posterior segments*.

Fibrous Layer

The outermost coat of the eyeball, the **fibrous layer**, is composed of dense avascular connective tissue. It has two obviously different regions: the sclera and the cornea.

Sclera The **sclera** (skler′ah; "hard"), forming the posterior portion and the bulk of the fibrous layer, is glistening white and opaque. Seen anteriorly as the "white of the eye," the tough, tendonlike sclera protects and shapes the eyeball and provides a sturdy anchoring site for the extrinsic eye muscles. Posteriorly, where the sclera is pierced by the optic nerve (cranial nerve II), it is continuous with the dura mater of the brain.

Cornea The anterior sixth of the fibrous layer is modified to form the transparent **cornea**, which bulges anteriorly from its junction with the sclera. The crystal-clear cornea forms a window that lets light enter the eye, and is a major part of the light-bending apparatus of the eye.

Epithelial sheets cover both faces of the cornea. The external sheet, a stratified squamous epithelium that protects the cornea from abrasion, merges with the bulbar conjunctiva at the corneoscleral junction. Epithelial cells that continually renew the cornea are located here. The deep *corneal endothelium*, composed of simple squamous epithelium, lines the inner face of the cornea.

The cornea is well supplied with nerve endings, most of which are pain receptors. When the cornea is touched, blinking and increased tearing occur reflexively. Even so, the cornea is the most exposed part of the eye and is vulnerable to damage from dust, slivers, and the like. Luckily, its capacity for regeneration and repair is extraordinary.

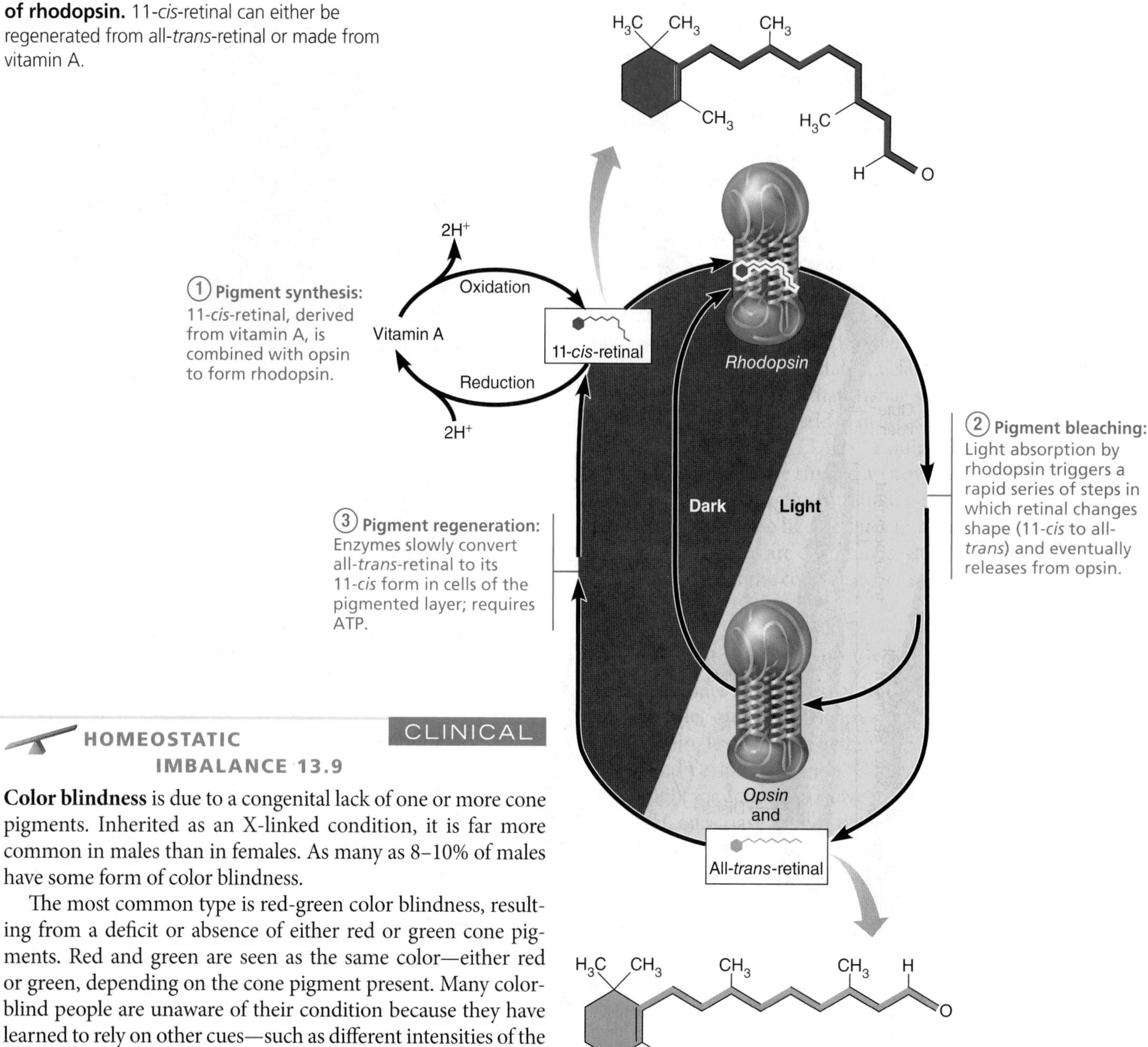

Figure 13.15 The formation and breakdown of rhodopsin. 11-*cis*-retinal can either be regenerated from all-*trans*-retinal or made from vitamin A.

HOMEOSTATIC IMBALANCE 13.9 CLINICAL

Color blindness is due to a congenital lack of one or more cone pigments. Inherited as an X-linked condition, it is far more common in males than in females. As many as 8–10% of males have some form of color blindness.

The most common type is red-green color blindness, resulting from a deficit or absence of either red or green cone pigments. Red and green are seen as the same color—either red or green, depending on the cone pigment present. Many color-blind people are unaware of their condition because they have learned to rely on other cues—such as different intensities of the same color—to distinguish something green from something red, such as traffic signals. ✚

Retinal is chemically related to vitamin A and is made from it. The cells of the pigmented layer of the retina absorb vitamin A from the blood and serve as the local vitamin A depot for rods and cones.

Retinal can assume a variety of three-dimensional forms, each form called an isomer. When bound to opsin, retinal has a bent shape called **11-*cis*-retinal**, as shown at the top of Figure 13.15. However, when the pigment absorbs a photon of light, retinal twists and snaps into a new configuration, **all-*trans*-retinal** (Figure 13.15, bottom). This change, in turn, causes opsin to change shape and assume its activated form.

The capture of light by visual pigments is the *only* light-dependent stage, and this simple photochemical event initiates a whole chain of chemical and electrical reactions in rods and cones that ultimately causes electrical impulses to be transmitted along the optic nerve. Let's look more closely at these events in rods and cones.

Phototransduction

Phototransduction is the process by which light energy is converted into a graded receptor potential. It begins when a visual pigment captures a photon of light.

Capturing Light The visual pigment of rods is a deep purple pigment called **rhodopsin** (ro-dop′sin; *rhodo* = rose, *opsis* =

vision). Rhodopsin molecules are arranged in a single layer in the membranes of each of the thousands of discs in the rods' outer segments. The formation and breakdown of rhodopsin follows the process shown in **Figure 13.15**.

1. **Pigment synthesis:** Rhodopsin forms and accumulates in the dark. Vitamin A is oxidized (and isomerized) to the 11-*cis*-retinal form and then combined with opsin to form rhodopsin.
2. **Pigment bleaching:** When rhodopsin absorbs light, retinal changes shape to its all-*trans* isomer, allowing the surrounding protein to quickly relax like an uncoiling spring into its light-activated form. Eventually, the retinal-opsin combination breaks down, allowing retinal and opsin to separate. The breakdown of rhodopsin to retinal and opsin is known as the **bleaching of the pigment**.
3. **Pigment regeneration:** Once the light-struck all-*trans*-retinal detaches from opsin, enzymes within the pigmented epithelium reconvert it to its 11-*cis* isomer. Then, retinal heads "homeward" again to the photoreceptor cells' outer segments. Rhodopsin is regenerated when 11-*cis*-retinal is rejoined to opsin.

The breakdown and regeneration of visual pigments in cones is essentially the same as for rhodopsin. However, cones are about a hundred times less sensitive than rods, which means that it takes higher-intensity (brighter) light to activate cones.

Light Transduction Reactions What happens when light changes opsin's shape? An enzymatic cascade occurs that ultimately results in closing cation channels that are normally kept open in the dark. **Figure 13.16** illustrates this process in detail, but in short, light-activated rhodopsin activates a G protein called **transducin**. Transducin, in turn, activates *PDE* (*phosphodiesterase*), the enzyme that breaks down **cyclic GMP (cGMP)**. In the dark, cGMP binds to cation channels in the outer segments of photoreceptor cells, holding them open. This allows Na^+ and Ca^{2+} to enter, depolarizing the cell to its *dark potential* of about −40 mV. In the light, PDE breaks down cGMP, the cation channels close, Na^+ and Ca^{2+} stop entering the cell, and the cell hyperpolarizes to about −70 mV.

This arrangement can seem bewildering, to say the least. Here we have receptors built to detect light that depolarize in the dark and hyperpolarize when exposed to light! However, all that is required is a signal and hyperpolarization is just as good a signal as depolarization.

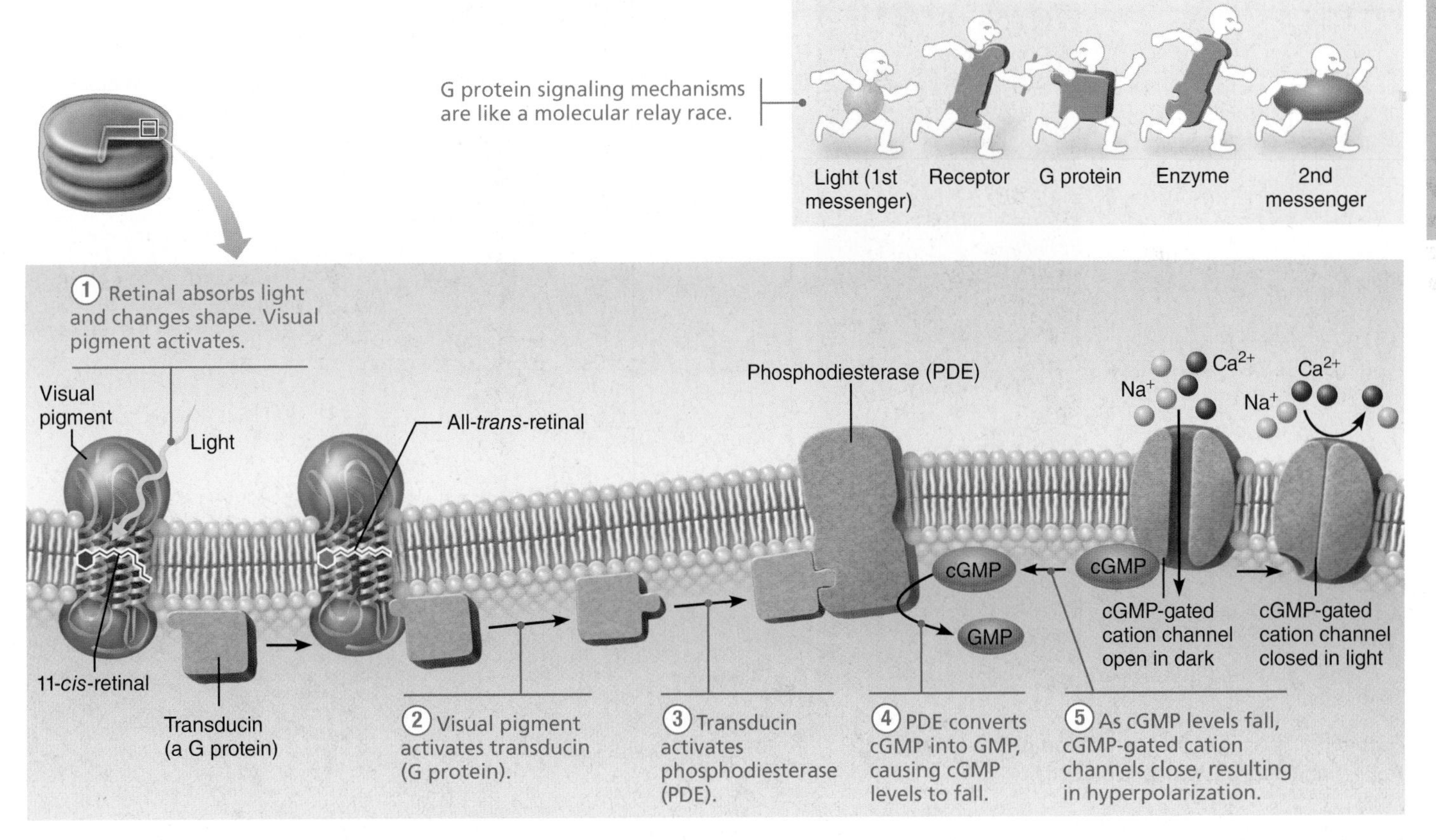

Figure 13.16 Events of phototransduction. A portion of photoreceptor disc membrane is shown. The G protein conversion of GTP to GDP has been omitted for clarity. For simplicity, the channels gated by cyclic GMP (cGMP) are shown on the same membrane as the visual pigment instead of in the plasma membrane. (For the basics of G protein signaling mechanisms, see *Focus on G Proteins*, Focus Figure 3.2 on p. 76.)

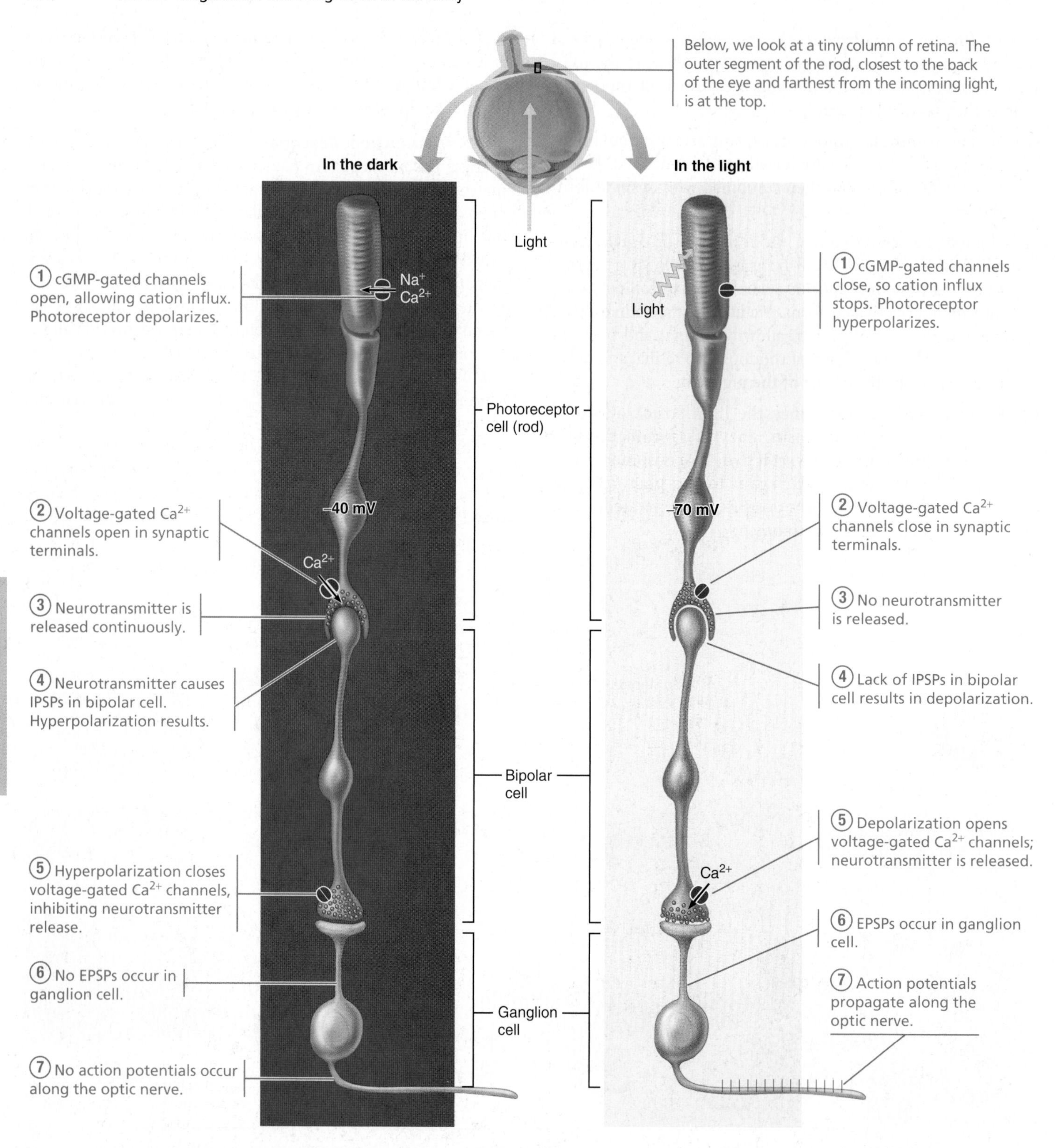

Figure 13.17 Signal transmission in the retina. (EPSP = excitatory postsynaptic potential; IPSP = inhibitory postsynaptic potential)

Information Processing in the Retina

How is the hyperpolarization of the photoreceptors transmitted through the retina and on to the brain? **Figure 13.17** illustrates this process. Notice that the photoreceptors do not generate action potentials (APs), and neither do the bipolar cells that are next in line. They only generate graded potentials. Photoreceptors generate receptor potentials, and bipolar cells generate excitatory or inhibitory postsynaptic potentials (EPSPs or IPSPs).

This is not surprising if you remember that the primary function of APs is to carry information rapidly over long distances. Retinal cells are small cells that are very close together. Graded potentials can serve quite adequately as signals that directly regulate neurotransmitter release at the synapse by opening or closing voltage-gated Ca^{2+} channels.

As shown in the right panel of Figure 13.17, for example, light hyperpolarizes photoreceptors, which then stop releasing their inhibitory neurotransmitter (glutamate). No longer inhibited, bipolar cells depolarize and release neurotransmitter onto ganglion cells. Once the signal reaches the ganglion cells, it is converted into an AP. This AP is transmitted to the brain along the ganglion cell axons that make up the optic nerve.

Light and Dark Adaptation

Rhodopsin is amazingly sensitive. Even starlight bleaches some of its molecules. As long as the light is low intensity, relatively little rhodopsin bleaches and the retina continues to respond to light stimuli. However, in high-intensity light, there is wholesale bleaching of the pigment, and rhodopsin bleaches as fast as it is re-formed. At this point, the rods are nonfunctional, but cones still respond. Hence, retinal sensitivity automatically adjusts to the amount of light present.

Light Adaptation **Light adaptation** occurs when we move from darkness into bright light, as when leaving a movie matinee. We are momentarily dazzled—all we see is white light—because the sensitivity of the retina is still "set" for dim light. Both rods and cones are strongly stimulated, and large amounts of the visual pigments break down almost instantaneously, producing a flood of signals that accounts for the glare.

Under such conditions, compensations occur. The rod system turns off—all of the transducins "pack up and move" to the inner segment, uncoupling rhodopsin from the rest of the transduction cascade. Without transducin in the outer segment, light hitting rhodopsin cannot produce a signal. At the same time, the less sensitive cone system and other retinal neurons rapidly adapt, and retinal sensitivity decreases dramatically. Within about 60 seconds, the cones, initially overexcited by the bright light, are sufficiently desensitized to take over. Visual acuity and color vision continue to improve over the next 5–10 minutes. Thus, during light adaptation, we lose retinal sensitivity (rod function) but gain visual acuity.

Dark Adaptation **Dark adaptation**, essentially the reverse of light adaptation, occurs when we go from a well-lit area into a dark one. Initially, we see nothing but velvety blackness because (1) our cones stop functioning in low-intensity light, and (2) the bright light bleached our rod pigments, and the rods are still turned off.

But once we are in the dark, rhodopsin accumulates, transducin returns to the outer segment, and retinal sensitivity increases. Dark adaptation is much slower than light adaptation and can go on for hours. However, there is usually enough rhodopsin within 20–30 minutes to allow adequate dim-light vision.

During both light and dark adaptation, reflexive changes occur in pupil size. Bright light shining in one or both eyes constricts both pupils (elicits the *pupillary* and *consensual light reflexes*). These pupillary reflexes are mediated by the pretectal nucleus of the midbrain and by parasympathetic fibers. In dim light, the pupils dilate, allowing more light to enter the eye.

HOMEOSTATIC IMBALANCE 13.10 CLINICAL

Night blindness, or *nyctalopia* (nic″tă-lo′pe-uh), is a condition in which rod function is seriously hampered, impairing one's ability to drive safely at night. In countries where malnutrition is common, the most common cause of night blindness is prolonged vitamin A deficiency, which leads to rod degeneration. Vitamin A supplements restore function if they are administered early.

In countries where nutrition isn't a problem, *retinitis pigmentosa*—a group of degenerative retinal diseases that destroy rods—are the most common causes of night blindness. Retinitis pigmentosa results from pigment epithelial cells that are unable to recycle the tips of the rods as they get sloughed off. ✚

Visual Pathways and Processing

The Visual Pathway to the Brain

As we described earlier, the axons of the retinal ganglion cells exit the eye in the **optic nerves**. At the X-shaped **optic chiasma** (*chiasm* = cross), fibers from the medial aspect of each eye cross over to the opposite side and then continue on via the **optic tracts** (Figure 13.18). As a result, each optic tract:

- Contains fibers from the lateral (temporal) aspect of the eye on the same side and fibers from the medial (nasal) aspect of the opposite eye
- Carries all the information from the same half of the visual field

Notice that, because the lens system of each eye reverses all images, the medial half of each retina receives light rays from the *temporal* (lateralmost) part of the visual field (from the far left or far right rather than from straight ahead), and the lateral half of each retina receives an image of the nasal (central) part of the visual field. Consequently, the left optic tract carries a complete representation of the right half of the visual field, and the opposite is true for the right optic tract.

The paired optic tracts sweep posteriorly around the hypothalamus and send most of their axons to synapse with neurons in the **lateral geniculate nuclei** (contained within the lateral geniculate bodies) of the thalamus. The lateral geniculate nuclei maintain the fiber separation established at the chiasma, but they balance and combine the retinal input for delivery to the visual cortex. Axons of these thalamic neurons project through the internal capsule to form the **optic radiation** of fibers in the cerebral white matter (Figure 13.18). These fibers project to the **primary visual cortex** in the occipital lobes, where conscious perception of visual images (seeing) occurs.

Some nerve fibers in the optic tracts send branches to the midbrain. One set of these fibers ends in the **superior colliculi**, visual reflex centers controlling the extrinsic muscles of the eyes. Another set comes from a small subset of ganglion cells in the retina that contain the visual pigment *melanopsin*, dubbed the circadian pigment. These ganglion cells respond directly to light stimuli and their fibers project to the **pretectal nuclei**, which mediate pupillary light reflexes, and to the **suprachiasmatic nucleus** of the hypothalamus, which functions as the "timer" to set our daily biorhythms.

Depth Perception

Each eye's visual field is about 170 degrees. The two visual fields overlap considerably, but each eye sees a slightly different view (Figure 13.18a). The visual cortex fuses the slightly different images delivered by the two eyes, providing us with **depth perception** (or **three-dimensional vision**), an accurate means of locating objects in space.

HOMEOSTATIC IMBALANCE 13.11

CLINICAL

Loss of an eye or destruction of one optic nerve eliminates true depth perception entirely, and peripheral vision on the damaged side.

If neural destruction occurs beyond the optic chiasma—in an optic tract, the thalamus, or visual cortex—then part or all of the opposite half of the visual field is lost. For example, a stroke affecting the left visual cortex leads to blindness in the right half of the visual field. +

Visual Processing

How does information received by rods and cones become vision? Visual processing begins in the retina.

Retinal cells simplify and condense the information from rods and cones, splitting it into a number of different "channels," each with its own type of ganglion cell. These "channels" include information about color and brightness, but also about more complex aspects of what we see—the angle, direction and speed of movement of *edges* (sudden changes in brightness or color). Edges are detected by a kind of contrast enhancement called *lateral inhibition*, which is the job of the amacrine and horizontal cells mentioned on pp. 442–443.

The ganglion cells pass the processed information to the lateral geniculate nuclei of the thalamus. There, information from each eye is combined in preparation for depth perception and input from cones is emphasized.

The *primary visual cortex* contains an accurate topographical map of the retina, with the left visual cortex receiving input from the right visual field and vice versa. Visual processing here occurs at a relatively basic level, with the processing neurons responding to dark and bright edges (contrast information) and object orientation. Surrounding areas process form, color, and movement.

Functional neuroimaging of humans has revealed that complex visual processing extends well forward into the temporal, parietal, and frontal lobes via two parallel streams, one that identifies objects in the visual field and another that assesses the location of objects in space. Output from these regions then passes to the frontal cortex for further processing.

☑ Check Your Understanding

6. What are tears and what structure secretes them?
7. What is the blind spot and why is it blind?
8. Arrange the following in the order that light passes through them to reach the photoreceptors (rods and cones): lens, bipolar cells, vitreous humor, cornea, aqueous humor, ganglion cells. (Hint: See Figure 13.8 if you need a reminder of where ganglion cells and bipolar cells are.)
9. You have been reading this book for a while now and your eyes are beginning to tire. Which intrinsic eye muscles are relaxing as you stare thoughtfully into the distance?
10. Why does your near point of vision move farther away as you age?
11. For each of the following, indicate whether it applies to rods or cones: vision in bright light; only one type of visual pigment; most abundant in the periphery of the retina; many feed into one ganglion cell; color vision; higher sensitivity; higher acuity.
12. Which part of the visual field would be affected by a tumor in the right visual cortex? By a tumor compressing the right optic nerve?

For answers, see Answers Appendix.

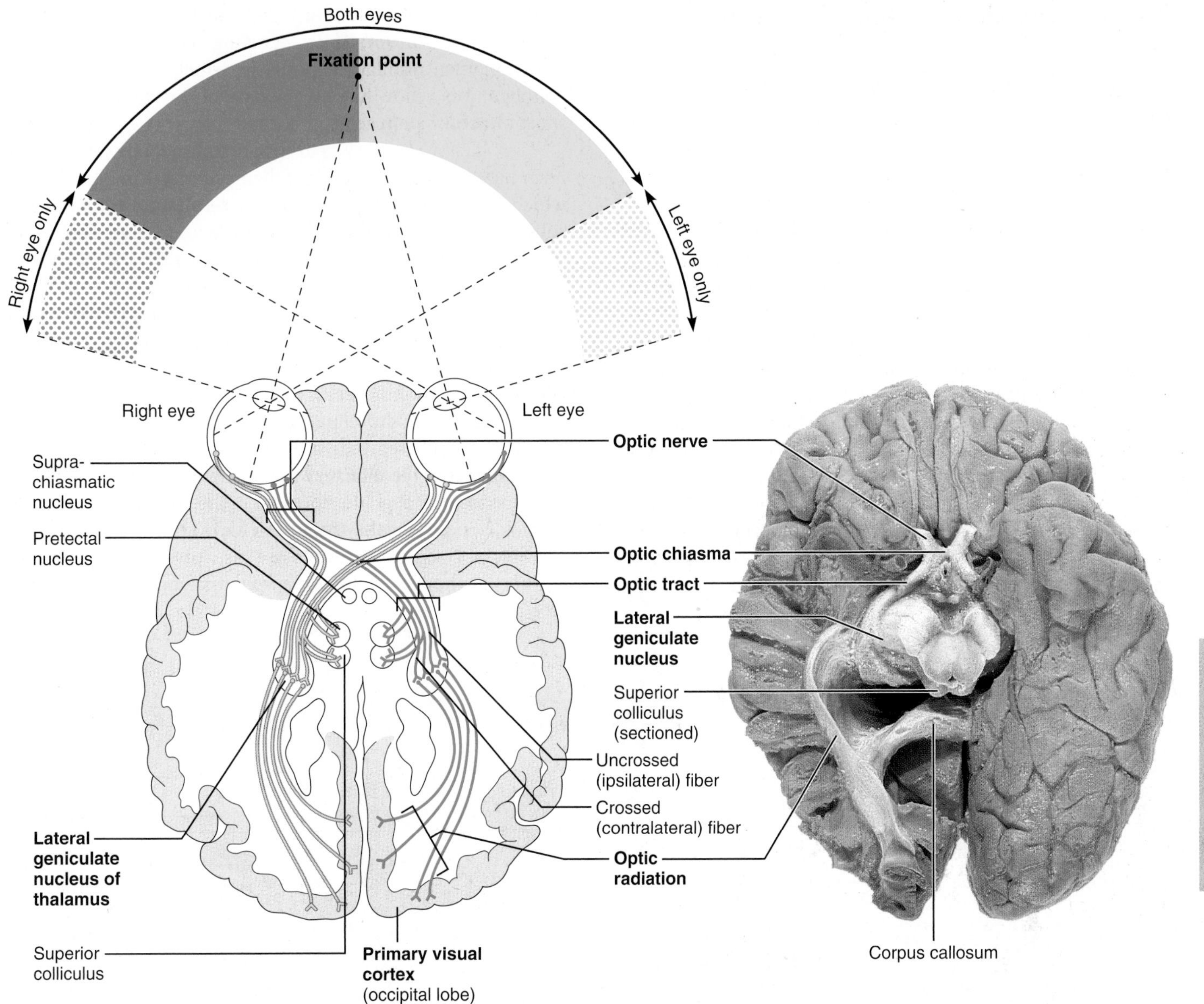

(a) The visual fields of the two eyes overlap considerably. Note that fibers from the lateral portion of each retinal field do not cross at the optic chiasma.

(b) Photograph of human brain, with the right side dissected to reveal internal structures.

Figure 13.18 Visual pathway to the brain and visual fields, inferior view.

Explore human cadaver
MasteringA&P®>Study Area>PAL

13.4 Receptors in the olfactory epithelium and taste buds detect chemicals, allowing us to smell and taste

→ Learning Objective

☐ **Describe the location, structure, and afferent pathways of smell and taste receptors, and explain how these receptors are activated.**

Smell and taste are gritty, primitive senses that alert us to whether that "stuff" nearby (or in our mouth) is to be savored or avoided. The receptors for smell (olfaction) and taste (gustation) are **chemoreceptors** (they respond to chemicals in an aqueous solution). They complement each other and respond to different classes of chemicals. Smell receptors are excited by airborne chemicals that dissolve in fluids coating nasal membranes, and taste receptors are excited by food chemicals dissolved in saliva.

Olfactory Epithelium and the Sense of Smell

Although our olfactory sense (*olfact* = to smell) is far less acute than that of many other animals, the human nose is still no slouch in picking up small differences in odors. Some people capitalize on this ability by becoming wine tasters.

Location and Structure of Olfactory Receptors

The organ of smell is a yellow-tinged patch (about 5 cm^2) of pseudostratified epithelium, called the **olfactory epithelium**, located in the roof of the nasal cavity (**Figure 13.19a**). Air entering the nasal cavity must make a hairpin turn to stimulate olfactory receptors before entering the respiratory passageway below, so the human olfactory epithelium is in a poor position for doing its job. (This is why sniffing, which draws more air superiorly across the olfactory epithelium, intensifies the sense of smell.)

The olfactory epithelium covers the superior nasal concha on each side of the nasal septum, and contains millions of bowling pin–shaped receptor cells—the **olfactory sensory neurons**. These are surrounded and cushioned by columnar **supporting cells**, which make up the bulk of the penny-thin epithelial membrane (Figure 13.19b). The supporting cells contain a yellow-brown pigment similar to lipofuscin, which gives the olfactory epithelium its yellow hue. At the base of the epithelium lie the short **olfactory stem cells**.

The olfactory sensory neurons are unusual bipolar neurons. Each has a thin apical dendrite that terminates in a knob from which several long cilia radiate. These **olfactory cilia**, which substantially increase the receptive surface area, typically lie flat on the nasal epithelium and are covered by a coat of thin mucus produced by the supporting cells and by olfactory glands in the underlying connective tissue. This mucus is a solvent that "captures" and dissolves airborne odorants. Unlike other cilia in the body, which beat rapidly in a coordinated manner, olfactory cilia are largely nonmotile.

The slender, nonmyelinated axons of the olfactory sensory neurons are gathered into small fascicles that collectively form the **filaments of the olfactory nerve** (cranial nerve I). They project superiorly through the openings in the cribriform plate of the ethmoid bone, where they synapse in the overlying olfactory bulbs.

Olfactory sensory neurons are also unusual because they are one of the few types of *neurons* that undergo noticeable

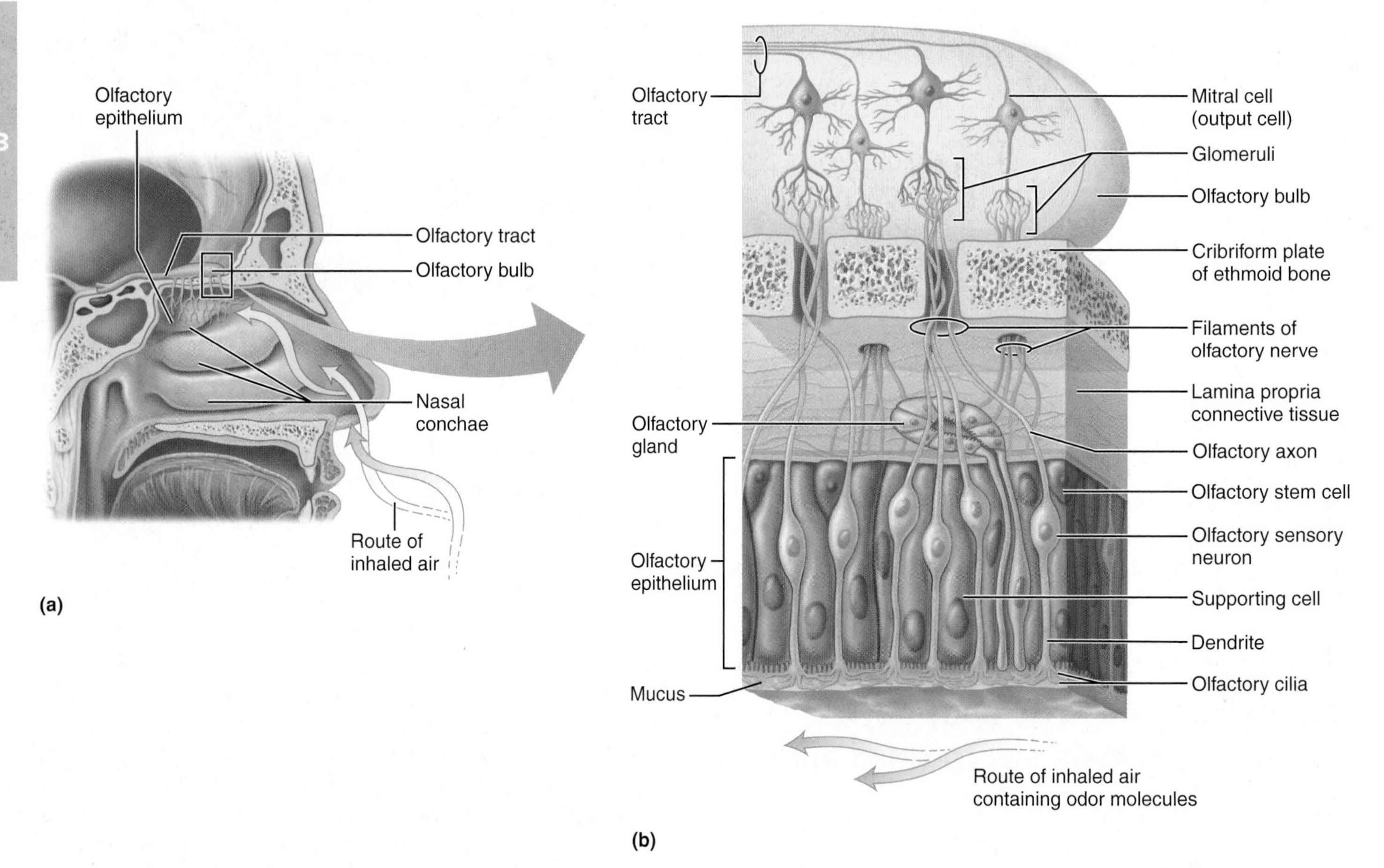

Figure 13.19 Olfactory receptors. (a) Site of olfactory epithelium in the superior nasal cavity. **(b)** An enlarged view of the olfactory epithelium showing the course of the fibers [filaments of the olfactory nerve (cranial nerve I)] through the ethmoid bone. These synapse in the glomeruli of the overlying olfactory bulb. The mitral cells are the output cells of the olfactory bulb.

turnover throughout adult life. Their superficial location puts them at risk for damage, and their typical life span is 30–60 days. Olfactory stem cells in the olfactory epithelium differentiate to replace them.

Specificity of Olfactory Receptors

Smell is difficult to research because any given odor (say, tobacco smoke) may be made up of hundreds of different chemicals (called *odorants*). Taste has been neatly packaged into five taste qualities as you will see, but science has yet to discover any similar means for classifying smell. Humans can distinguish 1 trillion or so odors, but our olfactory sensory neurons are stimulated by combinations of a more limited number of olfactory qualities.

There are about 400 "smell genes" in humans that are active only in the nose. Each gene encodes a unique receptor protein. It appears that each protein responds to one or more odors and each odor binds to several different receptor types. However, each receptor cell has only one type of receptor protein.

Olfactory neurons are exquisitely sensitive—in some cases, just a few molecules activate them. Some of what we call smell is really pain. The nasal cavities contain pain and temperature receptors that respond to irritants such as the sharpness of ammonia, the hotness of chili peppers, and the "chill" of menthol. Impulses from these receptors reach the central nervous system via afferent fibers of the trigeminal nerves.

Physiology of Smell

For us to smell a particular odorant, it must be *volatile*—that is, it must be in the gaseous state as it enters the nasal cavity. Additionally, it must dissolve in the fluid coating the olfactory epithelium.

Smell Transduction Olfactory transduction begins when an odorant binds to a receptor. This event activates G proteins, which activate enzymes (adenylate cyclases) that synthesize cyclic AMP (cAMP) as a second messenger. Cyclic AMP then acts directly on a plasma membrane cation channel, causing it to open, allowing Na^+ and Ca^{2+} to enter.

Na^+ influx leads to depolarization and impulse generation. Ca^{2+} influx causes the transduction process to adapt, decreasing its response to a sustained stimulus. This *olfactory adaptation* helps explain how a person working in a paper mill or sewage treatment plant can still enjoy lunch!

The Olfactory Pathway

As we have already noted, axons of the olfactory sensory neurons form the olfactory nerves that synapse in the overlying **olfactory bulbs**, the distal ends of the olfactory tracts (see Table 13.4 on p. 476). There, the filaments of the olfactory nerves synapse with **mitral cells** (mi′tral), which are second-order sensory neurons, in complex structures called **glomeruli** (glo-mer′u-li; "little balls") (Figure 13.19).

Axons from neurons bearing the same kind of receptor converge on a given type of glomerulus. That is, each glomerulus represents a single aspect of an odor (like one note in a chord) but each odor activates a unique set of glomeruli (the chord itself). Different odors activate different subsets of glomeruli (making different chords which may have some of the same notes). The mitral cells refine the signal, amplify it, and then relay it. The olfactory bulbs also house *amacrine granule cells*, GABA-releasing cells that inhibit mitral cells, so that only highly excitatory olfactory impulses are transmitted.

When the mitral cells are activated, impulses flow from the olfactory bulbs via the **olfactory tracts** (composed mainly of mitral cell axons) to the piriform lobe of the olfactory cortex. From there, two major pathways take information to various parts of the brain. One pathway brings information to part of the frontal lobe just above the orbit, where smells are consciously interpreted and identified. Only some of this information passes through the thalamus.

The other pathway flows to the hypothalamus, amygdaloid body, and other regions of the limbic system. There, emotional responses to odors are elicited. Smells associated with danger—smoke, cooking gas, or skunk scent—trigger the sympathetic fight-or-flight response. Appetizing odors stimulate salivation and the digestive tract, and unpleasant odors can trigger protective reflexes such as sneezing or choking.

HOMEOSTATIC IMBALANCE 13.12 CLINICAL

Most olfactory disorders, or *anosmias* (an-oz′me-ahz; "without smells"), result from head injuries that tear the olfactory nerves, the aftereffects of nasal cavity inflammation, and neurological disorders such as Parkinson's disease. Some people have olfactory hallucinations during which they experience a particular (usually unpleasant) odor, such as rotting meat. These are usually caused by temporal lobe epilepsy involving the olfactory cortex and can occur as olfactory auras before the seizure begins or as olfactory hallucinations during the seizure. ✚

Taste Buds and the Sense of Taste

The word *taste* comes from the Latin *taxare*, meaning "to touch, estimate, or judge." When we taste things, we are in fact intimately testing or judging our environment, and the sense of taste is considered by many to be the most pleasurable of the special senses.

Location and Structure of Taste Buds

Most of our 10,000 or so **taste buds**—the sensory organs for taste—are located on the tongue. A few taste buds are scattered on the soft palate, inner surface of the cheeks, pharynx, and epiglottis of the larynx, but most are found in **papillae** (pah-pil′e), peglike projections of the tongue mucosa that make the tongue surface slightly abrasive. Taste buds are located mainly on the tops of the mushroom-shaped **fungiform papillae** (fun′jĭ-form) scattered over the entire tongue surface and in the epithelium of the side walls of the **foliate papillae** and of the large round **vallate papillae** (val′āt). The vallate papillae are the

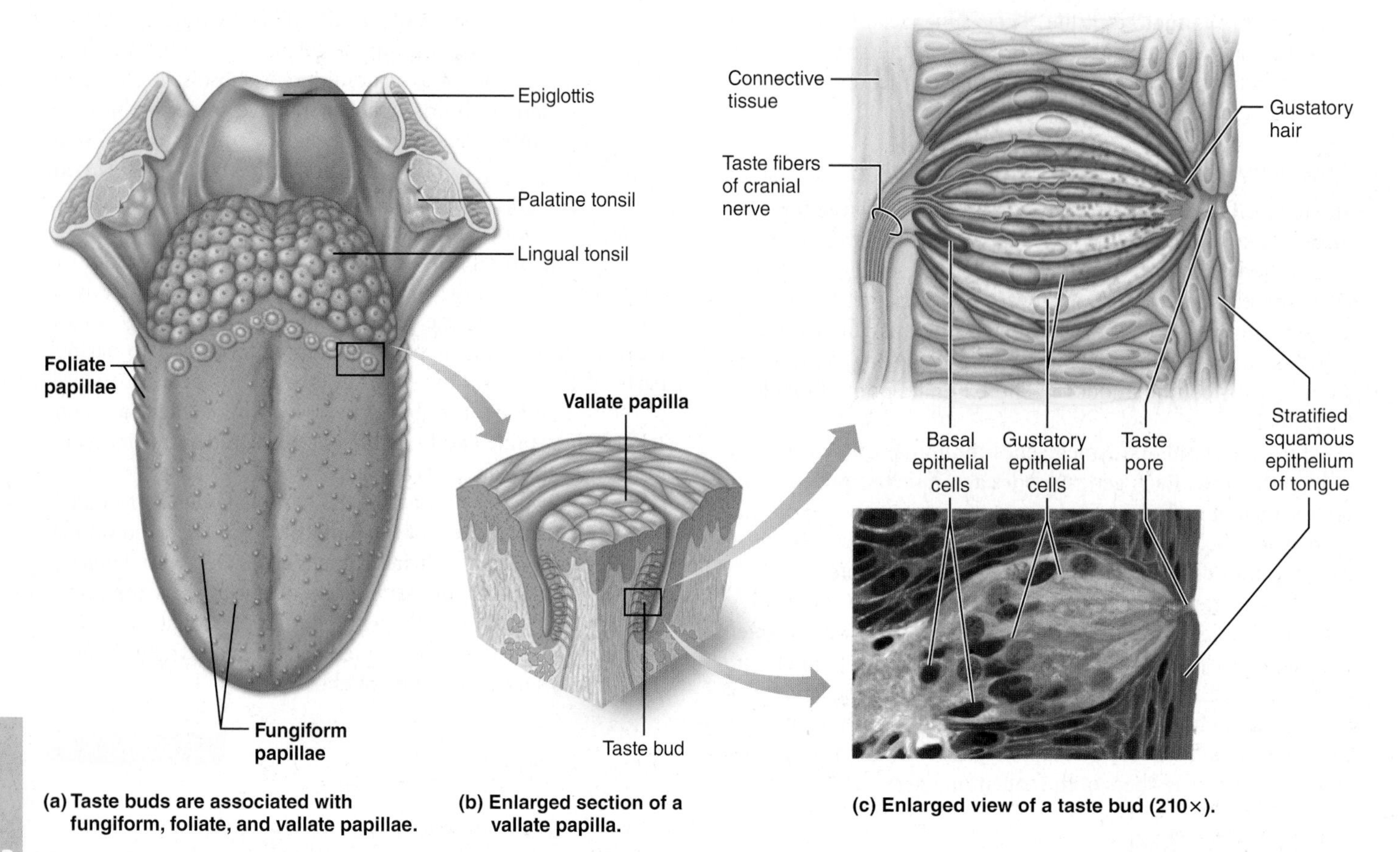

(a) Taste buds are associated with fungiform, foliate, and vallate papillae.

(b) Enlarged section of a vallate papilla.

(c) Enlarged view of a taste bud (210×).

13

Figure 13.20 Location and structure of taste buds on the tongue. (For a related image, see *A Brief Atlas of the Human Body*, Figure 62.)

View histology slides
MasteringA&P®>Study Area>PAL

largest and least numerous papillae, and 8 to 12 of them form an inverted V at the back of the tongue (**Figure 13.20a, b**).

Each flask-shaped taste bud consists of 50 to 100 *epithelial cells* of two major types: *gustatory epithelial cells* and *basal epithelial cells* (Figure 13.20c).

Gustatory Epithelial Cells The **gustatory epithelial cells** are the receptor cells for taste—the *taste cells.* Long microvilli called **gustatory hairs** project from the tips of all gustatory epithelial cells and extend through a **taste pore** to the surface of the epithelium, where they are bathed by saliva. The gustatory hairs are the sensitive portions (*receptor membranes*) of the gustatory epithelial cells. Coiling intimately around the gustatory epithelial cells are sensory dendrites that represent the initial part of the gustatory pathway to the brain. Each afferent fiber receives signals from several gustatory epithelial cells within the taste bud.

There are at least three kinds of gustatory epithelial cells. One kind forms traditional synapses with the sensory dendrites and releases the neurotransmitter serotonin. The other two kinds lack synaptic vesicles, but at least one releases ATP that acts as a neurotransmitter.

Basal Epithelial Cells Because of their location, taste bud cells are subjected to friction and are routinely burned by hot foods. Luckily, they are replaced every seven to ten days. **Basal epithelial cells** act as stem cells, dividing and differentiating into new gustatory epithelial cells.

Basic Taste Sensations

Normally, our taste sensations are complicated mixtures of qualities. However, all taste sensations can be grouped into one of five basic modalities: sweet, sour, salty, bitter, and umami.

- *Sweet* taste is elicited by many organic substances including sugars, saccharin, alcohols, some amino acids, and some lead salts (such as those found in lead paint).
- *Sour* taste is produced by acids, specifically their hydrogen ions (H^+) in solution.
- *Salty* taste is produced by metal ions (inorganic salts); table salt (sodium chloride) tastes the "saltiest."
- *Bitter* taste is elicited by alkaloids (such as quinine, nicotine, caffeine, morphine, and strychnine) as well as a number of nonalkaloid substances, such as aspirin.
- *Umami* (u-mam′e; "delicious"), a subtle taste discovered by the Japanese, is elicited by the amino acids glutamate and aspartate, which appear to be responsible for the "beef taste" of steak, the characteristic tang of aging cheese, and the flavor of the food additive monosodium glutamate.

In addition, there is growing evidence for our ability to taste long-chain fatty acids from lipids. This possible sixth modality may help explain our liking for fatty foods.

Keep in mind that many substances produce a mixture of these basic taste sensations, and taste buds generally respond to all five. However, it appears that a single taste cell has receptors for only one taste modality. Also, all areas of the tongue can detect all taste modalities.

Taste likes and dislikes have homeostatic value. Umami guides the intake of proteins, and a liking for sugar and salt helps satisfy the body's need for carbohydrates and minerals (as well as some amino acids). Many sour, naturally acidic foods (such as oranges, lemons, and tomatoes) are rich sources of vitamin C, an essential vitamin. On the other hand, intensely sour tastes warn us of spoilage. Likewise, many natural poisons and spoiled foods are bitter. Consequently, our dislike for sourness and bitterness is protective.

Physiology of Taste

For a chemical to be tasted it must dissolve in saliva, diffuse into a taste pore, and contact the gustatory hairs.

Activation of Taste Receptors Gustatory epithelial cells contain neurotransmitters. When a food chemical, or *tastant*, binds to receptors in the gustatory epithelial cell membrane, it induces a graded depolarizing potential that causes neurotransmitter release. Binding of the neurotransmitter to the associated sensory dendrites triggers generator potentials that elicit action potentials in these fibers.

The different gustatory epithelial cells have different thresholds for activation. In line with their protective nature, the bitter receptors detect substances present in minute amounts. The other receptors are less sensitive. Taste receptors adapt rapidly, with partial adaptation in 3–5 seconds and complete adaptation in 1–5 minutes.

Taste Transduction The mechanisms of taste transduction are only now beginning to become clear. Three different mechanisms underlie how we taste.

- Salty taste is due to Na^+ influx through Na^+ channels, which directly depolarizes gustatory epithelial cells.
- Sour is mediated by H^+, which acts intracellularly to open channels that allow other cations to enter.
- Bitter, sweet, and umami responses share a common mechanism, but each occurs in a different cell. Each taste's unique set of receptors is coupled to a common G protein called *gustducin*. Activation leads to the release of Ca^{2+} from intracellular stores, which causes cation channels in the plasma membrane to open, thereby depolarizing the cell and releasing the neurotransmitter ATP.

The Gustatory Pathway

Afferent fibers carrying taste information from the tongue are found primarily in two cranial nerve pairs. A branch of the **facial nerve** (VII), the *chorda tympani*, transmits impulses from taste receptors in the anterior two-thirds of the tongue. The lingual branch of the **glossopharyngeal nerve** (IX) services the posterior third and the pharynx just behind. Taste impulses from the few taste buds in the epiglottis and the lower pharynx are conducted primarily by the **vagus nerve** (X).

These afferent fibers synapse in the **solitary nucleus** of the medulla, and from there impulses stream to the thalamus and ultimately to the *gustatory cortex* in the insula. Fibers also project to the hypothalamus and limbic system structures, regions that determine our appreciation of what we are tasting.

HOMEOSTATIC IMBALANCE 13.13 CLINICAL

Taste disorders are less common than disorders of smell, in part because the taste receptors are served by three different nerves and thus are less likely to be "put out of business" completely. Causes of taste disorders include upper respiratory tract infections, head injuries, chemicals or medications, or head and neck radiation for cancer treatment. Zinc supplements may help some cases of radiation-induced taste disorders. +

☑ Check Your Understanding

13. How do the cilia of olfactory sensory neurons help these cells perform their function?

14. Name the five taste modalities. Name the three types of papillae that have taste buds.

For answers, see Answers Appendix.

13.5 Inner ear mechanoreceptors enable hearing and balance

→ Learning Objective

- ☐ **Describe the structure and general function of the outer, middle, and internal ears.**
- ☐ **Describe the sound conduction pathway to the fluids of the internal ear, and sound transduction.**
- ☐ **Describe the pathway of impulses traveling from the cochlea to the auditory cortex.**
- ☐ **Explain how we are able to differentiate pitch and loudness, and to localize the source of sounds.**
- ☐ **Explain how the balance organs of the semicircular canals and the vestibule help maintain equilibrium.**
- ☐ **List possible causes and symptoms of otitis media, deafness, and Ménière's syndrome.**

At first glance, the machinery for hearing and balance appears very crude—fluids must be stirred to stimulate the mechanoreceptors of the internal ear. Nevertheless, our hearing apparatus allows us to hear an extraordinary range of sound, and our equilibrium (balance) receptors continually inform the nervous system of head movements and position. Although the organs serving these two senses are structurally interconnected within the ear, their receptors respond to different stimuli and are activated independently of one another.

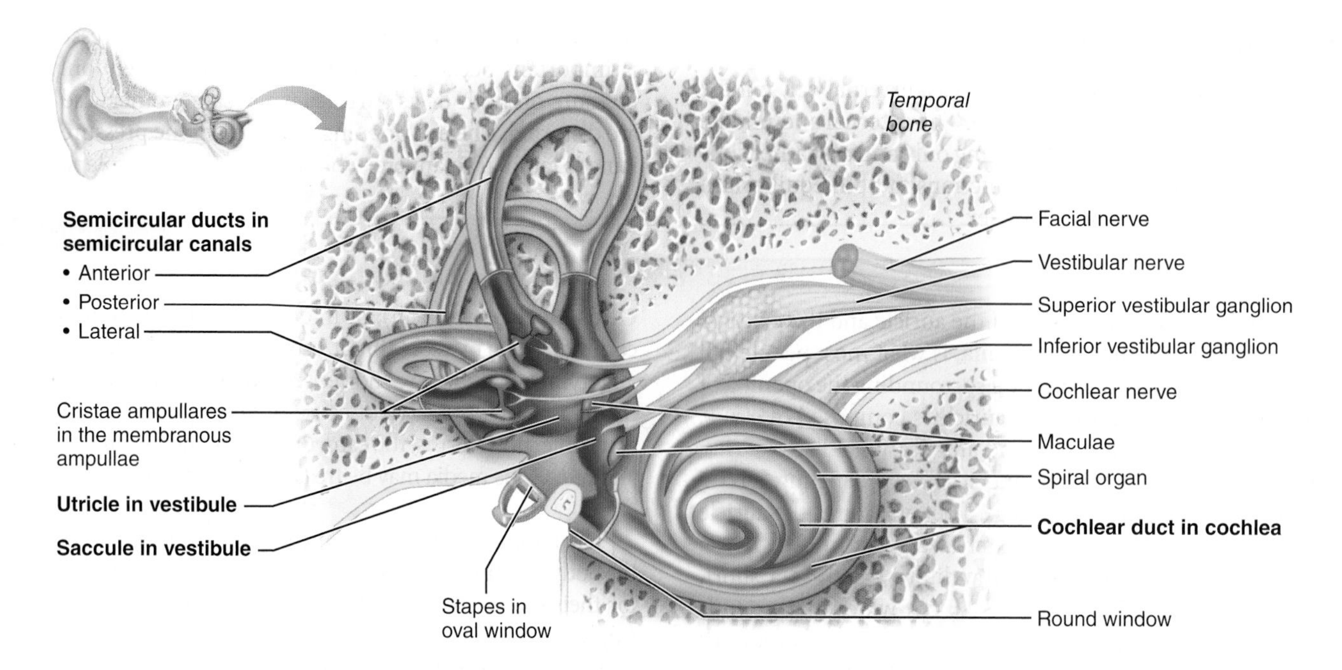

Figure 13.22 Membranous labyrinth of the internal ear. The membranous labyrinth (blue) lies within the chambers of the bony labyrinth (tan). The locations of the sensory organs for hearing (spiral organ) and equilibrium (maculae and cristae ampullares) are shown in purple.

intracellular fluid. These two fluids conduct the sound vibrations involved in hearing and respond to the mechanical forces occurring during changes in body position and acceleration.

The bony labyrinth has three regions: the *vestibule*, the *semicircular canals*, and the *cochlea*.

Vestibule The **vestibule** is the central egg-shaped cavity of the bony labyrinth. It lies posterior to the cochlea, anterior to the semicircular canals, and flanks the middle ear medially. In its lateral wall is the oval window.

Suspended in the vestibular perilymph and united by a small duct are two membranous labyrinth sacs, the **saccule** and **utricle** (u′trĭ-kl) (Figure 13.22). The smaller saccule is continuous with the membranous labyrinth extending anteriorly into the cochlea (the *cochlear duct*), and the utricle is continuous with the semicircular ducts extending into the semicircular canals posteriorly. The saccule and utricle house equilibrium receptor regions called *maculae* that respond to the pull of gravity and report on changes of head position.

Semicircular Canals The **semicircular canals** lie posterior and lateral to the vestibule. The cavities of the bony semicircular canals project from the posterior aspect of the vestibule, each oriented in one of the three planes of space. Accordingly, there is an *anterior*, a *posterior*, and a *lateral* semicircular canal in each internal ear. The anterior and posterior canals are oriented at right angles to each other in the vertical plane, whereas the lateral canal lies horizontally (Figure 13.22).

Snaking through each semicircular canal is a corresponding membranous **semicircular duct**, which communicates with the utricle anteriorly. Each of these ducts has an enlarged swelling at one end called an **ampulla**, which houses an equilibrium receptor region called a *crista ampullaris* (literally, crest of the ampulla). These receptors respond to rotational (angular) movements of the head.

Cochlea The **cochlea** (kok′le-ah), from the Latin "snail," is a spiral, conical, bony chamber about the size of a split pea. It extends from the anterior part of the vestibule and coils for about 2½ turns around a bony pillar called the **modiolus** (mo-di′o-lus) (**Figure 13.23a**). Running through its center like a wedge-shaped worm is the membranous **cochlear duct**, which ends blindly at the cochlear apex (Figure 13.22). The cochlear duct houses the receptor organ of hearing, called the **spiral organ** or the *organ of Corti* (Figure 13.23b).

The cochlear duct and the **osseous spiral lamina**, a thin shelflike extension of bone that spirals up the modiolus like the thread on a screw, together divide the cavity of the bony cochlea into three separate chambers or **scalae** (*scala* = ladder).

- The **scala vestibuli** (ska′lah ves-tĭ′bu-li) is continuous with the vestibule and begins at the oval window.
- The middle **scala media** is the cochlear duct itself.
- The **scala tympani** terminates at the membrane-covered round window.

Since the scala media is part of the membranous labyrinth, it is filled with endolymph. The scala vestibuli and the scala tympani, both part of the bony labyrinth, contain perilymph. The perilymph-containing chambers are continuous with each other at the cochlear apex, a region called the **helicotrema** (hel″ĭ-ko-tre′mah; "the hole in the spiral").

The "roof" of the cochlear duct, separating the scala media from the scala vestibuli, is the **vestibular membrane** (Figure 13.23b). The duct's external wall, the **stria vascularis**, is composed of an unusual richly vascularized mucosa that secretes endolymph.

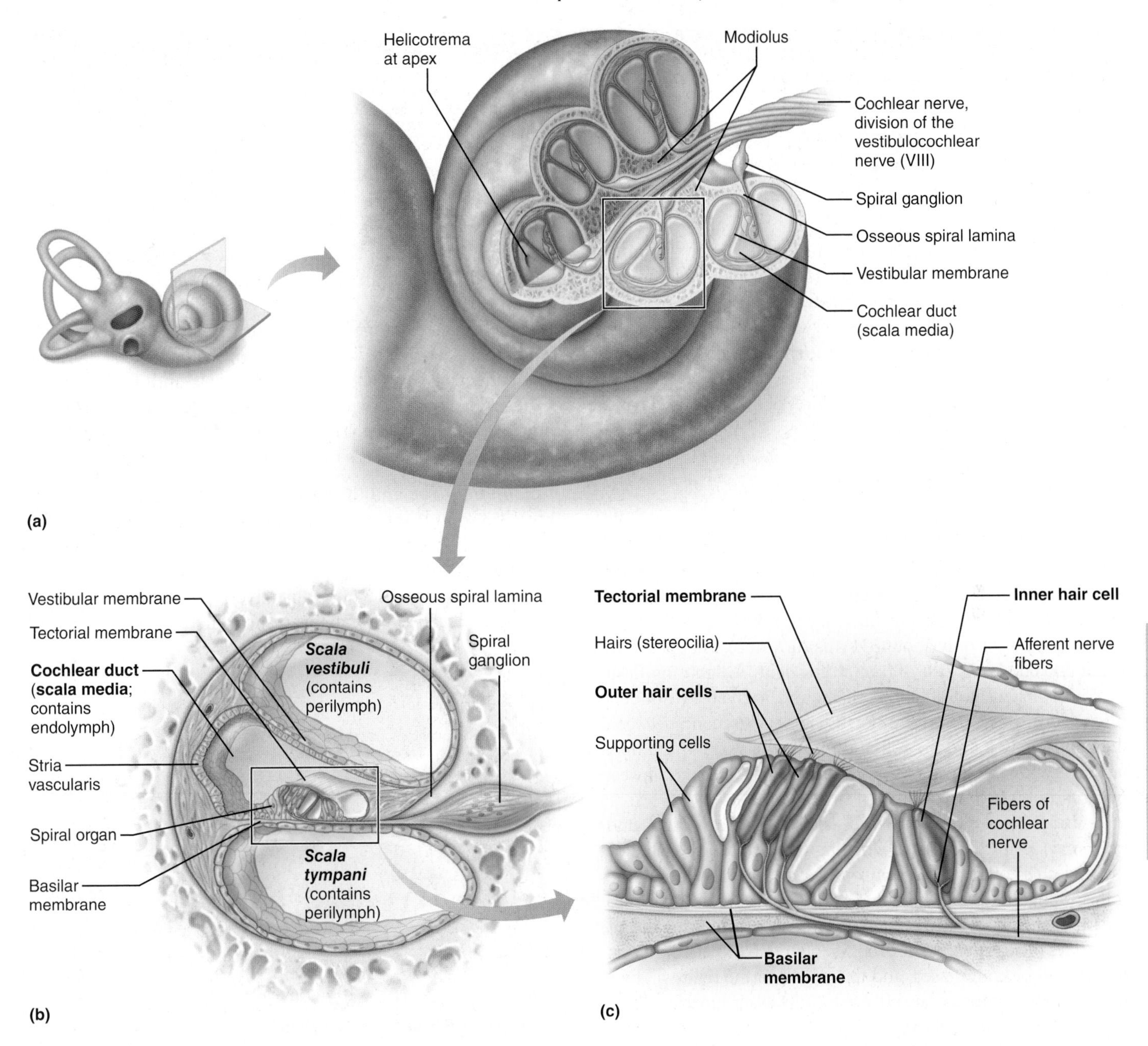

Figure 13.23 Anatomy of the cochlea. (a) Lateral view of part of the internal ear with a wedge-shaped section removed from the cochlea. **(b)** Magnified cross section of one turn of the cochlea, showing the relationship of the three scalae. This cross section has been rotated from its position in (a). **(c)** Detailed structure of the spiral organ. **(d)** Electron micrograph of cochlear hair cells (550×).

The "floor" of the cochlear duct is composed of the osseous spiral lamina and the flexible, fibrous basilar membrane, which supports the spiral organ. The **basilar membrane**, which plays a critical role in sound reception, is narrow and thick near the oval window and gradually widens and thins as it approaches the cochlear apex.

The spiral organ, which rests atop the basilar membrane, is composed of supporting cells and hearing receptor cells called *cochlear hair cells*. The hair cells are arranged functionally—

Table 13.3 Summary of the Internal Ear

BONY LABYRINTH	MEMBRANOUS LABYRINTH	FUNCTION	RECEPTOR REGION
Semicircular canals	Semicircular ducts	Equilibrium: rotational (angular) acceleration	Crista ampullaris
Vestibule	Utricle and saccule	Equilibrium: head position relative to gravity, linear acceleration	Macula
Cochlea	Cochlear duct (scala media)	Hearing	Spiral organ

specifically, one row of **inner hair cells** and three rows of **outer hair cells**—sandwiched between the tectorial and basilar membranes (Figure 13.23c). Afferent fibers of the **cochlear nerve** [a division of the vestibulocochlear nerve (VIII)] coil about the bases of the hair cells and run from the spiral organ through the modiolus to the brain.

Table 13.3 summarizes the structures of the internal ear and their functions.

Physiology of Hearing

Overview: Properties of Sound

Light can be transmitted through a vacuum (for instance, outer space), but sound depends on an *elastic* medium for its transmission. Sound also travels much more slowly than light. Its speed in dry air is only about 331 m/s (0.2 mi/s), as opposed to about 300,000 km/s (186,000 mi/s) for light. A lightning flash is almost instantly visible, but the sound it creates (thunder) reaches our ears much more slowly. (For each second between the lightning bolt and the roll of thunder, the storm is 1/5 mile farther away.) The speed of sound is fastest in solids and slowest in gases, but it is constant in a given medium.

Sound is a pressure disturbance—alternating areas of high and low pressure—produced by a vibrating object and propagated by the molecules of the medium. Consider a vibrating tuning fork. If the tuning fork is struck on the left, its prongs will move first to the right, creating an area of high pressure by compressing the air molecules there. Then, as the prongs rebound to the left, the air on the left will be compressed, and the region on the right will be a *rarefied*, or low-pressure, area (since most of its air molecules have been pushed farther to the right).

As the fork vibrates alternately from right to left, it produces a series of compressions and rarefactions, collectively called a *sound wave*, which moves outward in all directions. However, the individual air molecules just vibrate back and forth for short distances as they bump other molecules and rebound. Because the outward-moving molecules give up kinetic energy to the molecules they bump, energy is always transferred in the direction the sound wave is traveling. For this reason, the energy of the wave declines with time and distance.

We can illustrate a sound wave as an S-shaped curve, or *sine wave*, in which the compressed areas are crests and the rarefied areas are troughs. Sound can be described in terms of two physical properties: frequency and amplitude.

Frequency **Frequency** is defined as the number of waves that pass a given point in a given time. The sine wave of a pure tone has crests and troughs that repeat at specific intervals. The distance between two consecutive crests (or troughs) is the **wavelength** of the sound and is constant for a particular tone. The shorter the wavelength, the higher the frequency of the sound (Figure 13.24a).

The frequency range of human hearing is from 20 to 20,000 waves per second, or *hertz (Hz)*. Our ears are most sensitive to frequencies between 1500 and 4000 Hz and, in that range, we can distinguish frequencies differing by only 2–3 Hz. We perceive different sound frequencies as differences in **pitch**: the higher the frequency, the higher the pitch.

Amplitude The **amplitude**, or height, of the sine wave crests reveals a sound's intensity, which is related to its energy, or the pressure differences between its compressed and rarefied areas (Figure 13.24b).

Loudness refers to our subjective interpretation of sound intensity. Because we can hear such an enormous range of

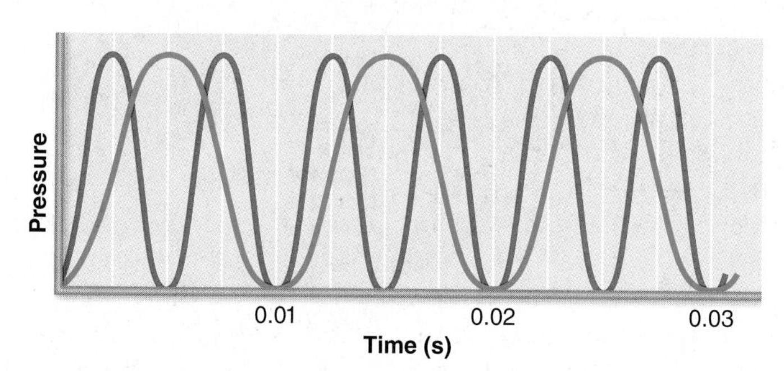

(a) Frequency is perceived as pitch.

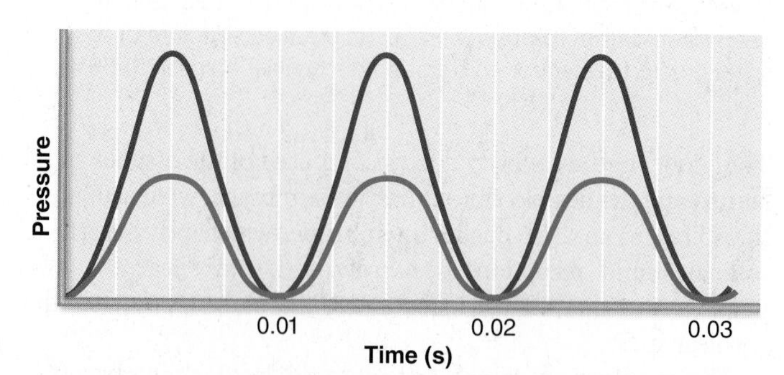

(b) Amplitude (size or intensity) is perceived as loudness.

Figure 13.24 **Frequency and amplitude of sound waves.**

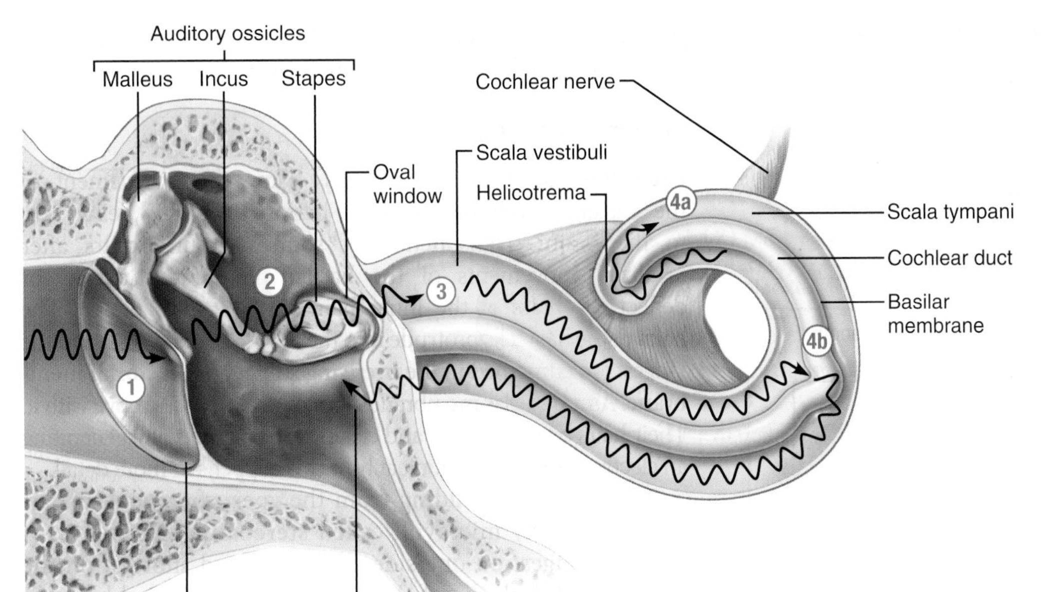

Figure 13.25 Pathway of sound waves. The cochlea is drawn as if uncoiled.

intensities, from the proverbial pin drop to a jet engine 10 million times more intense, sound intensity (and loudness) is measured in logarithmic units called **decibels (dB)** (des′ĭ-belz).

On a clinical audiometer, the decibel scale is arbitrarily set to begin at 0 dB, which is the threshold of hearing (barely audible sound) for normal ears. Each 10-dB increase represents a tenfold increase in sound intensity. A sound of 10 dB has 10 times more energy than one of 0 dB, and a 20-dB sound has 100 times (10×10) more energy. However, the same 10-dB increase represents only a twofold increase in loudness. In other words, most people would report that a 20-dB sound seems about twice as loud as a 10-dB sound. The normal range of hearing (from barely audible to the loudest sound we can process without excruciating pain) extends over a range of 120 dB. (The threshold of pain is 120 dB.)

Severe hearing loss occurs with frequent or prolonged exposure to sounds with intensities greater than 90 dB. That number becomes more meaningful when you realize that a normal conversation is in the 50-dB range, a noisy restaurant has 70-dB levels, and amplified concert music is often 120 dB or more, far above the 90-dB danger zone.

Transmission of Sound to the Internal Ear

Hearing occurs when the auditory area of the temporal lobe cortex is stimulated. However, before this can happen, sound waves must be propagated through air, membranes, bones, and fluids to stimulate receptor cells in the spiral organ (**Figure 13.25**).

(1) **Tympanic membrane.** Sound waves entering the external acoustic meatus strike the tympanic membrane and set it vibrating at the same frequency. The greater the intensity, the farther the membrane is displaced in its vibratory motion.

(2) **Auditory ossicles.** The motion of the tympanic membrane is amplified and transferred to the oval window by the ossicle lever system, which acts much like a piston to transfer the force striking the eardrum to the oval window. Because the tympanic membrane is 17–20 times larger than the oval window, the pressure (force per unit area) actually exerted on the oval window is about 20 times that on the tympanic membrane. This increased pressure overcomes the stiffness and inertia of cochlear fluid and sets it into wave motion.

This situation can be roughly compared to the difference in pressure relayed to the floor by the broad rubber heels of a man's shoes versus a woman's tiny spike heels. The man's weight is spread over several square inches, and his heels will not make dents in a pliable vinyl floor. But spike heels concentrate the same force in an area of about 2.5 cm^2 (1 square inch) and *will* dent the floor.

(3) **Scala vestibuli.** As the stapes rocks back and forth against the oval window, it sets the perilymph in the scala vestibuli into a similar back-and-forth motion. A pressure wave travels through the perilymph from the basal end toward the helicotrema, much as a piece of rope held horizontally can be set into wave motion by movements initiated at one end. Fluids cannot be compressed, so each time the stapes forces the fluid adjacent to the oval window medially, the membrane of the round window bulges laterally into the middle ear cavity and acts as a pressure valve.

(4a) **Helicotrema path.** Sounds of very low frequency (below 20 Hz) create pressure waves that take the complete route through the cochlea—up the scala vestibuli, around the helicotrema, and back toward the round window through the scala tympani (Figure 13.25). These low-frequency sounds do not activate the spiral organ and so are below the range of hearing.

(4b) **Basilar membrane path.** In contrast, sounds with frequencies high enough to hear create pressure waves that take a "shortcut" and are transmitted through the cochlear duct into the perilymph of the scala tympani. As a pressure wave descends through the flexible cochlear duct, it vibrates the basilar membrane. This vibration activates hair cells (receptor cells) at the site of this "shortcut" through the basilar membrane, causing action potentials to be sent to the brain.

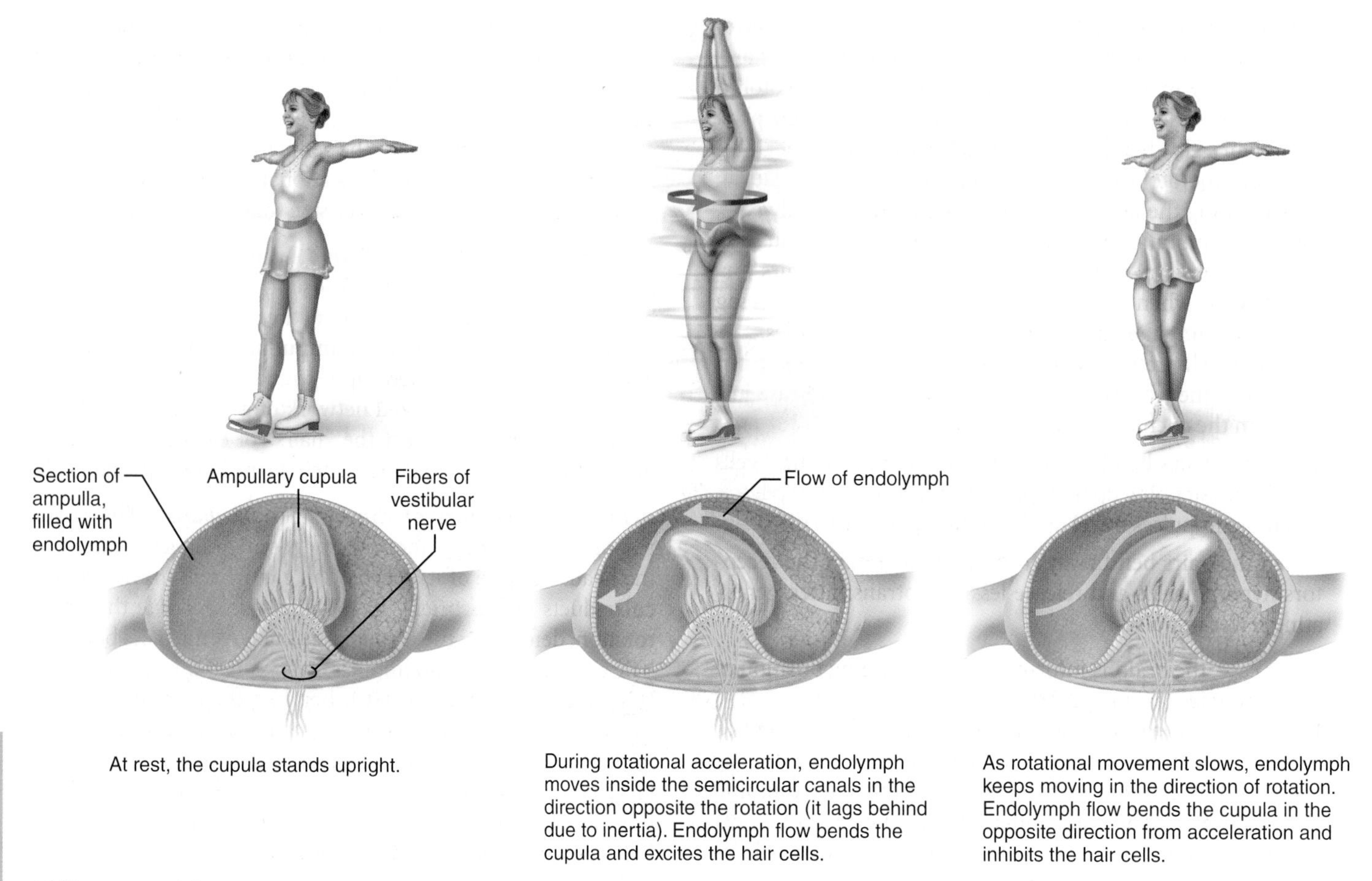

(c) Movement of the ampullary cupula during rotational acceleration and deceleration

Figure 13.29 *(continued)* **Location, structure, and function of a crista ampullaris in the internal ear.**

moving, the endolymph keeps on going, in effect reversing its direction within the canal. This sudden reversal in the direction of hair bending alters membrane voltage in the receptor cells and modifies the rate of impulse transmission, which tells the brain that we have slowed or stopped (Figure 13.29c, right).

The key point to remember when considering both types of equilibrium receptors is that the rigid bony labyrinth moves with the body, while the fluids (and gels) within the membranous labyrinth are free to move at various rates, depending on the forces (gravity, acceleration, and so on) acting on them.

Vestibular Nystagmus Impulses transmitted from the semicircular canals are particularly important to reflex movements of the eyes. **Vestibular nystagmus** is a complex of rather strange eye movements that occurs during and immediately after rotation.

As you rotate, your eyes slowly drift in the opposite direction, as though fixed on some object in the environment. This reaction relates to the backflow of endolymph in the semicircular canals. Then, because of CNS compensating mechanisms, the eyes jump rapidly toward the direction of rotation to establish a new fixation point. These alternating eye movements continue until the endolymph comes to rest.

When you stop rotating, at first your eyes continue to move in the direction of the previous spin, and then they jerk rapidly in the opposite direction. This sudden change is caused by the change in the direction in which the cristae bend after you stop. Nystagmus is often accompanied by vertigo, a false sensation of movement.

The Equilibrium Pathway to the Brain

Our responses to body imbalance, such as when we stumble, must be fast and reflexive. By the time we "thought about" correcting our fall, we would already be on the ground! Accordingly, information from equilibrium receptors goes directly to reflex centers in the brain stem, rather than to the cerebral cortex as with the other special senses.

The nerve pathways connecting the vestibular apparatus with the brain are complex. The transmission sequence begins when the hair cells in the vestibular apparatus are activated. Impulses travel initially to one of two destinations: the **vestibular nuclei** in the brain stem or the **cerebellum**. The vestibular nuclei, the major integrative center for balance, also receive inputs from the visual and somatic receptors, particularly from proprioceptors in neck muscles that report on the position of the head. These nuclei integrate this information and then send commands to brain stem motor centers that control the extrinsic eye muscles (cranial nerve nuclei III, IV, and VI) and reflex movements of the neck, limb, and trunk muscles (via the vestibulospinal tracts). The ensuing reflex movements of the eyes and body allow us to remain focused on the visual field and quickly adjust our body position to maintain or regain balance.

The cerebellum also integrates inputs from the eyes and somatic receptors (as well as from the cerebrum). It coordinates skeletal muscle activity and regulates muscle tone to maintain head position, posture, and balance, often in the face of rapidly changing inputs. Its "specialty" is fine control of delicate postural movements and timing.

Notice that the vestibular apparatus *does not automatically compensate* for forces acting on the body. Its job is to send warning signals to the CNS, which initiates the appropriate compensations.

CLINICAL

HOMEOSTATIC IMBALANCE 13.15

Equilibrium problems are usually obvious and unpleasant. Nausea, dizziness, and loss of balance are common and there may be nystagmus in the absence of rotational stimuli.

Motion sickness appears to be due to sensory input mismatches. For example, if you are inside a ship during a storm, visual inputs indicate that your body is fixed with reference to a stationary environment (your cabin). But as rough seas toss the ship about, your vestibular apparatus detects movement and sends impulses that disagree with the visual information. The brain thus receives conflicting signals, and its "confusion" somehow leads to motion sickness. Warning signals include excessive salivation, pallor, rapid deep breathing, and profuse sweating. Removal of the stimulus usually ends the symptoms. Over-the-counter drugs, such as meclizine (Bonine), depress vestibular inputs and help alleviate the symptoms. +

CLINICAL

Homeostatic Imbalances of Hearing and Equilibrium

Deafness

Any hearing loss, no matter how slight, is **deafness** of some sort. Deafness is classified as conduction or sensorineural deafness, according to its cause.

Conduction deafness occurs when something hampers sound conduction to the fluids of the internal ear. For example, compacted earwax can block the external acoustic meatus, or a *perforated* (*ruptured*) eardrum can prevent sound conduction from the eardrum to the ossicles. But the most common causes of conduction deafness are middle ear inflammations (otitis media) and **otosclerosis** (o″to-sklĕ-ro′sis) of the ossicles.

Otosclerosis ("hardening of the ear") occurs when overgrowth of bony tissue fuses the base of the stapes to the oval window or welds the ossicles to one another. In such cases, vibrations of the skull bones conduct sound to the receptors of that ear, which is far less satisfactory. Otosclerosis is treated surgically.

Sensorineural deafness results from damage to neural structures at any point from the cochlear hair cells to and including the auditory cortical cells. This type of deafness typically results from the gradual loss of hair cells throughout life. Hair cells can also be destroyed by a single explosively loud noise or prolonged exposure to high-intensity sounds, such as music or industrial noise, which literally tear off their cilia. Other causes of sensorineural deafness are degeneration of the cochlear nerve, strokes, and tumors in the auditory cortex.

Hair cells don't normally regenerate in mammals, but researchers are seeking ways to prod supporting cells into becoming hair cells. For congenital defects or age- or noise-related cochlear damage, cochlear implants (devices that convert sound energy into electrical signals) can be inserted into the temporal bone. Modern implants are so effective that even children born deaf can hear well enough to learn to speak well.

Tinnitus

Tinnitus (tĭ-ni′tus) is a ringing, buzzing, or clicking sound in the ears in the absence of auditory stimuli. It is usually a symptom rather than a disease. For example, tinnitus is one of the first symptoms of cochlear nerve degeneration. It may also signal inflammation of the middle or internal ears and is a side effect of some medications, such as aspirin.

Recent evidence suggests that tinnitus is analogous to phantom limb pain. In other words, it is "phantom cochlear noise" caused by destruction of some neurons in the auditory pathway and the subsequent ingrowth of nearby neurons whose signals are interpreted as noise by the CNS.

Ménière's Syndrome

Classic **Ménière's syndrome** (men″ē-ārz′) is a labyrinth disorder that seems to affect all three parts of the internal ear. The afflicted person has repeated attacks of vertigo, nausea, and vomiting. Balance is so disturbed that standing erect is nearly impossible. Hearing may be impaired or lost completely.

Mild cases can usually be managed by antimotion drugs or a low-salt diet and diuretics to decrease endolymph fluid volume. In severe cases, draining the excess endolymph from the internal ear may help. A last resort is removal of the entire malfunctioning labyrinth.

☑ Check Your Understanding

15. Apart from the bony boundaries, which structure separates the external from the middle ear? Which two (nonbone) structures separate the middle from the inner ear?
16. Which structure inside the spiral organ allows us to differentiate sounds of different pitch?
17. MAKING connections Opening cation channels causes depolarization. The influx of which ion usually causes this? (Hint: See the discussion on p. 369 of Chapter 11.) Which ions cause depolarization in hair cells? Why does K^+ move into these cells, rather than out?
18. If the brain stem did not receive input from both ears, what would you not be able to do?
19. For each of the following phrases, indicate whether it applies to a macula or a crista ampullaris: inside a semicircular canal; contains otoliths; responds to linear acceleration and deceleration; has a cupula; responds to rotational acceleration and deceleration; inside a saccule.

For answers, see Answers Appendix.

Table 13.4 Cranial Nerves *(continued)*

V Trigeminal Nerves

Largest cranial nerves; fibers extend from pons to face, and form three divisions (*trigemina* = threefold): ophthalmic (V_1), maxillary (V_2), and mandibular (V_3) divisions. As main general sensory nerves of face, transmit afferent impulses from touch, temperature, and pain receptors. Cell bodies of sensory neurons of all three divisions are located in large *trigeminal ganglion.*

The mandibular division also contains motor fibers that innervate chewing muscles.

Dentists desensitize upper and lower jaws by injecting local anesthetic (such as Novocain) into alveolar branches of maxillary and mandibular divisions, respectively. Since this blocks pain-transmitting fibers of teeth, the surrounding tissues become numb.

	Ophthalmic division (V_1)	Maxillary division (V_2)	Mandibular division (V_3)
Origin and course	Fibers run from face to pons via superior orbital fissure.	Fibers run from face to pons via foramen rotundum.	Fibers pass through skull via foramen ovale.
Function	Conveys sensory impulses from skin of anterior scalp, upper eyelid, and nose, and from nasal cavity mucosa, cornea, and lacrimal gland.	Conveys sensory impulses from nasal cavity mucosa, palate, upper teeth, skin of cheek, upper lip, lower eyelid.	Conveys sensory impulses from anterior tongue (except taste buds), lower teeth, skin of chin, temporal region of scalp. Supplies motor fibers to, and carries proprioceptor fibers from, muscles of mastication.
CLINICAL TESTING	Corneal reflex test: Touching cornea with wisp of cotton should elicit blinking.	Test sensations of pain, touch, and temperature with safety pin and hot and cold objects.	Assess motor branch by asking person to clench his teeth, open mouth against resistance, and move jaw side to side.

HOMEOSTATIC IMBALANCE *Trigeminal neuralgia* (nu-ral′je-ah), or *tic douloureux* (tik doo″loo-roo′; *tic* = twitch, *douloureux* = painful), caused by inflammation of trigeminal nerve, is widely considered to produce most excruciating pain known. The stabbing pain lasts for a few seconds to a minute, but it can be relentless, occurring a hundred times a day. Usually provoked by some sensory stimulus, such as brushing teeth or even a passing breeze hitting the face. Thought to be caused by a loop of artery or vein that compresses the trigeminal nerve near its exit from the brain stem. Analgesics and carbamazepine (an anticonvulsant) are only partially effective. In severe cases, surgery relieves the agony—either by moving the compressing vessel or by destroying the nerve. Nerve destruction results in loss of sensation on that side of face. ✚

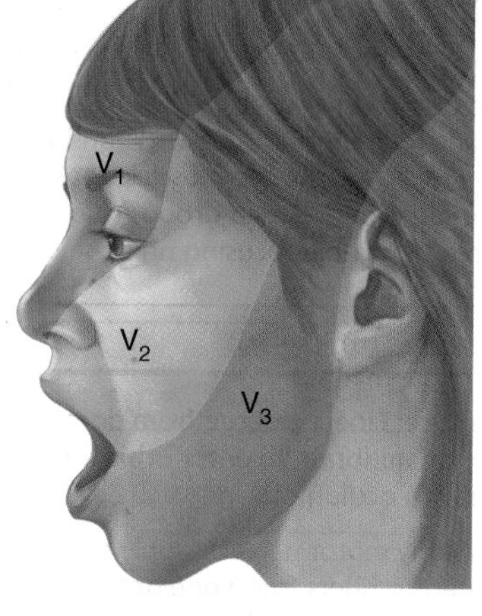

(b) Distribution of sensory fibers of each division

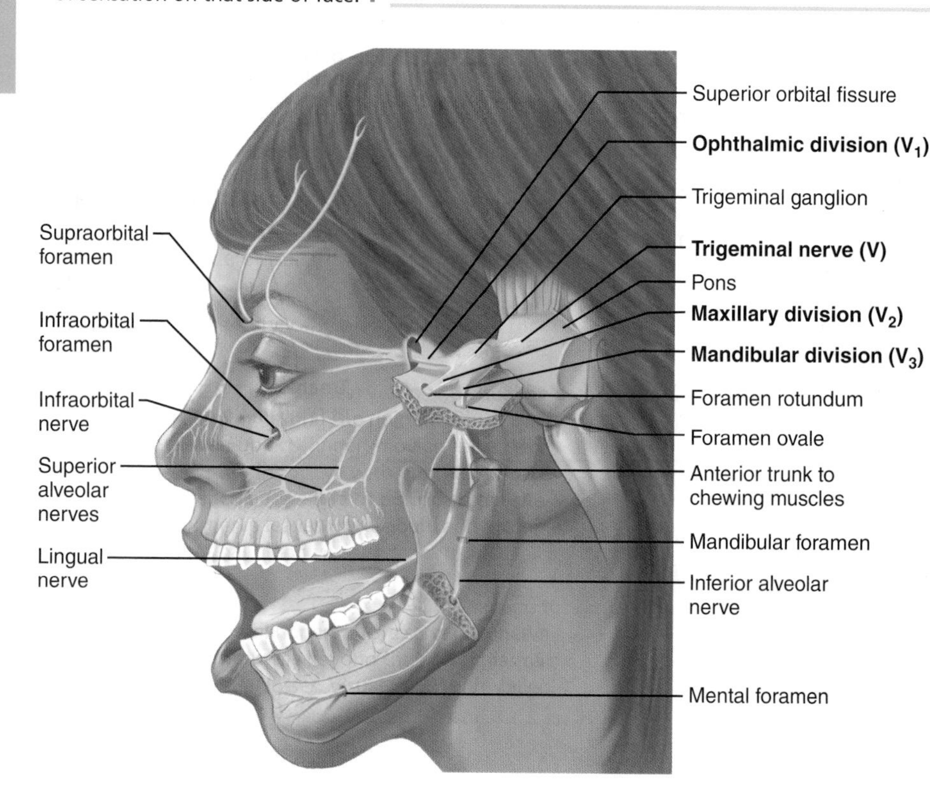

(a) Distribution of the trigeminal nerve

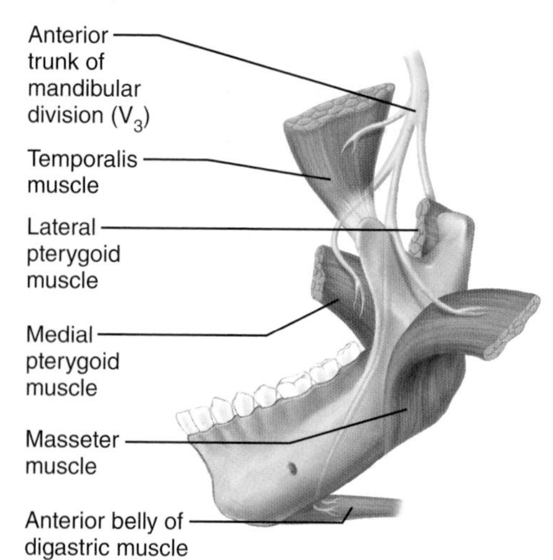

(c) Motor branches of the mandibular division (V_3)

Table 13.4 *(continued)*

VI Abducens Nerves (ab-du′senz)

Origin and course: Fibers leave inferior pons and enter orbit via superior orbital fissure to run to eye.

Function: Primarily motor; supply somatic motor fibers to lateral rectus muscle, an extrinsic muscle of the eye. Convey proprioceptor impulses from same muscle to brain.

CLINICAL TESTING: Test in common with cranial nerve III (oculomotor).

HOMEOSTATIC IMBALANCE In abducens nerve paralysis, eye cannot be moved laterally. At rest, eyeball rotates medially (*internal strabismus*). +

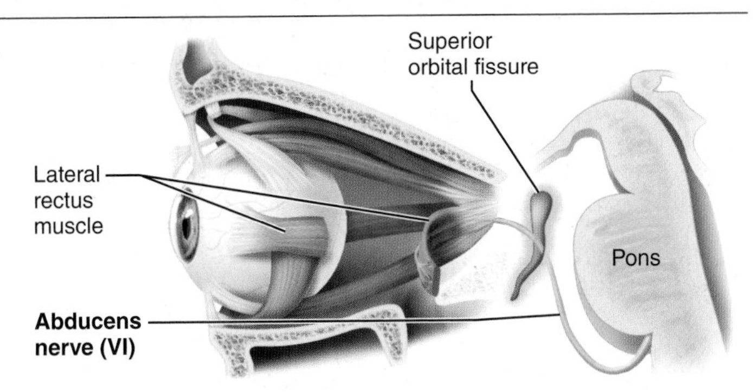

VII Facial Nerves

Origin and course: Fibers issue from pons, just lateral to abducens nerves (see Figure 13.32), enter temporal bone via *internal acoustic meatus*, and run within bone (and through inner ear cavity) before emerging through *stylomastoid foramen*. Nerve then courses to lateral aspect of face.

Function: Mixed nerves that are the chief motor nerves of face. Five major branches: temporal, zygomatic, buccal, mandibular, and cervical (see **c** on next page).

- Convey motor impulses to skeletal muscles of face (muscles of facial expression), except for chewing muscles served by trigeminal nerves, and transmit proprioceptor impulses from same muscles to pons (see **b** and pp. 290–292).
- Transmit parasympathetic (autonomic) motor impulses to lacrimal (tear) glands, nasal and palatine glands, and submandibular and sublingual salivary glands. Some of the cell bodies of these parasympathetic motor neurons are in *pterygopalatine* (ter″eh-go-pal′ah-tīn) and *submandibular ganglia* on the trigeminal nerve (see **a**).
- Convey sensory impulses from taste buds of anterior two-thirds of tongue; cell bodies of these sensory neurons are in *geniculate ganglion* (see **a**).

CLINICAL TESTING: Test anterior two-thirds of tongue for ability to taste sweet (sugar), salty, sour (vinegar), and bitter (quinine) substances. Check symmetry of face. Ask subject to close eyes, smile, whistle, and so on. Assess tearing with ammonia fumes.

HOMEOSTATIC IMBALANCE *Bell's palsy* is characterized by paralysis of facial muscles on affected side and partial loss of taste sensation. May develop rapidly (often overnight). Caused by inflamed and swollen facial nerve, possibly due to herpes simplex 1 viral infection. Lower eyelid droops, corner of mouth sags (making it difficult to eat or speak normally), tears drip continuously from eye and eye cannot be completely closed (conversely, dry-eye syndrome may occur). Treated with corticosteroids. Recovery is complete in 70% of cases. +

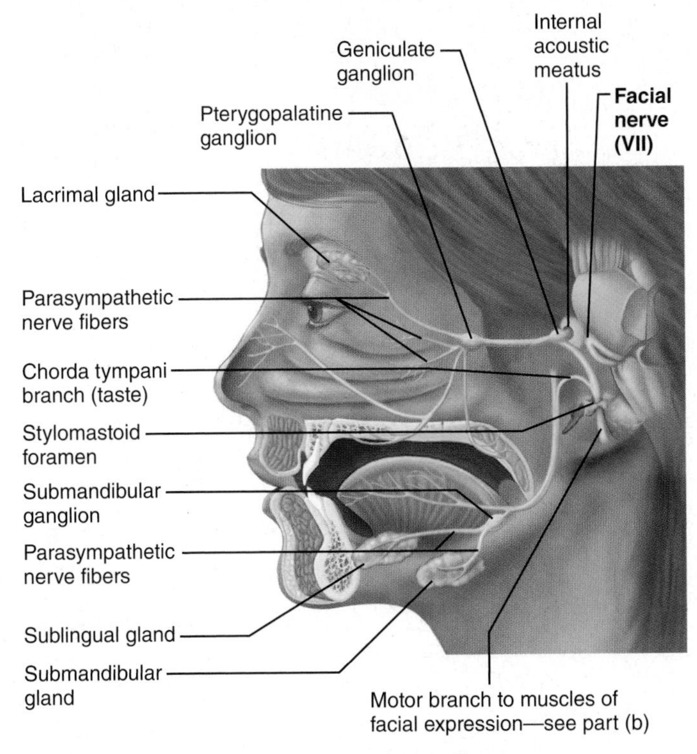

(a) Parasympathetic efferents and sensory afferents

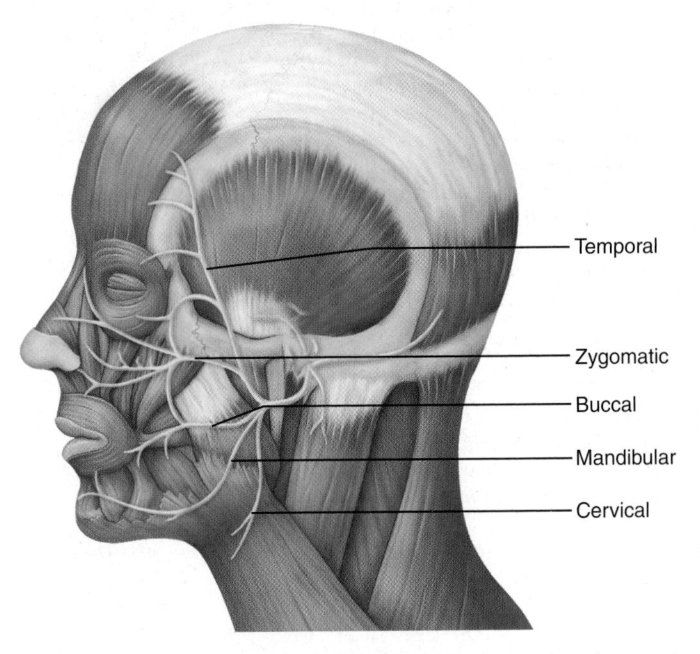

(b) Motor branches to muscles of facial expression and scalp muscles

➤

Table 13.4 Cranial Nerves *(continued)*

VII Facial Nerves *(continued)*

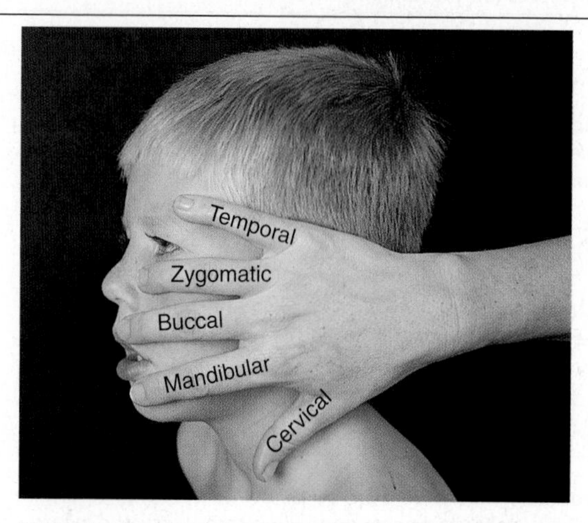

(c) A simple way to remember the courses of the five major branches of the facial nerve

VIII Vestibulocochlear Nerves (ves-tib″u-lo-kok′le-ar)

Origin and course: Fibers arise from hearing and equilibrium apparatus located within inner ear of temporal bone and pass through internal acoustic meatus to enter brain stem at pons-medulla border. Afferent fibers from hearing receptors in cochlea form the *cochlear division*; those from equilibrium receptors in semicircular canals and vestibule form the *vestibular division* (vestibular nerve). The two divisions merge to form vestibulocochlear nerve. See also Figure 13.22.

Function: Mostly sensory. Vestibular branch transmits afferent impulses for sense of equilibrium, and sensory nerve cell bodies are located in *vestibular ganglia*. Cochlear branch transmits afferent impulses for sense of hearing, and sensory nerve cell bodies are located in *spiral ganglion* within cochlea. Small motor component adjusts the sensitivity of sensory receptors. See also Figure 13.23.

CLINICAL TESTING: Check hearing by air and bone conduction using tuning fork.

Semicircular canals
Vestibular ganglia
Vestibular nerve
Internal acoustic meatus
Cochlear nerve
Vestibule
Pons
Cochlea (containing spiral ganglion)
Vestibulocochlear nerve (VIII)

HOMEOSTATIC IMBALANCE Lesions of cochlear nerve or cochlear receptors result in *central*, or *nerve*, *deafness*. Damage to vestibular division produces dizziness, rapid involuntary eye movements, loss of balance, nausea, and vomiting. +

IX Glossopharyngeal Nerves (glos″o-fah-rin′je-al)

Origin and course: Fibers emerge from medulla and leave skull via *jugular foramen* to run to throat.

Function: Mixed nerves that innervate part of tongue and pharynx. Provide somatic motor fibers to, and carry proprioceptor fibers from, a superior pharyngeal muscle called the *stylopharyngeus*, which elevates the pharynx in swallowing. Provide parasympathetic motor fibers to parotid salivary glands (some of the nerve cell bodies of these parasympathetic motor neurons are located in *otic ganglion*).

Sensory fibers conduct taste and general sensory (touch, pressure, pain) impulses from pharynx and posterior tongue, from chemoreceptors in the carotid body (which monitor O_2 and CO_2 levels in the blood and help regulate respiratory rate and depth), and from baroreceptors of carotid sinus (which monitor blood pressure). Sensory neuron cell bodies are located in *superior* and *inferior ganglia*.

CLINICAL TESTING: Check position of uvula; check gag and swallowing reflexes. Ask subject to speak and cough. Test posterior third of tongue for taste.

HOMEOSTATIC IMBALANCE Injured or inflamed glossopharyngeal nerves impair swallowing and taste. +

X Vagus Nerves (va′gus)

Origin and course: The only cranial nerves to extend beyond head and neck region. Fibers emerge from medulla, pass through skull via jugular foramen, and descend through neck region into thorax and abdomen. See also Figure 14.4.

Function: Mixed nerves. Nearly all motor fibers are parasympathetic efferents, except those serving skeletal muscles of pharynx and larynx (involved in swallowing). Parasympathetic motor fibers supply heart, lungs, and abdominal viscera and are involved in regulating heart rate, breathing, and digestive system activity. Transmit sensory impulses from thoracic and abdominal viscera, from the aortic arch baroreceptors (for blood pressure) and the carotid and aortic bodies (chemoreceptors for respiration), and taste buds on the epiglottis. Also carry general somatic sensory information from small area of skin on external ear. Carry proprioceptor fibers from muscles of larynx and pharynx.

CLINICAL TESTING: As for cranial nerve IX (glossopharyngeal); IX and X are tested in common, since they both innervate muscles of throat and mouth.

HOMEOSTATIC IMBALANCE Since laryngeal branches of the vagus innervate nearly all muscles of the larynx ("voice box"), vagal nerve paralysis can lead to hoarseness or loss of voice. Other symptoms are difficulty swallowing and impaired digestive system motility. These parasympathetic nerves are important for maintaining the normal state of visceral organ activity. Without their influence, the sympathetic nerves, which mobilize and accelerate vital body processes (and shut down digestion), would dominate. +

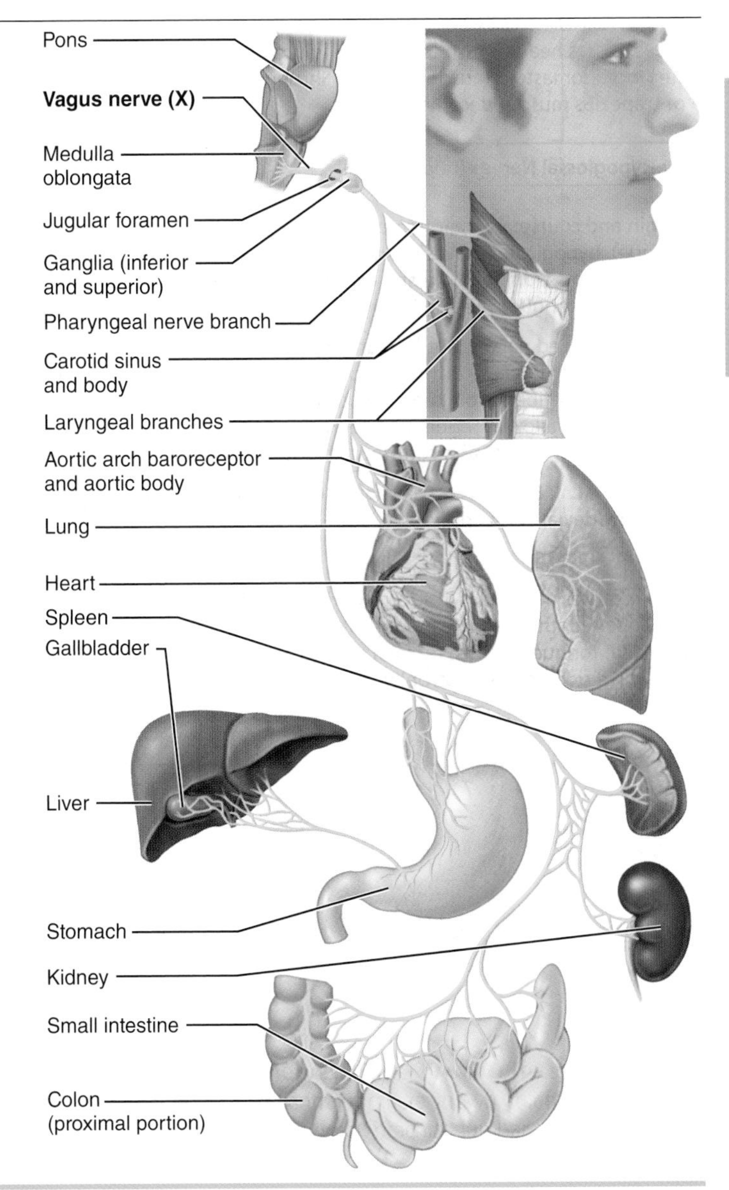

Musculocutaneous Nerve The **musculocutaneous nerve**, the major end branch of the lateral cord, courses inferiorly in the anterior arm, supplying motor fibers to the biceps brachii, brachialis, and coracobrachialis muscles. Distal to the elbow, it provides cutaneous sensation in the lateral forearm.

Median Nerve The **median nerve** descends through the arm to the anterior forearm, where it gives off branches to the skin and to most flexor muscles. On reaching the hand, it innervates five intrinsic muscles of the lateral palm. The median nerve activates muscles that pronate the forearm, flex the wrist and fingers, and oppose the thumb.

HOMEOSTATIC IMBALANCE 13.18 CLINICAL

Median nerve injury makes it difficult to use the pincer grasp (opposed thumb and index finger) to pick up small objects. Because this nerve runs down the midline of the forearm and wrist, it is a frequent casualty of wrist-slashing suicide attempts. In carpal tunnel syndrome (see p. 206), the median nerve is compressed. +

Ulnar Nerve The **ulnar nerve** branches off the medial cord of the plexus. It descends along the medial aspect of the arm toward the elbow, swings behind the medial epicondyle, and then follows the ulna along the medial forearm. There it supplies the flexor carpi ulnaris and the medial part of the flexor digitorum profundus (the flexors not supplied by the median nerve).

The ulnar nerve continues into the hand, where it innervates most intrinsic hand muscles and the skin of the medial aspect of the hand. It causes the wrist and fingers to flex, and (with the median nerve) adducts and abducts the medial fingers.

HOMEOSTATIC IMBALANCE 13.19 CLINICAL

Where it takes a superficial course, the ulnar nerve is very vulnerable to injury. Striking the "funny bone"—the spot where this nerve rests against the medial epicondyle—makes the little finger tingle. Severe or chronic damage can lead to sensory loss, paralysis, and muscle atrophy. Affected individuals have trouble making a fist and gripping objects. As the little and ring fingers become hyperextended at the knuckles and flexed at the distal interphalangeal joints, the hand contorts into a *clawhand*. +

Radial Nerve The **radial nerve**, the largest branch of the brachial plexus, is a continuation of the posterior cord. This nerve wraps around the humerus (in the radial groove), and then runs anteriorly around the lateral epicondyle at the elbow. There it divides into a superficial branch that follows the lateral edge of the radius to the hand and a deep branch (not illustrated) that runs posteriorly. It supplies the posterior skin of the limb along its entire course. Its motor branches innervate essentially all the extensor muscles of the upper limb. Muscles controlled by the radial nerve extend the elbow, supinate the forearm, extend the wrist and fingers, and abduct the thumb.

HOMEOSTATIC IMBALANCE 13.20 CLINICAL

Trauma to the radial nerve results in *wrist drop*, inability to extend the hand at the wrist. Improper use of a crutch or "Saturday night paralysis," in which an intoxicated person falls asleep with an arm draped over the back of a chair or sofa edge, can compress the radial nerve and impair its blood supply. +

Lumbosacral Plexus and Lower Limb

The sacral and lumbar plexuses overlap substantially. Because many fibers of the lumbar plexus contribute to the sacral plexus via the **lumbosacral trunk**, the two plexuses are often referred to as the **lumbosacral plexus**. Although the lumbosacral plexus serves mainly the lower limb, it also sends some branches to the abdomen, pelvis, and buttock.

Lumbar Plexus The **lumbar plexus** arises from spinal nerves L_1–L_4 and lies within the psoas major muscle (**Figure 13.37**). Its proximal branches innervate parts of the abdominal wall muscles and the psoas muscle, but its major branches descend to innervate the anterior and medial thigh.

Table 13.7 Branches of the Lumbar Plexus (See Figure 13.37)

NERVES	VENTRAL RAMI	STRUCTURES SERVED
Femoral	L_2–L_4	Skin of anterior and medial thigh via *anterior femoral cutaneous* branch; skin of medial leg and foot, hip and knee joints via *saphenous* branch; motor to anterior muscles (quadriceps and sartorius) of thigh and to pectineus, iliacus
Obturator	L_2–L_4	Motor to adductor magnus (part), longus, and brevis muscles, gracilis muscle of medial thigh, obturator externus; sensory for skin of medial thigh and for hip and knee joints
Lateral femoral cutaneous	L_2, L_3	Skin of lateral thigh; some sensory branches to peritoneum
Iliohypogastric	L_1	Skin on side of buttock and above pubis; muscles of anterolateral abdominal wall (internal obliques and transversus abdominis)
Ilioinguinal	L_1	Skin of external genitalia and proximal medial aspect of the thigh; inferior abdominal muscles
Genitofemoral	L_1, L_2	Skin of scrotum in males, of labia majora in females, and of anterior thigh inferior to middle portion of inguinal region; cremaster muscle in males

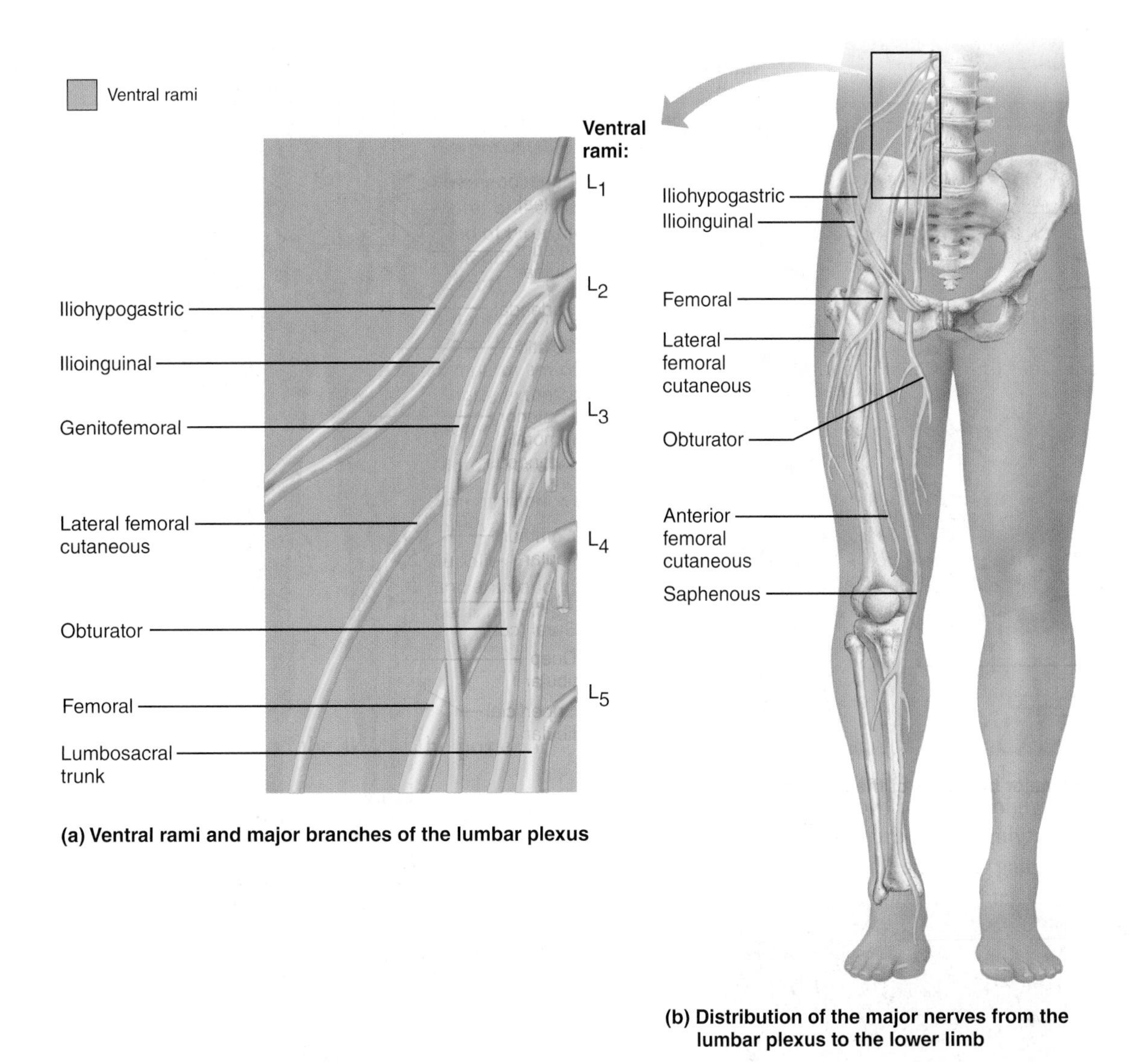

(a) Ventral rami and major branches of the lumbar plexus

(b) Distribution of the major nerves from the lumbar plexus to the lower limb

Figure 13.37 The lumbar plexus. (Anterior view.)

The **femoral nerve**, the largest terminal nerve of this plexus, runs deep to the inguinal ligament to enter the thigh and then divides into several large branches. The motor branches innervate anterior thigh muscles (quadriceps), which are the principal thigh flexors and knee extensors. The cutaneous branches serve the skin of the anterior thigh and the medial surface of the leg from knee to foot.

The **obturator nerve** (ob″tu-ra′tor) enters the medial thigh via the obturator foramen and innervates the adductor muscles. Table 13.7 summarizes the branches of the lumbar plexus.

HOMEOSTATIC IMBALANCE 13.21 CLINICAL

When the spinal roots of the lumbar plexus are compressed, as by a herniated disc, gait problems occur because the femoral nerve serves the prime movers that flex the hip and extend the knee. Other symptoms are pain or numbness of the anterior thigh and (if the obturator nerve is impaired) of the medial thigh. ✚

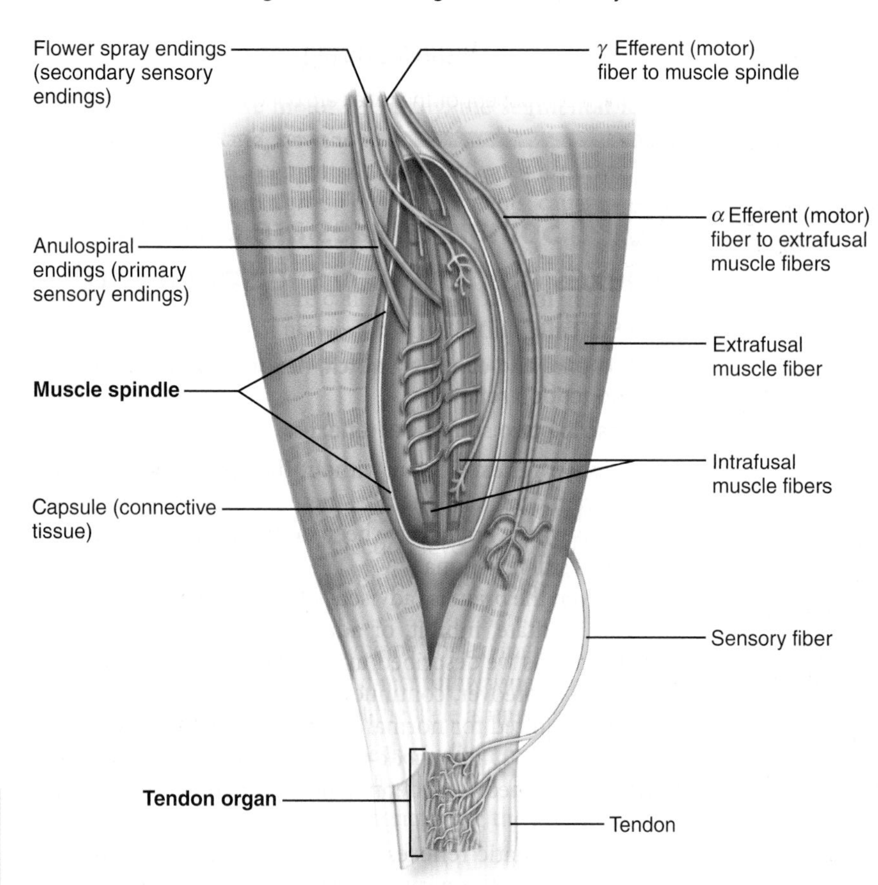

Figure 13.41 Anatomy of the muscle spindle and tendon organ. Myelin has been omitted from all nerve fibers for clarity.

(**Figure 13.41**). These fibers are less than one-quarter the size of the effector fibers of the muscle, called **extrafusal muscle fibers**.

The central regions of the intrafusal fibers lack myofilaments and are noncontractile. These regions are the receptive surfaces of the spindle. Two types of afferent endings send sensory inputs to the CNS:

- **Anulospiral endings** (also called *primary sensory endings*) are the endings of large axons that wrap around the spindle center. They are stimulated by both the rate and degree of stretch.
- **Flower spray endings** (also called *secondary sensory endings*) are formed by smaller axons that supply the spindle ends. They are stimulated only by degree of stretch.

The intrafusal muscle fibers have contractile regions at their ends, which are the only areas containing actin and myosin myofilaments. These regions are innervated by **gamma (γ) efferent fibers** that arise from small motor neurons in the ventral horn of the spinal cord. These γ motor fibers, which maintain spindle sensitivity (as described shortly), are distinct from the **alpha (α) efferent fibers** of the large **alpha (α) motor neurons** that stimulate the extrafusal muscle fibers to contract.

The muscle spindle is stretched (and excited) in one of two ways:

- By applying an external force that lengthens the entire muscle, such as when we carry a heavy weight or when antagonistic muscles contract (external stretch)
- By activating the γ motor neurons that stimulate the distal ends of the intrafusal fibers to contract, thereby stretching the middle of the spindle (internal stretch)

Whenever the muscle spindle is stretched, its associated sensory neurons transmit impulses at higher frequency to the spinal cord (**Figure 13.42a**).

During voluntary skeletal muscle contraction, the muscle shortens. If the intrafusal muscle fibers didn't contract along with the extrafusal fibers, the muscle spindle would go slack and cease generating action potentials (Figure 13.42b). At this point it would be unable to signal further changes in muscle length, so it would be useless.

Fortunately, **α-γ coactivation** prevents this from happening. Descending fibers of motor pathways synapse with both α and γ motor neurons, and motor impulses are simultaneously sent to the large extrafusal fibers and to muscle spindle intrafusal fibers. Stimulating the intrafusal fibers maintains the spindle's tension (and sensitivity) during muscle contraction, so that the brain continues to be notified of changes in the muscle length (Figure 13.42b). Without such a system, information on changes in muscle length would cease to flow from contracting muscles.

The Stretch Reflex

By sending commands to the motor neurons, the brain essentially sets a muscle's length. The **stretch reflex** makes sure that the muscle stays at that length. For example, the **patellar** (pah-tel′ar) or **knee-jerk reflex** is a stretch reflex that helps keep your knees from buckling when you are standing upright. As your knees begin to buckle and the quadriceps lengthens, the stretch reflex causes the quadriceps to contract without your having to think about it. *Focus on the Stretch Reflex* (**Focus Figure 13.1** on p. 498) shows the stretch reflex and a specific example—the knee-jerk reflex.

The stretch reflex is important for maintaining muscle tone and adjusting it reflexively. It is most important in the large extensor muscles that sustain upright posture and in postural muscles of the trunk. For example, stretch reflexes initiated first on one side of the spine and then on the other regulate contractions of the postural muscles of the spine almost continuously.

Let's look at how the stretch reflex works. As we've just seen in Figure 13.42, when stretch activates sensory neurons of muscle spindles, they transmit impulses at a higher frequency to the spinal cord. There the sensory neurons synapse directly with α motor neurons, which rapidly excite the extrafusal muscle fibers of the stretched muscle (Focus Figure 13.1). The reflexive

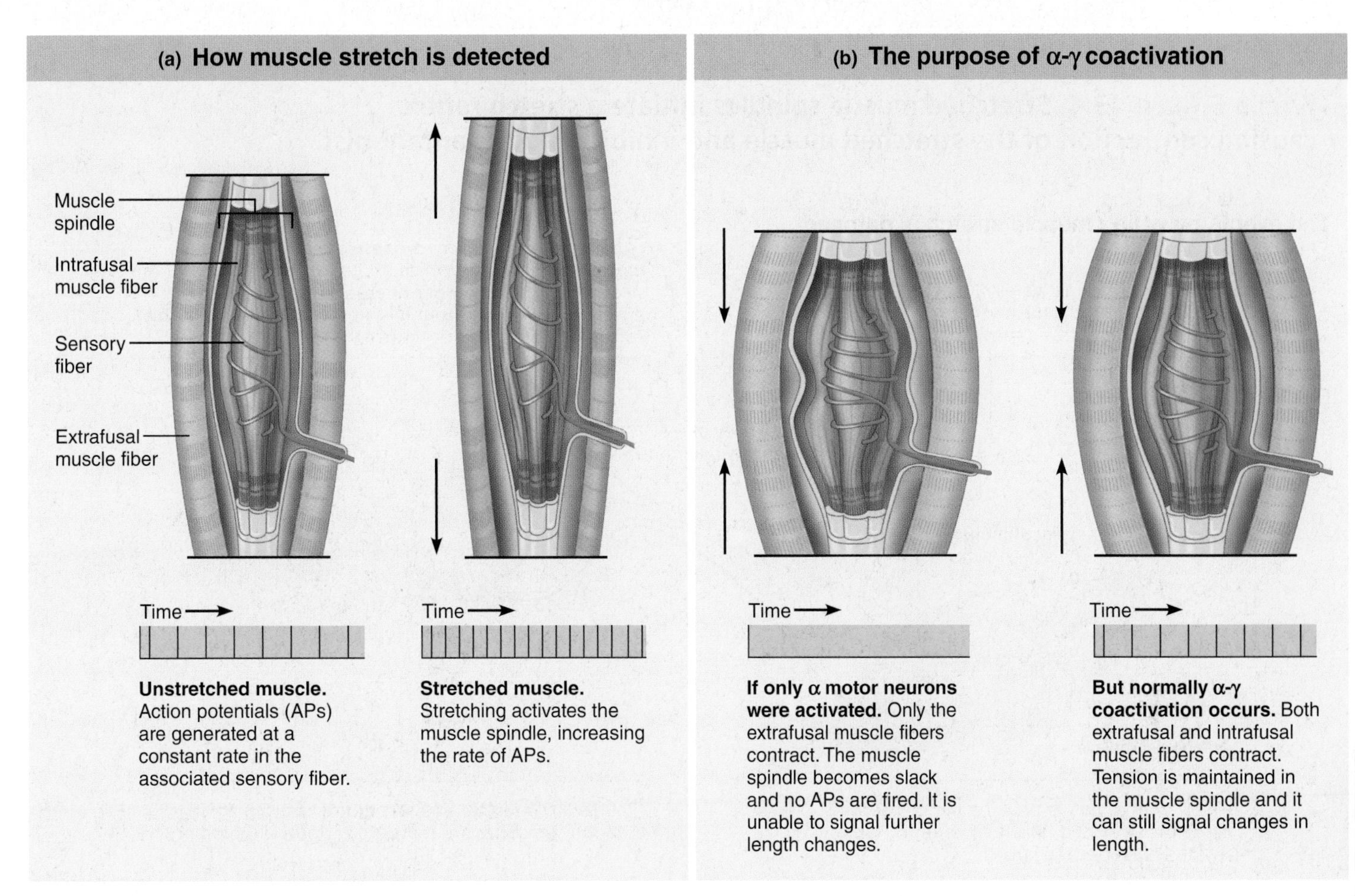

Figure 13.42 Operation of the muscle spindle. The action potentials generated in the sensory fibers are shown for each case as black lines in yellow bars.

muscle contraction that follows (an example of serial processing) resists further muscle stretching.

Branches of the afferent fibers also synapse with interneurons that inhibit motor neurons controlling antagonistic muscles (parallel processing), and the resulting inhibition is called **reciprocal inhibition**. Consequently, the stretch stimulus causes the antagonists to relax so that they cannot resist the shortening of the "stretched" muscle caused by the main reflex arc. While this spinal reflex is occurring, information on muscle length and the speed of muscle shortening is being relayed (mainly via the dorsal white columns) to higher brain centers (more parallel processing).

The most familiar clinical example of a stretch reflex is the knee-jerk reflex we have just described. Stretch reflexes can be elicited in any skeletal muscle by a sudden jolt to the tendon or the muscle itself. All stretch reflexes are **monosynaptic** and **ipsilateral**. In other words, they involve a single synapse and motor activity on the same side of the body. Stretch reflexes are the only monosynaptic reflexes in the body. However, even in these reflexes, the part of the reflex arc that inhibits the motor neurons serving the antagonistic muscles is polysynaptic.

A positive knee jerk (or a positive result for any other stretch reflex test) provides two important pieces of information. First, it proves that the sensory and motor connections between that muscle and the spinal cord are intact. Second, the vigor of the response indicates the degree of excitability of the spinal cord. When the spinal motor neurons are highly facilitated by impulses descending from higher centers, just touching the muscle tendon produces a vigorous reflex response. On the other hand, when inhibitory signals bombard the lower motor neurons, even pounding on the tendon may fail to trigger the reflex response.

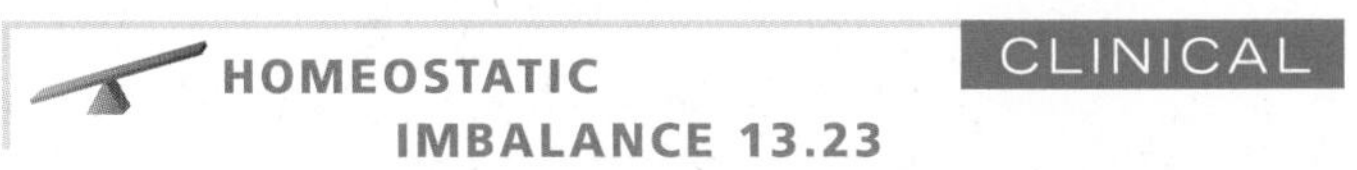

Stretch reflexes tend to be hypoactive or absent in cases of peripheral nerve damage or ventral horn injury involving the tested area. These reflexes are absent in those with chronic diabetes mellitus or neurosyphilis and during coma. However, they are hyperactive when lesions of the corticospinal tract reduce the inhibitory effect of the brain on the spinal cord (as in stroke patients). +

Adjusting Muscle Spindle Sensitivity

The motor supply to the muscle spindle allows the brain to voluntarily modify the stretch reflex response and the firing rate of α motor neurons. When the γ neurons are vigorously stimulated by impulses from the brain, the spindle is stretched and highly sensitive, and muscle contraction force is maintained or

Focus Figure 13.1 Stretched muscle spindles initiate a stretch reflex, causing contraction of the stretched muscle and inhibition of its antagonist.

The events by which muscle stretch is damped

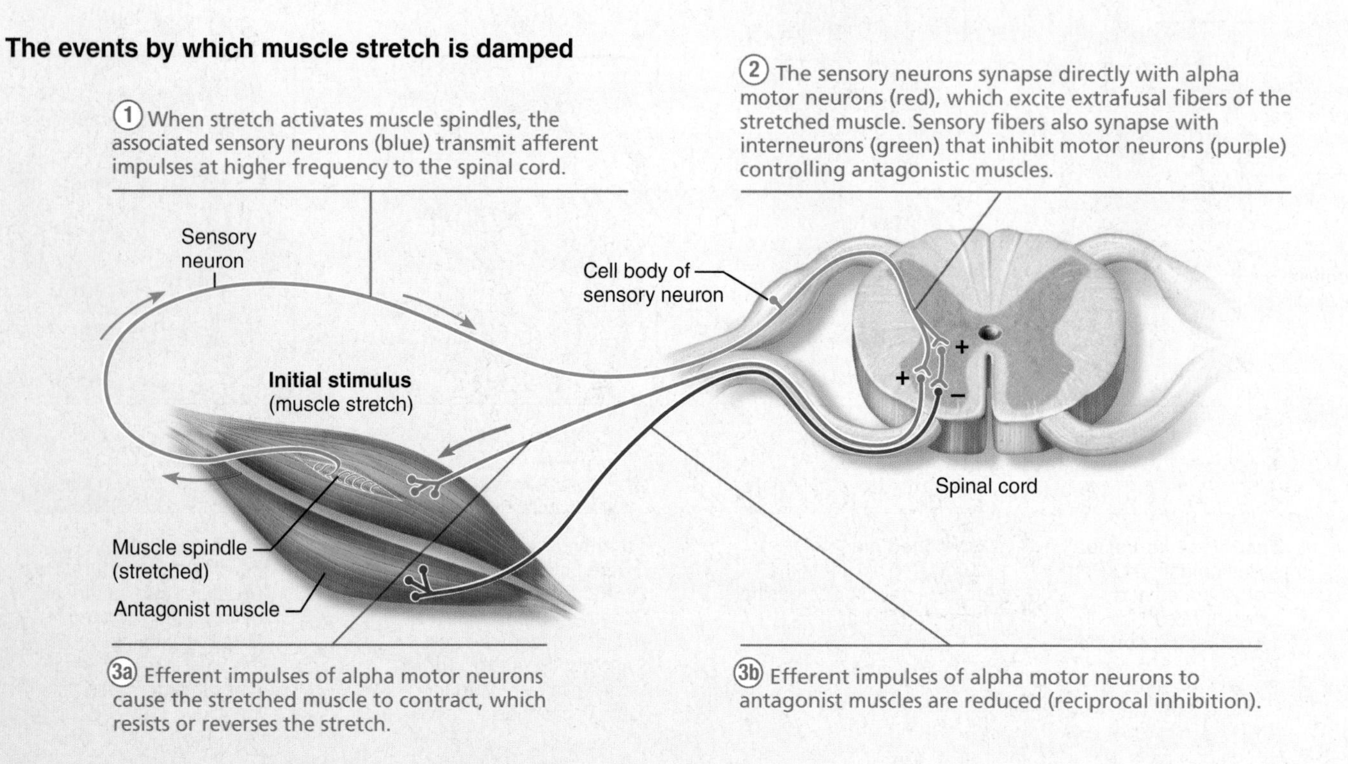

The patellar (knee-jerk) reflex—an example of a stretch reflex

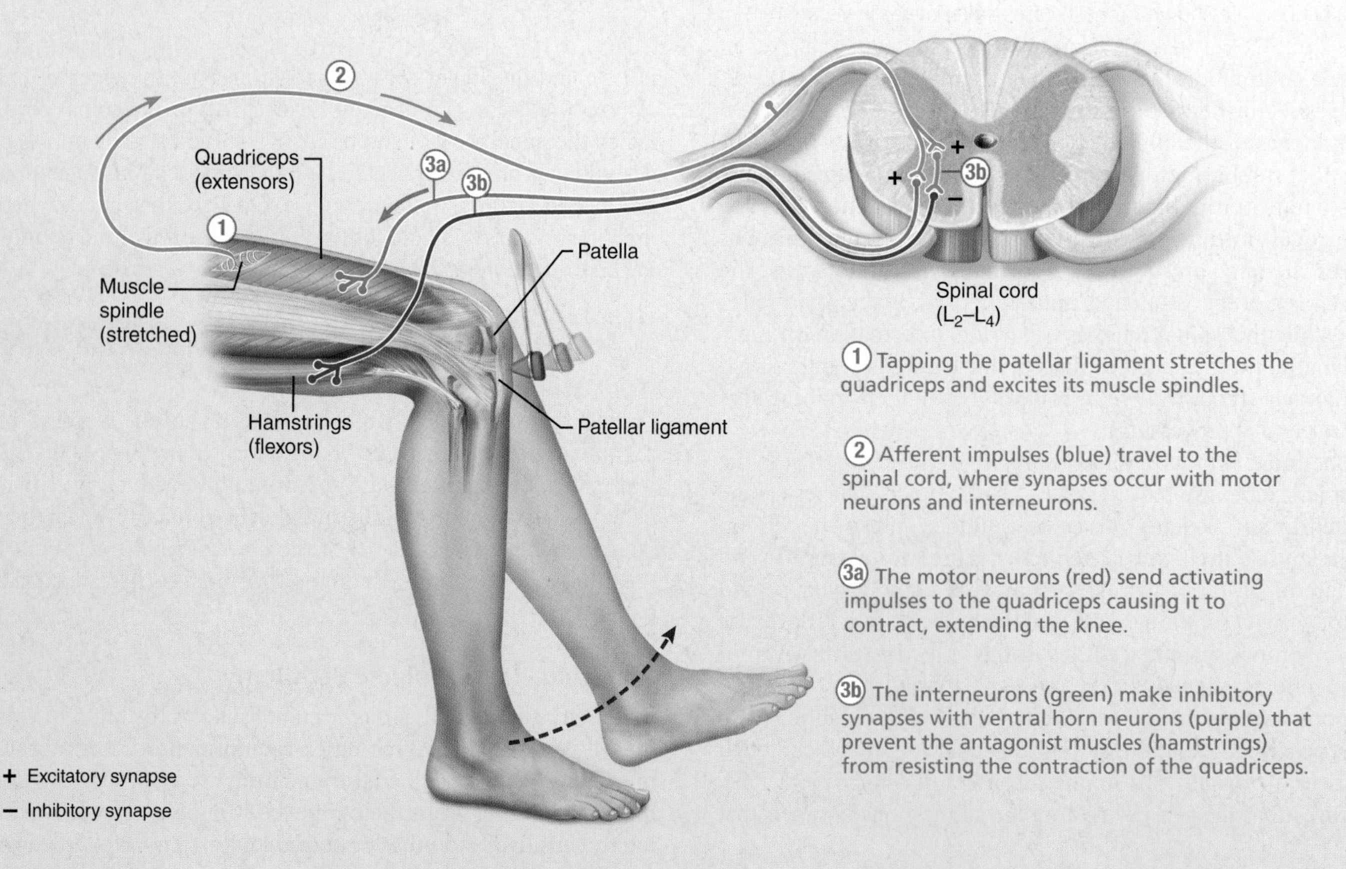

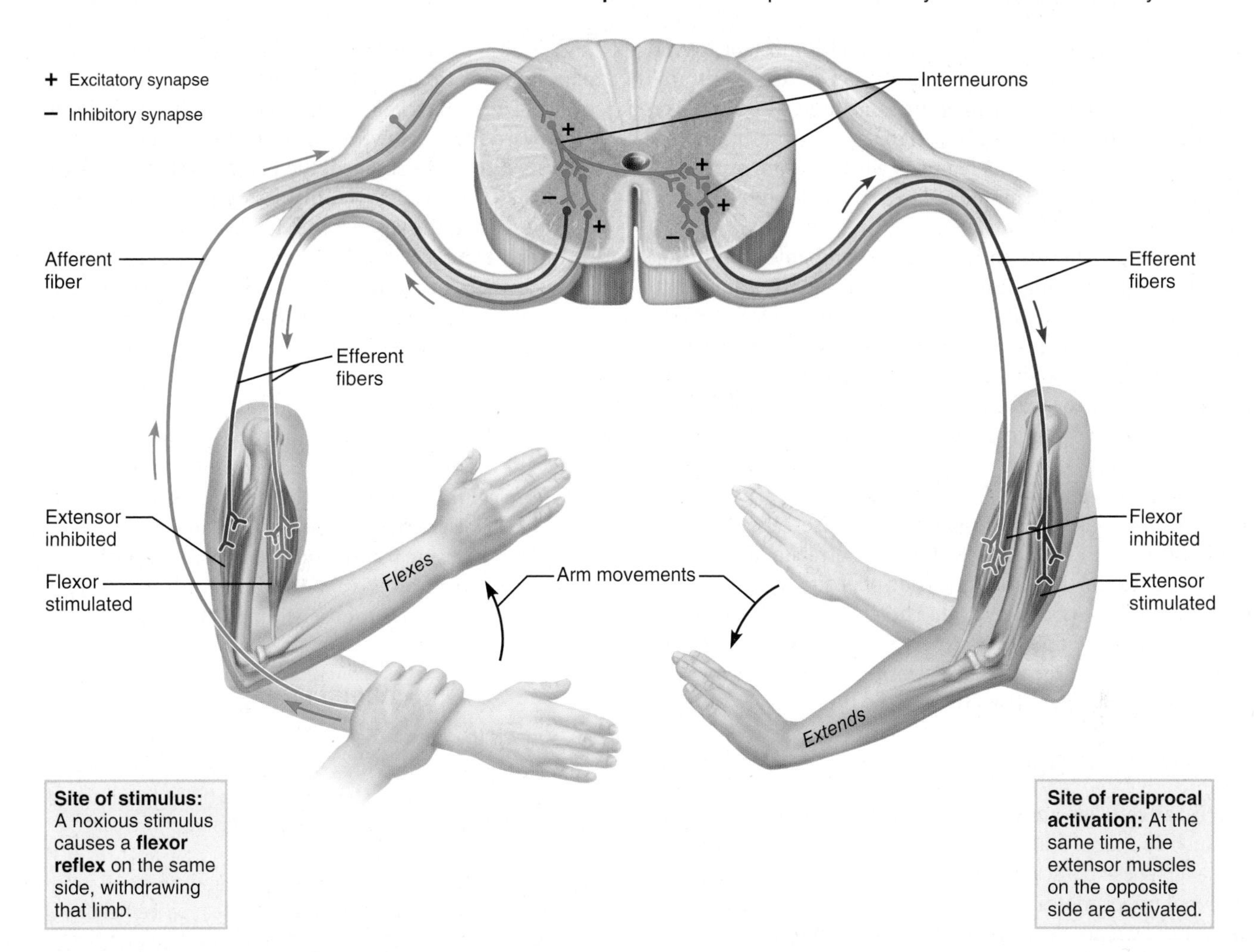

Figure 13.43 The crossed-extensor reflex. In this example, a stranger suddenly grasps the right arm, which is withdrawn reflexively while the opposite (left) arm reflexively extends and pushes the stranger away.

increased. When the γ motor neurons are inhibited, the spindle resembles a loose rubber band and is nonresponsive, and the extrafusal muscles relax.

The ability to modify the stretch reflex is important in many situations. As the speed and difficulty of a movement increase, the brain increases γ motor output to make the muscle spindles more sensitive. This sensitivity is highest when balance reflexes must be razor sharp, as for a gymnast on a balance beam. On the other hand, if you want to wind up to pitch a baseball, it is essential to suppress the stretch reflex so that your muscles can produce a large degree of motion (i.e., circumduct your pitching arm). Other athletes who require movements of maximum force learn to stretch muscles as much and as quickly as possible just before the movement. This advantage is demonstrated by the crouch that athletes assume just before jumping or running.

The Tendon Reflex

Stretch reflexes cause muscle contraction in response to increased muscle length (stretch). The polysynaptic **tendon reflexes**, on the other hand, produce exactly the opposite effect: Muscles relax and lengthen in response to tension.

When muscle tension increases substantially during contraction or passive stretching, high-threshold *tendon organs* may be activated. Afferent impulses are transmitted to the spinal cord, and then to the cerebellum, where the information is used to adjust muscle tension. Simultaneously, motor neurons in spinal cord circuits supplying the contracting muscle are inhibited and antagonist muscles are activated, a phenomenon called **reciprocal activation**. As a result, the contracting muscle relaxes as its antagonist is activated.

Tendon organs help to prevent muscles and tendons from tearing when they are subjected to potentially damaging stretching force. Tendon organs also function at normal muscle tensions, helping to ensure smooth onset and termination of muscle contraction.

The Flexor and Crossed-Extensor Reflexes

A painful stimulus initiates the **flexor**, or **withdrawal**, **reflex**, which causes automatic withdrawal of the threatened body part from the stimulus (**Figure 13.43**, left). Think of the response that occurs with a paper cut. Flexor reflexes are ipsilateral and polysynaptic, the latter a necessity when several muscles must be recruited to withdraw the injured body part.

Because flexor reflexes are protective and important to our survival, they override the spinal pathways and prevent any other reflexes from using them at the same time. However, like other spinal reflexes, descending signals from the brain can override flexor reflexes. This happens when you are expecting a painful stimulus, for example a skin prick as a lab technician prepares to draw blood from a vein.

The **crossed-extensor reflex** often accompanies the flexor reflex in weight-bearing limbs and is particularly important in maintaining balance. It is a complex spinal reflex consisting of an ipsilateral withdrawal reflex and a contralateral extensor reflex. Incoming afferent fibers synapse with interneurons that control the flexor withdrawal response on the same side of the body and with other interneurons that control the extensor muscles on the opposite side.

The crossed-extensor reflex is obvious when you step barefoot on broken glass. The ipsilateral response causes you to quickly lift your injured foot, while the contralateral response activates the extensor muscles of your opposite leg to support the weight suddenly shifted to it. The crossed-extensor reflex also occurs when someone unexpectedly grabs your arm. The grasped arm is withdrawn as the opposite arm pushes you away from the attacker (Figure 13.43).

Superficial Reflexes

Superficial reflexes are elicited by gentle cutaneous stimulation, such as that produced by stroking the skin with a tongue depressor. These clinically important reflexes depend both on functional upper motor pathways and on cord-level reflex arcs. The best known are the plantar and abdominal reflexes.

Plantar Reflex

The **plantar reflex** tests the integrity of the spinal cord from L_4 to S_2 and indirectly determines if the corticospinal tracts are functioning properly. To elicit the plantar reflex, draw a blunt object downward along the lateral aspect of the plantar surface (sole) of the foot. The normal response is for the toes to flex downward (curl). However, if the primary motor cortex or corticospinal tract is damaged, the plantar reflex is replaced by an abnormal reflex called **Babinski's sign**, in which the great toe dorsiflexes and the smaller toes fan laterally.

Infants exhibit Babinski's sign until they are about a year old because their nervous systems are incompletely myelinated. Despite its clinical significance, the physiological mechanism of Babinski's sign is not understood.

Abdominal Reflexes

Stroking the skin of the lateral abdomen above, to the side, or below the umbilicus induces a reflex contraction of the abdominal muscles in which the umbilicus moves toward the stimulated site. These reflexes, called **abdominal reflexes**, check the integrity of the spinal cord and ventral rami from T_8 to T_{12}.

Abdominal reflexes vary in intensity from one person to another. However, their absence indicates lesions in the corticospinal tract.

Check Your Understanding

29. What is the role of the stretch reflex? The flexor reflex?

30. Juan injured his back in a fall. When his ER physician stroked the bottom of Juan's foot, she noted that his big toe pointed up and his other toes fanned out. What is this response called and what does it indicate?

31. MAKING connections A technician is drawing blood from your arm for blood tests. As you feel the pain of the needle, you suppress your flexor reflex. Name the ascending pathway that carries pain signals and the region of the cortex that receives them (use Figure 12.32 on p. 424). Next, name the descending pathway you would use to inhibit the flexor reflex (use Figure 12.33 on p. 427).

For answers, see Answers Appendix.

REVIEW QUESTIONS

MAP For more chapter study tools, go to the Study Area of MasteringA&P®.

There you will find:

- Interactive Physiology iP
- A&PFlix A&PFlix
- Practice Anatomy Lab PAL
- PhysioEx PEx
- Videos, Practice Quizzes and Tests, MP3 Tutor Sessions, Case Studies, and much more!

Multiple Choice/Matching

(Some questions have more than one correct answer. Select the best answer or answers from the choices given.)

1. The large onion-shaped receptors that are found deep in the dermis and in subcutaneous tissue and that respond to deep pressure are **(a)** tactile discs, **(b)** lamellar corpuscles, **(c)** free nerve endings, **(d)** muscle spindles.
2. Proprioceptors include all of the following except **(a)** muscle spindles, **(b)** tendon organs, **(c)** tactile discs, **(d)** joint kinesthetic receptors.
3. The connective tissue sheath that surrounds a fascicle of nerve fibers is the **(a)** epineurium, **(b)** endoneurium, **(c)** perineurium, **(d)** epimysium.
4. Characterize each receptor activity described below by choosing the appropriate letter and number(s) from keys A and B.

____, ____ (1) You are enjoying an ice cream cone.
____, ____ (2) You have just scalded yourself with hot coffee.
____, ____ (3) The retinas of your eyes are stimulated.
____, ____ (4) You bump (lightly) into someone.
____, ____ (5) You are in a completely dark room and reaching toward the light switch.
____, ____ (6) You feel uncomfortable after a large meal.

Key A:
(a) exteroceptor
(b) interoceptor
(c) proprioceptor

Key B:
(1) chemoreceptor
(2) mechanoreceptor
(3) nociceptor
(4) photoreceptor
(5) thermoreceptor

5. Match the receptor type in column B to the correct description in column A.

Column A

____ **(1)** pain, itch, and temperature receptors
____ **(2)** contains intrafusal fibers and anulospiral and flower spray endings
____ **(3)** discriminative touch receptor in hairless skin (fingertips)
____ **(4)** contains receptor endings wrapped around thick collagen bundles
____ **(5)** rapidly adapting deep-pressure receptor
____ **(6)** slowly adapting deep-pressure receptor

Column B

(a) bulbous corpuscles
(b) tendon organ
(c) muscle spindle
(d) free nerve endings
(e) lamellar corpuscle
(f) tactile corpuscle

6. Match the names of the cranial nerves in column B to the appropriate description in column A.

Column A

____ **(1)** causes pupillary constriction
____ **(2)** the major sensory nerve of the face
____ **(3)** serves the sternocleidomastoid and trapezius muscles
____ **(4)** purely sensory (two nerves)
____ **(5)** serves the tongue muscles
____ **(6)** allows you to chew your food
____ **(7)** impaired in Bell's palsy
____ **(8)** helps regulate heart activity
____ **(9)** helps you hear and maintain your balance
____, ____, ____, ____, **(10)** contain parasympathetic motor fibers (four nerves)

Column B

(a) abducens
(b) accessory
(c) facial
(d) glossopharyngeal
(e) hypoglossal
(f) oculomotor
(g) olfactory
(h) optic
(i) trigeminal
(j) trochlear
(k) vagus
(l) vestibulocochlear

7. For each of the following muscles or body regions, identify the plexus and the peripheral nerve(s) (or branch of one) involved. Use choices from keys A and B.

____; ____ **(1)** the diaphragm
____; ____ **(2)** muscles of the posterior leg
____; ____ **(3)** anterior thigh muscles
____; ____ **(4)** medial thigh muscles
____; ____ **(5)** anterior arm muscles that flex the forearm
____; ____ **(6)** muscles that flex the wrist and digits (two nerves)
____; ____ **(7)** muscles that extend the wrist and digits
____; ____ **(8)** skin and extensor muscles of the posterior arm
____; ____ **(9)** fibularis muscles, tibialis anterior, and toe extensors
____; ____, ____, ____, ____ **(10)** elbow joint

Key A: Plexuses

(a) brachial
(b) cervical
(c) lumbar
(d) sacral

Key B: Nerves

(1) common fibular
(2) femoral
(3) median
(4) musculocutaneous
(5) obturator
(6) phrenic
(7) radial
(8) tibial
(9) ulnar

8. Accessory glands that produce an oily secretion are the **(a)** conjunctiva, **(b)** lacrimal glands, **(c)** tarsal glands.

9. The portion of the fibrous layer that is white and opaque is the **(a)** choroid, **(b)** cornea, **(c)** retina, **(d)** sclera.

10. Which sequence best describes a normal route for the flow of tears from the eyes into the nasal cavity? **(a)** lacrimal canaliculi, lacrimal sacs, nasolacrimal ducts; **(b)** lacrimal ducts, lacrimal canaliculi, nasolacrimal ducts; **(c)** nasolacrimal ducts, lacrimal canaliculi, lacrimal sacs.

11. Activation of the sympathetic nervous system causes **(a)** contraction of the sphincter pupillae muscles, **(b)** contraction of the dilator pupillae muscles, **(c)** contraction of the ciliary muscles, **(d)** a decrease in ciliary zonule tension.

12. Damage to the medial recti muscles would probably affect **(a)** accommodation, **(b)** refraction, **(c)** convergence, **(d)** pupil constriction.

13. The phenomenon of dark adaptation is best explained by the fact that **(a)** rhodopsin does not function in dim light, **(b)** rhodopsin breakdown occurs slowly, **(c)** rods exposed to intense light need time to generate rhodopsin, **(d)** cones are stimulated to function by bright light.

14. Blockage of the scleral venous sinus might result in **(a)** a sty, **(b)** glaucoma, **(c)** conjunctivitis, **(d)** a cataract.

15. Nearsightedness is more properly called **(a)** myopia, **(b)** hyperopia, **(c)** presbyopia, **(d)** emmetropia.

16. Of the neurons in the retina, the axons of which of these form the optic nerve? **(a)** bipolar cells, **(b)** ganglion cells, **(c)** cone cells, **(d)** horizontal cells.

17. Which reactions occur when a person looks at a distant object? **(a)** pupils constrict, ciliary zonule (suspensory ligament) relaxes, lenses become less convex; **(b)** pupils dilate, ciliary zonule becomes taut, lenses become less convex; **(c)** pupils dilate, ciliary zonule becomes taut, lenses become more convex; **(d)** pupils constrict, ciliary zonule relaxes, lenses become more convex.

18. The blind spot of the eye is **(a)** where more rods than cones are found, **(b)** where the macula lutea is located, **(c)** where only cones occur, **(d)** where the optic nerve leaves the eye.

19. Olfactory tract damage would probably affect your ability to **(a)** see, **(b)** hear, **(c)** feel pain, **(d)** smell.

20. Sensory impulses transmitted over the facial, glossopharyngeal, and vagus nerves are involved in the sensation of **(a)** taste, **(b)** touch, **(c)** equilibrium, **(d)** smell.

21. Taste buds are found on the **(a)** anterior part of the tongue, **(b)** posterior part of the tongue, **(c)** palate, **(d)** all of these.

22. Gustatory epithelial cells are stimulated by **(a)** movement of otoliths, **(b)** stretch, **(c)** substances in solution, **(d)** photons of light.

23. Olfactory nerve filaments are found **(a)** in the optic bulbs, **(b)** passing through the cribriform plate of the ethmoid bone, **(c)** in the optic tracts, **(d)** in the olfactory cortex.

24. Conduction of sound from the middle ear to the internal ear occurs via vibration of the **(a)** malleus against the tympanic membrane, **(b)** stapes in the oval window, **(c)** incus in the round window, **(d)** stapes against the tympanic membrane.

25. Which of the following statements does not correctly describe the spiral organ? **(a)** Sounds of high frequency stimulate hair cells at the basal end, **(b)** the "hairs" of the receptor cells are embedded in the tectorial membrane, **(c)** the basilar membrane acts as a resonator, **(d)** the more numerous outer hair cells are largely responsible for our perception of sound.

26. Pitch is to frequency of sound as loudness is to **(a)** quality, **(b)** intensity, **(c)** overtones, **(d)** all of these.

27. The structure that allows pressure in the middle ear to be equalized with atmospheric pressure is the **(a)** pinna, **(b)** pharyngotympanic tube, **(c)** tympanic membrane, **(d)** oval window.
28. Which of the following is important in maintaining the balance of the body? **(a)** visual cues, **(b)** semicircular canals, **(c)** the saccule, **(d)** proprioceptors, **(e)** all of these.
29. Equilibrium receptors that report the position of the head in space relative to the pull of gravity are **(a)** spiral organs, **(b)** maculae, **(c)** cristae ampullares, **(d)** otoliths.
30. Which of the following is not a possible cause of conduction deafness? **(a)** impacted cerumen, **(b)** middle ear infection, **(c)** cochlear nerve degeneration, **(d)** otosclerosis.
31. Which of the following are intrinsic eye muscles? **(a)** superior rectus, **(b)** orbicularis oculi, **(c)** smooth muscles of the iris and ciliary body, **(d)** levator palpebrae superioris.
32. Otoliths (ear stones) are **(a)** a cause of deafness, **(b)** a type of hearing aid, **(c)** important in equilibrium, **(d)** the rock-hard petrous temporal bones.
33. A reflex that causes reciprocal activation of the antagonist muscle is the **(a)** crossed-extensor, **(b)** flexor, **(c)** tendon, **(d)** muscle stretch.

Short Answer Essay Questions

34. List the structural components of the peripheral nervous system, and describe the function of each component.
35. How do rods and cones differ functionally?
36. Where is the fovea centralis, and why is it important?
37. Describe the response of rhodopsin to light stimuli. What is the outcome of this cascade of events?
38. Since there are only three types of cones, how can you explain the fact that we see many more colors?
39. Where are the olfactory sensory neurons, and why is that site poorly suited for their job?
40. Central pattern generators (CPGs) are found at the segmental level of motor control. (a) What is the job of the CPGs? (b) What controls them, and where is this control localized?
41. (a) Define plexus. (b) Indicate the spinal roots of origin of the four major nerve plexuses, and name the general body regions served by each.
42. What is the homeostatic value of flexor reflexes?
43. Compare and contrast flexor and crossed-extensor reflexes.
44. Explain how a crossed-extensor reflex exemplifies both serial and parallel processing.
45. What clinical information can be gained by conducting somatic reflex tests?
46. What is the structural and functional relationship between spinal nerves, skeletal muscles, and dermatomes?
47. Differentiate clearly between sensation and perception.
48. Why are the cerebellum and basal nuclei called precommand areas?

AT THE CLINIC

Clinical Case Studies

Peripheral Nervous System

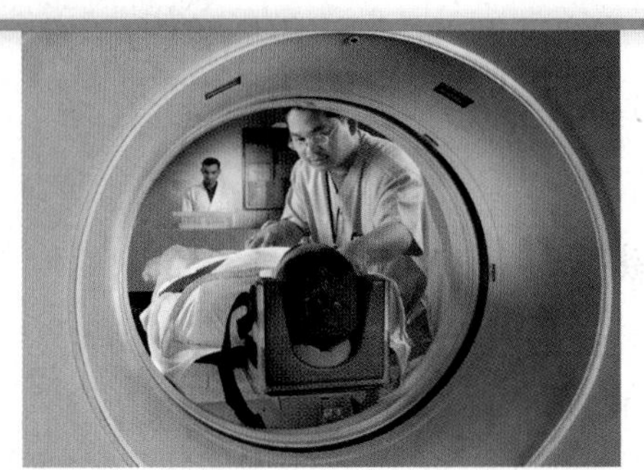

William Hancock, a 44-year-old male, was a passenger on the bus involved in the accident on Route 91. When emergency personnel arrived on the scene, they found Mr. Hancock unconscious, but with stable vital signs. As paramedics placed him on a backboard to stabilize his head, neck, and back, they noted watery blood leaking from his right ear. In the hospital, Mr. Hancock regained consciousness and was treated for deep lacerations on his scalp and face. Head CT scans revealed both longitudinal and transverse fractures of the right petrous temporal and sphenoid bones that extended through the foramen rotundum and foramen ovale.

The following observations were recorded on Mr. Hancock's chart on admission:

- Complete loss of hearing in the right ear.
- Paresthesia (sensation of "pins and needles") at the right corner of the mouth, extending to the lower lip and chin.
- Numbness of the right upper lip, lower eyelid, and cheek.
- Right eye turned slightly inward when looking straight ahead. Diplopia (double vision), particularly when looking to the right.

1. In addition to blood, which fluid was leaking from Mr. Hancock's right ear? Which structures must have been damaged to allow this to happen? Why would this lead Mr. Hancock's doctors to give him antibiotics? Why was the head of his bed elevated?
2. Each of the four observations on Mr. Hancock's chart indicates damage to a cranial nerve. Identify each cranial nerve involved. If applicable, identify which specific branch of that nerve is involved.

Mr. Hancock was given a course of antibiotics, the head of his bed was elevated by 30°, and he was placed under close observation. After 24 hours, doctors noted that the right side of Mr. Hancock's face showed signs of drooping, with incomplete eye closure and asymmetric facial expressions. Mr. Hancock's right eye showed minimal tear production. The weakness and asymmetry on the right side of his face began to subside after a few days, and the leak of fluid from his ear stopped, but he continued to complain of paresthesia, diplopia, and an inability to hear with his right ear.

3. The observations after 24 hours suggest that yet another cranial nerve has been damaged. Which one? How can you explain the lack of tear production in the right eye?

Special Senses

When emergency personnel arrived on the scene of the bus crash, they found Brian Rhen, 42, sitting on the side of the road holding his head in his hands. He complained of a severe headache and nausea, and was evaluated and treated for a concussion. Several days later, Mr. Rhen began to experience recurring episodes of vertigo and was referred to a neurologist. (Vertigo is a sensation of motion or movement while the person is stationary, and can be accompanied by nausea and vomiting.)

Mr. Rhen was diagnosed with *benign paroxysmal positional vertigo* (BPPV). With this condition, which can be caused by head trauma, vertigo can be provoked by specific changes in head position. Mr. Rhen reported that his vertigo usually occurred when rolling over in bed, or when turning his head from side to side while sitting up, and that these movements provoked the sensation of a spinning room, which led to nausea. The neurologist confirmed the diagnosis of BPPV by using a test called the Dix-Hallpike maneuver. During this test, the neurologist looks for nystagmus (involuntary, jerking eye movements) as he makes specific rotational changes to Mr. Rhen's head position.

1. The ear is divided into three major areas (compartments). What are these three areas, and which of these areas is involved in Mr. Rhen's BPPV?
2. What are the three main sources of sensory input that the body uses in order to control balance and equilibrium?
3. Name the two functional divisions of the vestibular apparatus. Identify the sensory receptor associated with each division, and state which aspect of equilibrium each receptor senses.
4. BPPV can be caused by otoliths that have been dislodged from the otolithic membrane of the maculae. Based on Mr. Rhen's symptoms and the head movements that provoke these symptoms, what part of the vestibular apparatus are the displaced otoliths now affecting?
5. Explain why nystagmus is associated with the Dix-Hallpike maneuver.

For answers, see Answers Appendix.

14 The Autonomic Nervous System

KEY CONCEPTS

The human body is exquisitely sensitive to changes in its internal environment, and engages in a lifelong struggle to balance competing demands for resources under ever-changing conditions. Although all body systems contribute, the stability of our internal environment depends largely on the **autonomic nervous system (ANS)**, the system of motor neurons that innervates smooth and cardiac muscle and glands (**Figure 14.1**).

At every moment, signals stream from visceral organs into the CNS, and autonomic nerves make adjustments as necessary to ensure optimal support for body activities. In response to changing conditions, the ANS shunts blood to "needy" areas, speeds or slows heart rate, adjusts blood pressure and body temperature, and increases or decreases stomach secretions.

Most of this fine-tuning occurs without our awareness or attention. Can you tell when your arteries are constricting or your pupils are dilating? Probably not—but if you've ever been stuck in a checkout line, and your full bladder was contracting as if it had a mind of its own, you've been very aware of visceral activity. The ANS controls all these functions, both those we're aware of and those we're not. Indeed, as the term *autonomic* (*auto* = self; *nom* = govern) implies, this motor subdivision of the peripheral nervous system has a certain amount of functional independence. The ANS is also called the **involuntary nervous system**, which reflects its subconscious control, or the **general visceral motor system**, which indicates the location of most of its effectors.

14.1 The ANS differs from the somatic nervous system in that it can stimulate or inhibit its effectors

→ **Learning Objectives**

☐ **Define autonomic nervous system and explain its relationship to the peripheral nervous system.**

☐ **Compare the somatic and autonomic nervous systems relative to effectors, efferent pathways, and neurotransmitters released.**

In our previous discussions of motor nerves, we have focused largely on the somatic nervous system. There are some key differences

Amanita muscaria mushrooms are the source of muscarine, a chemical that acts on targets of parasympathetic neurons.

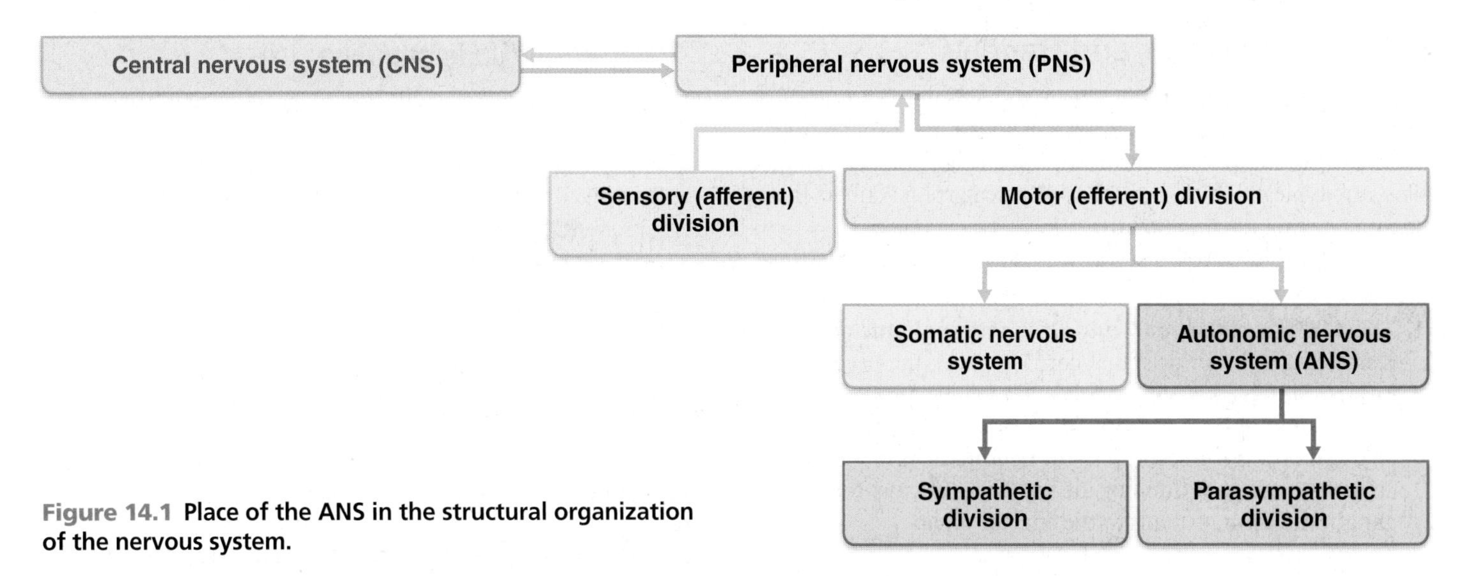

Figure 14.1 Place of the ANS in the structural organization of the nervous system.

between the somatic and autonomic systems as well as areas of functional overlap. Both systems have motor fibers, but the somatic and autonomic nervous systems differ in: (1) their effectors, (2) their efferent pathways and ganglia, and (3) target organ responses to their neurotransmitters. Consult **Figure 14.2** for a summary of the differences.

Effectors

The somatic nervous system stimulates skeletal muscles, whereas the ANS innervates cardiac and smooth muscle and glands. Differences in the physiology of the effector organs account for most of the remaining differences between somatic and autonomic effects on their target organs.

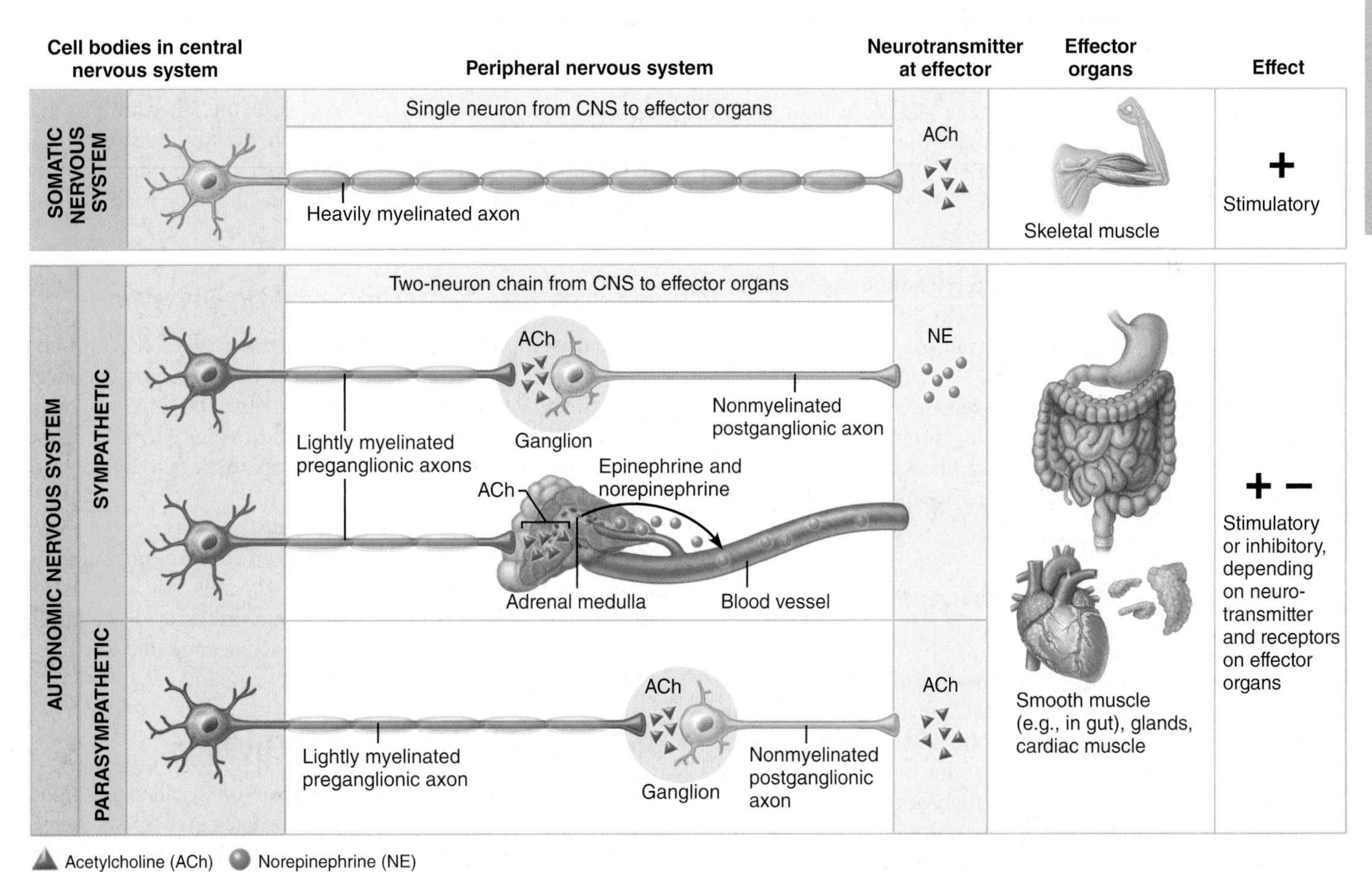

Figure 14.2 Comparison of motor neurons in the somatic and autonomic nervous systems.

Efferent Pathways and Ganglia

In the somatic nervous system, the motor neuron cell bodies are in the CNS, and their axons extend in spinal or cranial nerves all the way to the skeletal muscles they activate. Somatic motor fibers are typically thick, heavily myelinated group A fibers that conduct nerve impulses rapidly.

In contrast, the ANS uses a *two-neuron chain* to reach its effectors:

1. The cell body of the first neuron, the **preganglionic neuron**, resides in the brain or spinal cord. Its axon, the **preganglionic axon**, synapses with the second motor neuron.
2. The **postganglionic neuron** (sometimes called the *ganglionic neuron*), is the second motor neuron. Its cell body is in an **autonomic ganglion** outside the CNS. Its axon, the **postganglionic axon,** extends to the effector organ.

If you think about the meanings of all these terms while referring to Figure 14.2, understanding the rest of the chapter will be much easier.

Preganglionic axons are thin, lightly myelinated fibers, and postganglionic axons are even thinner and nonmyelinated. Consequently, conduction through the autonomic efferent chain is slower than conduction in the somatic motor system. For most of their course, many pre- and postganglionic fibers are incorporated into spinal or cranial nerves.

Keep in mind that autonomic ganglia are *motor* ganglia, containing the cell bodies of motor neurons. Technically, they are sites of synapse and information transmission from preganglionic to postganglionic neurons. Also, remember that the somatic motor division *lacks* ganglia entirely. The dorsal root ganglia are part of the sensory, not the motor, division of the PNS.

14

Neurotransmitter Effects

All somatic motor neurons release **acetylcholine (ACh)** at their synapses with skeletal muscle fibers. The effect is always *excitatory*, and if stimulation reaches threshold, the muscle fibers contract.

Autonomic postganglionic fibers release two neurotransmitters: **norepinephrine (NE)** secreted by most sympathetic fibers, and ACh secreted by parasympathetic fibers. Depending on the type of receptors on the target organ, the effect may be excitatory or inhibitory.

Overlap of Somatic and Autonomic Function

Higher brain centers regulate and coordinate both somatic and autonomic motor activities, and most spinal nerves (and many cranial nerves) contain both somatic and autonomic fibers. Moreover, most of the body's adaptations to changing internal and external conditions involve both skeletal muscles and visceral organs. For example, when skeletal muscles are working hard, they need more oxygen and glucose, so autonomic control mechanisms speed up heart rate and dilate airways to meet these needs and maintain homeostasis.

☑ Check Your Understanding

1. Name the three types of effectors of the autonomic nervous system.
2. Which relays instructions from the CNS to muscles more quickly, the somatic nervous system or the ANS? Explain why.
3. MAKING **connections** The cell bodies of autonomic postganglionic neurons are found in ANS ganglia. The cell bodies of another class of neuron are also found in ganglia (but not ANS ganglia). What are these other ganglia called? Determine the structural and functional classification of the neurons found in these other ganglia (use Table 11.1 on pp. 353–354).

For answers, see Answers Appendix.

14.2 The ANS consists of the parasympathetic and sympathetic divisions

→ Learning Objective

☐ **Compare and contrast the functions of the parasympathetic and sympathetic divisions.**

The ANS has two arms, parasympathetic and sympathetic. The *parasympathetic division* promotes maintenance functions and conserves body energy, whereas the *sympathetic division* mobilizes the body during activity.

Both divisions generally serve the same visceral organs but cause opposite effects: While one stimulates certain smooth muscles to contract or a gland to secrete, the other inhibits that action. Through this **dual innervation**, the two divisions counterbalance each other to keep body systems running smoothly. Let's focus briefly on extreme situations in which each division exerts primary control.

Role of the Parasympathetic Division

The **parasympathetic division**, sometimes called the "rest and digest" system, keeps body energy use as low as possible, even as it directs vital "housekeeping" activities like digesting food and eliminating feces and urine. (This explains why it is a good idea to relax after a heavy meal: so sympathetic activity does not interfere with digestion.)

Parasympathetic activity is best illustrated in a person who relaxes after a meal and reads a magazine. Blood pressure and heart rate are regulated at low normal levels, and the gastrointestinal tract is actively digesting food. In the eyes, the pupils are constricted and the lenses are accommodated for close vision to improve the clarity of the close-up image.

Role of the Sympathetic Division

The activity of the **sympathetic division** (often called the "fight-or-flight" system) is evident when we are excited or find ourselves in emergency or threatening situations, such as being frightened by street toughs late at night. A rapidly pounding heart; deep breathing; dry mouth; cold, sweaty skin; and dilated pupils are sure signs of sympathetic nervous system mobilization.

During any type of vigorous physical activity, the sympathetic division also promotes a number of other adjustments.

- It constricts visceral (and sometimes cutaneous) blood vessels, shunting blood to active skeletal muscles and the vigorously working heart
- It dilates the bronchioles in the lungs, increasing air flow (and thus increasing oxygen delivery to body cells)
- It stimulates the liver to release more glucose into the blood to accommodate the increased energy needs of body cells

At the same time, the sympathetic division temporarily reduces nonessential activities, such as gastrointestinal tract motility. If you are running from a mugger, digesting lunch can wait! It is far more important to give your muscles everything they need to get you out of danger. In such active situations, the sympathetic division enables the body to cope with potential threats to homeostasis. It provides the optimal conditions for an appropriate response, whether that response is to run, see distant objects better, or think more clearly.

We have just looked at two extreme situations in which one or the other branch of the ANS dominates. Think of the parasympathetic division as the **D** division [digestion, defecation, and diuresis (urination)], and the sympathetic division as the **E** division (exercise, excitement, emergency, embarrassment). Table 14.5 (p. 517) presents a more detailed summary of how each division affects various organs.

Remember, however, that the two ANS divisions rarely work in an all-or-none fashion as described above. A dynamic antagonism exists between the divisions, and both make continuous fine adjustments to maintain homeostasis.

Key Anatomical Differences

Before we explain the anatomy of the divisions of the ANS in detail, let's look at their key anatomical differences (**Figure 14.3**). The sympathetic and parasympathetic divisions differ in:

① **Sites of origin.** Parasympathetic fibers are craniosacral—they originate in the brain (cranium) and sacral spinal cord.

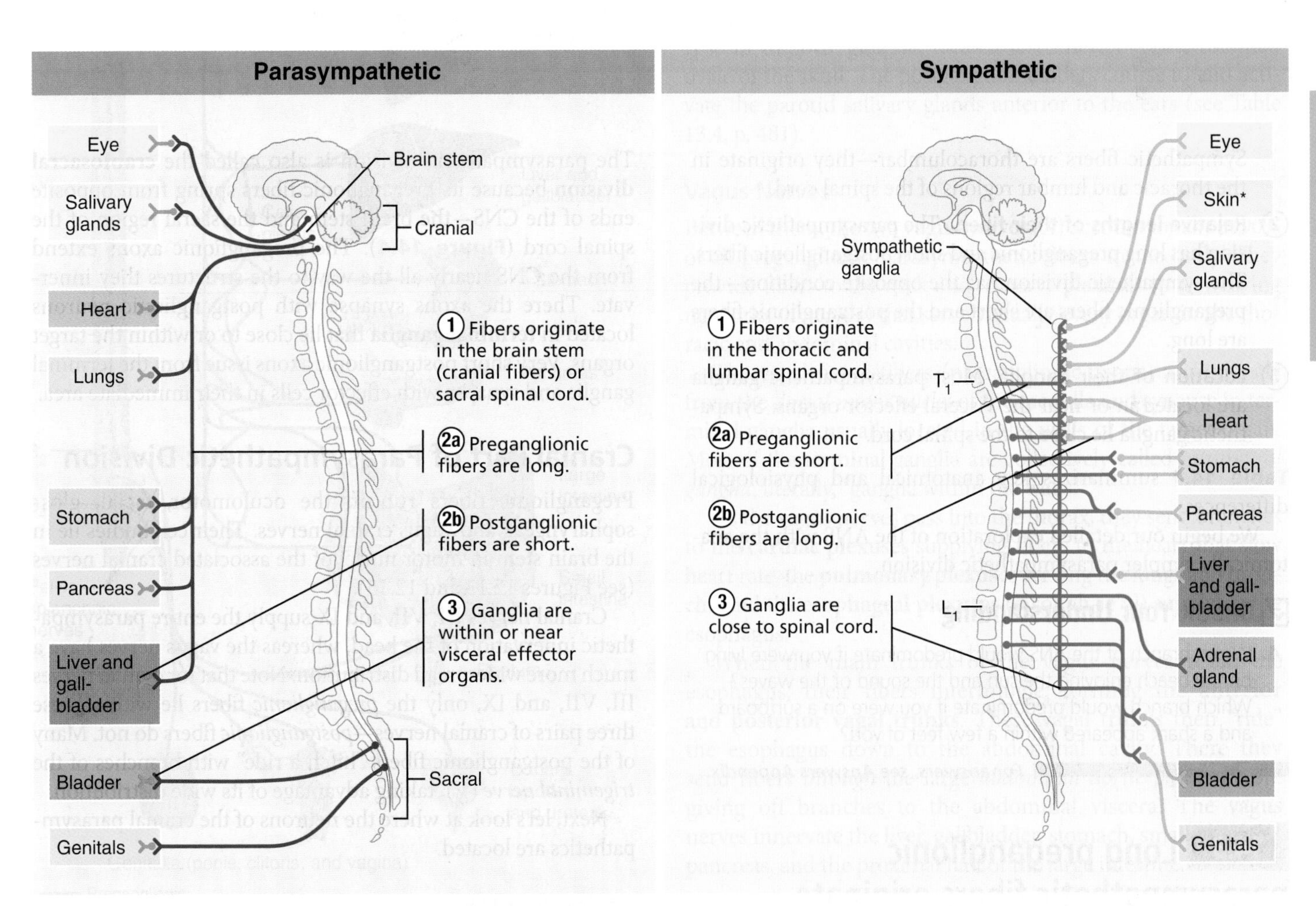

Figure 14.3 Key anatomical differences between ANS divisions.
*Although sympathetic innervation to the skin is mapped to the cervical region here, all nerves to the periphery carry postganglionic sympathetic fibers.

these neurons run in the ventral roots of the spinal nerves to the ventral rami and then branch off to form the **pelvic splanchnic nerves** (splank′nik; *splanchni* = viscera), which pass through the **inferior hypogastric (pelvic) plexus** in the pelvic floor (Figure 14.4). Some preganglionic fibers synapse with ganglia in this plexus, but most synapse in intramural ganglia in the walls of the following organs: distal half of the large intestine, urinary bladder, ureters, and reproductive organs.

☑ Check Your Understanding

5. In general terms, where are the cell bodies of preganglionic parasympathetic neurons that innervate the head? Where are the cell bodies of postganglionic parasympathetic neurons innervated by the vagus nerve?

For answers, see Answers Appendix.

14.4 Short preganglionic sympathetic fibers originate in the thoracolumbar CNS

→ Learning Objective

☐ **For the sympathetic division, describe the site of CNS origin, locations of ganglia, and general fiber pathways.**

The sympathetic division is anatomically more complex than the parasympathetic division, partly because it innervates more organs. Like the parasympathetic nervous system, the sympathetic nervous system supplies the visceral organs in the internal body cavities. In addition, it supplies all visceral structures in the superficial (somatic) part of the body. These superficial structures—some glands and smooth muscle—are innervated *only* by the sympathetic nervous system. These structures include:

- Sweat glands
- The hair-raising arrector pili muscles of the skin
- Smooth muscle in the walls of all arteries and veins, both deep and superficial (This will be a key point for you to remember when you study the cardiovascular system.)

We will explain exactly how this works later—let's get on with the basic anatomy of the sympathetic division.

All preganglionic fibers of the sympathetic division arise from cell bodies of preganglionic neurons in spinal cord segments T_1 through L_2 (Figure 14.3). For this reason, the sympathetic division is also referred to as the **thoracolumbar division** (tho-rah″ko-lum′bar).

The numerous cell bodies of preganglionic sympathetic neurons in the gray matter of the spinal cord form the **lateral horns** (see Figures 12.29b, p. 421, and 12.30, p. 422). The lateral horns are just posterolateral to the ventral horns that house somatic motor neurons. (Parasympathetic preganglionic neurons in the sacral cord are far less abundant than the comparable sympathetic neurons in the thoracolumbar regions, so there are no lateral horns in the sacral region of the spinal cord. This is a major anatomical difference between the two divisions.)

After leaving the cord via the ventral root, preganglionic sympathetic fibers pass through a **white ramus communicans**

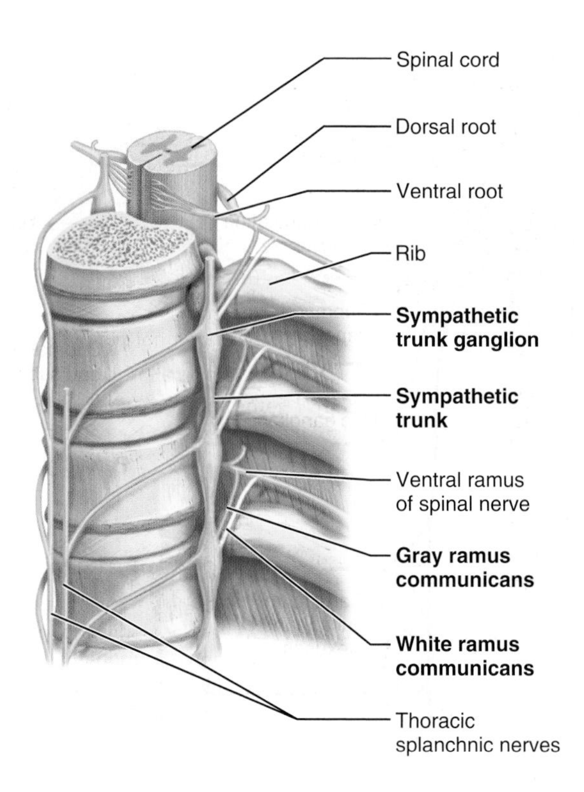

Figure 14.5 Location of the sympathetic trunk. The left sympathetic trunk in the posterior thorax.

[plural: **rami communicantes** (kom-mu″nĭ-kan′tēz)] to enter an adjoining **sympathetic trunk ganglion** forming part of the **sympathetic trunk** (or *sympathetic chain*, **Figure 14.5**). Looking like strands of glistening white beads, the sympathetic trunks flank each side of the vertebral column. They consist of the sympathetic ganglia and fibers running from one ganglion to another. The sympathetic trunk ganglia are also called *chain ganglia* or *paravertebral* ("near the vertebrae") *ganglia*.

Although the sympathetic *trunks* extend from neck to pelvis, sympathetic *fibers* arise only from the thoracic and lumbar cord segments, as shown in Figure 14.3. The ganglia vary in size, position, and number, but typically there are 23 in each sympathetic trunk—3 cervical, 11 thoracic, 4 lumbar, 4 sacral, and 1 coccygeal.

Once a preganglionic axon reaches a trunk ganglion, one of three things can happen (see **Figure 14.6**). The preganglionic and postganglionic neurons can:

1. **Synapse at the same level.** In this case, the synapse is in the same trunk ganglion.
2. **Synapse at a higher or lower level.** The preganglionic axon ascends or descends the sympathetic trunk to another trunk ganglion.
3. **Synapse in a distant collateral ganglion.** The preganglionic axon passes through the trunk ganglion and emerges from the sympathetic trunk without synapsing. These preganglionic fibers help form several *splanchnic nerves* and synapse in **collateral**, or *prevertebral*, **ganglia** located anterior to the vertebral column. Unlike sympathetic trunk ganglia, the

collateral ganglia are neither paired nor segmentally arranged. They occur only in the abdomen and pelvis.

Regardless of where the synapse occurs, all sympathetic ganglia are close to the spinal cord, and their postganglionic fibers are typically much longer than their preganglionic fibers. Recall that the opposite condition exists in the parasympathetic division, an important anatomical distinction. Autonomic ganglia are compared in **Table 14.2**.

Sympathetic Pathways with Synapses in Trunk Ganglia

When synapses are made in sympathetic trunk ganglia, the postganglionic axons enter the ventral (or dorsal) ramus of the adjoining spinal nerves by way of communicating branches called **gray rami communicantes** (Figure 14.6). From there they travel via branches of the rami to their effectors, including sweat glands and arrector pili muscles of the skin. Anywhere along their path, the postganglionic axons may transfer over to nearby blood vessels and innervate the vascular smooth muscle all the way to the final branches of the blood vessels.

Notice that the names of the rami communicantes—*white* or *gray*—reflect their appearance, revealing whether or not their fibers are myelinated. Preganglionic fibers composing the white rami are myelinated. Postganglionic axons forming the gray rami are not. (This has no relationship to the white and gray matter of the CNS.)

The white rami, which carry preganglionic axons to the sympathetic trunks, are found only in the T_1–L_2 cord segments, regions of sympathetic outflow. However, gray rami carrying postganglionic fibers headed for the periphery issue from every trunk ganglion from the cervical to

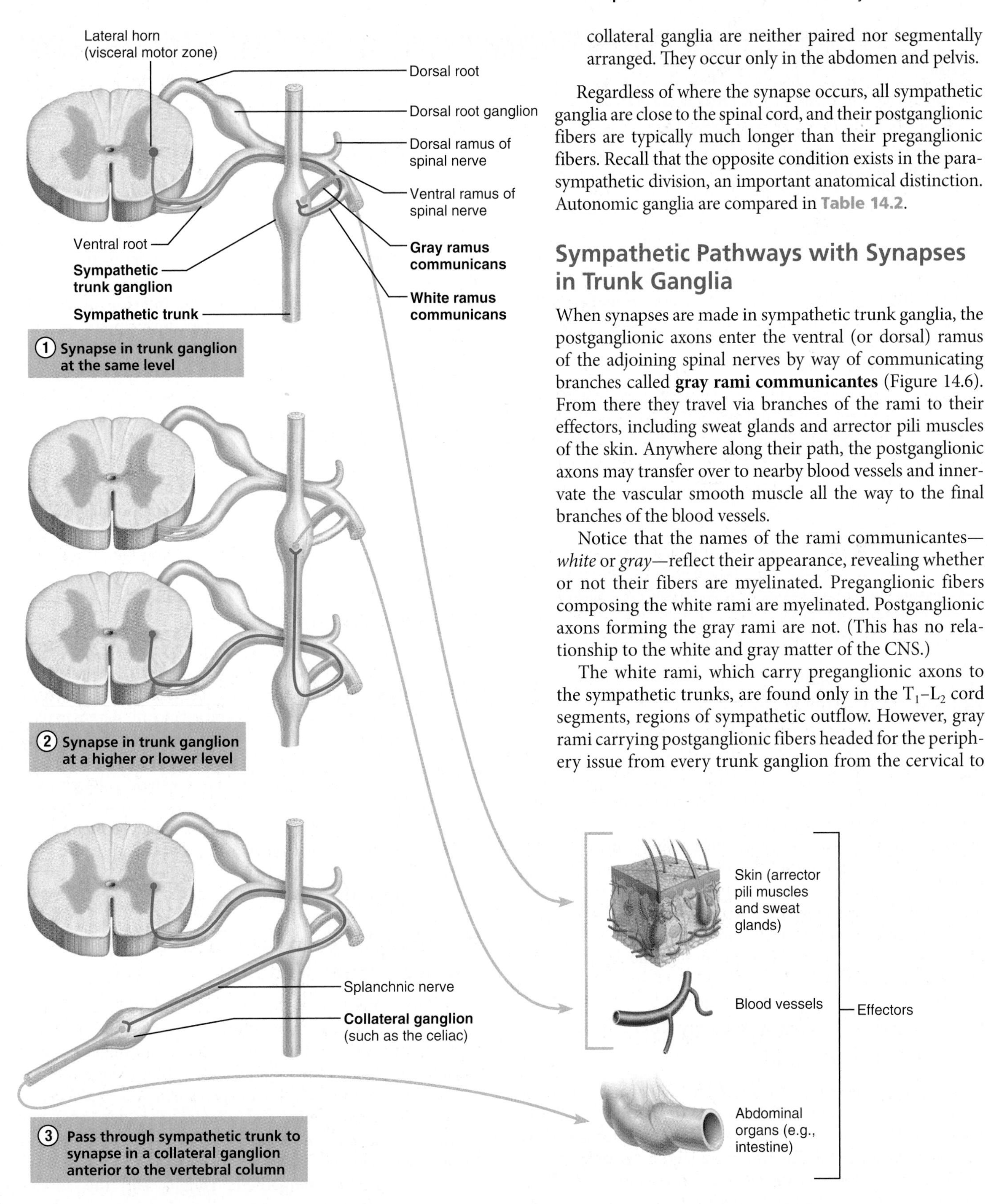

Figure 14.6 Three pathways of sympathetic innervation. Synapses between preganglionic and postganglionic sympathetic neurons can occur at three different locations.

Table 14.2 Summary of Autonomic Ganglia

NAME	DIVISION	LOCATION
Terminal ganglia	Parasympathetic nervous system	Within wall of organ served (*intramural ganglia*) or close to organ
Sympathetic trunk ganglia	Sympathetic nervous system	Paired, beside spinal cord
Collateral ganglia (prevertebral ganglia)	Sympathetic nervous system	Unpaired, anterior to spinal cord

the sacral region, allowing sympathetic output to reach all parts of the body.

Note that *rami communicantes are associated only with the sympathetic division*. They never carry parasympathetic fibers.

Pathways to the Head

Sympathetic preganglionic fibers serving the head emerge from spinal cord segments T_1–T_4 and ascend the sympathetic trunk to synapse with postganglionic neurons in the **superior cervical ganglion** (**Figure 14.7**). This ganglion contributes sympathetic fibers that run in several cranial nerves and in the upper three or four cervical spinal nerves.

Besides serving the skin and blood vessels of the head, fibers from the superior cervical ganglion stimulate the dilator muscles of the irises of the eyes, inhibit the nasal and salivary glands (the reason your mouth goes dry when you are scared), and innervate the smooth (tarsal) muscle that lifts the upper eyelid. The superior cervical ganglion also sends direct branches to the heart.

14

Pathways to the Thorax

Sympathetic preganglionic fibers innervating the thoracic organs originate at T_1–T_6. From there the preganglionic fibers run to synapse in the cervical trunk ganglia. Postganglionic fibers emerging from the **middle** and **inferior cervical ganglia** enter cervical nerves C_4–C_8 (Figure 14.7). Some of these fibers innervate the heart via the cardiac plexus, and some innervate the thyroid gland, but most serve the skin. Additionally, some T_1–T_6 preganglionic fibers synapse in the nearest trunk ganglion, and the postganglionic fibers pass directly to the organ served. Fibers to the heart, aorta, lungs, and esophagus take this direct route. Along the way, they run through the plexuses associated with those organs.

Sympathetic Pathways with Synapses in Collateral Ganglia

Most of the preganglionic fibers from T_5 down synapse in collateral ganglia, and so most of these fibers enter and leave the sympathetic trunks without synapsing. They form several nerves called **splanchnic nerves**, including the **greater**, **lesser**, and **least splanchnic nerves** (*thoracic splanchnic nerves*) and the **lumbar** and **sacral splanchnic nerves**.

The splanchnic nerves contribute to a number of interweaving nerve plexuses known collectively as the **abdominal aortic plexus**, which clings to the surface of the abdominal aorta. This complex plexus contains several ganglia that together serve the abdominopelvic viscera. From superior to inferior, the most important of these ganglia (and related subplexuses) are the **celiac**, **superior mesenteric**, and **inferior mesenteric**, named for the arteries with which they most closely associate (Figure 14.7). Postganglionic fibers issuing from these ganglia generally travel to their target organs in the company of the arteries serving these organs.

Pathways to the Abdomen

Sympathetic preganglionic fibers from T_5 to L_2 innervate the abdomen. They travel in the thoracic splanchnic nerves to synapse mainly at the celiac and superior mesenteric ganglia. Postganglionic fibers issuing from these ganglia serve the stomach, intestines (except the distal half of the large intestine), liver, spleen, and kidneys.

Pathways to the Pelvis

Preganglionic fibers innervating the pelvis originate from T_{10} to L_2 and then descend in the sympathetic trunk to the lumbar and sacral trunk ganglia. Some fibers synapse there and the postganglionic fibers run in lumbar and sacral splanchnic nerves to plexuses on the lower aorta and in the pelvis. Other preganglionic fibers pass directly to these autonomic plexuses and synapse in collateral ganglia, such as the inferior mesenteric ganglion.

Postganglionic fibers proceed from these plexuses to the pelvic organs (the urinary bladder and reproductive organs) and also the distal half of the large intestine. For the most part, sympathetic fibers *inhibit* the activity of muscles and glands in the abdominopelvic visceral organs.

Sympathetic Pathways with Synapses in the Adrenal Medulla

Some fibers traveling in the thoracic splanchnic nerves pass through the celiac ganglion without synapsing and terminate by synapsing with the hormone-producing medullary cells of the adrenal gland. When stimulated by preganglionic fibers, the medullary cells secrete *norepinephrine* and *epinephrine* (also called *noradrenaline* and *adrenaline*, respectively) into the blood, producing the excitatory effects we have all felt as a "surge of adrenaline."

Embryologically, sympathetic ganglia and the adrenal medulla arise from the same tissue. For this reason, the adrenal medulla is sometimes viewed as a "misplaced" sympathetic ganglion, and its hormone-releasing cells, although lacking nerve processes, are considered equivalent to postganglionic sympathetic neurons.

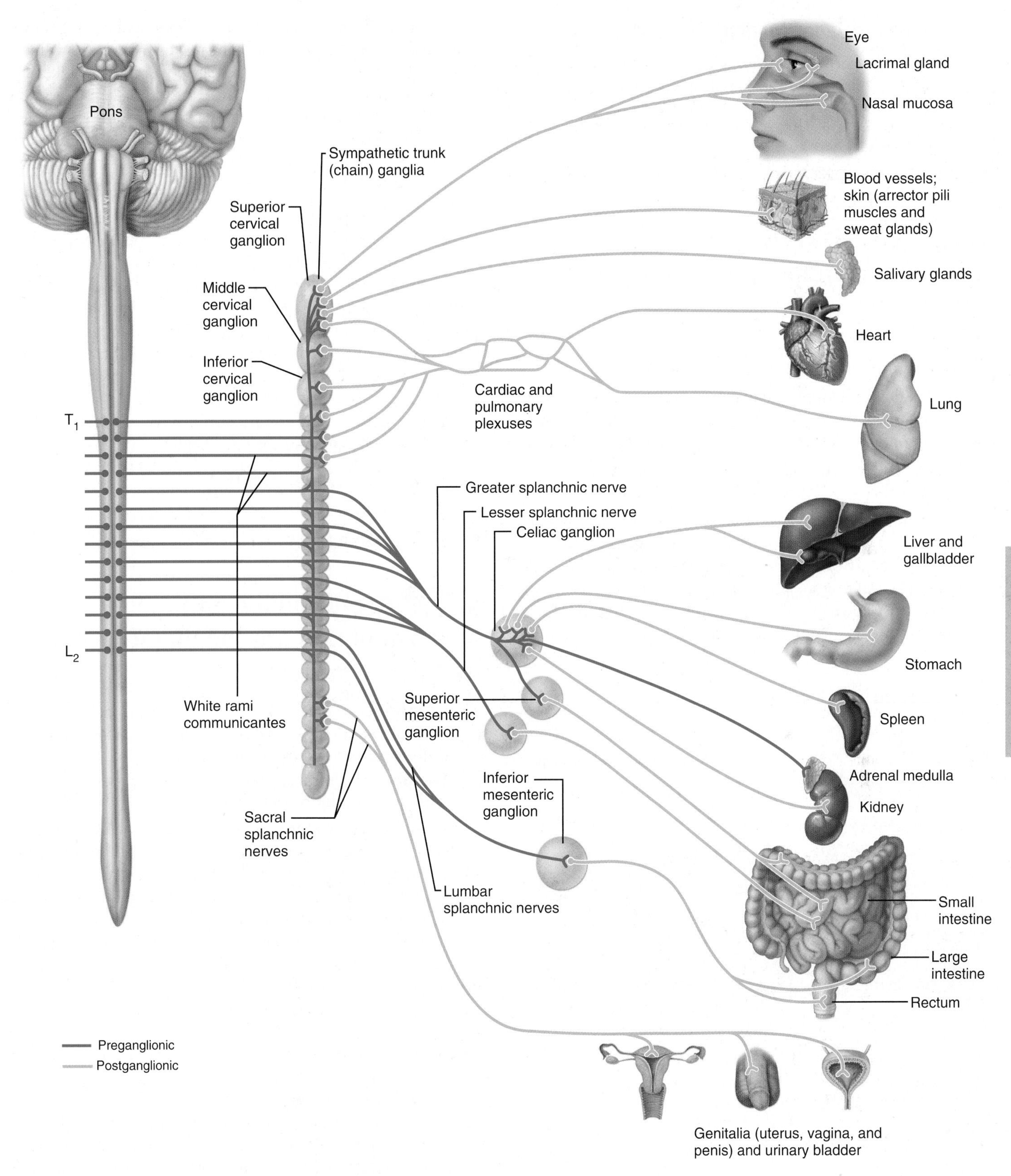

Figure 14.7 Sympathetic division of the ANS. Sympathetic innervation to peripheral structures (blood vessels, glands, and arrector pili muscles) occurs in all areas but is shown only in the cervical area.

☑ Check Your Understanding

6. State whether each of the following is a characteristic of the sympathetic or parasympathetic nervous system: short preganglionic fibers; origin from thoracolumbar region of spinal cord; terminal ganglia; collateral ganglia; innervates adrenal medulla.

For answers, see Answers Appendix.

14.5 Visceral reflex arcs have the same five components as somatic reflex arcs

→ Learning Objective

☐ **Compare visceral reflexes to somatic reflexes.**

Visceral reflex arcs have essentially the same components as somatic reflex arcs—receptor, sensory neuron, integration center, motor neuron, and effector. However, there are two key differences:

- A visceral reflex arc has *two consecutive* neurons in its motor component (**Figure 14.8**; compare with Figure 13.40).
- The afferent fibers are **visceral sensory neurons**, which send information about chemical changes, stretch, and irritation of the viscera. Because most anatomists consider the ANS to be a visceral motor system, the presence of sensory fibers (mostly visceral pain afferents) is often overlooked. However, they are the first link in autonomic reflexes.

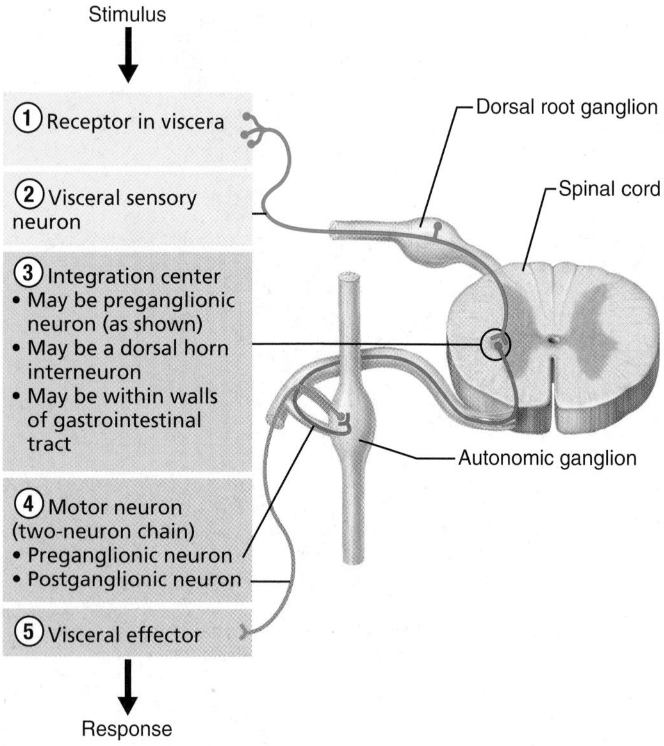

Figure 14.8 Visceral reflexes. Visceral reflex arcs have the same five elements as somatic reflex arcs. The visceral afferent (sensory) fibers are found both in spinal nerves (as depicted here) and in autonomic nerves.

Nearly all the sympathetic and parasympathetic fibers we have described so far are accompanied by the afferent fibers conducting sensory impulses from glands or muscles. Like sensory neurons serving somatic structures (skeletal muscles and skin), the cell bodies of visceral sensory neurons are located either in sensory ganglia of cranial nerves or in dorsal root ganglia of the spinal cord. Visceral sensory neurons are also found in sympathetic ganglia where preganglionic neurons synapse.

Two examples of visceral reflexes are the reflexes that empty the rectum and bladder, which we discuss in Chapters 22 and 24.

Complete three-neuron reflex arcs (with sensory neurons, interneurons, and motor neurons) exist entirely within the walls of the gastrointestinal tract. Neurons composing these reflex arcs make up the **enteric nervous system**, which plays an important role in controlling gastrointestinal tract activity. We will discuss the enteric nervous system in more detail in Chapter 22.

Visceral sensory neurons are also involved in the phenomenon of referred pain described in Chapter 13 (p. 434).

☑ Check Your Understanding

7. How does a visceral reflex differ from a somatic reflex?

For answers, see Answers Appendix.

14.6 Acetylcholine and norepinephrine are the major ANS neurotransmitters

→ Learning Objectives

☐ **Define cholinergic and adrenergic fibers, and list the different types of their receptors.**

☐ **Describe the clinical importance of drugs that mimic or inhibit adrenergic or cholinergic effects.**

The major neurotransmitters released by ANS neurons are *acetylcholine* (*ACh*) and *norepinephrine* (*NE*). ACh, the same neurotransmitter secreted by somatic motor neurons, is released by (1) all ANS preganglionic axons and (2) all parasympathetic postganglionic axons at synapses with their effectors. Fibers that release ACh are called **cholinergic fibers** (ko″lin-er′jik).

In contrast, most sympathetic postganglionic axons release NE and are called **adrenergic fibers** (ad″ren-er′jik). An exception is sympathetic postganglionic fibers that secrete ACh onto sweat glands.

The effects of ACh and NE on their effectors are not consistently excitatory or inhibitory. Why not? Because the action of any neurotransmitter depends on the receptor to which it binds. Each autonomic neurotransmitter binds with two or more kinds of receptors, allowing it to exert different effects (activation or inhibition) at different body targets. **Table 14.3** summarizes the receptor types that we introduce next, and **Table 14.4** describes some of the many drugs that act upon them.

Cholinergic Receptors

The two types of cholinergic (ACh-binding) receptors are named for drugs that bind to them and mimic acetylcholine's

Table 14.3 Cholinergic and Adrenergic Receptors

NEUROTRANSMITTER	RECEPTOR TYPE	MAJOR LOCATIONS*	EFFECT OF BINDING
Acetylcholine (ACh)	**Cholinergic**		
	Nicotinic	All postganglionic neurons; adrenal medullary cells (also neuromuscular junctions of skeletal muscle)	Excitation
	Muscarinic	All parasympathetic target organs	Excitation in most cases; inhibition of cardiac muscle
		Limited sympathetic targets (e.g., eccrine sweat glands[†])	Activation
Norepinephrine (NE) (and epinephrine released by adrenal medulla)	**Adrenergic**		
	β_1	Heart predominantly, but also kidneys and adipose tissue	Increases heart rate and force of contraction; stimulates kidneys to release renin
	β_2	Lungs and most other sympathetic target organs; abundant on blood vessels serving the heart, liver, and skeletal muscle	Effects mostly inhibitory; dilates blood vessels and bronchioles; relaxes smooth muscle walls of digestive and urinary visceral organs; relaxes uterus
	β_3	Adipose tissue	Stimulates lipolysis by fat cells
	α_1	Virtually all sympathetic target organs, especially blood vessels serving the skin, mucosae, abdominal viscera, kidneys, and salivary glands	Constricts blood vessels and visceral organ sphincters; dilates pupils of the eyes
	α_2	Membrane of adrenergic axon terminals; pancreas	Inhibits NE release from adrenergic terminals; inhibits insulin secretion

*Note that all of these receptor subtypes are also found in the CNS.
[†]Sympathetic cholinergic vasodilator fibers are found in other animals, but do not appear to be present in humans.

Table 14.4 Selected Drug Classes That Influence the Autonomic Nervous System

DRUG CLASS	RECEPTOR BOUND	EFFECTS	EXAMPLE	CLINICAL APPLICATION
Nicotinic agents	Nicotinic ACh receptors on all postganglionic neurons and in CNS	Typically stimulates sympathetic effects; blood pressure rises	Nicotine	Smoking cessation products
Parasympathomimetic agents (muscarinic agents)	Muscarinic ACh receptors	Enhance parasympathetic activity by mimicking effects of ACh	Pilocarpine	Glaucoma (opens aqueous humor drainage pores)
			Bethanechol	Difficulty urinating (increases bladder contraction)
Acetylcholinesterase inhibitors	None; bind to the enzyme (AChE) that degrades ACh	Indirect effect at all ACh receptors; prolong the effect of ACh	Neostigmine	Myasthenia gravis (increases availability of ACh)
			Sarin	Similar to widely used insecticides; used as chemical warfare agent
Sympathomimetic agents	Adrenergic receptors	Enhance sympathetic activity by binding to adrenergic receptors or increasing NE release	Albuterol (Ventolin)	Asthma (dilates bronchioles by binding to β_2 receptors)
			Phenylephrine	Colds (nasal decongestant, binds to α_1 receptors)
Sympatholytic agents	Adrenergic receptors	Decrease sympathetic activity by blocking adrenergic receptors	Propranolol	Hypertension (drugs called *beta-blockers* block β receptors, decreasing blood pressure)

effects. **Nicotinic receptors** (nik″o-tin′ik) respond to nicotine. **Muscarinic receptors,** the other set of ACh receptors, can be activated by the mushroom poison *muscarine* (mus′kah-rin). All ACh receptors are either nicotinic or muscarinic.

Nicotinic Receptors

Nicotinic receptors are found on:

- *All* postganglionic neurons (cell bodies and dendrites), both sympathetic and parasympathetic
- The hormone-producing cells of the adrenal medulla
- The sarcolemma of skeletal muscle cells at neuromuscular junctions (which are somatic and not autonomic targets)

When ACh binds to nicotinic receptors, the effect is *always* stimulatory. Just as at the sarcolemma of skeletal muscle (examined in Chapter 9), ACh binding to any nicotinic receptor directly opens ion channels, depolarizing the postsynaptic cell.

Muscarinic Receptors

Muscarinic receptors occur on all effector cells stimulated by postganglionic cholinergic fibers—that is, all parasympathetic target organs and a few sympathetic targets, such as eccrine sweat glands. When ACh binds to muscarinic receptors, the effect can be either inhibitory or stimulatory, depending on the subclass of muscarinic receptor on the target organ. For example, ACh binding to cardiac muscle receptors slows heart activity, whereas ACh binding to receptors on smooth muscle of the gastrointestinal tract increases its motility.

14

Adrenergic Receptors

There are also two major classes of adrenergic (NE-binding) receptors: **alpha (α)** and **beta (β)**. These receptors are further divided into subclasses (α_1 and α_2; β_1, β_2, and β_3). Organs that respond to NE (or to epinephrine) have one or more of these receptor subtypes.

NE or epinephrine can be either excitatory or inhibitory depending on which subclass of receptor predominates in the target organ. For example, NE binding to the β_1 receptors of cardiac muscle prods the heart into more vigorous activity, whereas epinephrine binding to β_2 receptors in bronchiole smooth muscle causes it to relax, dilating the bronchiole.

☑ Check Your Understanding

8. Would you find nicotinic receptors on skeletal muscle? Smooth muscle? Eccrine sweat glands? The adrenal medulla? CNS neurons?

For answers, see Answers Appendix.

14.7 The parasympathetic and sympathetic divisions usually produce opposite effects

→ Learning Objective

☐ **State the effects of the parasympathetic and sympathetic divisions on the following organs: heart, blood vessels, gastrointestinal tract, lungs, adrenal medulla, and external genitalia.**

As we mentioned earlier, most visceral organs receive *dual innervation* from both the parasympathetic and sympathetic divisions. Normally, both ANS divisions are partially active. Action potentials continually fire down both sympathetic and parasympathetic axons, producing a dynamic antagonism that precisely controls visceral activity. However, one division or the other usually predominates in given circumstances, and in a few cases, the two divisions actually cooperate with each other. Table 14.5 contains an organ-by-organ summary of their effects.

Antagonistic Interactions

Antagonistic effects are most clearly seen on the activity of the heart, respiratory system, and gastrointestinal organs. In a fight-or-flight situation, the sympathetic division increases heart rate, dilates airways, and inhibits digestion and elimination. When the emergency is over, the parasympathetic division restores heart rate and airway diameter to resting levels and then attends to processes that refuel body cells and discard wastes.

Sympathetic and Parasympathetic Tone

We have described the parasympathetic division as the "rest and digest" division, but the sympathetic division is the major actor in controlling blood pressure, even at rest. With few exceptions, blood vessels are entirely innervated by sympathetic fibers that keep the blood vessels in a continual state of partial constriction called **sympathetic**, or **vasomotor**, **tone**.

When blood pressure is too low to maintain blood flow, sympathetic fibers called **vasomotor fibers** fire more rapidly. This causes blood vessels to constrict and raises blood pressure. When blood pressure becomes too high, these sympathetic vasomotor fibers fire less rapidly and the vessels dilate.

During circulatory shock (inadequate blood flow to body tissues), or when more blood is needed to meet the soaring needs of working skeletal muscles, blood vessels serving the skin and abdominal viscera strongly constrict. This blood "shunting" helps maintain circulation to vital organs and skeletal muscles.

In contrast to the sympathetic division's dominance of blood vessels, parasympathetic effects normally dominate the heart and the smooth muscle of digestive and urinary tract organs. These organs exhibit **parasympathetic tone**. The parasympathetic division slows the heart and dictates the normal activity levels of the digestive and urinary tracts. However, the sympathetic division can override these parasympathetic effects during times of stress. Drugs that block parasympathetic responses increase heart rate and cause fecal and urinary retention. Parasympathetic fibers activate most glands, except for the adrenal glands and sweat glands of the skin.

Cooperative Effects

The best example of cooperative ANS effects occurs in the external genitalia. Parasympathetic stimulation dilates blood vessels in the external genitalia, producing the erection of the male penis or female clitoris during sexual excitement. (This may explain why anxiety can impair sexual performance—the sympathetic division is in charge.) Sympathetic stimulation then causes ejaculation of semen by the penis or reflex contractions of the vagina.

Table 14.5 Effects of the Parasympathetic and Sympathetic Divisions on Various Organs

TARGET ORGAN OR SYSTEM	PARASYMPATHETIC EFFECTS	SYMPATHETIC EFFECTS
Eye (iris)	Stimulates sphincter pupillae muscles; constricts pupils	Stimulates dilator pupillae muscles; dilates pupils
Eye (ciliary muscle)	Stimulates muscle, which makes lens bulge for close vision	Weakly inhibits muscle, which flattens lens for far vision
Glands (nasal, lacrimal, gastric, pancreas)	Stimulates secretory activity	Inhibits secretory activity; constricts blood vessels supplying the glands
Salivary glands	Stimulates secretion of watery saliva	Stimulates secretion of thick, viscous saliva
Sweat glands	No effect (no innervation)	Stimulates copious sweating (cholinergic fibers)
Adrenal medulla	No effect (no innervation)	Stimulates medulla cells to secrete epinephrine and norepinephrine
Arrector pili muscles attached to hair follicles	No effect (no innervation)	Stimulates contraction (erects hairs and produces "goosebumps")
Heart (muscle)	Decreases rate (slows heart)	Increases rate and force of heartbeat
Heart (coronary blood vessels)	No effect (no innervation)	Dilates blood vessels (vasodilation)*
Urinary bladder/urethra	Contracts smooth muscle of bladder wall; relaxes urethral sphincter; promotes voiding	Relaxes smooth muscle of bladder wall; constricts urethral sphincter; inhibits voiding
Lungs	Constricts bronchioles	Dilates bronchioles*
Digestive tract organs	Increases motility (peristalsis) and amount of secretion by digestive organs; relaxes sphincters to allow foodstuffs to move through tract	Decreases activity of glands and muscles of digestive system; constricts sphincters (e.g., anal sphincter)
Liver	Increases glucose uptake from blood	Stimulates release of glucose to blood*
Gallbladder	Excites (gallbladder contracts to expel bile)	Inhibits (gallbladder is relaxed)
Kidney	No effect (no innervation)	Promotes renin release; causes vasoconstriction; decreases urine output
Penis	Causes erection (vasodilation)	Causes ejaculation
Vagina/clitoris	Causes erection (vasodilation) of clitoris; increases vaginal lubrication	Causes vagina to contract
Blood vessels	Little or no effect	Constricts most vessels and increases blood pressure; constricts vessels of abdominal viscera and skin to divert blood to muscles, brain, and heart when necessary; epinephrine weakly dilates vessels of skeletal muscles during exercise*
Blood coagulation	No effect (no innervation)	Increases coagulation*
Cellular metabolism	No effect (no innervation)	Increases metabolic rate*
Adipose tissue	No effect (no innervation)	Stimulates lipolysis (fat breakdown)

*Effects are mediated by epinephrine release into the bloodstream from the adrenal medulla.

CLINICAL

HOMEOSTATIC IMBALANCE 14.1

Autonomic neuropathy (damage to autonomic nerves) is a common complication of diabetes mellitus. One of the earliest and most troubling symptoms is sexual dysfunction. Up to 75% of male diabetics experience erectile dysfunction, and female diabetics often experience reduced vaginal lubrication. Other frequent manifestations of autonomic neuropathy include dizziness after standing suddenly (poor blood pressure control), urinary incontinence, sluggish eye pupil reactions, and impaired sweating. Maintaining tight control of blood glucose levels is the best way to prevent diabetic neuropathy. +

Unique Roles of the Sympathetic Division

The adrenal medulla, sweat glands and arrector pili muscles of the skin, the kidneys, and most blood vessels receive only sympathetic fibers. It is easy to remember that the sympathetic system innervates these structures because most of us sweat under stress, our scalp "prickles" during fear, and our blood pressure skyrockets (from widespread constriction of blood vessels) when we get excited.

We have already described how sympathetic control of blood vessels regulates blood pressure and shunting of blood in the vascular system. We will now consider several other uniquely sympathetic functions.

- **Thermoregulatory responses to heat.** The sympathetic division mediates reflexes that regulate body temperature. For example, applying heat to the skin causes blood vessels in that area to dilate reflexively. When systemic body temperature rises, sympathetic nerves (1) dilate the skin's blood vessels, allowing heat to escape from skin flushed with warm blood, and (2) activate the sweat glands to help cool the body. When body temperature falls, skin blood vessels constrict, preventing heat loss from the skin.
- **Release of renin from the kidneys.** Sympathetic impulses stimulate the kidneys to release *renin*, an enzyme that causes the formation of potent blood pressure–increasing hormones (see Chapters 18 and 24).
- **Metabolic effects.** Through both direct neural stimulation and release of adrenal medullary hormones, the sympathetic division promotes a number of metabolic effects not reversed by parasympathetic activity. It (1) increases the metabolic rate of body cells; (2) raises blood glucose levels; and (3) mobilizes fats for use as fuels.

The medullary hormones also cause skeletal muscle to contract more strongly and quickly. As a side effect, this stimulates muscle spindles more often and, consequently, nerve impulses traveling to the muscles occur more synchronously. These neural bursts, which put muscle contractions on a "hair trigger," are great if you have to make a quick jump or run, but they can be disabling to the nervous musician or surgeon.

Localized versus Diffuse Effects

In the parasympathetic division, one preganglionic neuron synapses with one (or at most a few) postganglionic neurons. Additionally, all parasympathetic fibers release ACh, which is quickly destroyed (hydrolyzed) by acetylcholinesterase. Consequently, the parasympathetic division exerts short-lived, highly localized control over its effectors.

In contrast, in the sympathetic division, preganglionic axons branch profusely as they enter the sympathetic trunk, and they synapse with postganglionic neurons at several levels. As a result, when the sympathetic division is activated, it responds in a diffuse and highly interconnected way. Indeed, the literal translation of sympathetic (*sym* = together; *pathos* = feeling) relates to the bodywide mobilization this division provokes. Nevertheless, parts of the sympathetic nervous system can be activated individually. For example, just because your eye pupils dilate in dim light doesn't necessarily mean that your heart rate also speeds up.

Sympathetic activation produces much longer-lasting effects than parasympathetic activation. Adrenal medullary cells secrete NE and epinephrine into the blood when the sympathetic division is mobilized. These hormones reinforce and prolong the effects of the sympathetic nervous system. They have essentially the same effects as NE released by sympathetic neurons, although epinephrine is more potent at increasing heart rate and raising blood glucose levels and metabolic rate. In fact, circulating adrenal medullary hormones produce 25–50% of all the sympathetic effects acting on the body at a given time. These effects continue for several minutes until the liver destroys the hormones.

In short, sympathetic nerve impulses act only briefly, but the hormonal effects they provoke linger. The widespread and prolonged effect of sympathetic activation explains why we need time to "come down" after an extremely stressful experience.

☑ Check Your Understanding

9. Name the division of the ANS that does each of the following: increases digestive activity; increases blood pressure; dilates bronchioles; decreases heart rate; stimulates the adrenal medulla to release its hormones; causes ejaculation.

For answers, see Answers Appendix.

14.8 The hypothalamus oversees ANS activity

→ Learning Objective

☐ **Describe autonomic nervous system controls.**

Although the ANS is not usually considered to be under voluntary control, its activity is regulated by CNS controls in the spinal cord, brain stem, hypothalamus, and cerebral cortex (**Figure 14.9**). In general, the hypothalamus is the integrative center at the top of the ANS control hierarchy. From there, orders flow to lower and lower CNS centers for execution.

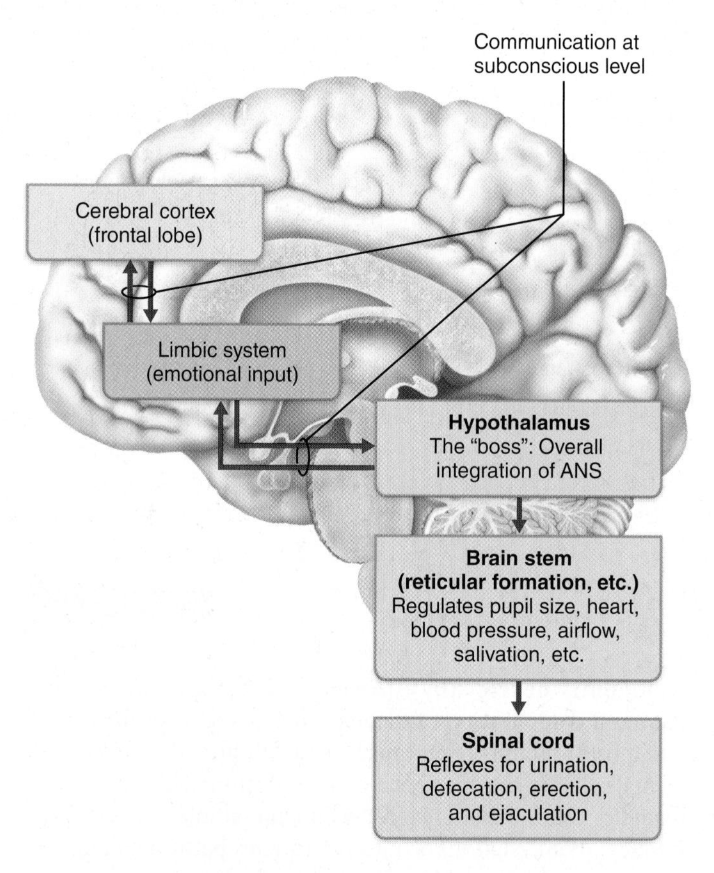

Figure 14.9 Levels of ANS control. The hypothalamus stands at the top of the control hierarchy as the integrator of ANS activity, but it is influenced by subconscious cerebral inputs via limbic system connections.

Although the cerebral cortex may modify the workings of the ANS, it does so at the subconscious level and by acting through limbic system structures on hypothalamic centers.

- **Brain stem and spinal cord controls.** The hypothalamus is the "boss," but the brain stem reticular formation appears to exert the most *direct* influence over autonomic functions (see Figure 12.15 on p. 400). For example, certain motor centers in the ventrolateral medulla (*cardiac* and *vasomotor centers*) reflexively regulate heart rate and blood vessel diameter. Other medullary regions oversee gastrointestinal activities. Most sensory impulses involved in these autonomic reflexes reach the brain stem via vagus nerve afferents. Midbrain centers (*oculomotor nuclei*) control the muscles concerned with pupil diameter and lens focus.

  Defecation and micturition (urination) reflexes that empty the rectum and urinary bladder are integrated at the spinal cord level but are subject to conscious inhibition. We will describe all of these autonomic reflexes in later chapters in relation to the organ systems they serve.
- **Hypothalamic controls.** As we noted, the hypothalamus is the main integration center of the autonomic nervous system. In general, anterior hypothalamic regions direct parasympathetic functions, and posterior areas direct sympathetic functions. Hypothalamic centers exert their effects both directly and via relays through the *reticular formation*, which in turn influences the preganglionic motor neurons in the brain stem and spinal cord (Figure 14.9). The hypothalamus, acting through the ANS, coordinates heart activity, blood pressure, body temperature, water balance, and endocrine activity.

  The hypothalamus also mediates our reactions to fear via its associations with the amygdala and the periaqueductal gray matter. Emotional responses of the limbic system of the cerebrum to danger and stress signal the hypothalamus to activate the sympathetic system to fight-or-flight status. In this way, the hypothalamus serves as the keystone of the emotional and visceral brain. Through its centers, emotions influence ANS function and behavior.
- **Cortical controls.** Originally, scientists believed the ANS was not subject to voluntary controls. However, we have all had occasions when just remembering a frightening event made our heart race (sympathetic response) or the thought of a favorite food, pecan pie for example, made our mouth water (parasympathetic response). These inputs converge on the hypothalamus through its connections to the limbic system.

  Additionally, biofeedback studies have shown that voluntary cortical control of visceral activities is possible—a capability untapped by most people.

☑ Check Your Understanding

10. Which part of the brain is the main integration center of the ANS? Which part exerts the most direct influence over autonomic functions?

For answers, see Answers Appendix.

CLINICAL

14.9 Most ANS disorders involve abnormalities in smooth muscle control

→ Learning Objective

- ☐ **Explain the relationship of some types of hypertension, Raynaud's disease, and autonomic dysreflexia to disorders of autonomic function.**

The ANS is involved in nearly every important process that goes on in the body, so it is not surprising that abnormalities of autonomic function can have far-reaching effects. Most autonomic disorders reflect exaggerated or deficient controls of smooth muscle activity. The most devastating involve blood vessels and include conditions such as hypertension, Raynaud's disease, and autonomic dysreflexia.

- *Hypertension*, or high blood pressure, may result from an overactive sympathetic vasoconstrictor response promoted by continuous high levels of stress. Hypertension is always serious because it forces the heart to work harder, which may precipitate heart disease, and increases the wear and tear on artery walls. Hypertension is sometimes treated with adrenergic receptor–blocking drugs that counteract the effects of the sympathetic nervous system on the cardiovascular system. We will discuss hypertension in more detail in Chapter 18.
- *Raynaud's disease* is characterized by intermittent attacks causing the skin of the fingers and toes to become pale, then cyanotic (bluish) and painful. Commonly provoked by exposure to cold or emotional stress, it is an exaggerated vasoconstriction response. The severity of Raynaud's disease ranges from merely uncomfortable to severe blood vessel constriction that causes ischemia and gangrene (tissue death).
- *Autonomic dysreflexia* is a life-threatening condition involving uncontrolled activation of autonomic neurons. It occurs in a majority of individuals with quadriplegia and in others with spinal cord injuries above the T_6 level, usually in the first year after injury. The usual trigger is a painful stimulus to the skin or an overfilled visceral organ, such as the urinary bladder. Arterial blood pressure skyrockets to life-threatening levels, which may rupture a blood vessel in the brain, precipitating stroke. Symptoms include headache, flushed face, sweating above the level of the injury, and cold, clammy skin below. The precise mechanism of autonomic dysreflexia is not yet clear.

☑ Check Your Understanding

11. Jackson works long, stress-filled shifts as an air traffic controller at a busy airport. His doctor has prescribed a beta-blocker. Why might his doctor have done this? What does a beta-blocker do?

For answers, see Answers Appendix.

In this chapter, we have described the structure and function of the ANS, one arm of the motor division of the peripheral nervous system. Because virtually every organ system still to be considered depends on autonomic controls, you will be hearing more about the ANS in chapters that follow.

REVIEW QUESTIONS

MAP For more chapter study tools, go to the Study Area of MasteringA&P®.

There you will find:

- Interactive Physiology iP
- A&PFlix A&PFlix
- Practice Anatomy Lab PAL
- PhysioEx PEx
- Videos, Practice Quizzes and Tests, MP3 Tutor Sessions, Case Studies, and much more!

Multiple Choice/Matching

(Some questions have more than one correct answer. Select the best answer or answers from the choices given.)

1. All of the following characterize the ANS except (a) a two-neuron efferent chain, (b) presence of neuron cell bodies in the CNS, (c) presence of neuron cell bodies in the ganglia, (d) innervation of skeletal muscles.
2. Relate each of the following terms or phrases to either the sympathetic (S) or parasympathetic (P) division of the autonomic nervous system:

____ (1) short preganglionic, long postganglionic fibers
____ (2) intramural ganglia
____ (3) craniosacral part
____ (4) adrenergic fibers
____ (5) cervical ganglia
____ (6) otic and ciliary ganglia
____ (7) generally short-duration action
____ (8) increases heart rate and blood pressure
____ (9) increases gastric motility and secretion of lacrimal, salivary, and digestive juices
____ (10) innervates blood vessels
____ (11) most active when you are relaxing in a hammock
____ (12) active when you are running in the Boston Marathon

3. The white rami communicantes contain what kind of fibers? (a) preganglionic parasympathetic, (b) postganglionic parasympathetic, (c) preganglionic sympathetic, (d) postganglionic sympathetic.
4. Collateral sympathetic ganglia are involved with innervating (a) abdominal organs, (b) thoracic organs, (c) head, (d) arrector pili, (e) all of these.

Short Answer Essay Questions

5. Briefly explain why the following terms are sometimes used to refer to the autonomic nervous system: involuntary nervous system and emotional-visceral system.
6. Describe the anatomical relationship of the white and gray rami communicantes to the spinal nerve, and indicate the kind of fibers found in each ramus type.
7. Indicate the results of sympathetic activation of the following structures: sweat glands, eye pupils, adrenal medullae, heart, bronchioles of the lungs, liver, blood vessels of vigorously working skeletal muscles, blood vessels of digestive viscera, salivary glands.
8. Which of the effects listed in response to question 7 would be reversed by parasympathetic activity?
9. Which ANS fibers release acetylcholine? Which release norepinephrine?
10. Describe the meaning and importance of sympathetic tone and parasympathetic tone.
11. Which area of the brain is most directly involved in mediating autonomic reflexes?
12. Describe the importance of the hypothalamus in controlling the autonomic nervous system.
13. Postganglionic neurons are also called ganglionic neurons. Why is the latter term more accurate?

AT THE CLINIC

Clinical Case Study

Autonomic Nervous System

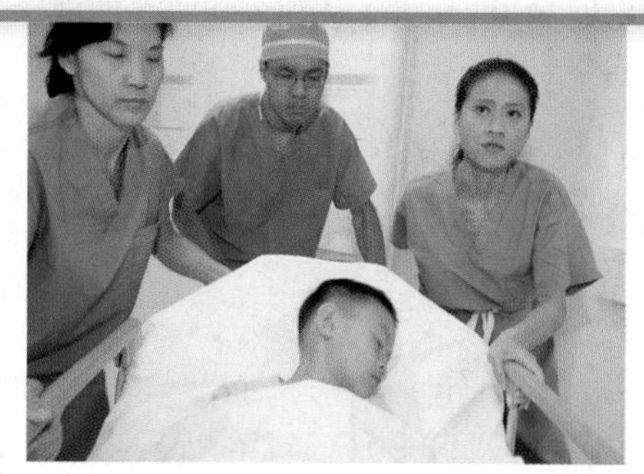

On arrival at Holyoke Hospital, Jimmy Chin, a 10-year-old boy, is immobilized on a rigid stretcher so that he is unable to move his head or trunk. The paramedics report that when they found him some 50 feet from the bus, he was awake and alert, but crying and complaining that he couldn't "get up to find his mom" and he had a "wicked headache." He has severe bruises on his upper back and head, and lacerations of his back and scalp. His blood pressure is low, body temperature is below normal, lower limbs are paralyzed, and he is insensitive to painful stimuli below the nipples. Although still alert on arrival, Jimmy soon begins to drift in and out of unconsciousness.

Jimmy is immediately scheduled for a CT scan, and an operating room is reserved.

Relative to Jimmy's condition:

1. Why were his head and torso immobilized for transport to the hospital?
2. What do his worsening neurological signs (drowsiness, incoherence, etc.) probably indicate? Relate this to the type of surgery that will be performed.
3. Assuming that Jimmy's sensory and motor deficits are due to a spinal cord injury, at what level do you expect to find a spinal cord lesion?
4. Two days after his surgery, Jimmy is alert and his MRI scan shows no residual brain injury, but pronounced swelling and damage to the spinal cord at T_4. On physical examination, Jimmy shows no reflex activity below the level of the spinal cord injury. His blood pressure is still low. Why are there no reflexes in his lower limbs and abdomen?
5. Over the next few days, his reflexes return in his lower limbs and become exaggerated. He is incontinent. Why is Jimmy hyperreflexive and incontinent?

On one occasion, Jimmy complains of a massive headache and his blood pressure is way above normal. On examination, he is sweating intensely above the nipples but has cold, clammy skin below the nipples and his heart rate is very slow.

6. What is this condition called and what precipitates it?
7. How does Jimmy's excessively high blood pressure put him at risk?

For answers, see Answers Appendix.

15 The Endocrine System

KEY CONCEPTS

You don't have to watch a blockbuster to experience action-packed drama. Molecules and cells inside your body have dynamic adventures on microscopic levels all the time.

For instance, when insulin molecules carried along in the blood attach to protein receptors of nearby cells, the response is dramatic: Glucose molecules disappear from the blood into the cells, and cellular activity accelerates. Such is the power of the second great control system of the body, the **endocrine system**, which interacts with the nervous system to coordinate and integrate the activity of body cells.

15.1 The endocrine system is one of the body's two major control systems

→ Learning Objectives

- ☐ **Indicate important differences between hormonal and neural controls of body functioning.**
- ☐ **List the major endocrine organs, and describe their body locations.**
- ☐ **Distinguish between hormones, paracrines, and autocrines.**

As we have seen, the nervous system regulates the activity of muscles and glands via electrochemical impulses delivered by neurons, and those organs respond within milliseconds. The means of control and speed of the endocrine system are very different: The endocrine system influences metabolic activity by means of *hormones* (*hormone* = to excite). Hormones are chemical messengers secreted by cells into the extracellular fluids. These messengers travel through the blood and regulate the metabolic function of other cells in the body. Binding of a hormone to cellular receptors initiates responses that typically occur after a lag period of seconds or even days. But, once initiated, those responses tend to last much longer than those induced by the nervous system.

Hormones ultimately target most cells of the body, producing widespread and diverse effects. The major processes that these "mighty molecules" control and integrate include:

- Reproduction
- Growth and development
- Maintenance of electrolyte, water, and nutrient balance of the blood
- Regulation of cellular metabolism and energy balance
- Mobilization of body defenses

As you can see, the endocrine system orchestrates processes that go on for relatively long periods, in some instances continuously. The scientific study of hormones and the endocrine organs is called **endocrinology**.

A man with diabetes testing his blood sugar before eating.

Compared with other organs, those of the endocrine system are small and unimpressive, but their influence is powerful. Unlike most organ systems, the endocrine organs are not grouped together but are widely scattered about the body.

As we explained in Chapter 4, there are two kinds of glands:

- *Exocrine glands* produce nonhormonal substances, such as sweat and saliva, and have ducts that carry these substances to a membrane surface.
- **Endocrine glands**, also called *ductless glands*, produce hormones and lack ducts. They release their hormones into the surrounding tissue fluid (*endo* = within; *crine* = to secrete), and typically have a rich vascular and lymphatic drainage that receives their hormones. Most of the hormone-producing cells in endocrine glands are arranged in cords and branching networks, which maximizes contact between them and the surrounding capillaries.

The endocrine glands include the pituitary, thyroid, parathyroid, adrenal, and pineal glands (**Figure 15.1**). The hypothalamus, along with its neural functions, produces and releases hormones, so we can consider the hypothalamus a **neuroendocrine organ**. In addition, several organs, such as the pancreas, gonads (ovaries and testes), and placenta, contain endocrine tissue. Most other organs also contain scattered endocrine cells or small clusters of endocrine cells.

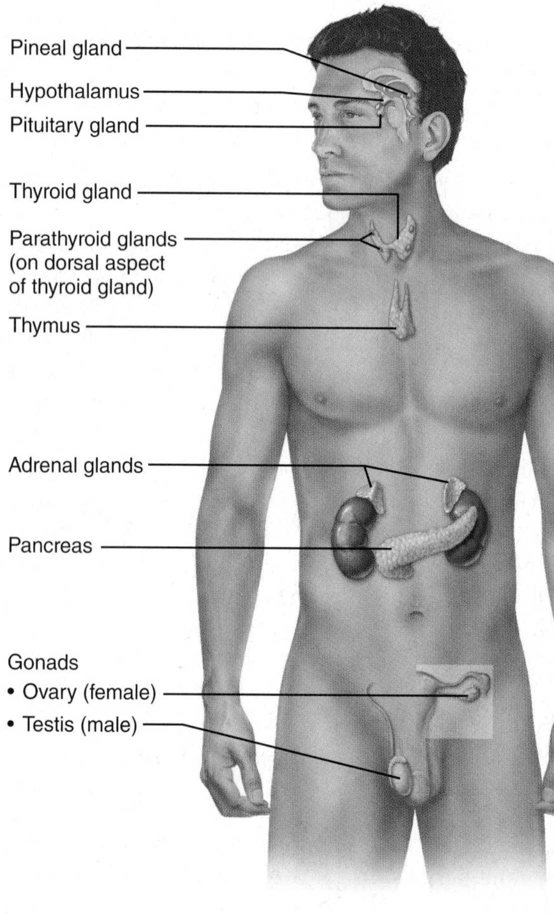

Figure 15.1 Location of selected endocrine organs of the body.

Practice art labeling
MasteringA&P®>Study Area>Chapter 15

Some physiologists include local chemical messengers—autocrines and paracrines—as part of the endocrine system, but that is not the consensus. Hormones are long-distance chemical signals that travel in blood or lymph throughout the body. Autocrines and paracrines, on the other hand, are short-distance signals. **Autocrines** are chemicals that exert their effects on the same cells that secrete them. For example, certain prostaglandins released by smooth muscle cells cause those smooth muscle cells to contract. **Paracrines** also act locally (within the same tissue) but affect cell types other than those releasing the paracrine chemicals. For example, somatostatin released by one population of pancreatic cells inhibits the release of insulin by a different population of pancreatic cells.

☑ Check Your Understanding

1. For each of the following statements, indicate whether it applies more to the endocrine system or the nervous system: rapid; discrete responses; controls growth and development; long-lasting responses.
2. Which two endocrine glands are found in the neck?
3. What is the difference between a hormone and a paracrine?

For answers, see Answers Appendix.

15.2 The chemical structure of a hormone determines how it acts

→ Learning Objective

☐ **Describe how hormones are classified chemically.**

Its chemical structure determines one critical property of a hormone: its solubility in water. Its water solubility in turn affects how the hormone is transported in the blood, how long it lasts before it is degraded, and what receptors it can act upon. Although a large variety of hormones are produced, nearly all of them can be classified chemically as either amino acid based or steroids.

- **Amino acid based:** Most hormones are amino acid based. Molecular size varies widely in this group—from simple amino acid derivatives [which include biogenic amines (e.g., epinephrine), and thyroxine], to peptides (short chains of amino acids), to proteins (long polymers of amino acids). These hormones are usually water soluble and cannot cross the plasma membrane.
- **Steroids:** Steroid hormones are synthesized from cholesterol. Of the hormones produced by the major endocrine organs, only gonadal and adrenocortical hormones are steroids. These hormones are all lipid soluble and can cross the plasma membrane.

Some researchers add a third class, **eicosanoids** (i-ko′să-noyds), which include *leukotrienes* and *prostaglandins.* Nearly all cell membranes release these biologically active lipids (made from arachidonic acid). Leukotrienes are signaling chemicals that mediate inflammation and some allergic reactions. Prostaglandins have multiple targets and effects, ranging from raising blood pressure and increasing the expulsive uterine contractions of birth to enhancing blood clotting, pain, and inflammation.

Because the effects of eicosanoids are typically highly localized, affecting only nearby cells, they generally act as paracrines and autocrines and do not fit the definition of true hormones, which influence distant targets. For this reason, we will not consider these hormonelike chemicals here, but will discuss them in later chapters as appropriate.

☑ Check Your Understanding

4. MAKING connections Where in the cell are steroid hormones synthesized? (Hint: Recall cell components from Chapter 3.) Where are peptide hormones synthesized? Which of these two types of hormone could be stored in vesicles and released by exocytosis?

For answers, see Answers Appendix.

15.3 Hormones act through second messengers or by activating specific genes

→ Learning Objective

☐ **Describe the two major mechanisms by which hormones bring about their effects on their target tissues.**

All major hormones circulate to virtually all tissues, but a hormone influences the activity of only those tissue cells that have receptors for it. These cells are its **target cells**. Hormones bring about their characteristic effects by *altering* target cell activity, increasing or decreasing the rates of normal cellular processes.

The precise response depends on the target cell type. For example, when the hormone epinephrine binds to certain smooth muscle cells in blood vessel walls, it stimulates them to contract. Epinephrine binding to cells other than muscle cells may have a totally different effect, but it does not cause those cells to contract.

A hormone typically produces one or more of the following changes:

- Alters plasma membrane permeability or membrane potential, or both, by opening or closing ion channels
- Stimulates synthesis of enzymes and other proteins within the cell
- Activates or deactivates enzymes
- Induces secretory activity
- Stimulates mitosis

How does a hormone communicate with its target cell? In other words, how is hormone receptor binding harnessed to the intracellular machinery needed for hormone action? The answer depends on the chemical nature of the hormone and the cellular location of the receptor. In general, hormones act at receptors in one of two ways.

- *Water-soluble hormones* (all amino acid–based hormones except thyroid hormone) act on *receptors in the plasma membrane*. These receptors are usually coupled via regulatory molecules called G proteins to one or more intracellular second messengers which mediate the target cell's response.
- *Lipid-soluble hormones* (steroid and thyroid hormones) act on *receptors inside the cell*, which directly activate genes.

This will be easy for you to remember if you think about why the hormones must bind where they do. Receptors for water-soluble hormones must be in the plasma membrane since these hormones *cannot* enter the cell, and receptors for lipid-soluble steroid and thyroid hormones are inside the cell because these hormones *can* enter the cell.

Plasma Membrane Receptors and Second-Messenger Systems

With the exception of thyroid hormone, amino acid–based hormones exert their signaling effects through intracellular **second messengers** generated when a hormone binds to a receptor in the plasma membrane. You are already familiar with one of these second messengers, **cyclic AMP (cAMP)**, which is used by neurotransmitters (Chapter 11) and olfactory receptors (Chapter 13).

The Cyclic AMP Signaling Mechanism

As you recall, this mechanism involves the interaction of three plasma membrane components—a hormone receptor, a G protein, and an effector enzyme (adenylate cyclase)—to determine intracellular levels of cyclic AMP. **Figure 15.2** illustrates these steps:

1. **Hormone binds receptor.** The hormone, acting as the **first messenger**, binds to its receptor in the plasma membrane.
2. **Receptor activates G protein.** Hormone binding causes the receptor to change shape, allowing it to bind a nearby inactive **G protein**. The G protein is activated as the guanosine diphosphate (GDP) bound to it is displaced by the high-energy compound *guanosine triphosphate* (*GTP*). The G protein behaves like a light switch: It is "off" when GDP is bound to it, and "on" when GTP is bound.
3. **G protein activates adenylate cyclase.** The activated G protein (moving along the membrane) binds to the effector enzyme **adenylate cyclase**. Some G proteins (G_s) *stimulate* adenylate cyclase (as shown in Figure 15.2), but others (G_i) *inhibit* adenylate cyclase. Eventually, the GTP bound to the G protein is hydrolyzed to GDP and the G protein becomes inactive once again. (The G protein cleaves the terminal phosphate group off GTP in much the same way that ATPase enzymes hydrolyze ATP.)
4. **Adenylate cyclase converts ATP to cyclic AMP.** For as long as activated G_s is bound to it, adenylate cyclase generates the *second messenger* cAMP from ATP.
5. **Cyclic AMP activates protein kinases.** cAMP, which is free to diffuse throughout the cell, triggers a cascade of chemical reactions by activating protein kinases. **Protein kinases** are enzymes that *phosphorylate* (add a phosphate group to) various proteins, many of which are other enzymes. Because phosphorylation activates some of these proteins and inhibits others, it may affect a variety of processes in the same target cell at the same time.

This type of intracellular enzymatic cascade has a huge amplification effect. Each activated adenylate cyclase generates large numbers of cAMP molecules, and a single kinase enzyme can catalyze hundreds of reactions. As the reaction cascades

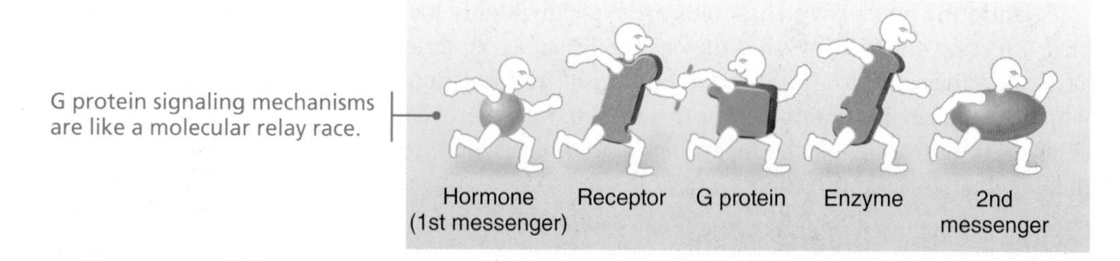

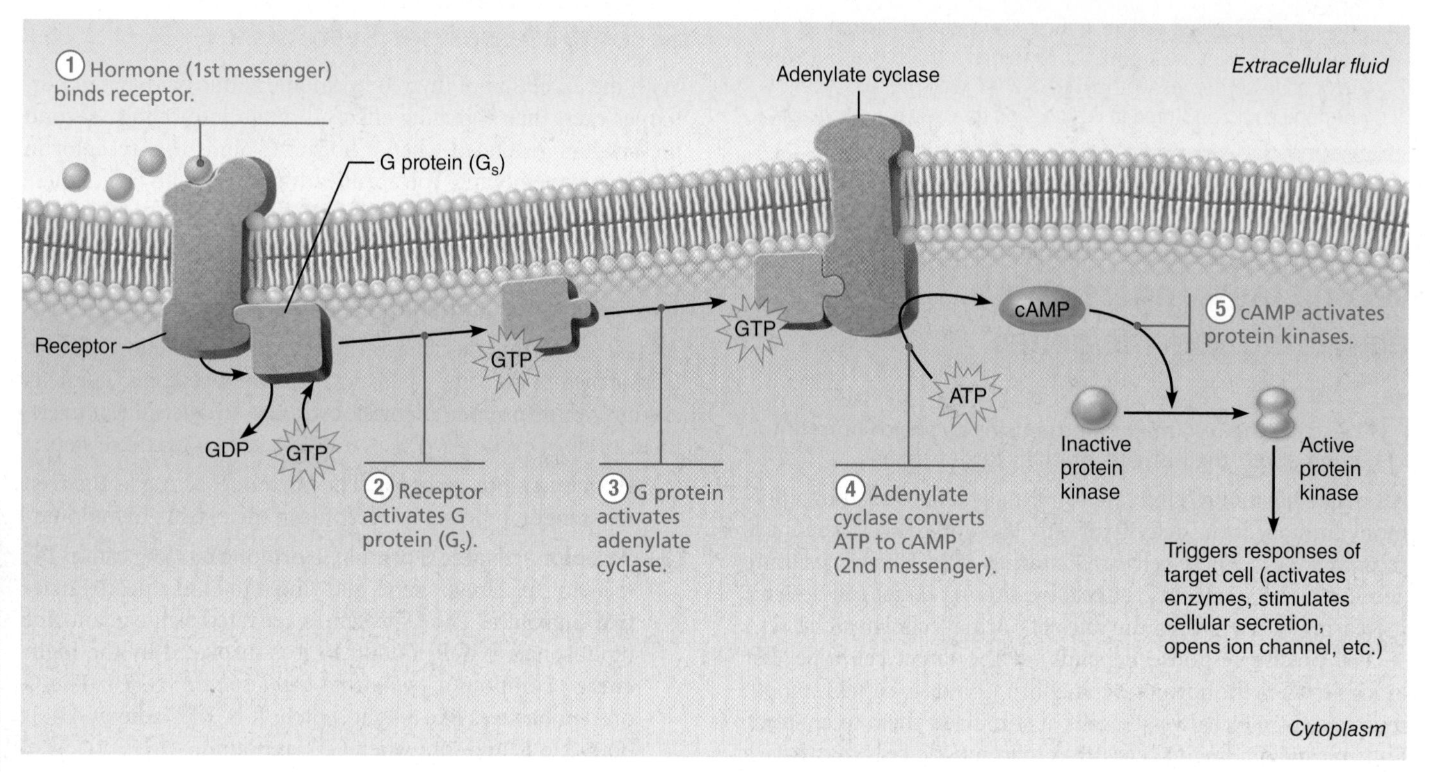

Figure 15.2 Cyclic AMP second-messenger mechanism of water-soluble hormones. (For the basics of G protein signaling mechanisms, see Focus Figure 3.2 on p. 76.)

through one enzyme intermediate after another, the number of product molecules increases dramatically at each step. A single hormone molecule binding to a receptor can generate millions of final product molecules!

The sequence of reactions set into motion by cAMP depends on the type of target cell, the specific protein kinases it contains, and the substrates within that cell available for phosphorylation. For example, in thyroid cells, binding of thyroid-stimulating hormone promotes synthesis of the thyroid hormone thyroxine; in liver cells, binding of glucagon activates enzymes that break down glycogen, releasing glucose to the blood. Since some G proteins inhibit rather than activate adenylate cyclase, thereby reducing the cytoplasmic concentration of cAMP, even slight changes in levels of antagonistic hormones can influence a target cell's activity.

The action of cAMP persists only briefly because the molecule is rapidly degraded by the intracellular enzyme **phosphodiesterase**. While at first glance this may appear to be a problem, it is quite the opposite. Because of the amplification effect, most hormones need to be present only briefly to cause results, and the quick work of phosphodiesterase means that no extracellular controls are necessary to stop the activity.

The PIP_2-Calcium Signaling Mechanism

Certain hormones use different second messengers. For example, in the PIP_2-calcium signaling mechanism, intracellular calcium ions act as a second messenger.

Like the cAMP signaling mechanism, the PIP_2-calcium signaling mechanism involves a G protein (G_q) and a membrane-bound effector, in this case an enzyme called **phospholipase C**. Phospholipase C splits a plasma membrane phospholipid called **PIP_2** (**p**hosphatidyl **i**nositol bis**p**hosphate) into two second messengers: **diacylglycerol (DAG)** and **inositol trisphosphate (IP_3)**. DAG, like cAMP, activates a protein kinase enzyme, which triggers responses within the target cell. In addition, IP_3 releases Ca^{2+} from intracellular storage sites.

The liberated Ca^{2+} also takes on a second-messenger role, either by directly altering the activity of specific enzymes and channels or by binding to the intracellular regulatory protein **calmodulin**. Once Ca^{2+} binds to calmodulin, it activates enzymes that amplify the cellular response.

Other Signaling Mechanisms

Other hormones that bind plasma membrane receptors act on their target cells through different signaling mechanisms. For example, cyclic guanosine monophosphate (cGMP) is a second messenger for selected hormones.

Still other hormones, such as insulin and certain growth factors, work without second messengers. The insulin receptor is a *tyrosine kinase* enzyme that is activated by adding phosphates to several of its own tyrosines when insulin binds. The activated insulin receptor provides docking sites for intracellular *relay proteins* that, in turn, initiate a series of protein phosphorylations that trigger specific cell responses.

Intracellular Receptors and Direct Gene Activation

Being lipid soluble, steroid hormones and thyroid hormone diffuse into their target cells where they bind to and activate an intracellular receptor (**Figure 15.3**). The activated receptor-hormone complex then makes its way to the nuclear chromatin and binds to a specific region of DNA. (There are exceptions to these generalizations. For example, thyroid hormone receptors are always bound to DNA even in the absence of thyroid hormone.)

When the receptor-hormone complex binds to DNA, it "turns on" a gene; that is, it prompts transcription of DNA to produce a messenger RNA (mRNA). The mRNA is then translated on the cytoplasmic ribosomes, producing specific proteins. These proteins include enzymes that promote the metabolic activities induced by that particular hormone and, in some cases,

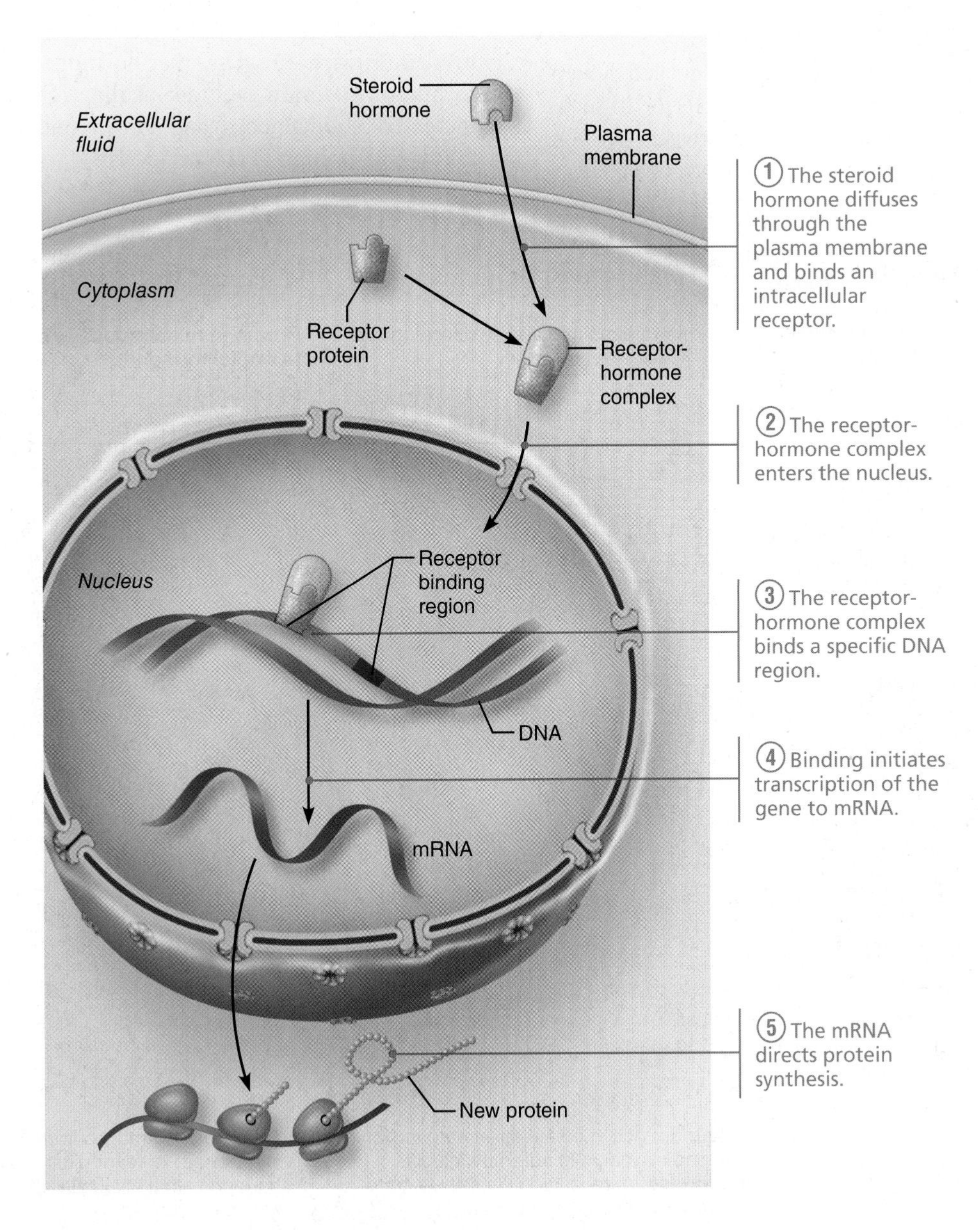

Figure 15.3 Direct gene activation mechanism of lipid-soluble hormones. Receptors may be located in the nucleus, or in the cytoplasm as shown.

promote synthesis of either structural proteins or proteins to be exported from the target cell.

Check Your Understanding

5. Which class of hormones consists entirely of lipid-soluble hormones? Name the only hormone in the other chemical class that is lipid soluble.
6. Consider the signaling mechanisms of water-soluble and lipid-soluble hormones. In each case, where are the receptors found and what is the result of the hormone binding to the receptor?

For answers, see Answers Appendix.

15.4 Three types of stimuli cause hormone release

→ Learning Objective

☐ **Explain how hormone release is regulated.**

The synthesis and release of most hormones are regulated by some type of **negative feedback mechanism** (see Chapter 1). In such a mechanism, some internal or external stimulus triggers hormone secretion. As levels of a hormone rise, it causes target organ effects, which then feed back to inhibit further hormone release. As a result, blood levels of many hormones vary only within a narrow range.

Endocrine Gland Stimuli

Three types of stimuli trigger endocrine glands to manufacture and release their hormones: *humoral* (hu′mer-ul), *neural*, and *hormonal stimuli*. Some endocrine organs respond to more than one type of stimulus.

Humoral Stimuli

Some endocrine glands secrete their hormones in direct response to changing blood levels of certain critical ions and nutrients. These stimuli are called *humoral stimuli* (from the Latin term *humor*, which refers to moisture or bodily fluids).

Humoral stimuli are the simplest endocrine controls. For example, cells of the parathyroid glands monitor the body's crucial blood Ca^{2+} levels and release parathyroid hormone as needed (**Figure 15.4a**). Other hormones released in response to humoral stimuli include insulin (released in response to increased blood glucose) and aldosterone (released in response to low Na^{+} or high K^{+} blood levels).

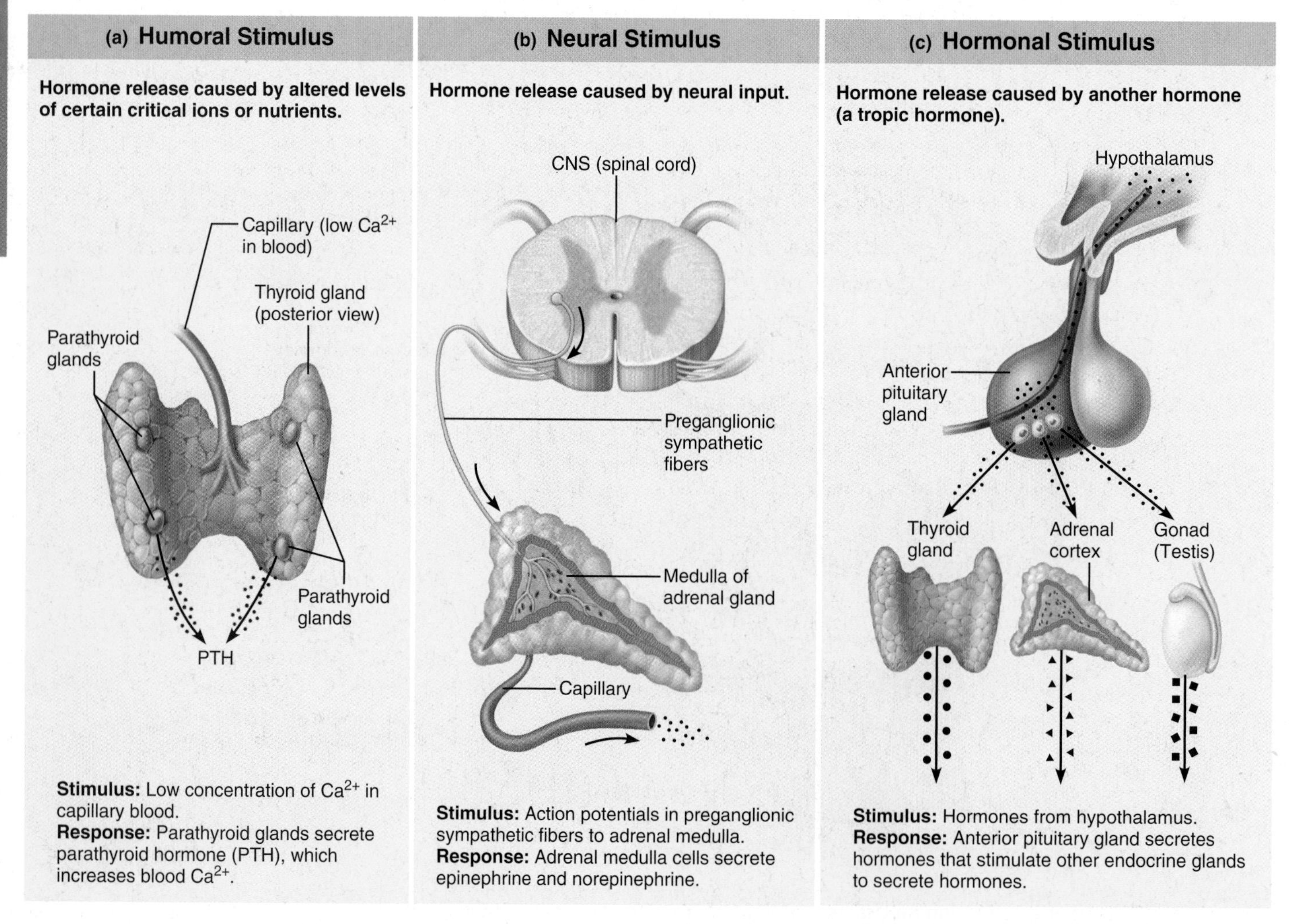

Figure 15.4 Three types of endocrine gland stimuli.

Neural Stimuli

In a few cases, nerve fibers stimulate hormone release. The classic example of neural stimuli is the response to stress, in which the sympathetic nervous system stimulates the adrenal medulla to release norepinephrine and epinephrine (Figure 15.4b).

Hormonal Stimuli

Many endocrine glands release their hormones in response to hormones produced by other endocrine organs. For example, releasing and inhibiting hormones produced by the hypothalamus regulate the secretion of most anterior pituitary hormones, and many anterior pituitary hormones in turn stimulate other endocrine organs to release their hormones (Figure 15.4c). As blood levels of the hormones produced by the final target glands increase, they inhibit the release of anterior pituitary hormones and thus their own release.

This hypothalamic–pituitary–target endocrine organ feedback loop lies at the very core of endocrinology, and it will come up many times in this chapter. Hormonal stimuli promote rhythmic hormone release, with hormone blood levels rising and falling in a specific pattern.

Nervous System Modulation

The nervous system can modify both "turn-on" factors (hormonal, humoral, and neural stimuli) and "turn-off" factors (feedback inhibition and others) that affect the endocrine system. Without this added safeguard, endocrine system activity would be strictly mechanical, much like a household thermostat. A thermostat can maintain the temperature at or around its set value, but it cannot sense that your grandmother visiting from Florida feels cold at that temperature and reset itself accordingly. You must make that adjustment. In your body, it is the nervous system that makes certain adjustments to maintain homeostasis by overriding normal endocrine controls.

For example, the action of insulin and several other hormones normally keeps blood glucose levels in the range of 90–110 mg/100 ml of blood. However, when your body is under severe stress, blood glucose levels rise because the hypothalamus and sympathetic nervous system centers are strongly activated. In this way, the nervous system ensures that body cells have sufficient fuel in case vigorous activity is required.

☑ Check Your **Understanding**

7. What are the three types of stimuli that control hormone release?

For answers, see Answers Appendix.

15.5 Cells respond to a hormone if they have a receptor for that hormone

→ Learning Objectives

- ☐ **Identify factors that influence activation of a target cell by a hormone.**
- ☐ **List three kinds of interaction of different hormones acting on the same target cell.**

In order for a target cell to respond to a hormone, the cell must have *specific* receptor proteins on its plasma membrane or in its interior to which that hormone can bind. For example, receptors for adrenocorticotropic hormone (ACTH) are normally found only on certain cells of the adrenal cortex. By contrast, thyroxine is the principal hormone stimulating cellular metabolism, and nearly all body cells have thyroxine receptors.

A hormone receptor responds to hormone binding by prompting the cell to perform, or turn on, some gene-determined "preprogrammed" function. As such, hormones are molecular triggers rather than informational molecules. Although binding of a hormone to a receptor is the crucial first step, target cell activation depends equally on three other factors:

- Blood levels of the hormone
- Relative numbers of receptors for that hormone on or in the target cells
- *Affinity* (strength) of the binding between the hormone and the receptor

The first two factors change rapidly in response to various stimuli and changes within the body. As a rule, for a given level of hormone in the blood, having a large number of high-affinity receptors produces a pronounced hormonal effect, and having a smaller number of low-affinity receptors reduces the target cell response or causes outright endocrine dysfunction.

Receptors are dynamic structures. For example, persistently low levels of a hormone can cause its target cells to form additional receptors for that hormone. This is called **up-regulation**. Likewise, prolonged exposure to high hormone concentrations can decrease the number of receptors for that hormone. This **down-regulation** desensitizes the target cells, so they respond less vigorously to hormonal stimulation, preventing them from overreacting to persistently high hormone levels. Receptors can also be uncoupled from their signaling mechanism, altering the sensitivity of the response.

Hormones influence not only the number of their own receptors but also the number of receptors that respond to other hormones. For example, progesterone down-regulates estrogen receptors in the uterus, thus antagonizing estrogens' actions. On the other hand, estrogens cause the same cells to produce more progesterone receptors, enhancing their ability to respond to progesterone.

Half-Life, Onset, and Duration of Hormone Activity

Hormones are potent chemicals, and they exert profound effects on their target organs even at very low concentrations. Hormones circulate in the blood in two forms—free or bound to a protein carrier. In general, lipid-soluble hormones (steroids and thyroid hormone) travel in the bloodstream attached to plasma proteins. Most others circulate without carriers.

The concentration of a circulating hormone in blood at any time reflects (1) its rate of release and (2) the speed at which it is inactivated and removed from the body. Some hormones are rapidly degraded by enzymes in their target cells. However, most hormones are removed from the blood by the kidneys or

Table 15.1 Comparison between Lipid- and Water-Soluble Hormones

	LIPID-SOLUBLE HORMONES	WATER-SOLUBLE HORMONES
Examples	All steroid hormones and thyroid hormone	All amino acid–based hormones except thyroid hormone
Sources	Adrenal cortex, gonads, and thyroid gland*	All other endocrine glands
Can be stored in secretory vesicles	No	Yes
Transport in blood	Bound to plasma proteins	Usually free in plasma
Half-life in blood	Long (most need to be metabolized by liver)	Short (most can be removed by kidneys)
Location of receptors	Usually inside cell	On plasma membrane
Mechanism of action at target cell	Activate genes, causing synthesis of new proteins	Usually act through second-messenger systems

*Skin is a source of cholecalciferol (an inactive form of vitamin D).

liver, and the body excretes their breakdown products in urine or, to a lesser extent, in feces. As a result, the length of time for a hormone's blood level to decrease by half, referred to as its **half-life**, varies from a fraction of a minute to a week. Water-soluble hormones have the shortest half-lives.

How long does it take for a hormone to have an effect? It varies. Some hormones provoke target organ responses almost immediately, while others, particularly steroid hormones, require hours to days before their effects are seen. Additionally, some hormones are secreted in a relatively inactive form and must be activated in the target cells.

The duration of hormone action is limited, ranging from 10 seconds to several hours. Effects may disappear rapidly as blood levels drop, or they may persist for hours even at very low levels. Because of these many variations, hormonal blood levels must be precisely and individually controlled to meet the continuously changing needs of the body.

As you can see, many characteristics of a hormone (such as its half-life and the time it takes to have an effect) depend on its solubility in water or lipids. **Table 15.1** compares the characteristics of lipid- and water-soluble hormones.

Interaction of Hormones at Target Cells

Understanding hormonal effects is a bit more complicated than you might expect because multiple hormones may act on the same target cells at the same time. In many cases the result of such an interaction is not predictable, even when you know the effects of the individual hormones. Here we will look at three types of hormone interaction—permissiveness, synergism, and antagonism.

- **Permissiveness** is the situation in which one hormone cannot exert its full effects without another hormone being present. For example, reproductive system hormones largely regulate the development of the reproductive system, as we might expect. However, thyroid hormone is also necessary (has a permissive effect) for normal *timely* development of reproductive structures. Lack of thyroid hormone delays reproductive development.
- **Synergism** occurs when more than one hormone produces the same effects at the target cell and their combined effects are amplified. For example, both glucagon and epinephrine cause the liver to release glucose to the blood. When they act together, the amount of glucose released is about 150% of what is released when each hormone acts alone.
- **Antagonism** occurs when one hormone opposes the action of another. For example, insulin, which lowers blood glucose levels, is antagonized by glucagon, which raises blood glucose levels. How does antagonism occur? Antagonists may compete for the same receptors, act through different metabolic pathways, or even cause down-regulation of the receptors for the antagonistic hormone.

☑ Check Your Understanding

8. Which type of hormone generally stays in the blood longer following its secretion, lipid soluble or water soluble?

For answers, see Answers Appendix.

15.6 The hypothalamus controls release of hormones from the pituitary gland in two different ways

→ Learning Objectives

- ☐ **Describe structural and functional relationships between the hypothalamus and the pituitary gland.**
- ☐ **Discuss the structure of the posterior pituitary, and describe the effects of the two hormones it releases.**
- ☐ **List and describe the chief effects of anterior pituitary hormones.**

Securely seated in the sella turcica of the sphenoid bone, the tiny **pituitary gland**, or **hypophysis** (hi-pof′ĭ-sis; "to grow under"), secretes at least eight hormones. This gland is the size and shape of a pea on a stalk. Its stalk, the funnel-shaped **infundibulum**, connects the gland to the hypothalamus superiorly as shown in *Focus on Hypothalamus and Pituitary Interactions* (**Focus Figure 15.1** on pp. 530–531).

In humans, the pituitary gland has two major lobes.

- The **posterior pituitary** (lobe) is composed largely of neural tissue such as pituicytes (glia-like supporting cells) and nerve fibers. It releases **neurohormones** (hormones secreted by neurons) received ready-made from the hypothalamus. Consequently, this lobe is a hormone-storage area and not a true endocrine gland that manufactures hormones. The posterior lobe plus the infundibulum make up the region called the **neurohypophysis** (nu″ro-hi-pof′ĭ-sis).
- The **anterior pituitary** (lobe), or **adenohypophysis** (ad″ĕ-no-hi-pof′ĭ-sis), is composed of glandular tissue (*adeno* = gland). It manufactures and releases a number of hormones (Table 15.2 on pp. 532–533).

Hypophyseal branches of the internal carotid arteries deliver arterial blood to the pituitary. The veins leaving the pituitary drain into the dural sinuses.

Pituitary-Hypothalamic Relationships

The contrasting histology of the two pituitary lobes reflects the dual origin of this tiny gland. The posterior lobe is actually part of the brain. It derives from a downgrowth of hypothalamic tissue and maintains its neural connection with the hypothalamus via a nerve bundle called the **hypothalamic-hypophyseal tract**, which runs through the infundibulum (Focus Figure 15.1).

This tract arises from neurons in the **paraventricular** and **supraoptic nuclei** of the hypothalamus. These neurosecretory cells synthesize one of two neurohormones and transport them along their axons to the posterior pituitary. When these hypothalamic neurons fire, they release the stored hormones into a capillary bed in the posterior pituitary for distribution throughout the body.

The glandular anterior lobe originates from epithelial tissue as a superior outpocketing of the oral mucosa. After touching the posterior lobe, the anterior lobe adheres to the neurohypophysis and loses its connection with the oral mucosa. There is no direct neural connection between the anterior lobe and hypothalamus, but there is a vascular connection. Specifically, the **primary capillary plexus** in the infundibulum communicates inferiorly via the small **hypophyseal portal veins** with a **secondary capillary plexus** in the anterior lobe. The primary and secondary capillary plexuses and the intervening hypophyseal portal veins make up the *hypophyseal portal system* (Focus Figure 15.1). Note that a *portal system* is an unusual arrangement of blood vessels in which a capillary bed feeds into veins, which in turn feed into a second capillary bed.

Via the hypophyseal portal system, **releasing** and **inhibiting hormones** secreted by neurons in the ventral hypothalamus circulate to the anterior pituitary, where they regulate secretion of its hormones. The portal system ensures that the minute quantities of hormones released by the hypothalamus arrive rapidly at the anterior pituitary without being diluted by the systemic circulation. All these hypothalamic regulatory hormones are amino acid based, but they vary in size from a single amine to peptides to proteins.

The Posterior Pituitary and Hypothalamic Hormones

The posterior pituitary consists largely of axon terminals of hypothalamic neurons whose cell bodies are located in the supraoptic or paraventricular nuclei. The paraventricular neurons primarily make oxytocin (ok″sĭ-to′sin), and the supraoptic neurons mainly produce antidiuretic hormone (ADH). Axon terminals in the posterior pituitary release these hormones "on demand" in response to action potentials that travel down the axons of these same hypothalamic neurons.

Oxytocin and ADH, each composed of nine amino acids, are almost identical. They differ in only two amino acids, and yet they have dramatically different physiological effects, summarized in Table 15.2 and described next.

Oxytocin

A strong stimulant of uterine contraction, **oxytocin** is released in significantly higher amounts during childbirth (*oxy* = rapid; *tocia* = childbirth) and in nursing women. The number of oxytocin receptors in the uterus peaks near the end of pregnancy, and uterine smooth muscle becomes more and more sensitive to the hormone's stimulatory effects. Stretching of the uterus and cervix as birth nears dispatches afferent impulses to the hypothalamus. The hypothalamus responds by synthesizing oxytocin and triggering its release from the posterior pituitary. Oxytocin acts via the PIP_2-Ca^{2+} second-messenger system to mobilize Ca^{2+}, allowing stronger contractions. As blood levels of oxytocin rise, the expulsive contractions of labor gain momentum and finally end in birth.

Oxytocin also acts as the hormonal trigger for milk ejection (the "letdown" reflex) in women whose breasts are producing milk in response to prolactin. Suckling causes a reflex-initiated release of oxytocin, which targets specialized myoepithelial cells surrounding the milk-producing glands. These cells contract and force milk from the breast into the infant's mouth. Both childbirth and milk ejection result from *positive feedback mechanisms*.

Both natural and synthetic oxytocic drugs are used to induce labor or to hasten labor that is progressing slowly. Less frequently, oxytocic drugs are used to stop postpartum bleeding.

Oxytocin also acts as a neurotransmitter in the brain. There, it is involved in sexual and affectionate behavior (as the "cuddle hormone"), and promotes nurturing, couple bonding, and trust.

Antidiuretic Hormone (ADH)

Diuresis (di″u-re′sis) is urine production, so an *antidiuretic* is a substance that inhibits or prevents urine formation. **Antidiuretic hormone (ADH)** prevents wide swings in water balance, helping the body avoid dehydration and water overload.

Hypothalamic neurons called *osmoreceptors* continually monitor the solute concentration (and thus the water concentration) of the blood. When solutes threaten to become too concentrated (as might follow excessive perspiration or inadequate fluid intake), the osmoreceptors transmit excitatory impulses to the hypothalamic neurons, which release ADH. ADH targets the kidney tubule cells, which respond by reabsorbing more water

15

Prolactin (PRL)

Prolactin (PRL) is a protein hormone structurally similar to GH. Produced by **prolactin cells**, PRL's only well-documented effect in humans is to stimulate milk production by the breasts (*pro* = for; *lact* = milk). The role of prolactin in males is not well understood.

Unlike other anterior pituitary hormones, PRL release is controlled primarily by an inhibitory hormone, **prolactin-inhibiting hormone (PIH)**, now known to be **dopamine**, which prevents prolactin secretion. Decreased PIH secretion leads to a surge in PRL release. There are a number of *prolactin-releasing factors*, including TRH, but their exact roles are not well understood.

In females, prolactin levels rise and fall in rhythm with estrogen blood levels. Estrogens stimulate prolactin release, both directly and indirectly. A brief rise in prolactin levels just before the menstrual period partially accounts for the breast swelling and tenderness some women experience at that time, but because this PRL stimulation is so brief, the breasts do not produce milk. In pregnant women, PRL blood levels rise dramatically toward the end of pregnancy, and milk production becomes possible. After birth, the infant's suckling stimulates release of prolactin-releasing factors in the mother, encouraging continued milk production.

HOMEOSTATIC IMBALANCE 15.3 — CLINICAL

Hypersecretion of prolactin is more common than hyposecretion (which is not a problem in anyone except women who choose to nurse). In fact, *hyperprolactinemia* is the most frequent abnormality of anterior pituitary tumors. Clinical signs include inappropriate lactation, lack of menses, infertility in females, and impotence in males. +

☑ Check Your Understanding

9. What is the key difference between the way the hypothalamus communicates with the anterior pituitary and the way it communicates with the posterior pituitary?
10. Zoe drank too much alcohol one night and suffered from a headache and nausea the next morning. What caused these "hangover" effects?
11. List the four anterior pituitary hormones that are tropic hormones and name their target glands.

For answers, see Answers Appendix.

15.7 The thyroid gland controls metabolism

→ **Learning Objectives**

- ☐ **Describe the effects of the two groups of hormones produced by the thyroid gland.**
- ☐ **Follow the process of thyroxine formation and release.**

Location and Structure

The butterfly-shaped **thyroid gland** is located in the anterior neck, on the trachea just inferior to the larynx (Figure 15.1 and **Figure 15.8a**). A median tissue mass called the *isthmus*

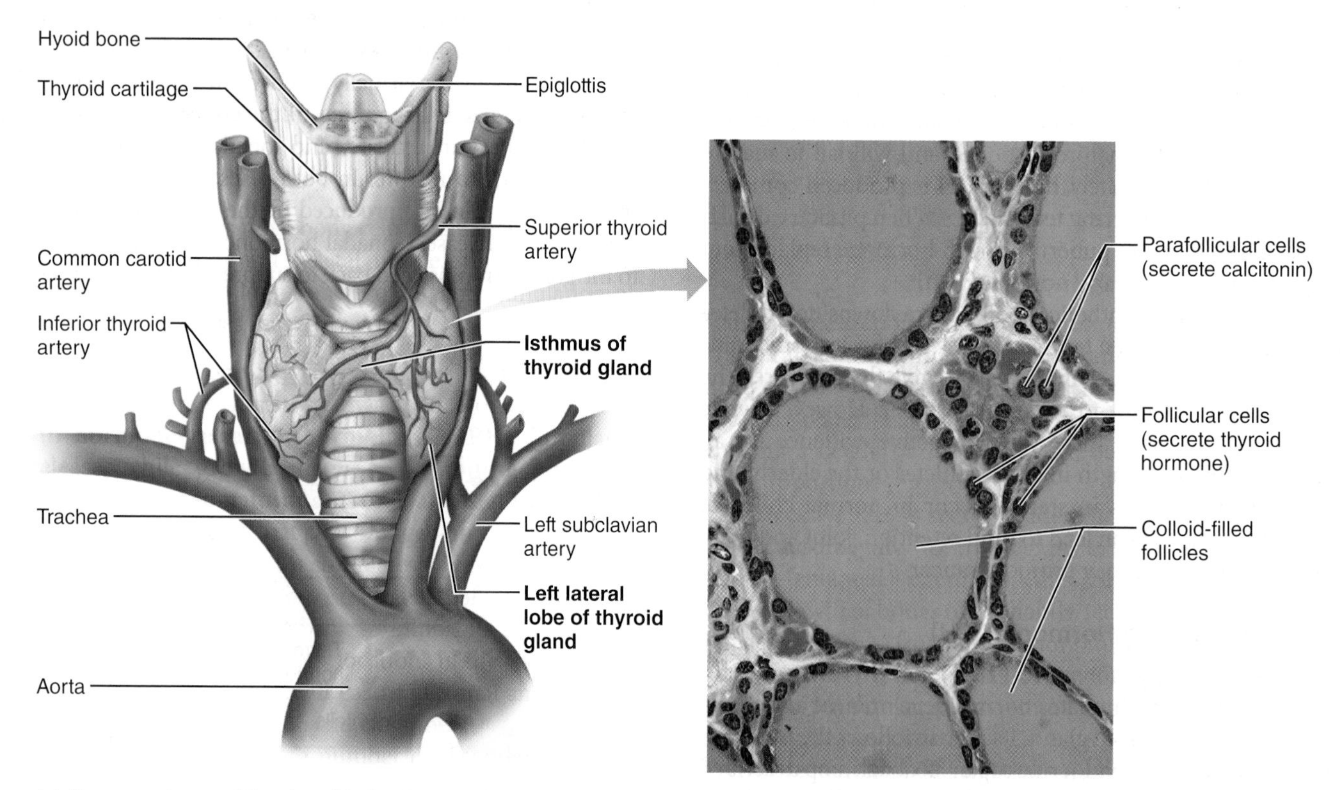

(a) Gross anatomy of the thyroid gland, anterior view

(b) Photomicrograph of thyroid gland follicles (315×)

Figure 15.8 The thyroid gland.

View histology slides
MasteringA&P®>Study Area>PAL

(is′mus) connects its two lateral *lobes*. The thyroid gland is the largest pure endocrine gland in the body. Its prodigious blood supply (from the *superior* and *inferior thyroid arteries*) makes thyroid surgery a painstaking (and bloody) endeavor.

Internally, the gland is composed of hollow, spherical **follicles** (Figure 15.8b). The walls of each follicle are formed largely by cuboidal or squamous epithelial cells called *follicular cells*, which produce the glycoprotein **thyroglobulin** (thi″ro-glob′u-lin). The central cavity, or lumen, of the follicle stores **colloid**, an amber-colored, sticky material consisting of thyroglobulin molecules with attached iodine atoms. *Thyroid hormone* is derived from this iodinated thyroglobulin.

The *parafollicular cells*, another population of endocrine cells in the thyroid gland, produce *calcitonin*. The parafollicular cells lie in the follicular epithelium but protrude into the soft connective tissue that separates and surrounds the thyroid follicles.

Thyroid Hormone (TH)

Often referred to as the body's major metabolic hormone, **thyroid hormone (TH)** is actually two iodine-containing amine hormones, **thyroxine** (thi-rok′sin), or $\mathbf{T_4}$, and **triiodothyronine** (tri″i-o″do-thi′ro-nēn), or $\mathbf{T_3}$. T_4 is the major hormone secreted by the thyroid follicles. Most T_3 is formed at the target tissues by conversion of T_4 to T_3. Both T_4 and T_3 are constructed from two linked tyrosine amino acids, but T_4 has four bound iodine atoms, and T_3 has three (thus, T_4 and T_3).

TH affects virtually every cell in the body (Table 15.3). Like steroids, TH enters a target cell, binds to intracellular receptors within the cell's nucleus, and initiates transcription of mRNA for protein synthesis. Effects of thyroid hormone include:

- Increasing basal metabolic rate and body heat production, by turning on transcription of genes concerned with glucose

Table 15.3 Major Effects of Thyroid Hormone (T_4 and T_3) in the Body

PROCESS OR SYSTEM AFFECTED	NORMAL PHYSIOLOGICAL EFFECTS	EFFECTS OF HYPOSECRETION	EFFECTS OF HYPERSECRETION
Basal metabolic rate (BMR)/ temperature regulation	Promotes normal oxygen use and BMR; calorigenesis; enhances effects of sympathetic nervous system	BMR below normal; decreased body temperature, cold intolerance; decreased appetite; weight gain; reduced sensitivity to catecholamines	BMR above normal; increased body temperature, heat intolerance; increased appetite; weight loss
Carbohydrate/lipid/protein metabolism	Promotes glucose catabolism; mobilizes fats; essential for protein synthesis; enhances liver's synthesis of cholesterol	Decreased glucose metabolism; elevated cholesterol/triglyceride levels in blood; decreased protein synthesis; edema	Enhanced catabolism of glucose, proteins, and fats; weight loss; loss of muscle mass
Nervous system	Promotes normal development of nervous system in fetus and infant; promotes normal adult nervous system function	In infant, slowed/deficient brain development, intellectual disability; in adult, mental dulling, depression, paresthesias, memory impairment, hypoactive reflexes	Irritability, restlessness, insomnia, personality changes, exophthalmos (in Graves' disease)
Cardiovascular system	Promotes normal functioning of the heart	Decreased efficiency of heart's pumping action; low heart rate and blood pressure	Increased sensitivity to catecholamines can lead to rapid heart rate, palpitations, high blood pressure, and ultimately heart failure
Muscular system	Promotes normal muscular development and function	Sluggish muscle action; muscle cramps; myalgia	Muscle atrophy and weakness
Skeletal system	Promotes normal growth and maturation of the skeleton	In child, growth retardation, skeletal stunting and retention of child's body proportions; in adult, joint pain	In child, excessive skeletal growth initially, followed by early epiphyseal closure and short stature; in adult, demineralization of skeleton
Gastrointestinal (GI) system	Promotes normal GI motility and tone; increases secretion of digestive juices	Depressed GI motility, tone, and secretory activity; constipation	Excessive GI motility; diarrhea
Reproductive system	Promotes normal female reproductive ability and lactation	Depressed ovarian function; sterility; depressed lactation	In females, depressed ovarian function; in males, impotence
Integumentary system	Promotes normal hydration and secretory activity of skin	Skin pale, thick, and dry; facial edema; hair coarse and thick	Skin flushed, thin, and moist; hair fine and soft; nails soft and thin

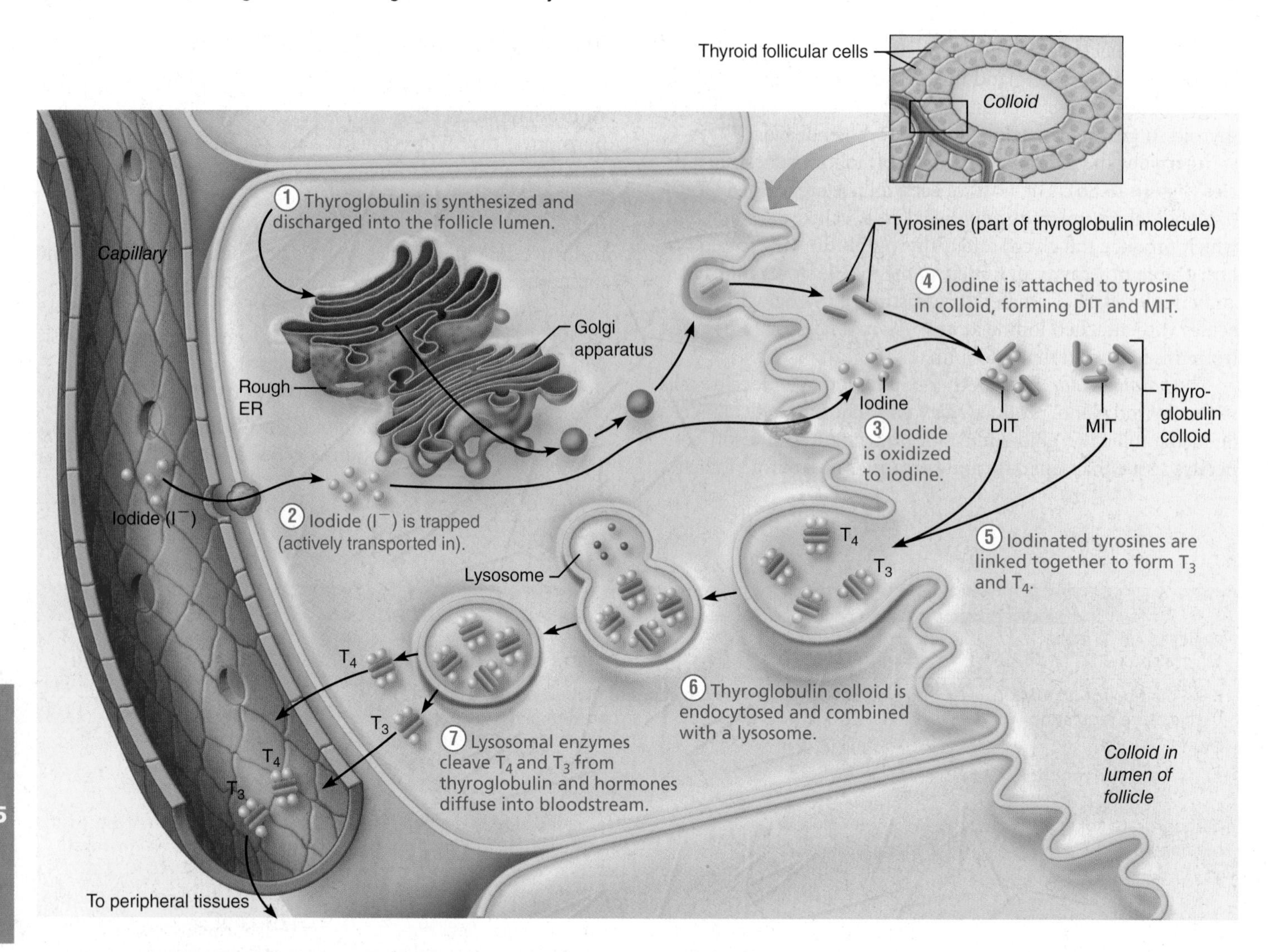

Figure 15.9 Synthesis of thyroid hormone. Only a few tyrosines of the thyroglobulins in the colloid are illustrated. The unstructured yellow substance in the follicle lumen is colloid. (MIT = monoiodotyrosine; DIT = diiodotyrosine)

oxidation. This is the hormone's **calorigenic effect** (*calorigenic* = heat producing).

- Regulating tissue growth and development. TH is critical for normal skeletal and nervous system development and maturation and for reproductive capabilities.
- Maintaining blood pressure by increasing the number of adrenergic receptors in blood vessels.

Synthesis

The thyroid gland is unique among the endocrine glands in its ability to store its hormone extracellularly and in large quantities. A normal thyroid gland stores enough colloid to provide normal levels of hormone for two to three months.

When TSH from the anterior pituitary binds to receptors on follicular cells, their *first* response is to secrete stored thyroid hormone. Their *second* response is to begin synthesizing more colloid to "restock" the follicle lumen. As a general rule, TSH levels are lower during the day, peak just before sleep, and remain high during the night. Consequently, thyroid hormone release and synthesis follows a similar pattern.

Let's examine how follicular cells synthesize thyroid hormone (**Figure 15.9**):

① **Thyroglobulin is synthesized and discharged into the follicle lumen.** After being synthesized on the ribosomes of the thyroid's follicular cells, thyroglobulin is transported to the Golgi apparatus, where sugar molecules are attached and the thyroglobulin is packed into transport vesicles. These vesicles move to the apex of the follicular cell, where they discharge their contents into the follicle lumen to become part of the stored colloid.

② **Iodide is trapped.** To produce the functional iodinated hormones, the follicular cells must accumulate iodides (anions of iodine, I^-) from the blood. Iodide trapping depends on active transport. (The concentration of I^- is over 30 times higher inside the cell than in blood.) Once trapped inside the follicular cell, iodide then moves into the follicle lumen by facilitated diffusion.

③ **Iodide is oxidized to iodine.** At the border of the follicular cell and colloid, iodides are oxidized (by removal of electrons) and converted to iodine (I_2).

④ **Iodine is attached to tyrosine.** Once formed, iodine is attached to tyrosine amino acids that form part of the thyroglobulin colloid. This iodination reaction, mediated by peroxidase enzymes, occurs at the junction of the follicular cell and the colloid. Attachment of one iodine to a tyrosine produces **monoiodotyrosine (MIT)**, and attachment of two iodines produces **diiodotyrosine (DIT)**.

⑤ **Iodinated tyrosines are linked together to form T_3 and T_4.** Enzymes in the colloid link MIT and DIT together. Two linked DITs result in T_4, and coupling of MIT and DIT produces T_3. At this point, the hormones are still part of the thyroglobulin colloid.

⑥ **Thyroglobulin colloid is endocytosed.** To secrete the hormones, the follicular cells must reclaim iodinated thyroglobulin by endocytosis and combine the vesicles with lysosomes.

⑦ **Lysosomal enzymes cleave T_4 and T_3 from thyroglobulin and the hormones diffuse from the follicular cell into the bloodstream.** The main hormonal product secreted is T_4. Some T_4 is converted to T_3 before secretion, but most T_3 is generated in the peripheral tissues.

Transport and Regulation

Most released T_4 and T_3 immediately binds to *thyroxine-binding globulins (TBGs)* and other transport proteins produced by the liver. Both T_4 and T_3 bind to target tissue receptors, but T_3 binds more avidly and is about 10 times more active. Most peripheral tissues have the enzymes needed to convert T_4 to T_3 by removing one iodine atom.

Figure 15.7 shows the negative feedback loop that regulates blood levels of TH. Falling TH blood levels trigger release of *thyroid-stimulating hormone (TSH)*, and ultimately of more TH. Rising TH levels feed back to inhibit the hypothalamic–anterior pituitary axis, temporarily shutting off the stimulus for TSH release.

In infants, exposure to cold stimulates the hypothalamus to secrete *thyrotropin-releasing hormone (TRH)*, which triggers TSH release. The thyroid gland then releases larger amounts of thyroid hormones, enhancing body metabolism and heat production. Factors that inhibit TSH release include GHIH, dopamine, and rising levels of glucocorticoids. Excessively high blood iodide concentrations also inhibit TH release.

HOMEOSTATIC IMBALANCE 15.4 — CLINICAL

Both overactivity and underactivity of the thyroid gland can cause severe metabolic disturbances. Hypothyroid disorders may result from some thyroid gland defects or secondarily from inadequate TSH or TRH release. They also occur when the thyroid gland is removed surgically and when dietary iodine is inadequate.

In adults, the full-blown hypothyroid syndrome is called **myxedema** (mik″sĕ-de′mah; "mucous swelling"). Symptoms include a low metabolic rate; feeling chilled; constipation; thick, dry skin and puffy eyes; edema; lethargy; and mental sluggishness. A **goiter** (an enlarged protruding thyroid gland) occurs if myxedema results from lack of iodine (Figure 15.10a). The follicular cells produce colloid but cannot iodinate it and make functional hormones. The pituitary gland secretes increasing amounts of TSH in a futile attempt to stimulate the thyroid to produce TH, but the only result is that the follicles accumulate more and more *unusable* colloid. Depending on the cause, iodine supplements or hormone replacement therapy can reverse myxedema.

Before iodized salt became available, the midwestern United States was called the "goiter belt." Goiters were common because this area had iodine-poor soil and no access to iodine-rich seafood. In places where goiters are especially common, these goiters are called *endemic goiters*.

Like many other hormones, the important effects of TH depend on a person's age and development. Severe hypothyroidism in infants is called *cretinism* (kre′tĭ-nizm) when it is due to iodine deficiency, and *congenital hypothyroidism* when it is due to a congenital abnormality of the thyroid gland. The child is intellectually disabled, and has a short, disproportionately sized body and a thick tongue and neck. Cretinism may reflect a genetic deficiency of the fetal thyroid gland or maternal factors, such as lack of dietary

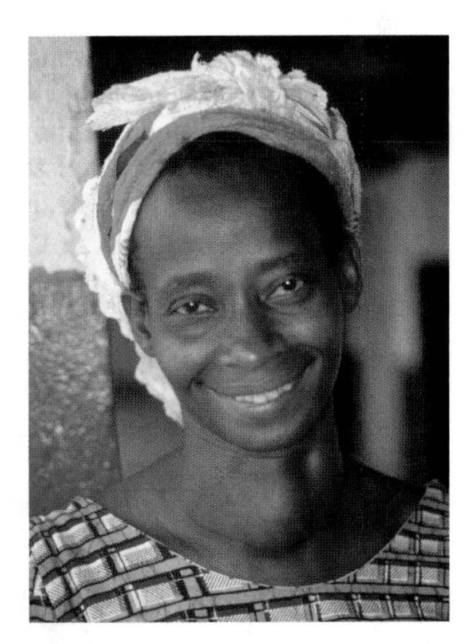

(a) An enlarged thyroid (goiter); due to iodine deficiency

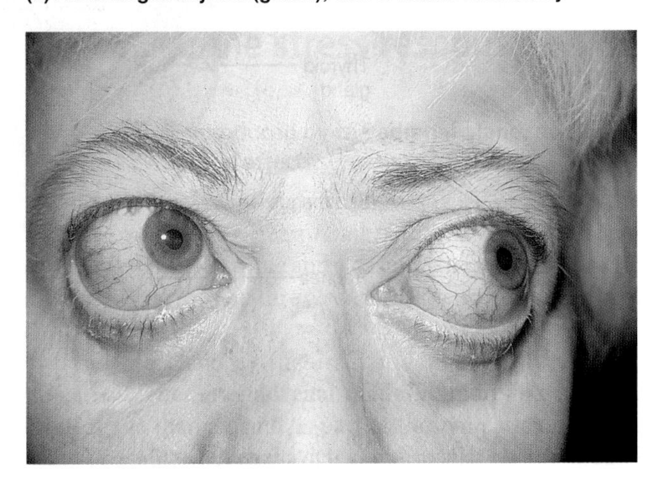

(b) Bulging eyes (exophthalmos) of Graves' disease

Figure 15.10 Thyroid disorders.

15

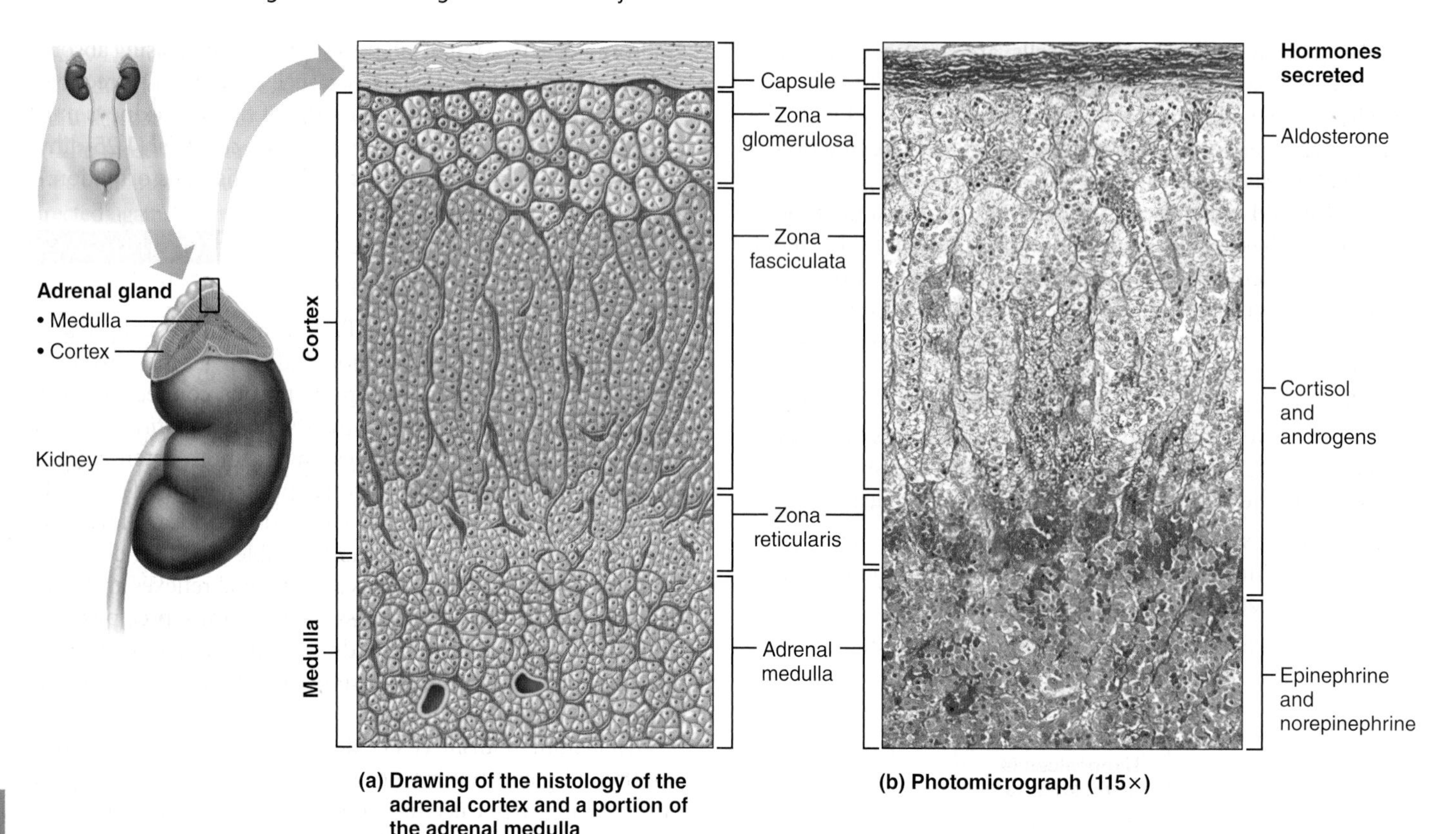

(a) Drawing of the histology of the adrenal cortex and a portion of the adrenal medulla

(b) Photomicrograph (115×)

Figure 15.13 Microscopic structure of the adrenal gland.

View histology slides
MasteringA&P®>Study Area>PAL

of hormones summarized in Table 15.4, but all adrenal hormones help us cope with stressful situations.

The Adrenal Cortex

The adrenal cortex synthesizes well over two dozen steroid hormones, collectively called **corticosteroids**. The multistep steroid synthesis pathway begins with cholesterol, and involves varying intermediates depending on the hormone being formed. Unlike the amino acid–based hormones, steroid hormones are not stored in cells. Consequently, their rate of release depends on their rate of synthesis.

The large, lipid-laden cortical cells are arranged in three layers or zones (Figure 15.13). From the outside in, they are:

- **Zona glomerulosa** (zo′nah glo-mer″u-lo′sah). The cell clusters forming this superficial layer produce mineralocorticoids, hormones that help control the balance of minerals and water in the blood.
- **Zona fasciculata** (fah-sik″u-la′tah). The cells of this middle layer, arranged in more or less linear cords, mainly produce the metabolic hormones called glucocorticoids.
- **Zona reticularis** (rĕ-tik″u-lar′is). The cells of this innermost layer, next to the adrenal medulla, have a netlike arrangement. They mainly produce small amounts of adrenal sex hormones, or gonadocorticoids.

Note, however, that the two innermost layers of the adrenal cortex share production of glucocorticoids and gonadocorticoids, although each layer predominantly produces one type.

Mineralocorticoids

The essential function of **mineralocorticoids** is to regulate the electrolyte (mineral salt) concentrations in extracellular fluids, particularly of Na^+ and K^+. The single most abundant cation in extracellular fluid is Na^+, and the amount of Na^+ in the body largely determines the volume of the extracellular fluid—where Na^+ goes, water follows. Changes in Na^+ concentration lead to changes in blood volume and blood pressure. Moreover, the regulation of Na^+ is coupled to the regulation of many other ions, including K^+, H^+, HCO_3^- (bicarbonate), and Cl^- (chloride).

The extracellular concentration of K^+ is also critical—it sets the resting membrane potential of all cells and determines how easily action potentials are generated in nerve and muscle. Not surprisingly, Na^+ and K^+ regulation are crucial to overall body homeostasis. Their regulation is the primary job of **aldosterone** (al-dos′ter-ōn), the most potent mineralocorticoid. Aldosterone accounts for more than 95% of the mineralocorticoids produced and is essential for life.

Aldosterone reduces excretion of Na^+ from the body. Its primary target is the distal parts of the kidney tubules, where it:

- Stimulates Na^+ reabsorption (increasing blood volume and blood pressure)
- Causes K^+ secretion into the tubules for elimination from the body

In some instances, aldosterone can alter the acid-base balance of the blood (by increasing H^+ excretion). Aldosterone also enhances Na^+ reabsorption from perspiration, saliva, and gastric juice.

Table 15.4 Adrenal Gland Hormones: Summary of Regulation and Effects

HORMONE	REGULATION OF RELEASE	TARGET ORGAN AND EFFECTS	EFFECTS OF HYPERSECRETION ↑ AND HYPOSECRETION ↓
Adrenocortical Hormones			
Mineralocorticoids (chiefly aldosterone)	**Stimulated** by renin-angiotensin-aldosterone mechanism (activated by decreasing blood volume or blood pressure), elevated blood K^+ levels, and ACTH (minor influence) **Inhibited** by increased blood volume and pressure, and decreased blood K^+ levels	Kidneys: increase blood levels of Na^+ and decrease blood levels of K^+; since water reabsorption usually accompanies sodium retention, blood volume and blood pressure rise	↑ Aldosteronism ↓ Addison's disease
Glucocorticoids (chiefly cortisol)	**Stimulated** by ACTH **Inhibited** by feedback inhibition exerted by cortisol	Body cells: promote gluconeogenesis and hyperglycemia; mobilize fats for energy metabolism; stimulate protein catabolism; assist body to resist stressors; depress inflammatory and immune responses	↑ Cushing's syndrome ↓ Addison's disease
Gonadocorticoids (chiefly androgens, converted to testosterone or estrogens after release)	**Stimulated** by ACTH; mechanism of inhibition incompletely understood, but feedback inhibition not seen	Insignificant effects in males; contributes to female libido; development of pubic and axillary hair in females; source of estrogens after menopause	↑ Masculinization of females (adrenogenital syndrome) ↓ No effects known
Adrenal Medullary Hormones			
Catecholamines (epinephrine and norepinephrine)	**Stimulated** by preganglionic fibers of the sympathetic nervous system	Sympathetic nervous system target organs: effects mimic sympathetic nervous system activation; increase heart rate and metabolic rate; increase blood pressure by promoting vasoconstriction	↑ Prolonged fight-or-flight response; hypertension ↓ Unimportant

Aldosterone's regulatory effects are brief (approximately 20 minutes), allowing plasma electrolyte balance to be precisely controlled and continuously modified. The mechanism of aldosterone activity involves the synthesis and activation of proteins required for Na^+ transport such as Na^+-K^+ ATPase, the pump that exchanges Na^+ for K^+.

Decreasing blood volume and blood pressure, and rising blood levels of K^+, stimulate aldosterone secretion. The reverse conditions inhibit its secretion. Four mechanisms regulate aldosterone secretion, but two of them—the renin-angiotensin-aldosterone mechanism and plasma concentrations of potassium—are by far the most important (**Figure 15.14** and Table 15.4).

The Renin-Angiotensin-Aldosterone Mechanism The renin-angiotensin-aldosterone mechanism (re′nin an″je-o-ten′-sin) influences both blood volume and blood pressure by regulating the release of aldosterone and therefore Na^+ and water reabsorption by the kidneys.

1. When blood pressure (or blood volume) falls, specialized cells of the *juxtaglomerular complex* in the kidneys are excited.
2. These cells respond by releasing **renin** into the blood.
3. Renin cleaves off part of the plasma protein **angiotensinogen** (an″je-o-ten′sin-o-jen), triggering an enzymatic cascade that forms **angiotensin II**, which stimulates the glomerulosa cells to release aldosterone.

However, the renin-angiotensin-aldosterone mechanism does much more than trigger aldosterone release, and all of its effects ultimately raise blood pressure. We describe these additional effects in Chapters 18 and 25.

Plasma Concentrations of Potassium Fluctuating blood levels of K^+ directly influence the zona glomerulosa cells in the adrenal cortex. Increased K^+ stimulates aldosterone release, whereas decreased K^+ inhibits it.

ACTH Under normal circumstances, ACTH released by the anterior pituitary has little or no effect on aldosterone release. However, when a person is severely stressed, the hypothalamus secretes more corticotropin-releasing hormone (CRH), and the resulting rise in ACTH blood levels steps up the rate of aldosterone secretion to a small extent. The resulting increase

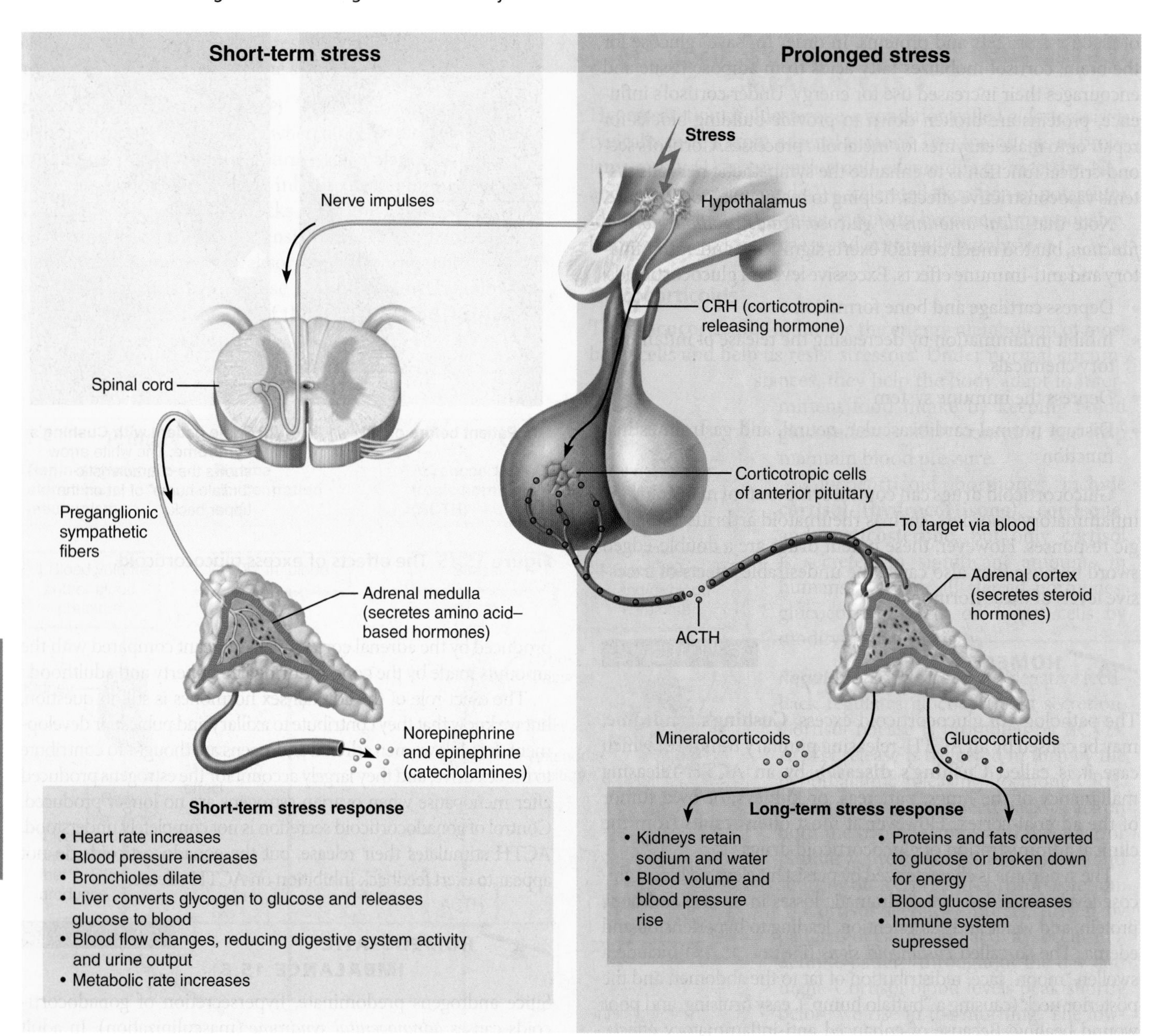

Figure 15.16 Stress and the adrenal gland. Stressful stimuli cause the hypothalamus to activate the adrenal medulla via sympathetic nerve impulses and the adrenal cortex via hormonal signals.

vessels constrict and the heart beats faster (together raising the blood pressure), and blood is diverted from temporarily nonessential organs to the heart and skeletal muscles. Blood glucose levels rise, and preganglionic sympathetic nerve endings weaving through the adrenal medulla signal for release of catecholamines, which reinforce and prolong the fight-or-flight response.

Unequal amounts of the two hormones are stored and released. Approximately 80% is epinephrine and 20% norepinephrine. With a few exceptions, the two hormones exert the same effects (see Table 14.3, p. 515). Epinephrine is the more potent stimulator of metabolic activities and bronchial dilation, but norepinephrine has a greater influence on peripheral vasoconstriction and blood pressure. Epinephrine is used clinically as a heart stimulant and to dilate the bronchioles during acute asthmatic attacks.

Unlike hormones from the adrenal cortex, which promote long-lasting body responses to stressors, catecholamines cause fairly brief responses. **Figure 15.16** depicts the interrelationships of the hypothalamus, the "director" of the stress response, with adrenal hormones.

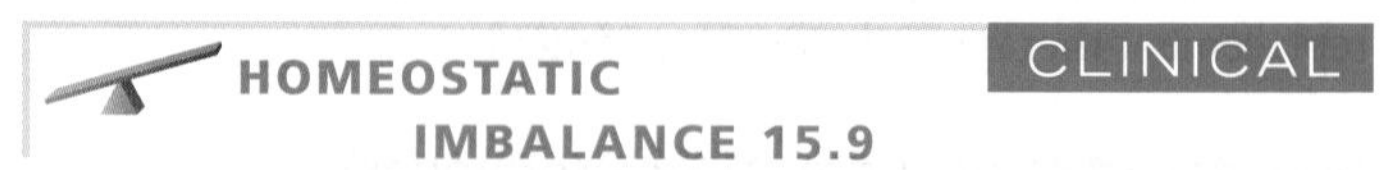

A deficiency in adrenal medulla hormones is not a problem because these hormones merely intensify activities set into motion by the sympathetic nervous system neurons. Unlike glucocorticoids and mineralocorticoids, adrenal catecholamines are not essential for life.

On the other hand, hypersecretion of catecholamines, sometimes arising from a medullary chromaffin cell tumor called a *pheochromocytoma* (fe-o-kro″mo-si-to′mah), produces symptoms of uncontrolled sympathetic nervous system activity—**hyperglycemia** (elevated blood glucose), increased metabolic rate, rapid heartbeat and palpitations, hypertension, intense nervousness, and sweating. +

Check Your Understanding

15. List the three classes of hormones released from the adrenal cortex and briefly state the major effect(s) of each.

For answers, see Answers Appendix.

15.10 The pineal gland secretes melatonin

→ Learning Objective

☐ **Briefly describe the importance of melatonin.**

The tiny, pinecone-shaped **pineal gland** hangs from the roof of the third ventricle in the diencephalon (see Figure 15.1 and Figure 12.11 on p. 395). Its secretory cells, called **pinealocytes**, are arranged in compact cords and clusters. Lying between pinealocytes in adults are dense particles containing calcium salts. These salts are radiopaque, making the pineal gland a handy landmark for determining brain orientation in X rays.

Although many peptides and amines have been isolated from this minute gland, its only major secretory product is **melatonin** (mel″ah-to′nin), an amine hormone derived from serotonin. Melatonin concentrations in the blood rise and fall in a diurnal (daily) cycle. Peak levels occur during the night and make us drowsy, and lowest levels occur around noon. Recent evidence suggests that melatonin also controls the production of protective antioxidant and detoxification molecules within cells.

The pineal gland indirectly receives input from the visual pathways (retina → suprachiasmatic nucleus of hypothalamus → superior cervical ganglion → pineal gland) concerning the intensity and duration of daylight. In some animals, mating behavior and gonadal size vary with relative lengths of light and dark periods, and melatonin mediates these effects.

In children, melatonin may have an antigonadotropic effect. In other words, it may affect the timing of puberty and inhibit precocious (too early) sexual maturation.

The *suprachiasmatic nucleus* of the hypothalamus, an area referred to as our "biological clock," is richly supplied with melatonin receptors, and exposure to bright light (known to suppress melatonin secretion) can reset the clock timing. As a result, changing melatonin levels may influence rhythmic variations in physiological processes such as body temperature, sleep, and appetite.

Check Your Understanding

16. Synthetic melatonin supplements are available, although their safety and efficacy have not been proved. What do you think they might be used for?

For answers, see Answers Appendix.

15.11 The pancreas, gonads, and most other organs secrete hormones

→ Learning Objectives

☐ **Compare and contrast the effects of the two major pancreatic hormones.**

☐ **Describe the functional roles of hormones of the testes, ovaries, and placenta.**

☐ **State the location of enteroendocrine cells.**

☐ **Briefly explain the hormonal functions of the heart, kidney, skin, adipose tissue, bone, and thymus.**

So far, we've examined the endocrine role of the hypothalamus and of glands dedicated solely to endocrine function. We will now consider a set of organs that contain endocrine tissue but also have other major functions. These include the pancreas, gonads, and placenta.

The Pancreas

Located partially behind the stomach in the abdomen, the soft, tadpole-shaped **pancreas** is a mixed gland composed of both endocrine and exocrine gland cells (see Figure 15.1). Along with the thyroid and parathyroids, it develops as an outpocketing of the epithelial lining of the gastrointestinal tract. *Acinar cells*, forming the bulk of the gland, produce an enzyme-rich juice that is carried by ducts to the small intestine during digestion.

Scattered among the acinar cells are approximately a million **pancreatic islets** (also called *islets of Langerhans*), tiny cell clusters that produce pancreatic hormones (**Figure 15.17**). The islets contain two major populations of hormone-producing cells, the glucagon-synthesizing **alpha (α) cells** and the more

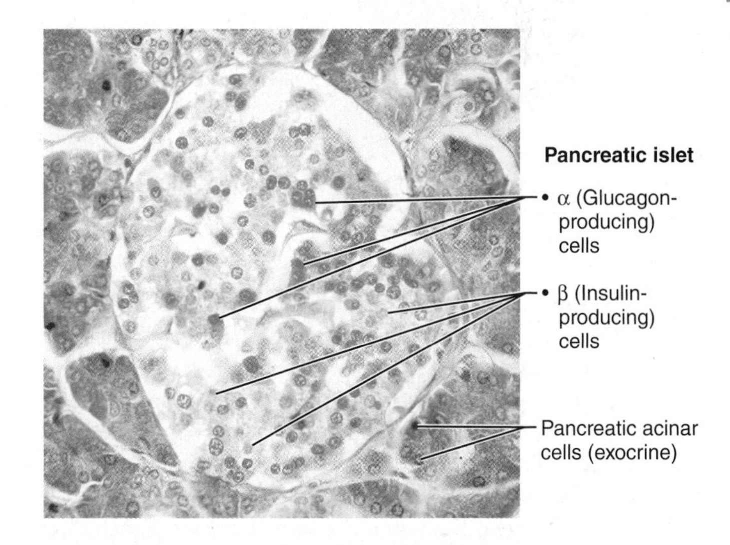

Figure 15.17 Photomicrograph of differentially stained pancreatic tissue. A pancreatic islet is surrounded by acinar cells, which produce the exocrine product (enzyme-rich pancreatic juice) (190×).

View histology slides
MasteringA&P®>Study Area>PAL

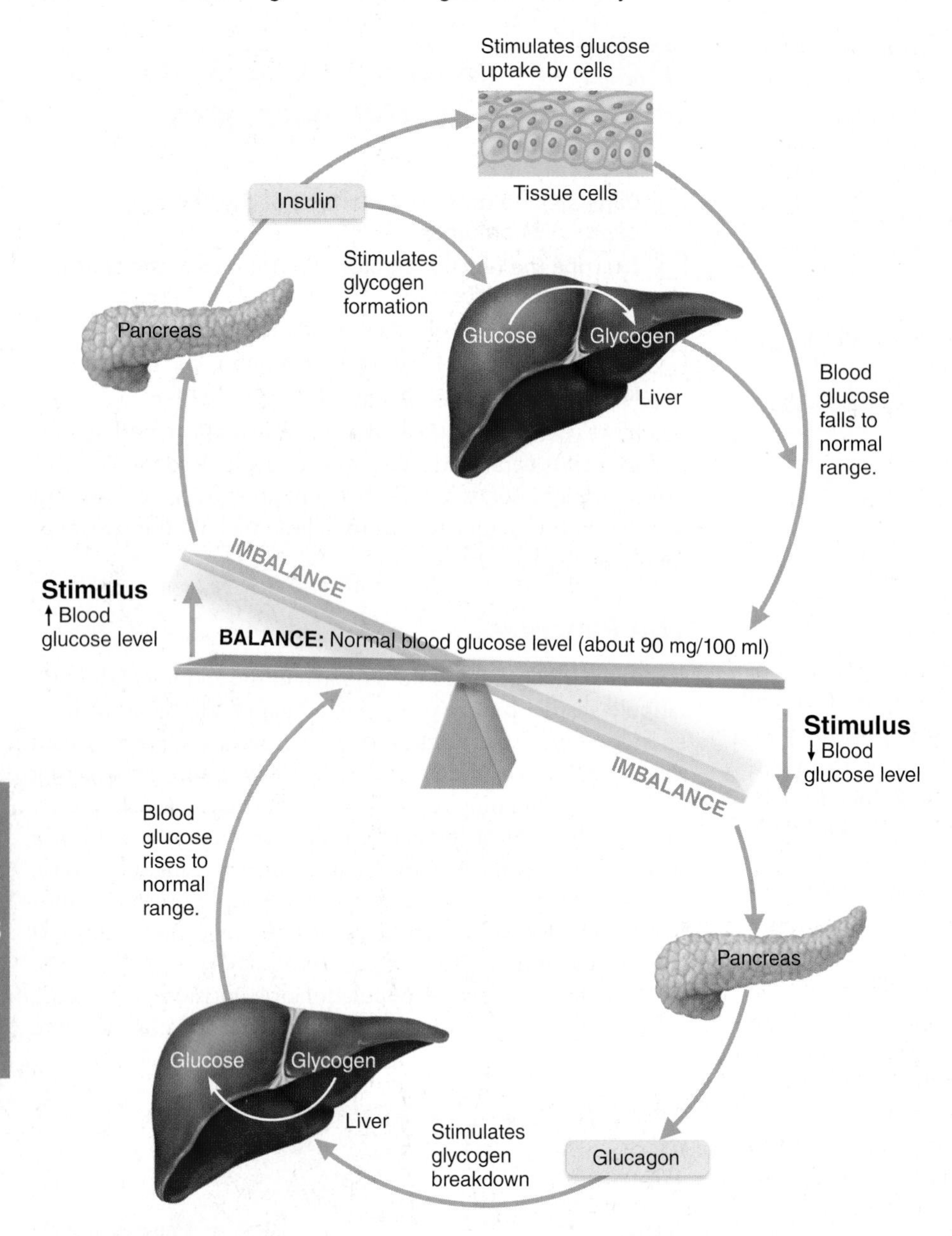

Figure 15.18 Insulin and glucagon from the pancreas regulate blood glucose levels.

numerous insulin-synthesizing **beta (β) cells**. These cells act as tiny fuel sensors, secreting glucagon and insulin appropriately during the fasting and fed states.

Insulin and glucagon are intimately but independently involved in regulating blood glucose levels. Their effects are antagonistic: Glucagon is a *hyperglycemic* hormone, whereas insulin is a *hypoglycemic* hormone (**Figure 15.18**). Some islet cells also synthesize other peptides in small amounts, including *somatostatin, pancreatic polypeptide* (*PP*), and others. However, here we will focus on glucagon and insulin.

Glucagon

Glucagon (gloo′kah-gon), a 29-amino-acid polypeptide, is an extremely potent hyperglycemic agent: One molecule can cause the release of 100 million glucose molecules into the blood! The major target of glucagon is the liver, where it promotes the following actions:

- Breakdown of glycogen to glucose (*glycogenolysis*) (Figure 15.18)
- Synthesis of glucose from lactic acid and from noncarbohydrate molecules (*gluconeogenesis*)
- Release of glucose to the blood by liver cells, causing blood glucose levels to rise

A secondary effect is to lower blood levels of amino acids as the liver cells sequester these molecules to make new glucose molecules.

Humoral stimuli, mainly falling blood glucose levels, prompt the alpha cells to secrete glucagon. However, sympathetic nervous system stimulation and rising amino acid levels (as might follow a protein-rich meal) are also stimulatory. Glucagon release is suppressed by rising blood glucose levels, insulin, and somatostatin.

Insulin

Insulin is a small (51-amino-acid) protein consisting of two amino acid chains linked by disulfide (–S–S–) bonds. It is synthesized as part of a larger polypeptide chain called **proinsulin**. Enzymes then excise the middle portion of this chain, releasing functional insulin. This "clipping" process occurs in the secretory vesicles just before the beta cell releases insulin.

Insulin's effects are most obvious when we have just eaten. Its main effect is to lower blood glucose levels (Figure 15.18), but it also promotes protein synthesis and fat storage. Circulating insulin lowers blood glucose levels in three ways.

- It enhances membrane transport of glucose (and other simple sugars) into most body cells, especially muscle and fat cells.
- It inhibits the breakdown of glycogen to glucose.
- It inhibits the conversion of amino acids or fats to glucose. These inhibiting effects counter any metabolic activity that would increase plasma levels of glucose.

Insulin is *not* needed for glucose entry into liver, kidney, and brain tissue, all of which have easy access to blood glucose regardless of insulin levels. However, insulin does have important roles in the brain—it participates in neuronal development, feeding behavior, and learning and memory.

Insulin activates its receptor (a tyrosine kinase enzyme), which phosphorylates specific proteins, beginning the cascade that promotes glucose uptake and insulin's other effects. After glucose enters a target cell, insulin binding triggers enzymatic activities that:

- Catalyze the oxidation of glucose for ATP production
- Join glucose molecules together to form glycogen
- Convert glucose to fat (particularly in adipose tissue)

As a rule, energy needs are met first, followed by glycogen formation. Finally, if excess glucose is still available, it is converted to fat. Insulin also stimulates amino acid uptake and protein synthesis in muscle tissue.

Factors That Influence Insulin Release Pancreatic beta cells secrete insulin when stimulated by:

- Elevated blood glucose levels. This is the chief controlling factor.
- Rising blood levels of amino acids and fatty acids.
- Acetylcholine released by parasympathetic nerve fibers.
- Hyperglycemic hormones (such as glucagon, epinephrine, growth hormone, thyroxine, or glucocorticoids). This effect is indirect and occurs because all of these hormones increase blood glucose levels.

Somatostatin and sympathetic nervous system activation depress insulin release.

As you can see, blood glucose levels represent a balance of humoral, neural, and hormonal influences. Insulin is the major hypoglycemic factor that counterbalances the many hyperglycemic hormones.

HOMEOSTATIC IMBALANCE 15.10 CLINICAL

Diabetes mellitus (DM) results from either hyposecretion or hypoactivity of insulin. When insulin is absent, the result is *type 1 diabetes mellitus*. If insulin is present, but its effects are deficient, the result is *type 2 diabetes mellitus*. In either case, blood glucose levels remain high after a meal because glucose is unable to enter most tissue cells. Ordinarily, when blood glucose levels rise, hyperglycemic hormones are not released, but when hyperglycemia becomes excessive, the person begins to feel nauseated, which precipitates the fight-or-flight response. This response results, inappropriately, in all the reactions that normally occur in the hypoglycemic (fasting) state to make glucose available—that is, glycogenolysis, lipolysis (breakdown of fat), and gluconeogenesis. Consequently, high blood glucose levels soar even higher, and excess glucose begins to be lost from the body in urine (*glycosuria*).

The three cardinal signs of diabetes mellitus are:

- **Polyuria.** Excessive glucose in the blood leads to excessive glucose in the kidney filtrate where it acts as an osmotic diuretic (that is, it inhibits water reabsorption by the kidney tubules), resulting in **polyuria**, a huge urine output that decreases blood volume and causes dehydration.
- **Polydipsia.** Dehydration stimulates hypothalamic thirst centers, causing **polydipsia**, or excessive thirst.
- **Polyphagia. Polyphagia** refers to excessive hunger and food consumption, a sign that the person is "starving in the land of plenty." Although plenty of glucose is available, the body cannot use it. Instead, the body breaks down protein and fat to supply energy, and this is thought to stimulate appetite.

When sugars cannot be used as cellular fuel, more fats are mobilized, resulting in high fatty acid levels in the blood, a condition called lipidemia. In severe cases of diabetes mellitus, blood levels of fatty acids and their metabolites (acetoacetic acid, acetone, and others) rise dramatically. The fatty acid metabolites, collectively called **ketones** (ke′tōnz) or **ketone bodies**, are organic acids. When they accumulate in the blood, the blood pH drops, resulting in **ketoacidosis**, and ketone bodies begin to spill into the urine (*ketonuria*).

Severe ketoacidosis is life threatening. The nervous system responds by initiating rapid deep breathing (hyperpnea) to blow off carbon dioxide from the blood and increase blood pH. (We will explain the physiological basis of this mechanism in Chapter 21.) Serious electrolyte losses also occur as the body rids itself of excess ketone bodies. Ketone bodies are negatively charged and carry positive ions out with them, so sodium and potassium ions are also lost. The electrolyte imbalance leads to abdominal pain and possibly vomiting. If untreated, ketoacidosis disrupts heart activity and oxygen transport, and severe depression of the nervous system leads to coma and death.

Figure 15.19 summarizes the consequences of insulin deficiency.

Hyperinsulinism, or excessive insulin secretion, results in low blood glucose levels, or **hypoglycemia**. This condition triggers the release of hyperglycemic hormones, which cause anxiety, nervousness, tremors, and weakness. Insufficient glucose delivery to the brain causes disorientation, progressing to convulsions, unconsciousness, and even death. In rare cases, hyperinsulinism results from an islet cell tumor. More commonly, it is caused by an overdose of insulin and is easily treated by ingesting sugar. +

The Gonads and Placenta

The male and female **gonads** produce steroid sex hormones, identical to those produced by adrenal cortical cells (see Figure 15.1). The major distinction is the source and relative amounts produced. As described earlier, gonadotropins regulate the release of gonadal hormones.

The paired *ovaries* are small, oval organs located in the female's abdominopelvic cavity. Besides producing ova, or eggs, the ovaries produce several hormones, most importantly **estrogens** and **progesterone** (pro-jes′tĕ-rōn). Estrogens are responsible for maturation of the reproductive organs and the appearance of the secondary sex characteristics of females at puberty. Acting with progesterone, estrogens promote breast development and cyclic changes in the uterine mucosa (the menstrual cycle).

The male *testes*, located in an extra-abdominal skin pouch called the scrotum, produce sperm and male sex hormones, primarily **testosterone** (tes-tos′tĕ-rōn). During puberty, testosterone initiates the maturation of the male reproductive organs and the appearance of secondary sex characteristics and sex drive. In addition, testosterone is necessary for normal sperm production and maintains the reproductive organs in their mature functional state in adult males.

The *placenta* is a temporary endocrine organ. Besides sustaining the fetus during pregnancy, it secretes several steroid and protein hormones that influence the course of pregnancy.

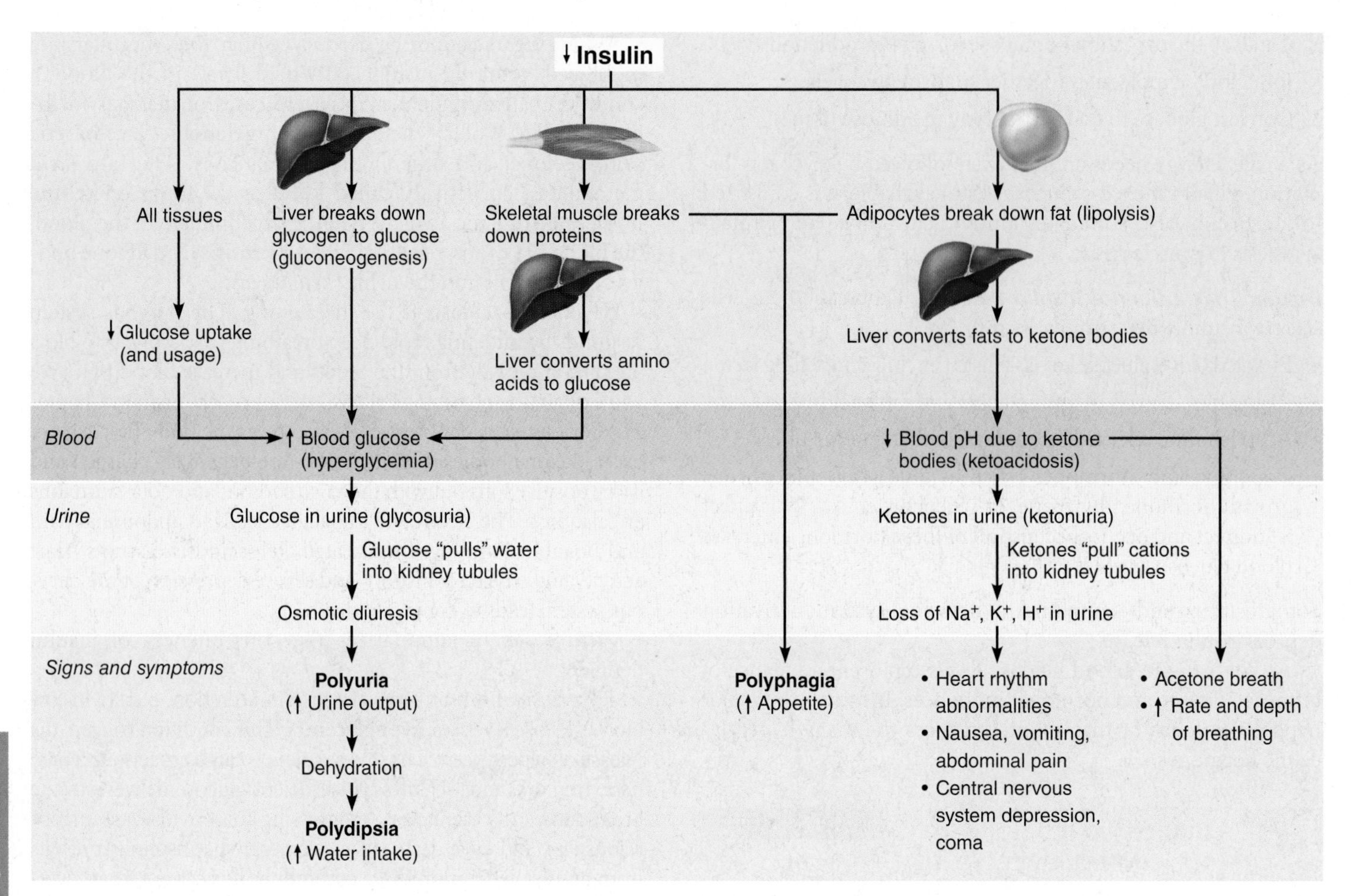

Figure 15.19 Consequences of insulin deficit (diabetes mellitus).

Placental hormones include estrogens, progesterone, and human chorionic gonadotropin (hCG).

We will discuss the roles of the gonadal, placental, and gonadotropic hormones in Chapter 26, where we consider the reproductive system.

Hormone Secretion by Other Organs

Other hormone-producing cells occur in various organs including the heart, gastrointestinal tract, kidneys, skin, adipose tissue, skeleton, and thymus (Table 15.5).

Adipose Tissue

Adipose cells release **leptin**, which serves to tell your body how much stored energy (as fat) you have. The more fat you have, the more leptin there will be in your blood. As we describe in Chapter 23, leptin binds to CNS neurons concerned with appetite control, producing a sensation of satiety. It also appears to stimulate increased energy expenditure.

Two other hormones released by adipose cells affect the sensitivity of cells to insulin. *Resistin* is an insulin antagonist, while *adiponectin* enhances sensitivity to insulin.

Gastrointestinal Tract

Enteroendocrine cells are hormone-secreting cells sprinkled in the mucosa of the gastrointestinal tract. These scattered cells release several peptide hormones that help regulate a wide variety of digestive functions, some of which are summarized in Table 15.5. Enteroendocrine cells also release amines such as serotonin, which act as paracrines, diffusing to and influencing nearby target cells without first entering the bloodstream. Enteroendocrine cells have been referred to as *paraneurons* because they are similar in certain ways to neurons and many of their hormones and paracrines are chemically identical to neurotransmitters.

Heart

The atria contain specialized cardiac muscle cells that secrete **atrial natriuretic peptide (ANP).** As noted on p. 544, ANP decreases the amount of sodium in the extracellular fluid, thereby reducing blood volume and blood pressure.

Kidneys

Interstitial cells in the kidneys secrete **erythropoietin** (ĕ-rith″ro-poi′ĕ-tin; "red-maker"), a glycoprotein hormone that signals the bone marrow to increase production of red blood cells. The kidneys also release **renin**, which acts as an enzyme to initiate the renin-angiotensin-aldosterone mechanism of aldosterone release described earlier.

Table 15.5 Selected Examples of Hormones Produced by Organs Other than the Major Endocrine Organs

SOURCE	HORMONE	CHEMICAL COMPOSITION	TRIGGER	TARGET ORGAN AND EFFECTS
Adipose tissue	Leptin	Peptide	Secretion proportional to fat stores; increased by nutrient uptake	Brain: suppresses appetite; increases energy expenditure
Adipose tissue	Resistin, adiponectin	Peptides	Secretion proportional to fat stores for resistin, inversely proportional for adiponectin	Fat, muscle, liver: resistin antagonizes insulin's action and adiponectin enhances it
Gastrointestinal (GI) tract mucosa				
• Stomach	Gastrin	Peptide	Secreted in response to food	Stomach: stimulates glands to release hydrochloric acid (HCl)
• Stomach	Ghrelin	Peptide	Secreted in response to fasting	Hypothalamus and pituitary: stimulates food intake and GH release
• Duodenum (of small intestine)	Secretin	Peptide	Secreted in response to food	Pancreas and liver: stimulates release of bicarbonate-rich juice Stomach: inhibits secretory activity
• Duodenum	Cholecystokinin (CCK)	Peptide	Secreted in response to food	Pancreas: stimulates release of enzyme-rich juice Gallbladder: stimulates expulsion of stored bile Hepatopancreatic sphincter: causes sphincter to relax, allowing bile and pancreatic juice to enter duodenum
• Duodenum (and other gut regions)	Incretins [glucose-dependent insulinotropic peptide (GIP) and glucagon-like peptide 1 (GLP-1)]	Peptide	Secreted in response to glucose in intestinal lumen	Pancreas: enhances glucose-dependent release of insulin and inhibition of glucagon release
Heart (atria)	Atrial natriuretic peptide (ANP)	Peptide	Secreted in response to stretching of atria (by rising blood pressure)	Kidney: inhibits sodium ion reabsorption and renin release Adrenal cortex: inhibits secretion of aldosterone; decreases blood pressure
Kidney	Erythropoietin (EPO)	Glycoprotein	Secreted in response to hypoxia	Red bone marrow: stimulates production of red blood cells
Skeleton	Osteocalcin	Peptide	Unknown; insulin promotes its activation	Increases insulin production and insulin sensitivity
Skin (epidermal cells)	Cholecalciferol (provitamin D_3)	Steroid	Activated by the kidneys to active vitamin D_3 (calcitriol) in response to parathyroid hormone	Intestine: stimulates active transport of dietary calcium across cell membranes of small intestine
Thymus	Thymulin, thymopoietins, thymosins	Peptides	Unknown	Mostly act locally as paracrines; involved in T lymphocyte development and in immune responses

Skeleton

Osteoblasts in bone secrete *osteocalcin*, a hormone that prods pancreatic beta cells to divide and secrete more insulin. It also restricts fat storage by adipocytes, and triggers the release of adiponectin. This improves glucose handling and reduces body fat.

Interestingly, insulin promotes the conversion of inactive osteocalcin to active osteocalcin in bone, forming a two-way conversation between bone and the pancreas. Osteocalcin levels are low in type 2 diabetes, and increasing its level may offer a new treatment approach.

Skin

The skin produces **cholecalciferol**, an inactive form of vitamin D_3, when modified cholesterol molecules in epidermal cells are exposed to ultraviolet radiation. This compound then enters the blood via the dermal capillaries, is modified in the liver, and becomes fully activated in the kidneys. The active form of vitamin D_3, **calcitriol**, is an essential regulator of the carrier system that intestinal cells use to absorb Ca^{2+} from food. Without this vitamin, bones become soft and weak. In addition, most cells throughout the body have vitamin D receptors. Vitamin D modulates immune functions, decreases inflammation, and may act as an anticancer agent.

Thymus

Located deep to the sternum in the thorax is the **thymus** (see Figure 15.1). Large and conspicuous in infants and children, the thymus shrinks throughout adulthood. By old age, it is composed largely of adipose and fibrous connective tissues.

Thymic epithelial cells secrete several different families of peptide hormones, including **thymulin, thymopoietins,** and **thymosins** (thi′mo-sinz). These hormones are thought to be involved in the normal development of *T lymphocytes* and the immune response, but their roles are not well understood. Although called hormones, they mainly act locally as paracrines. We describe the thymus in Chapter 19 in our discussion of lymphoid organs and tissues.

15

☑ Check Your Understanding

17. Which hormone does the heart produce and what is its function?

18. What is the main function of the hormone produced by the skin?

19. MAKING connections Diabetes mellitus and diabetes insipidus are both due to lack of a hormone. Which hormone causes which? What symptom do they have in common? What would you find in the urine of a patient with one but not the other?

20. MAKING connections Which of the two chemical classes of hormones introduced at the beginning of this chapter do the gonadal hormones belong to? Which major endocrine gland secretes hormones of this same chemical class?

For answers, see Answers Appendix.

In this chapter, we have covered the general mechanisms of hormone action and have provided an overview of the major endocrine organs, their chief targets, and their most important physiological effects. However, every one of the hormones discussed here comes up in at least one other chapter, where its actions are described as part of the functional framework of a particular organ system. For example, we described the effects of PTH and calcitonin on bone mineralization in Chapter 6 along with the discussion of bone remodeling.

REVIEW QUESTIONS

MAP® For more chapter study tools, go to the Study Area of MasteringA&P®.

There you will find:

- Interactive Physiology iP
- A&PFlix A&PFlix
- Practice Anatomy Lab PAL
- PhysioEx PEx
- Videos, Practice Quizzes and Tests, MP3 Tutor Sessions, Case Studies, and much more!

Multiple Choice/Matching

(Some questions have more than one correct answer. Select the best answer or answers from the choices given.)

1. The major stimulus for release of parathyroid hormone is **(a)** hormonal, **(b)** humoral, **(c)** neural.
2. The anterior pituitary secretes all but **(a)** antidiuretic hormone, **(b)** growth hormone, **(c)** gonadotropins, **(d)** TSH.
3. A hormone not involved in glucose metabolism is **(a)** glucagon, **(b)** cortisone, **(c)** aldosterone, **(d)** insulin.
4. Parathyroid hormone **(a)** increases bone formation and lowers blood calcium levels, **(b)** increases calcium excretion from the body, **(c)** decreases calcium absorption from the gut, **(d)** demineralizes bone and raises blood calcium levels.
5. Choose from the following key to identify the hormones described.

Key: **(a)** aldosterone, **(b)** antidiuretic hormone, **(c)** growth hormone, **(d)** luteinizing hormone, **(e)** oxytocin, **(f)** prolactin, **(g)** T_4 and T_3, **(h)** TSH

____ **(1)** important anabolic hormone; many of its effects mediated by IGFs

____ **(2)** cause the kidneys to conserve water and/or salt (two choices)

____ **(3)** stimulates milk production

____ **(4)** tropic hormone that stimulates the gonads to secrete sex hormones

____ **(5)** increases uterine contractions during birth

____ **(6)** major metabolic hormone(s) of the body

____ **(7)** causes reabsorption of sodium ions by the kidneys

____ **(8)** tropic hormone that stimulates the thyroid gland to secrete thyroid hormone

____ **(9)** secreted by the posterior pituitary (two choices)
____ **(10)** the only steroid hormone in the list

6. A hypodermic injection of epinephrine would **(a)** increase heart rate, increase blood pressure, dilate the bronchi of the lungs, and increase peristalsis, **(b)** decrease heart rate, decrease blood pressure, constrict the bronchi, and increase peristalsis, **(c)** decrease heart rate, increase blood pressure, constrict the bronchi, and decrease peristalsis, **(d)** increase heart rate, increase blood pressure, dilate the bronchi, and decrease peristalsis.
7. Testosterone is to the male as which hormone is to the female? **(a)** luteinizing hormone, **(b)** progesterone, **(c)** estrogen, **(d)** prolactin.
8. If anterior pituitary secretion is deficient in a growing child, the child will **(a)** develop acromegaly, **(b)** become a dwarf but have fairly normal body proportions, **(c)** mature sexually at an earlier than normal age, **(d)** be in constant danger of becoming dehydrated.
9. If there is adequate carbohydrate intake, secretion of insulin results in **(a)** lower blood glucose levels, **(b)** increased cell utilization of glucose, **(c)** storage of glycogen, **(d)** all of these.
10. Hormones **(a)** are produced by exocrine glands, **(b)** are carried to all parts of the body in blood, **(c)** remain at constant concentration in the blood, **(d)** affect only non-hormone-producing organs.
11. Some hormones act by **(a)** increasing the synthesis of enzymes, **(b)** converting an inactive enzyme into an active enzyme, **(c)** affecting only specific target organs, **(d)** all of these.
12. Absence of thyroid hormone would result in **(a)** increased heart rate and increased force of heart contraction, **(b)** depression of the CNS and lethargy, **(c)** exophthalmos, **(d)** high metabolic rate.
13. Medullary chromaffin cells are found in the **(a)** parathyroid gland, **(b)** anterior pituitary gland, **(c)** adrenal gland, **(d)** pineal gland.
14. Atrial natriuretic peptide secreted by the heart has exactly the opposite function of this hormone secreted by the zona glomerulosa: **(a)** antidiuretic hormone, **(b)** epinephrine, **(c)** calcitonin, **(d)** aldosterone, **(e)** androgens.

Short Answer Essay Questions

15. Define hormone.
16. Which type of hormone receptor—plasma membrane bound or intracellular—would be expected to provide the most long-lived response to hormone binding and why?
17. (a) Describe the body location of each of the following endocrine organs: anterior pituitary, pineal gland, pancreas, ovaries, testes, and adrenal glands. (b) List the hormones produced by each organ.
18. Name two endocrine glands (or regions) that are important in the stress response, and explain why they are important.
19. The anterior pituitary is often referred to as the master endocrine organ, but it, too, has a "master." What controls the release of anterior pituitary hormones?
20. The posterior pituitary is not really an endocrine gland. Why not? What is it?
21. Endemic goiter is not really the result of a malfunctioning thyroid gland. What does cause it?
22. How are the hyperglycemia and lipidemia of insulin deficiency linked?
23. Name a hormone secreted by a muscle cell and two hormones secreted by neurons.

AT THE CLINIC

Clinical Case Study

Endocrine System

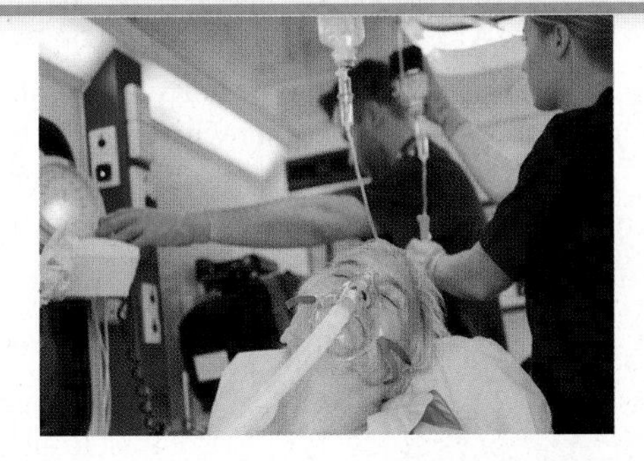

We have a new patient to consider today. Mr. Gutteman, a 70-year-old male, was brought into the ER in a comatose state and has yet to come out of it. It is obvious that he suffered severe head trauma—his scalp was badly lacerated, and he has an impacted skull fracture. His initial lab tests (blood and urine) were within normal limits. His fracture was repaired and the following orders (and others) were given:

- Check qh (every hour) and record: spontaneous behavior, level of responsiveness to stimulation, movements, pupil size and reaction to light, speech, and vital signs.
- Turn patient q4h and maintain meticulous skin care and dryness.

1. Explain the rationale behind these orders.

On the second day of his hospitalization, the aide reports that Mr. Gutteman is breathing irregularly, his skin is dry and flaccid, and that she has emptied his urine reservoir several times during the day. Upon receiving this information, the physician ordered:

- Blood and urine tests for presence of glucose and ketones
- Strict I&O (fluid intake and output recording)

Mr. Gutteman is found to be losing huge amounts of water in urine and the volume lost is being routinely replaced (via IV line). Mr. Gutteman's blood and urine tests are negative for glucose and ketones.

Relative to these findings:

2. What would you say Mr. Gutteman's hormonal problem is, and what do you think caused it?
3. Is it life threatening? (Explain your answer.)

For answers, see Answers Appendix.

16 Blood

KEY CONCEPTS

Blood is the river of life that surges within us, transporting nearly everything that must be carried from one place to another. Long before modern medicine, blood was viewed as magical—an elixir that held the mystical force of life—because when it drained from the body, life departed as well. Today, blood still has enormous importance in the practice of medicine. Clinicians examine it more often than any other tissue when trying to determine the cause of disease in their patients.

In this chapter, we describe the composition and functions of this life-sustaining fluid that serves as a transport "vehicle" for the organs of the cardiovascular system (*cardio* = heart, *vasc* = blood vessels). To get started, we need a brief overview of blood circulation, which is initiated by the pumping action of the heart. Blood exits the *heart* via *arteries*, which branch repeatedly until they become tiny *capillaries*. By diffusing across the capillary walls, oxygen and nutrients leave the blood and enter the body tissues, and carbon dioxide and wastes move from the tissues to the bloodstream. As oxygen-deficient blood leaves the capillary beds, it flows into *veins*, which return it to the heart. The returning blood then flows from the heart to the lungs, where it picks up oxygen and then returns to the heart to be pumped throughout the body once again.

16.1 The functions of blood are transport, regulation, and protection

→ Learning Objective

☐ **List eight functions of blood.**

Blood performs a number of functions, all concerned in one way or another with transporting substances, regulating blood levels of particular substances, or protecting the body.

Transport

Transport functions of blood include:

- Delivering oxygen from the lungs and nutrients from the digestive tract to all body cells.
- Transporting metabolic waste products from cells to elimination sites (to the lungs to eliminate carbon dioxide, and to the kidneys to dispose of nitrogenous wastes in urine).
- Transporting hormones from the endocrine organs to their target organs.

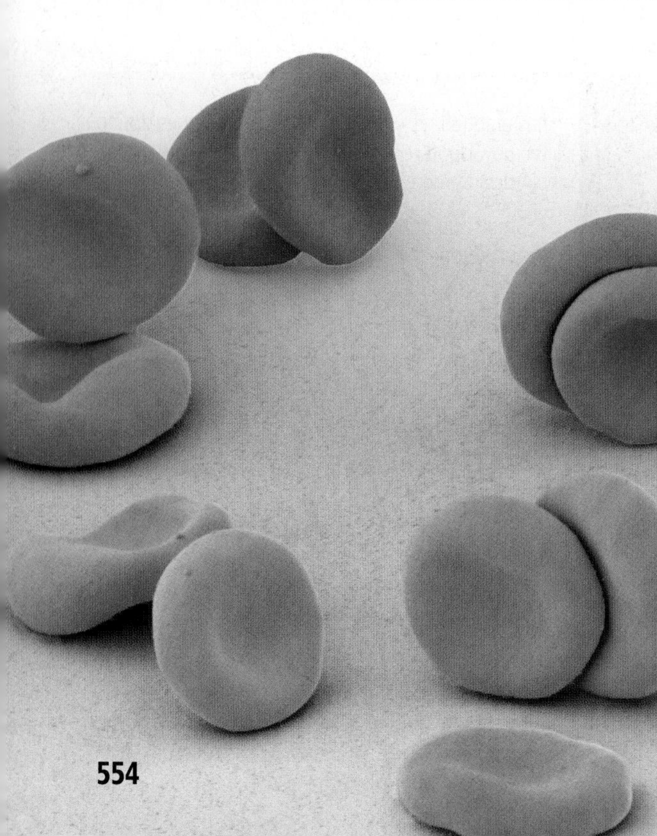

Electron micrograph of erythrocytes (red blood cells).

Regulation

Regulatory functions of blood include:

- Maintaining appropriate body temperature by absorbing and distributing heat throughout the body and to the skin surface to encourage heat loss.
- Maintaining normal pH in body tissues. Many blood proteins and other bloodborne solutes act as buffers to prevent excessive or abrupt changes in blood pH that could jeopardize normal cell activities. Blood also acts as the reservoir for the body's "alkaline reserve" of bicarbonate ions.
- Maintaining adequate fluid volume in the circulatory system. Blood proteins prevent excessive fluid loss from the bloodstream into the tissue spaces. As a result, the fluid volume in the blood vessels remains ample to support efficient blood circulation to all parts of the body.

Protection

Protective functions of blood include:

- Preventing blood loss. When a blood vessel is damaged, platelets and plasma proteins initiate clot formation, halting blood loss.
- Preventing infection. Drifting along in blood are antibodies, complement proteins, and white blood cells, all of which help defend the body against foreign invaders such as bacteria and viruses.

☑ Check Your **Understanding**

1. List two protective functions of blood.

For answers, see Answers Appendix.

16.2 Blood consists of plasma and formed elements

→ Learning Objectives

- ☐ **Describe the composition and physical characteristics of whole blood. Explain why it is classified as a connective tissue.**
- ☐ **Discuss the composition and functions of plasma.**

Blood is the only fluid tissue in the body. It appears to be a thick, homogeneous liquid, but the microscope reveals that it has both cellular and liquid components. Blood is a specialized connective tissue in which living blood cells, called the *formed elements*, are suspended in a nonliving fluid matrix called *plasma* (plaz′mah). Blood lacks the collagen and elastic fibers typical of other connective tissues, but dissolved fibrous proteins become visible as fibrin strands during blood clotting.

If we spin a sample of blood in a centrifuge, centrifugal force packs down the heavier formed elements and the less dense plasma remains at the top (**Figure 16.1**). Most of the reddish mass at the bottom of the tube is *erythrocytes* (ĕ-rith′ro-sīts; *erythro* = red), the red blood cells that transport oxygen. A thin, whitish layer called the **buffy coat** is present at the erythrocyte-plasma junction. This layer contains *leukocytes* (*leuko* = white), the white blood cells that act in various ways to protect the body, and *platelets*, cell fragments that help stop bleeding.

Erythrocytes normally constitute about 45% of the total volume of a blood sample, a percentage known as the **hematocrit** (he-mat′o-krit; "blood fraction"). Normal hematocrit values vary. In healthy males the norm is 47% ± 5%; in females it is 42% ± 5%. Leukocytes and platelets contribute less than 1% of blood volume. Plasma makes up most of the remaining 55% of whole blood.

Physical Characteristics and Volume

Blood is a sticky, opaque fluid with a characteristic metallic taste. As children, we discover its saltiness the first time we stick a cut finger into our mouth. Depending on the amount of oxygen it is carrying, the color of blood varies from scarlet (oxygen rich) to dark red (oxygen poor). It is slightly alkaline, with a pH between 7.35 and 7.45. Blood is more dense than water and about five times more viscous, largely because of its formed elements.

Erythrocytes are the major factor contributing to blood viscosity. Women typically have a lower red blood cell count than men [4.2–5.4 million cells per microliter (1 μl = 1 mm^3) of blood versus 4.7–6.1 million cells/μl respectively]. When the number of red blood cells increases beyond the normal range, blood becomes more viscous and flows more slowly. Similarly, as the number of red blood cells drops below the lower end of the range, the blood thins and flows more rapidly.

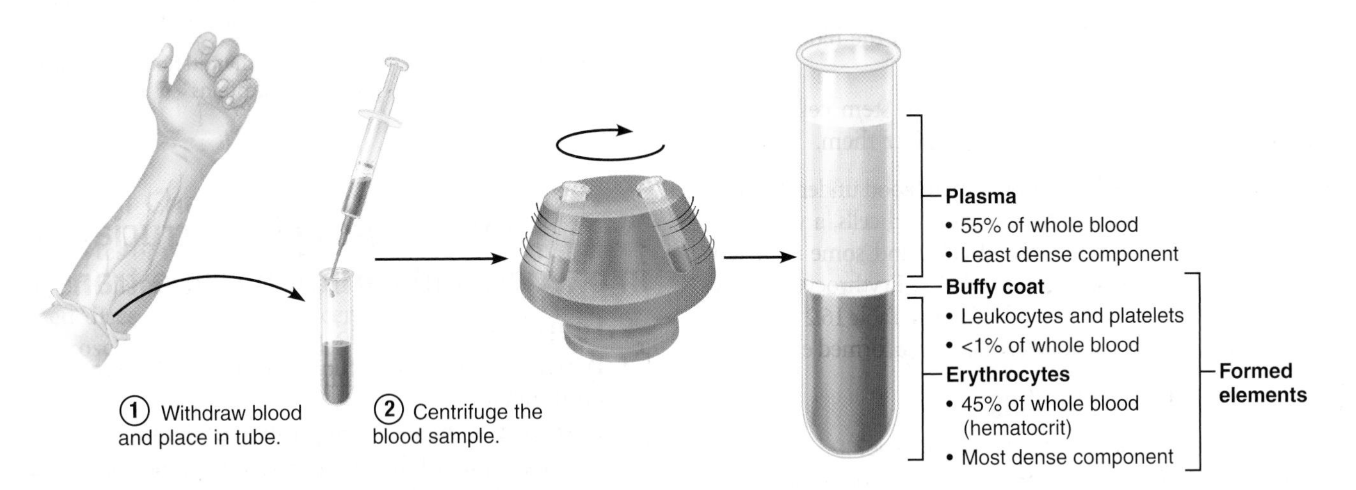

Figure 16.1 The major components of whole blood.

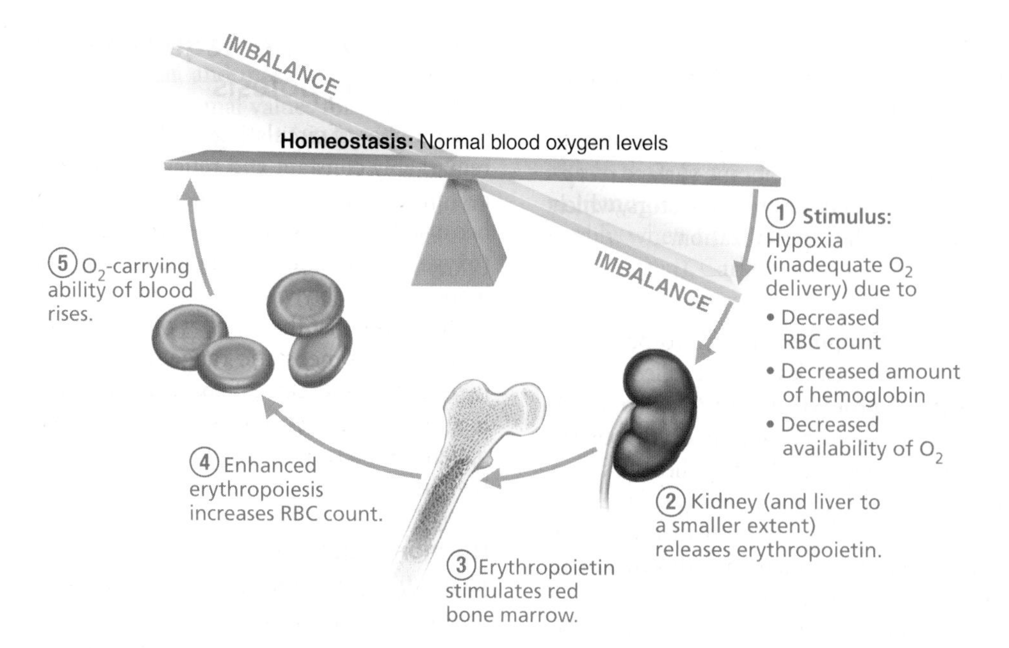

Figure 16.6 Erythropoietin mechanism for regulating erythropoiesis.

of erythropoiesis. Instead, control is based on their ability to transport enough oxygen to meet tissue demands.

Bloodborne erythropoietin stimulates red marrow cells that *are already committed* to becoming erythrocytes, causing them to mature more rapidly. Two to three days after erythropoietin levels rise in the blood, the rate of reticulocyte release and the reticulocyte count rise markedly. Notice that hypoxia does not activate the bone marrow directly. Instead it stimulates the kidneys, which in turn provide the hormonal stimulus that activates the bone marrow.

Some athletes abuse recombinant EPO—particularly professional bike racers and marathon runners seeking increased stamina and performance. However, the consequences can be deadly. By injecting EPO, healthy athletes increase their normal hematocrit from 45% to as much as 65%. Then, with the dehydration that occurs in a long race, the blood concentrates even further, becoming a thick, sticky "sludge" that can cause clotting, stroke, or heart failure. ✚

The male sex hormone *testosterone* also enhances the kidneys' production of EPO. Because female sex hormones do not have similar stimulatory effects, testosterone may be at least partially responsible for the higher RBC counts and hemoglobin levels seen in males.

Dietary Requirements

The raw materials required for erythropoiesis include the usual nutrients and structural materials—amino acids, lipids, and carbohydrates. Iron is essential for hemoglobin synthesis. Iron is available from the diet, and intestinal cells precisely control its absorption into the bloodstream in response to changing body stores of iron.

Approximately 65% of the body's iron supply (about 4000 mg) is in hemoglobin. Most of the remainder is stored in the liver, spleen, and (to a much lesser extent) bone marrow. Free iron ions (Fe^{2+}, Fe^{3+}) are toxic, so iron is stored inside cells as protein-iron complexes such as **ferritin** (fer′ĭ-tin) and **hemosiderin** (he″mo-sid′er-in). In blood, iron is transported loosely bound to a transport protein called **transferrin**, and developing erythrocytes take up iron as needed to form hemoglobin (**Figure 16.7**). Small amounts of iron are lost each day in feces, urine, and perspiration. The average daily loss of iron is 1.7 mg in women and 0.9 mg in men. In women, the menstrual flow accounts for the additional losses.

Two B-complex vitamins—vitamin B_{12} and folic acid—are necessary for normal DNA synthesis. Even slight deficits jeopardize rapidly dividing cell populations, such as developing erythrocytes.

Fate and Destruction of Erythrocytes

Red blood cells have a useful life span of 100 to 120 days. Their anucleate condition carries with it some important limitations. Red blood cells are unable to synthesize new proteins, grow, or divide. Erythrocytes become "old" as they lose their flexibility, become increasingly rigid and fragile, and their hemoglobin begins to degenerate. They become trapped and fragment in smaller circulatory channels, particularly in those of the spleen. For this reason, the spleen is sometimes called the "red blood cell graveyard."

Macrophages engulf and destroy dying erythrocytes. The heme of their hemoglobin is split off from globin (Figure 16.7). Its core of iron is salvaged, bound to protein (as ferritin or hemosiderin), and stored for reuse. The balance of the heme group is degraded to **bilirubin** (bil″ĭ-roo′bin), a yellow pigment that is released to the blood and binds to albumin for transport. Liver cells pick up bilirubin and in turn secrete it (in bile) into the intestine, where it is metabolized to *urobilinogen*. Most of this degraded pigment

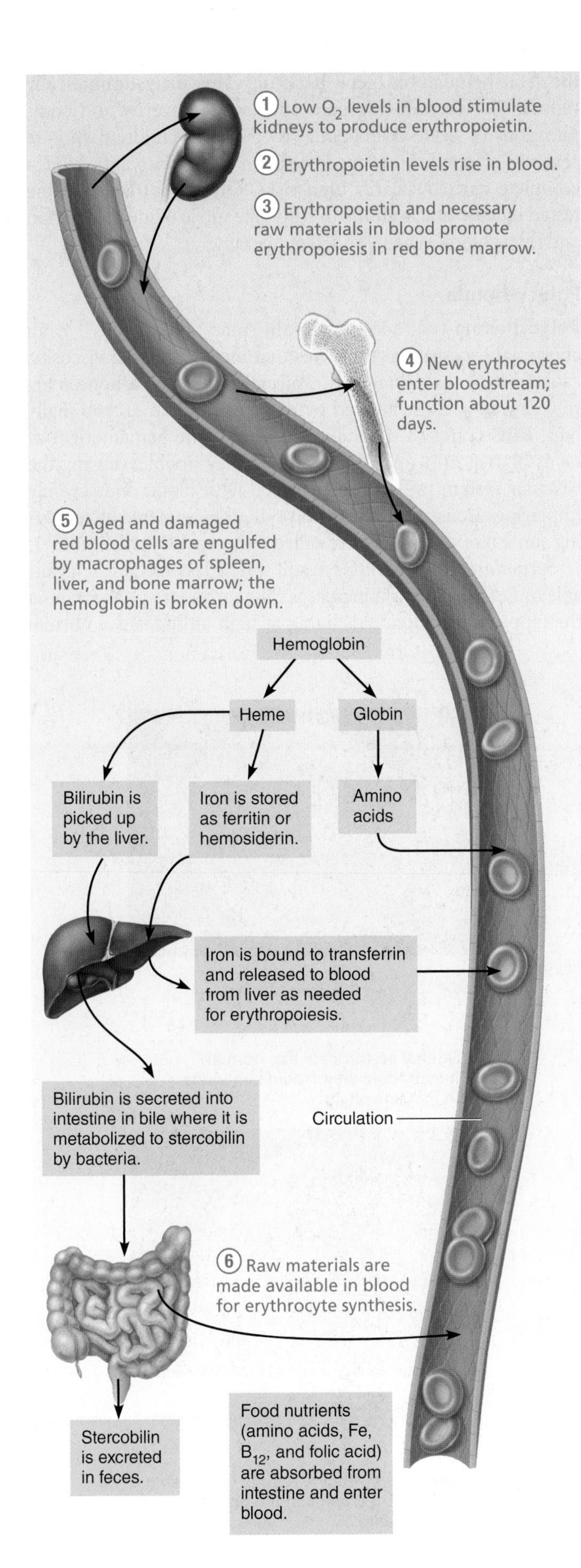

Figure 16.7 Life cycle of red blood cells.

leaves the body in feces, as a brown pigment called *stercobilin*. The protein (globin) part of hemoglobin is metabolized or broken down to amino acids, which are released to the circulation.

CLINICAL

Erythrocyte Disorders

Most erythrocyte disorders can be classified as anemia or polycythemia.

Anemia

Anemia (ah-ne′me-ah; "lacking blood") is a condition in which the blood's oxygen-carrying capacity is too low to support normal metabolism. It is a *sign* of some disorder rather than a disease in itself. Its hallmark is blood oxygen levels that are inadequate to support normal metabolism. Anemic individuals are fatigued, often pale, short of breath, and chilled.

The causes of anemia can be divided into three groups: blood loss, not enough red blood cells produced, or too many of them destroyed.

Blood Loss *Hemorrhagic anemia* (hem″o-raj′ik) is caused by blood loss. In acute hemorrhagic anemia, blood loss is rapid (as might follow a severe stab wound); it is treated by replacing the lost blood. Slight but persistent blood loss (due to hemorrhoids or an undiagnosed bleeding ulcer, for example) causes chronic hemorrhagic anemia. Once the primary problem is resolved, normal erythropoietic mechanisms replace the lost blood cells.

Not Enough Red Blood Cells Produced A number of problems can decrease erythrocyte production. These problems range from lack of essential raw materials (such as iron) to complete failure of the red bone marrow.

Iron-deficiency anemia is generally a secondary result of hemorrhagic anemia, but it also results from inadequate intake of iron-containing foods and impaired iron absorption. The erythrocytes produced, called **microcytes**, are small and pale because they cannot synthesize their normal complement of hemoglobin. The treatment is to increase iron intake in diet or through iron supplements.

Pernicious anemia is an autoimmune disease that most often affects the elderly. The immune system of these individuals destroys cells of their own stomach mucosa. These cells produce a substance called **intrinsic factor** that must be present for vitamin B_{12} to be absorbed by intestinal cells. Without vitamin B_{12}, the developing erythrocytes grow but cannot divide, and large, pale cells called **macrocytes** result. Treatment involves regular intramuscular injections of vitamin B_{12} or application of a B_{12}-containing gel to the nasal lining once a week.

As you might expect, lack of vitamin B_{12} in the diet also leads to anemia. However, this is usually a problem only in strict vegetarians because meats, poultry, and fish provide ample vitamin B_{12}.

Renal anemia is caused by the lack of EPO, the hormone that controls red blood cell production. Renal anemia frequently accompanies renal disease because damaged or diseased kidneys cannot produce enough EPO. Fortunately, it can be treated with synthetic EPO.

Aplastic anemia may result from destruction or inhibition of the red marrow by certain drugs and chemicals, ionizing radiation,

16

or viruses. In most cases, though, the cause is unknown. Because marrow destruction impairs formation of *all* formed elements, anemia is just one of its signs. Defects in blood clotting and immunity are also present. Blood transfusions provide a stopgap treatment until stem cells harvested from a donor's blood, bone marrow, or umbilical cord blood can be transplanted.

Too Many Red Blood Cells Destroyed In *hemolytic anemias* (he″mo-lit′ik), erythrocytes rupture, or lyse, prematurely. Hemoglobin abnormalities, transfusion of mismatched blood, and certain bacterial and parasitic infections are possible causes. Here we focus on the hemoglobin abnormalities.

Production of abnormal hemoglobin usually has a genetic basis. Two such examples, thalassemia and sickle-cell anemia, can be serious, incurable, and sometimes fatal diseases. In both diseases the globin part of hemoglobin is abnormal and the erythrocytes produced are fragile and rupture prematurely.

Thalassemias (thal″ah-se′me-ahs; "sea blood") typically occur in people of Mediterranean ancestry. One of the globin chains is absent or faulty, and the erythrocytes are thin, delicate, and deficient in hemoglobin. There are many subtypes of thalassemia, classified according to which hemoglobin chain is affected and where. They range in severity from mild to so severe that monthly blood transfusions are required.

In **sickle-cell anemia**, the havoc caused by the abnormal hemoglobin, *hemoglobin S* (*HbS*), results from a change in just one of the 146 amino acids in a beta chain of the globin molecule! (See **Figure 16.8**.) This alteration causes the beta chains to link together under low-oxygen conditions, forming stiff rods so that hemoglobin S becomes spiky and sharp. This, in turn, causes the red blood cells to become crescent shaped when they unload oxygen molecules or when the oxygen content of the blood is lower than normal, as during vigorous exercise and other activities that increase metabolic rate.

The stiff, deformed erythrocytes rupture easily and tend to dam up in small blood vessels. These events interfere with oxygen delivery, leaving the victims gasping for air and in extreme pain. Bone and chest pain are particularly severe, and infection and stroke often follow. Blood transfusion is still the standard treatment for an acute sickle-cell crisis, but preliminary results using inhaled nitric oxide to dilate blood vessels are promising.

Sickle-cell anemia occurs chiefly in black people who live in the malaria belt of Africa and among their descendants. It strikes nearly one of every 500 African-American newborns.

Why would such a dangerous genetic trait persist in a population? Globally, about 250 million people are infected with malaria and about a million die each year. While individuals with two copies of the sickle-cell gene have sickle-cell anemia, individuals with only one copy of the gene (sickle-cell trait) have a better chance of surviving malaria. Their cells only sickle under abnormal circumstances, most importantly when they are infected with malaria. Sickling reduces the malaria parasites' ability to survive and enhances macrophages' ability to destroy infected RBCs and the parasites they contain.

Several treatment approaches for sickle-cell anemia focus on preventing RBCs from sickling. Fetal hemoglobin (HbF) does not "sickle," even in those destined to have sickle-cell anemia. *Hydroxyurea*, a drug used to treat chronic leukemia, switches the fetal hemoglobin gene back on. This drug dramatically reduces the excruciating pain and overall severity and complications of sickle-cell anemia (by 50%). In children who are severely affected, bone marrow stem cell transplants offer a complete cure, but carry high risks. Other approaches being tested include oral arginine to stimulate nitric oxide production and dilate blood vessels, and gene therapy.

Polycythemia

Polycythemia (pol″e-si-the′me-ah; "many blood cells") is an abnormal excess of erythrocytes that increases blood viscosity, causing it to flow sluggishly. *Polycythemia vera*, a bone marrow cancer, is characterized by dizziness and an exceptionally high RBC count (8–11 million cells/μl). The hematocrit may be as high as 80% and blood volume may double, causing the vascular system to become engorged with blood and severely impairing circulation. Severe polycythemia is treated by removing some blood (a procedure called a therapeutic phlebotomy).

Secondary polycythemias result when less oxygen is available or EPO production increases. The secondary polycythemia that appears in individuals living at high altitudes is a normal

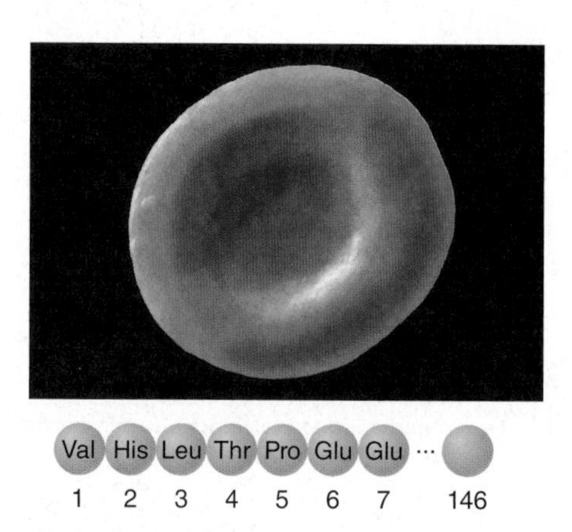

(a) Normal erythrocyte has normal hemoglobin amino acid sequence in the beta chain.

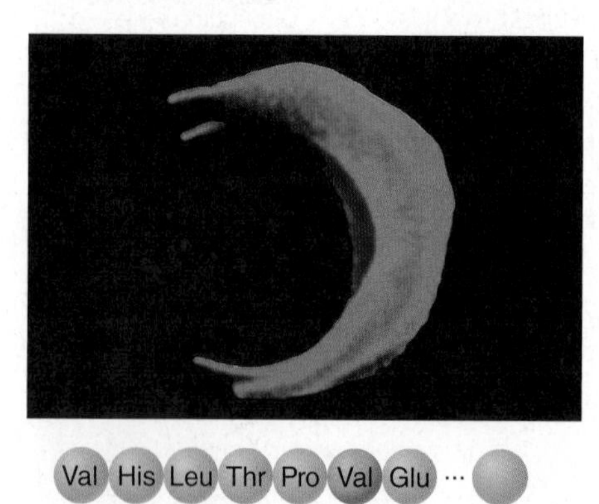

(b) Sickled erythrocyte results from a single amino acid change in the beta chain of hemoglobin.

Figure 16.8 Sickle-cell anemia. Scanning electron micrographs (4950×).

physiological response to the reduced atmospheric pressure and lower oxygen content of the air in such areas. RBC counts of 6–8 million/μl are common in such people.

Blood doping, practiced by some athletes competing in aerobic events, is artificially induced polycythemia. Some of the athlete's red blood cells are drawn off and stored. The body quickly replaces these erythrocytes because removing blood triggers the erythropoietin mechanism. Then, when the stored blood is reinfused a few days before the athletic event, a temporary polycythemia results.

Since red blood cells carry oxygen, the additional infusion should translate into increased oxygen-carrying capacity due to a higher hematocrit, and hence greater endurance and speed. Other than the risk of stroke and heart failure due to high hematocrit and high blood viscosity, blood doping seems to work. However, the practice is considered unethical and has been banned from the Olympic Games.

☑ Check Your Understanding

4. How many molecules of oxygen can each hemoglobin molecule transport? What part of the hemoglobin binds the oxygen?
5. Patients with advanced kidney disease often have anemia. Explain the connection.

For answers, see Answers Appendix.

16.4 Leukocytes defend the body

→ Learning Objectives

- ☐ **List the classes, structural characteristics, and functions of leukocytes.**
- ☐ **Describe how leukocytes are produced.**
- ☐ **Give examples of leukocyte disorders, and explain what goes wrong in each disorder.**

General Structural and Functional Characteristics

Leukocytes (*leuko* = white), or **white blood cells (WBCs)**, are the only formed elements that are complete cells, with nuclei and the usual organelles. Accounting for less than 1% of total blood volume, leukocytes are far less numerous than red blood cells. On average, there are 4800–10,800 WBCs/μl of blood.

Leukocytes are crucial to our defense against disease. They form a mobile army that helps protect the body from damage by bacteria, viruses, parasites, toxins, and tumor cells. As such, they have special functional characteristics. Red blood cells are confined to the bloodstream, and they carry out their functions in the blood. But white blood cells are able to slip out of the capillary blood vessels—a process called **diapedesis** (di″ah-pĕ-de′sis; "leaping across"). The circulatory system is simply their means of transport to areas of the body (mostly loose connective tissues or lymphoid tissues) where they mount inflammatory or immune responses.

As we explain in more detail in Chapter 20, the signals that prompt WBCs to leave the bloodstream at specific locations are cell adhesion molecules displayed by endothelial cells forming the capillary walls at sites of inflammation. Once out of the bloodstream, leukocytes move through the tissue spaces by **amoeboid motion** (they form flowing cytoplasmic extensions that move them along). By following the chemical trail of molecules released by damaged cells or other leukocytes, a phenomenon called **positive chemotaxis**, they pinpoint areas of tissue damage and infection and gather there in large numbers to destroy foreign substances and dead cells.

Whenever white blood cells are mobilized for action, the body speeds up their production and their numbers may double within a few hours. A *white blood cell count* of over 11,000 cells/μl is **leukocytosis**. This condition is a normal response to an infection in the body.

Leukocytes are grouped into two major categories on the basis of structural and chemical characteristics (**Figure 16.9**, and **Table 16.2** on p. 564). *Granulocytes* contain obvious membrane-bound cytoplasmic granules, and *agranulocytes* lack obvious granules.

Students are often asked to list the leukocytes in order from most abundant to least abundant. The following phrase may help you with this task: **N**ever **l**et **m**onkeys **e**at **b**ananas (neutrophils, lymphocytes, monocytes, eosinophils, basophils).

Granulocytes

Granulocytes (gran′u-lo-sīts), which include neutrophils, eosinophils, and basophils, are all roughly spherical in shape. They are larger and much shorter-lived (in most cases) than erythrocytes. They characteristically have lobed nuclei (rounded nuclear masses connected by thinner strands of nuclear material), and their membrane-bound cytoplasmic granules stain quite specifically with Wright's stain.

Neutrophils

Neutrophils (nu′tro-filz), the most numerous white blood cells, account for 50–70% of the WBC population. Neutrophils are about twice as large as erythrocytes.

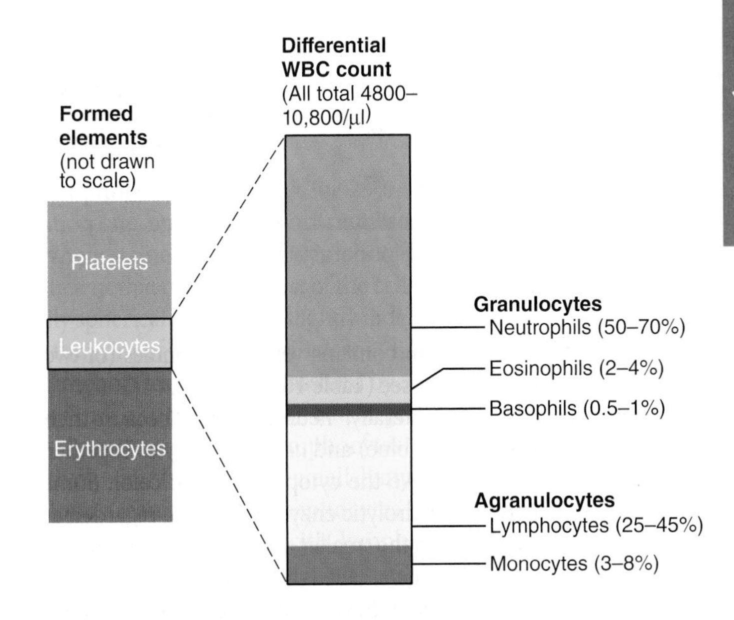

Figure 16.9 Types and relative percentages of leukocytes in normal blood. Erythrocytes comprise nearly 98% of the formed elements, and leukocytes and platelets together account for the remaining 2+%.

excessive numbers of lymphocytes. Many of these lymphocytes are so large and atypical that they were originally misidentified as monocytes, and the disease was mistakenly named mononucleosis. The affected individual complains of being tired and achy, and has a chronic sore throat and a low-grade fever. There is no cure, but with rest the condition typically runs its course to recovery in four to six weeks.

☑ Check Your Understanding

6. Which WBCs turn into macrophages in tissues? Which other WBC is a voracious phagocyte?
7. Amos has leukemia. Even though his WBC count is abnormally high, Amos is prone to severe infections, bleeding, and anemia. Explain.
8. MAKING connections Because of the blood brain barrier, the brain is largely inaccessible to circulating macrophages. Instead (as we discussed in Chapter 11), one type of CNS cell (related to macrophages) can become phagocytic. Which type of cell is this?

For answers, see Answers Appendix.

16.5 Platelets are cell fragments that help stop bleeding

→ Learning Objective

☐ **Describe the structure and function of platelets.**

Platelets are not cells in the strict sense. About one-fourth the diameter of a lymphocyte, they are cytoplasmic fragments of extraordinarily large cells (up to 60 μm in diameter) called **megakaryocytes** (meg″ah-kar′e-o-sītz). In blood smears, each platelet exhibits a blue-staining outer region and an inner area containing granules that stain purple. The granules contain an impressive array of chemicals that act in the clotting process, including serotonin, Ca^{2+}, a variety of enzymes, ADP, and platelet-derived growth factor (PDGF).

Platelets are essential for the clotting process that occurs in plasma when blood vessels are ruptured or their lining is injured. By sticking to the damaged site, platelets form a temporary plug that helps seal the break. (We explain this process shortly.) Because they are anucleate, platelets age quickly and degenerate in about 10 days if they are not involved in clotting. In the meantime, they circulate freely, kept mobile but inactive by molecules (nitric oxide, prostacyclin) secreted by endothelial cells lining the blood vessels.

A hormone called **thrombopoietin** regulates the formation of platelets. Their immediate ancestral cells, the megakaryocytes, are progeny of the hematopoietic stem cell and the myeloid stem cell, but their formation is quite unusual (**Figure 16.12**). In this line, repeated mitoses of the **megakaryoblast** (also called a stage I megakaryocyte) occur, but cytokinesis does not. The final result is the mature (stage IV) megakaryocyte (literally "big nucleus cell"), a bizarre cell with a huge, multilobed nucleus and a large cytoplasmic mass.

After it forms, the megakaryocyte presses against a sinusoid (the specialized type of capillary in the red marrow) and sends cytoplasmic extensions through the sinusoid wall into the bloodstream. These extensions rupture, releasing platelet fragments like leaves blowing off a tree, seeding the blood with platelets. The plasma membranes associated with each fragment quickly seal around the cytoplasm to form the grainy, roughly disc-shaped platelets (see Table 16.2), each with a diameter of 2–4 μm. Each microliter of blood contains 150,000 to 400,000 tiny platelets.

☑ Check Your Understanding

9. What is a megakaryocyte? What does its name mean?

For answers, see Answers Appendix.

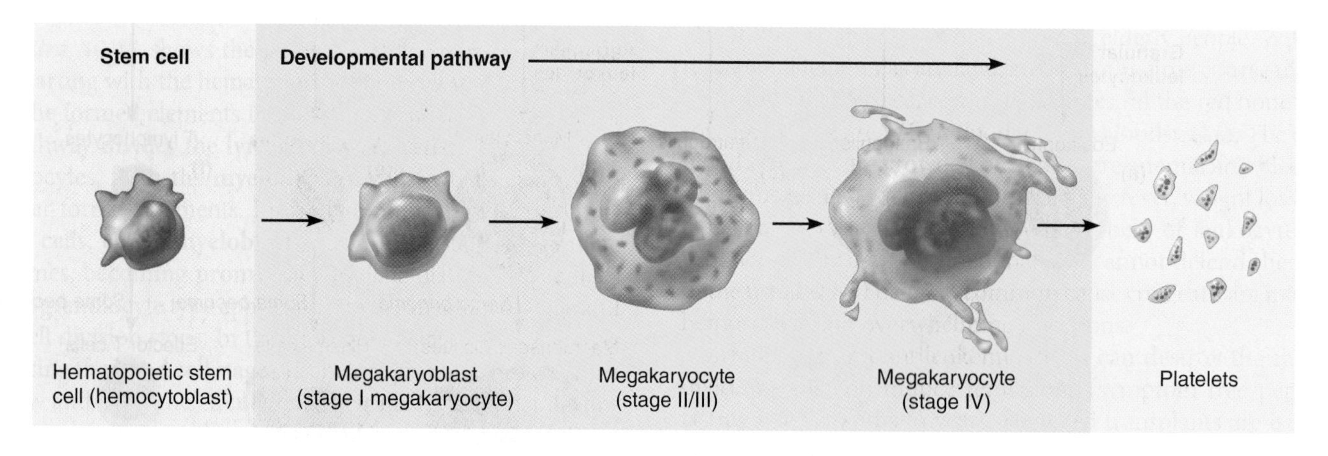

Figure 16.12 Formation of platelets. The hematopoietic stem cell gives rise to cells that undergo several mitotic divisions unaccompanied by cytoplasmic division to produce megakaryocytes. The plasma membrane of the megakaryocyte fragments, liberating the platelets. (Intermediate stages between the hematopoietic stem cell and megakaryoblast are not illustrated.)

16.6 Hemostasis prevents blood loss

→ Learning Objectives

- ☐ **Describe the process of hemostasis. List factors that limit clot formation and prevent undesirable clotting.**
- ☐ **Give examples of hemostatic disorders. Indicate the cause of each condition.**

Normally, blood flows smoothly past the intact blood vessel lining (endothelium). But if a blood vessel wall breaks, a whole series of reactions is set in motion to accomplish **hemostasis** (he″mo-sta′sis), which stops the bleeding (*stasis* = halting). Without this plug-the-hole defensive reaction, we would quickly bleed out our entire blood volume from even the smallest cuts.

The hemostasis response is fast, localized, and carefully controlled. It involves many *clotting factors* normally present in plasma as well as several substances that are released by platelets and injured tissue cells. During hemostasis, three steps occur in rapid sequence (**Figure 16.13**): ① vascular spasm, ② platelet plug formation, and ③ coagulation (blood clotting). Following hemostasis, the clot retracts. It then dissolves as it is replaced by fibrous tissue that permanently prevents blood loss.

Step 1: Vascular Spasm

In the first step, the damaged blood vessels respond to injury by constricting (vasoconstriction) (Figure 16.13 ①). Factors that trigger this **vascular spasm** include direct injury to vascular smooth muscle, chemicals released by endothelial cells and platelets, and reflexes initiated by local pain receptors. The spasm mechanism becomes more and more efficient as the amount of tissue damage increases, and is most effective in the smaller blood vessels. The spasm response is valuable because a strongly constricted artery can significantly reduce blood loss for 20–30 minutes, allowing time for the next two steps to occur.

Step 2: Platelet Plug Formation

In the second step, platelets play a key role in hemostasis by aggregating (sticking together), forming a plug that temporarily seals the break in the vessel wall (Figure 16.13 ②). They also help orchestrate subsequent events that form a blood clot.

As a rule, platelets do not stick to each other or to the smooth endothelial linings of blood vessels. Intact endothelial cells release nitric oxide and a prostaglandin called **prostacyclin** (or *PGI_2*). Both chemicals prevent platelet aggregation in undamaged tissue and restrict aggregation to the site of injury.

However, when the endothelium is damaged and the underlying collagen fibers are exposed, platelets adhere tenaciously to the collagen fibers. A large plasma protein called *von Willebrand factor* stabilizes bound platelets by forming a bridge between collagen and platelets. Platelets become activated: They swell, form spiked processes, and become stickier. In addition, they release chemical messengers including the following:

- **Adenosine diphosphate (ADP)**—a potent aggregating agent that causes more platelets to stick to the area and release their contents

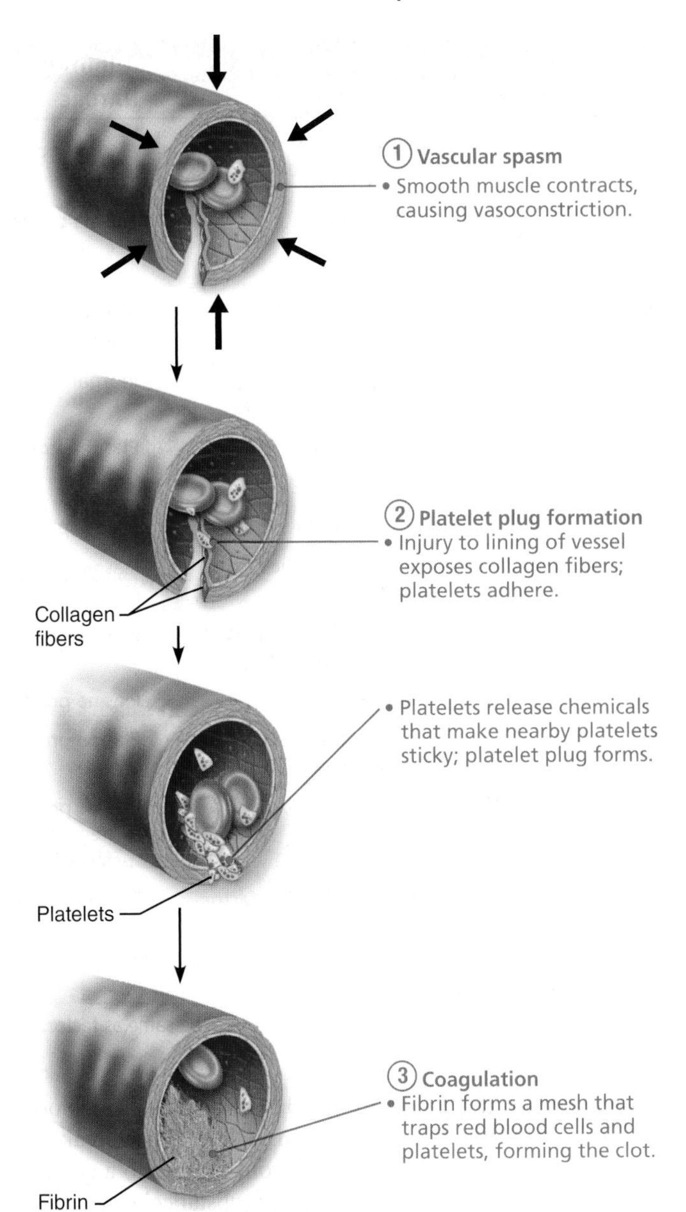

Figure 16.13 **Events of hemostasis.**

- **Serotonin** and **thromboxane A_2** (throm-boks′ān; a short-lived prostaglandin derivative)—messengers that enhance vascular spasm and platelet aggregation

As more platelets aggregate, they release more chemicals, aggregating more platelets, and so on, in a positive feedback cycle (see Figure 1.6 on p. 11). Within one minute, a platelet plug is built up, further reducing blood loss. Platelets alone are sufficient for sealing the thousands of minute rips and holes that occur unnoticed as part of the daily wear and tear in your smallest blood vessels. Because platelet plugs are loosely knit, larger breaks need additional reinforcement.

Step 3: Coagulation

The third step, **coagulation** or **blood clotting**, reinforces the platelet plug with fibrin threads that act as a "molecular glue" for

- Triggered by negatively charged surfaces such as activated platelets, collagen, or glass. (This is why this pathway can initiate clotting in a test tube.)
- Slower because it has many intermediate steps.

The *extrinsic pathway* is:

- Called *extrinsic* because the tissue factor it requires is *outside* of blood.
- Triggered by exposing blood to a factor found in tissues underneath the damaged endothelium. This factor is called **tissue factor (TF)** or **factor III**.
- Faster because it bypasses several steps of the intrinsic pathway. In severe tissue trauma, it can form a clot in 15 seconds.

Phase 1 ends with the formation of a complex substance called *prothrombin activator*.

Phase 2: Common Pathway to Thrombin

Prothrombin activator catalyzes the conversion of a plasma protein called **prothrombin** into the active enzyme **thrombin**.

Phase 3: Common Pathway to the Fibrin Mesh

The end point of phase 3 is a *fibrin mesh* that traps blood cells and effectively seals the hole until the blood vessel can be permanently repaired. Thrombin catalyzes the transformation of the *soluble* clotting factor **fibrinogen** into **fibrin**. The fibrin molecules then polymerize (join together) to form long, hair-like, *insoluble* fibrin strands. (Notice that, unlike other clotting factors, activating fibrinogen does not convert it into an enzyme, but instead allows it to polymerize.) The fibrin strands glue the platelets together and make a web that forms the structural basis of the clot. Fibrin makes the liquid plasma become gel-like and traps formed elements that try to pass through it (Figure 16.15).

In the presence of calcium ions, thrombin also activates **factor XIII (fibrin stabilizing factor)**, a cross-linking enzyme that binds the fibrin strands tightly together, forming a fibrin mesh. Cross-linking further strengthens and stabilizes the clot, effectively sealing the hole until the blood vessel can be permanently repaired.

Role of Anticoagulants

Factors that inhibit clotting are called **anticoagulants**. Whether or not blood clots depends on a delicate balance between clotting factors and anticoagulants. Normally, anticoagulants dominate and prevent clotting, but when a vessel is ruptured, clotting factor activity in that area increases dramatically and a clot begins to form. Clot formation is normally complete within 3 to 6 minutes after blood vessel damage.

Clot Retraction and Fibrinolysis

Although the process of hemostasis is complete when the fibrin mesh is formed, there are still things that need to be done to stabilize the clot and then remove it when the injury is healed.

Clot Retraction

Within 30 to 60 minutes, a platelet-induced process called **clot retraction** further stabilizes the clot. Platelets contain contractile proteins (actin and myosin), and they contract in much the same manner as smooth muscle cells. As the platelets contract, they pull on the surrounding fibrin strands, squeezing **serum** (plasma minus the clotting proteins) from the mass, compacting the clot and drawing the ruptured edges of the blood vessel more closely together.

Even as clot retraction is occurring, the vessel is healing. **Platelet-derived growth factor (PDGF)** released by platelets stimulates smooth muscle cells and fibroblasts to divide and rebuild the vessel wall. As fibroblasts form a connective tissue patch in the injured area, endothelial cells, stimulated by vascular endothelial growth factor (VEGF), multiply and restore the endothelial lining.

Fibrinolysis

A clot is not a permanent solution to blood vessel injury, and a process called **fibrinolysis** removes unneeded clots when healing has occurred. This cleanup detail is crucial because small clots form continually in vessels throughout the body. Without fibrinolysis, blood vessels would gradually become completely blocked.

The critical natural "clot buster" is a fibrin-digesting enzyme called **plasmin**, which is produced when the plasma protein **plasminogen** is activated. Large amounts of plasminogen are incorporated into a forming clot, where it remains inactive until appropriate signals reach it. The presence of a clot in and around the blood vessel causes the endothelial cells to secrete **tissue plasminogen activator (tPA).** Activated factor XII and

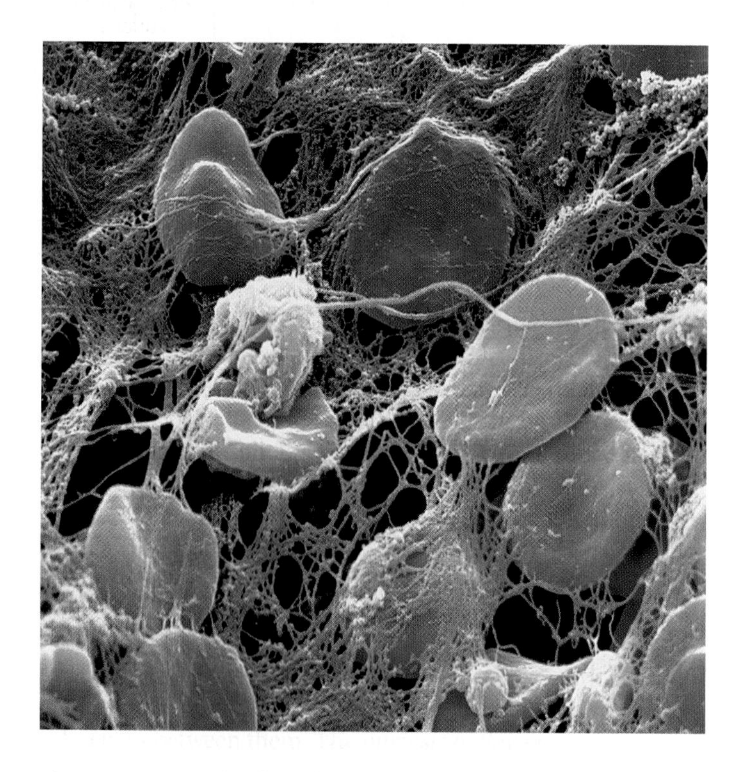

Figure 16.15 Scanning electron micrograph of erythrocytes trapped in a fibrin mesh. (2700×).

thrombin released during clotting also activate plasminogen. As a result, most plasmin activity is confined to the clot, and circulating enzymes quickly destroy any plasmin that strays into the plasma. Fibrinolysis begins within two days and continues slowly over several days until the clot finally dissolves.

Factors Limiting Clot Growth or Formation

Factors Limiting Normal Clot Growth

Once the clotting cascade has begun, it continues until a clot forms. Normally, two homeostatic mechanisms prevent clots from becoming unnecessarily large: (1) swift removal of clotting factors, and (2) inhibition of activated clotting factors. For clotting to occur, the concentration of activated clotting factors must reach certain critical levels. Clots do not usually form in rapidly moving blood because the activated clotting factors are washed away and diluted. For the same reasons, a clot stops growing when it contacts blood flowing normally.

Other mechanisms block the final step in which fibrinogen is polymerized into fibrin. They work by restricting thrombin to the clot or by inactivating it if it escapes into the general circulation. As a clot forms, almost all of the thrombin produced is bound onto the fibrin threads. This is an important safeguard because thrombin also exerts positive feedback effects on the coagulation process prior to the common pathway. Not only does it speed up the production of prothrombin activator by acting indirectly through factor V, but it also accelerates the earliest steps of the intrinsic pathway by activating platelets. By binding thrombin, fibrin effectively acts as an anticoagulant, preventing the clot from enlarging and thrombin from acting elsewhere.

Antithrombin III, a protein present in plasma, quickly inactivates any thrombin not bound to fibrin. Antithrombin III and **protein C**, another protein produced in the liver, also inhibit the activity of other intrinsic pathway clotting factors.

Heparin, the natural anticoagulant contained in basophil and mast cell granules, is also found on the surface of endothelial cells. It inhibits thrombin by enhancing the activity of antithrombin III. Like most other clotting inhibitors, heparin also inhibits the intrinsic pathway.

Factors Preventing Undesirable Clotting

As long as the endothelium is smooth and intact, platelets are prevented from clinging and piling up. Also, antithrombic substances—nitric oxide and prostacyclin—secreted by the endothelial cells normally prevent platelet adhesion. Additionally, vitamin E quinone, a molecule formed in the body when vitamin E reacts with oxygen, is a potent anticoagulant.

CLINICAL

Disorders of Hemostasis

Blood clotting is one of nature's most elegant creations, but it sometimes goes awry. The two major disorders of hemostasis are at opposite poles. **Thromboembolic disorders** result from conditions that cause undesirable clot formation. **Bleeding disorders** arise from abnormalities that prevent normal clot formation. **Disseminated intravascular coagulation (DIC)**, which has characteristics of both types of disorder, involves both widespread clotting and severe bleeding.

Thromboembolic Disorders

Despite the body's many safeguards, undesirable intravascular clotting sometimes occurs.

Thrombi and Emboli A clot that develops and persists in an *unbroken* blood vessel is called a **thrombus**. If the thrombus is large enough, it may block circulation to the cells beyond the occlusion and lead to death of those tissues. For example, if the blockage occurs in the coronary circulation of the heart (coronary thrombosis), the consequences may be death of heart muscle and a fatal heart attack.

If the thrombus breaks away from the vessel wall and floats freely in the bloodstream, it becomes an **embolus** (plural: *emboli*). An embolus ("wedge") is usually no problem until it encounters a blood vessel too narrow for it to pass through. Then it becomes an **embolism**, obstructing the vessel. For example, emboli that become trapped in the lungs (pulmonary embolisms) dangerously impair the body's ability to obtain oxygen. A cerebral embolism may cause a stroke.

Conditions that roughen the vessel endothelium, such as atherosclerosis or inflammation, cause thromboembolic disease by allowing platelets to gain a foothold. Slowly flowing blood or blood stasis is another risk factor, particularly in bedridden patients and those taking a long flight without moving around. In this case, clotting factors are not washed away as usual and accumulate, allowing clots to form.

Anticoagulant Drugs A number of drugs—most importantly aspirin, heparin, and warfarin—are used clinically to prevent undesirable clotting. **Aspirin** is an antiprostaglandin drug that inhibits thromboxane A_2 formation (blocking platelet aggregation and platelet plug formation). Clinical studies of men taking low-dose aspirin over several years demonstrated a 50% reduction in incidence of heart attack.

Other medications prescribed as anticoagulants are heparin (see above) and warfarin. Administered in injectable form, heparin is the anticoagulant most used in the hospital (for preoperative and postoperative heart patients and for those receiving blood transfusions). Taken orally, **warfarin** (Coumadin) is a mainstay of outpatient treatment to reduce the risk of stroke in those prone to atrial fibrillation, a condition in which blood pools in the heart. Warfarin works via a different mechanism than heparin—it interferes with the action of vitamin K in the production of some clotting factors (see Impaired Liver Function below). Because treatment with warfarin is difficult to manage, the introduction of new oral anticoagulants using other mechanisms has been welcomed.

Other drugs, including tPA, can dissolve blood clots, and there are innovative medical technologies for treating clots.

Bleeding Disorders

Anything that interferes with the clotting mechanism can result in abnormal bleeding. The most common causes are platelet deficiency (thrombocytopenia) and deficits of some clotting

7. The blood cell that can become an antibody-secreting cell is the **(a)** lymphocyte, **(b)** megakaryocyte, **(c)** neutrophil, **(d)** basophil.
8. Which of the following does not promote multiple steps in the clotting pathway? **(a)** platelet phospholipids, **(b)** factor XI, **(c)** thrombin, **(d)** Ca^{2+}.
9. The normal pH of the blood is about **(a)** 8.4, **(b)** 7.8, **(c)** 7.4, **(d)** 4.7.
10. Suppose your blood is AB positive. This means that **(a)** agglutinogens A and B are present on your red blood cells, **(b)** there are no anti-A or anti-B antibodies in your plasma, **(c)** your blood is Rh^+, **(d)** all of the above.

Short Answer Essay Questions

11. (a) Define formed elements and list their three major categories. (b) Which is least numerous? (c) Which comprise(s) the buffy coat in a hematocrit tube?
12. Discuss hemoglobin relative to its chemical structure, its function, and the color changes it undergoes during loading and unloading of oxygen.
13. If you had a high hematocrit, would you expect your hemoglobin determination to be low or high? Why?
14. What nutrients are needed for erythropoiesis?
15. (a) Describe the process of erythropoiesis. (b) What name is given to the immature cell type released to the circulation? (c) How does it differ from a mature erythrocyte?
16. Besides the ability to move by amoeboid motion, what other physiological attributes contribute to the function of white blood cells in the body?
17. (a) If you had a severe infection, would you expect your WBC count to be closest to 5000, 10,000, or 15,000/μl? (b) What is this condition called?
18. (a) Describe the appearance of platelets and state their major function. (b) Why should platelets not be called "cells"?
19. (a) Define hemostasis. (b) List the three major phases of coagulation. Explain what initiates each phase and what the phase accomplishes. (c) In what general way do the intrinsic and extrinsic mechanisms of clotting differ? (d) Which ion is essential to virtually all stages of coagulation?
20. (a) Define fibrinolysis. (b) What is the importance of this process?
21. (a) How is clot overgrowth usually prevented? (b) List two conditions that may lead to unnecessary (and undesirable) clot formation.
22. How can liver dysfunction cause bleeding disorders?
23. (a) What is a transfusion reaction and why does it happen? (b) What are its possible consequences?
24. How can poor nutrition lead to anemia?

AT THE CLINIC

Clinical Case Study

Blood

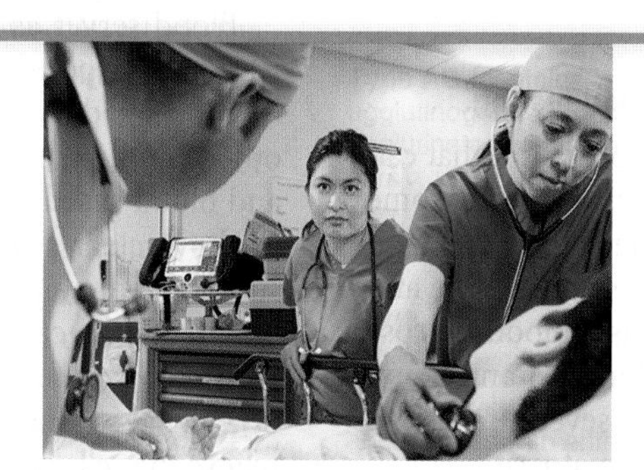

Earl Malone is a 20-year-old passenger on the bus that crashed on Route 91. Upon arrival at the scene, paramedics make the following observations:

- Right upper quadrant (abdominal) pain
- Cyanotic
- Cool and clammy skin
- Blood pressure 100/60 and falling, pulse 100

Paramedics start an IV to rapidly infuse a 0.9% sodium chloride solution (normal saline). They transport him to a small rural hospital where Mr. Malone's blood pressure continues to fall and his cyanosis worsens. The local physician begins infusing O negative packed red blood cells (PRBCs) and arranges transport by helicopter to a trauma center. She sends additional PRBC units in the helicopter for transfusion en route. After arrival at the trauma center, the following notes were added to Mr. Malone's chart:

- Abdomen firm and distended
- Blood drawn for typing and cross matching; packed A positive blood cells infused
- Emergency FAST (focused assessment with sonography for trauma) ultrasound is positive for intraperitoneal fluid

A positive FAST scan indicates intra-abdominal bleeding. Mr. Malone's condition continues to deteriorate, so he is prepared for surgery, which reveals a lacerated liver. The laceration is repaired, and Mr. Malone's vital signs stabilize.

1. Mr. Malone was going into shock because of blood loss, so paramedics infused a saline solution. Why would this help?
2. Mr. Malone was switched from saline to PRBCs. What problem does infusion of PRBCs address that the saline solution could not?
3. Why was the physician able to use O negative blood before the results of the blood type tests were obtained?
4. Mr. Malone's blood type was determined to be A positive. What plasma antibodies (agglutinins) does he have, and what type of blood can he receive?
5. What would happen if doctors had infused type B PRBCs into Mr. Malone's circulation?

For answers, see Answers Appendix.

17 The Cardiovascular System: The Heart

KEY CONCEPTS

Our ceaselessly beating heart has intrigued people for centuries. The ancient Greeks believed the heart was the seat of intelligence. Others thought it was the source of emotions. While these ideas have proved false, we do know that emotions affect heart rate. When your heart pounds or skips a beat, you become acutely aware of how much you depend on this dynamic organ for your very life.

Despite its vital importance, the heart does not work alone. Indeed, it is only part of the cardiovascular system, which includes the miles of blood vessels that run through your body. Day and night, tissue cells take in nutrients and oxygen and excrete wastes. Cells can make such exchanges only with their immediate environment, so some means of changing and renewing that environment is necessary to ensure a continual supply of nutrients and prevent a buildup of wastes. The cardiovascular system provides the transport system "hardware" that keeps blood continuously circulating to fulfill this critical homeostatic need.

17.1 The heart has four chambers and pumps blood through the pulmonary and systemic circuits

→ **Learning Objectives**

- ☐ **Describe the size, shape, location, and orientation of the heart in the thorax.**
- ☐ **Name the coverings of the heart.**
- ☐ **Describe the structure and function of each of the three layers of the heart wall.**
- ☐ **Describe the structure and functions of the four heart chambers. Name each chamber and provide the name and general route of its associated great vessel(s).**

The Pulmonary and Systemic Circuits

Stripped of its romantic cloak, the **heart** is no more than the transport system pump, and the blood vessels are the delivery routes. In fact, the heart is actually two pumps side by side (**Figure 17.1**).

- The *right side* of the heart receives oxygen-poor blood from body tissues and then pumps this blood to the lungs to pick up oxygen and dispel carbon dioxide. The blood vessels that carry blood to and from the lungs form the **pulmonary circuit** (*pulmo* = lung).
- The *left side* of the heart receives the oxygenated blood returning from the lungs and pumps this blood throughout the body to supply oxygen and nutrients to body tissues. The blood vessels that carry blood to and from all body tissues form the **systemic circuit**.

The heart has two receiving chambers, the *right atrium* and *left atrium*, that receive blood returning from the systemic and pulmonary circuits. The heart also has two main pumping chambers, the *right ventricle* and *left ventricle*, that pump blood around the two circuits.

Electron micrograph of a mitral valve of the human heart.

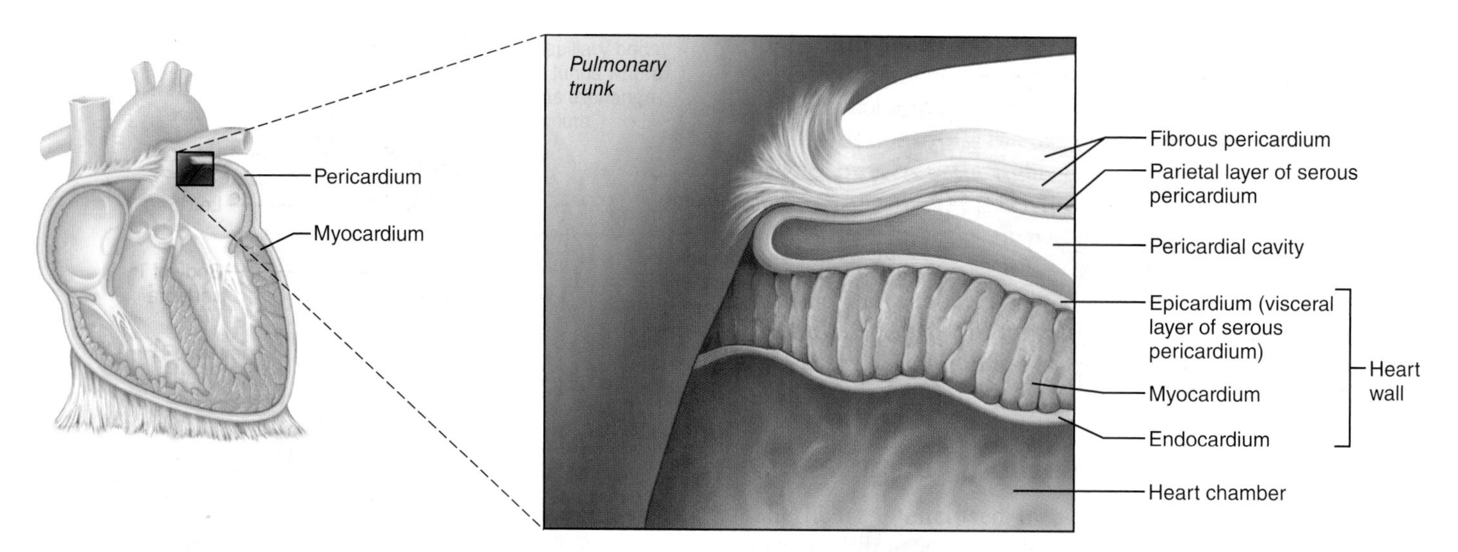

Figure 17.3 The layers of the pericardium and of the heart wall.

The third layer of the heart wall, the **endocardium** ("inside the heart"), is a glistening white sheet of endothelium (squamous epithelium) resting on a thin connective tissue layer. Located on the inner myocardial surface, it lines the heart chambers and covers the fibrous skeleton of the valves. The endocardium is continuous with the endothelial linings of the blood vessels leaving and entering the heart.

Chambers and Associated Great Vessels

The heart has four chambers (**Figure 17.5e** on p. 585)—two superior **atria** (a′tre-ah) and two inferior **ventricles** (ven′trĭ-klz). The internal partition that divides the heart longitudinally is called the **interatrial septum** where it separates the atria, and the **interventricular septum** where it separates the ventricles.

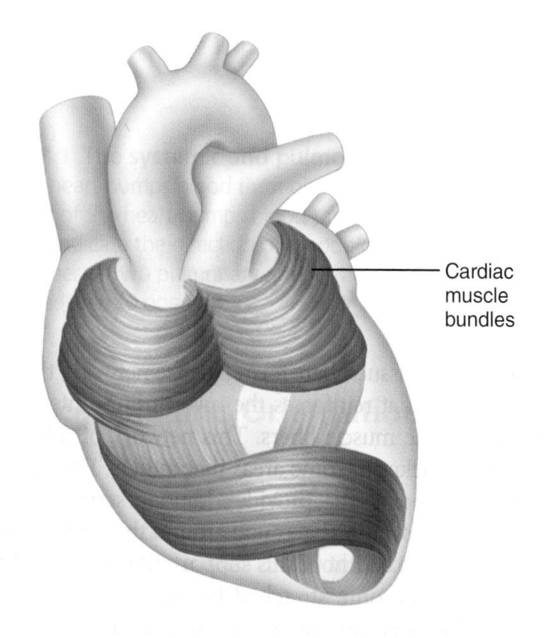

Figure 17.4 The circular and spiral arrangement of cardiac muscle bundles in the myocardium of the heart.

The right ventricle forms most of the anterior surface of the heart. The left ventricle dominates the inferoposterior aspect of the heart and forms the heart apex.

Two grooves visible on the heart surface indicate the boundaries of its four chambers and carry the blood vessels supplying the myocardium. The **coronary sulcus** (Figure 17.5b, d), or *atrioventricular groove*, encircles the junction of the atria and ventricles like a crown (*corona* = crown). The **anterior interventricular sulcus**, cradling the anterior interventricular artery, marks the anterior position of the septum separating the right and left ventricles. It continues as the **posterior interventricular sulcus**, which provides a similar landmark on the heart's posteroinferior surface.

Atria: The Receiving Chambers

Except for small, wrinkled, protruding appendages called **auricles** (or′ĭ-klz; *auricle* = little ear), which increase the atrial volume somewhat, the right and left atria are remarkably free of distinguishing surface features. Internally, the right atrium has two basic parts (Figure 17.5c): a smooth-walled posterior part and an anterior portion in which bundles of muscle tissue form ridges in the walls. These muscle bundles are called **pectinate muscles** because they look like the teeth of a comb (*pectin* = comb). The posterior and anterior regions of the right atrium are separated by a C-shaped ridge called the *crista terminalis* ("terminal crest").

In contrast, the left atrium is mostly smooth and pectinate muscles are found only in the auricle. The interatrial septum bears a shallow depression, the **fossa ovalis** (o-vă′lis), that marks the spot where an opening, the *foramen ovale*, existed in the fetal heart (Figure 17.5c, e).

Functionally, the atria are receiving chambers for blood returning to the heart from the circulation (*atrium* = entryway). The atria are relatively small, thin-walled chambers because they need to contract only minimally to push blood "downstairs" into the ventricles. They contribute little to the propulsive pumping activity of the heart.

Blood enters the *right atrium* via three veins (Figure 17.5c–e):

- The **superior vena cava** returns blood from body regions superior to the diaphragm.

(Text continues on p. 586.)

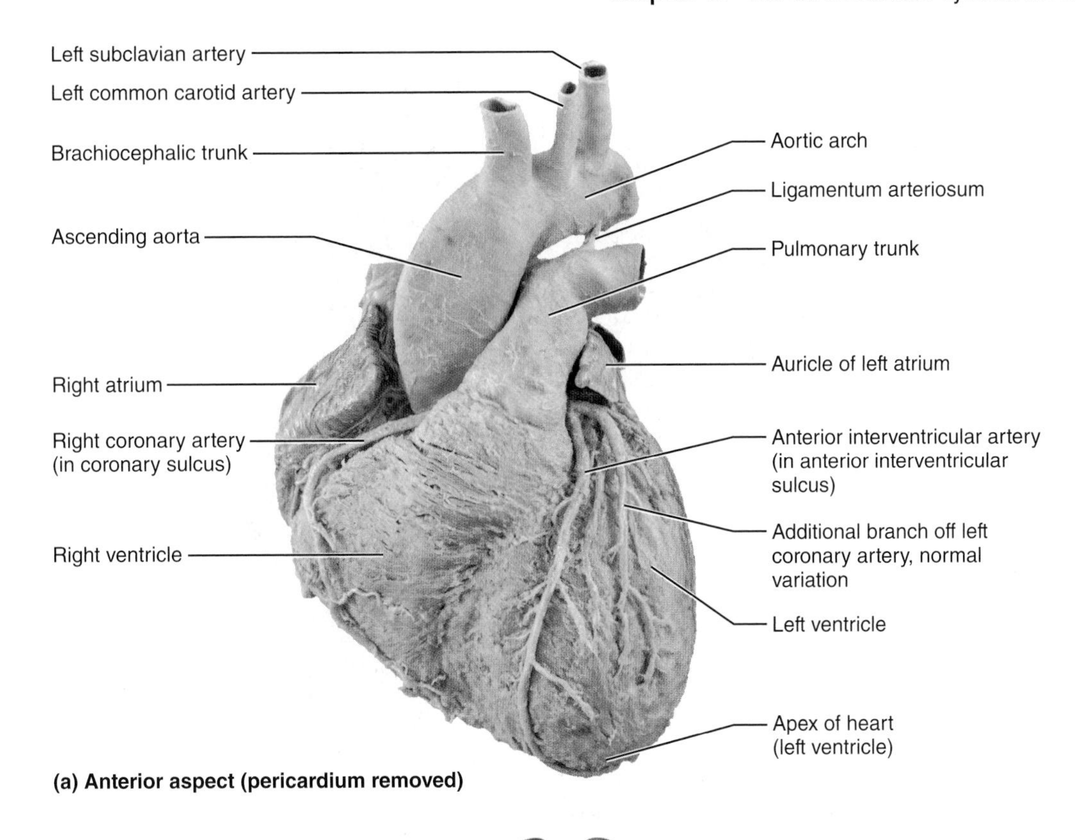

(a) Anterior aspect (pericardium removed)

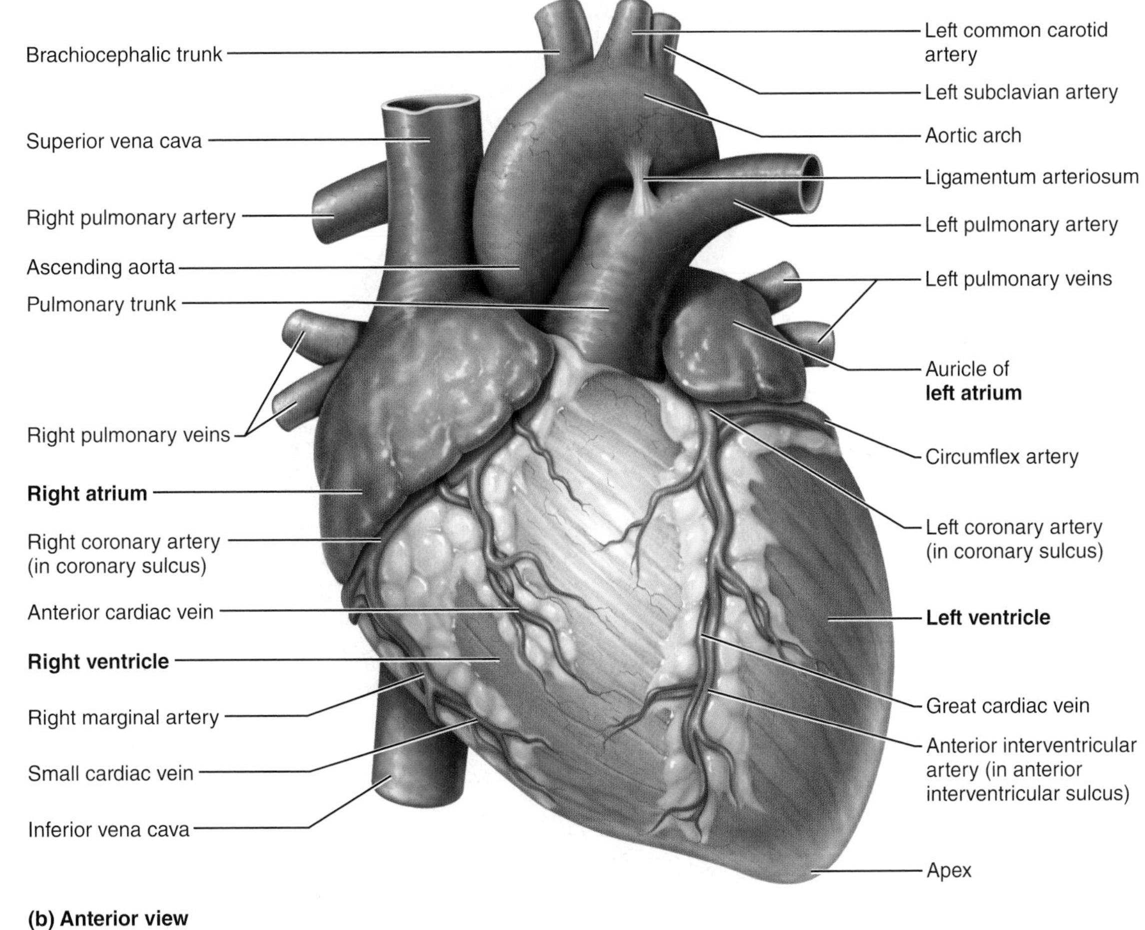

(b) Anterior view

Figure 17.5 Gross anatomy of the heart. In diagrammatic views, vessels transporting oxygen-rich blood are red; those transporting oxygen-poor blood are blue.

Practice art labeling
MasteringA&P®>Study Area>Chapter 17

17

- The **inferior vena cava** returns blood from body areas below the diaphragm.
- The **coronary sinus** collects blood draining from the myocardium.

Four **pulmonary veins** enter the *left atrium*, which makes up most of the heart's base. These veins, which transport blood from the lungs back to the heart, are best seen in a posterior view (Figure 17.5d).

Ventricles: The Discharging Chambers

Together the ventricles (*ventr* = underside) make up most of the volume of the heart. As already mentioned, the right ventricle forms most of the heart's anterior surface and the left ventricle dominates its posteroinferior surface. Irregular ridges of muscle called **trabeculae carneae** (trah-bek′u-le kar′ne-e; "crossbars of flesh") mark the internal walls of the ventricular chambers. Other muscle bundles, the **papillary muscles**, which play a role in valve function, project into the ventricular cavity (Figure 17.5e).

The ventricles are the discharging chambers, the actual pumps of the heart. Their walls are much more massive than the atrial walls, reflecting the difference in function between the atria and ventricles (Figure 17.5e and f). When the ventricles contract, they propel blood out of the heart into the circulation. The right ventricle pumps blood into the **pulmonary trunk**, which routes the blood to the lungs where gas exchange occurs. The left ventricle ejects blood into the **aorta** (a-or′tah), the largest artery in the body.

☑ Check Your Understanding

1. The heart is in the mediastinum. Just what is the mediastinum?
2. From inside to outside, list the layers of the heart wall and the coverings of the heart.
3. What is the purpose of the serous fluid inside the pericardial cavity?

For answers, see Answers Appendix.

17

17.2 Heart valves make blood flow in one direction

→ Learning Objective

☐ **Name the heart valves and describe their location, function, and mechanism of operation.**

Blood flows through the heart in one direction: from atria to ventricles and out the great arteries leaving the superior aspect of the heart. Four valves enforce this one-way traffic (Figure 17.5e and **Figure 17.6**). They open and close in response to differences in blood pressure on their two sides.

Atrioventricular (AV) Valves

The two **atrioventricular (AV) valves**, one located at each atrial-ventricular junction, prevent backflow into the atria when the ventricles contract.

- The right AV valve, the **tricuspid valve** (tri-kus′pid), has three flexible cusps (flaps of endocardium reinforced by connective tissue cores).
- The left AV valve, with two cusps, is called the **mitral valve** (mi′tral) because it resembles the two-sided bishop's miter (tall, pointed hat). It is sometimes called the *bicuspid valve*.

Attached to each AV valve flap are tiny white collagen cords called **chordae tendineae** (kor′de ten″dĭ′ne-e; "tendinous cords"), "heart strings" which anchor the cusps to the papillary muscles protruding from the ventricular walls (Figure 17.6c, d).

When the heart is completely relaxed, the AV valve flaps hang limply into the ventricular chambers below. During this time, blood flows into the atria and then through the open AV valves into the ventricles (**Figure 17.7a**). When the ventricles contract, compressing the blood in their chambers, the intraventricular pressure rises, forcing the blood superiorly against the valve flaps. As a result, the flap edges meet, closing the valve (Figure 17.7b).

The chordae tendineae and the papillary muscles serve as guy-wires that anchor the valve flaps in their *closed* position. If the cusps were not anchored, they would be blown upward (everted) into the atria, in the same way an umbrella is blown inside out by a gusty wind. The papillary muscles contract with the other ventricular musculature so that they take up the slack on the chordae tendineae as the full force of ventricular contraction hurls the blood against the AV valve flaps.

Semilunar (SL) Valves

The **aortic** and **pulmonary (semilunar, SL) valves** guard the bases of the large arteries issuing from the ventricles (aorta and pulmonary trunk, respectively) and prevent backflow into the associated ventricles. Each SL valve is fashioned from three pocketlike cusps, each shaped roughly like a crescent moon (*semilunar* = half-moon).

Like the AV valves, the SL valves open and close in response to differences in pressure. When the ventricles contract and intraventricular pressure rises above the pressure in the aorta and pulmonary trunk, the SL valves are forced open and their cusps flatten against the arterial walls as blood rushes past them (**Figure 17.8a**, p. 589). When the ventricles relax, and the blood flows backward toward the heart, it fills the cusps and closes the valves (Figure 17.8b).

We complete the valve story by noting what seems to be an important omission—there are no valves guarding the entrances of the venae cavae and pulmonary veins into the right and left atria, respectively. Small amounts of blood *do* spurt back into these vessels during atrial contraction, but backflow is minimal because of the inertia of the blood and because as it contracts, the atrial myocardium compresses (and collapses) these venous entry points.

HOMEOSTATIC IMBALANCE 17.2 — CLINICAL

Heart valves are simple devices, and the heart—like any mechanical pump—can function with "leaky" valves as long as the impairment is not too great. However, severe valve deformities can seriously hamper cardiac function.

An *incompetent*, or *insufficient*, *valve* forces the heart to repump the same blood over and over because the valve does not close properly and blood backflows. In valvular *stenosis* ("narrowing"), the valve flaps become stiff (typically due to calcium

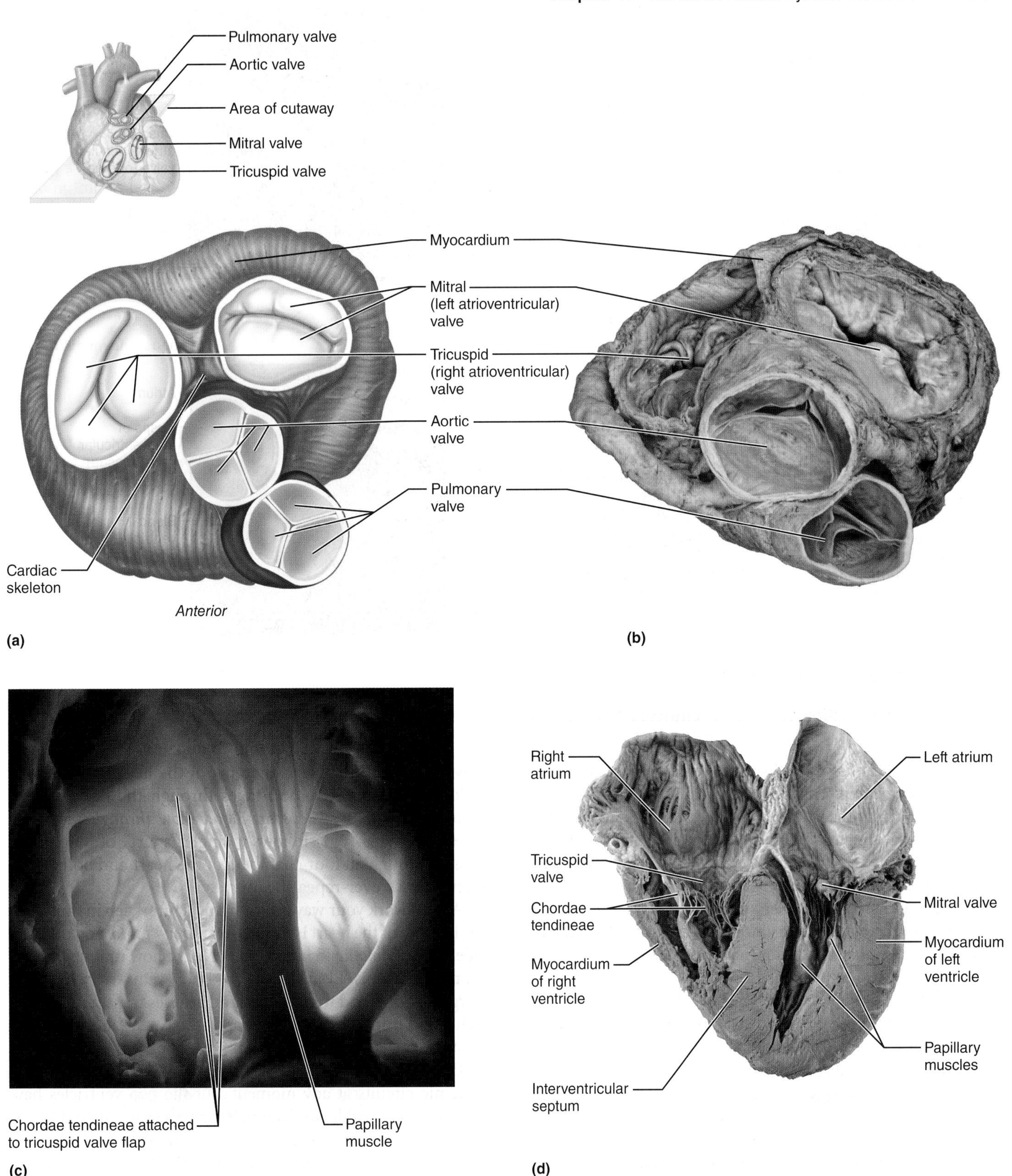

Figure 17.6 Heart valves. (a) Superior view of the two sets of heart valves (atria removed). The paired atrioventricular valves are located between atria and ventricles; the two semilunar valves are located at the junction of the ventricles and the arteries issuing from them. **(b)** Photograph of the heart valves, superior view. **(c)** Photograph of the tricuspid valve. This bottom-to-top view shows the valve as seen from the right ventricle. **(d)** Frontal section of the heart. (For related images, see *A Brief Atlas of the Human Body*, Figures 58, 60, and 61.)

Explore human cadaver
MasteringA&P®Study Area>PAL

FOCUS Blood Flow through the Heart

Focus Figure 17.1 The heart is a double pump, each side supplying its own circuit.

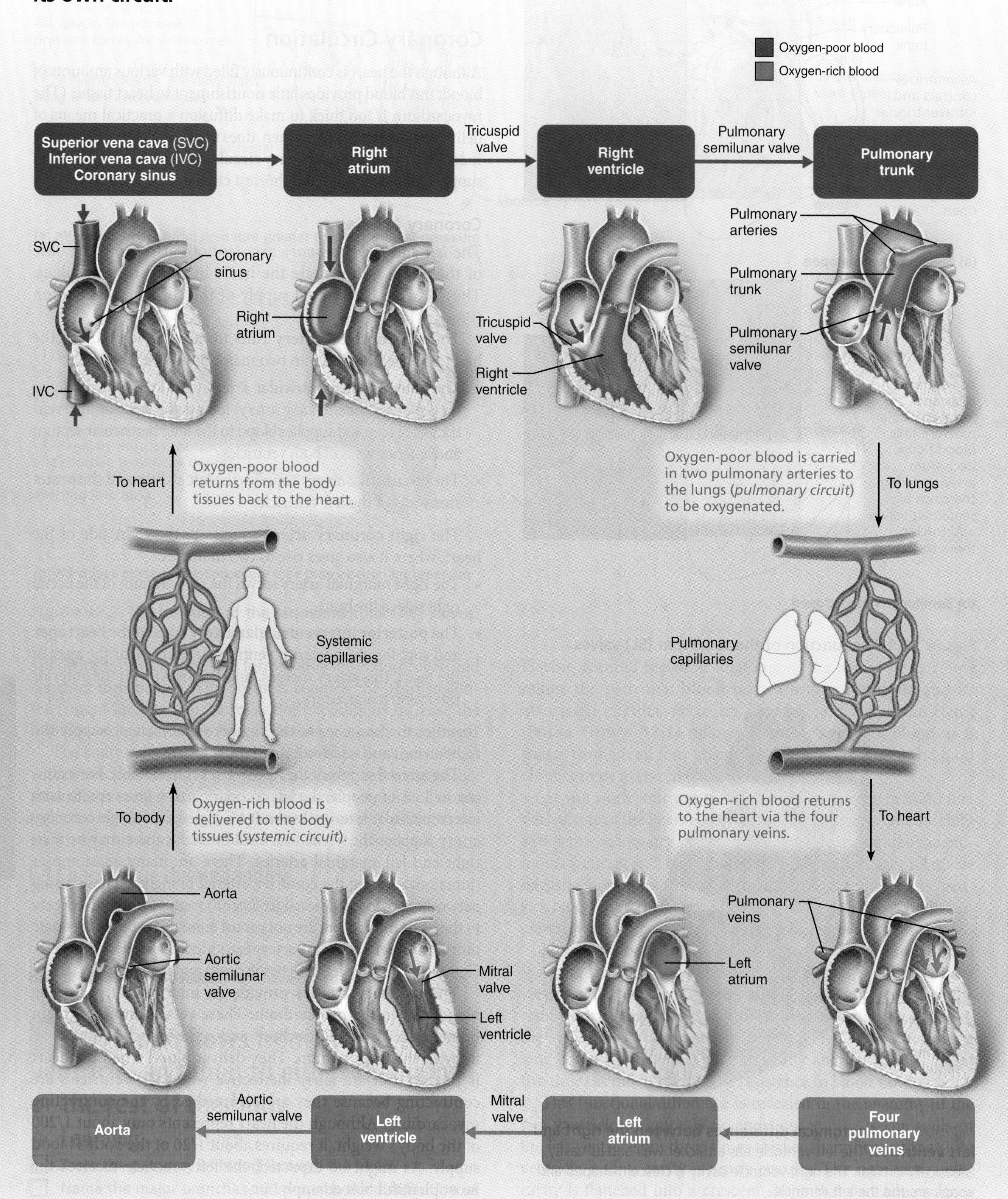

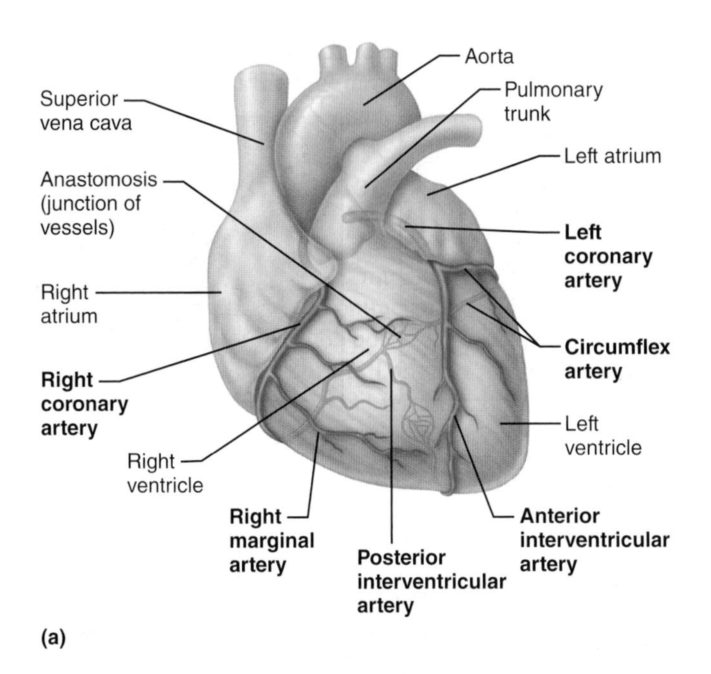

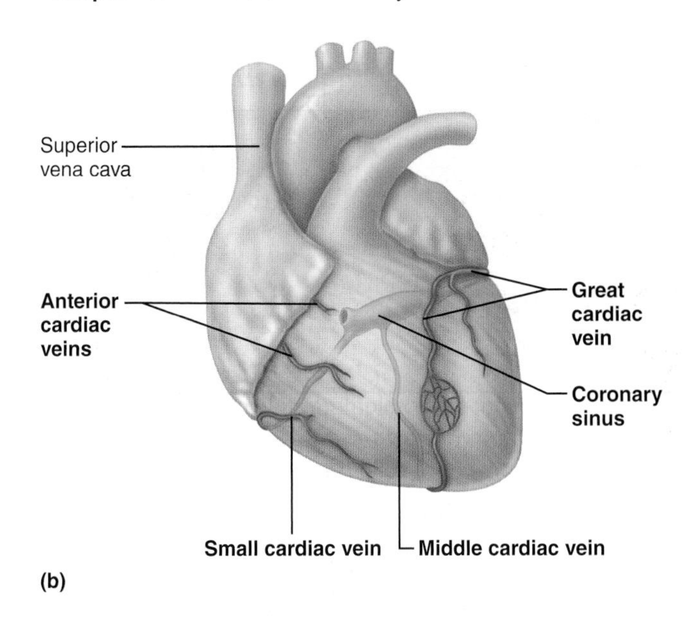

Figure 17.10 Coronary circulation. (a) The major coronary arteries. Lighter-tinted vessels are more posterior in the heart. **(b)** The major cardiac veins. Lighter-tinted vessels are more posterior in the heart.

Coronary Veins

After passing through the capillary beds of the myocardium, the venous blood is collected by the **cardiac veins**, whose paths roughly follow those of the coronary arteries. These veins join to form an enlarged vessel called the **coronary sinus**, which empties the blood into the right atrium. The coronary sinus is obvious on the posterior aspect of the heart (Figure 17.10b).

The sinus has three large tributaries: the **great cardiac vein** in the anterior interventricular sulcus; the **middle cardiac vein** in the posterior interventricular sulcus; and the **small cardiac vein**, running along the heart's right inferior margin. Additionally, several **anterior cardiac veins** empty directly into the right atrium anteriorly.

HOMEOSTATIC IMBALANCE 17.3 — CLINICAL

Blockage of the coronary arterial circulation can be serious and sometimes fatal. **Angina pectoris** (an-ji′nah pek′tor-is; "choked chest") is thoracic pain caused by a fleeting deficiency in blood delivery to the myocardium. It may result from stress-induced spasms of the coronary arteries or from increased physical demands on the heart. The myocardial cells are weakened by the temporary lack of oxygen but do not die.

Prolonged coronary blockage is far more serious because it can lead to a **myocardial infarction (MI)**, commonly called a **heart attack**, in which cells *do* die. Since adult cardiac muscle is essentially amitotic, most of the dead tissue is replaced with noncontractile scar tissue. Whether or not a person survives a myocardial infarction depends on the extent and location of the damage. Damage to the left ventricle—the systemic pump—is most serious. ✚

☑ Check Your Understanding

6. Which side of the heart acts as the pulmonary pump? The systemic pump?
7. Which of the following statements are true? (a) The left ventricle wall is thicker than the right ventricle wall. (b) The left ventricle pumps blood at a higher pressure than the right ventricle. (c) The left ventricle pumps more blood with each beat than the right ventricle. Explain.
8. Name the two main branches of the right coronary artery.

For answers, see Answers Appendix.

17.4 Intercalated discs connect cardiac muscle fibers into a functional syncytium

→ Learning Objectives

- ☐ **Describe the structural and functional properties of cardiac muscle, and explain how it differs from skeletal muscle.**
- ☐ **Briefly describe the events of excitation-contraction coupling in cardiac muscle cells.**

Although similar to skeletal muscle, cardiac muscle displays some special anatomical features that reflect its unique blood-pumping role.

Microscopic Anatomy

Like skeletal muscle, **cardiac muscle** is striated and contracts by the sliding filament mechanism. However, in contrast to the long, cylindrical, multinucleate skeletal muscle fibers, cardiac cells are short, fat, branched, and interconnected. Each fiber contains one or at most two large, pale, *centrally* located nuclei (**Figure 17.11a**). The intercellular spaces are filled with a loose connective tissue matrix (the *endomysium*) containing numerous capillaries. This delicate matrix is connected to the fibrous cardiac skeleton, which acts both as a tendon and as an insertion, giving the cardiac cells something to pull or exert their force against.

Check Your Understanding

9. For each of the following, state whether it applies to skeletal muscle, cardiac muscle, or both: (a) refractory period is almost as long as the contraction; (b) source of Ca^{2+} for contraction is *only* SR; (c) has troponin; (d) has triads.

For answers, see Answers Appendix.

17.5 Pacemaker cells trigger action potentials throughout the heart

→ Learning Objectives

- ☐ **Describe and compare action potentials in cardiac pacemaker and contractile cells.**
- ☐ **Name the components of the conduction system of the heart, and trace the conduction pathway.**
- ☐ **Draw a diagram of a normal electrocardiogram tracing. Name the individual waves and intervals, and indicate what each represents. Name some abnormalities that can be detected on an ECG tracing.**

Although the ability of the heart to depolarize and contract is intrinsic (no nerves required), the healthy heart *is* supplied with autonomic nerve fibers that alter its basic rhythm. In this module, we examine how the basic rhythm is generated and modified.

Setting the Basic Rhythm: The Intrinsic Conduction System

The independent, but coordinated, activity of the heart is a function of (1) the presence of gap junctions, and (2) the activity of the heart's "in-house" conduction system. The **intrinsic cardiac conduction system** consists of noncontractile cardiac cells specialized to initiate and distribute impulses throughout the heart, so that it depolarizes and contracts in an orderly, sequential manner. Let's look at how this system works.

Action Potential Initiation by Pacemaker Cells

Unstimulated contractile cells of the heart (and neurons and skeletal muscle fibers) maintain a stable resting membrane potential. However, about 1% of cardiac fibers are autorhythmic ("self-rhythmic") **cardiac pacemaker cells**, having the special ability to depolarize spontaneously and thus pace the heart. Pacemaker cells are a part of the intrinsic conduction system. They have an *unstable resting potential* that continuously depolarizes, drifting slowly toward threshold. These spontaneously changing membrane potentials, called **pacemaker potentials** or **prepotentials**, initiate the action potentials that spread throughout the heart to trigger its rhythmic contractions. Let's look at the three parts of an action potential in typical pacemaker cells as shown in **Figure 17.12**.

① **Pacemaker potential.** The pacemaker potential is due to the special properties of the ion channels in the sarcolemma. In these cells, hyperpolarization at the end of an action potential both closes K^+ channels and opens slow Na^+ channels. The Na^+ influx alters the balance between K^+ loss and Na^+ entry, and the membrane interior becomes less and less negative (more positive).

② **Depolarization.** Ultimately, at threshold (approximately −40 mV), **Ca^{2+} channels** open, allowing explosive entry of Ca^{2+} from the extracellular space. As a result, in pacemaker cells, it is the influx of Ca^{2+} (rather than Na^+) that produces the rising phase of the action potential and reverses the membrane potential.

③ **Repolarization.** Ca^{2+} channels inactivate. As in other excitable cells, the falling phase of the action potential and repolarization reflect opening of K^+ channels and K^+ efflux from the cell.

Once repolarization is complete, K^+ channels close, K^+ efflux declines, and the slow depolarization to threshold begins again.

Sequence of Excitation

Typical cardiac pacemaker cells are found in the sinoatrial (si″no-a′tre-al) and atrioventricular nodes (**Figure 17.13**).

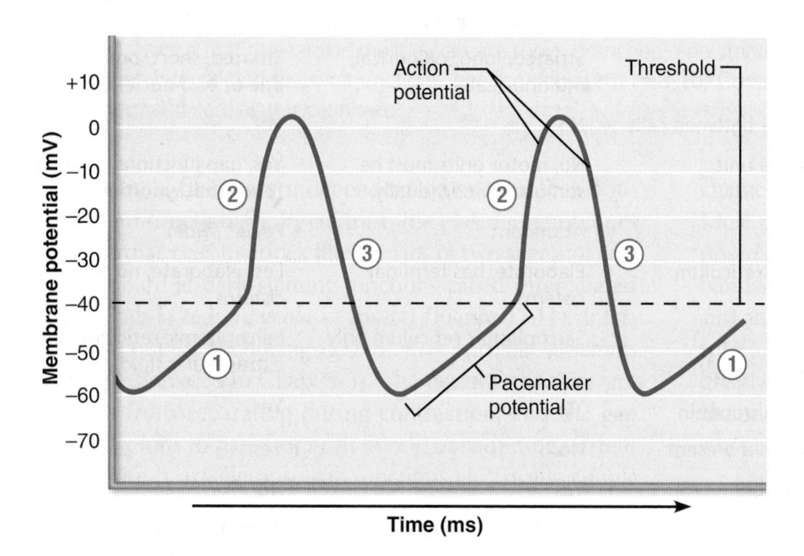

① **Pacemaker potential** This slow depolarization is due to both opening of Na^+ channels and closing of K^+ channels. Notice that the membrane potential is never a flat line.

② **Depolarization** The action potential begins when the pacemaker potential reaches threshold. Depolarization is due to Ca^{2+} influx through Ca^{2+} channels.

③ **Repolarization** is due to Ca^{2+} channels inactivating and K^+ channels opening. This allows K^+ efflux, which brings the membrane potential back to its most negative voltage.

Figure 17.12 Pacemaker and action potentials of typical cardiac pacemaker cells.

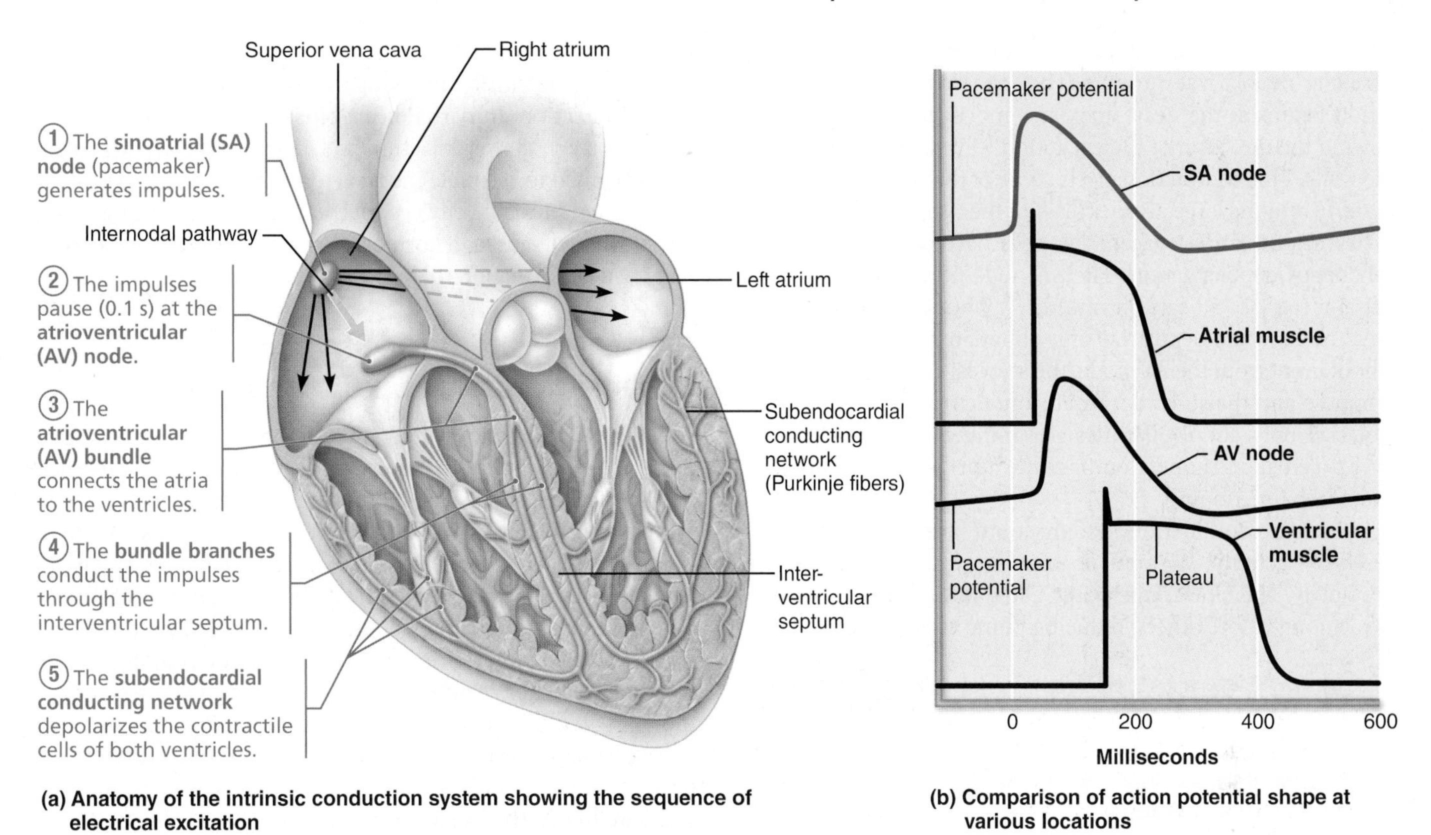

(a) Anatomy of the intrinsic conduction system showing the sequence of electrical excitation

(b) Comparison of action potential shape at various locations

Figure 17.13 Intrinsic cardiac conduction system and action potential succession during one heartbeat.

Interact with physiology
MasteringA&P*>Study Area>iP2

In addition, cells of the atrioventricular bundle, right and left bundle branches, and subendocardial conducting network (Purkinje fibers) can sometimes act as pacemakers. Impulses pass across the heart in order from ① to ⑤ following the yellow pathway in Figure 17.13a.

① **Sinoatrial (SA) node.** The crescent-shaped **sinoatrial node** is located in the right atrial wall, just inferior to the entrance of the superior vena cava. A minute cell mass with a mammoth job, the SA node typically generates impulses about 75 times every minute. The SA node sets the pace for the heart as a whole because no other region of the conduction system or the myocardium has a faster depolarization rate. For this reason, it is the heart's **pacemaker**, and its characteristic rhythm, called **sinus rhythm**, determines heart rate.

② **Atrioventricular (AV) node.** From the SA node, the depolarization wave spreads via gap junctions throughout the atria and via the *internodal pathway* to the **atrioventricular node**, located in the inferior portion of the interatrial septum immediately above the tricuspid valve. At the AV node, the impulse is delayed for about 0.1 second, allowing the atria to respond and complete their contraction before the ventricles contract. This delay reflects the smaller diameter of the fibers here and the fact that they have fewer gap junctions for current flow. Consequently, the AV node conducts impulses more slowly than other parts of the system, just as traffic slows when cars are forced to merge from four lanes into two. Once through the AV node, the signaling impulse passes rapidly through the rest of the system.

③ **Atrioventricular (AV) bundle.** From the AV node, the impulse sweeps to the **atrioventricular bundle** (also called the **bundle of His**) in the superior part of the interventricular septum. Although the atria and ventricles are adjacent to each other, they are *not* connected by gap junctions. The AV bundle is the *only* electrical connection between them. The fibrous cardiac skeleton is nonconducting and insulates the rest of the AV junction.

④ **Right and left bundle branches.** The AV bundle persists only briefly before splitting into two pathways—the **right** and **left bundle branches**, which course along the interventricular septum toward the heart apex.

⑤ **Subendocardial conducting network.** Essentially long strands of barrel-shaped cells with few myofibrils, the **subendocardial conducting network**, also called **Purkinje fibers** (purkin′je), completes the pathway through the interventricular septum, penetrates into the heart apex, and then turns superiorly into the ventricular walls. The bundle branches excite the septal cells, but the bulk of ventricular depolarization depends on the large fibers of the conducting network and, ultimately, on cell-to-cell transmission of the impulse via gap junctions between the ventricular muscle cells. Because the left ventricle is much larger than the right, the subendocardial conducting network is more elaborate in that side of the heart.

The total time between initiation of an impulse by the SA node and depolarization of the last of the ventricular muscle cells is approximately 0.22 s (220 ms) in a healthy human heart.

Ventricular contraction almost immediately follows the ventricular depolarization wave. The wringing motion of contraction begins at the heart apex and moves toward the atria, following the direction of the excitation wave through the ventricle walls. This contraction ejects some of the contained blood *superiorly* into the large arteries leaving the ventricles.

The various cardiac pacemaker cells have different rates of spontaneous depolarization. The SA node normally drives the heart at a rate of 75 beats per minute. Without SA node input, the AV node would depolarize only about 50 times per minute. Without input from the AV node, the atypical pacemakers of the AV bundle and the subendocardial conducting network would depolarize only about 30 times per minute. Note that these slower pacemakers cannot dominate the heart unless faster pacemakers stop functioning.

The cardiac conduction system coordinates and synchronizes heart activity. Without it, impulses would travel much more slowly. This slower rate would allow some muscle fibers to contract long before others, reducing pump effectiveness.

HOMEOSTATIC IMBALANCE 17.4 CLINICAL

Defects in the intrinsic conduction system can cause irregular heart rhythms, or **arrhythmias** (ah-rith′me-ahz). They may also cause uncoordinated atrial and ventricular contractions, or even **fibrillation**, a condition of rapid and irregular or out-of-phase contractions in which control of heart rhythm is taken away from the SA node by rapid activity in other heart regions. The heart in fibrillation has been compared with a squirming bag of worms. Fibrillating ventricles are useless as pumps; and unless the heart is defibrillated quickly, circulation stops and brain death occurs.

Defibrillation is accomplished by electrically shocking the heart, which interrupts its chaotic twitching by depolarizing the entire myocardium. The hope is that "with the slate wiped clean" the SA node will begin to function normally and sinus rhythm will be reestablished. Implantable cardioverter defibrillators (ICDs) can continually monitor heart rhythms and slow an abnormally fast heart rate or emit an electrical shock if the heart begins to fibrillate.

A defective SA node may have several consequences. An **ectopic focus** (ek-top′ik) (an abnormal pacemaker) may appear and take over the pacing of heart rate, or the AV node may become the pacemaker. The pace set by the AV node (**junctional rhythm**) is 40 to 60 beats per minute, slower than sinus rhythm but still adequate to maintain circulation.

Occasionally, ectopic pacemakers appear even when the SA node is operating normally. A small region of the heart becomes hyperexcitable, sometimes as a result of too much caffeine or nicotine, and generates impulses more quickly than the SA node. This leads to a *premature contraction* or **extrasystole** (ek″strah-sis′to-le) before the SA node initiates the next contraction. Then, because the heart has a longer time to fill, the next (normal) contraction is felt as a thud. As you might guess, premature *ventricular* contractions (PVCs) are most problematic.

The only route for impulse transmission from atria to ventricles is through the AV node, AV bundle, and bundle branches. Damage to any of these structures interferes with the ability of the ventricles to receive pacing impulses, and may cause **heart block**. In total heart block, no impulses get through and the ventricles beat at their intrinsic rate, which is too slow to maintain adequate circulation. In partial heart block, only some of the atrial impulses reach the ventricles. In both cases, artificial pacemakers are implanted to recouple the atria to the ventricles as necessary. These programmable devices speed up in response to increased physical activity just as a normal heart would, and many can send diagnostic information to the patient's doctor. ✚

Modifying the Basic Rhythm: Extrinsic Innervation of the Heart

Although the intrinsic conduction system sets the basic heart rate, fibers of the autonomic nervous system modify the marchlike beat and introduce a subtle variability from one beat to the next. The sympathetic nervous system (the "accelerator") increases both the rate and the force of the heartbeat. Parasympathetic activation (the "brakes") slows the heart. We explain these neural controls later—here we discuss the anatomy of the nerve supply to the heart.

The cardiac centers are located in the medulla oblongata. The **cardioacceleratory center** projects to sympathetic neurons in the T_1–T_5 level of the spinal cord. These preganglionic neurons, in turn, synapse with postganglionic neurons in the cervical and upper thoracic sympathetic trunk (**Figure 17.14**). From there, postganglionic fibers run through the cardiac plexus to the heart where they innervate the SA and AV nodes, heart muscle, and coronary arteries.

The **cardioinhibitory center** sends impulses to the parasympathetic dorsal vagus nucleus in the medulla, which in turn sends inhibitory impulses to the heart via branches of the vagus nerves. Most parasympathetic postganglionic motor neurons lie in ganglia in the heart wall and their fibers project most heavily to the SA and AV nodes.

Action Potentials of Contractile Cardiac Muscle Cells

The bulk of heart muscle is composed of *contractile muscle* fibers responsible for the heart's pumping activity. As we have seen, the sequence of events leading to contraction of these cells is similar to that in skeletal muscle fibers. However, the action potential has a characteristic "hump" or *plateau* as shown in **Figure 17.15**.

1. Depolarization opens a few **fast voltage-gated Na^+ channels** in the sarcolemma, allowing extracellular Na^+ to enter. This influx initiates a positive feedback cycle that causes the rising phase of the action potential (and reversal of the membrane potential from −90 mV to nearly +30 mV). The period of Na^+ influx is very brief, because the sodium channels quickly inactivate and the Na^+ influx stops.
2. When Na^+-dependent membrane depolarization occurs, the voltage change also opens channels that allow Ca^{2+} to enter from the extracellular fluid. These channels are called **slow Ca^{2+} channels** because their opening is delayed a bit. The Ca^{2+} surge across the sarcolemma prolongs the depolarization, producing a **plateau** in the action potential tracing. Not many voltage-gated K^+ channels are open yet, so the plateau is prolonged. As long as Ca^{2+} is entering, the cells continue to

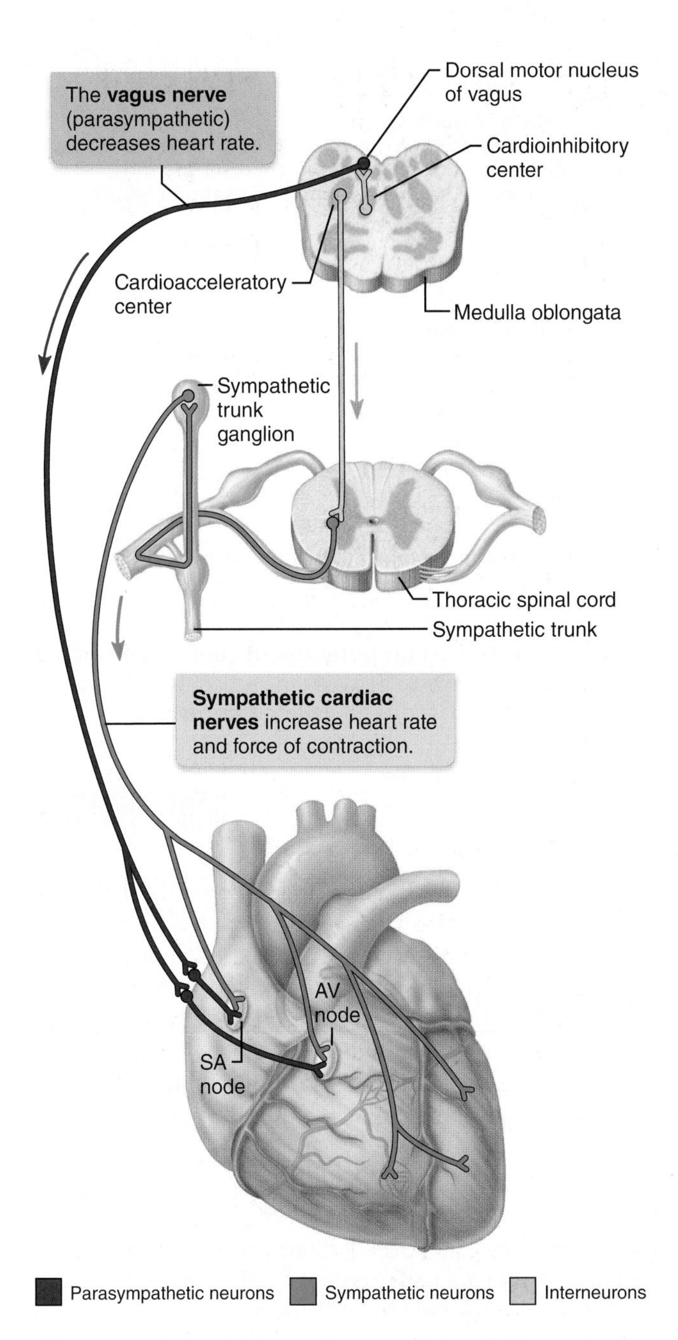

Figure 17.14 Autonomic innervation of the heart.

contract. Notice in Figure 17.15 that muscle tension develops during the plateau, and peaks just after the plateau ends.

③ After about 200 ms, the slope of the action potential tracing falls rapidly. This repolarization results from inactivation of Ca^{2+} channels and opening of voltage-gated K^+ channels. The rapid loss of potassium from the cell through K^+ channels restores the resting membrane potential. During repolarization, Ca^{2+} is pumped back into the SR and the extracellular space.

Notice that the action potential and contractile phase lasts much longer in cardiac muscle than in skeletal muscle. In skeletal muscle, the action potential typically lasts 1–2 ms and the contraction (for a single stimulus) 15–100 ms. In cardiac muscle, the action potential lasts 200 ms or more (because of the plateau), and tension development persists for 200 ms or more. This long plateau in cardiac muscle has two consequences:

- It ensures that the contraction is sustained so that blood is ejected efficiently from the heart.
- It ensures that there is a long refractory period, so that tetanic contractions cannot occur and the heart can fill again for the next beat.

Electrocardiography

The electrical currents generated in and transmitted through the heart spread throughout the body and can be detected with a device called an **electrocardiograph**. An **electrocardiogram (ECG)** is a graphic record of heart activity. An ECG is a composite of all the action potentials generated by nodal and contractile cells at a given time (**Figure 17.16**)—*not*, as sometimes assumed, a tracing of a single action potential.

To record an ECG, recording electrodes are placed at various sites on the body surface. In a typical 12-lead ECG, three electrodes form bipolar leads that measure the voltage difference either between the arms or between an arm and a leg, and nine form unipolar leads. Together the 12 leads provide a comprehensive picture of the heart's electrical activity.

A typical ECG has three almost immediately distinguishable waves or *deflections*: the P wave, the QRS complex, and the T wave (Figure 17.16). The first, the small **P wave**, lasts about 0.08 s and results from movement of the depolarization wave from the SA node through the atria. Approximately 0.1 s after the P wave begins, the atria contract.

The large **QRS complex** results from ventricular depolarization and precedes ventricular contraction. It has a complicated shape because the paths of the depolarization waves through the ventricular walls change continuously, producing corresponding changes in current direction. Additionally, the time required for each ventricle to depolarize depends on its size relative to the other ventricle. Average duration of the QRS complex is 0.08 s.

The **T wave,** caused by ventricular repolarization, typically lasts about 0.16 s. Repolarization is slower than depolarization, so the T wave is more spread out and has a lower amplitude (height) than the QRS complex. Because atrial repolarization takes place during the period of ventricular excitation, the wave representing atrial repolarization is normally obscured by the large QRS complex being recorded at the same time.

The **P-R interval** is the time (about 0.16 s) from the beginning of atrial excitation to the beginning of ventricular excitation. If the Q wave is visible (which is often not the case), it marks the beginning of ventricular excitation, and for this reason this interval is sometimes called the **P-Q interval**. The P-R interval includes atrial depolarization (and contraction) as well as the passage of the depolarization wave through the rest of the conduction system.

During the **S-T segment** of the ECG, when the action potentials of the ventricular myocytes are in their plateau phases, the entire ventricular myocardium is depolarized. The **Q-T interval**, lasting about 0.38 s, is the period from the beginning of ventricular depolarization through ventricular repolarization.

Figure 17.17 relates the parts of an ECG to the sequence of depolarization and repolarization in the heart.

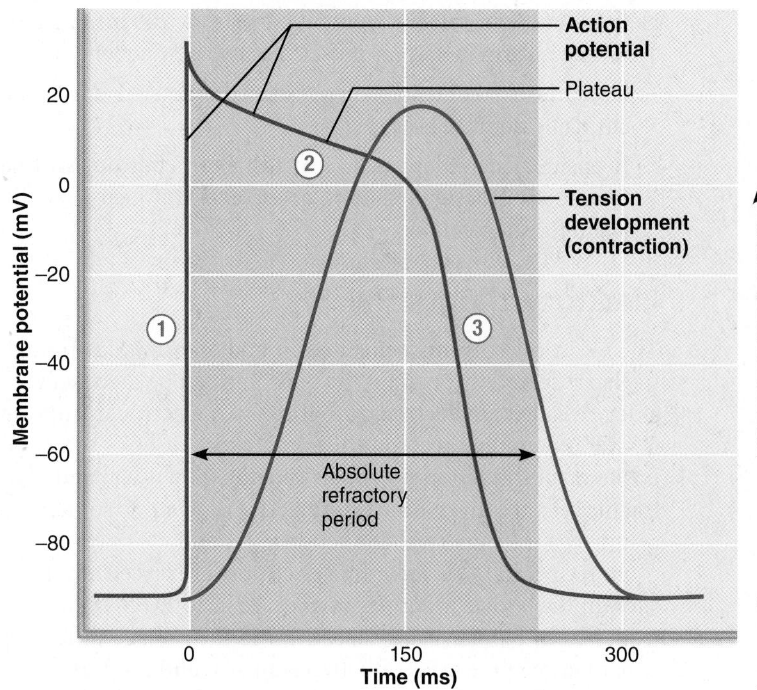

① **Depolarization** is due to Na^+ influx through fast voltage-gated Na^+ channels. A positive feedback cycle rapidly opens many Na^+ channels, reversing the membrane potential. Channel inactivation ends this phase.

② **Plateau phase** is due to Ca^{2+} influx through slow Ca^{2+} channels. This keeps the cell depolarized because most K^+ channels are closed.

③ **Repolarization** is due to Ca^{2+} channels inactivating and K^+ channels opening. This allows K^+ efflux, which brings the membrane potential back to its resting voltage.

Figure 17.15 The action potential of contractile cardiac muscle cells. Relationship between the action potential, period of contraction, and absolute refractory period in a single ventricular cell.

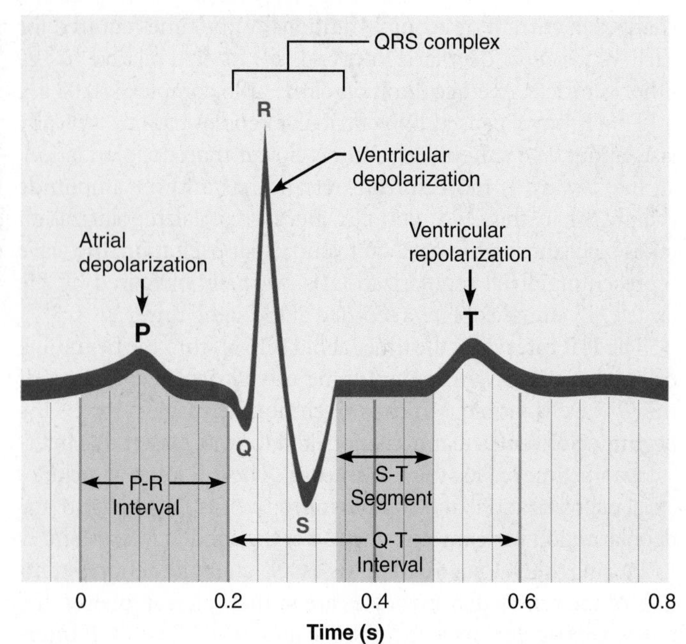

Figure 17.16 An electrocardiogram (ECG) tracing. The labels identify the three normally recognizable deflections (waves) and the important intervals.

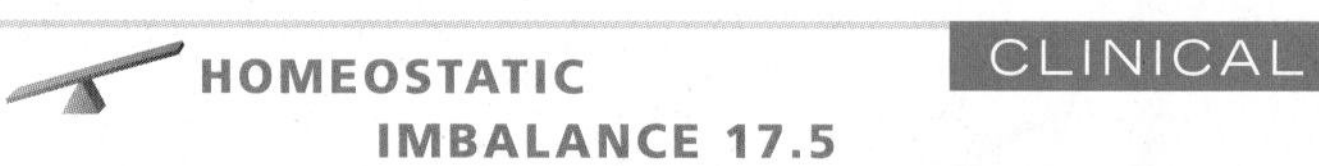

In a healthy heart, the size, duration, and timing of the deflection waves tend to be consistent. Changes in the pattern or timing of the ECG may reveal a diseased or damaged heart or problems with the heart's conduction system (**Figure 17.18**). For example, an enlarged R wave hints of enlarged ventricles, an S-T segment that is elevated or depressed indicates cardiac ischemia, and a prolonged Q-T interval reveals a repolarization abnormality that increases the risk of ventricular arrhythmias. +

Check Your Understanding

10. Cardiac muscle cannot go into tetany. Why?

11. Which part of the intrinsic conduction system directly excites ventricular myocardial cells? In which direction does the depolarization wave travel across the ventricles?

12. Describe the electrical event in the heart that occurs during each of the following: (a) the QRS wave of the ECG; (b) the T wave of the ECG; (c) the P-R interval of the ECG.

13. MAKING connections Below are drawings of three different action potentials. Two of these occur in the heart, and one occurs in skeletal muscle (as you learned in Chapter 9).

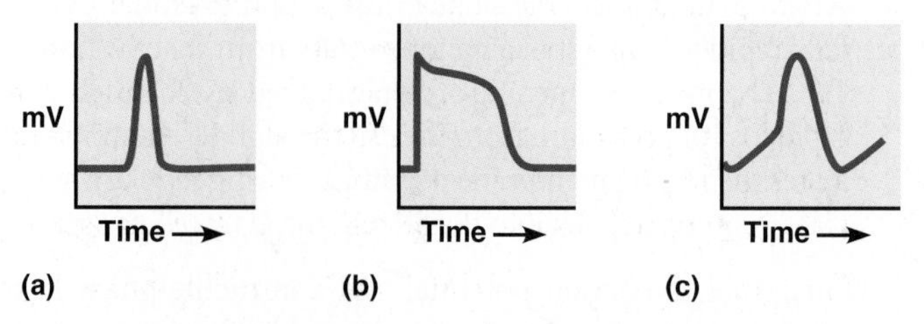

Which one comes from a contractile cardiac muscle cell? A skeletal muscle cell? A cardiac pacemaker cell? For each one, state which ion is responsible for the depolarization phase and which ion is responsible for the repolarization phase.

For answers, see Answers Appendix.

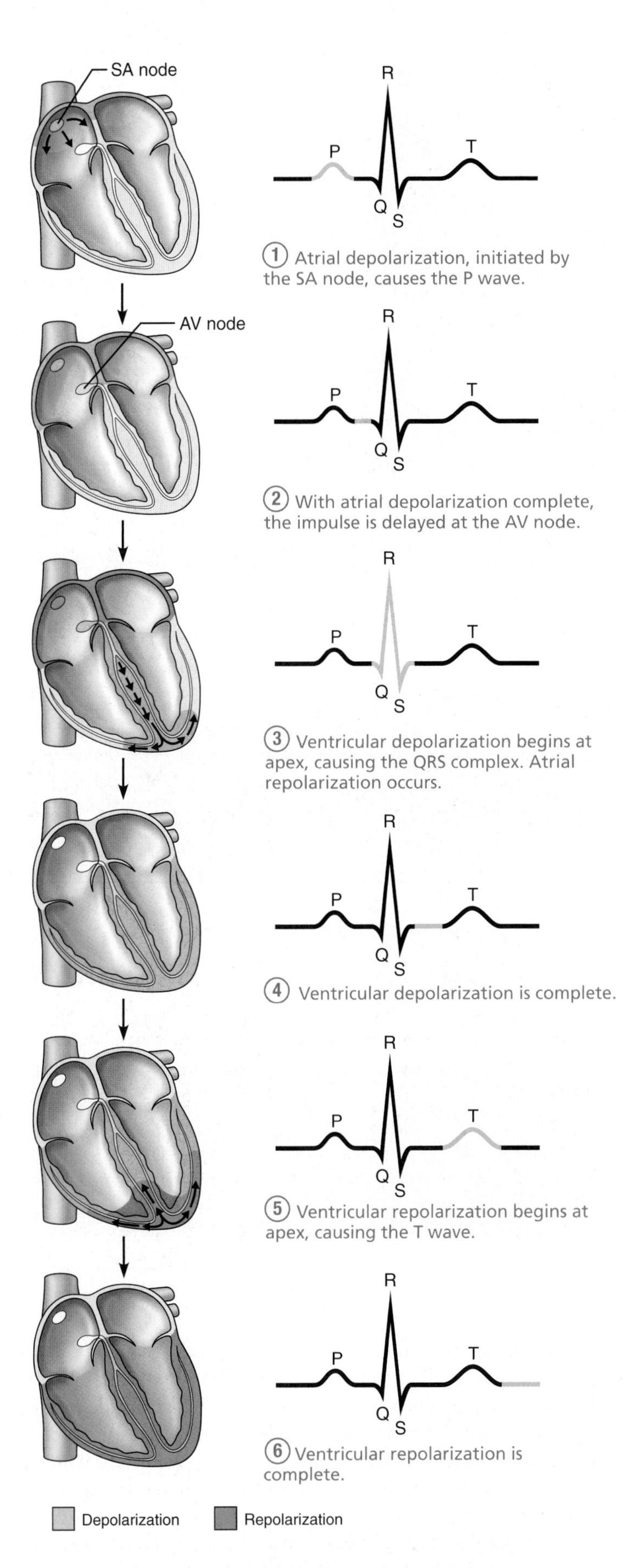

Figure 17.17 The sequence of depolarization and repolarization of the heart related to the deflection waves of an ECG tracing.

17.6 The cardiac cycle describes the mechanical events associated with blood flow through the heart

→ Learning Objectives

- ☐ **Describe the timing and events of the cardiac cycle.**
- ☐ **Describe normal heart sounds, and explain how heart murmurs differ.**

The heart undergoes some dramatic writhing movements as it alternately contracts, forcing blood out of its chambers, and then relaxes, allowing its chambers to refill with blood. The term **systole** (sis′to-le) refers to these periods of contraction, and **diastole** (di-as′to-le) refers to those of relaxation. The **cardiac cycle** includes *all* events associated with the blood flow through the heart during one complete heartbeat—atrial systole and diastole followed by ventricular systole and diastole. These mechanical events always *follow* the electrical events seen in the ECG.

The cardiac cycle is marked by a succession of pressure and blood volume changes in the heart. Because blood circulates continuously, we must choose an arbitrary starting point for one turn of the cardiac cycle. As shown in **Figure 17.19**, which outlines what happens in the left side of the heart, we begin with the heart in total relaxation: Atria and ventricles are quiet, and it is mid-to-late diastole.

1. **Ventricular filling: mid-to-late diastole.** Pressure in the heart is low, blood returning from the circulation is flowing passively through the atria and the open AV valves into the ventricles, and the aortic and pulmonary valves are closed. More than 80% of ventricular filling occurs during this period, and the AV valve flaps begin to drift toward the closed position. (The remaining 20% is delivered to the ventricles when the atria contract toward the end of this phase.)

   Now the stage is set for atrial systole. Following depolarization (P wave of ECG), the atria contract, compressing the blood in their chambers. This causes a sudden slight rise in atrial pressure, which propels residual blood out of the atria into the ventricles. At this point the ventricles are in the last part of their diastole and have the maximum volume of blood they will contain in the cycle, an amount called the *end diastolic volume* (*EDV*). Then the atria relax and the ventricles depolarize (QRS complex). Atrial diastole persists through the rest of the cycle.

2. **Ventricular systole (atria in diastole).** As the atria relax, the ventricles begin contracting. Their walls close in on the blood in their chambers, and ventricular pressure rises rapidly and sharply, closing the AV valves. The split-second period when the ventricles are completely closed chambers and the blood volume in the chambers remains constant as the ventricles contract is the **isovolumetric contraction phase** (i″so-vol″u-met′rik).

   Ventricular pressure continues to rise. When it finally exceeds the pressure in the large arteries issuing from the ventricles, the isovolumetric stage ends as the SL valves are forced open and blood rushes from the ventricles into the aorta and pulmonary trunk. During this ventricular ejection phase, the pressure in the aorta normally reaches about 120 mm Hg.

(a) Infant undergoing an electrocardiogram (ECG)

(b) Normal sinus rhythm

Normal ECG trace (sinus rhythm)

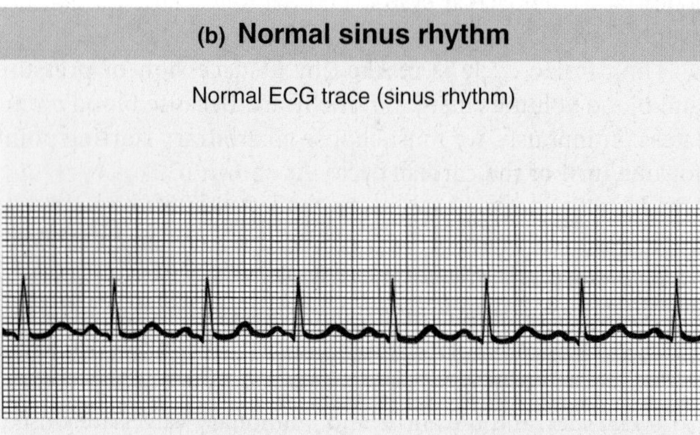

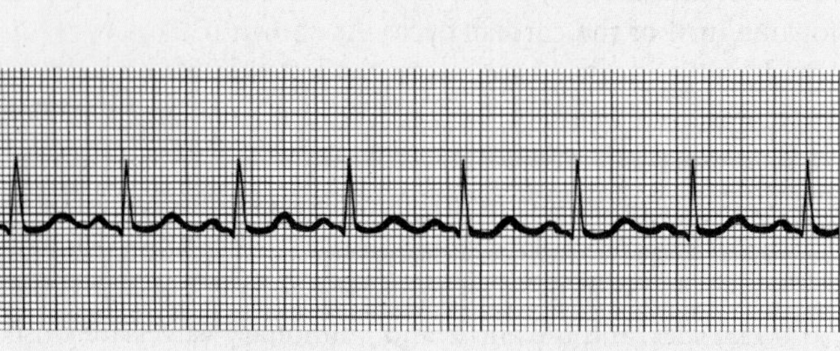

(c) Junctional rhythm

The SA node is nonfunctional. As a result:
- P waves are absent.
- The AV node paces the heart at 40–60 beats per minute.

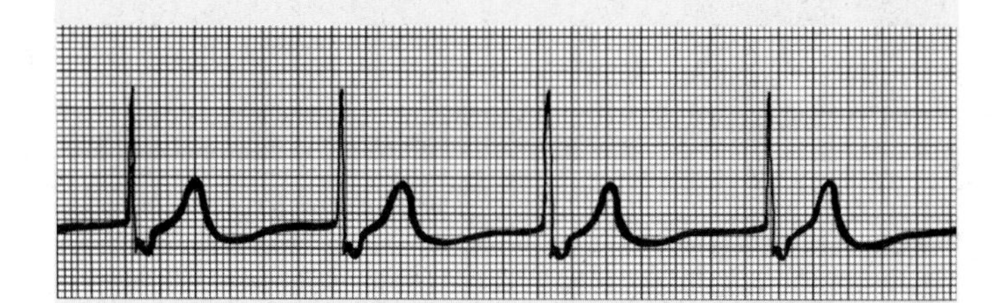

(d) Second-degree heart block

The AV node fails to conduct some SA node impulses.
- As a result, there are more P waves than QRS waves.
- In this tracing, there are usually two P waves for each QRS wave.

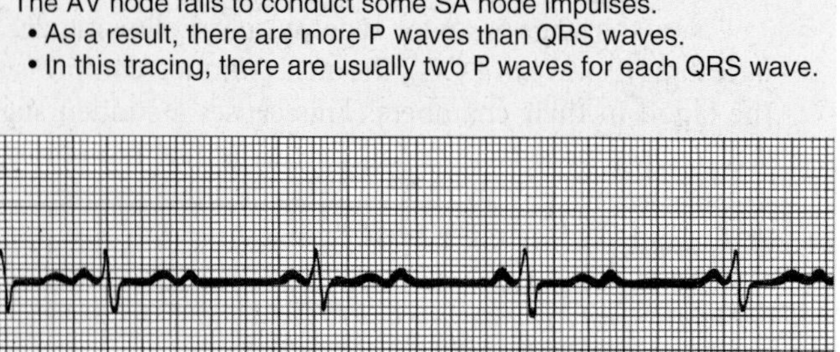

(e) Ventricular fibrillation

Electrical activity is disorganized. Action potentials occur randomly throughout the ventricles.
- Results in chaotic, grossly abnormal ECG deflections.
- Seen in acute heart attack and after an electrical shock.

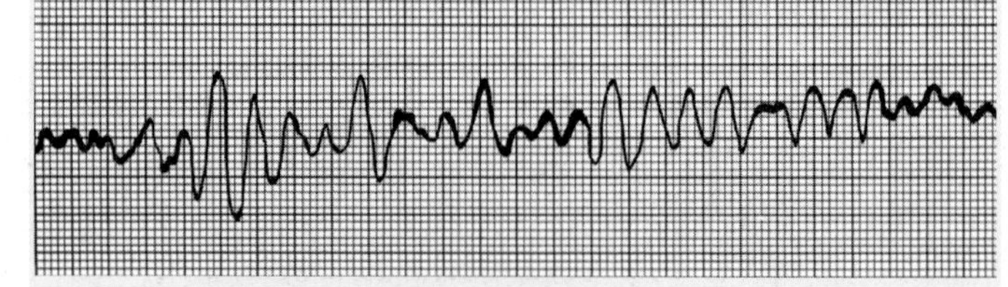

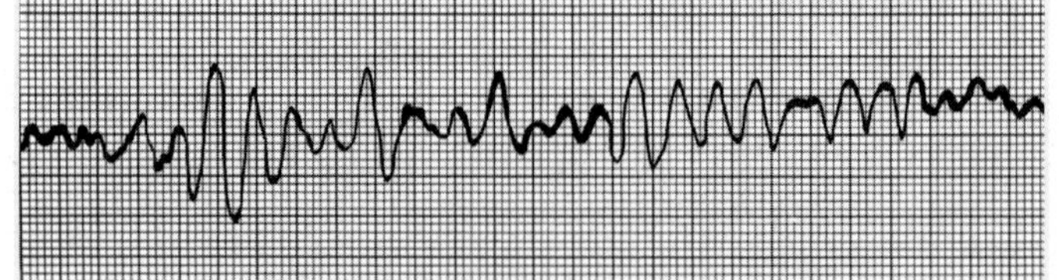

Figure 17.18 Normal and abnormal ECG tracings.

③ **Isovolumetric relaxation: early diastole.** During this brief phase following the T wave, the ventricles relax. Because the blood remaining in their chambers, referred to as the *end systolic volume* (*ESV*), is no longer compressed, ventricular pressure drops rapidly and blood in the aorta and pulmonary trunk flows back toward the heart, closing the SL valves. Closure of the aortic valve raises aortic pressure briefly as backflowing blood rebounds off the closed valve cusps, an event beginning at the **dicrotic notch** shown on the pressure graph. Once again the ventricles are totally closed chambers.

All during ventricular systole, the atria have been in diastole. They have been filling with blood and the intra-atrial pressure has been rising. When blood pressure on the atrial side of the AV valves exceeds that in the ventricles, the AV valves are forced open and ventricular filling, phase ①, begins again. Atrial pressure drops to its lowest point and ventricular pressure begins to rise, completing the cycle.

Assuming the average heart beats 75 times each minute, the cardiac cycle lasts about 0.8 s, with atrial systole accounting for 0.1 s and ventricular systole 0.3 s. The remaining 0.4 s is a period of total heart relaxation, the **quiescent period**.

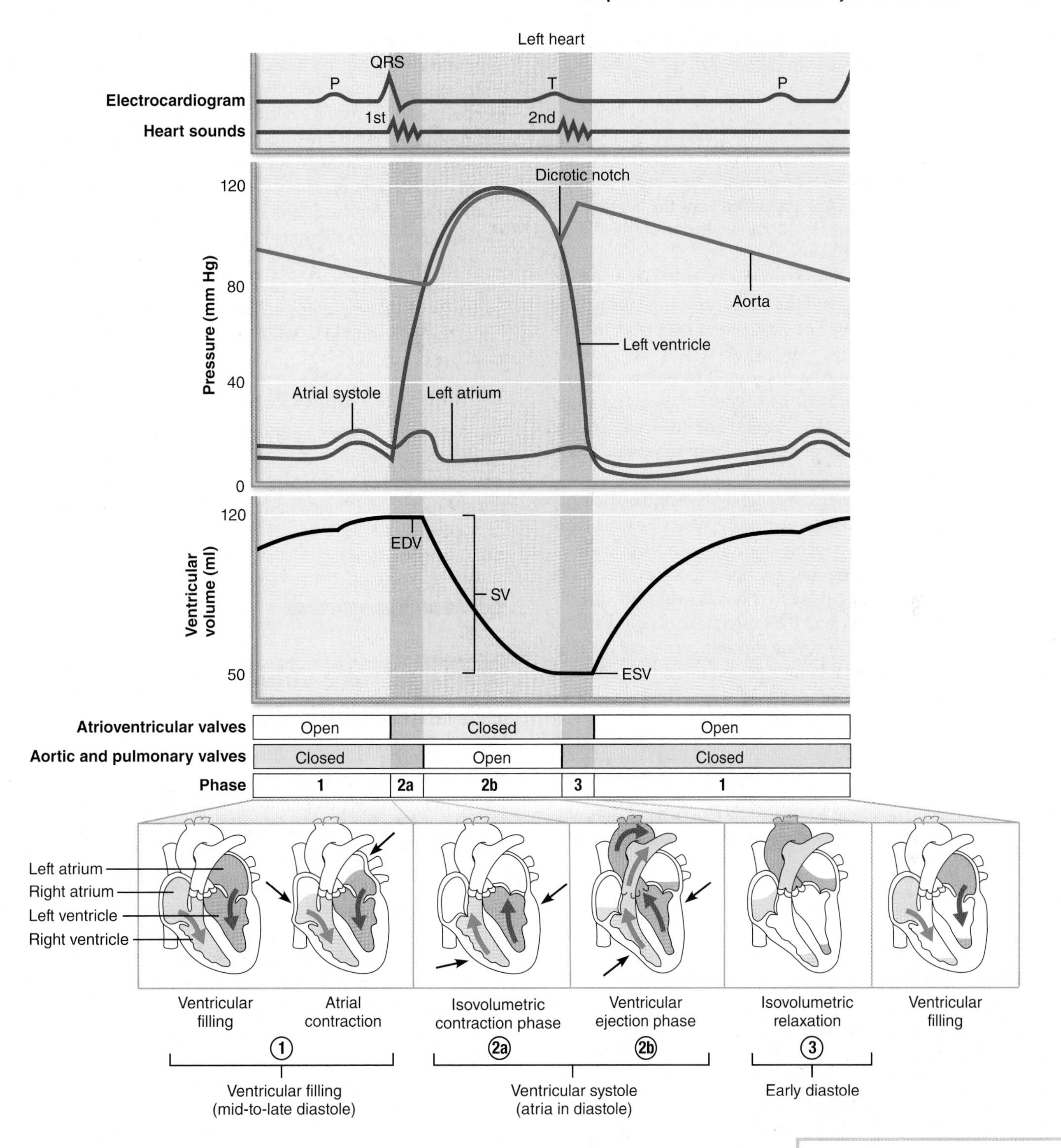

Figure 17.19 Summary of events during the cardiac cycle. An ECG tracing (*top*) correlated with graphs of pressure and volume changes (*center*) in the left side of the heart. Pressures are lower in the right side of the heart. Timing of heart sounds is also indicated. (*Bottom*) Events of phases 1 through 3 of the cardiac cycle. (EDV = end diastolic volume, ESV = end systolic volume, SV = stroke volume)

Interact with physiology
MasteringA&P®>Study Area>iP2

Notice two important points: (1) Blood flow through the heart is controlled entirely by pressure changes, and (2) blood flows down a pressure gradient through any available opening. The pressure changes, in turn, reflect the alternating contraction and relaxation of the myocardium and cause the heart valves to open, which keeps blood flowing in the forward direction.

The situation in the right side of the heart is essentially the same as in the left side *except* for pressure. The pulmonary circulation is a low-pressure circulation as evidenced by the much thinner myocardium of its right ventricle. So, typical systolic and diastolic pressures for the pulmonary artery are 24 and 10 mm Hg, compared to systolic and diastolic pressures of 120 and 80 mm Hg, respectively, for the aorta. However, the two sides of the heart eject the same blood volume with each heartbeat.

Heart Sounds

Auscultating (listening to) the thorax with a stethoscope will reveal two sounds during each heartbeat. These **heart sounds**, often described as lub-dup, are associated with the heart valves closing. (The top of Figure 17.19 shows the timing of heart sounds in the cardiac cycle.)

The basic rhythm of the heart sounds is lub-dup, pause, lub-dup, pause, and so on, with the pause indicating the period when the heart is relaxing. The first sound occurs as the AV valves close. It signifies the point when ventricular pressure rises above atrial pressure (the beginning of ventricular systole). The first sound tends to be louder, longer, and more resonant than the second. The second sound occurs as the SL valves snap shut at the beginning of ventricular relaxation (diastole), resulting in a short, sharp sound.

Because the mitral valve closes slightly before the tricuspid valve does, and the aortic SL valve generally snaps shut just before the pulmonary valve, it is possible to distinguish the individual valve sounds by auscultating four specific regions of the thorax (**Figure 17.20**). Notice that these four points, while not directly superficial to the valves (because the sounds take oblique paths to reach the chest wall), handily define the four corners of the normal heart. Knowing normal heart size and location is essential for recognizing an enlarged (and often diseased) heart.

HOMEOSTATIC IMBALANCE 17.6 CLINICAL

Blood flows silently as long as its flow is smooth and uninterrupted. If blood strikes obstructions, however, its flow becomes turbulent and generates abnormal heart sounds, called **heart murmurs**, that can be heard with a stethoscope. Heart murmurs are fairly common in young children (and some elderly people) with perfectly healthy hearts, probably because their heart walls are relatively thin and vibrate with rushing blood.

Most often, however, murmurs indicate valve problems. An *insufficient* or *incompetent* valve fails to close completely. There is a swishing sound as blood backflows or regurgitates through the partially open valve *after* the valve has (supposedly) closed.

A *stenotic* valve fails to open completely and its narrow opening restricts blood flow *through* the valve. In a stenotic aortic valve, for instance, a high-pitched sound or click can be detected when the valve should be wide open during ventricular contraction, but is not. +

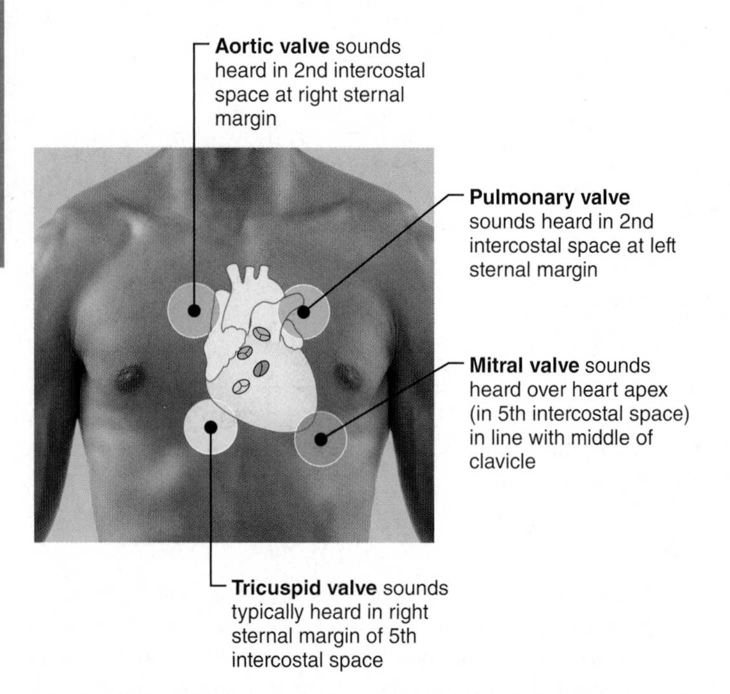

Figure 17.20 Areas of the thoracic surface where the sounds of individual valves are heard most clearly.

Check Your Understanding

14. The second heart sound is associated with the closing of which valve(s)?

15. If the mitral valve were insufficient, would you expect to hear the murmur (of blood flowing through the valve that should be closed) during ventricular systole or diastole?

16. During the cardiac cycle, there are two periods when all four valves are closed. Name these two periods.

For answers, see Answers Appendix.

17.7 Stroke volume and heart rate are regulated to alter cardiac output

Learning Objectives

- ☐ **Name and explain the effects of various factors regulating stroke volume and heart rate.**
- ☐ **Explain the role of the autonomic nervous system in regulating cardiac output.**

Cardiac output (CO) is the amount of blood pumped out by *each* ventricle in 1 minute. It is the product of heart rate (HR) and stroke volume (SV). **Stroke volume** is defined as the volume of blood pumped out by one ventricle with each beat. In general, stroke volume correlates with the force of ventricular contraction.

Using normal resting values for heart rate (75 beats/min) and stroke volume (70 ml/beat), the average adult cardiac output can be computed:

$$\text{CO} = \text{HR} \times \text{SV} = \frac{75\ \cancel{\text{beats}}}{\text{min}} \times \frac{70\ \text{ml}}{\cancel{\text{beat}}} = \frac{5250\ \text{ml}}{\text{min}} = \frac{5.25\ \text{L}}{\text{min}}$$

The normal adult blood volume is about 5 L (a little more than 1 gallon). As you can see, the entire blood supply passes through each side of the heart once each minute.

Notice that cardiac output varies directly with SV and HR. This means that CO increases when the stroke volume increases or the heart beats faster or both, and it decreases when either or both of these factors decrease.

Cardiac output is highly variable and increases markedly in response to special demands, such as running to catch a bus. **Cardiac reserve** is the difference between resting and maximal

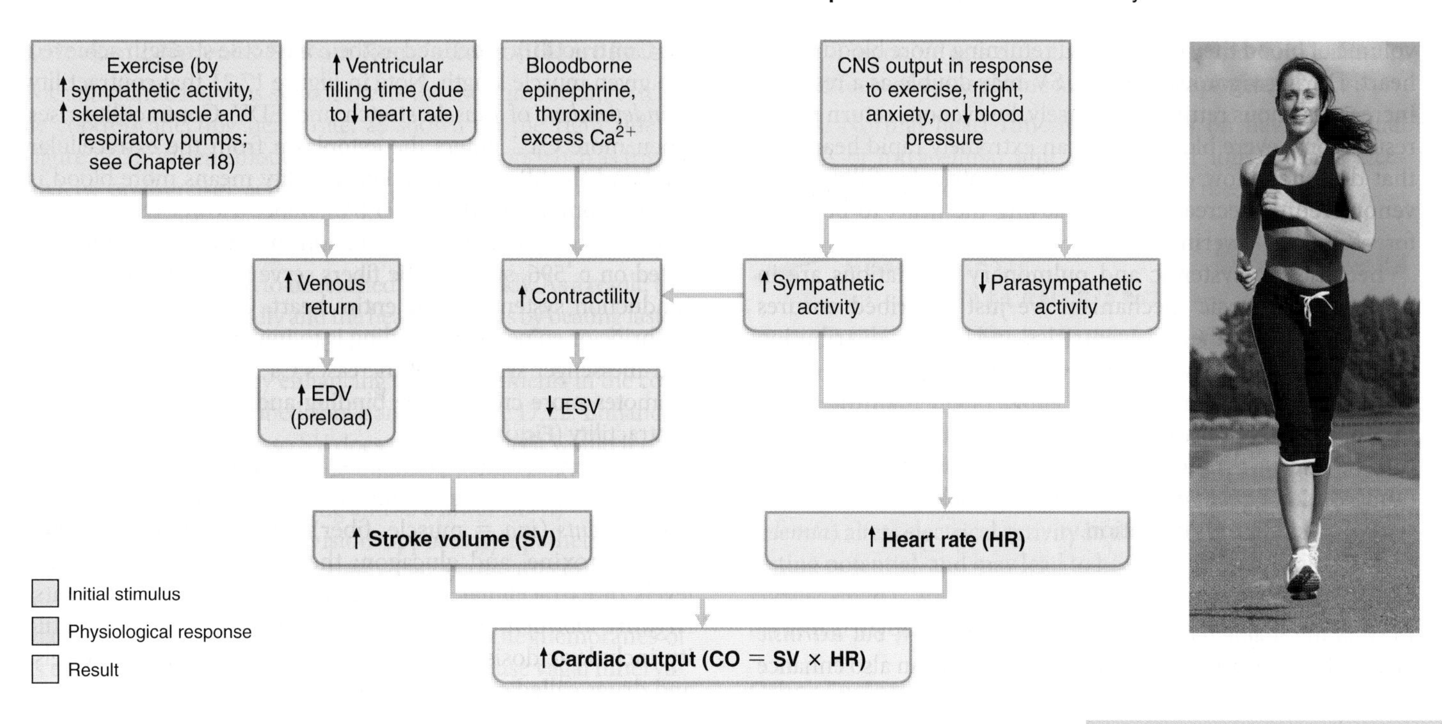

Figure 17.21 Factors involved in determining cardiac output. (EDV = end diastolic volume, ESV = end systolic volume)

Interact with physiology
MasteringA&P®>Study Area>iP2

CO. In nonathletic people, cardiac reserve is typically four to five times resting CO (20–25 L/min), but CO in trained athletes during competition may reach 35 L/min (seven times resting CO).

How does the heart accomplish such tremendous increases in output? To understand this feat, let's look at how stroke volume and heart rate are regulated. See **Figure 17.21** for an overview of the factors that affect stroke volume and heart rate, and consequently, cardiac output.

Regulation of Stroke Volume

Mathematically, stroke volume (SV) represents the difference between **end diastolic volume (EDV)**, the amount of blood that collects in a ventricle during diastole, and **end systolic volume (ESV)**, the volume of blood remaining in a ventricle *after* it has contracted. The EDV, determined by how long ventricular diastole lasts and by venous pressure, is normally about 120 ml. (An increase in either factor *raises* EDV.) The ESV, determined by arterial blood pressure and the force of ventricular contraction, is approximately 50 ml. (The higher the arterial blood pressure, the higher the ESV.) To figure normal stroke volume, simply plug these values into this equation:

$$SV = EDV - ESV = \frac{120 \text{ ml}}{\text{beat}} - \frac{50 \text{ ml}}{\text{beat}} = \frac{70 \text{ ml}}{\text{beat}}$$

As you can see, each ventricle pumps out about 70 ml of blood with each beat, which is about 60% of the blood in its chambers.

So what is important here—how do we make sense out of this alphabet soup (SV, ESV, EDV)? Although many factors affect SV by altering EDV or ESV, the three most important are *preload, contractility*, and *afterload*. As we describe in detail next, preload affects EDV, whereas contractility and afterload affect the ESV.

Preload: Degree of Stretch of Heart Muscle

The degree to which cardiac muscle cells are stretched just before they contract, called the **preload**, controls stroke volume. In a normal heart, the higher the preload, the higher the stroke volume. This relationship between preload and stroke volume is called the **Frank-Starling law of the heart**. Recall that at an *optimal length* of muscle fibers (and sarcomeres) (1) the maximum number of active cross bridge attachments is possible between actin and myosin, and (2) the force of contraction is maximal (see Figure 9.19, p. 270). Cardiac muscle, like skeletal muscle, exhibits a *length-tension relationship*.

Resting skeletal muscle fibers are kept near optimal length for developing maximal tension while resting cardiac cells are normally *shorter* than optimal length. As a result, stretching cardiac cells can produce dramatic increases in contractile force. The most important factor stretching cardiac muscle is **venous return**, the amount of blood returning to the heart and distending its ventricles.

Anything that increases venous return increases EDV and, consequently, SV and contraction force (Figure 17.21). Basically:

↑ Venous return → ↑ EDV (preload) → ↑ SV → ↑ Cardiac output

(↑ EDV → ↑ SV: Frank-Starling law)

Both exercise and increased filling time increase EDV. Exercise increases venous return because both increased sympathetic nervous system activity and the squeezing action of the skeletal muscles compress the veins, decreasing the

17

18 The Cardiovascular System: Blood Vessels

KEY CONCEPTS

Blood vessels are sometimes compared to a system of pipes with blood circulating in them, but this analogy is only a starting point. Unlike rigid pipes, blood vessels are dynamic structures that pulsate, constrict, relax, and even proliferate. In this chapter we examine the structure and function of these important circulatory passageways.

The **blood vessels** of the body form a closed delivery system that begins and ends at the heart. The idea that blood circulates in the body dates back to the 1620s with the inspired experiments of William Harvey, an English physician. Prior to that time, people thought, as proposed by the ancient Greek physician Galen, that blood moved through the body like an ocean tide, first moving out from the heart and then ebbing back in the same vessels.

PART 1

BLOOD VESSEL STRUCTURE AND FUNCTION

The three major types of blood vessels are *arteries, capillaries*, and *veins*. As the heart contracts, it forces blood into the large arteries leaving the ventricles. The blood then moves into successively smaller arteries, finally reaching their smallest branches, the *arterioles* (ar-te′re-ōlz; "little arteries"), which feed into the capillary beds of body organs and tissues. Blood drains from the capillaries into *venules* (ven′ūlz), the smallest veins, and then on into larger and larger veins that merge to form the large veins that ultimately empty into the heart. Altogether, the blood vessels in the adult human stretch for about 100,000 km (60,000 miles) through the internal body landscape!

Arteries carry blood *away from* the heart, so they are said to "branch," "diverge," or "fork" as they form smaller and smaller divisions. **Veins**, by contrast, carry blood *toward* the heart and so are said to "join," "merge," and "converge" into the successively larger vessels approaching the heart. In the systemic circulation, arteries always carry oxygenated blood and veins always carry oxygen-poor blood. The opposite is true in the pulmonary circulation, where the arteries, still defined as the vessels leading away from the heart, carry oxygen-poor blood to the lungs, and the veins carry oxygen-rich blood from the lungs to the heart. The special umbilical vessels of a fetus also differ in the roles of veins and arteries.

Of all the blood vessels, only the capillaries have intimate contact with tissue cells and directly serve cellular needs. Exchanges between the blood and tissue cells occur primarily through the gossamer-thin capillary walls.

Electron micrograph of a resin cast of blood vessels.

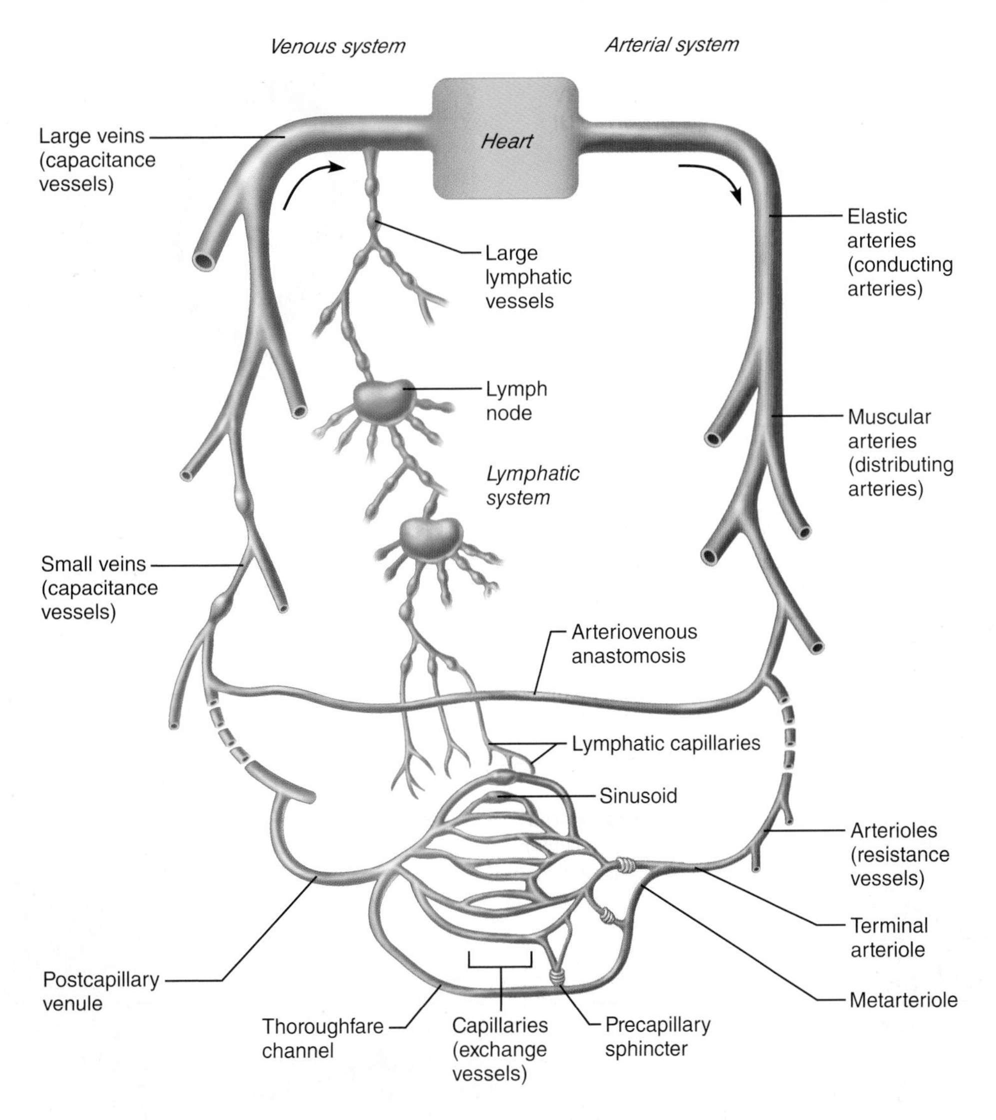

Figure 18.1 The relationship of blood vessels to each other and to lymphatic vessels. Lymphatic vessels recover excess tissue fluid and return it to the blood.

Figure 18.1 summarizes how these vascular channels relate to one another and to vessels of the lymphatic system. The lymphatic system recovers fluids that leak from the circulation and is described in Chapter 19.

18.1 Most blood vessel walls have three layers

→ Learning Objectives

- ☐ **Describe the three layers that typically form the wall of a blood vessel, and state the function of each.**
- ☐ **Define vasoconstriction and vasodilation.**

The walls of all blood vessels, except the very smallest, have three distinct layers, or *tunics* ("coverings"), that surround a central blood-containing space, the vessel **lumen** (**Figure 18.2**).

The innermost tunic is the **tunica intima** (in′tĭ-mah). The name is easy to remember once you know that this tunic is in *intimate* contact with the blood in the lumen. The tunica intima contains the **endothelium**, the simple squamous epithelium that lines the lumen of all vessels. The endothelium is continuous with the endocardial lining of the heart, and its flat cells fit closely together, forming a slick surface that minimizes friction as blood moves through the lumen. In vessels larger than 1 mm in diameter, a *subendothelial layer*, consisting of a basement membrane and loose connective tissue, supports the endothelium.

The middle tunic, the **tunica media** (me′de-ah), is mostly circularly arranged smooth muscle cells and sheets of elastin. The activity of the smooth muscle is regulated by sympathetic *vasomotor nerve fibers* of the autonomic nervous system and a whole battery of chemicals. Depending on the body's needs at any given moment, regulation causes either **vasoconstriction** (lumen diameter decreases as the smooth muscle contracts) or **vasodilation** (lumen diameter increases as the smooth muscle relaxes). The activities of the tunica media are critical in regulating circulatory dynamics because small changes in vessel diameter greatly influence blood flow and blood pressure. Generally, the tunica media is

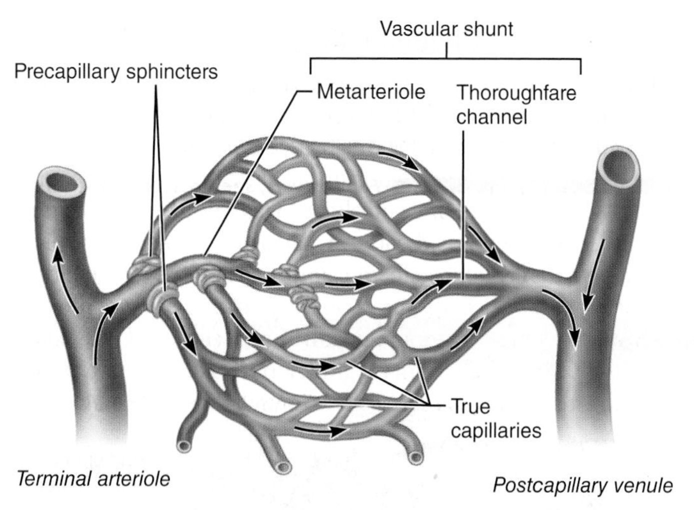

(a) **Sphincters open**—blood flows through true capillaries.

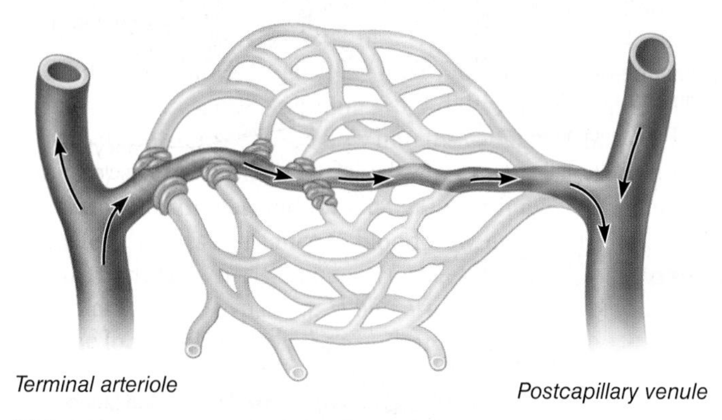

(b) **Sphincters closed**—blood flows through metarteriole–thoroughfare channel and bypasses true capillaries.

Figure 18.4 Anatomy of a capillary bed.

18.4 Veins are blood reservoirs that return blood toward the heart

→ **Learning Objective**

☐ **Describe the structure and function of veins, and explain how veins differ from arteries.**

Veins carry blood from the capillary beds toward the heart. Along the route, the diameter of successive venous vessels increases, and their walls gradually thicken as they progress from venules to larger and larger veins.

Venules

Capillaries unite to form **venules**, which range from 8 to 100 μm in diameter. The smallest venules, the *postcapillary venules*, consist entirely of endothelium around which pericytes congregate. Postcapillary venules are extremely porous (more like capillaries than veins in this way), and fluid and white blood cells move easily from the bloodstream through their walls. Indeed, a well-recognized sign of inflammation is adhesion of white blood cells to the postcapillary venule endothelium, followed by their migration through the wall into the inflamed tissue.

Larger venules have one or two layers of smooth muscle cells (a scanty tunica media) and a thin tunica externa as well.

Veins

Venules join to form *veins*. Veins usually have three distinct tunics, but their walls are always thinner and their lumens larger than those of corresponding arteries (see Figure 18.2 and Table 18.1). Consequently, in histological preparations, veins are usually collapsed and their lumens appear slitlike.

There is relatively little smooth muscle or elastin in the tunica media, which is poorly developed and tends to be thin even in the largest veins. The tunica externa is the heaviest wall layer. Consisting of thick longitudinal bundles of collagen fibers and elastic networks, it is often several times thicker than the tunica media. In the largest veins—the venae cavae, which return blood directly to the heart—longitudinal bands of smooth muscle make the tunica externa even thicker.

With their large lumens and thin walls, veins can accommodate a fairly large blood volume. Veins are called **capacitance vessels** and **blood reservoirs** because they can hold up to 65% of the body's blood supply at any time (**Figure 18.5**). Even so, these distensible vessels are usually not filled to capacity.

The walls of veins can be much thinner than arterial walls without danger of bursting because the blood pressure in veins is low. However, the low-pressure condition demands several structural adaptations to ensure that veins return blood to the heart at the same rate it was pumped into the circulation. One such adaptation is their large-diameter lumens, which offer relatively little resistance to blood flow.

Venous Valves

Venous valves prevent blood from flowing backward in veins just as valves do in the heart, and represent another adaptation to compensate for low venous pressure. They are formed from folds of the tunica intima and resemble the semilunar valves of the heart (see Figure 18.2). Venous valves are most abundant in the veins of the limbs, where gravity opposes the upward flow of blood. They are usually absent in veins of the thoracic and abdominal body cavities.

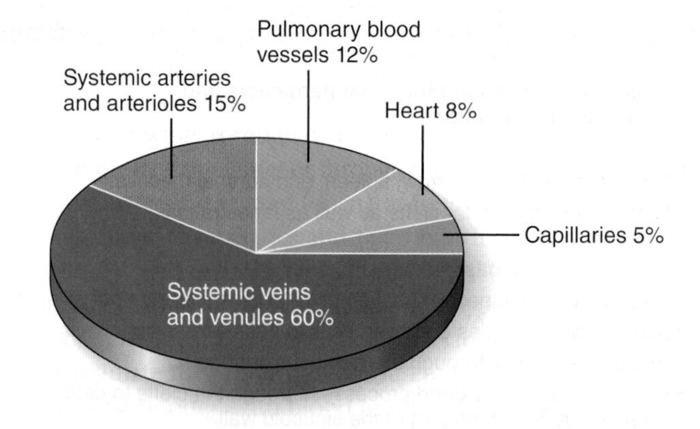

Figure 18.5 Relative proportion of blood volume throughout the cardiovascular system. The systemic veins are called capacitance vessels because they are distensible and contain a large proportion of the blood volume. Pulmonary blood vessels supply the lungs; systemic blood vessels supply the rest of the body.

The effectiveness of venous valves is demonstrated by this simple experiment: Hang one hand by your side until the blood vessels on its dorsal aspect distend with blood. Next place two fingertips against one of the distended veins, and pressing firmly, move the superior finger proximally along the vein and then release that finger. The vein will remain collapsed (flat) despite the pull of gravity. Finally, remove your distal fingertip and watch the vein refill with blood.

HOMEOSTATIC IMBALANCE 18.1 CLINICAL

Varicose veins are veins that are tortuous and dilated because of incompetent (leaky) valves. More than 15% of adults suffer from varicose veins, usually in the lower limbs.

Several factors contribute, including heredity and conditions that hinder venous return, such as prolonged standing in one position, obesity, or pregnancy. Both the "potbelly" of an overweight person and the enlarged uterus of a pregnant woman exert downward pressure on vessels of the groin, restricting return of blood to the heart. Consequently, blood pools in the lower limbs, and with time, the valves weaken and the venous walls stretch. Superficial veins, which receive little support from surrounding tissues, are especially susceptible.

Elevated venous pressure can also cause varicose veins. For example, straining to deliver a baby or have a bowel movement raises intra-abdominal pressure, preventing blood from draining from anal veins. The resulting varicosities in the anal veins are called *hemorrhoids* (hem′ŏ-roidz). +

Venous Sinuses

Venous sinuses, such as the *coronary sinus* of the heart and the *dural venous sinuses* of the brain, are highly specialized, flattened veins with extremely thin walls composed only of endothelium. They are supported by the tissues that surround them, rather than by any additional tunics. The dural venous sinuses, which receive cerebrospinal fluid and blood draining from the brain, are reinforced by the tough dura mater that covers the brain surface.

☑ Check Your Understanding

5. What is the function of venous valves? What forms the valves?
6. In the systemic circuit, which contains more blood—arteries or veins—or is it the same?

For answers, see Answers Appendix.

18.5 Anastomoses are special interconnections between blood vessels

→ Learning Objective

☐ **Explain the importance of vascular anastomoses.**

Blood vessels form special interconnections called **vascular anastomoses** (ah-nas″to-mo′sēz; "coming together"). Most organs receive blood from more than one arterial branch, and arteries supplying the same territory often merge, forming **arterial anastomoses**. These anastomoses provide alternate pathways, called **collateral channels**, for blood to reach a given body region. If one branch is cut or blocked by a clot, the collateral channel can often provide sufficient blood to the area.

Arterial anastomoses occur around joints, where active movement may hinder blood flow through one channel. They are also common in abdominal organs, the heart, and the brain (for example, the *cerebral arterial circle* in Figure 18.22d on p. 641). Arteries that supply the retina, kidneys, and spleen either do not anastomose or have a poorly developed collateral circulation. If their blood flow is interrupted, cells supplied by such vessels die.

The metarteriole–thoroughfare channel shunts of capillary beds that connect arterioles and venules are examples of **arteriovenous anastomoses**. Veins interconnect much more freely than arteries, and **venous anastomoses** are common. (You may be able to see venous anastomoses through the skin on the dorsum of your hand.) Because venous anastomoses are abundant, an occluded vein rarely blocks blood flow or leads to tissue death.

☑ Check Your Understanding

7. Which have more anastomoses, arteries or veins?

For answers, see Answers Appendix.

PART 2 PHYSIOLOGY OF CIRCULATION

Have you ever climbed a mountain? Well, get ready to climb a hypothetical mountain as you learn about circulatory dynamics. Like scaling a mountain, tackling blood pressure regulation and other topics of cardiovascular physiology is challenging while you're doing it, and exhilarating when you succeed. Let's begin the climb.

To sustain life, blood must be kept circulating. By now, you are aware that the heart is the pump, the arteries are pressure reservoirs and conduits, the arterioles are resistance vessels that control distribution, the capillaries are exchange sites, and the veins are conduits and blood reservoirs. Now for the dynamics of this system.

18.6 Blood flows from high to low pressure against resistance

→ Learning Objective

☐ **Define blood flow, blood pressure, and resistance, and explain the relationships between these factors.**

First we need to define three physiologically important terms—blood flow, blood pressure, and resistance—and examine how these factors relate to the physiology of blood circulation.

Definition of Terms

Blood Flow

Blood flow is the volume of blood flowing through a vessel, an organ, or the entire circulation in a given period (ml/min). If we consider the entire vascular system, blood flow is equivalent to cardiac output (CO), and under resting conditions, it is relatively constant. At any given moment, however, blood flow through *individual* body organs may vary widely according to their immediate needs.

Blood Pressure (BP)

Blood pressure (BP), the force per unit area exerted on a vessel wall by the contained blood, is expressed in millimeters of mercury (mm Hg). For example, a blood pressure of 120 mm Hg is equal to the pressure exerted by a column of mercury 120 mm high.

Unless stated otherwise, the term *blood pressure* means systemic arterial blood pressure in the largest arteries near the heart. The pressure gradient—the *differences* in blood pressure within the vascular system—provides the driving force that keeps blood moving, always from an area of higher pressure to an area of lower pressure, through the body.

Resistance

Resistance is opposition to flow and is a measure of the amount of friction blood encounters as it passes through the vessels. Because most friction is encountered in the peripheral (systemic) circulation, well away from the heart, we generally use the term **peripheral resistance**.

There are three important sources of resistance: blood viscosity, vessel length, and vessel diameter.

Blood Viscosity The internal resistance to flow that exists in all fluids is *viscosity* (vis-kos′ĭ-te) and is related to the thickness or "stickiness" of a fluid. The greater the viscosity, the less easily molecules slide past one another and the more difficult it is to get and keep the fluid moving. Blood is much more viscous than water. Because it contains formed elements and plasma proteins, it flows more slowly under the same conditions.

Blood viscosity is fairly constant, but conditions such as polycythemia (excessive numbers of red blood cells) can increase blood viscosity and, hence, resistance. On the other hand, if the red blood cell count is low, as in some anemias, blood is less viscous and peripheral resistance declines.

Total Blood Vessel Length The relationship between total blood vessel length and resistance is straightforward: the longer the vessel, the greater the resistance. For example, an infant's blood vessels lengthen as he or she grows to adulthood, and so both peripheral resistance and blood pressure increase.

Blood Vessel Diameter Because blood viscosity and vessel length are normally unchanging in the short term, the influence of these factors can be considered constant. However, blood vessel diameter changes frequently and significantly alters peripheral resistance. How so? The answer lies in principles of fluid flow. Fluid close to the wall of a tube or channel is slowed by friction as it passes along the wall, whereas fluid in the center of the tube flows more freely and faster. You can verify this by watching the flow of water in a river. Water close to the bank hardly seems to move, while that in the middle of the river flows quite rapidly.

In a tube of a given size, the relative speed and position of fluid in the different regions of the tube's cross section remain constant, a phenomenon called *laminar flow* or *streamlining*. The smaller the tube, the greater the friction, because relatively more of the fluid contacts the tube wall, where its movement is impeded.

Resistance varies *inversely* with the *fourth power* of the vessel radius (one-half the diameter). This means, for example, that if the radius of a vessel doubles, the resistance drops to one-sixteenth of its original value ($r^4 = 2 \times 2 \times 2 \times 2 = 16$ and $1/r^4 = 1/16$). For this reason, the large arteries close to the heart, which do not change dramatically in diameter, contribute little to peripheral resistance. Instead, the small-diameter arterioles, which can enlarge or constrict in response to neural and chemical controls, are the major determinants of peripheral resistance.

When blood encounters either an abrupt change in vessel diameter or rough or protruding areas of the tube wall (such as the fatty **plaques** on the tunica intima of the blood vessel that can intrude into the vessel lumen and narrow it, a condition called **atherosclerosis**), the smooth laminar blood flow is replaced by *turbulent flow*, that is, irregular fluid motion where blood from the different laminae (different layers of the tube's cross section) mixes. Turbulence dramatically increases resistance.

Relationship between Flow, Pressure, and Resistance

Now that we have defined these terms, let's summarize the relationships between them.

- Blood flow (F) is *directly* proportional to the difference in blood pressure (ΔP) between two points in the circulation, that is, the blood pressure, or hydrostatic pressure, gradient. Thus, when ΔP increases, blood flow speeds up, and when ΔP decreases, blood flow declines.
- Blood flow is *inversely* proportional to the peripheral resistance (R) in the systemic circulation; if R increases, blood flow decreases.

We can express these relationships by the formula

$$F = \frac{\Delta P}{R}$$

Of these two factors influencing blood flow, R is far more important than ΔP in influencing local blood flow because R can easily be changed by altering blood vessel diameter. For example, when the arterioles serving a particular tissue dilate (decreasing the resistance), blood flow to that tissue increases, even though the systemic pressure is unchanged or may actually be falling.

☑ Check Your **Understanding**

8. List three factors that determine resistance in a vessel. Which of these factors is physiologically most important?
9. Suppose vasoconstriction decreases the diameter of a vessel to one-third its size. What happens to the rate of flow through that vessel? Calculate the expected size of the change.

For answers, see Answers Appendix.

18.7 Blood pressure decreases as blood flows from arteries through capillaries and into veins

→ Learning Objective

☐ **Describe how blood pressure differs in the arteries, capillaries, and veins.**

Any fluid driven by a pump through a circuit of closed channels operates under pressure, and the nearer the fluid is to the pump, the greater the pressure exerted on the fluid. Blood flow in blood vessels is no exception, and blood flows through the blood vessels along a pressure gradient, always moving from higher- to lower-pressure areas. Fundamentally, *the pumping action of the heart generates blood flow. Pressure results when flow is opposed by resistance.*

As illustrated in **Figure 18.6**, systemic blood pressure is highest in the aorta and declines throughout the pathway to finally reach 0 mm Hg in the right atrium. The steepest drop in blood pressure occurs in the arterioles, which offer the greatest resistance to blood flow. However, as long as a pressure gradient exists, no matter how small, blood continues to flow until it completes the circuit back to the heart.

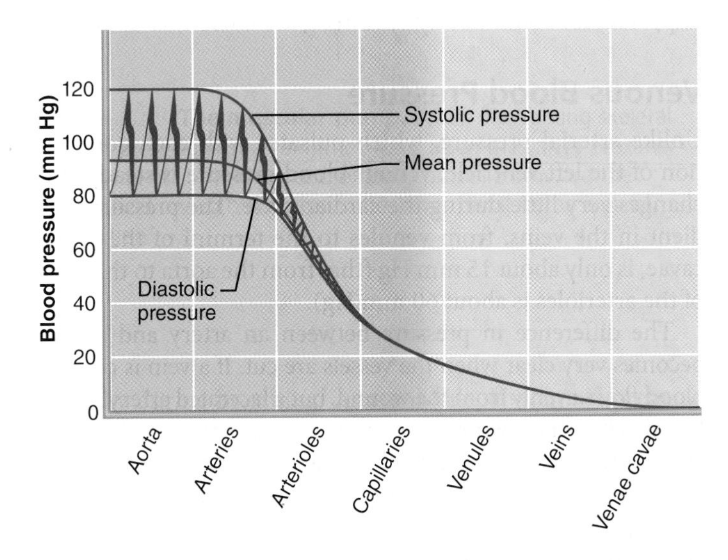

Figure 18.6 Blood pressure in various blood vessels of the systemic circulation.

Arterial Blood Pressure

Arterial blood pressure reflects two factors: (1) how much the elastic arteries close to the heart can stretch (their *compliance* or *distensibility*) and (2) the volume of blood forced into them at any time. If the amounts of blood entering and leaving the elastic arteries in a given period were equal, arterial pressure would be constant. Instead, as Figure 18.6 reveals, blood pressure is *pulsatile*—it rises and falls in a regular fashion—in the elastic arteries near the heart.

As the left ventricle contracts and expels blood into the aorta, it imparts kinetic energy to the blood, which stretches the elastic aorta as aortic pressure reaches its peak. Indeed, if the aorta were opened during this period, blood would spurt upward 5 or 6 feet! This pressure peak generated by ventricular contraction is called the **systolic pressure** (sis-tah′lik) and averages 120 mm Hg in healthy adults. Blood moves forward into the arterial bed because the pressure in the aorta is higher than the pressure in the more distal vessels.

During diastole, the aortic valve closes, preventing blood from flowing back into the heart. The walls of the aorta (and other elastic arteries) recoil, maintaining sufficient pressure to keep the blood flowing forward into the smaller vessels. During this time, aortic pressure drops to its lowest level (approximately 70 to 80 mm Hg in healthy adults), called the **diastolic pressure** (di-as-tah′lik). You can picture the elastic arteries as pressure reservoirs that operate as auxiliary pumps to keep blood circulating throughout the period of diastole, when the heart is relaxing. Essentially, the volume and energy of blood stored in the elastic arteries during systole are given back during diastole.

The difference between the systolic and diastolic pressures is called the **pulse pressure**. It is felt as a throbbing pulsation in an artery (a **pulse**) during systole as ventricular contraction forces blood into the elastic arteries and expands them. Increased stroke volume and faster blood ejection from the heart (a result of increased contractility) raise pulse pressure *temporarily*. Atherosclerosis chronically increases pulse pressure because the elastic arteries become less stretchy.

Because aortic pressure fluctuates up and down with each heartbeat, the important pressure to consider is the **mean arterial pressure (MAP)**—the pressure that propels the blood to the tissues. Diastole usually lasts longer than systole, so MAP is not simply the value halfway between systolic and diastolic pressures. Instead, it is roughly equal to the diastolic pressure plus one-third of the pulse pressure.

$$\text{MAP} = \text{diastolic pressure} + \frac{\text{pulse pressure}}{3}$$

For a person with a systolic blood pressure of 120 mm Hg and a diastolic pressure of 80 mm Hg:

$$\text{MAP} = 80 \text{ mm Hg} + \frac{40 \text{ mm Hg}}{3} = 93 \text{ mm Hg}$$

MAP and pulse pressure both decline with increasing distance from the heart. The MAP loses ground to the never-ending friction between the blood and the vessel walls, and the pulse pressure is gradually phased out in the less elastic muscular arteries, where elastic rebound of the vessels ceases to occur. At the end of the arterial tree, blood flow is steady and the pulse pressure has disappeared.

Clinical Monitoring of Circulatory Efficiency

Clinicians can assess the efficiency of a person's circulation by measuring pulse and blood pressure. These values, along with measurements of respiratory rate and body temperature, are referred to collectively as the body's **vital signs**. Let's examine how vital signs are determined or measured.

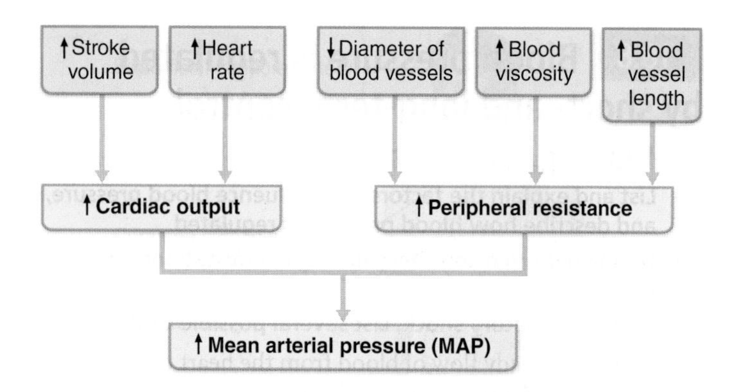

Figure 18.9 Major factors determining MAP. In addition, cardiac output increases as blood volume increases (not shown).

peripheral resistance and CO. *Long-term regulation* alters blood volume via the kidneys. Figure 18.12 (p. 624) summarizes the influence of nearly all of the important factors.

Short-Term Regulation: Neural Controls

Neural controls alter both cardiac output and peripheral resistance. We discussed neural control of cardiac output in Chapter 17, so we will focus on peripheral resistance here. Neural controls of peripheral resistance are directed at two main goals:

- Maintaining adequate MAP by altering blood vessel diameter on a moment-to-moment basis. (Remember, very small changes in blood vessel diameter cause substantial changes in peripheral resistance, and hence in systemic blood pressure.) Under conditions of low blood volume, all vessels except those supplying the heart and brain are constricted to allow as much blood as possible to flow to those two vital organs.
- Altering blood distribution to respond to specific demands of various organs. For example, during exercise blood is shunted temporarily from the digestive organs to the skeletal muscles.

Most neural controls operate via reflex arcs involving *baroreceptors* and associated afferent fibers. These reflexes are integrated in the cardiovascular center of the medulla, and their output travels via autonomic fibers to the heart and vascular smooth muscle. Occasionally, inputs from *chemoreceptors* and higher brain centers also influence the neural control mechanism.

Role of the Cardiovascular Center

Several clusters of neurons in the medulla oblongata act together to integrate blood pressure control by altering cardiac output and blood vessel diameter. This **cardiovascular center** consists of the *cardiac centers* (the cardioacceleratory and cardioinhibitory centers discussed in Chapter 17), and the **vasomotor center** that controls the diameter of blood vessels.

The vasomotor center transmits impulses at a fairly steady rate along sympathetic efferents called **vasomotor fibers**. These fibers exit from the T_1 through L_2 levels of the spinal cord and innervate the smooth muscle of blood vessels, mainly arterioles. As a result, the arterioles are almost always in a state of moderate constriction, called **vasomotor tone**.

The degree of vasomotor tone varies from organ to organ. Generally, arterioles of the skin and digestive viscera receive vasomotor impulses more frequently and tend to be more strongly constricted than those of skeletal muscles. Any increase in sympathetic activity produces generalized vasoconstriction and raises blood pressure. Decreased sympathetic activity allows the vascular muscle to relax somewhat and lowers blood pressure to basal levels.

Cardiovascular center activity is modified by inputs from (1) baroreceptors (pressure-sensitive mechanoreceptors that respond to changes in arterial pressure and stretch), (2) chemoreceptors (receptors that respond to changes in blood levels of carbon dioxide, H^+, and oxygen), and (3) higher brain centers. Let's take a look.

Baroreceptor Reflexes

When arterial blood pressure rises, it activates **baroreceptors**. These stretch receptors are located in the *carotid sinuses* (dilations in the internal carotid arteries, which provide the major blood supply to the brain), in the *aortic arch*, and in the walls of nearly every large artery of the neck and thorax. When stretched, baroreceptors send a rapid stream of impulses to the cardiovascular center, inhibiting the vasomotor and cardioacceleratory centers and stimulating the cardioinhibitory center. The result is a decrease in blood pressure (**Figure 18.10**).

Two mechanisms bring this about:

- **Vasodilation.** Decreased output from the vasomotor center allows arterioles and veins to dilate. **Arteriolar vasodilation** reduces peripheral resistance, so MAP falls. **Venodilation** shifts blood to the venous reservoirs, which decreases venous return and CO.
- **Decreased cardiac output.** Impulses to the cardiac centers inhibit sympathetic activity and stimulate parasympathetic activity, reducing heart rate and contractile force. As CO falls, so does MAP.

In the opposite situation, a decline in MAP initiates reflex vasoconstriction and increases cardiac output, bringing blood pressure back up. In this way, peripheral resistance and cardiac output are regulated in tandem to minimize changes in blood pressure.

Rapidly responding baroreceptors protect the circulation against short-term (acute) changes in blood pressure. For example, blood pressure falls (particularly in the head) when you stand up after reclining. Baroreceptors taking part in the **carotid sinus reflex** protect the blood supply to your brain, whereas those activated in the **aortic reflex** help maintain adequate blood pressure in your systemic circuit as a whole.

Baroreceptors are relatively *ineffective* in protecting us against sustained pressure changes, as evidenced by the fact that many people develop chronic hypertension. In such cases, the baroreceptors are "reprogrammed" (adapt) to monitor pressure changes at a higher set point.

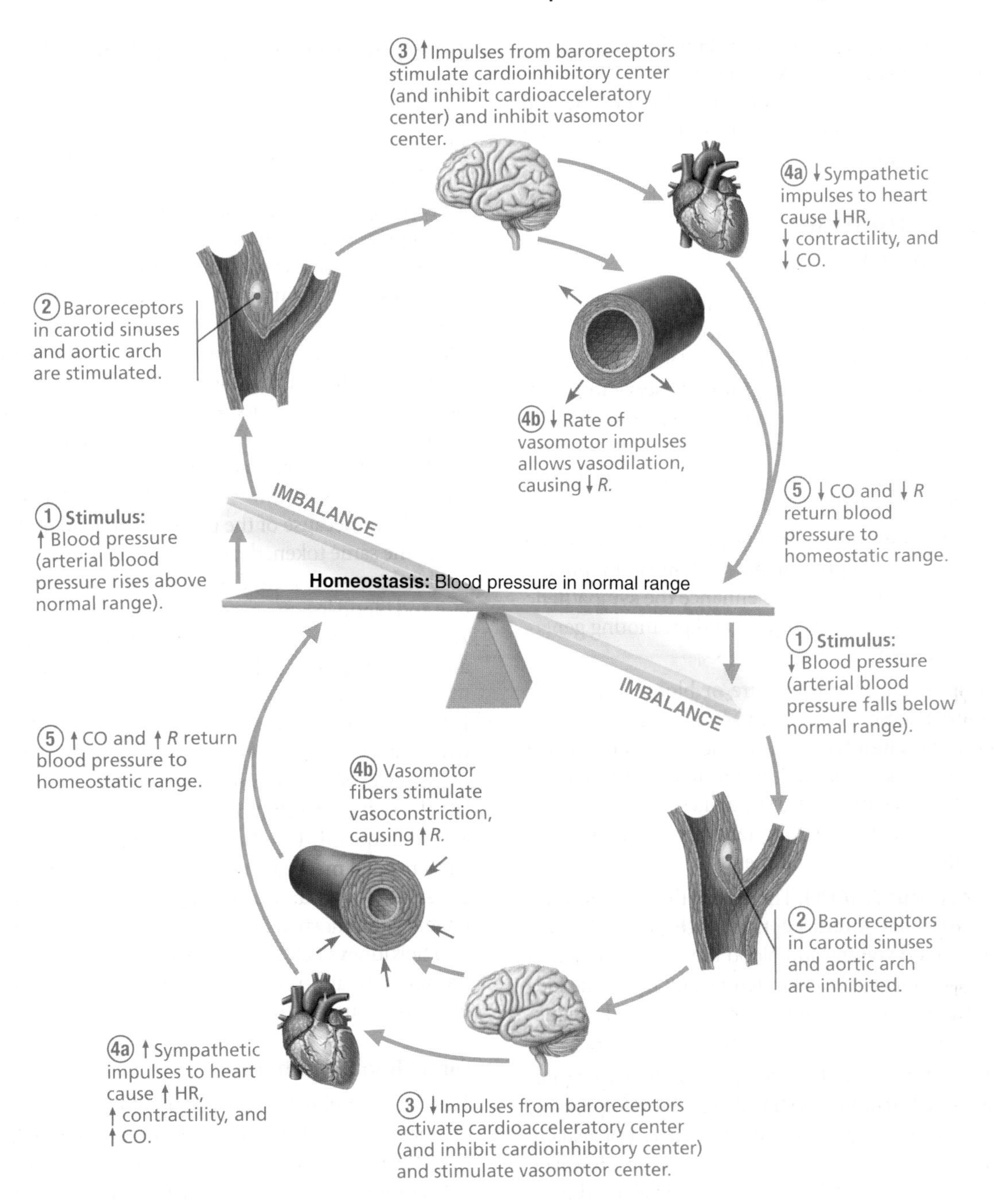

Figure 18.10 Baroreceptor reflexes that help maintain blood pressure homeostasis. (CO = cardiac output; *R* = peripheral resistance; HR = heart rate; BP = blood pressure)

Chemoreceptor Reflexes

When the carbon dioxide levels rise, or the pH falls, or oxygen content of the blood drops sharply, **chemoreceptors** in the aortic arch and large arteries of the neck transmit impulses to the cardioacceleratory center, which then increases cardiac output. Chemoreceptors also activate the vasomotor center, which causes reflex vasoconstriction. The rise in blood pressure that follows speeds the return of blood to the heart and lungs.

The most prominent chemoreceptors are the *carotid* and *aortic bodies* located close by the baroreceptors in the carotid sinuses and aortic arch. Chemoreceptors play a larger role in regulating respiratory rate than blood pressure, so we consider their function in Chapter 21.

Influence of Higher Brain Centers

Reflexes that regulate blood pressure are integrated in the medulla oblongata of the brain stem. Although the cerebral cortex and hypothalamus are not involved in routine controls of blood pressure, these higher brain centers can modify arterial pressure via relays to the medullary centers.

For example, the fight-or-flight response mediated by the hypothalamus has profound effects on blood pressure. (Even

18

Vascular Shock In **vascular shock**, blood volume is normal, but circulation is poor as a result of extreme vasodilation. A huge drop in peripheral resistance follows, as revealed by rapidly falling blood pressure.

A common cause of vascular shock is loss of vasomotor tone due to anaphylaxis (*anaphylactic shock*), a systemic allergic reaction in which the massive release of histamine triggers bodywide vasodilation. Two other common causes are failure of autonomic nervous system regulation (*neurogenic shock*), and septicemia (*septic shock*), a severe systemic bacterial infection (bacterial toxins are notorious vasodilators).

Cardiogenic Shock **Cardiogenic shock**, or pump failure, occurs when the heart is so inefficient that it cannot sustain adequate circulation. Its usual cause is myocardial damage, as might follow numerous myocardial infarctions (heart attacks).

Check Your Understanding

11. Describe the baroreceptor reflex changes that occur to maintain blood pressure when you rise from a lying-down to a standing position.
12. The kidneys play an important role in maintaining MAP by influencing which variable? Explain how renal artery obstruction could cause secondary hypertension.
13. Your neighbor, Bob, calls you because he thinks he is having an allergic reaction to a medication. You find Bob on the verge of losing consciousness and having trouble breathing. When paramedics arrive, they note his blood pressure is 63/38 and he has a rapid, thready pulse. Explain Bob's low blood pressure and rapid heart rate.
14. MAKING connections You have just learned that hypertension can be treated with a variety of different drugs including diuretics, beta-blockers, and calcium channel blockers. Using your knowledge of the autonomic nervous system (Chapter 14), smooth muscle (Chapter 9), and cardiac muscle (Chapter 17), explain how these drugs work to decrease blood pressure.

For answers, see Answers Appendix.

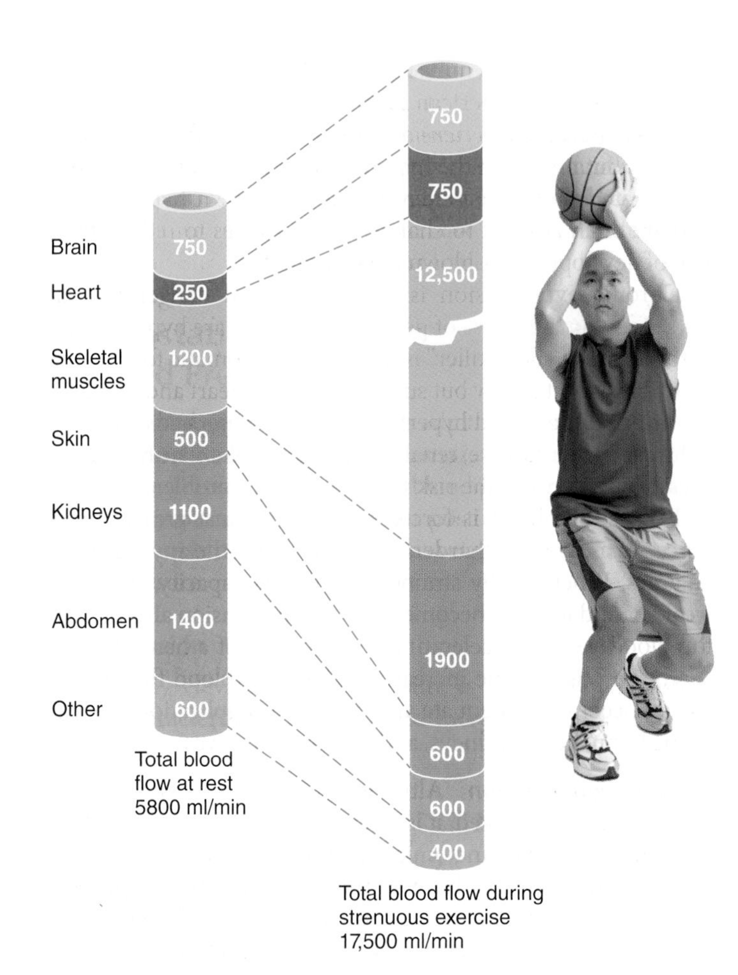

Figure 18.13 Distribution of blood flow at rest and during strenuous exercise.

18.9 Intrinsic and extrinsic controls determine blood flow through tissues

Learning Objective

- Explain how blood flow through tissues is regulated in general and in specific organs.

Blood flow through body tissues, or **tissue perfusion**, is involved in (1) delivering oxygen and nutrients to tissue cells, and removing wastes, (2) exchanging gases in the lungs, (3) absorbing nutrients from the digestive tract, and (4) forming urine in the kidneys. The rate of blood flow to each tissue and organ is almost exactly the right amount to provide for proper function—no more, no less. This is achieved by **intrinsic controls** (*autoregulation*) acting automatically on the smooth muscle of arterioles that feed any given tissue. We will examine these intrinsic mechanisms in the next section.

First, let's step back and look at the big picture. What do you think would happen if all of the arterioles in your body dilated at once? Because there is only a finite amount of blood, blood pressure would fall. Critical tissues, such as the brain, would be deprived of the oxygen and nutrients they need and would stop functioning. **Extrinsic controls** keep this from happening by acting on arteriolar smooth muscle to maintain blood pressure. The extrinsic controls act via the nerves (sympathetic nervous system) and hormones of the nervous and endocrine systems, the two major control systems of the body. They reduce blood flow to regions that need it the least, maintaining a constant MAP and allowing intrinsic mechanisms to direct blood flow to where it is most needed.

The redistribution of blood during exercise provides an example of how this works (**Figure 18.13**). When the body is at rest, the brain receives about 13% of total blood flow, the heart 4%, kidneys 20%, and abdominal organs 24%. Skeletal muscles, which make up almost half of body mass, normally receive about 20% of total blood flow. During exercise, however, nearly all of the increased cardiac output flushes into the skeletal muscles as intrinsic autoregulatory controls dilate skeletal muscle arterioles. To maintain blood pressure in spite of the widespread dilation of arterioles in skeletal muscle, the extrinsic controls act to decrease blood flow to the kidneys and digestive organs.

Autoregulation: Intrinsic (Local) Regulation of Blood Flow

As our activities change throughout the day, how does each organ or tissue manage to get the blood flow it needs? The answer is **autoregulation**. Local conditions regulate blood flow independent of control by nerves or hormones. Changes in blood flow through individual organs are controlled *intrinsically* by modifying the diameter of local arterioles feeding the capillaries.

You can compare blood flow autoregulation to water use in your home. Whether you have several taps open or none, the pressure in the main water pipe in the street remains relatively constant, as it does in the even larger water lines closer to the pumping station. Similarly, local conditions in the arterioles feeding the capillary beds of an organ have little effect on pressure in the muscular artery feeding that organ, or in the large elastic arteries. The pumping station is, of course, the heart. As long as the water company (circulatory feedback mechanisms) maintains a relatively constant water pressure (MAP), local demand regulates the amount of fluid (blood) delivered to various areas.

Organs regulate their own blood flows by varying the resistance of their arterioles. These intrinsic control mechanisms may be classed as *metabolic* (chemical) or *myogenic* (physical). Generally, both metabolic and myogenic factors determine the final autoregulatory response of a tissue. For example, **reactive hyperemia** (hi″per-e′me-ah) refers to the dramatically increased blood flow into a tissue that occurs after the blood supply to the area has been temporarily blocked. It results both from the myogenic response and from the metabolic wastes that accumulated during occlusion. Figure 18.14 summarizes the various intrinsic (local) and extrinsic controls of arteriolar diameter.

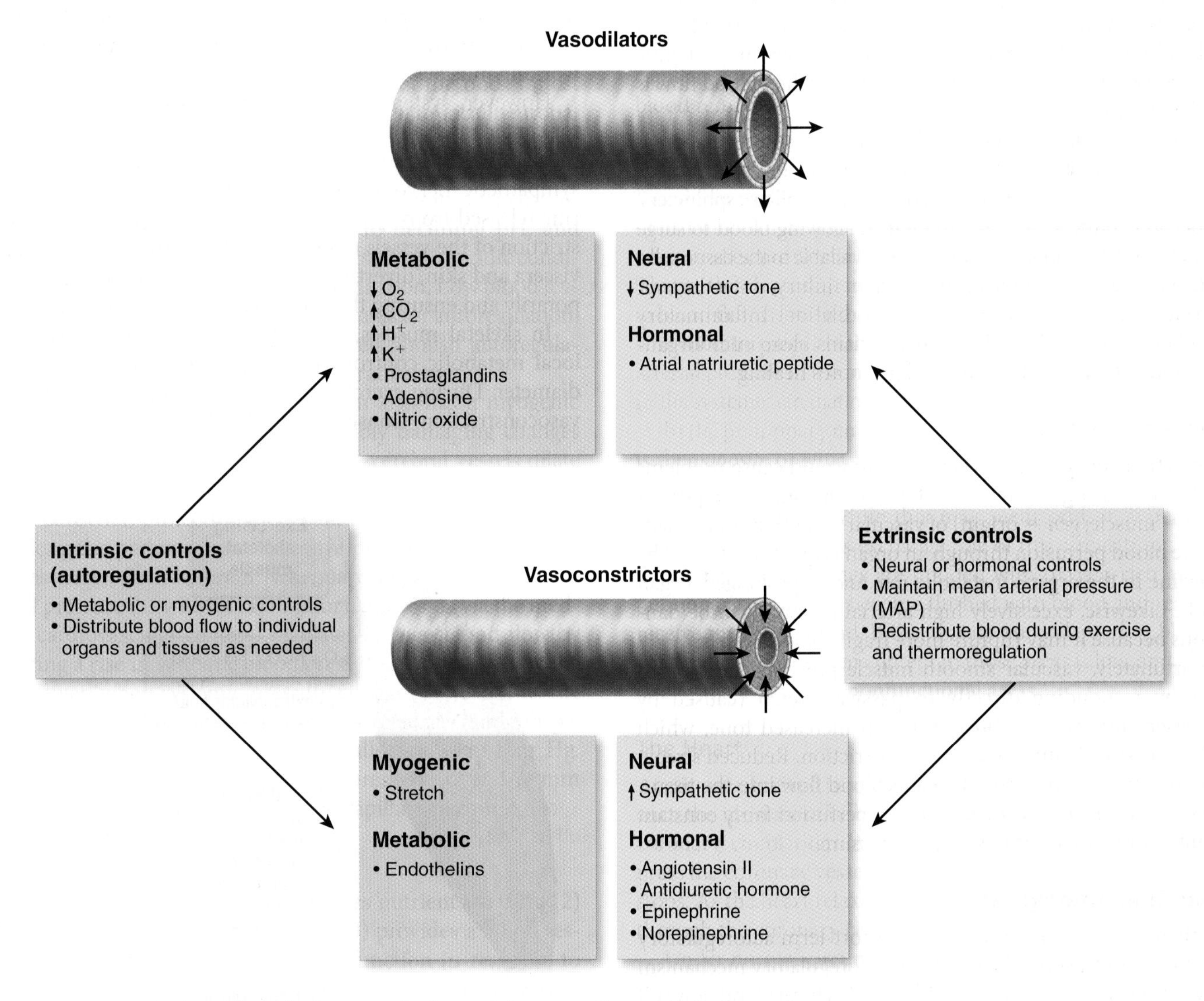

Figure 18.14 Intrinsic and extrinsic control of arteriolar smooth muscle in the systemic circulation. Controls are listed in the boxes below the arterioles. Epinephrine and norepinephrine constrict arteriolar smooth muscle by acting at α-adrenergic receptors. β-adrenergic receptors (causing vasodilation) are present in arterioles supplying skeletal and heart muscle, but their physiological relevance is minimal.

18

Table 18.7 Arteries of the Abdomen

The arterial supply to the abdominal organs arises from the abdominal aorta (Figure 18.24a). Under resting conditions, about half of the entire arterial flow moves through these vessels. Except for the celiac trunk, the superior and inferior mesenteric arteries, and the median sacral artery, all are paired vessels. These arteries supply the abdominal wall, diaphragm, and visceral organs of the abdominopelvic cavity. We discuss the branches in the order of their issue.

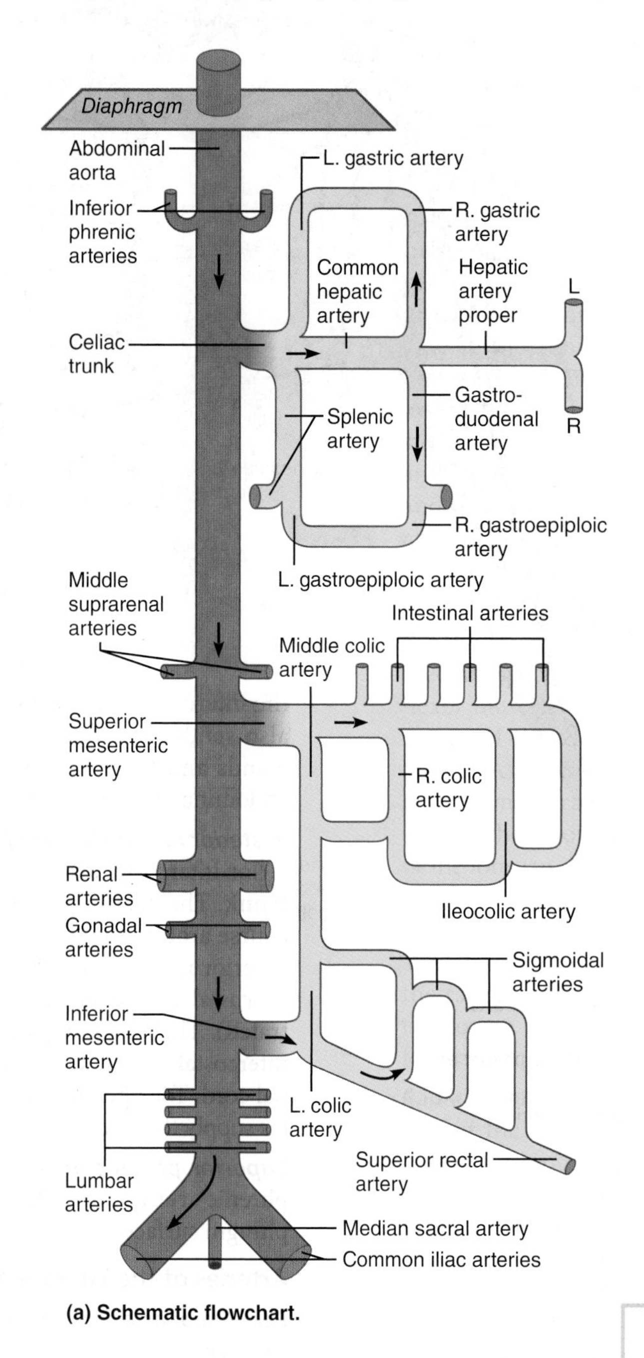

(a) Schematic flowchart.

Practice art labeling
MasteringA&P®>Study Area>Chapter 18

Figure 18.24 Arteries of the abdomen.

Table 18.7 *(continued)*

Description and Distribution

Inferior phrenic arteries. The inferior phrenics emerge from the aorta at T_{12}, just inferior to the diaphragm (Figure 18.24c). They serve the inferior diaphragm surface.

Celiac trunk. This very large unpaired branch of the abdominal aorta divides almost immediately into three branches (Figure 18.24b):

- **Common hepatic artery.** The **common hepatic artery** (hĕ-pat′ik) gives off branches to the stomach, duodenum, and pancreas. Where the **gastroduodenal artery** branches off, the common hepatic becomes the **hepatic artery proper**, which splits into right and left branches that serve the liver.
- **Splenic artery.** As the **splenic artery** (splen′ik) passes deep to the stomach, it sends branches to the pancreas and stomach and terminates in branches to the spleen.
- **Left gastric artery.** The **left gastric artery** (*gaster* = stomach) supplies part of the stomach and the inferior esophagus.

The **right** and **left gastroepiploic arteries** (gas″tro-ep″ĭ-plo′ik)—branches of the gastroduodenal and splenic arteries, respectively—serve the greater curvature of the stomach. A **right gastric artery**, which supplies the stomach's lesser curvature, may arise from the common hepatic artery or from the hepatic artery proper.

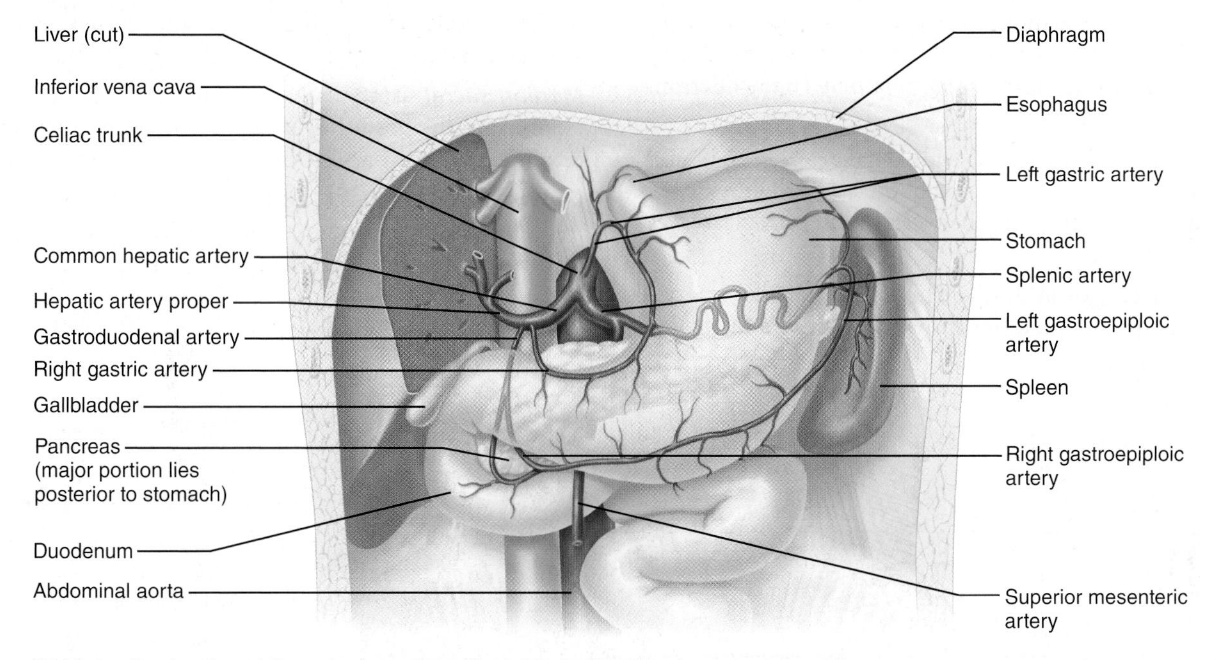

(b) The celiac trunk and its major branches. The left half of the liver has been removed.

Figure 18.24 *(continued)*

18

Table 18.7 Arteries of the Abdomen *(continued)*

Superior mesenteric artery (mes-en-ter′ik). This large, unpaired artery arises from the abdominal aorta at the L_1 level immediately below the celiac trunk (Figure 18.24d). It runs deep to the pancreas and then enters the mesentery (a drapelike membrane that supports the small intestine), where its numerous anastomosing branches serve virtually all of the small intestine via the **intestinal arteries**, and most of the large intestine—the appendix, cecum, ascending colon (via the **ileocolic** and **right colic arteries**), and part of the transverse colon (via the **middle colic artery**).

Suprarenal arteries (soo″prah-re′nal). The **middle suprarenal arteries** flank the origin of the superior mesenteric artery as they emerge from the abdominal aorta (Figure 18.24c). They supply blood to the adrenal (suprarenal) glands overlying the kidneys. The adrenal glands also receive two sets of branches not illustrated: *superior suprarenal* branches from the nearby inferior phrenic arteries, and *inferior suprarenal* branches from the nearby renal arteries.

Renal arteries. The short but wide renal arteries, right and left, issue from the lateral surfaces of the aorta slightly below the superior mesenteric artery (between L_1 and L_2). Each serves the kidney on its side.

Gonadal arteries (go-nă′dul). The paired gonadal arteries are called the **ovarian arteries** in females and the **testicular arteries** in males. The ovarian arteries extend into the pelvis to serve the ovaries and part of the uterine tubes. The much longer testicular arteries descend through the pelvis and inguinal canals to enter the scrotum, where they serve the testes.

Inferior mesenteric artery. This final major branch of the abdominal aorta is unpaired and arises from the anterior aortic surface at the L_3 level. It serves the distal part of the large intestine—from the midpart of the transverse colon to the midrectum—via its **left colic**, **sigmoidal**, and **superior rectal branches** (Figure 18.24d). Looping anastomoses between the superior and inferior mesenteric arteries help ensure that blood will continue to reach the digestive viscera in cases of trauma to one of these abdominal arteries.

Lumbar arteries. Four pairs of lumbar arteries arise from the posterolateral surface of the aorta in the lumbar region. These segmental arteries supply the posterior abdominal wall.

Median sacral artery. The unpaired median sacral artery issues from the posterior surface of the abdominal aorta at its terminus. This tiny artery supplies the sacrum and coccyx.

Common iliac arteries. At the L_4 level, the aorta splits into the right and left common iliac arteries, which supply blood to the lower abdominal wall, pelvic organs, and lower limbs (Figure 18.24c).

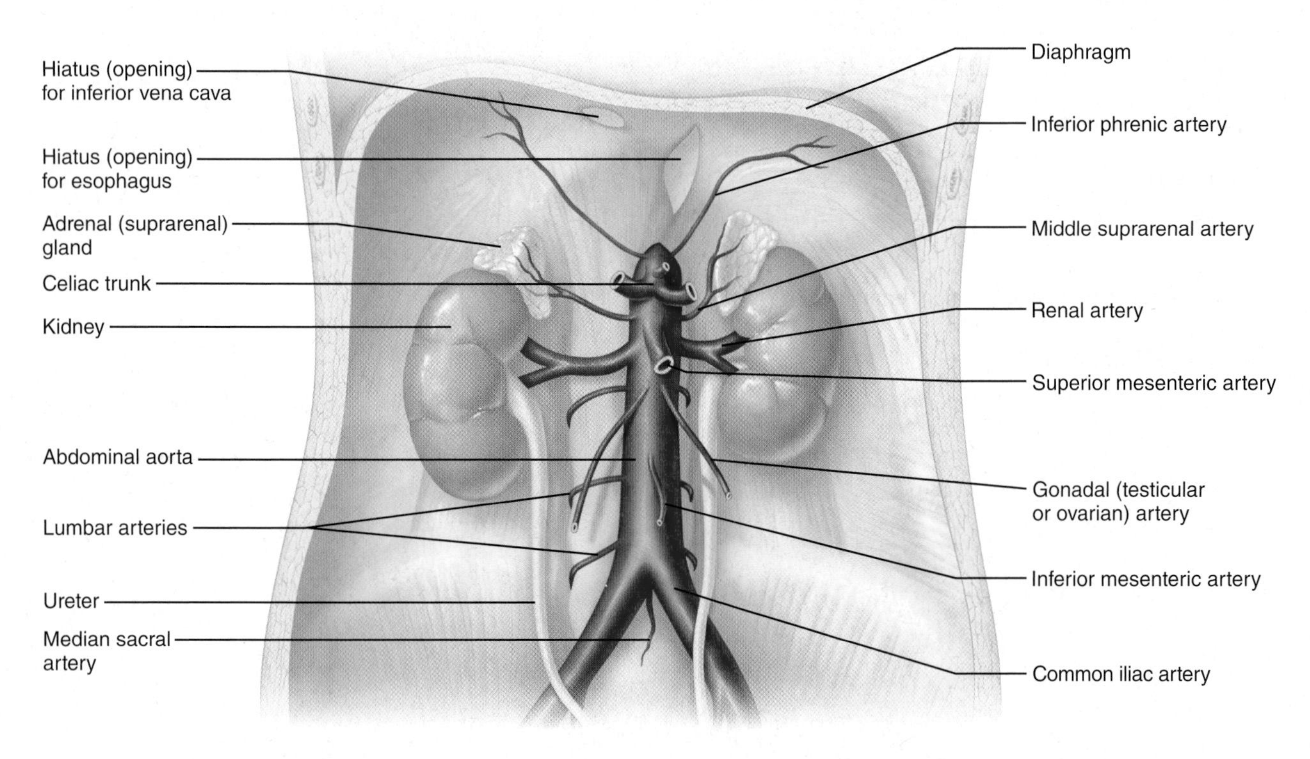

(c) Major branches of the abdominal aorta.

Figure 18.24 *(continued)* **Arteries of the abdomen.**

Table 18.7 *(continued)*

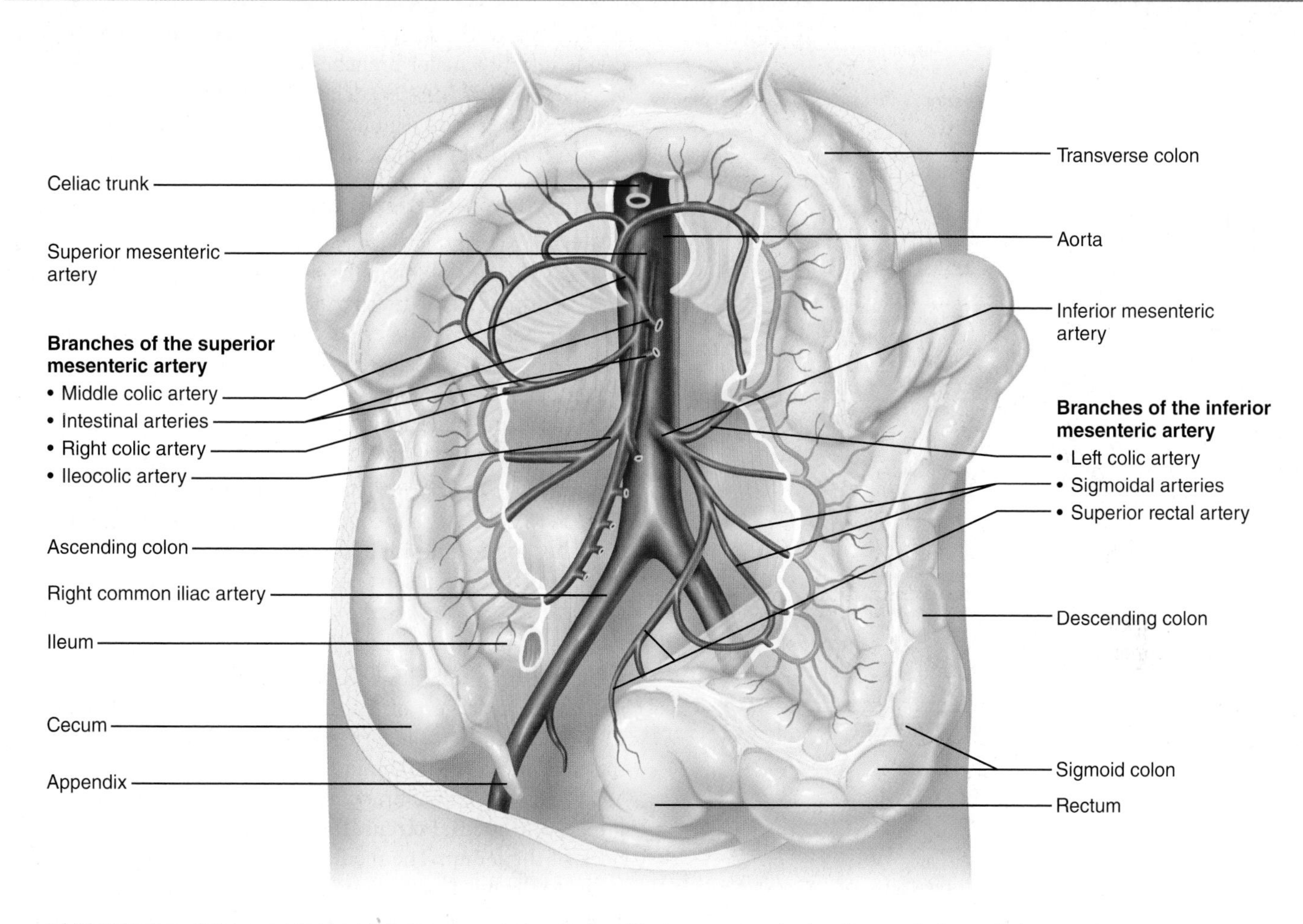

(d) Distribution of the superior and inferior mesenteric arteries. The transverse colon has been pulled superiorly.

Figure 18.24 *(continued)*

☑ Check Your Understanding

18. Which paired artery supplies most of the tissues of the head except for the brain and orbits?

19. Name the arterial anastomosis at the base of the cerebrum.

20. Name the four unpaired arteries that emerge from the abdominal aorta.

For answers, see Answers Appendix.

18

Table 18.8 Arteries of the Pelvis and Lower Limbs

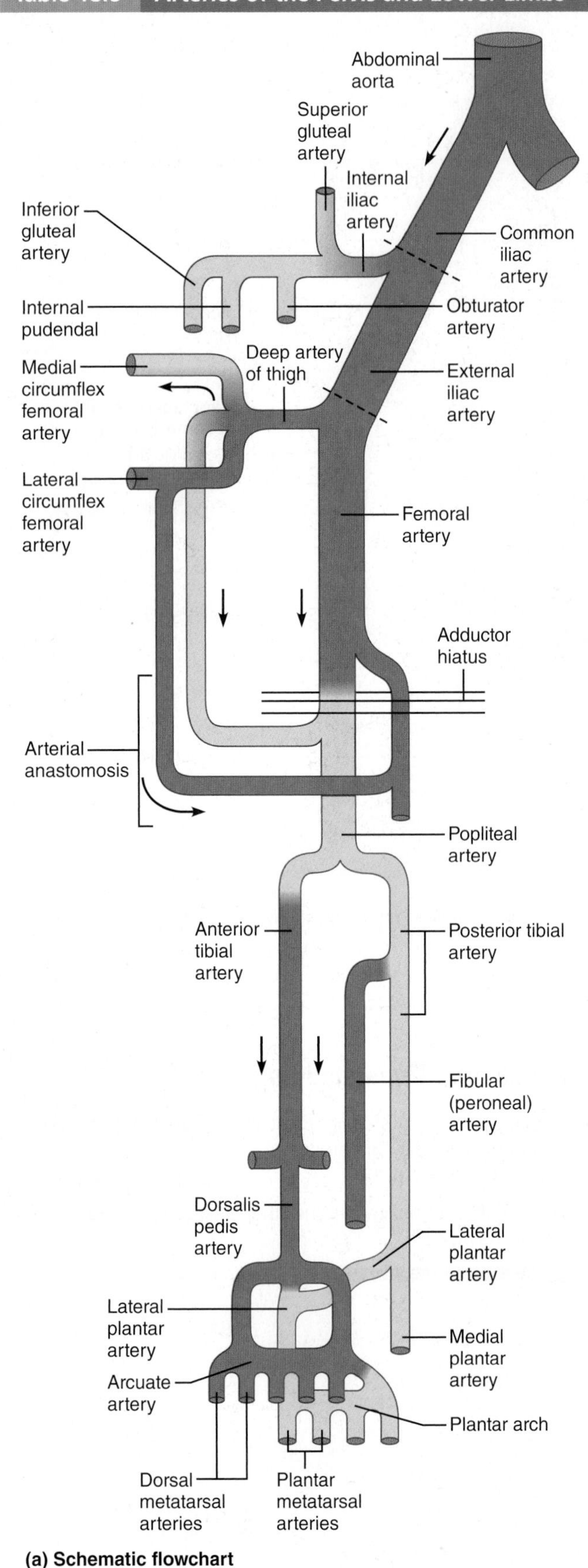

(a) Schematic flowchart

At the level of the sacroiliac joints, the **common iliac arteries** divide into two major branches, the internal and external iliac arteries (**Figure 18.25a**). The internal iliacs distribute blood mainly to the pelvic region. The external iliacs primarily serve the lower limbs but also send branches to the abdominal wall.

Description and Distribution

Internal iliac arteries. These paired arteries run into the pelvis and distribute blood to the pelvic walls and viscera (bladder and rectum, plus the uterus and vagina in the female and the prostate and ductus deferens in the male). Additionally they serve the gluteal muscles via the **superior** and **inferior gluteal arteries**, adductor muscles of the medial thigh via the **obturator artery**, and external genitalia and perineum via the **internal pudendal artery** (not illustrated).

External iliac arteries. These arteries supply the lower limbs (Figure 18.25b). As they course through the pelvis, they give off branches to the anterior abdominal wall. After passing under the inguinal ligaments to enter the thigh, they become the femoral arteries.

Femoral arteries. As each of these arteries passes down the anteromedial thigh, it gives off several branches to the thigh muscles. The largest of the deep branches is the **deep artery of the thigh** (also called the *deep femoral artery*), which is the main supply to the thigh muscles (hamstrings, quadriceps, and adductors). Proximal branches of the deep femoral artery, the **lateral** and **medial circumflex femoral arteries**, encircle the neck of the femur. The medial circumflex artery is the major vessel to the head of the femur. If it is torn in a hip fracture, the bone tissue of the head of the femur dies. A long descending branch of the lateral circumflex artery supplies the vastus lateralis muscle. Near the knee the femoral artery passes posteriorly and through a gap in the adductor magnus muscle, the *adductor hiatus*, to enter the popliteal fossa, where its name changes to popliteal artery.

Popliteal artery. This posterior vessel contributes to an arterial anastomosis that supplies the knee region and then splits into the anterior and posterior tibial arteries of the leg.

Anterior tibial artery. The anterior tibial artery runs through the anterior compartment of the leg, supplying the extensor muscles along the way. At the ankle, it becomes the **dorsalis pedis artery**, which supplies the ankle and dorsum of the foot, and gives off a branch, the **arcuate artery**, which issues the **dorsal metatarsal arteries** to the metatarsus of the foot. The superficial dorsalis pedis ends by penetrating into the sole where it forms the medial part of the **plantar arch**. The dorsalis pedis artery provides a clinically important pulse point, the pedal pulse. If the pedal pulse is easily felt, it is fairly certain that the blood supply to the leg is good.

Figure 18.25 Arteries of the right pelvis and lower limb.

18

Table 18.8 *(continued)*

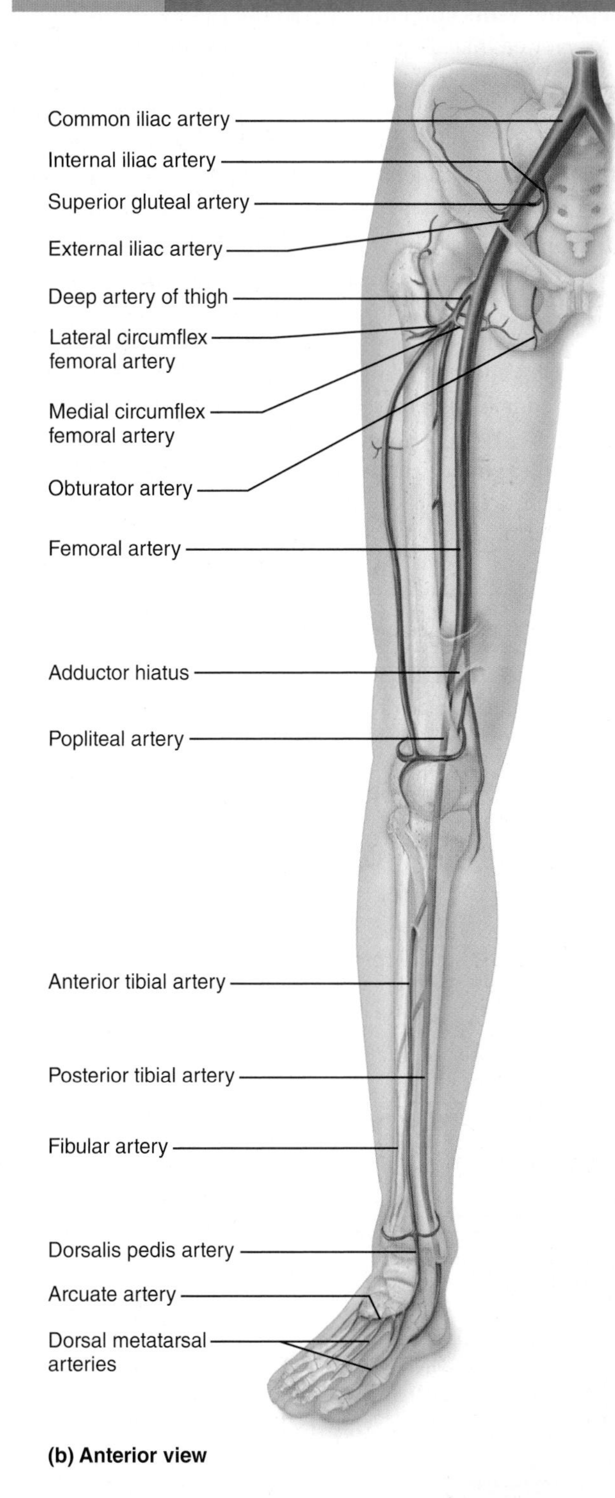

(b) Anterior view

Posterior tibial artery. This large artery courses through the posteromedial part of the leg and supplies the flexor muscles. Proximally, it gives off a large branch, the **fibular (peroneal) artery**, which supplies the lateral fibularis muscles of the leg. On the medial side of the foot, the posterior tibial artery divides into **lateral** and **medial plantar arteries** that serve the plantar surface of the foot. The lateral plantar artery forms the lateral end of the plantar arch. **Plantar metatarsal arteries** and digital arteries to the toes arise from the plantar arch.

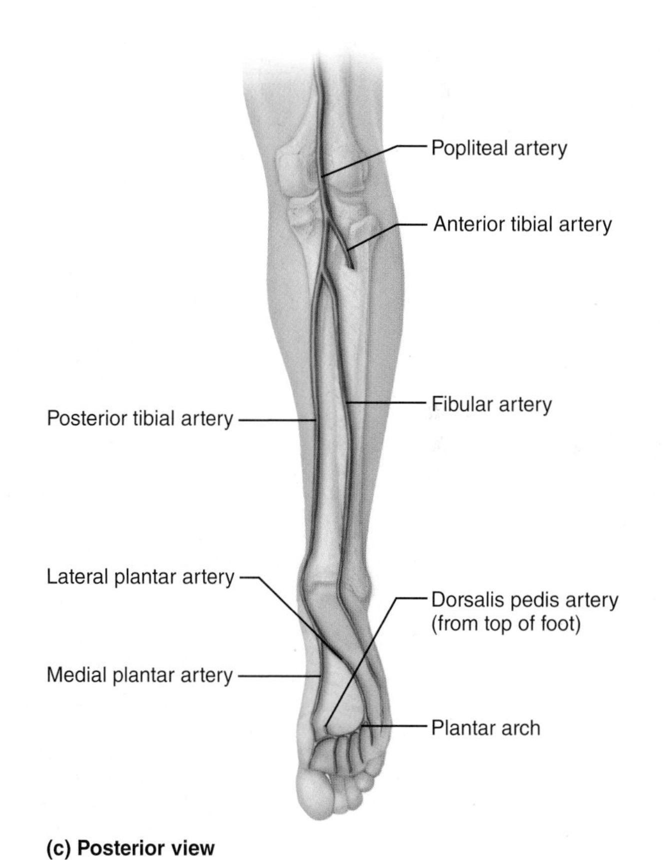

(c) Posterior view

Figure 18.25 *(continued)*

☑ Check Your Understanding

21. You are assessing the circulation in the leg of a diabetic patient at the clinic. Name the artery you palpate in each of these three locations: behind the knee, behind the medial malleolus of the tibia, on the dorsum of the foot.

For answers, see Answers Appendix.

18

Table 18.9 The Venae Cavae and the Major Veins of the Systemic Circulation

In our survey of the systemic veins, the major tributaries (branches) of the venae cavae are noted first in **Figure 18.26**, followed by a description in Tables 18.10 through 18.13 of the venous pattern of the various body regions. Because veins run toward the heart, the most distal veins are named first and those closest to the heart last. Deep veins generally drain the same areas served by their companion arteries, so they are not described in detail.

Description and Areas Drained

Superior vena cava. This great vein receives systemic blood draining from all areas superior to the diaphragm, except the heart wall. It is formed by the union of the **right** and **left brachiocephalic veins** and empties into the right atrium (Figure 18.26b). Notice that there are two brachiocephalic veins, but only one brachiocephalic artery (trunk). Each brachiocephalic vein is formed by the joining of the **internal jugular** and **subclavian veins** on its side. In most of the flowcharts that follow, only the vessels draining blood from the right side of the body are shown (except for the azygos circulation of the thorax).

Inferior vena cava. The widest blood vessel in the body, this vein returns blood to the heart from all body regions below the diaphragm. The abdominal aorta lies directly to its left. The paired **common iliac veins** join at L_5 to form the distal end of the inferior vena cava. From this point, it courses superiorly along the anterior aspect of the spine, receiving venous blood from the abdominal walls, gonads, and kidneys. Immediately above the diaphragm, the inferior vena cava ends as it enters the inferior aspect of the right atrium.

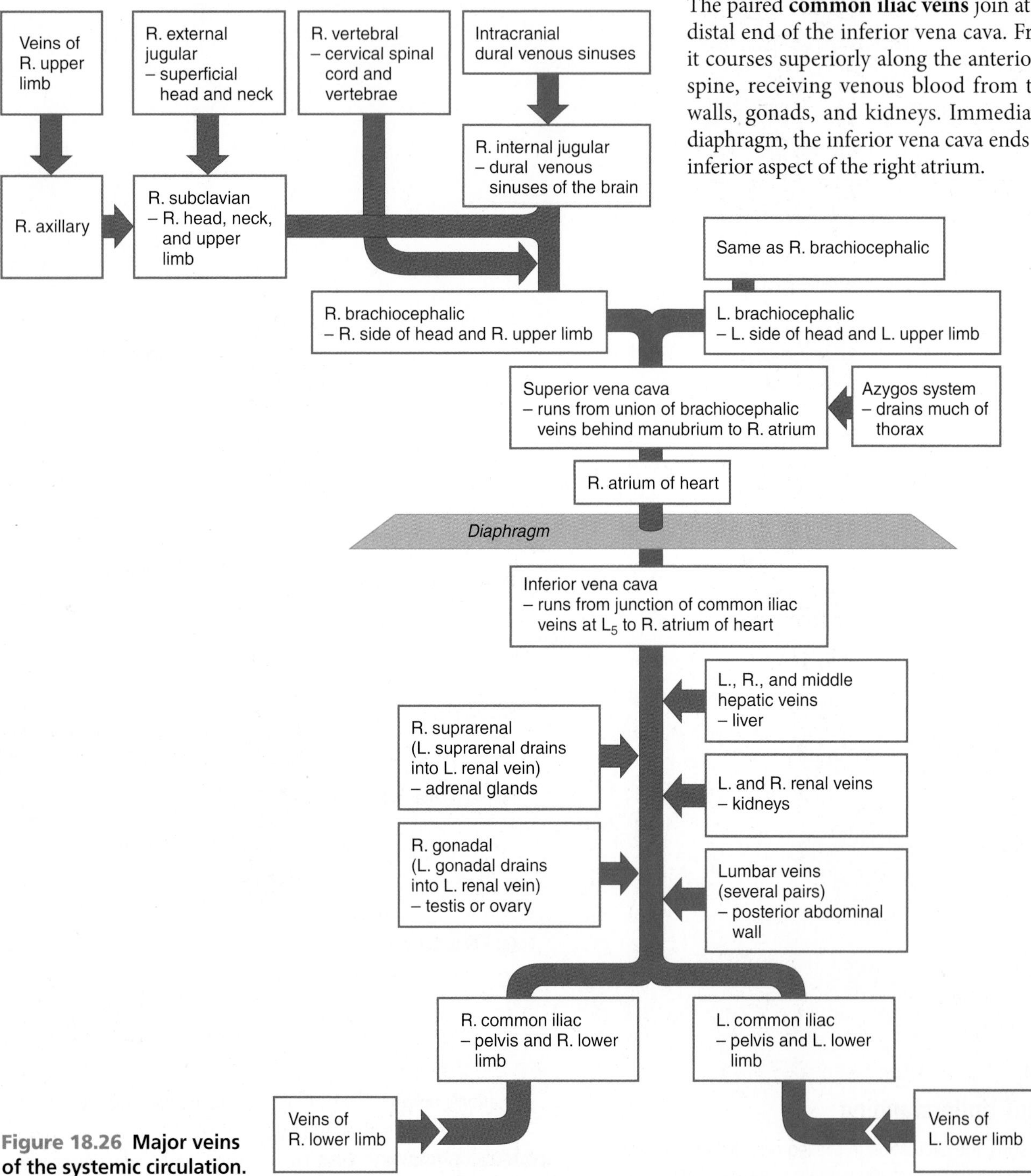

Figure 18.26 Major veins of the systemic circulation.

(a) Schematic flowchart

Table 18.9 *(continued)*

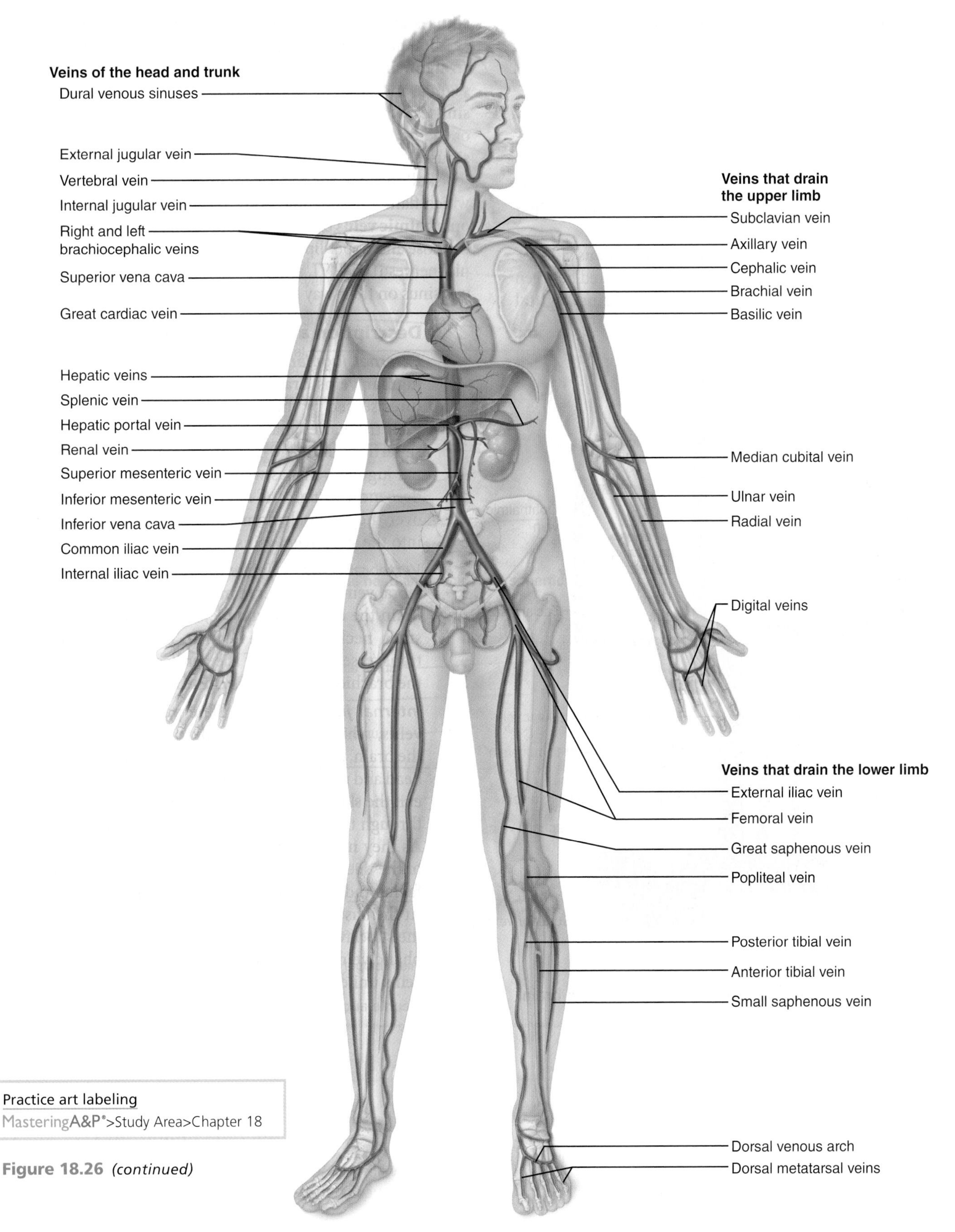

Practice art labeling
MasteringA&P®>Study Area>Chapter 18

Figure 18.26 *(continued)*

(b) Illustration, anterior view. The vessels of the pulmonary circulation are not shown.

Table 18.13 Veins of the Pelvis and Lower Limbs

As in the upper limbs, most deep veins of the lower limbs have the same names as the arteries they accompany and many are double. Poorly supported by surrounding tissues, the two superficial saphenous veins (great and small) are common sites of varicosities. The great saphenous (*saphenous* = obvious) vein is frequently excised and used as a coronary bypass vessel.

Description and Areas Drained

Deep veins. After being formed by the union of the **medial** and **lateral plantar veins**, the **posterior tibial vein** ascends deep in the calf muscle and receives the **fibular (peroneal) vein** (Figure 18.30). The **anterior tibial vein**, which is the superior continuation of the **dorsalis pedis vein** of the foot, unites at the knee with the posterior tibial vein to form the **popliteal vein**, which crosses the back of the knee. As the popliteal vein emerges from the knee, it becomes the **femoral vein**, which drains the deep structures of the thigh. The femoral vein becomes the **external iliac vein** as it enters the pelvis. In the pelvis, the external iliac vein unites with the **internal iliac vein** to form the **common iliac vein**. The distribution of the internal iliac veins parallels that of the internal iliac arteries.

Superficial veins. The **great** and **small saphenous veins** (sah-fe′nus) issue from the **dorsal venous arch** of the foot (Figure 18.30b and c). These veins anastomose frequently with each other and with the deep veins along their course. The great saphenous vein is the longest vein in the body. It travels superiorly along the medial aspect of the leg to the thigh, where it empties into the femoral vein just distal to the inguinal ligament. The small saphenous vein runs along the lateral aspect of the foot and then through the deep fascia of the calf muscles, which it drains. At the knee, it empties into the popliteal vein.

(a) Schematic flowchart of the anterior and posterior veins

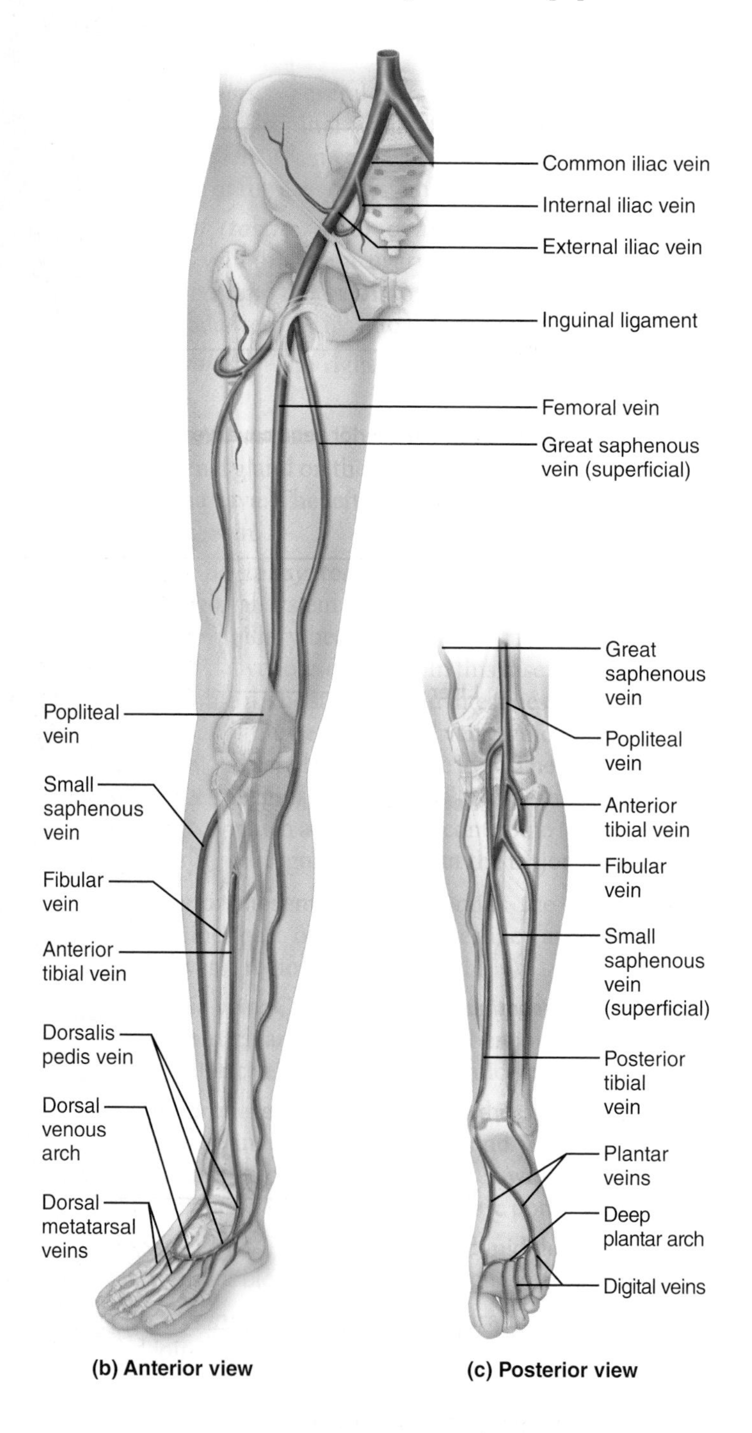

(b) Anterior view

(c) Posterior view

Figure 18.30 Veins of the right lower limb.

Check Your Understanding

24. What is a portal system? What is the function of the hepatic portal system?

25. Name the leg veins that often become varicosed.

For answers, see Answers Appendix.

Now that we have described the structure and function of blood vessels, our survey of the cardiovascular system is complete. The pump, the plumbing, and the circulating fluid form a dynamic organ system that ceaselessly services every other organ system of the body. However, our study of the *circulatory system* is still unfinished because we have yet to examine the lymphatic system, which acts with the cardiovascular system to ensure continuous circulation and to provide sites from which lymphocytes can police the body and provide immunity. These are the topics of Chapter 19.

REVIEW QUESTIONS

For more chapter study tools, go to the Study Area of MasteringA&P®.

There you will find:

- Interactive Physiology iP
- A&PFlix A&PFlix
- Interactive Physiology 2.0 iP2
- PhysioEx PEx
- Practice Anatomy Lab PAL
- Videos, Practice Quizzes and Tests, MP3 Tutor Sessions, Case Studies, and much more!

Multiple Choice/Matching

(Some questions have more than one correct answer. Select the best answer or answers from the choices given.)

1. Which statement does not accurately describe veins? **(a)** Have less elastic tissue and smooth muscle than arteries, **(b)** contain more fibrous tissue than arteries, **(c)** most veins in the extremities have valves, **(d)** always carry deoxygenated blood.
2. Smooth muscle in the blood vessel wall **(a)** is found primarily in the tunica intima, **(b)** is mostly circularly arranged, **(c)** is most abundant in veins, **(d)** is usually innervated by the parasympathetic nervous system.
3. Peripheral resistance **(a)** is inversely proportional to the length of the vascular bed, **(b)** increases in anemia, **(c)** decreases in polycythemia, **(d)** is inversely related to the diameter of the arterioles.
4. Which of the following can lead to decreased venous return of blood to the heart? **(a)** an increase in blood volume, **(b)** an increase in venous pressure, **(c)** damage to the venous valves, **(d)** increased muscular activity.
5. Arterial blood pressure increases in response to **(a)** increasing stroke volume, **(b)** increasing heart rate, **(c)** atherosclerosis, **(d)** rising blood volume, **(e)** all of these.
6. Which of the following would *not* result in the dilation of the feeder arterioles and opening of the precapillary sphincters in systemic capillary beds? **(a)** a decrease in local tissue O_2 content, **(b)** an increase in local tissue CO_2, **(c)** a local increase in histamine, **(d)** a local increase in pH.
7. The structure of a capillary wall differs from that of a vein or an artery because **(a)** it has two tunics instead of three, **(b)** there is less smooth muscle, **(c)** it has a single tunic—only the tunica intima, **(d)** none of these.
8. The baroreceptors in the carotid sinus and aortic arch are sensitive to **(a)** a decrease in CO_2, **(b)** changes in arterial pressure, **(c)** a decrease in O_2, **(d)** all of these.
9. The myocardium receives its blood supply directly from the **(a)** aorta, **(b)** coronary arteries, **(c)** coronary sinus, **(d)** pulmonary arteries.
10. Blood flow in the capillaries is steady despite the rhythmic pumping of the heart because of the **(a)** elasticity of the large arteries, **(b)** small diameter of capillaries, **(c)** thin walls of the veins, **(d)** venous valves.
11. Using the letters from column B, match the artery descriptions in column A. (Note that some require more than a single choice.)

Column A	Column B
____ (1) unpaired branch of abdominal aorta	(a) right common carotid
	(b) superior mesenteric
	(c) left common carotid
____ (2) second branch of aortic arch	(d) external iliac
	(e) inferior mesenteric
____ (3) branch of internal carotid	(f) superficial temporal
	(g) celiac trunk
____ (4) branch of external carotid	(h) facial
	(i) ophthalmic
____ (5) origin of femoral arteries	(j) internal iliac

12. Tracing the blood from the heart to the right hand, we find that blood leaves the heart and passes through the aorta, the right subclavian artery, the axillary and brachial arteries, and through either the radial or ulnar artery to arrive at the hand. Which artery is missing from this sequence? **(a)** coronary, **(b)** brachiocephalic, **(c)** cephalic, **(d)** right common carotid.
13. Which of the following do not drain directly into the inferior vena cava? **(a)** inferior phrenic veins, **(b)** hepatic veins, **(c)** inferior mesenteric vein, **(d)** renal veins.
14. Suppose that at a given point along a capillary, the following forces exist: capillary hydrostatic pressure (HP_c) = 30 mm Hg, interstitial fluid hydrostatic pressure (HP_{if}) = 0 mm Hg, capillary colloid osmotic pressure (OP_c) = 25 mm Hg, and interstitial fluid colloid osmotic pressure (OP_{if}) = 2 mm Hg. The net filtration pressure at this point in the capillary is **(a)** 3 mm Hg, **(b)** −3 mm Hg, **(c)** −7 mm Hg, **(d)** 7 mm Hg.

Short Answer Essay Questions

15. How is the anatomy of capillaries and capillary beds well suited to their function?
16. Distinguish between elastic arteries, muscular arteries, and arterioles relative to location, histology, and functional adaptations.
17. Write an equation showing the relationship between peripheral resistance, blood flow, and blood pressure.

18

18. (a) Define blood pressure. Differentiate between systolic and diastolic blood pressure. (b) What is the normal blood pressure value for an adult?
19. Describe the neural mechanisms responsible for controlling blood pressure.
20. Explain the reasons for the observed changes in blood flow velocity in the different regions of the circulation.
21. How does the control of blood flow to the skin for the purpose of regulating body temperature differ from the control of nutrient blood flow to skin cells?
22. Describe neural and chemical (both systemic and local) effects exerted on the blood vessels when you are fleeing from a mugger. (Be careful, this is more involved than it appears at first glance.)
23. How are nutrients, wastes, and respiratory gases transported to and from the blood and tissue spaces?
24. (a) What blood vessels contribute to the formation of the hepatic portal circulation? (b) Why is a portal circulation a "strange" circulation?
25. Physiologists often consider capillaries and postcapillary venules together. (a) What functions do these vessels share? (b) Structurally, how do they differ?

AT THE CLINIC

Clinical Case Study

Cardiovascular System: Blood Vessels

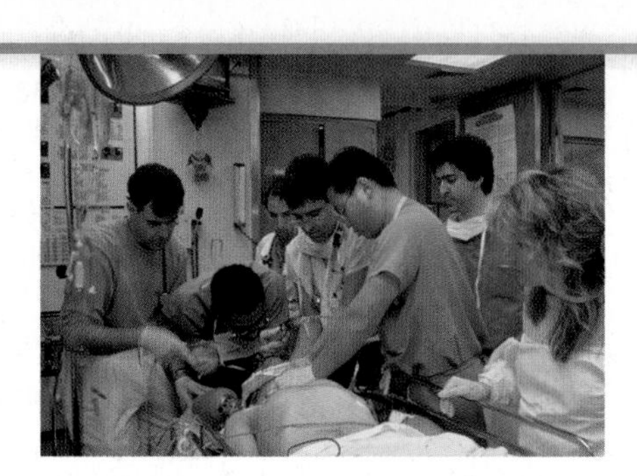

Mr. Hutchinson, another middle-aged victim of the collision on Route 91, has a tourniquet around his thigh when admitted in an unconscious state to Noble Hospital. The emergency technician who brings him in states that his right lower limb was pinned beneath the bus for at least 30 minutes. He is immediately scheduled for surgery. Admission notes include the following:

- Multiple contusions of lower limbs
- Compound fracture of the right tibia; bone ends covered with sterile gauze
- Right leg blanched and cold, no pulse
- Blood pressure 90/48; pulse 140/min and thready; patient diaphoretic (sweaty)

1. Relative to what you have learned about tissue requirements for oxygen, what is the condition of the tissues in the right lower limb?
2. Will the fracture be attended to, or will Mr. Hutchinson's other homeostatic needs take precedence? Explain your answer choice and predict his surgical treatment.
3. What do you conclude regarding Mr. Hutchinson's cardiovascular measurements (pulse and BP), and what measures do you expect will be taken to remedy the situation before commencing surgery?

For answers, see Answers Appendix.

19 The Lymphatic System and Lymphoid Organs and Tissues

KEY CONCEPTS

They can't all be superstars! When we mentally tick off the names of the body's organ systems, the lymphatic (lim-fat′ik) system and the lymphoid organs and tissues are probably not the first to come to mind. Yet if they failed their quiet background work, our cardiovascular system would stop working and our immune system would be hopelessly impaired.

In this chapter, we will explore two functionally different but structurally overlapping systems, the *lymphatic system* and the *lymphoid organs and tissues*. The **lymphatic system** returns fluids that have leaked from the vascular system back to the blood. It consists of three parts:

- A meandering network of *lymphatic vessels*
- *Lymph*, the fluid contained in those vessels
- *Lymph nodes* that cleanse the lymph as it passes through them

The **lymphoid organs and tissues** provide the structural basis of the immune system. These organs and tissues play essential roles in the body's defense mechanisms and its resistance to disease. These structures include the spleen, thymus, tonsils, and other lymphoid tissues scattered throughout the body. The lymph nodes are also part of this system, and like a keystone, they have important roles to play in both the lymphoid organs and tissues and the lymphatic system.

Let's begin by looking at the lymphatic system.

19.1 The lymphatic system includes lymphatic vessels, lymph, and lymph nodes

→ Learning Objectives

- ☐ **List the functions of the lymphatic vessels.**
- ☐ **Describe the structure and distribution of lymphatic vessels.**
- ☐ **Describe the source of lymph and mechanism(s) of lymph transport.**

As blood circulates through the body, nutrients, wastes, and gases are exchanged between the blood and the interstitial fluid. As we explained in *Focus on Bulk Flow across Capillary Walls* (Focus Figure 18.1 on pp. 632–633), the hydrostatic and colloid osmotic pressures operating at capillary beds force fluid out of the blood at the arterial ends of the beds ("upstream") and cause most of it to be reabsorbed at the venous ends ("downstream"). The fluid that remains behind in the tissue spaces, as much as 3 L daily, becomes part of the interstitial fluid.

This leaked fluid, plus any plasma proteins that escape from the bloodstream, must somehow be returned to the blood to ensure that the cardiovascular system has sufficient blood volume. This problem of circulatory dynamics is resolved by the **lymphatic vessels**, or **lymphatics**, elaborate networks of drainage vessels that collect the excess protein-containing interstitial fluid and return it to the bloodstream. Once interstitial fluid enters the lymphatic vessels, it is called **lymph** (*lymph* = clear water).

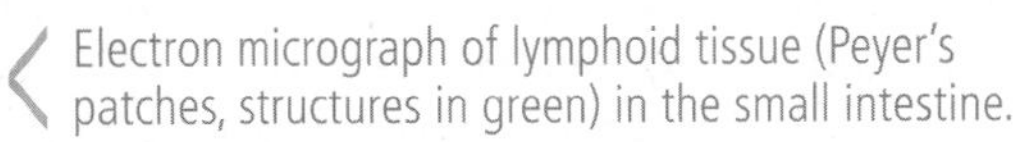

Electron micrograph of lymphoid tissue (Peyer's patches, structures in green) in the small intestine.

Distribution and Structure of Lymphatic Vessels

The lymphatic vessels form a one-way system in which lymph flows only toward the heart.

Lymphatic Capillaries

The transport of lymph begins in microscopic blind-ended **lymphatic capillaries** (Figure 19.1a). These capillaries weave between the tissue cells and blood capillaries in the loose connective tissues of the body. Lymphatic capillaries are widespread, but they are absent from bones and teeth, bone marrow, and the entire central nervous system (where the excess tissue fluid drains into the cerebrospinal fluid).

Although similar to blood capillaries, lymphatic capillaries are so remarkably permeable that they were once thought to be open at one end like a straw. We now know that they owe their permeability to two unique structural modifications:

- The endothelial cells forming the walls of lymphatic capillaries are not tightly joined. Instead, the edges of adjacent cells overlap each other loosely, forming easily opened, flaplike *minivalves* (Figure 19.1b).
- Collagen filaments anchor the endothelial cells to surrounding structures so that any increase in interstitial fluid volume opens the minivalves, rather than causing the lymphatic capillaries to collapse.

So, what we have is a system analogous to one-way swinging doors in the lymphatic capillary wall. When fluid pressure in the interstitial space is greater than the pressure in the lymphatic capillary, the minivalve flaps gape open, allowing fluid to enter the lymphatic capillary. However, when the pressure is greater *inside* the lymphatic capillary, it forces the endothelial minivalve flaps shut, preventing lymph from leaking back out as the pressure moves it along the vessel.

Proteins in the interstitial space are unable to enter blood capillaries, but they enter lymphatic capillaries easily. In addition, when tissues become inflamed, lymphatic capillaries develop openings that permit uptake of even larger particles such as cell debris, pathogens (disease-causing microorganisms such as bacteria and viruses), and cancer cells. The pathogens can then use the lymphatics to travel throughout the body. This threat to the body is partly offset by the lymph traveling through the lymph nodes, where it is cleansed of debris and "examined" by cells of the immune system.

A special set of lymphatic capillaries called **lacteals** (lak′te-alz) transports absorbed fat from the small intestine to the bloodstream. Lacteals are so called because of the milky white lymph that drains through them (*lact* = milk). This fatty lymph, called **chyle** ("juice"), drains from the fingerlike villi of the intestinal mucosa.

Larger Lymphatic Vessels

From the lymphatic capillaries, lymph flows through successively larger and thicker-walled channels—first collecting vessels, then trunks, and finally the largest of all, the ducts (Figure 19.1). The **collecting lymphatic vessels** have the same three tunics as veins, but the collecting vessels have thinner walls and

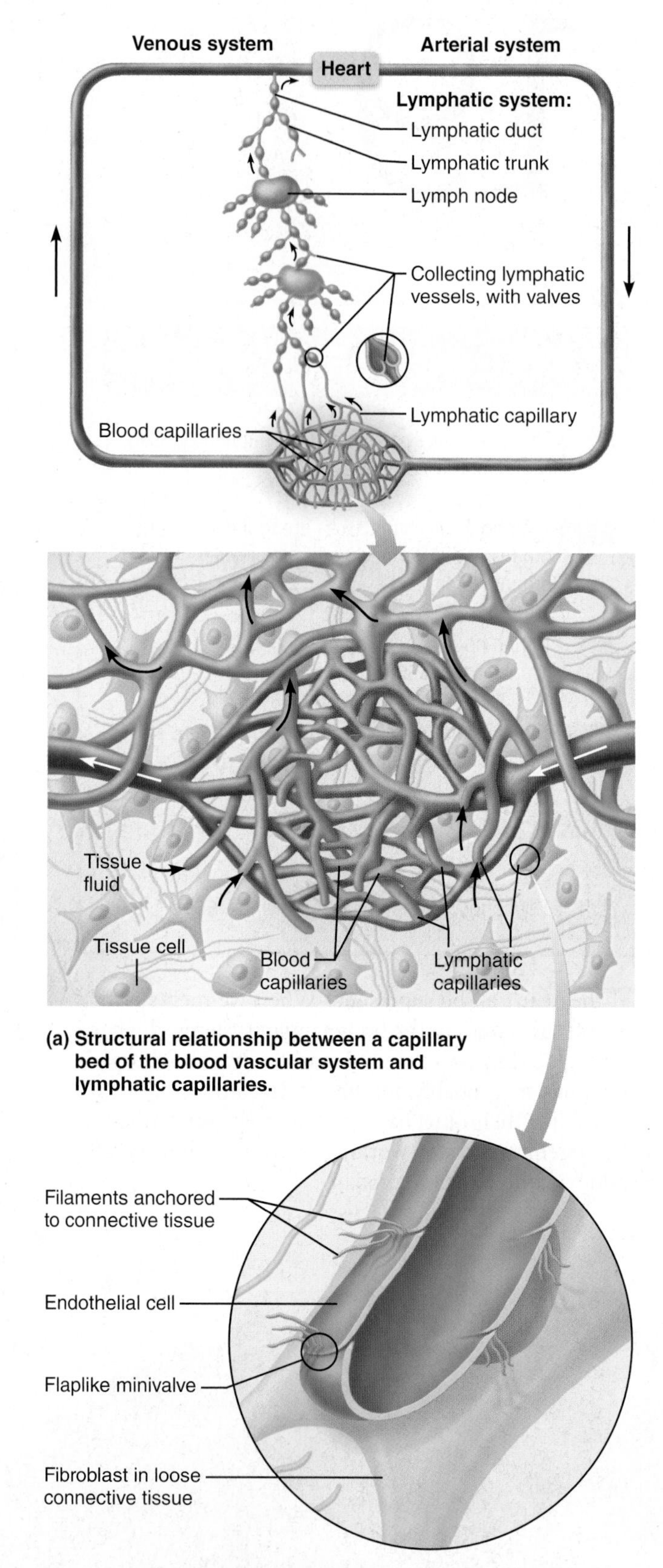

(a) Structural relationship between a capillary bed of the blood vascular system and lymphatic capillaries.

(b) Lymphatic capillaries are blind-ended tubes in which adjacent endothelial cells overlap each other, forming flaplike minivalves.

Figure 19.1 Distribution and special features of lymphatic capillaries. Arrows in (a) indicate direction of fluid movement.

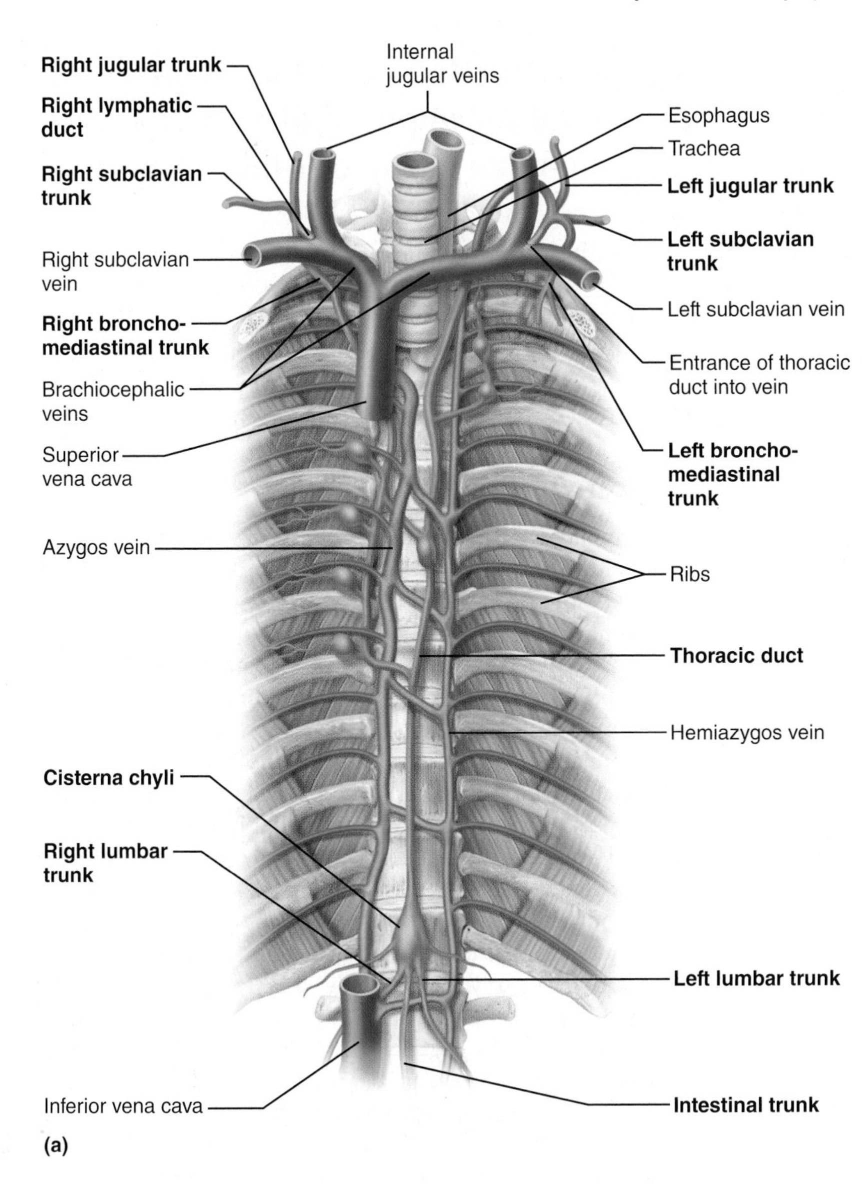

Figure 19.2 The lymphatic system. (a) Major lymphatic trunks and ducts in relation to veins and surrounding structures. Anterior view of thoracic and abdominal wall.

more internal valves, and they anastomose more. In general, lymphatics in the skin travel along with superficial *veins*, while the deep lymphatic vessels of the trunk and digestive viscera travel with the deep *arteries*. The exact anatomical distribution of lymphatic vessels varies greatly between individuals, even more than it does for veins.

The largest collecting vessels unite to form **lymphatic trunks**, which drain fairly large areas of the body. The major trunks, named mostly for the regions from which they drain lymph, are the paired **lumbar**, **bronchomediastinal**, **subclavian**, and **jugular trunks**, and the single **intestinal trunk** (Figure 19.2a).

Lymph is eventually delivered to one of two large *ducts* in the thoracic region. The **right lymphatic duct** drains lymph from the right upper limb and the right side of the head and thorax (Figure 19.2b). The much larger **thoracic duct** receives lymph from the rest of the body. It collects lymph from the two large lumbar trunks that drain the lower limbs and from the intestinal trunk that drains the digestive organs. In about half of individuals, the thoracic duct begins as an enlarged sac, the **cisterna chyli** (sis-ter′nah ki′li), located in the region between the last thoracic and second lumbar vertebrae. As the thoracic duct runs superiorly, it receives lymphatic drainage from the left side of the thorax, left upper limb, and the left side of the head. Each terminal duct empties its lymph into the venous circulation at the junction of the internal jugular vein and subclavian vein on its own side of the body (Figure 19.2b).

HOMEOSTATIC IMBALANCE 19.1 CLINICAL

Like the larger blood vessels, the larger lymphatics receive their nutrient blood supply from a branching vasa vasorum. When lymphatic vessels are severely inflamed, the related vessels of the vasa vasorum become congested with blood. As a result, the pathway of the associated superficial lymphatics becomes visible through the skin as red lines that are tender to the touch. This unpleasant condition is called *lymphangitis* (lim″fan-ji′tis; *angi* = vessel). ✚

Lymph Transport

The lymphatic system lacks an organ that acts as a pump. Under normal conditions, lymphatic vessels are low-pressure conduits, and the same mechanisms that promote venous return in blood vessels act here as well—the milking action of active skeletal muscles, pressure changes in the thorax during breathing, and valves to prevent backflow. Lymphatic vessels are usually bundled together in connective tissue sheaths along with blood vessels, and pulsations of nearby arteries also promote lymph flow. In addition to these mechanisms, smooth muscle in the walls of all but the smallest lymphatic vessels contracts rhythmically, helping to pump the lymph along.

Even so, lymph transport is sporadic and slow. Movement of adjacent tissues is extremely important in propelling lymph through the lymphatics. When physical activity or passive movements increase, lymph flows much more rapidly (balancing the greater rate of fluid loss from the blood in such situations). For this reason, it is a good idea to immobilize a badly infected body part to hinder flow of inflammatory material from that region.

HOMEOSTATIC IMBALANCE 19.2 CLINICAL

Anything that prevents the normal return of lymph to the blood—such as when tumors block the lymphatics or lymphatics are removed during cancer surgery—results in short-term but severe localized edema (*lymphedema*). In some cases, the lymphedema improves if some lymphatic pathways remain and can enlarge. ✚

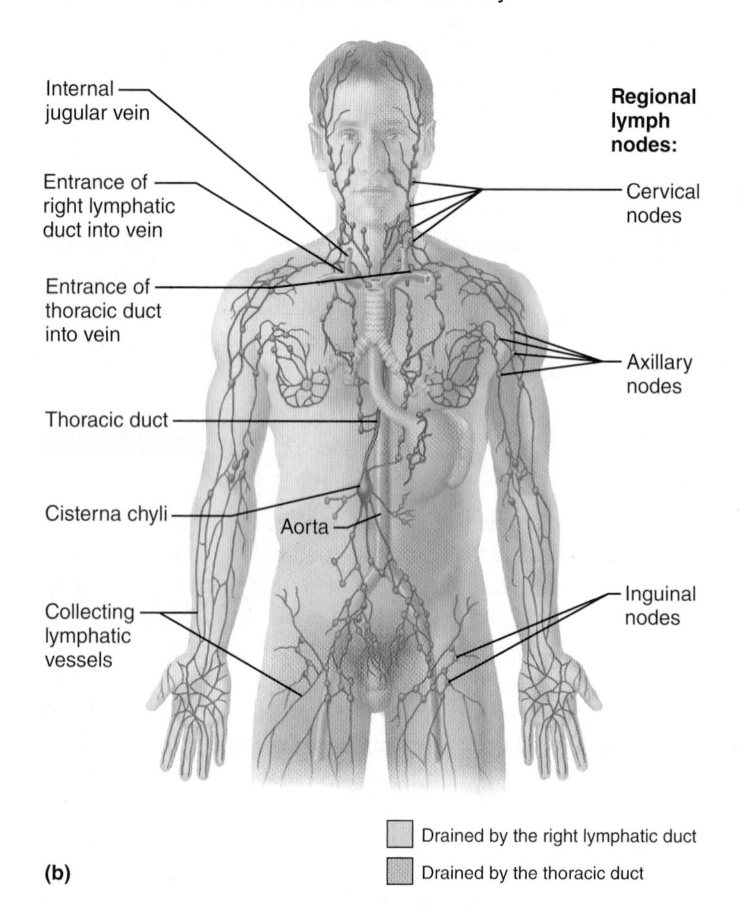

Figure 19.2 *(continued)* **The lymphatic system. (b)** General distribution of collecting lymphatic vessels and regional lymph nodes.

Practice art labeling
MasteringA&P®>Study Area>Chapter 19

19

To summarize, the lymphatic vessels:

- Return excess tissue fluid to the bloodstream
- Return leaked proteins to the blood
- Carry absorbed fat from the intestine to the blood (through lacteals)

Check Your Understanding

1. What is lymph? Where does it come from?
2. Name two lymphatic ducts and indicate the body regions usually drained by each.
3. What is the driving force for lymph movement?
4. MAKING connections A tumor in the left groin is blocking lymphatic drainage from Mr. Thomas's left leg, causing obvious edema. Which two of the functions of lymphatic vessels listed above no longer work in that leg? Think about the pressures driving bulk flow (Chapter 18) and state which two pressures are affected by the loss of these two lymphatic vessel functions and how the pressures are affected.

For answers, see Answers Appendix.

19.2 Lymphoid cells and tissues are found in lymphoid organs and in connective tissue of other organs

→ Learning Objective

☐ **Describe the basic structure and cellular population of lymphoid tissue. Differentiate between diffuse and follicular lymphoid tissues.**

To understand the role of the lymphoid organs in the body, we need to investigate their components—lymphoid cells and lymphoid tissues—before considering the organs themselves.

Lymphoid Cells

The lymphoid cells consist of immune system cells found in lymphoid tissues together with the supporting cells that form the "scaffolding" of those tissues.

Lymphocytes are the main warriors of the immune system. There are two main varieties of lymphocytes—**T cells (T lymphocytes)** and **B cells (B lymphocytes)**—that protect the body against antigens. (*Antigens* are anything that provokes an immune response, such as bacteria and their toxins, viruses, mismatched RBCs, or cancer cells.) Activated T cells manage the immune response, and some of them directly attack and destroy infected cells. B cells protect the body by producing **plasma cells**, daughter cells that secrete antibodies into the blood (or other body fluids). Antibodies mark antigens for destruction. Chapter 20 explores the roles of the lymphocytes in immunity.

Macrophages play a crucial role in body protection and the immune response by phagocytizing foreign substances and by helping to activate T cells. So, too, do the spiny-looking **dendritic cells** that capture antigens and bring them back to the lymph nodes.

Last but not least are the **reticular cells**, fibroblast-like cells that produce the reticular fiber **stroma** (stro′mah), which is the network that supports the other cell types in lymphoid organs and tissues (**Figure 19.3**).

Lymphoid Tissue

Lymphoid tissue is an important component of the immune system, mainly because it:

- Houses and provides a proliferation site for lymphocytes
- Furnishes an ideal surveillance vantage point for lymphocytes and macrophages

Lymphoid tissue, largely composed of loose connective tissue called **reticular connective tissue**, dominates all the lymphoid organs except the thymus. Macrophages live on the fibers of the reticular connective tissue network. Huge numbers of lymphocytes squeeze through the walls of postcapillary venules coursing through this network. The lymphocytes temporarily occupy the spaces in the network before leaving to patrol the body again (Figure 19.3). The cycling of lymphocytes between the circulatory vessels, lymphoid tissues, and loose connective tissues of the body ensures that lymphocytes reach infected or damaged sites quickly.

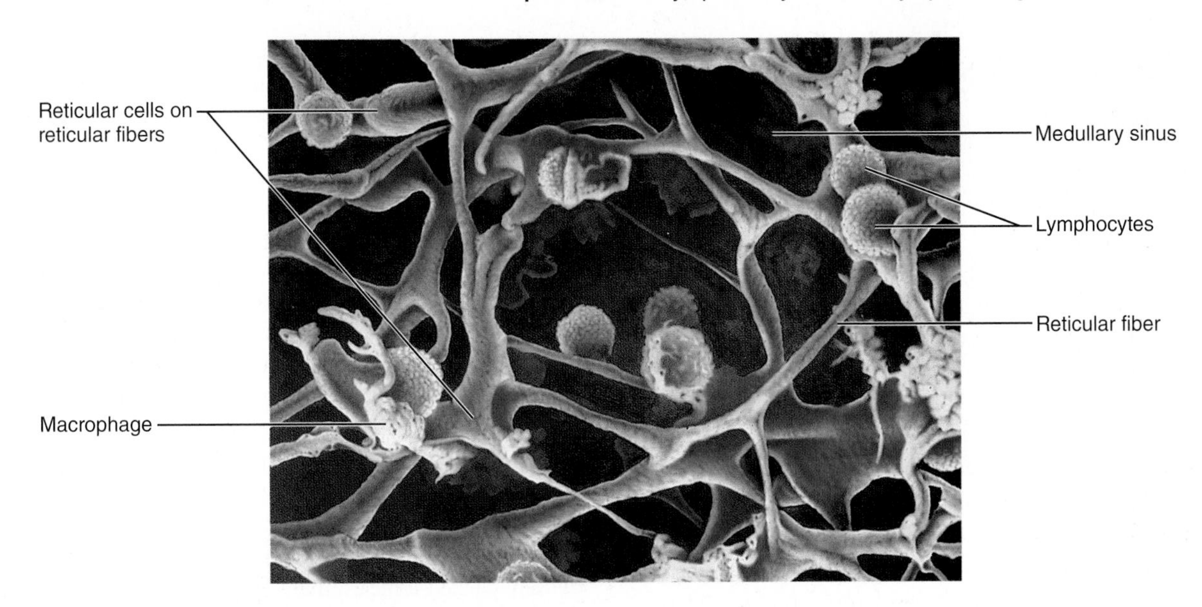

Figure 19.3 Reticular connective tissue in a human lymph node. Scanning electron micrograph (780×).

Lymphoid tissue comes in various "packages":

- **Diffuse lymphoid tissue**—a loose arrangement of lymphoid cells and some reticular fibers—is found in virtually every body organ. Larger collections appear in the lamina propria of mucous membranes such as those lining the digestive tract.
- **Lymphoid follicles (lymphoid nodules)** are solid, spherical bodies consisting of tightly packed lymphoid cells and reticular fibers. Follicles often have lighter-staining **germinal centers** where proliferating B cells predominate. These centers enlarge dramatically when the B cells are dividing rapidly and producing plasma cells. In many cases, the follicles form part of larger lymphoid organs, such as lymph nodes. However, isolated aggregations of lymphoid follicles occur in the intestinal wall as Peyer's patches (aggregated lymphoid nodules) and in the appendix (see p. 669).

Lymphoid Organs

The **lymphoid organs** (Figure 19.4) are grouped into two functional categories.

- The **primary lymphoid organs** are where B and T cells mature—the *red bone marrow* and the *thymus.* While both B and T cells originate in the red bone marrow, B cells mature in the red bone marrow and T cells mature in the thymus. We described the structure of bone marrow and its role in blood cell formation in Chapter 16. We will describe the thymus shortly.
- The **secondary lymphoid organs** are where mature lymphocytes first encounter their antigens and are activated. They include the lymph nodes, the spleen, and the collections of mucosa-associated lymphoid tissue (MALT) that form the tonsils, Peyer's patches (aggregated lymphoid nodules) in the small intestine, and the appendix. Lymphocytes also encounter antigens and are activated in the diffuse lymphoid tissues.

Although all lymphoid organs help protect the body, only the lymph nodes filter lymph. The other secondary lymphoid organs typically have efferent lymphatics draining them, but lack afferent lymphatics.

☑ Check Your Understanding

5. What are the primary lymphoid organs and what makes them special?

For answers, see Answers Appendix.

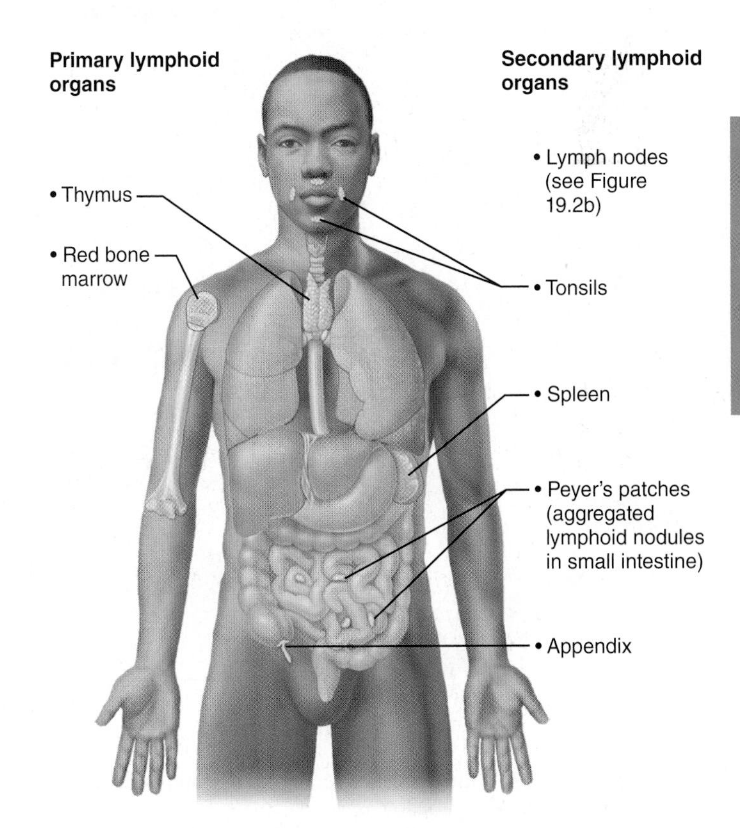

Figure 19.4 Lymphoid organs.

19.3 Lymph nodes filter lymph and house lymphocytes

→ Learning Objective

☐ **Describe the general location, histological structure, and functions of lymph nodes.**

The most important of the secondary lymphoid organs in the body are the **lymph nodes**, which cluster along the lymphatic vessels of the body. There are hundreds of these small organs, but because they are usually embedded in connective tissue, they are not ordinarily visible. Large clusters of lymph nodes occur near the body surface in the inguinal, axillary, and cervical regions, places where the collecting lymphatic vessels converge to form trunks (see Figure 19.2b).

Lymph nodes have two basic protective functions:

- **Cleansing the lymph.** As lymph is transported back to the bloodstream, the lymph nodes act as lymph "filters." Macrophages in the nodes remove and destroy microorganisms and other debris that enter the lymph from the loose connective tissues, preventing them from being delivered to the blood and spreading to other parts of the body.
- **Immune system activation.** Lymph nodes and other lymphoid organs are strategically located sites where lymphocytes encounter antigens and are activated to mount an attack against them.

Structure of a Lymph Node

The structure of a lymph node supports its defensive functions. Lymph nodes vary in shape and size, but most are bean shaped and less than 2.5 cm (1 inch) in length. Each node is surrounded by a dense fibrous **capsule** from which connective tissue strands called **trabeculae** extend inward to divide the node into a number of compartments (**Figure 19.5**). The node's internal framework, or stroma, of reticular fibers physically supports its ever-changing population of lymphocytes.

A lymph node has two histologically distinct regions, the **cortex** and the **medulla**. The superficial part of the cortex contains densely packed follicles, many with germinal centers heavy with dividing B cells. The deeper part of the cortex primarily houses T cells in transit. T cells circulate continuously between the blood, lymph nodes, and lymph, performing their surveillance role. Dendritic cells are abundant in the cortex and intimately associated with both B and T cells. The dendritic cells in the cortex are critical for preparing B and T cells to become effective defensive cells.

Medullary cords are thin inward extensions from the cortical lymphoid tissue, and contain both types of lymphocytes. Throughout the node are **lymph sinuses** (e.g., the subcapsular and medullary

Afferent lymphatic vessels
Cortex
• Lymphoid follicle
• Germinal center
• Subcapsular sinus
Efferent lymphatic vessels
Hilum
Medulla
• Medullary cord
• Medullary sinus
Trabeculae
Capsule

(a) Longitudinal view of the internal structure of a lymph node and associated lymphatics

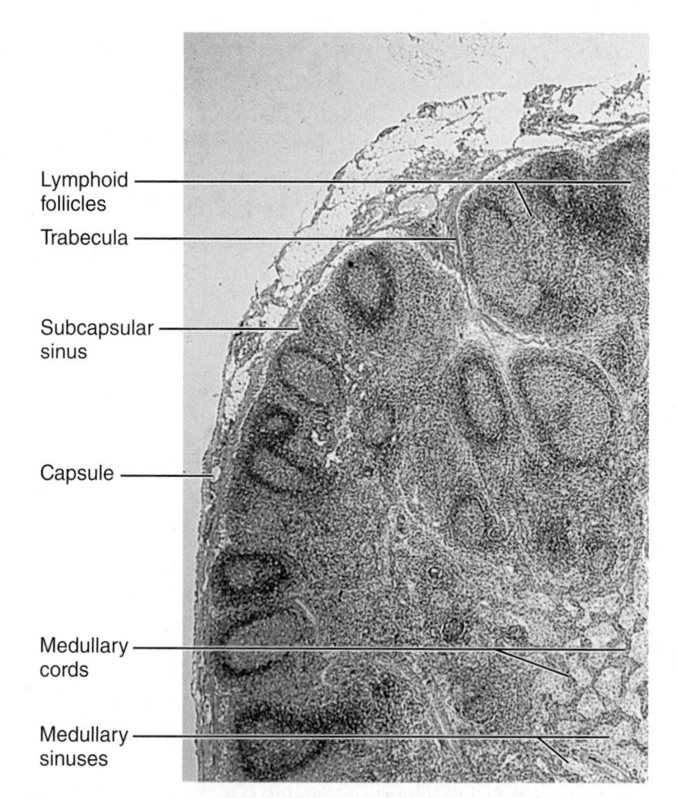

(b) Photomicrograph of part of a lymph node (72×)

Figure 19.5 Lymph node. In (a), notice that several afferent lymphatics converge on its convex side, whereas fewer efferent lymphatics exit at its hilum.

Practice art labeling
MasteringA&P®>Study Area>Chapter 19

sinuses described below). These sinuses are large lymphatic capillaries spanned by crisscrossing reticular fibers. Numerous macrophages reside on these reticular fibers and phagocytize foreign matter in the lymph as it flows by in the sinuses. Additionally, some of the lymph-borne antigens in the percolating lymph leak into the surrounding lymphoid tissue, where they activate lymphocytes to mount an immune attack against them.

Circulation in the Lymph Nodes

Lymph enters the convex side of a lymph node through a number of **afferent lymphatic vessels**. It then moves through a large, baglike sinus, the **subcapsular sinus**, into a number of smaller sinuses that cut through the cortex and enter the medulla. The lymph meanders through these **medullary sinuses** and finally exits the node at its **hilum** (hi′lum), the indented region on the concave side, via **efferent lymphatic vessels**.

There are fewer efferent vessels draining the node than afferent vessels feeding it, so the flow of lymph through the node stagnates somewhat, allowing time for the lymphocytes and macrophages to carry out their protective functions. Lymph passes through several nodes before it is completely cleansed.

HOMEOSTATIC IMBALANCE 19.3 — CLINICAL

Sometimes lymph nodes are overwhelmed by the agents they are trying to destroy. For example, when large numbers of bacteria are trapped in the nodes, the nodes become inflamed, swollen, and tender to the touch, a condition often referred to (erroneously) as swollen "glands." Such infected lymph nodes (often pus-filled) are called *buboes* (bu′bōz). (The bubonic plague was named for these buboes.)

Lymph nodes can also become secondary cancer sites, particularly when metastasizing cancer cells enter lymphatic vessels and become trapped there. Cancer-infiltrated lymph nodes are swollen but usually not painful, a fact that helps distinguish cancerous nodes from those infected by microorganisms. +

☑ Check Your Understanding

6. What is a lymphoid follicle? What type of lymphocyte predominates in follicles, especially in their germinal centers?
7. What is the benefit of having fewer efferent than afferent lymphatics in lymph nodes?

For answers, see Answers Appendix.

19.4 The spleen removes bloodborne pathogens and aged red blood cells

→ Learning Objective

☐ **Compare and contrast the structure and function of the spleen and lymph nodes.**

The soft, blood-rich **spleen** is about the size of a fist and is the largest lymphoid organ. Located in the left side of the abdominal cavity just beneath the diaphragm, it curls around the anterior aspect of the stomach (Figure 19.4 and **Figure 19.6**). It is served by the large *splenic artery* and *vein*, which enter and exit the *hilum* on its slightly concave anterior surface.

The spleen provides a site for lymphocyte proliferation and immune surveillance and response. But perhaps even more important are its blood-cleansing functions. Besides extracting aged and defective blood cells and platelets from the blood, its macrophages remove debris and foreign matter. The spleen also performs three additional, and related, functions. The spleen:

- Recycles the breakdown products of red blood cells for later reuse. It releases the breakdown products to the blood for processing by the liver and stores some of the iron salvaged from hemoglobin.
- Stores blood platelets and monocytes for release into the blood when needed.
- May be a site of erythrocyte production in the fetus.

Like lymph nodes, the spleen is surrounded by a fibrous capsule and has trabeculae that extend inward. Histologically, the spleen consists of two components: white pulp and red pulp.

- **White pulp** is where immune functions take place, so it is composed mostly of lymphocytes suspended on reticular fibers. The white pulp clusters or forms "cuffs" around central arteries (small branches of the splenic artery). These clusters of white pulp look like islands in a sea of red pulp.
- **Red pulp** is where worn-out red blood cells and bloodborne pathogens are destroyed, so it contains huge numbers of erythrocytes and the macrophages that engulf them. It is essentially all splenic tissue that is not white pulp. It consists of **splenic cords**, regions of reticular connective tissue, that separate the blood-filled **splenic sinusoids** (venous sinuses).

The names of the pulp regions reflect their appearance in fresh spleen tissue rather than their staining properties. Indeed, as you can see in Figure 19.6d, the white pulp sometimes appears darker than the red pulp due to the darkly staining nuclei of the densely packed lymphocytes.

HOMEOSTATIC IMBALANCE 19.4 — CLINICAL

Because the spleen's capsule is relatively thin, a direct blow or severe infection may cause it to rupture, spilling blood into the peritoneal cavity. Once, *splenectomy* (surgical removal of the ruptured spleen) was the standard treatment and thought necessary to prevent life-threatening hemorrhage and shock. However, surgeons have discovered that, if left alone, the spleen can often repair itself and so the frequency of emergency splenectomies has decreased dramatically. If the spleen must be removed, the liver and bone marrow take over most of its functions. In children younger than 12, the spleen will regenerate if a small part of it is left in the body. +

☑ Check Your Understanding

8. List several functions of the spleen.

For answers, see Answers Appendix.

19

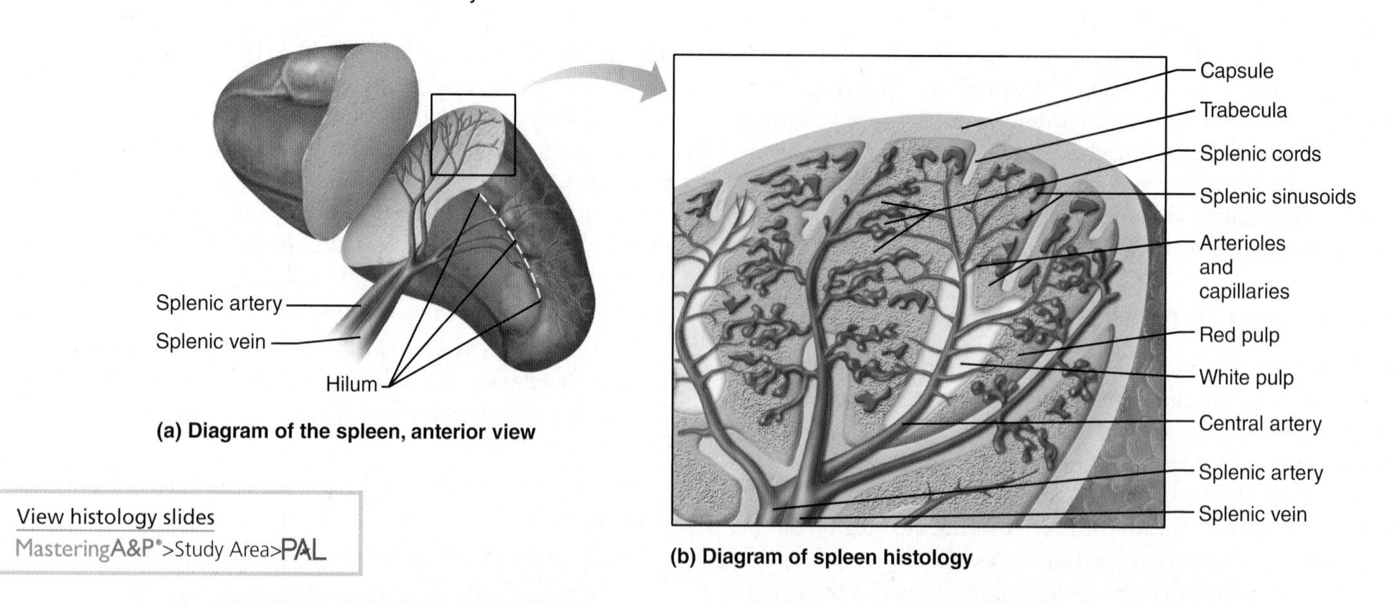

(a) Diagram of the spleen, anterior view

(b) Diagram of spleen histology

View histology slides
MasteringA&P®>Study Area>PAL

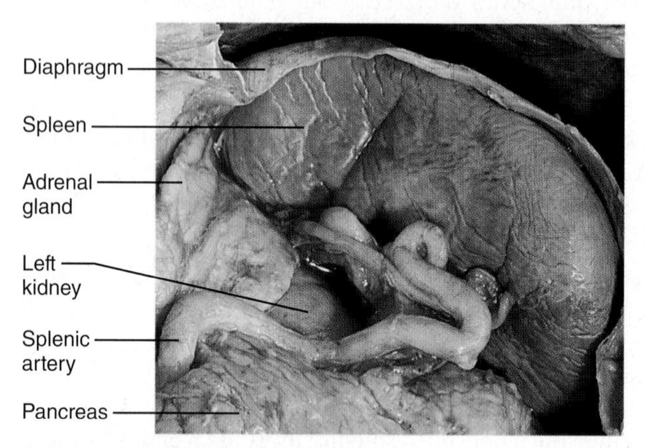

(c) Photograph of the spleen in its normal position in the abdominal cavity, anterior view.

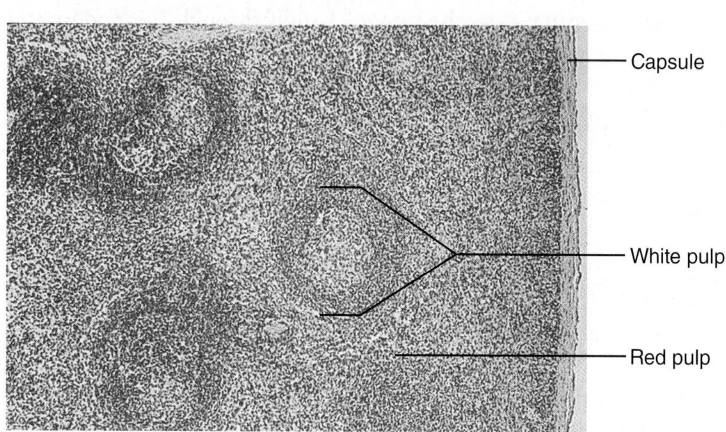

(d) Photomicrograph of spleen tissue (30×). The white pulp, a lymphoid tissue with many lymphocytes, is surrounded by red pulp containing abundant erythrocytes.

Figure 19.6 The spleen. (For a related image, see *A Brief Atlas of the Human Body*, Plate 39.)

19

19.5 MALT guards the body's entryways against pathogens

→ **Learning Objective**

☐ **Define MALT and list its major components.**

Mucosa-associated lymphoid tissues (MALT) are a set of distributed lymphoid tissues strategically located in mucous membranes throughout the body (see Figure 4.11 on p. 130 to review mucous membranes). MALT helps protect us from the never-ending onslaught of pathogens that seek to enter our bodies. Here we will consider the largest collections of MALT—the tonsils, Peyer's patches, and appendix. In addition to these large named collections, MALT also occurs in the mucosa of the respiratory and genitourinary organs as well as the rest of the digestive tract.

Tonsils

The **tonsils** form a ring of lymphoid tissue around the entrance to the pharynx (throat), where they appear as swellings of the mucosa (**Figure 19.7** and Figure 21.4). The tonsils are named according to location.

- The paired **palatine tonsils** are located on either side at the posterior end of the oral cavity. These are the largest tonsils and the ones most often infected.
- The **lingual tonsil** is the collective term for a lumpy collection of lymphoid follicles at the base of the tongue.
- The **pharyngeal tonsil** (referred to as the *adenoids* if enlarged) is in the posterior wall of the nasopharynx.
- The tiny **tubal tonsils** surround the openings of the auditory tubes into the pharynx.

The tonsils gather and remove many of the pathogens entering the pharynx in food or in inhaled air.

The lymphoid tissue of the tonsils contains follicles with obvious germinal centers surrounded by diffusely scattered lymphocytes. The tonsils are not fully encapsulated, and the epithelium overlying them invaginates deep into their interior, forming blind-ended **tonsillar crypts** (Figure 19.7). The crypts trap bacteria and particulate matter, and the bacteria work their way

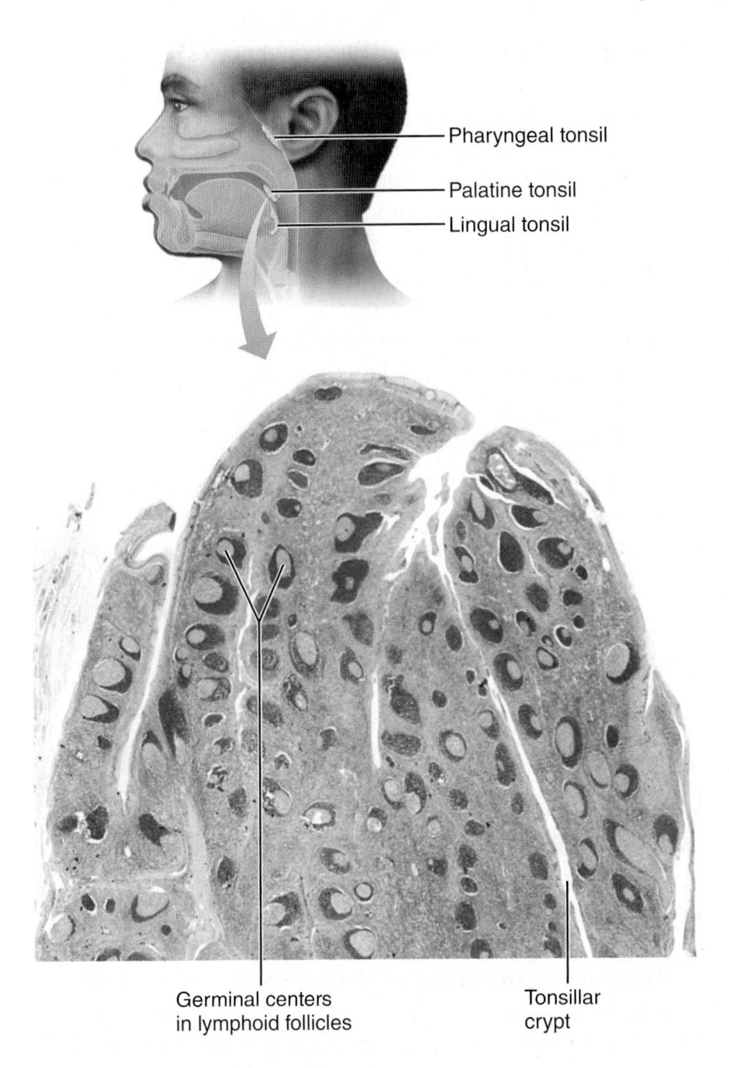

Figure 19.7 Histology of the palatine tonsil. The exterior surface of the tonsil is covered by stratified squamous epithelium, which invaginates deeply to form tonsillar crypts (10×).

View histology slides
MasteringA&P®>Study Area>PAL

through the mucosal epithelium into the lymphoid tissue, where most are destroyed. It seems a bit dangerous to "invite" infection this way, but this strategy produces a wide variety of immune cells that have a "memory" for the trapped pathogens. In other words, the body takes a calculated risk early on (during childhood) for the benefits of heightened immunity and better health later.

Peyer's Patches

Peyer's patches (pi′erz), or **aggregated lymphoid nodules**, are large clusters of lymphoid follicles, structurally similar to the tonsils. They are located in the wall of the distal portion of the small intestine (Figure 19.4 and **Figure 19.8**).

Appendix

The **appendix** is a tubular offshoot of the first part of the large intestine and contains a high concentration of lymphoid

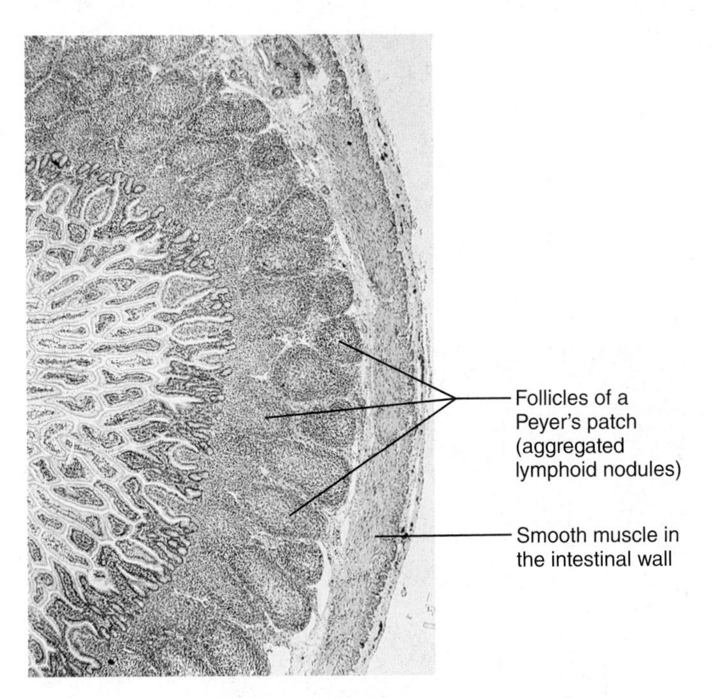

Figure 19.8 Peyer's patch (aggregated lymphoid nodules). Cross section of wall of the ileum of the small intestine (20×).

View histology slides
MasteringA&P®>Study Area>PAL

follicles. Like Peyer's patches, the appendix is in an ideal position (1) to prevent bacteria (present in large numbers in the intestine) from breaching the intestinal wall, and (2) to generate many "memory" lymphocytes for long-term immunity.

☑ Check Your Understanding

9. What is MALT? List several components of MALT.

For answers, see Answers Appendix.

19.6 T lymphocytes mature in the thymus

→ Learning Objective

☐ **Describe the structure and function of the thymus.**

The bilobed **thymus** (thi′mus) has important functions primarily during the early years of life. It is found in the inferior neck and extends into the superior thorax, where it partially overlies the heart deep to the sternum (see Figure 19.4 and **Figure 19.9**). In the thymus, T lymphocyte precursors mature to become immunocompetent lymphocytes. In other words, the thymus is where T lymphocytes become able to defend us against specific pathogens in the immune response.

Prominent in newborns, the thymus continues to increase in size during the first year, when it is highly active. After puberty, it gradually atrophies and by old age it has been replaced almost entirely by fibrous and fatty tissue and is difficult to distinguish from surrounding connective tissue. Even though it atrophies,

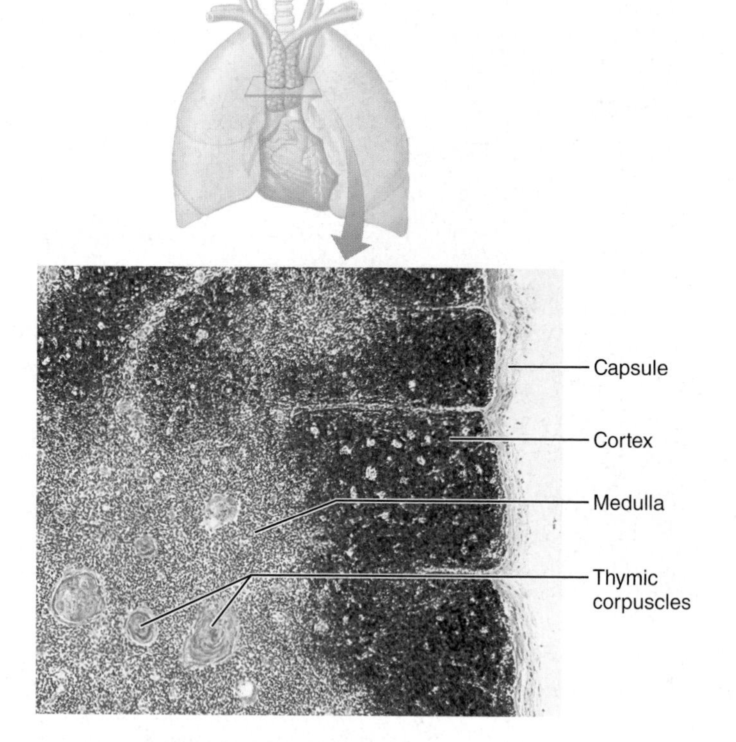

Figure 19.9 The thymus. The photomicrograph of a portion of the thymus shows part of a lobule with cortical and medullary regions (85×).

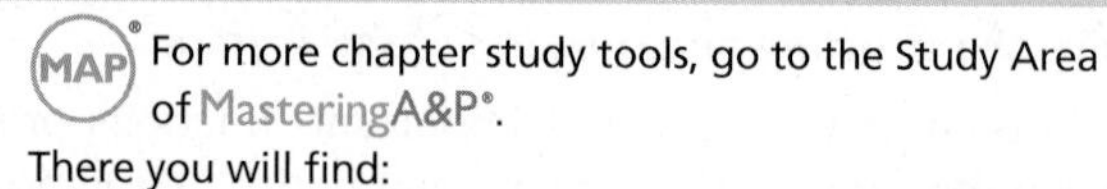

the thymus continues to produce immunocompetent cells as we age, although at a declining rate.

To understand thymic histology, it helps to compare the thymus to a cauliflower head—the flowerets represent *thymic lobules*, each containing an outer cortex and an inner medulla (Figure 19.9). Most thymic cells are lymphocytes. In the cortical regions the rapidly dividing lymphocytes are densely packed, with a few macrophages scattered among them.

The lighter-staining medullary areas contain fewer lymphocytes plus some bizarre structures called **thymic corpuscles**. Consisting of concentric whorls of keratinized epithelial cells, they were thought to be sites of T cell destruction. Recent evidence suggests that thymic corpuscles are involved in the development of *regulatory T cells*, a class of T lymphocytes that are important for preventing autoimmune responses.

The thymus is a primary lymphoid organ and differs from secondary lymphoid organs in three important ways:

- The thymus has no follicles because it lacks B cells.
- The thymus does not *directly* fight antigens. Instead, the thymus functions strictly as a maturation site for T lymphocyte precursors. These precursors must be kept isolated from foreign antigens to prevent their premature activation. In fact, there is a *blood thymus barrier* that keeps bloodborne antigens out of the thymus.
- The stroma of the thymus consists of epithelial cells rather than reticular fibers. These epithelial cells provide the physical and chemical environment in which T lymphocytes mature.

☑ Check Your Understanding

10. MAKING connections T cells mature in the thymus. Where do B cells mature?

For answers, see Answers Appendix.

Although the functions of the lymphatic vessels and lymphoid organs overlap, each helps maintain homeostasis in unique ways. The lymphatic vessels help maintain blood volume. The macrophages of lymphoid organs remove and destroy foreign matter in lymph and blood. Additionally, lymphoid organs and tissues provide sites from which the immune system can be mobilized. In Chapter 20, we continue this story as we examine the inflammatory and immune responses that allow us to resist a constant barrage of pathogens.

19

REVIEW QUESTIONS

MAP® For more chapter study tools, go to the Study Area of MasteringA&P®.
There you will find:

- Interactive Physiology iP
- A&PFlix A&PFlix
- Practice Anatomy Lab PAL
- PhysioEx PEx
- Videos, Practice Quizzes and Tests, MP3 Tutor Sessions, Case Studies, and much more!

Multiple Choice/Matching

(Some questions have more than one correct answer. Select the best answer or answers from the choices given.)

1. Lymphatic vessels **(a)** serve as sites for immune surveillance, **(b)** filter lymph, **(c)** transport leaked plasma proteins and fluids to the cardiovascular system, **(d)** are represented by vessels that resemble arteries, capillaries, and veins.
2. The sac that often forms the initial portion of the thoracic duct is the **(a)** lacteal, **(b)** right lymphatic duct, **(c)** cisterna chyli, **(d)** lymph sac.
3. Entry of lymph into the lymphatic capillaries is promoted by which of the following? **(a)** one-way minivalves formed by overlapping endothelial cells, **(b)** the respiratory pump, **(c)** the skeletal muscle pump, **(d)** greater fluid pressure in the interstitial space.
4. The structural framework of lymphoid organs is **(a)** areolar connective tissue, **(b)** hematopoietic tissue, **(c)** reticular tissue, **(d)** adipose tissue.
5. Lymph nodes are densely clustered in all of the following body areas *except* **(a)** the brain, **(b)** the axillae, **(c)** the groin, **(d)** the cervical region.

6. The germinal centers in lymph nodes are largely sites of **(a)** macrophages, **(b)** proliferating B lymphocytes, **(c)** T lymphocytes, **(d)** all of these.
7. The red pulp areas of the spleen are sites of **(a)** splenic sinusoids, macrophages, and red blood cells, **(b)** clustered lymphocytes, **(c)** connective tissue septa.
8. The lymphoid organ that functions primarily during youth and then begins to atrophy is the **(a)** spleen, **(b)** thymus, **(c)** palatine tonsils, **(d)** bone marrow.
9. Collections of lymphoid tissue (MALT) that guard mucosal surfaces include all of the following except **(a)** appendix follicles, **(b)** the tonsils, **(c)** Peyer's patches, **(d)** the thymus.

Short Answer Essay Questions

10. Compare and contrast blood, interstitial fluid, and lymph.
11. Compare the structure and functions of a lymph node to those of the spleen.
12. (a) Which anatomical characteristic ensures that the flow of lymph through a lymph node is slow? (b) Why is this desirable?
13. There are no lymphatic arteries. Why isn't this a problem?

AT THE CLINIC

Clinical Case Study

Lymphatic System/Immunity

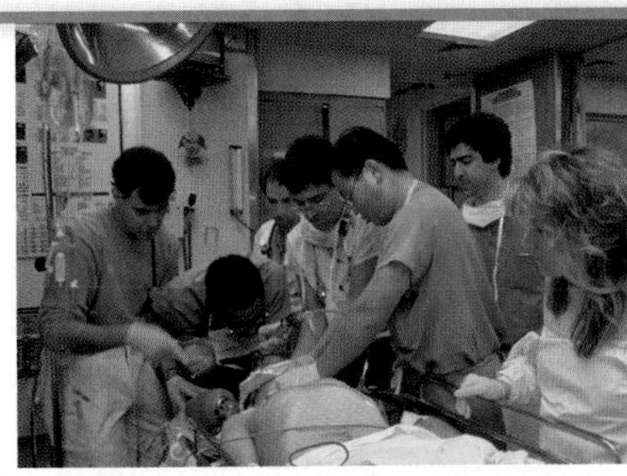

Back to following the progress of Mr. Hutchinson, we learn that the routine complete blood count (CBC) performed on admission reveals both a dangerously low total leukocyte count and a low proportion of lymphocytes. One day postsurgery, he complains of pain in his right ring finger (that hand had a crush injury). When examined, the affected finger and the dorsum of the right hand are edematous, and red streaks radiate superiorly on his right forearm. Higher-than-normal doses of antibiotics are prescribed, and a sling is applied to the affected arm. Nurses are instructed to wear gloves and gown when giving Mr. Hutchinson his care.

Relative to these observations:

1. What do the red streaks emanating from the bruised finger indicate? What would you conclude his problem was if there were no red streaks but the right arm was very edematous?
2. Why is it important that Mr. Hutchinson not move the affected arm excessively (i.e., why was the sling ordered)?
3. How might the low lymphocyte count, megadoses of antibiotics, and orders for additional clinical staff protection be related?
4. Do you predict that Mr. Hutchinson's recovery will be uneventful or problematic? Why?

For answers, see Answers Appendix.

20 The Immune System: Innate and Adaptive Body Defenses

KEY CONCEPTS

Every second of every day, armies of hostile bacteria, fungi, and viruses swarm on our skin and yet we stay amazingly healthy most of the time. The body has evolved a single-minded approach to such foes—if you're not with us, you're against us! To implement that stance, it relies heavily on two intrinsic defense systems that act both independently and cooperatively to provide resistance to disease, or **immunity** (*immun* = free).

1. The **innate (nonspecific) defense system**, like a lowly foot soldier, is always prepared, responding within minutes to protect the body from foreign substances. This system has two "barricades." The *first line of defense* is the external body membranes—intact skin and mucosae. The *second line of defense*, called into action whenever the first line has been penetrated, relies on internal defenses such as antimicrobial proteins, phagocytes, and other cells to inhibit the invaders' spread throughout the body. The hallmark of the second line of defense is inflammation.
2. The **adaptive (specific) defense system** functions like an elite fighting force equipped with high-tech weapons to attack *particular* foreign substances. The adaptive defense response, which provides the body's *third line of defense*, takes considerably longer to mount than the innate defense response.

Although we consider them separately, the innate and adaptive systems always work hand in hand. An overview of these two systems is shown in **Figure 20.1**. Small portions of this diagram will reappear in subsequent figures to let you know which part of the immune system we're dealing with.

Although certain organs of the body (notably lymphoid organs) are intimately involved in the immune response, the **immune system** is a *functional system* rather than an organ system in an anatomical sense. Its "structures" are a diverse array of molecules plus trillions of immune cells (especially lymphocytes) that inhabit lymphoid tissues and circulate in body fluids.

Once, the term *immune system* was equated with the adaptive defense system only. However, we now know that the innate and adaptive defenses are deeply intertwined. Specifically:

- The innate and adaptive systems release and recognize many of the same defensive molecules.
- The innate responses are not as nonspecific as once thought. Indeed, they have specific pathways to target certain foreign substances.
  - Proteins released during innate responses alert cells of the adaptive system to the presence of specific foreign molecules in the body.

When the immune system is operating effectively, it protects the body from most infectious microorganisms, cancer cells, and (unfortunately) transplanted organs and grafts. It does this both directly, by cell attack, and indirectly, by releasing mobilizing chemicals and protective antibody molecules.

Vaccination prevents many childhood illnesses that were once fatal.

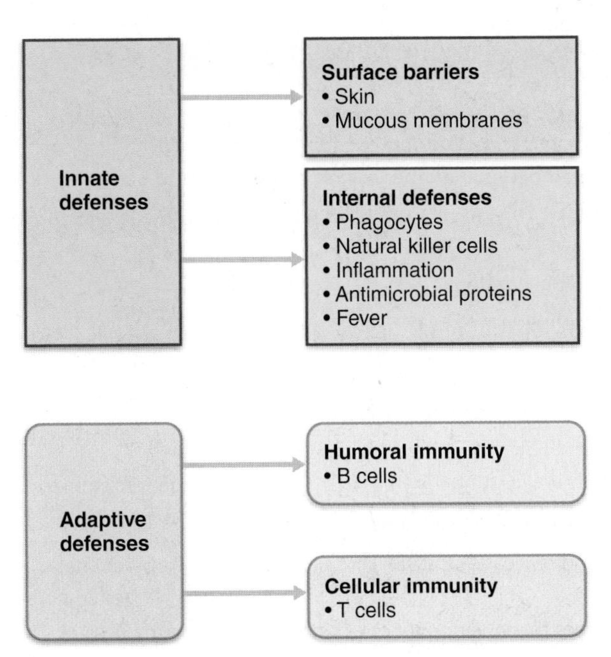

Figure 20.1 Simplified overview of innate and adaptive defenses. Humoral immunity (primarily involving B lymphocytes) and cellular immunity (involving T lymphocytes) are distinct but overlapping areas of adaptive immunity.

PART 1

INNATE DEFENSES

You could say we come fully equipped with innate defenses. The mechanical barriers that cover body surfaces and the cells and chemicals that act on the initial internal battlefronts are in place at birth, ready to ward off invading **pathogens** (harmful or disease-causing microorganisms).

Many times, our innate defenses alone ward off infection. In other cases, the adaptive immune system is called into action to reinforce and enhance the innate defenses. Either way, the innate defenses reduce the workload of the adaptive system by preventing the entry and spread of microorganisms in the body.

20.1 Surface barriers act as the first line of defense to keep invaders out of the body

→ Learning Objective

☐ **Describe surface membrane barriers and their protective functions.**

The body's first line of defense—the *skin* and the *mucous membranes*, along with the secretions these membranes produce—is highly effective. As long as the epidermis is unbroken, this heavily keratinized epithelial membrane is a formidable physical barrier to most microorganisms. Keratin is also resistant to most weak acids and bases and to bacterial enzymes and toxins. Intact mucosae provide similar mechanical barriers within the body. Recall that mucous membranes line all body cavities that open to the exterior: the digestive, respiratory, urinary, and reproductive tracts.

Besides serving as physical barriers, skin and mucous membranes produce a variety of protective chemicals:

- **Acid.** The acidity of skin, vaginal, and stomach secretions—the *acid mantle*—inhibits bacterial growth.
- **Enzymes.** *Lysozyme*—found in saliva, respiratory mucus, and lacrimal fluid of the eye—destroys bacteria. Protein-digesting enzymes in the stomach kill many different microorganisms.
- **Mucin.** *Mucin* dissolved in water forms thick, sticky mucus that lines the digestive and respiratory passageways. This mucus traps many microorganisms. Likewise, the mucin in watery saliva traps microorganisms and washes them out of the mouth into the stomach where they are digested.
- **Defensins.** Mucous membranes and skin secrete small amounts of broad-spectrum antimicrobial peptides called *defensins*. Defensin output increases dramatically in response to inflammation when surface barriers are breached. Using various mechanisms, such as disruption of microbial membranes, defensins help to control bacterial and fungal colonization in the exposed areas.
- **Other chemicals.** In the skin, some lipids in sebum and *dermcidin* in eccrine sweat are toxic to bacteria.

The respiratory tract mucosae also have structural modifications that counteract potential invaders. Tiny mucus-coated hairs inside the nose trap inhaled particles, and cilia on the mucosa of the upper respiratory tract sweep dust- and bacteria-laden mucus toward the mouth, preventing it from entering the lower respiratory passages.

Although these surface barriers (summarized in Table 20.1) are quite effective, they are breached by everyday nicks and cuts, for example, when you brush your teeth or shave. When this happens and microorganisms invade deeper tissues, your *internal* innate defenses—the second line of defense—come into play.

☑ Check Your Understanding

1. What distinguishes the innate defense system from the adaptive defense system?
2. What is the first line of defense against disease?

For answers, see Answers Appendix.

20.2 Innate internal defenses are cells and chemicals that act as the second line of defense

→ Learning Objectives

☐ **Explain the importance of phagocytosis, natural killer cells, and fever in innate body defense.**

☐ **Describe the inflammatory process. Identify several inflammatory chemicals and indicate their specific roles.**

☐ **Name the body's antimicrobial substances and describe their function.**

The body uses an enormous number of nonspecific cellular and chemical means to protect itself, including phagocytes, natural killer cells, antimicrobial proteins, and fever. The inflammatory response enlists macrophages, mast cells, all types of white

Table 20.1 The First Line of Defense: Surface Membrane Barriers

CATEGORY/ASSOCIATED ELEMENTS	PROTECTIVE MECHANISM
Intact skin epidermis	Forms mechanical barrier that prevents entry of pathogens and other harmful substances into body
• Acid mantle of skin	Skin secretions (sweat and sebum) make epidermal surface acidic, which inhibits bacterial growth; also contain various bactericidal chemicals
• Keratin	Provides resistance against acids, alkalis, and bacterial enzymes
Intact mucous membranes	Form mechanical barrier that prevents entry of pathogens
• Mucus	Traps microorganisms in respiratory and digestive tracts
• Nasal hairs	Filter and trap microorganisms in nasal passages
• Cilia	Propel debris-laden mucus away from nasal cavity and lower respiratory passages
• Gastric juice	Contains concentrated hydrochloric acid and protein-digesting enzymes that destroy pathogens in stomach
• Acid mantle of vagina	Inhibits growth of most bacteria and fungi in female reproductive tract
• Lacrimal secretion (tears); saliva	Continuously lubricate and cleanse eyes (tears) and oral cavity (saliva); contain lysozyme, an enzyme that destroys microorganisms
• Urine	Normally acid pH inhibits bacterial growth; cleanses the lower urinary tract as it flushes from the body

blood cells, and dozens of chemicals that kill pathogens and help repair tissue. These protective tactics identify potentially harmful substances by recognizing (binding tightly to) molecules with specific shapes that are part of infectious organisms (bacteria, viruses, and fungi) but not normal human cells. The receptors that do this are called **pattern recognition receptors**.

Phagocytes

Pathogens that get through the skin or mucosae into the underlying connective tissue are confronted by *phagocytes* (*phago* = eat). **Neutrophils**, the most abundant type of white blood cell, become phagocytic on encountering infectious material in the tissues. However, the most voracious phagocytes are **macrophages** ("big eaters"), which derive from white blood cells called **monocytes** that leave the bloodstream, enter the tissues, and develop into macrophages.

Free macrophages wander throughout the tissue spaces in search of cellular debris or "foreign invaders." *Fixed macrophages*, like *stellate macrophages* in the liver, are permanent residents of particular organs.

Phagocytosis

A phagocyte engulfs particulate matter much the way an amoeba ingests a food particle. Flowing cytoplasmic extensions bind to the particle and then pull it inside, enclosed within a membrane-lined vesicle (Figure 20.2a). The resulting **phagosome** then fuses with a *lysosome* to form a **phagolysosome** (steps ①–③ in Figure 20.2b).

Neutrophils and macrophages generally kill ingested prey by acidifying the phagolysosome and digesting its contents with lysosomal enzymes. However, some pathogens such as the tuberculosis bacillus and certain parasites are resistant to lysosomal enzymes and can even multiply within the phagolysosome. In this case, other immune cells called helper T cells release chemicals that stimulate the macrophage, activating additional enzymes that produce a lethal **respiratory burst**. The respiratory burst promotes killing of pathogens by:

- Liberating a deluge of highly destructive free radicals (including superoxide)
- Producing oxidizing chemicals (hydrogen peroxide and a substance identical to household bleach)
- Increasing the phagolysosome's pH and osmolarity, which activates other protein-digesting enzymes that digest the invader

Neutrophils also pierce the pathogen's membrane by using *defensins*, the antimicrobial peptides we mentioned earlier.

Phagocytic attempts are not always successful. In order for a phagocyte to ingest a pathogen, the phagocyte must first *adhere* to that pathogen, a feat made possible by recognizing the pathogen's carbohydrate "signature." Many bacteria have external capsules that conceal their carbohydrate signatures, allowing them to elude capture because phagocytes cannot bind to them.

Our immune systems get around this problem by coating pathogens with **opsonins**. Opsonins are complement proteins (discussed shortly) or antibodies. Both provide "handles" to which phagocyte receptors can bind. Any pathogen can be coated with opsonins, a process called **opsonization** ("to make tasty"), which greatly accelerates phagocytosis of that pathogen.

When phagocytes are unable to ingest their targets (because of size, for example), they can release their toxic chemicals into the extracellular fluid. Whether killing ingested or extracellular targets, neutrophils rapidly destroy themselves in the process. In contrast, macrophages are more robust and can survive to kill another day.

Natural Killer (NK) Cells

Natural killer (NK) cells, which "police" the body in blood and lymph, are a unique group of defensive cells that can lyse and kill cancer cells and virus-infected body cells before the adaptive

Innate defenses ⟶ Internal defenses

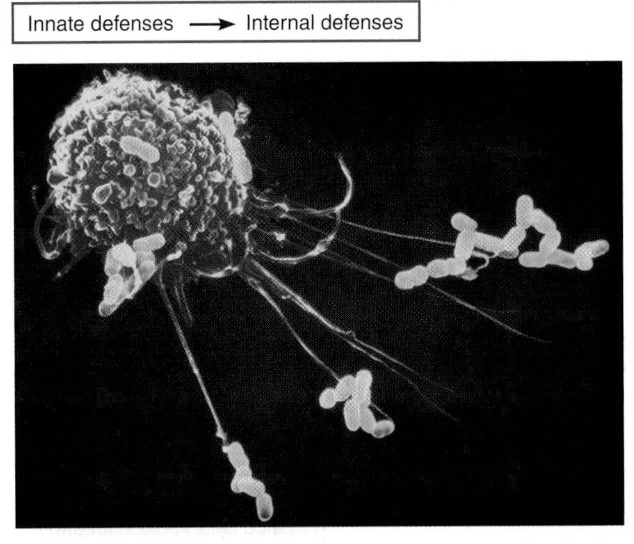

(a) A macrophage (purple) uses its cytoplasmic extensions to pull rod-shaped bacteria (green) toward it. Scanning electron micrograph (4800×).

Practice art labeling
MasteringA&P®>Study Area>Chapter 20

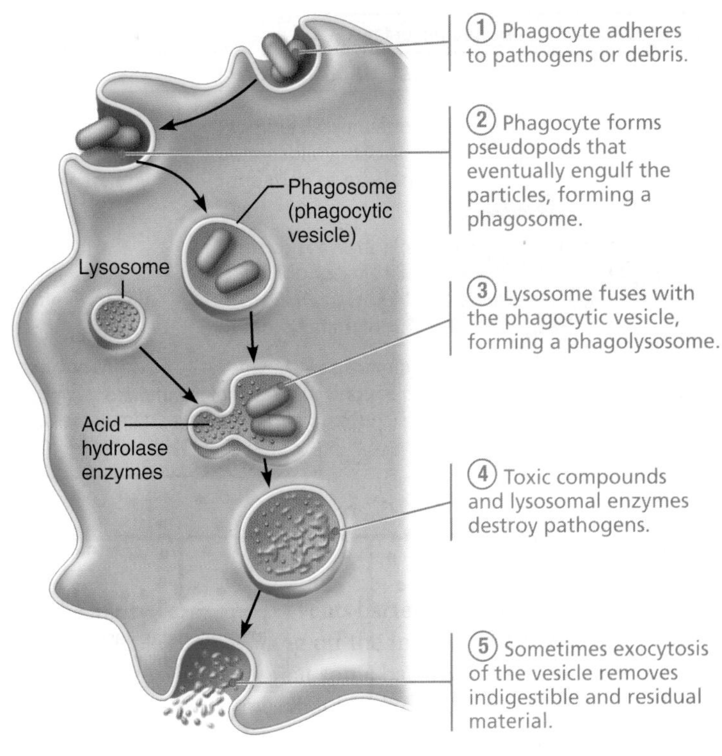

(b) Events of phagocytosis.

Figure 20.2 Phagocytosis.

immune system is activated. NK cells are part of a small group of *large granular lymphocytes*.

Unlike lymphocytes of the adaptive immune system, which only recognize and react against *specific* virus-infected or tumor cells, NK cells are far less picky. They can eliminate a variety of infected or cancerous cells by detecting general abnormalities such as the lack of "self" cell-surface proteins called MHC, described on p. 682. The name "natural" killer cells reflects their nonspecificity.

NK cells are not phagocytic. They kill by directly contacting the target cell, inducing it to undergo apoptosis (programmed cell death). This is the same method used by cytotoxic T cells (described on p. 697). NK cells also secrete potent chemicals that enhance the inflammatory response.

Inflammation: Tissue Response to Injury

Inflammation is triggered whenever body tissues are injured by physical trauma, intense heat, irritating chemicals, or infection by viruses, fungi, or bacteria. This inflammatory response to injury is summarized in Figure 20.3. Inflammation has several beneficial effects:

- It prevents the spread of damaging agents to nearby tissues.
- It disposes of cell debris and pathogens.
- It alerts the adaptive immune system.
- It sets the stage for repair.

The four *cardinal signs* of short-term, or acute, inflammation are *redness, heat* (*inflam* = set on fire), *swelling*, and *pain*. Some authorities consider *impaired function* to be a fifth cardinal sign. For instance, movement in an inflamed joint may be hampered temporarily, forcing it to rest, which aids healing.

Inflammatory Chemical Release

The inflammatory process begins with a chemical "alarm"—a flood of inflammatory chemicals released into the extracellular fluid. Inflammatory chemicals are released by injured or stressed tissue cells, and immune cells. For example, **mast cells**, a key component of the inflammatory response, release the potent inflammatory chemical **histamine** (his′tah-mēn). Inflammatory chemicals can also be formed from chemicals circulating in the blood (Table 20.2).

Macrophages (and cells of certain boundary tissues such as epithelial cells lining the gastrointestinal and respiratory tracts) have special pattern recognition receptors that allow them to recognize invaders and sound a chemical alarm. One class of these receptors, called **Toll-like receptors (TLRs)**, plays a central role in triggering immune responses. There are 11 types of human TLRs, each recognizing a particular class of attacking microbe. For example, one type responds to a glycolipid in cell walls of the tuberculosis bacterium and another to a component of gram-negative bacteria such as *Salmonella*. Once activated, a TLR triggers the release of inflammatory chemicals called *cytokines*.

Other inflammatory chemicals include **kinins** (ki′ninz), **prostaglandins** (pros″tah-glan′dinz), and **complement**. All inflammatory chemicals dilate local arterioles and make local capillaries leakier. In addition, many attract leukocytes to the injured area and some have individual inflammatory roles as well (Table 20.2).

③ **Diapedesis.** Continued chemical signaling prompts the neutrophils to flatten and squeeze between the endothelial cells of the capillary walls—a process called **diapedesis**.

④ **Chemotaxis.** Inflammatory chemicals act as homing devices, or more precisely **chemotactic agents**. Neutrophils and other WBCs migrate up the gradient of chemotactic agents to the site of injury. Within an hour after the inflammatory response has begun, neutrophils have collected at the site and are devouring any foreign material present.

As the body's counterattack continues, monocytes follow neutrophils into the injured area. Monocytes are fairly poor phagocytes, but within 12 hours of leaving the blood and entering the tissues, they swell and develop large numbers of lysosomes, becoming macrophages with insatiable appetites. These late-arriving macrophages replace the neutrophils on the battlefield.

Macrophages are the central actors in the final disposal of cell debris as acute inflammation subsides, and they predominate at sites of chronic inflammation. The ultimate goal of an inflammatory response is to clear the injured area of pathogens, dead tissue cells, and any other debris so that tissue can be repaired. Once this is accomplished, healing usually occurs quickly.

CLINICAL

HOMEOSTATIC IMBALANCE 20.1

In severely infected areas, the battle takes a considerable toll on both sides, and creamy yellow **pus** (a mixture of dead or dying neutrophils, broken-down tissue cells, and living and dead pathogens) may accumulate in the wound. If the inflammatory mechanism fails to clear the area of debris, collagen fibers may be laid down, which walls off the sac of pus, forming an *abscess*. The abscess may need to be surgically drained before healing can occur.

Some bacteria, such as tuberculosis bacilli, resist digestion by the macrophages that engulf them. They escape the effects of prescription antibiotics by remaining snugly enclosed within their macrophage hosts. In such cases, *granulomas* form. These tumor-like growths contain a central region of infected macrophages surrounded by uninfected macrophages and an outer fibrous capsule.

A person may harbor pathogens walled off in granulomas for years without displaying any symptoms. However, if the person's resistance to infection is ever compromised, the bacteria may be activated and break free, leading to clinical disease symptoms. ✚

Antimicrobial Proteins

A variety of **antimicrobial proteins** enhance our innate defenses by attacking microorganisms directly or by hindering their ability to reproduce. The most important antimicrobial proteins are interferons and complement proteins (Table 20.3).

Interferons

Viruses—essentially nucleic acids surrounded by a protein envelope—lack the cellular machinery to generate ATP or synthesize proteins. They do their "dirty work" in the body by invading tissue cells and taking over the cellular metabolic machinery needed to reproduce themselves.

Infected cells can do little to save themselves, but some can secrete small proteins called **interferons (IFNs)** (in″ter-fēr′onz) to help protect cells that have not yet been infected. The IFNs diffuse to nearby cells, which they stimulate to synthesize proteins that "interfere" with viral replication in still-healthy cells by blocking protein synthesis and degrading viral RNA (Figure 20.5). Because IFN protection is *not* virus-specific, IFNs produced against a particular virus protect against other viruses, too.

The IFNs are a family of immune modulating proteins produced by a variety of body cells, each having a slightly different physiological effect. IFN alpha (α) and beta (β) have the antiviral effects that we've just described and also activate NK cells.

20

Table 20.3 The Second Line of Defense: Innate Cellular and Chemical Defenses

CATEGORY/ASSOCIATED ELEMENTS	PROTECTIVE MECHANISM
Phagocytes	Engulf and destroy pathogens that breach surface membrane barriers; macrophages also contribute to adaptive immune responses
Natural killer (NK) cells	Promote apoptosis (cell suicide) by directly attacking virus-infected or cancerous body cells; recognize general abnormalities rather than specific antigens; do not form memory cells
Inflammatory response	Prevents injurious agents from spreading to adjacent tissues, disposes of pathogens and dead tissue cells, and promotes tissue repair; released inflammatory chemicals attract phagocytes (and other immune cells) to the area
Antimicrobial proteins	
• Interferons (α, β, γ)	Proteins released by virus-infected cells and certain lymphocytes; act as chemical messengers to protect uninfected tissue cells from viral takeover; mobilize immune system
• Complement	A group of bloodborne proteins that, when activated, lyse microorganisms, enhance phagocytosis by opsonization, and intensify inflammatory and other immune responses
Fever	Systemic response initiated by pyrogens; high body temperature inhibits microbes from multiplying and enhances body repair processes

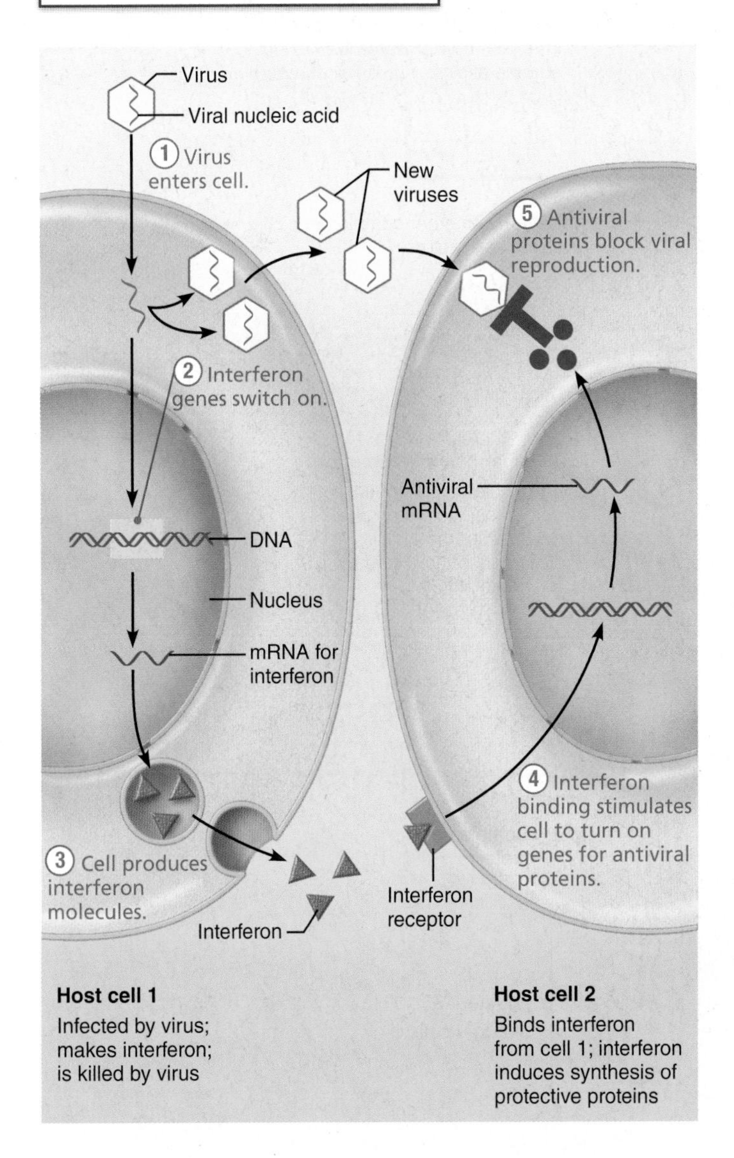

Figure 20.5 The interferon mechanism against viruses.

Another interferon, IFN gamma (γ), or immune interferon, is secreted by lymphocytes and has widespread immune mobilizing effects, such as activating macrophages. Because both macrophages and NK cells can also act directly against cancerous cells, the interferons have an indirect role in fighting cancer. Genetically engineered IFNs are used to treat several disorders including hepatitis C, genital warts, and multiple sclerosis.

Complement

The term **complement system**, or simply **complement**, refers to a group of at least 20 plasma proteins that normally circulate in the blood in an inactive state. These proteins include C1 through C9, factors B, D, and P, plus several regulatory proteins.

Complement provides a major mechanism for destroying foreign substances in the body. Its activation unleashes inflammatory chemicals that amplify virtually all aspects of the inflammatory process. Activated complement also lyses and kills certain bacteria and other cell types. (Luckily our own cells are equipped with proteins that normally inhibit complement activation.) Although complement is a nonspecific defensive mechanism, it "complements" (enhances) the effectiveness of *both* innate and adaptive defenses.

Figure 20.6 outlines the three pathways by which complement can be activated.

- The **classical pathway** involves *antibodies*, water-soluble protein molecules that the adaptive immune system produces to fight off foreign invaders. When antibodies bind to pathogens, they can also bind complement components. This double binding, called *complement fixation*, is the first step in this complement activation pathway. (We describe this in more detail on p. 690.)
- The **lectin pathway** involves *lectins*, water-soluble protein molecules that the innate immune system produces to recognize foreign invaders. When lectins bind specific sugars on the surface of microorganisms, they can then bind and activate complement.
- The **alternative pathway** is triggered when spontaneously activated C3 and other complement factors interact on the surface of microorganisms. These microorganisms lack the complement activation inhibitors our own cells have.

Like the blood clotting cascade, complement activation by any of these pathways involves a cascade in which proteins are activated in an orderly sequence—each step catalyzing the next. The three pathways converge at C3, which is split into C3a and C3b. Splitting C3 initiates a common terminal pathway that enhances inflammation, promotes phagocytosis, and can cause cell lysis.

Cell lysis begins when C3b binds to the target cell's surface and triggers the insertion of a group of complement proteins called **MAC (membrane attack complex)** into the cell's membrane. MAC forms and stabilizes a hole in the membrane that allows a massive influx of water, lysing the target cell.

The C3b molecules act as *opsonins*. As previously described, opsonins coat the microorganism, providing "handles" that receptors on macrophages and neutrophils can adhere to. This allows them to engulf the particle more rapidly. C3a and other cleavage products formed during complement fixation amplify the inflammatory response by stimulating mast cells and basophils to release histamine and by attracting neutrophils and other inflammatory cells to the area.

Fever

Inflammation is a localized response to infection, but sometimes the body's response to the invasion of microorganisms is more widespread. **Fever**, an abnormally high body temperature, is a systemic response to invading microorganisms.

When leukocytes and macrophages are exposed to foreign substances in the body, they release chemicals called **pyrogens** (*pyro* = fire). These pyrogens act on the body's thermostat—a cluster of neurons in the hypothalamus—raising the body's temperature above normal [37°C (98.6°F)].

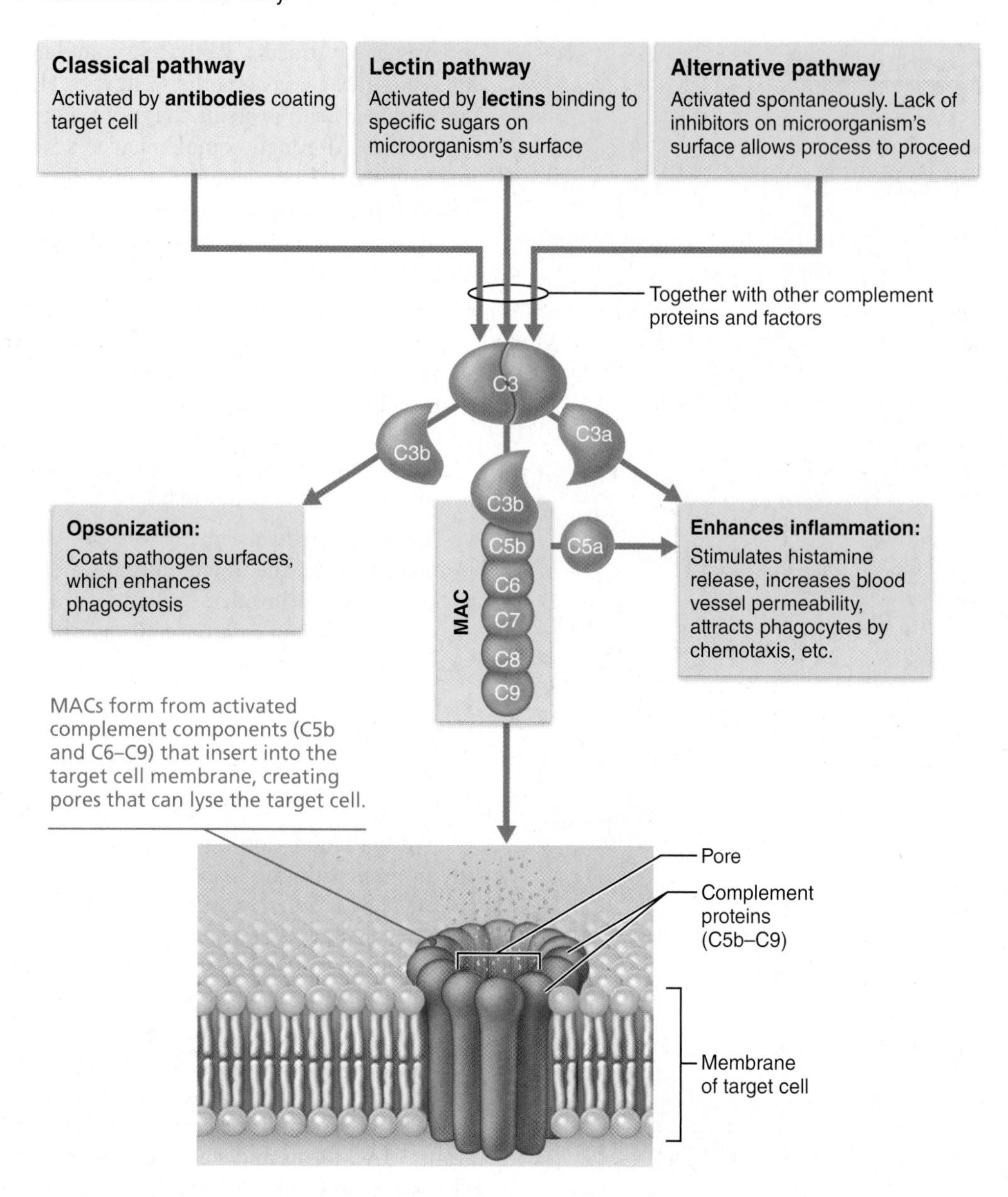

Figure 20.6 Complement activation. All three pathways that activate complement converge at C3. C3 splits into two active pieces: C3a and C3b, which enhance inflammation and act as opsonins. In certain target cells (mostly bacteria), C3b also activates other complement proteins that can form a membrane attack complex (MAC).

Fever is an adaptive response that seems to benefit the body, but exactly how it does so is unclear. Fever causes the liver and spleen to sequester iron and zinc, which may make them less available to support bacterial growth. Additionally, fever increases the metabolic rate of tissue cells, and may speed up repair processes.

☑ Check Your Understanding

3. What is opsonization and how does it help phagocytes? Give an example of a molecule that acts as an opsonin.
4. Under what circumstances might NK cells kill our own cells?
5. What are the cardinal signs of inflammation and what causes them?

For answers, see Answers Appendix.

PART 2

ADAPTIVE DEFENSES

Most of us would find it wonderfully convenient if we could walk into a single clothing store and buy a complete wardrobe—hat to shoes—that fits perfectly regardless of any special figure problems. We know that such a service would be next to impossible to find. And yet, we take for granted our **adaptive immune system**, the body's built-in *specific defensive system* that stalks and eliminates with nearly equal precision almost any type of pathogen that intrudes into the body.

When it operates effectively, the adaptive immune system protects us from a wide variety of infectious agents, as well as from abnormal body cells. When it fails, or is disabled, devastating diseases such as cancer and AIDS result. The activity of the adaptive immune system tremendously amplifies the inflammatory response and is responsible for most complement activation.

At first glance, the adaptive system seems to have a major shortcoming. Unlike the innate system, which is always ready and able to react, the adaptive system must "meet" or be primed by an initial exposure to a specific foreign substance (antigen). Only then can it protect the body against that substance, and this priming takes precious time.

Experiments in the late 1800s revealed the basis of this specific immunity. Researchers demonstrated that animals surviving a serious bacterial infection have protective factors (the proteins we now call *antibodies*) in their blood that defend against future attacks by the same pathogen. Furthermore, researchers found that if antibody-containing serum from the surviving animals was injected into animals that had not been exposed to the pathogen, the injected animals would also be protected. These landmark experiments were exciting because they revealed three important aspects of the adaptive immune response:

- **It is specific.** It recognizes and targets *particular* pathogens or foreign substances that initiate the immune response.
- **It is systemic.** Immunity is not restricted to the initial infection site.
- **It has "memory."** After an initial exposure, it recognizes and mounts even stronger attacks on previously encountered pathogens.

At first antibodies were thought to be the sole artillery of the adaptive immune system. Then, in the mid-1900s researchers discovered that injecting antibody-containing serum did *not* always protect the recipient from diseases the serum donor had survived. In such cases, however, injecting the donor's lymphocytes *did* provide immunity. As the pieces fell into place, researchers recognized two separate but overlapping arms of adaptive immunity, each using a variety of attack mechanisms that vary with the intruder.

Humoral immunity (hu′mor-ul), also called **antibody-mediated immunity**, is provided by antibodies present in the body's "humors," or fluids (blood, lymph, etc.). Though they are produced by lymphocytes, antibodies circulate freely in the blood and lymph, where they bind primarily to *extracellular* targets—bacteria, bacterial toxins, and free viruses—inactivating them temporarily and marking them for destruction by phagocytes or complement.

When lymphocytes themselves rather than antibodies defend the body, the immunity is called **cellular** or **cell-mediated immunity** because living cells provide the protection. Cellular immunity also has *cellular* targets—virus-infected or parasite-infected tissue cells, cancer cells, and cells of foreign grafts. The lymphocytes act against such targets either *directly*, by killing the infected cells, or *indirectly*, by releasing chemicals that enhance the inflammatory response or activate other lymphocytes or macrophages.

Before describing the humoral and cellular responses, we will first consider the central role of *antigens*.

20.3 Antigens are substances that trigger the body's adaptive defenses

→ Learning Objectives

- ☐ **Define antigen and describe how antigens affect the adaptive defenses.**
- ☐ **Define complete antigen, hapten, and antigenic determinant.**

Antigens (an′tĭ-jenz) are substances that can mobilize the adaptive defenses. They are the ultimate targets of all adaptive immune responses. (*Antigen* is a contraction of "*anti*body *gen*erating.") Most antigens are large, complex molecules (natural or synthetic) that are not normally present in the body. Consequently, as far as our immune system is concerned, they are intruders, or **nonself**.

Complete Antigens and Haptens

Antigens can be *complete* or *incomplete*. **Complete antigens** have two important functional properties:

- **Immunogenicity**, which is the ability to stimulate specific lymphocytes to proliferate (multiply).
- **Reactivity**, which is the ability to react with the activated lymphocytes and the antibodies released by immunogenic reactions.

An almost limitless variety of foreign molecules can act as complete antigens, including virtually all foreign proteins, many large polysaccharides, and some lipids and nucleic acids. Of these, proteins are the strongest antigens. Pollen grains and microorganisms—such as bacteria, fungi, and virus particles—are all immunogenic because their surfaces bear many different foreign macromolecules.

As a rule, small molecules—such as peptides, nucleotides, and many hormones—are not immunogenic. But if they link up with the body's own proteins, the adaptive immune system may recognize the *combination* as foreign and mount an attack that is harmful rather than protective. (We describe these reactions, called *hypersensitivities*, later in the chapter.) In such cases, the troublesome small molecule is called a **hapten** (hap′ten; *hapein* = grasp) or **incomplete antigen**. Unless attached to protein carriers, haptens have reactivity but not immunogenicity. Besides certain drugs (particularly penicillin), chemicals that act as haptens are found in poison ivy, animal dander, detergents, cosmetics, and a number of common household and industrial products.

Antigenic Determinants

The ability of a molecule to act as an antigen depends on both its size and its complexity. Only certain parts of the antigen, called **antigenic determinants**, are immunogenic. Antibodies or lymphocyte receptors bind to these antigenic determinants in much the same manner that an enzyme binds to a substrate.

Most naturally occurring antigens have a variety of antigenic determinants on their surfaces, some more potent than others

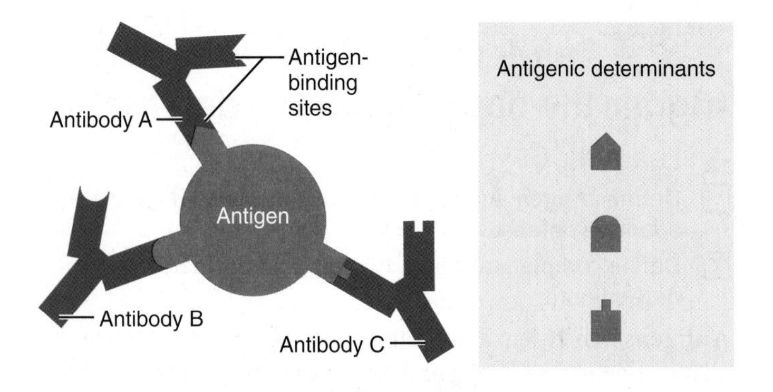

Figure 20.7 Most antigens have several different antigenic determinants. Antibodies (and related receptors on lymphocytes) bind to small areas on the antigen surface called antigenic determinants. In this example, three different types of antibodies react with different antigenic determinants on the same antigen molecule.

in provoking an immune response (**Figure 20.7**). Different lymphocytes "recognize" different antigenic determinants, so a single antigen may mobilize several lymphocyte populations and stimulate formation of many kinds of antibodies.

Large proteins have hundreds of chemically different antigenic determinants, which accounts for their high immunogenicity and reactivity. However, large simple molecules such as plastics, which have many identical, regularly repeating units, have little or no immunogenicity. Such substances are used to make artificial implants because the substances are not seen as foreign and rejected by the body.

Self-Antigens: MHC Proteins

A huge variety of protein molecules dot the external surfaces of all our cells. Assuming your immune system has been properly "programmed," your **self-antigens** are not foreign or antigenic to you, but they are strongly antigenic to other individuals. (This is the basis of transfusion reactions and graft rejection.)

Among the cell surface proteins that identify a cell as *self* is a group of glycoproteins called **MHC proteins**. Genes of the **major histocompatibility complex (MHC)** code for these proteins. Because millions of combinations of these genes are possible, it is unlikely that any two people except identical twins have the same MHC proteins. Each MHC protein has a deep groove that holds a peptide, either a self-antigen or a foreign antigen. As we will describe shortly, T lymphocytes can only bind antigens that are presented (displayed to them) on MHC proteins.

☑ Check Your Understanding

6. Name three key characteristics of adaptive immunity.
7. What is the difference between a complete antigen and a hapten?
8. What marks a cell as "self" as opposed to "nonself"?

For answers, see Answers Appendix.

20.4 B and T lymphocytes and antigen-presenting cells are cells of the adaptive immune response

→ Learning Objectives

- ☐ **Compare and contrast the origin, maturation process, and general function of B and T lymphocytes.**
- ☐ **Define immunocompetence and self-tolerance, and describe their development in B and T lymphocytes.**
- ☐ **Name several antigen-presenting cells and describe their roles in adaptive defenses.**

The adaptive immune system involves three crucial types of cells: two distinct populations of lymphocytes, plus *antigen-presenting cells (APCs)*.

- **B lymphocytes (B cells)** oversee humoral immunity.
- **T lymphocytes (T cells)** are non-antibody-producing lymphocytes that constitute the cellular arm of adaptive immunity.
- APCs do not respond to specific antigens as lymphocytes do. Instead, they play essential auxiliary roles. As we will see, T cells cannot recognize their antigens without APCs.

Lymphocytes

Despite their differences, B and T lymphocytes share a common pattern of development and common steps in their life cycles. Let's take a look.

Lymphocyte Development, Maturation, and Activation

The development, maturation, and activation of B and T cells share the five general steps shown in **Figure 20.8**.

Origin (Figure 20.8 ①) Like all blood cells, lymphocytes originate in red bone marrow from hematopoietic stem cells.

Maturation (Figure 20.8 ②) Lymphocytes are "educated" (go through a rigorous selection process) as they mature. The aim of this education is twofold:

- **Immunocompetence.** Each lymphocyte must become able (competent) to recognize its one specific antigen by binding to it. This ability is called **immunocompetence**. When B or T cells become immunocompetent, they display a unique type of receptor on their surface. These receptors (some 10^5 per cell) enable the lymphocyte to recognize and bind a specific antigen. Once these receptors appear, the lymphocyte is committed to react to one (and only one) distinct antigenic determinant because *all* of its antigen receptors are the same. The receptors on B cells are in fact membrane-bound antibodies. The receptors on T cells are not antibodies but are products of the same gene superfamily and have similar functions.
- **Self-tolerance.** Each lymphocyte must be relatively unresponsive to self-antigens so that it does not attack the body's own cells. This is called **self-tolerance**.

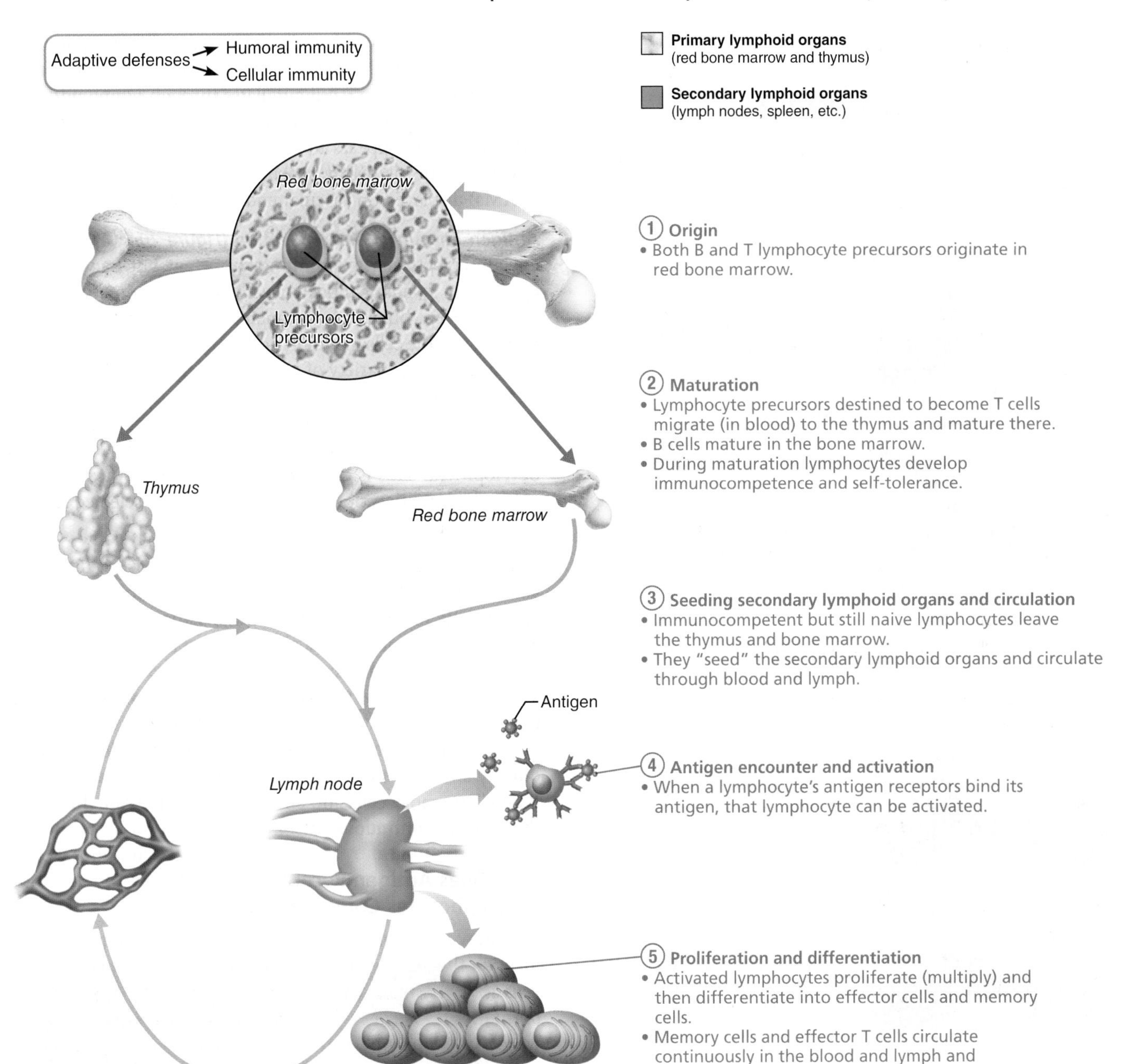

Figure 20.8 Lymphocyte development, maturation, and activation.

20

Maturation is a two- to three-day process that occurs in the bone marrow for B cells and in the thymus for T cells. Recall from Chapter 19 that the lymphoid organs where lymphocytes become immunocompetent—thymus and bone marrow—are called **primary lymphoid organs**. All other lymphoid organs are referred to as **secondary lymphoid organs**.

The selection process (education) that lymphocytes undergo is best understood in T cells. T cell education consists of positive and negative selection in the thymus (**Figure 20.9**).

1. **Positive selection** is the first of two tests a developing T lymphocyte must pass. It ensures that *only* T cells that are able to recognize self-MHC proteins survive. Remember that T cells cannot bind antigens unless the antigens are presented on self-MHC proteins. T cells that are unable to recognize self-MHC are eliminated by apoptosis.
2. **Negative selection**, the second test, ensures that T cells do not recognize self-antigens displayed on self-MHC. If they do, they are eliminated by apoptosis. Negative selection is the basis for immunological self-tolerance, making sure that T cells don't attack the body's own cells, which would cause autoimmune disorders. Because the self-reactive lymphocyte and all of its potential progeny are eliminated, this is called *clonal deletion*.

This education of T cells is expensive indeed—only about 2% of T cells survive it and continue to become successful immunocompetent, self-tolerant T cells.

Table 20.4 Overview of B and T Lymphocytes

	B LYMPHOCYTES	T LYMPHOCYTES
Type of immune response	Humoral	Cellular
Antibody secretion	Yes	No
Primary targets	Extracellular pathogens (e.g., bacteria, fungi, parasites, some viruses in extracellular fluid)	Intracellular pathogens (e.g., virus-infected cells) and cancer cells
Site of origin	Red bone marrow	Red bone marrow
Site of maturation	Red bone marrow	Thymus
Effector cells	Plasma cells	Cytotoxic T (T_C) cells Helper T (T_H) cells Regulatory T (T_{Reg}) cells
Memory cell formation	Yes	Yes

☐ **Describe the structure and functions of antibodies and name the five antibody classes.**

Now that you understand the common steps in lymphocyte maturation and activation, let's examine how this basic pattern applies to B lymphocytes. When a B cell encounters its antigen, that antigen provokes the *humoral immune response*, in which antibodies specific for that antigen are made.

Activation and Differentiation of B Cells

An immunocompetent but naive B lymphocyte is *activated* when matching antigens bind to its surface receptors and cross-link adjacent receptors together. Antigen binding is quickly followed by receptor-mediated endocytosis of the cross-linked antigen-receptor complexes. As we described previously, this is called *clonal selection* and is followed by proliferation and differentiation into effector cells (**Figure 20.11**). (As we will see shortly, interactions with T cells are usually required to help B cells achieve full activation.)

Most cells of the clone differentiate into **plasma cells**, the antibody-secreting *effector cells* of the humoral response. Plasma cells develop the elaborate internal machinery (largely rough endoplasmic reticulum) needed to secrete antibodies at the unbelievable rate of about 2000 molecules per second. Each plasma cell functions at this breakneck pace for 4 to 5 days and then dies. The secreted antibodies, each with the same antigen-binding properties as the receptor molecules on the surface of the parent B cell, circulate in the blood or lymph. There they bind to free antigens and mark them for destruction by other innate or adaptive mechanisms.

Clone cells that do not become plasma cells become long-lived **memory cells**. They can mount an almost immediate humoral response if they encounter the same antigen again in the future (Figure 20.11, bottom).

Immunological Memory

The cellular proliferation and differentiation we have just described constitute the **primary immune response**, which occurs on first exposure to a particular antigen. The primary response typically has a lag period of 3 to 6 days after the antigen encounter. This lag period mirrors the time required for the few B cells specific for that antigen to proliferate (about 12 generations) and for their offspring to differentiate into plasma cells. After the mobilization period, plasma antibody levels rise, reach peak levels in about 10 days, and then decline (**Figure 20.12**).

If (and when) someone is reexposed to the same antigen, whether it's the second or twenty-second time, a **secondary immune response** occurs. Secondary immune responses are faster, more prolonged, and more effective, because the immune system has already been primed to the antigen, and sensitized memory cells are already "on alert." These memory cells provide **immunological memory**.

Within hours after recognizing the "old enemy" antigen, a new army of plasma cells is being generated. Within 2 to 3 days the antibody concentration in the plasma, called the *antibody titer*, rises steeply to reach much higher levels than in the primary response. Secondary response antibodies not only bind with greater affinity (more tightly), but their blood levels remain high for weeks to months. (When the appropriate chemical signals are present, plasma cells can keep functioning for much longer than the 4 to 5 days seen in primary responses.) Memory cells persist for long periods and many retain their capacity to produce powerful secondary humoral responses for life.

The same general events occur in the cellular immune response: A primary response sets up a pool of effector cells (in this case, T cells) and generates memory cells that can then mount secondary responses.

Active and Passive Humoral Immunity

When your B cells encounter antigens and produce antibodies against them, you are exhibiting **active humoral immunity**. Active immunity is acquired in two ways (**Figure 20.13**). It is (1) *naturally acquired* when you get a bacterial or viral infection, during which time you may develop symptoms of the disease and suffer a little (or a lot), and (2) *artificially acquired* when you receive a **vaccine**. Indeed, once researchers realized that

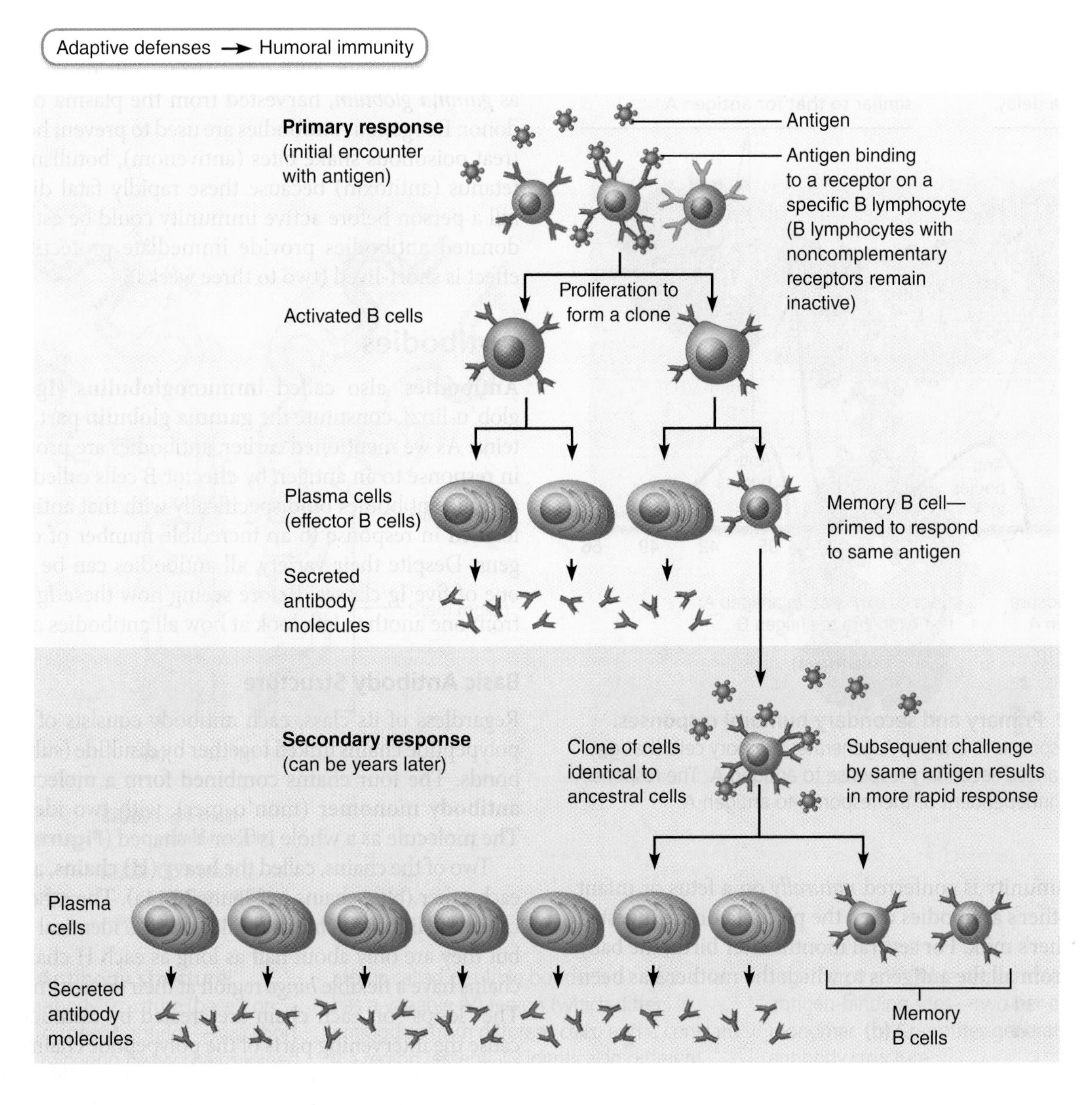

Figure 20.11 Clonal selection of a B cell.

secondary responses are so much more vigorous than primary responses, the race was on to develop vaccines to "prime" the immune response by providing a first encounter with the antigen.

Most vaccines contain pathogens that are dead or *attenuated* (living, but extremely weakened), or their components. Vaccines provide two benefits:

- Their weakened antigens provide functional antigenic determinants that are both immunogenic and reactive.
- They spare us most of the symptoms and discomfort of the disease that would otherwise occur during the primary response.

Vaccine *booster shots* are used in some cases to intensify the immune response at later encounters with the same antigen.

Vaccines have wiped out smallpox and have substantially lessened the illness caused by such former childhood killers as whooping cough, polio, and measles. Although vaccines have dramatically reduced hepatitis B, tetanus, influenza, and pneumonia in adults, immunization of adults in the U.S. has a much lower priority than that of children. As a result more than 65,000 Americans die each year from preventable infections.

Conventional vaccines have shortcomings. In extremely rare cases, vaccines have caused the very disease they are trying to prevent because the attenuated virus wasn't weakened enough. In some individuals, contaminating proteins (for example, egg albumin) cause allergic responses to the vaccine. The new "naked DNA" antiviral vaccines, blasted into the skin with a gene gun, and vaccines taken orally appear to circumvent these problems, but are not always effective.

Passive humoral immunity differs from active immunity, both in the antibody source and in the degree of protection it provides (Figure 20.13). Instead of being made by your plasma cells, ready-made antibodies are introduced into your body. As a result, your B cells are not challenged by antigens, immunological memory does not occur, and the protection provided by the "borrowed" antibodies ends when they naturally degrade in the body.

Table 20.5 Immunoglobulin Classes*

Class	Characteristics
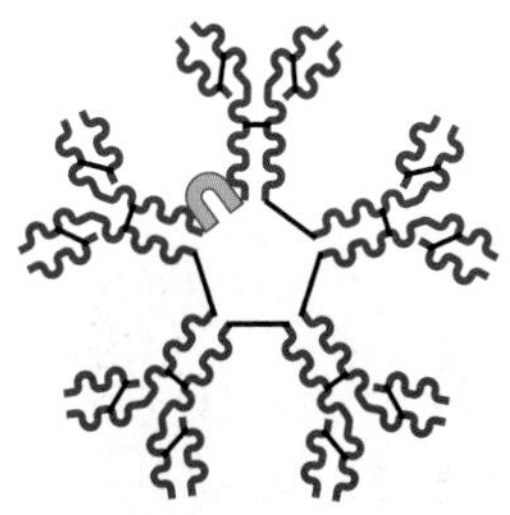 IgM (pentamer)	• The first immunoglobulin class secreted by plasma cells during the primary response. (This fact is diagnostically useful because presence of IgM in plasma usually indicates current infection by the pathogen eliciting IgM's formation.) • Readily fixes and activates complement. • Exists in monomer and pentamer (five united monomers) forms. • The monomer serves as an antigen receptor on the B cell surface. • The pentamer circulates in blood plasma. • Numerous antigen-binding sites make it a potent agglutinating agent.
 IgA (dimer)	• The dimer, referred to as **secretory IgA**, is found in body secretions such as saliva, sweat, intestinal juice, and milk. • Secretory IgA helps stop pathogens from attaching to epithelial cell surfaces (including mucous membranes and the epidermis). • The monomer exists in limited amounts in plasma.
 IgD (monomer)	• Found on the B cell surface. • Functions as a B cell antigen receptor (as does IgM).
 IgG (monomer)	• The most abundant antibody in plasma, accounting for 75–85% of circulating antibodies. • The main antibody of both secondary and late primary responses. • Readily fixes and activates complement. • Protects against bacteria, viruses, and toxins circulating in blood and lymph. • Crosses the placenta and confers passive immunity from the mother to the fetus.
 IgE (monomer)	• Stem end binds to mast cells or basophils. Antigen binding to its receptor end triggers these cells to release histamine and other chemicals that mediate inflammation and an allergic reaction. • Secreted by plasma cells in skin, mucosae of the gastrointestinal and respiratory tracts, and tonsils. • Only traces of IgE are found in plasma. • Levels rise during severe allergic attacks or chronic parasitic infections of the gastrointestinal tract.

*Key characteristics are listed in blue type.

from Chapter 16 that agglutination occurs when mismatched blood is transfused (the foreign red blood cells clump) and is the basis of tests used for blood typing.

Precipitation In **precipitation**, soluble molecules (instead of cells) are cross-linked into large complexes that settle out of solution. Like agglutinated bacteria, precipitated antigen molecules are much easier for phagocytes to capture and engulf than are freely moving antigens.

Complement Fixation and Activation Complement fixation and activation is the chief antibody defense used against cellular antigens, such as bacteria or mismatched red blood cells. When several antibodies bind close together on the same cell, the complement-binding sites on their stem regions align. This triggers complement fixation into the antigenic cell's surface, followed by cell lysis.

Additionally, as we described earlier, molecules released during complement activation tremendously amplify the inflammatory response and promote phagocytosis via opsonization. This sets into motion a positive feedback cycle that enlists more and more defensive elements.

A quick and dirty way to remember how antibodies work is to remember they have a PLAN of action—**p**recipitation, **l**ysis (by complement), **a**gglutination, and **n**eutralization.

CLINICAL

Around the world, billions of people are infected by parasitic worms such as *Ascaris* and *Schistosoma*. These large pathogens are difficult for our immune systems to deal with and "PLAN" is insufficient.

Nevertheless, antibodies still play a critical role in the worm's destruction. IgE antibodies coat the surface of parasitic worms, marking them for destruction by eosinophils. When eosinophils encounter antibody-coated worms, they bind to the exposed stems of the IgE. This triggers the eosinophils to release the toxic contents of their large cytoplasmic granules all over their prey. +

Monoclonal Antibodies as Clinical and Research Tools

Monoclonal antibodies are pure antibody preparations specific for a single antigenic determinant. They are produced by descendants of a single cell. Commercially prepared monoclonal antibodies are essential in research, clinical testing, and treatment.

Monoclonal antibodies are used to diagnose pregnancy, certain sexually transmitted infections, some cancers, hepatitis, and rabies. These monoclonal antibody tests are more specific, sensitive, and rapid than other tests. Monoclonal antibodies are also used to treat leukemia and lymphomas, cancers that are present in the circulation and so are easily accessible to injected antibodies. They also serve as "guided missiles" to deliver anti-cancer drugs only to cancerous tissue, and to treat certain autoimmune diseases (as we will discuss later).

Summary of Antibody Actions

At the most basic level, the race between antibody production and pathogen multiplication determines whether or not

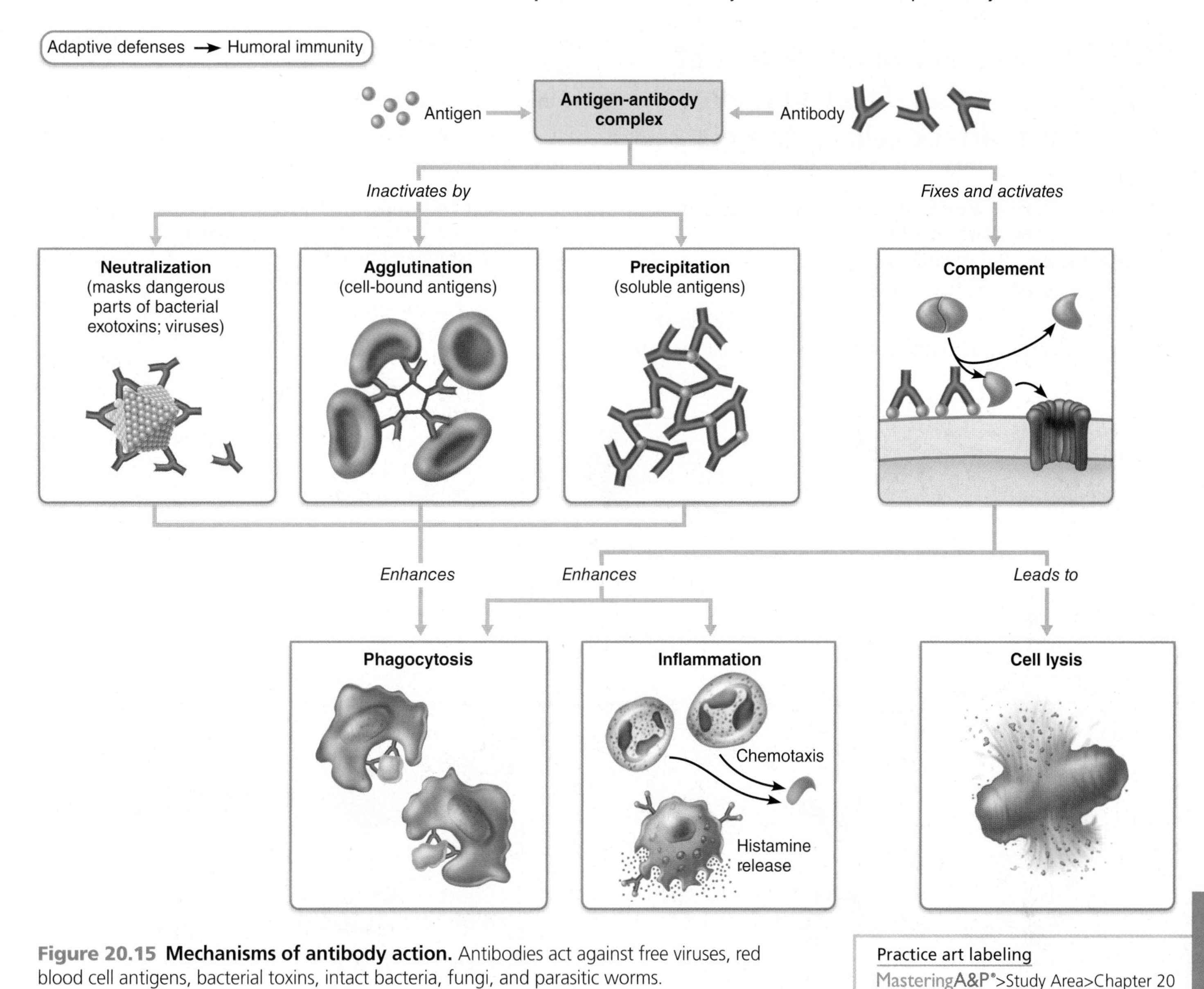

Figure 20.15 Mechanisms of antibody action. Antibodies act against free viruses, red blood cell antigens, bacterial toxins, intact bacteria, fungi, and parasitic worms.

Practice art labeling
MasteringA&P®>Study Area>Chapter 20

you become sick. Remember, however, that forming antigen-antibody complexes does *not* destroy the antigens. Instead, it prepares them for destruction by innate defenses.

Antibodies produced by plasma cells are in many ways the simplest, most versatile ammunition of the immune response. Nevertheless, they provide only partial immunity. Their prey are *extracellular pathogens*—intact bacteria, free viruses, and soluble foreign molecules—in other words, pathogens that are free in body secretions and tissue fluid and circulating in blood and lymph. Antibodies never invade solid tissues unless a lesion is present.

Until recently, the accepted dogma was that antibodies *only* act extracellularly. Remarkably, we now know that antibodies can act intracellularly as well. Antibodies that are attached to a virus before that virus infects a cell can "hang on" to the virus as it slips inside the cell. There, the antibodies activate intracellular mechanisms that destroy the virus. Even so, antibodies are not very effective against pathogens like viruses and tuberculosis bacilli that quickly slip inside body cells to multiply there. For these intracellular pathogens, the cellular arm of adaptive immunity comes into play.

☑ Check Your Understanding

13. Why is the secondary response to an antigen so much faster than the primary response?

14. How do vaccines protect against common childhood illnesses such as chicken pox, measles, and mumps?

15. Which class of antibody is most abundant in blood? Which is secreted first in a primary immune response? Which is most abundant in secretions?

16. List four ways in which antibodies can bring about destruction of a pathogen.

17. MAKING connections What is the function of the abundant endoplasmic reticulum in plasma cells? What other organelle(s) (described in Chapter 3) would be especially abundant in plasma cells? Why?

For answers, see Answers Appendix.

20.6 Cellular immunity consists of T lymphocytes that direct adaptive immunity or attack cellular targets

→ Learning Objectives

- ☐ **Define cellular immunity and describe the process of activation and clonal selection of T cells.**
- ☐ **Describe the roles of different types of T cells.**
- ☐ **Describe T cell functions in the body.**

T cells are best suited for cell-to-cell interactions. When antigens are presented to a T lymphocyte, they provoke a cellular immune response. Some activated T cells directly kill body cells if they are:

- Infected by viruses or bacteria
- Cancerous or abnormal
- Foreign cells (e.g., transplanted cells)

Other T cells release chemicals that regulate the immune response.

T cells are a diverse lot, much more complex than B cells in both classification and function. There are two major populations of T cells based on which of two structurally related *cell differentiation glycoproteins*—CD4 or CD8—a mature T cell displays. The CD4 and CD8 glycoproteins are surface receptors but are distinct from the T cell antigen receptors. They play a role in interactions between T cells and other cells.

When activated, CD4 and CD8 cells differentiate into the three major kinds of *effector cells* of cellular immunity (**Figure 20.16**).

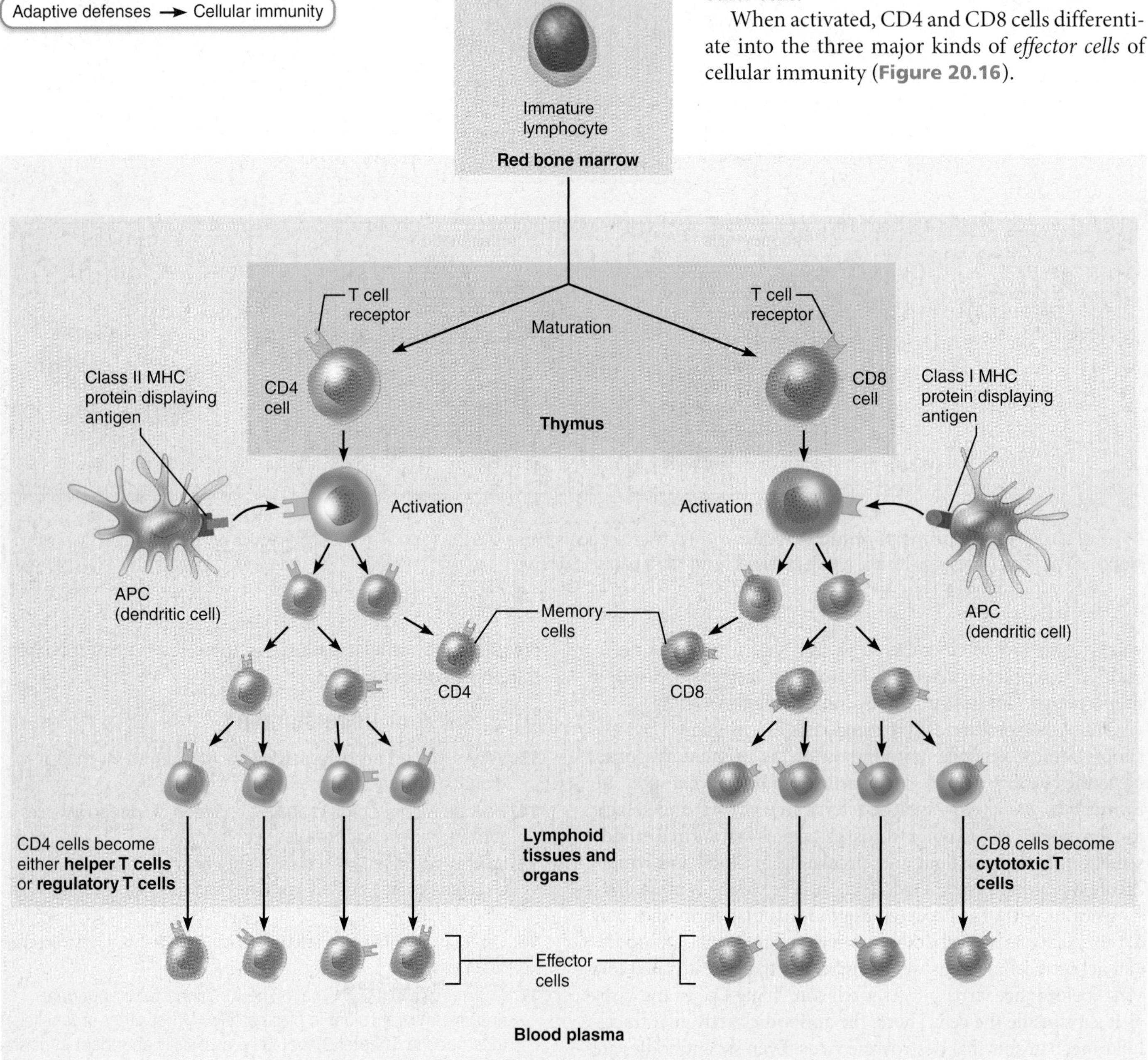

Figure 20.16 **Major types of T cells.**

- **CD4 cells** usually become *helper T* (T_H) *cells* that help activate B cells, other T cells, and macrophages, and direct the adaptive immune response.
- **CD8 cells** become *cytotoxic T* (T_C) *cells* that destroy cells in the body that harbor anything foreign.
- Some CD4 cells become *regulatory T* (T_{Reg}) *cells*, which moderate the immune response.

Activated CD4 and CD8 cells can also become memory T cells. Note that the names of the effector cells (helper, cytotoxic, regulatory) are reserved for *activated* T cells, while naive T cells are simply called CD4 or CD8 cells.

MHC Proteins and Antigen Presentation

Unlike B cells and antibodies, T cells cannot "see" either free antigens or antigens that are in their natural state. T cells can recognize and respond only to *processed* fragments of protein antigens displayed on surfaces of body cells (APCs and others).

Antigen presentation is necessary for both activation of naive T cells and the normal functioning of effector T cells. As previously noted, the cell surface proteins on which antigens are presented to T cells are the major histocompatibility complex (MHC) proteins. It is important to understand how these critical players in antigen presentation work before we can understand how T cells work. There are two classes of MHC proteins—class I and class II—summarized in **Table 20.6**.

Class I MHC Proteins

Class I MHC proteins are found on the surface of virtually *all* body cells except red blood cells. Each class I MHC protein has a groove that holds an antigen—a protein fragment 8 or 9 amino acids long.

Where do these protein fragments come from? All antigens displayed on class I MHC proteins are **endogenous antigens**—fragments of proteins synthesized *inside the cell.* In a healthy cell, endogenous antigens are all self-antigens, generally bits of digested cellular proteins. But in an infected cell, endogenous antigens may also include fragments of foreign antigens that are synthesized within the infected cell but "belong to" the pathogen. In a cancerous cell, endogenous antigens can include altered (cancer) proteins.

As proteases (protein-digesting enzymes) degrade cytoplasmic proteins as part of their natural recycling, a random sample of the resulting protein fragments is transported into the endoplasmic reticulum. Inside the ER, these peptides bind to newly synthesized class I MHC proteins. Transport vesicles then export the "loaded" class I MHC proteins to the cell surface.

Class I MHC proteins are crucial for both activating naive CD8 cells and "informing" cytotoxic T cells that infectious microorganisms are hiding in body cells. Without them, viruses and certain bacteria that thrive in cells could multiply unnoticed and unbothered.

When class I MHC proteins display fragments of our own proteins (self-antigens), cytotoxic T cells passing by get the signal "Leave this cell alone, it's ours!" and ignore them. But when class I MHC proteins display foreign antigens, they "sound a molecular alarm" that signals invasion. In this signaling, the class I MHC proteins both (1) act as antigen holders and (2) form the self part of the self-nonself complexes that cytotoxic T cells must recognize in order to kill.

Table 20.6 Role of MHC Proteins in Cellular Immunity

	CLASS I MHC PROTEINS	CLASS II MHC PROTEINS
Displayed by	All nucleated cells 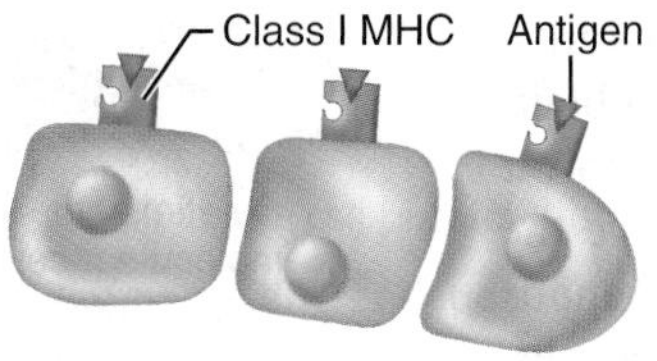	APCs (dendritic cells, macrophages, B cells) 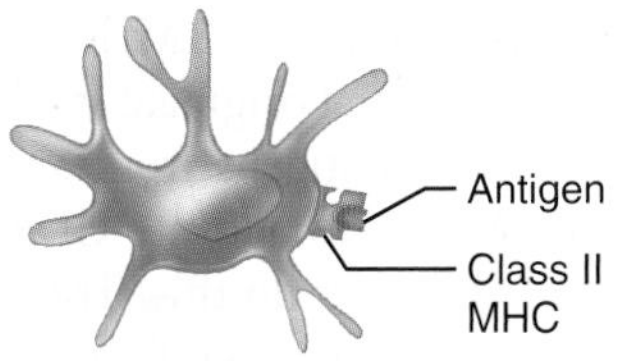
Recognized by	Naive CD8 cells and cytotoxic T cells 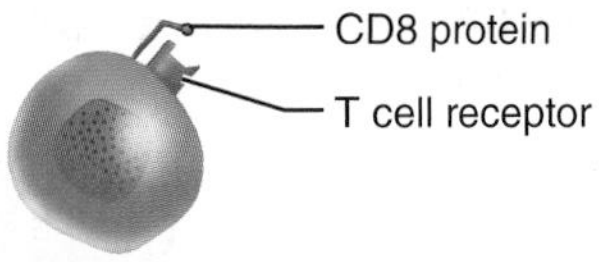	Naive CD4 cells and helper T cells 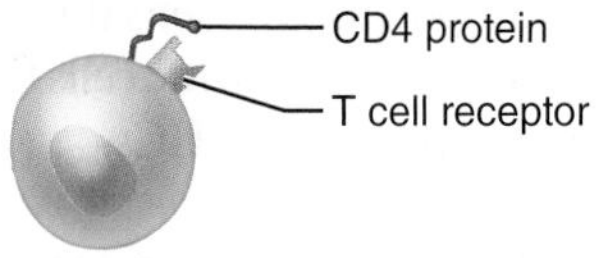
Foreign antigens on MHC are	Endogenous (intracellular pathogens or proteins made by cancerous cells)* 	Exogenous (phagocytized extracellular pathogens) 
Cells displaying foreign antigens on MHC send this message	**If the cell is an APC:** "I belong to self, but have captured a foreign invader. This is what it looks like. Kill any cell that displays it." **If the cell is not an APC:** "I belong to self, but have been invaded or become cancerous. Kill me!"	"I belong to self, but have captured a foreign invader. This is what it looks like. Help me mount a defense against it."

*Dendritic cells are an exception because they can present *another cell's* endogenous antigens on their class I MHC proteins to activate CD8 cells.

Regulatory T Cells

While T_H cells help activate adaptive immune responses, related T cells called **regulatory T (T_{Reg}) cells** dampen the immune response. They act either by direct contact or by releasing inhibitory cytokines.

T_{Reg} cells are important in preventing autoimmune reactions because they suppress self-reactive lymphocytes in the periphery—that is, outside the lymphoid organs. T_{Reg} cells and their subpopulations are currently hot research topics. For example, researchers hope to use them to induce tolerance to transplanted tissue and to lessen the severity of autoimmune diseases.

• • •

Table 20.8 summarizes the cells and molecules of the adaptive immune response. Figure 20.20 gives an overview of the entire primary immune response, both innate and adaptive.

CLINICAL

Organ Transplants and Prevention of Rejection

The goal of organ transplantion is to provide patients with a functional organ from a living or deceased donor. This is often the only viable option left in end-stage cardiac or renal disease. Immune rejection presents a particular problem and so transplant success depends in part on the similarity of the donor and recipient tissues. NK cells, macrophages, antibodies, and especially T cells act vigorously to destroy any tissue they recognize as foreign.

The most common type of transplant is the *allograft*. Allografts are grafts transplanted from different individuals of the same species. Before an allograft is attempted, the ABO and other blood group antigens of donor and recipient must be determined because these antigens are also present on most body cells and mismatches lead to immediate rejection. Matching donor and recpient MHCs helps minimize long-term rejection.

Following surgery the patient is treated with *immunosuppressive therapy*. Many of the drugs used to suppress rejection kill rapidly dividing cells (such as activated lymphocytes), and all of them have severe side effects. The major problem with immunosuppressive therapy is that the patient's suppressed immune system cannot protect the body against other foreign agents. As a result, overwhelming bacterial and viral infection remains the most frequent cause of death in transplant patients. Even under the best conditions, by ten years after receiving a transplant, roughly 50% of patients have rejected the donor organ.

☑ Check Your Understanding

18. Class II MHC proteins display what kind of antigens? What class of T cell recognizes antigens bound to class II MHC? What types of cells display these proteins?

Table 20.8 Cells and Molecules of the Adaptive Immune Response

ELEMENT	FUNCTION IN IMMUNE RESPONSE
Cells	
B cell	Lymphocyte that matures in bone marrow. Its progeny (clone members) form plasma cells and memory cells.
Plasma cell	Antibody-producing "machine"; produces huge numbers of antibodies. An effector B cell.
Helper T (T_H) cell	An effector CD4 T cell central to both humoral and cellular immunity. It stimulates production of cytotoxic T cells and plasma cells, activates macrophages, and acts both directly and indirectly by releasing cytokines.
Cytotoxic T (T_C) cell	An effector CD8 T cell that kills virus-invaded body cells and cancer cells.
Regulatory T (T_{Reg}) cell	Slows or stops activity of immune system. Important in controlling autoimmune diseases; several different kinds.
Memory cell	Descendant of activated B cell or any class of activated T cell; generated during initial immune response. May exist in body for years, enabling it to respond quickly and efficiently to subsequent encounters with same antigen.
Antigen-presenting cell (APC)	Any of several cell types (dendritic cell, macrophage, B cell) that engulfs and digests antigens that it encounters, then presents parts of them on its plasma membrane (bound to an MHC protein) for recognition by T cells bearing receptors for the same antigen. This function, antigen presentation, is essential for activation of T cells.
Molecules	
Antigen	Substance capable of provoking an immune response. Typically a large, complex molecule (e.g., protein or modified protein) not normally present in the body.
Antibody (immunoglobulin or Ig)	Protein produced by B cell or by plasma cell. Antibodies produced by plasma cells are released into body fluids (blood, lymph, saliva, mucus, etc.), where they attach to antigens. This causes complement fixation, neutralization, precipitation, or agglutination, which "marks" the antigens for destruction by phagocytes or complement.
Perforins, granzymes	Released by T_C cells. Perforins create large pores in the target cell's membrane, allowing entry of apoptosis-inducing granzymes.
Complement	Group of bloodborne proteins activated after binding to antibody-covered antigens or certain molecules on the surface of microorganisms; enhances inflammatory response and lyses some microorganisms.
Cytokines	Small proteins that act as chemical messengers between various parts of the immune system. See Table 20.7.

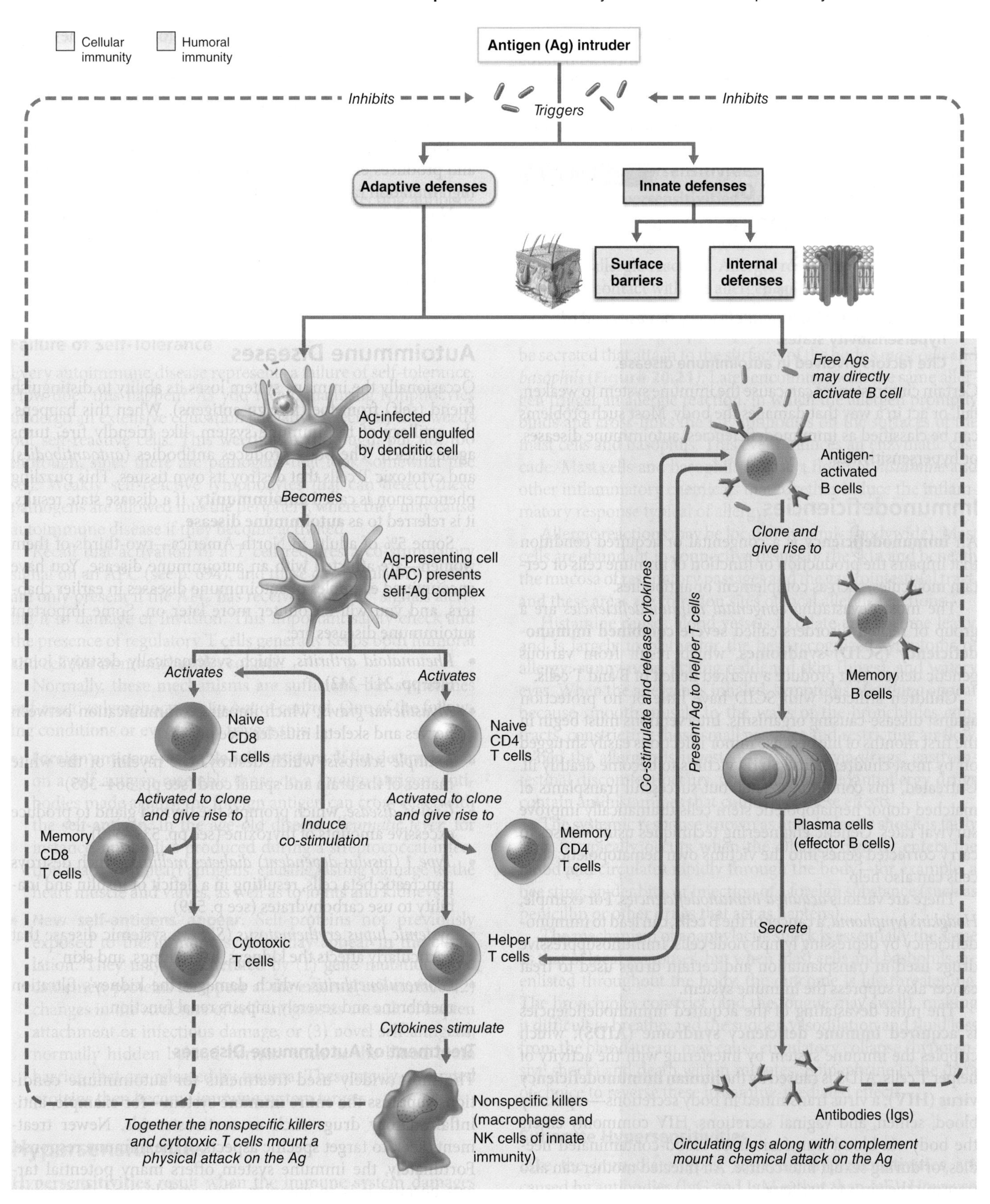

Figure 20.20 Simplified summary of the primary immune response. Co-stimulation usually requires direct cell-cell interactions; cytokines enhance these and many other events. Although complement, NK cells, and phagocytes are innate defenses, they are enlisted in the fight by cytokines. (For simplicity, only B cell receptors are illustrated.)

AT THE CLINIC

Clinical Case Study

Immune System

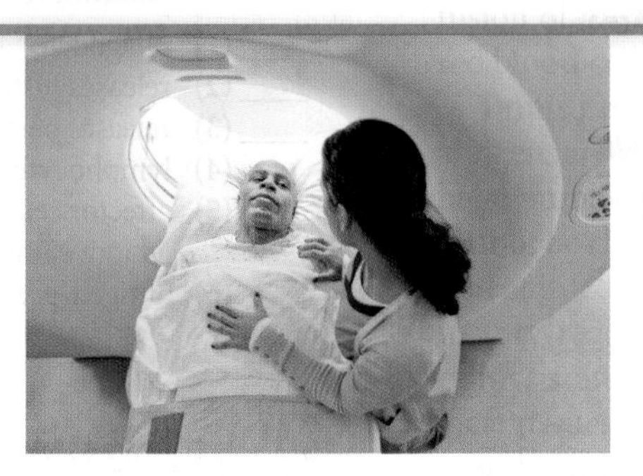

Remember Mr. Ayers, the bus driver from Chapter 17? When we last saw him, he was headed for surgery. Although his dissected aorta was repaired, by the time surgical exposure and blood vessel clamping had been achieved, the dissection had extended up into the origin of his left common carotid artery. As a result, a clot formed that caused a massive stroke. Unfortunately, this left him with severe and permanent brain damage, and he was declared brain dead.

A discussion of Mr. Ayers's situation with his family confirmed his status as an organ donor. The organ recovery coordinator evaluated Mr. Ayers's suitability as a candidate for organ donation. Tissue typing (histocompatibility) tests were conducted, and the results were entered into the UNOS (United Network for Organ Sharing) database. Two potential recipients were identified. Mr. Ayers's right kidney was given in transplantation to a 35-year-old man, and his left kidney was given to a 27-year-old woman. Following surgery, both recipients were placed on immunosuppressive drug therapy.

1. In organ transplants, the transplanted organ is referred to as a graft. What type of graft is represented by the two kidneys that Mr. Ayers has donated?
2. Tissue typing characterizes the class I and II MHC proteins. What is an MHC protein?
3. What is the difference between class I and class II MHC proteins?
4. Why is the matching of the MHC molecules and the tissue compatibility so important in this case?
5. Why were the recipients of the two kidneys put on immunosuppressive drug therapy?

For answers, see Answers Appendix.

21 The Respiratory System

KEY CONCEPTS

Far from self-sustaining, our bodies depend on the external environment, both as a source of substances we need to survive and as a catch basin for wastes. Our trillions of cells require a continuous supply of oxygen to carry out their vital functions. We can live without food or water for days, but we cannot do without oxygen for even a little while.

As cells use oxygen, they give off carbon dioxide, a waste product the body must get rid of. They also generate dangerous free radicals, the inescapable by-products of living in a world full of oxygen.

The major function of the **respiratory system** is to supply the body with oxygen and dispose of carbon dioxide. To accomplish this function, at least four processes, collectively called **respiration**, must happen:

1. **Pulmonary ventilation** (commonly called breathing): Air is moved into and out of the lungs (during *inspiration* and *expiration*) so the gases there are continuously changed and refreshed.
2. **External respiration:** Oxygen diffuses from the lungs to the blood, and carbon dioxide diffuses from the blood to the lungs.
3. **Transport of respiratory gases:** Oxygen is transported from the lungs to the tissue cells of the body, and carbon dioxide is transported from the tissue cells to the lungs. The cardiovascular system accomplishes this transport using blood as the transporting fluid.
4. **Internal respiration:** Oxygen diffuses from blood to tissue cells, and carbon dioxide diffuses from tissue cells to blood.

The respiratory system is responsible for only the first two processes (**Figure 21.1**), but it cannot accomplish its primary goal of obtaining oxygen and eliminating carbon dioxide unless the third and fourth processes also occur. As you can see, the respiratory and circulatory systems are closely coupled, and if either system fails, the body's cells begin to die from oxygen starvation.

The actual use of oxygen and production of carbon dioxide by tissue cells, known as cellular respiration, is the cornerstone of all energy-producing chemical reactions in the body. We discuss *cellular respiration*, which is not a function of the respiratory system, in the metabolism section of Chapter 23.

Because it moves air, the respiratory system is also involved with the sense of smell and with speech.

PART 1
FUNCTIONAL ANATOMY

The respiratory system (Figure 21.1) includes the *nose, nasal cavity*, and *paranasal sinuses*; the *pharynx*; the *larynx*; the *trachea*; the *bronchi* and their smaller branches; and the *lungs*, which contain tiny air

A spirometer, which measures airflow from the lungs, is a valuable tool for assessing lung function.

protective stratified squamous epithelium. This structural adaptation accommodates the increased friction and chemical trauma (characteristic of hot and spicy foods) accompanying food passage.

The paired **palatine tonsils** lie embedded in the lateral walls of the oropharyngeal mucosa just posterior to the oral cavity. The **lingual tonsil** covers the posterior surface of the tongue.

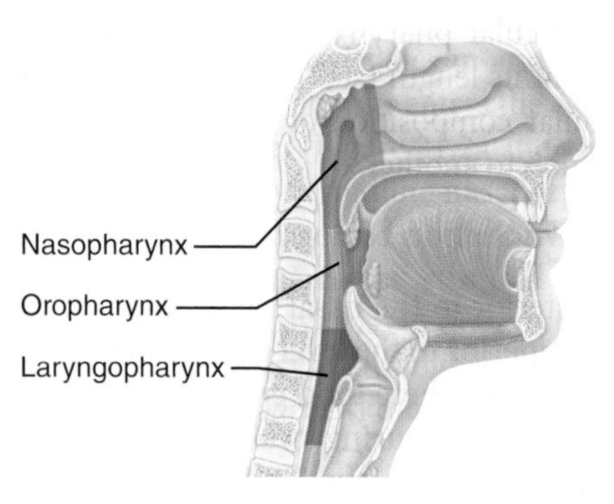

(a) Regions of the pharynx

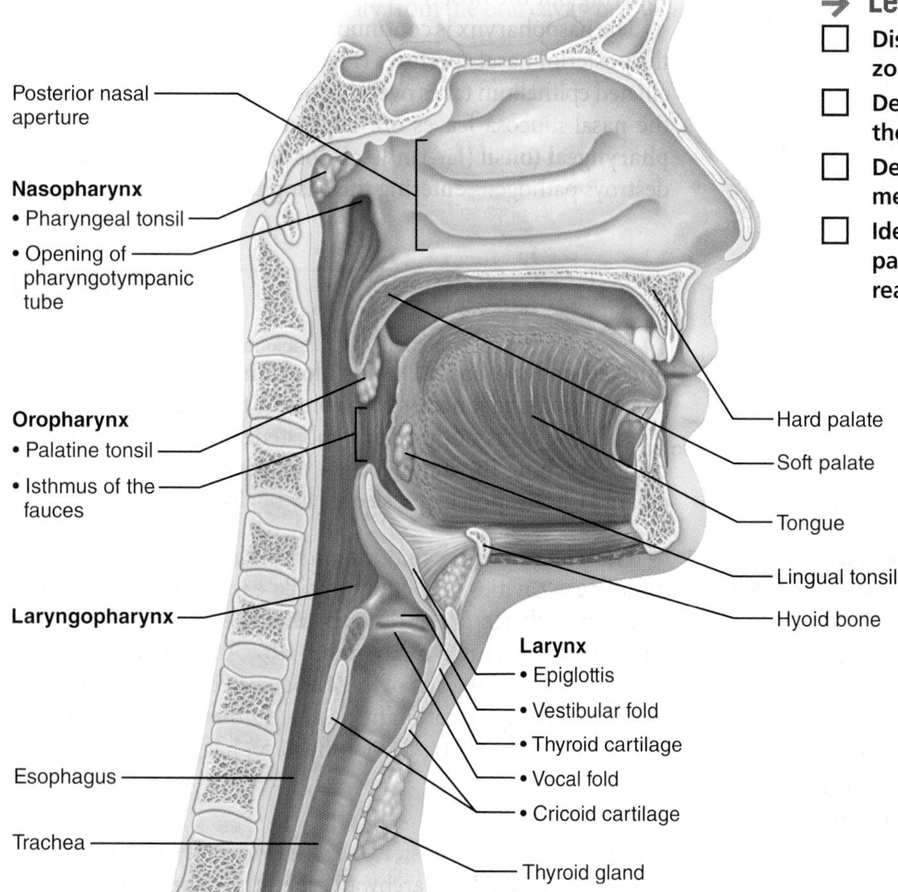

(b) Structures of the pharynx and larynx

Figure 21.4 The pharynx, larynx, and upper trachea. Midsagittal section of the head and neck. (For a related image, see *A Brief Atlas of the Human Body*, Figures 46 and 47.)

The Laryngopharynx

Like the oropharynx above it, the **laryngopharynx** (lah-ring″go-far′ingks) serves as a passageway for food and air and is lined with a stratified squamous epithelium. It lies directly posterior to the larynx, where the respiratory and digestive pathways diverge, and extends to the inferior edge of the cricoid cartilage. The laryngopharynx is continuous with the esophagus posteriorly.

The esophagus conducts food and fluids to the stomach; air enters the larynx anteriorly. During swallowing, food has the "right of way," and air passage temporarily stops.

Check Your Understanding

1. Air moving from the nose to the larynx passes by a number of structures. List (in order) as many of these structures as you can.
2. Which part of the pharynx houses the pharyngeal tonsil?

For answers, see Answers Appendix.

21.2 The lower respiratory system consists of conducting and respiratory zone structures

Learning Objectives

- ☐ **Distinguish between conducting and respiratory zone structures.**
- ☐ **Describe the structure, function, and location of the larynx, trachea, and bronchi.**
- ☐ **Describe the makeup of the respiratory membrane, and relate structure to function.**
- ☐ **Identify the organs forming the respiratory passageway(s) in descending order until you reach the alveoli.**

Anatomically, the lower respiratory system consists of the *larynx*, *trachea*, *bronchi*, and *lungs*. Functionally, the respiratory system as a whole consists of two zones:

- The **respiratory zone**, the actual site of gas exchange, is composed of the respiratory bronchioles, alveolar ducts, and alveoli, all microscopic structures.
- The **conducting zone** consists of all of the respiratory passageways from the nose to the respiratory bronchioles. These provide fairly rigid conduits for air to reach the gas exchange sites. The conducting zone organs also cleanse, humidify, and warm incoming air. As a result, air reaching the lungs has fewer irritants (dust, bacteria, etc.) than when it entered the body, and it is warm and damp, like the air of the tropics.

The Larynx

Basic Anatomy

The **larynx** (lar′ingks), or voice box, extends for about 5 cm (2 inches) from the level of the third to the sixth cervical vertebra. Superiorly it attaches to the hyoid bone and opens into the laryngopharynx. Inferiorly it is continuous with the trachea (Figure 21.4b).

The larynx has three functions:

- Provide a *patent* (open) airway
- Act as a switching mechanism to route air and food into the proper channels
- Voice production [because it houses the vocal folds (vocal cords)]

The framework of the larynx is an intricate arrangement of nine cartilages connected by membranes and ligaments (**Figure 21.5**). Except for the epiglottis, all laryngeal cartilages are hyaline cartilages.

The large, shield-shaped **thyroid cartilage** is formed by the fusion of two cartilage plates. The midline **laryngeal prominence** (lah-rin′je-al), which marks the fusion point, is obvious externally as the *Adam's apple* (Figure 21.5a). The thyroid cartilage is typically larger in males than in females because male sex

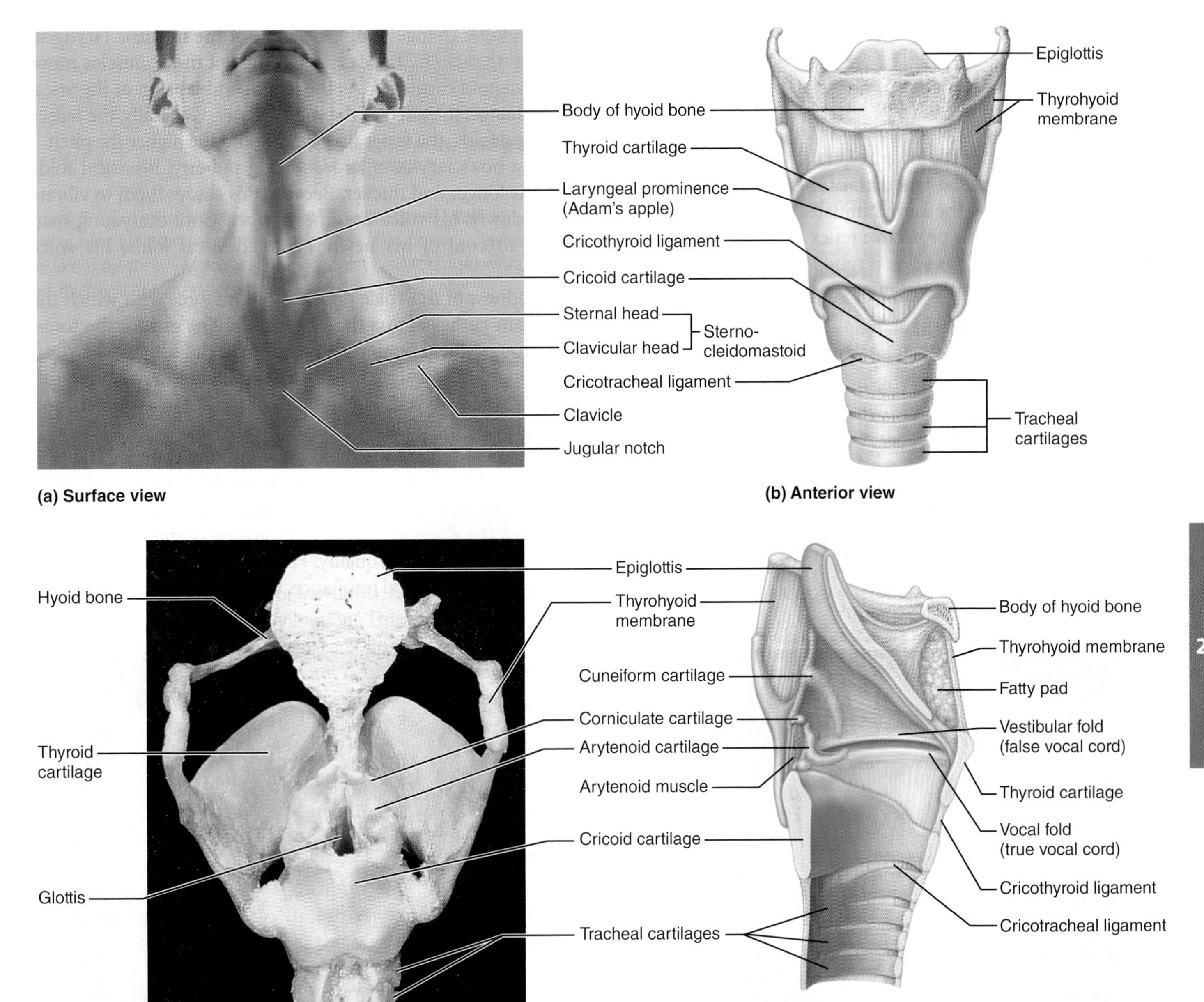

Figure 21.5 The larynx.

Practice art labeling
MasteringA&P®>Study Area>Chapter 21

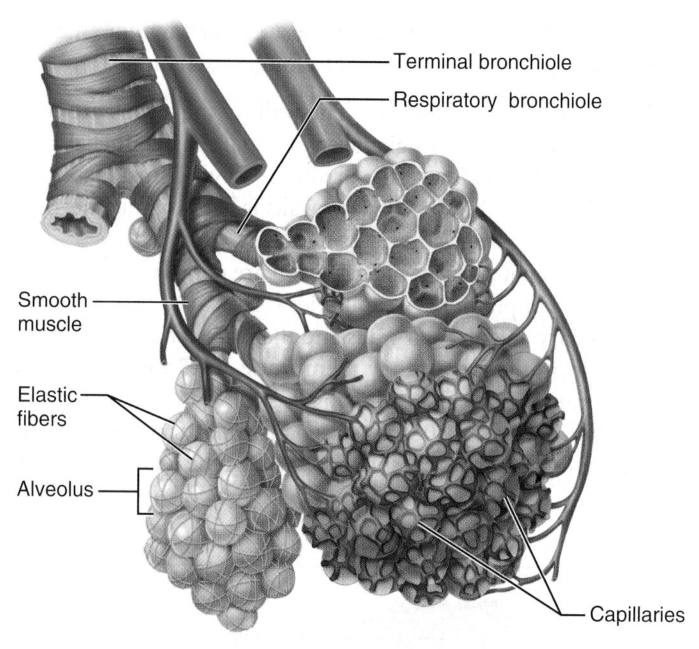

(a) Diagrammatic view of capillary-alveoli relationships

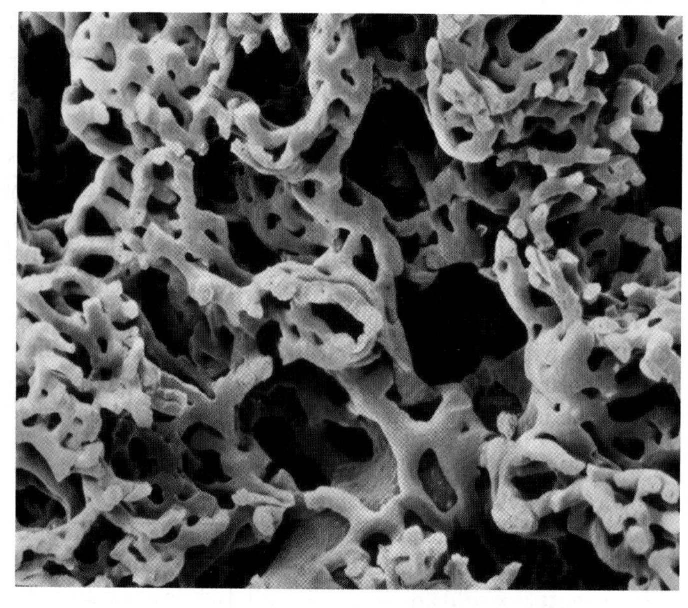

(b) Scanning electron micrograph of pulmonary capillary casts (300×)

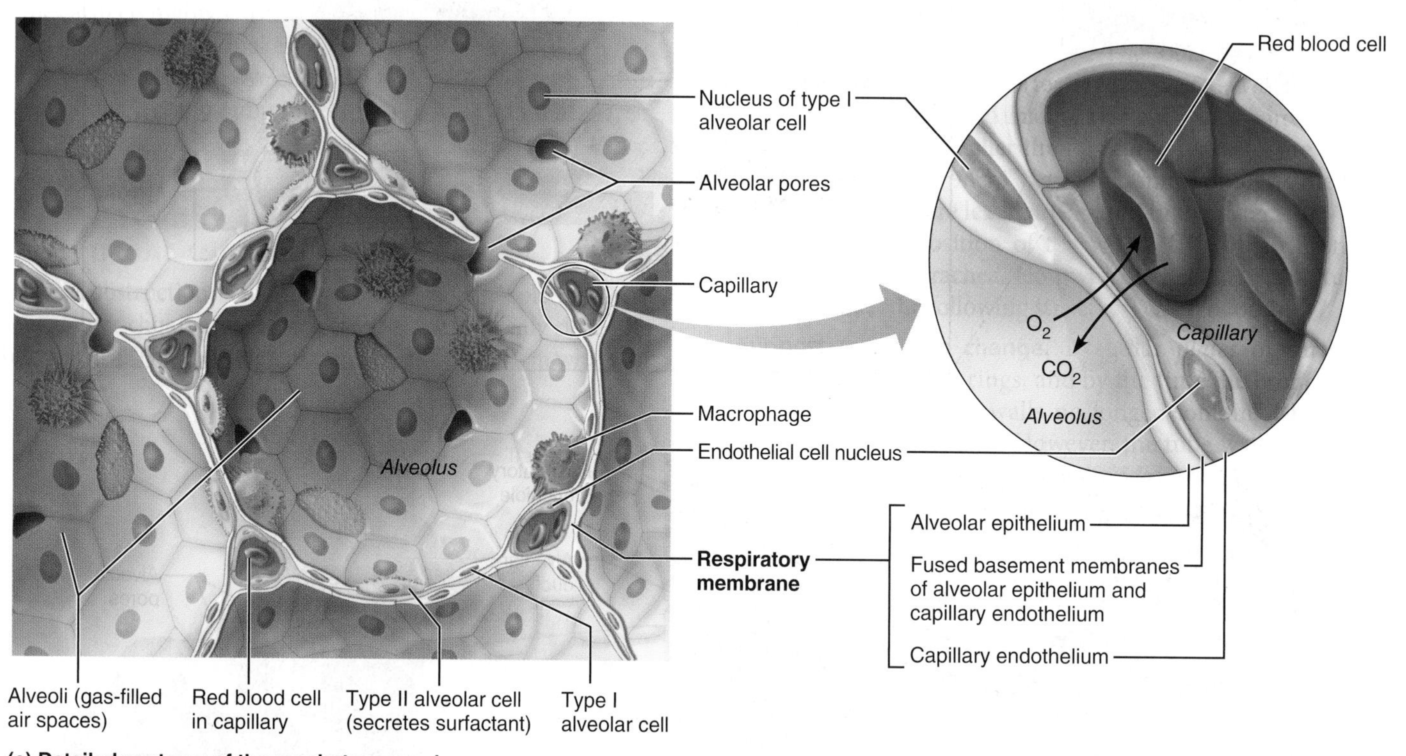

(c) Detailed anatomy of the respiratory membrane

Figure 21.10 Alveoli and the respiratory membrane. Elastic fibers and capillaries surround all alveoli, but for clarity they are shown only on some alveoli in (a). In (b), the tissue forming the alveoli has been removed. Only the capillary network remains.

the alveolus into the blood, and CO_2 leaves the blood to enter the gas-filled alveolus.

Scattered amid the squamous type I alveolar cells that form the major part of the alveolar walls are cuboidal type II alveolar cells (Figure 21.10c). **Type II alveolar cells** secrete a fluid containing a detergent-like substance called *surfactant* that coats the gas-exposed alveolar surfaces. (We describe surfactant's role in reducing the surface tension of the alveolar fluid later in this chapter.) Type II alveolar cells also secrete a number of antimicrobial proteins that are important elements of innate immunity.

The alveoli have three other significant features: (1) They are surrounded by fine elastic fibers of the same type that surround the entire bronchial tree. (2) Open **alveolar pores** connecting adjacent alveoli allow air pressure throughout the lung to be equalized and provide alternate air routes to any alveoli whose bronchi have collapsed due to disease. (3) Remarkably efficient **alveolar macrophages** crawl freely along the internal alveolar surfaces.

Although huge numbers of infectious microorganisms are continuously carried into the alveoli, alveolar surfaces are usually sterile. Because the alveoli are "dead ends," aged and dead macrophages must be prevented from accumulating in them. Most macrophages simply get swept up by the ciliary current of superior regions and carried to the pharynx. In this manner, we clear and swallow over 2 million alveolar macrophages per hour!

☑ Check Your Understanding

3. Which structure seals the larynx when we swallow?
4. Which structural features of the trachea allow it to expand and contract, yet keep it from collapsing?
5. What features of the alveoli and their respiratory membranes suit them to their function of exchanging gases by diffusion?
6. A 3-year-old boy is brought to the emergency department after aspirating (inhaling) a peanut. Bronchoscopy confirms the suspicion that the peanut is lodged in a bronchus and then it is successfully extracted. Which main bronchus was the peanut most likely to be in? Why?

For answers, see Answers Appendix.

21.3 Each multilobed lung occupies its own pleural cavity

→ Learning Objective

☐ **Describe the gross structure of the lungs and pleurae.**

The paired **lungs** occupy all of the thoracic cavity except the mediastinum, which houses the heart, great blood vessels, bronchi, esophagus, and other organs (**Figure 21.11**).

Gross Anatomy of the Lungs

Each cone-shaped lung is surrounded by pleurae and connected to the mediastinum by vascular and bronchial attachments, collectively called the lung **root**. The anterior, lateral, and posterior lung surfaces lie in close contact with the ribs and form the continuously curving **costal surface**. Just deep to the clavicle is the **apex**, the narrow superior tip of the lung. The concave, inferior surface that rests on the diaphragm is the **base**.

On the mediastinal surface of each lung is an indentation, the **hilum**, through which pulmonary and systemic blood vessels, bronchi, lymphatic vessels, and nerves enter and leave the lungs. Each main bronchus plunges into the hilum on its own side and begins to branch almost immediately. All conducting and respiratory passageways distal to the main bronchi are found in the lungs.

The two lungs differ slightly in shape and size because the apex of the heart is slightly to the left of the median plane. The left lung is smaller than the right, and the **cardiac notch**—a concavity in its medial aspect—is molded to and accommodates the heart (Figure 21.11a). The left lung is subdivided into superior and inferior **lobes** by the *oblique fissure*, whereas the right lung is partitioned into superior, middle, and inferior lobes by the *oblique* and *horizontal fissures*.

Each lobe contains a number of pyramid-shaped **bronchopulmonary segments** separated from one another by connective tissue septa. The right lung has 10 bronchopulmonary segments, but the left lung is more variable and consists of 8 to 10 segments (**Figure 21.12**). Each segment is served by its own artery and vein and receives air from an individual segmental (tertiary) bronchus.

The bronchopulmonary segments are clinically important because pulmonary disease is often confined to one or a few segments. Their connective tissue partitions allow diseased segments to be surgically removed without damaging neighboring segments or impairing their blood supply.

The smallest subdivisions of the lung visible with the naked eye are the **lobules**, which appear at the lung surface as hexagons ranging from the size of a pencil eraser to the size of a penny (Figure 21.11b). A large bronchiole and its branches serve each lobule. In most city dwellers and in smokers, the connective tissue that separates the individual lobules is blackened with carbon.

As we mentioned earlier, the lungs consist largely of air spaces. The balance of lung tissue, or its **stroma** ("mattress" or "bed"), is mostly elastic connective tissue. As a result, the lungs are soft, spongy, elastic organs that together weigh just over 1 kg (2.2 lb). The elasticity of healthy lungs reduces the work of breathing, as we will describe shortly.

Blood Supply and Innervation of the Lungs

The lungs are perfused by two circulations, the pulmonary and the bronchial, which differ in size, origin, and function.

Pulmonary Circulation of the Lungs

Systemic venous blood that is to be oxygenated in the lungs is delivered by the **pulmonary arteries**, which lie anterior to the main bronchi (Figure 21.11c). In the lungs, the pulmonary arteries branch profusely along with the bronchi and finally feed into the **pulmonary capillary networks** surrounding the alveoli (see Figure 21.10a).

The **pulmonary veins** convey the freshly oxygenated blood from the respiratory zone of the lungs to the heart. Their tributaries course back to the hilum both with the corresponding bronchi and in the connective tissue septa separating the bronchopulmonary segments.

The pulmonary circuit is a low-pressure, high-volume circulation. Because *all* of the body's blood passes through the lungs about once each minute, the lung capillary endothelium is an ideal location for enzymes that act on materials in the blood. Examples include *angiotensin converting enzyme*, which activates an important blood pressure hormone, and enzymes that inactivate certain prostaglandins.

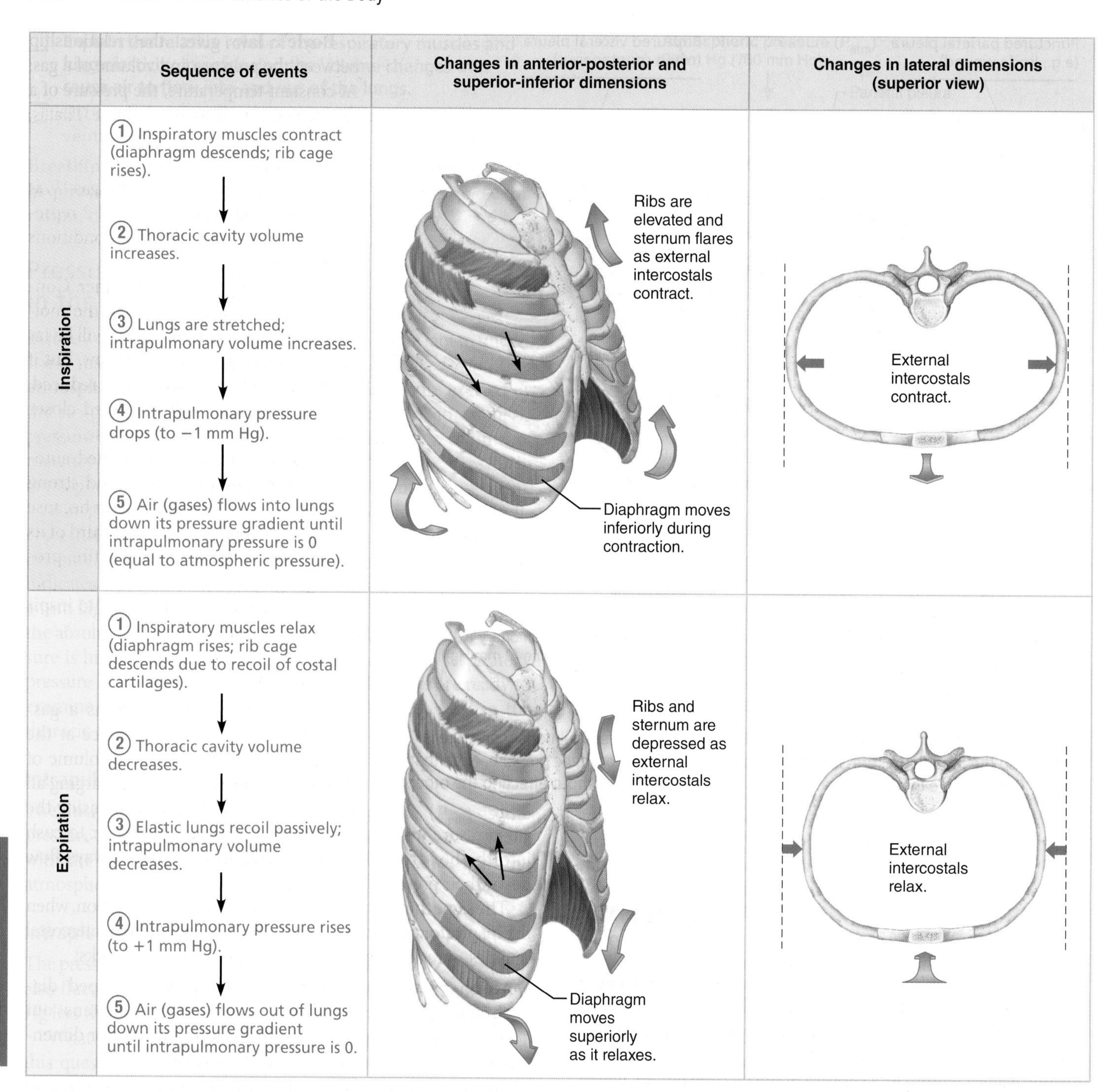

Figure 21.15 Changes in thoracic volume and sequence of events during inspiration and expiration. The sequence of events in the left column includes volume changes during inspiration (top) and expiration (bottom). The lateral views in the middle column show changes in the superior-inferior dimension (as the diaphragm alternately contracts and relaxes, see black arrows) and in the anterior-posterior dimension (as the external intercostal muscles alternately contract and relax). The superior views of transverse thoracic sections in the right column show lateral dimension changes resulting from alternate contraction and relaxation of the external intercostal muscles.

increase thoracic volume by almost 500 ml—the usual volume of air that enters the lungs during a normal quiet inspiration. Of the two types of inspiratory muscles, the diaphragm is far more important in producing these volume changes that lead to normal quiet inspiration.

As the thoracic dimensions increase during inspiration, the lungs are stretched and the intrapulmonary volume increases. As a result, P_{pul} drops about 1 mm Hg relative to P_{atm}. Anytime the intrapulmonary pressure is less than the atmospheric pressure ($P_{pul} < P_{atm}$), air rushes into the lungs along the pressure gradient.

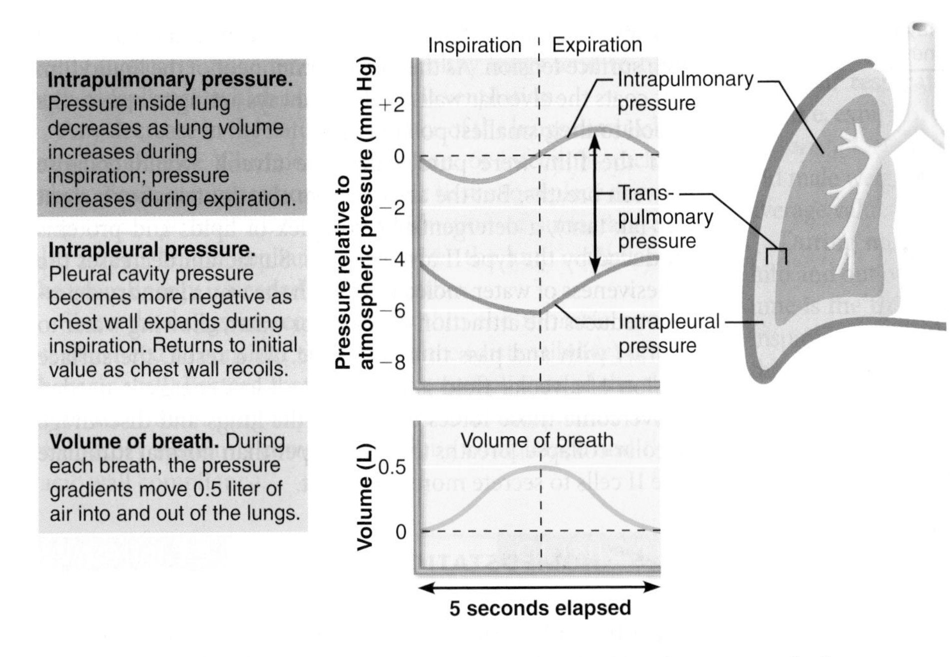

Figure 21.16 Changes in intrapulmonary and intrapleural pressures during inspiration and expiration. Notice that normal atmospheric pressure (760 mm Hg) is given a value of 0 on the scale.

Inspiration ends when $P_{pul} = P_{atm}$. During the same period, P_{ip} declines to about −6 mm Hg relative to P_{atm} (**Figure 21.16**).

During the *deep* or *forced inspirations* that occur during vigorous exercise and in some chronic obstructive pulmonary diseases, accessory muscles further increase thoracic volume. Several muscles, including the scalenes and sternocleidomastoid muscles of the neck and the pectoralis minor of the chest, raise the ribs even more than during quiet inspiration. Additionally, the back extends as the erector spinae muscles straighten the thoracic curvature.

Expiration

In healthy individuals, quiet expiration is a passive process that depends more on lung elasticity than on muscle contraction. As the inspiratory muscles relax and resume their resting length, the rib cage descends and the lungs recoil (Figure 21.15, bottom). As a result, both the thoracic and intrapulmonary volumes decrease. This volume decrease compresses the alveoli, and P_{pul} rises to about 1 mm Hg above atmospheric pressure (Figure 21.16). When $P_{pul} > P_{atm}$, the pressure gradient forces gases to flow out of the lungs.

Forced expiration is an active process produced by contracting abdominal wall muscles, primarily the oblique and transversus muscles. These contractions (1) increase the intra-abdominal pressure, which forces the abdominal organs superiorly against the diaphragm, and (2) depress the rib cage. The internal intercostal muscles also help depress the rib cage and decrease thoracic volume.

To precisely regulate air flow from the lungs, it is necessary to control the accessory muscles of expiration. For instance, the ability of a trained vocalist to hold a musical note depends on the coordinated activity of several muscles normally used in forced expiration.

Nonrespiratory Air Movements

Many processes other than breathing move air into or out of the lungs, altering the normal respiratory rhythm. These **nonrespiratory air movements** occur whenever you cough, sneeze, cry, laugh, hiccup, or yawn. Some can be produced voluntarily, but some (such as sneezing and hiccups) are reflexive.

Physical Factors Influencing Pulmonary Ventilation

As we have seen, the lungs are stretched during inspiration and recoil passively during expiration. The inspiratory muscles consume energy to enlarge the thorax. Energy is also used to overcome various factors that hinder air passage and pulmonary ventilation. We examine these factors next.

Airway Resistance

The major *nonelastic* source of resistance to gas flow is friction, or drag, encountered in the respiratory passageways. The following equation gives the relationship between gas flow (*F*), pressure (*P*), and resistance (*R*):

$$F = \frac{\Delta P}{R}$$

Notice that the factors determining gas flow in the respiratory passages and blood flow in the cardiovascular system are equivalent. The amount of gas flowing into and out of the alveoli is directly proportional to ΔP, the *difference* in pressure, or pressure gradient, between the external atmosphere and the alveoli.

Normally, very small differences in pressure produce large changes in gas flow. The average pressure gradient during normal quiet breathing is 2 mm Hg or less, and yet it is sufficient to move 500 ml of air in and out of the lungs with each breath.

But, as the equation also indicates, gas flow changes *inversely* with resistance. In other words, gas flow decreases as resistance increases. As in the cardiovascular system, resistance in the respiratory tree is determined mostly by the diameters of the conducting tubes. However, as a rule, airway resistance is insignificant for two reasons:

- Airway diameters in the first part of the conducting zone are huge, relative to the low viscosity of air.
- As the airways get progressively smaller, there are progressively more branches. As a result, although individual bronchioles are tiny, there are an enormous number of them in parallel, so the total cross-sectional area is huge.

Consequently, the greatest resistance to gas flow occurs in the medium-sized bronchi (**Figure 21.17**). At the terminal bronchioles, gas flow stops and diffusion takes over as the main force driving gas movement, so resistance is no longer an issue.

Table 21.4 Comparison of Gas Partial Pressures and Approximate Percentages in the Atmosphere and in the Alveoli

	ATMOSPHERE (SEA LEVEL)		ALVEOLI	
GAS	APPROXIMATE PERCENTAGE	PARTIAL PRESSURE (mm Hg)	APPROXIMATE PERCENTAGE	PARTIAL PRESSURE (mm Hg)
N_2	78.6	597	74.9	569
O_2	20.9	159	13.7	104
CO_2	0.04	0.3	5.2	40
H_2O	0.46	3.7	6.2	47
	100.0%	760	100.0%	760

phase, some of the dissolved gas molecules will reenter the gaseous phase. So the direction and amount of movement of a gas are determined by its partial pressure in the two phases. This flexible situation is exactly what occurs when gases are exchanged in the lungs and tissues. For example, when P_{CO_2} in the pulmonary capillaries is higher than in the lungs, CO_2 diffuses out of the blood and enters the air in the alveoli.

How much of a gas will dissolve in a liquid at any given partial pressure also depends on the *solubility* of the gas in the liquid and the *temperature* of the liquid. The gases in air have very different solubilities in water (and in blood plasma). Carbon dioxide is most soluble. Oxygen is only 1/20 as soluble as CO_2, and N_2 is only half as soluble as O_2. For this reason, at a given partial pressure, much more CO_2 than O_2 dissolves in water, and practically no N_2 goes into solution.

When a liquid's temperature rises, gas solubility decreases. Think of club soda, which is produced by forcing CO_2 gas to dissolve in water under high pressure. If you take the cap off a bottle of club soda and leave it in the fridge, it will slowly go flat. But if you leave it at room temperature, it will very quickly go flat. In both cases, you end up with plain water—all the CO_2 gas has escaped from solution.

Hyperbaric oxygen chambers provide clinical applications of Henry's law. These chambers contain O_2 gas at pressures higher than 1 atm and are used to force greater-than-normal amounts of O_2 into the blood of patients suffering from carbon monoxide poisoning (see p. 734) or tissue damage following radiation therapy. Hyperbaric therapy is also used to treat individuals with gas gangrene, because the anaerobic bacteria causing this infection cannot live in the presence of high O_2 levels.

Scuba diving provides another illustration of Henry's law. If divers rise rapidly from the depths, dissolved nitrogen forms bubbles in their blood, causing "the bends."

HOMEOSTATIC IMBALANCE 21.11 CLINICAL

Although breathing O_2 gas at 2 atm is not a problem for short periods, **oxygen toxicity** develops rapidly when P_{O_2} is greater than 2.5–3 atm. Excessively high O_2 concentrations generate huge amounts of harmful free radicals, resulting in profound CNS disturbances, coma, and death. +

Composition of Alveolar Gas

As shown in Table 21.4, the gaseous makeup of the atmosphere is quite different from that in the alveoli. The atmosphere is almost entirely O_2 and N_2; the alveoli contain more CO_2 and water vapor and much less O_2. These differences reflect the effects of:

- Gas exchanges occurring in the lungs (O_2 diffuses from the alveoli into the pulmonary blood and CO_2 diffuses in the opposite direction).
- Humidification of air by conducting passages.
- The mixing of alveolar gas that occurs with each breath. Because only 500 ml of air enter with each tidal inspiration, gas in the alveoli is actually a mixture of newly inspired gases and gases remaining in the respiratory passageways between breaths.

The alveolar partial pressures of O_2 and CO_2 are easily changed by increasing breathing depth and rate. A high AVR brings more O_2 into the alveoli, increasing alveolar P_{O_2} and rapidly eliminating CO_2 from the lungs.

External Respiration

During external respiration (pulmonary gas exchange), dark red blood flowing through the pulmonary circuit is transformed into the scarlet river that is returned to the heart for distribution by systemic arteries to all body tissues. This color change is due to O_2 uptake and binding to hemoglobin in red blood cells (RBCs), but CO_2 exchange (unloading) is occurring equally fast.

The following three factors influence external respiration:

- Partial pressure gradients and gas solubilities
- Thickness and surface area of the respiratory membrane
- Ventilation-perfusion coupling (matching alveolar ventilation with pulmonary blood perfusion)

Let's look at these factors one by one.

Partial Pressure Gradients and Gas Solubilities

Partial pressure gradients of O_2 and CO_2 drive the diffusion of these gases across the respiratory membrane. A steep oxygen partial pressure gradient exists across the respiratory membrane because the P_{O_2} of deoxygenated blood in the pulmonary arteries is only 40 mm Hg, as opposed to a P_{O_2} of approximately 104 mm Hg in the alveoli. As a result, O_2 diffuses rapidly from the alveoli into the pulmonary capillary blood (**Figure 21.19**).

Equilibrium—that is, a P_{O_2} of 104 mm Hg on both sides of the respiratory membrane—usually occurs in 0.25 second, which is about one-third of the time a red blood cell spends in a pulmonary capillary (**Figure 21.20**). The lesson here is that blood can flow through the pulmonary capillaries three times as quickly and still be adequately oxygenated.

Carbon dioxide diffuses in the opposite direction along a much gentler partial pressure gradient of about 5 mm Hg (45 mm Hg to 40 mm Hg) until equilibrium occurs at 40 mm Hg. Expiration then gradually expels carbon dioxide from the alveoli.

Even though the O_2 pressure gradient for oxygen diffusion is much steeper than the CO_2 gradient, equal amounts of these gases are exchanged. Why? The reason is that CO_2 is 20 times more soluble in plasma and alveolar fluid than O_2.

Thickness and Surface Area of the Respiratory Membrane

In healthy lungs, the respiratory membrane is only 0.5 to 1 μm thick, and gas exchange is usually very efficient.

The effective thickness of the respiratory membrane increases dramatically if the lungs become waterlogged and edematous, as in pneumonia or left heart failure (see p. 606). Under such conditions, even the 0.75 s that red blood cells spend in transit through the pulmonary capillaries may not be enough for adequate gas exchange, and body tissues suffer from oxygen deprivation. +

The greater the surface area of the respiratory membrane, the more gas can diffuse across it in a given time period. In healthy lungs, the alveolar surface area is enormous. Spread flat, the total gas exchange surface of these tiny sacs in an adult male's lungs is about 90 m^2—approximately 40 times greater than the surface area of his skin!

Certain pulmonary diseases drastically reduce the alveolar surface area. For instance, in emphysema the walls of adjacent alveoli break down and the alveolar chambers enlarge. Tumors, mucus, or inflammatory material also reduce surface area by blocking gas flow into the alveoli. +

Ventilation-Perfusion Coupling

For optimal gas exchange, there must be a close match, or coupling, between *ventilation* (the amount of gas reaching the alveoli) and *perfusion* (the blood flow in pulmonary capillaries). Both are controlled by local autoregulatory mechanisms that continuously respond to local conditions. For the most part:

- P_{O_2} controls perfusion by changing *arteriolar* diameter.
- P_{CO_2} controls ventilation by changing *bronchiolar* diameter.

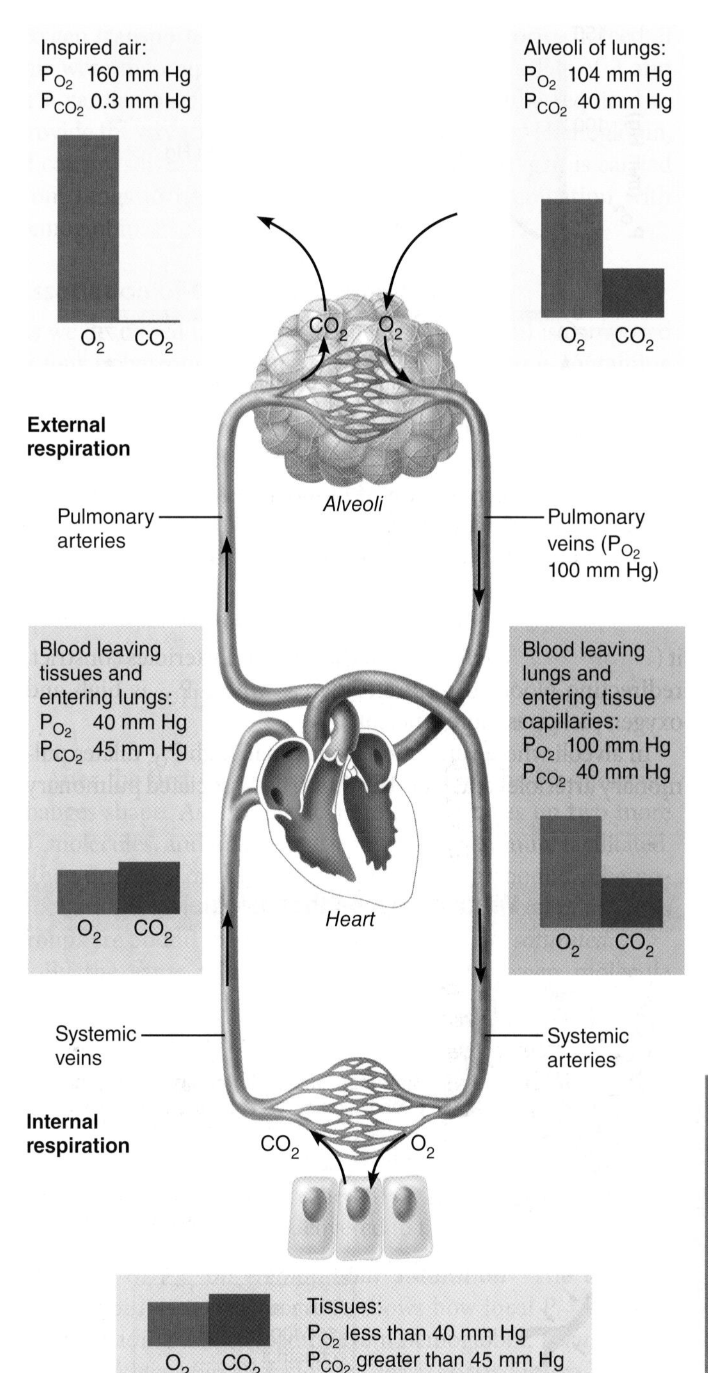

Figure 21.19 Partial pressure gradients promoting gas movements in the body. Gradients promoting O_2 and CO_2 exchange across the respiratory membrane in the lungs (top) and across systemic capillary membranes in body tissues (bottom). (The small decrease in P_{O_2} in blood leaving lungs is due to partial dilution of pulmonary capillary blood with less oxygenated blood.)

Influence of Local P_{O_2} on Perfusion We begin with perfusion because we introduced its autoregulatory control in Chapter 18. If alveolar ventilation is inadequate, local P_{O_2} is low because blood takes O_2 away more quickly than ventilation can replenish

of medullary neurons. For example, pontine centers appear to smooth out the transitions from inspiration to expiration, and vice versa. When lesions are made in its superior region, inspirations become very prolonged, a phenomenon called *apneustic breathing*.

The **pontine respiratory group** and other pontine centers transmit impulses to the VRG of the medulla (Figure 21.24). This input modifies and fine-tunes the breathing rhythms generated by the VRG during certain activities such as vocalization, sleep, and exercise. As you would expect from these functions, the pontine respiratory centers, like the DRG, receive input from higher brain centers and from various sensory receptors in the periphery.

Generation of the Respiratory Rhythm

There is little question that breathing is rhythmic, but we still cannot fully explain the origin of its rhythm. One hypothesis is that there are *pacemaker neurons,* which have intrinsic (automatic) rhythmicity like the pacemaker cells found in the heart. Pacemaker-like activity has been demonstrated in certain VRG neurons, but suppressing their activity does not abolish breathing.

This leads us to the second (and more widely accepted) hypothesis: Normal respiratory rhythm results from reciprocal inhibition of interconnected neuronal networks in the medulla. Rather than a single set of pacemaker neurons, there are two sets that inhibit each other and cycle their activity to generate the rhythm.

Factors Influencing Breathing Rate and Depth

Inspiratory depth is determined by how actively the respiratory centers stimulate the motor neurons serving the respiratory muscles. The greater the stimulation, the greater the number of motor units excited and the greater the force of respiratory muscle contractions. Respiratory rate is determined by how long the inspiratory center is active or how quickly it is switched off.

Changing body demands can modify depth and rate of breathing. The respiratory centers in the medulla and pons are sensitive to both excitatory and inhibitory stimuli, as summarized in **Figure 21.25**.

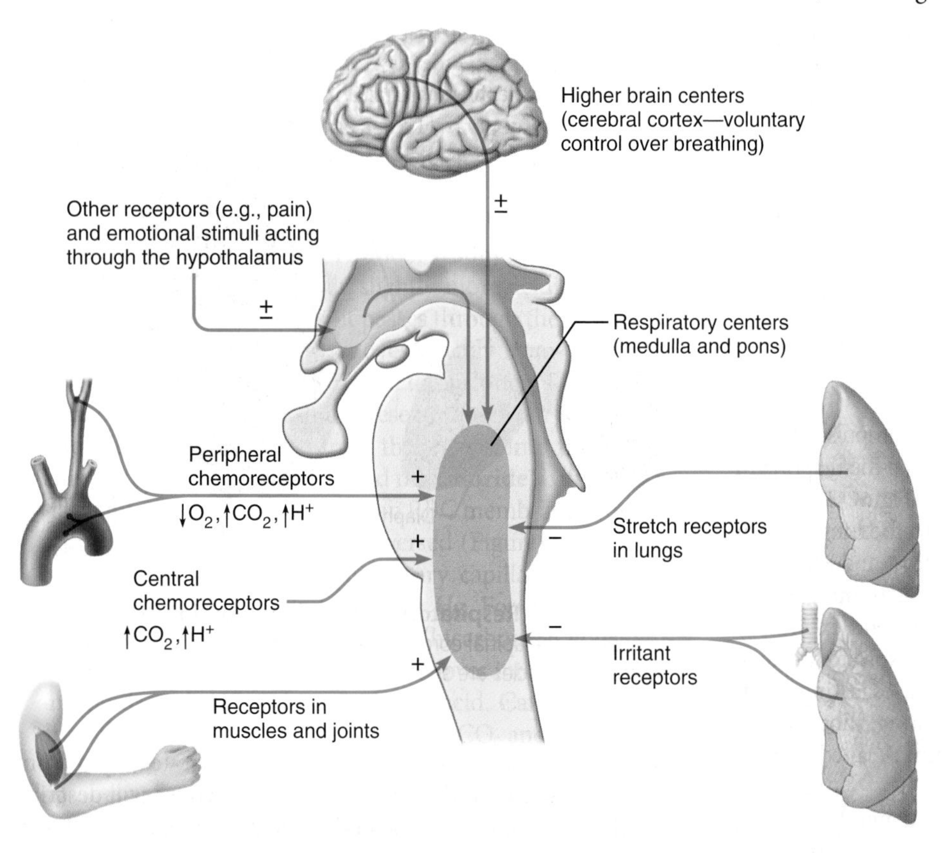

Figure 21.25 Neural and chemical influences on brain stem respiratory centers. Excitatory influences (+) increase the frequency of impulses sent to the muscles of respiration and recruit additional motor units, resulting in deeper, faster breathing. Inhibitory influences (−) have the reverse effect. In some cases, the influences may be excitatory or inhibitory (±), depending on which receptors or brain regions are activated. The cerebral cortex also directly innervates respiratory muscle motor neurons (not shown).

Chemical Factors

Among the factors that influence breathing rate and depth, the most important are changing levels of CO_2, O_2, and H^+ in arterial blood. Sensors responding to such chemical fluctuations, called **chemoreceptors**, are found in two major body locations:

- **Central chemoreceptors** are located throughout the brain stem, including the ventrolateral medulla.
- **Peripheral chemoreceptors** are found in the aortic arch and carotid arteries.

Influence of P_{CO_2} Of all the chemicals influencing respiration, CO_2 is the most potent and the most closely controlled. Normally, arterial P_{CO_2} is 40 mm Hg and is maintained within ±3 mm Hg of this level by an exquisitely sensitive homeostatic mechanism that is mediated mainly by the effect of rising CO_2 levels on the central chemoreceptors of the brain stem (**Figure 21.26**).

As P_{CO_2} levels rise in the blood, a condition referred to as **hypercapnia** (hi″per-kap′ne-ah), CO_2 accumulates in the brain. As CO_2 accumulates, it is hydrated to form carbonic acid. The acid dissociates, H^+ is liberated, and the pH drops. This is the same reaction that occurs when CO_2 enters RBCs (see pp. 735–736).

The increase in H^+ excites the central chemoreceptors, which make abundant synapses with the respiratory regulatory centers. As a result, the depth and rate of breathing increase. This enhanced alveolar ventilation quickly flushes CO_2 out of the blood, raising blood pH.

An elevation of only 5 mm Hg in arterial P_{CO_2} doubles alveolar ventilation, even when arterial O_2 levels and pH haven't changed. When P_{O_2} and pH are below normal, the response to elevated P_{CO_2}

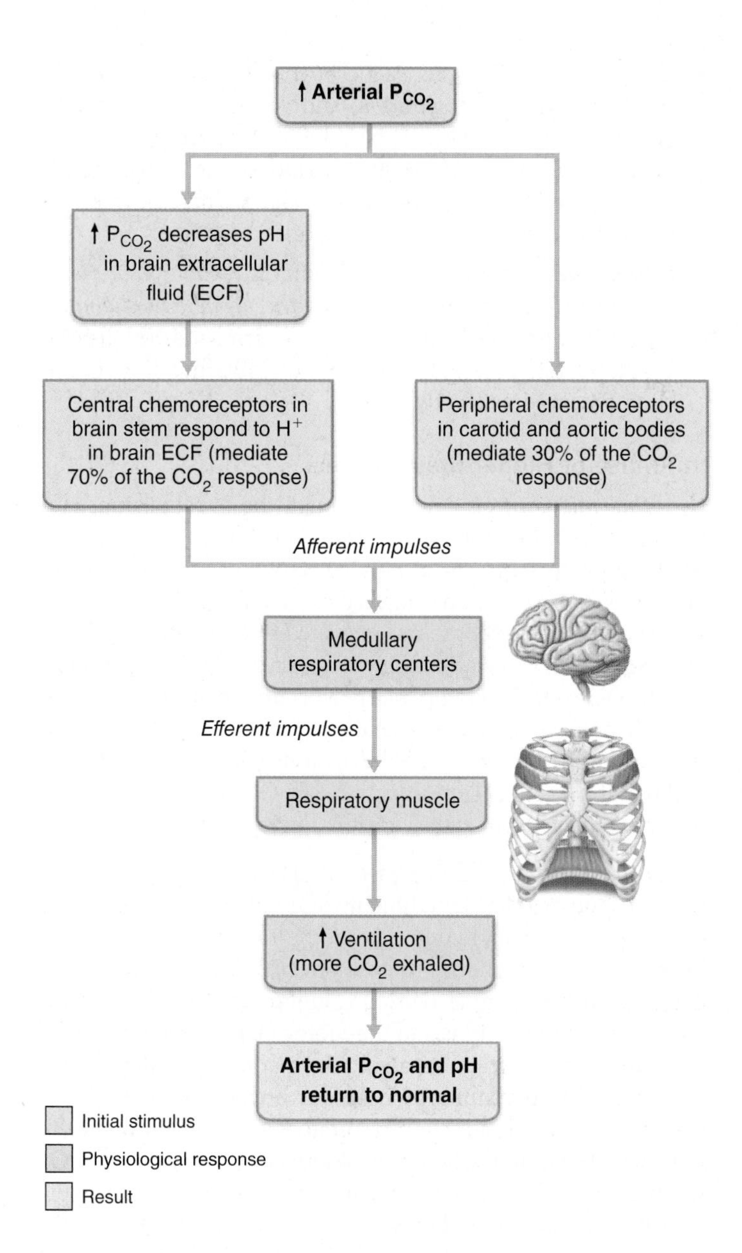

Figure 21.26 Changes in P_{CO_2} regulate ventilation by a negative feedback mechanism.

is even greater. Increased ventilation is normally self-limiting, ending when homeostatic blood P_{CO_2} levels are restored.

Notice that while rising blood CO_2 levels act as the initial stimulus, it is rising levels of H^+ generated within the brain that prod the central chemoreceptors into increased activity. (CO_2 readily diffuses across the blood brain barrier between the brain and the blood, but H^+ does not.) In the final analysis, control of breathing during rest is aimed primarily at *regulating the H^+ concentration in the brain.*

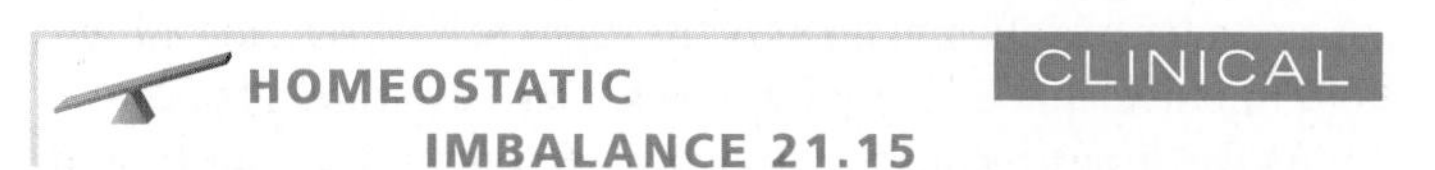

Hyperventilation is an increase in the rate and depth of breathing that exceeds the body's need to remove CO_2. A person experiencing an anxiety attack may hyperventilate involuntarily. As they blow off CO_2, the low CO_2 levels in the blood (**hypocapnia**) constrict cerebral blood vessels. This reduces brain perfusion, producing cerebral ischemia that causes dizziness or fainting. Earlier symptoms of hyperventilation are tingling and involuntary muscle spasms (tetany) in the hands and face caused by blood Ca^{2+} levels falling as pH rises.

The symptoms of hyperventilation may be averted by breathing into a paper bag. The air being inspired from the bag is expired air, rich in carbon dioxide, which causes carbon dioxide to be retained in the blood. ✚

When P_{CO_2} is abnormally low, respiration is inhibited and becomes slow and shallow. In fact, periods of **apnea** (breathing cessation) may occur until arterial P_{CO_2} rises and again stimulates respiration.

Sometimes swimmers voluntarily hyperventilate so they can hold their breath longer during swim meets. This is dangerous. Blood O_2 content rarely drops much below 60% of normal during regular breath-holding, because as P_{O_2} drops, P_{CO_2} rises enough to make breathing unavoidable. However, strenuous hyperventilation can lower P_{CO_2} so much that a lag period occurs before P_{CO_2} rebounds enough to stimulate respiration again. This lag may allow oxygen levels to fall well below 50 mm Hg, causing the swimmer to black out (and perhaps drown) before he or she has the urge to breathe.

Influence of P_{O_2} The peripheral chemoreceptors—found in the **aortic bodies** of the aortic arch and in the **carotid bodies** at the bifurcation of the common carotid arteries—contain cells sensitive to arterial O_2 levels (**Figure 21.27**). The main oxygen sensors are in the carotid bodies.

Under normal conditions, declining P_{O_2} has only a slight effect on ventilation, mostly limited to enhancing the sensitivity of peripheral receptors to increased P_{CO_2}. Arterial P_{O_2} must drop *substantially*, to at least 60 mm Hg, before O_2 levels become a major stimulus for increased ventilation.

This is not as strange as it may appear. Remember, there is a huge reservoir of O_2 bound to Hb, and Hb remains almost entirely saturated unless or until the P_{O_2} of alveolar gas and arterial blood falls below 60 mm Hg. The brain stem centers then begin to suffer from O_2 starvation, and their activity is depressed. At the same time, the peripheral chemoreceptors become excited and stimulate the respiratory centers to increase ventilation, even if P_{CO_2} is normal. In this way, the peripheral chemoreceptor system can maintain ventilation even though the brain stem centers are depressed by hypoxia.

Influence of Arterial pH Changes in arterial pH can modify respiratory rate and rhythm even when CO_2 and O_2 levels are normal. Because H^+ does not cross the blood brain barrier, the increased ventilation that occurs in response to falling arterial pH is mediated through the peripheral chemoreceptors.

Although changes in P_{CO_2} and H^+ concentration are interrelated, they are distinct stimuli. A drop in blood pH may reflect CO_2 retention, but it may also result from metabolic causes, such as accumulation of lactic acid during exercise or of fatty acid metabolites (ketone bodies) in patients with poorly controlled diabetes mellitus. Regardless of cause, as arterial pH declines, respiratory system controls attempt to compensate

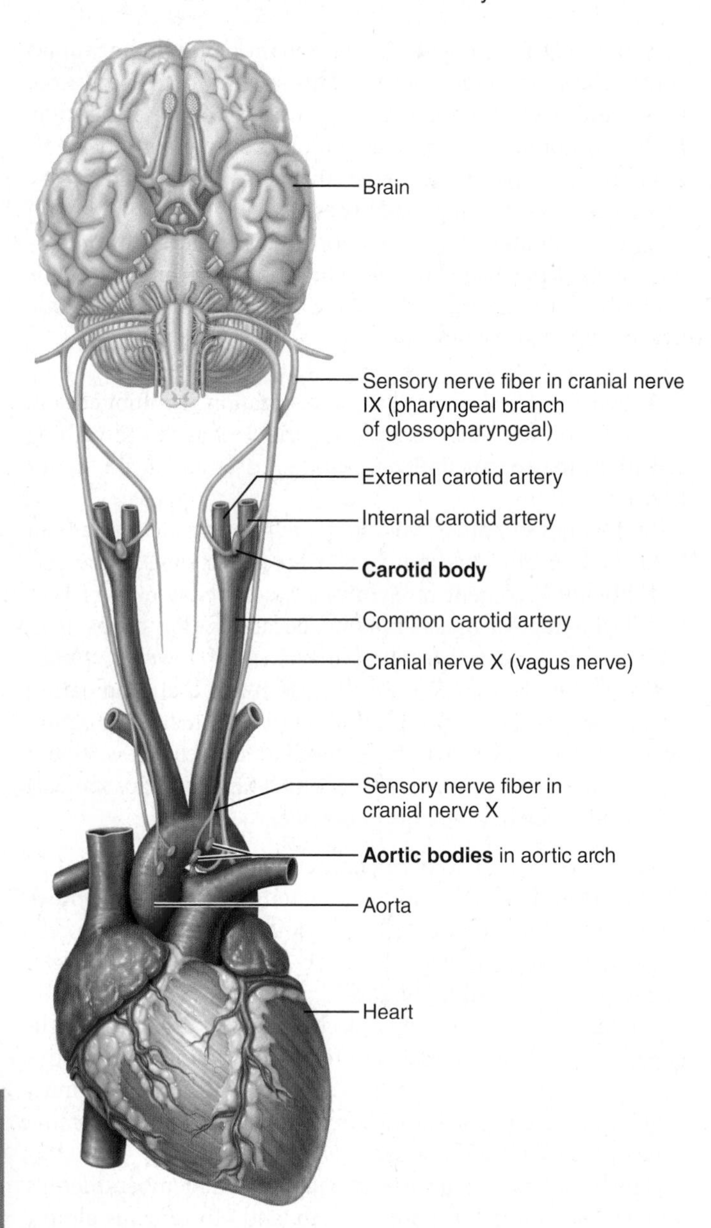

Figure 21.27 Location and innervation of the peripheral chemoreceptors in the carotid and aortic bodies.

21

and raise the pH. They do this by increasing respiratory rate and depth to eliminate CO_2 (and carbonic acid) from the blood.

Summary of Interactions of P_{CO_2}, P_{O_2}, and Arterial pH The body's need to rid itself of CO_2 is the most important stimulus for breathing in a healthy person. However, CO_2 does not act in isolation, and various chemical factors enforce or inhibit one another's effects. These interactions are summarized here:

- *Rising CO_2 levels are the most powerful respiratory stimulant.* As CO_2 is hydrated in brain tissue, liberated H^+ acts directly on the central chemoreceptors, causing a reflexive increase in breathing rate and depth. Low P_{CO_2} levels depress respiration.
- *Under normal conditions, blood P_{O_2} affects breathing only indirectly* by influencing peripheral chemoreceptor sensitivity to changes in P_{CO_2}. Low P_{O_2} augments P_{CO_2} effects, and high P_{O_2} levels diminish the effectiveness of CO_2 stimulation.
- *When arterial P_{O_2} falls below 60 mm Hg, it becomes the major stimulus for respiration*, and ventilation is increased via reflexes initiated by the peripheral chemoreceptors. This may increase O_2 loading into the blood, but it also causes hypocapnia (low P_{CO_2} blood levels) and an increase in blood pH, both of which inhibit respiration.
- *Changes in arterial pH resulting from CO_2 retention or metabolic factors act indirectly through the peripheral chemoreceptors to alter ventilation*, which in turn modifies arterial P_{CO_2} and pH. Arterial pH does not influence the central chemoreceptors directly.

Influence of Higher Brain Centers

Hypothalamic Controls Acting through the hypothalamus and the rest of the limbic system, strong emotions and pain send signals to the respiratory centers, modifying respiratory rate and depth. For example, have you ever touched something cold and clammy and gasped? That response was mediated through the hypothalamus. So too is the breath-holding that occurs when we are angry and the increased respiratory rate that occurs when we are excited. A rise in body temperature raises the respiratory rate, while a drop in body temperature produces the opposite effect. Sudden chilling (a dip in the North Atlantic Ocean in late October) can stop your breathing (apnea)—or at the very least, leave you gasping.

Cortical Controls Although the brain stem respiratory centers normally regulate breathing involuntarily, we can also exert conscious (voluntary) control over the rate and depth of our breathing. We can choose to hold our breath or take an extra-deep breath, for example. During voluntary control, the cerebral motor cortex sends signals to the motor neurons that stimulate the respiratory muscles, bypassing the medullary centers.

Our ability to voluntarily hold our breath is limited, however, because the brain stem respiratory centers automatically reinitiate breathing when the blood concentration of CO_2 reaches critical levels. That explains why drowning victims typically have water in their lungs.

Pulmonary Irritant Reflexes

The lungs contain receptors that respond to an enormous variety of irritants. When activated, these receptors communicate with the respiratory centers via vagal nerve afferents. Accumulated mucus, inhaled debris such as dust, or noxious fumes stimulate receptors in the bronchioles that promote reflex constriction of those air passages. The same irritants stimulate a cough in the trachea or bronchi, and a sneeze in the nasal cavity.

The Inflation Reflex

The visceral pleurae and conducting passages in the lungs contain numerous stretch receptors that are vigorously stimulated when the lungs are inflated. These receptors signal the medullary respiratory centers via afferent fibers of the vagus nerves, sending inhibitory impulses that end inspiration and allow expiration to occur.

As the lungs recoil, the stretch receptors become quiet, and inspiration is initiated once again. This reflex, called the **inflation reflex**, or **Hering-Breuer reflex** (her′ing broy′er), is thought to be more a protective response (to prevent the lungs from being stretched excessively) than a normal regulatory mechanism.

Check Your Understanding

21. Which brain stem respiratory area is thought to generate the respiratory rhythm?

22. Which chemical factor in blood normally provides the most powerful stimulus to breathe? Which chemoreceptors are most important for this response?

For answers, see Answers Appendix.

21.9 Exercise and high altitude bring about respiratory adjustments

→ Learning Objectives

- ☐ **Compare and contrast the hyperpnea of exercise with hyperventilation.**
- ☐ **Describe the process and effects of acclimatization to high altitude.**

Exercise

Respiratory adjustments during exercise are geared to both the intensity and duration of the exercise. Working muscles consume tremendous amounts of O_2 and produce large amounts of CO_2, so ventilation can increase 10- to 20-fold during vigorous exercise. Increased ventilation in response to metabolic needs is called **hyperpnea** (hi″perp-ne′ah).

How does hyperpnea differ from hyperventilation? The respiratory changes in hyperpnea do not alter blood O_2 and CO_2 levels significantly. In contrast, hyperventilation is excessive ventilation, and is characterized by low P_{CO_2} and alkalosis.

Exercise-enhanced ventilation does *not* appear to be prompted by rising P_{CO_2} and declining P_{O_2} and pH in the blood for two reasons.

- Ventilation increases abruptly as exercise begins, followed by a gradual increase, and then reaches a steady state. When exercise stops, there is a small but abrupt decline in ventilation rate, followed by a gradual decrease to the pre-exercise value.
- Although venous levels change, arterial P_{CO_2} and P_{O_2} levels remain surprisingly constant during exercise. In fact, P_{CO_2} may fall below normal and P_{O_2} may rise slightly because the respiratory adjustments are so efficient.

The most widely accepted explanation for the abrupt increase in ventilation that occurs as exercise begins reflects interaction of three neural factors:

- Psychological stimuli (our conscious anticipation of exercise)
- Simultaneous cortical motor activation of skeletal muscles and respiratory centers
- Excitatory impulses reaching respiratory centers from proprioceptors in moving muscles, tendons, and joints

The subsequent gradual increase and then plateauing of respiration probably reflect the rate of CO_2 delivery to the lungs (the "CO_2 flow").

The rise in lactic acid levels during exercise results from anaerobic respiration. However, it is *not* a result of inadequate respiratory function, because alveolar ventilation and pulmonary perfusion are as well matched during exercise as during rest (hemoglobin remains fully saturated). Rather, it reflects cardiac output limitations or inability of the skeletal muscles to further increase their oxygen consumption.

In light of this fact, the practice of inhaling pure O_2 by mask, used by some football players to replenish their "oxygen-starved" bodies as quickly as possible, is useless. The panting athlete *does* need more oxygen, but inspiring extra oxygen will not help, because the shortage is in the muscles—not the lungs.

High Altitude

Most people live between sea level and an altitude of approximately 2400 m (8000 feet). In this range, differences in atmospheric pressure are not great enough to cause healthy people any problems when they spend brief periods in higher-altitude areas.

However, if you travel quickly from sea level to elevations above 8000 ft, where atmospheric pressure and P_{O_2} are lower, your body responds with symptoms of *acute mountain sickness (AMS)*—headaches, shortness of breath, nausea, and dizziness. AMS is sometimes seen in travelers to ski resorts such as Vail, Colorado (8120 ft). In severe cases of AMS, lethal pulmonary and cerebral edema may occur.

When you move on a *long-term* basis from sea level to the mountains, your body makes respiratory and hematopoietic adjustments via an adaptive response called **acclimatization**. As we have already explained, decreases in arterial P_{O_2} cause the peripheral chemoreceptors to become more responsive to increases in P_{CO_2}, and a substantial decline in P_{O_2} directly stimulates them. As a result, ventilation increases as the brain attempts to restore gas exchange. Increased ventilation also reduces arterial CO_2 levels, so the P_{CO_2} of individuals living at high altitudes is typically below 40 mm Hg (its value at sea level).

High-altitude conditions always result in lower-than-normal hemoglobin saturation levels because less O_2 is available to be loaded. For example, at about 19,000 ft above sea level, O_2 saturation of arterial blood is only 67% (compared to nearly 98% at sea level). But Hb unloads only 20–25% of its oxygen at sea level, which means that even at the reduced saturations at high altitudes, the O_2 needs of the tissues are still met under resting conditions.

Additionally, at high altitudes hemoglobin's affinity for O_2 is reduced because BPG concentrations increase. This releases more O_2 to the tissues during each circulatory round.

When blood O_2 levels decline, the kidneys produce more erythropoietin, which stimulates bone marrow production of RBCs (see Chapter 16, p. 559). This phase of acclimatization, which occurs slowly, provides long-term compensation for living at high altitudes.

Check Your Understanding

23. An injured soccer player arrives by ambulance in the emergency room. She is in obvious distress, breathing rapidly. Her blood P_{CO_2} is 26 mm Hg and pH is 7.5. Is she suffering from hyperventilation or hyperpnea? Explain.

24. What long-term adjustments does the body make when living at high altitude?

For answers, see Answers Appendix.

CLINICAL

21.10 Lung diseases are major causes of disability and death

→ Learning Objective

☐ **Compare the causes and consequences of chronic bronchitis, emphysema, asthma, tuberculosis, and lung cancer.**

The respiratory system is particularly vulnerable to infectious diseases because it is wide open to airborne pathogens. However, some of the most disabling respiratory disorders are *chronic obstructive pulmonary disease* (COPD), *asthma, tuberculosis,* and *lung cancer*. COPD and lung cancer are living proof of the devastating effects of tobacco smoke on the body. Long known to promote cardiovascular disease, smoking is perhaps even more effective at destroying the lungs.

Chronic Obstructive Pulmonary Disease (COPD)

The **chronic obstructive pulmonary diseases (COPD)**, exemplified best by emphysema and chronic bronchitis, are a major cause of disability and death in North America. The key physiological feature of these diseases is an irreversible decrease in the ability to force air out of the lungs. Other features they share in common (**Figure 21.28**):

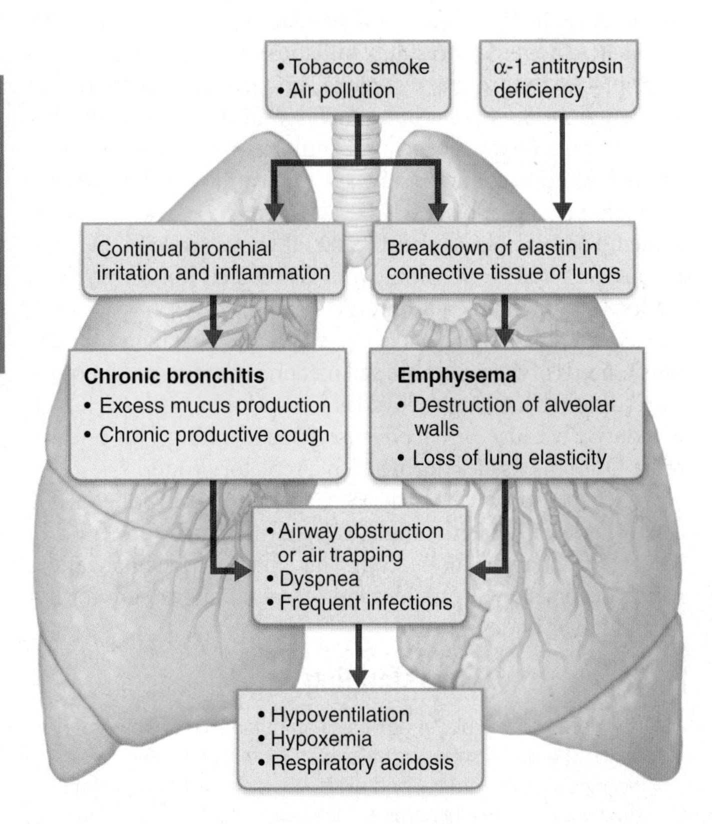

Figure 21.28 The pathogenesis of COPD.

- More than 80% of patients have a history of smoking.
- **Dyspnea** (disp-ne′ah), difficult or labored breathing often referred to as "air hunger," gets progressively worse.
- Coughing and frequent pulmonary infections are common.
- Most COPD victims develop respiratory failure manifested as **hypoventilation** (insufficient ventilation in relation to metabolic needs, causing them to retain CO_2), respiratory acidosis, and hypoxemia.

Emphysema

Emphysema is distinguished by permanent enlargement of the alveoli, accompanied by destruction of the alveolar walls. Invariably the lungs lose their elasticity. This has three important consequences:

- Accessory muscles must be enlisted to breathe, and victims are perpetually exhausted because breathing requires 15–20% of their total body energy supply (as opposed to 5% in healthy individuals).
- For complex reasons, the bronchioles open during inspiration but collapse during expiration, trapping huge volumes of air in the alveoli. This hyperinflation leads to development of a permanently expanded "barrel chest" and flattens the diaphragm, thus reducing ventilation efficiency.
- Damage to the pulmonary capillaries as the alveolar walls disintegrate increases resistance in the pulmonary circuit, forcing the right ventricle to overwork and consequently become enlarged.

Emphysema is usually caused by smoking, but hereditary factors (e.g., alpha-1 antitrypsin deficiency) cause emphysema in some patients.

Chronic Bronchitis

In **chronic bronchitis**, inhaled irritants lead to chronic production of excessive mucus. The mucosae of the lower respiratory passageways become inflamed and fibrosed. These responses obstruct the airways, severely impairing lung ventilation and gas exchange. Pulmonary infections are frequent because bacteria thrive in the stagnant pools of mucus. Smoking is a major risk factor. Environmental pollution also promotes chronic bronchitis.

COPD: Symptoms and Treatments

In the clinical setting you might see two very different patterns that represent the extremes of patients with COPD. One pattern has traditionally been called the "pink puffer": These patients work so hard to maintain adequate ventilation that they lose weight, becoming thin but still having nearly normal blood gases. In contrast, "blue bloaters," commonly of stocky build, become sufficiently hypoxic that they are obviously cyanotic. The hypoxia causes constriction of pulmonary blood vessels, leading to pulmonary hypertension and right-sided heart failure.

Traditionally, "pink puffers" were associated with emphysema while "blue bloaters" were associated with chronic bronchitis. As usual, things are not that clear-cut. It turns out that patients with the same underlying disease can display either of these clinical patterns, and this may depend on a third factor—the strength of their innate respiratory drive. Most COPD patients fall between these two clinical extremes.

COPD is routinely treated with inhaled bronchodilators and corticosteroids. Severe dyspnea and hypoxia mandate oxygen use. For a few patients, surgical treatment for COPD, called *lung volume reduction surgery*, may be beneficial. In this procedure, some lung tissue is removed, allowing the remaining lung tissue to expand. While this surgery does not prolong life, it can offer certain patients better quality of life.

COPD patients in acute respiratory distress are commonly given oxygen. Oxygen must be administered with care, however. In some of these patients, giving pure oxygen can increase the blood P_{CO_2} (and lower blood pH) to life-threatening levels. The solution is to use the minimum concentration of oxygen that relieves the patient's hypoxia.

Asthma

Asthma is characterized by episodes of coughing, dyspnea, wheezing, and chest tightness—alone or in combination. A sense of panic accompanies most acute attacks. Although sometimes classed with COPD because it is an obstructive disorder, asthma is marked by acute episodes followed by symptom-free periods—that is, the obstruction is *reversible*.

The cause of asthma has been hard to pin down. Initially it was viewed as a consequence of bronchospasms triggered by various factors such as cold air, exercise, or allergens. However, researchers have found that in allergic asthma (the most common kind), active inflammation of the airways comes first. The inflammation is an immune response controlled by a subset of T lymphocytes that stimulate the production of IgE and recruit inflammatory cells to the site.

Once someone has allergic asthma, the inflammation persists even during symptom-free periods and makes the airways hypersensitive. (The most common triggers are in the home—the allergens from dust mites, cockroaches, cats, dogs, and fungi.) Once the airway walls are thickened with inflammatory exudate, the effect of bronchospasm is vastly magnified and can dramatically reduce air flow.

About one in ten people in North America suffer from asthma—children more than adults. Over the past 20 years, the number of cases has risen dramatically, an increase which may now be plateauing.

While asthma remains a major health problem, better treatment options have reduced the number of asthma-related deaths. Instead of merely treating the symptoms with fast-acting bronchodilators, we now treat the underlying inflammation using inhaled corticosteroids. Newer approaches limit airway inflammation by using antileukotrienes and antibodies against the patient's own IgE class of antibodies.

Tuberculosis (TB)

Tuberculosis (TB), the infectious disease caused by the bacterium *Mycobacterium tuberculosis*, is spread by coughing and primarily enters the body in inhaled air. TB mostly affects the lungs but can spread through the lymphatics to other organs.

One-third of the world's population is infected, but most people never develop active TB because a massive inflammatory and immune response usually contains the primary infection in fibrous, or calcified, nodules (tubercles) in the lungs. However, the bacteria survive in the nodules and when the person's immunity is weakened, they may break out and cause symptomatic TB. Symptoms include fever, night sweats, weight loss, racking cough, and coughing up blood.

Deadly strains of drug-resistant (even multidrug-resistant) TB can develop when treatment is incomplete or inadequate. Resistant strains are found elsewhere in the world and have appeared in North America.

Homeless shelters, with their densely packed populations, are ideal breeding grounds for drug-resistant strains. The TB bacterium grows slowly and drug therapy entails a 12-month course of antibiotics. The transient nature of shelter populations makes it difficult to track TB patients and ensure they take their medications for the full 12 months. The threat of TB epidemics is so real that health centers in some cities are detaining such patients against their will for as long as it takes to complete a cure.

Lung Cancer

Lung cancer is the leading cause of cancer death for both men and women in North America, killing more people every year than breast, prostate, and colorectal cancer combined. This is tragic, because lung cancer is largely preventable—nearly 90% of cases result from smoking.

The cure rate for lung cancer is notoriously low, with most victims dying within one year of diagnosis. The five-year survival rate is about 17%. Because lung cancer is aggressive and metastasizes rapidly and widely, most cases are not diagnosed until they are well advanced.

Lung cancer, like many other cancers, results from a series of mutations that activate **oncogenes** (Greek *onco* = tumor), or cancer-causing genes, and inactivate tumor suppressor genes. The products of **tumor suppressor genes** inhibit cell growth and division, so the inactivation of these genes leads to uncontrolled cell growth. Ordinarily, nasal hairs, sticky mucus, and cilia do a fine job of protecting the lungs from chemical and biological irritants, but when a person smokes, these defenses are overwhelmed and eventually stop functioning. In particular, smoking paralyzes the cilia that clear mucus from the airways, allowing irritants and pathogens to accumulate. The "cocktail" of free radicals and other carcinogens in tobacco smoke eventually translates into lung cancer.

The most common types of lung cancer are:

- **Adenocarcinoma** (about 40% of cases), which originates in peripheral lung areas as solitary nodules that develop from bronchial glands and alveolar cells.
- **Squamous cell carcinoma** (25–30%), which arises in the epithelium of the bronchi or their larger subdivisions and tends to form masses that may cavitate (hollow out) and bleed.
- **Small cell carcinoma** (about 20%), round lymphocyte-sized cells that originate in the main bronchi and grow aggressively in small grapelike clusters within the mediastinum. Metastasis from the mediastinum is especially rapid. Some small cell

carcinomas cause additional problems because they produce certain hormones. For example, some secrete antidiuretic hormone (ADH), resulting in the syndrome of inappropriate ADH secretion (see p. 531).

Because lung cancers metastasize aggressively and early, the key to survival is early detection. If the cancer has not metastasized before it is discovered, complete removal of the diseased lung has the greatest potential for prolonging life and providing a cure. With metastatic lung cancer, radiation therapy and chemotherapy are the only options, but these have low success rates.

Fortunately, there are several new therapies on the horizon and as clinical trials progress, we will learn which of these approaches is most effective.

☑ Check Your Understanding

25. What distinguishes the obstruction in asthma from that in chronic bronchitis?

For answers, see Answers Appendix.

Lungs, bronchial tree, heart, and connecting blood vessels—together, these organs fashion a remarkable system that oxygenates blood, removes carbon dioxide, and ensures that all tissue cells have access to these services. Although the cooperation of the respiratory and cardiovascular systems is obvious, all organ systems depend on the functioning of the respiratory system.

REVIEW QUESTIONS

MAP For more chapter study tools, go to the Study Area of MasteringA&P®.
There you will find:
- Interactive Physiology iP
- A&PFlix A&PFlix
- Practice Anatomy Lab PAL
- PhysioEx PEx
- Videos, Practice Quizzes and Tests, MP3 Tutor Sessions, Case Studies, and much more!

Multiple Choice/Matching

(Some questions have more than one correct answer. Select the best answer or answers from the choices given.)

1. Cutting the phrenic nerves will result in **(a)** air entering the pleural cavity, **(b)** paralysis of the diaphragm, **(c)** stimulation of the diaphragmatic reflex, **(d)** paralysis of the epiglottis.

2. Which of the following laryngeal cartilages is/are not paired? **(a)** epiglottis, **(b)** arytenoid, **(c)** cricoid, **(d)** cuneiform, **(e)** corniculate.

3. Under ordinary circumstances, the inflation reflex is initiated by **(a)** noxious chemicals, **(b)** the ventral respiratory group, **(c)** overinflation of the alveoli and bronchioles, **(d)** the pontine respiratory centers.

4. The detergent-like substance that keeps the alveoli from collapsing between breaths because it reduces the surface tension of the water film in the alveoli is called **(a)** lecithin, **(b)** bile, **(c)** surfactant, **(d)** reluctant.

5. Which of the following determines the direction of gas movement? **(a)** solubility in water, **(b)** partial pressure gradient, **(c)** temperature, **(d)** molecular weight and size of the gas molecule.

6. When the inspiratory muscles contract, **(a)** the size of the thoracic cavity increases in diameter, **(b)** the size of the thoracic cavity increases in length, **(c)** the volume of the thoracic cavity decreases, **(d)** the size of the thoracic cavity increases in both length and diameter.

7. The nutrient blood supply of the lungs is provided by **(a)** the pulmonary arteries, **(b)** the aorta, **(c)** the pulmonary veins, **(d)** the bronchial arteries.

8. Oxygen and carbon dioxide are exchanged in the lungs and through all cell membranes by **(a)** active transport, **(b)** diffusion, **(c)** filtration, **(d)** osmosis.

9. Which of the following would not normally be treated by 100% oxygen therapy? (Choose all that apply.) **(a)** anoxia, **(b)** carbon monoxide poisoning, **(c)** respiratory crisis in an emphysema patient, **(d)** eupnea.

10. Most oxygen carried in the blood is **(a)** in solution in the plasma, **(b)** combined with plasma proteins, **(c)** chemically combined with the heme in red blood cells, **(d)** in solution in the red blood cells.

11. Which of the following has the greatest stimulating effect on the respiratory centers in the brain? **(a)** oxygen, **(b)** carbon dioxide, **(c)** calcium, **(d)** willpower.

12. In mouth-to-mouth artificial respiration, the rescuer blows air from his or her own respiratory system into that of the victim. Which of the following statements are correct?
(1) Expansion of the victim's lungs is brought about by blowing air in at higher than atmospheric pressure (positive-pressure breathing).
(2) During inflation of the lungs, the intrapleural pressure increases.
(3) This technique will not work if the victim has a hole in the chest wall, even if the lungs are intact.
(4) Expiration during this procedure depends on the elasticity of the alveolar and thoracic walls.
(a) all of these, **(b)** 1, 2, 4, **(c)** 1, 2, 3, **(d)** 1, 4.

13. A baby holding its breath will **(a)** have brain cells damaged because of low blood oxygen levels, **(b)** automatically start to breathe again when the carbon dioxide levels in the blood reach a high enough value, **(c)** suffer heart damage because of increased pressure in the carotid sinus and aortic arch areas, **(d)** be called a "blue baby."

14. Under ordinary circumstances, which of the following blood components is of no physiological significance? **(a)** bicarbonate ions, **(b)** carbaminohemoglobin, **(c)** nitrogen, **(d)** chloride.

15. Damage to which of the following would most likely result in cessation of breathing? **(a)** the pontine respiratory group, **(b)** the ventral respiratory group of the medulla, **(c)** the stretch receptors in the lungs, **(d)** the dorsal respiratory group of the medulla.

16. The bulk of carbon dioxide is carried **(a)** chemically combined with the amino acids of hemoglobin as carbaminohemoglobin in the red blood cells, **(b)** as the ion HCO_3^- in the plasma after first entering the red blood cell, **(c)** as carbonic acid in the plasma, **(d)** chemically combined with the heme portion of Hb.

Short Answer Essay Questions

17. Trace the route of air from the nares to an alveolus. Name subdivisions of organs where applicable, and differentiate between conducting and respiratory zone structures.
18. (a) Why is it important that the trachea is reinforced with cartilage rings? (b) Why is it advantageous that the rings are incomplete posteriorly?
19. Briefly explain the anatomical "reason" why most men have deeper voices than boys or women.
20. The lungs are mostly passageways and elastic tissue. (a) What is the role of the elastic tissue? (b) Of the passageways?
21. Describe the functional relationships between volume changes and gas flow into and out of the lungs.
22. Discuss how airway resistance, lung compliance, and alveolar surface tension influence pulmonary ventilation.
23. (a) Differentiate clearly between minute ventilation and alveolar ventilation rate. (b) Which provides a more accurate measure of ventilatory efficiency, and why?
24. State Dalton's law of partial pressures and Henry's law.
25. (a) Define hyperventilation. (b) If you hyperventilate, do you retain or expel more carbon dioxide? (c) What effect does hyperventilation have on blood pH?

AT THE CLINIC

Clinical Case Study

Respiratory System

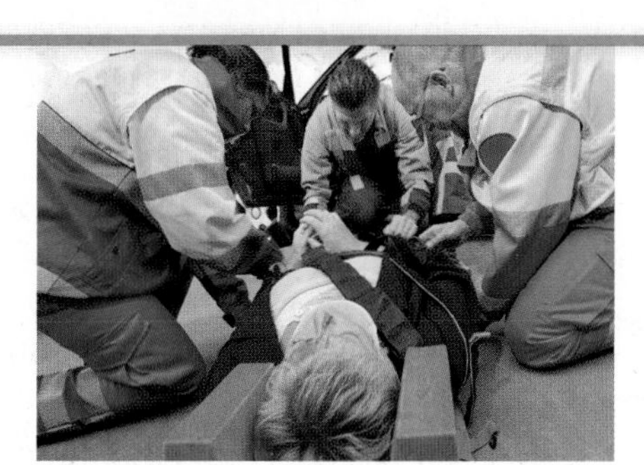

Barbara Joley was in the bus that was hit broadside. When she was freed from the wreckage, she was deeply cyanotic and her respiration had stopped. Her heart was still beating, but her pulse was fast and thready. The emergency medical technician reported that when Barbara was found, her head was cocked at a peculiar angle and it looked like she had a fracture at the level of the C_2 vertebra. The following questions refer to these observations.

1. How might the "peculiar" head position explain Barbara's cessation of breathing?
2. What procedures (do you think) the emergency personnel should have initiated immediately?
3. Why is Barbara cyanotic? Explain cyanosis.
4. Assuming that Barbara survives, how will her accident affect her lifestyle in the future?

Barbara survived transport to the hospital and notes recorded at admission included the following observations.

- Right thorax compressed; ribs 7 to 9 fractured
- Right lung atelectasis

Relative to these notes:

5. What is atelectasis and why is only the right lung affected?
6. How do the recorded injuries relate to the atelectasis?
7. What treatment will be done to reverse the atelectasis? What is the rationale for this treatment?

For answers, see Answers Appendix.

22 The Digestive System

KEY CONCEPTS

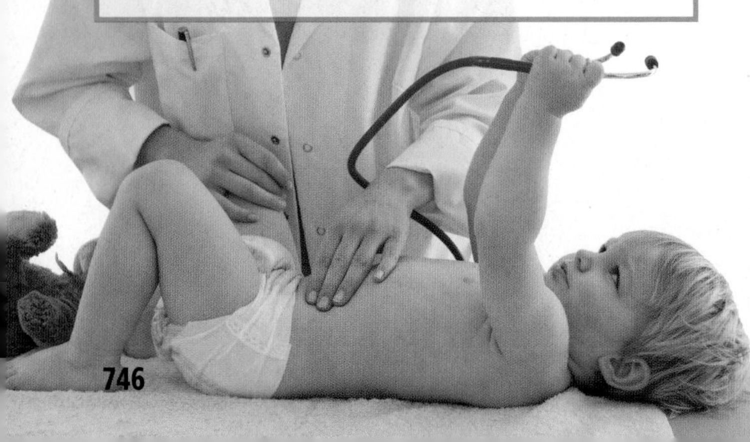

Some digestive system organs can be assessed by palpating the abdomen.

Children are fascinated by the workings of the digestive system. They relish crunching a potato chip, delight in making "mustaches" with milk, and giggle when their stomach "growls." As adults, we know that a healthy digestive system is essential to life, because it converts foods into the raw materials that build and fuel our body's cells. Specifically, the **digestive system** takes in food, breaks it down into nutrient molecules, absorbs these molecules into the bloodstream, and then rids the body of the indigestible remains.

PART 1

OVERVIEW OF THE DIGESTIVE SYSTEM

→ **Learning Objective**

☐ **Describe the functions of the digestive system, and differentiate between organs of the alimentary canal and accessory digestive organs.**

The organs of the digestive system fall into two main groups: (1) those of the *alimentary canal* (al″ĭ-men′tar-e; *aliment* = nourish) and (2) *accessory digestive organs* (**Figure 22.1**).

The **alimentary canal**, also called the **gastrointestinal (GI) tract** or gut, is the continuous muscular tube that winds through the body from the mouth to the anus. It **digests** food—breaks it down into smaller fragments (*digest* = dissolve)—and **absorbs** the digested fragments through its lining into the blood.

The organs of the alimentary canal are the *mouth*, *pharynx*, *esophagus*, *stomach*, *small intestine*, and *large intestine*. The large intestine leads to the terminal opening, or *anus*. In a cadaver, the alimentary canal is approximately 9 m (about 30 ft) long, but in a living person, it is considerably shorter because of its muscle tone. Food material in this tube is technically outside the body because the canal is open to the external environment at both ends.

The **accessory digestive organs** are the *teeth*, *tongue*, *gallbladder*, and a number of large digestive glands—the *salivary glands*, *liver*, and *pancreas*. The teeth and tongue are in the mouth, or oral cavity, while the digestive glands and gallbladder lie outside the GI tract and connect to it by ducts. The accessory digestive glands produce a variety of secretions that help break down foodstuffs.

22.1 What major processes occur during digestive system activity?

→ **Learning Objective**

☐ **List and define the major processes occurring during digestive system activity.**

Practice art labeling
MasteringA&P®>Study Area>Chapter 22

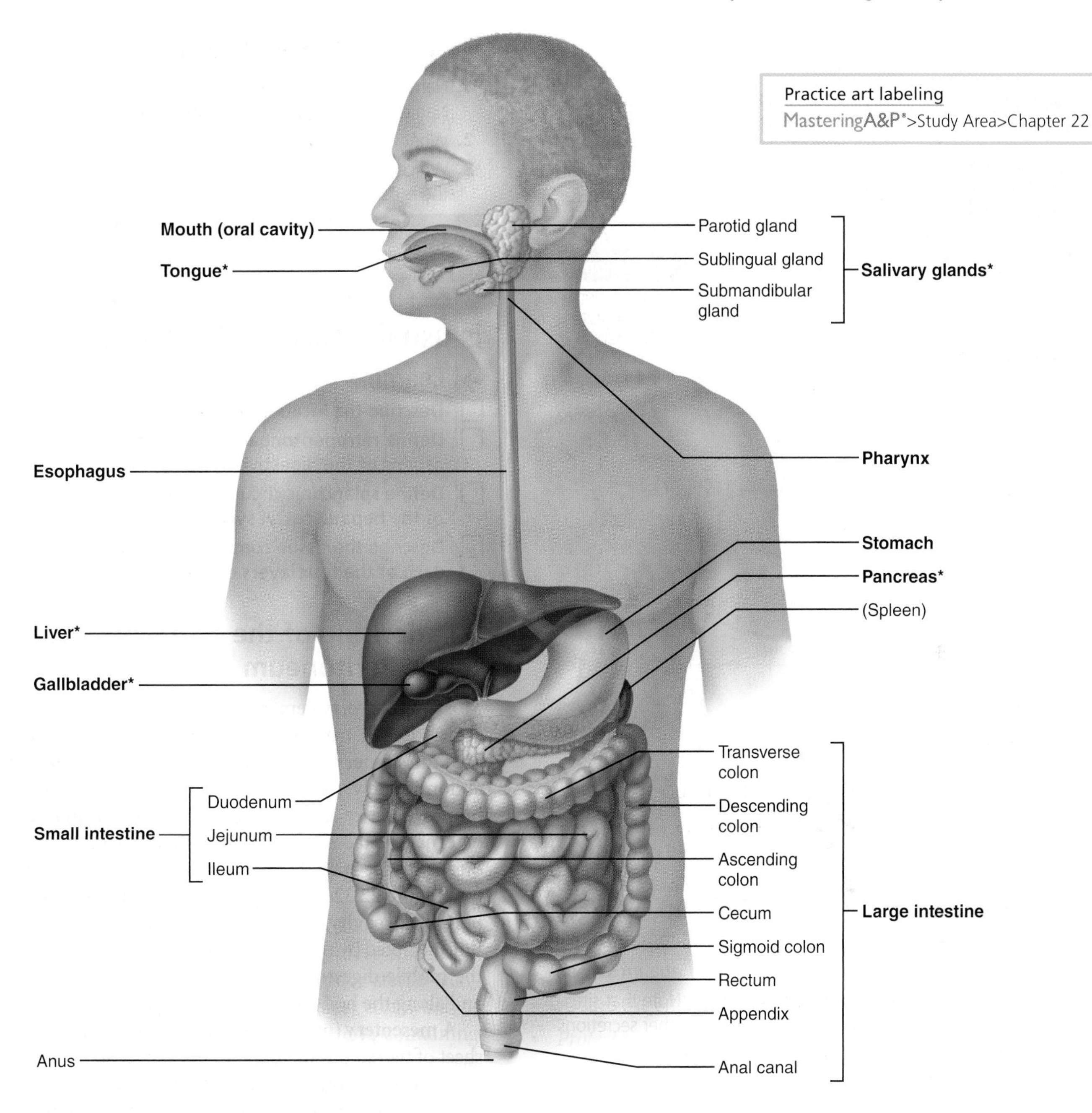

Figure 22.1 Alimentary canal and related accessory digestive organs. Organs with asterisks are accessory organs. Those without asterisks are alimentary canal organs (except the spleen, a lymphoid organ). (For a related image, see *A Brief Atlas of the Human Body*, Figure 64a.)

We can view the digestive tract as a "disassembly line" in which food becomes less complex at each step of processing and its nutrients become available to the body. The processing of food by the digestive system involves six essential activities (**Figure 22.2**):

- **Ingestion** is taking food into the digestive tract (eating).
- **Propulsion**, which moves food through the alimentary canal, includes *swallowing*, which is initiated voluntarily, and *peristalsis* (per″ĭ-stal′sis), an involuntary process. **Peristalsis** (*peri* = around; *stalsis* = constriction), the major means of propulsion, involves alternating waves of contraction and relaxation of muscles in the organ walls (**Figure 22.3a**). Its main effect is to squeeze food along the tract, but some mixing occurs as well. In fact, peristaltic waves are so powerful that, once swallowed, food and fluids will reach your stomach even if you stand on your head.
- **Mechanical breakdown** increases the surface area of ingested food, physically preparing it for digestion by enzymes. Mechanical processes include chewing, mixing food with saliva by the tongue, churning food in the stomach, and **segmentation** (rhythmic local constrictions of the small intestine, Figure 22.3b). Segmentation mixes food with digestive juices and makes absorption more efficient by repeatedly moving different parts of the food mass over the intestinal wall.

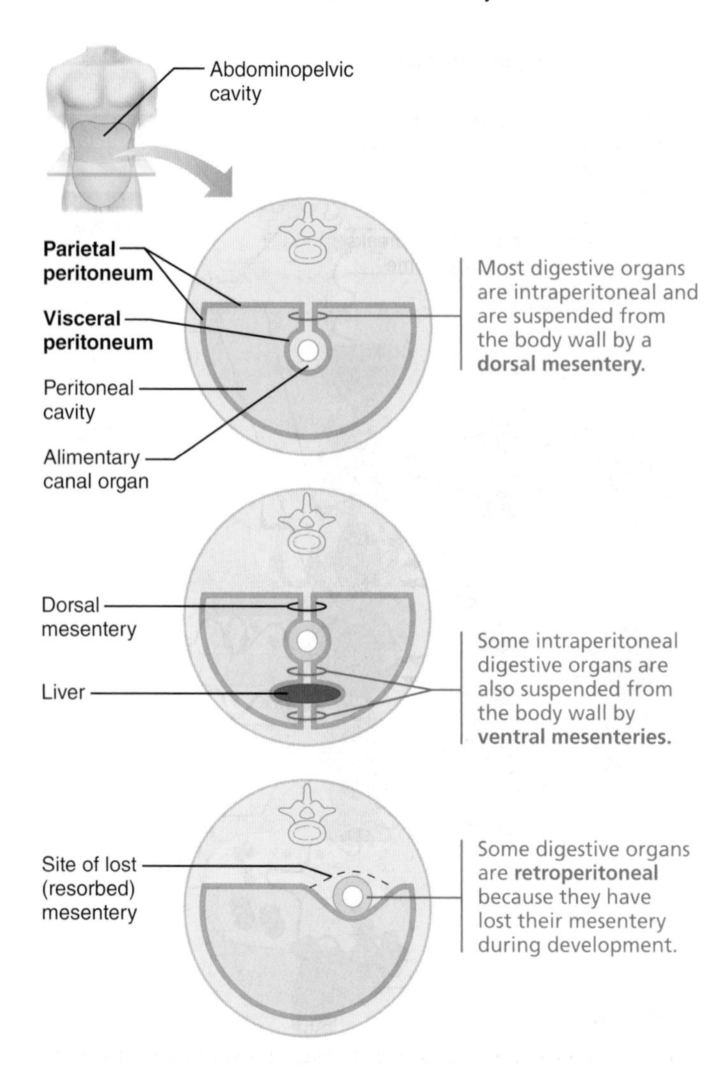

Figure 22.4 The peritoneum and the peritoneal cavity. Note that the peritoneal cavity is much smaller than depicted here.

and other pathogens, which have rather free access to our digestive tract. Particularly large collections of lymphoid follicles occur within the pharynx (as the tonsils) and in the appendix.

External to the lamina propria is the **muscularis mucosae**, a scant layer of smooth muscle cells that produces local movements of the mucosa that can enhance absorption and secretion.

The Submucosa

The **submucosa**, just external to the mucosa, is areolar connective tissue containing a rich supply of blood and lymphatic vessels, lymphoid follicles, and nerve fibers which supply the surrounding tissues of the GI tract wall. Its abundant elastic fibers enable the stomach, for example, to regain its normal shape after temporarily storing a large meal.

The Muscularis Externa

Surrounding the submucosa is the **muscularis externa**, also simply called the **muscularis**. This layer is responsible for segmentation and peristalsis. It typically has an inner *circular layer* and an outer *longitudinal layer* of smooth muscle cells (see Figure 9.22 on p. 273 and Figure 22.5). In several places along the tract, the circular layer thickens, forming *sphincters* that act as valves to control food passage from one organ to the next and prevent backflow.

The Serosa

The **serosa**, the outermost layer of the intraperitoneal organs, is the *visceral peritoneum*. In most alimentary canal organs, it is formed of areolar connective tissue covered with *mesothelium*, a single layer of squamous epithelial cells (see Figures 4.8a and 4.3a, respectively).

In the esophagus, which is located in the thoracic instead of the abdominopelvic cavity, the serosa is replaced by an **adventitia** (ad″ven-tish′e-ah), ordinary dense connective tissue that binds the esophagus to surrounding structures. Retroperitoneal organs have *both* an adventitia (on the side facing the dorsal body wall) and a serosa (on the side facing the peritoneal cavity).

Blood Supply: The Splanchnic Circulation

The **splanchnic circulation** includes those arteries that branch off the abdominal aorta to serve the digestive organs and the *hepatic portal circulation*. The arterial supply—the branches of the celiac trunk that serve the spleen, liver, and stomach, and the mesenteric arteries that serve the small and large intestines (see pp. 644 and 647)—normally receives one-quarter of the cardiac output. This percentage increases after a meal. The hepatic portal circulation (pp. 656–657) collects nutrient-rich venous blood draining from the digestive viscera and delivers it to the liver.

☑ Check Your Understanding

4. How does the location of the visceral peritoneum differ from that of the parietal peritoneum?
5. Of the following organs, which is/are retroperitoneal? Stomach, pancreas, liver.
6. Name the layers of the alimentary canal from the inside out.
7. What name is given to the venous portion of the splanchnic circulation?
8. MAKING connections The two types of smooth muscle are unitary and multi unit (see Chapter 9). Which type would you expect to find in the muscularis externa, and what characteristics make it well suited for this location?

For answers, see Answers Appendix.

22.3 The GI tract has its own nervous system called the enteric nervous system

→ Learning Objective

☐ **Describe stimuli and controls of digestive activity.**

A theme we have stressed in this book is the body's efforts to maintain a constant internal environment. Most organ systems respond to changes in that environment either by attempting to restore some plasma variable to its former levels or by changing their own function.

The digestive system, however, creates an optimal environment for its functioning in the lumen of the GI tract, an area that

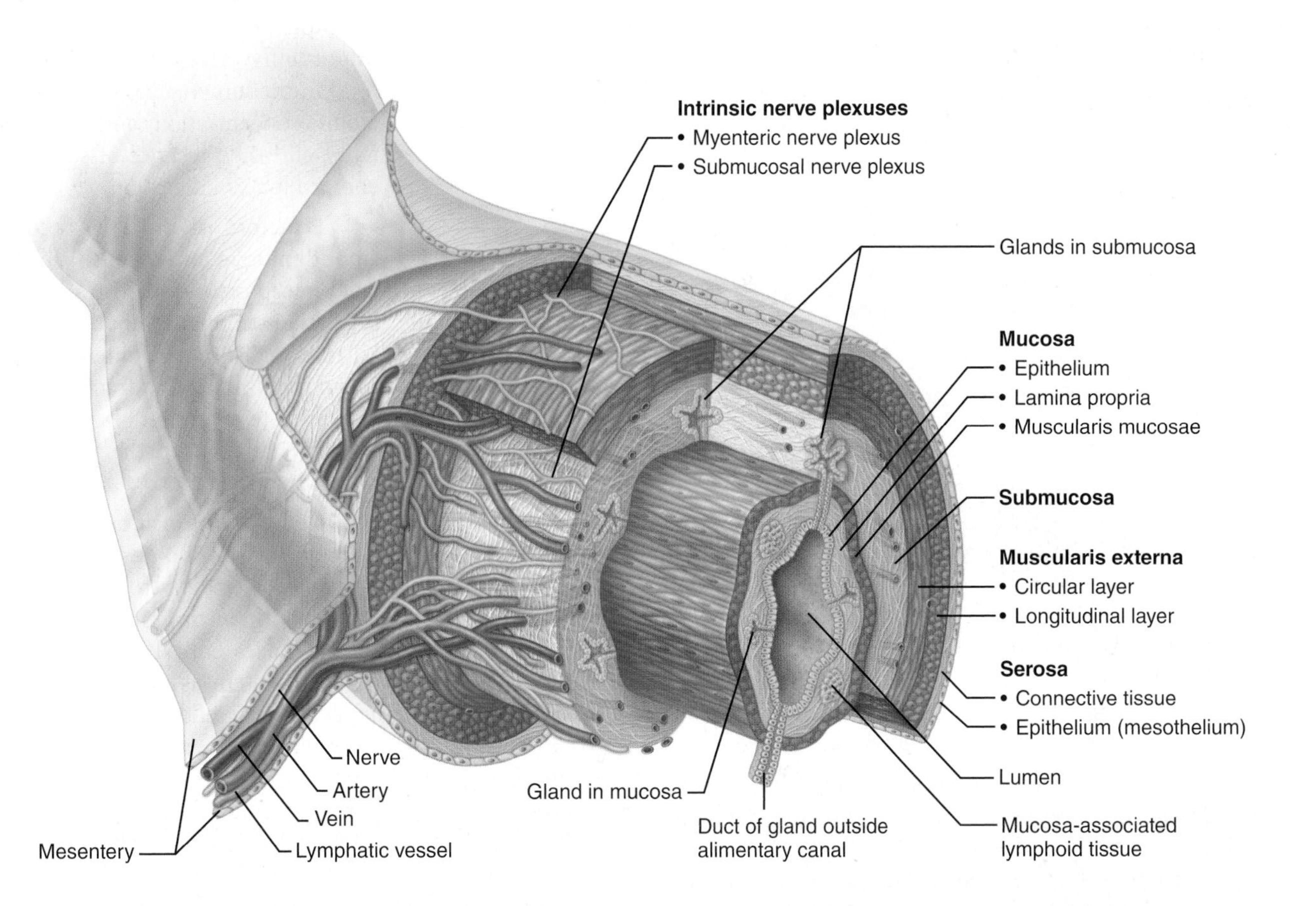

Figure 22.5 Basic structure of the alimentary canal. Its four basic layers are the mucosa, submucosa, muscularis externa, and serosa.

is actually *outside* the body. Essentially all digestive tract regulatory mechanisms control luminal conditions so that food breakdown and absorption can occur there as effectively as possible.

In order to accomplish this, the GI tract has its own *enteric nervous system* (sometimes also called the *gut brain*), which consists of over 100 million neurons. You could truly say that the gut, with more neurons than the entire spinal cord, has a mind of its own! We will describe the enteric nervous system next, and then outline key concepts that govern regulation of digestive activity.

Enteric Nervous System

The **enteric nervous system** (*enter* = gut) is the in-house nerve supply of the alimentary canal. It is staffed by **enteric neurons** that communicate widely with one another to regulate digestive system activity. These semiautonomous enteric neurons constitute the bulk of the two major *intrinsic nerve plexuses* (ganglia interconnected by unmyelinated fiber tracts) found in the walls of the alimentary canal: the submucosal and myenteric nerve plexuses (Figure 22.5). These plexuses interconnect like chicken wire all along the GI tract and regulate digestive activity throughout its length (**Figure 22.6**).

The **submucosal nerve plexus** occupies the submucosa, and the large **myenteric nerve plexus** (mi-en-ter′ik; "intestinal muscle") lies between the circular and longitudinal muscle

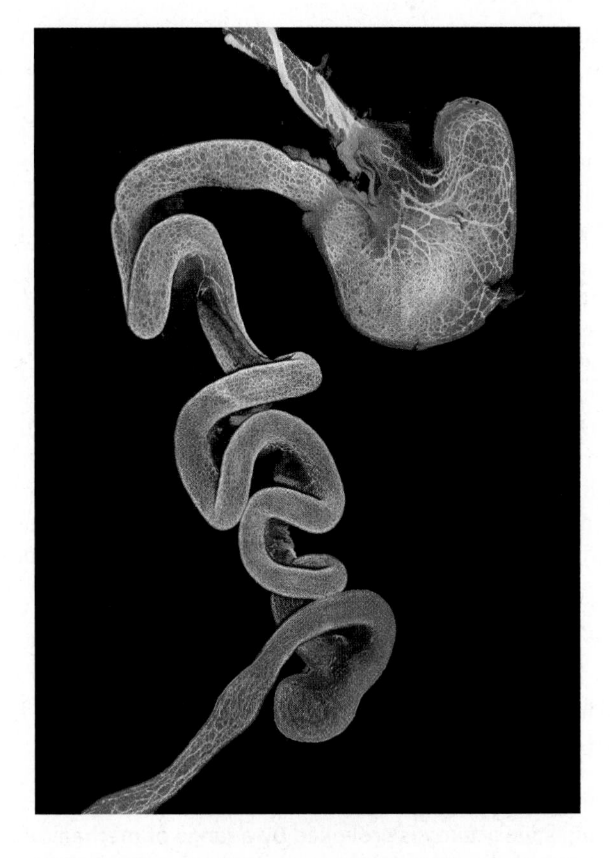

Figure 22.6 The enteric nervous system. The nerves of the enteric nervous sytem (yellow) extend throughout the alimentary canal. (Photo of immature mouse gut.)

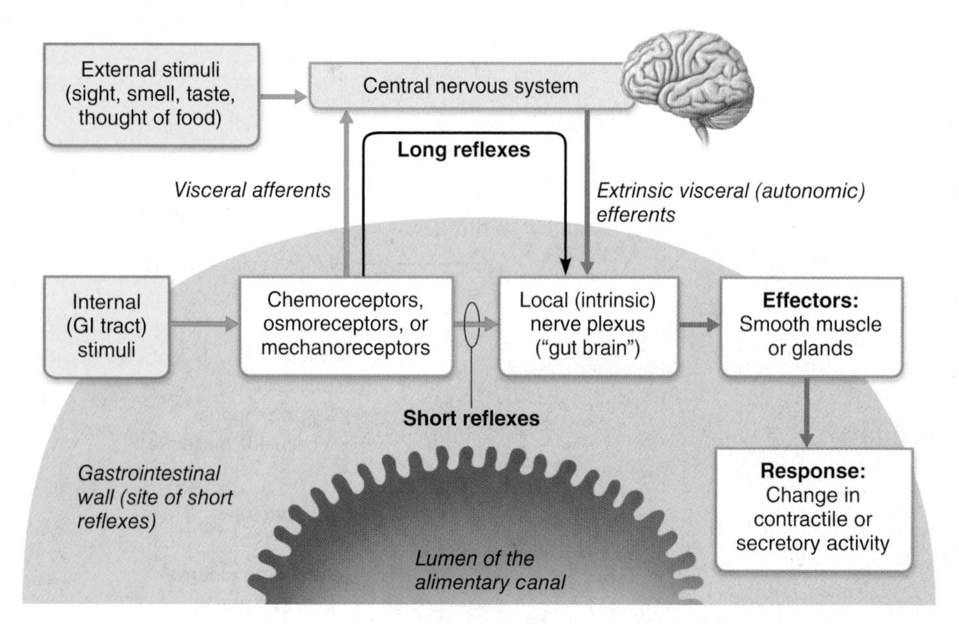

Figure 22.7 Neural reflex pathways initiated by stimuli inside or outside the gastrointestinal tract.

layers of the muscularis externa. Enteric neurons of these plexuses provide the major nerve supply to the GI tract wall and control GI tract motility (motion).

The enteric nervous system participates in both short and long reflex arcs (**Figure 22.7**).

- **Short reflexes** are mediated entirely by enteric nervous system plexuses in response to stimuli within the GI tract. Control of the patterns of segmentation and peristalsis is largely automatic, involving pacemaker cells and reflex arcs between enteric neurons in the same or different organs.
- **Long reflexes** involve CNS integration centers and extrinsic autonomic nerves. The enteric nervous system sends information to the central nervous system via afferent visceral fibers. It receives sympathetic and parasympathetic branches (motor fibers) of the autonomic nervous system that enter the intestinal wall to synapse with neurons in the intrinsic plexuses. Long reflexes can be initiated by stimuli arising inside or outside of the GI tract. In these reflexes, the enteric nervous system acts as a way station for the autonomic nervous system, allowing extrinsic controls to influence digestive activity (Figure 22.7). Generally speaking, parasympathetic inputs enhance digestive activity and sympathetic inpulses inhibit them.

Basic Concepts of Regulating Digestive Activity

Three key concepts govern regulation of digestive activity:

- **Digestive activity is provoked by a range of mechanical and chemical stimuli.** Receptors involved in controlling GI tract activity are located in the walls of the tract's organs. These receptors respond to several stimuli, most importantly stretching of the organ by food in the lumen, changes in osmolarity (solute concentration) and pH of the contents, and the presence of substrates and end products of digestion.
- **Effectors of digestive activity are smooth muscle and glands.** When stimulated, receptors in the GI tract initiate reflexes that stimulate smooth muscle of the GI tract walls to mix lumen contents and move them along the tract. Reflexes can also activate or inhibit glands that secrete digestive juices into the lumen or hormones into the blood.
- **Neurons (intrinsic and extrinsic) and hormones control digestive activity.** The nervous system controls digestive activity via both *intrinsic controls* (involving short reflexes entirely within the enteric nervous system as described above) and *extrinsic controls* (involving long reflexes).

The stomach and small intestine also contain hormone-producing cells. When stimulated, these cells release their products to the interstitial fluid in the extracellular space. Blood and interstitial fluid distribute these hormones to their target cells in the same or different digestive tract organs, where they affect secretion or contraction.

Check Your Understanding

9. When sensors in the GI tract are stimulated, they respond via reflexes. What types of digestive activity may be put into motion via those reflexes?
10. The term "gut brain" does not really mean there is a brain in the digestive system. What does it refer to?
11. Jerry has been given a drug that inhibits parasympathetic stimulation of his digestive tract. Should he "eat hearty" or temporarily refrain from eating, and why?

For answers, see Answers Appendix.

PART 2

FUNCTIONAL ANATOMY OF THE DIGESTIVE SYSTEM

Now that we have summarized some points that unify the digestive system organs, let's take a tour down the alimentary canal and examine the special structural and functional capabilities of each organ of this system. Figure 22.1 shows most of these organs in their normal body positions, so you may find it helpful to refer back to that illustration from time to time as you read the following modules.

As we tour the digestive system organs, we will examine how each participates in the six basic digestive processes. (This information is summarized in Table 22.2 on p. 767.) Since we cover digestion and absorption in a special physiology section later in the chapter, we will focus on the other digestive processes on our tour.

22.4 Ingestion occurs only at the mouth

→ Learning Objectives

- ☐ Describe the gross and microscopic anatomy and the basic functions of the mouth and its associated organs.
- ☐ Describe the composition and functions of saliva, and explain how salivation is regulated.
- ☐ Explain the dental formula and differentiate clearly between deciduous and permanent teeth.

Aside from ingestion, the digestive functions associated with the mouth mostly reflect the activity of the related accessory organs, such as teeth, salivary glands, and tongue. In the mouth we chew food and mix it with saliva containing enzymes that begin the process of digestion. The mouth also begins the propulsive process of swallowing, which carries food through the pharynx and esophagus to the stomach.

The Mouth

The **mouth** is also called the **oral cavity**, or *buccal cavity* (buk′al). Its boundaries are the lips anteriorly, cheeks laterally, palate superiorly, and tongue inferiorly (**Figure 22.8**). Its anterior opening is the **oral orifice**. Posteriorly, the oral cavity is continuous with the *oropharynx*.

The walls of the mouth are lined with a thick stratified squamous epithelium (see Figure 4.3e) which withstands considerable friction. The epithelium on the gums, hard palate, and dorsum of the tongue is slightly keratinized for extra protection against abrasion during eating.

The Lips and Cheeks

The **lips (labia)** and the **cheeks**, which help keep food between the teeth when we chew, are composed of a core of skeletal muscle covered externally by skin. The *orbicularis oris muscle* forms the fleshy lips; the cheeks are formed largely by the *buccinators*. The recess bounded externally by the lips and cheeks and internally by the gums and teeth is the **oral vestibule** ("porch"). The area that lies within the teeth and gums is the **oral cavity proper**. The **labial frenulum** (fren′u-lum) is a median fold that joins the internal aspect of each lip to the gum (Figure 22.8b).

The Palate

The **palate**, forming the roof of the mouth, has two distinct parts: the hard palate anteriorly and the soft palate posteriorly (Figure 22.8). The **hard palate** is underlain by the palatine bones and the palatine processes of the maxillae, and it forms a rigid surface against which the tongue forces food during chewing. The mucosa on either side of its *raphe* (ra′fe), a midline ridge, is slightly corrugated, which helps create friction.

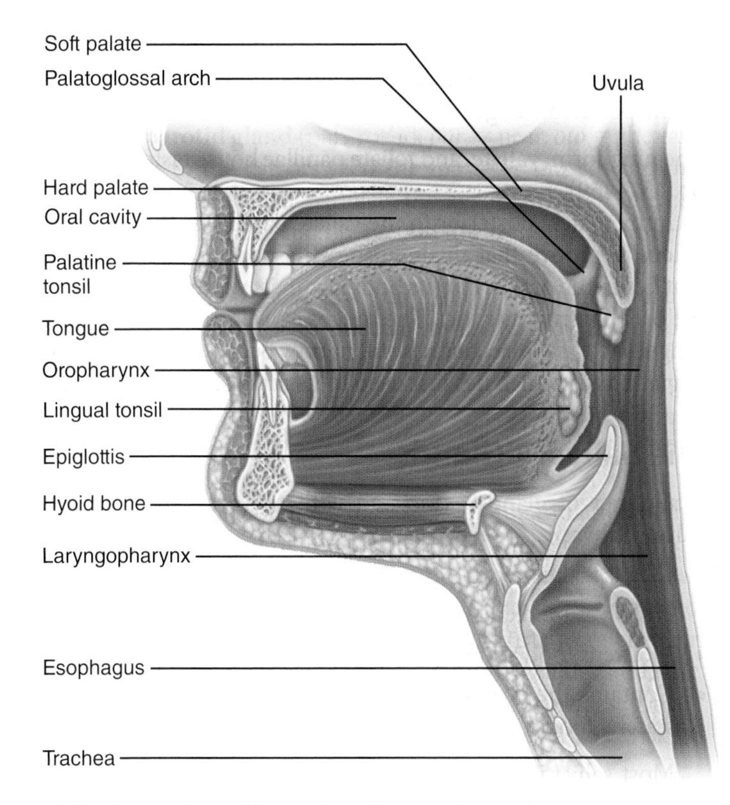

(a) Sagittal section of the oral cavity and pharynx

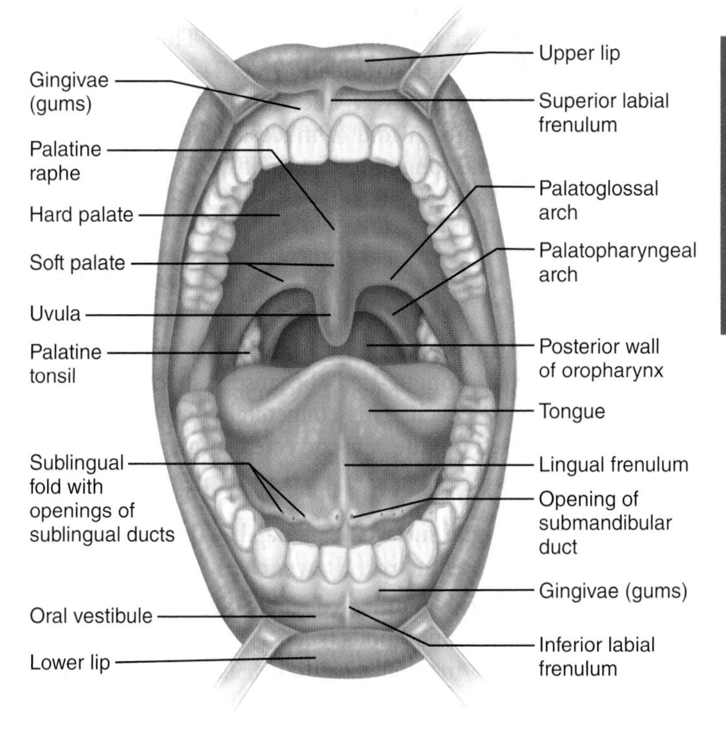

(b) Anterior view

Figure 22.8 Anatomy of the oral cavity (mouth).

The superior and inferior alveolar arteries, branches of the maxillary artery (see Figure 18.22b, p. 641), supply blood.

Dentin contains unique radial striations called *dentinal tubules* (Figure 22.12). Each tubule contains an elongated process of an **odontoblast** (o-don′to-blast; "tooth former"), the cell type that secretes and maintains the dentin. The odontoblasts line the pulp cavity just deep to the dentin. Dentin forms throughout adult life and gradually encroaches on the pulp cavity. New dentin can also be laid down fairly rapidly to compensate for tooth damage or decay.

Enamel, dentin, and cement are all calcified and resemble bone (to differing extents), but they differ from bone because they are avascular. Enamel differs from cement and dentin because it lacks collagen and is almost entirely mineral.

Tooth and Gum Disease

CLINICAL

Dental caries (kār′ēz; "rottenness"), or **cavities**, result from bacterial action that gradually demineralizes enamel and underlying dentin. Decay begins when **dental plaque** (a film of sugar, bacteria, and other mouth debris) adheres to the teeth. Bacterial metabolism of the trapped sugars produces acids, which dissolve the calcium salts of the teeth. Once the salts are leached out, enzymes released by the bacteria readily digest the remaining organic matrix of the tooth. Frequent brushing and daily flossing help prevent caries by removing plaque.

More serious than tooth decay is the effect of unremoved plaque on the gums. As dental plaque accumulates, it calcifies, forming **calculus** (kal′ku-lus; "stone") or tartar. These stony-hard deposits disrupt the seal between gingivae and teeth, deepening the sulcus and putting the gums at risk for infection by pathogenic anaerobic bacteria. In the early stages of such an infection, called **gingivitis** (jin″jĭ-vi′tis), the gums are red, sore, swollen, and may bleed.

Gingivitis is reversible if the calculus is removed, but if it is neglected the bacteria eventually form pockets of infection which become inflamed. Neutrophils and other immune cells attack not only the intruders but also body tissues, carving deep pockets around the teeth, destroying the periodontal ligament, and activating osteoclasts which dissolve the bone. This serious condition, **periodontal disease** or **periodontitis**, affects up to 95% of all people over age 35 and accounts for 80–90% of tooth loss in adults.

Tooth loss from periodontitis is not inevitable. Various treatments can alleviate the bacterial infestations and encourage the surrounding tissues to reattach to the teeth and bone.

Periodontal disease may jeopardize more than just teeth. Some contend that it increases the risk of heart disease and stroke in at least two ways: (1) the chronic inflammation promotes atherosclerotic plaque, and (2) bacteria entering the blood from infected gums stimulate the formation of clots that clog coronary and cerebral arteries. Risk factors for periodontal disease include smoking, diabetes mellitus, and oral (tongue or lip) piercing.

Digestive Processes of the Mouth

The mouth and its accessory digestive organs are involved in four of the six digestive processes described earlier. The mouth (1) ingests, (2) begins mechanical breakdown by chewing, (3) initiates propulsion by swallowing, and (4) starts the digestion of polysaccharides. Absorption does not occur in the mouth except for a few drugs that are absorbed through the oral mucosa (for example, nitroglycerine used to alleviate the pain of angina).

Chewing and swallowing are the mechanical processes that promote mechanical breakdown and propulsion, respectively. We describe chewing next, but because the mouth participates in only the first phase of swallowing, we will postpone its discussion until the end of the next module.

Mastication (Chewing)

As food enters the mouth, its mechanical breakdown begins with **mastication**, or chewing. The cheeks and closed lips hold food between the teeth, the tongue mixes food with saliva to soften it, and the teeth cut and grind solid foods into smaller morsels.

Mastication is partly voluntary and partly reflexive. We voluntarily put food into our mouths and contract the muscles that close our jaws. The pattern and rhythm of continued jaw movements are controlled mainly by stretch reflexes and in response to pressure inputs from receptors in the cheeks, gums, and tongue, but they can also be voluntary if desired.

☑ Check Your Understanding

12. Which structure forms the roof of the mouth?
13. Besides preparing food for swallowing, the tongue has another role. What is it?
14. Name three antimicrobial substances found in saliva.
15. Which tooth substance is harder than bone? Which tooth region includes nervous tissue and blood vessels?

For answers, see Answers Appendix.

22.5 The pharynx and esophagus move food from the mouth to the stomach

→ Learning Objectives

- ☐ **Describe the anatomy and basic functions of the pharynx and esophagus.**
- ☐ **Describe the mechanism of swallowing.**

The Pharynx

From the mouth, food passes posteriorly into the **oropharynx** and then the **laryngopharynx** (see Figures 21.4 and 22.8a), both common passageways for food, fluids, and air. (The nasopharynx has no digestive role.)

The histology of the pharyngeal wall resembles that of the oral cavity. The mucosa contains a friction-resistant stratified squamous epithelium well supplied with mucus-producing glands. The external muscle layer consists of two *skeletal muscle* layers. The cells of the inner layer run longitudinally. Those of the outer layer, the *pharyngeal constrictor* muscles, encircle the wall like three stacked fists (see Figure 10.9c). Contractions of these muscles propel food into the esophagus below.

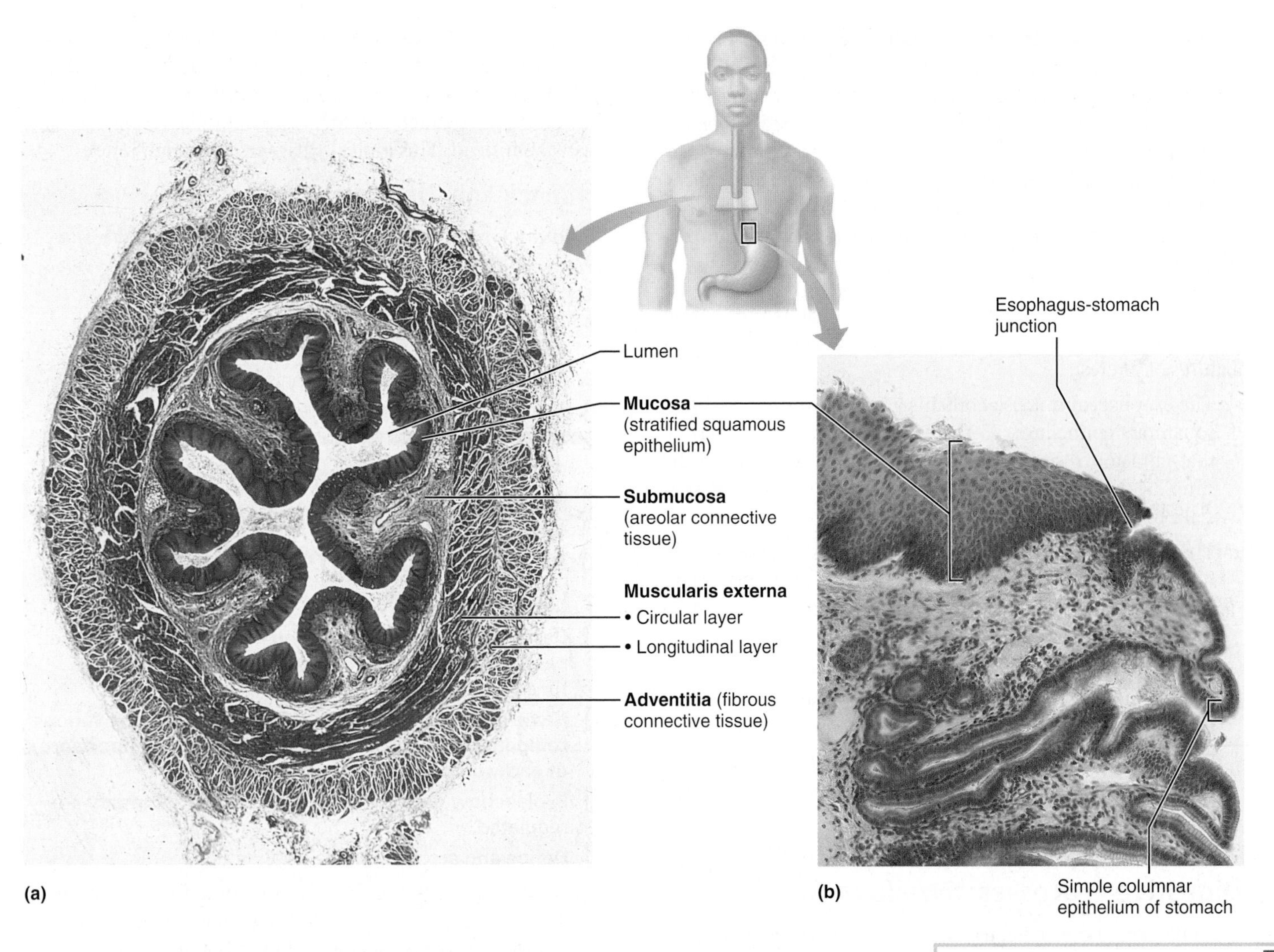

Figure 22.13 Microscopic structure of the esophagus. (a) Cross-sectional view of the esophagus taken from the region close to the stomach junction (10×). The muscularis is composed of smooth muscle. **(b)** Longitudinal section through the esophagus-stomach junction (130×). Notice the abrupt transition from the stratified squamous epithelium of the esophagus (top) to the simple columnar epithelium of the stomach (bottom).

View histology slides
MasteringA&P®>Study Area>PAL

The Esophagus

The **esophagus** (ĕ-sof′ah-gus; "carry food") is a muscular tube about 25 cm (10 inches) long and is collapsed when not involved in food propulsion (**Figure 22.13**). As food moves through the laryngopharynx, it is routed into the esophagus posteriorly because the epiglottis closes off the larynx to incoming food.

As shown in Figure 22.1, the esophagus takes a fairly straight course through the mediastinum of the thorax. It pierces the diaphragm at the **esophageal hiatus** (hi-a′tus; "gap") to enter the abdomen. It joins the stomach at the **cardial orifice** within the abdominal cavity. The cardial orifice is surrounded by the **gastroesophageal** or **cardiac sphincter** (gas″tro-ĕ-sof″ah-je′al), which is a *physiological* sphincter (see Figure 22.14). That is, it acts as a sphincter, but the only structural evidence of this sphincter is a slight thickening of the circular smooth muscle at that point. The muscular diaphragm, which surrounds this sphincter, helps keep it closed when food is not being swallowed. Mucous cells on both sides of the sphincter help protect the esophagus from reflux of stomach acid.

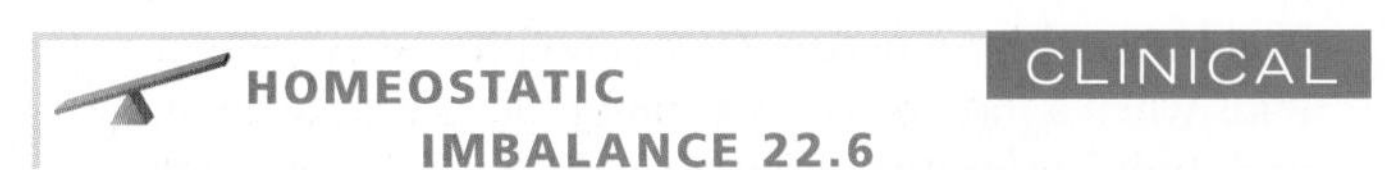

HOMEOSTATIC IMBALANCE 22.6 CLINICAL

Heartburn, the first symptom of *gastroesophageal reflux disease* (*GERD*), is the burning, radiating substernal pain that occurs when stomach acid regurgitates into the esophagus. Symptoms are so similar to those of a heart attack that many first-time sufferers of heartburn are rushed to the emergency room. Heartburn is most likely when a person has eaten or drunk to excess, and in conditions that force abdominal contents superiorly, such as extreme obesity, pregnancy, and running, which splashes stomach contents upward with each step.

Heartburn is also common in those with a **hiatal hernia**, a structural abnormality (most often due to abnormal relaxation or weakening of the gastroesophageal sphincter) in which the superior part of the stomach protrudes slightly above the diaphragm. Since the diaphragm no longer reinforces the sphincter, gastric juice may enter the esophagus, particularly when lying down. If the episodes are frequent and prolonged, *esophagitis* (inflammation of the esophagus) and *esophageal ulcers* may result. An even more threatening sequel is esophageal cancer. Treatment varies, but GERD is usually addressed with lifestyle and dietary modifications, along with antacids and certain prescription drugs. +

Unlike the mouth and pharynx, the esophagus wall has all four of the basic alimentary canal layers described earlier. Some features of interest:

- The esophageal mucosa contains a *nonkeratinized* stratified squamous epithelium. At the esophagus-stomach junction, that abrasion-resistant epithelium changes abruptly to the simple columnar epithelium of the stomach, which is specialized for secretion (Figure 22.13b).
- The submucosa contains mucus-secreting *esophageal glands.* As a bolus moves through the esophagus, it compresses these glands, causing them to secrete mucus that "greases" the esophageal walls and aids food passage.
- The muscularis externa is skeletal muscle in its superior third, a mixture of skeletal and smooth muscle in its middle third, and entirely smooth muscle in its inferior third.
- Instead of a serosa, the esophagus has a fibrous adventitia composed entirely of connective tissue, which blends with surrounding structures along its route.

Digestive Processes: Swallowing

The pharynx and esophagus merely serve as conduits to pass food from the mouth to the stomach. Their single digestive system function is food propulsion, accomplished by **deglutition** (deg″loo-tish′un), or swallowing.

To send food on its way from the mouth, it is first compacted by the tongue into a bolus and is then swallowed. This complicated process involves the coordinated activity of over 22 separate muscle groups. Before we examine the steps of this process in detail, let's outline the two major phases involved in deglutition.

- The **buccal phase** occurs in the mouth and is voluntary. It ends when a food bolus or a "bit of saliva" leaves the mouth and stimulates tactile receptors in the posterior pharynx, initiating the next phase.
- The **pharyngeal-esophageal phase** is involuntary and is controlled by the swallowing center in the brain stem (medulla and lower pons). Various cranial nerves, most importantly the vagus nerves, transmit motor impulses from the swallowing center to the muscles of the pharynx and esophagus. Once food enters the pharynx, respiration is momentarily inhibited and all routes except the desired one into the digestive tract are blocked off. Solid foods pass from the oropharynx to the stomach in about 8 seconds, and fluids, aided by gravity, pass in 1 to 2 seconds.

22

Figure 22.14 illustrates each phase of deglutition and the protective mechanisms that prevent food from being inhaled. If we talk or inhale while swallowing, these protective mechanisms may be short-circuited and food may enter the respiratory passageways instead. This typically triggers the cough reflex.

☑ Check Your Understanding

16. To which two organ systems does the pharynx belong?
17. How is the muscularis externa of the esophagus unique in the body?
18. What is the functional significance of the epithelial change seen at the esophagus-stomach junction?
19. What role does the tongue play in swallowing?
20. How are the respiratory passages blocked during swallowing?

For answers, see Answers Appendix.

22.6 The stomach temporarily stores food and begins protein digestion

→ Learning Objectives

- ☐ **Describe stomach structure and indicate changes in the basic alimentary canal structure that aid its digestive function.**
- ☐ **Name the cell types responsible for secreting the various components of gastric juice and indicate the importance of each component in stomach activity.**
- ☐ **Explain how gastric secretion and stomach motility are regulated.**
- ☐ **Define and account for the alkaline tide.**

Below the esophagus, the GI tract expands to form the **stomach** (see Figure 22.1), a temporary "storage tank" where chemical breakdown of proteins begins and food is converted to a paste called **chyme** (kīm; "juice"). The stomach lies in the upper left quadrant of the peritoneal cavity, nearly hidden by the liver and diaphragm.

Gross Anatomy of the Stomach

The adult stomach varies from 15 to 25 cm (6 to 10 inches) long, but its diameter and volume depend on how much food it contains. An empty stomach has a volume of about 50 ml and a cross-sectional diameter only slightly larger than the large intestine, but when it is really distended it can hold about 4 L (1 gallon) of food and may extend nearly to the pelvis! When empty, the stomach collapses inward, throwing its mucosa (and submucosa) into large, longitudinal folds called **rugae** (roo′ge; *ruga* = wrinkle, fold).

Figure 22.15a shows the major regions of the stomach. The small **cardial part**, or **cardia** ("near the heart"), surrounds the cardial orifice through which food enters the stomach from the esophagus. The **fundus** is the stomach's dome-shaped part, tucked beneath the diaphragm, that bulges superolaterally to the cardia. The **body**, or the midportion of the stomach, is continuous inferiorly with the funnel-shaped **pyloric part**. The wider and more superior area of the pyloric part, the **pyloric antrum** (*antrum* = cave), narrows to form the **pyloric canal**,

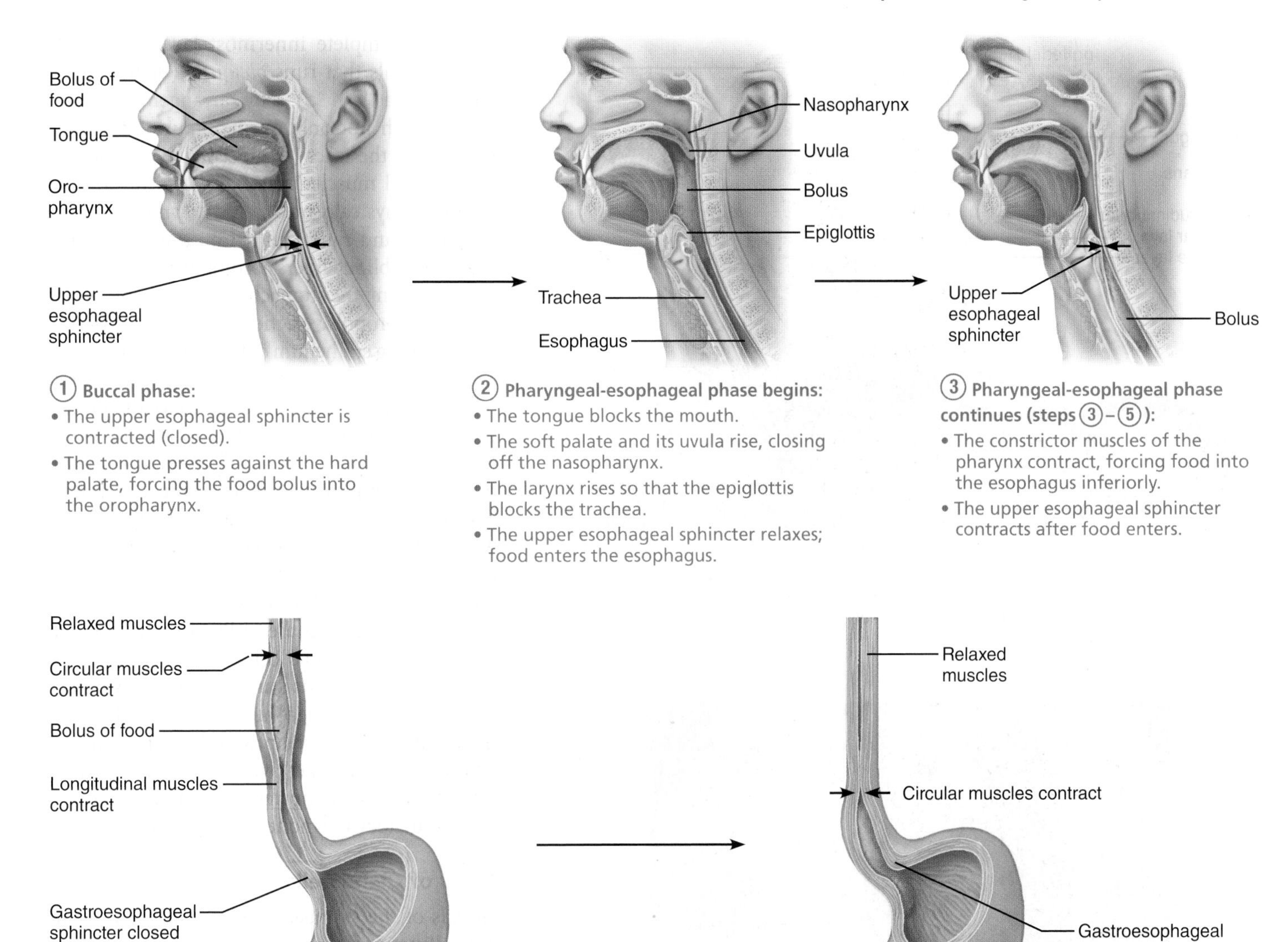

Figure 22.14 Deglutition (swallowing). The process of swallowing consists of a buccal (voluntary) phase (step ①) and a pharyngeal-esophageal (involuntary) phase (steps ②–⑤).

22

which terminates at the **pylorus**. The pylorus is continuous with the duodenum through the **pyloric sphincter** or **valve**, which controls stomach emptying (*pylorus* = gatekeeper).

The convex lateral surface of the stomach is its **greater curvature**, and its concave medial surface is the **lesser curvature**. Extending from these curvatures are two mesenteries, called *omenta* (o-men′tah), that help tether the stomach to other digestive organs and the body wall (see Figure 22.32, p. 784). The **lesser omentum** runs from the liver to the lesser curvature of the stomach, where it becomes continuous with the visceral peritoneum covering the stomach. The **greater omentum** drapes inferiorly from the greater curvature of the stomach to cover the coils of the small intestine. It then runs dorsally and superiorly, wrapping the spleen and the transverse portion of the large intestine before blending with the *mesocolon*, a dorsal mesentery that secures the large intestine to the parietal peritoneum of the posterior abdominal wall.

The greater omentum is riddled with fat deposits (*oment* = fatty skin) that give it the appearance of a lacy apron. It also contains large collections of lymph nodes. The immune cells and macrophages in these nodes "police" the peritoneal cavity and intraperitoneal organs.

The stomach is served by the autonomic nervous system. Sympathetic fibers from thoracic splanchnic nerves are relayed through the celiac plexus. Parasympathetic fibers are supplied by the vagus nerve. The arterial supply of the stomach is provided by branches (gastric and splenic) of the celiac trunk (see Figure 18.24, p. 644). The corresponding veins are

several digestive system target organs (Table 22.1). **Gastrin**, a hormone, plays essential roles in regulating stomach secretion and motility, as we will describe shortly.

The Mucosal Barrier

The stomach mucosa is exposed to some of the harshest conditions in the entire digestive tract. Gastric juice is corrosively acidic (the H^+ concentration in the stomach can be 100,000 times that found in blood), and its protein-digesting enzymes can digest the stomach itself.

However, the stomach protects itself by producing the **mucosal barrier**. Three factors create this barrier:

- *A thick coating of bicarbonate-rich mucus* builds up on the stomach wall.
- *The epithelial cells of the mucosa are joined together by tight junctions* that prevent gastric juice from leaking into underlying tissue layers.
- *Damaged epithelial mucosal cells are shed and quickly replaced* by division of *undifferentiated stem cells* that reside where the gastric pits join the gastric glands. The stomach surface epithelium of mucous cells is completely renewed every three to six days, but the more sheltered glandular cells deep within the gastric glands have a much longer life span.

HOMEOSTATIC IMBALANCE 22.7 — CLINICAL

Anything that breaches the gel-like mucosal barrier causes inflammation of the stomach wall, a condition called *gastritis*. Persistent damage to the underlying tissues can promote **peptic ulcers**, specifically called **gastric ulcers** when they are erosions of the stomach wall (Figure 22.17). The most distressing symptom of gastric ulcers is gnawing epigastric pain that seems to bore through to your back. The pain typically occurs 1–3 hours after eating and is often relieved by eating again. The danger posed by ulcers is perforation of the stomach wall, leading to peritonitis and, perhaps, massive hemorrhage.

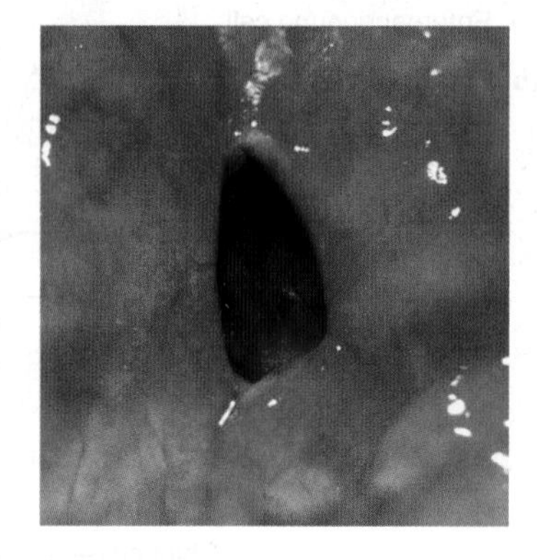

Figure 22.17 **A gastric ulcer.**

For years, ulcers were blamed on factors that increased HCl production or reduced mucus secretion, including aspirin and nonsteroidal anti-inflammatory drugs (NSAIDs such as ibuprofen), smoking, spicy food, alcohol, coffee, and stress. Although acidic conditions *are* necessary for ulcers to form, acidity in itself is not sufficient to cause them. Ninety percent of recurrent ulcers are the work of a strain of acid-resistant, corkscrew-shaped *Helicobacter pylori* bacteria (Figure 22.18), which burrow like a drill bit through the mucus and destroy the protective mucosal layer. Even more troubling are studies that link this bacterium to some stomach cancers.

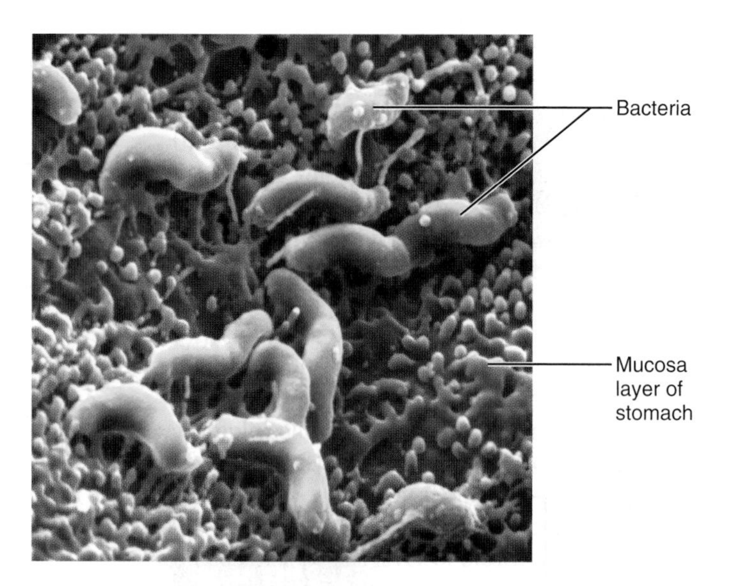

Figure 22.18 **Photomicrograph of *H. pylori*, the bacteria that most commonly cause gastric ulcers.**

More than half of the population harbors *H. pylori*, but these pathological effects occur in only 10–20% of infected individuals. The antimicrobial activity of gastric mucin appears to protect most of us from *H. pylori*'s invasive attacks.

A breath test can easily detect the presence of *H. pylori*. A two-week-long course of antibiotics kills the bacteria, promotes healing of the ulcers, and prevents recurrence. For active ulcers, a blocker for H_2 (histamine) receptors may also help because it inhibits HCl secretion by blocking histamine's effects.

The relatively few peptic ulcers not caused by *H. pylori* generally result from long-term use of NSAIDs. In such noninfectious cases, blocking HCl secretion either directly (with pump inhibitors) or indirectly [with H_2 (histamine) receptor blockers] is the therapy of choice. ✚

Digestive Processes in the Stomach

Except for ingestion and defecation, the stomach is involved in the whole "menu" of digestive activities. Besides serving as a holding area for ingested food, the stomach continues the demolition job begun in the oral cavity by further degrading food both physically and chemically. It then delivers chyme, the product of its activity, into the small intestine.

Protein digestion begins in the stomach and is the main type of enzymatic breakdown that occurs there. HCl produced by stomach glands denatures dietary proteins in preparation for enzymatic digestion. (The unfolded amino acid chain is more accessible to the enzymes.) The most important protein-digesting enzyme produced by the gastric mucosa is pepsin. In infants, however, the stomach glands also secrete **rennin**, an enzyme that acts on milk protein (casein), converting it to a curdy substance that looks like soured milk.

Fat digestion occurs primarily in the small intestine, but gastric and lingual lipases acting in the acidic pH of the stomach also contribute.

Table 22.1 Hormones and Paracrines That Act in Digestion*

HORMONE	SITE OF PRODUCTION	STIMULUS FOR PRODUCTION	TARGET ORGAN	ACTIVITY
Cholecystokinin (CCK)	Duodenal mucosa	Fatty chyme (also partially digested proteins)	Stomach	• Inhibits stomach's secretory activity
			Liver/pancreas	• Potentiates secretin's actions on these organs
			Pancreas	• Increases output of enzyme-rich pancreatic juice
			Gallbladder	• Stimulates organ to contract and expel stored bile
			Hepatopancreatic sphincter	• Relaxes sphincter to allow entry of bile and pancreatic juice into duodenum
Glucose-dependent insulinotropic peptide (GIP) (or gastric inhibitory peptide)	Duodenal mucosa	Fatty chyme	Stomach	• Inhibits HCl production (minor effect)
			Pancreas (beta cells)	• Stimulates insulin release
Gastrin	Stomach mucosa (G cells)	Food (particularly partially digested proteins) in stomach (chemical stimulation); acetylcholine released by nerve fibers	Stomach (parietal cells)	• Increases HCl secretion • Stimulates gastric emptying (minor effect)
			Small intestine	• Stimulates contraction of intestinal muscle
			Ileocecal valve	• Relaxes ileocecal valve
			Large intestine	• Stimulates mass movements
Histamine	Stomach mucosa	Food in stomach	Stomach	• Activates parietal cells to release HCl
Intestinal gastrin	Duodenal mucosa	Acidic and partially digested foods in duodenum	Stomach	• Stimulates gastric glands and motility
Motilin	Duodenal mucosa	Fasting; periodic release every 1½–2 hours by neural stimuli	Proximal duodenum	• Stimulates migrating motor complex
Secretin	Duodenal mucosa	Acidic chyme (also partially digested proteins and fats)	Stomach	• Inhibits gastric gland secretion and gastric motility
			Pancreas	• Increases output of pancreatic juice rich in bicarbonate ions; potentiates CCK's action
			Liver	• Increases bile output
Serotonin	Stomach mucosa	Food in stomach	Stomach	• Causes contraction of stomach muscle
Somatostatin	Stomach mucosa; duodenal mucosa	Food in stomach; stimulation by sympathetic nerve fibers	Stomach	• Inhibits gastric secretion of all products
			Pancreas	• Inhibits secretion
			Small intestine	• Inhibits GI blood flow; thus inhibits intestinal absorption
			Gallbladder and liver	• Inhibits contraction and bile release
Vasoactive intestinal peptide (VIP)	Enteric neurons	Chyme containing partially digested foods	Small intestine	• Stimulates buffer secretion • Dilates intestinal capillaries • Relaxes intestinal smooth muscle
			Pancreas	• Increases secretion
			Stomach	• Inhibits acid secretion

*Except for somatostatin, all of these polypeptides also stimulate the growth (particularly of the mucosa) of the organs they affect.

22

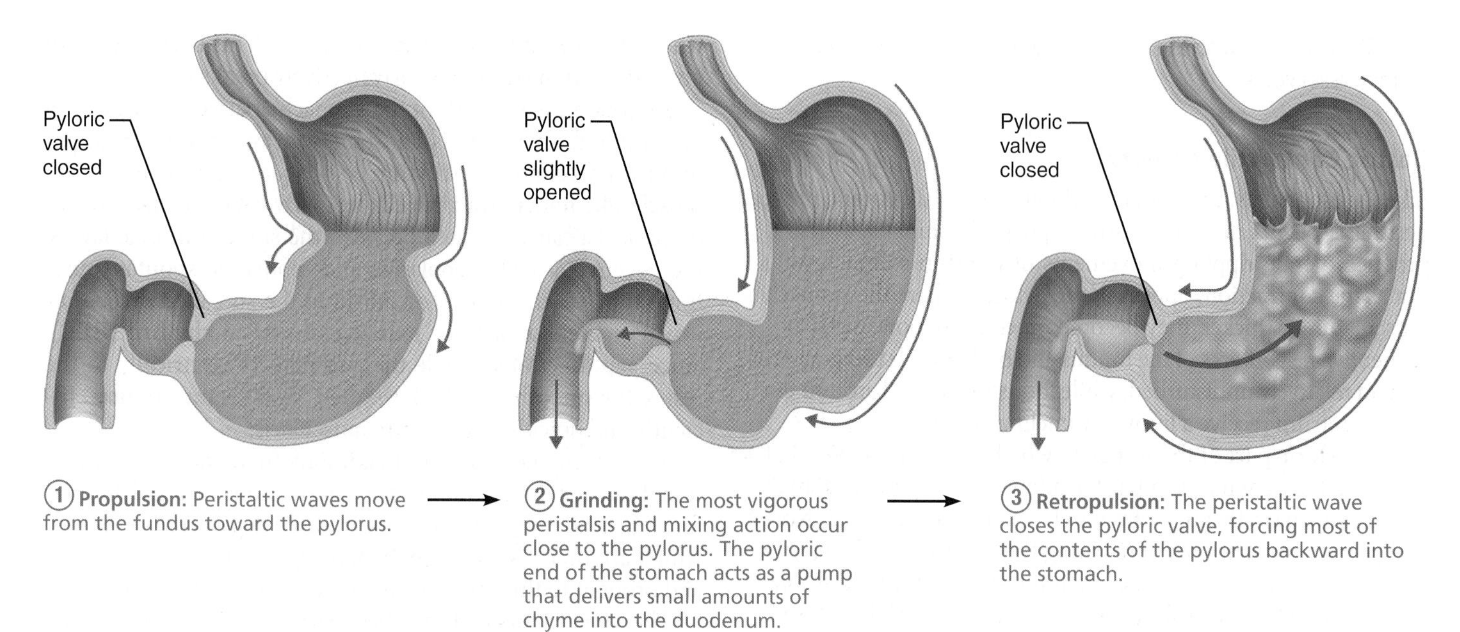

Figure 22.21 Peristaltic waves in the stomach.

empties. Fluids pass quickly through the stomach. Solids linger, remaining until they are well mixed with gastric juice and converted to the liquid state.

The rate of gastric emptying also depends as much—and perhaps more—on the contents of the duodenum as on what is happening in the stomach. The stomach and duodenum act in tandem. As chyme enters the duodenum, receptors in its wall respond to chemical signals and to stretch, initiating the enterogastric reflex and the hormonal (enterogastrone) mechanisms that inhibit acid and pepsin secretion as we described earlier. These mechanisms also prevent further duodenal filling by reducing the force of pyloric contractions (**Figure 22.22**).

A carbohydrate-rich meal moves through the duodenum rapidly, but fats form an oily layer at the top of the chyme and are digested more slowly by enzymes acting in the intestine. For this reason, when chyme entering the duodenum is fatty, reflexes slow stomach emptying, and food may remain in the stomach six hours or more.

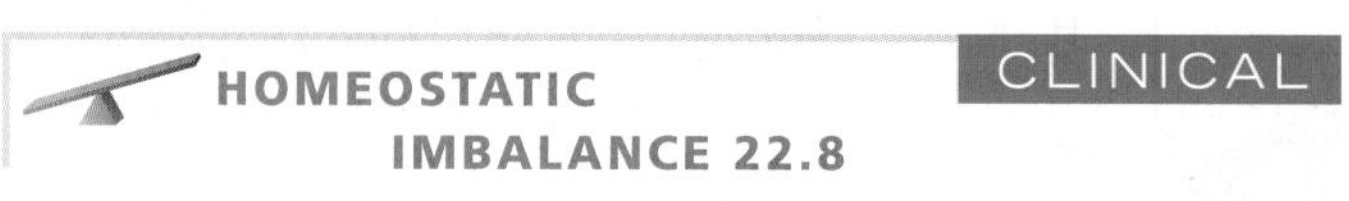

HOMEOSTATIC IMBALANCE 22.8

Vomiting, or **emesis**, is an unpleasant experience that empties the stomach by a different route. Many factors signal the stomach to "launch lunch," but the most common are extreme stretching of the stomach or intestine or irritants such as bacterial toxins, excessive alcohol, spicy foods, and certain drugs.

Bloodborne molecules and sensory impulses stream from the irritated sites to the **emetic center** (e-met′ik) of the medulla where they initiate a number of motor responses. Before vomiting, an individual typically feels nauseated, is pale, and salivates excessively. A deep inspiration directly precedes vomiting. The diaphragm and abdominal wall muscles contract, increasing intra-abdominal pressure, the gastroesophageal sphincter relaxes, and the soft palate rises to close off the nasal passages. As

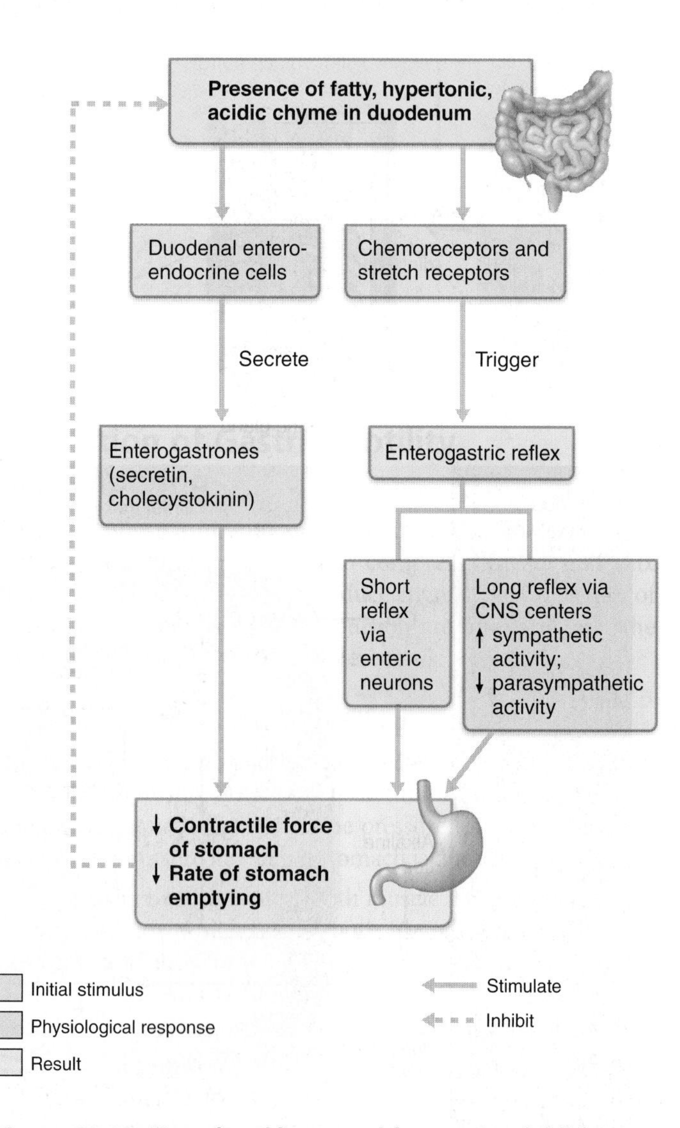

Figure 22.22 Neural and hormonal factors that inhibit gastric emptying. These controls ensure that the food is well liquefied in the stomach and prevent the small intestine from being overwhelmed.

a result, the stomach (and perhaps duodenal) contents are forced upward through the esophagus and pharynx and out the mouth.

Excessive vomiting can cause dehydration and severely disrupt the body's electrolyte and acid-base balance. Since large amounts of HCl are lost in vomitus, the blood becomes alkaline as the stomach attempts to replace its lost acid. +

Check Your Understanding

21. What structural modification of the stomach wall underlies the stomach's ability to mechanically break down food?
22. Two substances secreted by cells of the gastric glands are needed to produce the active protein-digesting enzyme pepsin. What are these substances and which cells secrete them?
23. Name the three phases of gastric secretion.
24. How does the presence of food in the small intestine inhibit gastric secretion and motility?

For answers, see Answers Appendix.

22.7 The liver secretes bile; the pancreas secretes digestive enzymes

Learning Objectives

- **Describe the histologic anatomy of the liver and pancreas.**
- **State the roles of bile and pancreatic juice in digestion.**
- **Describe the role of the gallbladder.**
- **Describe how bile and pancreatic juice secretion into the small intestine are regulated.**

The *liver*, *gallbladder*, and *pancreas* are accessory organs associated with the small intestine. We will take a side trip to these organs before we continue our journey down the digestive tract.

The liver has many metabolic and regulatory roles. However, its *digestive* system function is to produce bile for export to the duodenum (the first part of the small intestine). Bile is a fat emulsifier that breaks fats into tiny particles to make them more readily digestible. Although the liver also processes nutrient-laden venous blood delivered to it from the digestive organs, this is a metabolic rather than a digestive role (see Chapter 23). The gallbladder is chiefly a storage organ for bile. The pancreas supplies most of the enzymes that digest chyme as well as bicarbonate that neutralizes stomach acid.

The Liver

Gross Anatomy of the Liver

The ruddy, blood-rich **liver** is the largest gland in the body, weighing about 1.4 kg (3 lb) in the average adult. Shaped like a wedge, it occupies most of the right hypochondriac and epigastric regions (see Figure 1.12), extending farther to the right of the body midline than to the left. Located under the diaphragm, the liver lies almost entirely within the rib cage, which provides some protection (see Figure 22.1 and **Figure 22.23**).

The liver has four primary lobes. The largest, the *right lobe*, is visible on all liver surfaces and separated from the smaller *left lobe* by a deep fissure (Figure 22.23a). The posteriormost

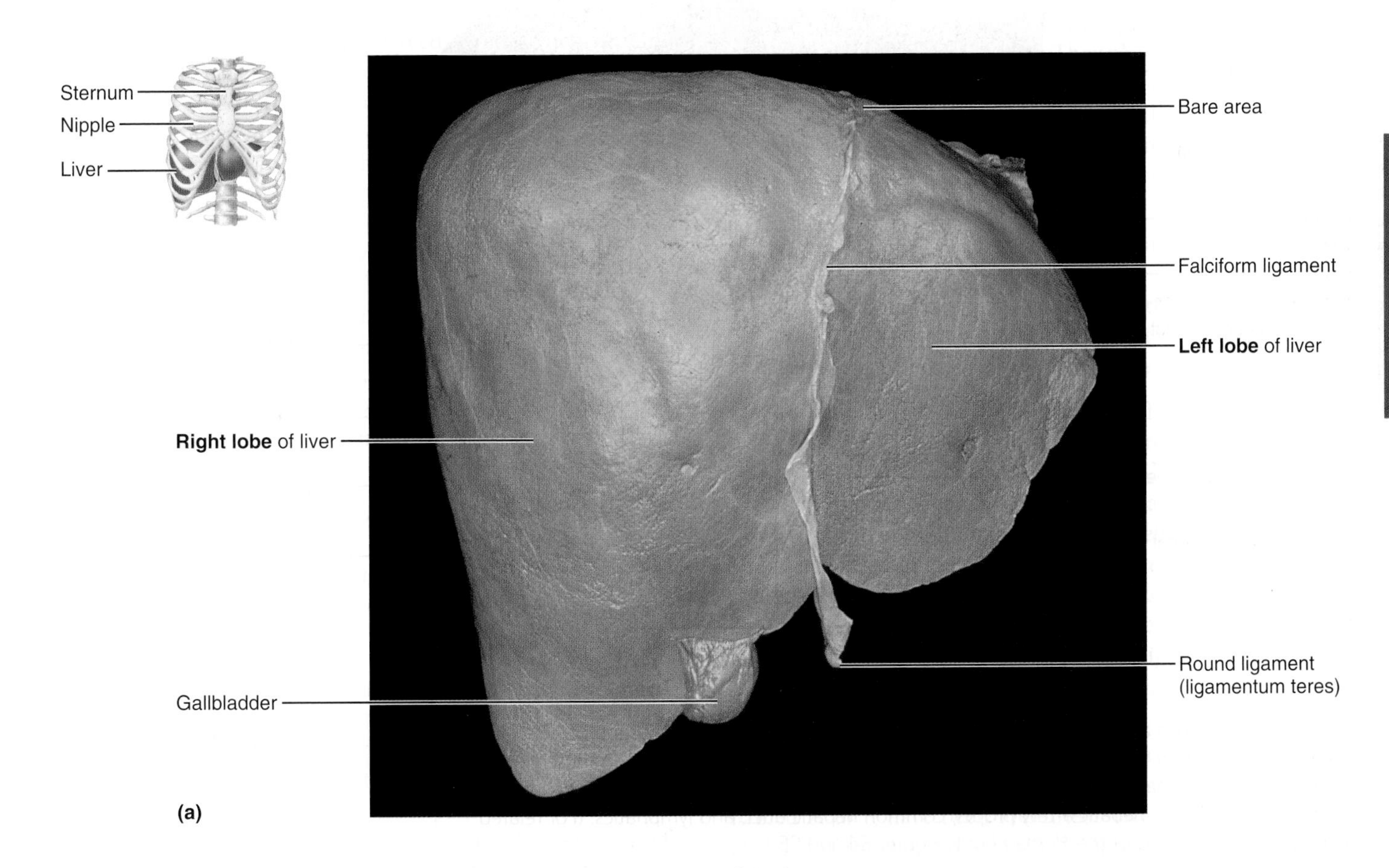

Figure 22.23 Gross anatomy of the human liver. (a) Anterior view of the liver.

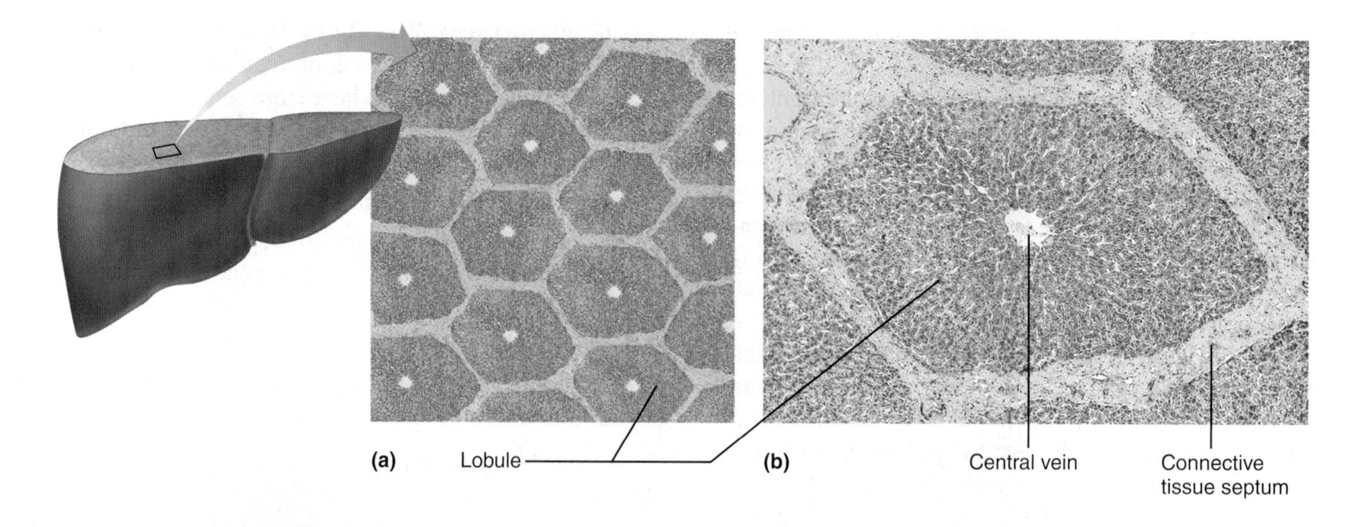

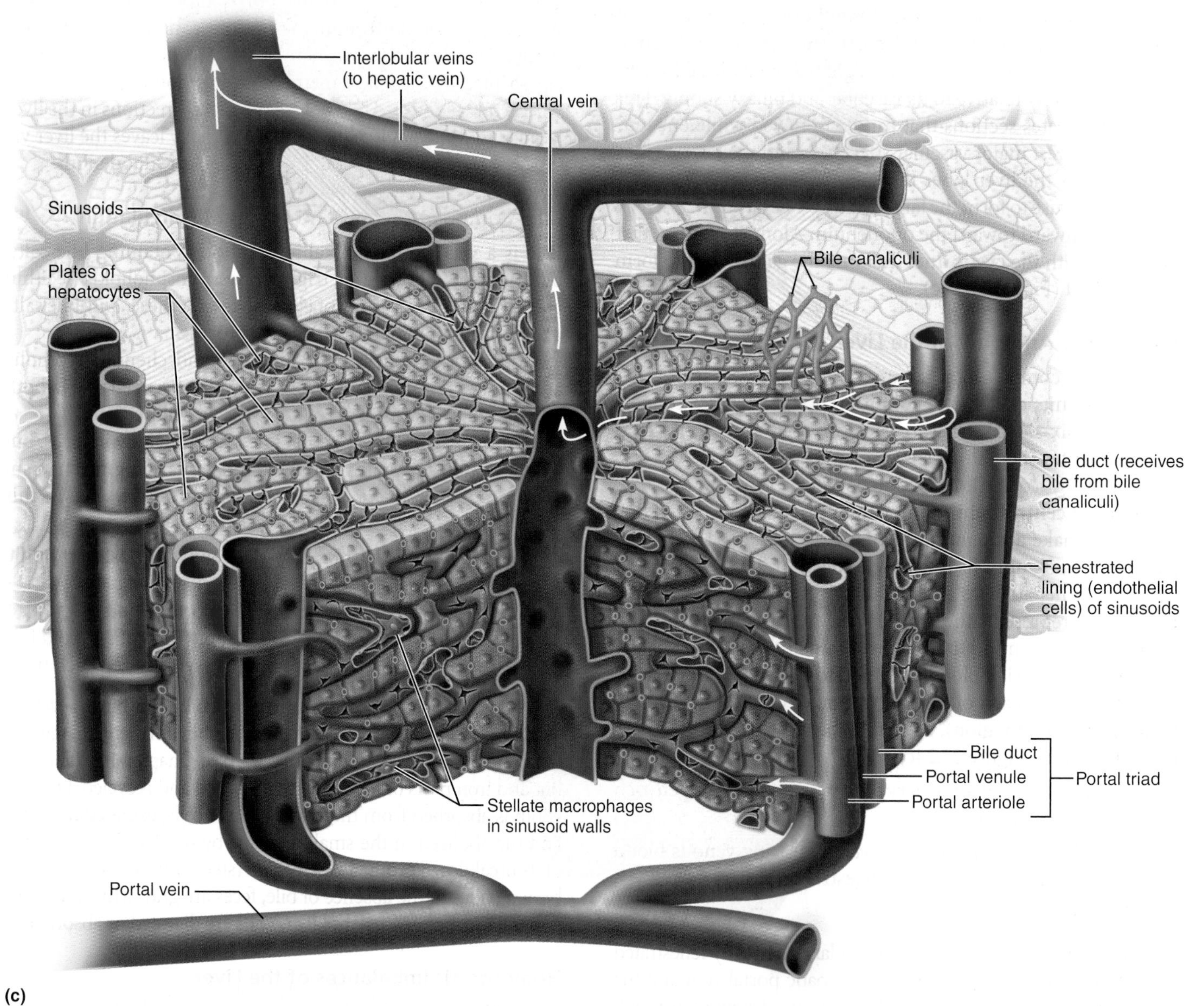

Figure 22.24 Microscopic anatomy of the liver. (a) Classic lobular pattern of a pig liver. **(b)** Enlarged view of one liver lobule. **(c)** Three-dimensional representation of a small portion of one liver lobule, showing the structure of sinusoids. Arrows indicate the direction of blood flow.

View histology slides
MasteringA&P®>Study Area>PAL

22

self-limiting. Those transmitted via blood—most importantly HVB and HVC—are linked to chronic hepatitis, liver cirrhosis, and cancer. Nonviral causes of acute hepatitis include alcohol- and drug-induced toxicity, and wild mushroom poisoning.

Hepatitis C has emerged as the most important infectious liver disease in the United States because it produces persistent or chronic liver infections (as opposed to acute infections). More than 4 million Americans are infected and over 10,000 die annually due to sequels of HVC infection. However, the life-threatening C form of hepatitis is now being successfully treated by a 12-week combination drug therapy consisting of sofosbuvir, ribavirin, and sometimes interferon.

Outpacing even hepatitis C and alcohol-associated liver damage, **non-alcoholic fatty liver disease (NAFLD)** has become the most common liver disease in North America. It affects about 30% of the general population, but 70% of the obese. Obesity and increased insulin resistance (in which a greater than normal amount of insulin is required to maintain normal blood glucose levels) are associated with abnormal lipid metabolism and liver inflammation, which cause NAFLD. There are usually no symptoms associated with NAFLD, but it predisposes the patient to develop full-blown cirrhosis or even liver cancer.

Cirrhosis (sĭr-ro′sis; "orange colored") is the last stage of progressive chronic inflammation of the liver. It typically results from severe chronic hepatitis due to chronic alcoholism, NAFLD, or infectious hepatitis. While damaged hepatocytes can regenerate, the liver's connective (scar) tissue regenerates faster. Liver activity is depressed and the liver becomes fibrous with scar tissue. The scar tissue obstructs blood flow throughout the hepatic portal system, causing **portal hypertension**.

Liver transplants are the only clinically proven effective treatment for patients with end-stage liver disease. The one- and five-year survival rate of such transplants is approximately 90% and 75%, respectively. The regenerative capacity of a healthy liver is exceptional. It can regenerate to its former size in 6–12 months even after surgical removal or loss of 80% of its mass. This means that part of a living donor's liver can be removed for transplant without long-term harm to the donor.

The Gallbladder

The **gallbladder** is a thin-walled green muscular sac about 10 cm (4 inches) long. The size of a kiwi fruit, it snuggles in a shallow fossa on the inferior surface of the liver (see Figures 22.1 and 22.23) from which its rounded fundus protrudes.

The gallbladder stores bile that is not immediately needed for digestion and concentrates it by absorbing some of its water and ions. When empty, its mucosa is thrown into honeycomb-like folds (see Figure 22.27) that, like the rugae of the stomach, allow the organ to expand as it fills. Its muscular wall contracts to expel bile into the *cystic duct*. From there bile flows into the bile duct. The gallbladder, like most of the liver, is covered by visceral peritoneum.

CLINICAL

Bile is the major vehicle for excreting cholesterol from the body, and bile salts keep the cholesterol dissolved within bile. Too much cholesterol or too few bile salts allows the cholesterol to crystallize, forming **gallstones** or *biliary calculi* (bil′e-a″re kal′ku-li), which obstruct the flow of bile from the gallbladder. When the gallbladder or its duct contracts, the sharp crystals cause agonizing pain that radiates to the right thoracic region.

Gallstones are easy to diagnose because they show up well with ultrasound imaging. Treatments for gallstones include dissolving the crystals with drugs, pulverizing them with ultrasound vibrations (lithotripsy), vaporizing them with lasers, and the classical treatment, surgically removing the gallbladder. When the gallbladder is removed, the bile duct enlarges to assume the bile-storing role.

Bile duct blockage prevents both bile salts and bile pigments from entering the intestine. As a result, yellow bile pigments accumulate in blood and eventually are deposited in the skin, causing it to become yellow, or *jaundiced*. Jaundice caused by blocked ducts is called *obstructive jaundice*, but jaundice may also reflect liver disease (in which the liver is unable to carry out its normal metabolic duties). ✚

The Pancreas

The **pancreas** (pan′kre-as; *pan* = all, *creas* = flesh, meat) is important to the digestive process because it produces enzymes that break down all categories of foodstuffs. The pancreas is a soft, tadpole-shaped gland that extends across the abdomen from its *tail* (next to the spleen) to its *head*, which is encircled by the C-shaped duodenum (see Figures 22.1 and 22.27). Most of the pancreas is retroperitoneal and lies deep to the greater curvature of the stomach.

The pancreas contains exocrine and endocrine parts. The exocrine part of the pancreas produces **pancreatic juice** and consists of the following (**Figure 22.25**):

- **Acini.** Acini (as′ĭ-ni; singular: acinus) are clusters of secretory acinar cells that produce the enzyme-rich component of pancreatic juice. Acinar cells are full of rough endoplasmic reticulum and exhibit deeply staining **zymogen granules** (zi′mo-jen; "fermenting"). These granules contain inactive digestive enzymes (proenzymes).
- **Ducts.** A system of ducts transports the secretions of the acinar cells. In addition, the epithelial cells of the smallest ducts secrete the water that makes up the bulk of the pancreatic juice and the bicarbonate that makes this secretion alkaline (about pH 8).

The endocrine part of the pancreas is a scattering of mini-endocrine glands called *pancreatic islets*. As we saw in Chapter 15, these islets release insulin and glucagon, hormones that play an important role in carbohydrate metabolism.

Composition of Pancreatic Juice

Approximately 1200 to 1500 ml of clear pancreatic juice is produced daily. It consists mainly of water, and contains enzymes and electrolytes (primarily bicarbonate ions). The high pH of pancreatic fluid helps neutralize acidic chyme entering the duodenum and provides the optimal environment for intestinal and pancreatic enzymes. The pancreatic enzymes include:

- **Proteases** (for proteins)

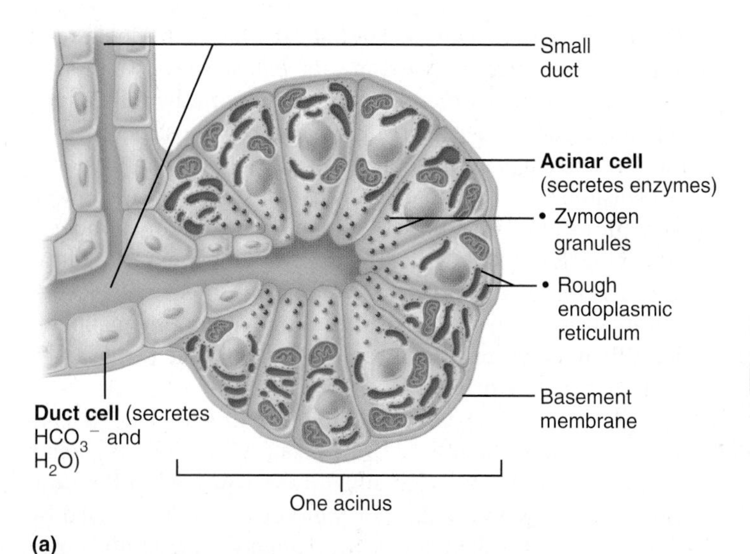

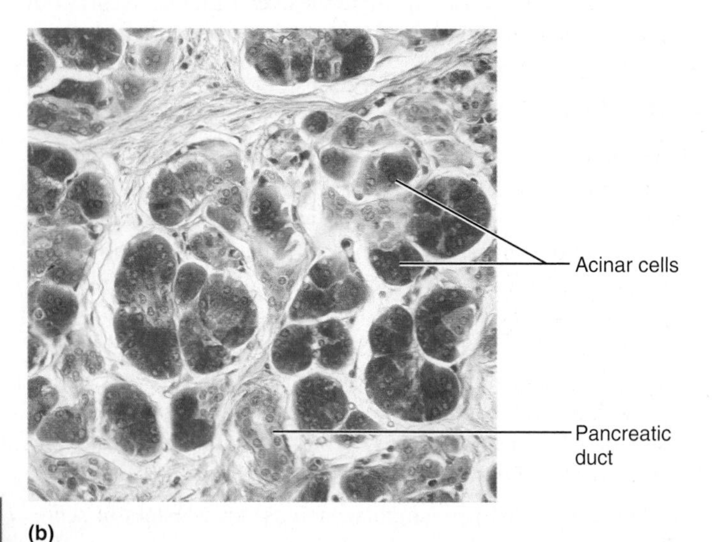

Figure 22.25 Structure of the enzyme-producing tissue of the pancreas. (a) Schematic view of one acinus (a secretory unit). The acinar cells contain abundant zymogen (enzyme-containing) granules and dark-staining rough ER (typical of gland cells producing large amounts of protein for export). **(b)** Photomicrograph of pancreatic acinar tissue (155×).

- **Amylase** (for starch)
- **Lipases** (for fats)
- **Nucleases** (for nucleic acids)

Like pepsin of the stomach, pancreatic proteases are produced and released in inactive forms that are activated in the duodenum, where they do their work. This protects the pancreas from digesting itself.

For example, within the duodenum, **enteropeptidase** (formerly called *enterokinase*), an enzyme bound to the plasma membrane of duodenal epithelial cells, activates *trypsinogen* to **trypsin**. Trypsin, in turn, activates more trypsinogen and two other pancreatic proteases (*procarboxypeptidase* and *chymotrypsinogen*) to their active forms, **carboxypeptidase** (kar-bok″se-pep′tĭ-dās) and **chymotrypsin** (ky″mo-trip′sin), respectively (**Figure 22.26**).

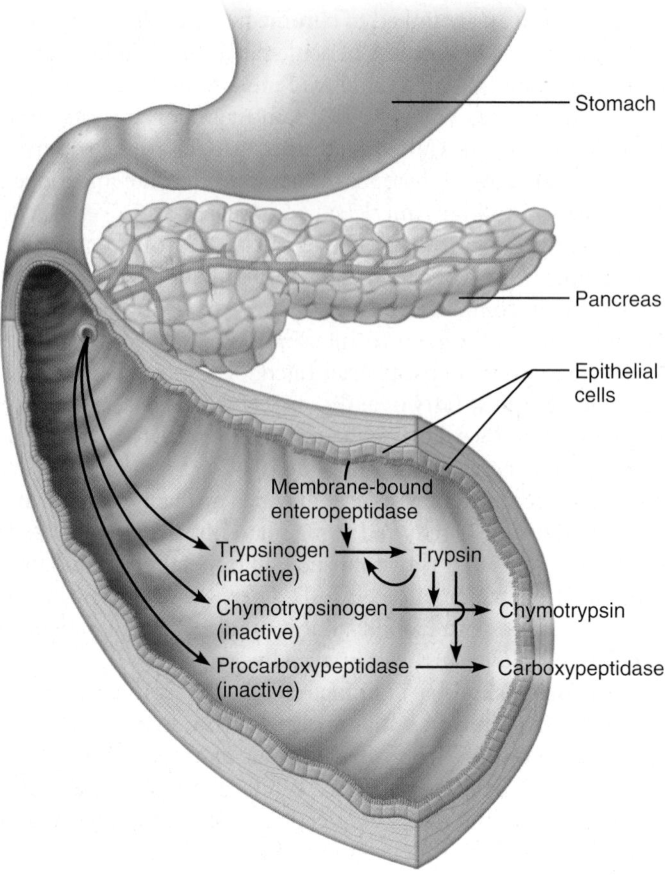

Figure 22.26 Activation of pancreatic proteases in the small intestine. Pancreatic proteases are secreted in an inactive form and are activated in the duodenum.

Bile and Pancreatic Secretion into the Small Intestine

Anatomy of Duct Systems

The bile duct, delivering bile from the liver, and the **main pancreatic duct**, carrying pancreatic juice from the pancreas, unite in the wall of the duodenum, the first section of the small intestine (**Figure 22.27**). They fuse together at a bulblike structure called the **hepatopancreatic ampulla** (hep″ah-to-pan″kre-at′ik am-pul′ah; *ampulla* = flask). The ampulla opens into the duodenum via the volcano-shaped **major duodenal papilla**. A smooth muscle valve called the **hepatopancreatic sphincter** controls the entry of bile and pancreatic juice. A smaller *accessory pancreatic duct* empties directly into the duodenum just proximal to the main duct.

Regulation of Bile and Pancreatic Secretion

Hormones and neural stimuli regulate both the secretion of bile and pancreatic juice and their release into the small intestine. The hormones include two *enterogastrones* that you are already familiar with—*cholecystokinin* and *secretin*. **Figure 22.28** summarizes the hormonal and neural mechanisms that control the secretion and release of bile and pancreatic juice.

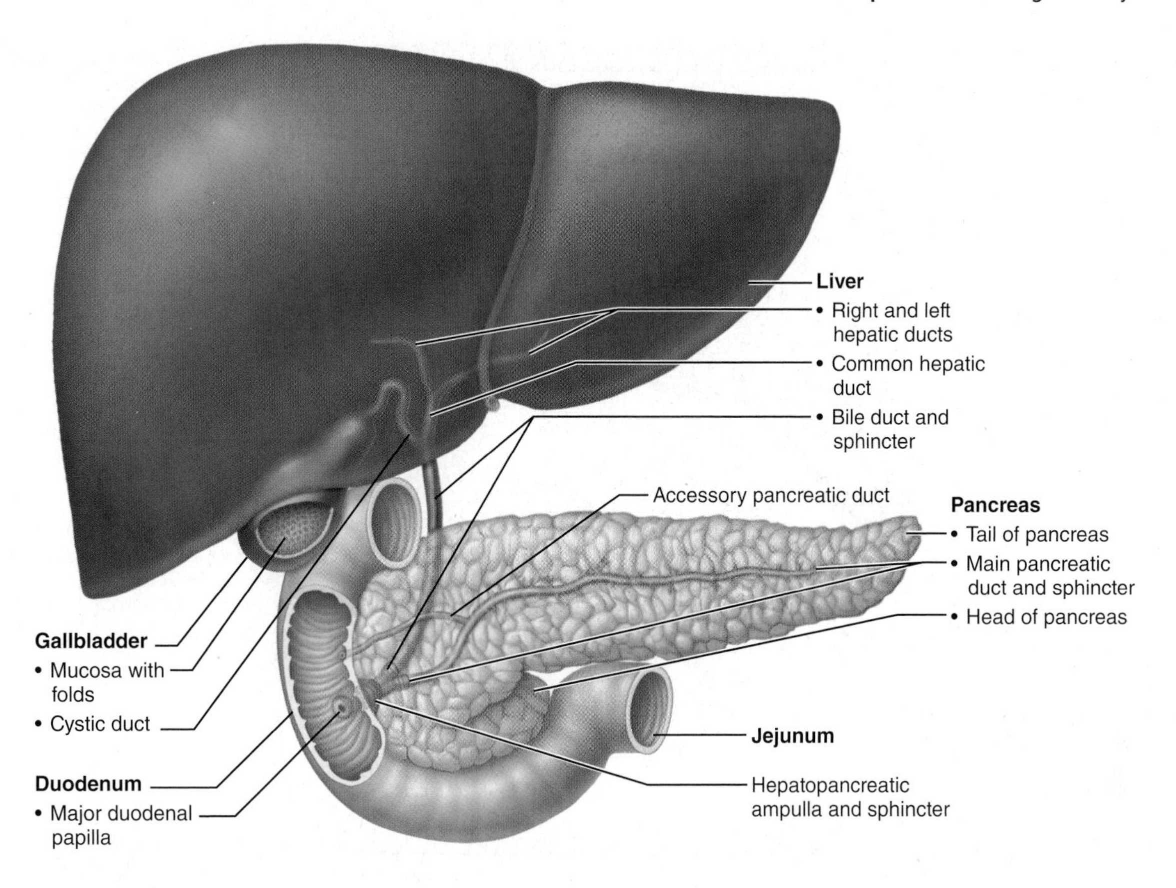

Figure 22.27 Relationship of the liver, gallbladder and pancreas to the duodenum. Ducts from the pancreas, gallbladder, and liver empty into the duodenum.

Bile salts themselves are the major stimulus for enhanced bile secretion (Figure 22.28). After a fatty meal, when the enterohepatic circulation is returning large amounts of bile salts to the liver, its output of bile rises dramatically. Secretin, released by intestinal cells exposed to fatty chyme, also stimulates liver cells to secrete bile.

When no digestion is occurring, the hepatopancreatic sphincter is closed and the released bile backs up the cystic duct into the gallbladder, where it is stored until needed. Although the liver makes bile continuously, bile does not usually enter the small intestine until the gallbladder contracts.

Check Your Understanding

25. What is a portal triad?
26. What is the importance of the enterohepatic circulation?
27. What is the functional difference between pancreatic acini and islets?
28. What is the makeup of the fluid in the pancreatic duct? In the cystic duct? In the bile duct?
29. What stimulates CCK release and what are its effects on the digestive process?

For answers, see Answers Appendix.

22.8 The small intestine is the major site for digestion and absorption

→ Learning Objectives

- ☐ **Identify and describe structural modifications of the wall of the small intestine that enhance the digestive process.**
- ☐ **Differentiate between the roles of the various cell types of the intestinal mucosa.**
- ☐ **Describe the functions of intestinal hormones and paracrines.**

The **small intestine** is the body's major digestive organ. Within its twisted passageway, digestion is completed (with the help of bile and pancreatic enzymes) and virtually all absorption occurs.

Gross Anatomy

The small intestine is a convoluted tube extending from the pyloric sphincter to the **ileocecal valve (sphincter)** (il″e-o-se′kal) where it joins the large intestine. It is the longest part of the alimentary canal, but is only about half the diameter of the large intestine, ranging from 2.5 to 4 cm (1–1.6 inches). Although

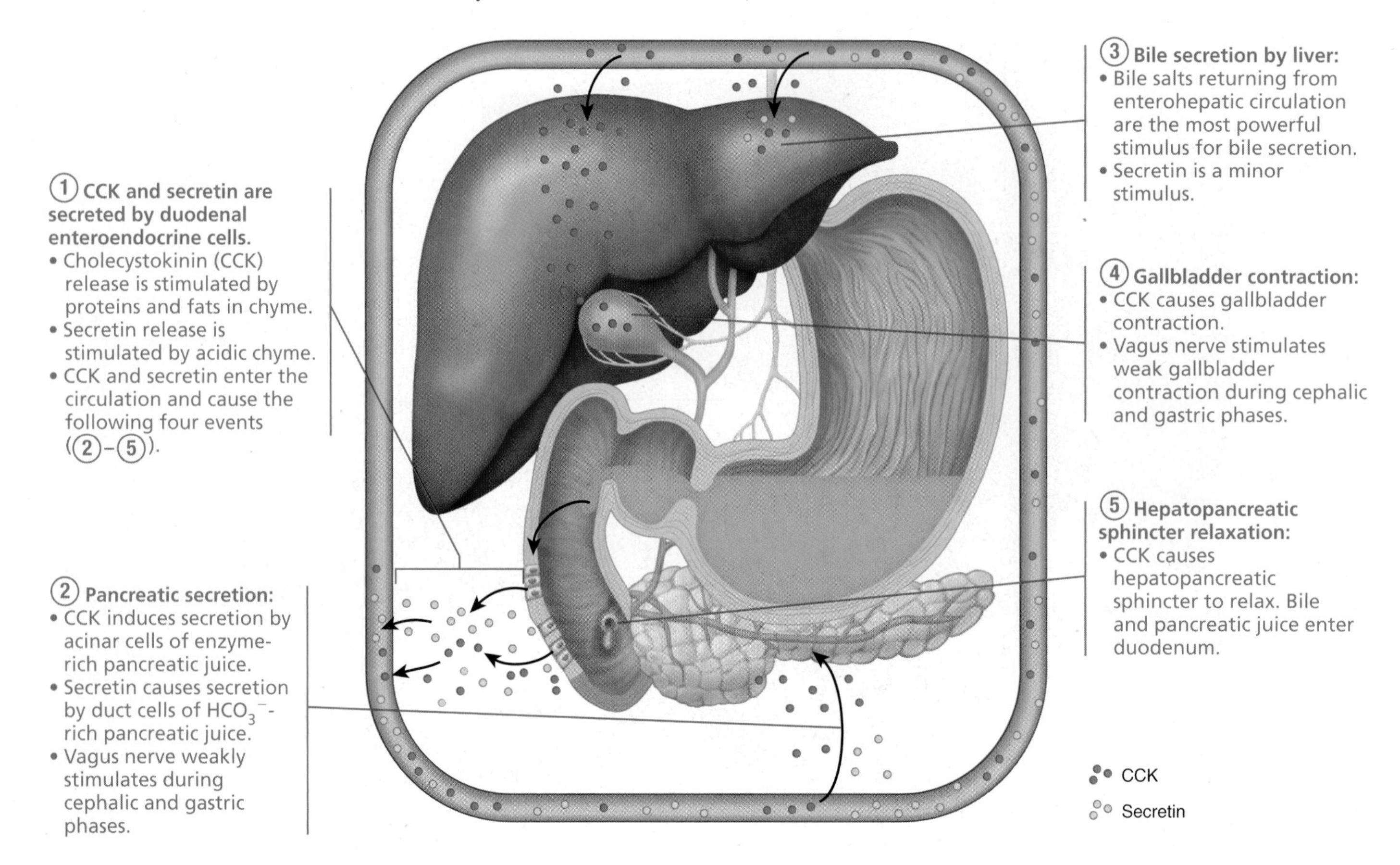

Figure 22.28 Mechanisms promoting secretion and release of bile and pancreatic juice.

6–7 m long (approximately 20 ft) in a cadaver, the small intestine is only 2–4 m (7–13 ft) long during life because of muscle tone.

The small intestine has three subdivisions: the duodenum, which is mostly retroperitoneal, and the jejunum and ileum, both intraperitoneal organs (see Figure 22.1). The relatively immovable **duodenum** (du″o-de′num; "twelve finger widths long"), which curves around the head of the pancreas, is about 25 cm (10 inches) long (Figure 22.27). Although it is the shortest intestinal subdivision, the duodenum has the most features of interest, including the *major duodenal papilla* mentioned earlier.

The **jejunum** (jĕ-joo′num; "empty"), about 2.5 m (8 ft) long, extends from the duodenum to the ileum. The **ileum** (il′e-um; "twisted"), approximately 3.6 m (12 ft) in length, joins the large intestine at the ileocecal valve. The jejunum and ileum hang in sausagelike coils in the central and lower part of the abdominal cavity, suspended from the posterior abdominal wall by a fan-shaped *mesentery* (see Figure 22.32). The large intestine encircles these more distal parts of the small intestine.

The arterial supply of the small intestine is primarily from the superior mesenteric artery (pp. 646–647). The veins parallel the arteries and typically drain into the superior mesenteric vein. From there, the nutrient-rich venous blood from the small intestine drains into the hepatic portal vein, which carries it to the liver.

Nerve fibers serving the small intestine include parasympathetics from the vagus and sympathetics from the thoracic splanchnic nerves, both relayed through the superior mesenteric (and celiac) plexus.

Microscopic Anatomy

Modifications of the Small Intestine for Absorption

The small intestine is highly adapted for absorbing nutrients. Its length alone provides a huge surface area, and its wall has three structural modifications—circular folds, villi, and microvilli—that amplify its absorptive surface enormously (by a factor of more than 600 times). In fact, the intestinal surface area is about equal to 200 square meters, the size of a singles tennis court!

- The **circular folds** are deep, permanent folds of the mucosa and submucosa (**Figure 22.29a**). Nearly 1 cm tall, these folds force chyme to spiral through the lumen, slowing its movement and allowing time for full nutrient absorption.
- **Villi** (vil′i; "tufts of hair") are fingerlike projections of the mucosa, over 1 mm high, that give it a velvety texture, much like the soft nap of a towel (Figure 22.29). The villi are large and leaflike in the duodenum (the intestinal site of most active absorption) and gradually narrow and shorten along the length of the small intestine. In the core of each villus is a dense capillary bed and a wide lymphatic capillary called a **lacteal** (lak′te-al). Digested foodstuffs are absorbed through the epithelial cells into both the capillary blood and the lacteal.
- **Microvilli** are long, densely packed cytoplasmic extensions of the absorptive cells of the mucosa that give the mucosal surface a fuzzy appearance called the **brush border** (Figure 22.29b enlargement and **Figure 22.30**). The plasma membranes

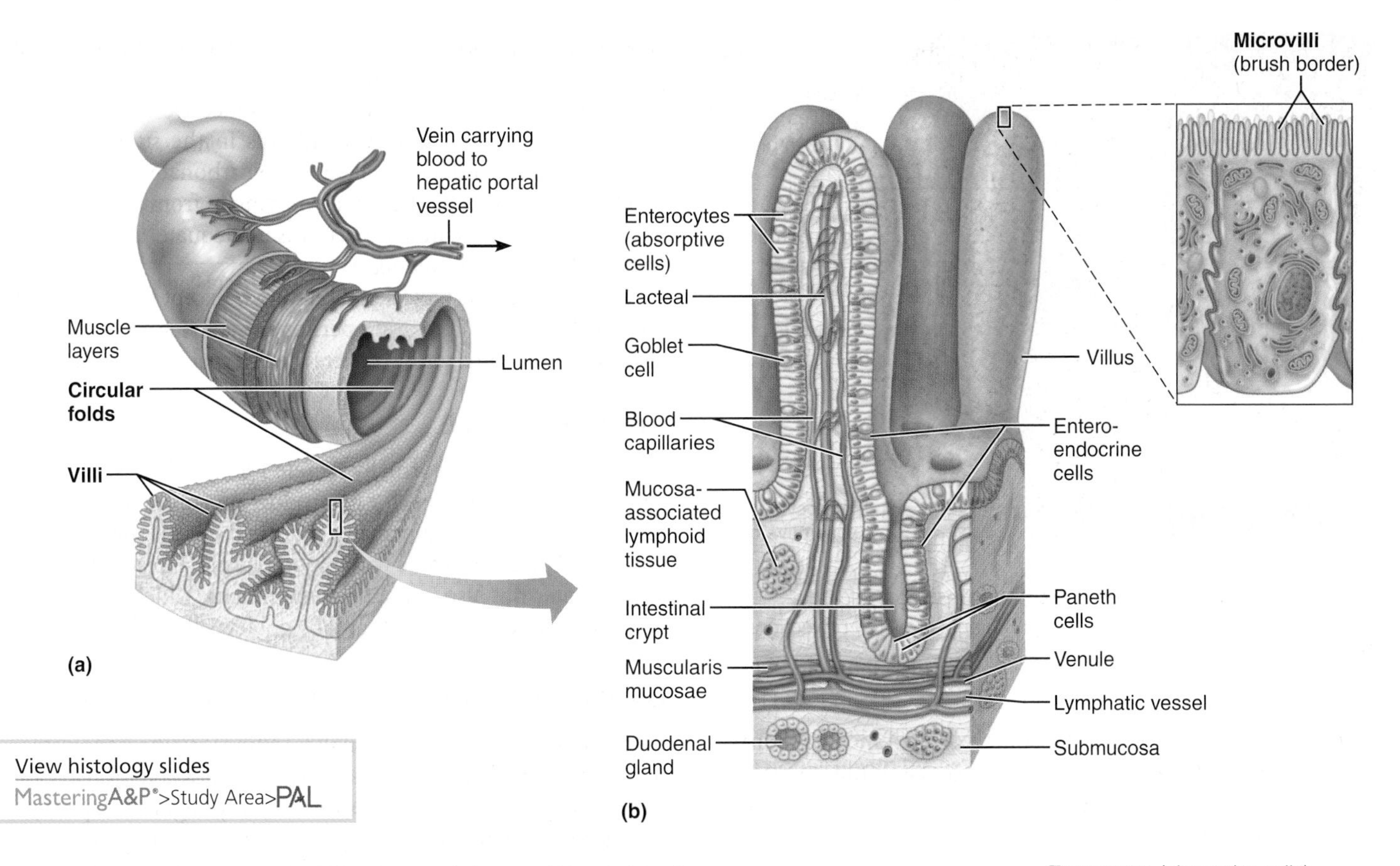

View histology slides
MasteringA&P®>Study Area>PAL

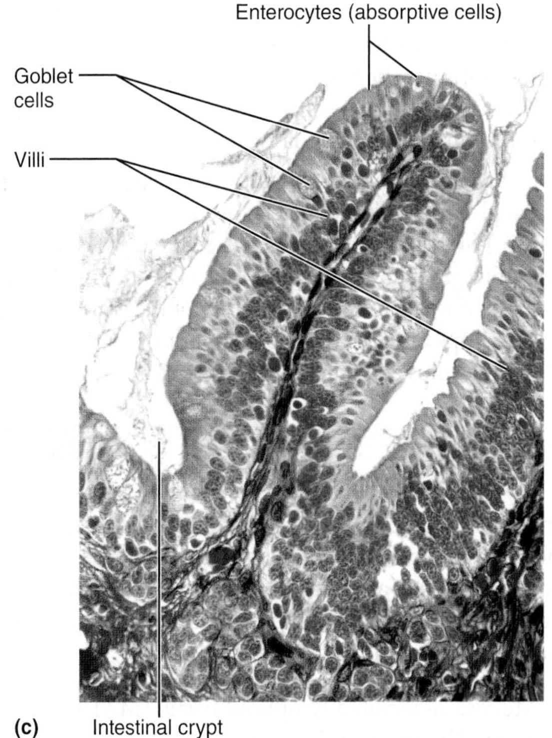

Figure 22.29 Structural modifications of the small intestine that increase its surface area for digestion and absorption. (a) Enlargement of a few circular folds, showing associated fingerlike villi (muscularis and serosa layers not indicated). **(b)** Structure of a villus. Enlargement shows absorptive enterocytes that exhibit microvilli on their free (apical) surface. **(c)** Photomicrograph of the mucosa, showing villi (250×). (For a related image, see *A Brief Atlas of the Human Body*, Figure 69b.)

of the microvilli bear enzymes referred to as **brush border enzymes**, which complete the digestion of carbohydrates and proteins in the small intestine.

Histology of the Small Intestine Wall

The four layers typical of the GI tract are also seen in the small intestine, but the mucosa and submucosa are modified to reflect the intestine's functions in the digestive pathway.

Between the villi, which are specialized for absorption, the small intestine mucosa is studded with tubular glands called **intestinal crypts** (see Figure 22.29b, c). The crypts decrease in number along the length of the small intestine.

Five major types of cells are found in the mucosal epithelium of the villi and crypts:

- *Enterocytes* form the bulk of the epithelium. They are simple columnar absorptive cells bound by tight junctions and richly endowed with microvilli. These cells bear the primary responsibility for absorbing nutrients and electrolytes in the villi. In the crypts, enterocytes are primarily secretory cells that secrete *intestinal juice*, a watery mixture that contains mucus and serves as a carrier fluid for absorbing nutrients from chyme.
- *Goblet cells* are mucus-secreting cells found in the epithelia of the villi and crypts.

The **large intestine** frames the small intestine on three sides and extends from the ileocecal valve to the anus (see Figure 22.1). Its diameter, at about 7 cm, is greater than that of the small intestine, but it is much shorter (1.5 m versus 6 m). Its major digestive functions are to absorb most of the remaining water from indigestible food residues, store the residues temporarily, and then eliminate them from the body as semisolid **feces** (fe′sēz), also called *stool.* It also absorbs metabolites produced by resident bacteria as they ferment carbohydrates not absorbed in the small intestine.

Gross Anatomy

The large intestine exhibits three features not seen elsewhere—teniae coli, haustra, and epiploic appendages. Except for its terminal end, the longitudinal muscle layer of its muscularis is mostly reduced to three bands of smooth muscle called **teniae coli** (ten′ne-e ko′li; "ribbons of the colon"). Their tone puckers the wall of the large intestine into pocketlike sacs called **haustra** (haw′strah; "to draw up"; singular: *haustrum*). Another obvious feature of the large intestine is its **epiploic appendages** (ep″ĭ-plo′ik; "membrane covered"), which are small fat-filled pouches of visceral peritoneum that hang from the surface of the large intestine (**Figure 22.31a**). Their significance is not known.

Subdivisions of the Large Intestine

The large intestine has the following subdivisions: cecum, appendix, colon, rectum, and anal canal. The saclike **cecum** (se′kum; "blind pouch"), which lies below the ileocecal valve in the right iliac fossa, is the first part of the large intestine (Figure 22.31a).

Attached to the posteromedial surface of the cecum is the blind, wormlike **appendix**. The appendix contains masses of lymphoid tissue, and as part of MALT (see p. 668) it plays an important role in body immunity. Additionally, it serves as a storehouse of bacteria and recolonizes the gut when needed. However, the appendix has an important structural shortcoming—its twisted structure makes it susceptible to blockage.

HOMEOSTATIC IMBALANCE 22.11 CLINICAL

Acute inflammation of the appendix, or **appendicitis**, results from a blockage (often by feces) that traps infectious bacteria in its lumen. Unable to empty its contents, the appendix swells, squeezing off venous drainage, which may lead to ischemia and necrosis (low blood flow and tissue death) of the appendix. If the appendix ruptures, feces containing bacteria spray over the abdominal contents, causing *peritonitis.*

The symptoms of appendicitis vary, but the first symptom is usually pain in the umbilical region. Loss of appetite, nausea and vomiting, and pain relocalization to the lower right abdominal quadrant follow. Immediate surgical removal of the appendix (appendectomy) is the accepted treatment. Appendicitis is most common during adolescence, when the entrance to the appendix is at its widest. +

The **colon** has several distinct regions. Proximally, as the **ascending colon**, it travels up the right side of the abdominal cavity to the level of the right kidney. Here it makes a right-angle turn—the **right colic (hepatic) flexure**—and travels across the abdominal cavity as the **transverse colon**. Directly anterior to the spleen, it bends acutely at the **left colic (splenic) flexure** and descends down the left side of the posterior abdominal wall as the **descending colon**. Inferiorly, it enters the pelvis, where it becomes the S-shaped **sigmoid colon**.

In the pelvis, at the level of the third sacral vertebra, the sigmoid colon joins the **rectum**, which runs posteroinferiorly just in front of the sacrum.

Despite its name (*rectum* = straight), the rectum has three lateral curves or bends, represented internally as three transverse folds called **rectal valves** (Figure 22.31b). These valves stop feces from being passed along with gas (flatus).

The **anal canal**, the last segment of the large intestine, lies in the perineum, entirely external to the abdominopelvic cavity. About 3 cm long, it begins where the rectum penetrates the levator ani muscle of the pelvic floor and opens to the body exterior at the **anus**. The anal canal has two sphincters, an involuntary **internal anal sphincter** composed of smooth muscle (part of the muscularis), and a voluntary **external anal sphincter** composed of skeletal muscle. The sphincters, which act rather like purse strings to open and close the anus, are ordinarily closed except during defecation.

The rectum and anal canal lack teniae coli and haustra. However, the rectum's muscularis muscle layers are complete and well developed, consistent with its role in generating strong contractions to expel feces.

Relationship of the Large Intestine to the Peritoneum

The cecum, appendix, and rectum are all retroperitoneal. The colon is also retroperitoneal, except for its transverse and sigmoid parts. These parts are intraperitoneal and anchored to the posterior abdominal wall by mesentery sheets called **mesocolons** (**Figure 22.32c, d**).

Microscopic Anatomy

The wall of the large intestine differs in several ways from that of the small intestine. The large intestine *mucosa* is simple columnar epithelium except in the anal canal. Because most food is absorbed before reaching the large intestine, there are no circular folds, villi, or brush border. However, its mucosa is thicker, its abundant crypts are deeper, and the crypts contain tremendous numbers of goblet cells. Mucus produced by goblet cells eases the passage of feces and protects the intestinal wall from irritating acids and gases released by resident bacteria.

The mucosa of the anal canal, a stratified squamous epithelium, merges with the true skin surrounding the anus and is quite different from the mucosa in the rest of the colon, reflecting the greater abrasion that this region receives. Superiorly, it hangs in long ridges or folds called **anal columns. Anal sinuses**, recesses between the anal columns, exude mucus when compressed by feces, which aids in emptying the anal canal (Figure 22.31b).

The horizontal, tooth-shaped line that parallels the inferior margins of the anal sinuses is called the *pectinate line*. Superior to this line, visceral sensory fibers innervate the mucosa, which is relatively insensitive to pain. The area inferior to

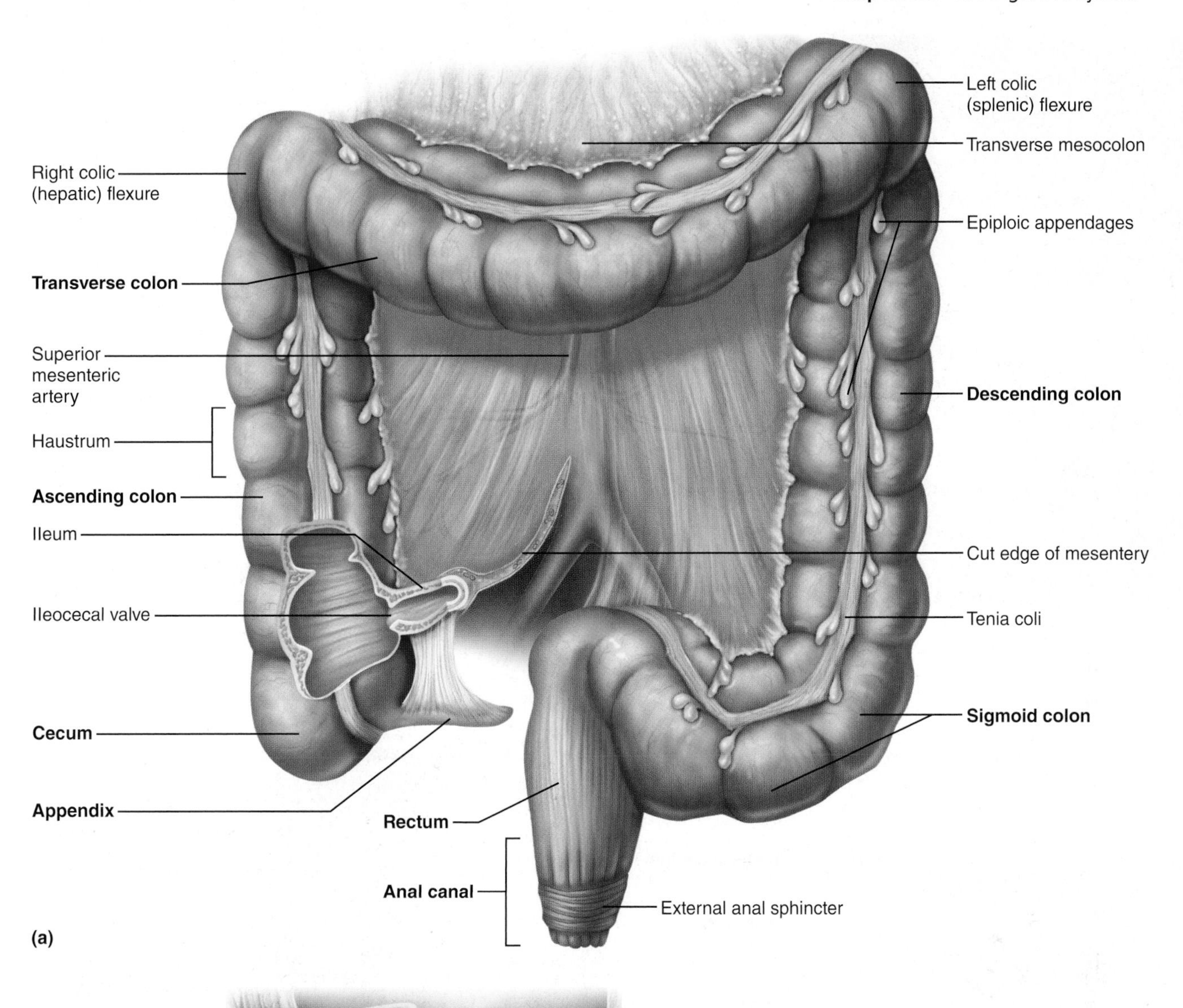

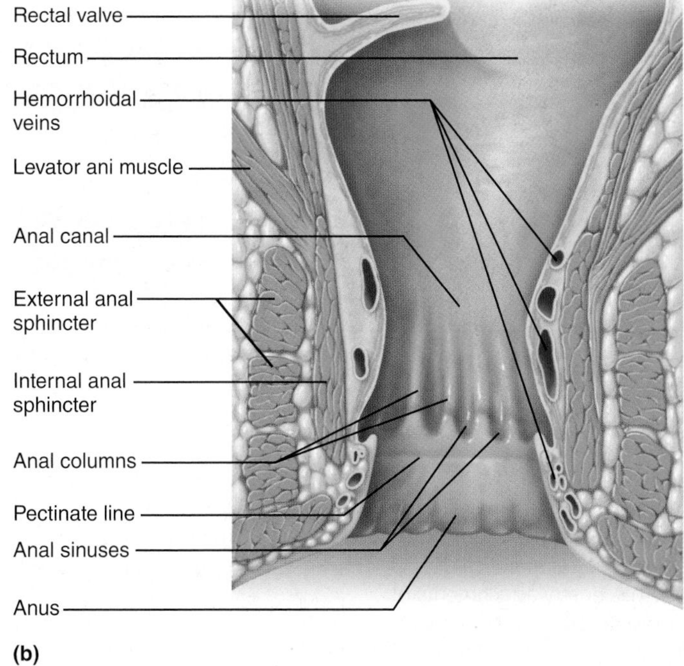

22

Practice art labeling
MasteringA&P®>Study Area>Chapter 22

Figure 22.31 Gross anatomy of the large intestine. **(a)** Diagrammatic view. **(b)** Structure of the anal canal.

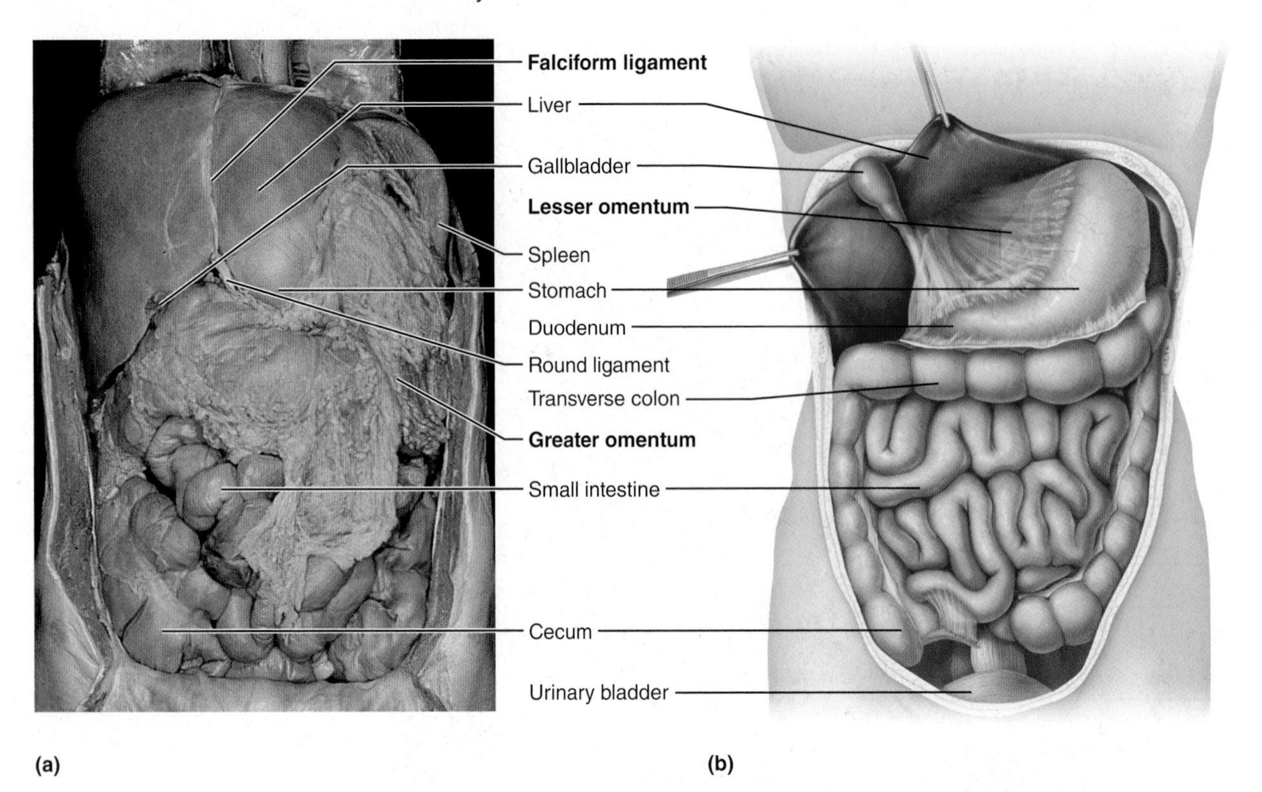

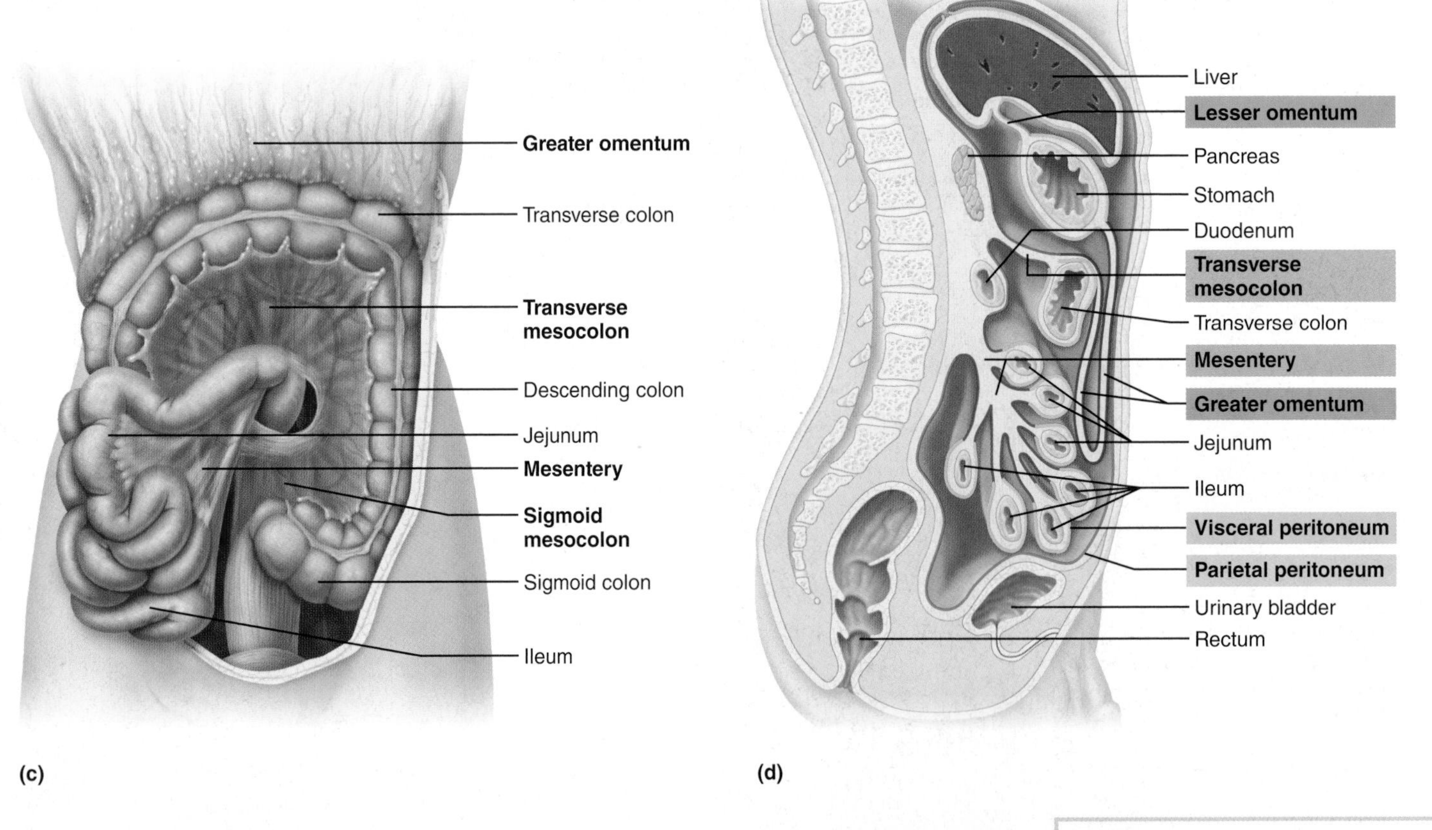

22

Figure 22.32 Mesenteries of the abdominal digestive organs. (a) The greater omentum, a dorsal mesentery, is shown in its normal position covering the abdominal viscera. **(b)** The liver and gallbladder have been reflected superiorly to reveal the lesser omentum, a ventral mesentery attaching the liver to the lesser curvature of the stomach. **(c)** The greater omentum has been reflected superiorly to reveal the mesentery attachments of the small and large intestine. **(d)** Sagittal section of the abdominopelvic cavity of a male. Mesentery labels appear in colored boxes.

Explore human cadaver
MasteringA&P®>Study Area>PAL

the pectinate line is very sensitive to pain, a reflection of the somatic sensory fibers serving it.

Two superficial venous plexuses are associated with the anal canal, one with the anal columns and the other with the anus itself. If these (hemorrhoidal) veins become dilated and inflamed, itchy varicosities called *hemorrhoids* result.

Bacterial Flora

The **bacterial flora** of the large intestine consists of over a thousand different types of bacteria. They outnumber the rest of our body's cells by 10 to 1 and account for a couple of pounds of our body weight. Some of these bacteria colonize the colon via the anus, but others enter from the small intestine still "alive and kicking" after running the gauntlet of antimicrobial defenses (lysozyme, defensins, HCl, and protein-digesting enzymes). We provide a home for these bacteria, but what's in it for us? We are just beginning to understand that we depend upon these bacteria just as much as they depend upon us.

Metabolic Functions

Our gut bacteria help us by recovering energy from otherwise indigestible foods and synthesizing some vitamins.

- **Fermentation.** Gut bacteria ferment some of the indigestible carbohydrates and mucin in gut mucus. The resulting short-chain fatty acids can be absorbed and used for fuel by the body's cells. Unfortunately, fermentation also produces a mixture of gases (including dimethyl sulfide, H_2, N_2, CH_4, and CO_2). Some of these gases, such as dimethyl sulfide, are quite odorous (smelly). About 500 ml of gas (flatus) is produced each day, much more when we eat foods (such as beans) rich in indigestible carbohydrates.
- **Vitamin synthesis.** B complex vitamins and some of the vitamin K the liver needs to produce several clotting proteins are synthesized by gut bacteria.

Keeping Pathogenic Bacteria in Check

The immune system and the gut flora live in a dynamic equilibrium. The immune system destroys any bacteria that threaten to breach the mucosal barrier. The gut bacteria, on the other hand, instruct the immune system not to overreact to their presence in the lumen.

Potentially harmful bacteria in our large intestine are kept in check in two ways. First, beneficial bacteria out-compete and actively suppress harmful bacteria, and as a result normally vastly outnumber them. Second, our immune system prevents bacteria from entering the body through the gut epithelium. An elegant system keeps the bacteria from breaching the mucosal barrier. Dendritic cells sample the microbial antigens in the lumen. They then migrate to the nearby lymphoid follicles within the gut mucosa (MALT) and trigger an IgA antibody–mediated response restricted to the gut lumen. This prevents the bacteria from straying into tissues deep to the mucosa where they might elicit a much more widespread systemic response.

Clostridium difficile, an anaerobic bacterium, is the most common cause of *antibiotic-associated diarrhea*, accounting for 14,000 deaths per year in the U.S. Where does it come from? For some people, *C. difficile* is a normal, but small, fraction of the gut's bacteria. Other people acquire *C. difficile* through the fecal-oral route (poor hand washing), particularly in hospital or long-term care settings. In either case, when other bacteria are wiped out by antibiotics, *C. difficile* flourishes in the gut and may cause pseudomembranous colitis (inflammation of the colon) that leads to bowel perforation and sepsis.

Because *C. difficile* infections are resistant to many antibiotics, they are notoriously difficult to treat and often recur. Instead of using ever more powerful antibiotics, a new treatment strategy seeks to restore competitive bacteria to the gut's ecosystem by perfoming a *fecal transplant*. Transferring fecal bacteria from an uninfected donor to the patient (e.g., by enema or during colonoscopy) cures *C. difficile* infections 90–100% of the time. In spite of the considerable "yuck factor," this treatment is becoming more and more mainstream. It is likely that optimized mixtures of cultured bacteria will become available for transplantation to treat *C. difficile* in the future. +

While the immune system keeps the gut bacteria in check, the gut bacteria also profoundly shape our immune system responses. For example, the type of bacteria present influences the balance between subtypes of T cells, and so affects the balance between pro- and anti-inflammatory responses.

The coexistence of enteric bacteria with our immune system does sometimes fail. When that happens, the painful and debilitating condition known as inflammatory bowel disease may result.

Gut Bacteria in Health and Disease

We are only just beginning to learn how gut bacteria affect the body. There is mounting evidence that the kinds and proportions of the bacteria in our gut can influence our body weight, our susceptibility to various diseases (including diabetes, atherosclerosis, fatty liver disease), and even our mood. Manipulating our various gut bacteria may become a routine health-care strategy in the near future.

Digestive Processes in the Large Intestine

What is finally delivered to the large intestine contains few nutrients, but it still has 12 to 24 hours more to spend there. Except for a small amount of digestion of that residue by the enteric bacteria, no further food breakdown occurs in the large intestine.

The large intestine harvests vitamins made by the bacterial flora and reclaims most of the remaining water and some of the electrolytes (particularly sodium and chloride). However, nutrient absorption is not its *major* function. As mentioned, the primary concerns of the large intestine are propulsive activities that force fecal material toward the anus and eliminate it from the body (defecation).

HOMEOSTATIC IMBALANCE 22.13 CLINICAL

The large intestine is important for our comfort, but it is not essential for life. If the colon is removed, the terminal ileum can be brought out to the abdominal wall in a procedure called an *ileostomy* (il″e-os′to-me). From there food residues are eliminated into a sac attached to the abdominal wall. ✚

Motility of the Large Intestine

When food residue enters the colon through the ileocecal valve, the colon becomes motile, but its contractions are sluggish or short-lived compared to those of the small intestine. The movements most seen in the colon are **haustral contractions**, slow segmenting movements that last about one minute and occur every 30 minutes or so.

These contractions, which occur mainly in the ascending and transverse colon, reflect local controls of smooth muscle within the walls of the individual haustra. As a haustrum fills with food residue, the distension stimulates its muscle to contract. These movements mix the residue, which aids in water absorption.

Mass movements (mass peristalsis) are long, slow-moving, but powerful contractile waves that move over large areas of the colon three or four times daily and force the contents toward the rectum. Typically, they occur during or just after eating. The presence of food in the stomach activates the gastroileal reflex in the small intestine and the propulsive **gastrocolic reflex** in the colon.

Segmenting movements in the descending and sigmoid colon promote the final drying out of the feces. This part of the colon also stores feces until mass movements propel the feces into the rectum. Fiber in the diet strengthens colon contractions and softens the feces, allowing the colon to act like a well-oiled machine.

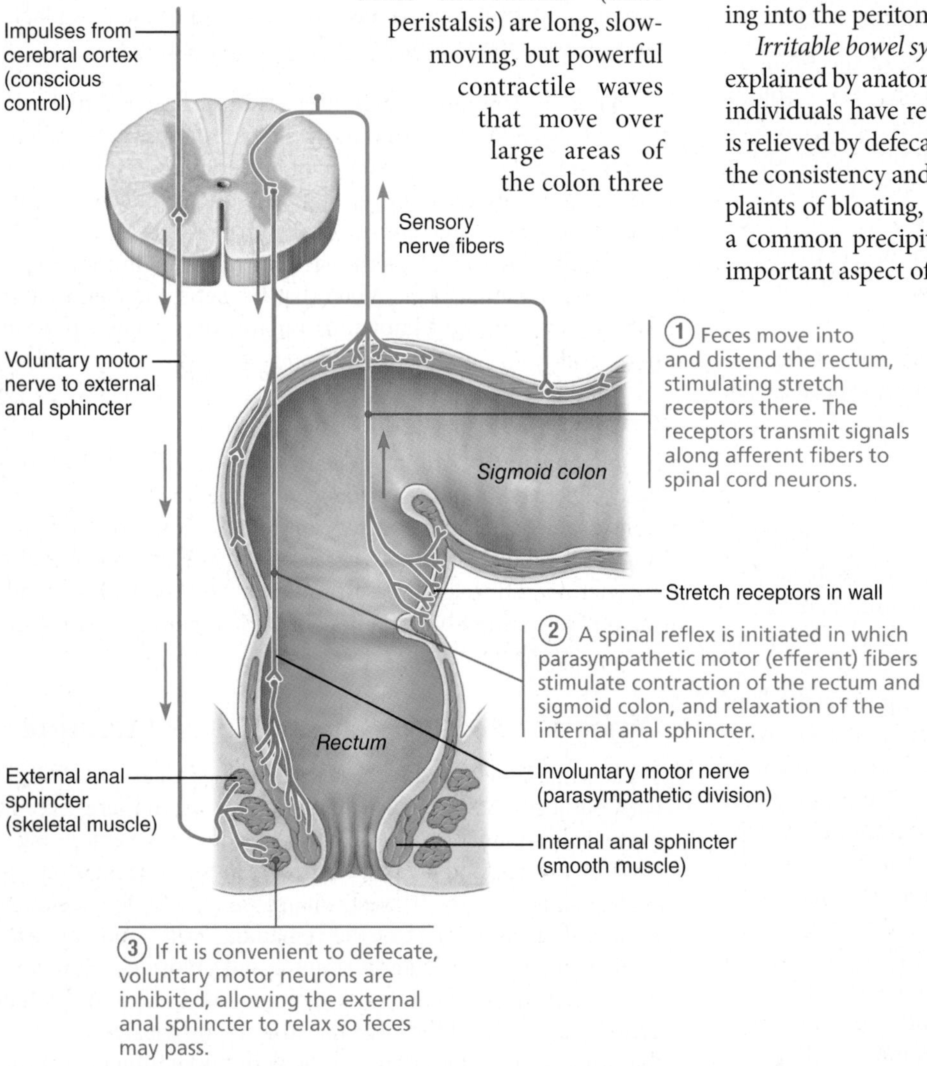

HOMEOSTATIC IMBALANCE 22.14 CLINICAL

When the diet lacks fiber and the volume of residues in the colon is small, the colon narrows and its contractions become more powerful, increasing the pressure on its walls. This promotes formation of **diverticula** (di″ver-tik′u-lah), small herniations of the mucosa through the colon walls.

This condition, called **diverticulosis**, most commonly occurs in the sigmoid colon, and affects over half of people over age 70. In 4–10% of cases, diverticulosis progresses to **diverticulitis**, in which the diverticula become inflamed and may rupture, leaking into the peritoneal cavity, which can be life threatening.

Irritable bowel syndrome (*IBS*) is a functional GI disorder not explained by anatomical or biochemical abnormalities. Affected individuals have recurring (or persistent) abdominal pain that is relieved by defecation. Additionally, they may have changes in the consistency and frequency of their stools, and varying complaints of bloating, flatulence, nausea, and depression. Stress is a common precipitating factor, and stress management is an important aspect of treatment. ✚

Figure 22.33 Defecation reflex.

The semisolid feces delivered to the rectum contain undigested food residues, mucus, sloughed-off epithelial cells, millions of bacteria, and just enough water to allow their smooth passage. Of the 500 ml or so of food residue entering the cecum daily, approximately 150 ml becomes feces.

Defecation

The rectum is usually empty, but when mass movements force feces into it, stretching of the rectal wall initiates the **defecation reflex**. This parasympathetic spinal reflex causes the sigmoid colon and the rectum to contract, and the internal anal sphincter to relax (**Figure 22.33** ① and ②). As feces are forced into the anal canal, messages reach the brain allowing us to decide whether the external (voluntary) anal sphincter should open or remain constricted to stop passage of feces temporarily (Figure 22.33 ③).

If defecation is delayed, the reflex contractions end within a few seconds and the rectal walls relax. The next mass movement initiates the defecation reflex again—and so on, until the person chooses to defecate or the urge becomes irresistible.

During defecation, the muscles of the rectum contract to expel the feces. We aid this process voluntarily by closing the glottis and contracting our diaphragm and abdominal wall muscles to increase the intra-abdominal pressure (a procedure called *Valsalva's maneuver*). We also contract the levator ani muscle (pp. 305–306), which lifts the anal canal superiorly. This lifting action leaves the feces below the anus—and outside the body. Involuntary or automatic defecation (fecal incontinence) occurs in infants because they have not yet gained control of their external anal sphincter. It also occurs in those with spinal cord transections.

HOMEOSTATIC IMBALANCE 22.15 CLINICAL

Watery stools, or **diarrhea**, result from any condition that rushes food residue through the large intestine before that organ has had sufficient time to absorb the remaining water. Causes include irritation of the colon by bacteria or, less commonly, prolonged physical jostling of the digestive viscera (occurs in marathon runners). Prolonged diarrhea may result in dehydration and electrolyte imbalance (acidosis and loss of potassium).

Conversely, when food remains in the colon for extended periods, too much water is absorbed and the stool becomes hard and difficult to pass. This condition, called **constipation**, may result from insufficient fiber or fluid in the diet, improper bowel habits (failing to heed the "call"), lack of exercise, or laxative abuse. +

☑ **Check Your Understanding**

35. Name and briefly describe the types of motility that occur in the large intestine.

36. What is the result of stimulation of stretch receptors in the rectal walls?

37. In what ways are enteric bacteria important to our nutrition?

For answers, see Answers Appendix.

PART 3

PHYSIOLOGY OF DIGESTION AND ABSORPTION

So far in this chapter, we have examined the structure and function of the organs that make up the digestive system. Now let's investigate the chemical processing (enzymatic breakdown) and absorption of each class of foodstuffs as it moves through the GI tract. As you read along, you may find it helpful to refer to the summary in **Figure 22.34**.

22.10 Digestion hydrolyzes food into nutrients that are absorbed across the gut epithelium

→ **Learning Objective**

☐ **Describe the general processes of digestion and absorption.**

After foodstuffs have spent even a short time in the stomach, they are unrecognizable, but mechanical breakdown has only changed their appearance. In contrast, digestion breaks down ingested foods into their chemical building blocks, which are very different molecules chemically. Only these molecules are small enough to be absorbed across the wall of the small intestine.

Mechanism of Digestion: Enzymatic Hydrolysis

Digestion is a catabolic process that breaks down large food molecules to *monomers* (chemical building blocks). Digestion is accomplished by enzymes secreted into the lumen of the alimentary canal by intrinsic and accessory glands. Recall from Chapter 2 that enzymatic breakdown of any food molecule is **hydrolysis** (hi-drol'ĭ-sis) because it involves adding a water molecule to each molecular bond to be broken (lysed).

Most digestion is done in the small intestine. Pancreatic enzymes break large chemicals (usually polymers) into smaller pieces that are, in turn, broken down into individual components by the intestinal (brush border) enzymes. Alkaline pancreatic juice neutralizes the acidic chyme that enters the small intestine from the stomach. This provides the proper environment for operation of the enzymes. Both pancreatic juice (the main source of lipases) and bile are necessary for fat breakdown.

Mechanisms of Absorption

Absorption is the process of moving substances from the lumen of the gut into the body. Because tight junctions join the epithelial cells of the intestinal mucosa at their apical surfaces, substances usually cannot move *between* cells. Instead, materials must pass *through* the epithelial cells. Materials enter an epithelial cell through its *apical membrane* from the lumen of the gut, and exit through the *basolateral membrane* into the interstitial fluid on the other side of the cell. Once in the interstitial fluid, substances diffuse into the blood capillaries. From the capillary blood in the villus they are transported in the hepatic portal vein to the liver. The exception is some lipid digestion products, which enter the lacteal in the villus to be carried via lymphatic fluid to the blood.

Remember that the structure of the plasma membrane means that nonpolar substances, which can dissolve in the lipid core of the membrane, can be absorbed passively. All other substances need a carrier mechanism. Most nutrients are absorbed by *active transport* processes driven directly or indirectly (secondarily) by metabolic energy (ATP).

There is much more flowing through the alimentary tube than food monomers. Indeed, up to 10 L of food, drink, and GI secretions enter the alimentary canal daily, but only 1 L or less reaches the large intestine. Virtually all of the foodstuffs, 80% of the electrolytes, and most of the water (remember water follows salt) are absorbed in the small intestine. Although absorption occurs all along the length of the small intestine, most of it is completed by the time chyme reaches the ileum. The major absorptive role of the ileum is to reclaim bile salts to be recycled back to the liver for resecretion. The absorptive capacity of the small intestine is truly remarkable and it is virtually impossible to exceed.

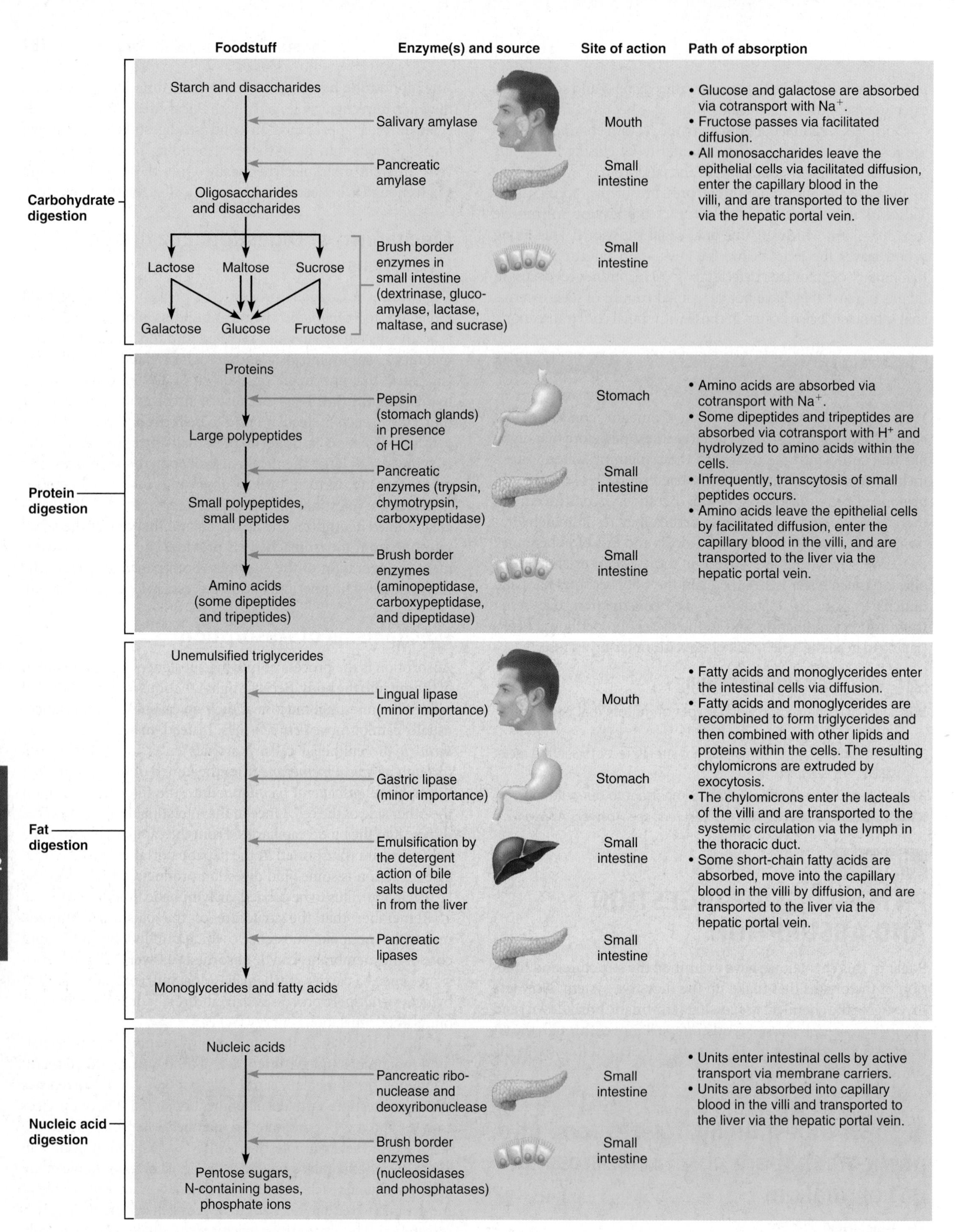

Figure 22.34 Flowchart of digestion and absorption of foodstuffs.

Check Your Understanding

38. In order to be absorbed, nutrients must pass through two plasma membranes. Name these membranes.

39. MAKING connections Name the layer and sublayer of the alimentary canal wall that houses the capillaries into which nutrients are absorbed.

For answers, see Answers Appendix.

22.11 How is each type of nutrient processed?

Learning Objectives

- **List the enzymes involved in digestion; name the foodstuffs on which they act.**
- **List the end products of protein, fat, carbohydrate, and nucleic acid digestion.**
- **Describe the process by which breakdown products of foodstuffs are absorbed in the small intestine.**

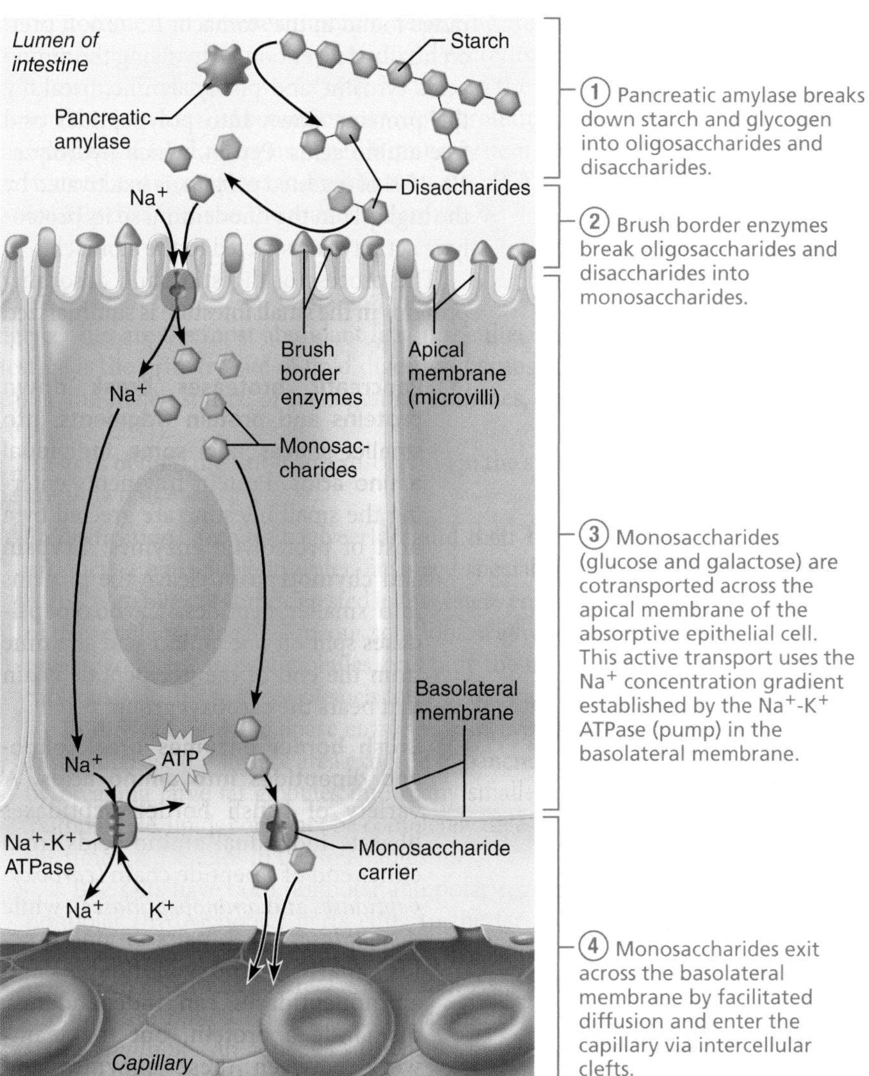

Figure 22.35 Carbohydrate digestion and absorption in the small intestine. Note that fructose enters epithelial cells via facilitated diffusion (not shown).

Carbohydrates

In the average diet, most (up to 60%) digestible carbohydrates are in the form of starch, with smaller amounts of disaccharides and monosaccharides. (See Figure 2.15 on p. 39 to review the structure of mono- and disaccharides.) Only three monosaccharides are common in our diet: *glucose*, *fructose*, and *galactose*. The more complex carbohydrates that our digestive system is able to break down to monosaccharides are the disaccharides *sucrose* (table sugar), *lactose* (milk sugar), and *maltose* (grain sugar), and the polysaccharides *glycogen* and *starch*.

Digestion of starch (and perhaps glycogen) begins in the mouth (Figure 22.34). **Salivary amylase**, present in saliva, splits starch into *oligosaccharides*, smaller fragments of two to eight linked glucose molecules. Starch digestion continues until salivary amylase is inactivated by stomach acid and broken apart by the stomach's protein-digesting enzymes. Generally speaking, the larger the meal, the longer salivary amylase continues to work in the stomach because foodstuffs in its relatively immobile fundus are poorly mixed with gastric juices.

The process of digesting and absorbing carbohydrates in the small intestine is summarized in **Figure 22.35**.

① **Pancreatic amylase breaks down starch and glycogen into oligosaccharides and disaccharides.** Starchy foods and other digestible carbohydrates that escape being broken down by salivary amylase are acted on by **pancreatic amylase** in the small intestine. About 10 minutes after entering the small intestine, starch is entirely converted to various oligosaccharides, mostly maltose.

② **Brush border enzymes break oligo- and disaccharides into monosaccharides.** Intestinal brush border enzymes further digest these products to monosaccharides. The most important brush border enzymes are **dextrinase** and **glucoamylase**, which act on oligosaccharides composed of more than three simple sugars, and **maltase**, **sucrase**, and **lactase**, which hydrolyze maltose, sucrose, and lactose respectively into their constituent monosaccharides. Because the intestine can absorb only monosaccharides, all dietary carbohydrates must be digested to monosaccharides to be absorbed.

③ **Monosaccharides are cotransported across the apical membrane of the absorptive epithelial cell.** Glucose and galactose, liberated by the breakdown of starch and disaccharides, are shuttled by secondary active transport with Na^+ into the epithelial cells. Fructose, on the other hand, enters the cells by facilitated

the hydrophobic core are cholesterol molecules and fat-soluble vitamins. Although micelles are similar to emulsion droplets, they are about 500 times smaller and easily diffuse between microvilli to come into close contact with the apical cell surface. Without micelles, the lipids would simply float on the surface of the chyme (like oil on water), inaccessible to the absorptive surfaces of the epithelial cells.

④ **Diffusion.** Upon reaching the epithelial cells, the various lipid substances leave the micelles and move through the lipid phase of the plasma membrane by simple diffusion.

⑤ **Chylomicron formation.** Once the free fatty acids and monoglycerides enter the epithelial cells, the smooth ER converts them back into triglycerides. The triglycerides are then combined with lecithin and other phospholipids and cholesterol, and coated with a "skin" of proteins to form water-soluble lipoprotein droplets called **chylomicrons** (ki″-lo-mi′kronz). This series of events is quite different from the absorption of amino acids and simple sugars, which pass through the epithelial cells unchanged.

⑥ **Chylomicron transport.** The milky-white chylomicrons are too large to pass through either the plasma membrane of the epithelial cell or the basement membrane of a blood capillary. Instead, the chylomicron-containing vesicles migrate to the basolateral membrane and are extruded by exocytosis. They then enter the more permeable lacteals. Thus, most fat enters the lymphatic stream for distribution in the lymph. Eventually the chylomicrons are emptied into the venous blood via the thoracic duct, which drains the lymphatics of the digestive viscera.

While in the bloodstream, the triglycerides of the chylomicrons are hydrolyzed to free fatty acids and glycerol by **lipoprotein lipase**, an enzyme associated with capillary endothelium. The fatty acids and glycerol can then pass through the capillary walls to be used by tissue cells for energy or stored as fats in adipose tissue. Liver cells then endocytose and process the residual chylomicron material.

Passage of short-chain fatty acids is quite different from what we have just described. These fat breakdown products do not depend on the presence of bile salts or micelles and are not recombined to form triglycerides within the intestinal cells. They simply diffuse into the portal blood for distribution.

Generally, fat absorption is completed in the ileum, but in the absence of bile (as might occur when a gallstone blocks the cystic duct), it happens so slowly that most of the fat passes into the large intestine and is lost in feces.

Nucleic Acids

The nuclei of the cells of ingested foods contain DNA and RNA. **Pancreatic nucleases** in pancreatic juice hydrolyze the nucleic acids to their **nucleotide** monomers. Intestinal brush border enzymes (**nucleosidases** and **phosphatases**) then break the nucleotides apart to release their nitrogenous bases, pentose sugars, and phosphate ions (see Figure 22.34).

Special carriers in the epithelium of the villi actively transport the breakdown products of nucleic acid digestion across the epithelium. These then enter the blood.

Absorption of Vitamins, Electrolytes, and Water

Vitamin Absorption

The small intestine absorbs dietary vitamins, and the large intestine absorbs some of the K and B vitamins made by its enteric bacterial "guests." As we already noted, fat-soluble vitamins (A, D, E, and K) dissolve in dietary fats, become incorporated into the micelles, and move across the villus epithelium passively (by diffusion). It follows that gulping pills containing fat-soluble vitamins without simultaneously eating some fat-containing food results in little or no absorption of these vitamins.

Most water-soluble vitamins (B vitamins and vitamin C) are absorbed via specific active or passive transporters. The exception is vitamin B_{12}, which is a very large, charged molecule. *Intrinsic factor*, produced by the stomach, binds to vitamin B_{12}. The vitamin B_{12}–intrinsic factor complex then binds to specific mucosal receptor sites in the terminal ileum, which trigger its active uptake by endocytosis.

Electrolyte Absorption

Absorbed electrolytes come from both ingested foods and gastrointestinal secretions. Most ions are actively absorbed along the entire length of the small intestine. But absorption of iron and calcium is largely limited to the duodenum.

As we mentioned earlier, absorption of sodium ions in the small intestine is coupled to active absorption of glucose and amino acids. For the most part, anions passively follow the electrical potential established by sodium transport. In other words, Na^+ is actively pumped out of the epithelial cells by a Na^+-K^+ pump after entering those cells. Usually, chloride ions passively follow Na^+. In the terminus of the small intestine, HCO_3^- is actively secreted into the lumen in exchange for Cl^-.

Potassium ions move across the intestinal mucosa passively by facilitated diffusion (or leaky tight junctions). As water is absorbed from the lumen, rising potassium levels in chyme create a concentration gradient for its absorption. Anything that interferes with water absorption (resulting in diarrhea) not only reduces potassium absorption but also "pulls" K^+ from the interstitial space into the intestinal lumen.

For most nutrients, the amount *reaching* the intestine is the amount absorbed, regardless of the nutritional state of the body. In contrast, absorption of iron and calcium is intimately related to the body's need for them at the time.

Ionic iron, essential for hemoglobin production, is actively transported into the mucosal cells, where it binds to the protein **ferritin** (fer′ĭ-tin). The intracellular iron-ferritin complexes then serve as local storehouses for iron. When body reserves of iron are adequate, only 10–20% is allowed to pass into the portal blood, and most of the stored iron is lost as the epithelial cells later slough off. However, when iron reserves are depleted (as during acute or chronic hemorrhage), iron uptake from the intestine and its release to the blood accelerate. In the blood, iron binds to **transferrin**, a plasma protein that transports it in the circulation.

Menstrual bleeding is a major route of iron loss in females, and premenopausal women require about 50% more iron in their diets. The intestinal epithelial cells of women have about four times as many iron transport proteins as do those of men, and little iron is lost from the body other than that lost in menses.

Calcium absorption is closely related to blood levels of ionic calcium. The active form of **vitamin D** promotes active calcium absorption. Decreased blood levels of ionic calcium prompt *parathyroid hormone* (*PTH*) release from the parathyroid glands. Besides facilitating the release of calcium ions from bone matrix and enhancing the reabsorption of calcium by the kidneys, PTH stimulates activation of vitamin D to calcitriol by the kidneys, which in turn accelerates calcium ion absorption in the small intestine.

Water Absorption

Approximately 9 L of water, mostly derived from GI tract secretions, enter the small intestine daily. Water is the most abundant substance in chyme, and 95% of it is absorbed in the small intestine by osmosis. Most of the rest is absorbed in the large intestine, leaving only about 0.1 L to soften the feces.

The normal rate of water absorption is 300 to 400 ml per hour. Water moves freely in both directions across the intestinal mucosa, but *net osmosis* occurs whenever a concentration gradient is established by the active transport of solutes (particularly Na^+) into the mucosal cells. In this way, water uptake is effectively coupled to solute uptake and, in turn, affects the absorption of substances that normally pass by diffusion. As water moves into mucosal cells, these substances follow along their concentration gradients.

Malabsorption, or impaired nutrient absorption, has many and varied causes. It can result from anything that interferes with the delivery of bile or pancreatic juice to the small intestine. Factors that damage the intestinal mucosa (severe bacterial infections and some antibiotics) or reduce its absorptive surface area are also common causes.

A common malabsorption syndrome is *gluten-sensitive enteropathy* or *celiac disease*, which affects one in 100 people. This chronic genetic condition is caused by an immune reaction to gluten, a protein plentiful in all grains but corn and rice. Breakdown products of gluten interact with molecules of the immune system in the GI tract, forming complexes. These complexes activate T cells, which then attack the intestinal lining, damaging intestinal villi and reducing the surface area of the brush border. Bloating, diarrhea, pain, and malnutrition result.

The usual treatment is to eliminate gluten-containing grains from the diet. In recent years, many gluten-free products have become available, although some are of questionable nutritional value. ✚

☑ Check Your Understanding

40. Fill in the blank: Amylase is to starch as ___ is to fats.

41. What is the role of bile salts in the digestive process? In absorption?

For answers, see Answers Appendix.

The digestive system keeps the blood well supplied with the nutrients needed by all body tissues to fuel their energy needs and to synthesize new proteins for growth and maintenance of health. Now we are ready to examine how body cells use these nutrients, the topic of Chapter 23.

REVIEW QUESTIONS

Multiple Choice/Matching

(Some questions have more than one correct answer. Select the best answer or answers from the choices given.)

1. The peritoneal cavity **(a)** is the same thing as the abdominopelvic cavity, **(b)** is filled with air, **(c)** like the pleural and pericardial cavities is a potential space containing serous fluid, **(d)** contains the pancreas and all of the duodenum.
2. Obstruction of the hepatopancreatic sphincter impairs digestion by reducing the availability of **(a)** bile and HCl, **(b)** HCl and intestinal juice, **(c)** pancreatic juice and intestinal juice, **(d)** pancreatic juice and bile.
3. The lamina propria forms part of the **(a)** muscularis externa, **(b)** submucosa, **(c)** serosa, **(d)** mucosa.
4. Carbohydrates are acted on by **(a)** peptidases, trypsin, and chymotrypsin, **(b)** amylase, maltase, and sucrase, **(c)** lipases, **(d)** peptidases, lipases, and galactase.
5. The parasympathetic nervous system influences digestion by **(a)** relaxing smooth muscle, **(b)** stimulating peristalsis and secretory activity, **(c)** constricting sphincters, **(d)** none of these.
6. The digestive juice product containing enzymes capable of digesting all four major foodstuff categories is **(a)** pancreatic, **(b)** gastric, **(c)** salivary, **(d)** biliary.
7. The vitamin associated with calcium absorption is **(a)** A, **(b)** K, **(c)** C, **(d)** D.
8. Someone has eaten a meal of buttered toast, cream, and eggs. Which of the following would you expect to happen? **(a)** Compared to the period shortly after the meal, gastric motility and secretion of HCl decrease when the food reaches the duodenum; **(b)** gastric motility increases even as the person is chewing the food (before swallowing); **(c)** fat will be emulsified in the duodenum by the action of bile; **(d)** all of these.
9. The site of production of cholecystokinin is **(a)** the stomach, **(b)** the small intestine, **(c)** the pancreas, **(d)** the large intestine.

10. Which of the following is not characteristic of the colon? **(a)** It is divided into ascending, transverse, and descending portions; **(b)** it contains abundant bacteria, some of which synthesize certain vitamins; **(c)** it is the main absorptive site; **(d)** it absorbs much of the water and salts remaining in the wastes.
11. The gallbladder **(a)** produces bile, **(b)** is attached to the pancreas, **(c)** stores and concentrates bile, **(d)** produces secretin.
12. The sphincter between the stomach and duodenum is **(a)** the pyloric sphincter, **(b)** the gastroesophageal sphincter, **(c)** the hepatopancreatic sphincter, **(d)** the ileocecal valve.

In items 13–17, trace the path of a single protein molecule that has been ingested.

13. The protein molecule will be digested by enzymes made by **(a)** the mouth, stomach, and colon, **(b)** the stomach, liver, and small intestine, **(c)** the small intestine, mouth, and liver, **(d)** the pancreas, stomach, and small intestine.
14. The protein molecule must be digested before it can be transported to and utilized by the cells because **(a)** protein is only useful directly, **(b)** protein has a low pH, **(c)** proteins in the circulating blood produce an adverse osmotic pressure, **(d)** the protein is too large to be readily absorbed.
15. The products of protein digestion enter the bloodstream largely through cells lining **(a)** the stomach, **(b)** the small intestine, **(c)** the large intestine, **(d)** the bile duct.
16. Before the blood carrying the products of protein digestion reaches the heart, it first passes through capillary networks in **(a)** the spleen, **(b)** the lungs, **(c)** the liver, **(d)** the brain.
17. Having passed through the regulatory organ selected above, the products of protein digestion are circulated throughout the body. They will enter individual body cells by **(a)** active transport, **(b)** diffusion, **(c)** osmosis, **(d)** phagocytosis.

Short Answer Essay Questions

18. Make a simple line drawing of the organs of the alimentary canal and label each organ. Then add three labels to your drawing—salivary glands, liver, and pancreas—and use arrows to show where each of these organs empties its secretions into the alimentary canal.
19. Lara was on a diet but she could not eat less and kept claiming her stomach had a mind of its own. She was joking, but indeed, there is a "gut brain" called the enteric nervous system. Is it part of the parasympathetic and sympathetic nervous system? Explain.
20. Name the layers of the alimentary canal wall. Note the tissue composition and major function of each layer.
21. What is a mesentery? Mesocolon? Greater omentum?
22. Name the six functional activities of the digestive system.
23. (a) Describe the boundaries of the oral cavity. (b) Why do you suppose its mucosa is stratified squamous epithelium rather than the more typical simple columnar epithelium?
24. (a) What is the normal number of permanent teeth? Of deciduous teeth? (b) What substance covers the tooth crown? Its root? (c) What substance makes up the bulk of a tooth? (d) What and where is pulp?
25. Describe the two phases of swallowing, noting the organs involved and the activities that occur.
26. Describe the role of these cells found in gastric glands: parietal, chief, mucous neck, and enteroendocrine.
27. Describe the regulation of the cephalic, gastric, and intestinal phases of gastric secretion.
28. (a) What is the relationship between the cystic, common hepatic, bile, and pancreatic ducts? (b) What is the point of fusion of the bile and pancreatic ducts called?
29. Explain why fatty stools result from the absence of bile or pancreatic juice.
30. Indicate the function of the stellate macrophages and the hepatocytes of the liver.
31. What are (a) brush border enzymes? (b) chylomicrons?
32. Explain why activation of pancreatic enzymes is delayed until they reach the small intestine.

AT THE CLINIC

Clinical Case Study

Digestive System

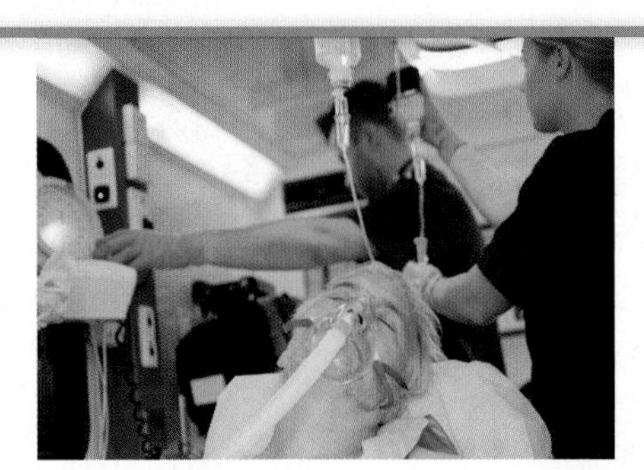

Remember Mr. Gutteman, the gentleman who was dehydrating? It seems that his tremendous output of urine was only one of his current problems. Today, he complains of a headache, gnawing epigastric pain, and "the runs" (diarrhea). To pinpoint the problem, he is asked the following questions.

- Have you had these symptoms previously? (Response: "Yes, but never this bad.")
- Are you allergic to any foods? (Response: "Shellfish doesn't like me and milk gives me the runs.")

As a result of his responses, a lactose-free diet is ordered for Mr. Gutteman instead of the regular diet originally prescribed.

1. Why is the new diet prescribed? (What is believed to be his problem?)

Mr. Gutteman's problem continues despite the diet change. In fact, the frequency of diarrhea increases and by the end of the next day, he is complaining of severe abdominal pain. Again, he is asked some questions to probe his condition. One is whether he has traveled outside the country recently. He has not, reducing the possibility of infection with *Shigella* bacteria, which is associated with poor sanitation. Other questions:

- Do you drink alcohol and how much? (Response: "Little or none.")
- Have you recently eaten raw eggs or a salad containing mayonnaise at a gathering? (Response: "No.")
- Are there certain foods that seem to precipitate these attacks? (Response: "Yes, when I have coffee and a sandwich.")

2. On the basis of these responses, what do you think Mr. Gutteman's diarrhea might stem from? How will it be diagnosed and treated?

For answers, see Answers Appendix.

23 Nutrition, Metabolism, and Energy Balance

KEY CONCEPTS

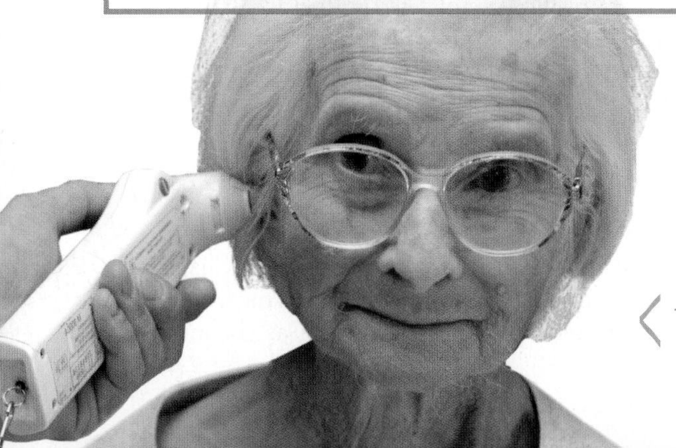

The ear is an easily accessible site for measuring body temperature.

Are you a food lover? We are too. In fact, most people fall into one of two camps—those who live to eat and those who eat to live. The saying "you are what you eat" is true in that part of the food we eat is converted to our living flesh. In other words, our bodies use some nutrients to build cell structures, replace worn-out parts, and synthesize functional molecules. However, most nutrients we ingest are used as metabolic fuel. That is, they are oxidized and transformed to **ATP**, the chemical energy form used by cells.

In Chapter 22, we talked about how foods are digested and absorbed, but what happens to these foods once they enter the blood? We will answer this question as we examine both the nature of nutrients and their metabolic roles.

PART 1

NUTRIENTS

Learning Objectives

- ☐ Define nutrient, essential nutrient, and calorie.
- ☐ List the five major nutrient categories. Note important sources and main cellular uses.

A **nutrient** is a substance in food the body uses to promote normal growth, maintenance, and repair. The nutrients needed for health divide into five categories. Three of these—carbohydrates, lipids, and proteins—are **macronutrients** that make up the bulk of what we eat. The fourth and fifth categories, vitamins and minerals, though equally crucial for health, are **micronutrients** required in only minute amounts.

Water, which accounts for about 60% by volume of the food we eat, is considered by some to also be a nutrient. We described its importance in the body in Chapter 2, so here we consider only the five nutrient categories listed above.

At least 45 and possibly 50 molecules, called **essential nutrients**, cannot be made fast enough to meet the body's needs, so our diet must provide them. As long as we ingest all the essential nutrients, the body can synthesize the hundreds of additional molecules required for life and good health. The ability of cells, especially liver cells, to convert one type of molecule to another is truly remarkable. These interconversions allow the body to use a wide range of foods and to adjust to varying food intakes. While "essential" is a standard way to describe the chemicals that must be obtained from outside sources, both essential and nonessential nutrients are equally vital for normal functioning.

Most foods offer a combination of nutrients. A balanced diet of foods from each of the different food groups normally guarantees adequate amounts of all the needed nutrients and adequate energy.

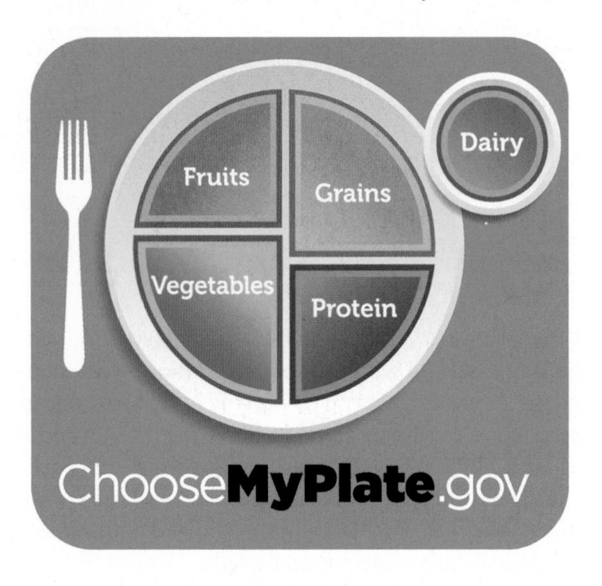

Figure 23.1 **USDA's MyPlate food guide.**

The energy value of foods is measured in **kilocalories** (kcal). One kilocalorie is the amount of heat energy needed to raise the temperature of 1 kilogram of water 1°C (1.8°F). This unit is the "calorie" (C) that dieters count so conscientiously.

Various organizations release dietary recommendations. Most of these emphasize eating more vegetables, whole grains, and fruits. For example, the U.S. Department of Agriculture (USDA) guidelines are represented as portions of a dinner plate (Figure 23.1). This image suggests how consumers might plan their meals relative to amounts and variety of foods from each food group. The MyPlate website provides details on healthy choices in each food group as well as personalized information according to age, sex, and activity level (www.choosemyplate.gov).

Nutrition advice is constantly in flux and often mired in the self-interest of food companies. Nonetheless, basic dietary principles have not changed in years and are not in dispute: Eat only what you need; eat plenty of fruits, vegetables, and whole grains; avoid junk food.

23.1 Carbohydrates, lipids, and proteins supply energy and are used as building blocks

→ Learning Objectives

- ☐ Distinguish between simple and complex carbohydrate sources.
- ☐ Distinguish between saturated, unsaturated, and trans fatty acid sources.
- ☐ Distinguish between nutritionally complete and incomplete proteins.
- ☐ Define nitrogen balance and indicate possible causes of positive and negative nitrogen balance.
- ☐ Indicate the major uses of carbohydrates, lipids, and proteins in the body.

Carbohydrates

Dietary Sources

Except for milk sugar (lactose) and negligible amounts of glycogen in meats, all the carbohydrates we ingest are derived from plants. Sugars (monosaccharides and disaccharides) come from fruits, sugar cane, sugar beets, honey, and milk. The polysaccharide starch is found in grains and vegetables.

Two varieties of polysaccharides provide fiber. Cellulose, plentiful in most vegetables, is not digested by humans but provides roughage, or *insoluble fiber*, which increases the bulk of the stool and facilitates defecation. *Soluble fiber*, such as pectin found in apples and citrus fruits, reduces blood cholesterol levels.

Uses in the Body

The monosaccharide **glucose** is *the* carbohydrate molecule ultimately used as fuel by body cells to produce ATP. Carbohydrate digestion also yields fructose and galactose, but the liver converts these monosaccharides to glucose before they enter the general circulation.

Many body cells also use fats as energy sources, but neurons and red blood cells rely almost entirely on glucose for their energy needs. Because even a temporary shortage of blood glucose can severely depress brain function and lead to neuron death, the body carefully monitors and regulates blood glucose levels. Any glucose in excess of what is needed for ATP synthesis is converted to glycogen or fat and stored for later use.

Other uses of monosaccharides are meager. Small amounts of pentose sugars are used to synthesize nucleic acids, and a variety of sugars are attached to externally facing plasma membrane proteins and lipids.

Dietary Requirements

The low-carbohydrate diet of the Inuit (Eskimos) and the high-carbohydrate diet of peoples in the Far East indicate that humans can be healthy even with wide variations in carbohydrate intake. The recommended intake to maintain health is 45–65% of total calorie intake, with the emphasis on *complex* carbohydrates (whole grains and vegetables), rather than simple carbohydrates (monosaccharides and disaccharides).

American adults typically consume about 46% of dietary food energy in the form of carbohydrates. Because starchy foods (rice, pasta, breads) cost less than meat and other high-protein foods, carbohydrates make up an even greater percentage of the diet in low-income groups. Highly processed carbohydrate foods such as candy and soft drinks only provide concentrated energy sources—so-called empty calories. Eating refined, sugary foods instead of more complex carbohydrates may cause nutritional deficiencies as well as obesity. Table 23.1 lists other possible consequences of excessive intake of carbohydrates.

Lipids

Dietary Sources

The most abundant dietary lipids are triglycerides (Chapter 2). We eat saturated fats in animal products such as meat and dairy foods, in a few tropical plant products such as coconut, and

Table 23.1 Summary of Carbohydrate, Lipid, and Protein Nutrients

FOOD SOURCES	RECOMMENDED DAILY ALLOWANCE (RDA) FOR ADULTS	PROBLEMS	
		EXCESSES	DEFICITS
Carbohydrates 			
Total Digestible • **Complex carbohydrates** (starches): bread, cereal, crackers, flour, pasta, rice, potatoes • **Simple carbohydrates (sugars):** carbonated drinks, candy, fruit, ice cream, pudding, young (immature) vegetables	130 g 45–65% of total caloric intake	Obesity; diabetes mellitus; nutritional deficits; dental caries; gastrointestinal irritation; elevated triglycerides in plasma	Tissue wasting (in extreme deprivation); metabolic acidosis resulting from accelerated fat use for energy
Total Fiber	25–30 g		
Lipids 			
Total: Animal sources (such as meat and dairy products) and plant sources (such as oils from nuts and seeds)	65 g Less than 30% of total caloric intake	Obesity and increased risk of cardiovascular disease (particularly with excesses of saturated and trans fat)	Weight loss; fat stores and tissue proteins catabolized to provide metabolic energy; problems controlling heat loss (due to depletion of subcutaneous fat)
• **Linoleic acid (an omega-6 fatty acid):** nuts, seeds, and vegetable oils (e.g., corn, soy, safflower)	11–17 g		
• **Linolenic acid (an omega-3 fatty acid):** fish oil, vegetable oils (e.g., canola, soy, flax), walnuts	1.1–1.6 g	Excess dietary intake of omega-3 fatty acid may increase risk of atherosclerosis	Poor growth; skin lesions (eczema-like); depression
• **Cholesterol:** organ meats (liver, kidneys, brains), egg yolks, fish roe; smaller concentrations in milk products and meat	As low as possible	Increased levels of blood cholesterol and low-density lipoproteins, correlated with increased risk of cardiovascular disease	Possible increased risk of stroke (CVA) in susceptible individuals
Proteins 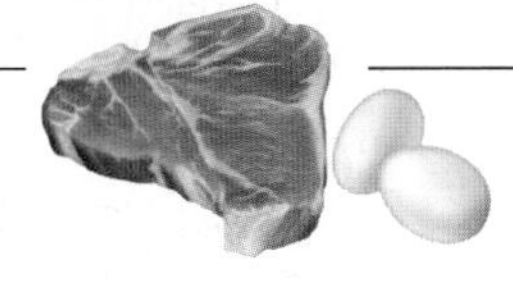			
• **Complete proteins:** eggs, milk, milk products, meat (fish, poultry, pork, beef, lamb), soybeans • **Incomplete proteins:** legumes (lima beans, kidney beans, lentils); nuts and seeds; grains and cereals; vegetables	0.8 g/kg body weight 12–20% of total caloric intake	Obesity; enhanced calcium excretion and bone loss; high cholesterol levels in blood; kidney stones	Profound weight loss and tissue wasting; retarded growth in children; anemia; edema (due to deficits of plasma proteins) During pregnancy: miscarriage or premature birth

in hydrogenated oils (trans fats) such as margarine and solid shortenings used in baking. Unsaturated fats are present in seeds, nuts, olive oil, and most vegetable oils.

Major sources of cholesterol are egg yolk, meats and organ meats, shellfish, and milk products. However, the liver produces about 85% of blood cholesterol regardless of dietary intake.

The liver is also adept at converting one fatty acid to another, but it cannot synthesize *linoleic acid* (lin″o-le′ik), a fatty acid component of *lecithin* (les′ĭ-thin). For this reason, linoleic acid, an omega-6 fatty acid, is an *essential fatty acid* that must be ingested. Linolenic acid, an omega-3 fatty acid, is also essential. Fortunately, most vegetable oils contain both linoleic and linolenic acids.

Uses in the Body

Fats have fallen into disfavor, particularly among those for whom the "battle of the bulge" is constant. But fats make foods tender, flaky, or creamy, and make us feel full and satisfied. Fats in the body *are* necessary for several reasons:

- Fatty deposits in adipose tissue provide (1) a protective cushion around body organs, (2) an insulating layer beneath the skin, and (3) an easy-to-store concentrated source of energy.
- Phospholipids are an integral component of myelin sheaths and cellular membranes.
- Cholesterol is a stabilizing component of plasma membranes and is the precursor from which bile salts, steroid hormones, and other essential molecules are formed. Unlike triglycerides, cholesterol is not used for energy.
- *Prostaglandins* (pros″tah-glan′dinz), regulatory molecules formed from linoleic acid via arachidonic acid (ah″rah-kĭ-don′ik), play a role in smooth muscle contraction, control of blood pressure, and inflammation.
- Triglycerides are the major energy fuel of skeletal muscle and hepatocytes.
- Fats help the body absorb fat-soluble vitamins.

Dietary Requirements

Fats represent over 40% of the calories in the typical American diet. There are no precise recommendations on amount or type of dietary fats, but the American Heart Association suggests:

- Fats should represent 30% or less of total caloric intake.
- Saturated fats should be limited to 10% or less of total fat intake.
- Daily cholesterol intake should be no more than 300 mg (the amount in 1½ egg yolks).

The goal of these recommendations is to keep total blood cholesterol below 200 mg/dl. Because a diet high in saturated fats and cholesterol may contribute to cardiovascular disease, these are wise guidelines. Table 23.1 summarizes sources of the various lipid classes and consequences of deficiency or excessive intake.

Proteins

Dietary Sources

Animal products contain the highest-quality proteins, in other words, those with the greatest amount and best ratios of *essential amino acids* (**Figure 23.2**). Proteins in eggs, milk, fish, and most meats are **complete proteins** that meet all the body's amino acid requirements for tissue maintenance and growth (Table 23.1). Legumes (beans and peas), nuts, and cereals are protein-rich, but their proteins are nutritionally incomplete because they are low in one or more of the essential amino acids. The exception to this generalization is soybeans, which provide plant-derived complete proteins.

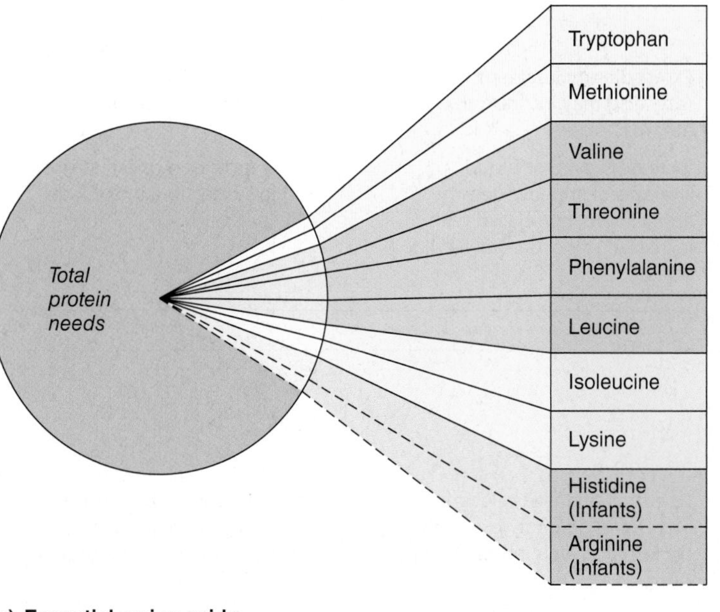

(a) Essential amino acids

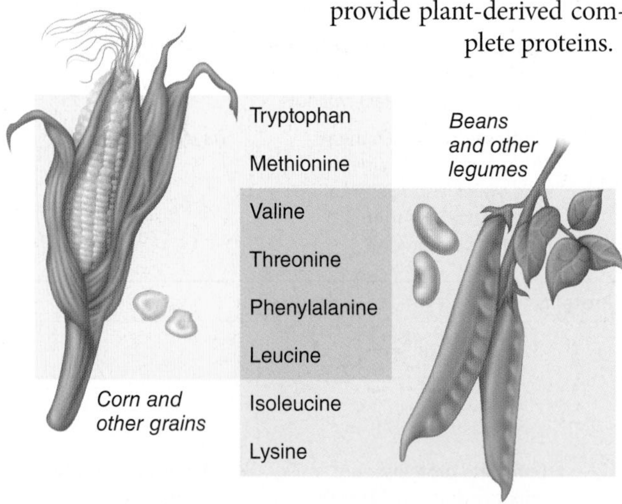

(b) Corn and beans together can provide all eight essential amino acids

Figure 23.2 Essential amino acids. (a) The essential amino acids represent only a small percentage of the total recommended protein intake. Histidine and arginine are essential in infants but not in adults. **(b)** Vegetarian diets must be carefully constructed to provide all essential amino acids.

Strict vegetarians must carefully plan their diets to obtain all the essential amino acids and prevent protein malnutrition. When ingested together, cereal grains and legumes provide all the essential amino acids (Figure 23.2b). Some combination of these foods is found in the diets of all cultures (for instance, the rice and beans seen on nearly every plate in a Mexican restaurant). For nonvegetarians, grains and legumes are useful as partial substitutes for more expensive animal proteins.

Uses in the Body

Proteins are important structural materials of the body, including, for example, keratin in skin, collagen and elastin in connective tissues, and muscle proteins. In addition, functional proteins such as enzymes and some hormones regulate an incredible variety of body functions. Whether amino acids are used to synthesize new proteins or burned for energy depends on a number of factors:

- **The all-or-none rule.** All amino acids needed to make a particular protein must be present in a cell at the same time and in sufficient amounts. If one is missing, the protein cannot be made. Because essential amino acids cannot be stored, those not used immediately to build proteins are oxidized for energy or converted to carbohydrates or fats.
- **Adequacy of caloric intake.** For optimal protein synthesis, the diet must supply sufficient carbohydrate or fat calories for ATP production. When it doesn't, dietary and tissue proteins are used for energy.
- **Hormonal controls.** Certain hormones, called *anabolic hormones*, accelerate protein synthesis and growth. The effects of these hormones vary continually throughout life. For example, pituitary growth hormone stimulates tissue growth during childhood and conserves protein in adults, and the sex hormones trigger the growth spurt of adolescence. Other hormones, such as the adrenal glucocorticoids released during stress, enhance protein breakdown and conversion of amino acids to glucose.

In healthy adults the rate of protein synthesis equals the rate of protein breakdown and loss, a homeostatic state called **nitrogen balance**. The body is in nitrogen balance when the amount of nitrogen ingested in proteins equals the amount excreted in urine and feces.

The body is in *positive nitrogen balance* when the amount of protein incorporated into tissue is greater than the amount being broken down and used for energy—the normal situation in growing children and pregnant women. A positive balance also occurs when tissues are being repaired following illness or injury.

In *negative nitrogen balance*, protein breakdown for energy exceeds the amount of protein being incorporated into tissues. This occurs during physical and emotional stress (for example, infection, injury, or burns), when the quality or quantity of dietary protein is poor, or during starvation.

Dietary Requirements

Besides supplying essential amino acids, dietary proteins furnish the raw materials for making nonessential amino acids and various nonprotein nitrogen-containing substances. The amount of protein a person needs to ingest reflects his or her age, size, metabolic rate, and the need to build new proteins (whether the body is in positive nitrogen balance). As a rule of thumb, nutritionists recommend a daily intake of 0.8 g per kilogram of body weight.

☑ Check Your Understanding

1. What are the five major nutrient categories?
2. Why is it important to include cellulose in a healthy diet even though we do not digest it?
3. How does the body use triglycerides? Cholesterol?
4. Jared eats nothing but baked bean sandwiches. Is he getting all the essential amino acids he needs in this restricted diet?

For answers, see Answers Appendix.

23.2 Most vitamins act as coenzymes; minerals have many roles in the body

→ Learning Objectives

- ☐ **Distinguish between fat- and water-soluble vitamins, and list the vitamins in each group.**
- ☐ **For each vitamin, list important sources, body functions, and important consequences of its deficit or excess.**
- ☐ **List minerals essential for health.**
- ☐ **Indicate important dietary sources of minerals and describe how each is used.**

Vitamins

Vitamins (*vita* = life) are organic compounds needed in minute amounts for growth and good health. Unlike other organic nutrients, vitamins do not serve as an energy source nor as building blocks, but they are crucial in helping the body use those nutrients that do. Without vitamins, all the carbohydrates, proteins, and fats we eat would be useless.

Most vitamins function as **coenzymes** (or parts of coenzymes), which act with an enzyme to accomplish a particular chemical task. For example, the B vitamins act as coenzymes when glucose is oxidized for energy.

Most vitamins are not made in the body, so we must ingest them in foods or vitamin supplements. The exceptions are vitamin D made in the skin, and small amounts of B vitamins and vitamin K synthesized by intestinal bacteria. In addition, the body can convert *beta-carotene* (kar′o-tēn), the orange pigment in carrots and other foods, to vitamin A. (For this reason, beta-carotene and substances like it are called *provitamins.*)

Vitamins are found in all major food groups, but no one food contains all the required vitamins. A balanced diet is the best way to ensure a full vitamin complement.

Initially vitamins were given letter designations that indicated the order of their discovery. Although more chemically descriptive names have been assigned to them, this earlier terminology is still commonly used.

Vitamins are either water soluble or fat soluble. **Water-soluble** vitamins—the B-complex vitamins and vitamin C—are absorbed along with water from the gastrointestinal tract. (The exception is vitamin B_{12}: To be absorbed, it must bind to *intrinsic factor*, a stomach secretion.) The body's lean tissue stores insignificant amounts of water-soluble vitamins, and any ingested amounts not taken up by cells within an hour or so are excreted in urine. Consequently, health problems resulting from excessive levels of these vitamins are rare.

Fat-soluble vitamins (A, D, E, and K) bind to ingested lipids and are absorbed along with their digestion products. Anything that interferes with fat absorption also interferes with the uptake of fat-soluble vitamins. Except for vitamin K, fat-soluble vitamins are stored in the body, and pathologies due to fat-soluble vitamin toxicity, particularly excess vitamin A, are well documented.

Metabolism uses oxygen, and during these reactions some potentially harmful free radicals are generated. Vitamins C, E, and A (in the form of its dimer beta-carotene) and the mineral selenium participate in *antioxidant reactions* that neutralize tissue-damaging free radicals. The whole story of how antioxidants interact in the body is still murky, but chemists propose that, much like a bucket brigade, they pass the dangerous free electron from one molecule to the next, until a chemical such as glutathione finally absorbs it and the body flushes it out in urine. Broccoli, cabbage, cauliflower, and brussels sprouts are all good sources of vitamins A and C.

The notion that megadoses of vitamin supplements are the road to eternal youth and glowing health is useless at best—and at worst, may cause serious health problems, particularly in the case of fat-soluble vitamins. Table 23.2 contains an overview of the roles of vitamins in the body.

Table 23.2 Vitamins

VITAMIN	MAJOR DIETARY SOURCES	MAJOR FUNCTIONS IN THE BODY	SYMPTOMS OF DEFICIENCY OR EXTREME EXCESS
Water-Soluble Vitamins			
Vitamin B_1 (thiamine)	Pork, legumes, peanuts, whole grains	Coenzyme used in removing CO_2 from organic compounds	Beriberi (nerve disorder—tingling, poor coordination, reduced heart function)
Vitamin B_2 (riboflavin)	Dairy products, meats, enriched grains, vegetables	Component of coenzymes FAD and FMN	Skin lesions such as cracks at corners of mouth
Vitamin B_3 (niacin)	Nuts, meats, grains	Component of coenzymes NAD^+ and $NADP^+$	Skin and gastrointestinal lesions, nervous disorders Liver damage
Vitamin B_5 (pantothenic acid)	Most foods: meats, dairy products, whole grains, etc.	Component of coenzyme A	Fatigue, numbness, tingling of hands and feet
Vitamin B_6 (pyridoxine)	Meats, vegetables, whole grains	Coenzyme used in amino acid metabolism	Irritability, convulsions, muscular twitching, anemia Unstable gait, numb feet, poor coordination
Vitamin B_7 (biotin)	Legumes, other vegetables, meats	Coenzyme in synthesis of fat, glycogen, and amino acids	Scaly skin inflammation, neuromuscular disorders
Vitamin B_9 (folic acid)	Green vegetables, oranges, nuts, legumes, whole grains	Coenzyme in nucleic acid and amino acid metabolism	Anemia, birth defects May mask deficiency of vitamin B_{12}
Vitamin B_{12}	Meats, eggs, dairy products	Coenzyme in nucleic acid metabolism; maturation of red blood cells	Anemia, nervous system disorders (numbness, loss of balance)
Vitamin C (ascorbic acid)	Fruits and vegetables, especially citrus fruits, broccoli, tomatoes	Used in collagen synthesis (such as for bone, cartilage, gums); antioxidant	Scurvy (degeneration of skin, teeth, blood vessels), weakness, delayed wound healing Gastrointestinal upset
Fat-Soluble Vitamins			
Vitamin A (retinol)	Provitamin A (beta-carotene) in deep green and orange vegetables and fruits; retinol in dairy products	Component of visual pigments; maintenance of epithelial tissues; antioxidant	Blindness, skin disorders, impaired immunity Headache, irritability, vomiting, hair loss, blurred vision, liver and bone damage
Vitamin D	Dairy products, egg yolk; also made in human skin in presence of sunlight	Aids in absorption and use of calcium and phosphorus	Rickets (bone deformities) in children, bone softening in adults Brain, cardiovascular, and kidney damage
Vitamin E (tocopherol)	Vegetable oils, nuts, seeds	Antioxidant; helps prevent damage to cell membranes	Degeneration of the nervous system
Vitamin K (phylloquinone)	Green vegetables, tea; also made by colon bacteria	Important in blood clotting	Defective blood clotting Liver damage and anemia

Source: From Jane B. Reece, CAMPBELL BIOLOGY, 10th Edition, © 2014. Reprinted by permission of Pearson Education, Inc., Upper Saddle River, N.J.

23

Minerals

The body requires moderate amounts of seven **minerals** (calcium, phosphorus, potassium, sulfur, sodium, chlorine, magnesium) and trace amounts of about a dozen others (Table 23.3). Minerals make up about 4% of the body by weight, with calcium and phosphorus (as bone salts) accounting for about three-quarters of this amount.

Minerals, like vitamins, are not used for fuel but work with other nutrients to ensure a smoothly functioning body. Incorporating minerals into structures makes them stronger. For example, calcium, phosphorus, and magnesium salts harden the teeth and strengthen the skeleton.

Most minerals are ionized in body fluids or bound to organic compounds to form phospholipids, hormones, and various proteins. For example, iron is essential to the oxygen-binding heme of hemoglobin, and sodium and chloride ions are the major electrolytes in blood. The amount of a particular mineral in the body gives very few clues to its importance in body function. For example, just a few milligrams of iodine (required for thyroid hormone synthesis) can make a critical difference to health.

Table 23.3 Minerals in the Body

MINERAL	MAJOR DIETARY SOURCES	MAJOR FUNCTIONS IN THE BODY	SYMPTOMS OF DEFICIENCY*
Greater than 200 mg per Day Required			
Calcium (Ca)	Dairy products, dark green vegetables, legumes	Bone and tooth formation, blood clotting, nerve and muscle function	Retarded growth, possibly loss of bone mass
Phosphorus (P)	Dairy products, meats, grains	Bone and tooth formation, acid-base balance, nucleotide synthesis	Weakness, loss of minerals from bone, calcium loss
Sulfur (S)	Proteins from many sources	Component of certain amino acids	Symptoms of protein deficiency
Potassium (K)	Meats, dairy products, many fruits and vegetables, grains	Nerve function, acid-base balance	Muscular weakness, paralysis, nausea, heart failure
Chlorine (Cl)	Table salt	Acid-base balance, formation of gastric juice, nerve function, osmotic balance	Muscle cramps, reduced appetite
Sodium (Na)	Table salt	Water balance, blood pressure, nerve function	Muscle cramps, reduced appetite
Magnesium (Mg)	Whole grains, green leafy vegetables	Cofactor; ATP bioenergetics	Nervous system disturbances
Trace Amounts Required			
Iron (Fe)	Meats, eggs, legumes, whole grains, green leafy vegetables	Component of hemoglobin and of electron carriers in energy metabolism; enzyme cofactor	Iron-deficiency anemia, weakness, impaired immunity
Fluorine (F)	Drinking water, tea, seafood	Maintenance of tooth (and probably bone) structure	Higher frequency of tooth decay
Zinc (Zn)	Meats, seafood, grains	Component of certain digestive enzymes and other proteins	Growth failure, skin abnormalities, reproductive failure, impaired immunity
Copper (Cu)	Seafood, nuts, legumes, organ meats	Enzyme cofactor in iron metabolism, melanin synthesis, electron transport	Anemia, cardiovascular abnormalities
Manganese (Mn)	Nuts, grains, vegetables, fruits, tea	Enzyme cofactor	Abnormal bone and cartilage
Iodine (I)	Seafood, iodized salt	Component of thyroid hormones	Goiter (enlarged thyroid)
Cobalt (Co)	Meats and dairy products	Component of vitamin B_{12}	None, except as B_{12} deficiency
Selenium (Se)	Seafood, meats, whole grains	Enzyme cofactor for antioxidant enzymes	Muscle pain, possibly heart muscle deterioration
Chromium (Cr)	Brewer's yeast, liver, seafood, meats, some vegetables	Involved in glucose and energy metabolism	Impaired glucose metabolism
Molybdenum (Mo)	Legumes, grains, some vegetables	Enzyme cofactor	Disorder in excretion of nitrogen-containing compounds

*All of these minerals are also harmful when consumed in excess.

Source: From Jane B. Reece, CAMPBELL BIOLOGY, 10th Edition, © 2014. Reprinted by permission of Pearson Education, Inc., Upper Saddle River, N.J.

A fine balance between uptake and excretion is crucial for retaining needed amounts of minerals while preventing toxic overload. Sodium present in virtually all natural and minimally processed foods poses little or no health risk. However, the large amounts added to processed foods and sprinkled on prior to eating may contribute to fluid retention and high blood pressure.

Fats and sugars are practically devoid of minerals, and highly refined cereals and grains are poor sources. The most mineral-rich foods are vegetables, legumes, milk, and some meats.

☑ Check Your Understanding

5. Vitamins are not used for energy fuels. What are they used for?
6. Which mineral is essential for thyroxine synthesis? For making bones hard? For hemoglobin synthesis?
7. MAKING **connections** Which B vitamin requires the help of a product made in the stomach to be absorbed? What is that gastric product and which cells in the gastric mucosa secrete it? What part of the small intestine ultimately absorbs this B vitamin? (Hint: See Chapter 22.) Lack of this B vitamin causes what kind of anemia? (Hint: See Chapter 16.)

For answers, see Answers Appendix.

PART 2

METABOLISM

Once inside body cells, nutrients become involved in an incredible variety of biochemical reactions known collectively as **metabolism** (*metabol* = change). During metabolism, substances are constantly built up and torn down. Cells use energy to extract more energy from foods, and then use some of this extracted energy to drive their activities. Even at rest, the body uses energy on a grand scale.

23

23.3 Metabolism is the sum of all biochemical reactions in the body

→ Learning Objectives

- ☐ **Define metabolism. Explain how catabolism and anabolism differ.**
- ☐ **Define oxidation and reduction and indicate the importance of these reactions in metabolism.**
- ☐ **Indicate the role of coenzymes used in cellular oxidation reactions.**
- ☐ **Explain the difference between substrate-level phosphorylation and oxidative phosphorylation.**

Anabolism and Catabolism

Metabolic processes are either *anabolic* (synthetic, building up) or *catabolic* (degradative, tearing down). **Anabolism** (ah-nab′o-lizm) is the general term for all reactions that build larger molecules or structures from smaller ones, such as the bonding together of amino acids to build proteins. **Catabolism** (kah-tab′o-lizm) refers to all processes that break down complex structures to simpler ones—for example, the hydrolysis of foods in the digestive tract.

In the group of catabolic reactions collectively called **cellular respiration**, food fuels, particularly glucose, are broken down in cells. Some of the energy released is captured to form ATP, the cells' energy currency that links energy-releasing catabolic reactions to cellular work.

Recall from Chapter 2 that reactions driven by ATP are coupled. As ATP is hydrolyzed, enzymes shift its high-energy phosphate groups to other molecules, which are then said to be **phosphorylated** (fos″for′ĭ-la-ted). Phosphorylation primes a molecule, changing it in a way that increases its activity, produces motion, or does work. For example, phosphorylation activates many regulatory enzymes that catalyze key steps in metabolic pathways.

Three major stages are involved in processing energy-containing nutrients in the body (**Figure 23.3**).

- *Stage 1* is digestion in the gastrointestinal tract. The absorbed nutrients are then transported in blood to the tissue cells.
- *Stage 2* occurs in the tissue cells. Newly delivered nutrients are either built into lipids, proteins, and glycogen by anabolic pathways or broken down by catabolic pathways to *pyruvic acid* (pi-roo′vik) and *acetyl CoA* (as′ĕ-til ko-a′) in the cell cytoplasm.
- *Stage 3*, which occurs in the mitochondria, is almost entirely catabolic. It requires oxygen, and completes the breakdown of foods, producing carbon dioxide and water and harvesting large amounts of ATP.

The primary function of *cellular respiration*, which consists of the glycolysis of stage 2 and all events of stage 3, is to generate ATP, which traps some of the chemical energy of the original food molecules in its own high-energy bonds. The body can also store energy in fuels, such as glycogen and fats, and mobilize these stores later to produce ATP for cellular use.

You do not need to memorize Figure 23.3, but you may want to refer to it often as a cohesive summary of nutrient processing and metabolism in the body.

Oxidation-Reduction Reactions and the Role of Coenzymes

Many of the reactions that take place within cells are **oxidation reactions**. *Oxidation* was originally defined as the combination of oxygen with other elements, seen in the rusting of iron (the slow formation of iron oxide) and the burning of wood. In burning, oxygen combines rapidly with carbon, releasing carbon dioxide, water, and an enormous amount of energy as heat and light.

Later it was discovered that oxidation also occurs when hydrogen atoms are *removed* from compounds, so the definition was expanded to its current form: *Oxidation is the gain of oxygen or the loss of hydrogen.* As explained in Chapter 2, whichever way oxidation occurs, the oxidized substance always *loses* (or nearly loses) electrons as they move to (or toward) a substance that more strongly attracts them.

To explain this loss of electrons, let's review the consequences of different electron-attracting abilities of atoms (see pp. 27–31). Consider a molecule made up of a hydrogen atom plus some other kinds of atoms. Hydrogen is very electropositive, so its lone electron usually spends more time orbiting the other

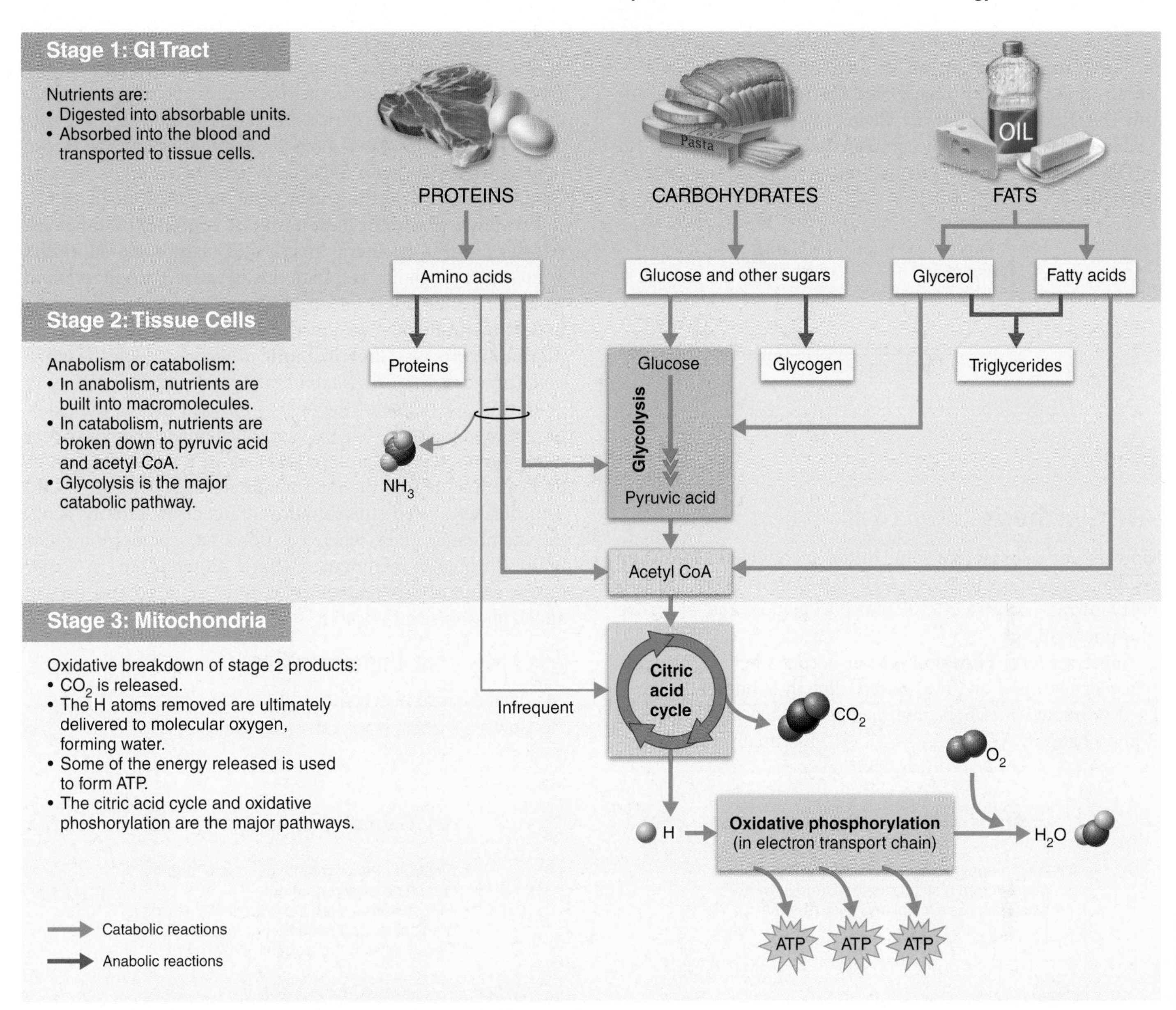

Figure 23.3 Three stages of metabolism of energy-containing nutrients.

atoms of the molecule. But when a hydrogen *atom* is removed, its electron goes with it, and the molecule as a whole loses that electron. Conversely, oxygen is very electron-hungry (electronegative), so when oxygen binds with other atoms the shared electrons spend more time in oxygen's vicinity. Again, the rest of the molecule loses electrons.

Essentially all oxidation of food fuels involves the step-by-step removal of pairs of hydrogen atoms (with their electrons) from the substrate molecules, eventually leaving only carbon dioxide (CO_2). Molecular oxygen (O_2) is the final electron acceptor. It combines with the removed hydrogen atoms at the very end of the process, to form water (H_2O).

Whenever one substance loses electrons (is oxidized), another substance gains them (is reduced). For this reason, oxidation and reduction are coupled reactions and we speak of **oxidation-reduction (redox) reactions**. The key understanding about redox reactions is that "oxidized" substances *lose* energy and "reduced" substances *gain* energy as energy-rich electrons are transferred from one substance to the next. Consequently, as food fuels are oxidized, their energy is transferred to a "bucket brigade" of other molecules and ultimately to ADP to form energy-rich ATP.

Like all other chemical reactions in the body, redox reactions are catalyzed by enzymes. Those that catalyze redox reactions in which hydrogen atoms are removed are called **dehydrogenases** (de-hi′dro-jen-ās″ez), while enzymes catalyzing the transfer of oxygen are **oxidases**.

Most of these enzymes require the help of a specific coenzyme, typically derived from one of the B vitamins. Although the enzymes catalyze the removal of hydrogen atoms to oxidize a substance, they cannot *accept* the hydrogen (hold on or bond to it). Their *coenzymes*, however, can act as hydrogen (or electron) acceptors, becoming reduced each time a substrate is oxidized.

Two very important coenzymes of the oxidative pathways are **nicotinamide adenine dinucleotide (NAD^+)** (nik″o-tin′ah-mīd), based on *niacin*, and **flavin adenine dinucleotide (FAD)**, derived from *riboflavin*. The oxidation of succinic acid to fumaric acid and the simultaneous reduction of FAD to $FADH_2$, an example of a coupled redox reaction, is shown on the right.

FAD (oxidized) → $FADH_2$ (reduced)

COOH–CH(H)–CH(H)–COOH (succinic acid) —oxidation (−2H)→ COOH–C(H)=C(H)–COOH (fumaric acid)

ATP Synthesis

How do our cells capture some of the energy liberated during cellular respiration to make ATP molecules? There are two mechanisms—substrate-level phosphorylation and oxidative phosphorylation.

Substrate-level phosphorylation occurs when high-energy phosphate groups are transferred directly from phosphorylated substrates (metabolic intermediates such as glyceraldehyde 3-phosphate) to ADP (**Figure 23.4a**). Essentially, this process occurs because the high-energy bonds attaching the phosphate groups to the substrates are even more unstable than those in ATP. ATP is synthesized by this route twice during glycolysis, and once during each turn of the citric acid cycle. The enzymes catalyzing substrate-level phosphorylations are located in both the cytosol (where glycolysis occurs) and in the watery matrix inside the mitochondria (where the citric acid cycle takes place) (**Figure 23.5**).

Oxidative phosphorylation is more complicated, but it also releases most of the energy that is eventually captured in ATP bonds during cellular respiration. Oxidative phosphorylation, which is carried out by electron transport proteins embedded in the inner mitochondrial membranes, is an example of a chemiosmotic process. **Chemiosmotic processes** couple the movement of substances across membranes to chemical reactions.

In this case, some of the energy released during the oxidation of food fuels (the "chemi" part of chemiosmotic) is used to pump (*osmo* = push) protons (H^+) across the inner mitochondrial membrane into the intermembrane space (Figure 23.4b). This creates a steep concentration gradient for protons across the membrane. Then, when H^+ flows back across the membrane (through a membrane channel protein called *ATP synthase*), some of this gradient energy is captured and used to attach phosphate groups to ADP.

☑ Check Your Understanding

8. What is a redox reaction?
9. How are anabolism and catabolism linked by ATP?

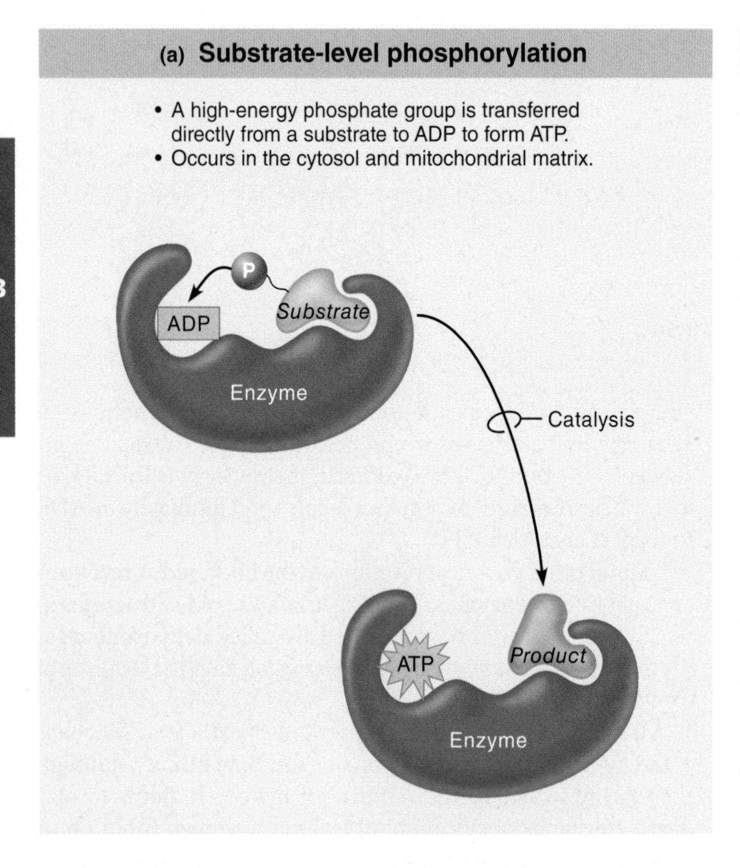

(b) Oxidative phosphorylation

- Electron transport proteins "pump" protons, creating a proton gradient.
- ATP synthase uses the energy of the proton gradient to bind phosphate groups to ADP.
- Occurs only in the mitochondrial matrix.

High H^+ concentration in intermembrane space

Inner mitochondrial membrane

H^+ H^+

Proton pumps (electron transport chain)

H^+

ATP synthase

Energy from food

ADP + P_i → ATP

Low H^+ concentration in mitochondrial matrix

Figure 23.4 Mechanisms of phosphorylation.

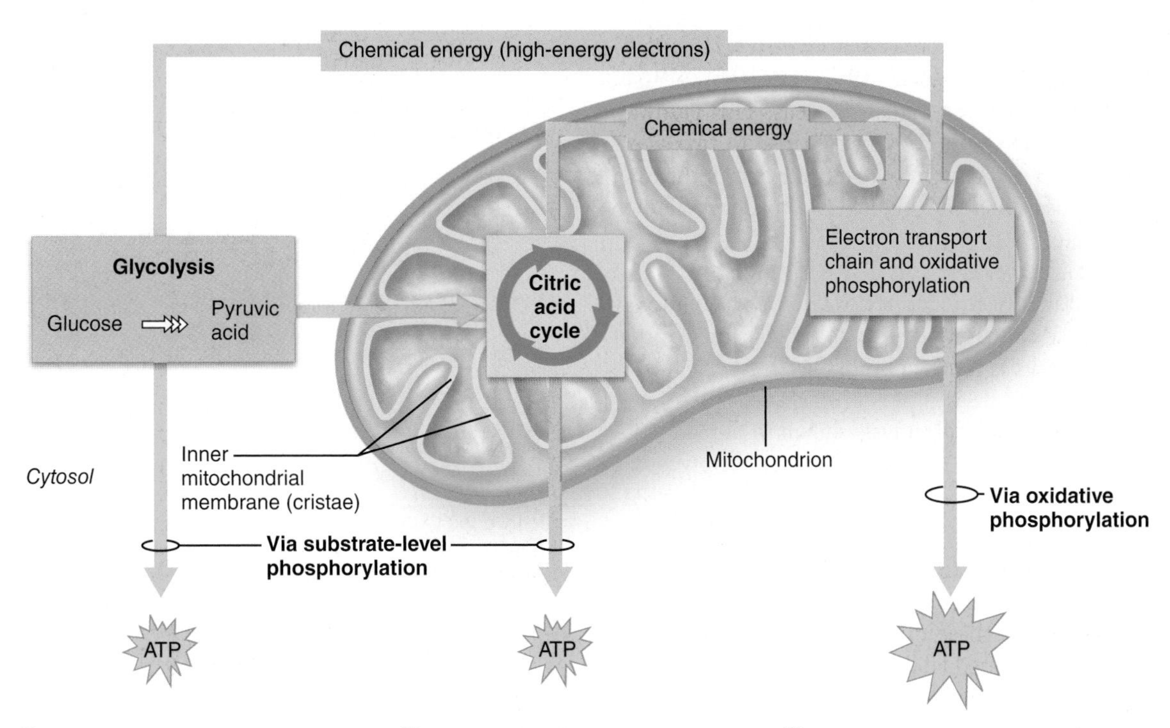

① Glycolysis, in the cytosol, breaks down each glucose molecule into two molecules of pyruvic acid.

② The pyruvic acid then enters the mitochondrial matrix, where the citric acid cycle oxidizes it to CO_2. During glycolysis and the citric acid cycle, substrate-level phosphorylation forms small amounts of ATP.

③ Energy-rich electrons picked up by coenzymes are transferred to the electron transport chain, built into the inner mitochondrial membrane. The electron transport chain carries out oxidative phosphorylation, which generates most of the ATP in cellular respiration.

Figure 23.5 During cellular respiration, ATP is formed in the cytosol and in the mitochondria.

10. What is the energy source for the proton pumps of oxidative phosphorylation?

For answers, see Answers Appendix.

23.4 Carbohydrate metabolism is the central player in ATP production

→ Learning Objectives

- ☐ **Summarize important events and products of glycolysis, the citric acid cycle, and electron transport.**
- ☐ **Define glycogenesis, glycogenolysis, and gluconeogenesis.**

The story of carbohydrate metabolism is really a tale of glucose metabolism because all food carbohydrates are eventually transformed to glucose. Glucose enters tissue cells by facilitated diffusion, a process that is greatly enhanced by insulin. Immediately after entering a cell, glucose is phosphorylated to *glucose-6-phosphate* by transfer of a phosphate group to its sixth carbon during a coupled reaction with ATP:

$$\text{Glucose} + \text{ATP} \rightarrow \text{glucose-6-PO}_4 + \text{ADP}$$

Most body cells lack the enzymes needed to reverse this reaction, so it effectively traps glucose inside the cells. Because glucose-6-phosphate is a *different* molecule from simple glucose, the reaction also keeps intracellular glucose levels low, maintaining a concentration gradient for glucose entry. Only intestinal epithelial cells, kidney tubule cells, and liver cells have the enzymes needed to reverse this phosphorylation reaction, which reflects their central roles in glucose uptake *and* release. The catabolic and anabolic pathways for carbohydrates all begin with glucose-6-phosphate.

Oxidation of Glucose

Glucose is the pivotal fuel molecule in the oxidative (ATP-producing) pathways. Glucose is catabolized via the reaction

$$\underset{\text{glucose}}{C_6H_{12}O_6} + \underset{\text{oxygen}}{6O_2} \rightarrow \underset{\text{water}}{6H_2O} + \underset{\text{carbon dioxide}}{6CO_2} + 32\ \text{ATP} + \text{heat}$$

This equation gives few hints that glucose breakdown is complex and involves three of the pathways featured in Figures 23.3 and 23.5:

1. Glycolysis (color-coded orange throughout the chapter)
2. The citric acid cycle (color-coded green)
3. The electron transport chain and oxidative phosphorylation (color-coded lavender)

These metabolic pathways occur sequentially.

Glycolysis

Also called the *glycolytic pathway*, **glycolysis** (gli-kol′ĭ-sis; "sugar splitting") occurs in the cytosol of cells. This pathway, a series of ten chemical steps, converts glucose to two *pyruvic acid* molecules. All steps are fully reversible except the first, during which glucose entering the cell is phosphorylated to glucose-6-phosphate.

Glycolysis is an *anaerobic process*. Although this term is sometimes mistakenly interpreted to mean the pathway occurs only in the absence of oxygen, it actually means that glycolysis *does not use oxygen and occurs whether or not oxygen is present.*

The three major phases of glycolysis shown in **Figure 23.6** are described next. Appendix A shows the complete glycolytic pathway.

Phase 1. Sugar activation. In phase 1, glucose is phosphorylated and converted to fructose-6-phosphate, which is then phosphorylated again. These three steps use two ATP molecules (which are recouped later) and yield fructose-1,6-bisphosphate. The two separate reactions of the sugar with ATP provide the *activation energy* needed to prime the later stages of the pathway, so phase 1 is sometimes called the *energy investment phase.*

Phase 2. Sugar cleavage. During phase 2, fructose-1,6-bisphosphate is split into two 3-carbon fragments that exist (interconvertibly) as one of two isomers: glyceraldehyde (glis″er-al′dĕ-hīd) 3-phosphate or dihydroxyacetone (di″hi-drok″se-as′ĕ-tōn) phosphate.

Phase 3. Sugar oxidation and ATP formation. In phase 3, actually consisting of six steps, two major events happen. First, the two 3-carbon fragments are oxidized by the removal of hydrogen, which NAD^+ picks up. In this way, some of glucose's energy is transferred to NAD^+. Second, inorganic phosphate groups (P_i) are attached to each oxidized fragment by high-energy bonds. Later, when these terminal phosphates are split off, enough energy is captured to form four ATP molecules. As we noted earlier, formation of ATP this way is called *substrate-level phosphorylation.*

The final products of glycolysis are two molecules of **pyruvic acid** and two molecules of reduced NAD^+ (which is $NADH + H^+$). There is a net gain of two ATP molecules per glucose molecule. Four ATPs are produced, but remember that two are consumed in phase 1 to "prime the pump." Each pyruvic acid molecule has the formula $C_3H_4O_3$, and glucose is $C_6H_{12}O_6$. Between them the two pyruvic acid molecules have lost four hydrogen atoms, whose electrons are now bound to two molecules of NAD^+. NAD carries a positive charge (NAD^+), so when it accepts a hydrogen pair, $NADH + H^+$ is the resulting reduced product. Although a small amount of ATP has been harvested, the other two products of glucose oxidation (H_2O and CO_2) have yet to appear.

The fate of pyruvic acid, which still contains most of glucose's chemical energy, depends on the availability of oxygen at the time the pyruvic acid is produced. Because the supply of NAD^+ is limited, glycolysis can continue only if the reduced coenzymes ($NADH + H^+$) formed during glycolysis are relieved of their extra hydrogen. Only then can they continue to act as hydrogen acceptors.

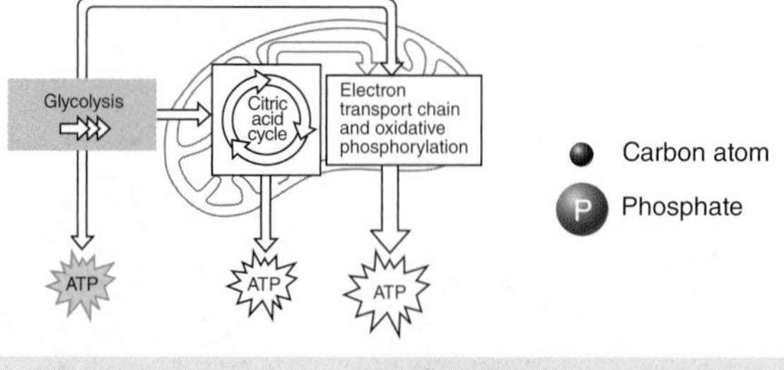

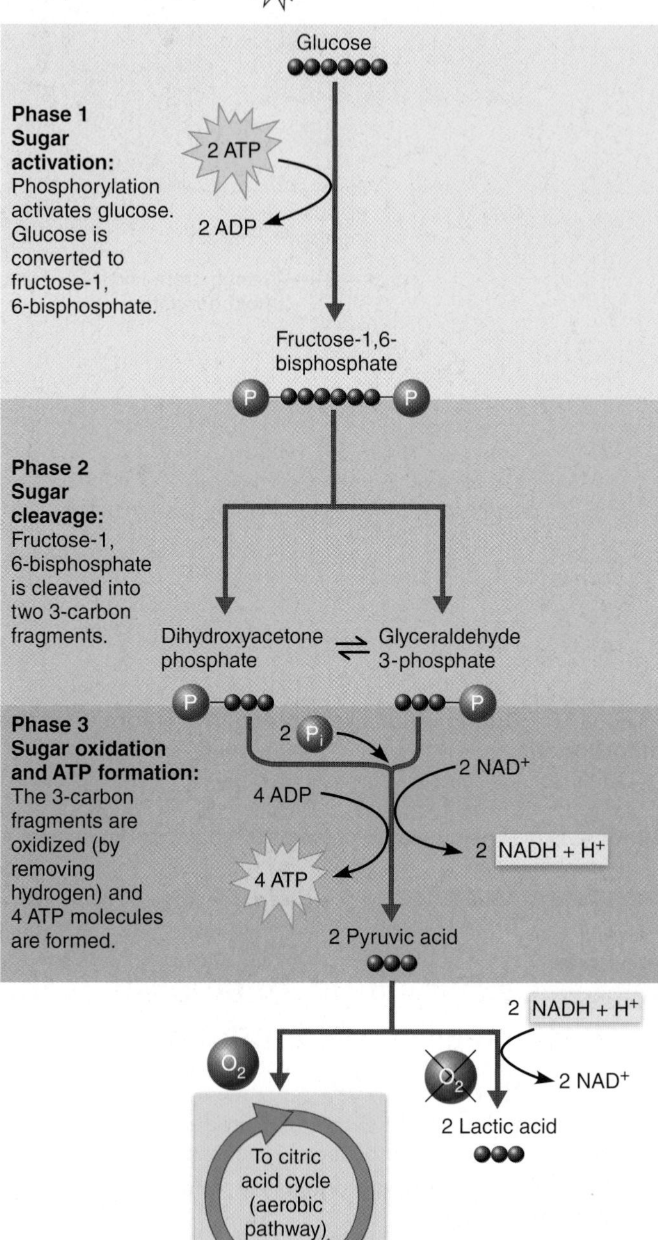

Figure 23.6 The three major phases of glycolysis. The fate of pyruvic acid depends on whether or not molecular O_2 is available.

When oxygen is readily available, this is no problem. NADH + H^+ delivers its burden of hydrogen atoms to the enzymes of the electron transport chain in the mitochondria, which deliver them to O_2, forming water. However, when oxygen is not present in sufficient amounts, as might occur during strenuous exercise, NADH + H^+ unloads its hydrogen atoms *back onto pyruvic acid*, reducing it. This addition of two hydrogen atoms to pyruvic acid yields **lactic acid** (see bottom right of Figure 23.6). Some of this lactic acid diffuses out of the cells and is transported to the liver for processing.

When oxygen is again available, lactic acid is oxidized back to pyruvic acid and enters the **aerobic pathways** (the oxygen-requiring citric acid cycle and electron transport chain within the mitochondria), and is completely oxidized to water and carbon dioxide. The liver may also convert lactic acid all the way back to glucose-6-phosphate (reverse glycolysis). Glucose-6-phosphate can either be stored as glycogen, or freed of its phosphate and released to the blood if blood sugar levels are low.

Except for red blood cells (which typically carry out *only* glycolysis), prolonged anaerobic metabolism ultimately results in acid-base problems. Consequently, *totally* anaerobic conditions resulting in lactic acid formation provide only a temporary route for rapid ATP production. Totally anaerobic conditions can go on without tissue damage for the longest periods in skeletal muscle, for much shorter periods in cardiac muscle, and almost not at all in the brain. Although glycolysis generates ATP rapidly, each glucose molecule yields only 2 ATP as compared to the 30 to 32 ATP when a glucose molecule is completely oxidized.

Citric Acid Cycle

The **citric acid cycle** (or Krebs cycle) is the next stage of glucose oxidation and is named for its first substrate. The citric acid cycle occurs in the mitochondrial matrix and is fueled largely by pyruvic acid produced during glycolysis and by fatty acids resulting from fat breakdown.

Because pyruvic acid is a charged molecule, it must enter the mitochondrion by active transport with the help of a transport protein. Once in the mitochondrion, the first order of business is a **transitional phase** that converts pyruvic acid to acetyl CoA. This occurs via a three-step process (**Figure 23.7**, top):

1. **Decarboxylation.** In this step, one of pyruvic acid's carbons is removed and released as carbon dioxide gas, a process called **decarboxylation.** CO_2 diffuses out of the cells into the blood to be expelled by the lungs. This is the first time that CO_2 is released during cellular respiration.
2. **Oxidation.** The remaining 2C fragment is oxidized to acetic acid by removing hydrogen atoms, which are picked up by NAD^+.
3. **Formation of acetyl CoA.** Acetic acid is combined with *coenzyme A* to produce the reactive final product, **acetyl coenzyme A (acetyl CoA)**. Coenzyme A is a sulfur-containing coenzyme derived from vitamin B_5.

Acetyl CoA is now ready to enter the citric acid cycle and be broken down completely by mitochondrial enzymes. Coenzyme A shuttles the 2-carbon acetic acid to an enzyme that joins it to a 4-carbon acid called **oxaloacetic acid** (ok″sah-lo″ah-sēt′ik) to produce the 6-carbon **citric acid.**

As the cycle moves through its eight successive steps, the atoms of citric acid are rearranged to produce different intermediate molecules, most called **keto acids** (Figure 23.7). The acetic acid that enters the cycle is broken apart carbon by carbon (decarboxylated) and oxidized, generating NADH + H^+ and $FADH_2$. At the end of the cycle, acetic acid has been totally disposed of and oxaloacetic acid, the *pickup molecule*, is regenerated.

For each turn of the cycle, we get:

- Two CO_2 molecules that come from two *decarboxylations.*
- Four molecules of reduced coenzymes (3 NADH + H^+ and 1 $FADH_2$). The addition of water at certain steps accounts for some of the released hydrogen.
- One molecule of ATP (via substrate-level phosphorylation).

The detailed events of each of the eight steps of the citric acid cycle are described in Appendix A.

Now let's back up and account for the pyruvic acid molecules entering the mitochondria. We need to consider the products of both the transitional phase and the citric acid cycle itself. Altogether, each pyruvic acid yields three CO_2 molecules and five molecules of reduced coenzymes—1 $FADH_2$ and 4 NADH + H^+ (equal to removing 10 hydrogen atoms). The products of glucose oxidation in the citric acid cycle are twice that (remember 1 glucose = 2 pyruvic acids): six CO_2, ten molecules of reduced coenzymes, and two ATP molecules.

Notice that it is these citric acid cycle reactions that produce the CO_2 released during glucose oxidation. The reduced coenzymes, which carry their extra electrons in high-energy linkages, must now be oxidized if the citric acid cycle and glycolysis are to continue.

Although glycolysis is exclusive to carbohydrate oxidation, breakdown products of carbohydrates, fats, and proteins can feed into the citric acid cycle to be oxidized for energy. On the other hand, some citric acid cycle intermediates can be siphoned off to make fatty acids and nonessential amino acids. Thus, the citric acid cycle is a source of building materials for anabolic reactions, as well as the final common pathway for oxidizing food fuels.

Electron Transport Chain and Oxidative Phosphorylation

Like glycolysis, none of the reactions of the citric acid cycle use oxygen directly. This is the exclusive function of the **electron transport chain**, which carries out the final catabolic reactions that occur on the inner mitochondrial membrane. However, because the reduced coenzymes produced in the citric acid cycle are the substrates for the electron transport chain, these two pathways are coupled, and both are *aerobic*, meaning they require oxygen.

In the electron transport chain, the hydrogens removed during the oxidation of food fuels are combined with O_2 to form water, and the energy released during those reactions is harnessed to attach P_i groups to ADP, forming ATP. As we noted earlier, this type of phosphorylation process is called *oxidative phosphorylation*. Let's peek under the hood of a cell's power plant and see how this rather complicated process works.

Just as in a car, what we see is a complicated structure with many parts. Why are there so many parts? In a car engine, we burn fuel for energy. If released all at once, that energy would

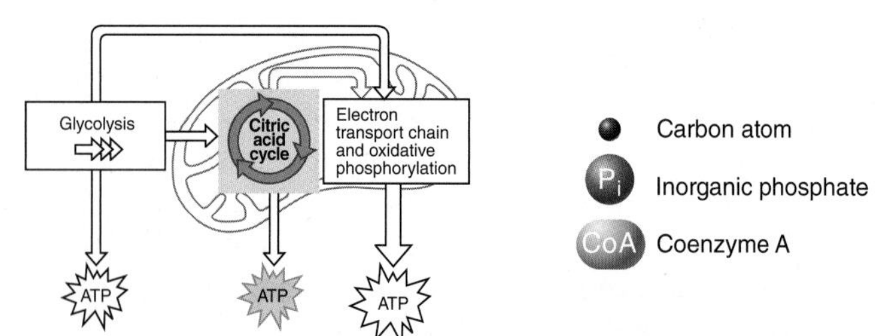

Figure 23.7 Simplified version of the citric acid (Krebs) cycle. During each turn of the cycle, two carbon atoms are removed from the substrates as CO_2 (decarboxylation reactions); four oxidations by removal of hydrogen atoms occur, producing four molecules of reduced coenzymes (3 NADH + H^+ and 1 $FADH_2$); and one ATP is synthesized by substrate-level phosphorylation. An additional decarboxylation and an oxidation reaction occur in the transitional phase (at top) that converts pyruvic acid, the product of glycolysis, to acetyl CoA, the molecule that enters the citric acid cycle pathway.

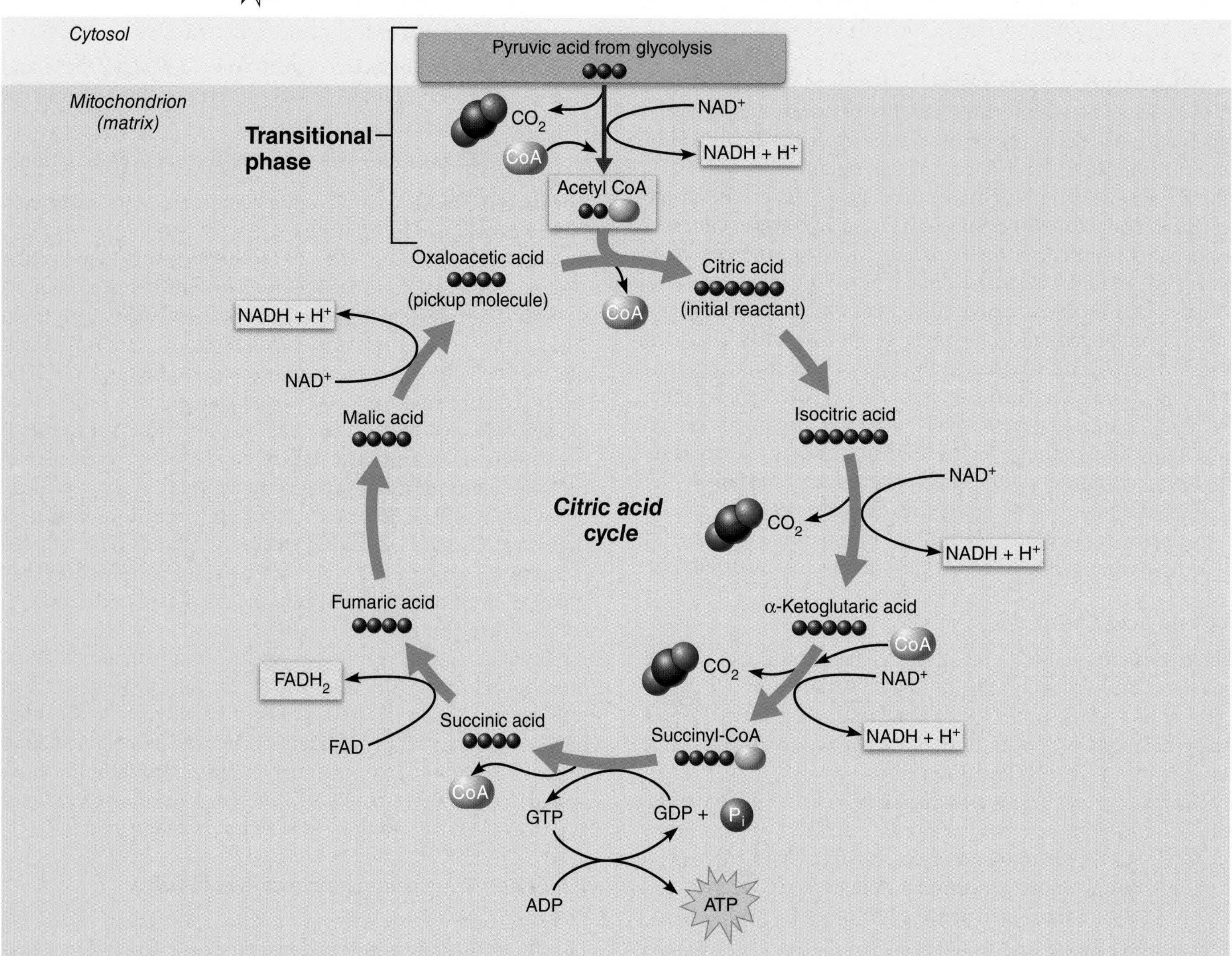

23

result in a big, hot explosion. All of the complex engine parts make it possible to capture some of this energy to do useful work. In the same way, the electron transport chain harvests the energy of food fuels a bit at a time in a step-by-step way so we can ultimately make ATP.

Most components of the electron transport chain are proteins that bind metal atoms (known as *cofactors*). These proteins vary in composition and form multiprotein complexes that are firmly embedded in the inner mitochondrial membrane as shown in *Focus on Oxidative Phosphorylation* (**Focus Figure 23.1**). For example, some of the proteins, the **flavins**, contain flavin mononucleotide (FMN) derived from the vitamin riboflavin, and others contain both sulfur (S) and iron (Fe). Most of these proteins, however, are brightly colored iron-containing pigments called **cytochromes** (si′to-krōmz; *cyto* = cell, *chrom* = color), including complexes III and IV depicted in Focus Figure 23.1. Neighboring carriers are clustered together to form four **respiratory enzyme complexes** that are alternately reduced and oxidized as they pick up electrons and pass them on to the next complex in the sequence.

As Focus Figure 23.1 shows, the first such complex accepts hydrogen atoms from NADH + H^+, oxidizing it to NAD^+. $FADH_2$ transfers its hydrogen atoms slightly farther along the chain to the small complex II. The hydrogen atoms that the reduced coenzymes deliver to the electron transport chain are quickly split into protons (H^+) plus electrons. The electrons are shuttled along the inner mitochondrial membrane from one complex to the next, losing energy with each transfer. The

FOCUS Oxidative Phosphorylation

Focus Figure 23.1 Oxidative phosphorylation has two phases:
Phase 1: The electron transport chain creates a proton (H^+) gradient across the inner mitochondrial membrane using high-energy electrons removed from food fuels.
Phase 2: Chemiosmosis uses the energy of the proton gradient to synthesize ATP.

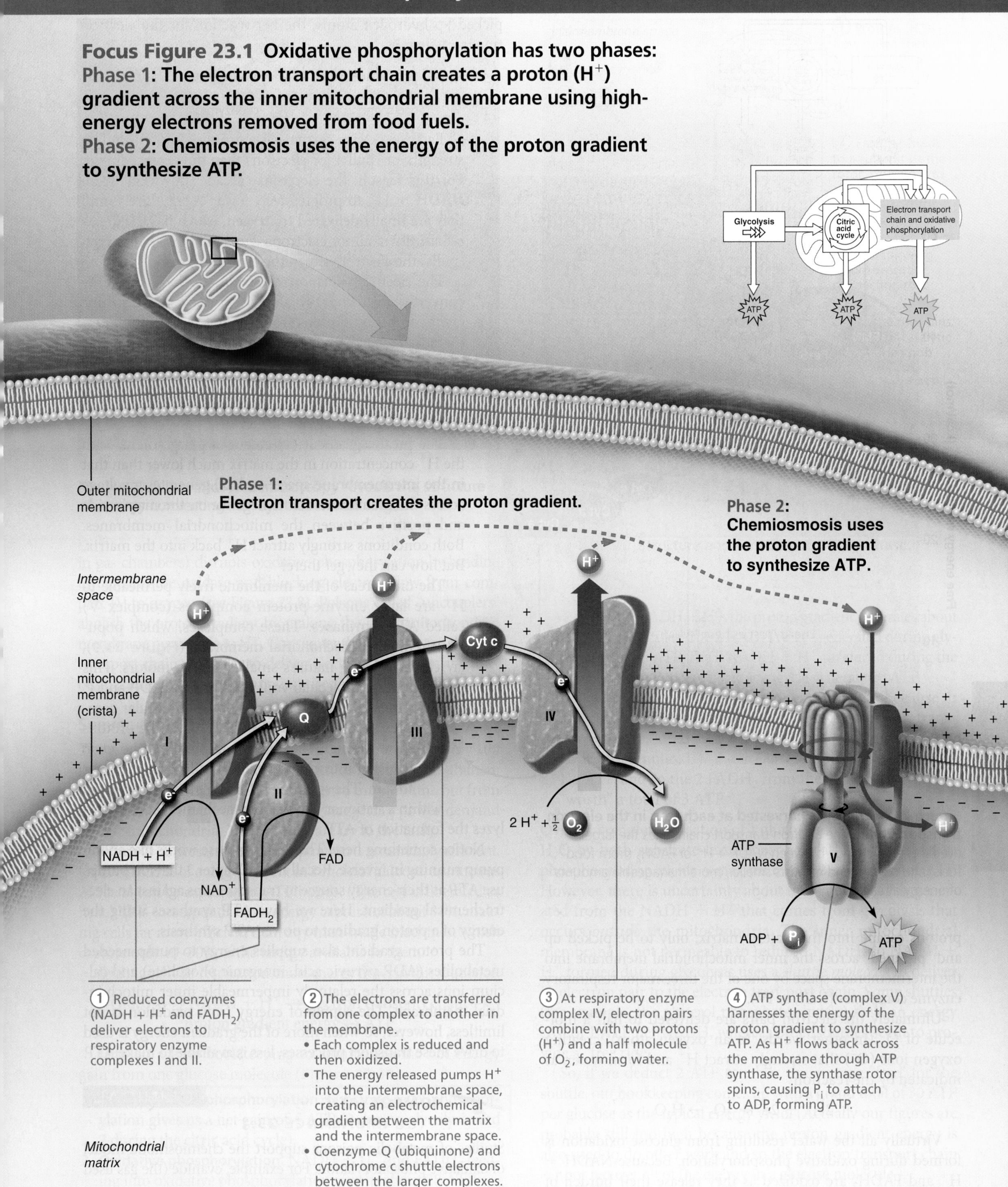

① Reduced coenzymes (NADH + H^+ and $FADH_2$) deliver electrons to respiratory enzyme complexes I and II.

② The electrons are transferred from one complex to another in the membrane.
- Each complex is reduced and then oxidized.
- The energy released pumps H^+ into the intermembrane space, creating an electrochemical gradient between the matrix and the intermembrane space.
- Coenzyme Q (ubiquinone) and cytochrome c shuttle electrons between the larger complexes.

③ At respiratory enzyme complex IV, electron pairs combine with two protons (H^+) and a half molecule of O_2, forming water.

④ ATP synthase (complex V) harnesses the energy of the proton gradient to synthesize ATP. As H^+ flows back across the membrane through ATP synthase, the synthase rotor spins, causing P_i to attach to ADP, forming ATP.

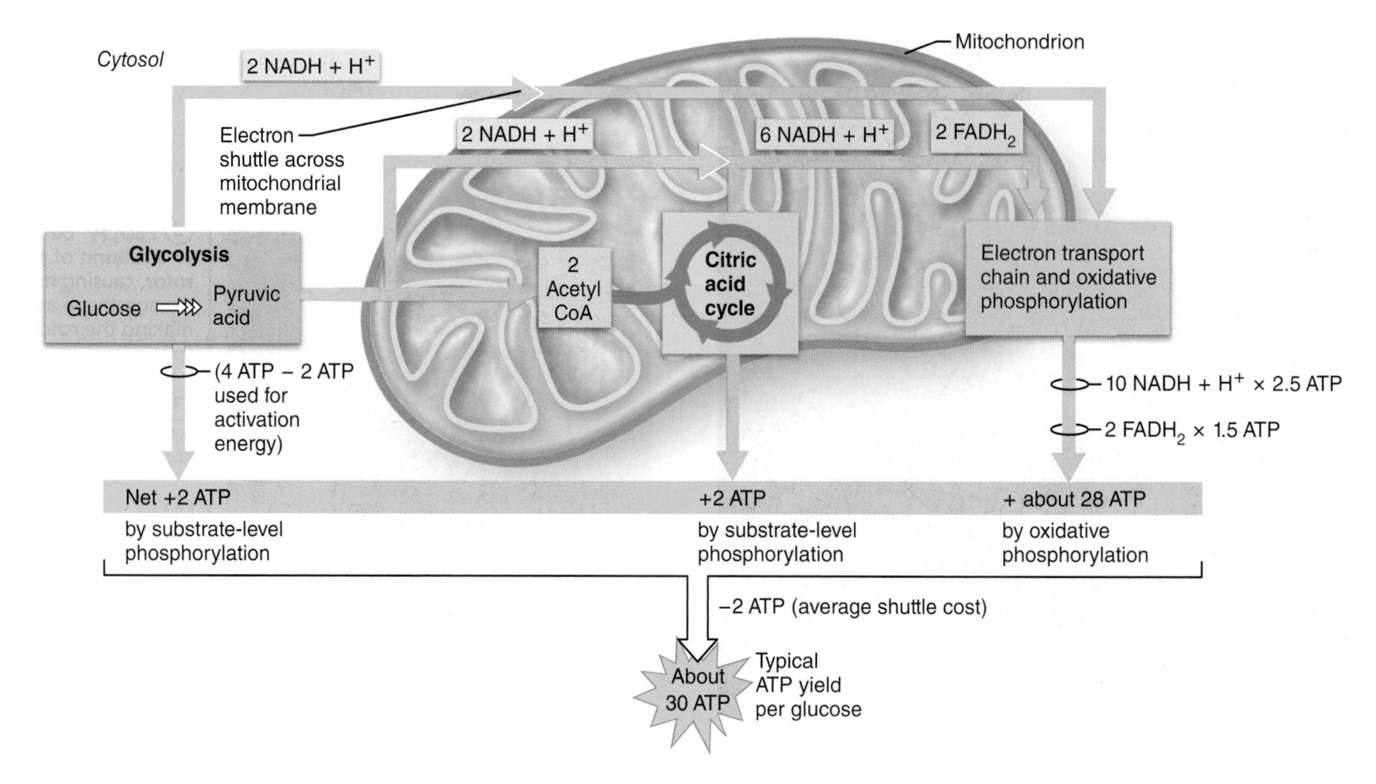

Figure 23.11 Energy yield during cellular respiration.

Glycogenesis, Glycogenolysis, and Gluconeogenesis

Although most glucose is used to generate ATP molecules, unlimited amounts of glucose do *not* result in unlimited ATP synthesis, because cells cannot store large amounts of ATP. Aside from cellular respiration, the goal of carbohydrate metabolism is to make sure that just the right amount of glucose is present in the blood. Three processes with similar sounding names are required. **Figure 23.12** will help you keep them straight.

When more glucose is available than can immediately be oxidized, rising intracellular ATP concentrations eventually inhibit glucose catabolism and cause glucose to be stored as glycogen or fat. Because the body can store much more fat than glycogen, fats account for 80–85% of stored energy.

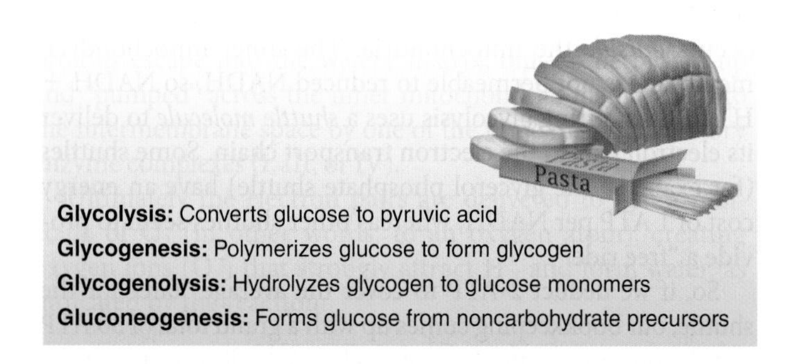

Figure 23.12 Quick summary of carbohydrate reactions.

Glycogenesis

When high ATP levels begin to "turn off" glycolysis, glucose molecules are combined in long chains to form glycogen, the animal carbohydrate storage product. This process is called **glycogenesis** (*glyco* = sugar; *genesis* = origin) (**Figure 23.13**, left side).

Glycogenesis begins as glucose entering cells is phosphorylated to glucose-6-phosphate and then converted to its isomer, *glucose-1-phosphate*. The terminal phosphate group is split off as the enzyme *glycogen synthase* catalyzes the attachment of glucose to the growing glycogen chain. Liver and skeletal muscle cells are most active in glycogen synthesis and storage.

Glycogenolysis

On the other hand, when blood glucose levels drop, glycogen lysis (splitting) occurs. This process is known as **glycogenolysis** (gli″ko-jĕ-nol′ĭ-sis) (Figure 23.13, right side). The enzyme *glycogen phosphorylase* oversees phosphorylation and splitting of glycogen to release glucose-1-phosphate, which is then converted to glucose-6-phosphate, a form that can enter the glycolysis pathway to be oxidized for energy.

In muscle cells and most other cells, the glucose-6-phosphate resulting from glycogenolysis is trapped because it cannot cross the cell membrane. However, hepatocytes (and some kidney and intestinal cells) contain *glucose-6-phosphatase*, an enzyme that removes the terminal phosphate, producing free glucose. Because glucose can then readily diffuse from the cell into the blood, the liver can use its glycogen stores to provide blood sugar for other organs when blood glucose levels drop. Liver glycogen is also an important energy source for skeletal muscles that have depleted their own glycogen reserves.

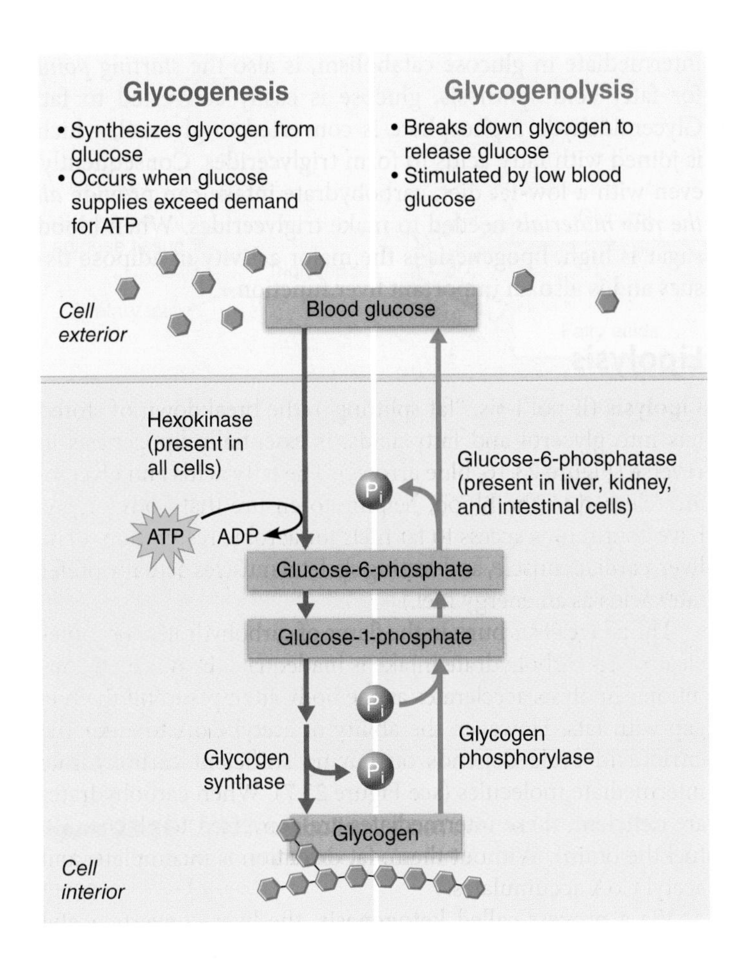

Figure 23.13 Glycogenesis and glycogenolysis.

Gluconeogenesis

When too little glucose is available to stoke the "metabolic furnace," glycerol and amino acids are converted to glucose. **Gluconeogenesis**, the process of forming new (*neo*) glucose from *noncarbohydrate* molecules, occurs in the liver.

Gluconeogenesis takes place when dietary sources and glucose reserves have been used up and blood glucose levels are beginning to drop. Gluconeogenesis protects the body, especially the nervous system, from the damaging effects of low blood sugar (*hypoglycemia*) by ensuring that ATP synthesis can continue.

☑ Check Your Understanding

11. Briefly, how do substrate-level and oxidative phosphorylation differ?
12. What happens in glycolysis if oxygen and pyruvic acid are absent and NADH + H^+ cannot transfer its "picked-up" hydrogen to pyruvic acid?
13. What two major kinds of chemical reactions occur in the citric acid cycle, and how are these reactions indicated symbolically?
14. What name is given to the chemical reaction in which glycogen is broken down to its glucose subunits?

For answers, see Answers Appendix.

23.5 Lipid metabolism is key for long-term energy storage and release

→ Learning Objectives

- ☐ **Describe the process by which fatty acids are oxidized for energy.**
- ☐ **Define ketone bodies, and indicate the stimulus for their formation.**

Fats are the body's most concentrated source of energy. They contain very little water, and the energy yield from fat catabolism is approximately twice that from either glucose or protein catabolism—9 kcal per gram of fat versus 4 kcal per gram of carbohydrate or protein. Most products of fat digestion are transported in lymph in the form of fatty-protein droplets called *chylomicrons* (see Chapter 22). Eventually, enzymes on capillary endothelium hydrolyze the lipids in the chylomicrons, and the resulting fatty acids and glycerol are taken up by body cells and processed in various ways. **Figure 23.14** summarizes the key metabolic reactions for lipids.

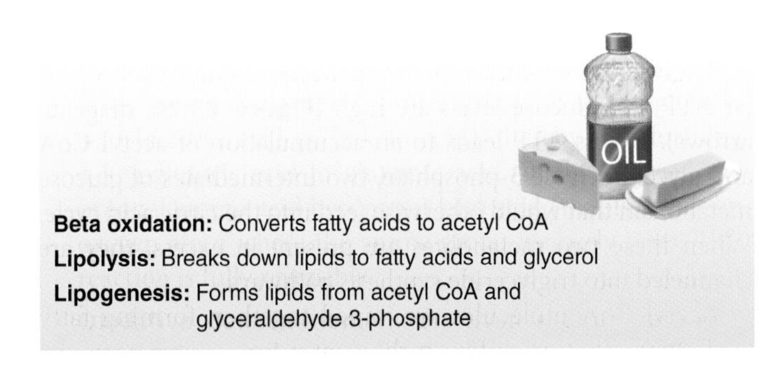

Figure 23.14 Quick summary of lipid reactions.

Oxidation of Glycerol and Fatty Acids

Of the various lipids, only triglycerides are routinely oxidized for energy. Their catabolism involves the separate oxidation of their two different building blocks: glycerol and fatty acid chains (**Figure 23.15**).

Most body cells easily convert glycerol to glyceraldehyde 3-phosphate (a glycolysis intermediate) and eventually to acetyl CoA that enters the citric acid cycle. Glyceraldehyde is equal to half a glucose molecule, and ATP energy harvest from its complete oxidation is approximately half that of glucose (15 ATP/glycerol).

Beta oxidation, the initial phase of fatty acid oxidation, occurs in the mitochondria. The net result is that the fatty acid chains are broken apart into two-carbon *acetic acid* fragments, and coenzymes (FAD and NAD^+) are reduced (Figure 23.15, right side). Each acetic acid molecule is fused to coenzyme A, forming acetyl CoA. The term "beta oxidation" reflects the fact that the carbon in the beta (third) position is oxidized each time a two-carbon fragment is broken off. Acetyl CoA then enters the citric acid cycle where it is oxidized to CO_2 and H_2O.

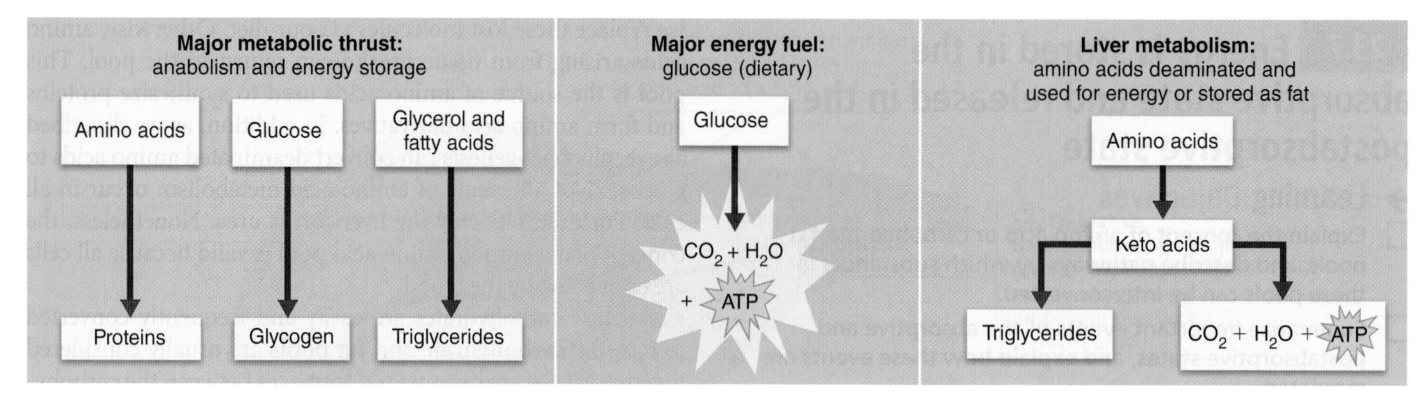

(a) Major events of the absorptive state

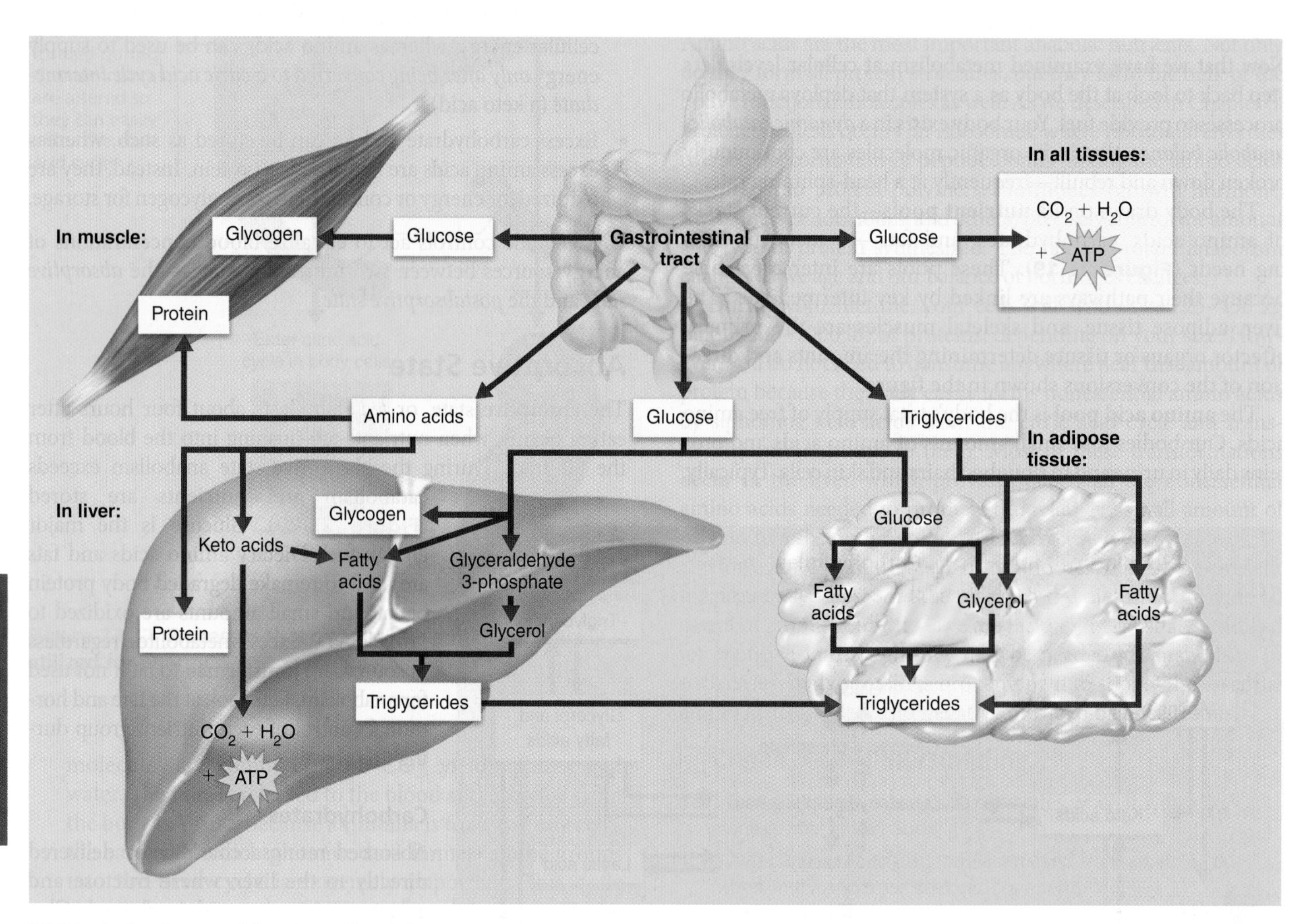

(b) Principal pathways of the absorptive state

Figure 23.20 **Major events and principal metabolic pathways of the absorptive state.** Although not indicated in (b), amino acids are also taken up by tissue cells and used for protein synthesis, and fats (triglycerides) are the primary energy fuel of muscle, liver cells, and adipose tissue.

Triglycerides

Nearly all products of fat digestion enter the lymph in the form of chylomicrons, which are hydrolyzed to fatty acids and glycerol before they can pass through the capillary walls. *Lipoprotein lipase*, the enzyme that catalyzes fat hydrolysis, is particularly active in the capillaries of muscle and fat tissues.

Adipose cells, skeletal and cardiac muscle cells, and liver cells use triglycerides as their primary energy source, but when dietary carbohydrates are limited, other cells begin to oxidize more fat for energy. Although some fatty acids and glycerol are used for anabolic purposes by tissue cells, most enter adipose tissue to be reconverted to triglycerides and stored.

Amino Acids

Absorbed amino acids are delivered to the liver, which deaminates some of them to keto acids. The keto acids may flow into the citric acid cycle to be used for ATP synthesis, or they may be converted to liver fat stores. The liver also uses some of the amino acids to synthesize plasma proteins, including albumin, clotting proteins, and transport proteins. However, most amino acids flushing through the liver sinusoids remain in the blood for uptake by other body cells, where they are used to synthesize proteins.

Hormonal Control of the Absorptive State

Insulin directs essentially all events of the absorptive state (**Figure 23.21**). After a meal, rising blood glucose and amino acid levels stimulate the beta cells of the pancreatic islets to secrete more insulin (see Figure 15.18). The GI tract hormone *glucose-dependent insulinotropic peptide* (*GIP*) and parasympathetic stimulation also promote the release of insulin.

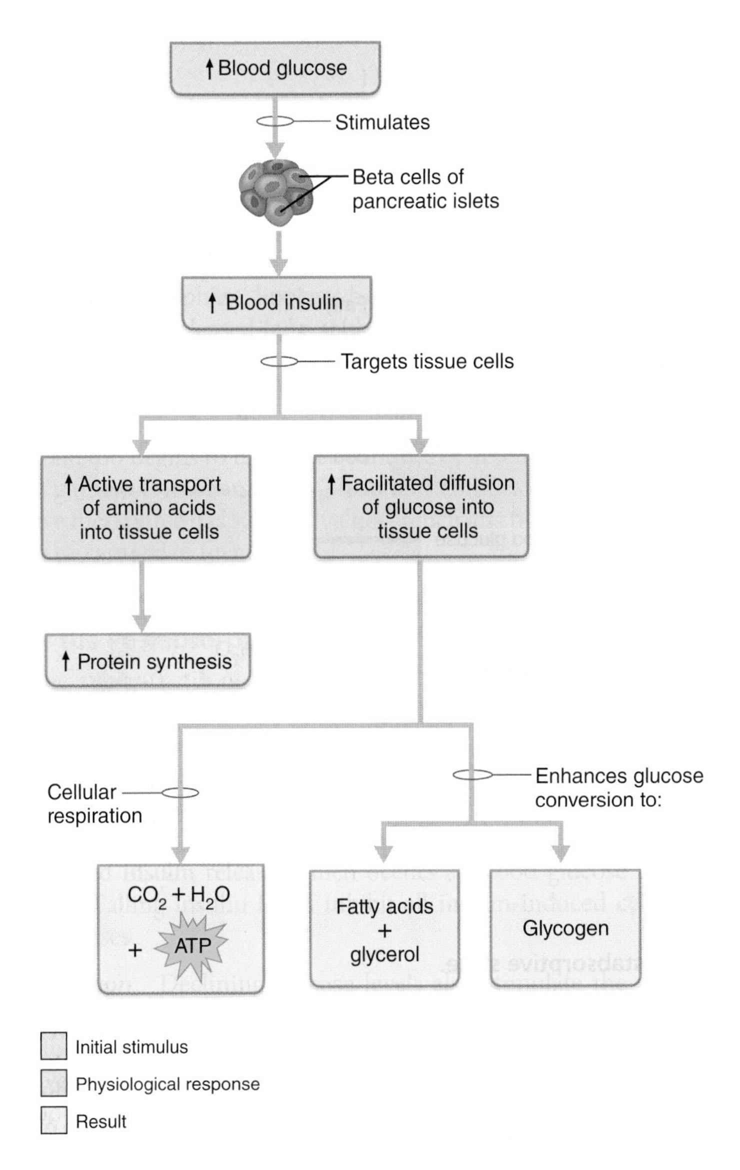

Figure 23.21 Insulin directs nearly all events of the absorptive state. (Note: Not all effects shown occur in all cells.)

Insulin binds to membrane receptors of its target cells. This stimulates the translocation of glucose transporters to the plasma membrane, which enhances the carrier-mediated facilitated diffusion of glucose into those cells. Within minutes, the rate of glucose entry into tissue cells (particularly muscle and adipose cells) increases about 20-fold. The exception is brain and liver cells, which take up glucose whether or not insulin is present.

Once glucose enters tissue cells, insulin enhances glucose oxidation for energy and stimulates its conversion to glycogen and, in adipose tissue, to triglycerides. Insulin also "revs up" the active transport of amino acids into cells, promotes protein synthesis, and inhibits liver export of glucose and virtually all liver enzymes that promote gluconeogenesis.

As you can see, insulin is a **hypoglycemic hormone** (hi″po-gli-se′mik). It sweeps glucose out of the blood into tissue cells, lowering blood glucose levels. It also enhances glucose oxidation or storage while inhibiting any process that might raise blood glucose levels.

Diabetes mellitus is a disorder of inadequate insulin production or abnormal insulin receptors. Without insulin or receptors that "recognize" it, glucose is unavailable to most body cells. Blood glucose levels remain high, and large amounts of glucose are excreted in urine. Metabolic acidosis, protein wasting, and weight loss occur as large amounts of fats and tissue proteins are used for energy. (Chapter 15 describes diabetes mellitus in more detail.) +

Postabsorptive State

The postabsorptive state, or *fasting state,* is the period when the GI tract is empty and body reserves are broken down to supply energy. Net synthesis of fat, glycogen, and proteins ends, and catabolism of these substances begins (**Figure 23.22a**).

The primary goal during the postabsorptive state is to maintain blood glucose levels within the homeostatic range (70–110 mg of glucose per 100 ml). Remember that constant blood glucose is important because the brain almost always uses glucose as its energy source. Most events of the postabsorptive state either (1) make glucose available to the blood or (2) make certain organs (such as skeletal muscle) switch over to using fats instead of glucose to spare glucose for organs that can't use fats.

Sources of Blood Glucose

So where does blood glucose come from in the postabsorptive state? Sources include stored glycogen in the liver and skeletal muscles, tissue proteins, and, in limited amounts, fats (Figure 23.22b).

1. **Glycogenolysis in the liver.** The liver's glycogen stores (about 100 g) are the first line of glucose reserves. They are mobilized quickly and can maintain blood sugar levels for about four hours during the postabsorptive state.
2. **Glycogenolysis in skeletal muscle.** Glycogen stores in skeletal muscle are approximately equal to those of the liver. Before liver glycogen is exhausted, glycogenolysis begins

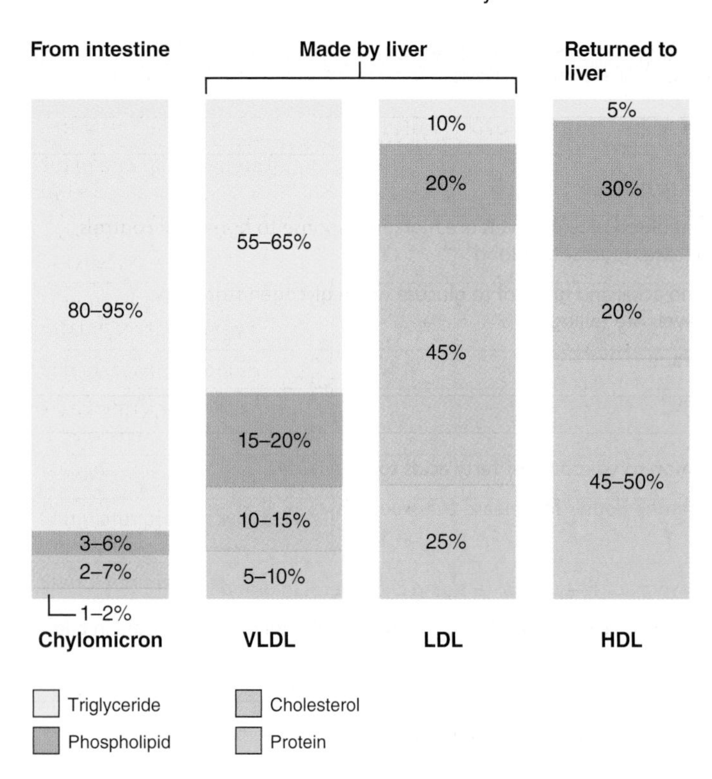

Figure 23.24 Approximate composition of lipoproteins that transport lipids in body fluids. (VLDL = very low-density lipoprotein, LDL = low-density lipoprotein, HDL = high-density lipoprotein)

this basis, there are **very low-density lipoproteins (VLDLs), low-density lipoproteins (LDLs),** and **high-density lipoproteins (HDLs).** *Chylomicrons*, which transport absorbed lipids from the GI tract, have the lowest density of all.

The liver is the primary source of VLDLs, which transport triglycerides from the liver to peripheral (nonliver) tissues, mostly to *adipose tissues*. Once the triglycerides are unloaded, the residues are converted to LDLs, which are cholesterol-rich. The job of the LDLs is to transport cholesterol *to peripheral tissues*, making it available to tissue cells to synthesize membranes or hormones, and to store it for later use.

Most cells other than liver and intestinal cells obtain the bulk of the cholesterol they need for membrane synthesis from the blood. When a cell needs cholesterol, it makes membrane receptor proteins for LDL. LDL binds to the receptors, is engulfed by endocytosis, and the endocytotic vesicles fuse with lysosomes, where the cholesterol is freed for use. When excessive cholesterol accumulates in a cell, it inhibits both the cell's own cholesterol synthesis and its synthesis of LDL receptors.

The major function of HDLs, which are particularly rich in phospholipids and proteins, is to scoop up and transport excess cholesterol *from peripheral tissues to the liver*, where it is broken down and becomes part of bile. The liver makes the protein envelopes of the HDL particles and then ejects them into the bloodstream in collapsed form, rather like deflated beach balls. Once in the blood, these still-incomplete HDL particles fill with cholesterol picked up from tissue cells and "pulled" from artery walls. HDL also provides the steroid-producing organs, like the ovaries and adrenal glands, with their raw material (cholesterol).

Recommended Total Cholesterol, HDL, and LDL Levels

For adults, the maximum recommended total cholesterol level is 200 mg/dl of blood. Blood cholesterol levels above 200 mg/dl have been linked to atherosclerosis, which clogs the arteries and causes strokes and heart attacks. However, it is not enough to simply measure total cholesterol. How cholesterol is packaged for transport in the blood is more important clinically.

As a rule, high levels of HDLs are considered *good* because the transported cholesterol is destined for degradation (think H for healthy). HDL levels above 60 mg/dl are thought to protect against heart disease, and levels below 40 are considered undesirable. In the United States, HDL levels average 40–50 in males and 50–60 in women.

High LDL levels (160 mg/dl or above) are considered *bad* (think L for lousy) because when LDLs are excessive, potentially lethal cholesterol deposits are laid down in the artery walls. The goal for LDL levels is 100 or less. A good rule of thumb is that HDL levels can't be too high and LDL levels can't be too low.

Factors Regulating Blood Cholesterol Levels

A negative feedback loop partially adjusts the amount of cholesterol produced by the liver according to the amount of cholesterol in the diet. A high cholesterol intake inhibits its synthesis by the liver, but it is not a one-to-one relationship because the liver produces a basal amount of cholesterol even when dietary intake is high. Conversely, severely restricting dietary cholesterol, although helpful, does not markedly reduce blood cholesterol levels.

However, the relative amounts of saturated and unsaturated fatty acids in the diet do have an important effect on blood cholesterol levels. Saturated fatty acids *stimulate liver synthesis* of cholesterol and *inhibit its excretion* from the body. In contrast, unsaturated fatty acids (found in olive and most other vegetable oils) *enhance excretion* of cholesterol and its catabolism to bile salts, thereby reducing total cholesterol levels.

Trans fats are "healthy" oils that have been hardened by hydrogenation to make them more solid, such as some margarines. Trans fats have a worse effect on blood cholesterol levels than saturated fats do. The trans fatty acids spark a greater increase in LDLs and a greater reduction in HDLs, producing the unhealthiest combination.

The unsaturated omega-3 fatty acids found in especially large amounts in some cold-water fish (such as salmon) lower the proportions of both saturated fats and cholesterol. The omega-3 fatty acids make blood platelets less sticky, thus helping prevent spontaneous clotting that can block blood vessels. They also appear to lower blood pressure.

Factors other than diet also influence blood cholesterol levels. For example, cigarette smoking and stress lower HDL levels, whereas regular aerobic exercise and estrogen lower LDL levels and increase HDL levels. Interestingly, body shape provides clues to risky blood levels of cholesterol and fats. "Apples" (people with upper body and abdominal fat, seen more often in men) tend to have higher levels of cholesterol and LDLs than "pears" (whose fat is localized in the hips and thighs, a pattern more common in women).

Previously, high cholesterol and LDL:HDL ratios were considered the most valid predictors of risk for atherosclerosis, cardiovascular disease, and heart attack. However, almost half of those who get heart disease have normal cholesterol levels, while others with poor lipid profiles remain free of heart problems. Presently, LDL levels and assessments of other cardiovascular disease risk factors are believed to be more accurate indicators of whether treatment is needed, and many physicians recommend dietary changes regardless of total cholesterol or HDL levels.

Cholesterol-lowering drugs such as *statins* are routinely prescribed for people with elevated LDL levels. It is estimated that more than 10 million Americans are now taking statins. +

☑ Check Your Understanding

24. If you had your choice, would you prefer to have high blood levels of HDLs or LDLs? Explain your answer.

25. What are trans fats and how do they affect LDL and HDL levels?

For answers, see Answers Appendix.

PART 3
ENERGY BALANCE

→ Learning Objective

☐ **Explain what is meant by body energy balance.**

When any fuel is burned, it consumes oxygen and liberates heat. The "burning" of food fuels by our cells is no exception. As we described in Chapter 2, energy can be neither created nor destroyed—only converted from one form to another. If we apply this principle (actually the *first law of thermodynamics*) to cell metabolism, it means that bond energy released as foods are catabolized (energy input) must be precisely balanced by the total energy output of the body. A dynamic balance exists between the body's energy intake and energy output:

$$\text{Energy intake} = \text{energy output} \quad (\text{heat} + \text{work} + \text{energy storage})$$

Energy intake is the energy liberated during food oxidation. **Energy output** includes energy (1) immediately lost as heat (about 60% of the total), (2) used to do work (driven by ATP), and (3) stored as fat or glycogen. Because losses of organic molecules in urine, feces, and perspiration are very small in healthy people, they are usually ignored in calculating energy output.

Nearly all the energy derived from foodstuffs is eventually converted to heat. Heat is lost during every cellular activity—when ATP bonds are formed and when they are broken to do work, as muscles contract, and through friction as blood flows through blood vessels. Though cells cannot use this energy to do work, the heat warms the tissues and blood and helps maintain the homeostatic body temperature that allows metabolic reactions to occur efficiently. Energy storage is an important part of the equation only during periods of growth and net fat deposit.

23.9 Neural and hormonal factors regulate food intake

→ Learning Objective

☐ **Describe several theories of food intake regulation.**

When energy intake and energy output are balanced, body weight remains stable. When they are not, weight is either gained or lost. Unhappily for many people, the body's weight-controlling systems appear to be designed more to protect us against weight loss than weight gain.

Obesity

How fat is too fat? What distinguishes a person who is obese from one who is merely overweight? Let's take a look.

The bathroom scale is an inaccurate guide because body weight tells little of body composition. Dense bones and well-developed muscles can make a fit, healthy person technically overweight. Arnold Schwarzenegger, for example, has tipped the scales at a hefty 257 lb.

Body mass index (BMI) is a formula for determining obesity based on a person's weight relative to height. To estimate BMI, multiply weight in pounds by 705 and then divide by your height in inches squared:

$$\text{BMI} = \text{wt(lb)} \times 705/\text{ht(inches)}^2$$

Overweight is defined by a BMI between 25 and 30 and carries some health risk. Obesity is a BMI greater than 30 and has a markedly increased health risk.

A body fat content of 18–20% of body weight (males and females respectively) is deemed normal for adults.

However it's defined, obesity is perplexing and poorly understood, and the economic toll of obesity-related disease is staggering. Chronic low-grade systemic inflammation accompanies obesity and contributes to insulin resistance and type 2 diabetes mellitus (in which a greater than normal amount of insulin is required to maintain normal blood glucose levels). People who are obese also have a higher incidence of atherosclerosis, hypertension, heart disease, and osteoarthritis.

The U.S. is big and getting bigger, at least around its middle. Two out of three adults are overweight, one out of three is obese, and one in twelve has diabetes. U.S. kids are getting fatter too: 20 years ago, 5% were overweight; today over 15% are and more are headed that way.

Regulation of Food Intake

Control of food intake poses difficult questions for researchers. For example, what type of receptor could sense the body's total calorie content and alert us to start eating or put down that fork? Despite heroic research efforts, no such single receptor type has been found.

It has been known for some time that the hypothalamus, particularly its *arcuate nucleus* (*ARC*) and two other areas—the *lateral hypothalamic area* (*LHA*) and the *ventromedial nucleus* (*VMN*)—release several peptides that influence feeding behavior. Most importantly, this influence ultimately

Convection

Convection is the process that occurs because warm air expands and rises and cool air, being denser, falls. Consequently, the warmed air enveloping the body is continually replaced by cooler air molecules. Convection substantially enhances heat transfer from the body surface to the air because the cooler air absorbs heat by conduction more rapidly than the already-warmed air.

Together, conduction and convection account for 15–20% of heat loss to the environment. These processes are enhanced by anything that moves air more rapidly across the body surface, such as wind or a fan, in other words, by *forced convection*.

Evaporation

The fourth mechanism by which the body loses heat is **evaporation**. Water evaporates because its molecules absorb heat from the environment and become energetic enough—in other words, vibrate fast enough—to escape as a gas, which we know as water vapor. The heat absorbed by water during evaporation is called **heat of vaporization**. The evaporation of water from body surfaces removes large amounts of body heat. Every gram of water that evaporates removes about 0.58 kcal of heat from the body.

There is a basal level of body heat loss due to the continuous evaporation of water from the lungs and oral mucosa, and through the skin. The unnoticeable water loss occurring via these routes is called **insensible water loss**, and the accompanying heat loss is **insensible heat loss**. Insensible heat loss dissipates about 10% of the basal heat production of the body and is a constant not subject to body temperature controls. When necessary, however, the body's control mechanisms do initiate heat-promoting activities to counterbalance this insensible heat loss.

Evaporative heat loss becomes an active or *sensible* process when body temperature rises and sweating produces increased amounts of water for vaporization. Extreme emotional states activate the sympathetic nervous system, causing body temperature to rise by one degree or so, and vigorous exercise can raise body temperature as much as 2–3°C (3.6–5.4°F). Vigorous muscular activity can produce and evaporate 1–2 L/h of perspiration, removing 600–1200 kcal of heat from the body each hour. This is more than 30 times the amount of heat lost via insensible heat loss!

HOMEOSTATIC IMBALANCE 23.6 CLINICAL

When sweating is heavy and prolonged, especially in untrained individuals, losses of water and NaCl may cause painful muscle spasms called *heat cramps*. The solution is simple: Drink fluids. +

Role of the Hypothalamus

Although other brain regions contribute, the hypothalamus, particularly its *preoptic* region, is the main integrating center for thermoregulation. Together the **heat-loss center** (located more anteriorly) and the **heat-promoting center** make up the brain's **thermoregulatory centers**.

The hypothalamus receives afferent input from (1) **peripheral thermoreceptors** located in the shell (the skin), and (2) **central thermoreceptors** sensitive to blood temperature and located in the body core including the anterior portion of the hypothalamus. Much like a thermostat, the hypothalamus responds to this input by reflexively initiating appropriate heat-promoting or heat-loss activities.

The central thermoreceptors have more influence than the peripheral ones, but varying inputs from the shell probably alert the hypothalamus to the need to prevent temperature changes in the core. In other words, they allow the hypothalamus to anticipate changes to be made.

Heat-Promoting Mechanisms

When the external temperature is low or blood temperature falls for any reason, the heat-promoting center is activated. It triggers one or more of the following mechanisms to maintain or increase core body temperature (Figure 23.28, bottom).

- **Constriction of cutaneous blood vessels.** Activation of the sympathetic vasoconstrictor fibers serving the blood vessels of the skin causes strong vasoconstriction. This restricts blood to deep body areas and largely bypasses the skin. Because a layer of insulating subcutaneous (fatty) tissue separates the skin from deeper organs, this reduces heat loss from the shell dramatically and lowers shell temperature toward that of the external environment.

HOMEOSTATIC IMBALANCE 23.7 CLINICAL

Restricting blood flow to the skin is not a problem for a brief period, but if it is prolonged (as during exposure to very cold weather), skin cells deprived of oxygen and nutrients begin to die. This extremely serious condition is *frostbite*. +

- **Shivering.** Shivering—involuntary shuddering contractions—is triggered when the hypothalamus activates brain centers that cause an increase in muscle tone. When muscle tone reaches sufficient levels, stretch receptors are alternately stimulated in antagonistic muscles. Shivering raises body temperature because skeletal muscle activity produces large amounts of heat.
- **Increase in metabolic rate.** Cold stimulates the adrenal medulla to release epinephrine and norepinephrine in response to sympathetic nerve stimuli, elevating the metabolic rate and enhancing heat production. This mechanism, called **chemical (nonshivering) thermogenesis**, occurs in infants. Recently, deposits of brown adipose tissue, a special kind of adipose tissue that dissipates energy by producing heat by this mechanism, have also been demonstrated in adult humans.
- **Enhanced release of thyroxine.** When environmental temperature decreases gradually, as in the transition from summer to winter, the hypothalamus of infants releases *thyrotropin-releasing hormone*. This hormone activates the anterior pituitary to release *thyroid-stimulating hormone*, which induces the thyroid to liberate more thyroid hormone to the blood. Because thyroid hormone raises metabolic rate, body heat production rises. Adults do not show a similar TSH response to cold exposure.

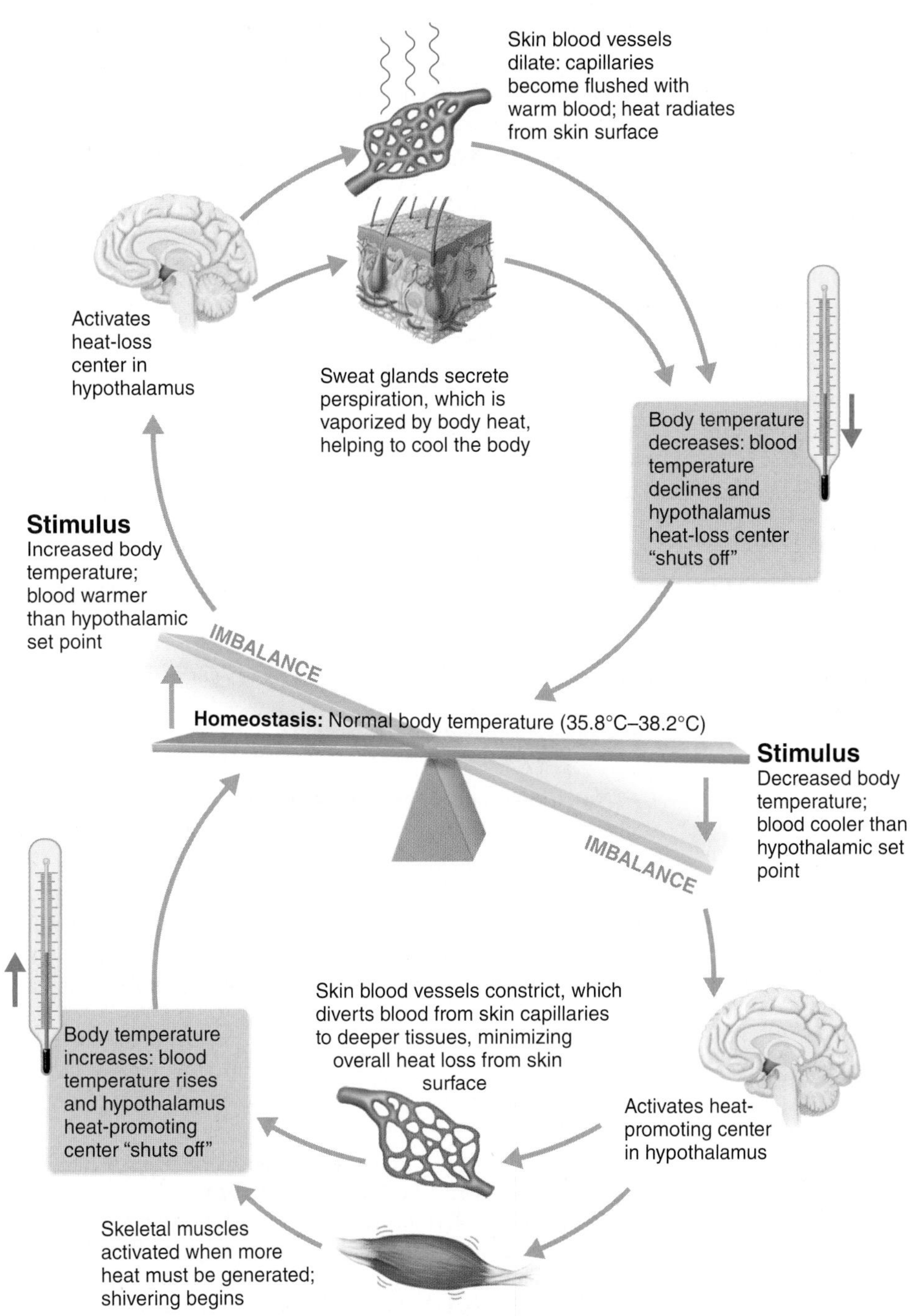

Figure 23.28 Mechanisms of body temperature regulation.

Besides these involuntary adjustments, we humans make a number of *behavioral modifications* to prevent overcooling of our body core:

- Putting on more or warmer clothing to restrict heat loss (hat, gloves, and insulated outer garments)
- Drinking hot fluids
- Changing posture to reduce exposed body surface area (hunching over or clasping the arms across the chest)
- Increasing physical activity to generate more heat (jumping up and down, clapping the hands)

Heat-Loss Mechanisms

How do heat-loss mechanisms protect the body from excessively high temperatures? Whenever core body temperature rises above normal, it inhibits the hypothalamic heat-promoting center. At the same time, it activates the heat-loss center and triggers one or both of the following (Figure 23.28, top):

- **Dilation of cutaneous blood vessels.** Inhibiting the vasomotor fibers serving blood vessels of the skin allows the vessels to dilate. As the blood vessels swell with warm blood, the shell loses heat by radiation, conduction, and convection.
- **Enhanced sweating.** If the body is extremely overheated or if the environment is so hot—over 33°C (about 92°F)—that heat cannot be lost by other means, evaporation becomes necessary. Sympathetic fibers activate the sweat glands to spew out large amounts of perspiration.

Evaporation of perspiration is an efficient means of ridding the body of surplus heat as long as the air is dry. However, when the relative humidity is high, evaporation occurs much more slowly. In such cases, the heat-liberating mechanisms cannot work well, and we feel miserable and irritable. Behavioral or voluntary measures commonly taken to reduce body heat in such circumstances include:

- Reducing activity ("laying low")
- Seeking a cooler environment (a shady spot) or using a device to increase convection (a fan) or cooling (an air conditioner)
- Wearing light-colored, loose clothing that reflects radiant energy. (This is actually cooler than being nude because bare skin absorbs most of the radiant energy striking it.)

Overexposure to a hot and humid environment makes normal heat-loss processes ineffective. The resulting **hyperthermia** (elevated body temperature) depresses the hypothalamus. At a core temperature of around 41°C (105°F), heat-control mechanisms are suspended, creating a vicious *positive feedback cycle*. Increasing temperatures increase the metabolic rate, which increases heat production. The skin becomes hot and dry and, as the temperature continues to spiral upward, multiple organ damage becomes a distinct possibility, including brain damage. This condition, called **heat stroke**, can be fatal unless corrective measures are initiated immediately (immersing the body in cool water and administering fluids).

The terms *heat exhaustion* and *exertion-induced heat exhaustion* are often used to describe the heat-associated extreme sweating and collapse of an individual during or following vigorous physical activity. This condition, evidenced by elevated body temperature and mental confusion and/or fainting, is due to dehydration and consequent low blood pressure. As heat-loss mechanisms struggling to function in heat exhaustion further give way, heat exhaustion merges into heat stroke. Death results if the body is not cooled and rehydrated promptly.

Hypothermia (hi″po-ther′me-ah) is low body temperature resulting from prolonged uncontrolled exposure to cold. Vital signs (respiratory rate, blood pressure, and heart rate) decrease as cellular enzymes become sluggish. Drowsiness sets in and, oddly, the person becomes comfortable even though previously he or she felt extremely cold. Shivering stops at a core temperature of 30–32°C (87–90°F) when the body has exhausted its heat-generating capabilities. Uncorrected, hypothermia progresses to coma and finally death (by cardiac arrest), when body temperatures approach 21°C (70°F). +

Fever

Fever is *controlled hyperthermia*. Most often, it results from infection somewhere in the body, but it may be caused by cancer, allergic reactions, or CNS injuries.

Whatever the cause, macrophages and other cells release cytokines that act as *pyrogens* (literally, "fire starters"). These chemicals act on the hypothalamus, causing release of prostaglandins which reset the hypothalamic thermostat to a higher-than-normal temperature, so that heat-promoting mechanisms kick in. As a result of vasoconstriction, heat loss from the body surface declines, the skin cools, and shivering begins to generate heat. These "chills" are a sure sign body temperature is rising.

The temperature rises until it reaches the new setting, and then is maintained at that setting until natural body defenses or medications reverse the disease process. Then, heat-loss mechanisms swing into action. Sweating begins and the skin becomes flushed and warm. Physicians have long recognized these signs as signals that body temperature is falling (aah, she has passed the crisis). As we explained in Chapter 20, fever speeds healing by increasing the metabolic rate, and it also appears to inhibit bacterial growth.

☑ Check Your Understanding

30. What is the body's core?

31. Andrea is flushed and her teeth are chattering even though her bedroom temperature is 72°F. Why do you think this is happening?

32. How does convection differ from conduction in causing heat loss?

For answers, see Answers Appendix.

Nutrition is one of the most overlooked areas in clinical medicine. Yet, what we eat and drink influences nearly every phase of metabolism and plays a major role in our overall health. Now that we have examined the fates of nutrients in body cells, we are ready to study the urinary system, the organ system that works tirelessly to rid the body of nitrogen wastes resulting from metabolism and to maintain the purity of our internal fluids.

REVIEW QUESTIONS

23

Multiple Choice/Matching

(Some questions have more than one correct answer. Select the best answer or answers from the choices given.)

1. Which of the following reactions would liberate the most energy? **(a)** complete oxidation of a molecule of sucrose to CO_2 and water, **(b)** conversion of a molecule of ADP to ATP, **(c)** respiration of a molecule of glucose to lactic acid, **(d)** conversion of a molecule of glucose to carbon dioxide and water.
2. The formation of glucose from glycogen is **(a)** gluconeogenesis, **(b)** glycogenesis, **(c)** glycogenolysis, **(d)** glycolysis.
3. The net gain of ATP from the complete metabolism (aerobic) of glucose is closest to **(a)** 2, **(b)** 30, **(c)** 3, **(d)** 4.
4. Which of the following best defines cellular respiration? **(a)** intake of carbon dioxide and output of oxygen by cells, **(b)** excretion of waste products, **(c)** inhalation of oxygen and exhalation of carbon dioxide, **(d)** oxidation of substances by which energy is released in usable form to the cells.
5. What is formed during aerobic respiration when electrons are passed down the electron transport chain? **(a)** oxygen, **(b)** water, **(c)** glucose, **(d)** NADH + H^+.
6. Metabolic rate is relatively low in **(a)** youth, **(b)** physical exercise, **(c)** old age, **(d)** fever.
7. In a temperate climate under ordinary conditions, the greatest loss of body heat occurs through **(a)** radiation, **(b)** conduction, **(c)** evaporation, **(d)** none of the above.
8. Which of the following is not a function of the liver? **(a)** glycogenolysis and gluconeogenesis, **(b)** synthesis of cholesterol, **(c)** detoxification of alcohol and drugs, **(d)** synthesis of glucagon, **(e)** deamination of amino acids.
9. Amino acids are essential (and important) to the body for all the following except **(a)** production of some hormones,

(b) production of antibodies, **(c)** formation of most structural materials, **(d)** as a source of quick energy.

10. A person has been on a hunger strike for seven days. Compared to normal, he has **(a)** increased release of fatty acids from adipose tissue, and ketosis, **(b)** elevated glucose concentration in the blood, **(c)** increased plasma insulin concentration, **(d)** increased glycogen synthase (enzyme) activity in the liver.
11. Transamination is a chemical process by which **(a)** protein is synthesized, **(b)** an amine group is transferred from an amino acid to a keto acid, **(c)** an amine group is split from the amino acid, **(d)** amino acids are broken down for energy.
12. Three days after removing the pancreas from an animal, the researcher finds a persistent increase in **(a)** acetoacetic acid concentration in the blood, **(b)** urine volume, **(c)** blood glucose, **(d)** all of the above.
13. Hunger, appetite, obesity, and physical activity are interrelated. Thus, **(a)** hunger sensations arise primarily from the stimulation of receptors in the stomach and intestinal tract in response to the absence of food in these organs; **(b)** obesity, in most cases, is a result of the abnormally high enzymatic activity of the fat-synthesizing enzymes in adipose tissue; **(c)** in all cases of obesity, the energy content of the ingested food has exceeded the energy expenditure of the body; **(d)** in a normal individual, increasing blood glucose concentration increases hunger sensations.
14. Body temperature regulation is **(a)** influenced by temperature receptors in the skin, **(b)** influenced by the temperature of the blood perfusing the heat regulation centers of the brain, **(c)** subject to both neural and hormonal control, **(d)** all of the above.
15. Which of the following yields the greatest caloric value per gram? **(a)** fats, **(b)** proteins, **(c)** carbohydrates, **(d)** all are equal in caloric value.

Short Answer Essay Questions

16. What is cellular respiration? What is the common role of FAD and NAD^+ in cellular respiration?
17. Describe the site, major events, and outcomes of glycolysis.
18. Pyruvic acid is a product of glycolysis, but it is not the substance that joins with the pickup molecule to enter the citric acid cycle. What is that substance?
19. Define glycogenesis, glycogenolysis, gluconeogenesis, and lipogenesis. Which is (are) likely to be occurring (a) shortly after a carbohydrate-rich meal, (b) just before waking up in the morning?
20. What is the harmful result when excessive amounts of fats are burned for energy? Name two conditions that might lead to this result.
21. Make a flowchart that indicates the pivotal intermediates through which glucose can be converted to fat.
22. Distinguish between the role of HDLs and that of LDLs.
23. List some factors that influence plasma cholesterol levels. Also list the sources and fates of cholesterol in the body.
24. What is meant by "body energy balance," and what happens if the balance is not precise?
25. Explain the effect of the following on metabolic rate: thyroxine levels, eating, body surface area, muscular exercise, emotional stress, starvation.
26. Explain the terms "core" and "shell" relative to body temperature balance. What serves as the heat-transfer agent from one to the other?
27. Compare and contrast mechanisms of heat loss with mechanisms of heat promotion, and explain how these mechanisms determine body temperature.

AT THE CLINIC

Clinical Case Study

Nutrition and Metabolism

Kyle Boulard, a 35-year-old male, is believed to be one of the primary causes of the accident on Route 91. Passengers on the bus reported that Mr. Boulard was clearly intoxicated when he boarded the bus. According to these reports, Mr. Boulard behaved erratically, appeared disoriented, and left his seat and staggered down the aisle. Just prior to the accident, he had stumbled into the driver's compartment. Paramedics found Mr. Boulard in a disoriented state when they arrived on the scene, but noted only minor injuries. A "fruity acetone combined with alcohol" smell was noted on his breath. What follows is a summary of the notable test results:

General	Blood	Urine
BP: 95/58	pH: 7.1	Odor: "fruity acetone"
HR: 110	Glucose: 345 mg/dl	pH: 4.3
	Ketone bodies: 22 mg/dl	Glucose: strongly positive
	Blood alcohol: 110 mg/dl	

1. Urine glucose is usually negative. Using a table of reference values for blood and urine tests, look up normal values for blood pH, blood glucose, blood ketone bodies, and urine pH, and identify whether each test result is normal or abnormal.
2. A "fruity acetone combined with alcohol" smell was detected in Mr. Boulard's breath, and in his urine. Which substance is producing the "fruity acetone" smell?
3. Where in the body are ketone bodies produced? What energy source does the body use to produce these substances?
4. The production of large amounts of ketone bodies is often seen when glucose is not readily available as an energy source (e.g., in starvation). In Mr. Boulard's case, large amounts of glucose are in the blood. Explain why his body is producing ketones in the presence of such large amounts of glucose.
5. Explain how the pH of Mr. Boulard's blood and urine is related to the ketone bodies measured in each of these fluids.

For answers, see Answers Appendix.

24 The Urinary System

KEY CONCEPTS

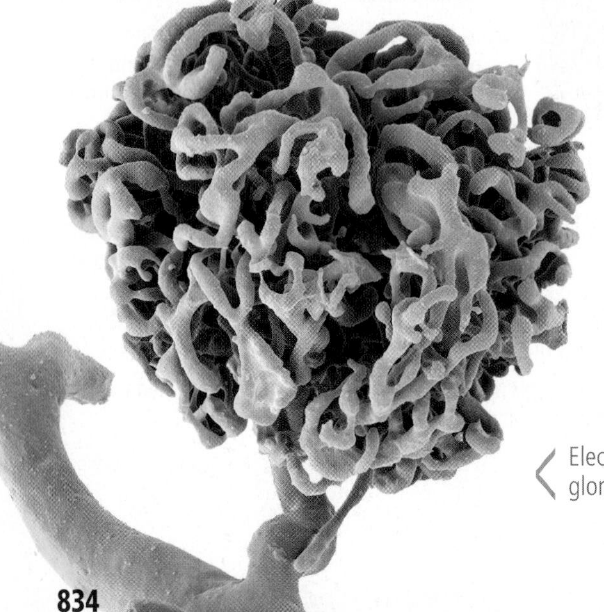

Electron micrograph of a cast of the blood vessels that form one glomerulus (a filtration unit) in a kidney.

Every day the kidneys filter nearly 200 liters of fluid from our bloodstream, allowing toxins, metabolic wastes, and excess ions to leave the body in urine while returning needed substances to the blood. Much like a water purification plant that keeps a city's water drinkable and disposes of its wastes, the kidneys are usually unappreciated until they malfunction and body fluids become contaminated.

The kidneys perform a chemical balancing act that would be tricky even for the best chemical engineer. They maintain the body's internal environment by:

- Regulating the total volume of water in the body and the total concentration of solutes in that water (osmolality).
- Regulating the concentrations of the various ions in the extracellular fluids. (Even relatively small changes in some ion concentrations such as K^+ can be fatal.)
- Ensuring long-term acid-base balance.
- Excreting metabolic wastes and foreign substances such as drugs or toxins.
- Producing *erythropoietin* and *renin* (re′nin; *ren* = kidney), important molecules for regulating red blood cell production and blood pressure, respectively.
- Converting vitamin D to its active form.
- Carrying out gluconeogenesis during prolonged fasting (see p. 821).

The urine-forming kidneys are crucial components of the **urinary system** (**Figures 24.1** and **24.2**). The urinary system also includes:

- *Ureters*—paired tubes that transport urine from the kidneys to the urinary bladder
- *Urinary bladder*—a temporary storage reservoir for urine
- *Urethra*—a tube that carries urine from the bladder to the body exterior

24.1 The kidneys have three distinct regions and a rich blood supply

→ Learning Objectives

- ☐ **Describe the gross anatomy of the kidney and its coverings.**
- ☐ **Trace the blood supply through the kidney.**

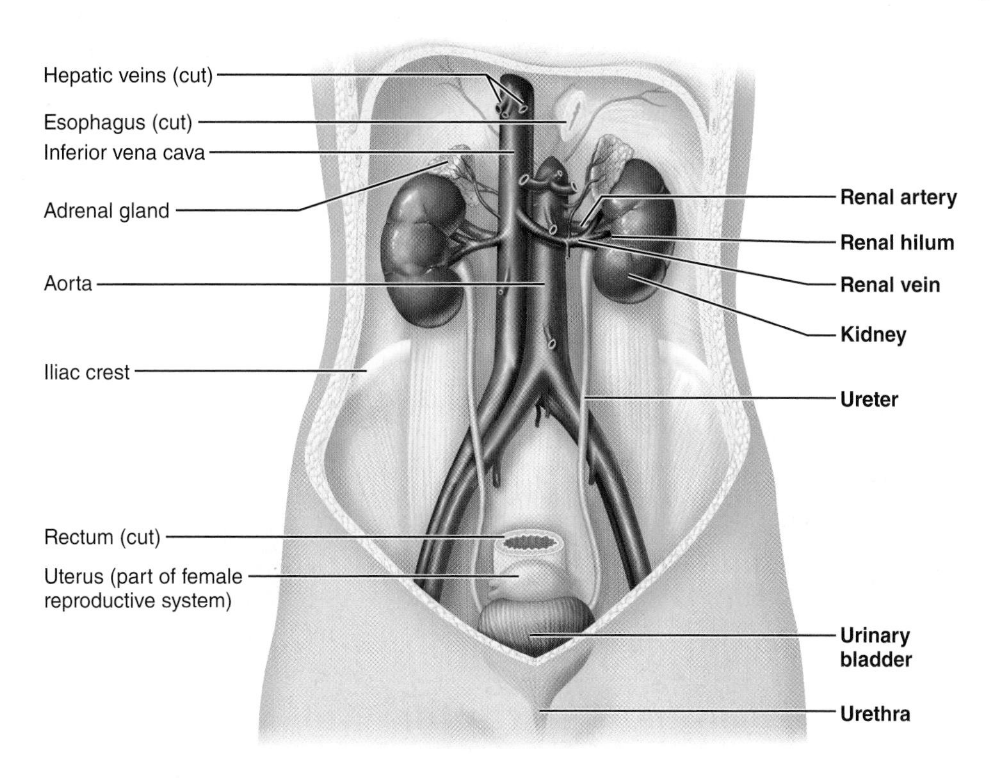

Figure 24.1 The urinary system. Anterior view of the urinary organs in a female. (Most unrelated abdominal organs have been omitted.)

Practice art labeling
MasteringA&P®>Study Area>Chapter 24

Location and External Anatomy

The bean-shaped kidneys lie in a retroperitoneal position (between the dorsal body wall and the parietal peritoneum) in the *superior* lumbar region (**Figure 24.3**). Extending approximately from T_{12} to L_3, the kidneys receive some protection from the lower part of the rib cage (Figure 24.3b). The right kidney is crowded by the liver and lies slightly lower than the left.

An adult's kidney has a mass of about 150 g (5 ounces) and its average dimensions are 11 cm long, 6 cm wide, and 3 cm thick—about the size of a large bar of soap. The lateral surface is convex. The medial surface is concave and has a vertical cleft called the **renal hilum** that leads into an internal space called the *renal sinus*. The ureter, renal blood vessels, lymphatics, and nerves all join each kidney at the hilum and occupy the sinus. Atop each kidney is an *adrenal* (or *suprarenal*) *gland*, an endocrine gland that is functionally unrelated to the kidney.

Three layers of supportive tissue surround each kidney (Figure 24.3a). From superficial to deep, these are:

- The **renal fascia**, an outer layer of dense fibrous connective tissue that anchors the kidney and the adrenal gland to surrounding structures
- The **perirenal fat capsule**, a fatty mass that surrounds the kidney and cushions it against blows
- The **fibrous capsule**, a transparent capsule that prevents infections in surrounding regions from spreading to the kidney

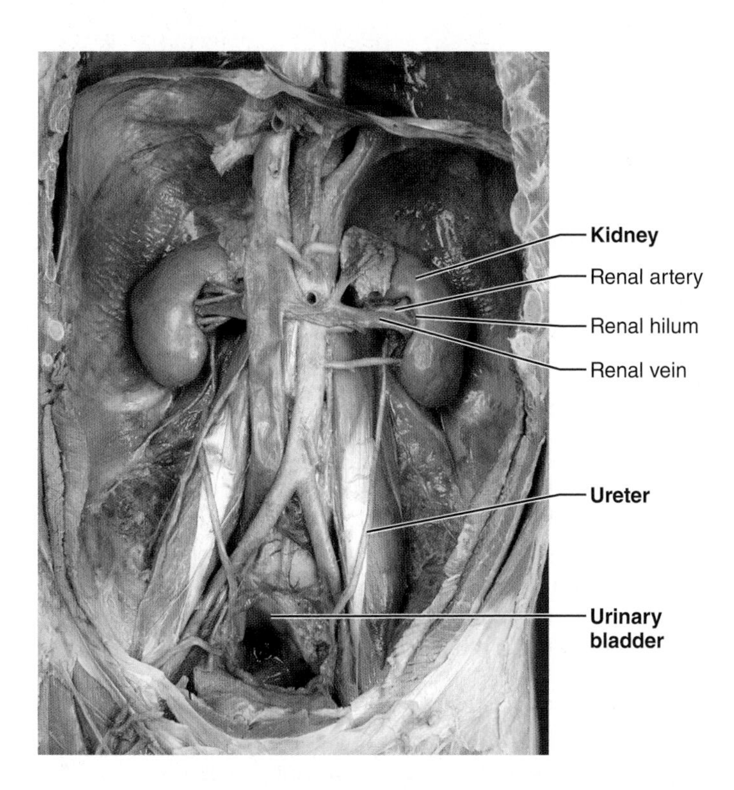

Figure 24.2 Dissection of urinary system organs (male).

The kidneys' fatty encasement holds them in their normal position. If the amount of fatty tissue dwindles (as with extreme emaciation or rapid weight loss), one or both kidneys may drop to a lower position, an event called *renal ptosis* (to′sis; "a fall"). Renal ptosis may cause a ureter to become kinked, causing urine to back up and exert pressure on kidney tissue. Backup of urine from ureteral obstruction or other causes is called *hydronephrosis* (hi″dro-nĕ-fro′sis; "water in the kidney"). Hydronephrosis can severely damage the kidney, leading to tissue death and renal failure. ✚

Internal Gross Anatomy

A frontal section through a kidney reveals three distinct regions: *cortex*, *medulla*, and *pelvis* (**Figure 24.4**). The most superficial region, the **renal cortex**, is light-colored and has a granular appearance. Deep to the cortex is the darker, reddish-brown **renal medulla**, which exhibits cone-shaped tissue masses called

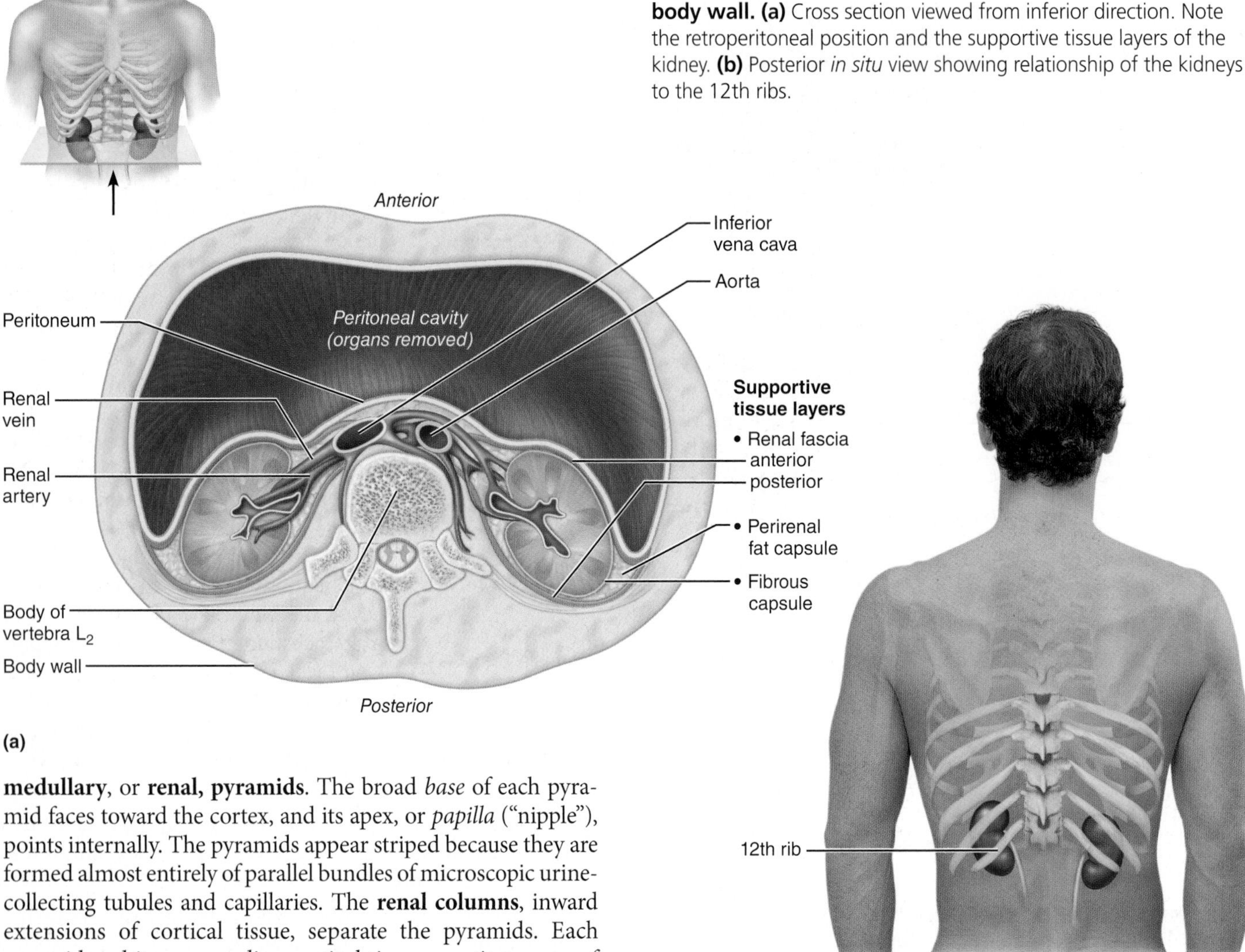

Figure 24.3 Position of the kidneys against the posterior body wall. (a) Cross section viewed from inferior direction. Note the retroperitoneal position and the supportive tissue layers of the kidney. **(b)** Posterior *in situ* view showing relationship of the kidneys to the 12th ribs.

medullary, or **renal, pyramids**. The broad *base* of each pyramid faces toward the cortex, and its apex, or *papilla* ("nipple"), points internally. The pyramids appear striped because they are formed almost entirely of parallel bundles of microscopic urine-collecting tubules and capillaries. The **renal columns**, inward extensions of cortical tissue, separate the pyramids. Each pyramid and its surrounding cortical tissue constitutes one of approximately eight **lobes** of a kidney.

The **renal pelvis**, a funnel-shaped tube, is continuous with the ureter leaving the hilum. Branching extensions of the pelvis form two or three **major calyces** (ka′lih-sēz; singular: calyx). Each major calyx subdivides to form several **minor calyces**, cup-shaped areas that enclose the papillae.

The calyces collect urine, which drains continuously from the papillae, and empty it into the renal pelvis. The urine then flows through the renal pelvis and into the ureter, which moves it to the bladder to be stored. The walls of the calyces, pelvis, and ureter contain smooth muscle that contracts rhythmically to propel urine by peristalsis.

Pyelitis (pi″ĕ-li′tis) is an infection of the renal pelvis and calyces. Infections or inflammations that affect the entire kidney are *pyelonephritis* (pi″ĕ-lo-nĕ-fri′tis). Kidney infections in females are usually caused by fecal bacteria that spread from the anal region to the urinary tract. Less often they result from blood-borne bacteria (traveling from other infected sites) that lodge and multiply in a kidney.

In severe cases of pyelonephritis, the kidney swells, abscesses form, and the pelvis fills with pus. Untreated, the kidney may be severely damaged, but antibiotic therapy can usually treat the infection successfully. ✚

Blood and Nerve Supply

The kidneys continuously cleanse the blood and adjust its composition, so it is not surprising that they have a rich blood supply. Under normal resting conditions, the large **renal arteries** deliver one-fourth of the total cardiac output to the kidneys—about 1200 ml each minute.

The renal arteries exit at right angles from the abdominal aorta, and the right renal artery is longer than the left because the aorta lies to the left of the midline. As each renal artery approaches a kidney, it divides into five **segmental arteries** (**Figure 24.5**). Within the renal sinus, each segmental artery branches further to form several **interlobar arteries**.

At the cortex-medulla junction, the interlobar arteries branch into the **arcuate arteries** (ar′ku-āt) that arch over the

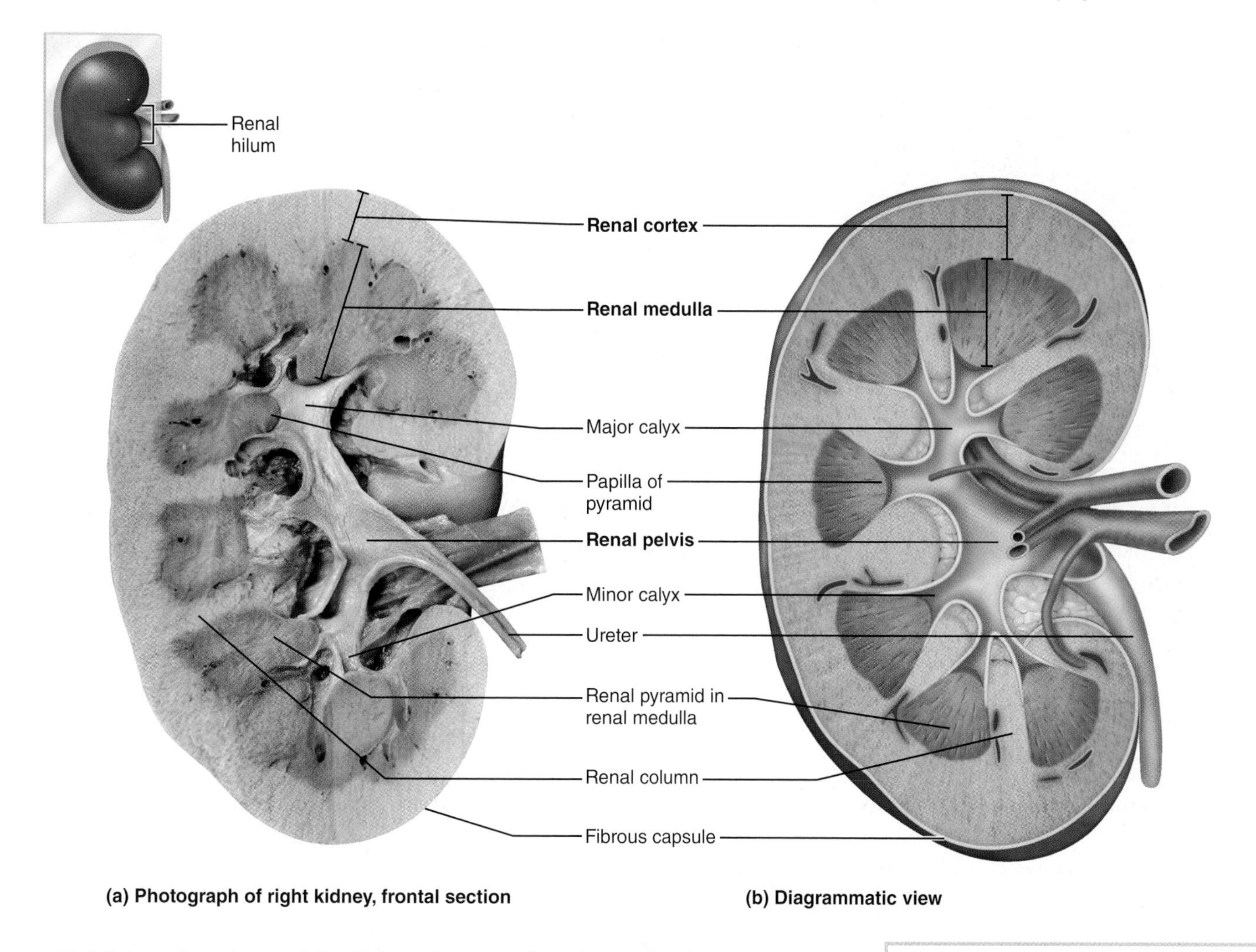

Figure 24.4 Internal anatomy of the kidney. Frontal sections. (For a related image, see *A Brief Atlas of the Human Body*, Figure 71.)

Practice art labeling
MasteringA&P®>Study Area>Chapter 24

bases of the medullary pyramids. Small **cortical radiate arteries** (also called *interlobular arteries*) radiate outward from the arcuate arteries to supply the cortical tissue. More than 90% of the blood entering the kidney perfuses the renal cortex.

Afferent arterioles branching from the cortical radiate arteries begin a complex arrangement of microscopic blood vessels. These vessels are key elements of kidney function, and we will examine them in the next module when we describe the nephron.

Veins pretty much trace the pathway of the arterial supply in reverse (Figure 24.5). Blood leaving the renal cortex drains sequentially into the **cortical radiate, arcuate, interlobar**, and finally **renal veins**. (There are no segmental veins.) The renal veins exit from the kidneys and empty into the inferior vena cava. Because the inferior vena cava lies to the right of the vertebral column, the left renal vein is about twice as long as the right.

The **renal plexus**, a variable network of autonomic nerve fibers and ganglia, provides the nerve supply of the kidney and its ureter. An offshoot of the celiac plexus, the renal plexus is largely supplied by sympathetic fibers from the most inferior thoracic and first lumbar splanchnic nerves, which course along with the renal artery to reach the kidney. These sympathetic vasomotor fibers regulate renal blood flow by adjusting the diameter of renal arterioles and also influence the formation of urine by the nephron.

☑ Check Your Understanding

1. Zach is hit in the lower back by an errant baseball. What protects his kidneys from this mechanical trauma?
2. From inside to outside, list the three layers of supportive tissue that surround each kidney. Where is the parietal peritoneum in relation to these layers?
3. The lumen of the ureter is continuous with a space inside the kidney. This space has branching extensions. What are the names of this space and its extensions?

For answers, see Answers Appendix.

24.2 Nephrons are the functional units of the kidney

→ Learning Objective

☐ **Describe the anatomy of a nephron.**

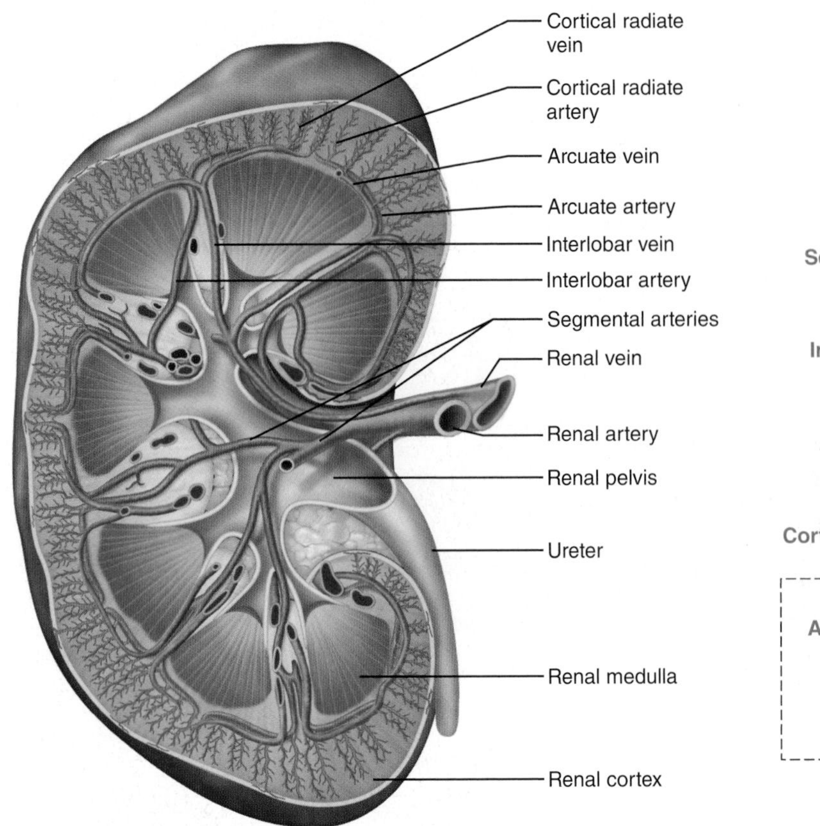

(a) Frontal section illustrating major blood vessels

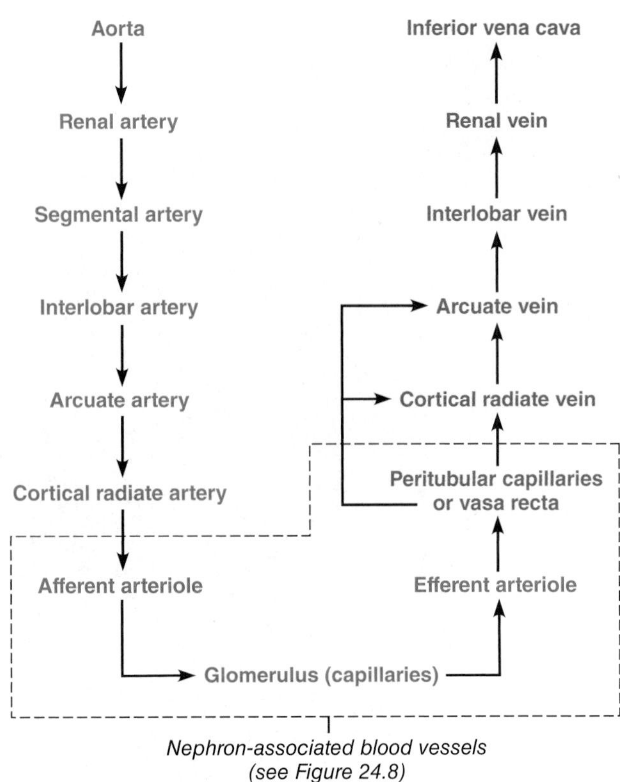

(b) Path of blood flow through renal blood vessels

Figure 24.5 **Blood vessels of the kidney.**

Nephrons (nef′ronz) are the structural and functional units of the kidneys. Each kidney contains over 1 million of these tiny blood-processing units, which carry out the processes that form urine (**Figure 24.6**). In addition, there are thousands of *collecting ducts*, each of which collects fluid from several nephrons and conveys it to the renal pelvis.

Each nephron consists of a *renal corpuscle* and a *renal tubule.* All of the renal corpuscles are located in the renal cortex, while the renal tubules begin in the cortex and then pass into the medulla before returning to the cortex.

Renal Corpuscle

Each **renal corpuscle** consists of a tuft of capillaries called a **glomerulus** (glo-mer′u-lus; *glom* = ball of yarn) and a cup-shaped hollow structure called the **glomerular capsule** (or **Bowman's capsule**). The glomerular capsule is continuous with its renal tubule and completely surrounds the glomerulus, much as a well-worn baseball glove encloses a ball.

Glomerulus

The endothelium of the glomerular capillaries is *fenestrated* (penetrated by many pores), which makes these capillaries exceptionally porous. This property allows large amounts of solute-rich but virtually protein-free fluid to pass from the blood into the glomerular capsule. This plasma-derived fluid or **filtrate** is the raw material that the renal tubules process to form urine.

Glomerular Capsule

The glomerular capsule has an external parietal layer and a visceral layer that clings to the glomerular capillaries.

- The *parietal layer* is simple squamous epithelium (Figures 24.6, 24.10, and 24.12a). This layer contributes to the capsule structure but plays no part in forming filtrate.
- The *visceral layer*, which clings to the glomerular capillaries, consists of highly modified, branching epithelial cells called **podocytes** (pod′o-sīts; "foot cells") (see Figure 24.12a–c). The octopus-like podocytes terminate in **foot processes**, which interdigitate as they cling to the basement membrane of the glomerulus. The clefts or openings between the foot processes are called **filtration slits**. Through these slits, filtrate enters the **capsular space** inside the glomerular capsule.

We describe the *filtration membrane*, the filter that lies between the blood in the glomerulus and the filtrate in the capsular space, on pp. 844–845.

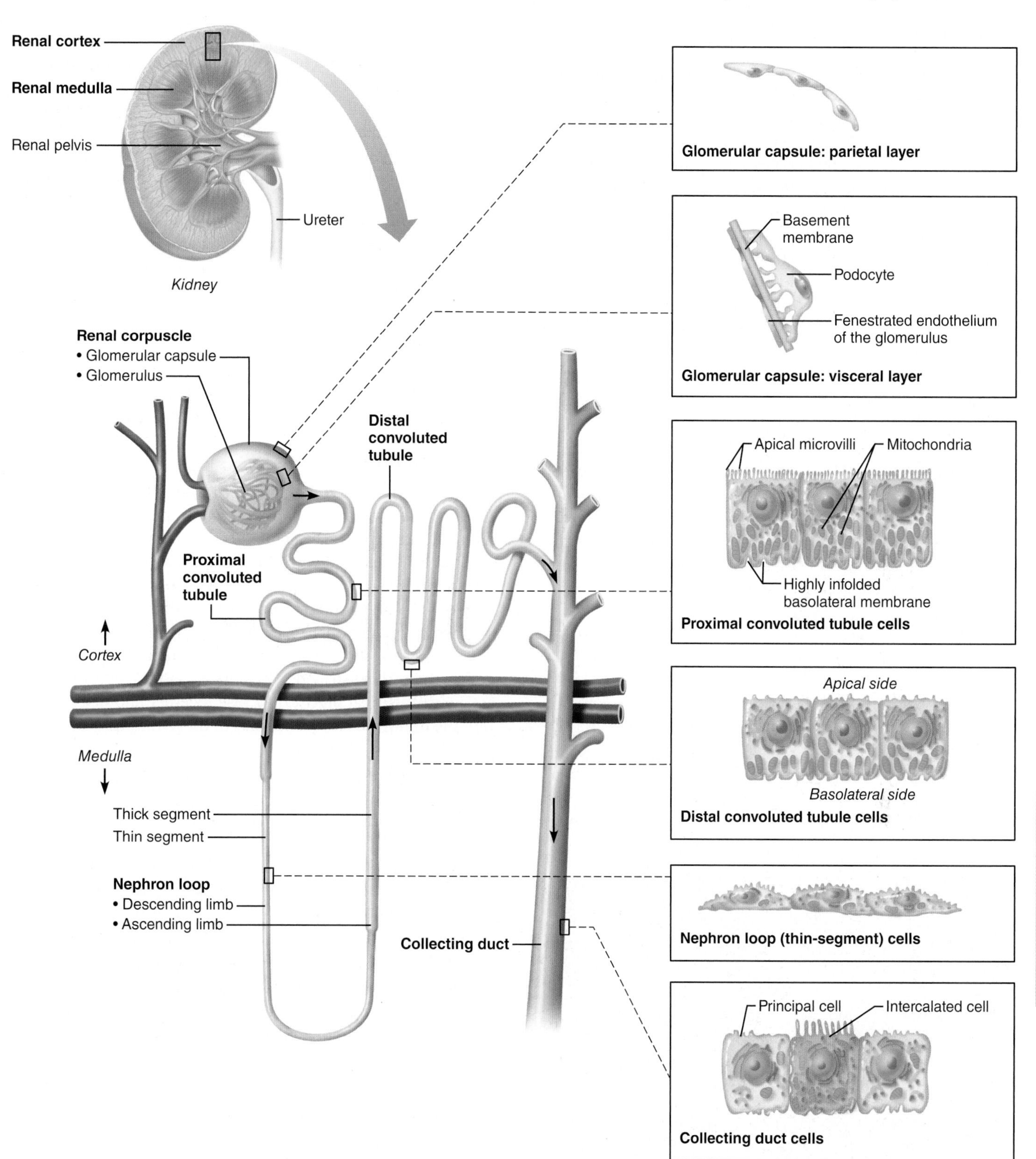

Figure 24.6 Location and structure of nephrons. Schematic view of a nephron and collecting duct depicting the structural characteristics of epithelial cells forming various regions.

Renal Tubule and Collecting Duct

The **renal tubule** is about 3 cm (1.2 inches) long and has three major parts. It leaves the glomerular capsule as the elaborately coiled *proximal convoluted tubule*, drops into a hairpin loop called the *nephron loop*, and then winds and twists again as the *distal convoluted tubule* before emptying into a collecting duct. The terms *proximal* and *distal* indicate the relationship of the convoluted tubules to the renal corpuscle—filtrate from the renal corpuscle passes through the proximal convoluted tubule first and then the distal convoluted tubule, which is thus "further away" from the renal corpuscle. The meandering nature of the renal tubule increases its length and enhances its filtrate processing capabilities.

Throughout their length, the renal tubule and collecting duct consist of a single layer of epithelial cells on a basement membrane. However, each region has a unique histology that reflects its role in processing filtrate.

Proximal Convoluted Tubule (PCT)

The walls of the **proximal convoluted tubule** are formed by cuboidal epithelial cells with large mitochondria, and their apical (luminal) surfaces bear dense microvilli (Figure 24.6 and **Figure 24.7**). Just as in the intestine, this *brush border* dramatically increases the surface area and capacity for reabsorbing water and solutes from the filtrate and secreting substances into it.

View histology slides
MasteringA&P®>Study Area>PAL

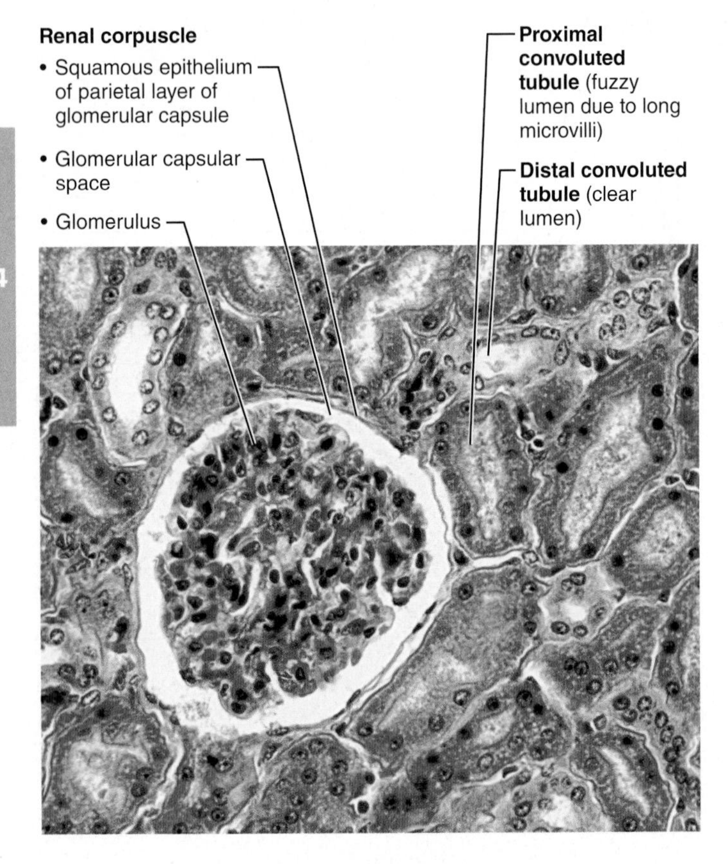

Figure 24.7 Renal cortical tissue. Photomicrograph (415×).

Nephron Loop

The U-shaped **nephron loop** (formerly called the *loop of Henle*) has **descending** and **ascending limbs**. The proximal part of the descending limb is continuous with the proximal tubule and its cells are similar. The rest of the descending limb, called the *descending thin limb*, consists of a simple squamous epithelium. The epithelium becomes cuboidal or even low columnar in the ascending part of the nephron loop, which is therefore called the *thick ascending limb*. In most nephrons, the entire ascending limb is thick, but in some nephrons, the thin segment extends around the bend as the *ascending thin limb*. The thick and thin parts of the nephron loop are also referred to as thick and thin segments.

Distal Convoluted Tubule (DCT)

The epithelial cells of the **distal convoluted tubule**, like those of the PCT, are cuboidal and confined to the cortex, but they are thinner and almost entirely lack microvilli (Figure 24.6).

Collecting Duct

Each **collecting duct** contains two cell types. The more numerous *principal cells* have sparse, short microvilli and are responsible for maintaining the body's water and Na^+ balance. The *intercalated cells* are cuboidal cells with abundant microvilli. There are two varieties of intercalated cells (types A and B), and each plays a role in maintaining the acid-base balance of the blood.

Each collecting duct receives filtrate from many nephrons. The collecting ducts run through the medullary pyramids, giving them their striped appearance. As the collecting ducts approach the renal pelvis, they fuse together and deliver urine into the minor calyces via papillae of the pyramids.

Classes of Nephrons

Nephrons are generally divided into two major groups, cortical and juxtamedullary (**Figure 24.8**).

- **Cortical nephrons** account for 85% of the nephrons in the kidneys. Except for small parts of their nephron loops that dip into the outer medulla, they are located entirely in the cortex.
- **Juxtamedullary nephrons** (juks″tah-mĕ′dul-ah-re) originate close to (*juxta* = near to) the cortex-medulla junction, and they play an important role in the kidneys' ability to produce concentrated urine. They have long nephron loops that deeply invade the medulla, and their ascending limbs have both thin and thick segments.

Nephron Capillary Beds

The renal tubule of every nephron is closely associated with two capillary beds. The first capillary bed (the *glomerulus*) produces the filtrate. The second (a combination of *peritubular capillaries* and *vasa recta*) reclaims most of that filtrate.

Glomerulus

The glomerulus, in which the capillaries run in parallel, is specialized for filtration. It differs from all other capillary beds in the body in that it is both fed and drained by arterioles—the

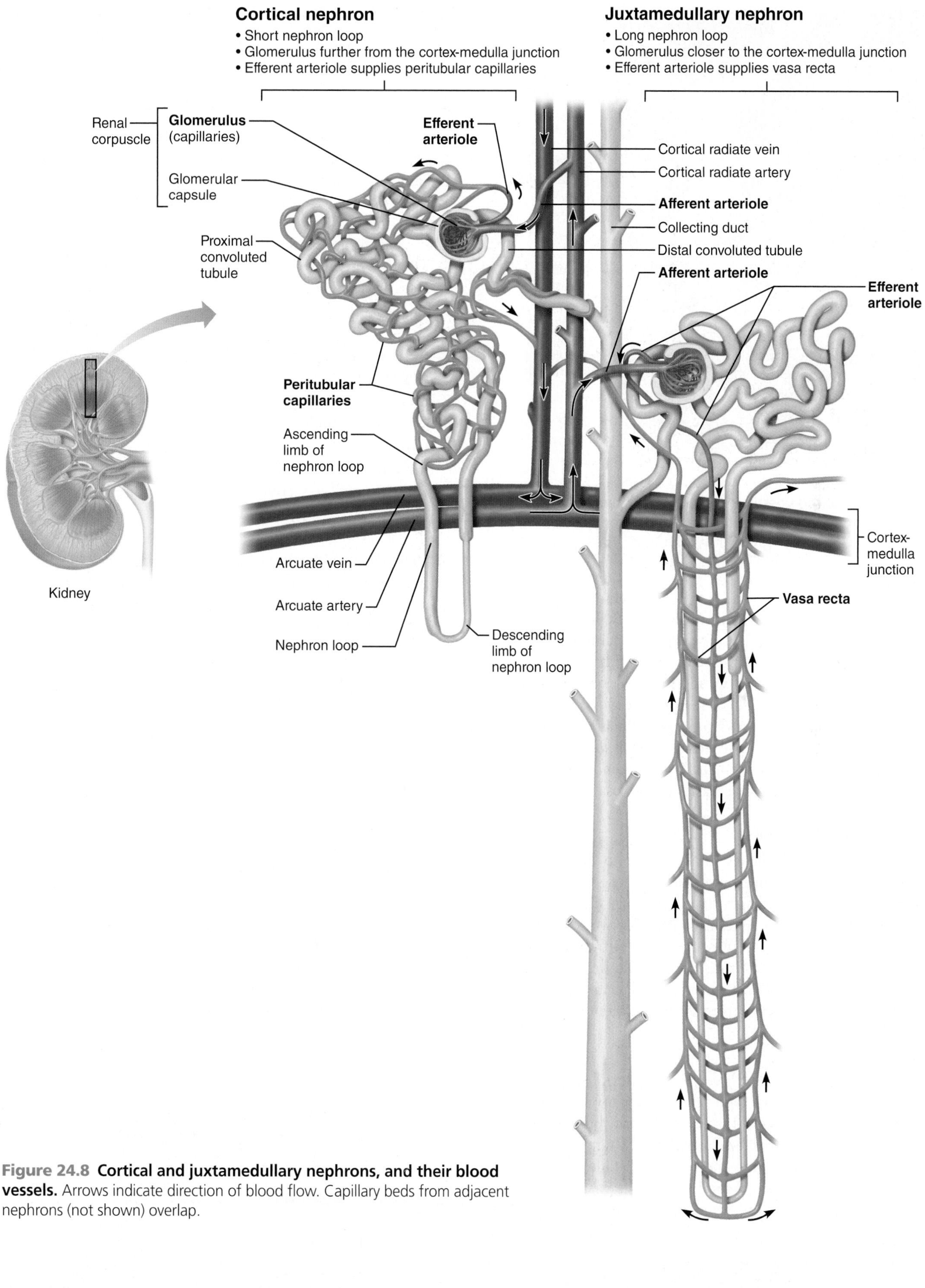

Cortical nephron

- Short nephron loop
- Glomerulus further from the cortex-medulla junction
- Efferent arteriole supplies peritubular capillaries

Juxtamedullary nephron

- Long nephron loop
- Glomerulus closer to the cortex-medulla junction
- Efferent arteriole supplies vasa recta

Renal corpuscle
Glomerulus (capillaries)
Glomerular capsule
Efferent arteriole
Proximal convoluted tubule
Peritubular capillaries
Ascending limb of nephron loop
Arcuate vein
Arcuate artery
Nephron loop
Descending limb of nephron loop
Kidney
Cortical radiate vein
Cortical radiate artery
Afferent arteriole
Collecting duct
Distal convoluted tubule
Afferent arteriole
Efferent arteriole
Cortex-medulla junction
Vasa recta

Figure 24.8 Cortical and juxtamedullary nephrons, and their blood vessels. Arrows indicate direction of blood flow. Capillary beds from adjacent nephrons (not shown) overlap.

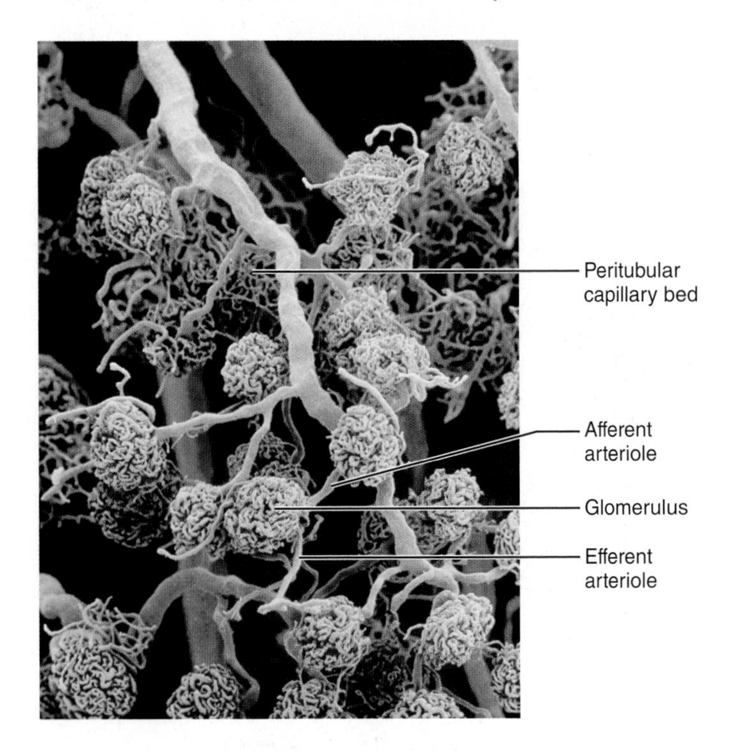

Figure 24.9 Blood vessels of the renal cortex. Scanning electron micrograph of a cast of blood vessels associated with nephrons (105×). View looking down onto the cortex.

afferent arteriole and **efferent arteriole**, respectively (Figure 24.8 and **Figure 24.9**). This arrangement maintains the high pressure in the glomerulus that is needed for filtration, a process we discuss in Module 24.4. Filtration produces a large amount of fluid, most (99%) of which is reabsorbed by the renal tubule cells and returned to the blood in the peritubular capillary beds.

The afferent arterioles arise from the *cortical radiate arteries* that run through the renal cortex. The efferent arterioles feed into either the peritubular capillaries or the vasa recta.

24

Peritubular Capillaries

The **peritubular capillaries** cling closely to adjacent renal tubules and empty into nearby venules. Because they arise from the efferent arterioles (which have high resistance), they only experience low pressure. As a result, these low-pressure, porous capillaries readily absorb solutes and water from the tubule cells as these substances are reclaimed from the filtrate. Renal tubules are closely packed together, so the peritubular capillaries of each nephron absorb substances from several adjacent nephrons.

Vasa Recta

Notice in Figure 24.8 that the efferent arterioles serving the juxtamedullary nephrons tend *not* to break up into meandering peritubular capillaries. Instead they form bundles of long straight vessels called **vasa recta** (va′sah rek′tah; "straight vessels") that extend deep into the medulla paralleling the longest nephron loops. Like all blood vessels, the vasa recta supply oxygen and nutrients to the tissue through which they pass (the renal medulla). However, the thin-walled vasa recta also play an important role in forming concentrated urine, as we will describe shortly.

Juxtaglomerular Complex (JGC)

Each nephron has a **juxtaglomerular complex (JGC)** (juks″-tah-glo-mer′u-lar), a region where the most distal portion of the ascending limb of the nephron loop lies against the afferent arteriole feeding the glomerulus (and sometimes the efferent arteriole) (**Figure 24.10**). Both the ascending limb and the afferent arteriole are modified at the point of contact.

The JGC includes three populations of cells that help regulate the rate of filtrate formation and systemic blood pressure.

- The **macula densa** (mak′u-lah den′sah; "dense spot") is a group of tall, closely packed cells in the ascending limb of the nephron loop that lies adjacent to the granular cells (Figure 24.10). The macula densa cells are chemoreceptors that monitor the NaCl content of the filtrate entering the distal convoluted tubule.
- **Granular cells** [also called *juxtaglomerular (JG) cells*] are in the arteriolar walls. They are enlarged smooth muscle cells with prominent secretory granules containing the enzyme *renin* (see p. 848). Granular cells act as mechanoreceptors that sense the blood pressure in the afferent arteriole.
- *Extraglomerular mesangial cells* lie between the arteriole and tubule cells, and are interconnected by gap junctions. These cells may pass regulatory signals between macula densa and granular cells.

We discuss the physiological role of the JGC on pp. 846–848.

☑ Check Your Understanding

4. Name the tubular components of a nephron in the order that filtrate passes through them.
5. What are the structural differences between juxtamedullary and cortical nephrons?
6. What type of capillaries are the glomerular capillaries? What is their function?
7. For the juxtamedullary nephron and collecting duct illustrated below, name each of the structures labeled a–f.

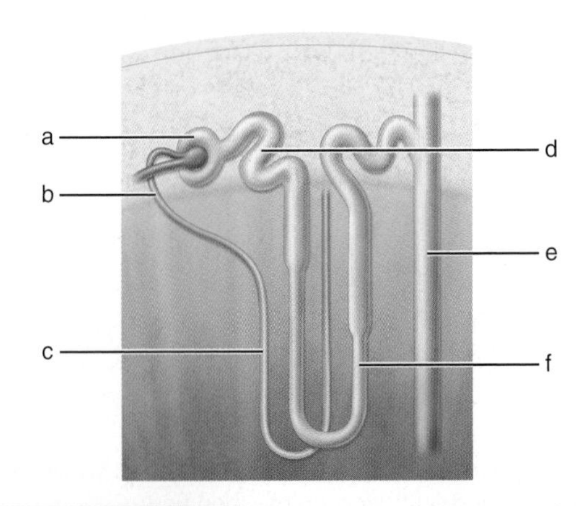

For answers, see Answers Appendix.

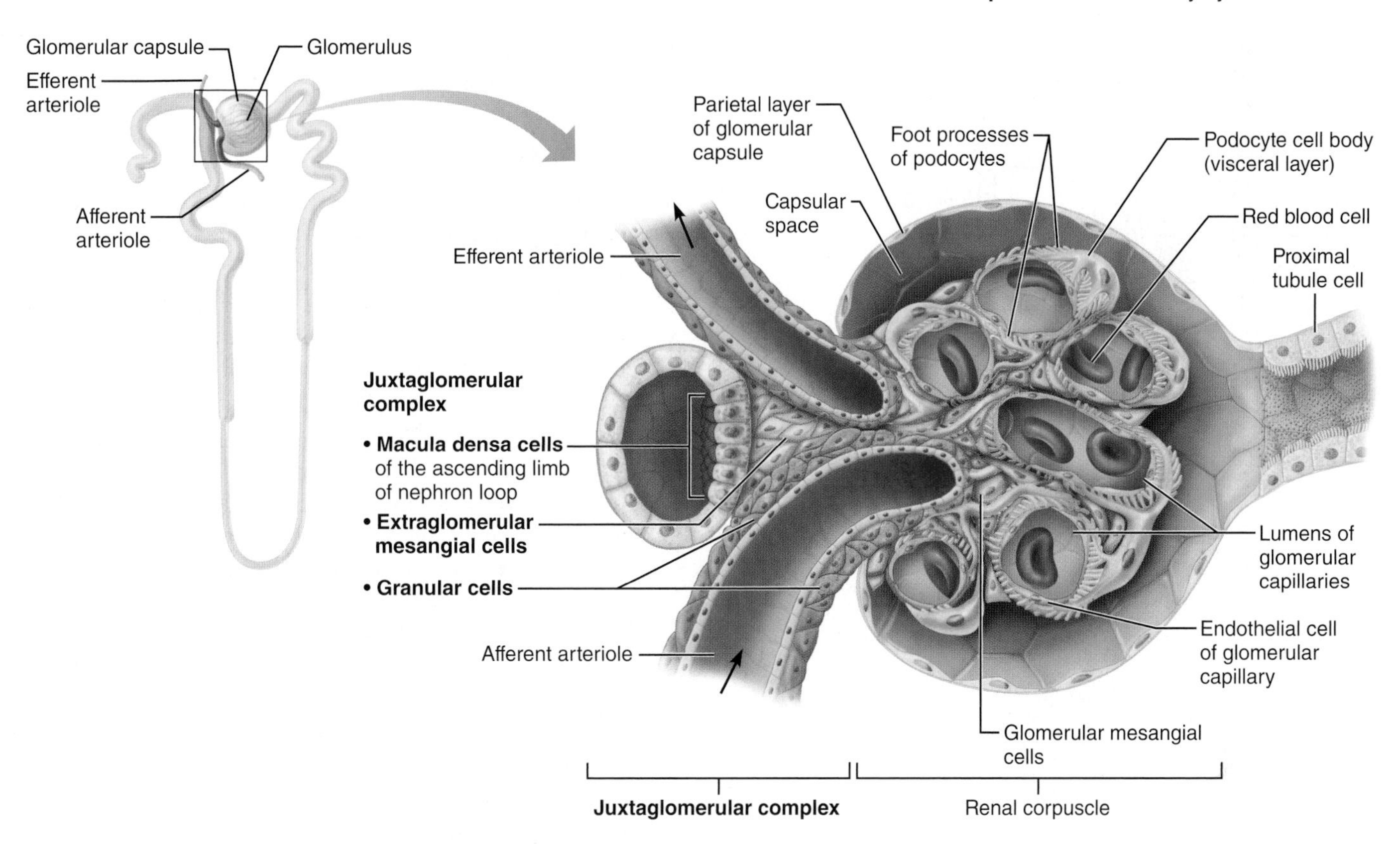

Figure 24.10 Juxtaglomerular complex (JGC) of a nephron. Mesangial cells that surround the glomerular capillaries (glomerular mesangial cells) are not part of the JGC.

24.3 Overview: Filtration, absorption, and secretion are the key processes of urine formation

→ Learning Objective

☐ **List and define the three major renal processes.**

If you had to design a system to chemically balance and cleanse the blood, how would you do it? Conceptually, it's really very simple. The body solves this problem in the following way. First, it "dumps" cell- and protein-free blood into a separate "waste container." From this container, it reclaims everything the body needs to keep (which is almost everything filtered). Finally, the kidney selectively adds specific things to the container, fine-tuning the body's chemical balance. Anything left in the container becomes urine. This is basically how nephrons work.

Urine formation and the adjustment of blood composition involve three processes (**Figure 24.11**):

1. **Glomerular filtration.** *Glomerular filtration* ("dumping into the waste container") takes place in the renal corpuscle and produces a cell- and protein-free filtrate.
2. **Tubular reabsorption.** *Tubular reabsorption* ("reclaiming what the body needs to keep") is the process of selectively moving substances from the filtrate back into the blood. It takes place in the renal tubules and collecting ducts. Tubular reabsorption reclaims almost everything filtered—all of the glucose and amino acids, and some 99% of the water, salt, and other components. Anything that is *not* reabsorbed becomes urine.
3. **Tubular secretion.** *Tubular secretion* ("selectively adding to the waste container") is the process of selectively moving substances from the blood into the filtrate. Like tubular reabsorption, it occurs along the length of the tubule and collecting duct.

The kidneys process an enormous volume of blood each day. Of the approximately 1200 ml of blood that passes through the glomeruli each minute, some 650 ml is plasma, and about one-fifth of this (120–125 ml) is forced into the glomerular capsules as filtrate. This is equivalent to filtering your entire plasma volume more than 60 times each day! Considering the magnitude of their task, it is not surprising that the kidneys (which account for only 1% of body weight) consume 20–25% of all oxygen used by the body at rest.

Filtrate and urine are quite different. Filtrate contains everything found in blood plasma except proteins. **Urine** contains unneeded substances such as excess salts and metabolic wastes. The kidneys process about 180 L (47 gallons!) of blood-derived fluid daily. Of this amount, less than 1% (1.5 L) typically leaves the body as urine; the rest returns to the circulation.

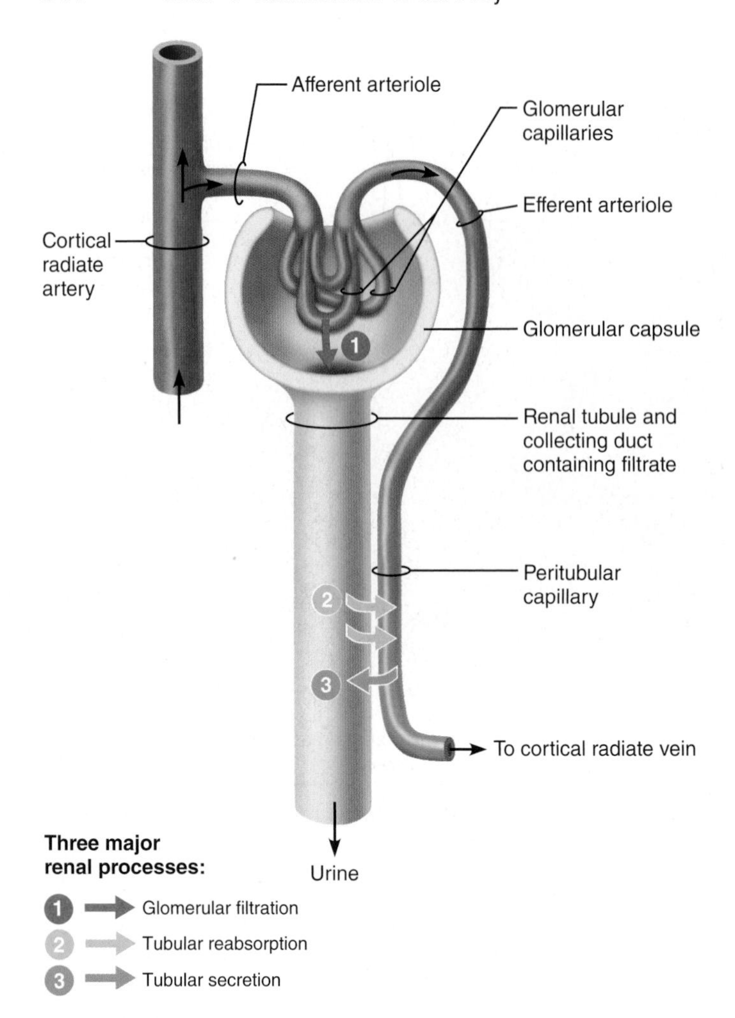

Figure 24.11 The three major renal processes. A single nephron is shown schematically, as if uncoiled. Each kidney actually has more than a million nephrons acting in parallel.

Check Your Understanding

8. MAKING connections In the kidneys, tubular secretion of a substance usually results in its excretion as well. Explain the difference between excretion (defined in Chapter 1) and tubular secretion.

For answers, see Answers Appendix.

24.4 Urine formation, step 1: The glomeruli make filtrate

Learning Objectives

- ☐ **Describe the forces (pressures) that promote or counteract glomerular filtration.**
- ☐ **Compare the intrinsic and extrinsic controls of the glomerular filtration rate.**

Glomerular filtration is a passive process in which hydrostatic pressure forces fluids and solutes through a membrane. The glomeruli can be viewed as simple mechanical filters because filtrate formation does not directly consume metabolic energy. Let's first look at the structure of the filtration membrane and then see how it works.

The Filtration Membrane

The **filtration membrane** lies between the blood and the interior of the glomerular capsule. It is a porous membrane that allows free passage of water and solutes smaller than plasma proteins. As Figure 24.12d shows, its three layers are:

- **Fenestrated endothelium of the glomerular capillaries.** The fenestrations (capillary pores) allow all blood components except blood cells to pass through.
- **Basement membrane.** The basement membrane lies between the other two layers and is composed of their fused basal laminae. It forms a physical barrier that blocks all but the smallest proteins while still permitting most other solutes to pass. The glycoproteins of the gel-like basement membrane give it a negative charge. As a result, the basement membrane electrically repels many negatively charged macromolecular anions such as plasma proteins, reinforcing the blockade based on molecular size.
- **Foot processes of podocytes of the glomerular capsule.** The visceral layer of the glomerular capsule is made of podocytes that have filtration slits between their foot processes. If any macromolecules manage to make it through the basement membrane, *slit diaphragms*—thin membranes that extend across the filtration slits—prevent almost all of them from traveling farther.

Macromolecules that get "hung up" in the filtration membrane are engulfed by specialized pericytes called *glomerular mesangial cells* (Figure 24.10).

Molecules smaller than 3 nm in diameter—such as water, glucose, amino acids, and nitrogenous wastes—pass freely from the blood into the glomerular capsule. As a result, these substances usually have similar concentrations in the blood and the glomerular filtrate. Larger molecules pass with greater difficulty, and those larger than 5 nm are generally barred from entering the tubule. Keeping the plasma proteins *in* the capillaries maintains the colloid osmotic (oncotic) pressure of the glomerular blood, preventing the loss of all its water to the capsular space. The presence of proteins or blood cells in the urine usually indicates a problem with the filtration membrane.

Pressures That Affect Filtration

The principles that govern filtration from the glomerulus are the same as those that govern filtration from any capillary bed. *Focus on Bulk Flow across Capillary Walls* (Focus Figure 18.1 on pp. 632–633 shows filtration in normal capillary beds. Figure 24.13 applies these principles to the glomerular capillaries of the nephron.

Outward Pressures

Outward pressures promote filtrate formation.

- The **hydrostatic pressure in glomerular capillaries (HP_{gc})** is essentially glomerular blood pressure. It is the chief force

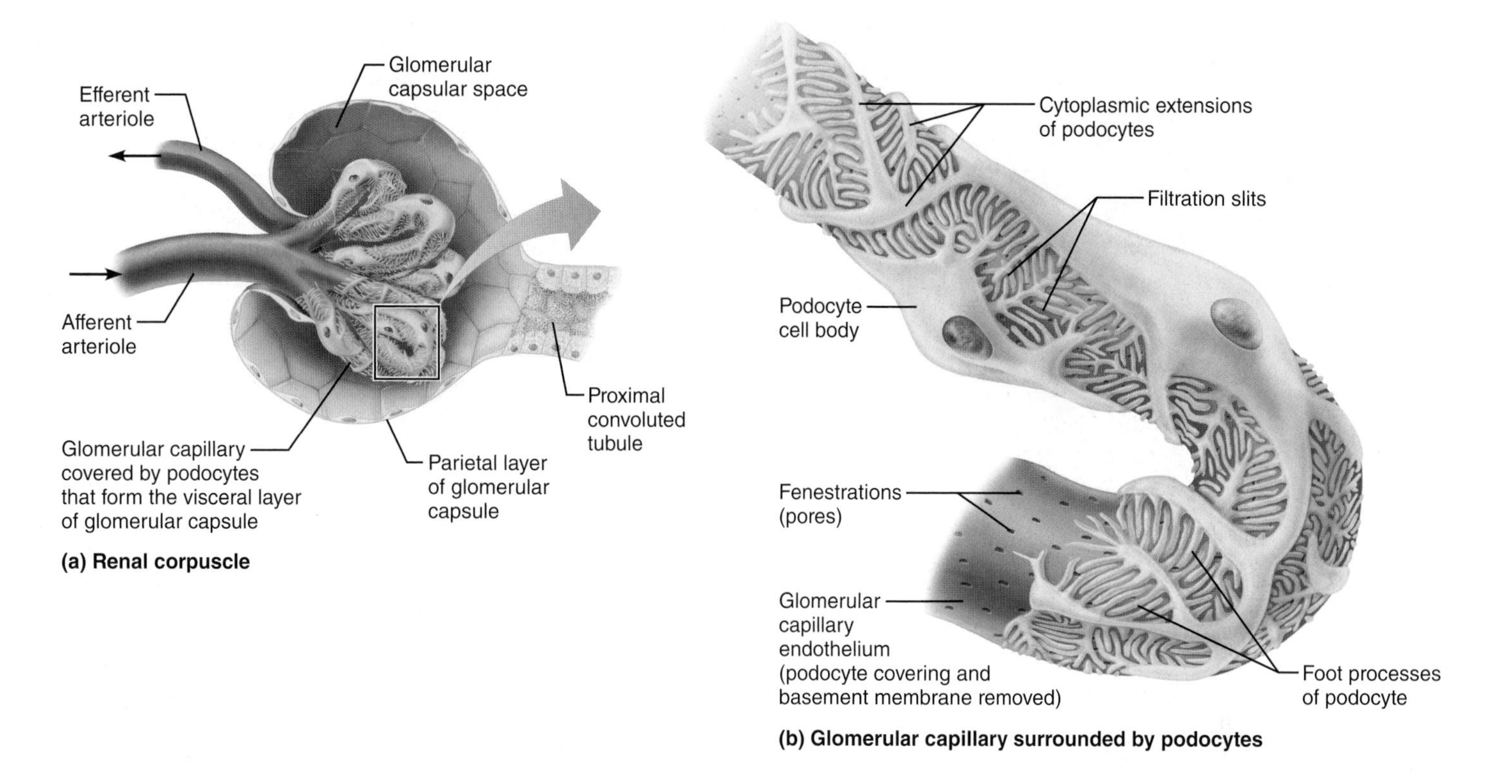

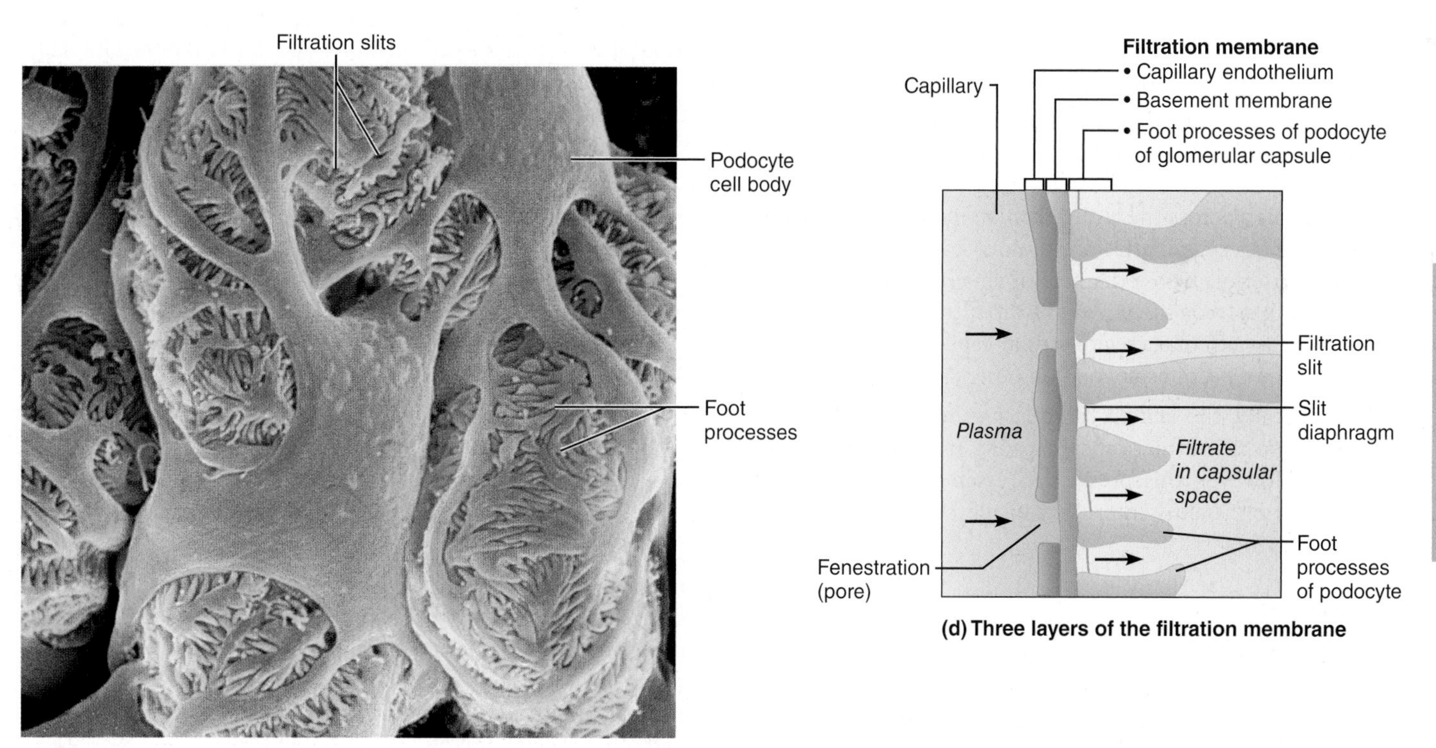

Figure 24.12 The filtration membrane. (a) The renal corpuscle consists of a glomerulus surrounded by a glomerular capsule. **(b)** Enlargement of a glomerular capillary covered by the visceral (inner) layer of the glomerular capsule consisting of podocytes. Some podocytes and the basement membrane have been removed to show the fenestrations (pores) in the underlying capillary wall. **(c)** Scanning electron micrograph of the visceral layer (6000×). **(d)** Diagram of a section through the filtration membrane showing its three layers.

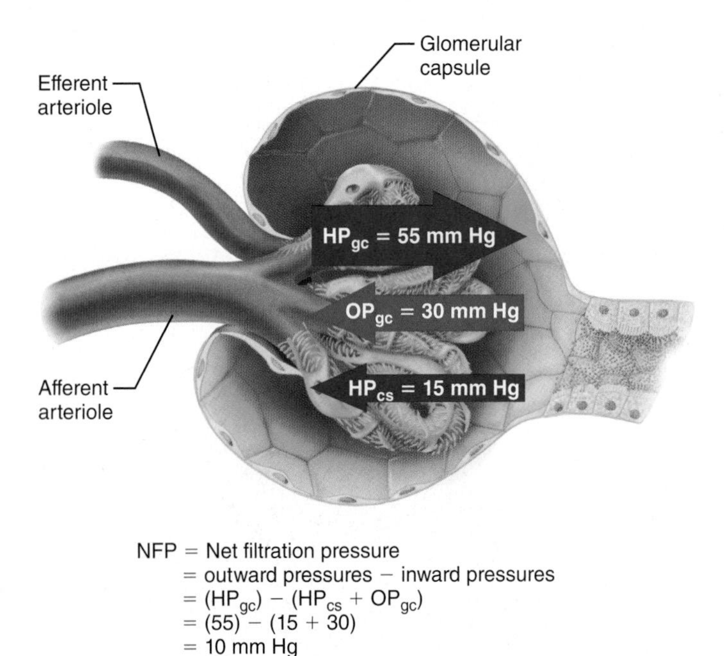

Figure 24.13 Forces determining net filtration pressure (NFP). The pressure values cited in the diagram are approximate. (HP_{gc} = hydrostatic pressure in glomerular capillaries, OP_{gc} = osmotic pressure in glomerular capillaries, HP_{cs} = hydrostatic pressure in capsular space)

pushing water and solutes out of the blood and across the filtration membrane. The blood pressure in the glomerulus is extraordinarily high (approximately 55 mm Hg compared to an average of 26 mm Hg or so in other capillary beds) and it remains high across the entire capillary bed. This is because the glomerular capillaries are drained by a high-resistance efferent arteriole whose diameter is smaller than the afferent arteriole that feeds them. As a result, filtration occurs along the entire length of each glomerular capillary and reabsorption does not occur as it would in other capillary beds.

- Theoretically, the *colloid osmotic pressure in the capsular space* of the glomerular capsule would "pull" filtrate into the tubule. However, this pressure is essentially zero because virtually no proteins enter the capsule, so we will not consider it further.

Inward Pressures

Two inward forces inhibit filtrate formation by opposing HP_{gc}.

- The **hydrostatic pressure in the capsular space (HP_{cs})** is the pressure exerted by filtrate in the glomerular capsule. HP_{cs} is much higher than hydrostatic pressure surrounding most capillaries because filtrate is confined in a small space with a narrow outlet.
- The **colloid osmotic pressure in glomerular capillaries (OP_{gc})** is the pressure exerted by the proteins in the blood.

As shown in Figure 24.13, the above pressures determine the **net filtration pressure (NFP)**. NFP largely determines the glomerular filtration rate, which we consider next.

Glomerular Filtration Rate (GFR)

The **glomerular filtration rate** is the volume of filtrate formed each minute by the combined activity of all 2 million glomeruli of the kidneys. GFR is directly proportional to each of the following factors:

- **Net filtration pressure.** NFP is the main controllable factor. Of the pressures determining NFP, the most important is hydrostatic pressure in the glomerulus. This pressure can be controlled by changing the diameter of the afferent (and sometimes the efferent) arterioles, as we will see shortly.
- **Total surface area available for filtration.** Glomerular capillaries have a huge surface area (collectively equal to the surface area of the skin). Glomerular mesangial cells surrounding these capillaries can fine-tune GFR by contracting to adjust the total surface area available for filtration.
- **Filtration membrane permeability.** Glomerular capillaries are thousands of times more permeable than other capillaries because of their fenestrations.

The huge surface area and high permeability of the filtration membrane explain how the relatively modest 10 mm Hg NFP can produce huge amounts of filtrate. Furthermore, the NFP in the glomerulus favors filtration over the entire length of the capillary, unlike other capillary beds where filtration occurs only at the arteriolar end and reabsorption occurs at the venous end. As a result, the adult kidneys produce about 180 L of filtrate daily, in contrast to the 2 to 4 L formed daily by all other capillary beds combined. This 180 L of filtrate per day translates to the normal GFR of 120–125 ml/min.

Regulation of Glomerular Filtration

GFR is tightly regulated to serve two crucial and sometimes opposing needs. The kidneys need a relatively constant GFR to make filtrate and do their job of maintaining extracellular homeostasis. On the other hand, the body as a whole needs a constant blood pressure, and this is closely tied to GFR in the following way: Assuming nothing else changes, an increase in GFR increases urine output, which reduces blood volume and blood pressure. The opposite holds true for a decrease in GFR.

Two types of controls serve these two different needs. *Intrinsic controls* (*renal autoregulation*) act locally within the kidney to maintain GFR, while *extrinsic controls* by the nervous and endocrine systems maintain blood pressure. In extreme changes of blood pressure (mean arterial pressure less than 80 or greater than 180 mm Hg), extrinsic controls take precedence over intrinsic controls in an effort to prevent damage to the brain and other crucial organs.

GFR can be controlled by changing a single variable—glomerular hydrostatic pressure. All major control mechanisms act primarily to change this one variable. If the glomerular hydrostatic pressure rises, NFP rises and so does GFR. If the glomerular hydrostatic pressure falls by as little as 18%, GFR drops to zero. Clearly hydrostatic pressure in the glomerulus must be tightly controlled. Let's see how the intrinsic and extrinsic mechanisms accomplish this feat.

Intrinsic Controls: Renal Autoregulation

By adjusting its own resistance to blood flow, a process called **renal autoregulation**, the kidney can maintain a nearly constant GFR despite fluctuations in systemic arterial blood pressure. Renal autoregulation uses two different mechanisms: (1) a *myogenic mechanism* and (2) a *tubuloglomerular feedback mechanism* (**Figure 24.14**, left side).

Myogenic Mechanism The **myogenic mechanism** (mi″o-jen′ik) reflects a property of vascular smooth muscle—it contracts when stretched and relaxes when not stretched. Rising systemic blood pressure stretches vascular smooth muscle in the arteriolar walls, causing the afferent arterioles to constrict. This constriction restricts blood flow into the glomerulus and prevents glomerular blood pressure from rising to damaging levels. Declining systemic blood pressure causes dilation of afferent arterioles and raises glomerular hydrostatic pressure. Both responses help maintain normal NFP and GFR.

Tubuloglomerular Feedback Mechanism Autoregulation by the flow-dependent **tubuloglomerular feedback mechanism** is "directed" by the *macula densa cells* of the *juxtaglomerular complex* (see Figure 24.10). These cells, located in the walls of the ascending limb of the nephron loop, respond to filtrate NaCl

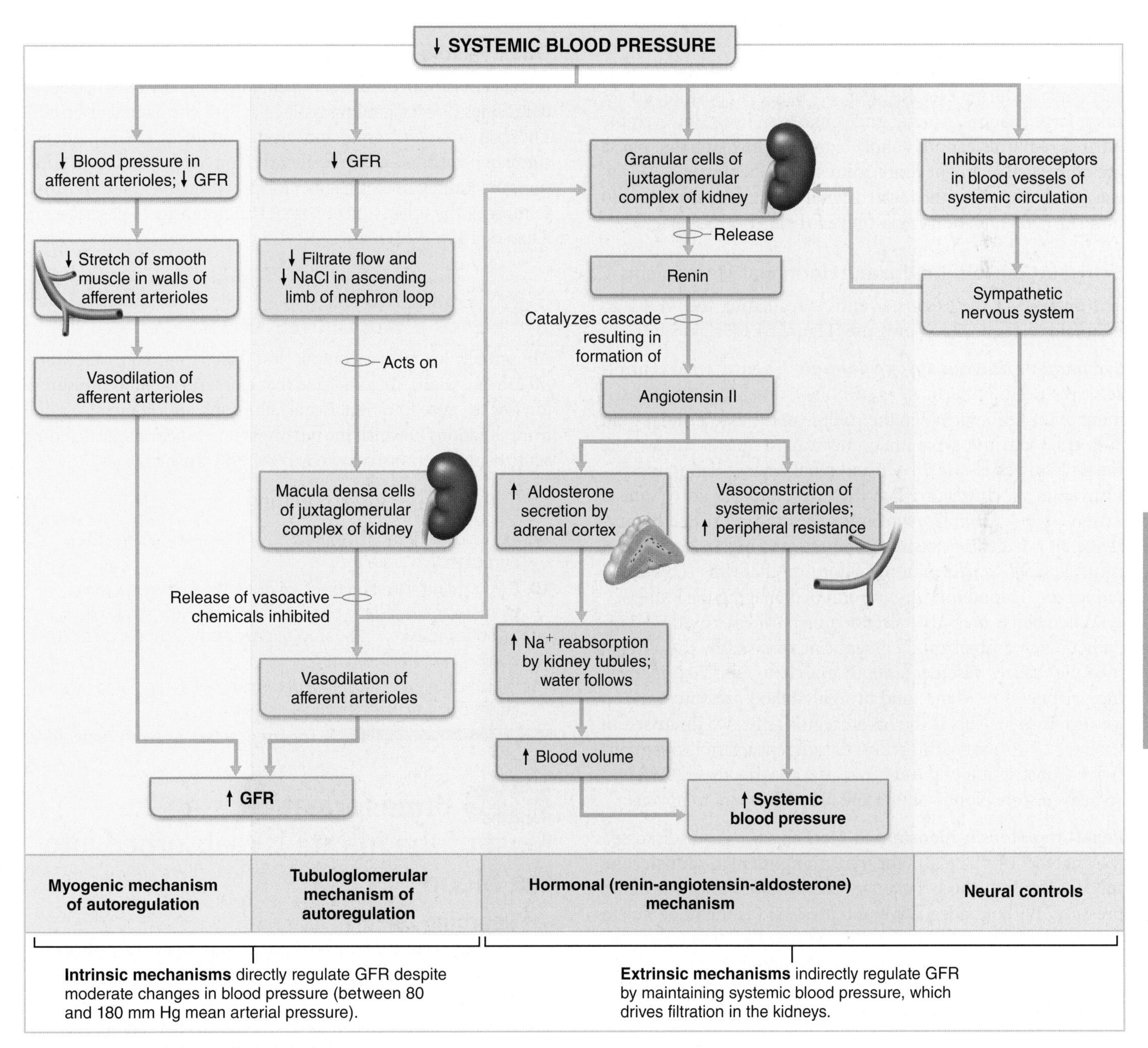

Figure 24.14 Regulation of glomerular filtration rate (GFR) in the kidneys. (Note that while the extrinsic controls are aimed at maintaining blood pressure, they also maintain GFR since bringing blood pressure back up allows the kidneys to maintain GFR.)

concentration (which varies directly with filtrate flow rate). When GFR increases, there is not enough time for reabsorption and the concentration of NaCl in the filtrate remains high. The macula densa cells respond to high levels of NaCl in filtrate by releasing vasoconstrictor chemicals (ATP and others) that cause intense constriction of the afferent arteriole, reducing blood flow into the glomerulus. This drop in blood flow decreases the NFP and GFR, slowing the flow of filtrate and allowing more time for filtrate processing (NaCl reabsorption).

In contrast, the low NaCl concentration of slowly flowing filtrate inhibits ATP release from macula densa cells, causing vasodilation of the afferent arterioles (Figure 24.14). This allows more blood to flow into the glomerulus, thus increasing NFP and GFR.

Autoregulatory mechanisms maintain a relatively constant GFR over an arterial pressure range from about 80 to 180 mm Hg. Consequently, normal day-to-day changes in our blood pressure (such as during exercise, sleep, or changes in posture) do not cause large changes in water and solute excretion. However, the intrinsic controls cannot handle extremely low systemic blood pressure, such as might result from serious hemorrhage (*hypovolemic shock*). Once the mean arterial pressure drops below 80 mm Hg, autoregulation ceases and extrinsic controls take over.

Extrinsic Controls: Neural and Hormonal Mechanisms

The purpose of the extrinsic controls regulating the GFR is to maintain systemic blood pressure (Figure 24.14, right side).

Sympathetic Nervous System Controls Neural renal controls serve the needs of the body as a whole—sometimes to the detriment of the kidneys. When the volume of the extracellular fluid is normal and the sympathetic nervous system is at rest, the renal blood vessels are dilated and renal autoregulation mechanisms prevail. However, when the extracellular fluid volume is extremely low (as in hypovolemic shock during severe hemorrhage), it is necessary to shunt blood to vital organs, and neural controls may override autoregulatory mechanisms. This could reduce renal blood flow to the point of damaging the kidneys.

When blood pressure falls, norepinephrine released by sympathetic nerve fibers (and epinephrine released by the adrenal medulla) causes vascular smooth muscle to constrict, increasing peripheral resistance and bringing blood pressure back up toward normal. This is the baroreceptor reflex we discussed in Chapter 18. As part of this reflex, the afferent arterioles also constrict. Constriction of the afferent arterioles decreases GFR and so helps restore blood volume and blood pressure to normal.

Renin-Angiotensin-Aldosterone Mechanism As we discussed in Chapter 18 (p. 623), the **renin-angiotensin-aldosterone mechanism** is the body's main mechanism for increasing blood pressure. Without adequate blood pressure (as might be due to hemorrhage, dehydration, etc.), glomerular filtration is not possible, so this mechanism regulates GFR indirectly.

Low blood pressure causes the granular cells of the juxtaglomerular complex to release **renin**. There are three pathways that stimulate granular cells:

- *Sympathetic nervous system.* As part of the baroreceptor reflex, renal sympathetic nerves activate β_1-adrenergic receptors that cause the granular cells to release renin.
- *Activated macula densa cells.* Low blood pressure or vasoconstriction of the afferent arterioles by the sympathetic nervous system reduces GFR, slowing down the flow of filtrate through the renal tubules. When macula densa cells sense the low NaCl concentration of this sluggishly flowing filtrate, they signal the granular cells to release renin. They may signal by releasing less ATP (also thought to be the tubuloglomerular feedback messenger), by releasing *more* of the prostaglandin PGE_2, or both.
- *Reduced stretch.* Granular cells act as mechanoreceptors. A drop in mean arterial blood pressure reduces the tension in the granular cells' plasma membranes and stimulates them to release more renin.

Other Factors Affecting GFR

Renal cells produce a battery of chemicals, many of which act as paracrines (local signaling molecules) affecting renal arterioles. These include *adenosine* and *prostaglandin E_2 (PGE_2)*; adenosine can be produced extracellularly from released ATP. In addition, the kidney makes its own locally acting *angiotensin II* that reinforces the effects of hormonal angiotensin II described in Chapter 18 (p. 622).

HOMEOSTATIC IMBALANCE 24.3 CLINICAL

Abnormally low urinary output (less than 50 ml/day), called *anuria* (ah-nu′re-ah), may indicate that glomerular blood pressure is too low to cause filtration. Renal failure and anuria can also result from situations in which the nephrons stop functioning, including acute nephritis, transfusion reactions, and crush injuries. +

☑ Check Your Understanding

9. Extrinsic and intrinsic controls of GFR serve two different purposes. What are they?
10. Calculate net filtration pressure given the following values: glomerular hydrostatic pressure = 50 mm Hg, blood colloid osmotic pressure = 25 mm Hg, capsular hydrostatic pressure = 20 Hg.
11. Which of the pressures that determine NFP is regulated by both intrinsic and extrinsic controls of GFR?

For answers, see Answers Appendix.

24.5 Urine formation, step 2: Most of the filtrate is reabsorbed into the blood

→ Learning Objectives

- ☐ **Describe the mechanisms underlying water and solute reabsorption from the renal tubules into the peritubular capillaries.**
- ☐ **Describe how sodium and water reabsorption are regulated in the distal tubule and collecting duct.**

Our total plasma volume filters into the renal tubules about every 22 minutes, so all our plasma would drain away as urine in less than 30 minutes were it not for **tubular reabsorption**, which

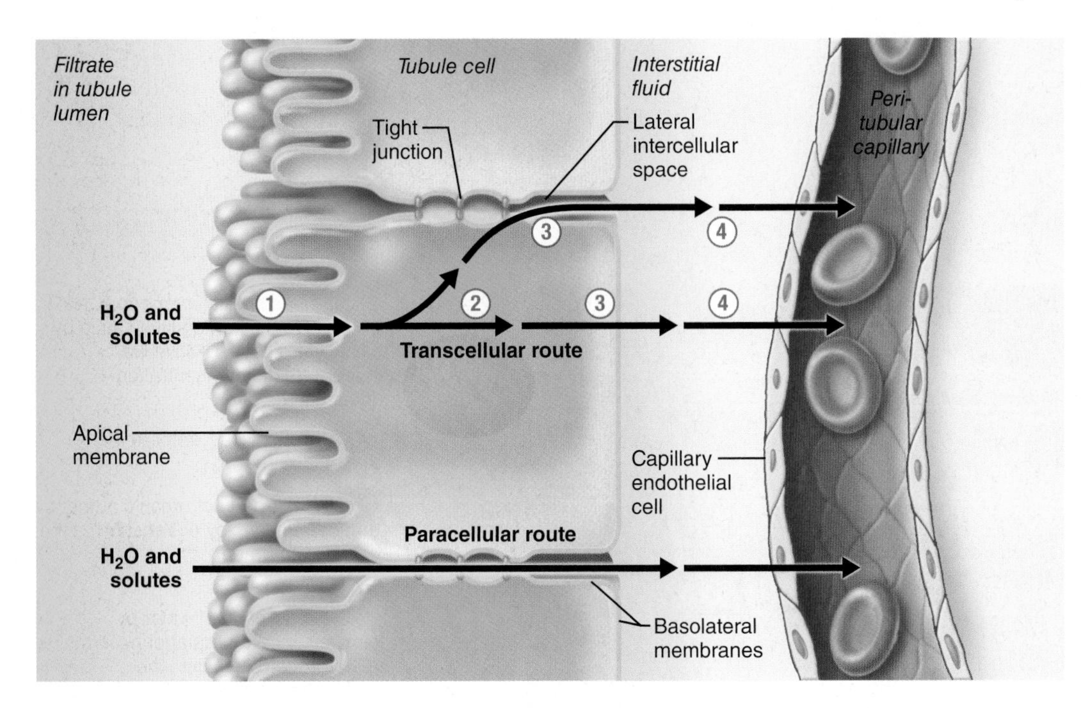

The transcellular route involves:

1. Transport across the apical membrane.
2. Diffusion through the cytosol.
3. Transport across the basolateral membrane. (Often involves the lateral intercellular spaces because membrane transporters transport ions into these spaces.)
4. Movement through the interstitial fluid and into the capillary.

The paracellular route involves:

- Movement through leaky tight junctions, particularly in the proximal convoluted tubule.
- Movement through the interstitial fluid and into the capillary.

Figure 24.15 Transcellular and paracellular routes of tubular reabsorption. Generally, water and solutes move into the peritubular capillaries through intercellular clefts. For simplicity, transporters, ion channels, intercellular clefts, and aquaporins are not depicted.

quickly reclaims most of the tubule contents and returns them to the blood. Tubular reabsorption is a selective *transepithelial process* that begins as soon as the filtrate enters the proximal tubules.

To reach the blood, reabsorbed substances follow either the *transcellular* or *paracellular route* (**Figure 24.15**). In the transcellular route, transported substances move through the *apical membrane*, the cytosol, and the *basolateral membrane* of the tubule cell and then the endothelium of the peritubular capillaries. Movement of substances in the paracellular route—*between* the tubule cells—is limited by the tight junctions connecting these cells. In the proximal nephron, however, these tight junctions are "leaky" and allow water and some important ions (Ca^{2+}, Mg^{2+}, K^+, and some Na^+) to pass through the paracellular route.

Given healthy kidneys, virtually all organic nutrients such as glucose and amino acids are completely reabsorbed to maintain or restore normal plasma concentrations. On the other hand, the reabsorption of water and many ions is continuously regulated and adjusted in response to hormonal signals. Depending on the substances transported, the reabsorption process may be *active* or *passive*. **Active tubular reabsorption** requires ATP either directly (primary active transport) or indirectly (secondary active transport) for at least one of its steps. **Passive tubular reabsorption** encompasses diffusion, facilitated diffusion, and osmosis—processes in which substances move down their electrochemical gradients. (You may wish to review these membrane transport processes in Chapter 3.)

Tubular Reabsorption of Sodium

Sodium ions are the single most abundant cation in the filtrate, and about 80% of the energy used for active transport is devoted to reabsorbing them. Sodium reabsorption is almost always active and via the transcellular route. Let's begin with the ATP-driven step.

Sodium Transport across the Basolateral Membrane

Na^+ is actively transported out of the tubule cell by *primary active transport*—a Na^+-K^+ ATPase pump in the basolateral membrane (**Figure 24.16** ①). From there, the bulk flow of water sweeps Na^+ into adjacent peritubular capillaries. This bulk flow of water and solutes into the peritubular capillaries is rapid because the blood there has low hydrostatic pressure and high osmotic pressure (remember, most proteins remain in the blood instead of filtering out into the tubule).

Sodium Transport across the Apical Membrane

Active pumping of Na^+ from the tubule cells results in a strong electrochemical gradient that favors its entry at the apical face via *secondary active transport* (*cotransport*) carriers (Figure 24.16 ②, ③) or via facilitated diffusion through channels (not illustrated). This occurs because (1) the pump maintains the intracellular Na^+ concentration at low levels, and (2) the K^+ pumped into the tubule cells almost immediately diffuses out into the interstitial fluid via leakage channels, leaving the interior of the tubule cell with a net negative charge.

Because each tubule segment plays a slightly different role in reabsorption, the precise mechanism by which Na^+ is reabsorbed at the apical membrane varies.

Tubular Reabsorption of Nutrients, Water, and Ions

The reabsorption of Na^+ by primary active transport provides the energy and the means for reabsorbing almost every other substance, including water.

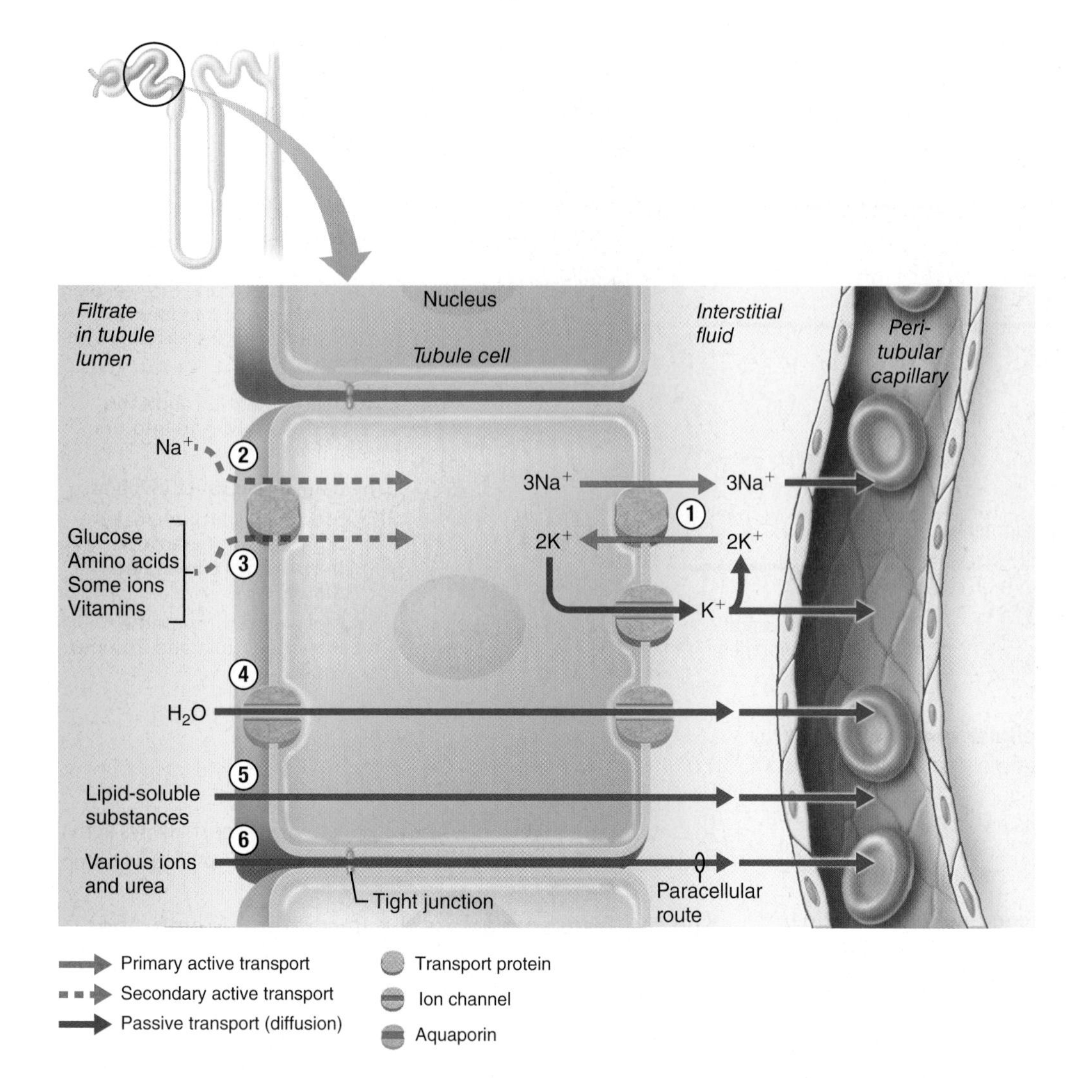

Figure 24.16 Reabsorption by PCT cells. Though not illustrated here, most organic nutrients reabsorbed in the PCT move through the basolateral membrane by facilitated diffusion. Microvilli have been omitted for simplicity.

24

Secondary Active Transport

Substances reabsorbed by *secondary active transport* (the "push" comes from the gradient created by Na^+-K^+ pumping at the basolateral membrane) include glucose, amino acids, some ions, and vitamins. In nearly all these cases, an apical carrier moves Na^+ down its concentration gradient as it cotransports another solute (Figure 24.16 ③). Cotransported solutes move across the basolateral membrane by facilitated diffusion via other transport proteins (not shown) before moving into the peritubular capillaries.

Passive Tubular Reabsorption of Water

The movement of Na^+ and other solutes establishes a strong osmotic gradient, and water moves by osmosis into the peritubular capillaries. Transmembrane proteins called **aquaporins** aid this process by acting as water channels across plasma membranes (Figure 24.16 ④).

In continuously water-permeable regions of the renal tubules, such as the PCT, aquaporins are always present in the tubule cell membranes. Their presence "obliges" the body to absorb water in the proximal nephron regardless of its state of over- or underhydration. This water flow is referred to as **obligatory water reabsorption**.

Aquaporins are virtually absent in the apical membranes of the collecting duct unless antidiuretic hormone (ADH) is present. Water reabsorption that depends on ADH is called **facultative water reabsorption**.

Passive Tubular Reabsorption of Solutes

As water leaves the tubules, the concentration of solutes in the filtrate increases and, if able, they too follow their concentration gradients into the peritubular capillaries. This phenomenon—solutes following solvent—explains the passive reabsorption of a number of solutes present in the filtrate, such as lipid-soluble substances, certain ions, and some urea (Figure 24.16 ⑤, ⑥). It also explains in part why lipid-soluble drugs and environmental pollutants are difficult to excrete: Since lipid-soluble compounds can generally pass through membranes, they will follow their concentration gradients and be reabsorbed, even if this is not "desirable."

As Na^+ ions move through the tubule cells into the peritubular capillary blood, they also establish an electrical gradient that favors passive reabsorption of anions (primarily Cl^-) to restore electrical neutrality in the filtrate and plasma.

Transport Maximum

The transcellular transport systems for the various solutes are quite specific and *limited*. There is a **transport maximum (T_m)** for nearly every substance that is reabsorbed using a transport protein in the membrane. The T_m (reported in mg/min) reflects the number of transport proteins in the renal tubules available to ferry a particular substance. In general, there are plenty of transporters for substances such as glucose that need to be retained, and few or no transporters for substances of no use to the body.

When the transporters are saturated—that is, all bound to the substance they transport—the excess is excreted in urine. This is what happens in individuals who become hyperglycemic because of uncontrolled diabetes mellitus. As plasma levels of glucose approach and exceed 180 mg/dl, the glucose T_m is exceeded and large amounts of glucose may be lost in the urine even though the renal tubules are still functioning normally.

Reabsorptive Capabilities of the Renal Tubules and Collecting Ducts

Table 24.1 (p. 852) compares the reabsorptive abilities of various regions of the renal tubules and collecting ducts.

Proximal Convoluted Tubule

The entire renal tubule is involved in reabsorption to some degree, but the PCT cells are by far the most active reabsorbers and the events just described occur mainly in this tubular segment. Normally, the PCT reabsorbs *all* of the glucose and amino acids in the filtrate and 65% of the Na^+ and water. The bulk of the electrolytes are reabsorbed by the time the filtrate reaches the nephron loop. Nearly all of the uric acid and about half of the urea are reabsorbed in the proximal tubule, but both are later secreted back into the filtrate.

Nephron Loop

Beyond the PCT, the permeability of the tubule epithelium changes dramatically. Here, for the first time, water reabsorption is not coupled to solute reabsorption. Water can leave the descending limb of the nephron loop but *not* the ascending limb, where aquaporins are scarce or absent in the tubule cell membranes. These permeability differences play a vital role in the kidneys' ability to form dilute or concentrated urine.

The rule for water is that it leaves the descending (but not the ascending) limb of the nephron loop. The opposite is true for solutes. Virtually no solute reabsorption occurs in the descending limb, but solutes are reabsorbed both actively and passively in the ascending limb.

In the thin segment of the ascending limb, Na^+ moves passively down the concentration gradient created by water reabsorption. In the thick ascending limb, a Na^+-K^+-$2Cl^-$ symporter is the main means of Na^+ entry at the apical surface. A Na^+-K^+ ATPase operates at the basolateral membrane to create the ionic gradient that drives the symporter. The thick ascending limb also has Na^+-H^+ antiporters. In addition, some 50% of Na^+ passes via the paracellular route in this region.

Distal Convoluted Tubule and Collecting Duct

While reabsorption in the PCT and nephron loop does not vary with the body's needs, hormones fine-tune reabsorption in the DCT and collecting duct (Figure 24.17). Because

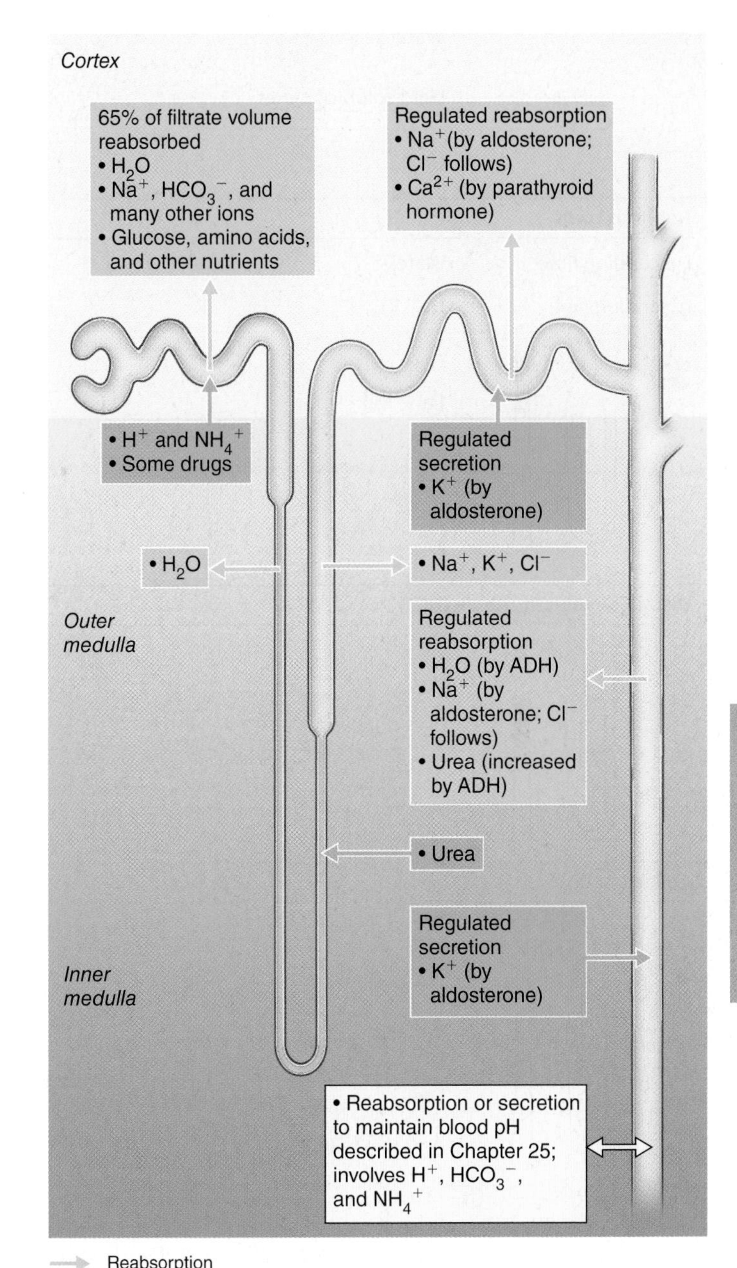

Figure 24.17 Summary of tubular reabsorption and secretion. The various regions of the renal tubule carry out reabsorption and secretion and maintain a gradient of osmolality within the medullary interstitial fluid. Color gradients represent varying osmolality at different points in the interstitial fluid.

24

Table 24.1 Reabsorption Capabilities of Different Segments of the Renal Tubules and Collecting Ducts

TUBULE SEGMENT	SUBSTANCE REABSORBED	MECHANISM
Proximal Convoluted Tubule (PCT)		
	Sodium ions (Na^+)	Primary active transport via basolateral Na^+-K^+ pump; crosses apical membrane through channels, symporters, or antiporters
	Virtually all nutrients (glucose, amino acids, vitamins, some ions)	Secondary active transport with Na^+
	Cl^-, K^+, Mg^{2+}, Ca^{2+}, and other ions	Passive paracellular diffusion driven by electrochemical gradient
	HCO_3^-	Secondary active transport linked to H^+ secretion and Na^+ reabsorption (see Chapter 25)
	Water	Osmosis; driven by solute reabsorption (obligatory water reabsorption)
	Lipid-soluble solutes	Passive diffusion driven by the concentration gradient created by reabsorption of water
	Urea	Primarily passive paracellular diffusion driven by chemical gradient
Nephron Loop		
Descending limb	Water	Osmosis
Ascending limb	Na^+, Cl^-, K^+	Secondary active transport of Cl^-, Na^+, and K^+ via Na^+-K^+-$2Cl^-$ cotransporter in thick portion; paracellular diffusion; Na^+-H^+ antiport
	Ca^{2+}, Mg^{2+}	Passive paracellular diffusion driven by electrochemical gradient
Distal Convoluted Tubule (DCT)		
	Na^+, Cl^-	Primary active Na^+ transport at basolateral membrane; secondary active transport at apical membrane via Na^+-Cl^- symporter and channels; aldosterone-regulated at distal portion
	Ca^{2+}	Passive uptake via PTH-modulated channels in apical membrane; primary and secondary active transport (antiport with Na^+) in basolateral membrane
Collecting Duct		
	Na^+, K^+, HCO_3^-, Cl^-	Primary active transport of Na^+ (requires aldosterone); passive paracellular diffusion of some Cl^-; cotransport of Cl^- and HCO_3^-; K^+ is both reabsorbed and secreted (aldosterone dependent), usually resulting in net K^+ secretion
	Water	Osmosis; controlled (facultative) water reabsorption; ADH required to insert aquaporins
	Urea	Facilitated diffusion in response to concentration gradient in the deep medulla region; recycles and contributes to medullary osmotic gradient

most of the filtered water and solutes have been reabsorbed by the time the DCT is reached, only a small amount of the filtered load is subject to this fine tuning (e.g., about 10% of the originally filtered NaCl and 25% of the water). We introduce hormones that act at the DCT and collecting duct here but discuss them in more detail in Chapter 25.

- **Antidiuretic hormone (ADH).** As its name reveals, ADH inhibits *diuresis* (di″u-re′sis), or urine output. ADH makes the principal cells of the collecting ducts more permeable to water by causing aquaporins to be inserted into their apical membranes. The amount of ADH determines the number of aquaporins, and thus the amount of water that is reabsorbed there. When the body is overhydrated, extracellular fluid osmolality decreases, decreasing ADH secretion by the posterior pituitary (see p. 529) and making the collecting ducts relatively impermeable to water. ADH also increases urea reabsorption by the collecting ducts, as we will describe later.

- **Aldosterone.** Aldosterone fine-tunes reabsorption of the remaining Na^+. Decreased blood volume or blood pressure, or high extracellular K^+ concentration (hyperkalemia), can cause the adrenal cortex to release aldosterone to the blood. Except for hyperkalemia (which *directly* stimulates the adrenal cortex to secrete aldosterone), these conditions promote the renin-angiotensin-aldosterone mechanism (see Figure 24.14).

  Aldosterone targets the principal cells of the collecting ducts and cells of the distal portion of the DCT (prodding them to synthesize and retain more apical Na^+ and K^+ channels, and more basolateral Na^+-K^+ ATPases). As a result, little or no Na^+ leaves the body in urine. In the absence of aldosterone, these segments reabsorb much less Na^+ and about 2% of Na^+ filtered daily can be lost—an amount incompatible with life.

  Physiologically, aldosterone's role is to increase blood volume, and therefore blood pressure, by enhancing Na^+ reabsorption. In general, water follows Na^+ if aquaporins are present. Aldosterone also reduces blood K^+ concentrations because aldosterone-induced reabsorption of Na^+ is coupled to K^+ secretion in the principal cells of the collecting duct. That is, as Na^+ enters the cell, K^+ moves into the lumen.
- **Atrial natriuretic peptide (ANP).** In contrast to aldosterone, which acts to conserve Na^+, ANP reduces blood Na^+, thereby decreasing blood volume and blood pressure. Released by cardiac atrial cells when blood volume or blood pressure is elevated, ANP exerts several effects that lower blood Na^+ content, including direct inhibition of Na^+ reabsorption at the collecting ducts.
- **Parathyroid hormone (PTH).** Acting primarily at the DCT, PTH increases the reabsorption of Ca^{2+}.

☑ Check Your Understanding

12. In which part of the nephron does most reabsorption occur?

13. How does the movement of Na^+ drive the reabsorption of water and solutes?

14. MAKING connections Primary and secondary active transport processes are shown in Figure 24.16 (and were introduced in Chapter 3). How do they differ?

For answers, see Answers Appendix.

24.6 Urine formation, step 3: Certain substances are secreted into the filtrate

→ Learning Objective

☐ **Describe the importance of tubular secretion and list several substances that are secreted.**

The most important way to clear plasma of unwanted substances is to simply not reabsorb them from the filtrate. Another way is **tubular secretion**—essentially, reabsorption in reverse. Tubular secretion moves *selected* substances (such as H^+, K^+, NH_4^+, creatinine, and certain organic acids and bases) from the peritubular capillaries through the tubule cells into the filtrate. Also, some substances (such as HCO_3^-) that are synthesized in the tubule cells are secreted.

The urine eventually excreted contains *both filtered and secreted substances.* With one major exception (K^+), the PCT is the main site of secretion, but the collecting ducts are also active (Figure 24.17).

Tubular secretion is important for:

- *Disposing of substances, such as certain drugs and metabolites, that are tightly bound to plasma proteins.* Because plasma proteins are generally not filtered, the substances they bind are not filtered and so must be secreted.
- *Eliminating undesirable substances or end products that have been reabsorbed by passive processes.* Urea and uric acid, two nitrogenous wastes, are both handled in this way. Urea handling in the nephron is complicated and will be discussed on p. 857, but the net effect is that 40–50% of the urea in the filtrate is excreted.
- *Ridding the body of excess K^+.* Because virtually all K^+ present in the filtrate is reabsorbed in the PCT and ascending nephron loop, nearly all K^+ in urine comes from aldosterone-driven active tubular secretion into the late DCT and collecting ducts.
- *Controlling blood pH.* When blood pH drops toward the acidic end of its homeostatic range, the renal tubule cells actively secrete more H^+ into the filtrate and retain and generate more HCO_3^- (a base). As a result, blood pH rises and the urine drains off the excess H^+. Conversely, when blood pH approaches the alkaline end of its range, Cl^- is reabsorbed instead of HCO_3^-, which is allowed to leave the body in urine. We will discuss the kidneys' role in pH homeostasis in more detail in Chapter 25.

☑ Check Your Understanding

15. List several substances that are secreted into the kidney tubules.

For answers, see Answers Appendix.

24

24.7 The kidneys create and use an osmotic gradient to regulate urine concentration and volume

→ Learning Objectives

☐ **Describe the mechanisms responsible for the medullary osmotic gradient.**

☐ **Explain how dilute and concentrated urine are formed.**

From day to day, and even hour to hour, our intake and loss of fluids can vary dramatically. For example, when you run on a hot summer day, you dehydrate as you rapidly lose fluid as sweat. On the other hand, if you drink a pitcher of lemonade while sitting on the porch, you may overhydrate. In response, the kidneys make adjustments to keep the solute concentration of body fluids

(Text continues on page 856.)

Focus Figure 24.1 Juxtamedullary nephrons create an osmotic gradient within the renal medulla that allows the kidney to produce urine of varying concentration.

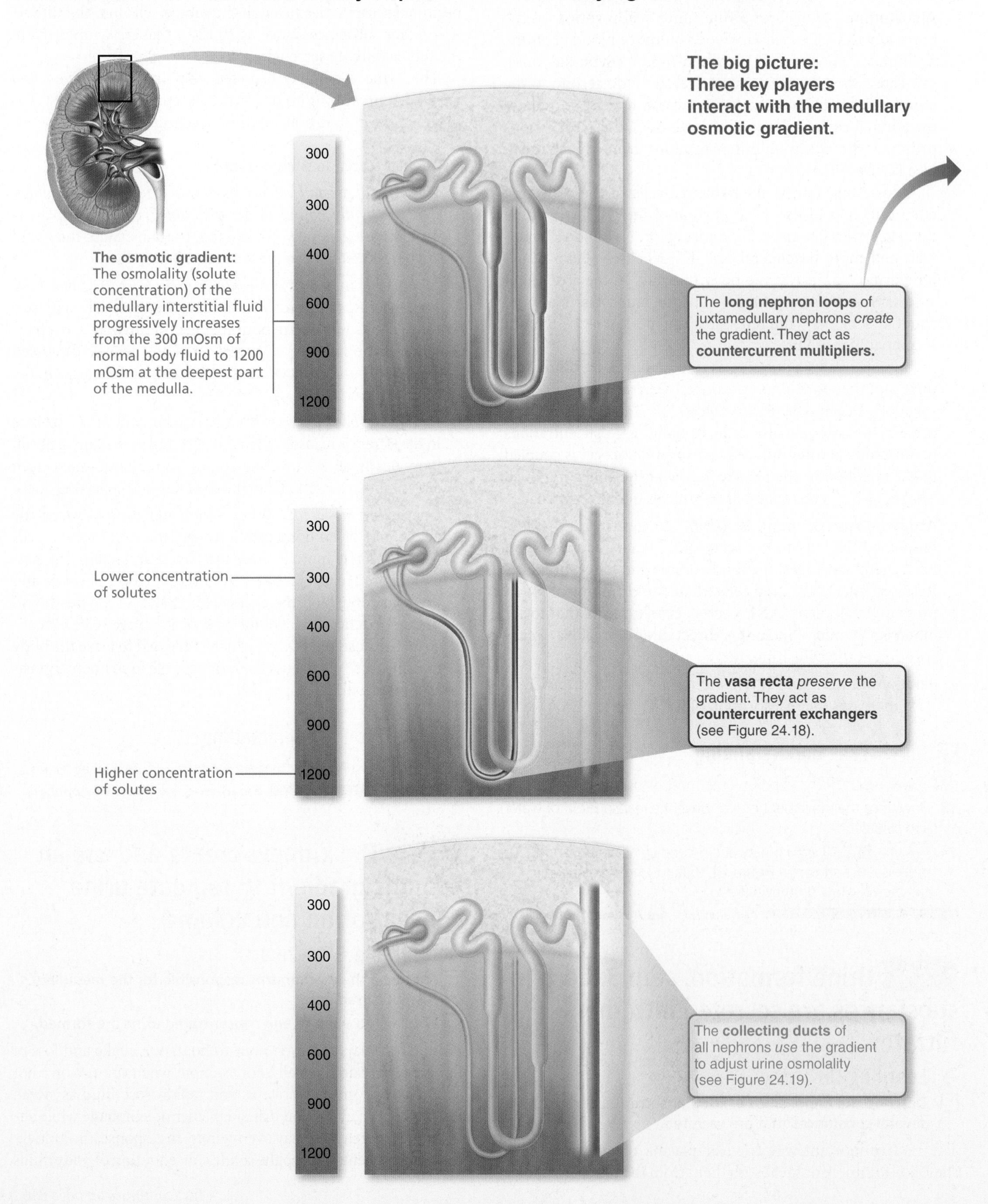

Long nephron loops of juxtamedullary nephrons create the gradient.

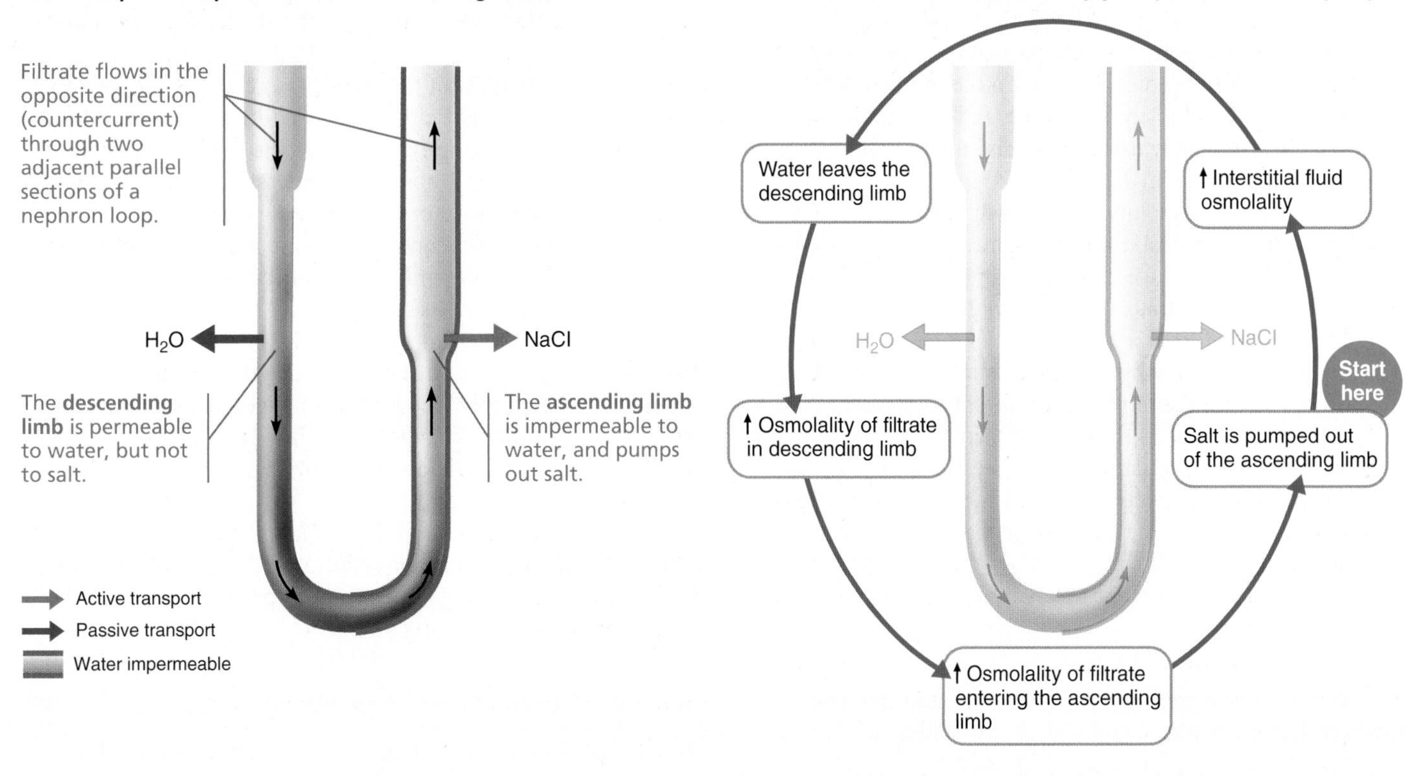

As water and solutes are reabsorbed, the loop first concentrates the filtrate, then dilutes it.

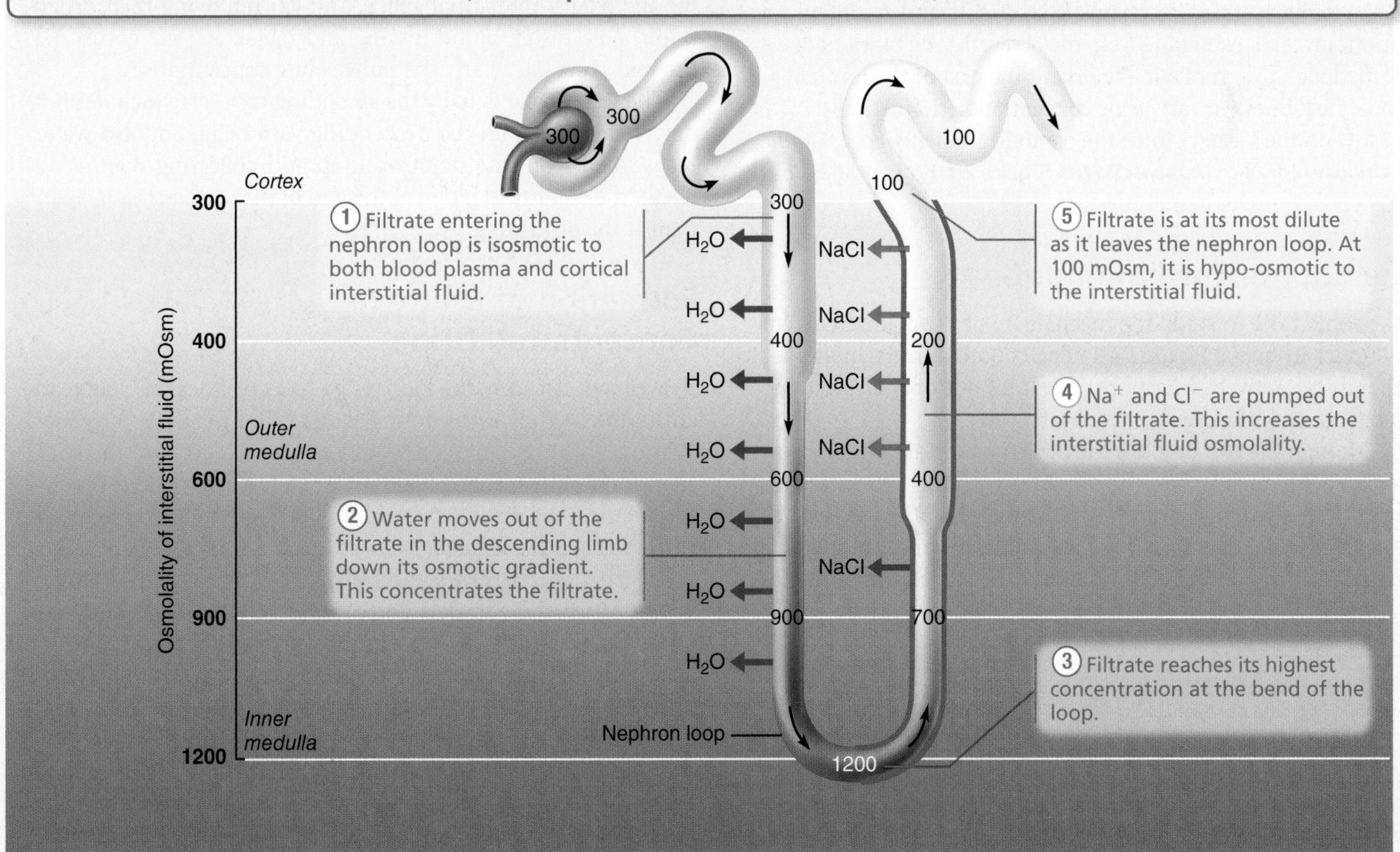

constant at about 300 mOsm, the normal osmotic concentration of blood plasma. Maintaining constant osmolality of extracellular fluids is crucial for preventing cells, particularly in the brain, from shrinking or swelling from the osmotic movement of water.

Recall from Chapter 3 (pp. 64–66) that a solution's osmolality is the concentration of solute particles per kilogram of water. Because 1 osmol (equivalent to 1 mole of particles) is a fairly large unit, the milliosmol (mOsm) (mil″e-oz′mōl), equal to 0.001 osmol, is generally used. In the discussion that follows, we use mOsm to indicate mOsm/kg.

The kidneys keep the solute load of body fluids constant by regulating urine concentration and volume. When you dehydrate, your kidneys produce a small volume of concentrated urine. When you overhydrate, your kidneys produce a large volume of dilute urine.

The kidneys accomplish this feat using countercurrent mechanisms. In the kidneys, the term *countercurrent* means that fluid flows in opposite directions through adjacent segments of the same tube connected by a hairpin turn* (see *Focus on the Medullary Osmotic Gradient*, Focus Figure 24.1, pp. 854–855). This arrangement makes it possible to exchange materials between the two segments.

Two types of countercurrent mechanisms determine urine concentration and volume:

- The **countercurrent multiplier** is the interaction between the flow of filtrate through the ascending and descending limbs of the long nephron loops of juxtamedullary nephrons.
- The **countercurrent exchanger** is the flow of blood through the ascending and descending portions of the vasa recta.

These countercurrent mechanisms establish and maintain an osmotic gradient extending from the cortex through the depths of the medulla. This gradient—the **medullary osmotic gradient**—allows the kidneys to vary urine concentration dramatically.

How do the kidneys form the osmotic gradient? *Focus on the Medullary Osmotic Gradient* (**Focus Figure 24.1** on pp. 854–855) explores the answer to this question.

24

The Countercurrent Multiplier

Take some time to study the mechanism of the countercurrent multiplier in Focus Figure 24.1. The countercurrent multiplier depends on actively transporting solutes out of the ascending limb ("Start" of the positive feedback cycle).

Although the two limbs of the nephron loop are not in direct contact with each other, they are close enough to influence each other's exchanges with the interstitial fluid they share. The more NaCl the ascending limb extrudes, the more water diffuses out of the descending limb and the saltier the filtrate in the descending limb becomes. The ascending limb then uses the increasingly "salty" filtrate left behind in the descending limb to raise the osmolality of the medullary interstitial fluid even further. This establishes a positive feedback cycle that produces the high osmolality of the fluids in the descending limb and interstitial fluid.

Notice at the bottom of the right page of Focus Figure 24.1 that there is a constant difference in filtrate concentration (200 mOsm) between the two limbs of the nephron loop, and between the ascending limb and the interstitial fluid. This difference reflects the power of the ascending limb's NaCl pumps, which are just powerful enough to create a 200 mOsm difference between the inside and outside of the ascending limb. A 200 mOsm gradient by itself would not be enough to allow excretion of very concentrated urine. The beauty of this system lies in the fact that, because of countercurrent flow, the nephron loop is able to "multiply" these small changes in solute concentration into a gradient change along the vertical length of the loop (both inside and outside) that is closer to 900 mOsm (1200 mOsm – 300 mOsm).

Notice also that while much of the Na^+ and Cl^- reabsorption in the ascending limb is active (via Na^+-K^+-$2Cl^-$ cotransporters in the thick ascending limb), some is passive (mostly in the thin portion of the ascending limb).

The Countercurrent Exchanger

The vasa recta act as countercurrent exchangers (**Figure 24.18**). Countercurrent exchange does not create the medullary gradient, but preserves it by (1) preventing rapid removal of salt from the medullary interstitial space, and (2) removing reabsorbed water. As a result, blood leaving and reentering the cortex via the vasa recta has nearly the same solute concentration.

The water picked up by the ascending vasa recta includes not only water lost from the descending vasa recta, but also water reabsorbed from the nephron loop and collecting duct. As a result, the volume of blood at the end of the vasa recta is greater than at the beginning.

Formation of Dilute or Concentrated Urine

As we have just seen, the kidneys go to a great deal of trouble to create the medullary osmotic gradient. But for what purpose? Without this gradient, you would not be able to raise the concentration of urine above 300 mOsm—the osmolality of interstitial fluid. As a result, you would not be able to conserve water when you are dehydrated.

Figure 24.19 shows the body's response to either overhydration or dehydration and ADH's role in controlling the production of dilute or concentrated urine. When we are overhydrated, ADH production decreases and the osmolality of urine falls as low as 100 mOsm. If aldosterone (not shown) is present, the DCT and collecting duct cells can remove Na^+ and selected other ions from the filtrate, making the urine that enters the renal pelvis even more dilute. The osmolality of urine can plunge as low as 50 mOsm, about one-sixth the concentration of glomerular filtrate or blood plasma.

---

*The term "countercurrent" is commonly misunderstood to mean that the direction of fluid flow in the nephron loops is opposite that of the blood in the vasa recta. In fact, there is no one-to-one relationship between individual nephron loops and capillaries of the vasa recta as might be suggested by two-dimensional diagrams such as Focus Figure 24.1. Instead, there are many tubules and capillaries packed together like a bundle of straws. Each tubule is surrounded by many blood vessels, whose flow is not necessarily counter to flow in that tubule (see Figure 24.8).

Vasa recta preserve the gradient.

- **The vasa recta are highly permeable to water and solutes.**
- **Countercurrent exchanges occur between each section of the vasa recta and its surrounding fluid. As a result:**
  - The blood within the vasa recta remains nearly isosmotic to the surrounding fluid.
  - The vasa recta are able to reabsorb water and solutes into the general circulation without undoing the osmotic gradient created by the countercurrent multiplier.

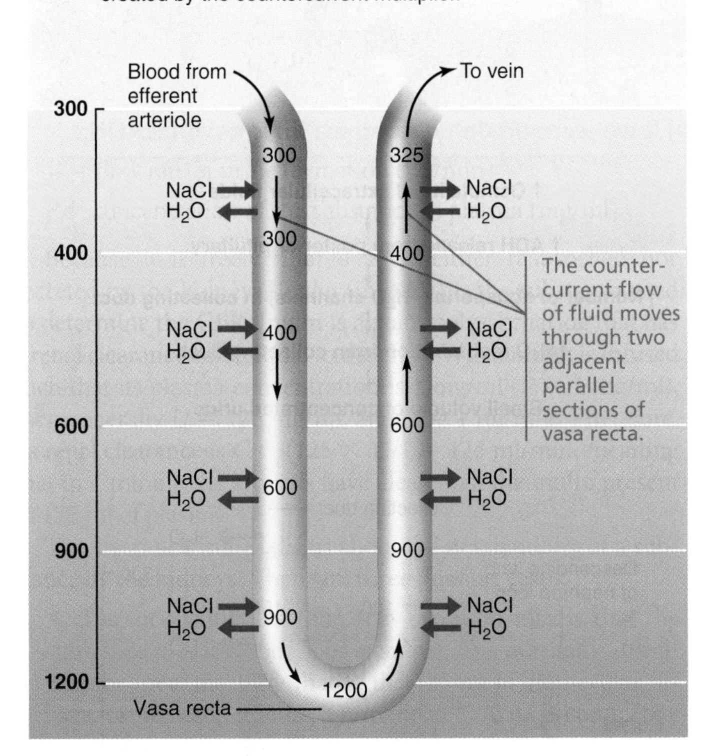

Figure 24.18 **Countercurrent exchange.**

When we are dehydrated, the posterior pituitary releases large amounts of ADH and the solute concentration of urine may rise as high as 1200 mOsm, the concentration of interstitial fluid in the deepest part of the medulla. With maximal ADH secretion, up to 99% of the water in the filtrate is reabsorbed and returned to the blood, and only half a liter per day of highly concentrated urine is excreted. The ability of our kidneys to produce such concentrated urine is critically tied to our ability to survive for a limited time without water.

Urea Recycling and the Medullary Osmotic Gradient

We're not quite done yet. There's one last piece of the puzzle left—urea. We usually think of urea as simply a metabolic waste product, but conserving water is so important that the kidneys actually use urea to help form the medullary gradient (Figure 24.19).

1. Urea enters the filtrate by facilitated diffusion in the ascending thin limb of the nephron loop.
2. As the filtrate moves on, the cortical collecting duct usually reabsorbs water, leaving urea behind.
3. When filtrate reaches the portion of the collecting duct in the deep medullary region, the now highly concentrated urea moves by facilitated diffusion out of the collecting duct into the interstitial fluid of the medulla. These movements form a pool of urea that recycles back into the ascending thin limb of the nephron loop. In this way, urea contributes substantially to the high osmolality in the medulla.

Antidiuretic hormone enhances urea transport out of the medullary collecting duct. When ADH is present, it increases urea recycling and strengthens the medullary osmotic gradient, allowing more concentrated urine to be formed.

Diuretics

There are several types of **diuretics**, chemicals that enhance urinary output. Alcohol encourages diuresis by inhibiting release of ADH. Other diuretics increase urine flow by inhibiting Na^+ reabsorption and the obligatory water reabsorption that normally follows. Examples include many drugs prescribed for hypertension or the edema of congestive heart failure. Most diuretics inhibit Na^+-associated symporters. "Loop diuretics" [like furosemide (Lasix)] are powerful because they inhibit formation of the medullary gradient by acting at the ascending limb of the nephron loop. Thiazides are less potent and act at the DCT. An *osmotic diuretic* is a substance that is not reabsorbed and that carries water out with it (for example, the high blood glucose of a diabetes mellitus patient).

☑ Check Your Understanding

16. Describe the special characteristics of the descending and ascending limbs of the nephron loop that cause the formation of the medullary osmotic gradient.

17. Under what conditions is ADH released from the posterior pituitary? What effect does ADH have on the collecting ducts?

For answers, see Answers Appendix.

24.8 Renal function is evaluated by analyzing blood and urine

→ Learning Objectives

- ☐ **Define renal clearance and explain how this value summarizes the way a substance is handled by the kidney.**
- ☐ **Describe the normal physical and chemical properties of urine.**
- ☐ **List several abnormal urine components, and name the condition characterized by the presence of detectable amounts of each.**

Since the time of the ancient Greek Hippocrates, physicians have examined their patients' urine for signs of disease. **Urinalysis**, the analysis of urine, can aid in the diagnosis of diseases or detect illegal substances. However, to fully understand renal function, we need to analyze *both blood and urine*. For example, renal function is often assessed by measuring levels of *nitrogenous wastes* in blood, whereas determination of renal clearance requires that both blood and urine be tested.

Table 24.2 Abnormal Urinary Constituents

SUBSTANCE	NAME OF CONDITION	POSSIBLE CAUSES
Glucose	Glycosuria	Diabetes mellitus
Proteins	Proteinuria, albuminuria	Nonpathological: excessive physical exertion, pregnancy Pathological (over 150 mg/day): glomerulonephritis, severe hypertension, heart failure, often an initial sign of renal disease
Ketone bodies	Ketonuria	Excessive formation and accumulation of ketone bodies, as in starvation and untreated diabetes mellitus
Hemoglobin	Hemoglobinuria	Various: transfusion reaction, hemolytic anemia, severe burns, etc.
Bile pigments	Bilirubinuria	Liver disease (hepatitis, cirrhosis) or obstruction of bile ducts from liver or gallbladder
Erythrocytes	Hematuria	Bleeding urinary tract (due to trauma, kidney stones, infection, or cancer)
Leukocytes (pus)	Pyuria	Urinary tract infection

Odor Fresh urine is slightly aromatic, but if allowed to stand, it develops an ammonia odor as bacteria metabolize its urea solutes. Some drugs and vegetables alter the odor of urine, as do some diseases. For example, in uncontrolled diabetes mellitus the urine smells fruity because of its acetone content.

pH Urine is usually slightly acidic (around pH 6), but changes in body metabolism or diet may cause the pH to vary from about 4.5 to 8.0. A predominantly *acidic* diet that contains large amounts of protein and whole wheat products produces acidic urine. A vegetarian (*alkaline*) diet, prolonged vomiting, and bacterial infection of the urinary tract all cause the urine to become alkaline.

Specific Gravity The ratio of the mass of a substance to the mass of an equal volume of distilled water is its **specific gravity**. Because urine is water plus solutes, a given volume has a greater mass than the same volume of distilled water. The specific gravity of distilled water is 1.0 and that of urine ranges from 1.001 to 1.035, depending on its solute concentration.

24

☑ Check Your Understanding

18. What would you expect the normal clearance value for amino acids to be? Explain.

19. What are the three major nitrogenous wastes excreted in the urine?

For answers, see Answers Appendix.

24.9 The ureters, bladder, and urethra transport, store, and eliminate urine

→ Learning Objectives

☐ **Describe the general location, structure, and function of the ureters, urinary bladder, and urethra.**

☐ **Compare the course, length, and functions of the male urethra with those of the female.**

☐ **Define micturition and describe its neural control.**

The kidneys form urine continuously and the ureters transport it to the bladder. It is usually stored in the bladder until its release through the urethra in a process called *micturition*.

Ureters

The **ureters** are slender tubes that convey urine from the kidneys to the bladder (Figures 24.1, 24.2, and 24.22). Each ureter begins at the level of L_2 as a continuation of the renal pelvis. From there, it descends behind the peritoneum and runs obliquely through the posterior bladder wall. This arrangement prevents backflow of urine because any increase in bladder pressure compresses and closes the distal ends of the ureters.

Histologically, the ureter wall has three layers (**Figure 24.20**). From the inside out:

- The *mucosa* contains a transitional epithelium that is continuous with the mucosae of the kidney pelvis superiorly and the bladder medially.

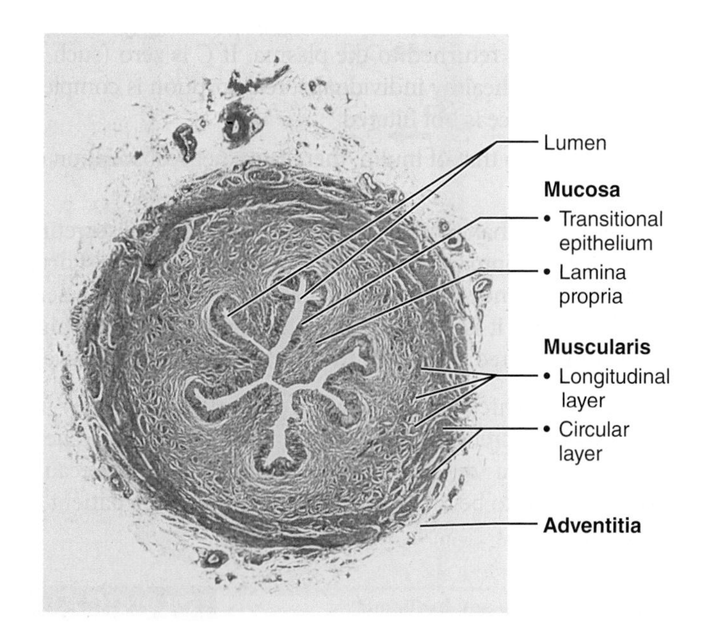

Figure 24.20 Cross-sectional view of the ureter wall (10×). The prominent mucosal folds seen in an empty ureter stretch and flatten to accommodate large pulses of urine.

View histology slides
MasteringA&P®>Study Area>PAL

- The *muscularis* is composed chiefly of two smooth muscle sheets: the internal longitudinal layer and the external circular layer. An additional smooth muscle layer, the external longitudinal layer, appears in the lower third of the ureter.
- The *adventitia* covering the ureter's external surface is typical fibrous connective tissue.

The ureter plays an active role in transporting urine. Incoming urine distends the ureter and stimulates its muscularis to contract, propelling urine into the bladder. (Urine does *not* reach the bladder through gravity alone.) The strength and frequency of the peristaltic waves are adjusted to the rate of urine formation. Both sympathetic and parasympathetic fibers innervate each ureter, but neural control of peristalsis appears to be insignificant compared to the way ureteral smooth muscle responds to stretch.

HOMEOSTATIC IMBALANCE 24.5 CLINICAL

On occasion, calcium, magnesium, or uric acid salts in urine may crystallize and precipitate in the renal pelvis, forming **renal calculi** (kal′ku-li; *calculus* = little stone), or kidney stones. Most calculi are under 5 mm in diameter and pass through the urinary tract without causing problems. However, larger calculi can obstruct a ureter and block urine drainage. Increasing pressure in the kidney causes excruciating pain, which radiates from the flank to the anterior abdominal wall on the same side. Pain also occurs during peristalsis when the contracting ureter wall closes in on the sharp calculi.

Predisposing conditions are frequent bacterial infections of the urinary tract, urine retention, high blood levels of calcium, and alkaline urine. Surgical removal of calculi has been almost entirely replaced by *shock wave lithotripsy*, a noninvasive procedure that uses ultrasonic shock waves to shatter the calculi. The pulverized, sandlike remnants of the calculi are then painlessly eliminated in the urine. People with a history of kidney stones are encouraged to drink enough water to keep their urine dilute. +

Urinary Bladder

The **urinary bladder** is a smooth, collapsible, muscular sac that stores urine temporarily.

Urinary Bladder Anatomy

The bladder is located retroperitoneally on the pelvic floor just posterior to the pubic symphysis. The prostate (part of the male reproductive system) lies inferior to the bladder neck, which empties into the urethra. In females, the bladder is anterior to the vagina and uterus (see Figure 26.13 on p. 905).

The interior of the bladder has openings for both ureters and the urethra (**Figure 24.21**). The smooth, triangular region of the bladder base outlined by these three openings is the **trigone** (tri′gōn; *trigon* = triangle), important clinically because infections tend to persist in this region.

The bladder wall has three layers: a mucosa containing transitional epithelium, a thick muscular layer, and a fibrous adventitia (except on its superior surface, where it is covered by the peritoneum). The muscular layer, called the **detrusor** (de-tru′sor; "to thrust out"), consists of intermingled smooth muscle fibers arranged in inner and outer longitudinal layers and a middle circular layer.

Urine Storage Capacity

The bladder is very distensible. When empty, the bladder collapses into its basic pyramidal shape and its walls are thick and thrown into folds (*rugae*). As urine accumulates, the bladder expands, becomes pear shaped, and rises superiorly in the abdominal cavity. The muscular wall stretches and thins, and rugae disappear. These changes allow the bladder to store more urine without a significant rise in internal pressure.

A moderately full bladder is about 12 cm (5 inches) long and holds approximately 500 ml (1 pint) of urine, but it can hold nearly double that if necessary. When tense with urine, it can be palpated well above the pubic symphysis. The maximum capacity of the bladder is 800–1000 ml and when it is overdistended, it may burst.

The urinary bladder and ureters can be seen in a special X ray called a pyelogram (**Figure 24.22**).

Urethra

The **urethra** is a thin-walled muscular tube that drains urine from the bladder and conveys it out of the body. The epithelium of its mucosal lining is mostly pseudostratified columnar epithelium. However, near the bladder it becomes transitional epithelium, and near the external opening it changes to a protective stratified squamous epithelium.

At the bladder-urethra junction, the detrusor smooth muscle thickens to form the **internal urethral sphincter** (Figure 24.21). This involuntary sphincter, controlled by the autonomic nervous system, keeps the urethra closed when urine is not being passed and prevents leaking between voiding.

The **external urethral sphincter** surrounds the urethra as it passes through the *urogenital diaphragm*. This sphincter is formed of skeletal muscle and is voluntarily controlled. The *levator ani* muscle of the pelvic floor also serves as a voluntary constrictor of the urethra (see Table 10.7, pp. 305–306).

The length and functions of the urethra differ in the two sexes. In females the urethra is only 3–4 cm (1.5 inches) long and fibrous connective tissue binds it tightly to the anterior vaginal wall. Its external opening, the **external urethral orifice**, lies anterior to the vaginal opening and posterior to the clitoris.

In males the urethra is approximately 20 cm (8 inches) long and has three regions.

- The **prostatic urethra**, about 2.5 cm (1 inch) long, runs within the prostate.
- The **intermediate part of the urethra** (or *membranous urethra*), which runs through the urogenital diaphragm, extends about 2 cm from the prostate to the beginning of the penis.
- The **spongy urethra**, about 15 cm long, passes through the penis and opens at its tip via the **external urethral orifice**.

The male urethra has a double function: It carries semen as well as urine out of the body. We discuss its reproductive function in Chapter 26.

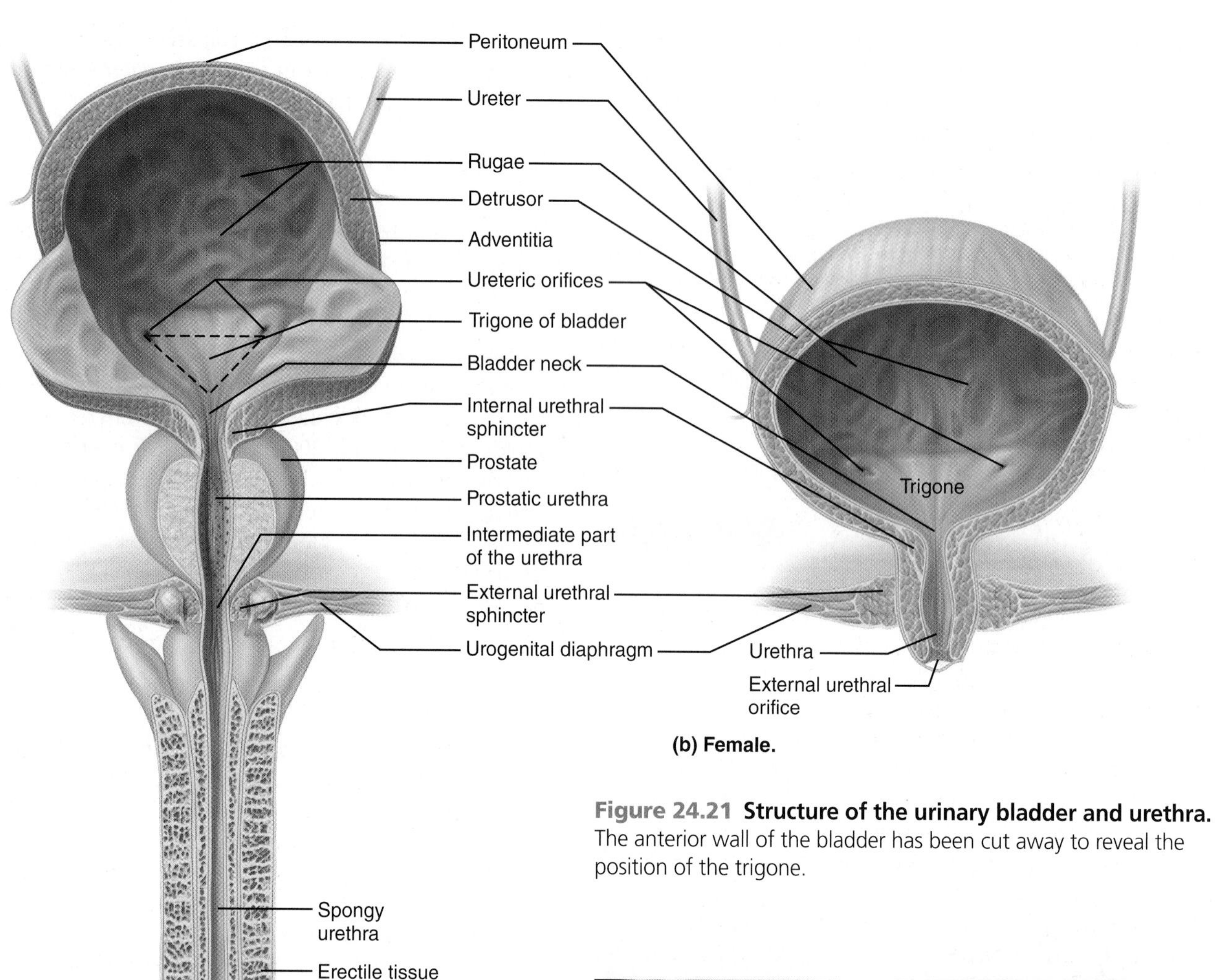

(a) Male. The long male urethra has three regions: prostatic, intermediate, and spongy.

(b) Female.

Figure 24.21 Structure of the urinary bladder and urethra. The anterior wall of the bladder has been cut away to reveal the position of the trigone.

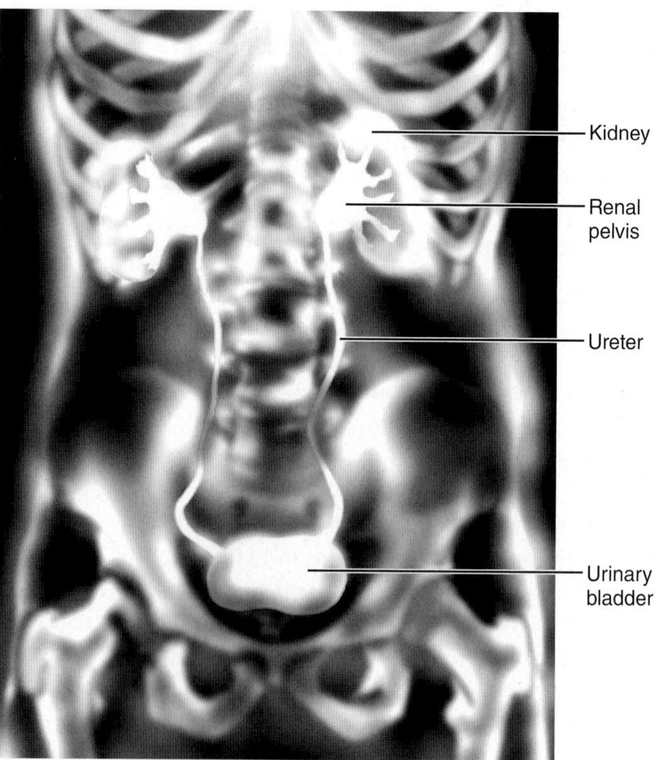

Figure 24.22 Pyelogram. This X-ray image was obtained using a contrast medium to show the ureters, kidneys, and urinary bladder.

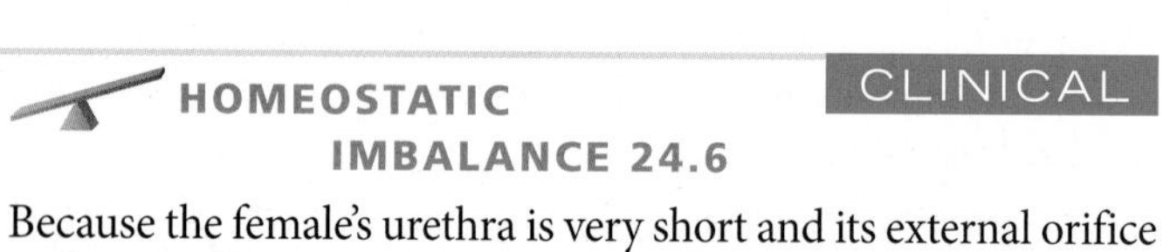

HOMEOSTATIC IMBALANCE 24.6

Because the female's urethra is very short and its external orifice is close to the anal opening, improper toilet habits (wiping back to front after defecation) can easily carry fecal bacteria into the urethra. Most *urinary tract infections* occur in sexually active women, because intercourse drives bacteria from the vagina and external genital region toward the bladder. The use of spermicides magnifies this problem, because the spermicide kills helpful bacteria, allowing infectious fecal bacteria to colonize the vagina. Overall, 40% of all women get urinary tract infections.

The urethral mucosa is continuous with that of the rest of the urinary tract, and an inflammation of the urethra (*urethritis*) can ascend the tract to cause bladder inflammation (*cystitis*) or even renal inflammations (*pyelitis* or *pyelonephritis*). Symptoms of urinary tract infection include dysuria (painful urination), urinary *urgency* and *frequency*, fever, and sometimes cloudy or blood-tinged urine. When the kidneys are involved, back pain and a severe headache often occur. Antibiotics can cure most urinary tract infections. +

Micturition

Micturition (mik″tu-rish′un; *mictur* = urinate), also called **urination** or *voiding*, is the act of emptying the urinary bladder. For micturition to occur, three things must happen simultaneously: (1) the detrusor must contract, (2) the internal urethral sphincter must open, and (3) the external urethral sphincter must open.

The detrusor and its internal urethral sphincter are composed of smooth muscle and are innervated by both the parasympathetic and sympathetic nervous systems, which have opposing actions. The external urethral sphincter, in contrast, is skeletal muscle, and therefore is innervated by the somatic nervous system.

How are the three events required for micturition coordinated? Micturition is most easily understood in infants where a spinal reflex coordinates the process. As urine accumulates, distension of the bladder activates stretch receptors in its walls. Impulses from the activated receptors travel via visceral afferent fibers to the sacral region of the spinal cord. Visceral afferent impulses, relayed by sets of interneurons, excite parasympathetic neurons and inhibit sympathetic neurons (**Figure 24.23**). As a result, the detrusor contracts and the internal sphincter opens. Visceral afferent impulses also decrease the firing rate of somatic efferents that normally keep the external urethral sphincter closed. This allows the sphincter to relax so urine can flow.

Between ages 2 and 3, descending circuits from the brain have matured enough to begin to override reflexive urination. The pons has two centers that participate in control of micturition. The *pontine storage center* inhibits micturition, whereas the *pontine micturition center* promotes this reflex. Afferent impulses from bladder stretch receptors are relayed to the pons, as well as to higher brain centers that provide the conscious awareness of bladder fullness.

Lower bladder volumes primarily activate the pontine storage center, which inhibits urination by suppressing parasympathetic and enhancing sympathetic output to the bladder. When a person chooses not to void, reflex bladder contractions subside within a

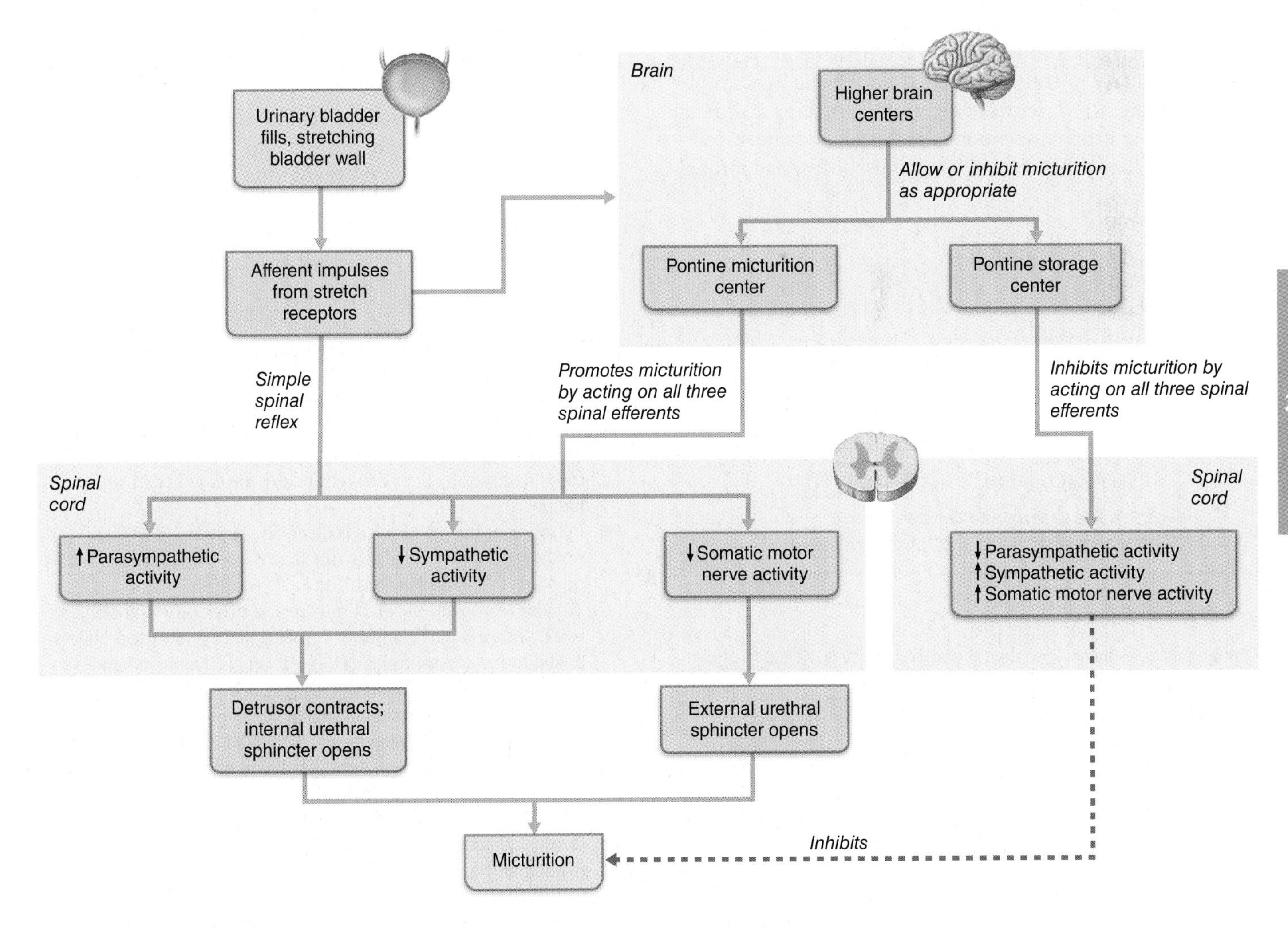

Figure 24.23 **Control of micturition.**

minute or so and urine continues to accumulate. Because the external sphincter is voluntarily controlled, we can choose to keep it closed and postpone bladder emptying temporarily. After additional urine has collected, the micturition reflex occurs again and, if urination is delayed again, is damped once more.

The urge to void gradually becomes greater and greater, and micturition usually occurs before urine volume exceeds 400 ml. After normal micturition, only about 10 ml of urine remains in the bladder.

CLINICAL

HOMEOSTATIC IMBALANCE 24.7

In adults, **urinary incontinence** (the inability to control urination) is usually a result of weakened pelvic muscles following childbirth or surgery, physical pressure during pregnancy, or nervous system problems. In *stress incontinence*, a sudden increase in intra-abdominal pressure (during laughing and coughing) forces urine through the external sphincter. This condition is common during pregnancy when the heavy uterus stretches the muscles of the pelvic floor and the urogenital diaphragm that support the external sphincter. In *overflow incontinence*, urine dribbles from the urethra whenever the bladder overfills.

In **urinary retention**, the bladder is unable to expel its contained urine. Urinary retention is common after general anesthesia (it takes a little time for the detrusor to regain its activity). Urinary retention in men often reflects hypertrophy of the prostate, which narrows the urethra, making it difficult to void. When urinary retention is prolonged, a slender drainage tube called a **catheter** (kath′ĕ-ter) must be inserted through the urethra to drain the urine and prevent bladder trauma from excessive stretching. +

☑ Check Your Understanding

20. A kidney stone blocking a ureter would interfere with urine flow to which organ? Why would the pain occur in waves?
21. What is the trigone of the bladder, and which landmarks define its borders?
22. Name the three regions of the male urethra.
23. How does the detrusor respond to increased firing of the parasympathetic fibers that innervate it? How does this affect the internal urethral sphincter?
24. MAKING **connections** Compare the structure and regulation of the sphincters that control micturition to those that control defecation (Chapter 22).

For answers, see Answers Appendix.

The ureters, urinary bladder, and urethra play important roles in transporting, storing, and eliminating urine from the body, but when the term "urinary system" is used, it is the kidneys that capture center stage. Other organ systems of the body contribute to the well-being of the urinary system in many ways. In turn, without continuous kidney function, the electrolyte and fluid balance of the blood is dangerously disturbed, and internal body fluids quickly become contaminated with nitrogenous wastes. No body cell can escape the harmful effects of such imbalances.

Now that we have described renal mechanisms, we are ready to integrate urinary system function into the larger topic of fluid and electrolyte balance in the body—the focus of Chapter 25.

REVIEW QUESTIONS

24

MAP For more chapter study tools, go to the Study Area of MasteringA&P®.
There you will find:
- Interactive Physiology iP
- A&PFlix A&PFlix
- Practice Anatomy Lab PAL
- PhysioEx PEx
- Videos, Practice Quizzes and Tests, MP3 Tutor Sessions, Case Studies, and much more!

Multiple Choice/Matching

(Some questions have more than one correct answer. Select the best answer or answers from the choices given.)

1. The lowest blood concentration of nitrogenous waste occurs in the **(a)** hepatic vein, **(b)** inferior vena cava, **(c)** renal artery, **(d)** renal vein.
2. The glomerular capillaries differ from other capillary networks in the body because they **(a)** have a larger area of anastomosis, **(b)** are derived from and drain into arterioles, **(c)** are not made of endothelium, **(d)** are sites of filtrate formation.
3. Damage to the renal medulla would interfere *first* with the functioning of the **(a)** glomerular capsules, **(b)** distal convoluted tubules, **(c)** collecting ducts, **(d)** proximal convoluted tubules.
4. Which is reabsorbed by the proximal convoluted tubule cells? **(a)** Na^+, **(b)** K^+, **(c)** amino acids, **(d)** all of the above.
5. Glucose is not normally found in the urine because it **(a)** does not pass through the walls of the glomerulus, **(b)** is kept in the blood by colloid osmotic pressure, **(c)** is reabsorbed by the tubule cells, **(d)** is removed by the body cells before the blood reaches the kidney.
6. Filtration at the glomerulus is inversely related to **(a)** water reabsorption, **(b)** capsular hydrostatic pressure, **(c)** arterial blood pressure, **(d)** acidity of the urine.
7. Tubular reabsorption **(a)** of glucose and many other substances is a T_m-limited active transport process, **(b)** of chloride is always linked to the passive transport of Na^+, **(c)** is the movement of substances from the blood into the nephron, **(d)** of sodium occurs only in the proximal tubule.
8. If a freshly voided urine sample contains excessive amounts of urochrome, it has **(a)** an ammonia-like odor, **(b)** a pH below normal, **(c)** a dark yellow color, **(d)** a pH above normal.
9. Conditions such as diabetes mellitus and starvation are closely linked to **(a)** ketonuria, **(b)** pyuria, **(c)** albuminuria, **(d)** hematuria.
10. Which of the following is/are true about ADH? **(a)** It promotes obligatory water reabsorption, **(b)** it is secreted in response to an

increase in extracellular fluid osmolality, **(c)** it causes insertion of aquaporins in the PCT, **(d)** it promotes Na^+ reabsorption.

Short Answer Essay Questions

11. What is the importance of the perirenal fat capsule that surrounds the kidney?
12. Trace the pathway a creatinine molecule takes from a glomerulus to the urethra. Name every microscopic or gross structure it passes through on its journey.
13. Explain the important differences between blood plasma and glomerular filtrate, and relate the differences to the structure of the filtration membrane.
14. Describe the mechanisms that contribute to renal autoregulation.
15. Describe the mechanisms of extrinsic regulation of GFR, and their physiological role.
16. Describe what is involved in active and passive tubular reabsorption.
17. Explain how the peritubular capillaries are adapted for receiving reabsorbed substances.
18. Explain the process and purpose of tubular secretion.
19. How does aldosterone modify the chemical composition of urine?
20. Explain why the filtrate becomes hypotonic as it flows through the ascending limb of the nephron loop. Also explain why the filtrate at the bend of the nephron loop (and the interstitial fluid of the deep portions of the medulla) is hypertonic.
21. How does urinary bladder anatomy support its storage function?
22. Define micturition and describe the micturition reflex.

AT THE CLINIC

Clinical Case Study

Urinary System

Let's return to Kyle Boulard, whom we met in the previous chapter. After two days in the hospital, Mr. Boulard has recovered from his acute diabetic crisis and his type 1 diabetes is once again under control. The last update on his chart before he is discharged includes the following:

- BP 150/95, HR 75, temperature 37.2°C
- Urine: pH 6.9, negative for glucose and ketones; 24-hour urine collection reveals 170 mg albumin in urine per day

Mr. Boulard is prescribed a thiazide diuretic and an angiotensin converting enzyme (ACE) inhibitor. He is counseled on the importance of keeping his diabetes under control, taking his medications regularly, and keeping his outpatient follow-up appointments.

1. What is albumin? Is it normally found in the urine? If not, what does its presence suggest?
2. Why were these medications prescribed for Mr. Boulard?
3. Where and how do thiazide diuretics act in the kidneys and how does this reduce blood pressure?

At his two-week appointment at the outpatient clinic, Mr. Boulard complains of fatigue, weakness, muscle cramps, and irregular heartbeats. A physical examination and lab tests produce the following observations:

- BP 133/90, HR 75
- Blood K^+ 2.9 mEq/L (normal 3.5–5.5 mEq/L); blood Na^+ 135 mEq/L (normal 135–145 mEq/L)
- Urine K^+ 55 mEq/L (normal $<$40 mEq/L); urine Na^+ 21 mEq/L (normal $>$20 mEq/L)

4. What is Mr. Boulard's main problem at this point?
5. Explain how the thiazide diuretic might have caused this problem.

When asked about his medications, Mr. Boulard admits that he did not fill his ACE inhibitor prescription because it was too expensive. He could only afford the thiazide medications along with his insulin.

6. How do ACE inhibitors reduce blood pressure?
7. Would taking ACE inhibitors and thiazides together have prevented Mr. Boulard's current symptoms? Explain.

For answers, see Answers Appendix.

25 Fluid, Electrolyte, and Acid-Base Balance

KEY CONCEPTS

Have you ever wondered why on some days you don't urinate for hours at a time, while on others it seems like you void every few minutes? Or why on occasion you cannot seem to quench your thirst? These situations reflect one of the body's most important functions: maintaining fluid, electrolyte, and acid-base balance.

Cell function depends not only on a continuous supply of nutrients and removal of metabolic wastes, but also on the physical and chemical homeostasis of the surrounding fluids. The French physiologist Claude Bernard recognized this truth with style in 1857 when he said, "It is the fixity of the internal environment which is the condition of free and independent life."

Maintaining water and salt balance is critical for runners participating in a week-long endurance race in the desert.

In this chapter, we first examine the composition and distribution of fluids in the internal environment and then consider the roles of various body organs and functions in establishing, regulating, and altering this balance.

25.1 Body fluids consist of water and solutes in three main compartments

→ Learning Objectives

- ☐ **List the factors that determine body water content and describe the effect of each factor.**
- ☐ **Indicate the relative fluid volume and solute composition of the fluid compartments of the body.**
- ☐ **Contrast the overall osmotic effects of electrolytes and nonelectrolytes.**
- ☐ **Describe factors that determine fluid shifts in the body.**

Body Water Content

Not all bodies contain the same amount of water. Total body water is a function not only of age and body mass, but also of sex and the relative amount of body fat. Infants, with their low body fat and low bone mass, are 73% or more water. (This high level of hydration accounts for their "dewy" skin, like that of a freshly picked peach.) After infancy total body water declines throughout life, accounting for only about 45% of body mass in old age.

A healthy young man is about 60% water, and a healthy young woman about 50%. This difference between the sexes reflects the fact that females have relatively more body fat and less skeletal muscle than males. Of all body tissues, adipose tissue is *least* hydrated (less than 20% water)—even bone contains more water than does fat. In contrast, skeletal muscle is about 75% water, so people with greater muscle mass have proportionately more body water.

Fluid Compartments

Water occupies two main **fluid compartments** within the body (**Figure 25.1**). Almost two-thirds by volume is in the **intracellular fluid (ICF) compartment**, which actually consists of trillions of tiny individual "compartments": the cells. In an adult male of average size (70 kg, or 154 lb), ICF accounts for about 25 L of the 40 L of body water.

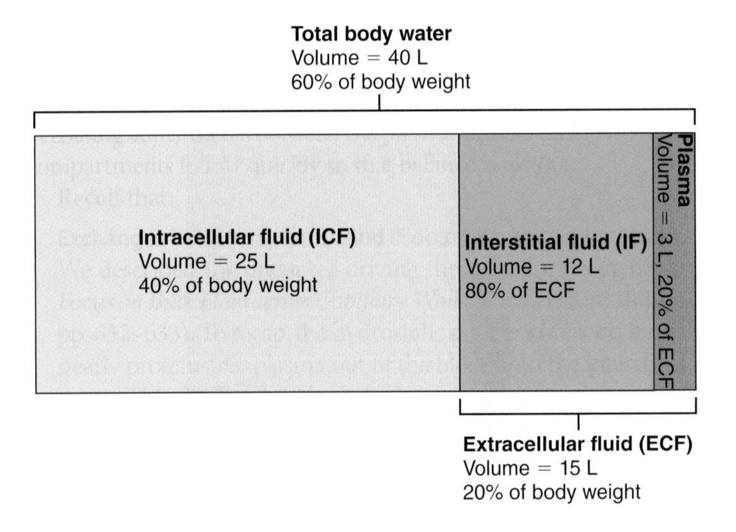

Figure 25.1 The major fluid compartments of the body. [Values are for a 70-kg (154-lb) male.]

The remaining one-third or so of body water is outside cells, in the **extracellular fluid (ECF) compartment**. The ECF constitutes the body's "internal environment" referred to by Claude Bernard and is the external environment of each cell. As Figure 25.1 shows, the ECF compartment is divisible into two subcompartments: (1) **plasma**, the fluid portion of blood, and (2) **interstitial fluid (IF)**, the fluid in the microscopic spaces between tissue cells. There are numerous other examples of ECF that are distinct from both plasma and interstitial fluid—lymph, cerebrospinal fluid, humors of the eye, synovial fluid, serous fluid, gastrointestinal secretions—but most of these are similar to IF and are usually considered part of it.

Composition of Body Fluids

Water serves as the *universal solvent* in which a variety of solutes are dissolved. Solutes may be classified broadly as *electrolytes* and *nonelectrolytes*.

Electrolytes and Nonelectrolytes

Nonelectrolytes have bonds (usually covalent bonds) that prevent them from dissociating in solution. For this reason, no electrically charged species are created when nonelectrolytes dissolve in water. Most nonelectrolytes are organic molecules—glucose, lipids, creatinine, and urea, for example.

In contrast, **electrolytes** are chemical compounds that *do* dissociate into ions in water. (See Chapter 2 if necessary to review these concepts of chemistry.) Because ions are charged particles, they can conduct an electrical current—and so have the name *electrolyte*. Typically, electrolytes include inorganic salts, both inorganic and organic acids and bases, and some proteins.

Although all dissolved solutes contribute to the osmotic activity of a fluid, electrolytes have much greater osmotic power than nonelectrolytes because each electrolyte molecule dissociates into at least two ions. For example, a molecule of sodium chloride (NaCl) contributes twice as many solute particles as glucose (which remains undissociated), and a molecule of magnesium chloride ($MgCl_2$) contributes three times as many:

$$NaCl \rightarrow Na^+ + Cl^- \quad \text{(electrolyte; two particles)}$$

$$MgCl_2 \rightarrow Mg^{2+} + 2Cl^- \quad \text{(electrolyte; three particles)}$$

$$\text{glucose} \rightarrow \text{glucose} \quad \text{(nonelectrolyte; one particle)}$$

Regardless of the type of solute particle, water moves according to osmotic gradients—from an area of lesser osmolality to an area of greater osmolality. For this reason, electrolytes have the greatest ability to cause fluid shifts.

Electrolyte concentrations of body fluids are usually expressed in **milliequivalents per liter (mEq/L)**, a measure of the number of electrical charges in 1 liter of solution. We can compute the concentration of any ion in solution using the equation

$$\text{mEq/L} = \frac{\text{ion concentration (mg/L)}}{\text{atomic weight of ion (mg/mmol)}} \times \text{no. of electrical charges on one ion}$$

(Recall from p. 26 that 1 millimole (mmol) = 0.001 mole.)

To calculate the mEq/L of sodium or calcium ions in solution in plasma, we would determine the normal concentration of these ions in plasma, look up their atomic weights in the periodic table, and plug these values into the equation:

$$Na^+: \quad \frac{3300 \text{ mg/L}}{23 \text{ mg/mmol}} \times 1 = 143 \text{ mEq/L}$$

$$Ca^{2+}: \quad \frac{100 \text{ mg/L}}{40 \text{ mg/mmol}} \times 2 = 5 \text{ mEq/L}$$

Notice that for ions with a single charge, 1 mEq is equal to 1 mmol, which, when dissolved in 1 kg of water, produces 1 mOsm (see p. 856). On the other hand, 1 mEq of ions with a double charge (like calcium) is equal to 1/2 In either case, 1 mEq provides the same amount of charge.

Comparison of Extracellular and Intracellular Fluids

A quick glance at the bar graphs in **Figure 25.2** reveals that each fluid compartment has a distinctive pattern of electrolytes. Except for the relatively high protein content in plasma, however, the extracellular fluids are very similar. Their chief cation is sodium, and their major anion is chloride. However, plasma contains somewhat fewer chloride ions than interstitial fluid, because the nonpenetrating plasma proteins are normally anions and plasma is electrically neutral.

In contrast to extracellular fluids, the ICF contains only small amounts of Na^+ and Cl^-. Its most abundant cation is potassium, and its major anion is HPO_4^{2-}. Cells also contain substantial quantities of soluble proteins (about three times the amount found in plasma).

Notice that sodium and potassium ion concentrations in ECF and ICF are nearly opposite (Figure 25.2). The characteristic distribution of these ions on the two sides of cellular membranes reflects the activity of cellular ATP-dependent sodium-potassium pumps, which keep intracellular Na^+ concentrations low and K^+ concentrations high.

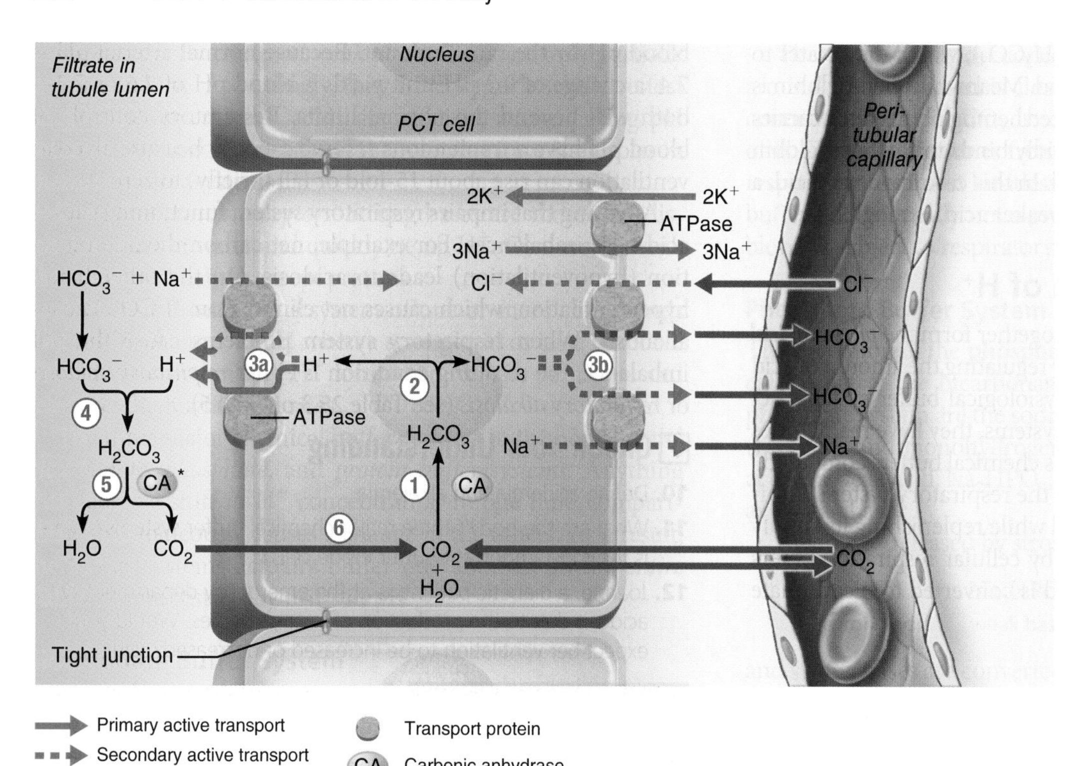

① CO_2 combines with water within the tubule cell, forming H_2CO_3.

② H_2CO_3 is quickly split, forming H^+ and bicarbonate ion (HCO_3^-).

③a H^+ is secreted into the filtrate.

③b For each H^+ secreted, a HCO_3^- enters the peritubular capillary blood either via symport with Na^+ or via antiport with Cl^-.

④ Secreted H^+ combines with HCO_3^- in the filtrate, forming carbonic acid (H_2CO_3). HCO_3^- disappears from the filtrate at the same rate that HCO_3^- (formed within the tubule cell) enters the peritubular capillary blood.

⑤ The H_2CO_3 formed in the filtrate dissociates to release CO_2 and H_2O.

⑥ CO_2 diffuses into the tubule cell, where it triggers further H^+ secretion.

Figure 25.12 Reabsorption of filtered HCO_3^- is coupled to H^+ secretion.

*The breakdown of H_2CO_3 to CO_2 and H_2O in the tubule lumen is catalyzed by carbonic anhydrase only in the proximal convoluted tubule (PCT).

Because the mechanisms for regulating acid-base balance depend on H^+ being secreted into the filtrate, we consider that process first. Secretion of H^+ occurs mainly in the PCT and in type A intercalated cells of the collecting duct. The H^+ secreted comes from the dissociation of carbonic acid, created from the combination of CO_2 and water within the tubule cells, a reaction catalyzed by *carbonic anhydrase* (**Figure 25.12** ①, ②). As H^+ is secreted into the lumen of the PCT, Na^+ is reabsorbed from the filtrate, maintaining the electrical balance (Figure 25.12 ③a).

The rate of H^+ secretion rises and falls with CO_2 levels in the ECF. The more CO_2 in the peritubular capillary blood, the faster the rate of H^+ secretion. Because blood CO_2 levels directly relate to blood pH, this system can respond to both rising and falling H^+ concentrations. Notice that secreted H^+ can combine with HCO_3^- in the filtrate, generating CO_2 and water (Figure 25.12 ④, ⑤). In this case, H^+ is bound in water. The rising concentration of CO_2 in the filtrate creates a steep diffusion gradient for its entry into the tubule cell, where it promotes still more H^+ secretion (Figure 25.12 ③a).

Conserving Filtered Bicarbonate Ions: Bicarbonate Reabsorption

Bicarbonate ions (HCO_3^-) are an important part of the bicarbonate buffer system, the most important inorganic blood buffer. If this reservoir of base, the *alkaline reserve*, is to be maintained, the kidneys must do more than just eliminate enough hydrogen ions to counter rising blood H^+ levels. Depleted stores of HCO_3^- have to be replenished. This task is more complex than it seems because the tubule cells are almost completely impermeable to the HCO_3^- in the filtrate—they cannot reabsorb it.

However, the kidneys can conserve filtered HCO_3^- in a rather roundabout way. As you can see, dissociation of carbonic acid liberates HCO_3^- as well as H^+ (Figure 25.12 ②). Although the tubule cells cannot reclaim HCO_3^- directly from the filtrate, they can and do shunt HCO_3^- generated within them (as a result of splitting H_2CO_3) into the peritubular capillary blood. HCO_3^- leaves the tubule cell either accompanied by Na^+ or in exchange for Cl^- (Figure 25.12 ③b). H^+ is actively secreted, mostly by a Na^+-H^+ antiporter, but also by a H^+ ATPase (Figure 25.12 ③a). In the filtrate, H^+ combines with filtered HCO_3^- (Figure 25.12 ④, ⑤). For this reason, reabsorption of HCO_3^- depends on the active secretion of H^+.

In short, for each filtered HCO_3^- that "disappears" from the filtrate, a HCO_3^- generated within the tubule cells enters the blood—a one-for-one exchange. When large amounts of H^+ are secreted, correspondingly large amounts of HCO_3^- enter the peri-tubular blood. The net effect is to remove HCO_3^- almost completely from the filtrate.

Generating New Bicarbonate Ions

Two renal mechanisms commonly carried out by cells of the PCT and collecting ducts generate *new* (as opposed to filtered) HCO_3^- that can be added to plasma. Both mechanisms involve renal excretion of acid, *via secretion and excretion* of either H^+ or ammonium ions in urine. Let's examine how these mechanisms differ.

As long as *filtered bicarbonate* is reclaimed, as we saw in Figure 25.12, the secreted H^+ is *not excreted or lost* from the body in urine. Instead, the H^+ is buffered by HCO_3^- in the filtrate

25

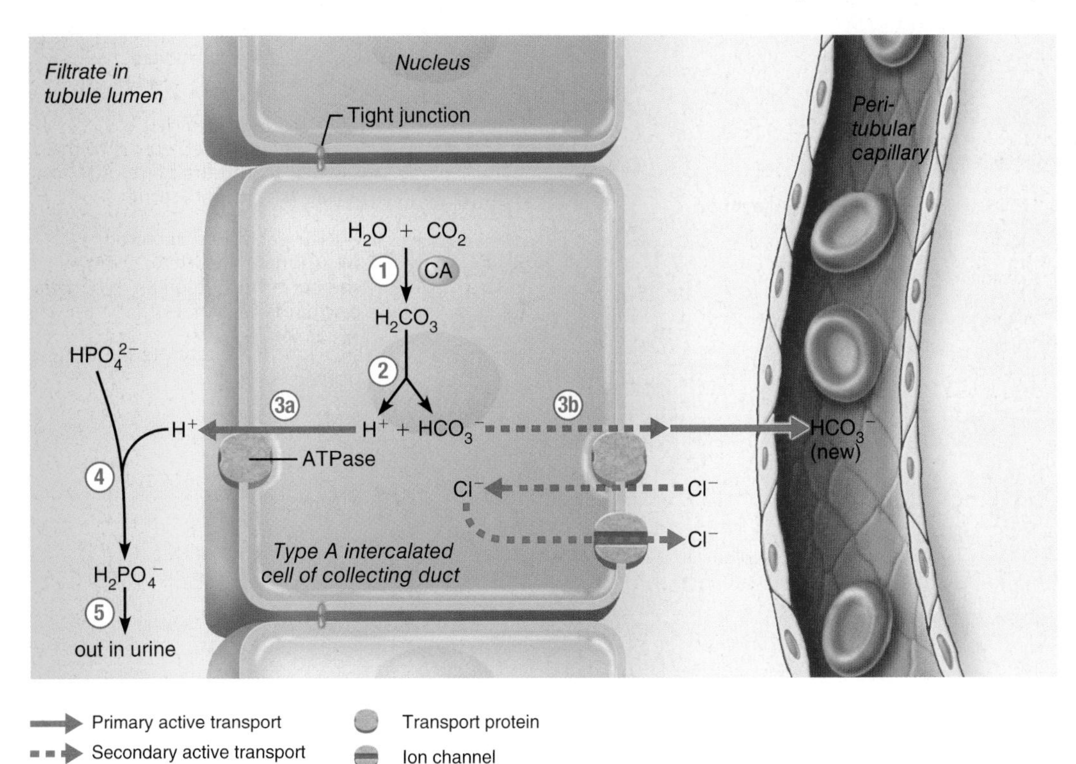

Figure 25.13 New HCO_3^- is generated via buffering of secreted H^+ by HPO_4^{2-} (monohydrogen phosphate).

and ultimately becomes part of water molecules (most of which are reabsorbed).

However, once the filtered HCO_3^- is "used up" (usually by the time the filtrate reaches the collecting ducts), any additional H^+ secreted is excreted in urine. More often than not, this is the case.

Reclaiming filtered HCO_3^- simply restores the bicarbonate concentration of plasma that exists at the time. However, metabolism of food normally releases new H^+ into the body. This additional H^+ uses up HCO_3^- and so must be balanced by generating *new* HCO_3^- that moves into the blood to counteract acidosis. This process of alkalinizing the blood is the way the kidneys compensate for acidosis.

Via Excretion of Buffered H^+

Binding H^+ to buffers in the filtrate minimizes the H^+ concentration gradient, allowing the proton pumps of the type A intercalated cells to secrete the large numbers of H^+ that the body must get rid of to prevent acidosis. (H^+ secretion ceases when urine pH falls to 4.5 because the proton pumps cannot pump against this large gradient.) The most important urine buffer is the *phosphate buffer system*, specifically its weak base *monohydrogen phosphate* (HPO_4^{2-}).

The components of the phosphate buffer system filter freely into the tubules, and about 75% of the filtered phosphate is reabsorbed. However, their reabsorption is inhibited during acidosis. As a result, the buffer pair becomes more and more concentrated as the filtrate moves through the renal tubules.

As shown in **Figure 25.13** (3a), the type A intercalated cells secrete H^+ actively via a H^+ ATPase pump and via a K^+-H^+ antiporter (not illustrated). The secreted H^+ combines with HPO_4^{2-}, forming $H_2PO_4^-$ which then flows out in urine (Figure 25.13 (4) and (5)).

Bicarbonate ions generated in the cells during the same reaction move into the interstitial space via a HCO_3^--Cl^- antiport process and then move passively into the peritubular capillary blood (Figure 25.13 (3b)). Notice again that when H^+ is being excreted, "brand new" bicarbonate ions are added to the blood—over and above those reclaimed from the filtrate. As you can see, in response to acidosis, the kidneys generate new HCO_3^- and add it to the blood (alkalinizing the blood) while adding an equal amount of H^+ to the filtrate (acidifying the urine).

Via NH_4^+ Excretion

The second and more important mechanism for excreting acid uses the ammonium ion (NH_4^+) produced by glutamine metabolism in PCT cells. Ammonium ions are weak acids that donate few H^+ at physiological pH.

As **Figure 25.14** step (1) shows, for each glutamine metabolized (deaminated, oxidized, and acidified by combining with H^+), two NH_4^+ and two HCO_3^- result. The HCO_3^- moves through the basolateral membrane into the blood (Figure 25.14 (2b)). The

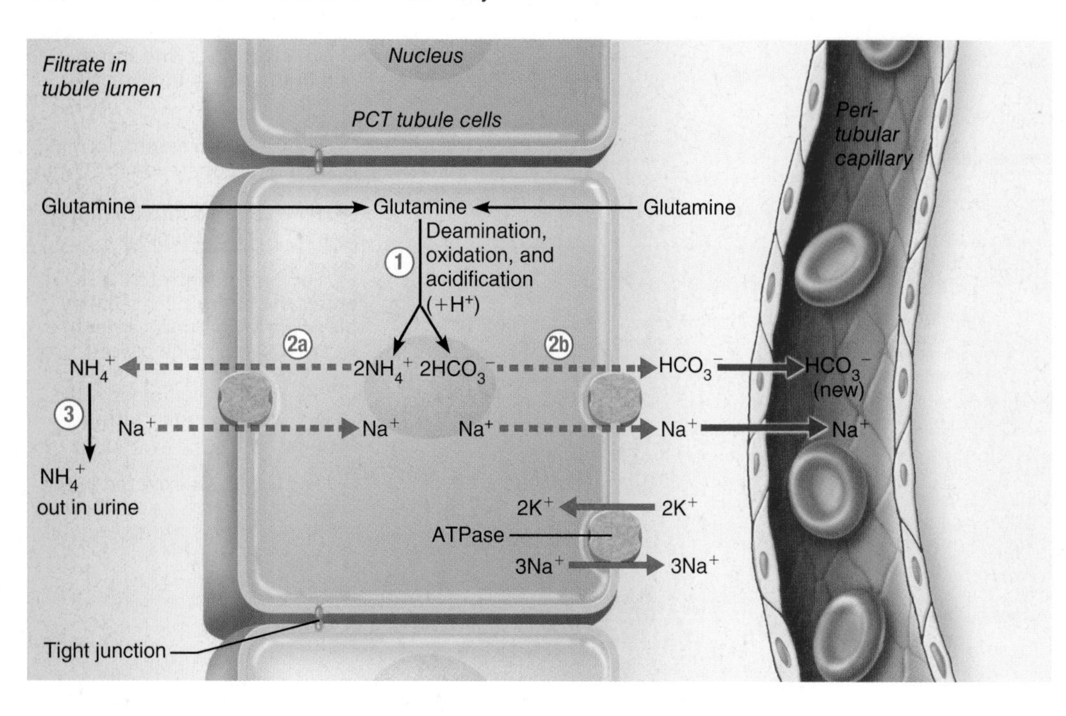

① PCT cells metabolize glutamine to NH_4^+ and HCO_3^-.

②a This weak acid NH_4^+ (ammonium) is secreted into the filtrate, taking the place of H^+ on a Na^+-H^+ antiport carrier.

②b For each NH_4^+ secreted, a bicarbonate ion (HCO_3^-) enters the peritubular capillary blood via a symport carrier.

③ The NH_4^+ is excreted in the urine.

Figure 25.14 New HCO_3^- is generated via glutamine metabolism and NH_4^+ secretion.

NH_4^+, in turn, is excreted and lost in urine (Figure 25.14 ②a, ③). As with the phosphate buffer system, this buffering mechanism replenishes the alkaline reserve of the blood, because the newly made HCO_3^- enters the blood as NH_4^+ is secreted.

Bicarbonate Ion Secretion

When the body is in alkalosis, another population of intercalated cells (type B) in the collecting ducts exhibit net HCO_3^- *secretion* while reclaiming H^+ to acidify the blood. Overall we can think of the type B cells as "flipped" type A cells, and we can visualize the HCO_3^- secretion process as the exact opposite of the HCO_3^- reabsorption process illustrated in Figure 25.12. However, the predominant process in the nephrons and collecting ducts is HCO_3^- reabsorption, and even during alkalosis, much more HCO_3^- is conserved than excreted.

Check Your Understanding

13. Reabsorption of HCO_3^- is always tied to the secretion of which ion?

14. What is the most important urinary buffer of H^+?

15. List the two mechanisms by which tubule and collecting duct cells generate new HCO_3^-.

16. MAKING connections Renal tubule cells acidify the urine and parietal cells acidify the stomach contents (see Chapter 22, p. 768). In each case (a) which intracellular enzyme is key, and (b) blood concentration of which ion increases?

For answers, see Answers Appendix.

CLINICAL

25.6 Abnormalities of acid-base balance are classified as metabolic or respiratory

Learning Objectives

- ☐ **Distinguish between acidosis and alkalosis resulting from respiratory and metabolic factors.**
- ☐ **Describe the importance of respiratory and renal compensations in maintaining acid-base balance.**

All cases of acidosis and alkalosis can be classed according to cause as *respiratory* or *metabolic* (**Table 25.3**).

Respiratory Acidosis and Alkalosis

Respiratory pH imbalances result from some failure of the respiratory system to perform its normal pH-balancing role. The partial pressure of carbon dioxide (P_{CO_2}) in the arteries is the single most important indicator of the adequacy of respiratory function. When respiratory function is normal, the P_{CO_2} fluctuates between 35 and 45 mm Hg. Generally speaking, values above 45 mm Hg indicate respiratory acidosis, and values below 35 mm Hg signal respiratory alkalosis.

Respiratory acidosis is a common cause of acid-base imbalance. It most often occurs when a person breathes shallowly or when gas exchange is hampered by diseases such as pneumonia, cystic fibrosis, or emphysema. Under such conditions, CO_2 accumulates in the blood. Thus, respiratory acidosis is characterized by falling blood pH and rising P_{CO_2}.

Respiratory alkalosis results when carbon dioxide is eliminated from the body faster than it is produced. This is called

Table 25.3 Causes and Consequences of Acid-Base Imbalances

CONDITION AND HALLMARK	POSSIBLE CAUSES; COMMENTS
Respiratory Acidosis (Hypoventilation)	
If uncompensated (uncorrected): P_{CO_2} >45 mm Hg; pH <7.35	**Impaired lung function** (e.g., chronic bronchitis, cystic fibrosis, emphysema): impaired gas exchange or alveolar ventilation **Impaired ventilatory movement:** paralyzed respiratory muscles, chest injury, extreme obesity **Narcotic or barbiturate overdose or injury to brain stem:** depression of respiratory centers, resulting in hypoventilation and respiratory arrest
Respiratory Alkalosis (Hyperventilation)	
If uncompensated: P_{CO_2} <35 mm Hg; pH >7.45	**Strong emotions:** pain, anxiety, fear, panic attack **Hypoxemia:** asthma, pneumonia, high altitude; represents effort to raise P_{O_2} at the expense of excessive CO_2 excretion **Brain tumor or injury:** abnormal respiratory controls
Metabolic Acidosis	
If uncompensated: HCO_3^- <22 mEq/L; pH <7.35	**Severe diarrhea:** bicarbonate-rich intestinal (and pancreatic) secretions rushed through digestive tract before their solutes can be reabsorbed; bicarbonate ions are replaced by renal mechanisms that generate new bicarbonate ions **Renal disease:** failure of kidneys to rid body of acids formed by normal metabolic processes **Untreated diabetes mellitus:** lack of insulin or inability of tissue cells to respond to insulin, resulting in inability to use glucose; fats are used as primary energy fuel, and ketoacidosis occurs **Starvation:** lack of dietary nutrients for cellular fuels; body proteins and fat reserves are used for energy—both yield acidic metabolites as they are broken down for energy **Excess alcohol ingestion:** results in excess acids in blood
Metabolic Alkalosis	
If uncompensated: HCO_3^- >26 mEq/L; pH >7.45	**Vomiting or gastric suctioning:** loss of stomach HCl requires that H^+ be withdrawn from blood to replace stomach acid; thus H^+ decreases and HCO_3^- increases proportionately **Selected diuretics:** cause K^+ depletion and H_2O loss. Low K^+ directly stimulates tubule cells to secrete H^+. Reduced blood volume elicits the renin-angiotensin-aldosterone mechanism, which stimulates Na^+ reabsorption and H^+ secretion. **Ingestion of excessive sodium bicarbonate (antacid):** bicarbonate moves easily into ECF, where it enhances natural alkaline reserve **Excess aldosterone** (e.g., adrenal tumors): promotes excessive reabsorption of Na^+, which pulls increased amount of H^+ into urine. Hypovolemia promotes the same relative effect because aldosterone secretion is increased to enhance Na^+ (and H_2O) reabsorption.

hyperventilation (deeper and faster breathing than needed to remove CO_2) (see p. 739), and results in the blood becoming more alkaline. While respiratory acidosis is frequently associated with respiratory system pathology, respiratory alkalosis is often due to stress or pain.

Metabolic Acidosis and Alkalosis

Metabolic pH imbalances include all abnormalities of acid-base imbalance *except* those caused by too much or too little carbon dioxide in the blood. Bicarbonate ion levels below or above the normal range of 22–26 mEq/L indicate a metabolic acid-base imbalance.

The second most common cause of acid-base imbalance, **metabolic acidosis**, is recognized by low blood pH and HCO_3^- levels. Typical causes are ingesting too much alcohol (which is metabolized to acetic acid) and excessive loss of HCO_3^-, as might result from persistent diarrhea. Other causes are accumulation of lactic acid during exercise or shock, the ketosis that occurs in diabetic crisis or starvation, and, infrequently, kidney failure.

Metabolic alkalosis, indicated by rising blood pH and HCO_3^- levels, is much less common than metabolic acidosis. Typical causes are vomiting the acidic contents of the stomach (or loss of those secretions through gastric suctioning) and intake of excess base (too many antacids, for example).

Effects of Acidosis and Alkalosis

The absolute blood pH limits for life are a low of 6.8 and a high of 7.8. When blood pH falls below 6.8, the central nervous system is so depressed that the person goes into coma and death soon follows. When blood pH rises above 7.8, the nervous system is overexcited, leading to muscle tetany, extreme nervousness, and convulsions. Death often results from respiratory arrest.

Respiratory and Renal Compensations

If one of the physiological buffer systems (lungs or kidneys) malfunctions and disrupts acid-base balance, the other system tries to compensate. The respiratory system attempts to compensate for metabolic acid-base imbalances, and the kidneys (although much slower) work to correct imbalances caused by respiratory disease. We can recognize these **respiratory** and **renal compensations** by the resulting changes in plasma P_{CO_2} and bicarbonate ion concentrations. Because the compensations act to restore normal blood pH, a patient may have a normal pH despite a significant medical problem.

Respiratory Compensations

As a rule, changes in respiratory rate and depth are evident when the respiratory system is attempting to compensate for metabolic acid-base imbalances. In metabolic acidosis, respiratory rate and depth are usually elevated—an indication that high H^+ levels are stimulating the respiratory centers. Blood pH is low (below 7.35) and the HCO_3^- level is below 22 mEq/L. As the respiratory system "blows off" CO_2 to rid the blood of excess acid, the P_{CO_2} falls below 35 mm Hg. In contrast, in respiratory acidosis, the respiratory rate is often depressed and *is the immediate cause of the acidosis* (with some exceptions such as pneumonia or emphysema where gas exchange is impaired).

Respiratory compensation for metabolic alkalosis involves slow, shallow breathing, which allows CO_2 to accumulate in the blood. Evidence of metabolic alkalosis being compensated by respiratory mechanisms includes a pH over 7.45 (at least initially), elevated bicarbonate levels (over 26 mEq/L), and a P_{CO_2} above 45 mm Hg.

Renal Compensations

When an acid-base imbalance is of respiratory origin, renal mechanisms are stepped up to compensate for the imbalance. For example, a hypoventilating individual will exhibit acidosis. When renal compensation is occurring, both the P_{CO_2} and the HCO_3^- levels are high. The high P_{CO_2} causes the acidosis, and the rising HCO_3^- level indicates that the kidneys are retaining bicarbonate to offset the acidosis.

Conversely, a person with renal-compensated respiratory alkalosis will have a high blood pH and a low P_{CO_2}. Bicarbonate ion levels begin to fall as the kidneys eliminate more HCO_3^- from the body by failing to reclaim it or by actively secreting it. Note that the kidneys cannot compensate for alkalosis or acidosis if that condition reflects a *renal* problem.

☑ Check Your Understanding

17. Which two abnormalities in plasma are key features of an uncompensated metabolic alkalosis? An uncompensated respiratory acidosis?

18. How do the kidneys compensate for respiratory acidosis?

For answers, see Answers Appendix.

In this chapter we have examined the chemical and physiological mechanisms that provide the optimal internal environment for survival. The kidneys are the superstars among homeostatic organs in regulating water, electrolyte, and acid-base balance, but they do not and cannot act alone. Rather, their activity is made possible by a host of hormones and enhanced both by bloodborne buffers, which give the kidneys time to react, and by the respiratory system, which shoulders a substantial responsibility for acid-base balance of the blood.

Now that we have discussed the topics relevant to renal functioning, the topics in Chapters 24 and 25 should draw together in an understandable way.

25

REVIEW QUESTIONS

For more chapter study tools, go to the Study Area of MasteringA&P®.

There you will find:

- Interactive Physiology iP
- A&PFlix A&PFlix
- Practice Anatomy Lab PAL
- PhysioEx PEx
- Videos, Practice Quizzes and Tests, MP3 Tutor Sessions, Case Studies, and much more!

Multiple Choice/Matching

(Some questions have more than one correct answer. Select the best answer or answers from the choices given.)

1. Body water content is greatest in **(a)** infants, **(b)** young adults, **(c)** elderly adults.
2. Potassium, magnesium, and phosphate ions are the predominant electrolytes in **(a)** plasma, **(b)** interstitial fluid, **(c)** intracellular fluid.
3. Sodium balance is regulated primarily by control of amount(s) **(a)** ingested, **(b)** excreted in urine, **(c)** lost in perspiration, **(d)** lost in feces.
4. Water balance is regulated by control of amount(s) (use choices in question 3).

Answer questions 5 through 10 by choosing responses from the following:

(a) ammonium ions	**(f)** magnesium
(b) bicarbonate	**(g)** phosphate
(c) calcium	**(h)** potassium
(d) chloride	**(i)** sodium
(e) hydrogen ions	**(j)** water

5. Two main substances regulated by the influence of aldosterone on the kidney tubules.
6. Two substances regulated by parathyroid hormone.
7. Two substances secreted into the proximal convoluted tubules in exchange for sodium ions.
8. Part of an important chemical buffer system in plasma.
9. Two ions produced during catabolism of glutamine.

10. Substance regulated by ADH's effects on the renal tubules.
11. Which of the following factors will enhance ADH release? **(a)** increase in ECF volume, **(b)** decrease in ECF volume, **(c)** decrease in ECF osmolality, **(d)** increase in ECF osmolality.
12. The pH of blood varies directly with **(a)** HCO_3^-, **(b)** P_{CO_2}, **(c)** H^+, **(d)** none of the above.
13. In an individual with metabolic acidosis, a clue that the respiratory system is compensating is provided by **(a)** high blood bicarbonate levels, **(b)** low blood bicarbonate levels, **(c)** rapid, deep breathing, **(d)** slow, shallow breathing.

Short Answer Essay Questions

14. Name the body fluid compartments, noting their locations and the approximate fluid volume in each.
15. Describe the thirst mechanism, indicating how it is triggered and terminated.
16. Explain why and how ECF osmolality is maintained.
17. Explain why and how sodium balance, ECF volume, and blood pressure are jointly regulated.
18. Describe the role of the respiratory system in controlling acid-base balance.
19. Explain how the chemical buffer systems resist changes in pH.
20. Explain the relationship of the following to renal secretion and excretion of hydrogen ions: (a) plasma carbon dioxide levels, (b) phosphate, and (c) sodium bicarbonate reabsorption.

AT THE CLINIC

Clinical Case Study
Fluid, Electrolyte, and Acid-Base Balance

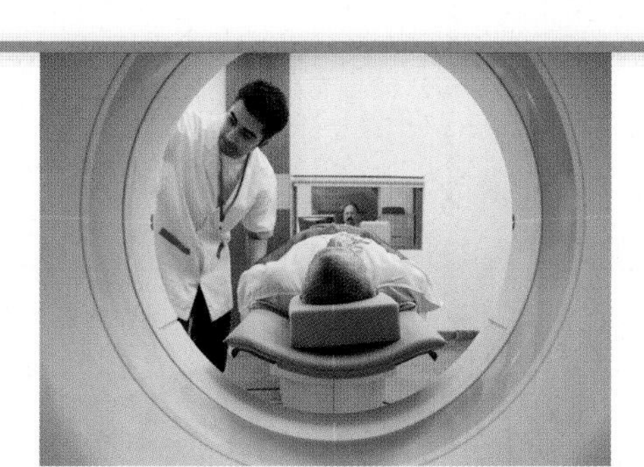

Mr. Heyden, a somewhat stocky 72-year-old man, is brought in to the emergency room (ER). The paramedics report that his left arm and the left side of his body trunk were pinned beneath some wreckage, and that when he was freed, his left hypogastric and lumbar areas appeared to be compressed and his left arm was blanched and without sensation. On admission, Mr. Heyden is alert, slightly cyanotic, and complaining of pain in his left side; he loses consciousness shortly thereafter. His vital signs are taken, blood is drawn for laboratory tests, and Mr. Heyden is catheterized and immediately scheduled for a CT scan of his left abdominal region.

Analyze the information that was subsequently recorded on Mr. Heyden's chart:

- Vital signs: Temperature 39°C (102°F); BP 90/50 mm Hg and falling; heart rate 116 beats/min and thready; 30 respirations/min

1. Given the values above and his attendant cyanosis, what would you guess is Mr. Heyden's immediate problem? Explain your reasoning.

- CT scan reveals a ruptured spleen and a large hematoma in the upper left abdominal quadrant. Splenic repair surgery is scheduled but unsuccessful; the spleen is removed.

2. Rupture of the spleen results in massive hemorrhage. Explain this observation. Which organs (if any) will compensate for the removal of Mr. Heyden's spleen?

- Hematology: Most blood tests yield normal results. However, renin, aldosterone, and ADH levels are elevated.

3. Explain the cause and consequence of each of the hematology findings.

- Urinalysis: Some granular casts (particulate cell debris) are noted, and the urine is brownish-red in color; other values are normal, but urine output is very low. An order is given to force fluids.

4. (a) What might account for the low volume of urine output? (Name at least two possibilities.) (b) What might explain the casts and abnormal color of his urine? Can you see any possible relationship between his crush injury and these findings?

The next day, Mr. Heyden is awake and alert. He says that he now has feeling in his arm, but he still complains of pain. However, the pain site appears to have moved from the left upper quadrant to his lumbar region. His urine output is still low. He is scheduled once again for a CT scan, this time of his lumbar region. The order to force fluids is renewed and some additional and more specific blood tests are ordered. We will visit Mr. Heyden again shortly, but in the meantime think about what these new findings may indicate.

For answers, see Answers Appendix.

26 The Reproductive System

KEY CONCEPTS

Most organ systems function almost continuously to maintain the well-being of the individual. The **reproductive system**, however, appears to "slumber" until puberty. The **primary sex organs**, or **gonads** (go′nadz; "seeds"), are the testes in males and the ovaries in females. The gonads produce sex cells, or **gametes** (gam′ēts; "spouses"), and secrete a variety of steroid hormones commonly called **sex hormones**. The remaining reproductive structures—ducts, glands, and external genitalia (jen-ĭ-ta′le-ah)—are **accessory reproductive organs**. Although male and female reproductive organs are quite different, their common purpose is to produce offspring.

The male's reproductive role is to manufacture male gametes called *sperm* and deliver them to the female reproductive tract, where fertilization can occur. The complementary role of the female is to produce female gametes, called *ova* or *eggs*. As a result of appropriately timed intercourse, a sperm and an egg may fuse to form a fertilized egg, or *zygote*. The zygote is the first cell of a new individual, from which all body cells will arise.

The male and female reproductive systems are equal partners in events leading up to fertilization, but once fertilization has occurred, the female partner's uterus provides the protective environment where the embryo develops until birth. Sex hormones—androgens in males and estrogens and progesterone in females—play vital roles both in the development and function of the reproductive organs and in sexual

A happy couple.

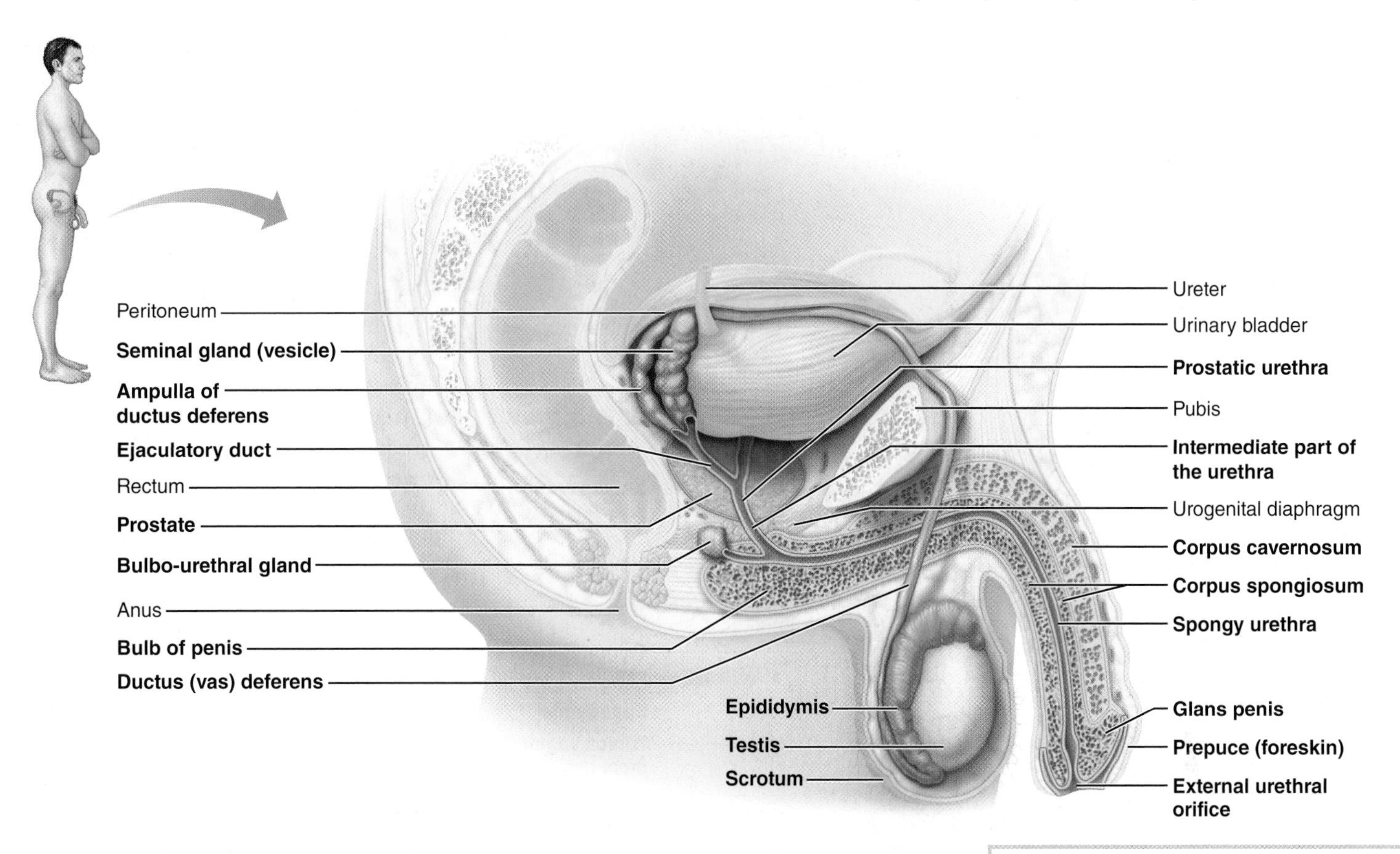

Figure 26.1 Reproductive organs of the male, sagittal view. A portion of the pubis of the hip bone has been left to show the relationship of the ductus deferens to the bony pelvis. (For related images, see *A Brief Atlas of the Human Body*, Figures 72 and 73.)

Practice art labeling
MasteringA&P®>Study Area>Chapter 26

behavior and drives. These hormones also influence the growth and development of many other organs and tissues of the body.

PART 1

ANATOMY OF THE MALE REPRODUCTIVE SYSTEM

The sperm-producing **testes** (tes′tez), or **male gonads**, lie within the *scrotum*. From the testes, the sperm are delivered to the body exterior through a system of ducts including (in order) the *epididymis*, the *ductus deferens*, the *ejaculatory duct*, and finally the *urethra*, which opens to the outside at the tip of the *penis*. The accessory sex glands, which empty their secretions into the ducts during ejaculation, are the *seminal glands, prostate*, and *bulbo-urethral glands*. Take a moment to trace the duct system in **Figure 26.1**, and identify the testis and accessory glands before continuing.

26.1 The testes are enclosed and protected by the scrotum

→ Learning Objective

☐ **Describe the structure and function of the testes, and explain the importance of their location in the scrotum.**

The Scrotum

The **scrotum** (skro′tum; "pouch") is a sac of skin and superficial fascia that hangs outside the abdominopelvic cavity at the root of the penis (Figure 26.1 and **Figure 26.2**). It is covered with sparse hairs, and contains paired oval testes. A midline *septum* divides the scrotum, providing a compartment for each testis.

This seems a rather vulnerable location for a man's testes, which contain his entire ability to father offspring. However, because viable sperm cannot be produced in abundance at core body temperature (37°C), the superficial location of the scrotum, which provides a temperature about 3°C lower, is an essential adaptation.

Furthermore, the scrotum is affected by temperature changes. When it is cold, the testes are pulled closer to the pelvic floor and the warmth of the body wall, and the scrotum becomes shorter and heavily wrinkled, decreasing its surface area and increasing its thickness to reduce heat loss. When it is warm, the scrotal skin is flaccid and loose to increase the surface area for cooling (sweating) and the testes hang lower, away from the body trunk.

These changes in scrotal surface area help maintain a fairly constant intrascrotal temperature and reflect the activity of two sets of muscles that respond to ambient temperature. The **dartos muscle** (dar′tos; "skinned"), a layer of smooth muscle in the superficial fascia, wrinkles the scrotal skin. The **cremaster muscles** (kre-mas′ter; "a suspender"), bands of skeletal muscle that arise from the internal oblique muscles of the trunk, elevate the testes.

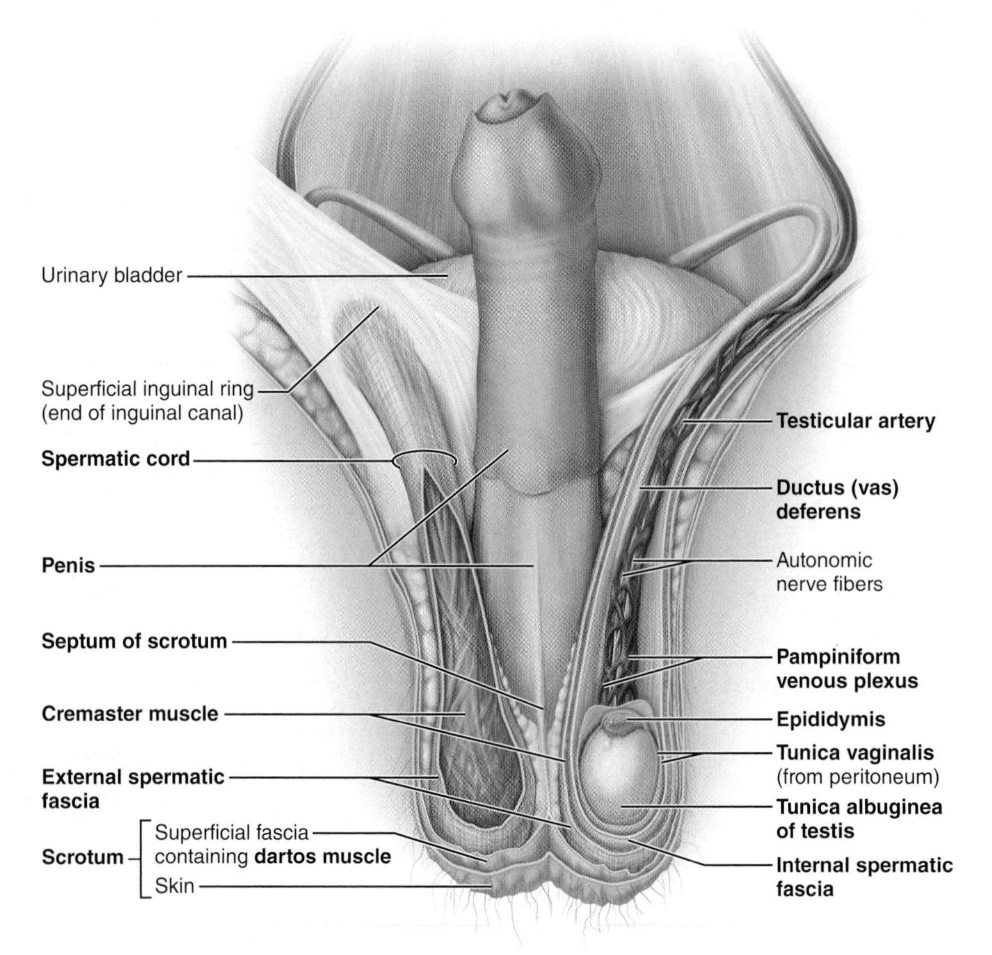

Figure 26.2 Relationships of the testis to the scrotum and spermatic cord. The scrotum has been opened and its anterior portion removed.

The Testes

Each plum-sized testis is approximately 4 cm (1.5 inches) long by 2.5 cm (1 inch) wide and is surrounded by two tunics. The outer tunic is the two-layered **tunica vaginalis** (vaj″ĭ-nal′is), derived from an outpocketing of the peritoneum (Figure 26.2 and **Figure 26.3a**). Deep to this serous layer is the **tunica albuginea** (al″bu-jin′e-ah; "white coat"), the fibrous capsule of the testis.

Septa extending inward from the tunica albuginea divide the testis into about 250 wedge-shaped *lobules*. Each contains one to four tightly coiled **seminiferous tubules** (sem″ĭ-nif′er-us; "sperm-carrying"), the actual "sperm factories" consisting of a thick stratified epithelium surrounding a central fluid-containing lumen (Figure 26.3a and c). The epithelium consists of spheroid *spermatogenic* ("sperm-forming") *cells* embedded in substantially larger columnar cells called *sustentocytes*. The sustentocytes are supporting cells that play several roles in sperm formation, as described shortly.

Surrounding each seminiferous tubule are three to five layers of smooth muscle–like **myoid cells** (Figure 26.3c). By contracting rhythmically, myoid cells may help to squeeze sperm and testicular fluids through the tubules and out of the testes.

The seminiferous tubules of each lobule converge to form a **straight tubule** that conveys sperm into the **rete testis** (re′te), a tubular network on the posterior side of the testis. From the rete testis, sperm leave the testis through the *efferent ductules* and enter the *epididymis* (ep″ĭ-did′ĭ-mis), which hugs the external testis surface posteriorly. The immature sperm pass through the head, the body, and then move into the tail of the epididymis, where they are stored until ejaculation.

Lying in the soft connective tissue surrounding the seminiferous tubules are the **interstitial endocrine cells**, also called *Leydig cells* (Figure 26.3c). These cells produce androgens (most importantly *testosterone*), which they secrete into the surrounding interstitial fluid. Thus, completely different cell populations carry out the sperm-producing and hormone-producing functions of the testis.

The long **testicular arteries**, which branch from the abdominal aorta superior to the pelvis (see *gonadal arteries* in Figure 18.24c, p. 646), supply the testes. The **testicular veins** draining the testes arise from a network called the **pampiniform venous plexus** (pam-pin′ĭ-form; "tendril-shaped") that surrounds the portion of each testicular artery within the scrotum like a climbing vine (see Figure 26.2). The cooler venous blood in each pampiniform plexus absorbs heat from the arterial blood, cooling it before it enters the testes. In this way, these plexuses help to keep the testes at their cool homeostatic temperature.

Both divisions of the autonomic nervous system serve the testes, and when the testes are hit forcefully, associated sensory nerves transmit impulses that result in agonizing pain and nausea. A connective tissue sheath encloses nerve fibers, blood vessels, and lymphatics. Collectively these structures make up the **spermatic cord**, which passes through the inguinal canal (see Figure 26.2).

HOMEOSTATIC IMBALANCE 26.1 CLINICAL

Although *testicular cancer* is relatively rare (affecting one of every 50,000 males), it is the most common cancer in young men ages 15 to 35. A history of mumps or orchitis (inflammation of the testis) and substantial maternal exposure to environmental toxins before birth increase the risk, but the most important risk factor for this cancer is *cryptorchidism* (nondescent of the testes, which also causes sterility).

Every male should examine his testes regularly. The most common sign of testicular cancer is a painless solid mass. If detected early, testicular cancer has an impressive cure rate. Over 90% of cases are cured by surgical removal of the cancerous testis (*orchiectomy*) alone or in combination with radiation therapy or chemotherapy. ✚

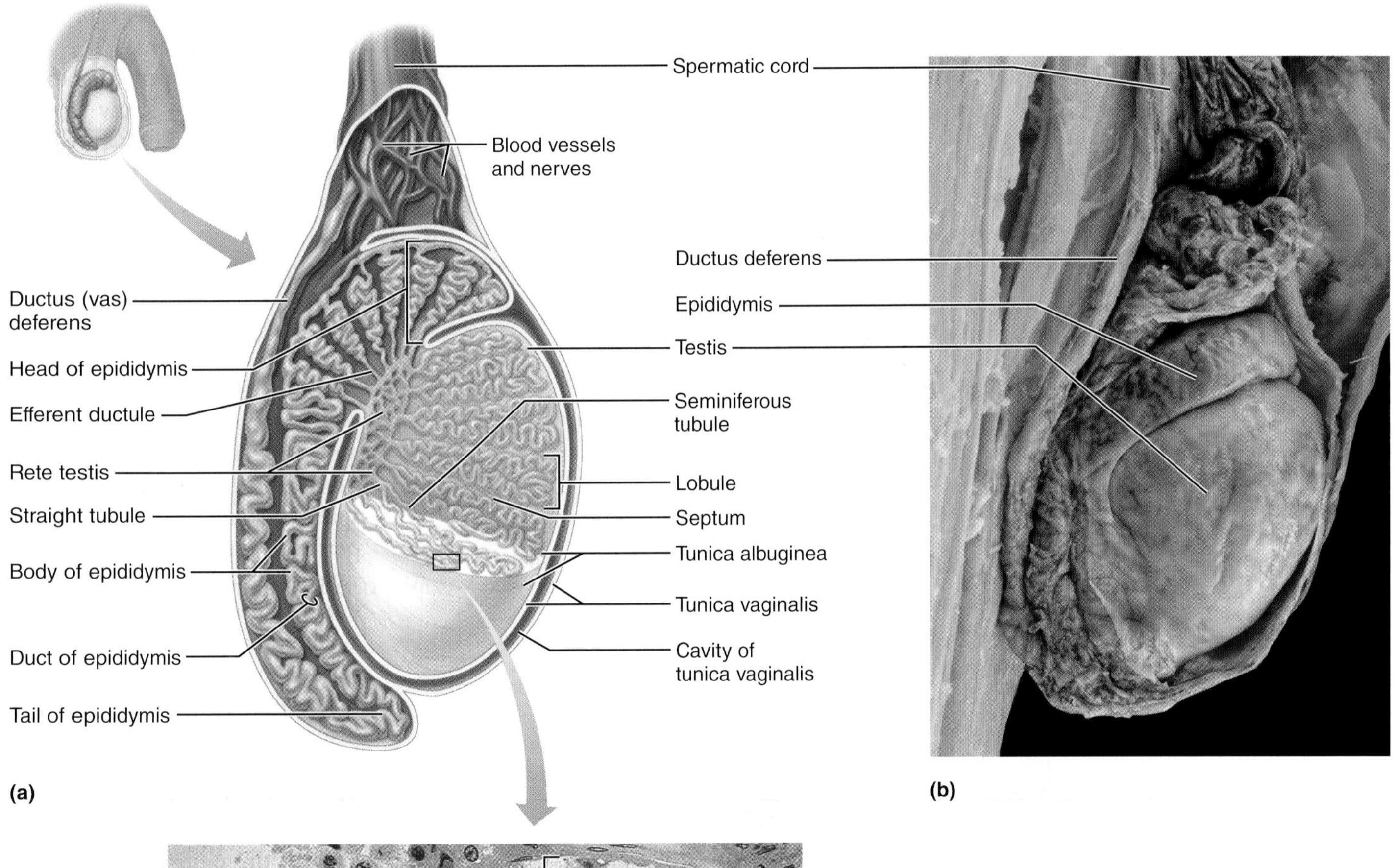

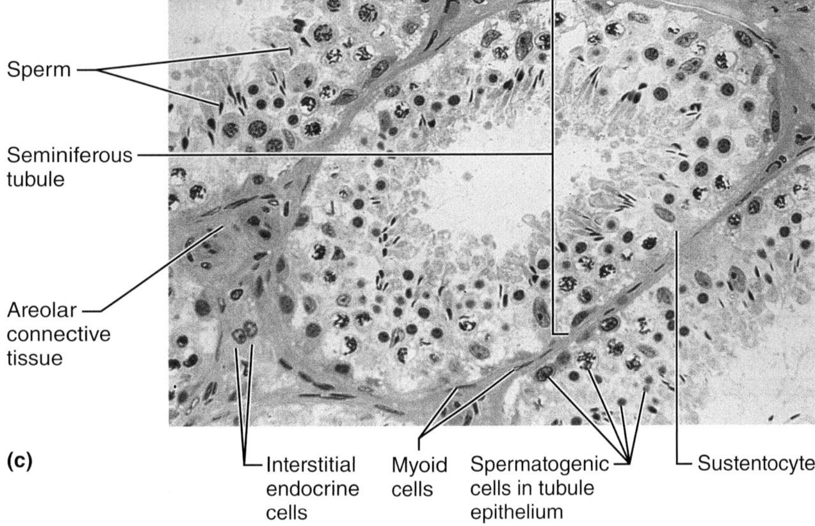

Figure 26.3 Structure of the testis. (a) Partial sagittal section through the testis and epididymis. The anterior aspect is to the right. (For a related image, see *A Brief Atlas of the Human Body*, Figure 73). **(b)** External view of a testis from a cadaver; same orientation as in (a). **(c)** Seminiferous tubule in cross section (270×). Note the spermatogenic (sperm-forming) cells in the tubule epithelium and the interstitial endocrine cells in the connective tissue between the tubules.

View histology slides
MasteringA&P®>Study Area>PAL

☑ Check Your Understanding

1. What are the two major functions of the testes?
2. Which of the tubular structures shown in Figure 26.3a are the sperm "factories"?
3. Muscle activity and the pampiniform venous plexus help to keep the temperature of the testes at homeostatic levels. How do they do that?

For answers, see Answers Appendix.

26.2 The penis is the copulatory organ of the male

→ Learning Objective

☐ **Describe the location, structure, and function of the penis.**

The **penis** ("tail") is a copulatory organ, designed to deliver sperm into the female reproductive tract (Figure 26.1 and Figure 26.4). The penis and scrotum, which hang suspended from the perineum, make up the external reproductive structures, or **external genitalia**, of the male.

The penis consists of an attached root and a free *body* or *shaft* that ends in an enlarged tip, the **glans penis**. The skin covering the penis is loose, and when it slides distally it forms a cuff called the **prepuce** (pre′pūs), or **foreskin**, around the glans. Frequently, the foreskin is surgically removed shortly after birth, a procedure called *circumcision* ("cutting around"). Interestingly, over 60% of

Because it is aligned with the second rib, it is a handy cue for finding that rib and then counting the ribs during a physical exam. **14.** The thoracic vertebrae also contribute to the thoracic cage. **15.** Each pectoral girdle is formed by a scapula and a clavicle. **16.** The pectoral girdle attaches to the sternal manubrium of the axial skeleton via the medial end of its clavicle. **17.** A consequence of its flexibility is that it is easily dislocated. **18.** The structures are (a) acromioclavicular joint, (b) coracoid process, (c) manubrium of sternum, (d) body of sternum, and (e) costal cartilage. **19.** Together the ulna and humerus form the elbow joint. **20.** The ulna and the radius each have a styloid process distally. **21.** Carpals are found in the proximal region of the palm. They are short bones. **22.** The third bone that forms the hip bone is the ischium. **23.** The pelvic girdle receives the weight of the upper body (trunk, head, and upper limbs) and transmits that weight to the lower limbs. **24.** The female pelvis is wider and has a shorter sacrum and a more movable coccyx. **25.** The tibia is the second largest bone in the body. **26.** The two bones shown are the tibia and fibula. They are from the right side of the body. The labeled structures are (a) medial condyle, (b) medial malleolus, and (c) head of fibula. **27.** The lateral condyles are not sites of muscle attachment, they are articular surfaces. **28.** Because of their springiness, the foot arches save energy during locomotion. **29.** The two largest tarsals are the talus and the calcaneus, which forms the heel.

Review Questions 1. (1)b, g; (2)h; (3)d; (4)d, f; (5)e; (6)c; (7)a, b, d, h; (8)i; **2.** (1)g, (2)f, (3)b, (4)a, (5)b, (6)c, (7)d, (8)e; **3.** (1)b, (2)c, (3)e, (4)a, (5)h, (6)e, (7)f

Clinical Case Study 1. The hemispherical socket at the point where the femur attaches is the *acetabulum*. **2.** The structure on the femur that forms the "ball" that fits into the "socket" named in question 1 is the *head* of the femur. **3.** The three bones in the pelvic girdle that fuse together at a point within the structure identified in question 1 are the *ilium, ischium,* and *pubis*. **4.** If you rest your hands on your hips, they are on the *iliac crests*. **5.** The structures on the femur where the large muscles of the buttocks and thigh attach are the *greater trochanter* and *lesser trochanter.* **6.** The structure of the pelvis that the sciatic nerve passes through as it travels into the upper thigh is the *greater sciatic notch* of the *ilium*.

Chapter 8

Check Your Understanding 1. The synarthroses are the least mobile of the joint types. **2.** In general, the more stable a joint, the less mobile it is. **3.** Most fibrous joints are synarthroses (immovable). **4.** Evan would not have synchondroses at the ends of his femur. By age 25, his epiphyseal plates have fused and become synostoses. **5.** Bursae and tendon sheaths help to reduce friction during joint movement. **6.** The muscle tendons that cross the joint are typically the most important factor in stabilizing synovial joints. **7.** John's hip joint was flexed and his knees extended and his thumb was in opposition (to his index finger). **8.** The hinge and pivot joints are uniaxial joints. **9.** The knee and temporomandibular joints have menisci. The elbow and knee each act mainly as a uniaxial hinge. The shoulder depends largely on muscle tendons for stability. **10.** Arthritis means inflammation of the joint. **11.** RA typically produces pain, swelling, and joint deformations that tend to be bilateral and crippling. OA patients tend to have pain, particularly on arising, which is relieved by gentle exercise, and enlarged bone ends (due to spurs) in affected joints. Affected joints may exhibit crepitus. **12.** Lyme disease is caused by spirochete bacteria and transmitted by a tick bite.

Review Questions 1. (1)c, (2)a, (3)a, (4)b, (5)c, (6)b, (7)b, (8)a, (9)c; **2.** b; **3.** d; **4.** d; **5.** b; **6.** d; **7.** d

Clinical Case Study 1. The hip joint would be structurally classified as a *synovial joint* and functionally classified as a *diarthrotic* (freely movable) joint. **2.** The six distinguishing features that define a synovial joint are: (1) articular cartilage, (2) joint (synovial) cavity, (3) articular capsule, (4) synovial fluid, (5) reinforcing ligaments, (6) nerves and blood vessels. **3.** The joint space in a synovial joint is normally filled with *synovial fluid.* **4.** The acetabular labrum is a piece of fibrocartilage that attaches to and extends the rim of the acetabulum. The diameter of the labrum is less than the diameter of the head of the femur, and this helps stabilize the joint to prevent it from dislocating. **5.** Mrs. Tanner's hip was bent (flexed) and her thigh was pulled toward the midline of her body (adducted) and turned in toward this midline (medially rotated). **6.** Mrs. Tanner suffered a posterior dislocation of the hip. When the head of the femur comes away from the acetabulum and then turns posteriorly, it causes the femur to rotate counterclockwise. This can be determined from the fact that her thigh was adducted and medially rotated. **7.** The hip movements include flexion, extension, abduction, adduction, rotation, and circumduction of the thigh.

Chapter 9

Check Your Understanding 1. Striated means "with stripes." **2.** He should respond "smooth muscle," which fits the description. **3.** "Epimysium" literally translates to "outside the muscle" and this connective tissue sheath is the outermost muscle sheath, enclosing the entire muscle. **4.** The thin myofilaments have binding sites for calcium on the troponin molecules forming part of those filaments. **5.** In a resting muscle fiber, the SR would have the highest concentration of calcium ions. The mitochondrion provides the ATP needed for muscle activity. **6.** The levels of organization are: (1) atom: phosphorus atom, (2) molecule: phospholipid molecule, (3) organelle: sarcoplasmic reticulum, (4) cell: skeletal muscle fiber, (5) tissue: muscle tissue, (6) organ: the biceps muscle, and (7) organ system: the muscular system. **7.** The components of the neuromuscular junction are the axon terminal, the synaptic cleft, and the junctional folds of the sarcolemma. **8.** The final trigger for contraction is a certain concentration of calcium ions in the cytosol. The initial trigger is depolarization of the sarcolemma. **9.** There are always some myosin cross bridges bound to the actin myofilament during the contraction phase. This prevents backward sliding of the actin filaments. **10.** Without ATP, rigor would occur because the myosin heads could not detach. **11.** A motor unit is an axon of a motor neuron and all the muscle fibers it innervates. **12.** During the latent period, events of excitation-contraction coupling are occurring. **13.** Immediately after Jacob grabs the bar, his biceps muscles are contracting isometrically. As his body moves upward toward the bar, they are contracting isotonically and concentrically. As he lowers his body toward the mat, the biceps are contracting isotonically and eccentrically. **14.** Eric was breathing heavily because it takes some time for his heart rate and overall metabolism to return to the resting state after exercise. Moreover, he had likely incurred an oxygen debt that required he take in extra oxygen, called EPOC, for the restorative processes. Although jogging is primarily an aerobic exercise, there is always some anaerobic respiration that occurs as well—the amount depends on exercise intensity. As fatigue occurs, potassium ions accumulate in the T tubules, and lactic acid and phosphate ions accumulate in the muscle cells. **15.** Factors that influence muscle contractile force include muscle fiber size, the number of muscle fibers stimulated, the frequency of stimulation, and the degree of muscle stretch. Factors that influence velocity of contraction include muscle fiber type, load, and the number of motor units contracting. **16.** Fast glycolytic fibers would provide for short periods of intense strength needed to lift and move furniture. **17.** To increase muscle size and strength, high-intensity resistance exercise (typically anaerobic) is best. Muscle endurance is enhanced by aerobic exercise. **18.** Both skeletal and smooth muscle fibers are elongated cells, but unlike smooth muscle cells, which are spindle shaped, uninucleate, and nonstriated, skeletal muscle cells are very large cigar-shaped, multinucleate, striated cells. **19.** Calcium binds to troponin on the thin filaments in skeletal muscle cells. In smooth muscle cells, it binds to a cytoplasmic protein called calmodulin. **20.** Hollow organs that have smooth muscle cells helping to form their walls often must temporarily store the organ's contents (urine, food residues, etc.), an ability ensured by the stress-relaxation response. **21.** Intracellular calcium is involved in exocytosis, including secretion of neurotransmitters and other chemical messengers, and acts as a second messenger.

Review Questions 1. c; **2.** b; **3.** (1)b, (2)a, (3)b, (4)a, (5)b, (6)a; **4.** c; **5.** a; **6.** a; **7.** d; **8.** a; **9.** (1)a, (2)a, c, (3)b, (4)c, (5)b, (6)b; **10.** a; **11.** c; **12.** c; **13.** c; **14.** b

Clinical Case Study 1. The first reaction to tissue injury is the initiation of the inflammatory response. The inflammatory chemicals increase the permeability of the capillaries in the injured area, allowing white blood cells, fluid, and other substances to reach the injured area. The next step in healing involves the formation of granulation tissue, in which the vascular supply for the injured area is regenerated and collagen fibers that knit the torn edges of the tissue together are formed. Skeletal muscle does not regenerate well, so the damaged areas of Mrs. DeStephano's muscle tissue will probably be repaired primarily by the formation of fibrous tissue, creating scar tissue. **2.** Healing is aided by good circulation of blood within the injured area. Vascular damage compromises healing because the supply of oxygen and nutrients to the tissue is reduced. **3.** Under normal circumstances, skeletal muscles receive electrical signals from the nervous system continuously. These signals help to maintain muscle tone and readiness. Severing of the sciatic nerve removes this continuous nervous input to the muscles and will lead to muscle atrophy. Immobility of muscles will lead to a replacement of contractile muscle tissue with noncontractile fibrous connective tissue. Distal to the point of transection, the muscle will begin to decrease in size within 3–7 days of becoming immobile. This process can be delayed by electrically stimulating the tissues. Passive range-of-motion exercises also help prevent loss of muscle tone and joint range, and improve circulation in the injured areas. **4.** Mrs. DeStephano's physician wants to supply her damaged tissues with the necessary building materials to encourage healing. A high-protein diet will provide plenty of amino acids to rebuild or replace damaged proteins, carbohydrates will provide the fuel molecules needed to generate the required ATP, and vitamin C is important for the regeneration of connective tissue.

Chapter 10

Check Your Understanding 1. The term "prime mover" refers to the muscle that bears the most responsibility for causing a particular movement. **2.** The iliacus overlies the iliac bone; the adductor brevis is a small (size) muscle that adducts (movement caused) the thigh; and the quadriceps (4 heads) femoris muscle follows the course of the femur. **3.** Of the muscles illustrated in Figure 10.1, the one with the parallel arrangement (sartorius) could shorten to the greatest degree. The stocky bipennate (rectus femoris) and multipennate (deltoid) muscles would be most powerful because they pack in the most fibers. **4.** Third-class levers are the fastest levers. **5.** A lever that operates at a mechanical advantage allows the muscle to exert less force than the load being moved. **6.** Mario was using the frontal belly of his epicranius to raise his eyebrows and the orbicularis oculi muscles to wink at Sarah. **7.** To make a sad clown's face you would contract your platysma, depressor anguli oris, and depressor labii inferioris muscles. **8.** The deltoid has a broad origin. When only its anterior fibers contract, it flexes and medially rotates the arm. When only its posterior fibers contract, it extends and laterally rotates the arm. **9.** The opponens pollicis does not have an insertion on the bones of the thumb. **10.** The muscles that cross the knee also reinforce the joint by providing increased stability.

Review Questions 1. c; **2.** c; **3.** (1)e, (2)c, (3)g, (4)f, (5)d; **4.** a; **5.** c; **6.** d; **7.** c; **8.** c; **9.** b; **10.** d; **11.** b; **12.** a; **13.** c; **14.** d; **15.** a, b; **16.** a

Clinical Case Study 1. A prime mover is a muscle that has primary or major responsibility for producing a specific movement. A synergist is a muscle that supports or helps the action of a prime mover by adding extra force, or providing stability so that the prime mover can perform its action. **2.** An antagonist is a muscle that opposes, resists, or reverses a particular movement. By mimicking the action of an antagonist, the therapist can test the strength of the agonist muscle and compare it with the same muscle in the other limb. **3.** Ideally, the therapist would assess each muscle individually. In reality, these assessments usually measure the function of a group of muscles because multiple muscles are often involved in similar actions. (a) This assessment focuses on the thigh adductors (magnus, longus, brevis), pectineus, and gracilis. Specifically, the adductor magnus is innervated by the damaged sciatic nerve. (b) This assessment addresses Mrs. Tanner's ability to dorsiflex her foot. Dorsiflexion involves all of the muscles in the anterior compartment of the lower leg: the tibialis anterior, extensor digitorum longus, fibularis (peroneus) tertius, and extensor hallucis longus. (c) This assessment addresses the function of the muscles of the posterior compartment of the thigh. The hamstrings (biceps femoris, semitendinosus, semimembranosus) are the prime movers involved in knee flexion. **4.** To assess these muscles, the therapist would apply resistance to the natural action of these muscles. (a) The extensor hallucis longus inserts on the distal phalanx of the great toe. The therapist can apply resistance to the top of the toe and ask Mrs. Tanner to extend the toe. (b) The fibularis longus is involved in eversion of the foot and plantar flexion. The therapist can apply resistance to the lateral aspect of the foot and ask Mrs. Tanner to evert (turn out) her ankle. In addition, the therapist can apply resistance to the bottom of the foot and ask Mrs. Tanner to push against that resistance. (c) The gastrocnemius, along with the soleus, is a powerful plantar flexor. The therapist can (1) ask Mrs. Tanner to raise her body up on her toes using her right foot, or (2) apply pressure to the bottom of the foot and ask Mrs. Tanner to push against that resistance.

Chapter 11

Check Your Understanding 1. Integration involves processing and interpreting sensory information, and making a decision about motor output. Integration occurs primarily in the CNS. **2.** (a) This "full stomach" feeling would be relayed by the sensory (afferent) division of the PNS (via its visceral afferent fibers). (b) The somatic nervous system, which is part of the motor (efferent) division of the PNS, controls movement of skeletal muscle. (c) The autonomic nervous system, which is part of the motor (efferent) division of the PNS, controls the heart rate. **3.** Astrocytes control the extracellular environment around neuron cell bodies in the CNS, whereas satellite cells perform this function in the PNS. **4.** Oligodendrocytes and Schwann cells form myelin sheaths in the CNS and PNS, respectively. **5.** A nucleus within the brain is a cluster of cell bodies, whereas the nucleus within each neuron is a large organelle that acts as the control center of the cell. **6.** In the CNS, a myelin sheath is formed by oligodendrocytes that wrap their plasma membranes around the axon. The myelin sheath protects and electrically insulates axons and increases the speed of transmission of nerve impulses. **7.** Burning a finger will first activate unipolar (pseudounipolar) neurons that are sensory (afferent) neurons. The impulse to move your finger away from the heat will be carried by multipolar neurons that are motor (efferent) neurons. **8.** A nerve fiber is a long axon, an extension of the cell. In connective tissue, fibers are extracellular proteins that provide support. In muscle tissue, a muscle fiber is a muscle cell. **9.** The concentration gradient and the electrical gradient—together called the electrochemical gradient—determine the direction in which ions flow through an open membrane channel.
10. There are more leakage channels for K^+ than for any other cation.
11. The size of a graded potential is determined by the strength of a stimulus. **12.** Action potentials are larger than graded potentials and travel further. Graded potentials generally initiate action potentials.
13. An action potential is regenerated anew at each membrane patch.
14. Conduction of action potentials is faster in myelinated axons because myelin allows the axon membrane between myelin sheath gaps to change

its voltage rapidly, and allows current to flow only at the widely spaced gaps. **15.** If a second stimulus occurs before the end of the absolute refractory period, no AP can occur because sodium channels are still inactivated. **16.** Voltage-gated ion (Ca^{2+}) channels are found in the presynaptic axon terminal and open when an action potential reaches the axon terminal. Chemically gated ion channels are found in the postsynaptic membrane and open when neurotransmitter binds to the receptor protein. **17.** At an electrical synapse, neurons are joined by gap junctions. **18.** IPSPs result from the flow of either K^+ or Cl^- through chemically gated channels. EPSPs result from the flow of both Na^+ and K^+ through chemically gated channels. **19.** Temporal summation is summation in time of graded potentials occurring in quick succession at the postsynaptic membrane. It can result from EPSPs arising from just one synapse. Spatial summation is summation in space—a postsynaptic neuron is stimulated by a large number of terminals at the same time. **20.** ACh interacts with more than one specific receptor type, and this explains how it can excite at some synapses and inhibit at others. **21.** Cyclic AMP (cAMP) is called a second messenger because it relays the message between the first messenger (the original chemical messenger) outside of the cell and effector molecules that will ultimately bring about the desired response within the cell. **22.** Reverberating circuits and parallel after-discharge circuits both result in prolonged output. **23.** The pattern of neural processing is serial processing. The response is a reflex arc. **24.** The pattern of neural processing is parallel processing.

Review Questions 1. b; **2.** (1)d, (2)b, (3)f, (4)c, (5)a; **3.** b; **4.** c; **5.** a; **6.** c; **7.** b; **8.** d; **9.** c; **10.** c; **11.** a; **12.** (1)d, (2)b, (3)a, (4)c

Clinical Case Study 1. The electrical signals generated by neurons are called *action potentials*. An action potential is a change in membrane potential that involves depolarization and repolarization phases. **2.** An inhibitory postsynaptic potential (IPSP) is a signal that makes it less likely that a postsynaptic neuron will be able to generate an action potential. This effect is usually produced when the signal causes the membrane potential of the postsynaptic neuron to become more negative, moving away from the axon's threshold potential. **3.** GABA falls into the amino acid class of neurotransmitters. This same class includes glycine, glutamate, and aspartate. **4.** To enhance the actions of GABA at a synapse, a drug could either (1) act presynaptically to increase the release of GABA at the synapse, (2) decrease the reuptake of GABA after it has been released, or (3) act postsynaptically to either increase the binding strength of GABA at its receptors, or increase the number of receptors. **5.** An influx of Cl^- into the postsynaptic cell causes the membrane potential to become more negative (hyperpolarize). When hyperpolarized, the cell is farther from its threshold potential and so is less likely to produce an action potential.

Chapter 12

Check Your Understanding 1. The third ventricle is surrounded by the diencephalon. **2.** The cerebral hemispheres and the cerebellum have an outside layer of gray matter in addition to central gray matter and its surrounding white matter. **3.** Convolutions increase surface area of the cortex, which allows more neurons to occupy the limited space within the skull. **4.** The central sulcus separates primary motor areas from somatosensory areas. **5.** Motor functions on the left side of the body are controlled by the right hemisphere of the brain because motor tracts from the right hemisphere cross over (in the medulla oblongata) to the left side of the spinal cord to go to the left side of the body. **6.** Commissural fibers (which form commissures) allow the cerebral hemispheres to "talk to each other." **7.** The caudate nucleus, putamen, and globus pallidus together form the basal nuclei. **8.** Virtually all inputs ascending to the cerebral cortex synapse in the thalamus en route. **9.** The hypothalamus oversees the autonomic nervous system. **10.** The pyramids of the medulla are the corticospinal (pyramidal) tracts, the large voluntary motor tracts descending from the motor cortex. The result of decussation (crossing over) is that each side of the motor cortex controls the opposite side of the body. **11.** The cerebral peduncles and the colliculi are associated with the midbrain. **12.** There are many possible answers to this question—here are a few: Structurally, the cerebellum and cerebrum are similar in that they both have a thin outer cortex of gray matter, internal white matter, and deep gray matter nuclei. Also, both have body maps (homunculi) and large fiber tracts connecting them to the brain stem. Both receive sensory input and influence motor output. A major difference is that the cerebellum is almost entirely concerned with motor output, whereas the cerebrum has much broader responsibilities. Also, while a cerebral hemisphere controls the opposite side of the body, a cerebellar hemisphere controls the same side of the body. **13.** The hypothalamus is part of the limbic system and also an autonomic (visceral) control center. **14.** Taylor is increasing the amount of sensory stimuli she receives, which will be relayed to the reticular activating system, which, in turn, will increase activation of the cerebral cortex. **15.** Transfer of memory from STM to LTM is enhanced by (1) rehearsal, (2) association (tying "new" information to "old" information), and (3) a heightened emotional state (for example, alert, motivated, surprised, or aroused). **16.** Delta waves are typically seen in deep sleep in normal adults. **17.** Drowsiness (or lethargy) and stupor are stages of consciousness between alertness and coma. **18.** Most skeletal muscles are actively inhibited during REM sleep. **19.** CSF, formed by the choroid plexuses as a filtrate of blood plasma, is a watery "broth" similar in composition to plasma. It protects the brain and spinal cord from blows and other trauma, helps nourish the brain, and carries chemical signals from one part of the brain to another. **20.** The brain surgeon cuts through (1) the skin of the scalp, (2) the periosteum, (3) skull bone, (4) dura mater, (5) arachnoid mater, and (6) pia mater to reach the brain. **21.** A TIA is a temporary loss of blood supply to brain tissue, and it differs from a stroke in that the resulting impairment is fully reversible. **22.** Mrs. Lee might have Parkinson's disease. **23.** The nerves serving the limbs arise in the cervical and lumbar enlargements of the spinal cord. **24.** A loss of motor function is called paralysis. Lower limb paralysis could be caused by a spinal cord injury in the thoracic region (between T_1 and L_1). If the spinal cord is transected, the result is paraplegia. If the cord is only bruised, he may regain function in the limbs. **25.** In the spinothalamic pathway, the cell bodies of first-order sensory neurons are outside the spinal cord in a ganglion, cell bodies of second-order sensory neurons are in the dorsal horn of the spinal cord, and cell bodies of third-order sensory neurons are in the thalamus. (See also Figure 12.32b.) **26.** The pyramidal cells controlling left big toe movement are in the right primary motor cortex in the frontal lobe. They synapse with the cell bodies of ventral horn neurons in the spinal cord. **27.** A nerve is a bundle of axons in the PNS, whereas a tract is a bundle of axons in the CNS. A nucleus is a collection of neuron cell bodies in the CNS, whereas a ganglion is a collection of neuron cell bodies in the PNS.

Review Questions 1. a; **2.** d; **3.** c; **4.** a; **5.** (1)d, (2)f, (3)e, (4)g, (5)b, (6)f, (7)i, (8)a; **6.** b; **7.** c; **8.** a; **9.** (1)a, (2)b, (3)a, (4)a, (5)b, (6)a, (7)b, (8)b, (9)a; **10.** d; **11.** (1)d, (2)e, (3)d, (4)a; **12.** c

Clinical Case Study 1. The four regions of the brain are the cerebral hemispheres, diencephalon, brain stem, and cerebellum. A cerebral hemisphere is involved in this case. Motor functions and language are both located in this cerebral hemisphere and have been affected by the injury. **2.** The left side of the brain has been affected in this case. The motor dysfunction on the right side of the body is the primary piece of evidence used to determine this. In addition, the brain areas serving speech are usually located in the left side. **3.** The motor dysfunction on the right side of the body suggests that the injury has affected the primary motor cortex and possibly the premotor cortex of the left cerebral hemisphere. In addition to the problems with motor function, Mrs. Bryans experienced difficulty with language. Two areas of the brain associated with language,

Broca's and Wernicke's areas, are usually located in the left cerebral cortex. The type of aphasia described suggests damage to Broca's area. **4.** From the surface of the brain to the skull, the three membranes that make up the meninges include the pia mater, arachnoid mater, and dura mater. **5.** The subarachnoid hemorrhage involves bleeding into the region below (*sub* = below) the arachnoid mater. The subdural hematoma involves blood collecting between the dura mater and the arachnoid mater.

Chapter 13

Check Your Understanding 1. The three levels of sensory integration are receptor level, circuit level, and perceptual level. **2.** Phasic receptors adapt, whereas tonic receptors exhibit little or no adaptation. Pain receptors are tonic so that we are reminded to protect the injured body part. **3.** Hot and cold are conveyed by different sensory receptors that are parts of separate "labeled lines." Cool and cold are two different intensities of the same stimulus, detected by frequency coding—the frequency of APs would be higher for a cold than a cool stimulus. Action potentials arising in the fingers and foot arrive at different locations in the somatosensory cortex via their own "labeled lines" and in this way the cortex can determine their origin. **4.** In addition to nerves, the PNS also consists of sensory receptors, motor endings, and ganglia. **5.** Nociceptors respond to painful stimuli. They are exteroceptors that are nonencapsulated (free nerve endings). **6.** Tears (lacrimal fluid) are a dilute saline secretion that contains mucus, antibodies, and lysozyme. They are secreted by the lacrimal glands. **7.** The blind spot of the eye is the optic disc. It is the part of the retina where the optic nerve exits the eye, and it is "blind" because it is a region of the retina that lacks photoreceptors. **8.** Light passes through the cornea, aqueous humor, lens, vitreous humor, ganglion cells, and bipolar cells before it reaches the photoreceptors. **9.** The ciliary muscles and sphincter pupillae relax for distant vision. (If you said the medial rectus muscles also relax, this is true, but remember that the rectus muscles are extrinsic eye muscles, not intrinsic.) **10.** The near point moves farther away as you age because the lens becomes less flexible (presbyopia), so that it is unable to assume the more rounded shape required for near vision. **11.** The following are characteristics of cones: "vision in bright light," "color vision," and "higher acuity." The following are characteristics of rods: "only one type of visual pigment," "most abundant in the periphery of the retina," "many feed into one ganglion cell," and "higher sensitivity." **12.** A tumor in the right visual cortex would affect the left visual field. A tumor compressing the right optic nerve would affect both the left and right visual fields from the right eye only. **13.** The cilia of these receptor cells greatly increase the surface area for sensory receptors. **14.** The five taste modalities are sweet, sour, bitter, salty, and umami. The fungiform, vallate, and foliate papillae contain taste buds. **15.** The tympanic membrane separates the external from the middle ear. The oval and round windows separate the middle from the inner ear. **16.** The basilar membrane allows us to differentiate sounds of different pitch. **17.** Influx of Na^+ ions is responsible for depolarization in most cells. K^+ and Ca^{2+} ions cause depolarization in hair cells. K^+ moves into hair cells (rather than out) because the extracellular fluid is endolymph, which is rich in K^+. As a result, the electrochemical gradient drives K^+ into these cells. **18.** You would not be able to locate the origin of a sound if the brain stem did not receive input from both ears. **19.** The following apply to a macula: "contains otoliths," "responds to linear acceleration or deceleration," and "inside a saccule." The following apply to a crista ampullaris: "inside a semicircular canal," "has a cupula," "responds to rotational acceleration and deceleration." **20.** Ganglia are collections of neuron cell bodies in the PNS. **21.** Nerves also contain connective tissue, blood vessels, lymphatic vessels, and the myelin surrounding the axons. **22.** Schwann cells, macrophages, and the neurons themselves were all important in healing the nerve. **23.** The oculomotor (III), trochlear (IV), and abducens (VI) nerves control eye movements. Sticking out your tongue involves the hypoglossal nerve (XII). The vagus nerve (X) influences heart rate and digestive activity. The accessory nerve (XI) innervates the trapezius muscle, which is involved in shoulder shrugging. **24.** Roots lie medial to spinal nerves, whereas rami lie lateral to spinal nerves. Dorsal roots are purely sensory, whereas dorsal rami carry both motor and sensory fibers. **25.** The spinal nerve roots were C_3–C_5, the spinal nerve was the phrenic nerve, the plexus was the cervical plexus. The phrenic nerve is the sole motor nerve supply to the diaphragm, the primary muscle for respiration. **26.** Varicosities are the series of knoblike swellings that are the axon endings of autonomic motor neurons. You would find them on axon endings serving smooth muscle or glands. **27.** The cerebellum and basal nuclei, which form the precommand level of motor control, plan and coordinate complex motor activities. **28.** The effector (muscle or gland) brings about the response. **29.** The stretch reflex is important for maintaining muscle tone and adjusting it reflexively by causing muscle contraction in response to increased muscle length (stretch). It maintains posture. The flexor or withdrawal reflex is initiated by a painful stimulus and causes automatic withdrawal of the painful body part from the stimulus. It is protective. **30.** This response is called Babinski's sign and it indicates damage to the corticospinal tract or primary motor cortex. **31.** The spinothalamic pathway carries pain signals to the somatosensory cortex in the parietal lobe. The pyramidal pathways carry voluntary motor information and would be involved in inhibiting the flexor reflex.

Review Questions 1. b; **2.** c; **3.** c; **4.** (1)a, 1 and 5; (2)a, 3 and 5; (3)a, 4; (4)a, 2; (5)c, 2; (6)b, 2; **5.** (1)d, (2)c, (3)f, (4)b, (5)e, (6)a; **6.** (1)f, (2)i, (3) b, (4)g, h, (5)e, (6)i, (7)c, (8)k, (9)l, (10)c, d, f, k; **7.** (1)b 6; (2)d 8; (3)c 2; (4)c 5; (5)a 4; (6)a 3, 9; (7)a 7; (8)a 7; (9)d 1; (10)a 3, 4, 7, 9; **8.** c; **9.** d; **10.** a; **11.** b; **12.** c; **13.** c; **14.** b; **15.** a; **16.** b; **17.** b; **18.** d; **19.** d; **20.** a; **21.** d; **22.** c; **23.** b; **24.** b; **25.** d; **26.** b; **27.** b; **28.** e; **29.** b; **30.** c; **31.** c; **32.** c; **33.** c

Clinical Case Studies: Peripheral Nervous System 1. Cerebrospinal fluid (CSF) is leaking out of Mr. Hancock's right ear. The fracture must have torn both the dura mater and arachnoid mater. In addition, the tympanic membrane must have ruptured. Antibiotics were administered to prevent infection by bacteria that might enter through the ruptured meninges, causing meningitis. Elevating the head of the bed decreases the CSF pressure in the skull. (This allows the torn meninges to heal spontaneously in the majority of cases.) **2.** The observations on Mr. Hancock's chart indicate: (a) Either damage to CN VIII (the vestibulocochlear nerve, which transmits afferent impulses for the sense of hearing) or destruction of the cochlea (the sensory organ for hearing). (b) Damage to CN V_3 (the mandibular division of the trigeminal nerve), which runs through the foramen ovale. This nerve conveys sensory information from the lower part of the face. (c) Damage to CN V_2 (the maxillary division of the trigeminal nerve), which runs through the foramen rotundum. This nerve conveys sensory information from the skin of the upper lip, lower eyelid, and cheek. (d) Damage to CN VI (the abducens nerve), which innervates the lateral rectus muscle of the eye. Because this muscle is responsible for pulling the eye laterally (abduction), loss of tone in this muscle at rest will cause the eye to turn inward. Diplopia will worsen when looking to the right because the eye cannot abduct. **3.** The facial nerve (cranial nerve VII) is the primary motor nerve associated with facial expression. The facial nerve also contains parasympathetic fibers that control secretion of tears from the lacrimal glands. Damage to this nerve explains both the motor symptoms and the dryness of his eye.

Special Senses 1. The ear is divided into three major areas: external ear, middle ear, and internal ear. The portion of the internal ear, or labyrinth, associated with balance and equilibrium is the part of the ear affected in Mr. Rhen's BPPV. **2.** Maintaining balance and equilibrium requires multiple sources of sensory input. The three main sources of sensory input are the vestibular apparatus of the ear, visual input, and input from the proprioceptors of the skin, muscles, and joints. **3.** The two functional divisions are the vestibule and semicircular canals. The vestibule's sensory receptors are the maculae, which sense linear (straight line) acceleration and deceleration. The semicircular canals' receptors are the cristae ampullares, which detect rotational acceleration and deceleration. **4.** Mr. Rhen's vertigo is brought on by rotational movements of the head, suggesting that the cristae ampullares of the semicircular canals are affected. **5.** The added mass of the displaced otoliths pushes on a cupula of a semicircular canal when the head is rotated during the Dix-Hallpike maneuver. The otoliths either stick to the gelatinous cupula, or swirl through the canals during head movement and drift against the cupula like snow. In both cases, the bending of the cupula is prolonged and vertigo persists. As a result, vestibular nystagmus is observed when it would ordinarily be absent.

Chapter 14

Check Your Understanding 1. The effectors of the autonomic nervous system are cardiac muscle, smooth muscle, and glands. **2.** The somatic motor system relays instructions to muscles more quickly because it involves only one motor neuron, whereas the ANS uses a two-neuron chain. Moreover, axons of somatic motor neurons are typically heavily myelinated, whereas preganglionic autonomic axons are lightly myelinated and postganglionic axons are nonmyelinated. **3.** Dorsal root ganglia also contain neuron cell bodies. These cell bodies belong to neurons that are structurally classified as unipolar (pseudounipolar) and functionally classified as sensory neurons (primary or first-order sensory neurons in this case). **4.** While you relax in the sun on the beach, the parasympathetic branch of the ANS would probably predominate. When you perceive danger (as in a shark), the sympathetic branch of the ANS predominates. **5.** Cell bodies of preganglionic parasympathetic neurons that innervate the head are in the brain stem. Cell bodies of postganglionic parasympathetic neurons innervated by the vagus nerve are in ganglia found mostly in the walls of visceral organs. **6.** "Short preganglionic fibers," "origin from thoracolumbar region of spinal cord," "collateral ganglia," and "innervates adrenal medulla" are all characteristic of the sympathetic nervous system. Terminal ganglia are found in the parasympathetic nervous system. **7.** The major differences are (1) the ANS has visceral afferents rather than somatic afferents, (2) the ANS has a two-neuron efferent chain, whereas the somatic nervous system has one, and (3) the effectors of the ANS are smooth muscles, cardiac muscle, and glands, whereas the effectors of the somatic nervous system are skeletal muscles. **8.** You would find nicotinic receptors on skeletal muscle and the hormone-producing cells of the adrenal medulla, but not on smooth muscle or glands. Virtually all types of receptors (including nicotinic receptors) are also found in the CNS (see Table 11.3 on pp. 374–375). **9.** The parasympathetic nervous system increases digestive activity and decreases heart rate. The sympathetic nervous system increases blood pressure, dilates bronchioles, stimulates the adrenal medulla to release its hormones, and causes ejaculation. **10.** The main integration center of the ANS is the hypothalamus, although the most direct influence is through the brain stem reticular formation and the reflex centers in the pons and medulla oblongata. **11.** Jackson's doctor may have prescribed a beta-blocker because Jackson has hypertension. (Chronic stress is a factor in causing hypertension.) The beta-blocker will decrease blood pressure by blocking beta-adrenergic receptors in the heart, thereby decreasing heart rate and force of contraction, and by decreasing renin release from the kidneys.

Review Questions 1. d; **2.** (1)S, (2)P, (3)P, (4)S, (5)S, (6)P, (7)P, (8)S, (9) P, (10)S, (11)P, (12)S; **3.** c; **4.** a

Clinical Case Study 1. The location of Jimmy's lacerations and bruises and his inability to rise led the paramedics to suspect a head, neck, or back injury. They immobilized his head and torso to prevent any further damage to the brain and spinal cord. **2.** The worsening neurological signs indicate a probable intracranial hemorrhage. The blood escaping from the ruptured blood vessel(s) will begin to compress Jimmy's brain and increase his intracranial pressure. Jimmy's surgery will involve repair of the damaged vessel(s) and removal of the mass of clotted blood pressing on his brain. **3.** Loss of motor and sensory function below the level of the nipples indicates a lesion at T_4. See Figure 13.39. **4.** Jimmy is suffering from spinal shock, which occurs as a result of injury to the spinal cord. Spinal shock is a temporary condition in which all reflex and motor activities caudal to the level of spinal cord injury are lost, so Jimmy's muscles are paralyzed. His blood pressure is low due to the loss of sympathetic tone in his vasculature. **5.** Jimmy's exaggerated reflexes are caused by damaged upper motor neuron axons in the spinal cord. These upper motor neurons normally inhibit spinal reflexes. He is incontinent because there are no longer pathways to support voluntary control of bowel and bladder emptying. **6.** This condition is called autonomic dysreflexia (or autonomic hyperreflexia). This is a condition in which a normal stimulus triggers a massive activation of autonomic neurons. **7.** Extremely high arterial blood pressure can cause a rupture of the cerebral blood vessels (as well as other blood vessels in the body) and put Jimmy's life at risk.

Chapter 15

Check Your Understanding 1. The endocrine system is more closely associated with growth and development, and its responses tend to be long-lasting, whereas nervous system responses tend to be rapid and discrete. **2.** The thyroid and parathyroid glands are found in the neck. **3.** Hormones are released into the blood and transported throughout the body, whereas paracrines act locally, generally within the same tissue. **4.** Steroid hormones are synthesized on the membrane of the smooth endoplasmic reticulum. Peptide hormones are synthesized on rough endoplasmic reticulum. Peptide hormones can be stored in vesicles and released by exocytosis. **5.** Steroids are all lipid soluble. Thyroid hormones are the only amino acid–based hormones that are lipid soluble. **6.** Water-soluble hormones act on receptors in the plasma membrane coupled most often via regulatory molecules called G proteins to intracellular second messengers. Lipid-soluble hormones act on intracellular receptors, directly activating genes and stimulating synthesis of specific proteins. **7.** Hormone release can be triggered by humoral, neural, or hormonal stimuli. **8.** Lipid-soluble hormones have longer half-lives, meaning that they stay in the blood longer. (They are not as readily excreted by the kidneys because they are bound to plasma proteins, and most need to be metabolized by the liver before they can be excreted.) **9.** The hypothalamus communicates with the anterior pituitary via hormones released into a special portal system of blood vessels. In contrast, it communicates with the posterior pituitary via action potentials traveling down axons that connect the hypothalamus to the posterior pituitary. **10.** Drinking alcoholic beverages inhibits ADH secretion from the posterior pituitary and causes copious urine output and dehydration. The dehydration causes the hangover effects. **11.** LH and FSH are tropic hormones that act on the gonads, TSH is a tropic hormone that acts on the thyroid, and ACTH is a tropic hormone that acts on the adrenal cortex. (If you said growth hormone, that's also a good answer, as GH causing the liver to release IGFs might also be considered a tropic effect.) **12.** T_4 has four bound iodine atoms, and T_3 has three. T_4 is the major hormone secreted, but T_3 is more potent. T_4 is referred to as thyroxine. **13.** Thyroid hormone increases basal metabolic rate (and heat production) in the body. Parathyroid hormone increases blood Ca^{2+} levels in a variety of ways. Calcitonin at high (pharmacological) levels has a Ca^{2+}-lowering, bone-sparing effect. (At normal blood levels its effects in humans are negligible.) **14.** Thyroid follicular cells release thyroid hormone, parathyroid cells in the parathyroid gland release parathyroid hormone, and parafollicular (C) cells in the thyroid gland release calcitonin. **15.** Glucocorticoids are stress hormones that, among many effects, increase blood glucose. Mineralocorticoids increase blood Na^{+} (and blood pressure) and decrease blood K^{+}. Gonadocorticoids are male and female sex hormones that are thought to have a variety of effects (for example, contribute to onset of puberty, sex drive in women, pubic and axillary hair development in women). **16.** Melatonin is used by some individuals as a sleep aid, particularly to counter jet lag. **17.** The heart produces atrial natriuretic peptide (ANP). ANP decreases blood volume and blood pressure by increasing the kidneys' production of salty urine. **18.** The major function of vitamin D_3, produced in inactive form by the skin, is to increase intestinal absorption of calcium. **19.** Diabetes mellitus is due to a lack of insulin production or action, whereas diabetes insipidus is due to a lack of ADH. Both conditions are characterized by production of copious amounts of urine. You would find glucose in the urine of a patient with diabetes mellitus, but not in the urine of a patient with diabetes insipidus. **20.** The gonadal hormones are steroid hormones. A major endocrine gland that also secretes steroid hormones is the adrenal cortex.

Review Questions 1. b; **2.** a; **3.** c; **4.** d; **5.** (1)c, (2)a and b, (3)f, (4)d, (5)e, (6)g, (7)a, (8)h, (9)b and e, (10)a; **6.** d; **7.** c; **8.** b; **9.** d; **10.** b; **11.** d; **12.** b; **13.** c; **14.** d

Clinical Case Study 1. Rationale for orders: As Mr. Gutteman is unconscious, the level of damage to his brain is unclear. Monitoring his responses and vital signs every hour will provide information for his care providers about the extent of his injuries. Turning him every 4 hours and providing careful skin care will prevent decubitus ulcers (bedsores) as well as stimulating his proprioceptive pathways. **2.** Mr. Gutteman's condition is termed diabetes insipidus, a condition in which insufficient quantities of antidiuretic hormone (ADH) are produced or released. Diabetes insipidus patients excrete large volumes of urine but do not have glucose or ketones present in the urine. The head trauma could have damaged Mr. Gutteman's hypothalamus, which produces the hormone, or injured his posterior pituitary gland, which releases ADH into the bloodstream. **3.** Diabetes insipidus is not life threatening for most individuals with normal thirst mechanisms, as they will be thirsty and drink to replenish the lost fluid. However, Mr. Gutteman is comatose, so his fluid output must be monitored closely so that the volume lost can be replaced by IV line. His subsequent recovery may be complicated if he has suffered damage to his hypothalamus, which houses the thirst center neurons.

Chapter 16

Check Your Understanding 1. Blood can prevent blood loss by forming clots when a blood vessel is damaged. Blood can prevent infection because it contains antimicrobial proteins and white blood cells. **2.** The hematocrit is the percentage of blood that is occupied by erythrocytes. It is normally about 45%. **3.** Plasma proteins are not used as fuel for body cells because their presence in blood is required to perform many key functions. **4.** Each hemoglobin molecule can transport four O_2. The heme portion of the hemoglobin binds the O_2. **5.** The kidneys' synthesis of erythropoietin is compromised in advanced kidney disease, so RBC production decreases, causing anemia. **6.** Monocytes become macrophages in tissues. Neutrophils are also voracious phagocytes. **7.** Amos's red bone marrow is spewing out many abnormal white blood cells, which are crowding out the production of normal bone marrow elements. The lack of normal white blood cells allows the infections, the low number of platelets fails to stop bleeding, and the lack of erythrocytes is anemia. **8.** Microglial cells can become phagocytes in the brain. **9.** A megakaryocyte is a cell that produces platelets. Its name means "big nucleus cell." **10.** The three steps of hemostasis are vascular spasm, platelet plug formation, and coagulation. **11.** Fibrinogen is water soluble, whereas fibrin is not. Prothrombin is an inactive precursor, whereas thrombin acts as an enzyme. Most factors are inactive in blood before activation and become enzymes upon activation. (There are exceptions, such as fibrinogen and calcium.) **12.** Thrombocytopenia (platelet deficiency) results in failure to plug the countless small tears in blood vessels, and so manifests as small purple spots. Hemophilia A results from the absence of clotting factor VIII. **13.** Nigel has anti-A antibodies in his blood and type B agglutinogens on his RBCs. He can donate blood to an AB recipient, but he should not receive blood from an AB donor because his anti-A antibodies will cause a transfusion reaction. **14.** If Emily has a bacterial meningitis, a differential WBC count would likely reveal an increase in neutrophils because neutrophils are a major body defense against bacteria.

Review Questions 1. c; **2.** c; **3.** d; **4.** b; **5.** d; **6.** a; **7.** a; **8.** b; **9.** c; **10.** d

Clinical Case Study 1. Saline infusion temporarily replaces the lost blood volume, thereby helping to restore Mr. Malone's circulation. **2.** The PRBCs contain oxygen-carrying hemoglobin. While the saline replaces lost blood volume, it cannot replace the hemoglobin in the lost RBCs. (In acute trauma, the rule of thumb is to use no more than 2 liters of normal saline before starting PRBCs, so that the hematocrit does not drop below 30%.) **3.** O negative blood cells bear neither the A nor the B nor the Rh agglutinogens (antigens). People with O negative blood are sometimes called "universal donors" because their cells lack the antigens responsible for most major transfusion reactions. **4.** Mr. Malone's blood would have anti-B antibodies (agglutinins) so he would not be able to receive B or AB blood. He can safely receive A and O blood. **5.** If doctors had transfused type B or AB blood into Mr. Malone's circulation, his anti-B antibodies would have "attacked" these foreign cells and caused them to agglutinate. This transfusion reaction can be dangerous because the agglutinated cells can clog small vessels. In addition the transfused cells would begin to hemolyze (rupture) or would be destroyed by phagocytes.

Chapter 17

Check Your Understanding 1. The mediastinum is the medial cavity of the thorax within which the heart, great vessels, thymus, and parts of the trachea, bronchi, and esophagus are found. **2.** The layers of the heart wall are the endocardium, the myocardium, and the epicardium. The epicardium is also called the visceral layer of the serous pericardium. This is surrounded by the parietal layer of the serous pericardium and the fibrous pericardium. **3.** The serous fluid decreases friction caused by movement of the layers against one another. **4.** The papillary muscles and chordae tendineae keep the AV valve flaps from everting into the atria as the ventricles contract. **5.** The mitral (left atrioventricular) valve has two cusps. **6.** The right side of the heart acts as the pulmonary pump, whereas the left acts as the systemic pump. **7.** (a) True. The left ventricle wall is thicker than the right. (b) True. The left ventricle pumps blood at much higher pressure than the right ventricle because the left ventricle supplies the whole body, whereas the right ventricle supplies only the lungs. (c) False. Each ventricle pumps the same amount of blood with each beat. If this were not true, blood would back up in either the systemic or pulmonary circulation (because the two ventricles are in series). **8.** The branches of the right coronary artery are the right marginal artery and the posterior interventricular artery. **9.** (a) The refractory period is almost as long as the contraction in cardiac muscle. (b) The source of Ca^{2+} for the contraction is only SR in skeletal muscle. (c) Both skeletal muscle and cardiac muscle have troponin. (d) Only skeletal muscle has triads. **10.** Cardiac muscle cannot go into tetany because the absolute refractory period is almost as long as the contraction. **11.** The subendocardial conducting network excites ventricular muscle fibers. The depolarization wave travels upward from the apex toward the atria. **12.** (a) The QRS wave occurs during ventricular depolarization. (b) The T wave of the ECG occurs during ventricular repolarization. (c) The P-R interval of the ECG occurs during atrial depolarization and the conduction of the action potential through the rest of the intrinsic conduction system. **13.** (b) Represents an action potential (AP) in a contractile cardiac muscle cell, (a) represents an AP in a skeletal muscle cell, and (c) represents an AP in cardiac pacemaker cells. The depolarization phase is due to Na^+ influx in skeletal muscle and contractile cardiac muscle cells, and it is due to Ca^{2+} entry in cardiac pacemaker cells. K^+ efflux is responsible for the repolarization in all action potentials. **14.** The second heart sound is associated with the closing of the semilunar valves. **15.** The murmur of mitral insufficiency occurs during ventricular systole (because this is when the valve should be closed, and the murmur is due to blood leaking through the incompletely closed valve into the atrium). **16.** The periods when all four valves are closed are the isovolumetric contraction phase and the isovolumetric relaxation phase. **17.** Exercise activates the sympathetic nervous system. Sympathetic nervous system activity increases heart rate. It also directly increases ventricular contractility, thereby increasing Josh's stroke volume. **18.** If the heart is beating very rapidly, the amount of time for ventricular filling between contractions is decreased. This decreases the end diastolic volume, decreases the stroke volume, and therefore decreases the cardiac output.

Review Questions 1. a; **2.** c; **3.** b; **4.** c; **5.** b; **6.** b; **7.** c; **8.** d; **9.** b

Clinical Case Study 1. The weak and thready pulse indicates a drop in stroke volume (SV). The pulse is felt as blood is ejected from the heart during ventricular contraction (systole). The weak and thready pulse suggests that less blood is being ejected during each contraction (a lower SV). **2.** An increase in heart rate leads to an increase in cardiac output (recall CO = HR × SV). Mr. Ayers's CO is abnormally low (as shown by his decreasing blood pressure). This is probably due to a decrease in SV. The increase in HR is an attempt to compensate for the decrease in SV in order to maintain CO as close to normal as possible. **3.** In cardiac tamponade, the fluid around the heart compresses the heart and prevents it from fully expanding as it relaxes (diastole). As a result of this restriction, less blood will flow into the heart (ventricular filling). With less blood flowing into the ventricles, the degree of stretch of the heart muscle (preload) will also be reduced. These events lead to a reduction in SV. **4.** Heart sounds are produced by the closing of heart valves during a normal cardiac cycle. When EDV is reduced, there is both reduced SV and reduced force of contraction, leading to slower, quieter valve closure. **5.** The enlarged mediastinum and pericardial effusions suggest that the bleeding is restricted within these compartments. The tear would most likely be located in the proximal portion of the ascending aorta (the part closest to the heart). This part of the aorta is located within the pericardium. (In addition, it is known that back pain can be caused by injury to the descending aorta. Mr. Ayers's back pain suggests that the tear proceeded distally. The descending aortic tear must have been contained within the aortic wall because an uncontained tear in the aortic arch or descending aorta would lead to bleeding in the thoracic and/or abdominal cavities, which was not observed in this case.) **6.** In the face of reduced SV, blood returning to the heart backs up, leading to a rise in venous pressure. This is a key sign of tamponade.

Chapter 18

Check Your Understanding 1. The sympathetic nervous system innervates blood vessels. The sympathetic nerves innervate the tunica media. The effector cells in the tunica media are smooth muscle cells. **2.** When vascular smooth muscle contracts, the diameter of the blood vessel becomes smaller. This is called vasoconstriction. **3.** Elastic arteries play a major role in dampening the pulsatile pressure of heart contractions. Dilation or constriction of arterioles determines blood flow to individual capillary beds. Muscular arteries have the thickest tunica media relative to their lumen size. **4.** If you were doing calf raises, your capillary bed would be in the condition depicted in part (a). The true capillaries would be flushed with blood to ensure that the working calf muscles could receive the needed nutrients and dispose of their metabolic wastes. **5.** Valves prevent blood from flowing backwards in veins. They are formed from folds of the tunica intima. **6.** In the systemic circuit, veins contain more blood than arteries (see Figure 18.5). **7.** Veins have more anastomoses than arteries. **8.** The three factors that determine resistance are blood viscosity, vessel length, and vessel diameter. Vessel diameter is physiologically most important. **9.** The rate of flow will decrease 81-fold from its original flow (3 × 3 × 3 × 3 = 81). **10.** Cole's pulse pressure is 60 mm Hg. His mean arterial pressure is 80 + 60/3 = 100 mm Hg. **11.** When you first stand up, mean arterial pressure (MAP) temporarily decreases and this is sensed by aortic and carotid baroreceptors. Medullary cardiac and vasomotor center reflexes increase sympathetic and decrease parasympathetic outflow to the heart. Heart rate and contractility increase, increasing cardiac output, and therefore MAP. Further, sympathetic constriction of arterioles increases peripheral resistance, also increasing MAP. (In addition, increased constriction of veins increases venous return, which increases end diastolic volume, increasing stroke volume, and therefore cardiac output and MAP.) See also Figure 18.10 (bottom). **12.** The kidneys help maintain MAP by influencing blood volume. In renal artery obstruction, the blood pressure in the kidney is lower than in the rest of the body (because it is downstream of the obstruction). Low renal blood pressure triggers both direct and indirect renal mechanisms to increase blood pressure by increasing blood volume. This can cause hypertension (called "secondary hypertension" because it is secondary to a defined cause—in this case the renal artery obstruction). **13.** Bob is in vascular shock due to anaphylaxis, a systemic allergic reaction to his medication. His blood pressure is low because of widespread vasodilation triggered by the massive release of histamine. Bob's rapid heart rate is a result of the baroreceptor reflex triggered by his low blood pressure. This activates the sympathetic nervous system, increasing heart rate, in an attempt to restore blood pressure. **14.** Diuretics cause a decrease in blood volume (because more fluid is lost in urine), decreasing cardiac output, which decreases blood pressure. Beta-blockers block beta-adrenergic receptors. Their antihypertensive effects are primarily due to their action in the heart, where they decrease contractility (and therefore stroke volume) and heart rate, resulting in lower cardiac output. Ca^{2+} channel blockers decrease Ca^{2+} entry into arteriolar smooth muscle, decreasing its contraction (causing vasodilation) and resulting in decreased peripheral resistance. By acting on Ca^{2+} channels in the heart, these drugs may also decrease contractility and heart rate. **15.** In a bicycle race, autoregulation by intrinsic metabolic controls causes arteriolar smooth muscle in your legs to relax, dilating the vessels and supplying more O_2 and nutrients to the exercising muscles. **16.** Extrinsic mechanisms, primarily the sympathetic nervous system, prevent blood pressure from plummeting by constricting arterioles elsewhere (such as the gut and kidneys). In addition, cardiac output increases, which also helps maintain MAP. **17.** (a) An increase in interstitial fluid osmotic pressure (OP_{if}) would tend to pull more fluid out of capillaries (causing localized swelling, or edema). (b) An increase of OP_{if} to 10 mm Hg would increase the outward pressure on both the arteriolar and venous ends of the capillary. The NFP at the venous end would become 1 mm Hg (27 mm Hg − 26 mm Hg). (c) Fluid would flow out of the venous end of the capillary rather than in. **18.** The external carotid arteries supply most of the tissues of the head except for the brain and orbits. **19.** The cerebral arterial circle (circle of Willis) is the arterial anastomosis at the base of the cerebrum. **20.** The four unpaired arteries that emerge from the abdominal aorta are the celiac trunk, the superior and inferior mesenteric arteries, and the median sacral artery. **21.** You would palpate the popliteal artery behind the knee, the posterior tibial artery behind the medial malleolus of the tibia, and the dorsalis pedis artery on the foot. (See also Figure 18.7.) **22.** The vertebral arteries help supply the brain, but the vertebral veins do not drain much blood from the brain. **23.** The internal jugular veins drain the dural venous sinuses. Each internal jugular vein joins a subclavian vein to form a brachiocephalic vein. **24.** A portal system is a system where two capillary beds occur in series. In other words, in a portal system, a capillary bed is drained by a vein that leads into a second capillary bed. The function of the hepatic portal system is to transport venous blood from the digestive organs to the liver for processing before it enters the rest of the systemic circulation. This plays an important role in defense against absorbed toxins or microorganisms and also allows direct delivery of absorbed nutrients to the liver for processing. **25.** The leg veins that often become varicosed are the great and small saphenous veins.

Review Questions 1. d; **2.** b; **3.** d; **4.** c; **5.** e; **6.** d; **7.** c; **8.** b; **9.** b; **10.** a; **11.** (1)b, e, g; (2)c; (3)i; (4)f, h; (5)d; **12.** b; **13.** c; **14.** d

Clinical Case Study 1. The tissues in Mr. Hutchinson's right leg were deprived of oxygen and nutrients for at least one-half hour. When tissues are deprived of oxygen, tissue metabolism decreases and eventually ceases, so these tissues may have died due to anoxia. **2.** Mr. Hutchinson's vital signs (low BP; rapid, thready pulse) indicate that he is facing a life-threatening problem that must be stabilized before other, less vital problems can be addressed. As for surgery, he may be scheduled for open reduction of his crushed bone, depending upon the condition of the tissues in his crushed right leg. If tissue death has occurred in his leg, he may undergo amputation of that limb. **3.** Mr. Hutchinson's rapid, thready pulse and falling blood pressure are indications of hypovolemic shock, a type of shock resulting from decreased blood volume. Because his blood volume is low, his heart rate is elevated to increase cardiac output in an effort to maintain the blood supply to his vital organs. Mr. Hutchinson's blood volume must be increased as quickly as possible with blood transfusions or intravenous saline. This will stabilize his condition and allow his physicians to continue with his surgery.

Chapter 19

Check Your Understanding 1. Lymph is the fluid inside lymphatic vessels. It enters lymphatic vessels from interstitial fluid. Interstitial fluid, in turn, is a filtrate of blood plasma. **2.** The right lymphatic duct receives lymph from the right upper arm and the right side of the head and thorax. The thoracic duct drains lymph from the rest of the body. **3.** Lymph movement is driven by the contraction of adjacent skeletal muscles, pressure changes in the thorax during breathing, the pulsations of nearby arteries, and contraction of smooth muscle in the lymphatic vessel walls. (Valves in lymphatic vessels prevent backflow of lymph.) **4.** The blocked lymphatic vessels in Mr. Thomas's left leg no longer return excess tissue fluid or proteins to the bloodstream. The lack of fluid return from the tissues into the blood will result in a higher interstitial fluid hydrostatic pressure. This is an inward pressure, and will tend to help limit the edema. However, the lack of proteins being returned from the interstitial fluid to the blood will result in a higher interstitial fluid osmotic pressure, which will tend to pull more fluid into the tissue. **5.** The primary lymphoid organs are the red bone marrow and thymus. Primary lymphoid organs are special because they are the organs where lymphocytes originate and mature. **6.** Lymphoid follicles are solid, spherical bodies consisting of tightly packed reticular fibers and lymphoid cells, often with a lighter-staining central region. They are regions where B cells predominate. **7.** Having fewer efferents causes lymph to accumulate in lymph nodes, allowing more time for its cleansing. **8.** The spleen cleanses the blood, recycles breakdown products of RBCs, stores iron, stores platelets and monocytes, and is thought to be a site of erythrocyte production in the fetus. **9.** MALT (mucosa-associated lymphoid tissue) is lymphoid tissue found in the mucosa of the digestive, respiratory, and genitourinary tracts. It includes tonsils, Peyer's patches, and the appendix. **10.** B cells mature in the bone marrow.

Review Questions 1. c; **2.** c; **3.** a, d; **4.** c; **5.** a; **6.** b; **7.** a; **8.** b; **9.** d

Clinical Case Study 1. The red streaks radiating from Mr. Hutchinson's finger indicate that his lymphatic vessels are inflamed. This inflammation may be caused by a bacterial infection. If Mr. Hutchinson's arm had exhibited edema without any accompanying red streaks, the problem would likely have been impaired lymph transport from his arm back to his trunk, due to injury or blockage of his lymphatic vessels. **2.** Mr. Hutchinson's arm was placed in a sling to immobilize it, slowing the drainage of lymph from the infected area in an attempt to limit the spread of the infection. **3.** Mr. Hutchinson's low lymphocyte count indicates that his body's ability to fight infection by bacteria or viruses is impaired. The antibiotics and additional staff protection will protect Mr. Hutchinson until his lymphocyte count increases again. Gloving and gowning also protect the staff caring for Mr. Hutchinson from any body infection he might have. **4.** Mr. Hutchinson's recovery may be problematic, as he probably already has an ongoing bacterial infection and his ability to raise a defense to this infection is impaired.

Chapter 20

Check Your Understanding 1. The innate defense system is always ready to respond immediately, whereas it takes considerable time to mount the adaptive defense system. The innate defenses consist of surface barriers and internal defenses, whereas the adaptive defenses consist of humoral and cellular immunity, which rely on B and T lymphocytes. **2.** Surface barriers (the skin and mucous membranes) constitute the first line of defense. **3.** Opsonization is the process of making pathogens more susceptible to phagocytosis by decorating their surface with molecules that phagocytes can bind. Antibodies and complement proteins are examples of molecules that act as opsonins. **4.** Our own cells are killed by NK cells when they have been infected by viruses or when they have become cancerous. **5.** Redness, heat, swelling, and pain are the cardinal signs of inflammation. Redness and local heat are both caused by vasodilation of arterioles, which increases the flow of blood (warmed by the body core) to the affected area. The swelling (edema) is due to the release of histamine and other chemical mediators of inflammation, which increase capillary permeability. This increased permeability allows proteins to leak into the interstitial fluid (IF), increasing the IF osmotic pressure and drawing more fluid out of blood vessels and into the tissues, thereby causing swelling. The pain is due to two things: (1) the actions of certain chemical mediators (kinins and prostaglandins) on nerve endings, and (2) the swelling, which can compress free nerve endings. **6.** Three key characteristics of adaptive immunity are that it is specific, it is systemic, and it has memory. **7.** A complete antigen has both immunogenicity and reactivity, whereas a hapten has reactivity but not immunogenicity. **8.** Self-antigens, particularly MHC proteins, mark a cell as self. **9.** Development of immunocompetence of a B or T cell is signaled by the appearance on its surface of specific and unique receptors for an antigen. In the case of a B cell, this receptor is a membrane-bound antibody. (In T cells, it is simply called the T cell receptor.) **10.** The T cell that would survive is (c), one that recognizes MHC but not self-antigen. **11.** Dendritic cells, macrophages, and B cells can all act as APCs. Dendritic cells are most important for T cell activation. **12.** In clonal selection, the antigen does the selecting. What is being selected is a particular clone of B or T cells that has antigen receptors corresponding to that antigen. **13.** The secondary response to an antigen is faster than the primary response because the immune system has already been "primed" and has memory cells that are specific for that particular antigen. **14.** Vaccinations protect by providing the initial encounter to an antigen—the primary response to that antigen. As a result, when the pathogen for that illness is encountered again, the pathogen elicits the much faster, more powerful secondary response, which is generally effective enough to prevent clinical illness. **15.** IgG antibody is most abundant in blood. IgM is secreted first in a primary immune response. IgA is most abundant in secretions. **16.** Antibodies can bring about destruction of pathogen via "PLAN"—**p**hagocytosis, **l**ysis (via complement), **a**gglutination, or **n**eutralization. **17.** Plasma cells make large amounts of antibodies—proteins that are exported from the cell. Rough endoplasmic reticulum is the site where proteins that are exported are synthesized. Ribosomes, the Golgi apparatus, and secretory vesicles are also required for protein synthesis and export, and so would also be abundant in these cells. **18.** Class II MHC proteins display exogenous antigens. Class II MHC proteins are recognized by CD4 T cells (which usually become helper T cells). APCs display class II MHC proteins. **19.** Helper T cells are central to both humoral and cellular immunity because they are required for activation of both cytotoxic T cells and most B cells. **20.** The cytotoxic T cell releases perforins and granzymes onto the identified target cell. Perforins form a pore in the target cell membrane, and granzymes enter through this pore, activating enzymes that trigger apoptosis (cell suicide). **21.** HIV is particularly hard for the immune system to defeat because (1) it destroys helper T cells, which are key players in adaptive immunity and (2) it has a high mutation rate and so it rapidly becomes resistant to drugs. **22.** Binding of an allergen onto specific IgE antibodies attached to mast cells triggers the mast cells to release histamine.

Review Questions 1. c; **2.** a; **3.** d; **4.** d, e; **5.** a; **6.** d; **7.** b; **8.** c; **9.** d; **10.** d; **11.** d; **12.** (1)b, g; (2)d, i; (3)a, e; (4)a, e, f, h; (5)e, h; (6)c, f, g

Clinical Case Study 1. The two kidneys transplanted in this case represent allografts. **2.** A major histocompatibility complex (MHC) protein is a type of cell surface protein that the human body uses to recognize *self* and to help coordinate the recognition of *nonself*, or foreign, antigens. These proteins are involved in the display of antigens to T cells. **3.** Class I MHC proteins are found on virtually all of the body's cells, while class II MHCs are found only on antigen-presenting cells (APCs). Class I MHCs display antigens for recognition by CD8 T cells (including cytotoxic T cells). Class II MHCs display antigens for recognition by CD4 T cells (including helper T cells). **4.** A foreign MHC protein will provoke an immune response, so the donor's and recipient's MHCs must match as closely as possible to minimize such an attack. Tissue typing dramatically reduces the risk of organ rejection due to attack by the recipient's immune system. **5.** In this case, the donor and recipients were not genetically identical. Even with very careful tissue typing and compatibility testing, there will still be some differences that the recipients' immune systems will recognize as foreign. To reduce the risk of organ rejection, the recipients are given drugs to suppress their immune systems.

Chapter 21

Check Your Understanding 1. The structures that air passes by are the nasal cavity (nares, nasal vestibule, nasal conchae), nasopharynx (with pharyngeal tonsil), oropharynx (with palatine tonsil), and laryngopharynx. **2.** The pharyngeal tonsil is in the nasopharynx. **3.** The epiglottis seals the larynx when we swallow. **4.** The incomplete, C-shaped cartilage rings of the trachea allow it to expand and contract and yet keep it from collapsing. **5.** The many tiny alveoli together have a large surface area. This and the thinness of their respiratory membranes make them ideal for gas exchange. **6.** The peanut was most likely in the right main bronchus because it is wider and more vertical than the left. **7.** The two circulations of the lungs are the pulmonary circulation, which delivers deoxygenated blood to the lungs for oxygenation and returns oxygenated blood to the heart, and the bronchial circulation, which provides systemic (oxygenated) blood to lung tissue. **8.** Angiotensin converting enzyme is found in the plasma membrane of lung capillary endothelial cells. This is a good location for it because all of the blood in the body passes through the lung capillaries about once every minute. Angiotensin converting enzyme is part of the renin-angiotensin-aldosterone hormone cascade, which increases blood pressure. **9.** The driving force for pulmonary ventilation is a pressure gradient created by changes in the thoracic volume. **10.** The intrapulmonary pressure decreases during inspiration because of the increase in thoracic cavity volume brought about by the muscles of inspiration. **11.** The partial vacuum (negative pressure) inside the pleural cavity is caused by the opposing forces acting on the visceral and parietal pleurae. The visceral pleurae are pulled inward by the lungs' natural tendency to recoil and the surface tension of the alveolar fluid. The parietal pleurae are pulled outward by the elasticity of the chest wall. If air enters the pleural cavity, the lung on that side will collapse. This condition is called pneumothorax. **12.** A lack of surfactant increases surface tension in the alveoli and causes them to collapse between breaths. (In other words, it markedly decreases lung compliance.) **13.** Slow, deep breaths ventilate the alveoli more effectively because a smaller fraction of the tidal volume of each breath is spent moving air into and out of the dead space. **14.** Respiratory capacities are combinations of two or more respiratory volumes. **15.** In a sealed container, the air and water would be at equilibrium. Therefore, the partial pressures of CO_2 and O_2 (P_{CO_2} and P_{O_2}) will be the same in the water as in the air: 100 mm Hg each. More CO_2 than O_2 molecules will be dissolved in the water (even though they are at the same partial pressure) because CO_2 is much more soluble than O_2 in water. **16.** The difference in P_{O_2} between inspired air and alveolar air can be explained by (1) the gas exchange occurring in the lungs (O_2 continuously diffuses out of the alveoli into the blood), (2) the humidification of inspired air (which adds water molecules that dilute the O_2 molecules), and (3) the mixing of newly inspired air with gases already present in the alveoli. **17.** The arterioles leading into the O_2-enriched alveoli would be dilated. This response allows matching of blood flow to availability of oxygen. **18.** About 70% of CO_2 is transported as bicarbonate ion (HCO_3^-) in plasma. Just over 20% is transported bound to hemoglobin in the RBCs, and 7–10% is dissolved in plasma. **19.** As blood CO_2 increases, blood pH decreases. This is because CO_2 combines with water to form carbonic acid. (However, the change in pH in blood for a given increase in CO_2 is minimized by other buffer systems.) **20.** The shift shown in graph (b) would allow more oxygen delivery to the tissues. Conditions that would cause the curve to shift this way are increased temperature, increased P_{CO_2}, decreased pH, or an increase in BPG levels. **21.** The ventral respiratory group of the medulla (VRG) is thought to be the rhythm-generating area. **22.** CO_2 in blood normally provides the most powerful stimulus to breathe. Central chemoreceptors are most important in this response (see Figure 21.26). **23.** The injured soccer player's P_{CO_2} is low. (Recall that normal P_{CO_2} = 40 mm Hg.) The low P_{CO_2} reveals that this is hyperventilation and not hyperpnea (which is not accompanied by changes in blood CO_2 levels). **24.** Long-term adjustments to altitude include an increase in erythropoiesis, resulting in a higher hematocrit; an increase in BPG, which decreases Hb affinity for oxygen; and an increase in minute respiratory volume. **25.** The obstruction in asthma is reversible, and acute exacerbations are typically followed by symptom-free periods. In contrast, the obstruction in chronic bronchitis is generally not reversible.

Review Questions 1. b; **2.** a and c; **3.** c; **4.** c; **5.** b; **6.** d; **7.** d; **8.** b; **9.** c, d; **10.** c; **11.** b; **12.** b; **13.** b; **14.** c; **15.** b; **16.** b

Clinical Case Study 1. Spinal cord injury from a fracture at the level of the C_2 vertebra would interrupt the normal transmission of signals from the brain stem down the phrenic nerve to the diaphragm, and Barbara would be unable to breathe due to paralysis of the diaphragm. **2.** Barbara's head, neck, and torso should have been immobilized to prevent further damage to the spinal cord. In addition, she required assistance to breathe, so her airway was probably intubated to permit ventilation of her lungs. **3.** Cyanosis is a decrease in the degree of oxygen saturation of hemoglobin. As Barbara's respiratory efforts cease, her alveolar P_{O_2} will fall, so there is less oxygen to load onto hemoglobin. In her peripheral tissues, what little oxygen hemoglobin carries will be consumed, leaving these tissues with a bluish tinge. **4.** Injury to the spinal cord at the level of the C_2 vertebra will cause quadriplegia (paralysis of all four limbs). **5.** Atelectasis is the collapse of a lung. Because it is the right thorax that is compressed, only her right lung is affected. Because the lungs are in separate pleural cavities, only the right lung collapsed. **6.** Barbara's fractured ribs probably punctured her lung tissue and allowed air within the lung to enter the pleural cavity. **7.** The atelectasis will be reversed by inserting a chest tube and removing the air from the pleural cavity. This will allow her lung to heal and reinflate.

Chapter 22

Check Your Understanding 1. The esophagus is found in the thorax. Three alimentary canal organs found in the abdominal cavity include the stomach, small intestine, and large intestine. **2.** The usual site of ingestion is the mouth. **3.** The process of absorption moves nutrients into the body. **4.** The visceral peritoneum is the outermost layer of the digestive organ; the parietal peritoneum is the serous membrane covering the wall of the abdominal cavity. **5.** The pancreas is retroperitoneal. **6.** From deep to superficial, the layers of the alimentary canal are the mucosa, submucosa, muscularis externa, and serosa. **7.** The hepatic portal circulation is the venous portion of the splanchnic circulation. **8.** The muscularis externa is unitary smooth muscle. Characteristics of unitary smooth muscle that make it well suited for this location are that it is electrically coupled by gap junctions and so contracts as a unit, is arranged in longitudinal and circular sheets, and exhibits rhythmic spontaneous action potentials. **9.** Reflexes associated with the GI tract promote muscle contraction and secretion of digestive juices or hormones. **10.** The term "gut brain" refers to the enteric nervous system, the web of neurons closely associated with the digestive organs. **11.** He should temporarily refrain from eating because the parasympathetic nervous system oversees digestive activities. **12.** The palate forms the roof of the mouth. The hard palate supported by bone is anterior to the soft palate (no bony support). **13.** The tongue is important for taste and for speech, particularly for uttering consonants. **14.** Antimicrobial substances found in saliva include lysozyme, defensins, and IgA antibodies. **15.** Enamel is harder than bone. Pulp consists of nervous tissue and blood vessels. **16.** The pharynx is part of the digestive and respiratory systems. **17.** The esophageal muscularis externa undergoes a transformation along its length from skeletal muscle superiorly to smooth muscle near the stomach. **18.** The esophagus is merely a chute for food passage and is subjected to a good deal of abrasion, which a stratified squamous epithelium can withstand. The stomach mucosa is a secretory mucosa served well by a simple columnar epithelium. **19.** The tongue mixes the chewed food with saliva, compacts the food into a bolus, and initiates swallowing. **20.** During swallowing the larynx rises and the epiglottis covers its lumen so that foodstuffs are diverted into the esophagus posteriorly. **21.** The stomach has an additional layer of smooth muscle—the oblique layer. The oblique layer allows the stomach to pummel food in addition to making its peristaltic movements. **22.** The chief cells produce pepsinogen, which is the inactive enzyme pepsin, and the parietal cells secrete HCl needed to activate pepsinogen. **23.** The three phases of gastric secretion are the cephalic, gastric, and intestinal phases. **24.** The presence of food in the duodenum inhibits gastric activity by triggering the enterogastric reflex and the secretion of enterogastrones (hormones). **25.** A portal triad is a region at the corner of a hepatic lobule that contains a branch of the hepatic portal vein, a branch of the hepatic artery proper, and a bile duct. **26.** The enterohepatic circulation is an important recycling mechanism for retaining bile salts needed for fat absorption. **27.** Pancreatic acini produce the exocrine products of the pancreas (digestive enzymes and bicarbonate-rich juice). The islets produce pancreatic hormones, most importantly insulin and glucagon. **28.** Fluid in the pancreatic duct is bicarbonate-rich, enzyme-rich pancreatic juice. Fluid in the cystic and bile ducts is bile. **29.** CCK is secreted in response to the entry into the duodenum of chyme rich in protein and fat. It causes the pancreatic acini to secrete digestive enzymes, stimulates the gallbladder to contract, and relaxes the hepatopancreatic sphincter. **30.** All of these modifications increase the surface area of the small intestine. **31.** Brush border enzymes are enzymes associated with the microvilli of the small intestine mucosal cells. **32.** Distension of stomach walls enhances stomach secretory activity. Distension of the walls of the small intestine reduces stomach secretory activity (to give the small intestine time to carry out its digestive and absorptive activities). **33.** Segmentation is the most common motility pattern after a meal. Segmentation mixes chyme with digestive enzymes and exposes the products of this digestion to the absorptive epithelium where brush border enzymes complete digestion and where absorption occurs. **34.** MMC is the migrating motor complex, a pattern of peristalsis seen in the small intestine that moves the last remnants of a meal plus bacteria and other debris into the large intestine. MMC is important to prevent the overgrowth of bacteria in the small intestine. **35.** Mass movements and haustral contractions occur in the large intestine. Mass movements are long, slow, powerful contractions that move over large areas of the colon three or four times a day, forcing the contents toward the rectum. Haustral contractions are a special type of segmentation. **36.** Activation of stretch receptors in the rectal wall initiates the defecation reflex. **37.** Enteric bacteria synthesize B vitamins and some of the vitamin K the liver needs to synthesize clotting proteins. **38.** To be absorbed, nutrients pass through the apical and then the basolateral membranes of absorptive epithelial cells. **39.** Capillaries that receive absorbed nutrients are in the lamina propria of the mucosa of the small intestine. **40.** Amylase is to starch as lipase is to fats. **41.** Bile salts emulsify fats so that they can be acted on efficiently by lipase enzymes, and form micelles that aid fat absorption.

Review Questions 1. c; **2.** d; **3.** d; **4.** b; **5.** b; **6.** a; **7.** d; **8.** d; **9.** b; **10.** c; **11.** c; **12.** a; **13.** d; **14.** d; **15.** b; **16.** c; **17.** a

Clinical Case Study 1. Mr. Gutteman's statement about the effects of milk on his digestive tract suggests that he may be deficient in lactase, a brush border enzyme that breaks down lactose (milk sugar). **2.** His responses to the questions reduced the possibility that he has gastric ulcers. Mr. Gutteman's diarrhea may be due to gluten-sensitive enteropathy. To verify this diagnosis, Mr. Gutteman should be screened for specific IgA antibodies in his blood. If these screening tests are positive, a biopsy of the intestinal mucosa would be performed. Observation of damaged intestinal villi and microvilli would confirm the diagnosis. A positive diagnosis of gluten-sensitive enteropathy would lead to a lifelong dietary restriction on all grains except rice and corn. Grains should not be restricted prior to the biopsy.

Chapter 23

Check Your Understanding 1. The five major nutrients are carbohydrates, proteins, fats, minerals, and vitamins. **2.** Cellulose provides fiber, which helps in elimination. **3.** Triglycerides are used for ATP synthesis, body insulation and protective padding, and to help the body absorb fat-soluble vitamins. Cholesterol is the basis of our steroid hormones and bile salts, and stabilizes cellular membranes. **4.** Beans (legumes) and grains (wheat) are good sources of protein, but neither is a complete one. However, together they provide all the essential amino acids. **5.** Vitamins serve as the basis for coenzymes, which work with enzymes to accomplish metabolic reactions. **6.** Iodine is essential for thyroxine synthesis. Calcium in the form of bone salts is needed to make bones hard. Iron is needed to make functional hemoglobin. **7.** Vitamin B_{12} needs intrinsic factor synthesized by parietal cells to be absorbed by the intestine. Vitamin B_{12} is absorbed in the ileum, and its lack causes pernicious anemia. **8.** A redox reaction is a combination of an oxidation and a reduction reaction. As one substance is oxidized, another is reduced. **9.** Some of the energy released during catabolism is captured in the bonds of ATP, which provides the energy needed to carry out the constructive activities of anabolism. **10.** The energy released during the oxidation of food fuels is used to pump protons across the inner mitochondrial membrane. **11.** In substrate-level phosphorylation, high-energy phosphate groups are transferred directly from phosphorylated intermediates to ADP to form ATP. In oxidative phosphorylation, electron transport proteins forming part of the inner mitochondrial membrane use energy released during oxidation of glucose to create a steep gradient for protons across this membrane. Then as protons flow back through the membrane, gradient energy is captured to attach phosphate to ADP. **12.** If oxygen is not available, glycolysis will stop because the supply of NAD^+ is limited and glycolysis can continue only if the reduced coenzymes ($NADH + H^+$) formed during glycolysis are relieved of their extra hydrogen. **13.** Oxidation (via removal of H) is common in the citric acid cycle; it is indicated by the reduction of a coenzyme (either NAD^+ or FAD). Decarboxylations are also common, and are indicated by the removal of CO_2 from the cycle. **14.** Glycogenolysis is the reaction in which glycogen is broken down to its glucose monomers. **15.** Glycerol, a breakdown product of fat metabolism, enters glycolysis. **16.** Acetyl CoA is the central molecule of fat metabolism. **17.** The products of beta oxidation are acetyl CoA (acetic acid + coenzyme A), $NADH^+ + H^+$, and $FADH_2$. **18.** The liver uses keto acids drained off the citric acid cycle and amino groups (from other nonessential amino acids) as substrates to make the nonessential amino acids that the body needs. **19.** The ammonia removed from amino acids is combined with carbon dioxide to form urea, which is then eliminated by the kidneys. **20.** The three organs or tissues that regulate the directions of interconversions in the nutrient pools are the liver, skeletal muscles, and adipose tissues. **21.** Anabolic reactions and energy storage typify the absorptive state. Catabolic reactions (to increase blood sugar levels) such as lipolysis and glycogenolysis, and glucose sparing occur in the postabsorptive state. **22.** The main antagonist of glucagon is insulin. **23.** A rise in amino acid levels in blood increases both insulin and glucagon release. **24.** High HDLs would be preferable because the cholesterol these particles transport is destined for the liver and elimination from the body. **25.** Trans fats are oils that have been hydrogenated (with H atoms). They are unhealthy because they cause LDLs to increase and HDLs to decrease—exactly the opposite of what is desirable. **26.** Among short-term stimuli influencing feeding behavior are neural signals from the digestive tract, nutrient signals related to energy stores, and GI tract hormones (CCK, insulin, glucagon, and ghrelin). **27.** Leptin is the most important long-term regulator of feeding behavior. **28.** Of the factors listed, breathing and kidney function contribute to BMR. **29.** Samantha is suffering from hypersecretion of thyroid hormone, which makes her feel restless and warm (because it raises her BMR), and gives her insomnia. The presence of autoantibodies suggests that she has Graves' disease, a condition in which antibodies bind to receptors for TSH (mimicking TSH) and stimulate continuous thyroid hormone release. **30.** The body's core is the organs within the skull and the thoracic and abdominal cavities. **31.** Andrea's body temperature is rising as heat-promoting mechanisms (shivering, chills) are activated. Something (an infection?) has caused the hypothalamic thermostat to be set to a higher level (fever) temporarily. **32.** In conduction, heat is transferred directly from one object to another (a hot surface to your palm). In convection, air warmed by body heat is continually removed (warm air rises) and replaced by cooler air (cool air falls), which in turn will absorb heat radiating from the body.

Review Questions 1. a; **2.** c; **3.** b; **4.** d; **5.** b; **6.** c; **7.** a; **8.** d; **9.** d; **10.** a; **11.** b; **12.** d; **13.** c; **14.** d; **15.** a

Clinical Case Study 1. Blood pH: Low (normal: 7.35–7.45); Blood glucose: High (normal: 70–120 mg/dl); Blood ketone bodies: High (normally negative); Urine pH: Low (normal: 4.5–8.0) **2.** Ketone bodies have an acetone smell that can be detected in the urine and the breath. Mr. Boulard is producing abnormal amounts of ketone bodies, which are accumulating in his blood. These substances are being excreted into the urine and diffusing out of the lungs and into the exhaled air. **3.** Ketogenesis, or the production of ketone bodies, is a process that occurs primarily in hepatocytes (liver cells) when carbohydrates are unavailable as an energy source. In this case, the hepatocytes metabolize fats. The fatty acids produced from fat breakdown (lipolysis) are converted to acetyl CoA by beta oxidation. In the absence of glucose, these acetyl CoA molecules cannot enter the citric acid cycle, and hepatocytes convert them into ketone bodies. **4.** Mr. Boulard's elevated blood glucose and his acidotic state indicate diabetes mellitus. In both type 1 and type 2 diabetes, the absence of insulin's actions means that the cells are unable to efficiently take up and utilize glucose from the blood. As a result, blood glucose levels are high, yet the cells are forced to switch to an alternative energy source such as lipids, which results in the production of ketone bodies in a diabetic patient. The stress of alcohol-induced dehydration can trigger ketoacidosis by causing the release of stress hormones that make diabetes worse. **5.** The pH of Mr. Boulard's blood and urine is abnormally low because ketone bodies are acidic (keto acids). As the concentration of ketone bodies in the blood rises, the blood pH falls (acidosis). The kidneys correct the blood pH by moving ketone bodies into the urine, making acidic urine.

Chapter 24

Check Your Understanding 1. The lower part of his rib cage and the perirenal fat capsule protect his kidneys from blows. **2.** The layers of supportive tissue around each kidney are the fibrous capsule, the perirenal fat capsule, and the renal fascia. The parietal peritoneum overlies the anterior renal fascia. **3.** The renal pelvis, which has extensions called calyces, is continuous with the ureter. **4.** Filtrate is formed in the glomerular capsule and then passes through the proximal convoluted tubule (PCT), the descending and ascending limbs of the nephron loop, and the distal convoluted tubule (DCT). **5.** The structural differences are (1) juxtamedullary nephrons have long nephron loops (with long thin segments) and renal corpuscles that are near the cortex-medulla junction, whereas cortical nephrons have short nephron loops and renal corpuscles that lie more superficially in the cortex; (2) efferent arterioles of juxtamedullary nephrons supply vasa recta, while efferent arterioles of cortical nephrons supply peritubular capillaries. **6.** The glomerular capillaries are fenestrated capillaries. (See Figure 18.3 on p. 613 to refresh your memory of capillary types.) Their function is to filter large amounts of plasma into the glomerular capsule. **7.** The structures shown are (a) glomerular capsule, (b) efferent arteriole, (c) vasa recta, (d) proximal convoluted tubule (PCT), (e) collecting duct, and (f) nephron loop (ascending thin limb). **8.** Excretion is a means of removing wastes from the body, whereas tubular secretion is the process of selectively moving substances from the blood into the filtrate. While secretion often leads to excretion, this is not always the case because something could be secreted in the proximal tubules and yet reabsorbed distally. **9.** Intrinsic controls serve to maintain a nearly constant GFR in spite of changes in systemic blood pressure. Extrinsic controls serve to maintain systemic blood pressure. **10.** Net filtration pressure is 5 mm Hg [50 mm Hg − (25 mm Hg + 20 mm Hg)]. **11.** Hydrostatic pressure in the glomerular capillaries (HP_{gc}) is regulated by intrinsic and extrinsic controls of GFR. **12.** The majority of reabsorption occurs in the proximal convoluted tubule. **13.** The reabsorption of Na^+ by primary active transport drives reabsorption of amino acids and glucose by secondary active transport. It also drives passive reabsorption of chloride, and reabsorption of water by osmosis. The reabsorption of water leaves behind other solutes, which become more concentrated and can therefore be reabsorbed by diffusion. **14.** In primary active transport, the energy for the process is provided directly by ATP. In secondary active transport, the energy for the process is provided by the Na^+ concentration gradient established by active pumping of Na^+ occurring elsewhere in the cell. As Na^+ moves down its own concentration gradient, it drives the movement of another substance (e.g., glucose) against that substance's concentration gradient. **15.** H^+, K^+, NH_4^+, creatinine, urea, and uric acid are all substances that are secreted into the kidney tubules. **16.** The descending limb of the nephron loop is permeable to water and impermeable to NaCl. The ascending limb is impermeable to water and permeable to NaCl. **17.** ADH is released from the posterior pituitary in response to hyperosmotic extracellular fluid (as sensed by hypothalamic osmoreceptors). ADH causes insertion of aquaporins into the apical membrane of the principal cells of the collecting ducts. **18.** The normal renal clearance value for amino acids is zero. You would expect this because amino acids are valuable as nutrients and as the building blocks for protein synthesis, so it would not be good to lose them in the urine. **19.** The three major nitrogenous wastes excreted in urine are urea, creatinine, and uric acid. **20.** A kidney stone blocking the ureter would interfere with urine flow to the bladder. The pain would occur in waves that coincide with the peristaltic contractions of the smooth muscle of the ureter. **21.** The trigone is a smooth triangular region at the base of the bladder. Its borders are defined by the openings for the ureters and the urethra. **22.** The prostatic urethra, the intermediate part of the urethra, and the spongy urethra are the three regions of the male urethra. **23.** The detrusor contracts in response to increased firing of parasympathetic nerves. The internal urethral sphincter opens. **24.** In both cases (micturition and defecation) the external sphincters are skeletal muscle under voluntary control, and the internal sphincters are smooth muscle controlled by the autonomic nervous system.

Review Questions 1. d; **2.** b; **3.** c; **4.** d; **5.** c; **6.** b; **7.** a; **8.** c; **9.** a; **10.** b

Clinical Case Study 1. Albumin is the smallest and most abundant plasma protein. More than trace amounts of albumin are not normally found in urine, so its presence indicates damage to the filtration membrane of the nephron. **2.** These medications were prescribed to treat Mr. Boulard's hypertension. Both diabetes and hypertension can cause kidney damage, and hypertension is a major cause of other cardiovascular diseases such as heart failure and stroke. Albuminuria indicates that Mr. Boulard already has damage to his kidneys, so it is important to protect his kidneys from further damage. **3.** Thiazide diuretics increase urine output by inhibiting Na^+ reabsorption in the DCT. This decreases blood volume, which decreases blood pressure. **4.** While Mr. Boulard still has hypertension, his main problem is that his blood K^+ is low. He is losing too much K^+ in his urine. This is the underlying cause of his irregular heartbeat, which could turn into a fatal arrhythmia if his hypokalemia is not corrected. **5.** Thiazide diuretics increase Na^+ excretion and decrease blood pressure. To compensate for these thiazide effects, Mr. Boulard's renin-angiotensin-aldosterone mechanism is activated. Aldosterone increases Na^+ reabsorption and K^+ secretion, resulting in hypokalemia (low blood K^+). **6.** ACE inhibitors decrease blood pressure by blocking the action of angiotensin converting enzyme and so reducing the amount of circulating angiotensin II. Because angiotensin II increases blood pressure in a number of ways, including by increasing aldosterone release and causing vasoconstriction (see Figure 18.11 on p. 623), ACE inhibitors are very effective at lowering blood pressure. (ACE inhibitors also help minimize kidney damage in diabetes.) **7.** Yes. The ACE inhibitors would have prevented the formation of excess angiotensin II and the resulting release of aldosterone. This would have lessened excess secretion of K^+.

Chapter 25

Check Your Understanding 1. You have more intracellular than extracellular fluid and more interstitial fluid than plasma. **2.** Na^+ is the major cation in the ECF and K^+ is the major cation in the ICF. The intracellular counterparts to extracellular Cl^- are HPO_4^{2-} and protein anions. **3.** If you eat salty pretzels, your extracellular fluid volume will expand even if you don't ingest fluids. This is because water will flow by osmosis from the intracellular fluid to the extracellular fluid. **4.** An increase in osmolality of the plasma is most important for triggering thirst. This change is sensed by osmoreceptors in the hypothalamus. **5.** ADH cannot add water—it can only conserve what is already there. In order to reduce an increase in osmolality of body fluids, the thirst mechanism is required. **6.** (a) A loss of plasma proteins causes edema. (b) Copious sweating causes dehydration. (c) Using ecstasy (together with drinking lots of fluids) could cause hypotonic hydration because ecstasy promotes ADH secretion, which interferes with the body's ability to get rid of extra water. **7.** Insufficient aldosterone would cause Nathan's plasma Na^+ to be decreased and his plasma K^+ to be elevated. The decrease in plasma Na^+ would cause a decrease in blood pressure, because plasma Na^+ is directly related to blood volume, which is a major determinant of blood pressure. **8.** Arrow 2 in the proximal convoluted tubule shows the site where most Ca^{2+} is reabsorbed. Arrow 5 corresponds to Na^+ reabsorption in the collecting duct, which is regulated by aldosterone. Arrow 4 corresponds to K^+ secretion in the collecting duct, which is regulated by aldosterone. **9.** The major regulator of calcium in the blood is parathyroid hormone. Hypercalcemia decreases excitability of neurons and muscle cells and may cause life-threatening cardiac arrhythmias. Hypocalcemia increases excitability and causes muscle tetany. **10.** Acidemia is an arterial pH below 7.35 and alkalemia is a pH above 7.45. **11.** The three major chemical buffer systems of the body are the bicarbonate buffer system, the phosphate buffer system, and the protein buffer system. The most important intracellular buffer is the protein buffer system. **12.** Joanne's ventilation would be increased. The acidosis caused by the accumulated ketone bodies will stimulate the peripheral chemoreceptors, and this will cause more CO_2 to be "blown off" in an attempt to restore pH to normal. **13.** Reabsorption of HCO_3^- is always linked with secretion of H^+. **14.** The most important urine buffer of H^+ is the phosphate buffer system. **15.** The tubule and collecting duct cells generate new HCO_3^- either by excreting ammonium ions (NH_4^+) or by excreting buffered H^+ ions. **16.** In both renal tubule cells and parietal cells, (a) the enzyme carbonic anhydrase is key, and (b) blood concentration of HCO_3^- increases. **17.** Key features of an uncompensated metabolic alkalosis are an increase in blood pH and an increase in blood HCO_3^-. Key features of an uncompensated respiratory acidosis are a decrease in blood pH and an increase in blood P_{CO_2}. **18.** The kidneys compensate for respiratory acidosis by excreting more H^+ and generating new HCO_3^- to buffer the acidosis.

Review Questions 1. a; **2.** c; **3.** b; **4.** a and b; **5.** h, i; **6.** c, g; **7.** a, e; **8.** b; **9.** a, b; **10.** j; **11.** b, d; **12.** a; **13.** c

Clinical Case Study 1. Mr. Heyden's vital signs suggest that he is in hypovolemic shock, which is probably due to an internal hemorrhage. **2.** The spleen is a highly vascular organ due to its role as a blood-filtering organ. The macrophages in Mr. Heyden's liver and bone marrow will help compensate for the loss of his spleen. **3.** Elevation of renin, aldosterone, and antidiuretic hormone indicate that Mr. Heyden's body is trying to compensate for his falling blood pressure and blood loss.

Renin: released when renal blood flow is diminished and blood pressure falls. The angiotensin-aldosterone response is initiated by renin. The formation of angiotensin II leads to vasoconstriction, which will increase blood pressure, and to the release of aldosterone. Aldosterone: increases Na^+ reabsorption by the kidney. The movement of this reabsorbed Na^+ into the bloodstream will promote the movement of water from the interstitial fluid, resulting in an increase in blood volume. Antidiuretic hormone (ADH): released when the hypothalamic osmoreceptors sense an increase in osmolality. ADH has two consequences: It is a potent vasoconstrictor and will increase blood pressure, and it promotes water retention by the kidney, increasing blood volume. **4.** (a) Mr. Heyden's urine production may be decreased for several reasons. The severe drop in his blood pressure would reduce renal blood flow, thus reducing glomerular blood pressure and decreasing his glomerular filtration rate. The elevation in his ADH levels can reduce urine output due to increased water reabsorption by the kidney. He may also have damage to the kidney due to his crush injury in the left lumbar region. (b) The presence of casts and a brownish-red color in his urine are probably due to damaged cells and blood. If he has suffered kidney damage due to being crushed, he may have nephron damage that would include disruption of the filtration membrane, allowing red blood cells to pass into the filtrate and therefore the urine. He may also have damaged renal tubules and peritubular capillaries, allowing entry of blood and damaged renal tubule epithelial cells into the filtrate.

Chapter 26

Check Your Understanding 1. The testes produce the male gametes (sperm) and testosterone. **2.** The sperm factories are the seminiferous tubules. **3.** When the ambient temperature is cold, the associated muscles contract, bringing the testes close to the warm body wall. When body temperature is high, the associated muscles relax, allowing the testes to hang away from the body wall. The pampiniform venous plexus absorbs heat from the arterial blood, cooling the blood before it enters the testes. **4.** The erectile tissue of the penis allows the penis to become stiff so that it may more efficiently enter the female vagina to deliver sperm. **5.** The male perineum is the external region where the penis and scrotum are suspended. **6.** The organs of the male duct system in order from the epididymis to the body exterior are the ductus deferens, ejaculatory duct, prostatic urethra, intermediate part of the urethra, and spongy urethra. **7.** These stereocilia pass nutrients to the sperm and absorb excess testicular fluid. **8.** The ductus deferens runs from the scrotum into the pelvic cavity. **9.** Arthur probably has a hypertrophied prostate, a condition which can be felt through the anterior wall of the rectum. **10.** Semen is sperm plus the secretions of the male accessory glands. **11.** The bulbo-urethral glands (c) secrete primarily mucus. The seminal glands (a) produce the largest fraction of semen volume. The prostate (b) is at the junction of the ejaculatory duct and urethra. **12.** Erection is the stiffening of the penis that occurs when blood in the cavernous tissue is prevented from leaving the penis. It is caused by the parasympathetic division of the autonomic nervous system. **13.** Resolution is a period of muscular and psychological relaxation that follows orgasm. It results as the sympathetic nervous system causes constriction of the internal pudendal arteries, reducing blood flow to the penis, and activates small muscles that force blood out of the penis. **14.** Meiosis reduces the chromosomal count from $2n$ to n and introduces variability. **15.** The sperm head is the compacted DNA-containing nucleus. The acrosome that caps the head is a lysosome-like sac of enzymes. The midpiece contains the energy-producing mitochondria. The tail, a flagellum fashioned by a centriole, is the propulsive structure. **16.** Sustentocytes provide nutrients and essential development signals to the developing sperm and form the blood testis barrier that prevents sperm antigens from escaping into the blood. Interstitial endocrine cells secrete testosterone. **17.** The HPG axis is the hormonal interrelationship between the hypothalamus, anterior pituitary, and gonads that regulates the production of gametes and sex hormones (e.g., sperm production and testosterone in the male). **18.** Follicle-stimulating hormone indirectly stimulates spermatogenesis by prompting the sustentocytes to secrete androgen-binding protein. Androgen-binding protein keeps the concentration of testosterone high in the vicinity of the spermatogenic cells, which directly stimulates spermatogenesis. **19.** Secondary sex characteristics of males include appearance of pubic, axillary, and facial hair, deepening of the voice, increased oiliness of the skin, and increased size (length and mass) of the bones and skeletal muscles. **20.** The female's internal genitalia include the ovaries and duct system (uterine tubes, uterus, and vagina). **21.** The ovaries produce the female gametes and secrete female sex hormones (estrogens and progesterone). **22.** The antrum is the fluid-filled cavity of a mature follicle. **23.** Women are more at risk for PID than men because the duct system of women is incomplete—there is no physical connection between the ovary and the uterine tubes, which are open to the pelvic cavity. In men, the duct system is continuous from the testes to the body exterior. **24.** The waving action of the fimbriae and currents created by the beating cilia help to direct the ovulated oocytes into the uterine tube. **25.** The usual site of fertilization is the uterine tube. The uterus serves as the incubator for fetal development. **26.** The greater vestibular glands are the female homologue of the male bulbo-urethral glands. **27.** Both the penis and clitoris are hooded by a skin fold and are largely erectile tissue. However, the clitoris lacks a corpus spongiosum containing a urethra, so the urinary and reproductive systems are completely separate in females. **28.** Developmentally, the mammary glands are modified sweat glands. **29.** Breast cancer usually arises from the epithelial cells of the small ducts. **30.** The products of meiosis in females are 3 polar bodies (tiny haploid cells with essentially no cytoplasm) and 1 haploid ovum (functional gamete). Meiosis in males yields 4 functional gametes, the haploid sperm. **31.** Identical twins develop from separation of a very young embryo (the result of fertilization of a single oocyte by a single sperm) into two parts. Fraternal twins develop when different oocytes are fertilized by different sperm. **32.** In the luteal phase, the follicle from which an oocyte has been ovulated develops into a corpus luteum, which then secretes progesterone (and some estrogens). **33.** Leptin is important in advising the brain of the girl's readiness (relative to energy stores) for puberty. **34.** FSH prompts follicle growth and LH prompts ovulation. **35.** Estrogens exert positive feedback on the anterior pituitary that leads to a burstlike release of LH. **36.** Estrogens are responsible for the secondary sex characteristics of females. **37.** Estrogens promote epiphyseal closure in both males and females. Long bones stop growing when the epiphyses are ossified. **38.** The vestibular glands help to lubricate the vestibule. **39.** The human papillomavirus (HPV) is most associated with cervical cancer. **40.** Chlamydia is the most common bacterial sexually transmitted infection in the U.S.

Review Questions 1. a and b; **2.** d; **3.** a; **4.** d; **5.** b; **6.** d; **7.** d; **8.** c; **9.** a, c, e, f; **10.** (1)c, f; (2)e, h; (3)g; (4)a; (5)b, e, and g; (6)f; **11.** a; **12.** c; **13.** d; **14.** b; **15.** b; **16.** c

Clinical Case Study 1. Carcinoma is the term for cancer originating from epithelial tissue. The primary source of Mr. Heyden's cancer is likely to be the prostate. **2.** Elevation of PSA levels suggests carcinoma of the prostate. (In addition, Mr. Heyden's age places him in a group that is at relatively higher risk for this type of cancer.) **3.** Digital examination of Mr. Heyden's prostate should detect the presence of carcinoma in this tissue. In addition, transrectal ultrasound imaging may be used, and biopsies of prostate tissue follow positive results of tests. **4.** Mr. Heyden's carcinoma has advanced to the point of metastasis. He will probably undergo a treatment that reduces the levels of androgens in his body, as androgens promote growth of the prostate-derived tissue. These treatments could include castration or administration of drugs that block the production and/or effects of androgens.

Two Important Metabolic Pathways

① Glucose enters the cell and is phosphorylated by the enzyme hexokinase, which catalyzes the transfer of a phosphate group, indicated as Ⓟ, from ATP to the number six carbon of the sugar, producing glucose-6-phosphate. The electrical charge of the phosphate group traps the sugar in the cell because the plasma membrane is impermeable to ions. Phosphorylation of glucose also makes the molecule more chemically reactive. Although glycolysis is supposed to *produce* ATP, ATP is actually consumed in step 1—an energy investment that will be repaid with dividends later in glycolysis.

② Glucose-6-phosphate is rearranged and converted to its isomer, fructose-6-phosphate. Isomers, remember, have the same number and types of atoms but in different structural arrangements.

③ In this step, still another molecule of ATP is used to add a second phosphate group to the sugar, producing fructose-1,6-bisphosphate. So far, the ATP ledger shows a debit of −2. With phosphate groups on its opposite ends, the sugar is now ready to be split in half.

④ This is the reaction from which glycolysis gets its name. An enzyme cleaves the sugar molecule into two different 3-carbon sugars: glyceraldehyde 3-phosphate and dihydroxyacetone phosphate. These two sugars are isomers of one another.

⑤ An isomerase enzyme interconverts the 3-carbon sugars, and if left alone in a test tube, the reaction reaches equilibrium. This does not happen in the cell, however, because the next enzyme in glycolysis uses only glyceraldehyde 3-phosphate as its substrate and not dihydroxyacetone phosphate. This pulls the equilibrium between the two 3-carbon sugars in the direction of glyceraldehyde 3-phosphate, which is removed as fast as it forms. Thus, the net result of steps 4 and 5 is cleavage of a 6-carbon sugar into two molecules of glyceraldehyde 3-phosphate; each will progress through the remaining steps of glycolysis.

⑥ An enzyme now catalyzes two sequential reactions while it holds glyceraldehyde 3-phosphate in its active site. First, the sugar is oxidized by the transfer of H from the number one carbon of the sugar to NAD^+, forming NADH + H^+. Here we see in metabolic context the oxidation-reduction reaction described in Chapter 23. This reaction releases substantial amounts of energy, and the enzyme capitalizes on this by coupling the reaction to the creation of a high-energy phosphate bond at the number one carbon of the oxidized substrate. The source of the phosphate is inorganic phosphate (P_i) always present in the cytosol. The enzyme releases NADH + H^+ and 1,3-bisphosphoglyceric acid as products. Notice in the figure that the new phosphate bond is symbolized with a squiggle (~), which indicates that the bond is at least as energetic as the high-energy phosphate bonds of ATP.

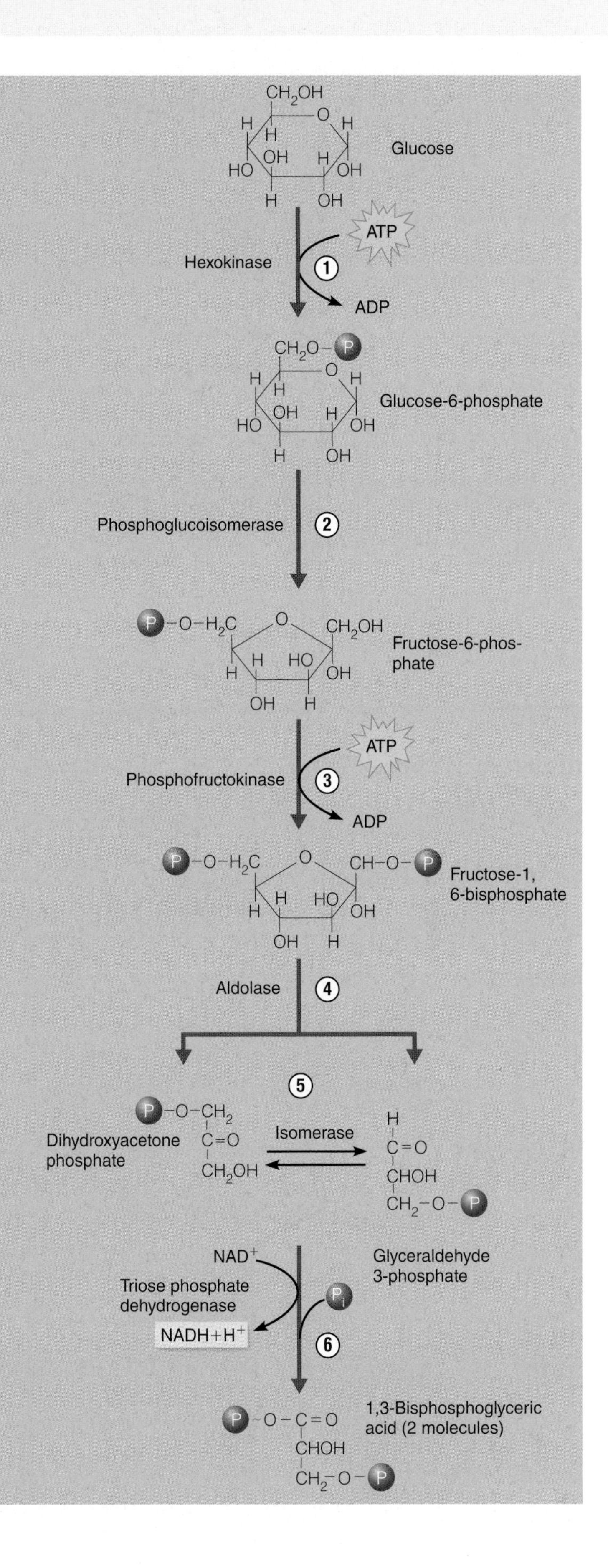

THE TEN STEPS OF GLYCOLYSIS Each of the ten steps of glycolysis is catalyzed by a specific enzyme found dissolved in the cytoplasm. All steps are reversible. An abbreviated version of the three major phases of glycolysis appears in the lower right-hand corner of the next page.

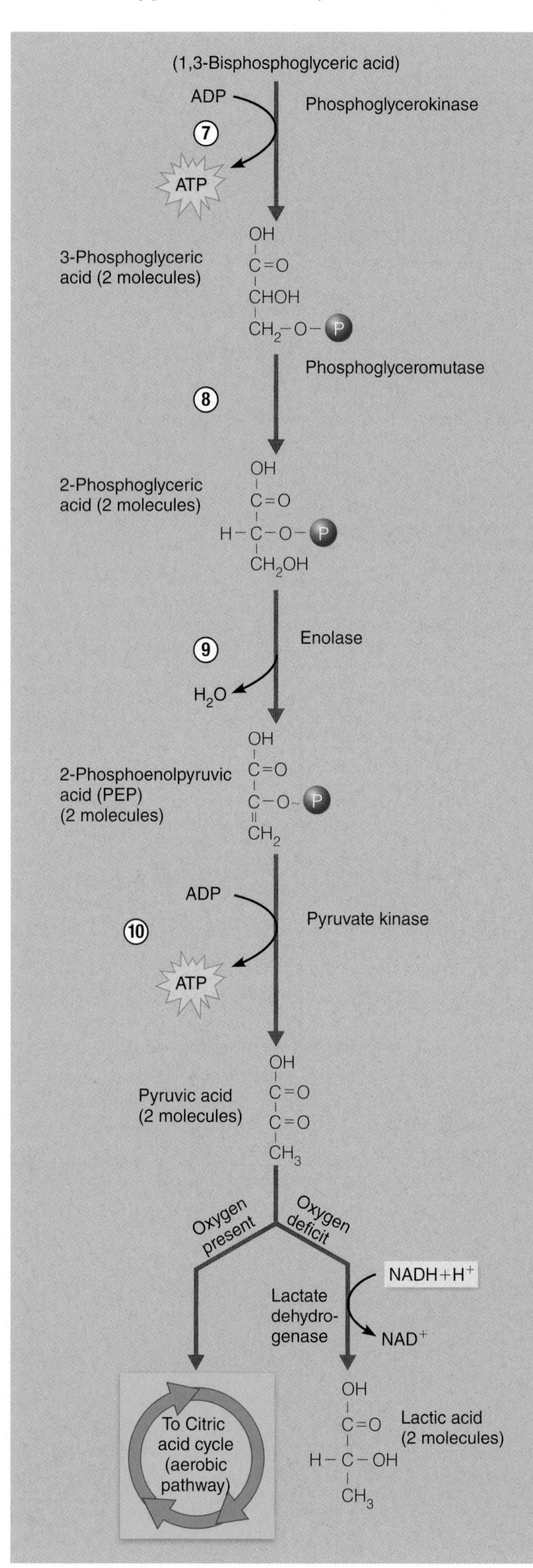

⑦ Finally, glycolysis produces ATP. The phosphate group, with its high-energy bond, is transferred from 1,3-bisphosphoglyceric acid to ADP. For each glucose molecule that began glycolysis, step 7 produces two molecules of ATP, because every product after the sugar-splitting step (step 4) is doubled. Of course, two ATPs were invested to get sugar ready for splitting. The ATP ledger now stands at zero. By the end of step 7, glucose has been converted to two molecules of 3-phosphoglyceric acid. This compound is not a sugar. The sugar was oxidized to an organic acid back in step 6, and now the energy made available by that oxidation has been used to make ATP.

⑧ Next, an enzyme relocates the remaining phosphate group of 3-phosphoglyceric acid to form 2-phosphoglyceric acid. This prepares the substrate for the next reaction.

⑨ An enzyme forms a double bond in the substrate by removing a water molecule from 2-phosphoglyceric acid to form phosphoenolpyruvic acid, or PEP. This results in the electrons of the substrate being rearranged in such a way that the remaining phosphate bond becomes very unstable; it has been upgraded to high-energy status.

⑩ The last reaction of glycolysis produces another molecule of ATP by transferring the phosphate group from PEP to ADP. Because this step occurs twice for each glucose molecule, the ATP ledger now shows a net gain of two ATPs. Steps 7 and 10 each produce two ATPs for a total credit of four, but a debt of two ATPs was incurred from steps 1 and 3. Glycolysis has repaid the ATP investment with 100% interest. In the meantime, glucose has been broken down and oxidized to two molecules of pyruvic acid, the compound produced from PEP in step 10.

Summary

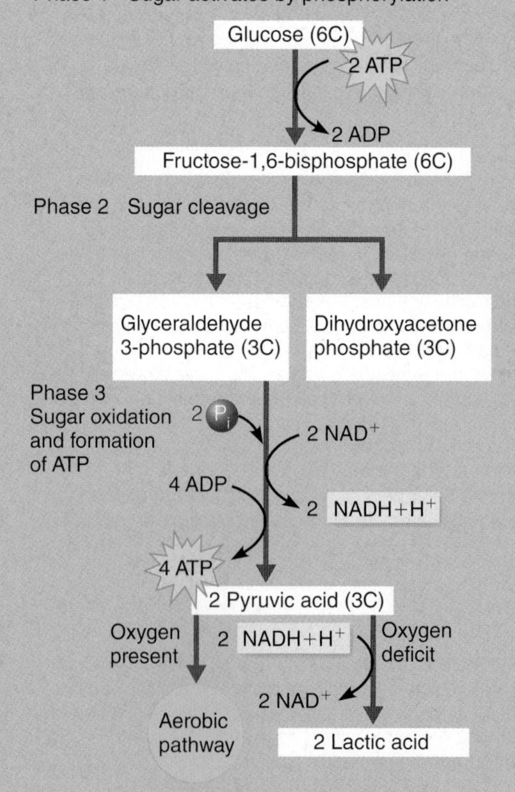

① Two-carbon acetyl CoA is combined with oxaloacetic acid, a 4-carbon compound. The unstable bond between the acetyl group and CoA is broken as oxaloacetic acid binds and CoA is freed to prime another 2-carbon fragment derived from pyruvic acid. The product is the 6-carbon citric acid, for which the cycle is named.

② A molecule of water is removed, and another is added back. The net result is the conversion of citric acid to its isomer, isocitric acid.

③ The substrate loses a CO_2 molecule, and the remaining 5-carbon compound is oxidized, forming an α-ketoglutaric acid and reducing NAD^+.

④ This step is catalyzed by a multienzyme complex very similar to the one that converts pyruvic acid to acetyl CoA. CO_2 is lost; the remaining 4-carbon compound is oxidized by the transfer of electrons to NAD^+ to form $NADH + H^+$ and is then attached to CoA by an unstable bond. The product is succinyl CoA.

⑤ Substrate-level phosphorylation occurs in this step. CoA is displaced by a phosphate group, which is then transferred to GDP to form guanosine triphosphate (GTP). GTP is similar to ATP, which is formed when GTP donates a phosphate group to ADP. The products of this step are succinic acid and ATP.

⑥ In another oxidative step, two hydrogens are removed from succinic acid (forming fumaric acid) and transferred to FAD to form $FADH_2$. The function of this coenzyme is similar to that of $NADH + H^+$, but $FADH_2$ stores less energy. The enzyme that catalyzes this oxidation-reduction reaction is the only enzyme of the cycle that is embedded in the mitochondrial membrane. All other enzymes of the citric acid cycle are dissolved in the mitochondrial matrix.

⑦ Bonds in the substrate are rearranged in this step by the addition of a water molecule. The product is malic acid.

⑧ The last oxidative step reduces another NAD^+ and regenerates oxaloacetic acid, which accepts a 2-carbon fragment from acetyl CoA for another turn of the cycle.

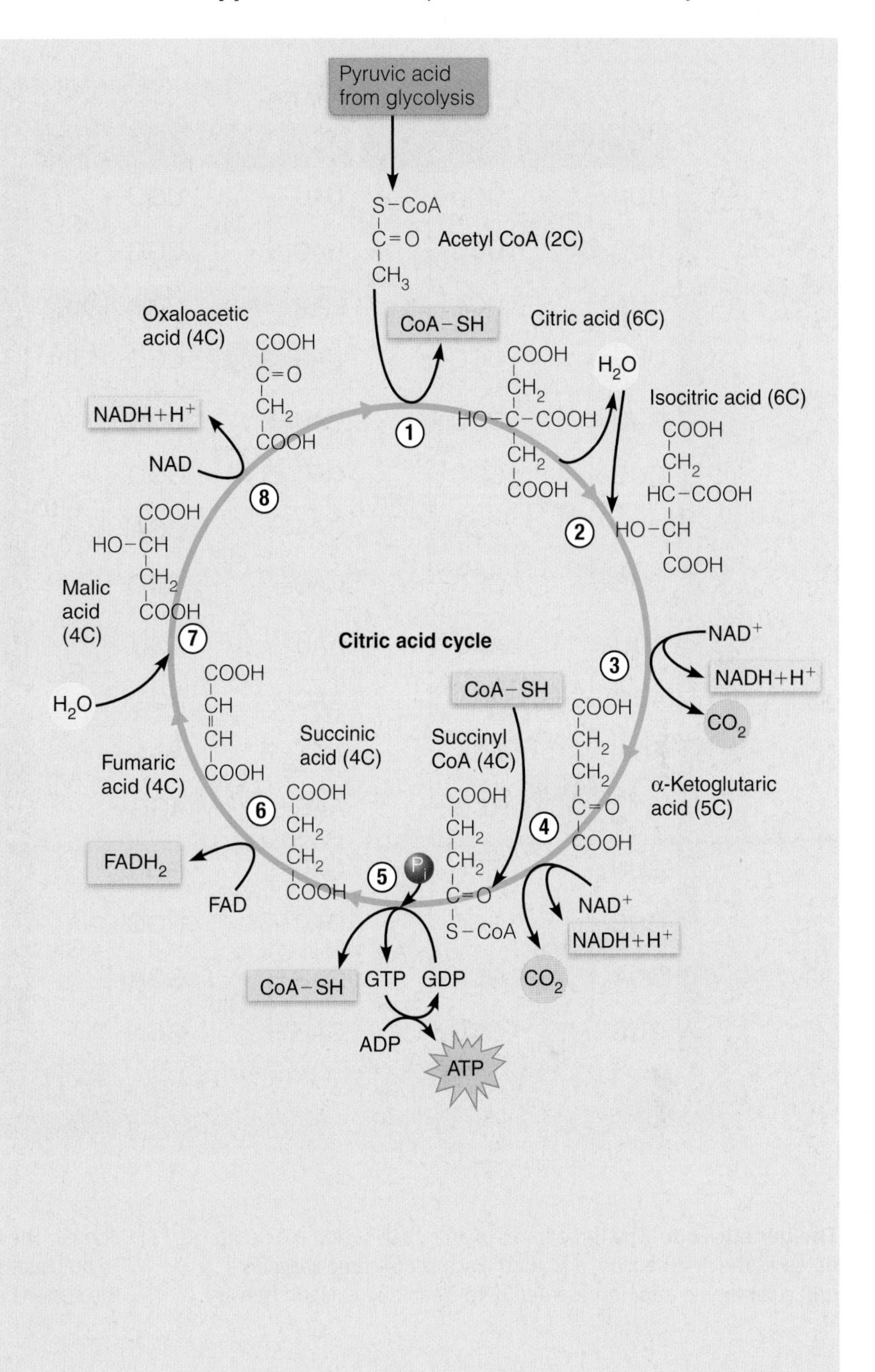

CITRIC ACID CYCLE (KREBS CYCLE) All but one of the steps (step 6) occur in the mitochondrial matrix. The preparation of pyruvic acid (by oxidation, decarboxylation, and reaction with coenzyme A) to enter the cycle as acetyl CoA is shown above the cycle. Acetyl CoA is picked up by oxaloacetic acid to form citric acid; and as it passes through the cycle, it is oxidized four more times [forming three molecules of reduced NAD ($NADH + H^+$) and one of reduced FAD ($FADH_2$)] and decarboxylated twice (releasing 2 CO_2). Energy is captured in the bonds of GTP, which then acts in a coupled reaction with ADP to generate one molecule of ATP by substrate-level phosphorylation.

Appendix B

The Genetic Code

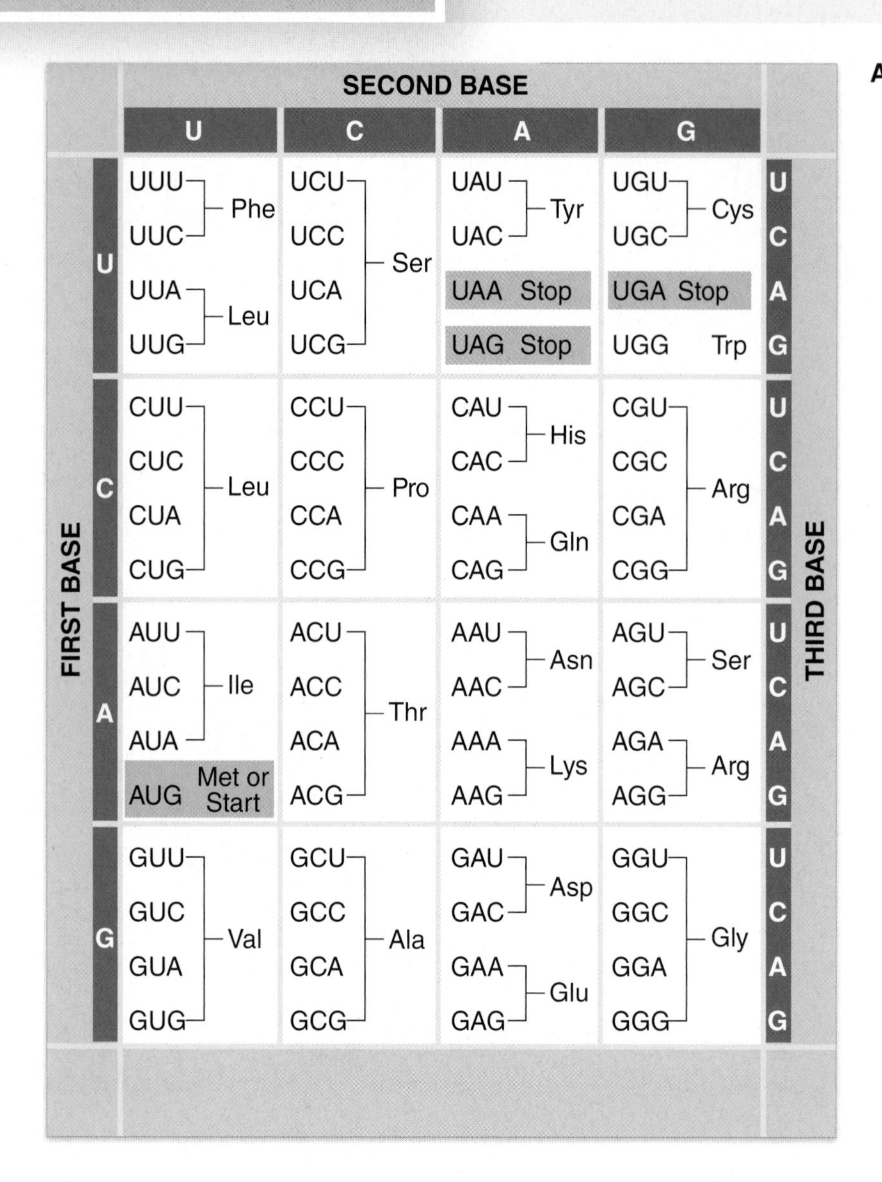

Abbreviation	Amino acid
Ala	Alanine
Arg	Arginine
Asn	Asparagine
Asp	Aspartic acid
Cys	Cysteine
Glu	Glutamic acid
Gln	Glutamine
Gly	Glycine
His	Histidine
Ile	Isoleucine
Leu	Leucine
Lys	Lysine
Met	Methionine
Phe	Phenylalanine
Pro	Proline
Ser	Serine
Thr	Threonine
Trp	Tryptophan
Tyr	Tyrosine
Val	Valine

The genetic code. The three bases in an mRNA codon are designated as the first, second, and third. Each set of three specifies a particular amino acid, represented here by an abbreviation (see list above). The codon AUG (which specifies the amino acid methionine) is the usual start signal for protein synthesis. The word *stop* indicates the codons that serve as signals to terminate protein synthesis.

Glossary

Pronunciation Key

′ = Primary accent

″ = Secondary accent

Pronounce:

a, fa, āt	as in	fate	o, no, ōt	as in	note
ă, hă, at		hat	ŏ, frŏ, og		frog
ah		father	oo		soon
ar		tar	or		for
e, stre, ēt		street	ow		plow
ĕ, hĕ, en		hen	oy		boy
er		her	sh		she
ew		new	u, mu, ūt		mute
g		go	ŭ, sŭ, un		sun
i, bi, īt		bite	z		zebra
ĭ, hĭ, im		him	zh		measure
ng		ring			

Abduct (ab-dukt′) To move away from the midline of the body.

Absolute refractory period Period following stimulation during which no additional action potential can be evoked.

Absorption Process by which the products of digestion pass through the alimentary tube mucosa into the blood or lymph.

Accessory digestive organs Organs that contribute to the digestive process but are not part of the alimentary canal; include the tongue, teeth, salivary glands, pancreas, liver.

Accommodation The process of increasing the refractive power of the lens of the eye; focusing.

Acetabulum (as″ĕ-tab′u-lum) Cuplike cavity on lateral surface of the hip bone that receives the femur.

Acetylcholine (ACh) (as″ĕ-til-ko′lēn) Chemical transmitter substance released by some nerve endings.

Acetylcholinesterase (AChE) (as″ĕ-til-ko″lin-es′ter-ās) Enzyme present at the neuromuscular junction and synapses that degrades acetylcholine and terminates its action.

Achilles tendon *See* Calcaneal tendon.

Acid A substance that releases hydrogen ions when in solution (compare with Base); a proton donor.

Acid-base balance Situation in which the pH of the blood is maintained between 7.35 and 7.45.

Acidosis (as″ĭ-do′sis) State of abnormally high hydrogen ion concentration in the extracellular fluid.

Actin (ak′tin) A contractile protein of muscle.

Action potential A large transient depolarization event, including polarity reversal, that is conducted along the membrane of a muscle cell or a nerve fiber.

Activation energy The amount of energy required to push a reactant to the level necessary for action.

Active immunity Immunity produced by an encounter with an antigen; provides immunological memory.

Active site Region on the surface of a functional (globular) protein where it binds and interacts chemically with other molecules of complementary shape and charge.

Active (transport) processes (1) Membrane transport processes for which ATP is directly or indirectly required, e.g., solute pumping and endocytosis. (2) "Active transport" also refers specifically to solute pumping.

Adaptation (1) Any change in structure or response to suit a new environment; (2) decline in the transmission of a sensory nerve when a receptor is stimulated continuously and at constant stimulus strength.

Adduct (a-dukt′) To move toward the midline of the body.

Adenine (A) (ad′ĕ-nēn) One of the two major purines found in both RNA and DNA; also found in various free nucleotides of importance to the body, such as ATP.

Adenohypophysis (ad″ĕ-no-hi-pof′ĭ-sis) Anterior pituitary; the glandular part of the pituitary gland.

Adenoids (ad′en-noids) Pharyngeal tonsil.

Adenosine triphosphate (ATP) (ah-den′o-sēn tri″fos′fāt) Organic molecule that stores and releases chemical energy for use in body cells.

Adenylate cyclase An enzyme, usually activated by a G protein, that converts ATP to the second messenger cyclic AMP.

Adipocyte (ad′ĭ-po-sīt) An adipose, or fat, cell.

Adipose tissue (ad′ĭ-pōs) Areolar connective tissue modified to store nutrients; a connective tissue consisting chiefly of fat cells.

Adrenal glands (uh-drē′nul) Hormone-producing glands located superior to the kidneys; each consists of medulla and cortex areas.

Adrenergic fibers (ad″ren-er′jik) Nerve fibers that release norepinephrine.

Word Roots, Prefixes, Suffixes, and Combining Forms

Prefixes and Combining Forms

a-, an- *absence, lack* acardia, lack of a heart; anaerobic, in the absence of oxygen
ab- *departing from, away from* abnormal, departing from normal
acou- *hearing* acoustics, the science of sound
ac-, acro- *extreme or extremity, peak* acrodermatitis, inflammation of the skin of the extremities
ad- *to, toward* adorbital, toward the orbit
aden-, adeno- *gland* adeniform, resembling a gland in shape
adren- *toward the kidney* adrenal gland, adjacent to the kidney
aero- *air* aerobic respiration, oxygen-requiring metabolism
af- *toward* afferent neurons, which carry impulses to the central nervous system
agon- *contest* agonistic and antagonistic muscles, which oppose each other
alb- *white* corpus albicans of the ovary, a white scar tissue
aliment- *nourish* alimentary canal, or digestive tract
allel- *of one another* alleles, alternative expressions of a gene
amphi- *on both sides, of both kinds* amphibian, an organism capable of living in water and on land
ana- *apart, up, again* anaphase of mitosis, when the chromosomes separate
anastomos- *come together* arteriovenous anastomosis, a connection between an artery and a vein
aneurysm *a widening* aortic aneurysm, a weak spot that causes enlargement of the blood vessel
angi- *vessel* angiitis, inflammation of a lymph vessel or blood vessel
angin- *choked* angina pectoris, a choked feeling in the chest due to dysfunction of the heart
ant-, anti- *opposed to, preventing, inhibiting* anticoagulant, a substance that prevents blood coagulation
ante- *preceding, before* antecubital, in front of the elbow
aort- *great artery* aorta
ap-, api- *tip, extremity* apex of the heart
append- *hang to* appendicular skeleton
aqua-, aque- *water* aqueous solutions
arbor *tree* arbor vitae of the cerebellum, the treelike pattern of white matter
areola- *open space* areolar connective tissue, a loose connective tissue
arrect- *upright* arrector pili muscles of the skin, which make the hairs stand erect
arthr-, arthro- *joint* arthropathy, any joint disease
artic- *joint* articular surfaces of bones, the points of connection
atri- *vestibule* atria, upper chambers of the heart
auscult- *listen* auscultatory method for measuring blood pressure
aut-, auto- *self* autogenous, self-generated
ax-, axi-, axo- *axis, axle* axial skeleton, axis of vertebral column
azyg- *unpaired* azygous vein, an unpaired vessel
baro- *pressure* baroreceptors for monitoring blood pressure
basal *base* basal lamina of epithelial basement membrane
bi- *two* bicuspid, having two cusps
bili- *bile* bilirubin, a bile pigment
bio- *life* biology, the study of life and living organisms
blast- *bud, germ* blastocyte, undifferentiated embryonic cell
brachi- *arm* brachial plexus of peripheral nervous system supplies the arm
brady- *slow* bradycardia, abnormally slow heart rate
brev- *short* fibularis brevis, a short leg muscle
broncho- *bronchus* bronchospasm, spasmodic contraction of bronchial muscle
bucco- *cheek* buccolabial, pertaining to the cheek and lip
calor- *heat* calories, a measure of energy
capill- *hair* blood and lymph capillaries
caput- *head* decapitate, remove the head
carcin- *cancer* carcinogen, a cancer-causing agent
cardi-, cardio- *heart* cardiotoxic, harmful to the heart
carneo- *flesh* trabeculae carneae, ridges of muscle in the ventricles of the heart
carot- *(1) carrot, (2) stupor* (1) carotene, an orange pigment, (2) carotid arteries in the neck, blockage of which causes fainting
cata- *down* catabolism, chemical breakdown
caud- *tail* caudal (directional term)
cec- *blind* cecum of large intestine, a blind-ended pouch
cele- *abdominal* celiac artery, in the abdomen
cephal- *head* cephalometer, an instrument for measuring the head
cerebro- *brain, especially the cerebrum* cerebrospinal, pertaining to the brain and spinal cord
cervic-, cervix *neck* cervix of the uterus
chiasm- *crossing* optic chiasma, where optic nerves cross
chole- *bile* cholesterol, cholecystokinin, a bile-secreting hormone
chondr- *cartilage* chondrogenic, giving rise to cartilage
chrom- *colored* chromosomes, so named because they stain darkly
cili- *small hair* ciliated epithelium
circum- *around* circumnuclear, surrounding the nucleus
clavic- *key* clavicle, a "skeleton key"
co-, con- *together* concentric, common center, together in the center
coccy- *cuckoo* coccyx, which is beak-shaped
cochlea *snail shell* the cochlea of the inner ear, which is coiled like a snail shell
coel- *hollow* coelom, the ventral body cavity
commis- *united* gray commissure of the spinal cord connects the two columns of gray matter
concha *shell* nasal conchae, coiled shelves of bone in the nasal cavity
contra- *against, opposite* contraceptive, agent preventing conception

corn-, cornu- *horn* stratum corneum, outer layer of the skin composed of (horny) cells
corona *crown* coronal suture of the skull
corp- *body* corpse; corpus luteum, hormone-secreting body in the ovary
cort- *bark* cortex, the outer layer of the brain, kidney, adrenal glands, and lymph nodes
cost- *rib* intercostal, between the ribs
crani- *skull* craniotomy, a skull operation
crypt- *hidden* cryptomenorrhea, a condition in which menstrual symptoms are experienced but no external loss of blood occurs
cusp- *pointed* bicuspid, tricuspid valves of the heart
cutic- *skin* cuticle of the nail
cyan- *blue* cyanosis, blue color of the skin due to lack of oxygen
cyst- *sac, bladder* cystitis, inflammation of the urinary bladder
cyt- *cell* cytology, the study of cells
de- *undoing, reversal, loss, removal* deactivation, becoming inactive
decid- *falling off* deciduous (milk) teeth
delta *triangular* deltoid muscle, roughly triangular in shape
den-, dent- *tooth* dentin of the tooth
dendr- *tree, branch* dendrites, branches of a neuron
derm- *skin* dermis, deep layer of the skin
desm- *bond* desmosome, which binds adjacent epithelial cells
di- *twice, double* dimorphism, having two forms
dia- *through, between* diaphragm, the wall through or between two areas
dialys- *separate, break apart* kidney dialysis, in which waste products are removed from the blood
diastol- *stand apart* cardiac diastole, between successive contractions of the heart
diure- *urinate* diuretic, a drug that increases urine output
dors- *the back* dorsal, dorsum, dorsiflexion
duc-, duct *lead, draw* ductus deferens, tube which carries sperm from the epididymis into the urethra during ejaculation
dura *hard* dura mater, tough outer meninx
dys- *difficult, faulty, painful* dyspepsia, disturbed digestion
ec-, ex-, ecto- *out, outside, away from* excrete, to remove materials from the body
ectop- *displaced* ectopic pregnancy; ectopic focus for initiation of heart contraction
edem- *swelling* edema, accumulation of water in body tissues
ef- *away* efferent nerve fibers, which carry impulses away from the central nervous system
ejac- *to shoot forth* ejaculation of semen
embol- *wedge* embolus, an obstructive object traveling in the bloodstream
en-, em- *in, inside* encysted, enclosed in a cyst or capsule
enceph- *brain* encephalitis, inflammation of the brain
endo- *within, inner* endocytosis, taking particles into a cell
entero- *intestine* enterologist, one who specializes in the study of intestinal disorders
epi- *over, above* epidermis, outer layer of skin
erythr- *red* erythema, redness of the skin; erythrocyte, red blood cell
eso- *within* esophagus
eu- *well* euesthesia, a normal state of the senses
excret- *separate* excretory system
exo- *outside, outer layer* exophthalmos, an abnormal protrusion of the eye from the orbit
extra- *outside, beyond* extracellular, outside the body cells of an organism
extrins- *from the outside* extrinsic regulation of the heart
fasci-, fascia- *bundle, band* superficial and deep fascia
fenestr- *window* fenestrated capillaries
ferr- *iron* transferrin, ferritin, both iron-storage proteins
flagell- *whip* flagellum, the tail of a sperm cell
flat- *blow, blown* flatulence
folli- *bag, bellows* hair follicle
fontan- *fountain* fontanelles of the fetal skull
foram- *opening* foramen magnum of the skull
foss- *ditch* fossa ovalis of the heart; mandibular fossa of the skull
gam-, gamet- *married, spouse* gametes, the sex cells
gangli- *swelling, or knot* dorsal root ganglia of the spinal nerves
gastr- *stomach* gastrin, a hormone that influences gastric acid secretion
gene *beginning, origin* genetics
germin- *grow* germinal epithelium of the gonads
gero-, geront- *old man* gerontology, the study of aging
gest- *carried* gestation, the period from conception to birth
glauc- *gray* glaucoma, which causes gradual blindness
glom- *ball* glomeruli, clusters of capillaries in the kidneys
glosso- *tongue* glossopathy, any disease of the tongue
gluco-, glyco- gluconeogenesis, the production of glucose from noncarbohydrate molecules
glute- *buttock* gluteus maximus, largest muscle of the buttock
gnost- *knowing* the gnostic sense, a sense of awareness of self
gompho- *nail* gomphosis, the term applied to the joint between tooth and jaw
gon-, gono- *seed, offspring* gonads, the sex organs
gust- *taste* gustatory sense, the sense of taste
hapt- *fasten, grasp* hapten, a partial antigen
hema-, hemato-, hemo- *blood* hematocyst, a cyst containing blood
hemi- *half* hemiglossal, pertaining to one-half of the tongue
hepat- *liver* hepatitis, inflammation of the liver
hetero- *different or other* heterosexuality, sexual desire for a person of the opposite sex
hiat- *gap* the hiatus of the diaphragm, the opening through which the esophagus passes
hippo- *horse* hippocampus of the brain, shaped like a seahorse
hirsut- *hairy* hirsutism, excessive body hair
hist- *tissue* histology, the study of tissues
holo- *whole* holocrine glands, whose secretions are whole cells
hom-, homo- *same* homeoplasia, formation of tissue similar to normal tissue; homocentric, having the same center
hormon- *to excite* hormones
humor- *a fluid* humoral immunity, which involves antibodies circulating in the blood
hyal- *glass, clear* hyaline cartilage, which has no visible fibers
hydr-, hydro- *water* dehydration, loss of body water
hyper- *excess* hypertension, excessive tension
hypno- *sleep* hypnosis, a sleeplike state
hypo- *below, deficient* hypodermic, beneath the skin; hypokalemia, deficiency of potassium
hyster-, hystero- *uterus, womb* hysterectomy, removal of the uterus; hysterodynia, pain in the womb
ile- *intestine* ileum, the last portion of the small intestine
im- *not* impermeable, not permitting passage, not permeable
inter- *between* intercellular, between the cells
intercal- *insert* intercalated discs, the end membranes between adjacent cardiac muscle cells
intra- *within, inside* intracellular, inside the cell
iso- *equal, same* isothermal, equal, or same, temperature
jugul- *throat* jugular veins, prominent vessels in the neck
juxta- *near, close to* juxtaglomerular complex, a cell cluster next to a glomerulus in the kidneys
karyo- *kernel, nucleus* karyotype, the assemblage of the nuclear chromosomes
kera- *horn* keratin, the water-repellent protein of the skin
kilo- *thousand* kilocalories, equal to 1000 calories
kin-, kines- *move* kinetic energy, the energy of motion

labi-, labri- *lip* labial frenulum, the membrane which joins the lip to the gum
lact- *milk* lactose, milk sugar
lacun- *space, cavity, lake* lacunae, the spaces occupied by cells of cartilage and bone tissue
lamell- *small plate* concentric lamellae, rings of bone matrix in compact bone
lamina *layer, sheet* basal lamina, part of the epithelial basement membrane
lat- *wide* latissimus dorsi, a broad muscle of the back
laten- *hidden* latent period of a muscle twitch
later- *side* lateral (directional term)
leuko- *white* leukocyte, white blood cell
leva- *raise, elevate* levator labii superioris, muscle that elevates upper lip
lingua- *tongue* lingual tonsil, adjacent to the tongue
lip-, lipo- *fat, lipid* lipophage, a cell that has taken up fat in its cytoplasm
lith- *stone* cholelithiasis, gallstones
luci- *clear* stratum lucidum, clear layer of the epidermis
lumen *light* lumen, center of a hollow structure
lut- *yellow* corpus luteum, a yellow, hormone-secreting structure in the ovary
lymph *water* lymphatic circulation, return of clear fluid to the bloodstream
macro- *large* macromolecule, large molecule
macula *spot* macula lutea, yellow spot on the retina
magn- *large* foramen magnum, largest opening of the skull
mal- *bad, abnormal* malfunction, abnormal functioning of an organ
mamm- *breast* mammary gland, breast
mast- *breast* mastectomy, removal of a mammary gland
mater *mother* dura mater, pia mater, membranes that envelop the brain
meat- *passage* external acoustic meatus, the ear canal
medi- *middle* medial (directional term)
medull- *marrow* medulla, the middle portion of the kidney, adrenal gland, and lymph node
mega- *large* megakaryocyte, large precursor cell of platelets
meio- *less* meiosis, nuclear division that halves the chromosome number
melan- *black* melanocytes, which secrete the black pigment melanin
men-, menstru- *month* menses, the cyclic menstrual flow
meningo- *membrane* meningitis, inflammation of the membranes of the brain
mer-, mero- *a part* merocrine glands, the secretions of which do not include the cell
meso- *middle* mesoderm, middle germ layer
meta- *beyond, between, transition* metatarsus, the part of the foot between the tarsus and the phalanges
metro- *uterus* endometrium, the lining of the uterus
micro- *small* microscope, an instrument used to make small objects appear larger
mictur- *urinate* micturition, the act of voiding the bladder
mito- *thread, filament* mitochondria, small, filament-like structures located in cells
mnem- *memory* amnesia
mono- *single* monospasm, spasm of a single limb
morpho- *form* morphology, the study of form and structure or organisms
multi- *many* multinuclear, having several nuclei
mur- *wall* intramural ganglion, a nerve junction within an organ
muta- *change* mutation, change in the base sequence of DNA
myelo- *spinal cord, marrow* myeloblasts, cells of the bone marrow
myo- *muscle* myocardium, heart muscle
nano- *dwarf* nanometer, one-billionth of a meter
narco- *numbness* narcotic, a drug producing stupor or numbed sensations
natri- *sodium* atrial natriuretic peptide, a sodium-regulating hormone
necro- *death* necrosis, tissue death
neo- *new* neoplasm, an abnormal growth
nephro- *kidney* nephritis, inflammation of the kidney
neuro- *nerve* neurophysiology, the physiology of the nervous system
noci- *harmful* nociceptors, receptors for pain
nom- *name* innominate artery; innominate bone
noto- *back* notochord, the embryonic structure that precedes the vertebral column
nucle- *pit, kernel, little nut* nucleus
nutri- *feed, nourish* nutrition
ob- *before, against* obstruction, impeding or blocking up
oculo- *eye* monocular, pertaining to one eye
odonto- *teeth* orthodontist, one who specializes in proper positioning of the teeth in relation to each other
olfact- *smell* olfactory nerves
oligo- *few* oligodendrocytes, neuroglial cells with few branches
onco- *a mass* oncology, study of cancer
oo- *egg* oocyte, precursor of female gamete
ophthalmo- *eye* ophthalmology, the study of the eyes and related disease
orb- *circular* orbicularis oculi, muscle that encircles the eye
orchi- *testis* cryptorchidism, failure of the testes to descend into the scrotum
org- *living* organism
ortho- *straight, direct* orthopedic, correction of deformities of the musculoskeletal system
osm- *smell* anosmia, loss of sense of smell
osmo- *pushing* osmosis
osteo- *bone* osteodermia, bony formations in the skin
oto- *ear* otoscope, a device for examining the ear
ov-, ovi- *egg* ovum; oviduct
oxy- *oxygen* oxygenation, the saturation of a substance with oxygen
pan- *all, universal* panacea, a cure-all
papill- *nipple* dermal papillae, projections of the dermis into the epidermal area
para- *beside, near* paranuclear, beside the nucleus
pect-, pectus *breast* pectoralis major, a large chest muscle
pelv- *a basin* pelvic girdle, which cradles the pelvic organs
peni- *a tail* penis; penile arteriole
penna- *feather* unipennate, bipennate muscles, whose fascicles have a feathered appearance
pent- *five* pentose, a 5-carbon sugar
pep-, peps-, pept- *digest* pepsin, a digestive enzyme of the stomach; peptic ulcer
per-, permea- *through* permeate; permeable
peri- *around* perianal, situated around the anus
phago- *eat* phagocyte, a cell that engulfs and digests particles or cells
pheno- *show, appear* phenotype, the physical appearance of an individual
phleb- *vein* phlebitis, inflammation of the veins
pia *tender* pia mater, delicate inner membrane around the brain and spinal cord
pili *hair* arrector pili muscles of the skin, which make the hairs stand erect
pin-, pino- *drink* pinocytosis, the engulfing of small particles by a cell
platy- *flat, broad* platysma, broad, flat muscle of the neck
pleur- *side, rib* pleural serosa, the membrane that lines the thoracic cavity and covers the lungs
plex-, plexus *net, network* brachial plexus, the network of nerves that supplies the arm
pneumo- *air, wind* pneumothorax, air in the thoracic cavity
pod- *foot* podiatry, the treatment of foot disorders
poly- *multiple* polymorphism, multiple forms
post- *after, behind* posterior, places behind (a specific) part
pre-, pro- *before, ahead of* prenatal, before birth
procto- *rectum, anus* proctoscope, an instrument for examining the rectum
pron- *bent forward* prone, pronate

propri- *one's own* proprioception, awareness of body parts and movement
pseudo- *false* pseudotumor, a false tumor
psycho- *mind, psyche* psychogram, a chart of personality traits
ptos- *fall* renal ptosis, a condition in which the kidneys drift below their normal position
pub- *of the pubis* puberty
pulmo- *lung* pulmonary artery, which brings blood to the lungs
pyo- *pus* pyocyst, a cyst that contains pus
pyro- *fire* pyrogen, a substance that induces fever
quad-, quadr- *four-sided* quadratus lumborum, a muscle with a square shape
re- *back, again* reinfect
rect- *straight* rectus abdominis, rectum
ren- *kidney* renal; renin, an enzyme secreted by the kidney
retin-, retic- *net, network* endoplasmic reticulum, a network of membranous sacs within a cell
retro- *backward, behind* retrogression, to move backward in development
rheum- *watery flow, change, flux* rheumatoid arthritis; rheumatic fever
rhin-, rhino- *nose* rhinitis, inflammation of the nose
ruga- *fold, wrinkle* rugae, the folds of the stomach, gallbladder, and urinary bladder
sagitt- *arrow* sagittal (directional term)
salta- *leap* saltatory conduction, the rapid conduction of impulses along myelinated neurons
sanguin- *blood* consanguineous, indicative of a genetic relationship between individuals
sarco- *flesh* sarcomere, unit of contraction in skeletal muscle
saphen- *visible, clear* great saphenous vein, superficial vein of the thigh and leg
sclero- *hard* sclerodermatitis, inflammatory thickening and hardening of the skin
seb- *grease* sebum, the oil of the skin
semen *seed, sperm* semen, the discharge of the male reproductive system
semi- *half* semicircular, having the form of half a circle
sens- *feeling* sensation, sensory
septi- *rotten* sepsis, infection; antiseptic
septum *fence* nasal septum
sero- *serum* serological tests, which assess blood conditions
serrat- *saw* serratus anterior, a muscle of the chest wall that has a jagged edge
sin-, sino- *a hollow* sinuses of the skull
soma- *body* somatic nervous system
somn- *sleep* insomnia, inability to sleep
sphin- *squeeze* sphincter
splanchn- *organ* splanchnic nerve, autonomic supply to abdominal viscera
spondyl- *vertebra* ankylosing spondylitis, rheumatoid arthritis affecting the spine
squam- *scale, flat* squamous epithelium, squamous suture of the skull
steno- *narrow* stenocoriasis, narrowing of the pupil
strat- *layer* strata of the epidermis; stratified epithelium
stria- *furrow, streak* striations of skeletal and cardiac muscle tissue
stroma *spread out* stroma, the connective tissue framework of some organs
sub- *beneath, under* sublingual, beneath the tongue
sucr- *sweet* sucrose, table sugar
sudor- *sweat* sudoriferous glands, the sweat glands
super- *above, upon* superior, quality or state of being above other parts
supra- *above, upon* supracondylar, above a condyle
sym-, syn- *together, with* synapse, the region of communication between two neurons
synerg- *work together* synergism
systol- *contraction* systole, contraction of the heart
tachy- *rapid* tachycardia, abnormally rapid heartbeat
tact- *touch* tactile sense
telo- *the end* telophase, the end of mitosis
templ-, tempo- *time* temporal summation of nerve impulses
tens- *stretched* muscle tension
terti- *third* fibularis tertius, one of three fibularis muscles
tetan- *rigid, tense* tetanus of muscles
therm- *heat* thermometer, an instrument used to measure heat
thromb- *clot* thrombocyte, thrombus
thyro- *a shield* thyroid gland
tissu- *woven* tissue
tono- *tension* tonicity, hypertonic
tox- *poison* toxicology, study of poisons
trab- *beam, timber* trabeculae, spicules of bone in spongy bone tissue
trans- *across, through* transpleural, through the pleura
trapez- *table* trapezius, the four-sided muscle of the upper back
tri- *three* trifurcation, division into three branches
trop- *turn, change* tropic hormones, whose targets are endocrine glands
troph- *nourish* trophoblast, from which develops the fetal portion of the placenta
tuber- *swelling* tuberosity, a bump on a bone
tunic- *covering* tunica albuginea, the covering of the testis
tympan- *drum* tympanic membrane, the eardrum
ultra- *beyond* ultraviolet radiation, beyond the band of visible light
vacc- *cow* vaccine
vagin- *a sheath* vagina
vagus *wanderer* the vagus nerve, which starts at the brain and travels into the abdominopelvic cavity
valen- *strength* valence shells of atoms
venter, ventr- *abdomen, belly* ventral (directional term), ventricle
vent- *the wind* pulmonary ventilation
vert- *turn* vertebral column
vestibul- *a porch* vestibule, the anterior entryway to the mouth and nose
vibr- *shake, quiver* vibrissae, hairs of the nasal vestibule
villus *shaggy hair* microvilli, which have the appearance of hair in light microscopy
viscero- *organ, viscera* visceroinhibitory, inhibiting the movements of the viscera
viscos- *sticky* viscosity, resistance to flow
vita- *life* vitamin
vitre- *glass* vitreous humor, the clear jelly of the eye
viv- *live* in vivo
vulv- *a covering* vulva, the female external genitalia
zyg- *a yoke, twin* zygote

Suffixes

-able *able to, capable of* viable, ability to live or exist
-ac *referring to* cardiac, referring to the heart
-algia *pain in a certain part* neuralgia, pain along the course of a nerve
-apsi *juncture* synapse, where two neurons communicate
-ary *associated with, relating to* coronary, associated with the heart
-asthen *weakness* myasthenia gravis, a disease involving paralysis
-bryo *swollen* embryo
-cide *destroy or kill* germicide, an agent that kills germs
-cipit *head* occipital
-clast *break* osteoclast, a cell that dissolves bone matrix
-crine *separate* endocrine organs, which secrete hormones into the blood
-dips *thirst, dry* polydipsia, excessive thirst associated with diabetes
-ectomy *cutting out, surgical removal* appendectomy, cutting out of the appendix
-ell, -elle *small* organelle
-emia *condition of the blood* anemia, deficiency of red blood cells
-esthesi *sensation* anesthesia, lack of sensation